T0143483

Lubricant Additives:
Chemistry and Applications,
Third Edition

CHEMICAL INDUSTRIES
A Series of Reference Books and Textbooks

Founding Editor

HEINZ HEINEMANN
Berkeley, California

Series Editor

JAMES G. SPEIGHT
CD & W, Inc.
Laramie, Wyoming

MOST RECENTLY PUBLISHED

Lubricant Additives: Chemistry and Applications, Third Edition

Edited by
Leslie R. Rudnick
Designed Materials LLC, Scottsdale, Arizona

CRC Press
Taylor & Francis Group
Boca Raton London New York

CRC Press is an imprint of the
Taylor & Francis Group, an **informa** business

Cover photographs courtesy of Leslie (Les) R. Rudnick, 2017

CRC Press
Taylor & Francis Group
6000 Broken Sound Parkway NW, Suite 300
Boca Raton, FL 33487-2742

First issued in paperback 2022

© 2017 by Taylor & Francis Group, LLC
CRC Press is an imprint of Taylor & Francis Group, an Informa business

No claim to original U.S. Government works

ISBN 13: 978-1-03-240216-1 (pbk)
ISBN 13: 978-1-4987-3172-0 (hbk)
ISBN 13: 978-1-315-12062-1 (ebk)

DOI: 10.1201/9781315120621

This book contains information obtained from authentic and highly regarded sources. Reasonable efforts have been made to publish reliable data and information, but the author and publisher cannot assume responsibility for the validity of all materials or the consequences of their use. The authors and publishers have attempted to trace the copyright holders of all material reproduced in this publication and apologize to copyright holders if permission to publish in this form has not been obtained. If any copyright material has not been acknowledged please write and let us know so we may rectify in any future reprint.

Except as permitted under U.S. Copyright Law, no part of this book may be reprinted, reproduced, transmitted, or utilized in any form by any electronic, mechanical, or other means, now known or hereafter invented, including photocopying, microfilming, and recording, or in any information storage or retrieval system, without written permission from the publishers.

For permission to photocopy or use material electronically from this work, please access www.copyright.com (http://www.copyright.com/) or contact the Copyright Clearance Center, Inc. (CCC), 222 Rosewood Drive, Danvers, MA 01923, 978-750-8400. CCC is a not-for-profit organization that provides licenses and registration for a variety of users. For organizations that have been granted a photocopy license by the CCC, a separate system of payment has been arranged.

Trademark Notice: Product or corporate names may be trademarks or registered trademarks, and are used only for identification and explanation without intent to infringe.

Publisher's Note
The publisher has gone to great lengths to ensure the quality of this reprint but points out that some imperfections in the original copies may be apparent.

Library of Congress Cataloging-in-Publication Data

Names: Rudnick, Leslie R., 1947-
Title: Lubricant Additives: Chemistry and Applications / [edited by] Leslie R. Rudnick.
Description: Third edition. | Boca Raton : CRC Press, Taylor & Francis Group, 2017.
Series: Chemical industries | Includes bibliographical references and index.
Identifiers: LCCN 2016043477 | ISBN 9781498731720 (hardback : alk. paper) | ISBN 9781498731744 (ebook)
Subjects: LCSH: Lubrication and lubricants--Additives.
Classification: LCC TJ1077.L815 2017 | DDC 621.8/9--dc22
LC record available at https://lccn.loc.gov/2016043477

Visit the Taylor & Francis Web site at
http://www.taylorandfrancis.com

and the CRC Press Web site at
http://www.crcpress.com

Contents

SECTION IV   Viscosity Control Additives

SECTION V   Miscellaneous Additives

SECTION VI Applications

SECTION VII Trends

SECTION VIII Methods and Resources

Preface

Lubricant additives continue to be developed to improve the properties and performance of modern lubricants. The market for passenger car and heavy-duty motor oils, industrial oils, and metalworking fluids is growing rapidly, and this growth is expected to continue. Additional additive technology needs to be developed to meet performance requirements and the increased demand, especially in the Asia-Pacific region. China and India, for example, represent highly populated markets that are continuing to grow in infrastructure, leading to a growth in industrial equipment and number of vehicles. Many U.S. and EU companies continue to retain their presence in the Asia-Pacific region through either new manufacturing facilities or sales and distribution offices.

This edition includes new chapters on chlorohydrocarbons, foaming chemistry and physics, antifoams for nonaqueous lubricants, hydrogenated styrene–diene viscosity modifiers, alkylated aromatics, and the impact of REACh (Registration, Evaluation, Authorisation and Restriction of Chemicals) and GHS (Global Harmonized System) on the lubricant industry. In addition, many chapters from the previous edition have new coauthors and have been updated since then.

Environmental issues and applications that require lubricants to operate under severe conditions cause an increase in the use of synthetics. Owing to performance and maintenance reasons, many applications that have historically relied on petroleum-derived lubricants are shifting to synthetic lubricant–based products. Cost issues, on the contrary, tend to shift the market toward group II and III base oils where hydrocarbons can be used. A shift to renewable and biodegradable fluids is also necessary, and this will require a greater need for new effective additives to meet the challenges of formulating lubricants for various applications.

There are several indications that the lubricant additive industry will grow and change.

Legislation is driving changes to fuel composition and lubricant components, and, therefore, future lubricant developments will be constrained compared to what has been done in the past. REACh in the EU is placing constraints on the incentive to develop new molecules that will serve as additives. The cost of introduction of new proprietary materials will be the burden of the company that develops the new material. The costs in generating any needed data on the toxicology or biodegradability of the materials for many common additives that are produced by several manufacturers will be shared among them.

Continued progress toward new engine oil requirements will require oils to provide improved fuel economy and to have additive chemistry that does not degrade emission system components. This will require a new test to evaluate the volatility of phosphorus in engine oils and to improve oil properties in terms of protecting the engine. Future developments and requirements will undoubtedly require new, more severe testing protocols.

More advanced technologies will require the application of new types of lubricants, containing new additive chemistries required for the exploration of space and oceans. Since these remote locations and extreme environments require low maintenance, they will place new demands on lubricant properties and performance.

This book would not have developed the way it has without the invaluable help and encouragement of many of my colleagues. I thank all of the authors of the chapters contained herein for responding to the deadlines. There is always a balance between job responsibilities and publishing projects like this one. I truly appreciate all of their effort toward this new edition. It is your contributions that have created this resource for our industry.

I especially thank Barbara (Glunn) Knott at Taylor & Francis Group, with whom I have worked earlier on *Synthetics, Mineral Oils, and Bio-Based Lubricants*, for her support to this project from its early stages through its completion. I also thank Cheryl Wolf, Editorial Assistant, Joette Lynch, Project Editor, and Vinithan Sethumadhavan, Account Manager at SPI, who have been invaluable in every way in the progress of this project and have been a tremendous asset to me as an editor and helpful to the many contributors of this book.

I also thank my wife, Paula, and our children, Eric and Rachel, for all their support during this project.

Dr. Leslie R. Rudnick
Designed Materials LLC, Scottsdale, Arizona

Contributors

Ewa A. Bardasz
ZUAL Associates in Lubrication, LLC
Mentor, Ohio

Girma Biresaw
Cereal Products and Food Science Research Unit
NCAUR-MWA-ARS-USDA
Peoria, Illinois

Thomas F. Bunemann
Uniqema
Gouda, the Netherlands

Christa M.A. Chilson
PolyMod® Technologies, Inc.
Fort Wayne, Indiana

Michael T. Costello
BASF Corporation
Tarrytown, New York

Michael J. Covitch
The Lubrizol Corporation
Wicklife, Ohio

Jun Dong
Chemtura Corporation
Middlebury, Connecticut

Alan C. Eachus
Independent Consultant
Villa Park, Illinois

Boris M. Eisenberg
Evonik Oil Additives
Darmstadt, Germany

Liehpao Oscar Farng
ExxonMobil Research and Engineering Company
Paulsboro, New Jersey

Achim Fessenbecker
Rhein Chemie Rheinau GmbH
Mannheim, Germany

Matthias Fies
BASF SE
Ludwigshafen, Germany

Alan Flamberg
Evonik Oil Additives USA, Inc.
Horsham, Pennsylvania

Ernest C. Galgoci
Münzing North America, LP
Bloomfield, New Jersey

Isabella Goldmints
Infineum USA L.P.
Linden, New Jersey

Lois J. Gschwender
University of Dayton Research Institute
Nonstructural Materials Division
Dayton, Ohio

John W. Harris (deceased)
Shell Global Solutions
Houston, Texas

Sandra Horstmann
Lanxess Deutschland GmbH
Cologne, Germany

Kevin Hughes
The Lubrizol Corporation
Wickliffe, Ohio

Maureen E. Hunter
King Industries, Inc.
Norwalk, Connecticut

Tze-Chi Jao
Afton Chemical Corporation
Richmond, Virginia

Tom E. Karis
Independent Consultant
Aromas, California

Dick Kenbeck
Uniqema
Gouda, the Netherlands

Bernard G. Kinker
Independent Consultant
Kintnersville, Pennsylvania

Don Knobloch
The Lubrizol Corporation
Wickliffe, Ohio

Kalman Koczo
Momentive Performance Materials
Tarrytown, New York

Gordon D. Lamb
Lubrizol Limited
Hazelwood, United Kingdom

Saurabh Lawate
The Lubrizol Corporation
Wickliffe, Ohio

and

CPI Fluid Engineering
Division of the Lubrizol Corporation
Berkshire Hathaway Company
Midland, Michigan

Mark D. Leatherman
Momentive Performance Materials
Tarrytown, New York

Fay Linn Lee
Shell Lubricants
Houston, Texas

Brian M. Lipowski
Avery Dennison Performance Tapes
Painesville, Ohio

Philip Ma
BASF Corporation
Tarrytown, New York

James MacNeil
Qualice LLC
Hamlet, North Carolina

Gino Mariani
National Starch and Chemical Company
Port Huron, Michigan

Randolf A. McDonald
Functional Products, Inc.
Cleveland, Ohio

Cyril A. Migdal
LanXess Solutions US, Inc.
Naugatuck, Connecticut

Mark Miller
RSC Bio Solutions
Indian Trail, North Carolina

Neal Milne
Chemtura Petroleum Additives
Manchester, United Kingdom

Sonia Oberoi
Infineum USA L.P.
Linden, New Jersey

W. David Phillips
Independent Consultant
Poynton
Cheshire, United Kingdom

Syed Q.A. Rizvi
Research and Development
Elevance Renewable Sciences, Inc.
Woodridge, Illinois

Thomas Rossrucker
Lanxess Deutschland GmbH
Cologne, Germany

Robert G. Rowland
LanXess Solutions US, Inc.
Naugatuck, Connecticut

Leslie R. Rudnick
Designed Materials LLC
Scottsdale, Arizona

Thomas Rühle
BASF SE
Ludwigshafen, Germany

William R. Schwingel
Masco Corporation
Taylor, Michigan

Luc Séguin
KMK Regulatory Services, Inc.
Quebec, Canada

Shashi Kant Sharma (Retired)
Air Force Research Laboratory
Materials and Manufacturing Directorate
Wright-Patterson Air Force Base Ohio
Ohio

Robert Silverstein
The Orelube Corporation
Bellport, New York

Carl E. Snyder, Jr.
University of Dayton Research Institute
Nonstructural Materials Division
Dayton, Ohio

Joan Souchik
Evonik Oil Additives USA, Inc.
Horsham, Pennsylvania

Daniel M. Vargo
Functional Products Inc.
Macedonia, Ohio

Ronald E. Zielinski
PolyMod® Technologies, Inc.
Fort Wayne, Indiana

Section I

Deposit Control Additives
Oxidation Inhibitors

1 Antioxidants

Robert G. Rowland, Jun Dong, and Cyril A. Migdal

CONTENTS

1.1 INTRODUCTION

Well before the mechanism of hydrocarbon oxidation was thoroughly investigated, researchers had come to understand that some oils provided greater resistance to oxidation than others. The difference was eventually identified as being due to naturally occurring antioxidants, which varied depending upon the nature of the crude oil, and refining techniques. Some of these natural antioxidants were found to contain sulfur- or nitrogen-bearing functional groups. Therefore, it is not surprising that certain additives that are used to impart special properties to the oil, such as sulfur-bearing chemicals, were found to provide additional antioxidant stability. The discovery of sulfurized additives providing oxidation stability was followed by the identification of similar properties with phenols, which led to the development of sulfurized phenols. Next, certain amines and metal salts of phosphorus- or sulfur-containing acids were identified as imparting oxidation stability. Numerous antioxidants for lubricating oils have been patented and described in the literature. Today, nearly all lubricants contain at least one antioxidant. Since oxidation is the primary cause of oil degradation, improving oxidative stability is critical to achieving extended drain intervals.

Oxidation produces various undesired reactive chemical species, which can degrade the lubricant. Further reactions of these oxidation products can lead to changes in oil viscosity, deposit formation, corrosion, and ultimately engine damage.

Lubricant oxidation is initiated by the exposure of hydrocarbons to oxygen and heat. Oxidation can be greatly accelerated by the presence of transition metals such as copper, iron, and nickel. The internal combustion engine is an excellent chemical reactor for catalyzing the process of oxidation with heat and engine metal parts acting as effective oxidation catalysts. Thus, in-service engine oils are probably exposed to more oxidative stress than are encountered in most other lubricant applications. Antioxidant additives protect the lubricant from oxidative degradation, allowing the fluid to meet the demanding requirements for use in engines and industrial applications.

Numerous effective antioxidant classes have been developed over the years and have seen use in engine oils, automatic transmission fluids, gear oils, turbine oils, compressor oils, greases, hydraulic fluids, and metal working fluids. The main classes include oil-soluble organic and organometallic antioxidants of the following types:

- Sulfur compounds
- Sulfur–nitrogen compounds
- Phosphorus compounds
- Sulfur–phosphorus compounds
- Aromatic amine compounds
- Hindered phenolic compounds
- Multifunctional amine and phenol derivatives
- Organocopper compounds
- Boron compounds
- Other organometallic compounds
- Controlled release antioxidants

These additive types will be discussed in the following. However, it is important to remember that while sulfur, phosphorus, and various metal-containing additives can impart antioxidant and antiwear benefits to engine oils, they are often at odds with recent environmental regulations requiring reduction in SO_2, NO_x, and soot emissions. Following the drastic regulatory reductions in the allowable sulfur content of fuels, limits are also set on the amounts of sulfated ash, phosphorus, and sulfur (SAPS) that can be present in an oil. This is done to protect emissions after-treatment devices such as catalytic converters and diesel particulate filters that can be damaged by SAPS [1]. Hence, the trend is toward employing ashless antioxidants, which do not contain sulfur or phosphorus. Various formulations have been proposed to address this problem [2–5].

1.2 SULFUR COMPOUNDS

The initial concepts of using antioxidants to inhibit oil oxidation date back to the 1800s. One of the earliest lubricant inventions described in the literature [6] is the heating of a mineral oil with elemental sulfur to produce a nonoxidizing oil. However, the major drawback to this approach is the high corrosivity of the sulfurized oil toward copper. An engine lubricant with antioxidant and corrosion inhibition characteristics was obtained from sulfurizing terpenes [7]. Paraffin wax has also been employed to prepare sulfur compounds [8–11].

Aromatic sulfides represent another class of sulfur additive used as oxidation and corrosion inhibitors. Examples of simple sulfides are dibenzyl sulfide and dixylyl disulfide. Mono- and dialkyldiphenyl sulfides obtained by reacting diphenyl sulfide with C_{10}–C_{18} α-olefins in the presence of aluminum chloride have been demonstrated to be powerful antioxidants for high-temperature lubricants, especially those utilizing synthetic base oils such as hydrogenated poly-α-olefins (PAOs), diesters, and polyol esters [12]. The hydroxyl groups of the alkyl phenol sulfides may also be treated with metals to form oil-soluble metal phenates. These metal phenates function as both detergents and antioxidants.

Representative structures of several sulfur compounds are illustrated in Figure 1.1. The actual additives can be chemically complex mixtures. Hindered phenols that contain sulfur will be discussed further in Section 1.6.2.

Multifunctional antioxidant and extreme pressure additives with heterocyclic structures were prepared by sulfurizing norbornene, 5-vinylnorbornene dicyclopentadiene, or methyl cyclopentadiene dimer [13]. Heterocyclic compounds such as n-alkyl 2-thiazoline disulfide in combination with ZDDP exhibited excellent antioxidant performance in laboratory engine tests [14]. Heterocyclic sulfur- and oxygen-containing compositions derived from mercaptobenzothiazole and β-thiodialkanol have been found to be excellent antioxidants in automatic transmission fluids [15]. Novel antioxidant and antiwear additives based on dihydrobenzothiophenes have been prepared via condensation of low-cost aryl thiols and carbonyl compounds in a high-yield one-step process [16].

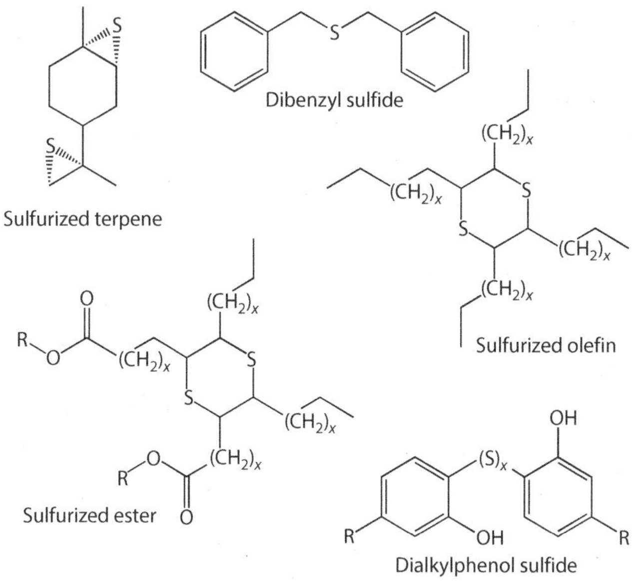

FIGURE 1.1 Examples of antioxidants containing sulfur.

1.3 SULFUR–NITROGEN COMPOUNDS

Dithiocarbamates were first introduced in the early 1940s as fungicides and pesticides [17]. Their potential use as antioxidants for lubricants was not realized until the mid-1960s [18], and since then, there has been continuous interest in this type of chemistry for lubricant applications [19]. Today, dithiocarbamates represent a main class of sulfur–nitrogen-bearing compounds being used as antioxidant, antiwear, and anticorrosion additives for lubricants.

Depending upon the type of adduct to the dithiocarbamate core, ashless and metal-containing dithiocarbamate derivatives can be formed. Typical examples of ashless materials are methylene bis(dialkyldithiocarbamate) and dithiocarbamate esters with general structures being illustrated in Figure 1.2. Both are synergistic with alkylated diphenylamine and organomolybdenum compounds in high-temperature deposit control [20]. In particular, methylene bis(dialkyldithiocarbamate) in combination with primary antioxidants such as arylamines or hindered phenolics and triazole derivatives is known to provide synergistic action in stabilizing mineral oils and synthetic lubricating oils [21–23]. This material has been used to improve antioxidation characteristics of internal combustion engine oils containing low levels (less than 0.1 wt%) of phosphorus [2].

It has been known that metal dithiocarbamates such as zinc, copper, lead, antimony, bismuth, and molybdenum dithiocarbamates possess desirable lubricating characteristics including antiwear and antioxidant properties. The associated metal ions affect the antioxidancy of the additives. Molybdenum dithiocarbamates are the most commercially important of the group and are widely used in engine crankcase lubricants. Certain molybdenum additives, which impart good oxidation resistance and have acceptable corrosion characteristics, are prepared by reacting water, an acidic molybdenum compound, a basic nitrogen complex, and a sulfur source [24,25]. Oil-soluble trinuclear molybdenum dithiocarbamates prepared by reacting ammonium polythiomolybdate with appropriate tetraalkylthiuram disulfides were found to be superior to dinuclear molybdenum compounds in terms of providing lubricant antioxidant, antiwear, and friction-reducing properties [26].

When combined with an appropriate aromatic amine, molybdenum dithiocarbamates can exhibit synergistic antioxidant effects in oxidation tests [27]. As a result, molybdenum dialkyldithiocarbamates (C_7–C_{24}) and alkylated diphenylamines are claimed broadly for lubricating oils [28]. More restrictive are claims for molybdenum dialkyldithiocarbamates (C_8–C_{23} and/or C_3–C_{18}) and alkylated diphenylamines in lubricating oils that contain less than 3 wt% of aromatic content and less than 50 ppm of sulfur and nitrogen [29]. Molybdenum dithiocarbamate was used to top treat engine oils formulated with Group I base oils (>300 ppm S) and an additive package designed for Group II base oils. The oils passed the Sequence IIIF oxidation test, in which the oils would otherwise fail without the molybdenum top treatment [30]. Further demonstrated is a combination of alkylated diphenylamines, sulfurized olefin or hindered phenolic, and oil-soluble molybdenum compounds including molybdenum dithiocarbamate. The mixture is highly effective in stabilizing lubricants, especially those formulated with highly saturated, low-sulfur base oils [31].

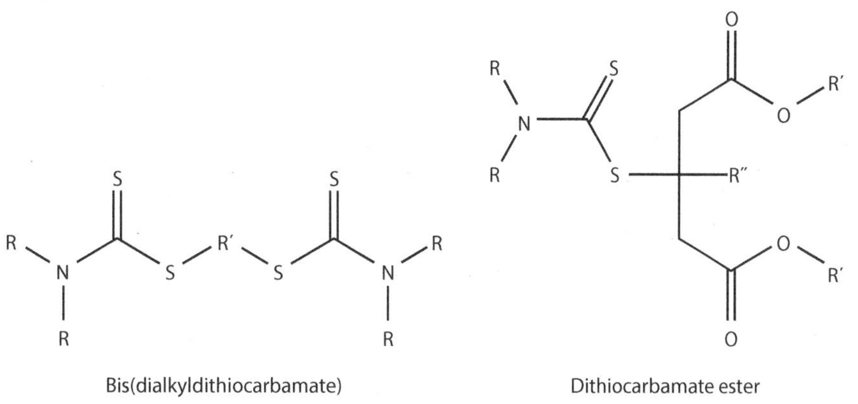

Bis(dialkyldithiocarbamate)

Dithiocarbamate ester

FIGURE 1.2 Ashless dithiocarbamates for lubricants.

Thiadiazole derivatives, particularly the monomers and dimers, represent another class of sulfur- and nitrogen-bearing multifunctional additives with antioxidant activity. For example, a monomeric 2-alkylesterthio-5-mercapto-1,3,4-thiadiazole has been reported to increase the oxidative stability of engine oils under thin-film oxidation conditions as measured by the Thin-Film Oxygen Uptake Test (TFOUT, ASTM D7098) [32]. Lithium 12-hydroxystearate grease containing the dimer 2,5-dithiobis(1,3,4-thiadiazole-2-thiol) exhibited superior oxidative stability in the ASTM D942 pressure vessel oxidation method [33]. When used in conjunction with alkylated diphenylamine and organomolybdenum compound, the thiadiazole derivative improved the Thermo-Oxidation Engine Oil Simulation Test (TEOST®, ASTM D7097) deposits relative to a control engine oil containing sulfurized isobutylene instead [34]. In addition to providing antioxidant benefit, the thiadiazole derivatives have been widely used as ashless antiwear and extreme pressure additives. Some of them can also provide corrosion inhibition and metal deactivation properties to nonferrous metals such as copper.

Alkylated phenothiazines are also well-known antioxidants containing sulfur and nitrogen, which have been used to stabilize ester-based aviation fluids [35] and engine oils [36]. They are usually prepared by sulfurization of alkylated diphenylamines [37] and are frequently used in combination with the parent alkylated diphenylamines [38]. A recent study found that in PAO, dioctylphenothiazine gave better control of total acid number and viscosity than dioctyldiphenylamine, but significantly worse deposits. A synergistic 1:1 mixture of the two components gave satisfactory results in all three tests [39]. N-substituted phenothiazines (especially N-alkylthioethyl) have been claimed as having improved antioxidant activities [40]. Functionalized phenothiazines together with aromatic amines can be attached to olefin copolymers resulting in a multifunctional antioxidant, antiwear agent, and viscosity index improver for lubricants [41].

Diamine sulfides, including diamine polysulfides, can also provide effective oxidation control when used in conjunction with oil-soluble copper. In demonstration, dimorpholine disulfide and di(dimethyl morpholine) disulfide were compared to primary alkyl ZDDP and found to be superior in controlling oil viscosity increase of engine crankcase lubricants at elevated temperatures [42].

1.4 PHOSPHORUS COMPOUNDS

The good performance of phosphorus as an oxidation inhibitor in oils was one of the earliest improvements in the development of additive formulated lubricants. The use of elemental phosphorus to reduce sludge formation in oils was described in 1917 [43]. However, elemental phosphorus, like elemental sulfur, has poor solubility in oil and may have corrosive side effects to many nonferrous metals and alloys, so it is rarely incorporated in oils in this form. Oil-soluble organic compounds of phosphorus are preferred instead. Naturally occurring phosphorus compounds such as lecithin have been utilized as antioxidants, and many patents have been issued on these materials for use alone or in combination with other additives [42–47]. Lecithin is a phosphatide that has been produced commercially as a by-product from the processing of crude soybean oil.

The antioxidant properties of synthetic neutral and acid phosphite esters have been known for some time. Alkyl and aryl phosphites, such as tributyl phosphite and triphenyl phosphite, are efficient antioxidants in some petroleum base oils, and many patents have been issued on such compositions [48,49]. Table 1.1 summarizes the patenting activities of the last 45 years on the stabilization of various lubricants with organophosphites.

TABLE 1.1
Applications of Organophosphites as Antioxidants for Lubricants

Applications	Phosphites	Supplementary Antioxidants	Reference
Compressor oils	Trinonylphenyl phosphite; tributyl phosphite; tridecylphosphite; triphenyl phosphite; trioctylphosphite; dilaurylphosphite	Secondary aminic and hindered phenolic	[53]
Automotive and industrial lubricants	Triaryl phosphites; trialkyl phosphites; alkyl aryl phosphites; acid dialkyl phosphites	Secondary aminic and hindered phenolic	[54]
Automotive and industrial lubricants	Triphenyl phosphite; diisodecyl pentaerythritol diphosphite; tri-isodecyl phosphite; dilauryl phosphite	Secondary aminic and hindered phenolic	[55]
Hydraulic fluids, steam turbine oils, compressor oils, and heat transfer oil	Steric hindered tributyl phosphite; bis(butylphenyl pentaerythritol) diphosphite	(3,5-Di-tert-butyl)4-hydroxybenzyl isocyanurate	[50]
Steam turbine oils, gas turbine oils	Triphenyl phosphite; trialkyl-substituted phenyl phosphite	Alkylated diphenylamine; phenyl-α-naphthylamine	[56]
Hydraulic fluids, automatic transmission fluids	Trialkyl phosphites	Secondary aminic and hindered phenolic including bis-phenol	[57]
Lubricant base oils, polymers	4-Oxo-1,3-dioxa-2-phosphanaphthalenes	Alkylated diphenylamines and hindered phenolics	[58]
Internal combustion engines, steam and gas turbines.	Diisodecyl pentaerythritol diphosphite	Alkylated diphenylamines and/or hindered phenolics	[59]

For optimum antioxidant performance, phosphites are customarily blended with aminic or hindered phenolic antioxidants that can lead to synergistic effect. For better hydrolytic stability, trisubstituted phosphites with sterically hindered structures such as tris(2,4-di-*tert*-butylphenyl) phosphite and those based on pentaerythritol, as described in the U.S. Patent 5,124,057 [50], are preferred. The aluminum, calcium, or barium salts of alkyl phosphoric acids are another type of phosphorus compound that displays antioxidant properties [51,52].

1.5 SULFUR–PHOSPHORUS COMPOUNDS

The most widely used class of sulfur–phosphorus additive is the metal dialkyldithiophosphates. Metal dithiophosphates are typically prepared by the reaction of phosphorus pentasulfide with alcohols to form dithiophosphoric acids, followed by neutralization of the acids with an appropriate metal oxide compound. Most readily available simple C_2–C_{18} alcohols, such as 1-hexanol, 2-ethylhexanol, 2-octanol, 2-butanol, cyclohexanol, phenol, and 4-nonylphenol, have been used in these reactions [60,61]. For the second reaction step, zinc, molybdenum, or barium oxides are usually chosen.

Providing both antioxidant and antiwear/extreme pressure activity, the zinc diorganodithiophosphates [62] (ZDDPs) have collectively been the most important multifunctional lubricant additive since the 1940s. (In historical usage, ZDDP is a broad term that is generally understood to include zinc dialkyldithiophosphates, zinc diaryldithiophosphates, and any alkaryl permutations thereof.) Functioning as peroxide decomposers [1], ZDDPs have been some of the most cost-effective antioxidants and therefore since their discovery have been included as a key component in many oxidation inhibitor packages for engine oils and transmission fluids. While zinc diorganodithiophosphates provide a valuable antioxidant credit, they are far more important to lubricant formulators for the very cost-effective antiwear protection that they offer.

Unfortunately, zinc diorganodithiophosphates can decompose to form volatile phosphorus species, which over time can coat and clog the supporting media in catalytic converters. When phosphorus was first implicated as causing this problem, a great deal of effort was expended toward the goal of entirely removing zinc diorganodithiophosphates from engine oils [1]. Because of concerns about retaining critical antiwear protection, most engine manufacturers have preferred to compromise by using oils with reduced phosphorus levels while designing zinc diorganodithiophosphates that form lesser amounts of volatile phosphorus upon decomposition [60,62,63]. The zinc diorganodithiophosphates prepared from 4-methyl-2-pentanol or 2-ethylhexanol have been identified as being particularly nonvolatile [60].

Many descriptions have recently appeared of organo-molybdenum phosphorodithioate complexes that impart excellent oxidation stability to lubricants. In certain circumstances, oil-soluble molybdenum compounds are preferred additives owing to their multifunctional characteristics such as antiwear, extreme pressure, antioxidant, antipitting, and antifriction properties. For instance, several molybdenum dialkylphosphorodithioate complexes with varying alkyl chains (e.g., pentyl, octyl, 2-ethylhexyl, and isodecyl) were reported to exhibit appreciable antioxidation, antiwear, and antifriction properties [64]. Novel trinuclear molybdenum dialkyldithiophosphates prepared by reacting an ammonium polythiomolybdate and an appropriate bisalkyldithiophosphoric acid possess excellent antioxidant as well as antiwear and friction-reducing properties [26]. Some molybdenum compounds have been used commercially in engine oils and metal working fluids as well as in a variety of industrial and automotive lubricating oils, greases, and specialties [65]. The combination of ZDDP with a molybdenum-containing adduct, prepared by reacting a phosphosulfurized polyisoalkylene or α-olefin with a molybdenum salt, has been described [66]. In this case, the molybdenum adduct alone gave poor performance in oxidation tests, but the mixture with ZDDP provided good oxidation stability. Novel organomolybdenum complexes prepared with vegetable oil have been identified as being synergistic with alkylated diphenylamines and ZDDPs in lubricating oils [67].

Due to increasing concerns about the use of metal dithiophosphates that are related to toxicity, waste disposal, filter clogging, and pollution, etc., there have been extensive research activities on the use of ashless technologies for both industrial and automotive applications. A number of ashless compounds based on derivatives of dialkylphosphorodithioic acids had been reported as multifunctional additives. Upon reacting diisoamylphosphorodithioic acid with various primary and secondary amines, eight alkylamino phosphorodithioates with varying chain length from C_5 to C_{18} were obtained and found to possess excellent antiwear and antioxidant properties as compared to ZDDP [68]. Alkylamino phosphorodithioates obtained from reacting heptylated, octylated, or nonylated phosphorodithioic acids with ethylene diamine, morpholine, or *tert*-alkyl (C_{12}–C_{14}) amines have been demonstrated to impart similar antioxidant and antiwear efficacy and superior hydrolytic stability over ZDDP [69]. Phosphorodithioate ester derivatives containing a hindered phenol moiety are also known to have antioxidant potency. This type of chemistry can be obtained by reacting metal salts of phosphorodithioic acids with hindered phenol halides [70] or with hindered phenol aldehydes [71]. Substituting the phenol aldehydes with hindered cyclic aldehydes, in which the carbon atom attached to the carbonyl carbon contains no hydrogen atoms, may also result in products having excellent antioxidant and thermal stability characteristics [72].

1.6 AMINE AND PHENOL DERIVATIVES

Alkylated aromatic amines and hindered phenols are the most important single-purpose, ashless antioxidants in many lubricant applications today. It is highly unlikely that there is any engine oil, industrial turbine oil, or transmission fluid in use today that meets current industry specifications (SAE, ACEA, JASO, etc.) without containing one or more of these additives.

The introduction of aminic and phenolic antioxidants for engine oils began in the 1930s. In 1932, Standard Oil applied for a patent that claims the use of alkylated phenolics as stabilizers for highly refined and hence readily oxidizable kerosene, mineral, and vegetable oils and fats [73]. In 1938, Gulf claimed dialkyl-cyclohexyl-phenols as stabilizers for hydrocarbon products, including gasoline, lubricating oils, and turbine oils [74]. Around the same time, alkylated aromatic amines were being developed as antioxidants for rubber [75]; and by the end of the decade, the utility of alkylated diphenylamines and phenyl-naphthylamines as stabilizers in rubber, gasoline, and oils was disclosed [76]. Aromatic amines and phenols were reacted together to form stabilizers for "gasoline inhibitors, high pressure lubricants, rubber antioxidants, etc." [77].

With the intervention of World War II, many subsequent early advancements in lubricant antioxidant technology were undoubtedly made under the cloak of military secrecy. B.F. Goodrich received U.S. Patent 2,530,769 in 1950 for a liquid alkylated diphenylamine antioxidant for rubber, fatty oils, and petroleum oils [78]. While utility was only demonstrated for rubber in this patent, after 65 years this type of octylated, styrenated diphenylamine is still used as a lubricant antioxidant in certain markets.

1.6.1 Aromatic Amine Compounds

Alkylated diphenylamines are one of the most important classes of amine antioxidants being used today. While diphenylamine is an effective antioxidant by itself, only alkylated forms are suitable for use in lubricants. Alkylation improves the oil solubility of the fresh antioxidant and highly polar, oxidized products that are produced as the antioxidant is

consumed. Increasing molecular weight reduces additive volatility, which is critical for retention of the antioxidant in the oil. Compared to diphenylamine, alkylated diphenylamines have markedly improved toxicological and environmental profiles.

Figure 1.3 illustrates a commercial route to the preparation of diphenylamine and the typical synthetic routes to some commonly used alkylated diphenylamines. The diphenylamine synthesis reactions start with benzene, which is first converted to nitrobenzene [79], followed by a high-temperature reduction to aniline [80]. Under very high-temperature (400°C–500°C) and high-pressure (50–150 psi) conditions, aniline can undergo a catalytic vapor-phase conversion to form diphenylamine [81].

To make alkylated diphenylamines, diphenylamine is reacted with an appropriate alkylating agent in the presence of an acidic catalyst. The alkylating agent is usually an olefin but can also be an alcohol, alkyl halide, or aliphatic carbonyl compound. From an economic viewpoint, the olefins are generally preferred. The most commonly used olefins are isobutylene (C_4), diisobutylene (C_8), propylene trimer ("nonenes," C_9), and propylene tetramer (C_{12}).

Depending on the acidic catalyst, olefin, stoichiometry, and other reaction conditions, such as time and temperature, the degree of alkylation may progress from mono- to dialkylation, and even some trialkylation may occur. Since the alkyl groups dilute the amount of functional diphenylamine that is present in the additive, overalkylation is to be avoided.

A number of commercial diphenylamine antioxidants have been prepared using diisobutylene as the alkylating olefin. Within a certain mole ratio range, diphenylamine can

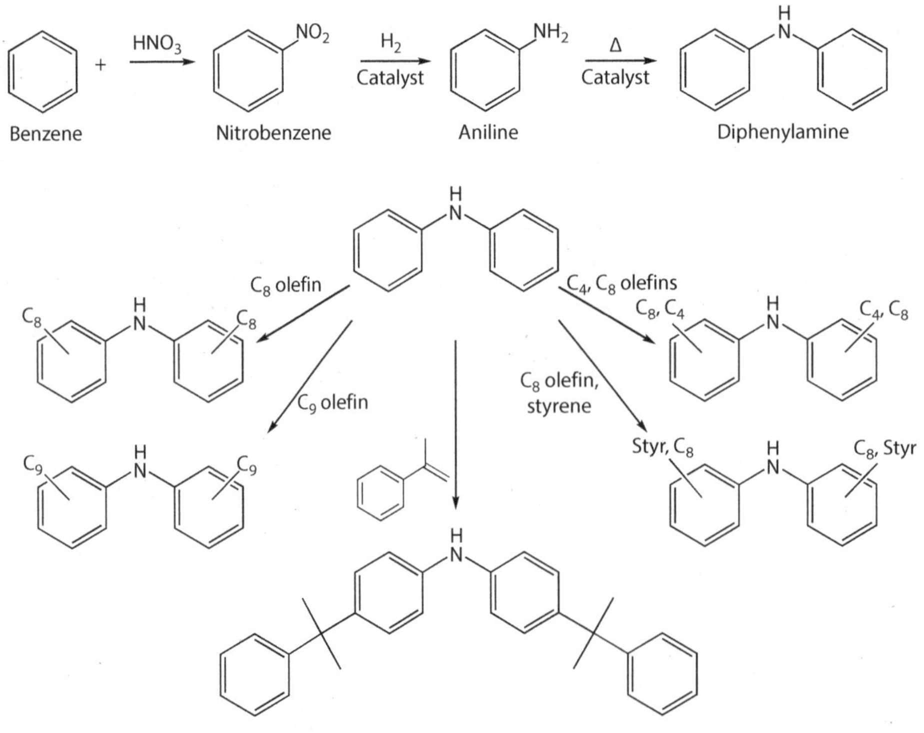

FIGURE 1.3 Synthesis of diphenylamine and some alkylated diphenylamine antioxidants.

be reacted with diisobutylene at a temperature of 160°C or higher to facilitate chain scission of diisobutylene [82]. In the presence of an acid clay catalyst, the resulting product has less than 25% of 4,4'-dioctyldiphenylamine, which yields a liquid at room temperatures. In another process that involves two-step reactions [83], a light-colored liquid product is obtained by first reacting diphenylamine with diisobutylene, followed by reaction with a second more reactive olefin, for example, styrene, α-methylstyrene, or isobutylene. Specific mole ratios, reaction temperatures, and reaction durations are critical to obtain the desired alkylated diphenylamines [84]. U.S. Patent 6,355,839 [85] discloses a one-step process using highly reactive polyisobutylene oligomers having an average molecular weight of about 160–280 Da and at least 25% of 2-methylvinylidene isomers as the alkylating agents to make alkylated diphenylamines and other types of alkylated diarylamine. The resulting products are liquid at ambient temperatures.

Diphenylamine has been alkylated with PAOs of 120–600 Da. These PAOs are a by-product recovered from commercial PAO lubricant manufacture [86].

It was found that monosubstituted diphenylamines more readily oligomerize under a variety of conditions to produce higher-molecular-weight linear oligomers. Oligomers with 2–10 degrees of polymerization are desirable antioxidants especially for high-temperature applications. Disubstituted and polysubstituted diphenylamines, on the other hand, are more restricted from forming oligomers higher than dimers. Oligomeric versions of monosubstituted diphenylamine prepared from reacting diphenylamine with C_4–C_{16} olefins have been described for use in ester lubricants [87]. The products are claimed to be more effective than simple diphenylamines for extremely high-temperature applications.

Substituted benzylamines or substituted 1-amino-1,2,3,4-tetrahydronaphthalene is particularly useful for synthetic lubricants such as PAOs or polyol esters. Oils bearing these additives demonstrate very low metal corrosion, low viscosity increase, and low sludge buildup [88]. Closely related mixed isomers of di(1,2,3,4-tetrahydro-naphthyl)amines are effective antioxidants for engine oils in the TEOST MHT™ (ASTM D7097) and for turbine oils in the Rotating Pressure Vessel Oxidation Test (RPVOT, ASTM D2272) [89].

A group of acridines, where one of the aromatic rings bears a fused ring substituent, have been claimed as antioxidants for lubricating oils [90]. Another group of novel substituted diphenylamines, where each phenyl ring bears a fused ring substituent, and the nitrogen is substituted with an allyl, benzyl, or methallyl group, have also been claimed as lubricant antioxidants [91]. Further improved antioxidant performance has been claimed when these compounds are used in combination with benzeneamines such as dialkyl anilines or anthranilic acid esters [92].

Alkylated phenyl-naphthylamines are another very important group of aminic antioxidants. While both the α- and β-isomers are effective antioxidants, only the phenyl-α-naphthylamine (PANA) derivatives are in use today, as the intermediate β-naphthylamine is a known carcinogen.

Alkylated PANAs are significantly more expensive than alkylated diphenylamines. Therefore, PANAs are used primarily in particularly demanding high-temperature applications such as aviation lubricants [93,94].

Homo-oligomers of alkylated (C_4–C_8) diphenylamines, styrenated diphenylamines, or cross-oligomers of the alkylated diphenylamines with substituted PANA are claimed to possess superior antioxidant efficacy in synthetic ester lubricants for high-temperature applications [95]. Complex substituted PANA reaction products have been claimed to offer superior lubricant antioxidant performance over various closely related commercial alkylated diphenylamines and PANAs. The claimed products include materials prepared by using mixtures of propylene trimer with either α-methylstyrene or styrene to alkylate PANA, or a mixture of PANA and DPA [96].

A product made by reacting a polyalkenylsuccinic acid or anhydride first with an aromatic secondary amine and then with an alkanol amine was found to provide appreciable antioxidancy, dispersancy, and anticorrosion effects to engine oils as tested in a Caterpillar engine test [97]. A more recent U.S. Patent [98] discloses materials made from the reaction of alkyl or alkenyl succinic acid derivative with a diaminonaphthyl compound for use as antioxidant, antiwear, and soot dispersing agents for lubricating oils. U.S. Patent 5,075,383 [99] describes novel antioxidant–dispersant additives obtained by reacting amino-aromatic polyamine compound, including aromatic secondary amines, with ethylene-propylene copolymer grafted with maleic anhydride. Engine oils containing the additives displayed improved performance characteristics in laboratory oxidation and sludge dispersancy tests as well as in the Sequence VE engine test.

Cross-products and co-oligomers of N-alkylated p-phenylenediamines with N-phenyl-α-naphthylamine have been claimed as antioxidants for liquid hydrocarbons and polyol esters [100]. Oligomeric products derived from thermal and chemical condensation of alkylated diphenylamine and alkylated PANA in the presence of aldehyde can provide high-performance and nonsludging attributes, as evident in the RPVOT, and in the ASTM D4310 sludging tendency test designed for turbine oils [101]. An antioxidant for ester-based conveyor chain lubricants has been claimed for oligomers of alkylated diphenylamines, alkylated PANAs, and their cross-products. The preferred aromatic amines are di-t-octyl-diphenylamine and 4-t-octyl-phenyl-α-naphthylamine [102]. The oligomerization is preferably carried out using Braid's potassium permanganate method [103].

Diarylamines containing a tetrahydroquinoline, benzofuran, or benzodioxin moiety have been used as antioxidants in a fully formulated Group II⁺ oil. The N-aryl-N-tetrahydroquinolinamines gave the best performance in a bulk oil bench oxidation test [104]. Synergy was observed between one of these compounds and an alkylated diphenylamine [105].

A number of 4-nitro-4'-alkyldiphenylamines demonstrate significant antioxidant activity when used at 1.0–1.5 wt% in combination with 0.5 wt% alkylated diarylamines. The nitro compounds are inactive when used alone [106].

Aromatic diamines are a broad group of aminic antioxidants suitable for lubricants and fuels. *N,N'*-diphenyl-*p*-phenylenediamines where the phenyl groups may be substituted with methyl, ethyl, or methoxy groups have been claimed as effective antioxidants [107]. A broader range of substituted *p*-phenylenediamines have been claimed for crankcase lubricating oils for use in environments where iron-catalyzed oxidation reactions can take place [108]. Phenylenediamines have been claimed alone and in combination with other ashless antioxidants as a means of reducing soot-induced viscosity increase in diesel engines equipped with exhaust gas recycle (EGR) [109]. Substantial improvement in the pressurized differential scanning calorimetry (PDSC) oxidation induction time (OIT) was noted for gas-to-liquid (GTL) and PAO base oils stabilized with *N,N'*-bis(2,6-diisopropylphenyl)-*p*-phenylenediamine [110] and *N,N'*-bis(aryl)-*ortho*-phenylenediamines [111].

Michael reaction adducts of phenylenediamines [112] and tetraaryl hydrazines (Ar₁Ar₂N-N'Ar₃Ar₄) [113] gave good deposit control in fully formulated crankcase oils in the TEOST MHT test. Novel hydrazide antioxidants were prepared by treating phenylenediamine Michael adducts with hydrazine. Hydrazides were further maleated to prepare multifunctional dispersants [114].

Dispersants have been prepared incorporating aromatic amines, which might be expected to provide an antioxidant credit. Several patents report improvements in soot control and/or dispersancy, which may be due in part to better oxidation control [115–117]. However, a specific improvement in antioxidancy (PDSC OIT) is reported only in U.S. Patent 8,324,139, where the grafted amine is *p*-phenylenediamine [116].

Various 3,5-diethyltoluenediamines with the amino moieties being located on the 2,4- and 2,6-positions relative to the methyl group have been claimed to be effective in the prevention of oil viscosity increase and acid buildup [118]. The additives are relatively noncorrosive to copper and lead bearings and are compatible with seals at high temperatures and pressures. Tetraalkyl-naphthalene-1,8-diamines have been used in combination with alkylated diphenylamines or phenylnaphthylamines as a lubricant antioxidant [119].

Some 2,3-dihydroperimidines can be prepared from the condensation of 1,8-diaminonaphthalenes with ketones or aldehydes and show good oxidation inhibition in the RPVOT (ASTM D2272). Synergistic behavior of the amines was also observed when an appropriate phenolic antioxidant is present [120]. Oils containing *N,N'*-disubstituted-2,4-diaminodiphenyl ethers and imines of the same ethers have shown low viscosity increase, low acid buildup, and reduced metal corrosion in bench tests [121,122].

Several antioxidant patents have been issued based on alkylated benzotriazole compounds. This class of antioxidant also has additional activity in the reduction of copper corrosion. Examples are *N-t*-alkylated benzotriazoles obtained by reacting a benzotriazole with an olefin such as diisobutylene [123] and the reaction products of a benzotriazole with an alkyl vinyl ether or a vinyl ester of a carboxylic acid such as vinyl acetate [124]. Antioxidant and antiwear properties were reported for benzotriazole adducts of an amine phosphate [125] or of organophosphorodithioate [126]. The former type also exhibited rust prevention characteristics in the ASTM D665 corrosion test. The reaction product of a hydrocarbyl succinic anhydride and 5-amino-triazole demonstrated antioxidant efficacy in a railway diesel oil composition [127].

1.6.2 Hindered Phenolic Compounds

Phenols, especially the sterically hindered phenols, are another class of antioxidants that are extensively used in industrial and automotive lubricating oils and greases. Based on chemical structure, phenols may be customarily categorized into simple phenols such as 2,6-di-*tert*-butyl-4-methylphenol (also known as BHT) and complex phenols that are typically in polymeric forms having molecular weights of 1000 or higher. The structures, important physical properties, and typical applications of some commonly used hindered phenols are given in Table 1.2.

Hindered phenols that incorporate sulfur atoms have been widely reported. The simplest are alkyl phenol sulfides. Alkyl phenols, such as mono- or dialkylated butyl-, pentyl-, or octylphenol, have been reacted with sulfur mono- or dichloride to form either mono- or disulfides [13,128–132]. The structure of a commercial sulfur-containing hindered phenol that has found use in various lubricant formulations is shown in Figure 1.4.

The reaction products of simple phenols, such as 2,6-di-*tert*-butylphenol, with selected thioalkenes have shown effectiveness in the prevention of acid buildup and oil viscosity increase, without causing lead corrosion [133]. Another patent describes a process for preparing alkylthio-substituted hindered phenols by reacting substituted phenols with hydrocarbyl disulfides using an aluminum phenoxide catalyst [134]. Using a 4,4'-methylene bis(2,6-di-tertiarybutylphenol) as reference, the (alkythio)phenols were found to be superior in a bulk oil oxidation tests and bench corrosion test on bearings. Reaction products prepared by sulfurizing mixtures of mono- and dialkylated *sec*- and *tert*-butyl phenols [135] and closely related high oligomeric phenolic antioxidants have been developed [136]. These compounds have lower volatility, better thermal stability, and improved seal compatibility and corrosion properties.

In general, sulfur-bridged hindered phenolics are more effective than the conventional phenolics under high-temperature oxidation conditions and are considered particularly suitable for the lubricants formulated with highly refined base oils [137]. Thioalkene-bridged hemi-hindered phenols prepared from the reaction of hindered phenols with thioalkenes have also been reported to be active in the stabilization of mineral oils and synthetic oils [133]. A number of antioxidants have been prepared, which combine a hindered phenol moiety with an ester that contains a sulfide linkage either β- to the phenolic ring or in the alkyl side chain [138,139].

Other hydroxybenzene compounds have been investigated recently. Hydroxychromans, including the general (Markush) structure illustrated in Figure 1.5, have been claimed as lubricant antioxidants [140].

TABLE 1.2

Structure, Physical Properties, and Typical Applications of Commercial Hindered Phenols for Lubricants

Phenols	Structure	Melting Point (°C)	Solubility (Minerals Oils)	Applications
2,6-Di-*tert*-butyl-phenol	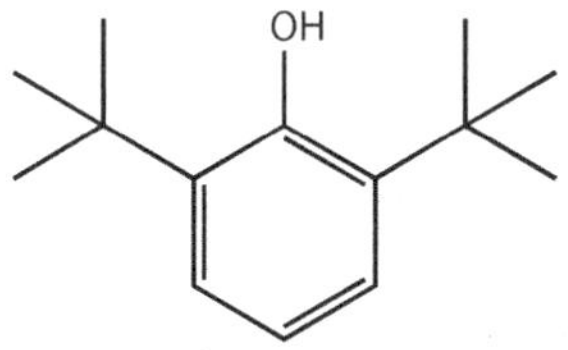	36–37	>5	Industrial oils, power transmission fluids, greases, fuels
2,6-Di-*tert*-butyl-4-methylphenol (BHT)	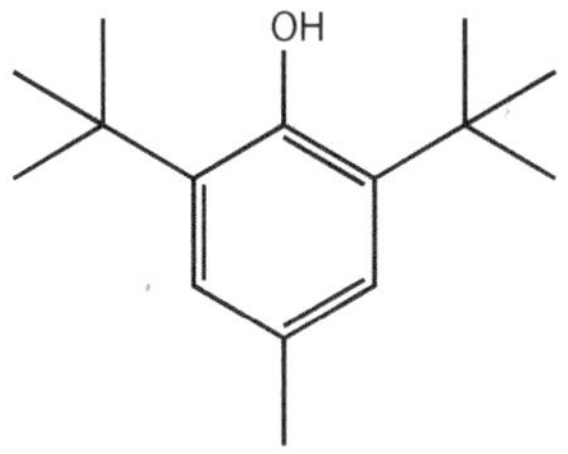	69	>5	Industrial oils, power transmission fluids, food-grade lubricants, greases
Tetrakis[methylene (3,5-di-*tert*-butyl-4-hydroxyphenyl) propionate] methane	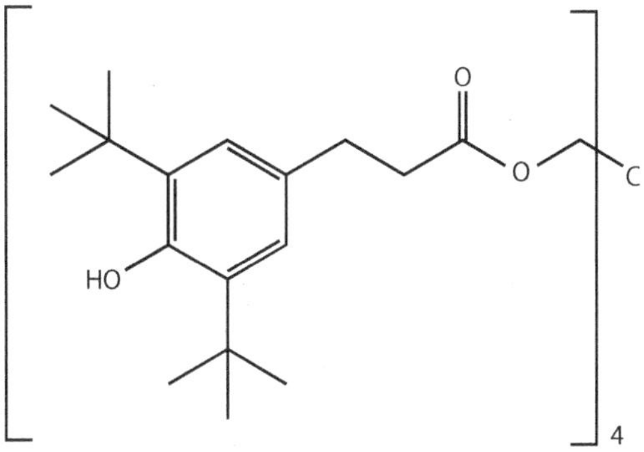	110–125	0.1–0.5	Industrial oils, food-grade lubricants, greases
Octadecyl-3-(3,5-di-*tert*-butyl-4-hydroxyphenyl) propionate	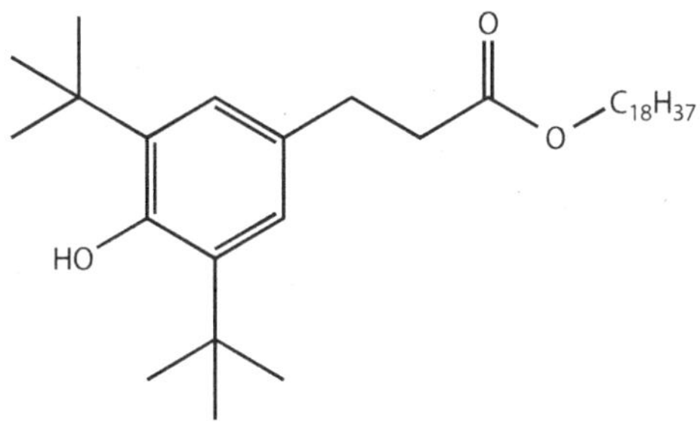	50–55	>5	Industrial oils, eco-friendly oils
3,5-Di-*tert*-butyl-4-hydroxy-hydrocinnamic acid, C_7–C_9 alkyl ester	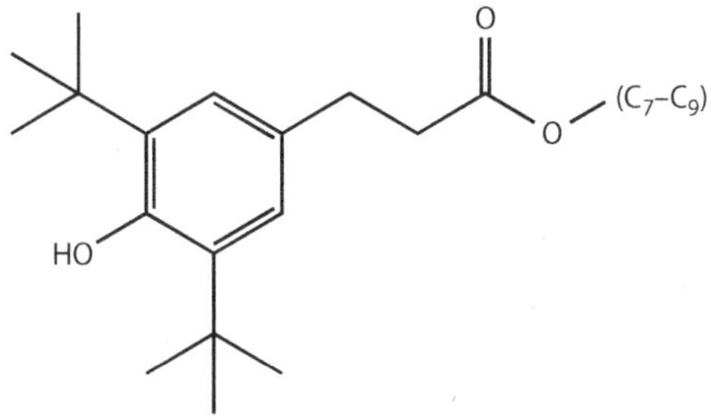	Liquid at 25°C	>5	Industrial oils, power transmission fluids, greases, fuels

(Continued)

TABLE 1.2 (*Continued*)
Structure, Physical Properties, and Typical Applications of Commercial Hindered Phenols for Lubricants

Phenols	Structure	Melting Point (°C)	Solubility (Minerals Oils)	Applications
3,5-Di-*tert*-butyl-4-hydroxy-hydrocinnamic acid, C_{13}–C_{15} alkyl ester	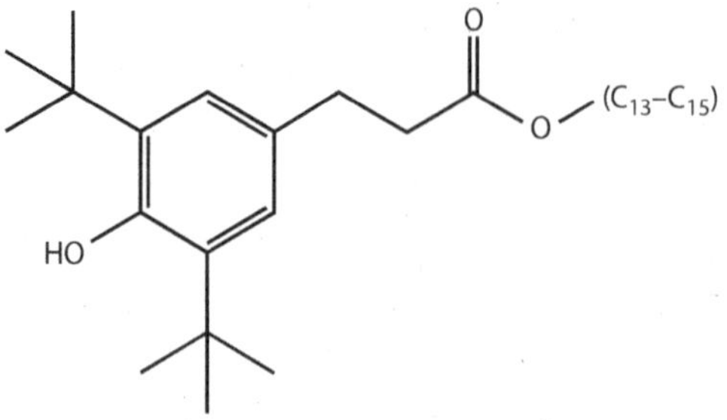	Liquid at 25°C	>5	Industrial oils, power transmission fluids, food-grade lubricants, greases
4,4′-Methylene bis(2,6-di-*tert*-butylphenol)	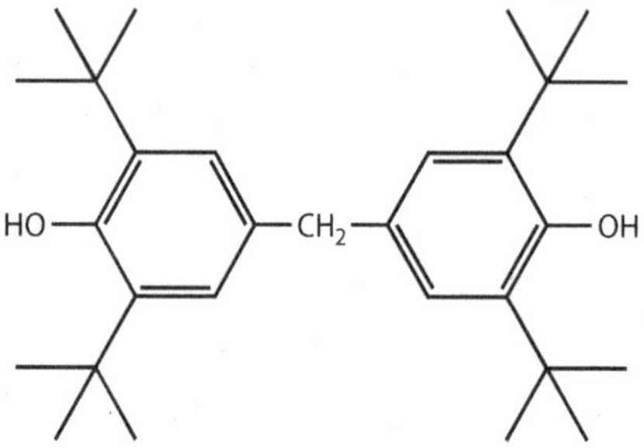	154	NA	Industrial oils, food-grade lubricants, greases
2,2′-Methylene bis(4-methyl-6-*tert*-butylphenol)	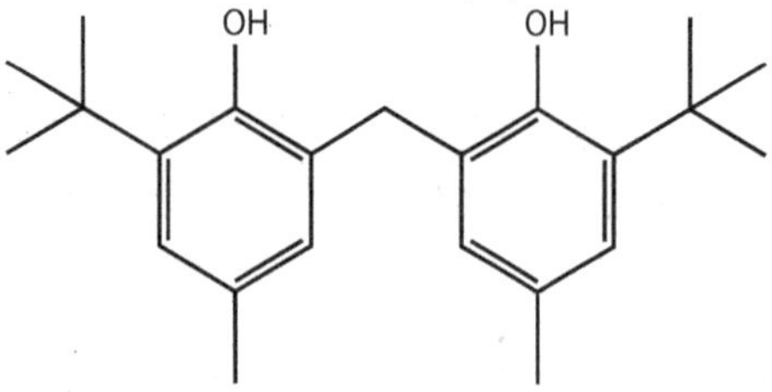	128	2–5	Industrial oils, eco-friendly oils
2-Propenoic acid, 3-[3,5-bis(1,1-dimethylethyl)-4-hydroxyphenyl]-1,6-hexanediyl ester	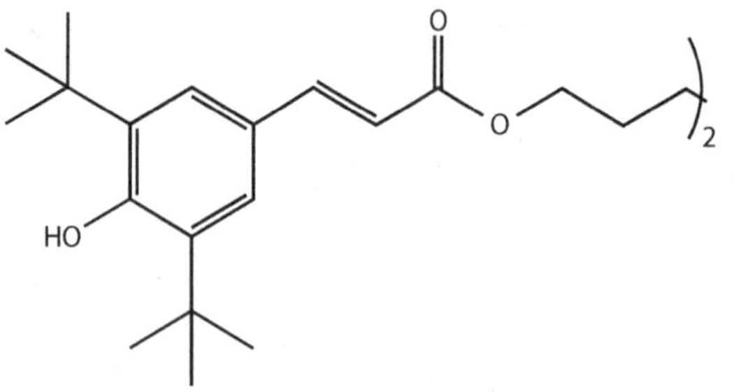	105	<1	Industrial oils, greases, food-grade lubricants

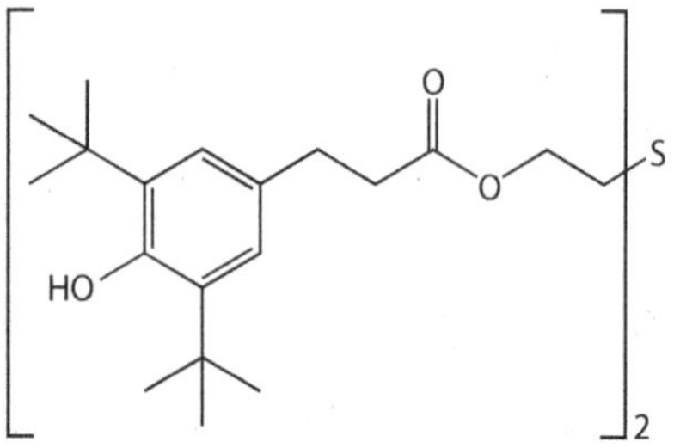

FIGURE 1.4 A commercial sulfur-bridged phenolic antioxidant used in lubricants.

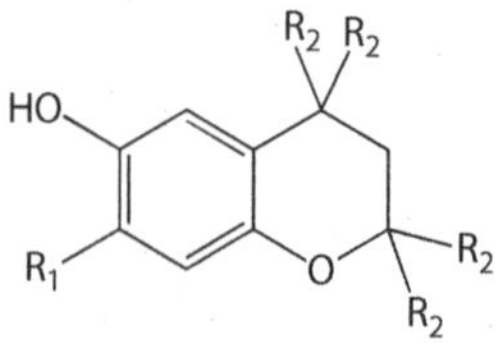

FIGURE 1.5 A generalized hydroxychroman antioxidant.

A group of C_{15}-substituted resorcinols derived from cashew nut shell liquid (CNSL) have been found to be effective antioxidants in crankcase formulations. The result is particularly intriguing, since the resorcinols are not hindered by substitution at the 2- and 6-positions [141]. A similar material, obtained by distillation of CNSL, followed by hydrogenation of the distillate, has been used as a stabilizer for trunk piston marine lubricants. When used in combination with overbased calcium salicylates, the composition also disperses asphaltene contaminants in the lubricant coming from the heavy fuel [142]. *Note*: CNSL has been reported in the literature as causing contact dermatitis [143].

1.6.3 Multifunctional Amine and Phenol Derivatives

Given the effectiveness of both amine and phenol derivatives as lubricant antioxidants and the synergy between them (see Section 1.12), combining both functionalities in one molecule is a logical way to achieve synergistic performance. Ingold and Pratt have recently challenged the merits of this approach based on both net antioxidant costs and the likelihood that the relative stoichiometry of the two functionalities will be less than optimal [144].

The Mannich reaction was used to couple an ethoxylated alkyl phenol with an alkyl arylamine. The resulting products had markedly improved antioxidant activity relative to an equal weight loading of the amine alone and were noted as having improved solubility in oils [145]. Phenolic imidazolines have been prepared from polyaminophenols and carbonyl compounds [146]. In addition to providing antioxidant activity, the products also have corrosion inhibition and metal deactivation properties owing to the cyclic imidazoline moiety.

Multifunctional additives containing sulfur, nitrogen, and phenolic moieties in one molecule have been reported. In one instance, mercaptobenzothiazoles or thiadiazoles are coupled by a Mannich reaction with hindered phenolic antioxidants to yield oil-soluble compounds with antioxidant and antiwear properties [147]. A more complex product having similar functionalities was obtained by reacting a sulfur-containing hindered phenolic ester with an alkylated diphenylamine [148]. Reaction products of polyisobutenyl succinimide dispersants with hindered cinnamic acid esters have been claimed as dispersants with antioxidant functionality. The product reduced the formation of sludge and varnish deposits in the Sequence VG test (ASTM D6593) [149].

1.7 ORGANOCOPPER ANTIOXIDANTS

The ability of copper compounds to function as oxidation inhibitors has been of interest to the lubricant industry for years. Copper is usually considered to be an oxidation promoter and its presence is of a concern in lubricants such as power transmission oils where fluid contact with copper-containing bearings, sintered bronze clutch plates, etc., takes place [150]. It has been suggested that copper corrosion products, originating from the surface attack of copper metal, are generally catalysts that accelerate the rate of oxidation [151], while oil-soluble copper salts are antioxidants [152]. To maximize the full antioxidant strength of a copper compound, the initial concentration needs to be maintained at an optimum range, normally from 100 to 200 ppm [150,152]. Below this range, the antioxidant effect of the copper compounds will not be fully realized, while above the range, interference with antiwear additives may occur, leading to pronounced increase in wear on high-stress contact points [153].

In early work, certain types of unsaturated hydrocarbons were sulfurized in the presence of copper. This produced a group of copper–sulfur complexes that are examples of oil-soluble copper antioxidants [154–156]. A more recent patent describes lubricant compositions that are stabilized with a zinc dihydrocarbyldithiophosphate (ZDDP) and 60–200 ppm of copper derived from oil-soluble copper compounds such as copper dihydrocarbyldithiophosphate or copper dithiocarbamates [153]. Oxidation data are given for fully formulated engine oils containing the ZDDP and a variety of supplemental antioxidants including amines, phenolics, a second ZDDP, and the copper salts. Only the blends with the copper salts passed the oxidation test. The viscosity increase was excessive with the other additives. Organocopper compounds including copper naphthenates, oleates, stearates, and polyisobutylene succinic anhydrides have been reported to be synergistic with multi-ring aromatic compounds in controlling high-temperature deposit formation in synthetic base oils [152]. Oil-soluble copper compounds such as copper oleates, naphthenates, dithiocarbamates, or thiophosphates have also been reported to be synergistic with phenothiazines in controlling viscosity increase in lubricating oils [157].

More complex compounds obtained from further reactions of copper salts have also been reported to be effective antioxidants in a variety of lubrication applications. For example, copper carboxylate or copper thiocyanate was reacted with a mono-oxazoline, bis-oxazoline, or lactone oxazoline dispersant to form coordination complexes wherein the nitrogen contained in the oxazoline moiety is the ligand that complexes with copper. The resulting products exhibit improved varnish control and oxidation inhibition capabilities [158]. Reaction products of a copper salt (acetate, carbonate, or hydroxide) with a substituted succinic anhydride derivative containing at least one free carboxylic acid group are effective high-temperature antioxidants and friction modifiers. When incorporated in an engine oil formulation, the oil passed rust, oxidation, and bearing corrosion engine tests [159]. In another patent [160], a hindered phenolic carboxylic acid was used as the coupling reagent. The resulting copper compounds are reported to be effective in the control of high-temperature sludge formation and oil viscosity increase when used alone or in synergistic mixtures with a conventional aminic or phenolic antioxidant.

1.8 BORON ANTIOXIDANTS

The search for more eco-friendly additives to replace ZDDP has led to renewed interest in boron esters owing to their ability to improve antiwear, antifriction, and oxidative stability properties of lubricants when used alone or in combination with other additives. A number of boron–oxygen-bearing compounds have been reported to be effective oxidation inhibitors in terms of the prevention of oil viscosity increase and acid formation at elevated temperature (163°C) [161–165]. Examples of these compounds include boron epoxides (especially 1,2-epoxyhexadecane) [161], borated single and mixed alkanediols [162], mixed hydroquinone-hydroxyester borates [159], phenol esters of hindered phenyl borates [164], and reaction products of boric acid with the condensates of phenols with aromatic or aliphatic aldehydes [165]. Appreciable oxidation inhibition effect has also been reported for borate esters of hydrocarbyl imidazolines [166], borates of mixed ethoxyamines and ethoxyamides [167], and borates of ether-diamines [168].

Borate esters with nitrogen (boron amides) are known for their improved antiwear properties, which are probably due to the formation of boron nitride films on the rubbing surfaces [169]. Borated adducts of alkyl diamines with long-chain hydrocarbylene alkoxides and low-molecular-weight carboxylic acids have been reported to have antifriction and antioxidant properties at elevated temperatures [170]. The treatment of inherently basic alkylamines with reactive boron compounds to form boron amides is known to improve the compatibility of the aminic compounds with engine seals [171,172].

Synergistic antioxidant effects of borate esters with alkylated diphenylamines or with zinc dithiophosphates have been established. When tested at 180°C in a PAO using a PDSC, strong synergistic antioxidant action was observed between borate esters and a dioctyldiphenylamine at a 1:1 (w/w) blending ratio [173]. A similar effect was observed for mixtures of borate esters and a ZDDP [174]. The synergy with ZDDP is of practical importance as it allows reduction in the phosphorus level in a finished lubricant without sacrifice of oxidative stability. The catalytic effect of boron in enhancing antioxidant performance has led to the development of phenolic-phosphorodithioate borates, obtained by borating a hindered phenol together with an alkyl phosphorodithioate-derived alcohol. The borates were found to possess exceptional antioxidant and antiwear activities. The synergy was proposed to arise from the formation of borate esters with both the hindered phenolic moiety and the phosphorodithioate alcohol within a single molecule [175].

Despite the many tribological and antioxidation benefits borate esters can offer, their use in lubricant applications has been limited. One serious drawback with most borate esters has been their high susceptibility to hydrolysis, a process that produces oil-insoluble and abrasive boric acid. The following approaches have been taken to address the issue:

1. Incorporation of hindered phenolic moiety to sterically inhibit the boron–oxygen bonds from hydrolytic attack. Commonly used hindered phenolics are 2,6-dialkyl phenols [176], 2,2′-thiobis(alkylphenols), and thiobis(alkylnaphthols) [177].

2. Incorporation of amines that have nonbonding pairs of electrons. The amines coordinate with the electron-deficient boron atom, thus preventing hydrolysis. U.S. Patents 4,975,211 [178] and 5,061,390 [179] disclose the stabilization of borated alkyl catechol against hydrolysis by forming a complex with diethylamine. Significant improvement in hydrolytic stability was reported for borate esters, which incorporate an N,N'-dialkylamino-ethyl moiety [180]. It was hypothesized that the formation of a stable intramolecular five-member ring structure involving coordination of nitrogen with boron substantially inhibited hydrolytic attack by water.

3. Use of certain hydrocarbon diols or tertiary amine diols to react with boric acid to form stable five-member ring structures [181].

1.9 MISCELLANEOUS ORGANOMETALLIC COMPOUNDS

A number of oil-soluble titanium, zirconium, and manganese organometallic compounds have been claimed to be effective stabilizers for lubricants [182,183]. These materials include organic acid salts, amine salts, oxygenates, phenates, and sulfonates. Some of the compounds are essentially devoid of sulfur and phosphorus and therefore candidates for use in low-SAPS automotive engine oils. In one example [182], lubricating oils having 25 to ~100 ppm of titanium (derived from titanium (IV) isopropoxide) exhibited excellent oxidative stability in the high-temperature (280°C) Komatsu hot tube test and in the ASTM D6618 test that evaluates engine oils for ring sticking, ring and cylinder wear, and the accumulation of piston deposits in a four-stroke cycle diesel engine. In another example [183], titanium (IV) isopropoxide was reacted with neodecanoic acid, glycerol mono-oleate, or polyisobutenyl bis-succinimide to form the respective titanium compounds. These compounds, when used as a top treatment in a SAE 5W-30 engine oil so as to result in 50 to about 800 ppm of titanium, improved the deposit control capability of the oil as tested by the mid-high-temperature Thermo-Oxidation Engine Oil Simulation Test method (TEOST MHT, ASTM D7097). Similar antioxidant effects were observed for zirconium and manganese neodecanoates in the same oil. Synergistic antioxidant performance, as measured by an extended OIT, was observed in the PDSC test when titanium isopropoxide was used in combination with a hindered amine {bis(1,2,2,6,6-pentamethyl-1-piperidinyl)sebacate}. The same patent also claims antioxidant formulations comprising similar hindered amines and molybdenum, tungsten, or boron compounds [184].

Amine tungstates and tungsten dithiocarbamates are soluble or dispersible in oil. They have been investigated as lubricant antioxidants and found to be synergistic with secondary diarylamines and/or alkylated phenothiazines. The mixtures are highly effective in controlling oil oxidation and deposit formation when they are added to an engine crankcase lubricant in an amount that results in ~20–1000 ppm of tungsten [185].

A formulation comprising an alkylated diphenylamine or an alkylated phenothiazine antioxidant and an ammonium tungstate gave improved PDSC performance. The ammonium tungstate can be prepared from a polyisobutylene mono-succinimide dispersant or dialkylamines [186]. This work has been extended to include organomolybdenum compounds in the formulation [187].

Sulfur-free molybdenum salts, such as molybdenum carboxylates, have been used as antioxidants and found to be synergistic with alkylated diphenylamines in lubricating oils [188]. The synergistic mixtures improved the oxidation stability of crankcase lubricants while providing additional friction modification characteristics. A sulfur- and phosphorus-free fatty acid/molybdenum complex demonstrated antioxidant synergy with di-*tert*-octyl diphenylamine in PAO, as measured by PDSC and Modified Penn State Micro-Oxidation Test [189].

1.10 CONTROLLED RELEASE ANTIOXIDANTS

As new specifications continue to extend drain intervals for lubricants, formulators want to guarantee that all additives are present at sufficient levels throughout the service life. Once the antioxidants in an oil have been exhausted, potentially catastrophic viscosity increase will occur very quickly. A recent approach to addressing this problem involves incorporating the additives into a controlled release carrier so that fresh additive becomes available across the service life.

An additive gel containing dispersant/detergent and antioxidant has been prepared. However, the additive gel must still be introduced into the lubrication system separately from the oil. This may be done through the use of a special gel-charged oil filter or various types of metering addition devices [190,191]. Another approach involves dispersing 0.01–50 μm microcapsules containing a polar core and polar additive(s) in an oil. The patent is prophetic for the use of antioxidants as microencapsulated additives [192].

1.11 MECHANISMS OF HYDROCARBON OXIDATION AND ANTIOXIDANT ACTION

Hydrocarbon-based lubricants undergo an oxidative process called autoxidation, which leads to the formation of acids and oil thickening. As the oil becomes increasingly degraded, oil-insoluble sludge, deposits, and varnish may be formed, causing poor lubrication, reduced fuel economy, and increased wear. Antioxidants are essential additives incorporated in lubricant formulations to delay the formation of oxidation products and keep the oil within specifications throughout the service life. Mechanisms of lubricant degradation and the stabilization of oils by antioxidants are discussed in the following.

1.11.1 Autoxidation of Lubricating Oil

The well-documented autoxidation mechanism involves a free-radical chain reaction [193–195]. It consists of four distinct reaction steps: chain initiation, propagation, branching, and termination.

1.11.1.1 Initiation

$$RH + O_2 \rightarrow R\cdot + HOO\cdot \qquad (1.1)$$

$$R-R + Energy \rightarrow R\cdot + R\cdot \qquad (1.2)$$

The initiation step is characterized as the formation of free alkyl radicals (R·) from the breakdown of hydrocarbon bonds by hydrogen abstraction and hemolytic cleavage of carbon–carbon bonds. These reactions take place when hydrocarbons are exposed to oxygen and/or energy in the form of heat, UV light, or mechanical shear stress [196]. The ease of homolytic cleavage of an R–H bond is determined by the C–H bond strength and the stability of the resulting radical. Empirically, the relative rates of hydrogen atom abstraction from carbon have been found to be benzylic > allylic > tertiary > secondary > primary > phenyl [197,198]. Thus, hydrocarbons containing tertiary hydrogen or hydrogen in an α-position to a carbon–carbon double bond or aromatic ring are most susceptible to oxidation. The reaction rate of chain initiation is generally slow under ambient conditions but can be greatly accelerated with temperature and catalyzed by the presence of transition metals (copper, iron, nickel, vanadium, manganese, cobalt, etc.).

1.11.1.2 Chain Propagation

$$R\cdot + O_2 \rightarrow ROO\cdot \qquad (1.3)$$

$$ROO\cdot + RH \rightarrow ROOH + R\cdot \qquad (1.4)$$

The first propagation step involves an alkyl radical reacting irreversibly with oxygen to form an alkyl peroxyl radical (ROO·). This reaction is extremely fast and the specific rate is dependent upon the radical's substituents [193]. Once formed, the peroxyl radical can readily abstract hydrogen from another hydrocarbon molecule to form a hydroperoxide (ROOH) and a new alkyl radical (R·). Therefore, each time a free alkyl radical is formed, a large number of hydrocarbon molecules may be oxidized to hydroperoxides.

1.11.1.3 Chain Branching
1.11.1.3.1 Radical Formation

$$ROOH \rightarrow RO\cdot + HO\cdot \qquad (1.5)$$

$$RO\cdot + RH \rightarrow ROH + R\cdot \qquad (1.6)$$

$$HO\cdot + RH \rightarrow H_2O + R\cdot \qquad (1.7)$$

1.11.1.3.2 Aldehyde or Ketone Formation

$$RR'HCO\cdot \rightarrow RCHO + R'\cdot \qquad (1.8)$$

$$RR'R''CO\cdot \rightarrow RR'CO + R''\cdot \qquad (1.9)$$

The chain branching steps begin with the cleavage of a hydroperoxide into an alkoxy radical (RO·) and a hydroxy radical (HO·). This reaction has a high activation energy and is only significant at temperatures greater than 150°C. Catalytic metal ions accelerate the process. The resulting radicals may undergo a number of possible reactions: (1) the alkoxy radical abstracts a hydrogen from a hydrocarbon to form a molecule of alcohol and a new alkyl radical according to Reaction 1.6, (2) the hydroxy radical follows the pathway of Reaction 1.7 to abstract hydrogen from a hydrocarbon molecule to form water and a new alkyl radical, (3) a secondary alkoxy radical (RR′HCO·) may decompose via reaction pathway (1.8) to form an aldehyde, and (4) a tertiary alkoxy radical (RR′R″CO·) may decompose to form a ketone (Reaction 1.9).

The chain branching reactions are very important to the subsequent condition of the oil. As a large number of alkyl radicals are formed, the oxidation process accelerates rapidly. The lower-molecular-weight alcohols, aldehydes, and ketones generated will immediately affect the physical properties of the lubricant by decreasing oil viscosity and increasing oil volatility and polarity. Under high-temperature oxidation conditions, the aldehydes and ketones can undergo further reactions to form acids and high molecular weight species that thicken the oil and contribute to the formation of sludge and varnish deposits.

1.11.1.4 Chain Termination

$$R \cdot + R' \cdot \rightarrow R - R' \qquad (1.10)$$

$$R \cdot + R'OO \cdot \rightarrow ROOR' \qquad (1.11)$$

As oxidation proceeds, oil viscosity will increase due to the formation of high-molecular-weight hydrocarbons. When the oil viscosity has reached a level where the diffusion of oxygen throughout the oil is significantly limited, chain termination reactions will dominate. As indicated by Reaction 1.10, two alkyl radicals can combine to form a hydrocarbon molecule. Alternatively, an alkyl radical can combine with an alkyl peroxyl radical to form a peroxide (Reaction 1.11). This peroxide, however, is not stable and can readily decompose to generate more alkyl peroxyl radicals. During the chain termination processes, formation of carbonyl compounds and alcohols may also take place via the peroxyl radicals that contain an abstractable α-hydrogen atom. This is illustrated in Figure 1.6 (Reaction 1.12).

1.11.2 METAL-CATALYZED LUBRICANT DEGRADATION

Metal ions are able to catalyze the initiation step as well as the hydroperoxide decomposition in the chain branching step [199] via a redox mechanism illustrated in the following section. The required activation energy is lower for this mechanism, and thus the initiation and propagation steps can commence at much lower temperatures.

1.11.2.1 Metal Catalysis
1.11.2.1.1 Initiation Step

$$M^{(n+1)+} + RH \rightarrow M^{n+} + H^+ + R \cdot \qquad (1.13)$$

$$M^{n+} + O_2 \rightarrow M^{(n+1)+} + O_2^{-\cdot} \qquad (1.14)$$

1.11.2.1.2 Propagation Step

$$M^{(n+1)+} + ROOH \rightarrow M^{n+} + H^+ + ROO \cdot \qquad (1.15)$$

$$M^{n+} + ROOH \rightarrow M^{(n+1)+} + HO^- + RO \cdot \qquad (1.16)$$

1.11.3 HIGH-TEMPERATURE LUBRICANT DEGRADATION

The preceding discussion describes the degradation of lubricants by autoxidation under both low- and high-temperature conditions. Low-temperature oxidation results in the formation of peroxides, alcohols, aldehydes, ketones, and water [200,201]. Under high-temperature oxidation conditions (>120°C), the decomposition of peroxides, including hydroperoxides, becomes predominant and the resulting carbonyl compounds (e.g., Reactions 1.8 and 1.9) will be oxidized to carboxylic acids. The carboxylic acids will start to consume the reserves of basic species provided by any overbased detergent/dispersant used in the formulation. As oxidation proceeds, acid- or base-catalyzed Aldol reactions take place [202]. Initially, α,β-unsaturated aldehydes or ketones are formed, and further condensation of these species leads to high-molecular-weight products. These products contribute to oil viscosity increase and eventually can combine with each other to form oil-insoluble polymeric products. These condensates can lead to sludge formation in a bulk oil oxidation environment or to varnish deposits on hot metal surface.

In a high-temperature study using the Penn State Micro-Oxidation test at 225°C, oxygenated polymeric species were observed to deposit on the test specimen in a single film. As soon as the film formed, it was observed to begin losing oxygen and hydrogen. Oxygenated deposits peaked early in the test, changing to black, largely carbon residue by the end [203]. Oil viscosity increase and deposit formation have been identified as being the principal oil-related causes of engine damage [204].

1.11.4 EFFECT OF BASE OIL COMPOSITION ON OXIDATIVE STABILITY

In the American Petroleum Institute (API) base oil classification system, mineral oils largely fall into the Groups I, II, III, and V categories, with some distinctions shown in Table 1.3 in terms of saturates, sulfur contents, and viscosity index. The market share

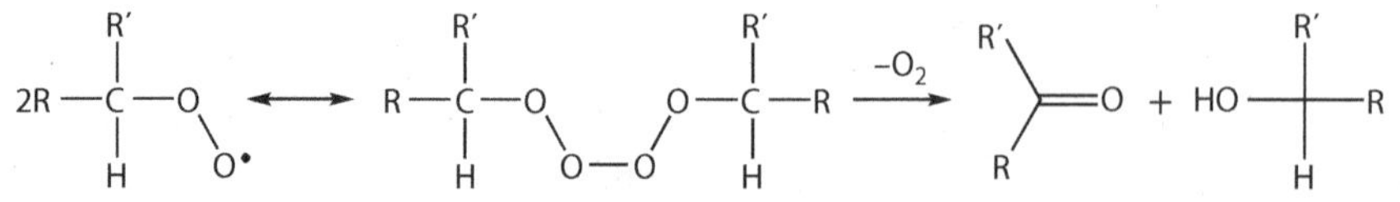

$$2R-\underset{\underset{H}{|}}{\overset{\overset{R'}{|}}{C}}-O-O\cdot \;\longleftrightarrow\; R-\underset{\underset{H}{|}}{\overset{\overset{R'}{|}}{C}}-O-O-O-\underset{\underset{H}{|}}{\overset{\overset{R'}{|}}{C}}-R \;\xrightarrow{-O_2}\; \underset{R}{\overset{R'}{\diagup}}C=O \;+\; HO-\underset{\underset{H}{|}}{\overset{\overset{R'}{|}}{C}}-R \qquad (1.12)$$

FIGURE 1.6 Abstraction of α-hydrogen atoms to form ketones and alcohols.

TABLE 1.3
API Base Oil Categories

API Category	Percent Saturates	Percent Sulfur	Viscosity Index
Group I	≤90	≥0.03	≥80 and ≤120
Group II	≥90	≤0.03	≥80 and ≤120
Group III	≥90	≤0.03	≥120
Group IV	PAOs		
Group V	All other base oils not included in the first four groups (primarily naphthenics and non-PAO synthetics).		

of Group I base oils has dropped significantly over the last decade and now accounts for about 43% of global capacity. Meanwhile, the more highly refined Group II and Group III base oils have grown to 33% and 12% of global capacity, respectively [205].

It has been widely recognized that base oil composition, for example, linear and branched hydrocarbons, saturates, unsaturates, monoaromatics, polyaromatics, together with traces of nitrogen-, sulfur-, and oxygen-containing heterocycles, etc., plays an important role in the oxidative stability of the oil. There have been quite extensive research activities attempting to establish correlations between base oil composition and oxidative stability [206–210]. However, due to the large variations in the origin of the oil samples, the test methods, test conditions, and the performance criteria employed, the conclusions are not always consistent and in some cases are contradictory to each other. In general, it has been agreed that saturated hydrocarbons are more stable than the unsaturated toward oxidation. Of the different saturated hydrocarbons found in mineral oils, paraffins are more stable than cycloparaffins. Aromatic compounds, due to their complex and large variation in the chemical makeup, play a more profound role. Monocyclic aromatics are relatively stable and resistant to oxidation, while bi- and polycyclic aromatics are unstable and susceptible to oxidation [211]. Alkylated aromatics oxidize more readily due to the presence of highly reactive benzylic hydrogen atoms. Kramer et al. [208] demonstrated that the oxidation rate of a hydrocracked 500 N base oil doubled when the aromatic content increased from 1 to 8.5 wt%. Naturally occurring sulfur compounds are known antioxidants for the inhibition of the early stage of oil oxidation. Laboratory experiments have shown that mineral oils containing as little as 0.03% of sulfur had better resistance to oxidation at 165°C than sulfur-free white oils and PAOs [150]. In hydrocracked oils that are essentially low in aromatics, better oxidative stability was found with elevated sulfur concentration (above 80 ppm) versus a level at 20 ppm or lower [207]. It has been proposed that sulfur compounds act as antioxidants by generating strong acids that catalyze the decomposition of peroxides via a nonradical route or by promoting the acid-catalyzed rearrangement of arylalkyl hydroperoxides to form phenols that are antioxidants [150,193]. Contrary to sulfur, nitrogen-bearing compounds, especially the heterocyclic components (also called "basic nitrogen"), accelerate oil oxidation even at relatively low concentrations [212]. In highly refined Group II and III base oils that are essentially devoid of heteroatom-containing molecules, aromatic and sulfur

contents are considered the main factors influencing the base oil oxidative stability [207,208]. It has been shown that the oxidative stability of a given base oil can be enhanced when the combinations and concentrations of base oil sulfur and aromatics have been optimized [209].

1.11.5 Oxidation Inhibition

There are two broad approaches to controlling lubricant oxidation: (1) the destruction of alkyl radicals, alkyl peroxyl radicals, and hydroperoxides and (2) the trapping of catalytic metal impurities. These approaches can be implemented through the use of an appropriate antioxidant with radical scavenging or peroxide decomposing functionality and by using a metal deactivator, respectively.

The radical scavengers are known as primary antioxidants. They function by donating hydrogen atoms to terminate alkoxy and alkyl peroxyl radicals, thus interrupting the radical chain mechanism of the autoxidation process. The key requirement for a compound to become a successful antioxidant is that peroxyl and alkoxy radicals abstract hydrogen from the compound much more readily than they do from hydrocarbons [213]. After hydrogen abstraction, the antioxidant becomes a stable radical and the alkyl radical becomes a hydrocarbon and the alkyl peroxyl radical becomes an alkyl hydroperoxide. Hindered phenolics and aromatic amines are the two main classes of primary antioxidants for lubricants.

The peroxide decomposers are also called secondary antioxidants [194]. They function by reducing alkyl hydroperoxides in the radical chain to nonradical, less reactive alcohols. Organosulfur and organophosphorus compounds and those containing both zinc and phosphorus, such as zinc dialkyldithiophosphates, are well-known secondary antioxidants.

Since transition metals are present in most lubrication systems, metal deactivators are usually added to lubricants to suppress the catalytic activities of the metals. Based on the mechanism of action, metal deactivators for petroleum products can be classified into two major types: chelators [194] and surface passivators [214]. The surface passivators act by attaching to a metal surface to form a protective layer, thereby preventing metal–hydrocarbon interaction. They can also minimize the corrosive attack of the metal surface by physically restricting access of the corrosive species to the metal surface. The chelators, on the other hand, function in the bulk of the lubricant by trapping metal ions to form an empirically inactive or much less active complex. With either mechanism, metal deactivators provide an antioxidant credit by effectively slowing the oxidation process catalyzed by the transition metals. Table 1.4 illustrates examples of metal deactivators that are commonly found in lubricant formulations.

1.11.6 Mechanisms of Primary Antioxidants

1.11.6.1 Hindered Phenolics

A representative example of a hindered phenolic antioxidant is 3,5-di-*t*-butyl-4-hydroxytoluene (2,6-di-*t*-butyl-4-methylphenol), also known as BHT. Figure 1.7 compares the reaction

TABLE 1.4

Metal Deactivators for Lubricants

Surface Passivators	Basic Structure
Triazole derivative	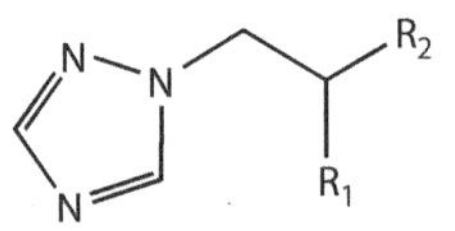
Benzotriazole	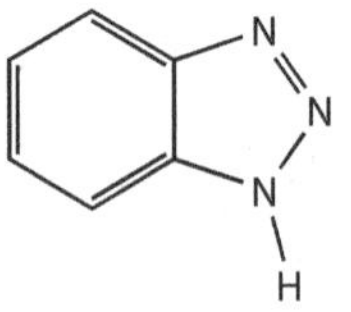
2-Mercaptobenzothiazole	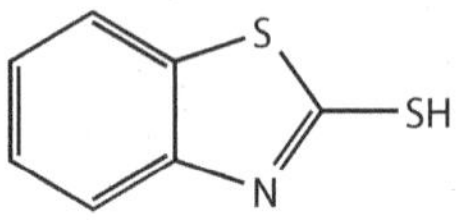
Tolyltriazole derivative	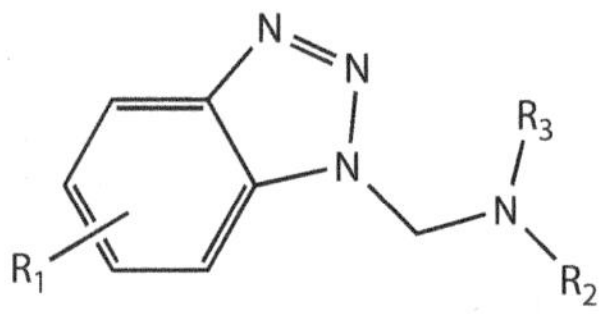
Chelator *N,N'*-Disalicylidene-1,2- diaminopropane	

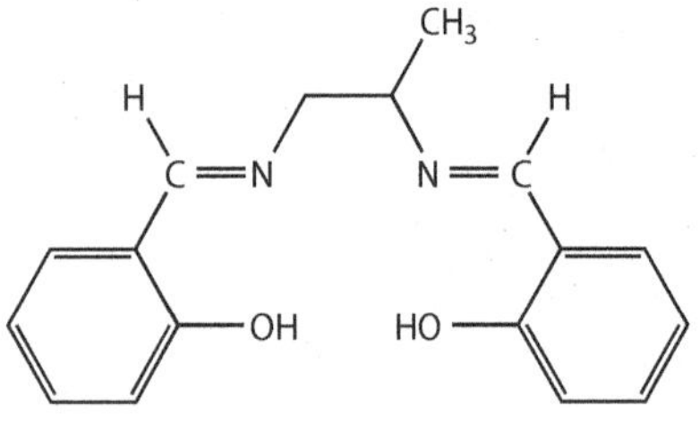

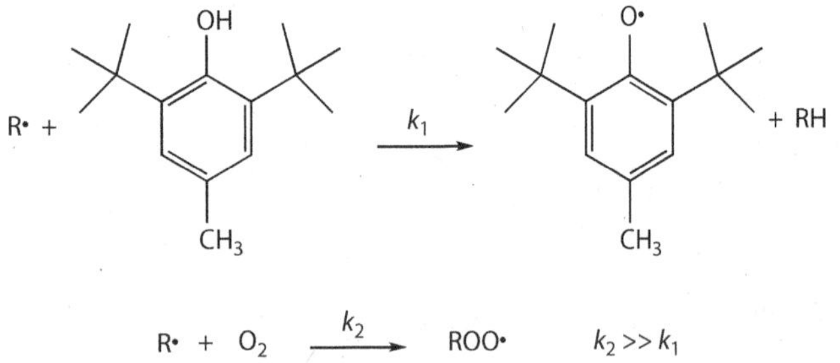

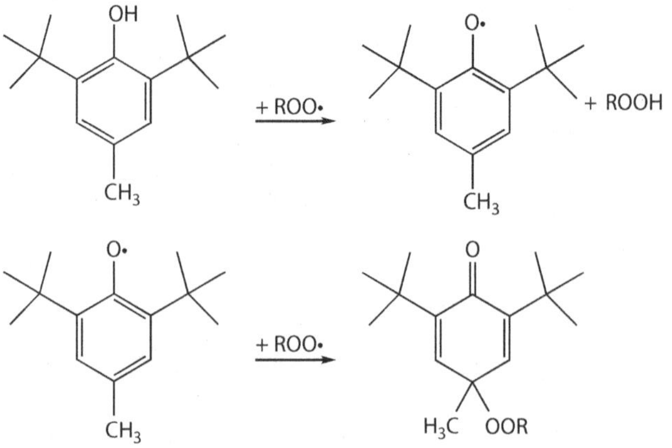

FIGURE 1.8 Hydrogen atom donation and peroxyl radical trapping mechanisms of BHT.

In this mechanism, a proton is transferred in a σ-bond fashion between heteroatoms (from one oxygen to another in this case), while an electron is simultaneously transferred from an orthogonal orbital on the phenol (generally perpendicular to the aromatic ring, allowing for 2p-π overlap) to a singly occupied molecular orbital (SOMO) on the radical [144]. The steric hindrance provided by the two butyl moieties on the *ortho*-positions effectively prevents the phenol radical from attacking other hydrocarbons [215]. The phenol radical can react with a second alkyl peroxyl radical to form a peroxycyclohexadienone, which is stable at temperatures up to about 100°C–120°C [216,217]. Since each phenol molecule is capable of reacting with two radicals, the stoichiometric factor $n = 2$ [144,216].

Under higher-temperature oxidation conditions, the peroxycyclohexadienone is no longer stable. As illustrated in Figure 1.9, it may decompose to form phenoxyl radical and an alkyl peroxyl radical, or alternatively to an alkyl radical, an alkoxy radical, and 2,6-di-*t*-butyl-1,4-benzoquinone. As can be expected, the generation of new radicals will deteriorate the overall effectiveness of the BHT under high-temperature oxidation conditions.

1.11.6.2 Aromatic Amines

Alkylated aromatic amines have been of critical industrial importance as antioxidants for lubricants and rubber for over 60 years. While phenolic antioxidants sacrificially quench radicals in a stoichiometric manner, at temperatures above about 120°C, aromatic amines consume radicals and can then be regenerated as part of a catalytic cycle involving nitroxide radicals. Stoichiometric factors of $n > 12$ have been reported [201,218,219].

While many mechanisms have been proposed for the activity of amine antioxidants, until recently there have been very few studies that rise to the rigor of modern mechanistic organic chemistry. Jensen, Korcek, and coworkers published the landmark paper on the mechanism of aromatic amine antioxidants in 1995 [219]. In the last few years, there has been an outpouring of excellent calculational and experimental mechanistic work through collaborations of the Pratt and Valgimigli [220,221] groups. Pratt et al. have proposed a novel refinement

FIGURE 1.7 Comparative reactivity of alkyl radical with BHT and oxygen.

rates of an alkyl radical with BHT and oxygen. The reaction rate constant (k_2) of alkyl radical with oxygen to form alkyl peroxyl radicals is much greater than that (k_1) of alkyl radical with BHT [193,215,216]. Hence with an ample supply of oxygen, the probability of BHT to react with alkyl radicals is low. As oxidation proceeds with more alkyl radicals being converted to alkyl peroxyl radicals, BHT starts to react by formally donating a hydrogen atom to the peroxyl radical as shown in Figure 1.8.

It has recently been proposed that this first step actually proceeds by a proton-coupled electron transfer (PCET) mechanism.

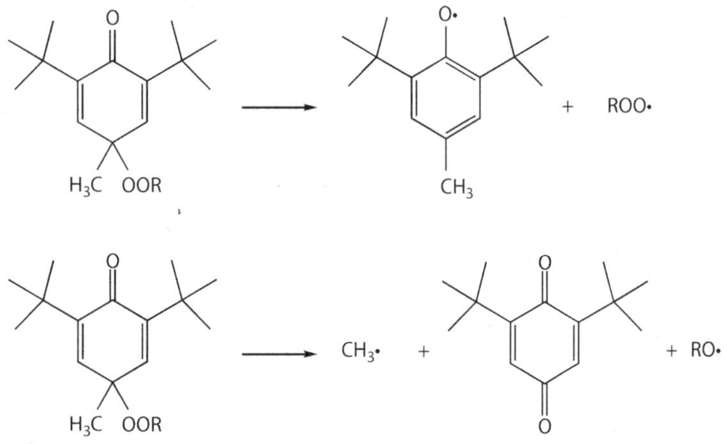

FIGURE 1.9 Decomposition of cyclohexadienone alkyl peroxide.

to the mechanism of amine antioxidants [218]. The very recent review by Ingold and Pratt discusses kinetics and mechanisms for aminic, phenolic, and other antioxidants [144].

The alkylated diphenylamines comprise a particularly effective class of aromatic amine primary antioxidants. The low-temperature (<100°C–120°C) reaction of the alkylated diphenylamine antioxidants begins with a formal hydrogen atom abstraction from the secondary amine by an alkyl peroxyl radical. The proposed overall mechanism for low-temperature reactions of aromatic amine antioxidants is illustrated in

Figure 1.10. Transition state calculations have shown that as with phenols, the first step actually proceeds by a PCET mechanism. In the amine case, a proton is transferred from nitrogen to oxygen, while an electron is simultaneously transferred from a π orbital on the amine to a singly occupied π orbital on the radical [144,216,218,221]. The resulting aminyl radical can react with a second alkyl peroxyl radical, but decomposition of the peroxide intermediate results in the formation of a stable nitroxide and a highly reactive, chain-propagating alkoxy radical. However, the nitroxide is also an effective antioxidant.

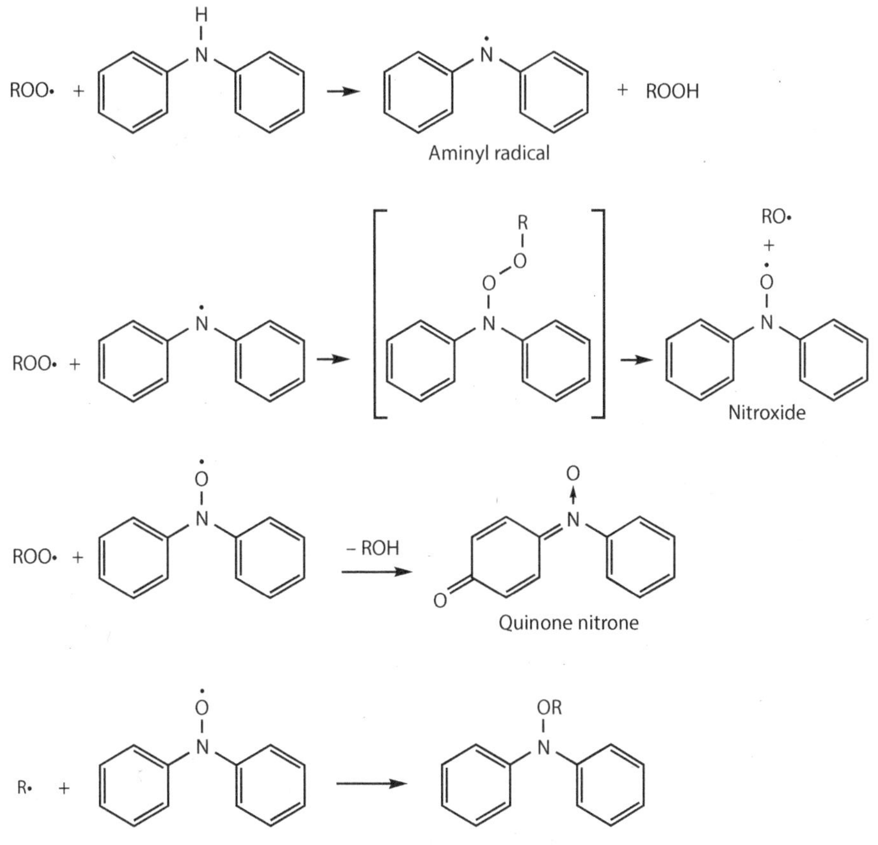

FIGURE 1.10 Proposed low-temperature reactions of aromatic amine antioxidants. (Reprinted with permission from Ingold, K.U. and Pratt, D.A., *Chem. Rev.*, 114, 9022. Copyright 2014, American Chemical Society.)

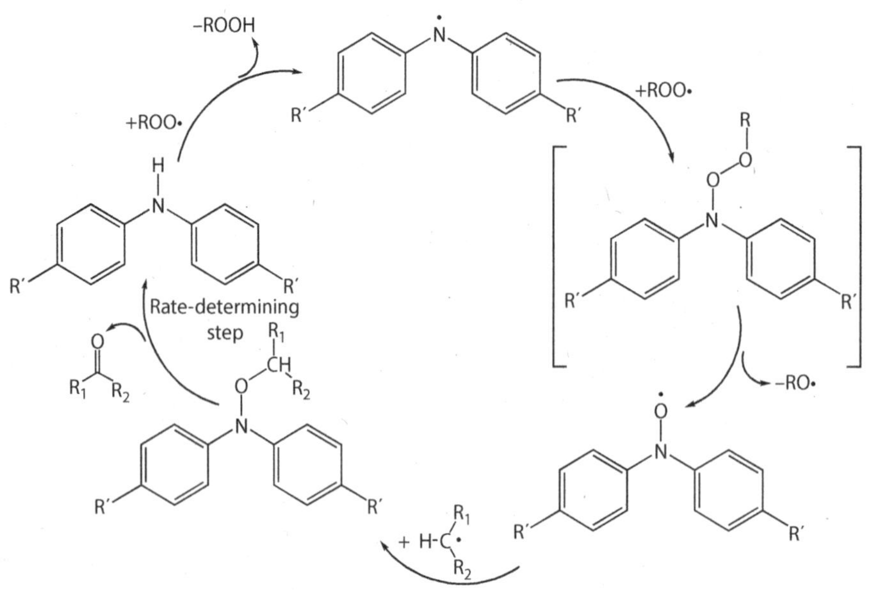

FIGURE 1.11 Proposed catalytic mechanism of alkylated diphenylamines at high temperature (>120°C). (Reprinted with permission from Haidasz, E.A., Shah, R., and Pratt, D.A., *J. Am. Chem. Soc.*, 136, 16643. Copyright 2014, American Chemical Society.)

It may react with an alkyl peroxyl radical to sacrificially form an alcohol and a quinone nitrone or potentially trap an alkyl radical. These termination steps combine to give a low-temperature stoichiometric factor n that is empirically about 2.

Reaction rates for N–H hydrogen atom abstraction by alkyl radicals have been determined at 100°C for DPA and several 4,4′-disubstituted alkylated diphenylamines [222]. N–H bond dissociation energies and relative reactivity toward alkyl, alkoxy, and peroxyl radicals (20°C–50°C) have been reported for DPA, alkylated diphenylamines, phenothiazines, and a few related structures. Trapping reactivity decreased RO· ≫ ROO· > R· [223].

At high temperatures (>120°C), the alkoxyamine becomes unstable, ultimately decomposing to form a ketone while regenerating the diphenylamine. The proposed catalytic mechanism for alkylated amines at high temperature is shown in Figure 1.11. Pratt and coworkers argue that the mechanism of alkoxyamine decomposition will vary depending upon whether the alkoxyamine R group is saturated (alkyl) or unsaturated (alkenyl). Alkyl radicals are proposed to decompose within a solvent cage by homolysis of the N–O bond, followed by disproportionation, while alkenyl radicals may undergo a novel retro-carbonyl-ene reaction [218]. Catalytic regeneration of the diphenylamine results in the high stoichiometric factors for ADPAs ($n > 12$) relative to phenolics ($n = 2$) cited previously. By-products such as the quinone nitrone (Figure 1.10) remove ADPA from the cycle and prevent the cycle from continuing indefinitely [144].

1.11.7 Mechanisms of Secondary Antioxidants

1.11.7.1 Organosulfur Compounds

Organosulfur compounds function as hydroperoxide decomposers via the formation of oxidation and decomposition products. The reaction mechanisms have only been studied in detail for a rather limited number of the many organosulfur functionalities.

Phenothiazines, unsubstituted at nitrogen, were found to react with dioxetanes by single electron transfer, generating a phenothiazine radical cation and a dioxetane radical anion. The dioxetane radical anion may fragment or may abstract a hydrogen atom from the radical cation. Fragmentation is favored for the more stable radical cations (phenothiazine has low oxidation potential), while hydrogen atom abstraction is favored for less stable radical ions (phenothiazine has higher oxidation potential). Electron-donating groups lower the oxidation potential of the phenothiazine, hence reducing hydrogen atom transfer. N-alkyl phenothiazines were oxidized to sulfoxides [224].

Sulfoxides RS(O)R can decompose by a Cope elimination to form a sulfenic acid, RSOH. Most sulfenic acids are too reactive to be isolated for direct use as an additive. The O–H bond of a sulfenic acid is very weak and readily donates H· to peroxyl radicals. The resulting RSO· radical is stabilized by the sulfur atom [144].

At elevated temperatures, sulfinic acid (RSO$_2$H) may decompose to form sulfur dioxide (SO$_2$), which is a particularly powerful Lewis acid for hydroperoxide decomposition through the formations of active sulfur trioxide and sulfuric acid. Previous work found that one equivalent of SO$_2$ was able to catalytically decompose up to 20,000 equivalents of cumene hydroperoxide [225].

1.11.7.2 Organophosphorus Compounds

Phosphites are important as ashless organophosphorus secondary antioxidants for lubricants. While phosphites can work synergistically with hindered phenolics, phosphites are less effective at chain termination. Phosphites can decompose hydroperoxides and peroxyl radicals following the reactions shown in the following. In these reactions, phosphite is oxidized to the corresponding phosphate, with the hydroperoxide and the peroxyl radical being reduced to alcohol and alkoxy

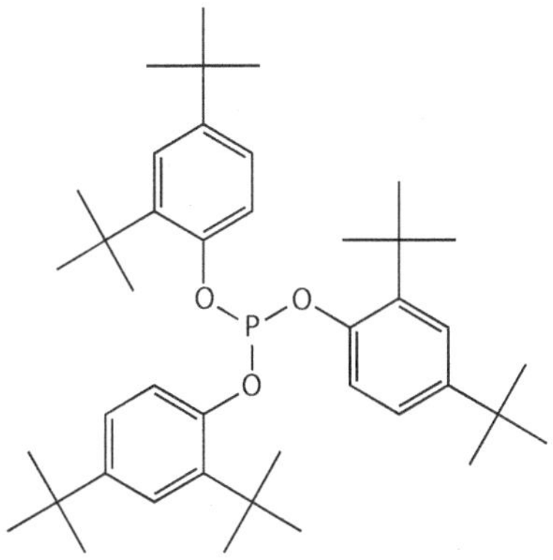

FIGURE 1.12 A hydrolytically stable commercial phosphite.

radical, respectively. The alkoxy radical can propagate the chain through an alkyl radical [226–228].

$$(ArO)_3P + R'OOH \rightarrow (ArO)_3 = O + R'OH$$

$$(ArO)_3P + R'OO\cdot \rightarrow (ArO)_3P = O + R'O\cdot$$

$$(ArO)_3P + R'O\cdot \rightarrow (ArO)_3P = O + R'\cdot$$

$$(ArO)_3P + R'O\cdot \rightarrow (ArO)_2P - OR' + ArO\cdot$$

Due to steric hindrance, aryl phosphites with alkyl groups on the *ortho*-positions tend to be more stable against hydrolysis. These phosphites are preferred for use in moist lubrication environments. An example of a hydrolytically stable commercial phosphite is shown in Figure 1.12.

1.11.8 ANTIOXIDANT SYNERGY*

Antioxidant synergy describes the combined use of two or more antioxidants to give better performance than can be obtained by an equivalent amount of any of the component antioxidants alone. Synergistic antioxidant systems offer practical solutions to problems where using a single antioxidant is inadequate to provide satisfactory results or where the treatment level has to be limited due to economic and/or environmental reasons. Synergies have been observed between primary antioxidants as well as between a primary and a secondary antioxidant.

The use of an alkylated diphenylamine in combination with a hindered phenolic is a classic example of two primary antioxidants working synergistically. Presumably, the synergy arises from the different selectivity of the aminic and phenolic radicals toward the various oxidation-derived radicals, thereby attacking different reactive pathways.

Primary aminic antioxidants are commonly used in combination with zinc dialkyldithiophosphate (ZDDP), a secondary antioxidant. The primary antioxidant scavenges radicals while the secondary antioxidant decomposes hydroperoxides by reducing them to more stable alcohols. Mixtures of alkylated diphenylamines with phenothiazines

may function in a similar manner, with the phenothiazine functioning primarily as a peroxide decomposer via the reactive sulfur atom.

1.12 OXIDATION BENCH TESTS

Oxidative degradation of lubricants can be classified into two main reactions: bulk oil oxidation and thin-film oxidation. Bulk oil oxidation usually takes place at a slower rate in a larger oil body, such as a crankcase sump. The exposure to air (oxygen) is regulated by the surface contact kinetics and the gas diffusion is limited. Bulk oxidation leads to increases in oil acidity, oil thickening, and eventually the formation of oil-insoluble polymers. The polymers may combine with other impurities (metals, ash, water, partially burnt fuel, etc.), with the mixture dropping out of the oil collectively as sludge. Thin-film oxidation describes a more rapid reaction in which a small amount of oil is exposed to severely elevated temperatures and air (oxygen). This may occur on the cylinder liner, for example. Under these conditions, hydrocarbons decompose much more quickly and the polar oil oxidation products formed at the oil–metal interface can rapidly build up on the metal surface, leading to the formation of varnish or deposits.

Fleet testing and fired engine tests are the best ways to evaluate lubricant performance. However, these tests are so expensive, labor intensive, and resource consuming that even the largest organizations use them selectively to validate leads drawn from bench tests.

Many oxidation bench tests have been developed over the years. These have proven to be valuable tools for lubricant formulators, particularly in the screening of new antioxidants and the development of new formulations. Most bench tests attempt to simulate the operating conditions of more expensive engine and field tests while taking into consideration either the bulk or thin-film oxidation conditions described previously. In addition, a third approach based on oxygen uptake in a closed system has been employed in some bench tests, such as the RPVOT (ASTM D2272) [229].

Due to the limitations of any laboratory setup, no single bench test can address all aspects of oxidation. The large variation in test conditions makes it rather difficult or even impossible to correlate one bench test with another. Furthermore, despite the best efforts of the test designers, the significance of proposed correlations between bench and engine tests is frequently the subject of spirited debate.

It is therefore common practice to run multiple tests at a time when characterizing a lubricant formulation and its additives. This section selectively reviews oxidation bench tests more closely related to the characterization of antioxidants. These tests have been standardized by some of the international standardization organizations such as ASTM International and the Co-ordinating European Council (CEC) and are widely used in the industry. It is important to note that there are a number of custom-tailored test methods designed for specific needs that have been proven to be advantageous in certain circumstances. The value of these tests should not be underestimated. A summary of the various tests is given in Table 1.5.

---

* With permission from Dong, J. and Migdal, C.A., Synergistic antioxidant systems for lubricants, *Twelfth Asia Fuels and Lubes Conference Proceedings*, Hong Kong, 2006.

TABLE 1.5
Conditions of Oxidation Test Methods

Test	Test Designation	Oxidation Regime	Temperature (°C)	Gas	Gas Flow or Initial Pressure	Catalyst	Sample Size	End of Test	Parameter(s) Measured
PDSC	D6186	Thin film	130, 155, 180, 210	O_2	500 psi, 100 mL/m	None	3.0 mg	Occurrence of oxidation exotherm	OIT
PDSC	CEC L-85	Thin film	210	Air	100 psi, no flow	None	3.0 mg	120 min max.	OIT
TEOST 33C	D6335	Bulk	100, 200–480	N_2O, moist air	3.6 mL/min	Fe naphthenate	116 mL	12 programmed cycles	Deposits
TEOST MHT	D7097	Thin film	285	Dry air	10 mL/m	Oil-soluble Fe, Pb, Sn	8.4 g	24 h	Deposits, volatile
TFOUT	D4742	O_2 uptake, thin film	160	O_2	90 psig	Fuel, naphthenates of Fe, Pb, Cu, Mg, and Sn, H_2O	1.5 g	Sharp pressure drop	OIT
TOST	D943 D4310	Bulk	95	O_2	3.0 L/H	Fe, Cu, H_2O	300 mL	ΔTAN = 2.0 1000 h	TAN, sludge, metal wt. loss
Dry TOST	D7873	Bulk	120	O_2	3.0 L/H	Fe, Cu	8 × 360 mL	D2272 residual <25%	Sludge, D2272 residual
IP 48	IP 48	Bulk	200	Air	15 L/H	None	40 mL	6 h × 2	Viscosity, carbon residue
IP 280/CIGRE	IP 280	Bulk	120	O_2	1 L/H	Cu, Fe naphthenates	25 g	164 h	Volatile acids, oil acidity, sludge
RPVOT	D2272	O_2 uptake	160	O_2	90 psig	Cu, H_2O	50 mL	ΔP = 25 psi	OIT

1.12.1 Thin-Film Oxidation Tests

1.12.1.1 Pressurized Differential Scanning Calorimetry (PDSC)

Differential scanning calorimetry, including PDSC, is a thermal technique for rapid and accurate determination of thermal-oxidative stability of base oils and performance of antioxidants. PDSC has been a more sought-after technique for two main reasons. First, high pressure elevates boiling points, thus effectively reducing experimental errors caused by volatilization losses of additives and the light fractions of base oil; second, it increases the saturation of the reacting gases in sample, allowing the use of lower test temperature or shorter test time at the same temperature [230].

PDSC experiments can be run in an isothermal mode to measure OIT corresponding to the onset of oil oxidation or in a programmed temperature mode to measure the onset temperature of oxidation. The temperature technique has been utilized to study the deposit-forming tendency of five engine oils, and the results obtained were consistent with their engine test ranking [231]. The OIT technique, however, is more commonly used for its simplicity and speed. Its early use can be traced back to the 1980s when Hsu et al. [232] tested a number of engine oils and found the induction periods of the samples to be indicative of the Sequence IIID viscosity break points. Soluble metals consisting of lead, iron, copper, manganese, and tin together with a synthetic oxidized fuel were included as catalysts to promote oil oxidation.

The CEC L-85 and the ASTM D6186 [233,234] are two standard methods that are based on OIT technique. Key test conditions of the methods are collected in Table 1.5. The CEC L-85 test method was originally developed for European ACEA (Association des Constructeurs Européens de l'Automobile) E5 specification for heavy duty diesel oils and has been incorporated in the current E7 and E9 specifications. The test is capable of differentiating between different quality base oils, additives, indicating antioxidant synergies and correlating with some bulk oil oxidation tests [235,236]. With appropriate modifications to the standard methods, PDSC has been successfully utilized in the characterization of a variety of lubricants in addition to automotive engine oils. These include, but are not limited to, base oils [237,238], greases [239], turbine oils [238], gear oils [240], synthetic ester lubricants [241], and biodegradable oils [242,243]. Using PDSC to study the kinetics of base oil oxidation [244] and antioxidant structure–performance relationship [245] has also been reported.

1.12.1.2 Thermo-Oxidation Engine Oil Simulation Test (TEOST, ASTM D6335; D7097)

The TEOST was originally developed to assess the high-temperature deposit forming characteristic of API SF quality engine oils under turbocharger operating conditions [246]. The original test conditions were specified as the 33C protocol and subsequently standardized in the ASTM D6335 method [247]. In this test, oil containing ferric naphthenate is in contact with nitrous oxide and moist air and is cyclically pumped to flow past a tared depositor rod. The rod is resistively heated through 12 temperature cycles, each going from 200°C to 480°C for 9.5 min. After the heating cycle is complete, the deposit formed on the depositor rod is determined by differential weighting. The 33C protocol was found capable of discriminating engine oils with known ability in resisting deposit formation in critical areas of engines [246].

The successful use of high-temperature deposition test to characterize engine oils has led to the development of a TEOST MHT protocol, a simplified procedure for the assessment of oil deposition tendency in the piston ring belt and under-crown areas of fired engines [248]. Thin-film oxidation condition was thought to be predominant in these areas, and accordingly, the depositor assembly was revised to allow the oil to flow down the rod in a slow and even manner to obtain a desired thin film. To better reflect the thermal-oxidative conditions of the engine zone of interest, a continuous depositor temperature of 285°C together with modified catalyst package and dry air are employed. The test runs for 24 h, and afterward, the amount of deposits formed on the tared depositor is gravimetrically determined [249]. Since introduction, the TEOST MHT has been incorporated in the ILSAC GF-3 through GF-6 engine oil specifications with an upper limit of 45 and 35 mg, respectively. Aside from being a thermo-oxidation test, TEOST can also be used to characterize neutral and overbased detergents of automotive engine oils [250].

1.12.1.3 Thin-Film Oxygen Uptake Test (TFOUT, ASTM D4742)

The TFOUT method was originally developed under the U.S. Congress mandate to monitor batch-to-batch variations in the oxidative stability of re-refined lubricating base oils [251]. The test heats a small amount of oil to 160°C in a high-pressure reactor pressurized with oxygen, along with a metal catalyst package, a fuel catalyst, and water to partially simulate the high-temperature oxidation conditions in automotive combustion engines [252]. Better oxidative stability of oil corresponds to a longer time taken to observe a sharp drop in oxygen pressure. TFOUT can be carried out in a RPVOT apparatus upon proper modification to the sampling accessories. Based on the results obtained from testing a limited number of reference engine oils, qualitative correlation between TFOUT and the Sequence IIID engine dynamometer test has been established [253]. Since being adopted as an ASTM standard method, there has been a wider utilization of the TFOUT to screen lubricants, base oils, and additive components prior to Sequence III engine testing [251].

1.12.2 Bulk Oil Oxidation Tests

1.12.2.1 Turbine Oil Stability Test (TOST, ASTM D943, D4310, Dry TOST D7873)

The turbine oil stability test (TOST) has been widely used in the industry to assess the oxidative stability of inhibited steam turbine oils under long-term service conditions. It can be used on other types of industrial lubricants such as hydraulic fluids and circulating oils and in particular on those that are prone to water contamination in service. The test runs at relatively low temperature (95°C) to represent the thermal-oxidative conditions of real steam turbine applications. Two versions of the TOST, namely, ASTM D943 and D4310, have been developed [254,255]. Both methods share some common test conditions including test apparatus, catalysts, sample size, temperature profile, and gas, with minor differences in the test duration and target oxidation parameters to be monitored. The ASTM D943 measures oxidation lifetime, which is the number of hours required for the test oil to reach an acid number of 2.0 mg KOH/g or above. The ASTM D4310 determines the sludging and corrosion tendencies of the test oil by gravimetrically measuring oil-insoluble products after 1000 h of thermal and oxidative stress. The total amount of copper in the oil, water, and sludge phases is also determined.

A modified TOST method that operates at higher temperature (120°C) and in the absence of water was proposed [256] and recently became ASTM D7873 (Dry TOST method) [257]. The procedure requires RPVOT as a monitoring tool and is specifically suitable for the determination of sludging tendencies of long-life steam and gas turbine oils formulated with the more stable Group II and III base oils and high-performance aminic antioxidants. The Dry TOST method is a potential alternative to the original methods that have found to be less discriminatory on such high-performance turbine oils.

1.12.2.2 IP 48 Method

The Energy Institute IP 48 is a high-temperature bulk oil oxidation test that was originally designed for the characterization of base oils [258]. The test stresses a 40 mL sample of oil in a glass tube at 200°C, along with air bubbling at 15 L per h, for two 6 h periods with a 15–30 h standby period in between. Oil viscosity increase and the formation of carbon residue are determined after the oxidation. The test is considered unsuitable for additized oils (other than those containing ashless additives) or those that form solid products or evaporate more than 10% by volume during the test. However, successful assessment of engine oils using a modified IP 48 method with four 6 h cycles has been reported [258,259].

1.12.2.3 IP 280/CIGRE

The IP 280, also known as the CIGRE test, was designed to assess the oxidative stability of inhibited mineral turbine oils, targeting formations of volatile acid products (via water absorption), sludge, and increase of oil acidity [260]. The IP 280 and the TOST D943 are similar to each other in terms of the oxidation regime employed. However, the test conditions are different. Therefore, it is common practice to conduct both tests, since the limits for each test are stipulated in some turbine oil specifications. The IP 280 was found to be more suitable for discriminating performance of additive packages, while the D943 is more suitable for comparative evaluation of base oils derived from different crude source and processing techniques [261].

1.12.3 Oxygen Uptake Test

1.12.3.1 Rotating Pressure Vessel Oxidation Test (RPVOT, ASTM D2272)

The RPVOT, originally known as the Rotating Bomb Oxidation Test, was designed to monitor the oxidative stability of new and in-service turbine oils having the same composition. It can also be used to characterize other types of industrial lubricants, for example, hydraulic fluids and circulating oils. The test utilizes a steel pressure vessel where sample oil is initially pressurized to 90 psi with oxygen and thermally stressed to 150°C in the presence of water and copper coil catalyst until a pressure drop of 25 psi is observed [262]. The test temperature was chosen in order to promote measurable oil breakdown in a relatively short time. However, this temperature means that the test is not representative of most steam turbines that operate below 100°C or to the combustion turbines that operate at much higher temperatures [263]. Due to its sensitivity to specific additive chemistries, RPVOT finds limited use in comparing differently formulated oils. In addition, the test is more suitable for the determination of remaining useful life of in-service turbine oils rather than the qualification of new oils. Efforts to correlate RPVOT to the lengthy TOST D943 on steam turbine oils have been successful, suggesting that the results from RPVOT may be used to estimate the relative lifetime of turbine oils in the TOST D943 [264].

1.12.4 Experimental Observations

The following two experiments demonstrate (1) the comparative performance of an alkylated aminic antioxidant (ADPA) versus a hindered phenolic that is in agreement with the mechanisms discussed previously and (2) how proper selection and combinations of antioxidants can lead to synergy that further enhances performance.

In the first experiment, two turbine oils each formulated with a base oil selected from an API Group I or Group IV base oil, a standard additive package of metal deactivator and corrosion inhibitor, and 0.8 wt% of the antioxidants of interest were tested by using the TOST D943 lifetime method. The aminic antioxidant was an alkylated diphenylamine containing a mixture of butylated and octylated diphenylamines. The hindered phenolic was a C_7–C_9 branched alkyl ester of 3,5-di-tert-butyl-4-hydroxyhydrocinnamic acid. As can be seen from the TOST results shown in Figure 1.13, in either oil the hindered phenolic significantly outperformed the alkylated diphenylamine by providing longer protection time against oxidation. Mixtures of the alkylated diphenylamine and the hindered phenolic at 0.4 wt% each in the oils provided even stronger protection, leading to an extended lifetime of about 5000 h for the Group I turbine oil and well over 8000 h for the Group IV turbine oil. Thus, under the low-temperature test conditions, the hindered phenolic was superior to the alkylated diphenylamine. Using the correct combination of the two additives produced synergy offering the maximum protection.

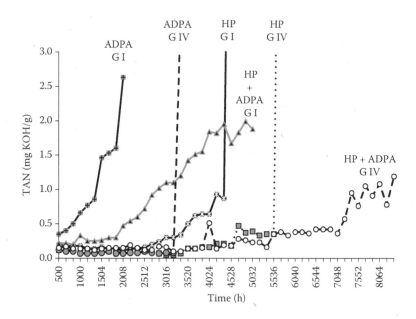

FIGURE 1.13 TOST results of turbine oils containing Group I or Group IV base oil and 0.8 wt% of antioxidant.

In the second experiment, a turbine oil formulated with an API Group I base oil, a metal deactivator, a corrosion inhibitor, and a 0.5 wt% of the same antioxidants as before was tested by using the RPVOT (ASTM D2272). The results are graphically presented in Figure 1.14. At the higher test temperature (150°C), the OIT of the blend containing alkylated diphenylamine was about 600 min, at which time the alkylated diphenylamine was depleted. The hindered phenolic protected the oil for about 300 min, indicating the hindered phenolic is only half as effective as the alkylated diphenylamine under the same test conditions. A mixture of alkylated diphenylamine and hindered phenolic with 0.25 wt% of each additive present provided a protection for over 700 min. Therefore, in contrast to the TOST results, under high-temperature conditions the alkylated diphenylamine was superior

to the hindered phenolic. Similar to what was observed in the TOST, a synergistic mixture of the two additives provided the maximum protection.

The superiority of alkylated diphenylamine over hindered phenolic and the benefit of antioxidant synergy for maximum oxidation protection have been further demonstrated in a GF-4 prototype passenger car motor oil (PCMO). The oil contained an API Group II base oil, a low level (0.05 wt%) of phosphorus derived from ZDDP, and a number of other additives (detergents, dispersant, viscosity index improvers, pour point depressant, etc.) that are commonly found in engine oil formulations. The alkylated diphenylamine, hindered phenolic, and their mixture were tested at 1.0 wt% in the oil on a TEOST MHT apparatus using the ASTM D7097 standard procedure. The results are presented in Figure 1.15.

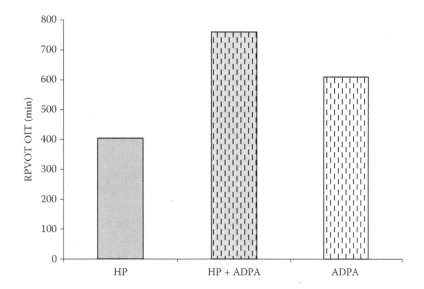

FIGURE 1.14 RPVOT results of turbine oil containing a Group I base oil and 0.5 wt% of antioxidant.

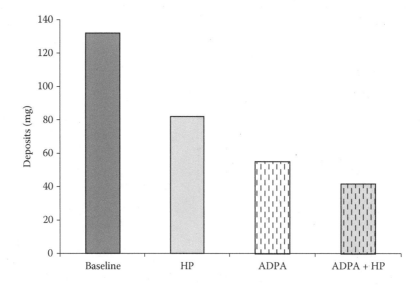

FIGURE 1.15 TEOST results of a prototype PCMO containing a Group II base oil and a total of 1.0 wt% of antioxidant.

The baseline blend, which contained all other additives except the antioxidant, produced a fairly high level (130 mg) of deposits. With the addition of the hindered phenolic, the deposit was substantially reduced to about 80 mg, and with the alkylated diphenylamine down to about 55 mg. By properly mixing the two antioxidants together while keeping the total level constantly at 1.0 wt%, the deposit was further reduced to about 40 mg. The TEOST results confirm the superior performance of alkylated diphenylamine and further demonstrate the benefit of antioxidant synergy for high-temperature oxidation conditions.

The antioxidant mechanisms discussed earlier well explain the experimental results and can serve as a foundation to guide lubricant formulators in the selection of correct antioxidant(s) for a particular end use. In order to obtain a successful formulation, other factors such as cost/performance, volatility, color, solubility, odor, physical form, toxicity, and compatibility with other additives need also be taken into consideration. From a performance standpoint, hindered phenolics are excellent primary antioxidants for their stoichiometric reactions with free radicals under lower-temperature conditions. In contrast, alkylated diphenylamines are excellent primary antioxidants for high-temperature conditions owing to their catalytic radical scavenging actions. As demonstrated, the synergy between the alkylated diphenylamine and the hindered phenolic is significant in the inhibition of oil oxidation. It is, however, important to note that the generation and the magnitude of an antioxidant synergy are dependent on the formulation, base oil, and test method used. The alkylated diphenylamine/hindered phenolic synergy is fairly robust as it was successfully reproduced in two oil formulations and tests that vastly differ from each other in terms of base oil makeup, additive type and complexity, test conditions, and oxidation regimes. In fact, this type of synergy has been used in a wide range of lubricants. In a more recent development, a methylene-bridged hindered phenol was utilized

and found to be synergistic with alkylated diphenylamine in low-phosphorus engine oils [265]. Several instances of other types of synergy have been demonstrated and discussed in greater depth elsewhere. These include, but are not limited to, the synergy between sulfur-bearing hindered phenolic and alkylated diphenylamine antioxidants for hydrotreated base oils [137,266], between different aminic antioxidants [267], and between primary antioxidants and organophosphites [55].

1.13 ANTIOXIDANT PERFORMANCE WITH BASE OIL SELECTION

Driven by ever-increasing environmental and performance requirements, the lubricant industry is rapidly changing for the better with the advances of additive and base oil technologies. One notable change from a formulation point of view is that the conventional solvent-extracted base oils (Group I) are increasingly being replaced by higher-quality, higher-performance Group II and III base oils made from hydrotreated (hydrocracked), hydrotreating, and hydrocatalytic dewaxing and GTL processes. These processes provide oils with low sulfur, high degree of saturation, and viscosity index (Table 1.3). Lubricants formulated with these base oils generally have improved performance characteristics such as superior oxidative stability, lower volatility, improved low-temperature properties, longer drain intervals, and improved fuel economy. Because of these benefits, the API Group III base oils are becoming a serious challenge to synthetic PAOs for top-tier oil formulations.

Many efforts have been made to understand the relationship between the base oil composition and the response to added antioxidants. Such knowledge is extremely important for lubricant formulators when it comes to the selection of an appropriate antioxidant system for a given oil. Figure 1.16 shows the RPVOT results of four base oils with and without the presence of an antioxidant. Each oil represents

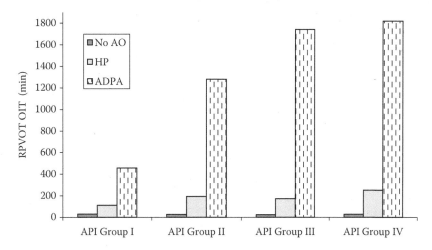

FIGURE 1.16 RPVOT results of hindered phenolic and alkylated diphenylamine in API Group I to IV base oils.

an API Group from I to IV. The hindered phenolic and the alkylated diphenylamine are the same as in Section 1.12. Clearly, without the protection of antioxidant, all oils performed equally poorly. A 0.5 wt% loading of the hindered phenolic antioxidant gave modest levels of protection that marginally increased from API Group I to API Group IV. When the base oils were treated with the same level of the alkylated diphenylamine, a dramatic performance boost is seen across the board. The improvement in performance of the highly refined Groups II, III, and IV to the added alkylated diphenylamine appears to be particularly strong. The superior antioxidant response of the Group II and III base oils over the conventional Group I base oils may be attributed to the removal of aromatic hydrocarbons and polar constituents and the large presence of saturated hydrocarbons in the oils [268,269].

ZDDP, another important class of antioxidant/antiwear agent, has been studied by others, and the results indicated that its antioxidant performance is dependent on the base oil aromatics, alkyl-substituted aromatics, average chain length of hydrocarbons, and the relative presence of normal paraffins and isoparaffins [211]. In Group I base oils, ZDDP gave good responses to highly saturated hydrocarbons characterized with normal paraffins having shorter chain length. Isoparaffins were found to decrease the antioxidant activity of ZDDP due to the steric hindrance of the side chains, which restricts the additive molecules from interacting with the hydrocarbons. In oils with higher monoaromatic hydrocarbons, ZDDP tends to perform better, which was believed to be related to improved solvency.

1.14 FUTURE REQUIREMENTS

The petroleum industry is currently working through the effects of nonconventional drilling and hydraulic fracturing, the disruptive new technologies commonly referred to as "fracking." The rapid development of the shale-gas industry has made possible the ready availability of higher-quality base oils from GTL processes [270]. With crude oil prices depressed, there is presently less incentive for investment in biobased lubricant and energy sources.

However, despite the new sourcing options for petrochemicals, the need to meet environmental regulations remains unchanged. Lubricant formulators face continued pressure to reduce the already low levels of sulfur, phosphorus, zinc, and other metals present in their oils, in order to safeguard and improve the operation of emissions control systems.

Engine and vehicle OEMs continue to press for ever extending oil drain intervals, for customer convenience, oil conservation, and waste minimization [270]. The amount of ashless antioxidants used in engine oils may increase. However, there can be a limit to the amount of antioxidant (and other additives) that can be added to a formulation, and the higher-quality base oils likely to be used in these formulations tend to be less effective in dissolving additives and oxidation products than lower-grade oils. Therefore, there will be a continuing need for developing more effective antioxidants and/or antioxidant combinations.

The most pressing challenge facing the automotive industry today comes in the area of fuel economy, with mandated CAFE requirements of 54.5 mpg by 2025 [271]. Lubricants must contribute to this goal. Fuel economy is measured in the proposed GF-6 standard by the Sequence VI-D test (ASTM D7589) or its equivalent [272] (Sequence VI-E, in development).

Over the last 30 years or so, the primary approach to improving the lubricant contribution to fuel economy has been to reduce viscosity. Whereas SAE 10W-30 was the primary PCMO viscosity grade in 2000, 5W-30 and 5W-20 grades dominate today [273]. The viscosity reduction trend continues, with the recent introduction of specifications for SAE 0W-16, 0W-12, and 0W-8 grades [274,275]. With the reduction in oil film thickness that accompanies lowered viscosity, we expect that these very low viscosity oils will behave in an increasingly boundary lubrication mode. These oils will require excellent antioxidant protection for deposit control and further protection with friction modifiers.

1.15 SOME COMMERCIAL ANTIOXIDANTS

Product	Company	Chemistry
Irganox® L 06	BASF	Octylated phenyl-α-naphthylamine
Irganox® L 57	BASF	Butylated, octylated diphenylamine
Irganox® L 64	BASF	Mixture of aminic and high-molecular-weight phenolic antioxidants
Irganox® L 67	BASF	Dinonyldiphenylamine
Irganox® L 101	BASF	Pentaerythritol tetrakis(3-(3,5-di-*tert*-butyl-4-hydroxyphenyl) propionate)
Irganox® L 107	BASF	Octadecyl 3-(3,5-di-*tert*-butyl-4-hydroxyphenyl)propionate
Irganox® L 109	BASF	Hindered bis-phenol: {2-Propenoic acid, 3-[3,5-bis(1,1-dimethylethyl)-4-hydroxyphenyl]-, 1,1′-(1,6-hexanediyl) ester}
Irganox® L 115	BASF	2,2′-Thiodiethylene bis-(3,5-di-*t*-butyl-4-hydroxyphenyl)propionate
Irganox® L 135	BASF	3,5-Di-*t*-butyl-4-hydroxy-hydrocinnamic acid, C_7–C_9 alkyl ester
Irgafos® 168	BASF	Tris(2,4-di-*tert*-butylphenyl) phosphite
Naugalube® 438	Chemtura	Dioctyldiphenylamine
Naugalube® 438L	Chemtura	Dinonyldiphenylamine
Naugalube® 750	Chemtura	Butylated-, octylated-diphenylamine
Naugalube® AMS	Chemtura	Alpha-methystyrenated DPA
Naugalube® APAN	Chemtura	Alkylated PANA
Naugard® PANA	Chemtura	Phenyl-alpha-naphthylamine
Additin® RC 7001	Rhein Chemie	*p,p′*-Dioctyldiphenylamine
Additin® RC 7010	Rhein Chemie	Polymeric trimethylol-dihydroquinoline
Additin® RC 7130	Rhein Chemie	Phenyl-alpha-naphthylamine
Additin® RC 7132	Rhein Chemie	Aminic antioxidant
Additin® RC 7110	Rhein Chemie	2,6-Di-*tert*-butyl-*p*-cresol
Additin® RC 7115	Rhein Chemie	Phenol derivative sterically hindered
Additin® RC 7120	Rhein Chemie	2,6-Di-*tert*-butyl phenol
Additin® RC 7201	Rhein Chemie	Tetrakis[methylene(3,5-di-*tert*-butyl-4-hydroxyphenyl) propionate] methane
Additin® RC 7207	Rhein Chemie	Octadecyl-3-(3,5-di-*tert*-butyl-4-hydroxyphenyl) propionate
Additin® RC 7209	Rhein Chemie	Hexamethylene bis[3-(3,5-di-*tert*-butyl-4-hydroxyphenyl) propionate]
Additin® RC 7215	Rhein Chemie	Thiodiethylene bis[3-(3,5-di-*tert*-butyl-4-hydroxyphenyl) propionate]
Additin® RC 7235	Rhein Chemie	Benzenepropanoic acid, 3,5-bis-(1,1-dimethylethyl)-4-hydroxy-, C_7–C_9-branched alkyl esters
Ethanox® 4701	SIGroup	2,6-Di-*t*-butyl phenol
Ethanox® 4702	SIGroup	4,4′-Methylene bis-(2,6-di-*t*-butyl phenol)
Ethanox® 4716	SIGroup	3,5-Di-t-butyl-4-hydroxy-hydrocinnamic acid, C_7–C_9 alkyl ester

(Continued)

Product	Company	Chemistry
Ethanox® 4727J	SIGroup	High-molecular-weight hindered phenol
Ethanox® 4782J	SIGroup	High-molecular-weight hindered phenol
Ethanox® 5057	SIGroup	Butylated, octylated diphenylamine
Molyvan® 855	RT Vanderbilt	Organomolybdenum complex
Molyvan® 3000	RT Vanderbilt	Molybdenum dialkyldithiocarbamate
Molyvan® L	RT Vanderbilt	Molybdenum di(2-ethylhexyl) phosphorodithioate
Vanlube® SS	RT Vanderbilt	Octylated diphenylamine
Vanlube® 81	RT Vanderbilt	Dioctyldiphenylamine
Vanlube® 869	RT Vanderbilt	Zinc dithiocarbamate/sulfurized olefin blend
Vanlube® 887	RT Vanderbilt	Tolutriazole compound in oil
Vanlube® 961	RT Vanderbilt	Butylated, octylated diphenylamine
Vanlube® 996E	RT Vanderbilt	Methylene bis(di-butyl-dithiocarbamate) and tolutriazole derivative
Vanlube® 1202	RT Vanderbilt	Octylated *N*-phenyl-1-naphthylamine
Vanlube® 7723	RT Vanderbilt	Methylene bis(dibutyldithiocarbamate)
Vanlube® 8610	RT Vanderbilt	Antimony dithiocarbamate/sulfurized olefin blend
Vanlube® 9317	RT Vanderbilt	Organic amine in synthetic ester
Vanlube® AZ	RT Vanderbilt	Zinc diamyldithiocarbamate in oil
Vanlube® BHC	RT Vanderbilt	3,5-Bis(1,1-Dimethylethyl)-4-hydroxybenzenepropanoic acid, branched alkyl (C = 7–9) ester
Vanlube® EZ	RT Vanderbilt	Zinc diamyldithiocarbamate and diamyl ammonium Diamyldithiocarbamate
Vanlube® NA	RT Vanderbilt	Nonylated, ethylated diphenylamine
Vanlube® RD	RT Vanderbilt	Oligomerized 1,2-dihydro-4-trimethylquinoline
Vanlube® SL	RT Vanderbilt	Octylated, styrenated diphenylamine
Vanlube® W-324	RT Vanderbilt	Dialkylammonium tungstate in oil

1.16 SOME COMMERCIAL METAL DEACTIVATORS

Product	Company	Chemistry
Irgamet® 30	BASF	Triazole derivative
Irgamet® 39	BASF	Tolutriazole derivative
Irgamet® BTZ	BASF	Benzotriazole
Irgamet® TTZ	BASF	Tolutriazole
Cuvan® 303	RT Vanderbilt	*N*,*N*-bis(2-ethylhexyl)-ar-methyl-1H-benzotriazole-1-methanamine
Cuvan® 484	RT Vanderbilt	2,5-Dimercapto-1,3,4-thiadiazole derivative
Cuvan® 826	RT Vanderbilt	2,5-Dimercapto-1,3,4-thiadiazole derivative
Vanlube® 601	RT Vanderbilt	Heterocyclic sulfur–nitrogen compound
Vanlube® 601E	RT Vanderbilt	Heterocyclic sulfur–nitrogen compound
Vanlube® 704	RT Vanderbilt	Proprietary blend

REFERENCES

1. Spikes, H. Low- and zero-sulphated ash, phosphorus and sulphur anti-wear additives for engine oils. *Lubrication Science*, 20(2), 103–136 (2008).

2. Nakazato, M., J. Magarifuchi, A. Mochizuki et al. Low phosphorus engine oil composition and additive compositions. U.S. Patent 6,351,428, Chevron Oronite Company LLC, USPTO, Washington, DC (March 11, 2003).

3. Khorramian, B.A. Phosphorus-free and ashless oil for aircraft and turbo engine application. U.S. Patent 5,726,135 (March 10, 1998).

4. Umehara, K. and W. Van Dam. Ultra-low SAPS lubricants for internal combustion engines. U.S. Patent Appl. 2014/0187455, Chevron-Oronite LLC, USPTO, Washington, DC (July 3, 2014).

5. Buck, W.H., L.O. Farng, D.E. Deckman, S. Kennedy, and A.G. Horodysky. Low-sulfur, low-phosphorus lubricating oil additives containing non-corrosive sulfur compounds and organic borates. PCT 2007120712 A2, ExxonMobil Research & Engineering, World Intellectual Property Organization, Geneva, Switzerland (October 25, 2007).

6. Baird, J. Great Britain Patent 1516 (1872).

7. Knowles, E.C., F.C. McCoy, and J.A. Patterson. Lubricating oil and method of lubricating. U.S. Patent 2,417,305, The Texas Company, USPTO, Washington, DC (March 11, 1947).

8. Lincoln, B.H., W.L. Steiner, and G.D. Byrkit. Sulphur containing lubricant. U.S. Patent 2,218,132, Continental Oil Company, USPTO, Washington, DC (October 15, 1940).

9. Lincoln, B.H., W.L. Steiner, and G.D. Byrkit. Sulphur containing lubricant. U.S. Patent 2,313,248, The Lubri-Zol Development Corporation, USPTO, Washington, DC (March 9, 1943).

10. Lincoln, B.H., W.L. Steiner, and G.D. Byrkit. Method for the synthesis of sulphur-bearing derivatives of high molecular weight. U.S. Patent 2,348,080, Continental Oil Company, USPTO, Washington, DC (May 2, 1944).

11. Farrington, B.B., V.M. Kostainsek, and G.H. Denison, Jr. Compounded lubricant. U.S. Patent 2,346,156, Standard Oil Company of California, USPTO, Washington, DC (April 11, 1944).

12. Hu, S.M., C.L. Gao, J.J. Tang et al. Properties of mono- and dialkyldiphenyl sulfides for high temperature lubricants and their molecular structures. *Acta Petrolei Sinica (Shiyou Xuebao)*, S1, 118–130 (1997).

13. Askew, H.F., G.J.J. Jayne, and J.S. Elliott. Lubricant compositions. U.S. Patent 3,882,031, Edwin Cooper & Company Ltd., USPTO, Washington, DC (May 6, 1975).

14. Spence, J.R. Lubricating compositions containing normal-alkyl substituted 2-thiazoline disulfide antioxidants. U.S. Patent 4,485,022, Phillips Petroleum Company, USPTO, Washington, DC (November 27, 1984).

15. Salomon, M.F. Antioxidant compositions. U.S. Patent 4,764,299, The Lubrizol Corporation, USPTO, Washington, DC (August 16, 1988).

16. Oumar-Mahamat, H., A.G. Horodysky, and A. Jeng. Dihydrobenzothiophenes as antioxidant and antiwear additives for lubricating oils. U.S. Patent 5,514,289, Mobil Oil Corporation, USPTO, Washington, DC (May 7, 1996).

17. Hester, W.F. Fungicidal composition. U.S. Patent 2,317,765, Rohm & Haas Company, USPTO, Washington, DC (April 27, 1943).

18. Denton, W.M. and S.A.M. Thompson. Screening compounds for antioxidant activity in motor oil. *Institute of Petroleum Review*, 20(230), 46–54 (1966).

19. Holubec, A.M. Lubricant compositions. U.S. Patent 3,876,550, The Lubrizol Corporation (April 8, 1975).

20. Karol, T.J., S.G. Donnelly, and R.J. Hiza. Improved antioxidant additive compositions and lubricating compositions containing the same. PCT 03/027215 A2, R.T. Vanderbilt Company, Inc., World Intellectual Property Organization, Geneva, Switzerland (2003).

21. Chesluk, R.P., J.D. Askew, Jr., and C.C. Henderson. Oxidation inhibited lubricating oil. U.S. Patent 4,125,479, Texaco, Inc., USPTO, Washington, DC (November 14, 1978).

22. Yao, Y.B. The application of ashless thiocarbamate as lubricant antioxidation and extreme pressure additive. *Lubricating Oil*, 20(6), 41–44 (2005).

23. Doe, L.A. Antioxidant synergists for lubricating compositions. U.S. Patent 4,880,551, R.T. Vanderbilt Company, Inc., USPTO, Washington, DC (November 14, 1989).

24. deVries, L. and J.M. King. Process of preparing molybdenum complexes, the complexes so-produced and lubricants containing same. U.S. Patent 4,263,152, Chevron Research Company, USPTO, Washington, DC (April 21, 1981).

25. deVries, L. and J.M. King. Process of preparing molybdenum complexes, the complexes so-produced and lubricants containing same. U.S. Patent 4,265,773, Chevron Research Company (May 5, 1981).

26. Stiefel, E.I., J.M. McConnachie, and D.P. Leta et al. Trinuclear molybdenum multifunctional additive for lubricating oils. U.S. Patent 6,232, 276 B1, Infineum USA L.P., USPTO, Washington, DC (May 15, 2001).

27. deVries, L. and J.M. King. Antioxidant combinations of molybdenum complexes and aromatic amine compounds. U.S. Patent 4,370,246, Chevron Research Company, USPTO, Washington, DC (January 25, 1983).

28. Shaub, H. Mixed antioxidant composition. European Patent 719,313 B1, Exxon Chemical Patents, Inc., European Patent Office, Munich (August 6, 1997)

29. Arai, K. and H. Tomizawa. Lubricating oil composition. U.S. Patent 5,605,880, Exxon Chemical Patents, Inc., USPTO, Washington, DC (February 25, 1997).

30. Kelly, J.C. Engine lubricant using molybdenum dithiocarbamate as an antioxidant top treatment in high sulfur base stocks. *IP.com Journal*, 1(6), 22 (2001).

31. Gatto, V.J. Antioxidant system for lubrication base oils. U.S. Patent 5,840,672, Ethyl Corporation, USPTO, Washington, DC (November 24, 1998).

32. Yao, J.B. Recent development of antiwear and extreme pressure-resistant additives for lubricating oils and greases. *Lubricating Oil*, 21(3), 29–37 (2006).

33. Hoffman, D.M., J.J. Feher, and H.H. Farmer. Lubricating compositions containing 5,5′-dithiobis(1,3,4-thiadiazole-2-thiol). U.S. Patent 4,517,103, R.T. Vanderbilt Company, Inc., Washington, DC (May 14, 1985).

34. Karol, T.J., S.G. Donnelly, and R.J. Hiza. Antioxidant additive compositions and lubricating compositions containing the same. U.S. Patent 6,806,241, R.T. Vanderbilt Company, Inc., USPTO, Washington, DC (October 19, 2004).

35. Oberender, F.G., A.W. Godfrey, and M.A. Wiley, Synthetic lubricating composition. U.S. Patent 3,476,685, Texaco, Inc., USPTO, Washington, DC (November 4, 1969).

36. Schumacher, R., S. Evans, and P. Dubs. Lubricant composition. U.S. Patent 5,273,669, Ciba-Geigy Corporation, USPTO, Washington, DC (December 28, 1993).

37. Evans, S. Mixtures and compositions containing phenothiazines. U.S. Patent 5,520,848, Ciba-Geigy Corporation, USPTO, Washington, DC (May 28, 1996).

38. Evans, S., S. Allenbach, and P. Dubs. Process for the preparation of a mixture of alkylated phenothiazines and diphenylamines. U.S. Patent 6,407,231 B1, Ciba Specialty Chemicals Corp., USPTO, Washington, DC (June 18, 2002).

39. Hui, W., W. Zuwang, W. Chaoliang, S. Dong, and W. Fuli, Antioxidant activity of 3,7-di-iso-octyl-phenothiazine and its synergistic effect with 4,4′-di-iso-octyldiphenylamine. *Tribology Transactions*, 50, 273–276, 2011.

40. Salomon, M.F. N-Substituted thio alkyl phenothiazines. U.S. Patent 5,034,019, The Lubrizol Corporation, USPTO, Washington, DC (July 23, 1991).

41. Kapuscinski, M.M. and R.T. Biggs. Dispersant and antioxidant VI improver based on olefin copolymers containing phenothiazine and aromatic amine groups. U.S. Patent 5,942,0471, Ethyl Corporation, USPTO, Washington, DC (August 24, 1999).

42. Colclough, T., M. Beltzer, and J. Habeeb. Lubricating compositions. U.S. Patent 5,558,805, Exxon Chemical Patents, Inc., USPTO, Washington, DC (September 24, 1996).

43. Brown, A.L. Treatment of hydrocarbon oils. U.S. Patent 1,234,862, Westinghouse and Electric Manufacturing Company, USPTO, Washington, DC (July 31, 1917).

44. Ashburn, H.V. and W.G. Alsop. Lubricating oil. U.S. Patent 2,221,162, The Texas Company, USPTO, Washington, DC (November 12, 1940).

45. Hall, F.W. and C.G. Towne. Method of lubrication. U.S. Patent 2,257,601, The Texas Company, USPTO, Washington, DC (September 30, 1941).

46. Musher, S. Lubricating oil and the method of making the same. U.S. Patent 2,223,941, The Musher Foundation, USPTO, Washington, DC (December 3, 1940).

47. Loane, C.M. and J.W. Gaynor. Lubricant. U.S. Patent 2,322,859, Standard Oil Company, USPTO, Washington, DC (June 29, 1943).

48. Moran, R.C., W.L. Evers, and E.W. Fuller. Petroleum product and method of making same. U.S. Patent 2,058,343, Socony-Vacuum Oil Company, Inc., USPTO, Washington, DC (October 20, 1936).

49. Moran, R.C. and A.P. Kozacik. Mineral oil composition. U.S. Patent 2,151,300, Socony-Vacuum Oil Company, Inc., USPTO, Washington, DC (March 21, 1939).

50. Cohen, S.C. Synergistic antioxidant system for severely hydrocracked lubricating oils. U.S. Patent 5,124,057, Petro-Canada, Inc., USPTO, Washington, DC (June 23, 1992).

51. Farrington, B.B. and J.O. Clayton. Compounded mineral oil. U.S. Patent 2,228,658, Standard Oil Company of California, USPTO, Washington, DC (January 14, 1941).

52. Farrington, B.B., J.O. Clayton, and J.T. Rutherford. Compounded mineral oil. U.S. Patent 2,228,659, Standard Oil Company of California, USPTO, Washington, DC (January 14, 1941).

53. Meyers, D. Method of lubricating compression cylinders used in the manufacture of high-pressure polyethylene. U.S. Patent 6,172,014 B1, Pennzoil-Quaker State, USPTO, Washington, DC (January 9, 2001).

54. Holt, A. and G. Mulqueen. Stabilizing compositions for lubricating oils. U.S. Patent Appl. 2003/0171227, Great Lakes Chemicals, USPTO, Washington, DC (September 11, 2003).

55. Dong, J. and C.A. Migdal. Stabilized lubricant compositions. U.S. Patent 7,829,511 B2, Chemtura Corporation, USPTO, Washington, DC (November 9, 2010).

56. Durr, S.M. and R.A. Krenowicz. Turbine oil compositions. U.S. Patent 3,923,672, Continental Oil Company, USPTO, Washington, DC (December 2, 1975).

57. Messina, N.V. and D.R. Senior. Stabilized fluids. U.S. Patent 3,556,999, Rohm and Haas Company, USPTO, Washington, DC (January 19, 1971).

58. Gelbin, M. and J. Dong. Phosphite stabilizer for lubricating base stocks and thermoplastic polymers, U.S. Patent 8,049,041 B2, Chemtura Corporation, USPTO, Washington, DC (November 1, 2011).

59. Dong, J. and C.A. Migdal. Stabilized lubricant compositions. U.S. Patent 7,799,101 B2, Chemtura Corporation, USPTO, Washington, DC (September 21, 2010).

60. Sheets, R.M., G.H. Guinther, and J.T. Loper. Lubricant formulations and method of lubricating a combustion system to achieve improved emissions catalyst durability. U.S. Patent Appl. 2010/0056407, Afton Chemical Corporation, USPTO, Washington, DC (March 4, 2010).

61. Hartley, R.J., S. Rea, and M. Waddoups, Lubricating oil composition for outboard engines. U.S. Patent 6,642,188, Infineum International LTD., USPTO, Washington, DC (November 4, 2003).

62. Selby, T.W., R.J. Bosch, and D.C. Fee. Phosphorus additive chemistry and its effects on the phosphorus volatility of engine oils. *Journal of ASTM International*, 2(9), Paper ID JAI12977, (Downloaded 07/07/15, no pp. given). (October 2005).

63. Bosch, R.J., D.C. Fee, and T.W. Selby. Continued studies of the causes of engine oil phosphorus volatility—Part 2. SAE International, Warrendale, PA, [Special Publication] SP (2007), SP-2090 (General Emissions), pp. 25–33 (2007).

64. Sarin, R., D.K. Tuli, A.V. Sureshbabu et al. Molybdenum dialkylphosphorodithioates: Synthesis and performance evaluation as multifunctional additives for lubricants. *Tribology International*, 27(6), 379–386 (1994).

65. Vanderbilt Chemicals LLC, Organo molybdenum compounds, http://www.vanderbiltchemicals.com/product_categories/product_listing/category/ga-ao-organo-molybdenum-compounds, accessed February 22, 2017.

66. Levine, S.A., R.C. Schlicht, H. Chafetz et al. Molybdenum derivatives and lubricants containing same. U.S. Patent 4,428,848, Texaco, Inc., USPTO, Washington, DC (January 31, 1984).

67. Nalesnik, T.E. and C.A. Migdal. Oil-soluble molybdenum multifunctional friction modifier additives for lubricant compositions. U.S. Patent 6,103,674, Uniroyal Chemical Company, Inc., USPTO, Washington, DC (August 15, 2000).

68. Sarin, R., D.K. Tuli, V. Martin et al. Development of N, P and S-containing multifunctional additives for lubricants. *Lubrication Engineering*, 53(5), 21–27 (1997).

69. Ripple, D.E. Zinc-free farm tractor fluid. PCT 2007005423 A2, The Lubrizol Corporation World Intellectual Property Organization, Geneva, Switzerland (January 11, 2007).

70. Schadenberg, H. Dithiophosphate ester derivatives and their use for stabilizing organic material. British Patent 1,506,917. Intellectual Property Office, London (September 30, 1975).

71. Davis, R.H., A. Okorodudu, and M. Sedlak. Lubricant compositions containing a dithiophosphoric acid ester-aldehyde reaction product. European Patent Appl. 00090506 A2, Mobil Oil Corporation European Patent Office, Munich (March 3, 1983).

72. Braid, M. Lubricating oils or fuels containing adducts of phosphorodithioate esters. U.S. Patent 3,644,206, Mobil Oil Corporation, USPTO, Washington, DC (February 22, 1972).

73. Buc, H.E. Stabilized refined mineral, vegetable, and animal oils. U.S. Patent 2,031,930, Standard Oil Development Company, USPTO, Washington, DC (February 25, 1936).

74. Stevens, D.R. and W.A. Gruse. Dialkyl-cyclohexyl-phenols. U.S. Patent 2,248,827, Gulf Research & Development Company, USPTO, Washington, DC (July 8, 1941).

75. Howland, L.H. Antioxidant. U.S. Patent 2,200,747, United States Rubber Company, USPTO, Washington, DC (May 14, 1940).

76. Craig, D. Method of preparing substituted diphenylamines. U.S. Patent 2,225,368, B.F. Goodrich Company, USPTO, Washington, DC (December 17, 1940).

77. Paul, P.T. Aryl-amino alkenyl phenols. U.S. Patent 2,246,942, United States Rubber Company, USPTO, Washington, DC (June 24, 1941).

78. Hollis, A.L. Diphenylamine derivatives. U.S. Patent 2,530,769, B.F. Goodrich Company, USPTO, Washington, DC (November 21, 1950).

79. Dubois, L.O. and R.N. Gartside. Continuous manufacture of nitrobenzene. U.S. Patent 2,773,911, E.I. du Pont de Nemours and Company, USPTO, Washington, DC (December 11, 1956).

80. Karkalits, O.C., Jr., C.M. Vanderwaart, and F.H. Megson. New catalyst for reducing nitrobenzene and the process of reducing nitrobenzene thereover. U.S. Patent 2,891,094, American Cyanamid Co., USPTO, Washington, DC (June 16, 1959).

81. Addis, G.I. Vapor phase process for the manufacture of diphenylamine. U.S. Patent 3,118,944, American Cyanamid Co., USPTO, Washington, DC (January 21, 1964).

82. Franklin, J. Liquid antioxidant produced by alkylating diphenylamine with a molar excess of diisobutylene. U.S. Patent 4,824,601, Ciba-Geigy Corporation, USPTO, Washington, DC (April 25, 1989).

83. Lai, J.T. Method of manufacturing alkylated diphenylamine compositions and products thereof. U.S. Patent 6,204,412, The B.F. Goodrich Company, USPTO, Washington, DC (March 12, 2001).

84. Lai, J.T. and D.S. Filla. Liquid alkylated diphenylamine antioxidant. U.S. Patent 5,672,752, B.F. Goodrich Company, USPTO, Washington, DC (September 30, 1997).

85. Onopchenko, A. Alkylation of diphenylamine with polyisobutylene oligomers. U.S. Patent 6,355,839, Chevron U.S.A., Inc., USPTO, Washington, DC (March 12, 2002).

86. Patil, A.O. Diphenylamine functionalization of poly-α-olefins. U.S. Patent 7,847,030 B2, ExxonMobil Research and Engineering Company, USPTO, Washington, DC (December 7, 2010).

87. Lai, J.T. Synthetic lubricant antioxidant from monosubstituted diphenylamines. U.S. Patent 5,489,711, The B.F. Goodrich Company, USPTO, Washington, DC (February 6, 1996).

88. Bandlish, B.K., F.C. Loveless, and W. Nudenberg. Amino compounds and use of amino compounds as antioxidants in lubricating oils. European Patent 022281 B1, Uniroyal, Inc., European Patent Office, Munich (September 14, 1983).

89. Nalesnik, T.E. Diaromatic amines. U.S. Patent 8,017,805 B2, Chemtura Corporation, USPTO, Washington, DC (September 13, 2011).

90. Ma, Q. and C.A. Migdal. Acridan derivatives as antioxidants. U.S. Patent 7,847,125 B2, Chemtura Corporation, USPTO, Washington, DC (December 7, 2010).

91. Ma, Q. and C.A. Migdal. Diaromatic amine derivatives as antioxidants. U.S. Patent 7,838,703 B2, Chemtura Corporation, USPTO, Washington, DC (November 23, 2010).

92. Galic Raguz, M. and V.A. Carrick. Lubricating composition with improved TBN retention. PCT 2012/166781A1, The Lubrizol Corporation, World Intellectual Property Organization, Geneva, Switzerland (December 6, 2012).

93. Nebzydoski, J.W., E.L. Patmore, and I.D. Rubin. Synthetic aircraft turbine oil. U.S. Patent 3,850,524, Texaco, Inc., USPTO, Washington, DC (November 26, 1974).

94. Odorisio, P.A., D.E. Chasan, and S.D. Pastor. Substituted 1-aminonaphthalenes and stabilized compositions. U.S. Patent 5,160,647, Ciba-Geigy Corporation, USPTO, Washington, DC (November 3, 1992).

95. Lai, J.T. Lubricant composition. U.S. Patent 6,426,324, Noveon IP Holdings Corp. and BP Exploration & Oil, Inc., USPTO, Washington, DC (July 30, 2002).

96. Aebli, B.M., S. Evans, M. Ribeaud, and D.E. Chasan. Alkylated PANA and DPA compositions. U.S. Patent 8,030,259 B2, Ciba Specialty Chemicals Corporation, USPTO, Washington, DC (October 4, 2011).

97. Andress, H.J. and H. Ashjian. Products of reaction involving alkenylsuccinic anhydrides with aminoalcohols and aromatic secondary amines and lubricants containing same. U.S. Patent 4,522,736, Mobil Oil Corporation, USPTO, Washington, DC (June 11, 1985).

98. Nelson, K.D. and E.A. Chiverton. Fused aromatic amine based wear and oxidation inhibitors for lubricants. U.S. Patent 8,138,130 B2, Chevron Texaco Corporation, USPTO, Washington, DC (March 20, 2012).

99. Migdal, C.A., T.E. Nalesnik, and C.S. Liu. Dispersant and antioxidant additive and lubricating oil composition containing same. U.S. Patent 5,075,383, Texaco Inc., USPTO, Washington, DC (December 24, 1991).

100. Rowland, R.G. Cross-products and co-oligomers of phenylenediamines and aromatic amines as antioxidants for lubricants. U.S. Patent 8,987,515 B2, Chemtura Corp., USPTO, Washington, DC (March 24, 2015).

101. Andress, H.J., Jr. and R.H. Davis. Arylamine-aldehyde lubricant antioxidants. European Patent Appl. 0083871 A2, Mobil Oil Corporation, European Patent Office, Munich (July 20, 1983).

102. Goujon, G., F. Severac, and M. Borel-Garlin. Use of an oligomer-based additive for stabilizing a lubricating composition for a conveyor chain. U.S. Patent 8,492,321 B2, NYCO S.A., USPTO, Washington, DC (July 23, 2013).

103. Braid, M. and D. Law. Oil soluble oxidized naphthylamine compositions. U.S. Patent 3,509,214, Mobil Oil Corporation, USPTO, Washington, DC (April 28, 1970).

104. Cherpeck, R.E. and C.Y. Chan. Synergistic lubricating oil compositions containing a mixture of a benzo[b]perhydroheterocyclic arylamine and a diarylamine. U.S. Patent 8,003,583 B2, Chevron Oronite Company LLC, USPTO, Washington, DC (August 23, 2011).

105. Cherpeck, R.E. and C.Y. Chan. Synergistic lubricating oil compositions containing a mixture of a benzo[b]perhydroheterocyclic arylamine and a diarylamine. U.S. Patent 7,501,386 B2, Chevron Oronite Company LLC, USPTO, Washington, DC (March 10, 2009).

106. Cherpeck, R.E. and C.Y. Chan. Synergistic lubricating oil composition containing a mixture of nitro-substituted arylamine and a diarylamine. U.S. Patent 7,683,017 B2, Chevron Oronite Company LLC, USPTO, Washington, DC (March 23, 2010).

107. Muller, R. and W. Hartmann. N,N'-diphenyl-p-phenylenediamines, method for their production and their use as stabilizers for organic materials. European Patent Appl. 072575 A1, Chemische Werke Lowi GmbH., European Patent Office, Munich (February 23, 1983).

108. Colclough, T. Lubricating oil components and additives for use therein. U.S. Patent 5,232,614, Exxon Chemical Patents, Inc., USPTO, Washington, DC (August 2, 1993).

109. Malandro, D.L. and M.L. Alessi. EGR equipped diesel engines and lubricating oil compositions. U.S. Patent 8,741,824 B2, Infineum International LTD., USPTO, Washington, DC (June 3, 2014).

110. Patil, A.O. and J.J. Habeeb. Lubricant compositions containing ashless catalytic antioxidant additives. U.S. Patent 7,586,007 B2, ExxonMobil Research and Engineering Company, USPTO, Washington, DC (September 8, 2009).

111. Patil, A.O., J.J. Habeeb, M.E. Landis, M.A. Francisco, and M. Varma-Nair. Lubricant compositions containing ashless catalytic antioxidant additives. U.S. Patent 7,977,286 B2, ExxonMobil Research and Engineering Company, USPTO, Washington, DC (July 12, 2009).

112. Bera, T.K., R.J. Hartley, J. Emert, J. Cheng, T.E. Nalesnik, and R.G. Rowland. Additive compositions. U.S. Patent 8,530,397 B2, Infineum International LTD, Chemtura Corporation, USPTO, Washington, DC (September 10, 2013).

113. Ma, Q. Tetraaromatic diamine compounds as antioxidants. U.S. Patent 7,608,568 B2, Chemtura Corporation, USPTO, Washington, DC (October 27, 2009).

114. Duyck, K., T.E. Nalesnik, and W. Batorewicz. Antioxidant hydrazides and derivatives thereof having multifunctional activity. U.S. Patent 7,375,061 B2, Chemtura Corporation, USPTO, Washington, DC (May 20, 2008).

115. Covitch, M.J., M.K. Pudelski, C. Friend, M.D. Gieselman, R.A. Eveland, M. Galic Raguz, and B.J. Schober. Dispersant viscosity modifiers containing aromatic amines. U.S. Patent 7,960,320 B2, The Lubrizol Corporation, USPTO, Washington, DC (June 14, 2011).

116. Gieselman, M.D. Mannich post-treatment of PIBSA dispersants for improved dispersion of EGR soot. U.S. Patent 8,324,139 B2, The Lubrizol Corporation, USPTO, Washington, DC (December 4, 2012).

117. Gieselman, M.D., C. Friend, and A.J. Preston. Lubricant composition containing a polymer. U.S. Patent 8,637,437 B2, The Lubrizol Corporation, USPTO, Washington, DC (January 28, 2014).

118. Wright, W.E. Antioxidant diamine. U.S. Patent 4,456,541, Ethyl Corporation, USPTO, Washington, DC (June 26, 1984).

119. Cherpeck, R.E., C.Y. Chan, and G. Bhalla. Lubricating oil compositions containing a tetraalkyl-napthalene-1,8 diamine antioxidant. U.S. Patent 8,623,798, Chevron Oronite Company LLC, USPTO, Washington, DC (January 7, 2014).

120. Malherbe, R.F. 2,3-Dihydroperimidines as antioxidants for lubricants. U.S. Patent 4,389,321, Ciba-Geigy Corporation, USPTO, Washington, DC (June 21, 1983).

121. Roberts, J.T. N,N'-Disubstituted 2,4'-diaminodiphenyl ethers as antioxidants. U.S. Patent 4,309,294, UOP, Inc., USPTO, Washington, DC (January 5, 1982).

122. Roberts, J.T. Imines of 2,4-diaminodiphenyl ethers as antioxidants for lubricating oils and greases. U.S. Patent 4,378,298, UOP, Inc., USPTO, Washington, DC (March 29, 1983).

123. Braid, M. Lubricant compositions containing N-tertiary alkyl benzotriazoles. U.S. Patent 4,519,928, Mobil Oil Corporation, USPTO, Washington, DC (May 28, 1985).

124. Shim, J. Lubricant compositions containing antioxidant mixtures comprising substituted thiazoles and substituted thiadiazole compounds. U.S. Patent 4,260,501, Mobil Oil Corporation, USPTO, Washington, DC (April 7, 1981).

125. Shim, J. Multifunctional additives. U.S. Patent 4,511,481, Mobil Oil Corporation, USPTO, Washington, DC (April 16, 1985).

126. Shim, J. Triazole-dithiophosphate reaction product and lubricant compositions containing same. U.S. Patent 4,456,539, Mobil Oil Corporation, USPTO, Washington, DC (June 26, 1984).

127. Sung, R.L. and B.H. Zoleski. Diesel lubricant composition containing 5-amino-triazole-succinic anhydride reaction product. U.S. Patent 4,256,595, Texaco, Inc., USPTO, Washington, DC (March 17, 1981).

128. Mikeska, L.A. and E. Lieber. Preparation of phenol sulfides. U.S. Patent 2,139,321, Standard Oil Development Company, USPTO, Washington, DC (December 6, 1938).

129. Mikeska, L.A. and C.A. Cohen. Mineral oil stabilizing agent and composition containing same. U.S. Patent 2,139,766, Standard Oil Development Company, USPTO, Washington, DC (December 13, 1938).

130. Mikeska, L.A. and E. Lieber. Stabilized lubricating composition. U.S. Patent 2,174,248, Standard Oil Development Company, USPTO, Washington, DC (September 26, 1939).

131. Mikeska, L.A. and E. Lieber. Polymerization and condensation products. U.S. Patent 2,239,534, Standard Oil Development Company, USPTO, Washington, DC (April 22, 1941).

132. Richardson, R.W. Oxidation inhibitor. U.S. Patent 2,259,861, Standard Oil Development Company, USPTO, Washington, DC (October 21, 1941).

133. Braid, M. Phenolic antioxidants and lubricants containing same. U.S. Patent 4,551,259, Mobil Oil Corporation, USPTO, Washington, DC (November 5, 1985).

134. McKinnie, B.G. and P.F. Ranken. (Hydrocarbylthio)phenols and their preparation. U.S. Patent 4,533,753, Ethyl Corporation, USPTO, Washington, DC (August 6, 1985).

135. Lam, W.Y. and V.J. Gatto. Sulfurized phenolic antioxidant composition, method of preparing same, and petroleum products containing same. U.S. Patent 6,096,695, Ethyl Corporation, USPTO, Washington, DC (August 1, 2000).

136. Gatto, V.J. and A. Kadkhodayan. Sulfurized phenolic antioxidant composition method of preparing same and petroleum products containing same. U.S. Patent 6,001,786, Ethyl Corporation, USPTO, Washington, DC (December 14, 1999).

137. Dong, J. and C.A. Migdal. Synergistic antioxidant systems for lubricants. *12th Asia Fuels and Lubes Conference Proceedings*, Hong Kong, 2006.

138. Rosenberger, S. and K. Schwarzenbach. Use of phenol-mercaptocarboxylic acid esters as stabilizers for lubricants. U.S. Patent 4,954,275, Ciba-Geigy Corporation, USPTO, Washington, DC (September 4, 1990).

139. Park, K.P. and A.P. Vellturo. Process for the production of hydroxyalkylphenyl derivatives. U.S. Patent 4,085,132, Ciba-Geigy Corporation, USPTO, Washington, DC (April 18, 1987).

140. Crawley, S.L. and V. Carrick. Hydroxychroman derivatives as engine oil antioxidants. U.S. Patent Appl. 2013/165355 A, The Lubizol Corporation, USPTO, Washington, DC (June 27, 2013).

141. Hartley, J.P. Lubricating oil composition. U.S. Patent 9,109,182 B2, Infineum International LTD., USPTO, Washington, DC (August 18, 2015).

142. Hartley, J.P. Marine engine lubrication. U.S. Patent 8,609,599 B2, Infineum International LTD., USPTO, Washington, DC (December 17, 2013).

143. Premi, T.H. and K. Jayaprakash. Histopathalogical observations of anacardic acid—A phytotoxin responsible for occupational dermatitis. *Chemical Science Review and Letters*, 3(10), 162–165 (2014).

144. Ingold, K.U. and Pratt, D.A. Advances in radical-tapping antioxidant chemistry in the 21st century: A kinetics and mechanisms perspective. *Chemical Reviews*, 114, 9022–9046 (2014).

145. Nelson, L.A. and R.R. Leslie. Mannich type compounds as antioxidants. U.S. Patent 5,338,469, Mobil Oil Corporation, USPTO, Washington, DC (August 16, 1994).

146. Oumar-Mahamat, H. and A.G. Horodysky. Phenolic imidazoline antioxidants. U.S. Patent 5,846,917, Mobil Oil Corporation, USPTO, Washington, DC (December 8, 1998).

147. Camenzind, H., A. Dratva, and P. Hanggi. Ash-free and phosphorus-free antioxidants and antiwear additives for lubricants. European Patent Appl. 894,793, Ciba Specialty Chemicals Holding, Inc., European Patent Office, Munich (February 3, 1999).

148. Hsu, S.Y. and A.G. Horodysky. Sulfur-containing ester derivative of arylamines and hindered phenols as multifunctional antiwear and antioxidant additives for lubricants. U.S. Patent 5,304,314, Mobil Oil Corporation., USPTO, Washington, DC (April 19, 1994).

149. Loper, J.T. Dispersant reaction product with antioxidant capability. U.S. Patent 7,645,726 B2, Afton Chemical Corporation, USPTO, Washington, DC (January 12, 2010).

150. Colclough, T. Lubricating oil oxidation and stabilization, in *Atmospheric Oxidation and Antioxidants*, G. Scott ed., Elsevier Science Publishers B.V. Amsterdam, The Netherlands, pp. 1–69 (1993).

151. Klaus, E.E., J.L. Duda, and J.C. Wang. Study of copper salts as high-temperature oxidation inhibitors. *Tribology Transactions*, 35(2), 316–324 (1992).

152. Holt, D.G.L. Multiring aromatics for enhanced deposit control. European Patent Appl. 0709447, Exxon Research and Engineering Company, European Patent Office, Munich (May 1, 1996).

153. Colclough, T., F.A. Gibson, and J.F. Marsh. Lubricating oil compositions containing ashless dispersant, zinc dihydrocarbyldithiophosphate, metal detergent and a copper compound. U.S. Patent 4,867,890 (September 19, 1989).

154. Downing, F.B. and H.M. Fitch. Lubricant. U.S. Patent 2,343,756, E.I. du Pont de Nemours & Company., USPTO, Washington, DC (March 7, 1944).

155. Fox, A.L. Solution of copper mercaptides from terpenes. U.S. Patent 2,349,820, E.I. du Pont de Nemours & Company, USPTO, Washington, DC (May 30, 1944).

156. Downing, F.B. and H.M. Fitch. Lubricating oil. U.S. Patent 2,356,661, E.I. du Pont de Nemours & Company, USPTO, Washington, DC (August 22, 1944).

157. Field, I.P. Lubricating compositions. U.S. Patent 5,731,273, Exxon Chemical Patents, Inc., USPTO, Washington, DC (March 24, 1988).

158. Gurierrez, A., D.W. Brownawell, and S. J. Brois. Copper complexes of oxazolines and lactone oxazolines as lubricating oil additives. U.S. Patent 4,486,326, Exxon Research & Engineering Co., USPTO, Washington, DC (December 4, 1984).

159. Hopkins, T.R. Copper salts of succinic anhydride derivatives. U.S. Patent 4,552,677, The Lubrizol Corporation, USPTO, Washington, DC (November 12, 1985).

160. Farng, L.O. and A.G. Horodysky. Copper salts of hindered phenolic carboxylates and lubricants and fuel containing same. U.S. Patent 4,828,733, Mobil Oil Corporation, USPTO, Washington, DC (May 9, 1989).

161. Horodysky, A.G. Borated epoxides and lubricants containing same. U.S. Patent 4,410,438, Mobil Oil Corporation, USPTO, Washington, DC (November 18, 1983).

162. Horodysky, A.G. Borated hydroxyl-containing compositions and lubricants containing same. U.S. Patent 4,788,340, Mobil Oil Corporation, USPTO, Washington, DC (November 29, 1998).

163. Farng, L.O. and A.G. Horodysky. Mixed hydroquinone-hydroxyester borates as antioxidants. U.S. Patent 4,828,740, Mobil Oil Corporation, USPTO, Washington, DC (May 9, 1989).

164. Braid, M. Phenol-hindered phenol borates and lubricant compositions containing same. U.S. Patent 4,530,770, Mobil Oil Corporation, USPTO, Washington, DC (July 23, 1985).

165. Koch, F.W. Boron-containing compositions and lubricants containing them. U.S. Patent 5,240,624, The Lubrizol Corporation, USPTO, Washington, DC (August 31, 1993).

166. Horodysky, A.G. and J.M. Kaminski. Friction reducing additives and compositions thereof. U.S. Patent 4,298,486, Mobil Oil Corporation, USPTO, Washington, DC (November 3, 1981).

167. Horodysky, A.G. and J.M. Kaminski. Friction reducing additives and compositions thereof. U.S. Patent 4,478,732, Mobil Oil Corporation, USPTO, Washington, DC (October 23, 1984).

168. Horodysky, A.G. and R.S. Herd. Etherdiamine borates and lubricants containing same. U.S. Patent 4,537,692, Mobil Oil Corporation, USPTO, Washington, DC (August 27, 1985).

169. Yao, J.B., W.L. Wang, S.Q. Chen et al. Borate esters used as lubricant additives. *Lubrication Science* 14(4), 415–423 (August 2002).

170. Horodysky, A.G. Borated adducts of diamines and alkoxides, as multifunctional lubricant additives, and compositions thereof. U.S. Patent 4,549,975, Mobil Oil Corporation, USPTO, Washington, DC (October 19, 1985).

171. Shough, A.M., D.J. Baillargeon, S.M. Jetter et al. Method for improving nitrile seal compatibility with lubricating oils. U.S. Patent Appl. 2014/0038864, ExxonMobil Research and Engineering Company, USPTO, Washington, DC (February 6, 2014).

172. Harrison, J.J. and W.R. Ruhe, Jr. Polyalkylene succinimides and post-treated derivatives thereof. U.S. Patent 5,716,912, Chevron Chemical Company, USPTO, Washington, DC (February 10, 1998).

173. Yao, J.B. and P. Ma. Interaction of organic borate ester containing nitrogen with other lubricant additives. *Lubricating Oil (Runhuayou)*, 21(2), 32–47 (2006).

174. Stanulov, K., H.N. Harbara, and G. Cholakov. Antioxidation properties of boron-containing lubricant additives and their mixtures with Zn dialkyldithiophosphates. *Oxidation Communications*, 22(3), 374–386 (1999).

175. Farng, L.O. and A.G. Horodysky. Lubricant composition containing phenolic/phosphorodithioate borates as multifunctional additives. U.S. Patent 4,956,105, Mobil Oil Corporation, USPTO, Washington, DC (September 11, 1990).

176. Hinkamp, J.B., J.D. Bartleson, and G.E. Irish. Boron esters and process of preparing same. U.S. Patent 3,356,707, Ethyl Corporation, USPTO, Washington, DC (December 5, 1967).

177. M. Braid. Borate esters and lubricant compositions containing such esters. U.S. Patent 4,547,302, Mobil Oil Corporation, USPTO, Washington, DC (October 10, 1985).

178. Small, V.R., Jr., T.V. Liston, and A. Onopchenko. Diethylamine complexes of borated alkyl catechols and lubricating oil compositions containing the same. U.S. Patent 4,975,211, Chevron Research Company, USPTO, Washington, DC (December 4, 1990).

179. Small, V.R., Jr., T.V. Liston, and A. Onopchenko. Diethylamine complexes of borated alkyl catechols and lubricating oil compositions containing the same. U.S. Patent 5,061,390, Chevron Research Company, USPTO, Washington, DC (October 29, 1991).

180. Yao, J.B. and J.X. Dong. Improvement of hydrolytic stability of borate esters used as lubricant additives. *Lubrication Engineering*, 51(6), 475–479 (1995).

181. Wright, W.E. and B.T. Davis. Haze-free boronated antioxidant. U.S. Patent 4,927,553, Ethyl Corporation, USPTO, Washington, DC (May 22, 1990).

182. Brown, J.R., P.E. Adams, V.A. Carrick et al. Titanium compounds and complexes as additives in lubricants. U.S. Patent 7,727,943 B2, The Lubrizol Corporation, USPTO, Washington, DC (June 1, 2010).

183. Esche, C.K., Jr. Additives and lubricant formulations for improved antioxidant properties. U.S. Patent 7,615,520, New Market Services Corporation, USPTO, Washington, DC (November 10, 2009).

184. Chase, K.J., J.M. DeMassa, B. Stunkel, G.A. Mazzamarro, and S.G. Donnelly. Lubricant antioxidant compositions containing a metal compound and a hindered amine. U.S. Patent 8,093,190 B2, R.T. Vanderbilt Company, Inc., USPTO, Washington, DC (January 10, 2012).

185. Ravichandran, R., F. Abi-Karam, A. Yermolenka et al. Amine tungstates and diarylamines in lubricant compositions. European Patent 1,907,517 B1, King Industries, Inc., European Patent Office, Munich (December 5, 2012).

186. Tynik, R.J., S.G. Donnelly, and G.A. Aguilar. Antioxidant additive for lubricant compositions, comprising organotungstate. U.S. Patent 7,858,565 B2, R.T. Vanderbilt Company, Inc., USPTO, Washington, DC (December 28, 2010).

187. Tynik, R.J., S.G. Donnelly, and G.A. Aguilar. Antioxidant additive for lubricant compositions, comprising organotungstate, diarylamine and organomolybdenum compounds. U.S. Patent 7,879,777 B2, R.T. Vanderbilt Company, Inc., USPTO, Washington, DC (February 1, 2011).

188. Gatto, V.J. and M.T. Devlin. Lubricant containing molybdenum compound and secondary diarylamine. U.S. Patent 5,650,381, Ethyl Corporation, USPTO, Washington, DC (July 22, 1997).

189. Hu, J.Q., X.Y. Wei, G.L. Dai, C.C. Liu, Y. Fu, Z.M. Zong, and J.B. Yao. Study demonstrating enhanced oxidation stability when arylamine antioxidants are combined with organic molybdenum complexes. Tribology Transactions, 50(2), 205–210 (2007).

190. Burrington, J.D. and B.P. Leffel. Viscosity modifiers in controlled release lubricant additive gels. U.S. Patent 7,833,955 B2 (November 16, 2010).

191. Burrington, J.D., B.H. Grasser, H.F. George, J.R. Martin, J.K. Pudelski, J.P. Roski, B.L. Soukup, and M.E. Bartlett. Slow release lubricant additives gel. U.S. Patent 6,843,916 B2, The Lubizol Corporation (January 18, 2005).

192. Calcavecchio, P., L.O. Farng, J.M. Krylowski, M.N. Webster, V. Minak-Bernero, and E.N. Drake. Microencapsulation of lubricant additives. U.S. Patent App. 2014/0087982 A1, ExxonMobil Research and Engineering Company (March 27, 2014).

193. Rasberger, M. Oxidative degradation and stabilisation of mineral oil based lubricants, in Chemistry and Technology of Lubricants, Motier, R.M. and S.T. Orszulik, eds., Blackie Academic & Professional, Blackie, London, U.K., pp. 98–143 (1997).

194. Paolino, P.R. Antioxidants, in Thermoplastic Polymer Additives, Lutz, J.T., Jr., ed., Marcel Dekker, Inc., New York, pp. 1–35 (1989).

195. Reyes-Gavilan, J.L. and R. Odorisio. A review of the mechanisms of action of antioxidants, metal deactivators and corrosion inhibitors, NLGI Spokesman, 64(11), 22–33 (2001).

196. Pospisil, J. Aromatic and heterocyclic amines in polymer stabilization. Advances in Polymer Science, 124, 87–190 (1995).

197. Lowry, T.H. and K.S. Richardson. Mechanism and Theory in Organic Chemistry, Harper and Row Publishers, New York, p. 472 (1976).

198. Golden, D.M. and S.W. Benson. Free-radical and molecule thermochemistry from studies of gas-phase iodine-atom reactions. Chemical Reviews, 69, 125–134 (1969).

199. Colclough, T. Role of additives and transition metals in lubricating oil oxidation, Industrial & Engineering Chemistry Research, 26, 1888–1895 (1987).

200. Maleville, X., D. Faure, A. Legros et al. Oxidation of mineral base oils of petroleum origin: The relationship between chemical composition, thickening, and composition of degradation, Lubrication Science, 9, 3–60 (1996).

201. Jensen, R.K., S. Korcek, L.R. Mahoney et al. Liquid-phase autoxidation of organic compounds at elevated temperatures. 1. The stirred flow reactor technique and analysis of primary products from n-hexadecane autoxidation at 120–180°C, Journal of the American Chemical Society, 101, 7574–7584 (1979).

202. March, J. Advanced Organic Chemistry: Reactions, Mechanisms and Structure, 4th Ed., John Wiley & Sons, Inc., New York, pp. 937–945 (1992).

203. Kouame, S.-D.B., R.L. Vander Wal, and J. Perez. Deposit formation from lubricant degradation: A uniform layer deposition model. Lubrication Science, 27, 1–13 (2015).

204. Hamblin, P.C. and P. Rohrbach. Piston deposit control using metal-free additives. Lubrication Science, 14(1), 3–22 (November 2001).

205. Lubes'N'Greases Supplement. 2015 Lubricants Industry Factbook, Persaud, M. ed., 21(8), 48 (2015).

206. Murray, D.W., C.T. Clarke, G.A. MacAlpine et al. The effect of base stock composition on lubricant performance, SAE Technical Paper 821236. Society of Automotive Engineers (SAE), Warrendale, PA.

207. Cerny, J., M. Pospisil, and G. Sebor. Composition and oxidative stability of hydrocracked base oils and comparison with a PAO. Journal of Synthetic Lubrication, 18(3), 199–213 (October 2001).

208. Kramer, D.C., J.N. Ziemer, M.T. Cheng et al. Influence of group II & III base oil composition on V.I. and oxidative stability. Presented at the 66th NLGI Annual Meeting, Tucson, AZ (October 1999).

209. Igarashi, J., T. Yoshida, and H. Watanabe. Concept of optimal aromaticity in base oil oxidative stability revisited. Symposium on Worldwide Perspectives on the Manufacture, Characterization and Application of Lubricant Base Oils: 213th Annual Meeting Preprint, American Chemical Society, San Francisco, CA (April 13–17, 1997).

210. Wang, H.D. and X.L. Hu. The study advance on effects of molecular structure of group II, III base oils on the oxidation stability. Lubricating Oil, 20(2), 10–14 (2005).

211. Adhvaryu, A., S.Z. Erhan, and I.D. Singh. The effect of molecular composition on the oxidative behavior of group I base oils in the presence of an antioxidant additive. Lubrication Science, 14(2), 119–129 (February 2002).

212. Yoshida, T., J. Igarashi, H. Watanabe et al. The impact of basic nitrogen compounds on the oxidative and thermal stability of base oils in automotive and industrial applications, SAE Paper 981405. Society of Automotive Engineers (SAE), Warrendale, PA, (1998).

213. MacFaul, P.A., K.U. Ingold, and J. Lusztyk. Kinetic solvent effects on hydrogen atom abstraction from phenol, aniline, and diphenylamine. The importance of hydrogen bonding on their radical-trapping (antioxidant) activities. Journal of Organic Chemistry, 61, 1316–1321 (1996).

214. Hamblin, P.C., D. Chasan, and U. Kristen. A review: Ashless antioxidants, copper deactivators and corrosion inhibitors: Their use in lubricating oils, in Fifth International Colloquium on Additives for Operational Fluids, Bartz, J., ed., Technische Akademie Esslingen Ostfildern, Germany (1986).

215. Foti, M.C. Antioxidant properties of phenols. Journal of Pharmacy and Pharmacology, 59(12), 1673–1685 (2007).

216. Gryn'ova, G., K.U. Ingold, and M.L. Coote. New insights into the mechanism of amine/nitroxide cycling during the hindered amine light stabilizer inhibited oxidative degradation of polymers. *Journal of the American Chemical Society*, 134, 12979–12988 (2012).

217. Boozer, E.B., G.S. Hammond, C.E. Hamilton et al. Air oxidation of hydrocarbons II. The stoichiometry and fate of inhibitors in benzene and chlorobenzene. *Journal of the American Chemical Society*, 77, 3233–3237 (1955).

218. Haidasz, E.A., R. Shah, and D.A. Pratt. The catalytic mechanism of diarylamine radical-trapping antioxidants. *Journal of the American Chemical Society*, 136, 16643–16650 (2014).

219. Jensen, R.K., S. Korcek, M. Zinbo et al. Regeneration of amine in catalytic inhibition of oxidation. *Journal of Organic Chemistry*, 60, 5396–5400 (1995).

220. Hanthorn, J.J., L. Valgimigli, and D.A. Pratt. Incorporation of ring nitrogens into diphenylamine antioxidants: Striking a balance between reactivity and stability. *Journal of the American Chemical Society*, 134, 8306–8309 (2012).

221. Hanthorn, J.J., R. Amorati, L. Valgimigli, and D.A. Pratt. The reactivity of air-stable pyridine- and pyrimidine-containing diarylamine antioxidants. *Journal of Organic Chemistry*, 77, 6895–6907 (2012).

222. Burton, A., K.U. Ingold, and J.C. Walton. Absolute rate constants for the reactions of primary alkyl radicals with aromatic amines. *Journal of Organic Chemistry*, 61, 3778–3782 (1996).

223. Lucarini, M., P. Pedrelli, G.F. Pedulli, L. Valgimigli, D. Gigmes, and P. Tordo. Bond dissociation energies of the N–H bond and rate constants for the reaction with alkyl, alkoxy, and peroxyl radicals of phenothiazines and related compounds. *Journal of the American Chemical Society*, 121, 11546–11553 (1999).

224. Adam, W., S. Hueckmann, and F. Vargas. Direct observation of electron transfer between phenothiazines and 1,2-dioxetanes. *Tetrahedron Letters*, 30(46), 6315–6318 (1989).

225. Bridgewater, A.J. and M.D. Sexton. Mechanism of antioxidant action: Reactions of alkyl and aryl sulphides with hydroperoxides. *Journal of the Chemical Society, Perkin Transactions 2: Physical Organic Chemistry*, 1972–1999 (6), 530 (1978).

226. Habicher, W.D., I. Bauer, and J. Pospisil. Organic phosphites as polymer stabilizers. *Macromolecular Symposia*, 225, 147–164 (2005).

227. Schwetlick, K. Chapter 2: Mechanisms of antioxidant action of phosphite and phosphonite esters, in *Mechanisms of Polymer Degradation and Stabilisation*, Scott, G., ed., Elsevier Applied Science, London, U.K. (1990).

228. Pospisil, J. Chapter 3: Antioxidants and related stabilizers, in *Oxidation Inhibition in Organic Materials*, Pospisil, J. and P.P. Klemchuk, eds., CRC Press, Boca Raton, FL (1990).

229. Hsu, S.M. Review of laboratory bench tests in assessing the performance of automotive crankcase oils. *Lubrication Engineering*, 37(12), 722–731 (1981).

230. Sharma, B.K. and A.J. Stipanovic. Development of a new oxidation stability test method for lubricating oils using high-pressure differential scanning calorimetry. *Thermochimica Acta*, 402, 1–18 (2003).

231. Zhang, Y., P. Pei, J.M. Perez et al. A new method to evaluate deposit-forming tendency of liquid lubricants by differential scanning calorimetry. *Lubrication Engineering*, 48(3), 189–195 (1992).

232. Hsu, S.M., A.L. Cummings, and D.B. Clark. Evaluation of automotive crankcase lubricants by differential scanning calorimetry, SAE Technical Paper 821252. Society of Automotive Engineers (SAE), Warrendale, PA (1982).

233. CEC L-85-T-99. Hot surface oxidation—Pressure differential scanning calorimeter (PDSC). http://www.cectests.org/disptestdocl.asp, accessed February 22, 2017.

234. ASTM Standard D6186-08. Standard test method for oxidation induction time of lubricating oils by pressure differential scanning calorimetry (PDSC). ASTM International, West Conshohocken, PA (2013).

235. Adamczewska, J.Z. and C. Love. Oxidative stability of lubricants measured by PDSC CEC L-85-T-99 test procedure. *Journal of Thermal Analysis and Calorimetry*, 80, 753–759 (2005).

236. Infineum LLC. ACEA 2012 oil sequences. https://www.infineum.com/media/16389/acea-2012-oil-sequences.pdf, accessed October 29, 2015. © Infineum International Limited, 2013.

237. Adhvaryu, A., S.Z. Erhan, S.K. Sahoo et al. Thermo-oxidative stability studies on some new generation API Group II and III base oils. *Fuel*, 81(6), 785–791 (2002).

238. Migdal, C.A. The influence of hindered phenolic and aromatic amine antioxidants on the stability of base oils. *213th ACS National Meeting Preprint*, San Francisco, CA (April 13–17, 1997).

239. Rohrbach, P., P.C. Hamblin, and M. Ribeaud. Benefits of antioxidants in lubricants and greases assessed by pressurized differential scanning calorimetry. *Tribotest Journal*, 11(3), 233–246 (March 2005).

240. Jain, M.R., R. Sawant, R.D.A. Paulmer et al. Evaluation of thermo-oxidative characteristics of gear oils by different techniques: Effect of antioxidant chemistry. *Thermochimica Acta*, 435(2), 172–175 (2005).

241. Nakanishi, H., K. Onodera, K. Inoue et al. Oxidative stability of synthetic lubricants. *Lubrication Engineering*, 29–37 (May 1997).

242. Sharma, B.K., J.M. Perez, and S.Z. Erhan. Soybean oil-based lubricants: A search for synergistic antioxidants. *Energy & Fuels*, 21, 2408–2414 (2007).

243. Cheenkachorn, K., J.M. Perez, and W.A. Lloyd. Use of pressurized differential scanning calorimetry (PDSC) to evaluate effectiveness of additives in vegetable oil lubricants. *ICE (American Society of Mechanical Engineers)*, 40, 197–206 (2003).

244. Gamlin, C.D., N.K. Dutta, N.R. Choudhury et al. Evaluation of kinetic parameters of thermal and oxidative decomposition of base oils by conventional, isothermal and modulated TGA and pressure DSC. *Thermochimica Acta*, 392–393, 357–369 (2002).

245. Gatto, V.J., H.Y. Elnagar, W.E. Moehle et al. Redesigning alkylated diphenylamine antioxidants for modern lubricants. *Lubrication Science*, 19(1), 25–40 (2007).

246. Florkowski, D.W. and T.W. Selby. The development of a thermo-oxidation engine oil simulation test (TEOST), SAE Technical Paper 932,837. Society of Automotive Engineers (SAE), Warrendale, PA (1993).

247. ASTM D6335-16. Standard test method for determination of high temperature deposits by thermo-oxidation engine oil simulation test., ASTM International, West Conshohocken, PA (2016). https://www.astm.org/Standards/D6335.htm, accessed February 22, 2017.

248. Selby, T.W. and D.W. Florkowski. The development of the TEOST protocol MHT bench test of engine oil piston deposit tendency. Presented at the *Twelfth Esslingen Colloquium*, Esslingen, Germany (January 11–13, 2000).

249. ASTM D7097-16a. Standard test method for determination of moderately high temperature piston deposits by thermo-oxidation engine oil simulation test—TEOST MHT, ASTM International, West Conshohocken, PA (2016). https://www.astm.org/Standards/D7097.htm, accessed February 22, 2017.

250. Anonymous. Correlation of TEOST performance with molar soap concentration for optimal deposit performance. *Research Disclosure*, 409, 531 (1998).

251. Sun, J.X., P.T. Pei, Z.S. Hu et al. A modified thin-film oxygen uptake test (TFOUT) for lubricant oxidative stability study. *Lubrication Engineering*, 54(5), 12–19 (1998).

252. ASTM D4742-16. Standard test method for oxidation stability of gasoline automotive engine oils by thin-film oxygen uptake (TFOUT), ASTM International, West Conshohocken, PA (2016). https://www.astm.org/Standards/D4742.htm, accessed February 22, 2017.

253. Ku, C.S. and S.M. Hsu. A thin-film oxygen uptake test for the evaluation of automotive crankcase lubricants. *Lubrication Engineering*, 40(2), 75–83 (1984).

254. ASTM D943-04a(2010)e1. Standard test method for oxidation characteristics of inhibited mineral oils, ASTM International, West Conshohocken, PA (2010). https://www.astm.org/Standards/D943.htm, accessed February 22, 2017.

255. ASTM D4310-10(2015). Standard test method for determination of the sludging and corrosion tendencies of inhibited mineral oils, ASTM International, West Conshohocken, PA (2015). https://www.astm.org/Standards/D4310.htm, accessed February 22, 2017.

256. Yano, A., S. Watanabe, Y. Miyazaki et al. Study on sludge formation during the oxidation process of turbine oils. *Tribology Transactions*, 47, 111–122 (2004).

257. ASTM D7873-13e2. Standard test method for determination of oxidation stability and insolubles formation of inhibited turbine oils at 120°C without the inclusion of water (dry TOST method), ASTM International, West Conshohocken, PA (2013). https://www.astm.org/Standards/D7873.htm, accessed February 22, 2017.

258. Cerny, J., D. Landtova, and G. Sebor. Development of a new laboratory oxidation test for engine oils. *Petroleum and Coal*, 44(1–2), 48–50 (2002).

259. Cerny, J., Z. Strnad, and G. Sebor. Composition and oxidation stability of SAE 15W-40 engine oils. *Tribology International*, 34(2), 127–134 (2001).

260. IP 280: Petroleum products and lubricants - Inhibited mineral turbine oils - Determination of oxidation stability, January 1999; Energy Institute, London, UK. http://publishing.energyinst.org/ip-test-methods/full-list-of-ip-test-methods-publications/ip-280-petroleum-products-and-lubricants-inhibited-mineral-turbine-oils-determination-of-oxidation-stability, accessed February 22, 2017.

261. Jayaprakash, K.C., S.P. Srivastava, K.S. Anand et al. Oxidation stability of steam turbine oils and laboratory methods of evaluation. *Lubrication Engineering*, 49(2), 89–95 (1984). https://www.astm.org/Standards/D2272.htm, accessed February 22, 2017.

262. ASTM D2272-14a. Standard test method for oxidation stability of steam turbine oils by rotating pressure vessel, ASTM International, West Conshohocken, PA (2014). https://www.astm.org/Standards/D2272.htm, accessed February 22, 2017.

263. Swift, S.T., K.D. Butler, and W. Dewald. Turbine oil quality and field application requirements, in *Turbine Lubrication in the 21st Century*, ASTM STP 1407, Herguth, W.R. and T.M. Warne, eds., American Society for Testing and Materials, West Conshohocken, PA (2001).

264. Mookken, R.T., D. Saxena, B. Basu et al. Dependence of oxidation stability of steam turbine oil on base oil composition. *Lubrication Engineering*, 53(10), 19–24 (1997).

265. Gatto, V.J. and W.E. Moehle. Lubricating oil composition with reduced phosphorus levels. U.S. Patent Application 2006/0223724 A1, Albemarle Corporation, USPTO, Washington, DC (October 5, 2006).

266. Gatto, V.J. and M.A. Grina. Effects of base oil type, oxidation test conditions and phenolic antioxidant structure on the detection and magnitude of hindered phenol/diphenylamine synergism. *Lubrication Engineering*, 11–20 (January 1999).

267. Dong, J. and C.A. Migdal. Lubricant compositions stabilized with multiple antioxidants. U.S. Patent 7,704,931 B2, Chemtura Corporation, USPTO, Washington, DC (April 27, 2010).

268. Niu, Q.S., H. Chui, and L.P. Yang. Effects of lube base oil composition on the lubricant oxidation. *Shiyou Xuebao Shiyou Jiagong*, 2(2), 61 (1986).

269. Adhvaryu, A., Y.K. Sharma, and I.D. Singh. Studies on the oxidative behavior of base oils and their effects of additives on oxidation stability. *Fuel*, 78, 1293 (1999).

270. Henderson, H.E. Gas-to-liquids, in *Synthetics, Mineral Oils, and Bio-Based Lubricants Chemistry and Technology*, 2nd edn., Rudnick, L.R., ed., CRC Press, Boca Raton, FL, pp. 332–346 (2013).

271. Henderson, H.E. Chemically modified mineral oils, in *Synthetics, Mineral Oils, and Bio-Based Lubricants Chemistry and Technology*, 2nd edn., Rudnick, L.R., ed., CRC Press, Boca Raton, FL, pp. 301–331 (2013).

272. International Lubricant Special Advisory Committee. ILSAC GF-6A recommendations for passenger car engine oils. Draft. http://www.pceo.com/2C-ILSAC_gf-6a_03-06-13_draft_rev_6.pdf, accessed October 12, 2015. (November 6, 2012).

273. 2015 Lubricants industry factbook. Persaud, M., ed., *Lubes'N'Greases* Supplement, 21(8), 10 (2015).

274. Swedberg, S. Volatility, front and center. *Lubes'N'Greases*, 21(2), 6–10 (2015).

275. Smolenski, D. Is fuel economy hiding in SAE J300? *Lubes'N'Greases*, 21(3), 28–32 (2015).

2 Zinc Dithiophosphates

Randolf A. McDonald

CONTENTS

2.1 INTRODUCTION

Zinc dialkyldithiophosphates (ZDDPs) have been used for more than 50 years in the lubricant industry as low-cost, multifunctional additives in engine oils, transmission fluids, hydraulic fluids, gear oils, greases, and other lubricant applications. The power of this particular compound is in its ability to simultaneously function as an excellent antiwear agent, a mild extreme-pressure (EP) agent, and an effective oxidation and corrosion inhibitor, all at a very low cost in comparison with the alternate chemistries available in the market. This is why it is still manufactured on a large scale by companies such as the ExxonMobil Corporation, Chevron Corporation, Ethyl Corporation, Lubrizol Corporation, and others. To date, as much as 300 million lb of ZDDP is still manufactured annually in the industrialized West.

2.2 SYNTHESIS AND MANUFACTURE

ZDDP was first patented on December 5, 1944, by Herbert C. Freuler of the Union Oil Company of California in Los Angeles [1]. The multifunctionality of ZDDP was immediately noticed as Freuler indicated a noticeable increase in both the oxidation and the corrosion resistance of the lubricants tested with the novel compound at a 0.1%–1.0% treatment level. The initial synthesis Freuler carried out involved the reaction of 4 mol of the intermediate dialkyldithiophosphate acid and 1 mol of hydrogen sulfide

$$4ROH + P_2S_5 \rightarrow 2(RO)_2 \overset{S}{\underset{|}{P}} SH + H_2S \tag{2.1}$$

followed by neutralization of the acid with 1 mol of zinc oxide

$$2(RO)_2 \overset{S}{\underset{|}{P}} SH + ZnO \rightarrow 2(RO)_2 \overset{S}{\underset{|}{P}} SZnS \overset{S}{\underset{|}{P}} (OR)_2 + H_2O \tag{2.2}$$

This synthetic route is still used today in the manufacturing of ZDDP. The P_2S_5, a flammable solid produced from the high-temperature reaction between elemental sulfur and phosphorus, is provided to the ZDDP manufacture in sealed aluminum bins containing 500–7200 lb P_2S_5. The P_2S_5 is hoppered into the reactor containing alcohol under a blanket of nitrogen. This is due to the ignitability of both the alcohol and the P_2S_5 when exposed to air. The hydrogen sulfide by-product, a highly toxic gas, is either converted to sodium sulfide solution in a caustic scrubber or thermally oxidized to sulfur dioxide. The heat of reaction and rate of hydrogen sulfide evolution are controlled by the addition rate of the P_2S_3 as well as the flow rate of the cooling water. The acid is then neutralized by zinc oxide; the reaction temperature is controlled by the addition rate of reaction depending on whether it is an acid to oxide or oxide to acid, addition scheme. Enough zinc oxide is used to neutralize the acid to a pH range, which will give a product suitably stable to thermal degradation and hydrogen sulfide evolution. The water formed from the reaction and the residual alcohol is vacuum-distilled. Any unreacted zinc oxide is then filtered, requiring a filtration system capable of removing particles as small as 0.1–0.8 μm. A larger molar excess of zinc oxide is often necessary to obtain the pH required for stability. The various manufactures have done much work to reduce the amount of zinc oxide used to obtain product stability (such as the addition of low-molecular-weight alcohols or carboxylic acids to lower the amount of residual sediment in the product before filtration) [2]. The filtered liquid product, with or without additional petroleum oil, is then provided to the customer in drums or in bulk.

2.3 CHEMICAL AND PHYSICAL NATURE

ZDDP is an organometallic compound having four sulfur atoms coordinated to the zinc atom, which is in a tetrahedral, sp^3 hybridized state. A Raman spectrum of ZDDP shows a strong P–S symmetric stretching band near 540 cm^{-1} and the absence of a strong Raman band near 660 cm^{-1}, indicating a symmetrical sulfur–zinc coordination arrangement as in the following structure

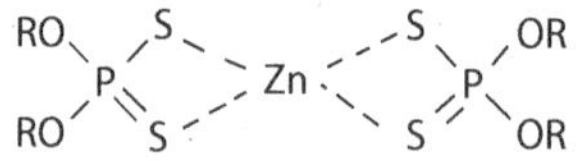

$$(2.3)$$

versus Structure 2.4

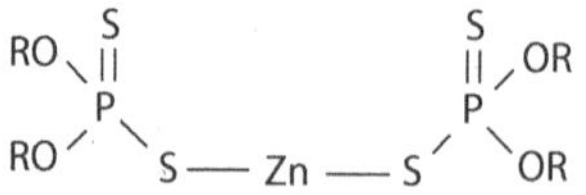

$$(2.4)$$

often given in the literature. The strong IR band at 600 cm^{-1} pointing to P=S stretching would be more consistent with PS$_2$ antisymmetrical stretching in light of the Raman spectrum [3]. The neutral ZDDP molecules as represented in Structure 2.3 actually exist as monomer, dimer, trimer, or oligomer depending on the state of the ZDDP, crystalline or liquid, the concentration of ZDDP in solvent, and the presence of additional compounds. The proposed structure for a tetramer in the case of a neutral zinc diisobutyldithiophosphate in hexane as determined by dynamic light scattering is shown in the following structure [4]:

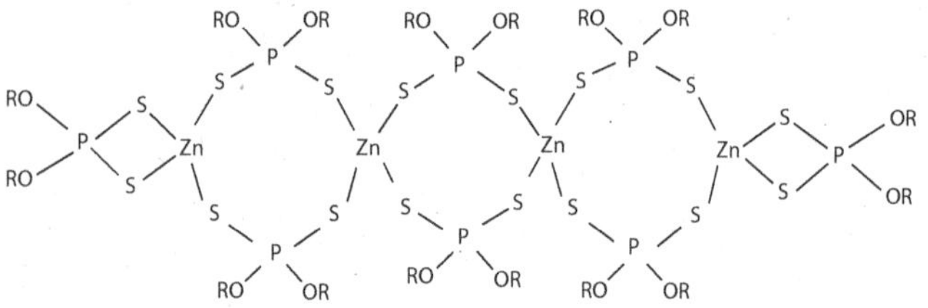

$$(2.5)$$

Under overbased condition, when the ratio of dialkyldithiophosphate acid to zinc oxide is less than 2:1, a basic zinc salt

$$(2.6)$$

will be synthesized along with the neutral salt. The basic salt is a tetrahedron of zinc atoms surrounding a central oxygen atom with (RO)$_2$PS$_2$ ligands along each edge of the tetrahedron. Crystallographic analysis of pure basic zinc salts has established the near equivalency of P–S–Zn bonds. Raman spectra have also shown symmetrical P–S stretching, supporting a symmetrical sulfur–zinc coordination arrangement for the basic ZDDPs [3]. In the presence of water, as one would encounter during commercial ZDDP manufacture, the basic zinc salt will be in equilibrium with the basic zinc double salt as seen in the following reaction:

R = alkyl, phenyl, or alkylphenyl

$$(2.7)$$

The stoichiometric excess of zinc oxide used in commercial ZDDP manufacture gives rise to a mixture of basic zinc salt (or zinc double salt) and neutral salt, the ratio depending on the amount of excess zinc oxide used and the molecular weight of the alkyl groups involved, where short alkyl groups tend to promote the formation

R = alkyl, phenyl, or alkylphenyl

$$(2.8)$$

but performance differences seen between the two salts with respect to wear would imply that a more complex situation may exist [5]. As reported in the literature, basic ZDDP salts spontaneously decompose in solution into neutral complexes and zinc oxide when the temperature is increased [6].

Pure ZDDPs, with alkyl groups of four carbons or less, are solid at ambient temperatures (with the exception of *sec*-butyl, which is a semisolid at room temperature) and tend to have limited or no solubility in petroleum base stocks. ZDDPs with aryl or alkyl groups with more than five carbons are liquid at ambient temperature. To utilize the less-expensive and more readily available low-molecular-weight alcohols and yet to produce oil-soluble products, commercial manufactures use mixtures of high- (i.e., more than four carbons) and low-molecular-weight alcohols to obtain a statistical distribution of products favoring lesser amounts of pure low-molecular-weight ZDDPs. Other methods have also been developed to increase the amount of lower-molecular-weight alcohols in ZDDPs. These include the addition of ammonium carboxylates to inhibit precipitation [7] and the use of alkyl succinimides as solubilizing-complexing agents [8].

2.4 THERMAL AND HYDROLYTIC STABILITY

The study of the thermal degradation of ZDDP is important in that much of the tribological characteristics of ZDDP can be explained by the effects of its decomposition products. The thermal decomposition of ZDDP in mineral oil has been found to be extremely complex. ZDDP in oil, upon heating to degradation, will give off volatile compounds such as olefin, alkyl disulfide, and alkyl mercaptan. A white precipitate will also form, which has been determined to be a low-sulfur-containing zinc pyrophosphate. The oil phase will contain varying amounts of *S,S,S*-trialkyltetrathiophosphate, *O,S,S*-trialkyltrithiophosphate, and *O,O,S*-trialkyldithiophosphate depending on the alkyl chain and the extent of degradation. The decomposition products of ZDDPs made from secondary alkyl alcohols, straight-chain primary alkyl alcohols, and branched primary alkyl alcohols appear similar in content but differ in proportions. This implies a similar mechanism for both primary and secondary ZDDP decomposition.

O-alkyl thiphosphate esters are powerful alkylating agents. The P–O–R group is susceptible to nucleophilic attack, thus producing an alkylated nucleophile and thiophosphate anion.

The incoming nucleophile initiates the reaction by an attack on the alpha carbon. This shows a kinetic dependence on alkyl structure

$$-\overset{\overset{\displaystyle S}{\parallel}}{\underset{\displaystyle |}{P}}-O-R \ + \ Nu^- \ \longrightarrow \ -\overset{\overset{\displaystyle S}{\parallel}}{\underset{\displaystyle |}{P}}-O^- \ + \ Nu\text{--}R \qquad (2.9)$$

Steric hindrance to the approach of the nucleophile will play a large rate-controlling factor here. The only nucleophile initially present is the dithiophosphate itself. The decomposition is initiated by one dithiophosphate anion attacking another, possibly on the same zinc atom:

$$2\left[\overset{\displaystyle OR}{\underset{\displaystyle OR}{S=P-S^-}} \right] \ \longrightarrow \ \overset{\displaystyle O^-}{\underset{\displaystyle OR}{S=P-S^-}} \ + \ \overset{\displaystyle OR}{\underset{\displaystyle OR}{S=P-SR}} \qquad (2.10)$$

The resulting di-anion then attacks the triester, producing O,S-dialkyldithiophosphate anion

$$\overset{\displaystyle O^-}{\underset{\displaystyle OR}{S-P-SR}} \qquad (2.11)$$

resulting in the migration of an alkyl group from oxygen to sulfur. This anion then, in a route analogous to the dialkyldithiophosphate anion, reacts with itself in a nucleophilic attack to effect another alkyl transfer from oxygen to sulfur producing O,S,S-trialkyldithiophosphate

$$\overset{\displaystyle OR}{\underset{\displaystyle O}{RS-P-SR}} \qquad (2.12)$$

The net effect of the above reactions is a double alkyl migration from oxygen to sulfur

$$\overset{\displaystyle OR}{\underset{\displaystyle OR}{{}^-S-P=S}} \ \longrightarrow \ \longrightarrow \ \overset{\displaystyle O}{\underset{\displaystyle O^-}{RS-P-SR}} \qquad (2.13)$$

The major gases associated with ZDDP decomposition are dialkylsulfide (RSR), alkyl mercaptan (RSH), and olefin. The relative amounts of each of these gases depend on whether the alkyl group in the ZDDP is primary, branched primary, or secondary [9]. In the presence of mercaptide anion (RS$^-$) from the intermediate zinc mercaptide (Zn[RS$_2$]), O,S,S-trialkyldithiophosphate will react with mercaptide to produce alkyl mercaptan and results in the following structure:

$$\overset{\displaystyle O^-}{\underset{\displaystyle O}{RS-P-SR}} \qquad (2.14)$$

The nucleophilic phosphoryl oxygen (P=O) will then attack another phosphorus atom to produce a P–O–P bond as in the following reaction:

$$-\overset{\displaystyle |}{\underset{\displaystyle |}{P}}-SR + O=\overset{\displaystyle |}{\underset{\displaystyle O}{P}}^- \ \longrightarrow \ -\overset{\displaystyle |}{\underset{\displaystyle |}{P}}-O-\overset{\displaystyle |}{\underset{\displaystyle |}{P}}- \ + \ {}^-SR \qquad (2.15)$$

A mercaptide anion subsequently cleaves the P–O–P bond at the original P–O site, giving rise to a net exchange of one atom of oxygen for one atom of sulfur between the two phosphorus atoms:

$$-\overset{\displaystyle |}{\underset{\displaystyle |}{P}}-O-\overset{\displaystyle |}{\underset{\displaystyle |}{P}}- \ + \ {}^-SR \ \longrightarrow \ -\overset{\displaystyle |}{\underset{\displaystyle |}{P}}-O^- \ + \ -\overset{\displaystyle +|}{\underset{\displaystyle |}{P}}-SR \qquad (2.16)$$

This gives rise to a net reaction for conversion of Structure 2.14 to S,S,S-trialkyltetrathiophosphate, dialkylsulfide and S-alkylthiophosphate di-anion as shown in the following reaction:

$$3\left[\overset{\overset{\displaystyle O}{\parallel}}{\underset{\displaystyle O^-}{RS-P-SR}} \right] + {}^-SR \ \longrightarrow \ 2\left[\overset{\overset{\displaystyle O}{\parallel}}{\underset{\displaystyle O^-}{RS-P-O^-}} \right] + R_2S + \overset{\overset{\displaystyle S}{\parallel}}{\underset{\displaystyle SR}{RS-P-SR}} \qquad (2.17)$$

The dialkylsulfide and S,S,S-trialkyltetrathiophosphate decomposition products are soluble in oil.

The S-alkylthiophosphate decomposition product can also react with itself by way of a phosphoryl nucleophilic attack and elimination of mercaptide anion as in Reaction 2.15. This process will continue until a zinc pyro- and polypyrophosphate molecule with low sulfur content is formed. The chain will continue to extend until the product precipitates out of solution.

The decomposition of primary alkyl ZDDPs can be accurately described as discussed earlier. ZDDPs made from branched primary alcohols will decompose in a similar fashion, although at a much slower rate. This can be explained by the fact that the alpha carbon of the branched primary alkyl group, being more sterically hindered than the unbranched primary alkyl group, will be less susceptible to nucleophilic attack, as described in Reaction 2.9. The increased steric hindrance from beta carbon branching will also decrease the amount of successful mercaptide anion attack on the branched alkyl P–O–R bond, resulting in less dialkylsulfide formation and a higher yield of mercaptan, an olefin by-product (through a competing protonation or elimination reaction with mercaptide anion). Lengthening the alkyl chain will have a much less pronounced effect on thermal stability than branching at the beta carbon due to the greater steric hindrance derived from the latter.

The decomposition of secondary alkyl ZDDPs, although similar to primary decomposition, shows that olefin formation is much more pronounced. The increase in elimination over nucleophilic substitution in secondary ZDDPs over primary ZDDPs is easily explained by the fact that elimination is accelerated by increasing the alkyl substitution around the double bond formed. Thus, secondary alkyl groups will favor a thermal decomposition into olefins and phosphate acids at the expense of the sulfur–oxygen interchange noted earlier. In a similar but much more pronounced way, tertiary ZDDP decomposition will be dominated by facile production of olefin through elimination. This occurs at even moderate temperatures, making their use in commercial applications prohibitive.

Aryl ZDDPs, due to the stability of the aromatic ring, are not susceptible to nucleophilic attack. Thus, the initial thermal decomposition reaction described in Reaction 2.9 cannot occur. Also the formation of olefin from an acid-catalyzed elimination reaction cannot occur. Aryl ZDDPs are, therefore, very thermally stable.

A rating of various ZDDPs in terms of thermal stability would, therefore, be aryl > branched primary alkyl > primary alkyl > secondary > tertiary. The varying amounts of decomposition products that depend on the heat history and the alkyl or aryl chain involved will directly control the amount of EP and wear protection the ZDDP will provide in a given circumstance [10].

Hydrolysis of ZDDP begins with cleavage of the carbon–oxygen bond of the thiophosphate ester, with the hydroxide anion displacing the thiophosphate-anion-leaving group. The stability of the intermediate alkyl cation determines the ease with which this cleavage takes place. The secondary alkyl cation is more stable and more easily formed than the primary alkyl cation; therefore, hydrolysis of secondary ZDDP occurs more easily than hydrolysis of primary ZDDP. For the case of an aryl ZDDP, the carbon–oxygen bond cannot be broken, and the site of hydrolytic attack is the phosphorus–oxygen bond with the displacement of phenoxide anion with hydroxide anion. The order of hydrolytic stability is, therefore, primary > secondary > aryl.

2.5 OXIDATION INHIBITION

Base oils used in lubricants degrade by an autocatalytic reaction known as *auto-oxidation*. The initial stages of oxidation are characterized by a slow, metal-catalyzed reaction with oxygen to form an alkyl-free radical and a hydroperoxy-free radical as seen in the following reaction:

$$RH + O_2 \xrightarrow{M^+} R^* + HOO^* \qquad (2.18)$$

This reaction is propagated by the reaction of the alkyl-free radical with oxygen to form an alkylperoxy radical. This radical further reacts with the base oil hydrocarbon to form alkyl hydroperoxide and another alkyl radical as seen in the following reaction:

$$R^* + O_2 \longrightarrow ROO^* \xrightarrow{RH} ROOH + R^* \qquad (2.19)$$

This initial sequence is followed by chain branching and termination reactions forming high-molecular-weight oxidation products [11].

The antioxidant functionality of ZDDP is ascribed to its affinity for peroxy radicals and hydroperoxides in a complex pattern of interaction.

The initial oxidation step of ZDDP by hydroperoxide is the rapid reaction involving the oxidative formation of the basic ZDDP salt as seen in the following reaction:

$$4\left[\underset{(RO)_2PS}{\overset{S}{\|}}\right]_2 Zn + R'OOH \longrightarrow \left[\underset{(RO)_2PS}{\overset{S}{\|}}\right]_6 Zn_4O + R'OH$$
$$+ \left[\underset{(RO)_2PS}{\overset{S}{\|}}\right]_2$$

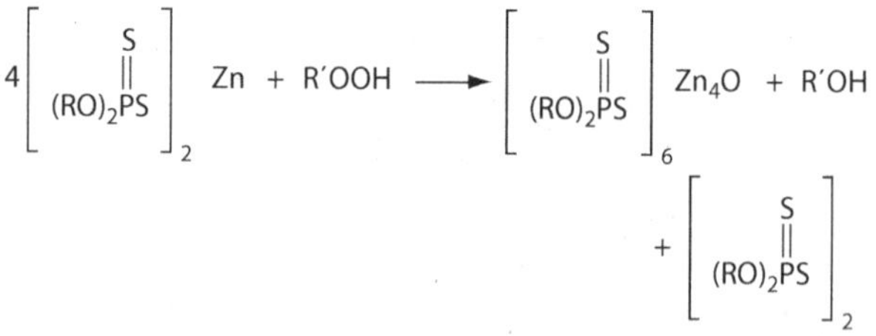

(2.20)

In this reaction, 1 mol of alkyl hydroperoxide converts 4 mol of neutral ZDDP to 1 mol of basic ZDDP and 2 mol of the dialkyldithiophosphoryl radical (which subsequently reacts to produce the disulfide) [12]. The rate of hydroperoxide decomposition slows during an induction period during which the basic zinc thermally breaks down into the neutral ZDDP and zinc oxide [6]. This is followed by the neutral ZDDP further reacting with hydroperoxide to produce more dialkyldithiophosphoryl disulfide and more basic ZDDP. When the concentration of the basic ZDDP becomes low enough, a final rapid neutral salt-induced decomposition of the hydroperoxide will occur in which the dialkyldithiophosphoryl radical will not react with itself to form the disulfide but will react with hydroperoxide to form the dialkyldithiophosphoric acid as seen in the following reaction [13]:

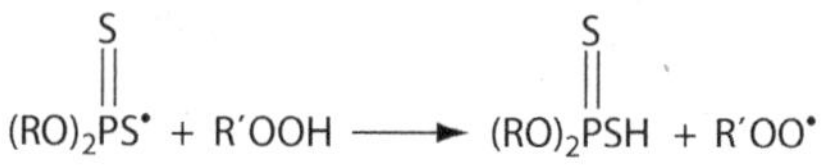

$$\underset{(RO)_2PS^*}{\overset{S}{\|}} + R'OOH \longrightarrow \underset{(RO)_2PSH}{\overset{S}{\|}} + R'OO^* \qquad (2.21)$$

The dialkyldithiophosphoric acid then rapidly reacts with alkyl hydroperoxide, producing oxidation products that are inactive in oxidation chain reactions. The simplest reaction scheme for the reduction of the hydroperoxide is seen in the following reaction:

$$2\underset{(RO)PSH}{\overset{S}{\|}} + R'OOH \longrightarrow \left[\underset{(RO)_2PS}{\overset{S}{\|}}\right]_2 + R'OH + H_2O$$

(2.22)

Oxidation products include the disulfide mentioned earlier, the analogous mono- and trisulfides, and compounds of the form $(RO)_n(RS)_{3-n}P{=}S$ and $(RO)_n(RS)_{3-n}P{=}O$ [3]. These products show little activity as either oxidation inhibitors or antiwear agents.

The literature also reveals an ionic process that will produce more dialkyl-dithiophosphoric acid as seen in the following reaction:

$$\left[\underset{(RO)PS}{\overset{S}{\|}}\right]_2 Zn + R'OO^* \longrightarrow R'OO^- + \underset{(RO)_2PS}{\overset{S}{\|}}Zn^+ + \underset{(RO)_2PS^*}{\overset{S}{\|}}$$

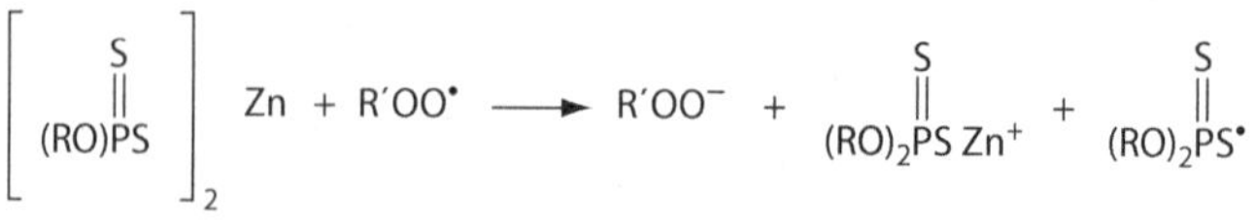

(2.23)

followed by Reaction 2.21 [14].

At low concentrations of ZDDP, hydrolysis of the ZDDP to the zinc basic double salt and dialkyldithiophosphoric acid becomes viable. At temperatures >125°C, the dialkyldithiophosphoryl disulfide decomposes into the dialkyldithiophosphoryl radicals, which further react with hydroperoxide to produce more dialkyldithiophosphoric acid [6]. Thus, many pathways are available to form the active dialkyldithiophosphoric acid.

The neutral ZDDP also reacts with alkyl peroxy radicals. This is an electron-transfer mechanism that involves the stabilization of a peroxy intermediate. An attack by a second peroxy radical leads to the intramolecular dimerization of the resulting dithiophosphate radical forming the inactive dialkyldithiophosphoryl disulfide as seen in the following reaction:

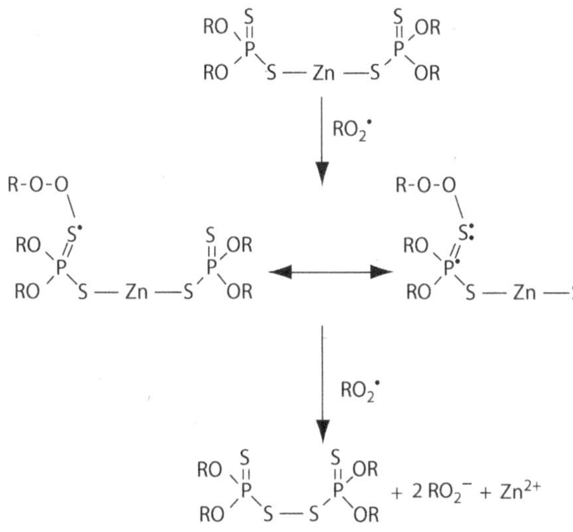

$$\text{(2.24)}$$

The zinc metal atom provides an easy route for heterolysis of the radical intermediate; thus, the disulfide, by itself, has little antioxidant functionality [15]. ZDDP acts as an oxidation inhibitor not only by trapping the alkyl radicals, thus slowing the chain reaction mechanism, but also by destroying alkyl hyperoxides and inhibiting the formulation of alkyl radicals. Empirical determination of the relative antioxidant capability of the three main classes of ZDDP shows secondary ZDDP > primary > aryl ZDDP. The relative performance of each ZDDP type may correlate with the stabilization of the dialkyl(aryl)dithiophosphoryl radical and its subsequent reactivity with alkyl hydroperoxide to produce the catalyzing acid.

Commercial ZDPs are a mixture of both neutral and basic salts. It has recently been determined that neutral and basic ZDDPs give essentially equivalent performance with respect to antioxidant behavior. This can be explained by the equilibrium shown in Reaction 2.8. At elevated temperatures, as would occur in an oxidation test, the basic ZDDP is converted into the neutral ZDDP. As the temperature is lowered, the equilibrium shifts back toward the formation of the basic ZDDP, indicating that the concentration of basic ZDDP as a function of temperature. The solvent used and the presence of other additives also play a role in this equilibrium. Thus, the exact composition of neutral versus basic salts at any time in an actual formulation is a complex function of many variables.

2.6 ANTIWEAR AND EXTREME-PRESSURE FILM FORMATION

ZDDPs operate mainly as antiwear agents but exhibit mild EP characteristics. As an antiwear agent, ZDDP operates under mixed lubrication conditions with a thin oil film separating the metal parts. Surface asperities, however, intermittently penetrate the liquid film, giving rise to metal-on-metal contact. The ZDDP reacts with these asperities to reduce the contact. Likewise, when the load is high enough to collapse the oil film, the ZDDP reacts with the entire metal surface to prevent welding and to reduce wear. A great deal of study has been done to determine the nature of this protective film and the mechanism of deposition, where the thermal degradation products of the ZDDP are the active antiwear agents.

The antiwear film thickness and composition are directly related to temperature and the extent of surface rubbing. Initially, ZDDP is reversibly absorbed onto the metal surface at low temperatures. As the temperature increases, catalytic decomposition of ZDDP to dialkyldithiophosphoryl disulfide occurs, with the disulfide absorbed onto the metal surface. From here, the thermal degradation products (as described in Section 2.3) are formed with increasing temperature and pressure until a film is formed on the surface [16]. The thickness and composition of this film have been studied using many different analytical techniques, but no analysis gives a concise description of the film size and composition for the various kinds of metal-to-metal contact found in industrial and automotive lubrication regimes. In general, the antiwear/EP ZDDP film can be said to be composed of various layers of ZDDP degradation products. Some of these degradation products are reacted with the metal making up the lubricated surface. The composition of the layers is temperature-dependent.

The first process that takes place is the reaction of sulfur (from the ZDDP thermal degradation products) with the exposed metal leading to the formation of a thin iron sulfide layer [17]. Next, phosphate reacts to produce an amorphous layer of short-chain ortho- and metaphosphates with minor sulfur incorporation. The phosphate chains become longer toward the surface, with the minimum chain length approaching 20 phosphate units. Some studies have indicated that this region is best described as a phosphate "glass" region in which zinc and iron cations act to stabilize the glass structure. At the outermost region of the antiwear film, the phosphate chains contain more and more organic ligands, eventually giving way to a region composed of organic ZDDP decomposition products and undegraded ZDDP itself. The thickness of the film has been analyzed to be as small as 20 nm using ultra thin film interferometry and as large as 1 μm using electrical capacitance [18–21].

Recent work has concluded that, although the rate of film formation is directly proportional to temperature, a stronger correlation exists between film formation and the extent of metal-to-metal rubbing as quantified by the actual distance that the metal slides during a given test period. The film reaches a maximum thickness at which point a steady state between formation and removal exists, the rate of formation being more temperature-dependent than the rate of removal. It was also found that the ZDDP reaction film has a "solid-like" nature (as opposed to be a highly viscous liquid) due to the lack of reduction of film thickness observed with time on a static test ball [22].

Another mechanism of wear found to be inhibited by ZDDP is wear produced from the reaction of alkyl hydroperoxides with metal surfaces. It was found that the wear rate of automobile engine cam lobes is directly proportional to alkyl hydroperoxide concentration. The mechanism proposes the direct attack of hydroperoxide (generally through fuel combustion and oil oxidation) on fresh metal, causing the oxidation of an iron atom from a neutral charge state to Fe^{+3} by reaction with 3 mol of alkyl hydroperoxide as described in the following reactions:

$$2ROOH + Fe \longrightarrow 2RO^* + 2OH^- + Fe^{+2} \qquad (2.25)$$

$$ROOH + Fe^{+2} \longrightarrow RO^* + OH^- + Fe^{+3} \qquad (2.26)$$

The ZDDP and its thermal degradation products neutralize the effect of the hydroperoxides by the mechanism described in Reactions 2.20 through 2.23 in Section 2.5. It was also shown that peroxy and alkoxy radicals were far less aggressive toward metal surfaces than hydroperoxides, indicating that free-radical scavengers such as hindered phenols would be ineffective in controlling this kind of engine wear. This may explain why the antiwear performance of ZDDP is directly related to its antioxidation performance in the order of secondary ZDDP > primary ZDDP > aryl ZDDP rather than correlating with the order of thermal stability (aryl > primary > secondary) [23].

A recent study has been conducted to investigate the difference in wear performance between neutral and basic ZDDPs in the sequence VE engine test. The neutral ZDDP performed better in value train wear protection than the basic ZDDP. The basic salt actually failed the sequence VE engine test, indicating that using commercial ZDDPs with lower basic salt content may be preferred when limited to 0.1% maximum phosphorus content (as mandated by the International Lubricant Standardization and Approval Committee [ILSAC] GF-3 motor oil specification). It was suggested that the increased wear protection by neutral ZDDP could be explained by the superior adsorption of the oligomeric structure of the neutral salt, leading to the formation of longer polyphosphate chains relative to the basic salt [5].

2.7 APPLICATIONS

ZDDPs are used in engine oils as antiwear and antioxidant agents. Primary and secondary ZDDPs are both used in engine oil formulations, but it has been determined that secondary ZDDPs perform better in cam lobe wear protection than primary ZDDPs. Secondary ZDDPs are generally used

when increased EP activity is required (i.e., during run-in to protect heavily loaded contacts such as valve trains). ZDDPs are generally used in combination with detergents and dispersants (alkaline earth sulfonate or phenate salts, polyalkenyl succine amides or Mannich-type dispersants), viscosity index improvers, additional organic antioxidants (hindered phenols, alkyl diphenyl amines), and pour point depressants. A typical lubricant additive package for engine oils can run in high at 25% in treatment level. The ILSAC has designated its GF-3 engine oil specification to include a maximum limit of 0.1% phosphorus to minimize the engine oil's negative impact on the emissions catalyst. For the GF-4 specification, the limit in phosphorus was reduced even further. As a result of the minimum phosphorus requirement, the treatment level for ZDDP in organic oils is limited to ~0.5% to 1.5%, depending on the alkyl chain length used.

The new challenge to motor oil formulators is in passing the required ILSAC tests while keeping the ZDDP level low. Yamaguchi et al. have shown that the antioxidant effect of ZDDP is significantly enhanced in API group II base stocks with as much as 50% increase noted for a basic ZDDP. An increase in antioxidancy was also noted when using ZDDPs in polyol ester [24]. Several studies have also shown that ZDDP oxidation by-products are in effective antiwear agents. The use of these base-stock effects to extend the oxidation life of the ZDDP may be a suitable method for the formulator to reduce the level of ZDDP needed to accommodate the GF-3 limits.

The synergistic effect between organic molybdenum compounds and ZDDP in wear reduction is currently being studied as a means of lowering phosphorus content in engine oils. In U.S. patent 5,736,491, molybdenum carboxylate is used with ZDDP to give a synergistic reduction in friction coefficient by as much as 30%, thus allowing a reduction in the total phosphorus content and an improvement in fuel economy [25]. The patent literature has sited other organic molybdenum compounds such as molybdenum dithiocarbamates (MoDTC) and dialkyldithiophosphates (MoDTP) as being useful, synergistic secondary antiwear agents [26].

ZDDPs are also used in hydraulic fluids as antiwear agents and antioxidants. The treatment level for ZDDP in hydraulic fluids is lower than that used for engine oils, typically running between 0.2% and 0.7% by weight. They are used in combustion with detergents, dispersants, additional organic antioxidants, viscosity index improvers, pour point depressants, corrosion inhibitors, defoamers, and demulsifiers for a total treatment level of between 0.5% and 1.25% [27]. Primary ZDDPs are preferred over secondary ZDDPs due to their better thermal and hydrolytic stability. One problem faced by hydraulic fluid formulators is the need for a fluid that will service both high-pressure rotary vane pumps and axial piston pumps, preferably out of the same sump. High-pressure vane pumps require a hydraulic fluid with antiwear properties and oxidative stability commonly achieved through the use of ZDDPs. High-pressure piston pumps need only rust and oxidation protection and do not require ZDDPs. ZDDPs can cause catastrophic failure to axial piston systems by adversely affecting the sliding steel–copper alloy interfaces. The patent literature has several examples of formulators trying to overcome this problem with the use of additional wear-moderating chemistries such as sulfurized olefins, polyol esters or borates of them, fatty acid imidazolines, aliphatic amines, and polyamines. Another problem faced by hydraulic fluid formulators is the interaction of ZDDPs with overbased alkaline earth detergent salts (as well as the interaction of carboxylic acid and alkenyl succinic anhydride rust preventatives with these detergents) in the presence of water to give filter-clogging byproducts. Formulators have tried to overcome this problem of poor "wet" filterability by using nonreactive rust inhibitors (i.e., alkenyl succinimides) and improving the hydrolytic stability of ZDDP antiwear agent [28].

ZDDPs are used in EP applications such as gear oils, greases, and metalworking fluids. Secondary ZDDPs are preferred due to their thermal instability resulting in quick film formation under high loads. In automotive gear oils, ZDDPs are used at 1.5%–4% in combination with EP agents (such as sulfurized olefins), corrosion inhibitors, foam inhibitors, demulsifiers, and detergents. Total multifunctional additive package treatment levels for automotive gear lubricant additives are from 5% to 12% by weight. Industrial gear oil formulators have generally gone to ashless systems using sulfur–phosphorus-based EP antiwear chemistries at total additive package treatment levels of 1.5%–3%. In general, the recent focus in gear oil technology improvement has centered on increased thermal stability and EP properties.

ZDDPs are used in greases in chemical systems that closely resemble gear oil formulations. Many gear oil lubricant additives are used in EP greases. In general, the ZDDP treatment level for greases is in the same range as that used for gear oils. ZDDP, usually secondary or a mixture of secondary and primary, is used in combination with sulfurized olefins, corrosion inhibitors, ashless antioxidants, and additional friction modifiers. A recent advancement in grease technology is the use of ZDDP/sulfurized olefin synergy to replace antimony and lead in high-EP grease formulations. This has generally been limited to the European market, having been pioneered in Germany.

ZDDPs, in combination with sulfurized olefins, are also used to replace chlorinated paraffins in medium- to heavy-duty metalworking fluids. This is due to the possible carcinogenicity of the low-molecular-weight analogs of chlorinated paraffin. European formulators, and to a certain extent Japanese formulators, use ZDDPs in this way. The use of ZDDPs in metalworking fluids in the United States is limited due to environmental concerns. The U.S. Environmental Protection Agency classifies them as marine pollutants.

In conclusion, after 50 years, ZDDPs still enjoy a wide variety of uses in the lubrication industry, with production volumes remaining at high levels. The majority of ZDDP production is used in automobile engine oil. The impact of the GF-2 and GF-3 phosphorus-level specification of 0.1%, however, was reduction of ZDDP production in the past 10 years. The Ford Motor Company is currently evaluating engine oils with 0%–0.6% phosphorus levels in fleet tests in preparation for the looming GF-4 standard in 2004, which will require engine oils to have minimal impact on emission system deterioration. This could further negatively impact ZDDP production. The need to understand clearly how ZDDPs function in terms of wear and oxidation protection is reinforced by the need to develop satisfactory phosphorus-free alternatives to ZDDP. The development of such chemistries, within the economic and functional limits that ZDDPs impose, will be a daunting task for future researchers. Until that time, the elimination of ZDDPs from various industrial lubricants will mandate either higher costs or less performance.

REFERENCES

1. Freuler, H.C. Modified lubricating oil. U.S. Patent 2,364,284 (December 5, 1944), Union Oil Co. of California.
2. Adams, D.R. Manufacture of dihydrocarbyl dithiophosphats. U.S. Patent 5,672,294 (May 6, 1997), Exxon Chemical Patents, Inc.
3. Paddy, J.L. et al. Zinc dialkyldithiophosphate oxidation by cumene hydroperoxide: Kinetic studies by Raman and ^{31}P NMR spectroscopy. Trib Trans 33(1):15–20, 1990.
4. Yamaguchi, E.S. et al. Dynamic light scattering studies of neutral diisobutyl zinc dithiophosphate. Trib Trans 40(2):330–337, 1997.
5. Yamaguchi, E.S. The relative wear performance of neutral and basic zinc dithiophosphates in engines. Trib Trans 42(1):90–94, 1999.
6. Bridgewater, A.J., J.R. Dever, M.D. Sexton. Mechanisms of antioxidant action, part 2. Reactions of zinc bis(O,O'-dialkyl(aryl)phosphorodithioates) and related compounds with hydroperoxides. J Chem Soc Perkin II 1006–1016, 1980.
7. Buckley, T.F. Methods for preventing the precipitation of mixed zinc dialkyldithiophosphates which contain high percentages of a lower alkyl group. U.S. Patent 4,577,037 (March 18, 1986), Chevron Research Co.
8. Yamaguchi, E.S. Oil soluble metal (lower) dialklyl dithiophosphate succinimide complex and lubricating oil composition containing same. U.S. Patent 4,306,984 (December 22, 1981), Chevron Research Co.
9. Luther, H., E. Baumgarten, K. Ul-Islam. Investigations by gas chromatography into the thermal decomposition of zinc dibutyldithiophosphates. Erdol und Kohle 26(9):501, 1973.
10. Coy, R.C., R.B. Jones. The chemistry of the thermal degradation of zinc dialkyldithiophosphate additives. ASLE Trans 24(1):91–97, 1979.
11. Rasberger, M. Oxidative degradation and stabilization of mineral based lubricants, in R.M. Moritier and S.T. Orszulik, eds., Chemistry and Technology of Lubricants, 2nd edn. London, U.K.: Blackie Academic & Professional, 1997, pp. 82–123.
12. Rossi, E., L. Imperoto. Chim Ind (Milan) 53:838–840, 1971.
13. Sexton, M.D. J Chem Soc Perkin Trans II 1771–1776, 1984.
14. Howard, S.A., S.B. Tong. Can J Chem 58:92–95, 1980.
15. Burn, A.J. The mechanism of the antioxidant action of zinc dialkyl dithiophosphates. Tetrahedron 22:2153–2161, 1966.
16. Bovington, C.H., B. Darcre. The adsorption and reaction of decomposition products of zinc dialkyldithiophosphate on steel. ASLE Trans 27:252–258, 1984.
17. Bell, J.C., K.M. Delargy. The composition and structure of model zinc dialkyldithiophosphate antiwear films, in M. Kozna, ed., Proceedings of the Sixth International Congress on Tribology Eurotrib '93, Budapest, Hungary, Vol. 2, pp. 328–332, 1993.
18. Willermet, P.A., R.O. Carter, E.N. Boulos. Lubricant-derived tribochemical films—An infra-red spectroscopic study. Trib Int 25:371–380, 1992.

19. Fuller, M. et al. Chemical characterization of tribochemical and thermal films generated from neutral and basic ZDDPs using x-ray absorption spectroscopy. *Trib Int* 30:305–315, 1997.

20. Allison-Greiner, A.F., J.A. Greenwood, A. Cameron. Thickness measurements and mechanical properties of reaction films formed by zinc dialkyldithiophosphate during running, in *Proceedings of IMechE International Conference on Tribology—Friction, Lubrication and Wear 50 Years on*, London, U.K., IMechE, Vol. 1, pp. 565–569, 1987.

21. Tripaldi, G., A. Vettor, H.A. Spikes. Friction behavior of ZDDP films in the mixed boundary/EHD regime. SAE Technical Paper 962036, 1996.

22. Taylor, L., A. Dratva, H.A. Spikes. Friction and wear behavior of zinc dialkyldithiophosphate additive. 43(3): 469–479, 2000.

23. Habeeb, J.J., W.H. Stover. The role of hydroperoxides in engine wear and the effect of zinc dialkyldithiophosphates. *ASLE Trans* 30(4):419–426, 1987.

24. Yamaguchi, E.S. et al. The relative oxidation inhibition performance of some neutral and basic zinc dithiophosphate salts. S.T.L.E. Preprint No. 99-AM-24, pp. 1–7, 1989.

25. Patel, J.A. Method of improving the fuel economy characteristics of a lubricant by friction reduction and compositions useful therein. U.S. Patent 5,736,491 (April 7, 1998), Texaco, Inc.

26. Naitoh, Y. Engine oil composition. U.S. Patent 6,063,741 (May 16, 2000), Japan Energy Corporation.

27. Brown, S.H. Hydraulic system using an improved antiwear hydraulic fluid. U.S. Patent 5,849,675 (December 15, 1998), Chevron Chemical Co.

28. Ryan, H.T. Hydraulic fluids. U.S. Patent 5,767,045 (June 16, 1998), Ethyl Petroleum Additives, Ltd.

3 Dispersants

Syed Q.A. Rizvi

CONTENTS

3.1 INTRODUCTION

Lubricants are composed of a base fluid and additives. The base fluid can be mineral, synthetic, or biological in origin. In terms of use, petroleum-derived (mineral) base fluids top the list, followed by synthetic fluids. Base oils of biological origin, that is, vegetable and animal oils, have not gained much popularity except in environmentally compatible lubricants. This is because of the inherent drawbacks these base oils have pertaining to their oxidation stability and low-temperature properties. Additives are added to the base fluid either to enhance an already-existing property, such as viscosity, of a base oil or to impart a new property, such as detergency, lacking in the base oil. The lubricants are designed to perform a number of functions, including lubrication, cooling, protection against corrosion, and keeping the equipment components clean by suspending ordinarily insoluble contaminants in the bulk lubricant [1]. Although for automotive applications all functions are important, suspending the insoluble contaminants and keeping the surfaces clean are the most critical. As mentioned in Chapter 4 on "detergents," this is achieved by the combined action of the detergents and the dispersants present in the lubricant. Dispersants differ from detergents in three significant ways:

1. Dispersants are metal-free, but detergents contain metals, such as magnesium, calcium, and sometimes barium [2]. This means that on combustion detergents will lead to ash formation and dispersants will not.

2. Dispersants have little or no acid-neutralizing ability, but detergents do. This is because dispersants have either no basicity, as is the case in ester dispersants, or low basicity, as is the case in imide/amide dispersants. The basicity of the imide/amide dispersants is due to the presence of the amine functionality. Amines are weak bases and therefore possess minimal acid-neutralizing ability. Conversely, detergents, especially basic detergents, contain reserve metal bases as metal hydroxides and metal carbonates. These are strong bases, with the ability to neutralize combustion and oxidation-derived inorganic acids, such as sulfuric and nitric acids, and oxidation-derived organic acids.

3. Dispersants are much higher in molecular weight, approximately 4–15 times higher, than the organic portion (soap) of the detergent. Because of this, dispersants are more effective in fulfilling the suspending and cleaning functions than detergents.

Dispersants, detergents, and oxidation inhibitors make up the general class of additives called *stabilizers* and *deposit control agents*. The goal of oxidation inhibitors is to minimize the formation of deposit precursors, such as hydroperoxides and radicals [3,4]. This is because these species are reactive, and they attack the hydrocarbon base oil and additives, which make up the lubricant, to form sludge, resin, varnish, and hard deposits. The goal of the dispersant and the soap portion of the detergent is to keep these entities suspended in the bulk lubricant. This not only results in deposit control but also minimizes particulate-related abrasive wear and viscosity increase. When the lubricant in the equipment is changed, the deposit precursors and the deposit-forming species are removed with the used oil.

The dispersants suspend deposit precursors in oil in various ways. These comprise the following:

Including the undesirable polar species into micelles.
Associating with colloidal particles, thereby preventing them from agglomerating and falling out of solution.
Suspending aggregates in the bulk lubricant, if they are formed.
Modifying soot particles so as to prevent their aggregation. The aggregation will lead to oil thickening, a typical problem in heavy-duty diesel engine oils [5,6].
Lowering the surface/interfacial energy of the polar species to prevent their adherence to metal surfaces.

3.2 NATURE OF DEPOSITS AND MODE OF THEIR FORMATION

A number of undesirable materials result from the oxidative degradation of various components of the lubricant. These are base oil, additives, and the polymeric viscosity modifier, if present. In engine oils, the starting point for the degradation is fuel combustion, which gives rise to hydroperoxides and free radicals [7]. The compounds in the fuel that are most likely to form peroxides, hydroperoxides, and radicals include highly branched aliphatics, unstaurates such as olefins, and aromatics such as alkylbenzenes. All these are present in both gasoline and diesel fuels. American Society for Testing and Materials (ASTM) test methods D4420 and D5186 are used to determine the aromatic content of gasoline and diesel fuels, respectively [8]. The fuel degradation products (peroxides, hydroperoxides, and radicals) go past the piston rings into the lubricant as blowby and, because they are highly energetic, attack largely the hydrocarbon lubricant. Again, the highly branched aliphatic, unsaturated, and aromatic structures are among those that are highly susceptible. ASTM Standard D5292 is commonly used to determine the aromatic content of the base oil [8]. The reaction between the contents of the blowby and these compounds results in the formation of the lubricant-derived peroxides and hydroperoxides that either oxidatively or thermally decompose to form aldehydes, ketones, and carboxylic acids [3,4,9]. Acids can also result from the high-temperature reaction of nitrogen and oxygen, both of which are present in the air–fuel mixture; the oxidation of the fuel sulfur; and the oxidation, hydrolysis, or thermal decomposition of additives such as zinc dialkyldithiophosphates. The reaction between nitrogen and oxygen to form NO_x is more prevalent in diesel engines and gasoline engines that are subjected to severe service, such as long-distance driving for extended periods. The NO_x formation initiates when the temperature reaches 137°C [10,11]. Zinc dialkyldithiophosphates are commonly used as oxidation

inhibitors in engine oils [12,13]. All these acids are neutralized by basic detergents to form inorganic metal salts and metal carboxylates. These compounds are of low hydrocarbon solubility and are likely to fall out of solution.

The aldehydes and ketones undergo aldol-type condensation in the presence of bases or acids to form oligomeric or polymeric compounds. These can further oxidize to highly oxygenated hydrocarbons, commonly referred to as *oxygenates*. The oxygenates are usually of sticky consistency, and the term *resin* is often used to describe them [14]. Resin is either the basic component in or the precursor to all types of deposits. Common types of deposits include varnish, lacquer, carbon, and sludge [15,16]. Varnish, lacquer, and carbon occur when resin separates on hot surfaces and dehydrates or polymerizes to make tenacious films. The quantity and the nature of deposits depend on the proximity of the engine parts to the combustion chamber. The parts closer to the combustion chamber, such as exhaust valve head and stem that experience approximate temperatures of 630°C–730°C [17,18], will develop carbon deposits. The same is true of the combustion chamber wall, piston crown, top land, and top groove, which are exposed to approximate temperatures of 200°C–300°C. Carbon deposits are more common in diesel engines than in gasoline engines and result from the burning of the liquid lubricating oil and the high-boiling fractions of the fuel that adhere to hot surfaces [19].

As we move away from these regions to the low-temperature regions, such as the piston skirt, the deposits are not heavy and form only a thin film. For diesel engine pistons, this type of deposit is referred to as *lacquer*; for gasoline engine pistons, this type of deposit is called *varnish*. The difference between lacquer and varnish is that lacquer is lubricant-derived and varnish is largely fuel-derived. In addition, the two differ in their solubility characteristics. That is, lacquer is water-soluble and varnish is acetone-soluble [15]. Lacquer usually occurs on piston skirts, on cylinder walls, and in the combustion chamber, whereas varnish occurs on valve lifters, piston rings, piston skirts, valve covers, and positive crankcase ventilation (PCV) valves.

The coolest parts of the engine, such as rocker arm covers, oil screen, and oil pan, that are exposed to temperatures of ≤200°C experience sludge deposits. Sludge can be watery or hard in consistency, depending on the severity of service. If the service is extremely mild and of short duration, as in the case of stop-and-go gasoline engine operation, the sludge is likely to be watery or mayonnaiselike [15]. This type of sludge is called low-temperature sludge, which occurs when the ambient temperature is <95°C. The high-temperature sludge is more common in diesel engines and gasoline engines with long, continuous operation. This type of sludge occurs when the ambient temperature is >120°C and is hard in consistency. In the former case, the engine does not get hot enough to expel combustion water, which stays mixed with oil, imparting sludge, a mayonnaise-like appearance. In the latter case, however, the ambient temperature is high enough to expel water, thereby resulting in hard sludge. Sludge is common in areas that experience low oil flow, such as crankcase bottoms and rocker boxes.

Another component of the combustion effluent that must be considered is soot. Soot not only contributes toward some types of deposits such as carbon and sludge, but it also leads to a viscosity increase. These factors can cause poor lubricant circulation and lubricating film formation, both of which will result in wear and catastrophic failure. Soot is particulate in nature and results from the incomplete combustion of the fuel and of the lubricating oil from the crankcase that might enter the combustion chamber by traveling past the piston rings [20]. Fuel-derived soot is a chronic problem in the case of diesel engines because diesel fuel contains high-boiling components that do not burn easily. In addition, diesel engine combustion is largely heterogeneous, with poor air–fuel mixing, hence poor combustion [20]. Soot is made of hydrocarbon fragments with some of the hydrogen atoms removed. The particles are charged and hence have the tendency to form aggregates. When aggregates occur on surfaces, such as those of the combustion chamber, soot deposits result. These deposits are soft and flaky in texture. If these occur in oil, lubricant experiences an increase in viscosity. A soot-related viscosity increase usually requires the presence of polar materials in oil that have the ability to associate with soot. These can be additives or polar lubricant oxidation and degradation products. Carbon deposits are lower in carbon content than soot and, in most cases, contain oily material and ash. This makes knowledge of the ash-forming tendency of a lubricant important to a formulator.

When soot associates with resin, one gets either resin-coated soot particles or soot-coated resin particles [16]. The first type of particles results when resin is in excess, and the second type results when soot is in excess. The amount of soot in resin determines the color of the deposits: the higher the soot, the darker the deposits. Sludge results when resin, soot, oil, and water mix [9].

Deposit formation in gasoline engines is initiated by NO_x and oxidation-derived hydroperoxides that react with hydrocarbons in the fuel and the lubricant to form organic nitrates and oxygenates [14,21]. Being thermally unstable, these species decompose and polymerize to form deposits. The deposits typically include resin, varnish, and low-temperature sludge. In diesel engines, however, soot is an important component of the deposits, which include lacquer, carbon deposits, and high-temperature sludge [16]. Typically, carbon deposits are of high metal content, which is mainly due to the presence of detergent additives in the lubricant [22,23].

Detailed mechanism of deposit formation in engines is described elsewhere [24,25]. The mechanism is based on the premise that both the lubricant and the fuel contribute toward deposit formation. The role of the blowby, NO_x, and high-temperature oxidative and thermal degradation of the lubricant, described earlier, are substantiated [24]. The importance of oxygenated precursors—their decomposition, condensation, and polymerization to form deposits—is also supported. The deposit precursors consist of approximately 15–50 carbon

atoms and contain multiple hydroxy and carboxy functional groups. Because of the polyfunctionality, these molecules have the ability to thermally polymerize to high-molecular-weight products [14,16]. As mentioned earlier, soot associates with polar oxidation products in oil to cause a viscosity increase. Viscosity increase can also occur in gasoline engine oils that have little or no soot. This happens when the oxygen content of the precursors is low and the resulting polymer is of low molecular weight and of good oil solubility [14]. This phenomenon is commonly referred to as *oil thickening* [6]. Conversely, if the oxygen content of the precursors is high, the polymerization results in the formation of high-molecular-weight products of low lubricant solubility. Such products constitute resin, which is of low oil solubility and separates on surfaces. If the surfaces are hot, subsequent dehydration and polymerization lead to the formation of varnish, lacquer, and carbon deposits. It is important to note that deposits are a consequence of lubricant oxidation that accelerates once the oxidation inhibitor package in the lubricant is exhausted.

Three other internal combustion engine problems—oil consumption, ring sticking, and corrosion and wear—are also related to lubricant degradation. Oil consumption is a measure of how much lubricant travels past piston rings into the combustion chamber and burns. A certain minimum amount of the lubricant is necessary in the vicinity of the piston rings to lubricate cylinder walls and cylinder liners and hence facilitate piston movement and minimize scuffing. However, if too much lubricant ends up in the combustion chamber, serious emission problems will result. Modern piston designs, such as articulated pistons and pistons with low crevice volume, allow just enough lubricant to minimize scuffing, but without adversely contributing to emissions [26,27]. Other parameters that affect oil consumption include the integrity of pistons and cylinders and the viscosity, volatility, and sealing characteristics of the lubricant. Pistons with stuck rings and out-of-square grooves and cylinders with increased wear will result in a poor seal between the crankcase and the combustion chamber [15]. As a consequence, a larger amount of blow-by will enter the crankcase and increase the rate of lubricant breakdown. This will complicate the situation further. Ring sticking occurs when sticky deposits form in the grooves behind the piston rings. This is a serious problem because it not only results in a poor seal but also leads to poor heat transfer from the cylinder to the wall. If not controlled, this will result in nonuniform thermal expansion of the pistons, loss of compression, and ultimately the failure of the engine [15]. The wear of pistons and the cylinders is undesired for the same reasons. Wear of engine parts is either corrosive or abrasive. *Corrosive wear* arises from the attack of fuel sulfur-derived products, such as sulfur oxides or sulfuric acid, or the acidic by-products of lubricant oxidation and degradation, such as carboxylic and sulfonic acids. Fuel sulfur–derived piston ring wear and cylinder wear are serious problems in large, slow-speed marine diesel engines that use a high-sulfur fuel. Corrosive wear is controlled by the use of lubricants with a base reserve, that is, those containing a large quantity of basic detergents. *Abrasive wear* results from the presence of the particulate matter, such as large soot particles, in the lubricant. Dispersants are crucial to the control of soot-related wear.

3.3 DEPOSIT CONTROL BY DISPERSANTS

Fuel and lubricant oxidation and degradation products, such as soot, resin, varnish, lacquer, and carbon, are of low lubricant (hydrocarbon) solubility, with a propensity to separate on surfaces. The separation tendency of these materials is a consequence of their particle size. Small particles are more likely to stay in oil than large particles. Therefore, resin and soot particles, which are the two essential components of all deposit-forming species, must grow in size through agglomeration before separation. Growth occurs either because of dipolar interactions, as is the case in resin molecules, or because of adsorbed polar impurities such as water and oxygen, as is the case in soot particles. Alternatively, soot particles are caught in the sticky resin. Dispersants interfere in agglomeration by associating with individual resin and soot particles. The particles with associated dispersant molecules are unable to coalesce because of either steric factors or electrostatic factors [28]. Dispersants consist of a polar group, usually oxygen- or nitrogen-based, and a large nonpolar group. The polar group associates with the polar particles, and the nonpolar group keeps such particles suspended in the bulk lubricant [16]. Neutral detergents, or soaps, operate by an analogous mechanism.

3.4 DESIRABLE DISPERSANT PROPERTIES

Dispersing soot, deposit precursors, and deposits is clearly the primary function of a dispersant. Dispersants, in addition, need other properties to perform effectively. These include thermal and oxidative stability and good low-temperature properties. If a dispersant has poor thermal stability, it will break down, thereby losing its ability to associate with and suspend potentially harmful products. Poor oxidative stability translates into the dispersant molecule contributing itself toward deposit formation. Good low-temperature properties of a lubricant are desired for many reasons: ease of cold cranking, good lubricant circulation, and fuel economy. Base oil suppliers have developed a number of ways to achieve these properties. The methods they use include isomerization of the base stock hydrocarbons through hydrocracking and the use of special synthetic oils as additives. Since dispersant is one of the major components of the engine oil formulations, its presence can adversely affect these properties, which must be preserved.

3.5 DISPERSANT STRUCTURE

A dispersant molecule consists of three distinct structural features: a *hydrocarbon group*, a *polar group*, and a connecting group or a *link* (see Figure 3.1). The hydrocarbon group is polymeric in nature, and depending on its molecular weight, dispersants can be classified into *polymeric dispersants* and *dispersant polymers*. Polymeric dispersants are of lower molecular weight than dispersant polymers. The molecular weight of polymeric dispersants ranges between 3,000 and 7,000 as compared to dispersant polymers, which have a molecular weight of 25,000 and higher. Although various olefins, such as polyisobutylene, polypropylene, poly-α-olefins (PAOs), and mixtures thereof, can be used to make polymeric dispersants, the polyisobutylene-derived dispersants are the most common. The number average molecular weight (Mn) of polyisobutylene ranges between 500 and 3000, with an Mn of 1000–2000 being typical [29]. In addition to Mn, other polyisobutylene parameters, such as molecular weight distribution and the length and degree of branching, are also important in determining the overall effectiveness of a dispersant.

Substances obtained through a polymerization reaction, especially those made by using an acid catalyst or a free-radical initiator, often contain molecules of different sizes. Molecular weight distribution, or polydispersity index, is commonly used to assess the heterogeneity in molecular size. Polydispersity index is the ratio of weight average molecular weight (Mw) and Mn, or Mw/Mn [30–32]. These molecular weights are determined by subjecting the polymer to gel permeation chromatography (GPC). The method separates molecules based on size [33]. The larger molecules come out first, followed by the next size. When the molecules are of the same size, Mw/Mn equals 1 and the polymer is called a *monodisperse polymer*. The polymers with an index >1 are called *polydisperse polymers*. For most applications, monodispersity is desired. Polyisobutylene, derived from acid-catalyzed polymerization reaction, typically has a polydispersity index between 2 and 3. This will impact many of the dispersant properties described below.

Dispersant polymers, also called dispersant viscosity modifiers (DVMs) and dispersant viscosity index improvers (DVIIs), are derived from hydrocarbon polymers of molecular weights between 25,000 and 500,000. Polymer substrates used to make DVMs include high-molecular-weight olefin copolymers (OCPs), such as ethylene–propylene copolymers (EPRs), ethylene–propylene–diene copolymers (EPDMs), polymethacrylates (PMAs), styrene–diene rubbers (SDRs) of both linear and star configurations, and styrene–ester polymers (SEs).

The polar group is usually nitrogen- or oxygen-derived. Nitrogen-based groups are derived from amines and

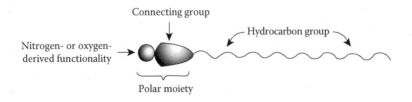

FIGURE 3.1 Graphic representation of a dispersant molecule.

are usually basic in character. Oxygen-based groups are alcohol-derived and are neutral. The amines commonly used to synthesize dispersants are polyalkylene polyamines such as diethylenetriamine and triethylenetetramine. In the case of DVMs or dispersant polymers, the polar group is introduced by direct grafting, copolymerization, or by introducing a reactable functionality. The compounds used for this purpose include monomers such as 2- or 4-vinylpyridine, N-vinylpyrrolidinone, and N,N-dialkylaminoalkyl acrylate and unsaturated anhydrides and acids such as maleic anhydride, acrylic acid, and glyoxylic acid. The details of these reactions are described in Section 3.6, which deals with the dispersant synthesis. Amine-derived dispersants are called nitrogen or amine dispersants, and those that are alcohol-derived are called oxygen or ester dispersants [28]. Oxygen-derived phosphonate ester dispersants were popular at one time, but their use in engine oils is now restrained because of the phosphorus limit. Phosphorus limit pertains to its tendency to poison noble metal catalysts used in catalytic converters. Formulators prefer to take advantage of the phosphorus limit by using zinc dialkyldithiophosphates, which are excellent oxidation inhibitors and antiwear agents. In the case of amine dispersants, it is customary to leave some of the amino groups unreacted to impart basicity to the dispersant. The reasons for this are described in Section 3.7.

3.6 DISPERSANT SYNTHESIS

Since it is not easy to attach the polar group directly to the hydrocarbon group, except in the case of olefins that are used to make DVMs, the need for a connecting group or a link arises. Although many such groups can be used, the two common ones are phenol and succinic anhydride. Olefin, such as polyisobutylene, is reacted either with phenol to form an alkylphenol or with maleic anhydride to form an alkenylsuccinic anhydride. The polar functionality is then introduced by reacting these substrates with appropriate reagents.

3.6.1 THE HYDROCARBON GROUP

Polyisobutylene is the most common source of the hydrocarbon group in polymeric dispersants. It is manufactured through acid-catalyzed polymerization of isobutylene [34,35]. Figure 3.2 depicts the mechanism of its formation. In Figure 3.2, polyisobutylene is shown as a terminal olefin, whereas in reality it is a mixture of various isomers. Those that predominate include geminally disubstituted (vinylidene), trisubstituted, and tetrasubstituted olefins. Figure 3.3 shows their structure and the possible mechanism of their formation. Polyisobutylenes of structures I and II result from the loss of a proton from carbon 1 and carbon 3 of the intermediate of structure V. Polyisobutylenes of structures III and IV result from the rearrangement of the initially formed carbocation, as shown in Figure 3.3. The reactivity of these olefins toward phenol and maleic anhydride varies. In general, the more substituted the olefin, the lower the reactivity, which is a consequence of the steric factors. Similarly, the larger the size of the polyisobutyl pendant group, that is, the higher the molecular weight, the lower the reactivity. This is due to the dilution effect, which results from low olefin-to-hydrocarbon ratio. As mentioned earlier, polyisobutylene is the most commonly used olefin. One of the reasons for its preference is its extensive branching. This makes the derived dispersants to possess excellent oil solubility, in both non-associated and associated forms. However, if the hydrocarbon chain in the dispersant is too small, its lubricant solubility greatly suffers. Because of this, the low-molecular-weight components in polyisobutylene are not desired. This is despite their higher reactivity. These must be removed, which is carried out through distillation. Alternatively, one can minimize the formation of these components by decreasing the amount of the catalyst during polymerization and by lowering the polymerization reaction temperature.

A new class of dispersants derived from ethylene/α-OCP with an Mn of 300–20,000 has also been reported, primarily by the Exxon scientists [36,37]. Such dispersants are claimed to have superior low- and high-temperature viscometrics than those of the polyisobutylene-derived materials.

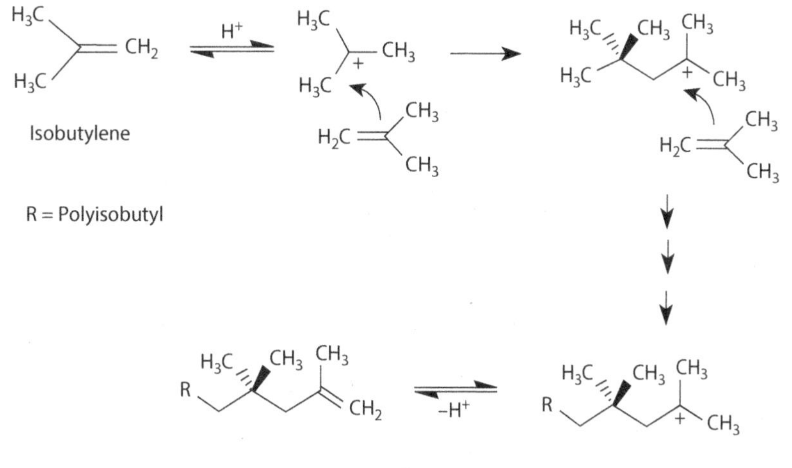

FIGURE 3.2 Acid-catalyzed polymerization of isobutylene.

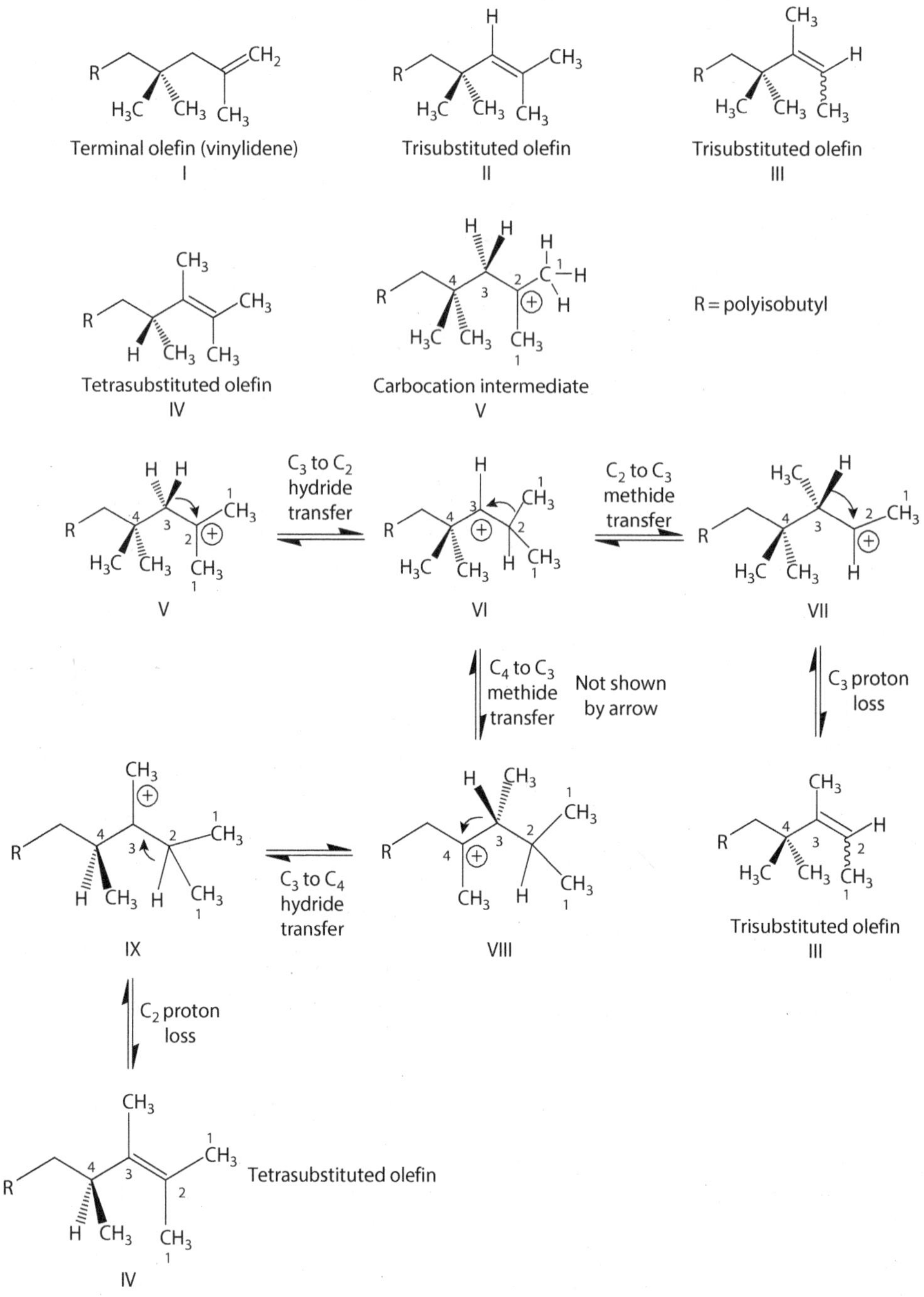

FIGURE 3.3 Polyisobutylene structures and the mode of their formation.

As mentioned earlier, dispersant polymers are derived from EPRs, styrene–butadiene copolymers, polyacrylates, PMAs, and styrene esters. The ethylene–propylene rubbers are synthesized by Ziegler–Natta catalysis [38]. The styrene–butadiene rubbers are synthesized through anionic polymerization [38]. Polyacrylates and PMAs are synthesized through polymerization of the monomers using free-radical initiators [38]. Styrene esters are made by reacting styrene–maleic anhydride copolymer or styrene–maleic anhydride–alkyl acrylate terpolymer with alcohols, usually in the presence of a protic acid, such as sulfuric or methanesulfonic acid, catalyst. Since complete esterification of the anhydride is hard to achieve, the neutralization of the residual carboxylic acid anhydride is carried out by alternative means [38–40].

3.6.2 The Connecting Group

As mentioned in Section 3.5, succinimide, phenol, and phosphonate are the common connecting groups used to make dispersants. Of these, succinimide and phenol are the most prevalent [2]. Succinimide group results when a cyclic carboxylic acid anhydride is reacted with a primary amino group. Alkenylsuccinic anhydride is the precursor for introducing the succinimide connecting group in dispersants.

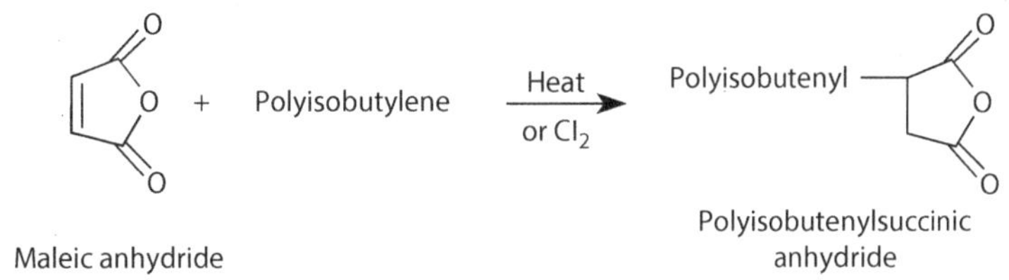

FIGURE 3.4 Alkenylsuccinic anhydride formation.

Alkenylsuccinic anhydride is synthesized by reacting an olefin, such as polyisobutylene, with maleic anhydride [2]. This is shown in Figure 3.4.

The reaction is carried out either thermally [29,41,42] or in the presence of chlorine [43]. The thermal process involves heating the two reactants together usually >200°C [29,41,42], whereas the chlorine-mediated reaction with a mixture is carried out by introducing chlorine to react containing polyisobutylene and maleic anhydride [43–48]. Depending on the manner in which chlorine is added, the procedure is either *one-step* or *two-step* [44]. If chlorine is first reacted with polyisobutylene before adding maleic anhydride, the procedure is considered two-step. If chlorine is added to a mixture of polyisobutylene and maleic anhydride, it is a one-step procedure. The one-step procedure is generally preferred.

The chlorine-mediated process has several advantages, which include having a low reaction temperature, having a faster reaction rate, and working well with internalized or highly substituted olefins. The low reaction temperature minimizes the chances of thermal breakdown of polyisobutylene and saves energy. The major drawback of the chlorine process is that the resulting dispersants contain residual chlorine as organic chlorides. Their presence in the environment is becoming a concern because they can lead to the formation of carcinogenic dioxins. A number of strategies are reported in the literature to decrease the chlorine content in dispersants [49–54]. The thermal process does not suffer from the presence of chlorine, although it is less energy-efficient and requires the use of predominantly a terminal olefin, that is, the polyisobutylene of high vinylidene content.

The mechanism by which the two processes proceed is also different [46,47,50–52]. The thermal process is postulated to occur through an ene reaction. The chlorine-mediated reaction is postulated to proceed through a Diels–Alder reaction. The mechanism of the diene formation is shown in Figure 3.5. Chlorine first reacts with polyisobutylene 1 to form allylic chloride II. By the loss of the chloride radical, this yields the intermediate III, which through C_4 to C_3 methyl radical transfer is converted into the intermediate IV. A C_3 to C_4 hydrogen shift in the intermediate results in the formation of the radical V. This radical can lose hydrogen either from C_4 to yield the diene VI or from C_5 to result in the diene VII. The resulting dienes then react with maleic anhydride through a 4 + 2 addition reaction, commonly called a Diels–Alder reaction [55], to form alkenyltetrahydrophthalic anhydrides [50,52]. These reactions are shown in Figure 3.6.

These anhydrides can be converted into phthalic anhydrides through dehydrogenation by using sulfur [50–52]. These compounds can then be transformed into dispersants by reacting with polyamines and polyhydric alcohols [51,52]. During the thermal reaction of polyisobutylene with maleic anhydride, that is, the ene reaction, the vinylidene double bond moves down the chain to the next carbon. Since thermal reaction requires a terminal olefin, further reaction of the new olefin with another mole of maleic anhydride will not occur if the double bond internalizes, and the reaction will stop at this stage. This is shown in Reaction 3.3 of Figure 3.6. If the new double bond is external, the reaction with another molecule of maleic anhydride is possible [45]. This is shown in Reaction 3.4.

For dispersants, polyisobutylphenol is the alkylphenol of choice. It is synthesized by reacting polyisobutylenes with phenol in the presence of an acid catalyst [56–58]. Lewis acid catalysts, such as aluminum chloride and boron trifluoride, are often employed. Boron trifluoride is preferred over aluminum chloride because the reaction can be carried out at low temperatures, which minimizes acid-mediated breakdown of polyisobutylene [58]. This is desired because dispersants derived from low-molecular-weight phenols are not very effective. Other catalysts, such as sulfuric acid, methanesulfonic acid, and porous acid catalysts of Amberlyst® type, can also be used to make alkylphenols [59,60]. Polyisobutylene also reacts with phosphorus pentasulfide through an ene reaction. The resulting adduct is hydrolyzed by the use of steam to alkenylphosphonic and alkenylthiophosphonic acids [2,3]. The methods to synthesize alkylphenols and alkenylphosphonic acids are shown in Figure 3.7.

A new carboxylate moiety derived from glyoxylic acid to make dispersants has been reported in the literature [61–65]. However, at present, no commercial products appear to be based on this chemistry.

3.6.3 THE POLAR MOIETY

The two common polar moieties in dispersants are based on polyamines and polyhydric alcohols. The structures of common amines and alcohols used to make dispersants are shown in Figure 3.8.

The polyamines are manufactured from ethylene through chlorination, followed by the reaction with ammonia [66]. The reaction scheme is given in Figure 3.9. As shown, polyamines contain piperazines as a by-product. Examining the structures of various amines, one can see that they contain

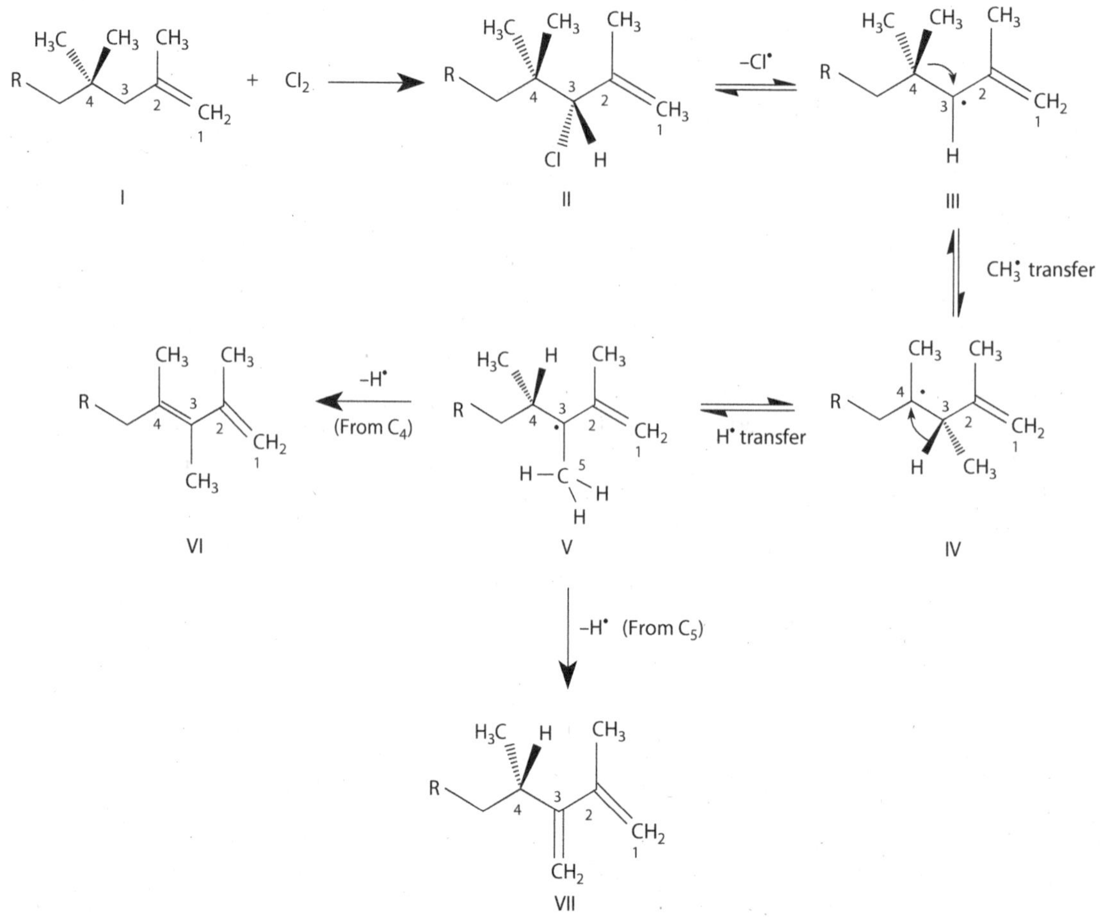

FIGURE 3.5 Mechanism of chlorine-assisted diene formation.

primary, secondary, and tertiary amino groups. Each type of amino group has different reactivity toward alkenylsuccinic anhydride. The primary amino group reacts with the anhydride to form a cyclic imide, the secondary amino group reacts with the anhydride to form an amide/carboxylic acid, and the tertiary amino group does not react with the anhydride at all [67].

However, it can make a salt if a free carboxylic acid functionality is present in the molecule, as is the case in amide/carboxylic acid. These reactions are shown in Figure 3.10. New high-molecular-weight amines derived from phosphoric acid–catalyzed condensation of polyhydroxy compounds, such as pentaerythritol, and polyalkylene polyamines, such as triethylenetetramine, are known [68]. These amines are claimed to form high total base number (TBN) dispersants with low free-amine content and better engine test performance than dispersants made from conventional polyamines.

Imide and ester dispersants are made by reacting polyamines and polyhydric alcohols with alkenylsuccinic anhydrides. The reaction typically requires a reaction temperature between 130°C and 200°C to remove the resulting water and complete the reaction [44]. As mentioned earlier, imide dispersants are made by the use of polyalkylene polyamines, such as diethylenetriamine and triethylenetetramine. Many

polyhydric alcohols can be used to make ester dispersants. These include trimethylolpropane, tris(hydroxymethyl)aminoethane, and pentaerythritol. When one uses tris (hydroxymethyl) aminoethane as the alcohol, one can obtain an ester dispersant with basicity. The reactions to make succinimide and succinate dispersants are depicted in Figure 3.11.

The alkylphenol-derived dispersants are made by reacting an alkylphenol, such as polyisobutylphenol, with formaldehyde and a polyamine [58,69]. The result is the formation of 2-aminomethyl-4-polyisobutylphenol. The reaction of ammonia or an amine, formaldehyde, and a compound with active hydrogen(s), such as a phenol, is called the *Mannich reaction* [70,71]. Hence, such dispersants are called Mannich dispersants. For making phosphonate dispersants, the common method is to react the free acid with an olefin epoxide, such as propylene oxide or butylene oxide, or an amine [2,72,73]. These reactions are shown in Figure 3.12. Salts derived from the direct reaction of amine and metal bases with olefin-phosphorus pentasulfide adduct are also known [74,75]. It is important to note that structures in figures are idealized structures. The actual structures will depend on the substrate (alkylphenol and alkenylsuccinic anhydride)-to-reactant (formaldehyde and polyamines) ratio.

Because of the polyfunctionality of the succinic anhydride group and of the amines and polyhydric alcohols,

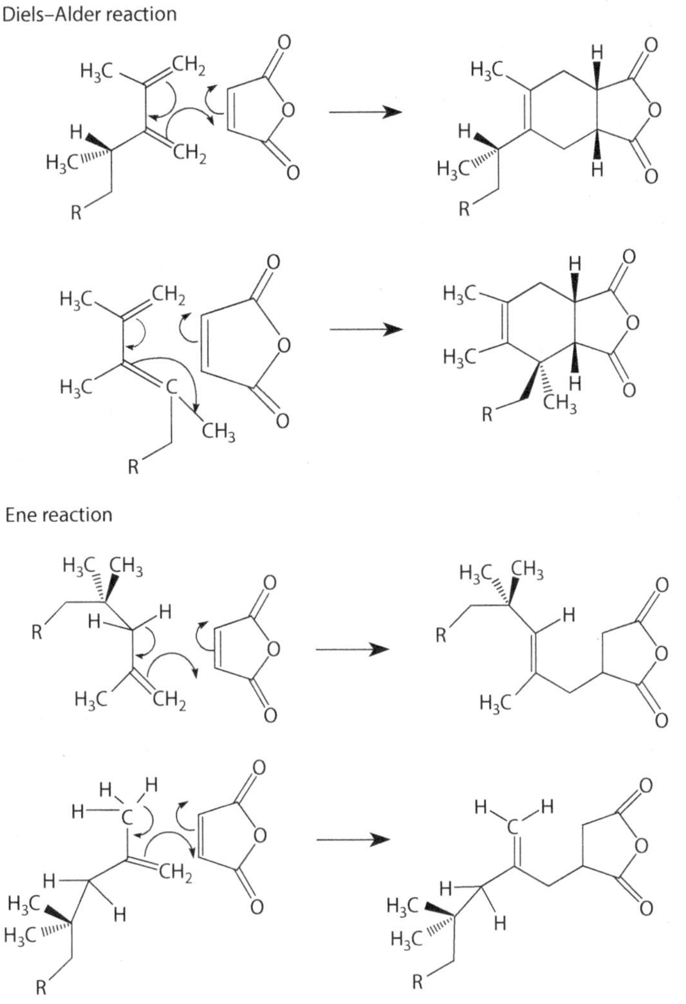

Diels–Alder reaction

(3.1)

(3.2)

Ene reaction

(3.3)

(3.4)

FIGURE 3.6 Mechanism of alkenylsuccinic anhydride formation.

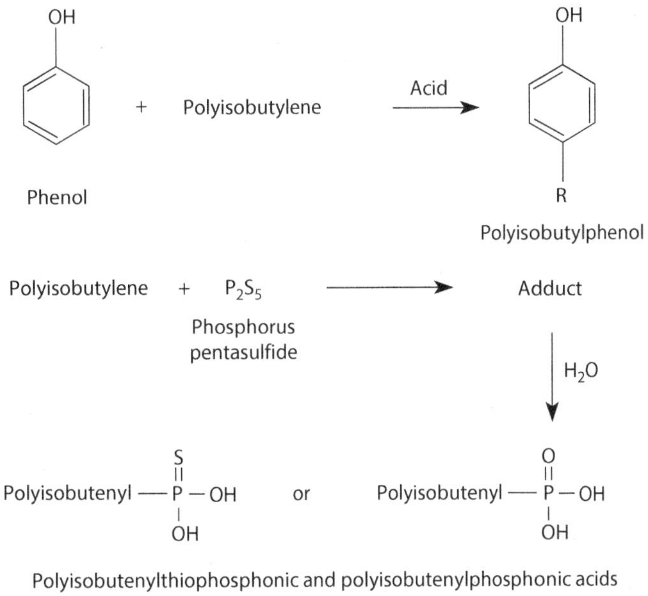

Phenol

Polyisobutylphenol

Polyisobutylene + P_2S_5 $\longrightarrow$ Adduct

Phosphorus
pentasulfide

H_2O

Polyisobutenyl — P — OH   or   Polyisobutenyl — P — OH

Polyisobutenylthiophosphonic and polyisobutenylphosphonic acids

FIGURE 3.7 Synthesis of alkylphenols and alkenylphosphonic acids.

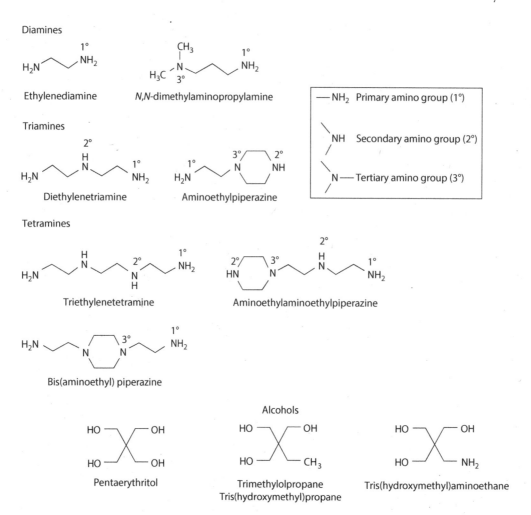

FIGURE 3.8   Amines and alcohols used to synthesize dispersants.

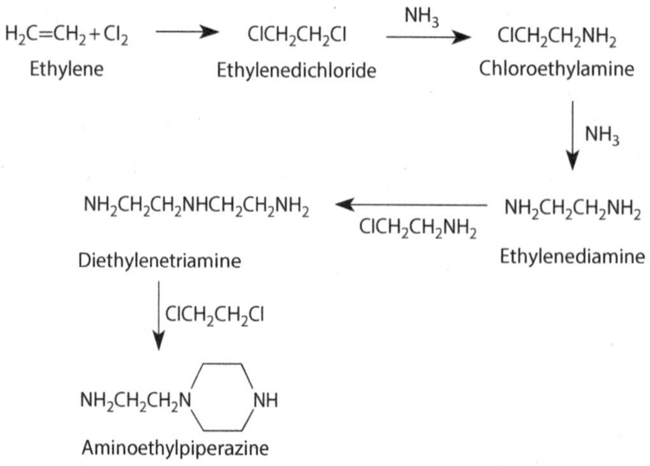

FIGURE 3.9   Manufacture of polyamines.

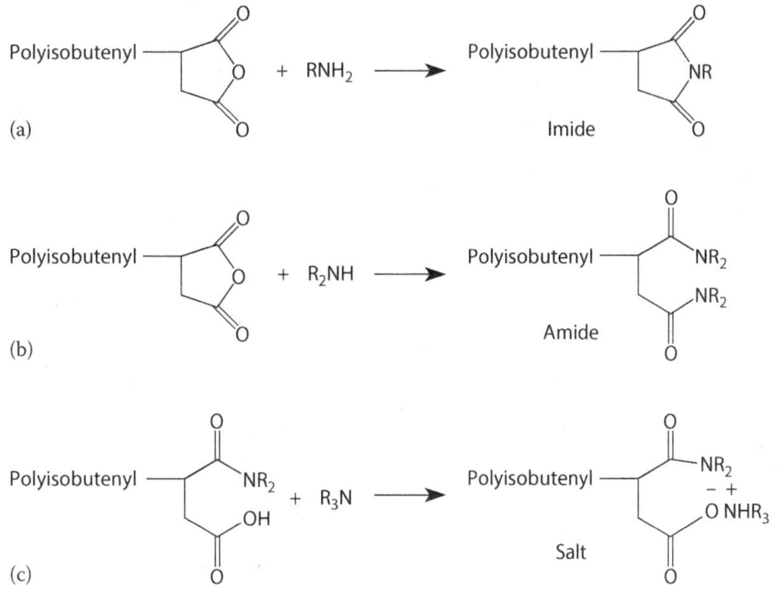

FIGURE 3.10 Amine-anhydride reaction products. (a) Primary amine, (b) secondary amine, and (c) tertiary amine. (Based on Harrison, J.J. et al., Two-step thermal process for the preparation of alkenylsuccinic anhydride, U.S. Patent 5,625,004, April 29, 1997.)

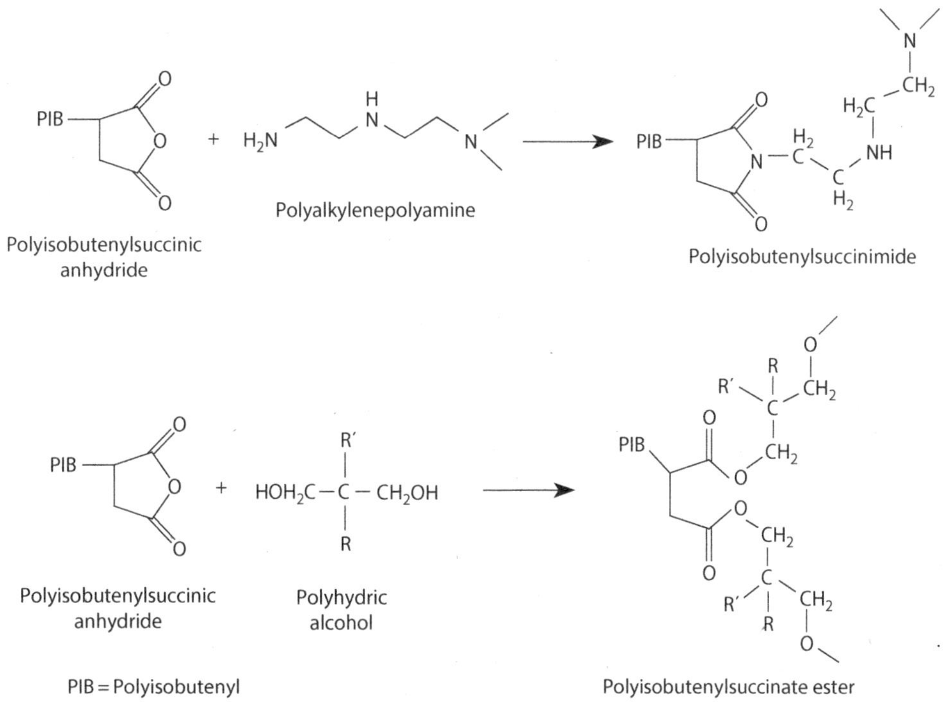

FIGURE 3.11 Synthesis of imide and ester dispersants.

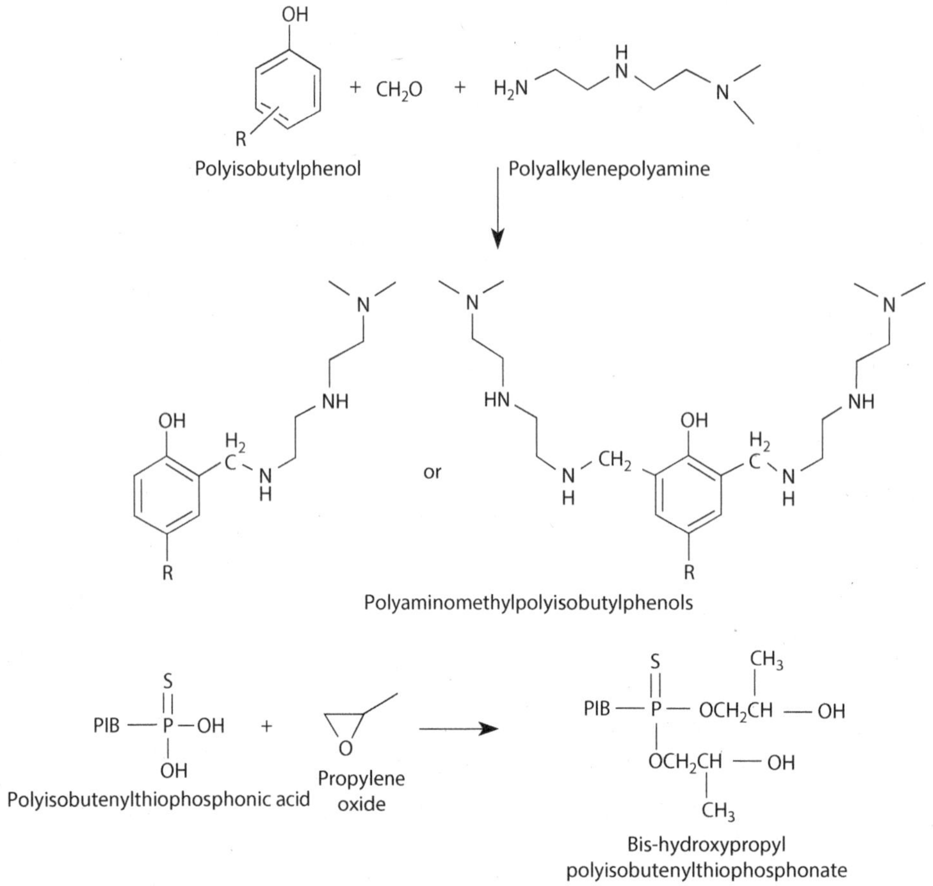

FIGURE 3.12 Synthesis of Mannich and phosphonate dispersants.

various dispersants can be made by altering the anhydride-to-amine or anhydride-to-alcohol ratios. These dispersants differ not only in their molecular weight but also in their properties. Polyfunctionality of the two reactants leads to dispersants, which have molecular weights that are three to seven times higher than expected if the two reactants were monofunctional.

The methods to make DVMs are shown in Figures 3.13 through 3.15. These are synthesized by

Grafting or reacting of a dispersancy-imparting monomer on an already-formed polymer, as in the case of EPRs and SDRs [76–84].

Including such a monomer during the polymerization process, as in the case of polyacrylates and PMAs [85].

Introducing a reactive functional group in the polymer that can be reacted with a reagent to impart dispersancy, as in the case of styrene–maleic anhydride copolymers [40,86–93].

Although most of the examples in Figures 3.13 through 3.15 pertain to the introduction of the basic nitrogen-containing moieties, neutral DVMs are also known in the literature. These are made by using nonbasic reactants, such as

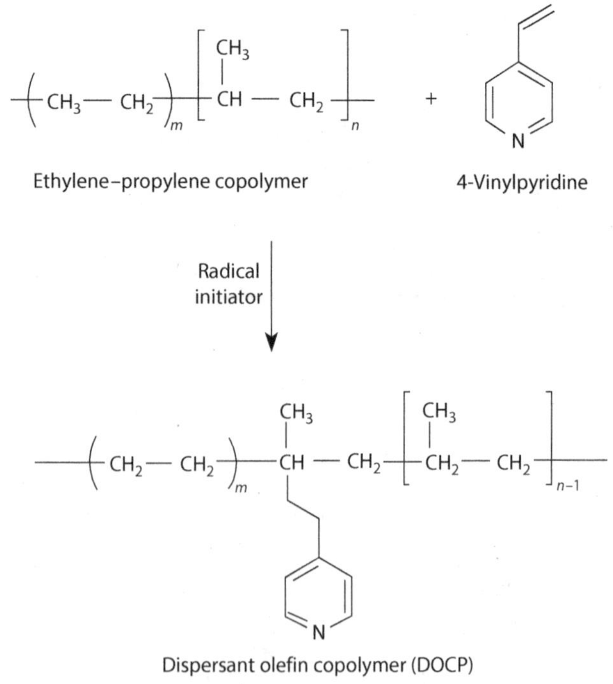

FIGURE 3.13 Dispersant viscosity modifier synthesis through grafting.

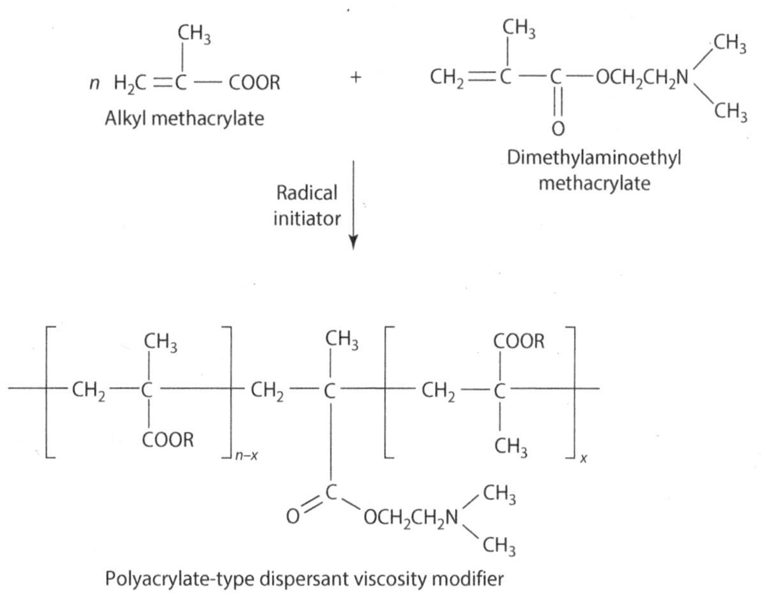

FIGURE 3.14 Dispersant viscosity modifier synthesis through copolymerization.

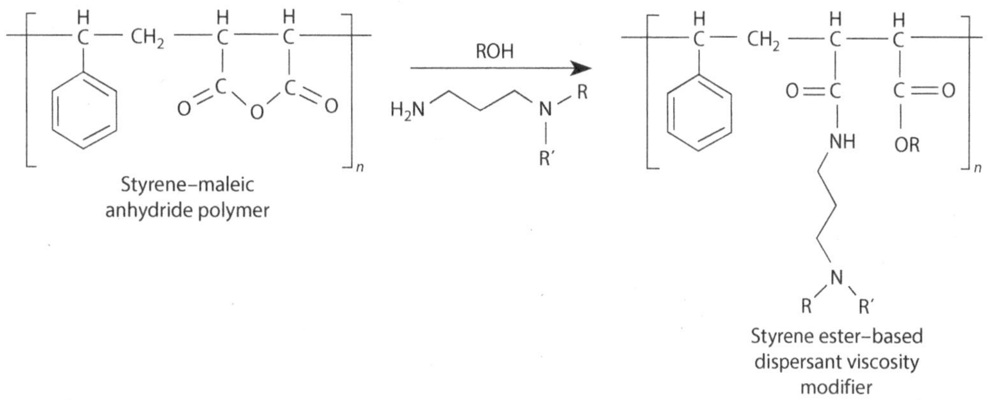

FIGURE 3.15 Dispersant viscosity modifier synthesis through chemical reaction.

N-vinylpyrrolidinone, alcohols, or polyether-derived methacrylate ester [79,94,95]. Recently, dispersant viscosity–improving additives with built-in oxidation inhibiting and antiwear moieties have been reported in the patent literature [77,96,97]. Dispersant polymers containing oxidation-inhibiting moieties are commercially available from Texaco Chemical Company now part of Ethyl Petroleum Additives Company. As the examples show, grafting usually allows the introduction of the connecting group in the dispersant polymers at the same time as the polar moiety.

3.7 DISPERSANT PROPERTIES

A dispersant consists of a hydrocarbon chain, a connecting group, and a polar functionality. Although each structural feature imparts unique properties to a dispersant, the dispersant's overall performance depends on all the three. The overall performance is assessed in terms of its dispersancy, thermal and oxidative stability, viscosity characteristics, and seal performance. These criteria primarily relate to engine oils, where dispersants find major use.

3.7.1 DISPERSANCY

As mentioned, dispersancy pertains to a dispersant's ability to suspend by-products of combustion, such as soot, and lubricant degradation, such as resin, varnish, lacquer, and carbon deposits. The overall performance of a dispersant depends on all the three of its structural features: the hydrocarbon chain, the connecting group, and the polar moiety. The molecular weight of the hydrocarbon group in a dispersant determines its ability to associate with undesirable polar species and suspend them in the bulk lubricant. For dispersants that have the same connecting group and the polar moiety, the lower the molecular weight, the higher the ability to associate with polar materials and the lower the ability to suspend them. Because of the trade-off between the two properties, the hydrocarbon chain must have the correct size and branching.

The size affects a dispersant's affinity toward polar materials, and branching affects its solubility, both before association and after association with the species, a dispersant is designed to suspend in oil. Experience has demonstrated that hydrocarbon groups containing 70–200 carbon atoms and extensive branching, as in the case of polyisobutylenes, are extremely suitable to design dispersants with good dispersancy. The hydrocarbon chains of larger size, even if the branching is similar, lead to dispersants with low affinity toward polar materials.

That is why dispersant polymers possess lower dispersancy than polymeric dispersants. However, since dispersant polymers have additional attributes, such as good thickening efficiency and in some cases good thermal and oxidative stability, their use is advantageous. They usually replace additives, called viscosity modifiers, in the package. Since they impart some dispersancy because of their structure, the amount of polymeric dispersant in engine oil formulations is somewhat decreased [79,98].

Both the connecting group and the polar moiety are important to the dispersancy of the dispersant molecule. They must be considered together since both contribute toward polarity. In Mannich dispersants, the phenol functional group, and in imide and ester dispersants, succinimide, succinate, and phosphonate functional groups are also polar, the same as the amine and the alcohol-derived portion of the molecule. The polarity is a consequence of the electronegativity difference between carbon, oxygen, nitrogen, and phosphorus atoms. The greater the electronegativity difference, the stronger the polarity. This implies that groups that contain phosphorus–oxygen bonds are more polar than those containing carbon–oxygen bonds, carbon–nitrogen bonds, and carbon–phosphorus bonds. The electronegativity difference for such bonds is 1.4, 1.0, 0.5, and 0.4, respectively [99]. However, since dispersants have many bonds with various combinations of atoms, the overall polarity in a dispersant and its ability to associate with polar materials are not easy to predict. Because some of the materials with which the dispersant associates are acidic, such as carboxylic acids derived from lubricant oxidation, the presence of an amine nitrogen is an advantage because of its basic character. Therefore, in certain gasoline engine tests, nitrogen dispersants are superior to ester dispersants. Ester dispersants are usually superior in diesel engine tests because of their higher thermo-oxidative stability. Mannich dispersants are good low-temperature dispersants; hence, they are typically used in gasoline engine oils.

As mentioned earlier, commercial polyisobutylenes have a molecular weight distribution. This will lead to dispersant structures of varying size, hence molecular weight. An optimum ratio between the molecular weight of the hydrocarbon chain and that of the polar functionality (polar/nonpolar ratio) is a prerequisite for good dispersancy. If a dispersant composition has an excessive amount of components with short hydrocarbon chains, that is, of low molecular weight, its associating ability increases, but its oil solubility suffers. This is likely to deteriorate its dispersancy, especially after associating with polar impurities. Such structures in dispersants are, therefore,

undesired. Their formation can be minimized by using polyolefins of low polydispersity index, controlling the formation of low-molecular-weight components, removing such components through distillation [100], or postreacting with another reagent, preferably of the hydrocarbon type. Polyolefins of low polydispersity index (≤2) are available from BP and Exxon Chemical Company. Controlling the formation of low-molecular-weight components is exemplified by the use of boron trifluoride catalyst for making alkylphenols instead of aluminum chloride, which tends to fragment polyisobutylene. Removing the lower-molecular-weight components, although not easy, is possible at the precursor stage, which is before reacting with the alcohol or the amine. A number of reagents can be used for the postreaction [101]. Hydrocarbon posttreatment agents include polyepoxides [102], polycarboxylic acid [103], alkylbenzenesulfonic acids [104], and alkenylnitriles [105]. Whenever postreacted dispersants are used in engine oils, improved dispersancy, viscosity index credit, improved fluorocarbon elastomer compatibility, hydrolytic stability, and shear stability are often claimed.

3.7.2 THERMAL AND OXIDATIVE STABILITY

All the three components of the dispersant structure determine its thermal and oxidative stability, the same as dispersancy. The *hydrocarbon group* can oxidize in the same manner as the lubricant hydrocarbons to form oxidation products that can contribute toward deposit-forming species [4,9]. (This is described in Section 3.2.) Although the rate of oxidation is quite slow for largely paraffinic hydrocarbon groups, such as polyisobutyl group, it is quite high for those that contain multiple bonds, such as polyisobutenyl, and the benzylic groups. The benzylic functional group is present in styrene butadiene and styrene ester–derived dispersant polymers. Purely paraffinic hydrocarbon groups that contain tertiary hydrogen atoms, such as EPRs, oxidize at a faster rate than those that contain only primary and secondary hydrogen atoms. Styrene isoprene–derived materials contain both benzylic and tertiary hydrogen atoms. This implies that highly branched alkyl groups, such as polyisobutyl and polyisobutenyl, have a higher susceptibility toward oxidation than linear or unbranched alkyl groups. Dispersant polymers with built-in oxidation-inhibiting moieties are known in the literature [77,78,96]. The polar moiety in an amine-derived dispersant is also likely to oxidize at a faster rate than the oxygen-derived moiety because of the facile formation of the amine oxide functional group on oxidation. Such groups are known to thermally undergo β-elimination [40], called the *cope reaction*, to form an olefin. This can oxidize at a faster rate as well as lead to deposit-forming polymeric products.

From a thermal stability perspective, the hydrocarbon group in the case of high-molecular-weight dispersant polymers, such as those derived from OCPs, is more likely to break down (unzip) than that derived from the low-molecular-weight polymers. Dispersants based on 1000–2000 molecular weight polyisobutylenes are relatively stable, except at very high temperatures that are experienced in some engine parts, such as near the top of the piston [17,18]. Thermal breakdown of the oxidized amine polar group is mentioned in the previous paragraph.

The chemical reactivity of certain dispersants toward water and other reactive chemicals present in the lubricant formulation is an additional concern. The most likely reaction site is the connecting group. The common connecting groups are amide and imide in amine-derived dispersants and ester in alcohol-derived dispersants. All three can hydrolyze in the presence of water [106], but at different rates. Esters are easier to hydrolyze than amides and imides. The hydrolysis is facilitated by the presence of bases and acids. Basic detergents are the source of the metal carbonate and metal hydroxide bases, which at high temperatures catalyze the hydrolysis reaction. Additives, such as zinc dialkyldithiophosphates, are a source of strong acids that result when these additives hydrolyze, thermally decompose, or oxidize. The fate of the ester-, amide-, and imide-type dispersant polymers, such as those derived from polyacrylates, PMAs, and styrene ester substrates, is the same. Some OCP-derived dispersant polymers, such as those obtained by grafting of monomers 2- or 4-vinylpyridine and 1-vinyl-2-pyrrolidinone [76,80], do not suffer from this problem since they do not contain easily hydrolyzable groups. Reactivity toward other chemicals present in the formulation is again prevalent in the case of ester-derived dispersants. Reaction with metal-containing additives, such as detergents and zinc dialkyldithiophosphates, can occur after hydrolysis to form metal salts. This can destroy the polymeric structure of a dispersant and hence its effectiveness. Some formulations contain amines or their salts as corrosion inhibitors or friction modifiers. Depending on the molecular weight and the ambient temperature, these can displace the polyol or sometimes the polyamine, thereby altering the dispersant structure, hence its properties.

3.7.3 Viscosity Characteristics

The amount of dispersant in automotive engine oils typically ranges between 3% and 7% by weight [79], making it the highest among additives. In addition, dispersant is the highest molecular-weight component except the viscosity improver [107]. Both of these factors can alter some physical properties, such as viscosity, of the lubricant. A boost in the viscosity of a lubricant at high temperatures is desired, but at low temperatures it is a disadvantage. At high temperatures, the lubricant loses some of its viscosity [108], hence its film-forming ability, resulting in poor lubrication. Maintaining good high-temperature viscosity of a lubricant is therefore imperative to minimize wear damage. This is usually achieved by the use of polymeric viscosity modifiers [3,109]. Some dispersants, especially those that are based on high-molecular-weight polyolefins and have been oversuccinated partly fulfill this need [44]. Therefore, the amount of polymeric viscosity modifier necessary to achieve specific high-temperature viscosity is reduced. Unfortunately, dispersants that provide a viscosity advantage lead to a viscosity increase at low temperatures as well. The low-temperature viscosity requirements for engine oils have two components: *cranking viscosity* and *pumping*

viscosity [110]. Cranking viscosity is an indication of how easily the engine will turn over in extremely cold weather conditions. Pumping viscosity is the ability of the lubricant to be pumped to reach various parts of the engine. For cold weather operation, low to moderate cranking and pumping viscosities are highly desirable. Although pumping viscosity and the pour point can be lowered by the use of additives, called pour point depressants [3,13], lowering cranking viscosity is not easy. In the case of base oils, this is usually achieved by blending carefully selected base stocks. An ideal polymeric dispersant must provide high-temperature viscosity advantage without adversely affecting the cold-cranking viscosity of the lubricant. Dispersant polymers have the same requirement. Good high-temperature viscosity to cranking viscosity ratio in polymeric dispersants can be achieved by

Carefully balancing the type and the molecular weight of the hydrocarbon chain [111]

Choosing the optimum olefin to maleic anhydride molar ratio [112]

Selecting the type and the amount of the polyamine used

In dispersant polymers this can be achieved by selecting (1) a polymer of correct molecular weight and branching and (2) a suitable pendant group. Dispersant polymers derived from mediummolecular-weight, highly branched structures, and ester-type pendant groups are best suited for use as additives. Examples include polyacrylate, PMA, and styrene ester–derived dispersants. These additives not only act as viscosity modifiers and dispersants but also act as pour point depressants, thereby improving the low-temperature properties of the lubricant.

A number of patents pertaining to dispersants with balanced high-temperature viscosity and low-temperature properties are reported in the patent literature [113–117]. A Mannich (alkylphenol) dispersant, derived from ethylene/1-butene polymers of Mn 1500–7500, has been claimed to possess improved dispersancy and pour point [113]. Another patent claiming the synthesis of a dispersant with superior dispersancy and pour point depressing properties has also been issued [114]. The dispersant is based on the reaction of maleic anhydride/lauryl methacrylate/stearyl methacrylate terpolymer with dimethylaminopropylamine, and a Mannich base was obtained by reacting N-aminoethylpiperazine, paraformaldehyde, and 2,6-di-t-butyl phenol. A number of patents describe the use of ethylene/α-olefin/diene interpolymers to make dispersants [115–117]. These dispersants are claimed to possess excellent high- and low-temperature viscosities, as defined by VR'/VR. Here VR' pertains to the dispersant and VR pertains to the precursor, such as alkylphenol or alkenylsuccinic anhydride. VR' is the ratio of the −20°C cold-cranking simulator (CCS) viscosity (cP) of a 2% solution of dispersant in a reference oil to the 100°C kinematic viscosity (cSt) of the dispersant. VR is the ratio of the −20°C CCS viscosity (cP) of a 2% solution of precursor in the reference oil to the 100°C kinematic viscosity (cSt) of the precursor. The values of 2.0–3.9 for *VR* and *VR′* and of <1.11 for *VR′/VR* are considered suitable for balanced low- and high-temperature viscosities.

3.7.4 SEAL PERFORMANCE

Seals in automotive equipment are used for many purposes, the most prominent of which are to have easy access to malfunctioning parts to perform repair and to minimize contamination and loss of lubricant. Various polymeric materials are used to make seals. These include fluoroelastomers, nitrile rubber, polyacrylates, and polysiloxanes (silicones). Maintaining the integrity of seals is critical; otherwise, the lubricant will be lost, and wear damage and equipment failure will occur. The seals fail in a number of ways. They can shrink, elongate, or become brittle and thus deteriorate. The damage to elastomer seals is assessed by examining volume, hardness, tensile strength change, and the tendency to elongate and rupture [118]. Two primary mechanisms by which seal damage can occur include abrasion due to particulate matter in the lubricant and the attack of various lubricant components on the seals. The lubricant-related damage can occur when some of its components diffuse into the seals. This will either cause a change in the seal's hardness, thereby leading to swelling and or elongation, or extract the plasticizer, an agent used to impart flexibility and strength to polymeric materials.

Abrasive damage is not common since most equipment has an installed lubricant filtration system. The lubricant-related damage, however, is of primary interest to us. The lubricant is a blend of base stocks and an additive package. Certain base stocks, such as those of high aromatics content or those that are of the ester type, have the tendency to extract the plasticizer because of their high polarity. Additives, however, have the ability to diffuse into the seal material and alter its properties as well as remove the plasticizer. Among additives, dispersants are the most implicated in causing seal damage, especially to fluoroelastomer (Viton®) seals. Although in many cases seal failure can be corrected by the use of additives, called the seal-swell agents, it is wise to eliminate such damage by prevention. Elastomer compatibility requirements are a part of the current United States, Association des Contsructeurs Européens de l'Automobile (ACEA), and Japanese standards for engine oils and worldwide automotive transmission and tractor hydraulic fluid specifications [119]. Damage to seals is prevalent in the case of nitrogen dispersants. In general, the higher the nitrogen content, the higher the seal problems [118]. Rationally, these problems occur due to the presence of low-molecular-weight molecules in the dispersant. These include free amine either as such or in a labile form, such as an alkylammonium salt, or low-molecular-weight succinimides and succinamides. Because of their high polarity and smaller size, these molecules are more likely to diffuse into the seal material and alter its physical and mechanical properties [120]. It is believed that in the case of Viton seals, the loss of fluoride ions is responsible for seal deterioration. Removal of the free amine

and of low-molecular-weight succinimides will improve seal performance. Alternatively, one can posttreat dispersants with reagents, such as boric acid and epoxides, which will either make such species innocuous or hinder their diffusion into the seal material. Many chemical treatments of dispersants, covered in Section 3.7.1, claim to improve seal performance of dispersants and crankcase lubricants that use them. These reagents react with seal-damaging amines and low-molecular-weight succinimides to make them harmless. Strategies other than those listed earlier are also reported in the patent literature [121–125].

3.8 PERFORMANCE TESTING

Engine oils account for almost 80% of the automatic transmission dispersant use. Other applications that use these additives include automatic transmission fluids, gear lubricants, hydraulic fluids, and refinery processes as antifoulants. Dispersants of relatively lower molecular weight are also used in fuels to control injector and combustion chamber deposits [126,127]. Such dispersants usually contain a polyether functionality [128].

Succinimide and succinate ester–type polymeric dispersants are used in gasoline and diesel engine oils, but the use of alkylphenol-derived dispersants, that is, of the Mannich type, is limited to gasoline engine oils. Dispersant polymers derived from ethylene–propylene rubbers, styrene–diene copolymers, and PMAs are also used in both gasoline and diesel engine oils. As mentioned earlier, dispersant polymers lack sufficient dispersancy to be used alone and hence are used in combination with polymeric dispersants. The PMA and styrene ester–derived dispersant polymers are used in automatic transmission fluids, in power-steering fluids, and, to a limited extent, in gear oils.

Additive manufacturers use various laboratory screen tests and engine tests to evaluate a dispersant's effectiveness. Many of the screen tests are proprietary, but all are developed around evaluating performance in terms of a dispersant's ability to disperse lamp black or used engine oil sludge. The laboratory engine tests are industry-required tests and include both gasoline engine and diesel engine tests. These are listed in International Lubricant Standardization and Approval Committee (ILSAC), American Petroleum Institute (API), ACEA 2002, Japanese Automobile Standards Organization (JASO), and Bureau of Indian Standards (BIS) standards. It is important to note that the U.S. military and original equipment manufacturers (OEMs) have their own performance requirements, which are over and above those of the API. Although the details of various tests are available in these standards and elsewhere [119], the important engine tests that evaluate a dispersant's performance are listed in Tables 3.1 through 3.4.

As mentioned earlier, soot-related viscosity increase and deposit-related factors are the primary criteria for evaluating a dispersant's performance. Neutral detergents (soaps) also help control deposits such as varnish, lacquer, sludge, and carbon. Therefore, besides the control of soot-related viscosity

TABLE 3.1
U.S. Gasoline Engine Tests

Engine Test	Engine Type	Evaluation Criteria
CRC L-38	CLR single-cylinder engine	Bearing corrosion, sludge, varnish, oil oxidation, and viscosity change
ASTM sequence IIIE	1987 Buick V6 engine	Sludge, varnish, wear, and viscosity change
ASTM sequence IIIF	1996 Buick V6 engine	Sludge, varnish, wear, and viscosity change
ASTM sequence VE	Ford Dual-Plug head four-cylinder engine	Sludge, varnish, and wear
ASTM sequence VG	Ford V8 engine	Sludge, varnish, and wear
TEOST	Bench test	Thermal and oxidative stability
High-temperature deposit test	Bench test	High-temperature deposits

TABLE 3.2
U.S. Diesel Engine Tests

Engine Test	Engine Type	Evaluation Criteria
Caterpillar 1K	Caterpillar single-cylinder engine	Piston deposits and oil consumption
Caterpillar 1M-PC	Caterpillar single-cylinder engine	Piston deposits and oil consumption
Caterpillar 1N	Caterpillar single-cylinder engine	Piston deposits and oil consumption
Caterpillar 1P	Caterpillar single-cylinder engine	Piston deposits and oil consumption
Mack T-6	Multicylinder engine	Piston deposits, wear, oil consumption, and oil thickening
Mack T-7	Multicylinder engine	Oil thickening
Mack T-8	Multicylinder engine	Oil thickening
Mack T-9	Multicylinder engine	Soot thickening

TABLE 3.3
European Gasoline Engine Tests

Engine Test	Engine Type	Evaluation Criteria
ASTM sequence IIIE	Six-cylinder engine	High-temperature oxidation (sludge, varnish, wear, and viscosity increase)
ASTM sequence VE	Four-cylinder engine	Low-temperature sludge, varnish, and wear
Peugeot TU-3M high temperature	Four-cylinder single-point injection engine	Piston deposits, ring sticking, viscosity increase
M-B M111 black sludge	Four-cylinder multipoint injection engine	Engine sludge and cam wear
VW 1302	Four-cylinder carbureted engine	Piston deposits, varnish, wear, and oil consumption
VW T-4	Four-cylinder multipoint injection engine	Extended drain capability

TABLE 3.4
Current European Diesel Engine Tests

Engine Test	Engine Type	Evaluation Criteria
VW 1.6TC diesel intercooler	Four-cylinder engine	Piston deposits, varnish, and ring sticking
VW D1	Four-cylinder direct-injection engine	Piston deposits, viscosity increase, and ring sticking
Peugeot XUD11ATE	Four-cylinder indirect-injection engine	Piston deposits and viscosity increase
Peugeot XUD11BTE	Four-cylinder indirect-injection engine	Piston deposits and viscosity increase
M-B OM 602A	Five-cylinder indirect-injection engine	Engine wear and cleanliness
M-B OM 364A/LA	Four-cylinder direct-injection engine	Bore polishing, piston deposits, varnish, sludge, wear, and oil consumption
M-B OM 441LA	Six-cylinder direct-injection engine	Piston deposits, bore polishing, wear, oil consumption, valve train condition, and turbo deposits
MAN 5305	Single-cylinder engine	Piston deposits, bore polishing, and oil consumption
Mack T-8	Multicylinder engine	Soot-related oil thickening

increase, which is the sole domain of dispersants, deposit control is the result of a joint performance of the detergent and the dispersant. However, in this regard, the dispersant plays a more prominent role.

Besides engine oils, transmission fluids are the primary users of dispersants. Certain parts of the transmission see very high temperatures, which lead to extensive lubricant oxidation. The oxidation products, such as sludge and varnish, appear on parts; for instance, clutch housing, clutch piston, control valve body, and oil screen components. This can impair the functioning of these parts. A turbohydramatic oxidation test (THOT) is used to determine a transmission fluid's oxidative stability.

Polymeric dispersants are useful in controlling sludge buildup [129]. When friction modification of the transmission fluid is the goal, either dispersants or their precursors, such as alkenylsuccinic acids or anhydrides, are used in combination with metal sulfonates [130–134]. In many such formulations, the borated dispersant and the borated detergent (metal sulfonate) are used.

Dispersants are used in gear oils to improve their properties also. Gear oils usually contain thermally labile extreme-pressure additives. Their decomposition by-products are highly polar, and dispersants are used to contain them to avoid corrosion and deposit formation [135,136]. Polymeric dispersants are used in hydraulic fluids to overcome wet filtration (Association Française de Normalisation [AFNOR]) problems, which is often required for HF-0-type fluids [137]. Filtration problems occur due to the interaction of water with metal sulfonate detergent and zinc dialkyldithiophosphate that are used as additives in hydraulic fluid formulations. Fouling is a common problem in many processes, including refinery processes. Fouling refers to the deposition of various inorganic and organic materials, such as salt, dirt, and asphaltenes, on heat-transfer surfaces and other processing equipment. This results in poor heat transfer, among other problems. Antifoulants are chemicals used in refinery operations to overcome fouling. Detergents and dispersants are often used for this purpose [138–140].

REFERENCES

1. Sieloff, F.X., J.L. Musser. What does the engine designer need to know about engine oils? Presented to Detroit Section of the Society of Automotive Engineers, Detroit, MI. March 16, 1982.
2. Colyer, C.C., W.C. Gergel. Detergents/dispersants. In R.M. Mortier, S.T. Orszulik, eds. *Chemistry and Technology of Lubricants*. New York: VCH Publishers, Inc., 1992, pp. 62–82.
3. Rizvi, S.Q.A. Lubricant additives and their functions. In S.D. Henry, ed. *American Society of Metals Handbook*, 10th edn., Vol. 18, 1992, pp. 8–112.
4. Ingold, K.U. Inhibition of autoxidation of organic substances in liquid phase. *Chemical Reviews* 61: 563–589, 1961.
5. Kornbrekke, R.E. et al. Understanding soot-mediated oil thickening—Part 6: Base oil effects. SAE Technical Paper 982, 665. Society of Automotive Engineers, October 1, 1998.

   Also see parts 1–5 by E. Bardasz et al., SAE Papers 952, 527 (October 1995), 961, 915 (October 1, 1996), 971, 692 (May 1, 1997), 976, 193 (May 1, 1997), and 972, 952 (October 1, 1997), Society of Automotive Engineers, Warrendale, PA.
6. Covitch, M.J., B.K. Humphrey, D.E. Ripple. Oil thickening in the Mack T-7 engine test—Fuel effects and the influence of lubricant additives on soot aggregation. Presented at SAE Fuels and Lubricants Meeting, Tulsa, OK, October 23, 1985.
7. Obert, E.F. *Internal Combustion Engines and Air Pollution*. New York: Intext Educational Publishing, 1968.
8. Petroleum products, lubricants, and fossil fuels. In *Annual Book of ASTM Standards*. Philadelphia, PA: American Society of Testing and Materials, 1998.
9. Cochrac, J., S.Q.A. Rizvi. Oxidation and oxidation inhibitors. In *ASTM Manual on Fuels and Lubricants*, to be published in 2003.
10. Gas and expansion turbines. In D.M. Considine, ed. *Van Nostrand's Scientific Encyclopedia*, 5th edn. New York: Van Nostrand Reinhold, 1976, pp. 1138–1148.
11. Zeldovich, Y.B., P.Y. Sadovnikov, D.A. Frank-Kamenetskii. *Oxidation of Nitrogen in Combustion*. Moscow-Leningrad, Russia: Academy of Sciences, U.S.S.R., 1947.
12. Ford, J.F. Lubricating oil additives—A chemist's eye view. *Journal of the Institute of Petroleum* 54: 188–210, 1968.
13. Rizvi, S.Q.A. Additives: Chemistry and testing. In E.R. Booser, ed. *Tribology Data Handbook—An Excellent Friction, Lubrication, and Wear Resource*. Boca Raton, FL: CRC Press, 1997, pp. 117–137.
14. Kreuz, K.L. Gasoline engine chemistry as applied to lubricant problems. *Lubrication* 55: 53–64, 1969.
15. Bouman, C.A. *Properties of Lubricating Oils and Engine Deposits*. London, U.K.: MacMillan and Co., 1950, pp. 69–92.
16. Kreuz, K.L. Diesel engine chemistry as applied to lubricant problems. *Lubrication* 56: 77–88, 1970.
17. Chamberlin, W.B., J.D. Saunders. Automobile engines. In R.E. Booser, ed. *CRC Handbook of Lubrication*, Vol. I: Theory and Practice of Tribology: Applications and Maintenance. Boca Raton, FL: CRC Press, 1983, pp. 3–44.
18. Obert, E.F. Basic engine types and their operation. In *Internal Combustion Engines and Air Pollution*. New York: Intext Educational Publishing, 1968, pp. 1–25.
19. Schilling, A. Antioxidant and anticorrosive additives. In *Motor Oils and Engine Lubrication*. London, U.K.: Scientific Publications, 1968, Section 2.8, p. 2.61.
20. Patterson, D.J., N.A. Henein. *Emissions from Combustion Engines and Their Control*. Ann Arbor, MI: Ann Arbor Science Publishers, 1972.
21. Lachowicz, D.R., K.L. Kreuz. Peroxynitrates. The unstable products of olefin nitration with dinitrogen tetroxide in the presence of oxygen. A new route to α-nitroketones. *Journal of the Organic Chemistry* 32: 3885–3888, 1967.
22. Covitch, M.J., R.T. Graf, D.T. Gundic. Microstructure of carbonaceous diesel engine piston deposits. *Lubricant Engineering* 44: 128, 1988.
23. Covitch, M.J., J.P. Richardson, R.T. Graf. Structural aspects of European and American diesel engine piston deposits. *Lubrication Science* 2: 231–251, 1990.
24. Nahamuck, W.M., C.W. Hyndman, S.A. Cryvoff. Development of the PV-2 engine deposit and wear test. An ASTM Task Force Progress Report, SAE Publication 872,123. Presented at International Fuels and Lubricants Meeting and Exposition, Toronto, Ontario, Canada, November 2–5, 1987.

25. Rasberger, M. Oxidative degradation and stabilization of mineral oil based lubricants. In R.M. Mortier, S.T. Orszulik, eds. *Chemistry and Technology of Lubricants*. New York: VCH Publishers, Inc., 1992, pp. 83–123.

26. Oliver, C.R., R.M. Reuter, J.C. Sendra. Fuel efficient gasoline-engine oils. *Lubrication* 67: 1–12, 1981.

27. Stone, R. *Introduction to Internal Combustion Engines*. Society of Automotive Engineers, Warrendale, PA, 1993.

28. Rizvi, S.Q.A. Additives and additive chemistry. In *ASTM Manual on Fuels and Lubricants*, West Conshohocken, PA, 2003.

29. Stuart, F.A., R.G. Anderson, A.Y. Drummond. Lubricating-oil compositions containing alkenyl succinimides of tetraethylene pentamine. U.S. Patent 3,361,673, January 2, 1968.

30. Cooper, A.R. Molecular weight determination. In J.I. Kroschwitz, ed. *Concise Encyclopedia of Polymer Science and Engineering*. New York: Wiley Interscience, 1990, pp. 638–639.

31. Ravve, A. Molecular weights of polymers. In *Organic Chemistry of Macromolecules*. New York: Marcel Dekker, 1967, pp. 39–54.

32. Deanin, R.D. *Polymer Structure, Properties, and Applications*. New York: Cahner Books, 1972, p. 53.

33. Randall, J.C. Microstructure. In J.I. Kroschwitz, ed. *Concise Encyclopedia of Polymer Science and Engineering*. New York: Wiley Interscience, 1990, p. 625.

34. Fotheringham, J.D. Polybutenes. In L.R. Rudnick, R.L. Shubkin, eds. *Synthetic Lubricants and High-Performance Functional Fluids*, 2nd edn. New York: Marcel Dekker, 1999.

35. Randles, S.J. et al. Synthetic base fluids. In R.M. Mortier, S.T. Orszulik, eds. *Chemistry and Technology of Lubricants*. New York: VCH Publishers, 1992, pp. 32–61.

36. Gutierrez, A., R.A. Kleist, W.R. Song, A. Rossi, H.W. Turner, H.C. Welborn, R.D. Lundberg. Ethylene alpha-olefin polymer substituted mono- and dicarboxylic acid dispersant additives. U.S. Patent 5,435,926, July 25, 1995.

37. Gutierrez, A., W.R. Song, R.D. Lundberg, R.A. Kleist. Novel ethylene alpha-olefin copolymersubstituted mannich base lubricant dispersant additives. U.S. Patent 5,017,299, May 21, 1991.

38. Stambaugh, R.L. Viscosity index improvers and thickeners. In R.M. Mortier, S.T. Orszulik, eds. *Chemistry and Technology of Lubricants*. New York: VCH Publishers, 1992.

39. Bryant, C.P., H.M. Gerdes. Nitrogen-containing esters and lubricants containing them. U.S. Patent 4,604,221, August 5, 1986.

40. Shanklin, J.R., Jr., N.C. Mathur. Lubricating oil additives. U.S. Patent 6,071,862, June 6, 2000.

41. Morris, J.R., R. Roach. Lubricating oils containing metal derivatives. U.S. Patent 2,628,942, February 17, 1953.

42. Sparks, W.J., D.W. Young, Roselle, J.D. Garber. Modified olefin–diolefin resin. U.S. Patent 2,634,256, April 7, 1953.

43. Le Suer, W.M., G.R. Norman. Reaction product of high molecular weight succinic acids and succinic anhydrides with an ethylene polyamine. U.S. Patent 3,172,893, March 9, 1965.

44. Meinhardt, N.A., K.E. Davis. Novel carboxylic acid acylating agents, derivatives thereof, concentrate and lubricant compositions containing the same, and processes for their preparation. U.S. Patent 4,234,435, November 18, 1980.

45. Rense, R.J. Lubricant. U.S. Patent 3,215,707, November 2, 1965.

46. Weill, J., B. Sillion. Reaction of chlorinated polyisobutene with maleic anhydride: Mechanism of catalysis by dichloromaleic anhydride. *Revue de l'Institut Francais du Petrole* 40(1): 77–89, 1985.

47. Weill, J. PhD dissertation, 1982.

48. Weill, J., J. Garapon, B. Sillion. Process for manufacturing anhydrides of alkenyl dicarboxylic acids. U.S. Patent 4,433,157, February 21, 1984.

49. Baumanis, C.K., M.M. Maynard, A.C. Clark, M.R. Sivik, C.P. Kovall, D.L. Westfall. Treatment of organic compounds to reduce chlorine level. U.S. Patent 5,708,097, January 13, 1998.

50. Pudelski, J.K., M.R. Sivik, K.F. Wollenberg, R. Yodice, J. Rutter, J.G. Dietz. Low chlorine polyalkylene substituted carboxylic acylating agent compositions and compounds derived therefrom. U.S. Patent 5,885,944, March 23, 1999.

51. Pudelski, J.K., C.J. Kolp, J.G. Dietz, C.K. Baumanis, S.L. Bartley, J.D. Burrington. Low chlorine content composition for use in lubricants and fuels. U.S. Patent 6,077,909, June 20, 2000.

52. Wollenberg, K.F., J.K. Pudelski. Preparation, NMR characterization and lubricant additive application of novel polyisobutenyl phthalic anhydrides. Symposium on Recent Advances in the Chemistry of Lubricant Additives, 218th National Meeting of the American Chemical Society, Paper presented before the Division of Petroleum Chemistry, Inc., New Orleans, LA, August 22–26, 1999.

53. Harrison, J.J., R. Ruhe, Jr., R. William. One-step process for the preparation of alkenyl succinic anhydride. U.S. Patent 5,319,030, June 7, 1994.

54. Harrison, J.J., R. Ruhe, Jr., R. William. Two-step thermal process for the preparation of alkenylsuccinic anhydride. U.S. Patent 5,625,004, April 29, 1997.

55. Morrison, R.T., R.N. Boyd. The Diels–Alder reaction. In *Organic Chemistry*, 3rd edn. Boston, MA: Allyn and Bacon, 1976, Section 27.8, pp. 876–878.

56. Alkylation of phenols. In K. Othmer, ed. *Kirk-Othmer Encyclopedia of Chemical Technology*, Vol. 1. New York: Interscience Publishers, 1963, pp. 894–895.

57. Ion exchange. In K. Othmer, ed. *Kirk-Othmer Encyclopedia of Chemical Technology*, Vol. 2. New York: Interscience Publishers, 1967, pp. 871–899.

58. McAtee, J.R. Aromatic Mannich compound-containing composition and process for making same. U.S. Patent 6,179,885, January 30, 2001.

59. Merger, F., G. Nestler. Manufacture of alkylphenol compounds. U.S. Patent 4,202,199, May 13, 1980.

60. Kolp, C.J. Methods for preparing alkylated hydroxyaromatics. U.S. Patent 5,663,457, September 2, 1997.

61. Adams, P.E., M.R. Baker, J.G. Dietz. Hydroxy-group containing acylated nitrogen compounds useful as additives for lubricating oil and fuel compositions. U.S. Patent 5,696,067, December 9, 1997.

62. Pudelski, J.K. Mixed carboxylic compositions and derivatives and use as lubricating oil and fuel. U.S. Patent 6,030,929, February 29, 2000.

63. Baker, M.R., J.G. Dietz, R. Yodice. Substituted carboxylic acylating agent compositions and derivatives thereof for use in lubricants and fuels. U.S. Patent 5,912,213, June 15, 1999.

64. Baker, M.R. Acylated nitrogen compounds useful as additives for lubricating oil and fuel compositions. U.S. Patent 5,856,279, January 5, 1999.

65. Baker, M.R., K.M. Hull, D.L. Westfall. Process for preparing condensation product of hydroxy-substituted aromatic compounds and glyoxylic reactants. U.S. Patent 6,001,781, December 14, 1999.

66. Ethylene amines. In K. Othmer, ed. *Kirk-Othmer Encyclopedia of Chemical Technology*, 2nd edn., Vol. 7. New York: Interscience Publishers, 1965, pp. 22–37.

67. Morrison, R.T., R.N. Boyd. *Amines II. Reactions. Organic Chemistry*, 3rd edn. Boston, MA: Allyn and Bacon, 1976, pp. 745–748.

68. Steckel, T.F. High molecular weight nitrogen-containing condensates and fuels and lubricants containing same. U.S. Patent 5,053,152, October 1, 1991.

69. Pindar, J.F., J.M. Cohen, C.P. Bryant. Dispersants and process for their preparation. U.S. Patent 3,980,569, September 14, 1976.

70. Harmon, J., F.M. Meigs. Artificial resins and method of making. U.S. Patent 2,098,869, November 9, 1937.

71. March, J. Aminoalkylation and amidoalkylation. In *Advanced Organic Chemistry, Reactions, Mechanisms, and Structure*, 4th edn. New York: Wiley Interscience, 1992, pp. 550–551.

72. Schallenberg, E.E., R.G. Lacoste. Ethylenediamine salts of thiphosphonic acids. U.S. Patent 3,185,728, May 25, 1965.

73. Schlicht, R.C. Friction reducing agents for lubricants. U.S. Patent 3,702,824, November 14, 1972.

74. Brois, S.J. Olefin-thionophosphine sulfide reaction products, their derivatives, and use thereof as oil and fuel additives. U.S. Patent 4,042,523, August 16, 1977.

75. Brois, S.J. Olefin-thionophosphine sulfide reaction products, their derivatives, and use thereof as oil and fuel additives. U.S. Patent 4,100,187, July 11, 1978.

76. Kapusciniski, M.M., B.J. Kaufman, C.S. Liu. Oil containing dispersant VII olefin copolymer. U.S. Patent 4,715,975, December 29, 1987.

77. Kapuscinski, M.M., R.E. Jones. Dispersant-antioxidant multifunction viscosity index improver. U.S. Patent 4,699,723, October 13, 1987.

78. Kapuscinski, M.M., T.E. Nalesnik, R.T. Biggs, H. Chafetz, C.S. Liu. Dispersant anti-oxidant VI improver and lubricating oil composition containing same. U.S. Patent 4,948,524, August 14, 1994.

79. Goldblatt, I., M. McHenry, K. Henderson, D. Carlisle, N. Ainscough, M. Brown, R. Tittel. Lubricant for use in diesel engines. U.S. Patent 6,187,721, February 13, 2001.

80. Lange, R.M., C.V. Luciani. Graft copolymers and lubricants containing such as dispersant-viscosity improvers. U.S. Patent 5,298,565, March 29, 1994.

81. Sutherland, R.J. Dispersant viscosity index improvers. U.S. Patent 6,083,888, July 4, 2000.

82. Stambaugh, R.L., R.D. Bakule. Lubricating oils and fuels containing graft copolymers. U.S. Patent 3,506,574, April 14, 1970.

83. Trepka, W.J. Viscosity index improvers with dispersant properties prepared by reaction of lithiated hydrogenated copolymers with 4-substituted aminopyridines. U.S. Patent 4,402,843, September 6, 1983.

84. Trepka, W.J. Viscosity index improvers with dispersant properties prepared by reaction of lithiated hydrogenated copolymers with substituted aminolactams. U.S. Patent 4,402,844, September 6, 1983.

85. Seebauer, J.G., C.P. Bryant. Viscosity improvers for lubricating oil composition. U.S. Patent 6,124,249, September 26, 2000.

86. Adams, P.E., R.M. Lange, R. Yodice, M.R. Baker, J.G. Dietz. Intermediates useful for preparing dispersant-viscosity improvers for lubricating oils. U.S. Patent 6,117,941, September 12, 2000.

87. Lange, R.M., C.V. Luciani. Dispersant-viscosity improves for lubricating oil composition. U.S. Patent 5,512,192, April 30, 1996.

88. Lange, R.M. Dispersant-viscosity improvers for lubricating oil compositions. U.S. Patent 5,540,851, July 30, 1996.

89. Hayashi, K., T.R. Hopkins, C.R. Scharf. Graft copolymers from solvent-free reactions and dispersant derivatives thereof. U.S. Patent 5,429,758, July 4, 1995.

90. Nalesnik, T.E. Novel VI improver, dispersant, and antioxidant additive and lubricating oil composition containing same. U.S. Patent 4,863,623, September 5, 1989.

91. Mishra, M.K., I.D. Rubin. Functionalized graft co-polymer as a viscosity index improver, dispersant, and anti-oxidant additive and lubricating oil composition containing same. U.S. Patent 5,409,623, April 25, 1995.

92. Kapuscinski, M.K., C.S. Liu, L.D. Grina, R.E. Jones. Lubricating oil containing dispersant viscosity index improver. U.S. Patent 5,520,829, May 28, 1996.

93. Sutherland, R.J. Process for making dispersant viscosity index improvers. U.S. Patent 5,486,563, January 23, 1996.

94. Bryant, C.P., B.A. Grisso, R. Cantiani. Dispersant-viscosity improvers for lubricating oil compositions. U.S. Patent 5,969,068, October 19, 1999.

95. Kiovsky, T.E. Star-shaped dispersant viscosity index improver. U.S. Patent 4,077,893, March 7, 1978.

96. Patil, A.O. Multifunctional viscosity index improver-dispersant antioxidant. U.S. Patent 5,439,607, August 8, 1995.

97. Baranski, J.R., C.A. Migdal. Lubricants containing ashless antiwear-dispersant additive having viscosity index improver credit. U.S. Patent 5,698,5000, December 16, 1997.

98. Sutherland, R.J., R.B. Rhodes. Dispersant viscosity index improvers. U.S. Patent 5,360,564, November 1, 1994.

99. Brady, J.E., G.E. Humiston. Chemical bonding: General concepts—Polar molecules and electronegativity. In *General Chemistry: Principles and Structure*, 2nd edn. New York: Wiley, 1978, pp. 114–117.

100. Diana, W.B., J.V. Cusumano, K.R. Gorda, J. Emert, W.B. Eckstrom, D.C. Dankworth, J.E. Stanat, J.P. Stokes. Dispersant additives and process. U.S. Patents 5,804,667, September 8, 1998 and 5,936,041, August 10, 1999.

101. Degonia, D.J., P.G. Griffin. Ashless dispersants formed from substituted acylating agents and their production and use. U.S. Patent 5,241,003, August 31, 1993.

102. Emert, J., R.D. Lundberg, A. Gutierrez. Oil soluble dispersant additives useful in oleaginous compositions. U.S. Patent 5,026,495, June 25, 1991.

103. Sung, R.L., B.J. Kaufman, K.J. Thomas. Middle distillate containing storage stability additive. U.S. Patent 4,948,386, August 14, 1990.

104. Ratner, H., R.F. Bergstrom. Non-ash containing lubricant oil composition. U.S. Patent 3,189,544, June 15, 1965.

105. Norman, G.R., W.M. Le Suer. Reaction products of hydrocarbon-substituted succinic acid-producing compound, an amine, and an alkenyl cyanide. U.S. Patents 3,278,550, October 11, 1966 and 3,366,569, June 30, 1968.

106. Morrison, R.T., R.N. Boyd. Hydrolysis of amides, pp. 671–672; Alkaline and acidic hydrolysis of esters, pp. 677–681. In *Organic Chemistry*, 3rd edn. Boston, MA: Allyn and Bacon, 1976.

107. Baczek, S.K., W.B. Chamberlin. Petroleum additives. In *Encyclopedia of Polymer Science and Engineering*, 2nd edn., Vol. 11. New York: Wiley, 1998, p. 22.

108. Klamann, D. Viscosity–temperature (VT) function. In *Lubricants and Related Products*. Weinheim, Germany: Verlag Chemie, 1984, pp. 7–12.

109. Schilling, A. Viscosity index improvers. In *Motor Oils and Engine Lubrication*. London, U.K.: Scientific Publications, 1968, pp. 2.28–2.43.

110. SAE Standard. Engine oil viscosity classification. SAE J300. Society of Automotive Engineers, Revised December 1999.

111. Adams, D.R., P. Brice. Multigrade lubricating compositions containing no viscosity modifier. U.S. Patent 5,965,497, October 12, 1999.

112. Emert, J., R.D. Lundberg. High functionality low molecular weight oil soluble dispersant additives useful in oleaginous compositions. U.S. Patent 5,788,722, August 4, 1998.

113. Emert, J., A. Rossi, S. Rea, J.W. Frederick, M.W. Kim. Polymers derived from ethylene and 1-butene for use in the preparation of lubricant dispersant additives. U.S. Patent 6,030,930, February 29, 2000.

114. Hart, W.P., C.S. Liu. Lubricating oil containing dispersant VII and pour depressant. U.S. Patent 4,668,412, May 26, 1987.

115. Song, W.R., A. Rossi, H.W. Turner, H.C. Welborn, R.D. Lundberg, A. Gutierrez, R.A. Kleist. Ethylene alpha-olefin/diene interpolymer-substituted carboxylic acid dispersant additives. U.S. Patents 5,759,967, June 2, 1998 and 5,681,799, October 28, 1997.

116. Song, W.R., A. Rossi, H.W. Turner, H.C. Welborn, R.D. Lundberg. Ethylene alpha-olefin polymer substituted mono- and dicarboxylic acid dispersant additives. U.S. Patent 5,433,757, July 18, 1995.

117. Song, W.R., R.D. Lundberg, A. Gutierrez, R.A. Kleist. Borated ethylene alpha-olefin copolymer substituted Mannich base lubricant dispersant additives. U.S. Patent 5,382,698, January 17, 1995.

118. Harrison, J.J., W.A. Ruhe, Jr. Polyalkylene polysuccinimides and post-treated derivatives thereof. U.S. Patent 6,146,431, November 14, 2000.

119. Ready reference for lubricant and fuel performance. Lubrizol Publication. Available at http://www.lubrizol.com.

120. Stachew, C.F., W.D. Abraham, J.A. Supp, J.R. Shanklin, G.D. Lamb. Engine oil having dithiocarbamate and aldehyde/epoxide for improved seal performance, sludge and deposit performance. U.S. Patent 6,121,211, September 9, 2000.

121. Viton seal compatible dispersant and lubricating oil composition containing same. U.S. Patent 5,188,745, February 23, 1993.

122. Nalesnik, T.E., C.M. Cusano. Dibasic acid lubricating oil dispersant and viton seal additives. U.S. Patent 4,663,064, May 5, 1987.

123. Nalesnik, T.E. Lubricating oil dispersant and viton seal additives. U.S. Patent 4,636,332, January 13, 1987.

124. Scott, R.M., R.W. Shaw. Dispersant additives. U.S. Patent 6,127,322, October 3, 2000.

125. Fenoglio, D.J., P.R. Vettel, D.W. Eggerding. Method for preparing engine seal compatible dispersant for lubricating oils comprising reacting hydrocarbyl substituted dicarboxylic compound with aminoguanidine or basic salt thereof. U.S. Patent 5,080,815, January 14, 1992.

126. Cunningham, L.J., D.P. Hollrah, A.M. Kulinowski. Compositions for control of induction system deposits. U.S. Patent 5,679,116, October 21, 1997.

127. Ashjian, H., M.P. Miller, D.-M. Shen, M.M. Wu. Deposit control additives and fuel compositions containing the same. U.S. Patent 5,334,228, August 2, 1994.

128. Mulard, P., Y. Labruyere, A. Forestiere, R. Bregent. Additive formulation of fuels incorporating ester function products and a detergent-dispersant. U.S. Patent 5,433,755, July 18, 1995.

129. Takigawa, S. Gear and transmission lubricant compositions of improved sludge-dispersibility, fluids comprising the same. U.S. Patent 5,665,685, September 9, 1997.

130. Otani, H., R.J. Hartley. Automatic transmission fluids and additives thereof. U.S. Patent 5,441,656, August 15, 1995.

131. O'Halloran, R. Hydraulic automatic transmission fluid with superior friction performance. U.S. Patent 4,253,977, March 3, 1981.

132. Ichihashi, T., H. Igarashi, J. Deshimaru, T. Ikeda. Lubricating oil composition for automatic transmission. U.S. Patent 5,972,854, October 26, 1999.

133. Kitanaka, M. Automatic transmission fluid composition. U.S. Patent, September 28, 1999.

134. Srinivasan, S., D.W. Smith, J.P. Sunne. Automatic transmission fluids having enhanced performance capabilities. U.S. Patent 5,972,851, October 26, 1999.

135. Conary, G.S., R.J. Hartley. Gear oil additive concentrates and lubricants containing them. U.S. Patent 6,096,691, August 1, 2000.

136. Ichihashi, T. Lubricating oil composition for high-speed gears. U.S. Patent 5,756,429, May 26, 1998.

137. Ryan, H.T. Hydraulic fluids. U.S. Patent 5,767,045, June 16, 1988.

138. Forester, D.R. Use of dispersant additives as process antifoulants. U.S. Patent 5,368,777, November 29, 1994.

139. Forester, D.R. Multifunctional antifoulant compositions. U.S. Patent 4,927,561, May 22, 1990.

140. Forester, D.R. Multifunctional antifoulant compositions and methods of use thereof. U.S. Patent 4,775,458, October 4, 1988.

4 Detergents

Philip Ma

CONTENTS

4.1 INTRODUCTION

Detergents, together with dispersants, antioxidants, antiwear/extreme-pressure (EP) agents, viscosity modifiers, friction modifiers, pour point depressants, metal deactivators, corrosion inhibitors, rust inhibitors, demulsifiers, defoamers, and so on are import lubricant additives for the modern lubricant industry.

A detergent molecule, also called "soap," refers to a type of oil-soluble surfactant with an amphiphilic structure that has a hydrophilic polar head and a hydrophobic nonpolar tail. Sulfonates, phenates, and salicylates are the three most common types of detergents. They are obtained from the neutralization of corresponding organic acidic molecules using various readily available alkaline or alkaline earth metals (sodium [Na], calcium [Ca], magnesium [Mg], and barium [Ba]). Detergents with acid/base neutralization close to the same equivalent are called "neutral" detergents. Ca salts are the most common detergent salts. Na salts are typically used in industrial oil as emulsifiers. Mg salts have less sulfated ash than Ca salts for the same alkaline reserve. This is desirable for low sulfated ash, phosphorous, sulfur (SAPS) engine oils. $MgCO_3$ is less basic and neutralizes acid slower than $CaCO_3$, and Mg salts maintain the alkaline reserve better than Ca salts [1]; Mg salts are usually used in combination with Ca salts as detergents in engine oils. Mg salts are also used as fuel additives for corrosion control for vanadium-containing fuels in gas turbines [2]. Ba is a heavy metal, which can cause health and environmental concerns; the usage of Ba salts is declining although there is some limited usage in industrial oil as corrosion inhibitor, where Ba salts have shown some unique performance against other salts.

Modern detergents are typically "overbased," that is, incorporated with excess amounts of alkaline reserve. The alkaline reserve mostly consists of the carbonate salt of Ca, Mg, or Ba. Overbased detergents (Figure 4.1) comprise the largest volume of nanomaterials [3] produced in the world, with the annual volume exceeding 100,000 tons. Overbased detergents contain stable nanoparticles [4] dispersed in diluent oil. The nanoparticle has a core–shell, inverse micelle [5] structure with the metal carbonate salt nanoparticle forming the hard core; the neutral, amphiphilic organic salt soap molecule self-assemblies anchor on the surface of the hard nanoparticle to form the shell. Overbased sulfonates, phenates, and salicylates all have similar structures [6]. For Ca carbonate salt, the $CaCO_3$ nanoparticle within the core could be amorphous [7] or crystalline [8]. There is a small

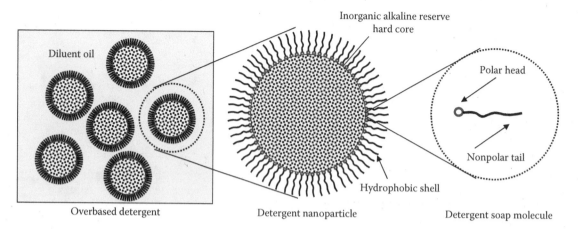

FIGURE 4.1 Overbased detergent.

amount of uncarbonated $Ca(OH)_2$ [9] in the amorphous $CaCO_3$ core. The amorphous nanoparticle has a diameter of 1–10 nm and the surfactant layer has a thickness of 1–5 nm [10]. The detergent nanoparticles are far smaller than the visible light wavelength (390–750 nm) in size and thus appear transparent. Detergent particles with crystalline $CaCO_3$ cores, which have a calcite or vaterite structure, are typically larger, with a diameter ranging from 40–60 nm up to more than 500–1000 nm, and have a hazy appearance. They are not desirable in engine oils, are typically used as rheology modifiers or made into greases [11], and have excellent EP/antiwear properties.

Overbased detergents are mostly used in high-temperature engine oil formulations, where the excess alkaline reserve is needed to provide permanent neutralization of the harmful effects of corrosive acids (SO_2, NO_x, and their products obtained by their reaction with water) that are formed by the combustion of diesel and gasoline fuels (sulfur content in gasoline/diesel is 30–2000 ppm). Overbased detergents reduce the corrosive effect of acids on metal parts and reduce/prevent acids from catalyzing further decomposition of lubricants [12]. The ability of detergents and engine oils to neutralize these acids is measured by their total base number (TBN). Passenger car motor oils (PCMOs) typically have a TBN of 6–10 mg KOH/g and contain 1%–3% of detergents. Heavy-duty diesel engine oils (HDDEOs) typically have a TBN of 8–16 mg KOH/g and contain 3%–5% of detergents due to a higher operating temperature, which generates more acidic by-products. Marine cylinder lubricants (MCLs) typically have a TBN of 40–100 mg KOH/g and contain 10%–30% of detergents due to the usage of high sulfur-containing bunker fuels (contain up to 5% of sulfur).

It was discovered in the 1940s that lubricants containing oil-soluble carboxylated metal salts could remove engine ring deposits and reduce ring stickiness. Since then, detergents have been widely used in lubricant formulations that are intended for cleaning metal surfaces and prolong the performance and service life of automotive engine oils. Lubricant oils subjected to higher-temperature application, especially engine oils, generate sludge, varnish, and resin due to thermal-oxidative

stress. They can block oil lines and oil passages and prevent the flow of engine oil from reaching parts that need to be lubricated. This in turn can result in increased wear, heat buildup, and eventual engine failure. Both detergents and dispersants are used to control deposit formation in engine oil formulation. The most common dispersant is polyisobutylene succinimide (PIBSI). PIBSI dispersants contain a higher number average molecular weight (Mn) oil-soluble polyisobutylene (PIB) tail (Mn 1000–2300) than detergents (oil-soluble alkylbenzene Mn ~500), thus dispersants have better dispersancy than detergents. Detergent soap can clean the varnish and deposit formed on the engine surface and disperse them from the high-temperature engine piston crown zone into the much cooler lubricant oil sump [13]. Detergents can effectively solubilize and disperse polar oxygenated species such as peroxides, alcohols, aldehydes, ketones, esters, organic acids, and so on, which form due to lubricant or fuel oxidation; they can also retard further oxidation, condensation, polymerization, and accumulation that might result in insoluble sludge, varnish, and resin, which can deposit on the surface of metal parts [14]. Detergent molecules form two kinds of barrier films that can disperse deposits. On small particles (generally less than 0.02 μm in size), they form an absorbed film that slows down coagulation of the particles. On much larger particles (ranging from 0.5 to 1.5 μm in size), they cause the particle surfaces to acquire an electrical charge of the same sign so they will repel each other [15].

Detergent soap molecules can also form a thin film [16] on engine metal parts to protect the metal from corrosion and reduce contact of metal to metal that could cause friction and wear [17] and thus improve energy efficiency.

Detergents are also widely used in other areas, such as industrial oil, oil field, mining, and so on, as emulsifiers, demulsifiers, rust inhibitors, dispersants, surfactants for enhanced oil recovery, ore-floatation agents, and wetting agents, among others.

The physical property and performance of a detergent are closely related to the type of soap molecules (sulfonate, phenate, salicylate, carboxylate, phosphonate, and so on [Figure 4.2]), raw material structure, purity, molecular weight, molecular weight

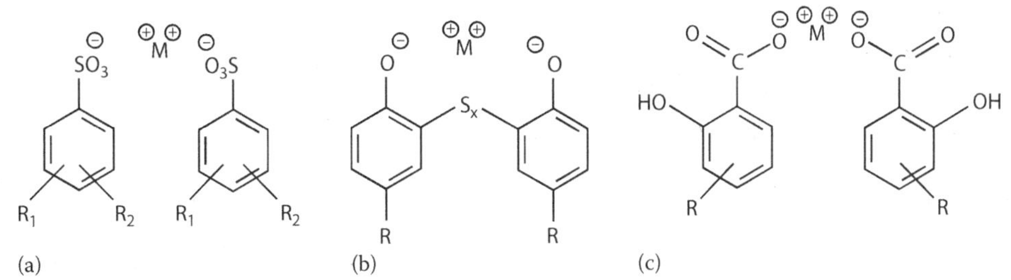

FIGURE 4.2 Sulfonate (a), phenate (b), salicylate (c) soap molecule structure. *x*: 1–3.

distribution [18], type of metal salt, particle size, crystallinity, and degree of carbonation.

This chapter focuses on the three most important modern detergent types: sulfonates (metallic salt of alkaryl sulfonic acid), phenates (metallic salt of alkyl phenol sulfide), and salicylates (metallic salt of alkyl salicylic acid).

4.2 DETERGENT TYPES

4.2.1 SULFONATES

Sulfonates are the most widely used detergents. Sulfonates can be classified as petroleum sulfonates (natural sulfonates) if the sulfonic acid raw materials are from sulfonation of petroleum oils, or they can be classified as synthetic sulfonates if the sulfonic acid raw materials are from synthetic alkylaryl compounds, such as alkyl benzene, alkyl toluene, alkyl xylene, alkyl naphthalene, and so on. Among all the salts, Ca salts are the most widely used, followed by Mg, Na, and Ba salts. Oil-soluble sulfonates have excellent high-temperature detergency and dispersancy in comparison to phenates and salicylates, and excellent alkaline reserve capability in comparison to dispersants. However, sulfonates are considered to be pro-oxidants [19], possibly due to the strong electron-withdrawing effect of the sulfonate group, which destabilizes the α-hydrogen on the carbon adjacent to the aromatic ring.

4.2.1.1 Petroleum (Natural) Sulfonates

Petroleum sulfonates used to be the by-product of "white oil" production and are the earliest sulfonates to be studied and used. Petroleum distillate cuts often contain aromatic components that could react with SO_3 to generate petroleum sulfonic acid containing oil. The sulfonic acids are usually neutralized with an aqueous NaOH solution in the downstream process and extracted and concentrated with polar solvents (typically low-molecular weight alcohols). After phase separation, the lower polar layer contains petroleum (natural) sulfonate Na salts, and the upper aromatic-depleted, nonpolar naphthenic oil is usually called "white oil" [20].

Depending on the boiling point range and the origins of the petroleum distillate cuts, for every ton of neat petroleum sulfonate Na salt produced, 3–10 tons of "white oil" are coproduced. This method of producing "white oil" is less economical in comparison to other methods such as catalytic hydrogenation. "Sulfonatable" petroleum distillate cuts with suitable molecular weight and molecular weight distribution are not consistent, thus the quality of petroleum sulfonates may be inconsistent. Shell closed the largest petroleum sodium sulfonate Martinez plant located in California in 2003. As a result, there is a shortage of high-quality petroleum sulfonates in the market. Only a few producers that have close ties with some refineries are still producing petroleum sulfonates [21].

Petroleum sulfonic acids have complicated chemical structures [22]. Petroleum sulfonic acid that is lower in molecular weight is "greenish" and more water soluble, and is also called "green acid"; petroleum sulfonic acid that is higher in molecular weight is "reddish brown" and more oil soluble and is also called "mahogany acid."

Petroleum sulfonate alkaline earth salts are typically produced from a metathesis reaction [23]. For example, petroleum sulfonate Ca salts are produced from the reaction of petroleum sulfonate Na salts with aqueous $CaCl_2$. Petroleum sulfonate Ca salts can also be produced from direct neutralization of the petroleum sulfonic acids with CaO or $Ca(OH)_2$. Petroleum sulfonates have very different molecular weight distributions and are more "continuous" than sulfonates from synthetic origin. In some applications, some end users still prefer to use petroleum sulfonates instead of synthetic sulfonates.

4.2.1.2 Synthetic Sulfonates

Synthetic sulfonates are obtained from neutralization of synthetic alkylaryl sulfonic acids, such as alkylbenzene sulfonic acids, alkyltoluene sulfonic acids, and alkylnaphthalene sulfonic acids, with alkaline and alkaline earth metals. Alkylaryl sulfonic acids are derived from sulfonation of the alkylaryl compounds with a sulfonating agent. Alkylaryl compounds are made from alkylation of the aromatic compounds with olefins.

Olefins can be sulfonated directly to form alkane sulfonic acids. However, the olefin sulfonation process generates a series of potent skin sensitizers, such as γ-sultones [24], and is accompanied by complex products, such as alkene sulfonic acid and hydroxy alkane sulfonic acid, which are thermal oxidatively not stable enough as detergents for higher-temperature application, such as for engine oil application.

For benzene alkylation, linear alkylbenzene (LAB) is obtained when linear olefin is used, and branched alkylbenzene (BAB) is obtained when branched olefin is used.

After World War II, water-soluble BAB sodium sulfonate household detergent formed from the alkylation of propylene tetramer and benzene using $AlCl_3$ as catalyst was very popular. In the 1960s, it was discovered that since BAB sulfonate is not biodegradable [25], residues in the environment can last for a long time. In places where large numbers of people live and use BAB sodium sulfonate detergent, wastewater treatment plants, rivers, and lakes accumulate a lot of persistent foam. Water-soluble LAB sulfonate detergent has excellent detergency and biodegradability [26], thus BAB sodium sulfonate is gradually being phased out. This trend may have impacted the oil-soluble detergents in the lubricant industry, and therefore oil-soluble LAB sulfonates are now more common than BAB sulfonates.

4.2.1.3 Synthesis of Alkylaryl Compounds

Alkylaryl compounds are synthesized from the Friedel–Crafts alkylation of olefins or alkyl halides with aromatic compounds in the presence of a catalyst [27] (Scheme 4.1).

The raw olefin material, the catalyst, and the alkylation process affect the alkylaryl compound structure and molecular weight and molecular weight distribution. All these affect the subsequent sulfonation process and the quality and performance of the resulting sulfonates.

There are various types of olefins, such as α-olefin, internal olefins, dimeric olefins, oligomeric olefins (polypropylenes, polybutenes, polyisobutylenes, etc.). Aromatics could be benzene [28], toluene, xylene [29], or naphthalene [30]. Alkylxylenes from xylene alkylation are tri-substituted benzene; certain isomers cannot be readily sulfonated due to steric hindrance in the subsequent sulfonation process, and thus the sulfonation yield is low. Naphthalene is more expensive than benzene/toluene/xylene, has a higher boiling point (218°C), and is more difficult/expensive to be removed by distillation in the process; furthermore, naphthalene is toxic, and residual naphthalene must be low in alkylnaphthalene. Thus, alkylnaphthalene sulfonate is typically more expensive and is mostly used for some specialty applications, such as rust inhibiting, due to its outstanding performance. Alkylbenzene, dialkylbenzene (DAB), and alkyltoluene sulfonates are the most common alkylates used for synthetic sulfonate detergents.

There are two types of alkylation catalysts: Brønsted acids, such as anhydrous HF, methane sulfonic acid, sulfuric acid, phosphoric acid, and sulfonic acid ion exchange resin; and Lewis acids, such as acid clay, BF_3, and $AlCl_3$ [31]. Different catalysts affect the yield, structure composition, and properties of the resulting alkylates, such as alkylbenzene positional

isomer distribution (such as 2-phenylalkane content), tetrahydronaphthalene and indan derivative content, bromine number, and subsequent sulfonatability.

In the presence of Brønsted acid catalysts such as HF [32], α-olefin is first protonated to form active carbon-centered cation at the 2-position, and through rapid proton migration [33] the carbon cation is distributed along the alkyl chain excluding the carbon atoms at the chain end, before electrophilic addition with benzene to form a series of positional phenylalkanes; Lewis acid catalysts such as $AlCl_3$ [34] can isomerize alkylbenzene more easily [35]. It is generally considered that among all the positional isomers, sulfonates from 2-phenylalkane have the best detergency and biodegradability [36]. For alkylation with α-olefin, the 2-phenylalkane isomer content ranges from 20% to 40% [37]. Tetrahydronaphthalene and indan derivative content is less than 1%, and their corresponding sulfonates are less biodegradable.

Linear α-olefins (LAOs) for oil-soluble linear alkylbenzene sulfonates typically have more than 14–16 carbons, up to 40, or higher. LAO is typically derived from the ethylene oligomerization process [38], such as Shell SHOP process [39], SABIC-Linde α-SABLIN process, CP chemicals process, and Ineos process. LAOs from these processes all have an even number of carbon atoms. Apart from Ineos process, LAOs from the other processes are seen in Anderson–Schulz–Flory distribution [40]; the content of LAOs decreases with molecular weight increases [41]. The C22 α-olefin content is less than 1% of the oligomeric polyethylene in the SHOP process, thus higher–molecular weight α-olefin and corresponding higher number average molecular weight linear alkylbenzenes are less available. Sulfonates from long LAOs tend to be waxy.

Heavy alkylbenzene (HAB) is the leftover high boiling point heavy residue after distillation of lower–molecular weight mono LAB; HAB is the by-product of linear alkylbenzene sodium sulfonate detergent production. Hydrotreated kerosene C_nH_{2n+2} with n typically lying between 10 and 16 (although generally supplied as a tighter cut, such as C12–C15, C12–C13, C10–C13, and C10–C14) is a typical feedstock for high-purity linear paraffins (n-paraffins). These are subsequently dehydrogenated over iron-containing solid catalysts under high temperature to form linear olefins, as well as a small amount of dienes. Linear olefin reacts with benzene in the presence of catalysts to form alkylbenzene, while diene reacts with benzene to form either diphenylalkane (DPA) or tetralin/indan compounds. DPAs could be sulfonated on both aromatic rings to form less oil-soluble sludges, affecting the resulting sulfonate production and quality. The most common

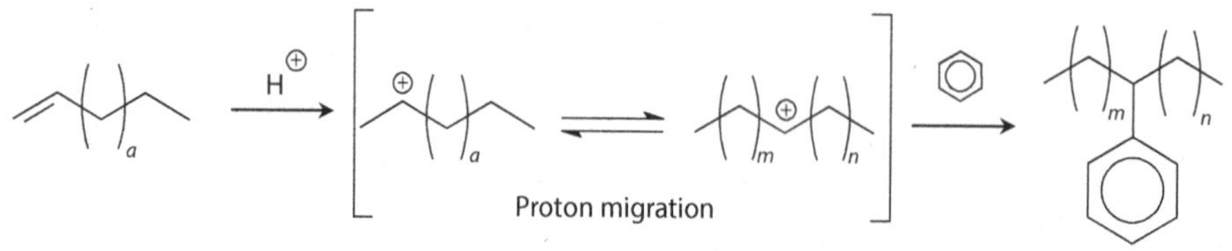

Proton migration

SCHEME 4.1 Brønsted acid–catalyzed Friedel–Crafts alkylation of α-olefin on benzene and positional isomers along the alkyl chain. a, m, n are integers.

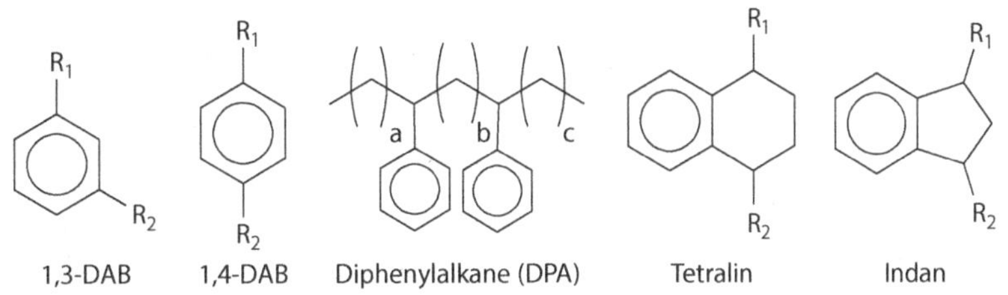

FIGURE 4.3 HAB compositions.

catalysts for LAB production are anhydrous HF, AlCl$_3$, and the solid Si–Al catalyst in Detal [42] process from UOP. HAB (Figure 4.3) consists mostly of DAB, the majority of which is 1,3-dialkylbenzene (with some 1,4-dialkylbenzene [43], diphenylalkanes, tetralin, and indan compound); and a small amount of lower–molecular weight mono LAB residue is left over from distillation. HAB is an important raw material for oil-soluble sulfonates. Different catalysts and processes affect HAB's yield, composition, and the subsequent sulfonation yield and product properties.

Anhydrous HF catalysts are commonly used for LAB production. Even though HF is highly toxic, it is highly effective and can be separated easily from reactants. HAB yield amounts to about 10% of the LAB production when using HF as a catalyst. However, HAB formed by the HF-catalyzed alkylation process contains more of 1,4-dialkylbenzene, and some of the isomers cannot be sulfonated easily due to steric hindrance. The sulfonation yield is also relatively low (55%–60%) [44], and therefore such HAB is a less desired raw material for oil-soluble sulfonate.

HAB from AlCl$_3$-catalyzed alkylation process yields about 20% of the LAB production, and the sulfonation yield for HAB from AlCl$_3$-catalyzed alkylation process is about 70%–90%. However, AlCl$_3$ is corrosive to the reactor, and the process generates plenty of waste when a caustic solution is used to wash the product. There is little LAB produced using AlCl$_3$ as a catalyst.

Solid catalysts can be easily separated from reactants, and the process is easy to operate with little waste. However, the problem with solid catalysts is that the catalyst bed tends to coke and lose activity. The UOP Detal process was able to solve this problem and is the only commercial process that uses solid catalysts. More than 90% of global LAB production capacity is based on UOP technology. HAB formed from this process has a high sulfonation yield.

Alkylate structure, molecular weight, and molecular weight distribution are all key to the production of sulfonate and its performance. Bromine index [45] (or iodine number [46]) is another key parameter for alkylate quality and is expressed as milligrams of bromine consumed per 100 g of sample. Bromine easily adds across to the double bond–containing impurities, and these impurities will lead to sludge formation during the subsequent sulfonation and overbasing process, affecting the quality of the sulfonates. A typical bromine index is required to be less than 5 for alkylbenzene used for oil-soluble sulfonates.

4.2.1.4 Sulfonation

Sulfonation reactions through electrophilic addition on the aromatic ring introduce a polar sulfonic acid group [47] (Scheme 4.2), transforming the oil-soluble alkylaryl compounds into amphiphilic molecules. A common sulfonating agent is SO$_3$ and its various complexes, such as oleum (pyrosulfuric acid or fuming sulfuric acid), concentrated sulfuric acid, chlorosulfuric acid, and so on. SO$_3$ is a highly reactive sulfonating agent and reacts with alkylbenzene violently; the heat of the reaction is 150–170 kJ/mol [48]. The sulfonation reaction is instantaneous, and the reaction rate is close to diffusion control. SO$_3$ is also a strong oxidizing agent. High temperatures during the sulfonation reaction result in SO$_3$ depriving hydrogen from alkylbenzene and lead to the formation of dark-colored species and excess amount of sulfuric acid by-product. For alkylbenzene sulfonation, the activation energy of the sulfonation reaction is less than oxidation reactions, and oxidation reactions occur more often at higher temperatures. In order to reduce the extent of oxidation reaction, the large amount of heat generated by the sulfonation process needs to be removed as soon as possible. Strict control of temperature is required to avoid over-sulfonation. SO$_3$ could be added to alkylbenzene in a traditional batch reactor, but because of the rapid increase in viscosity after alkylbenzene is sulfonated, heat transfer becomes difficult, hence it is more difficult to control the color. SO$_3$ needs to be added slowly.

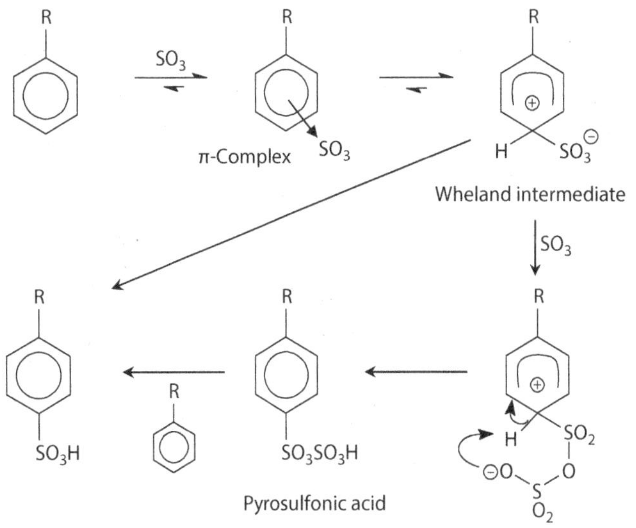

SCHEME 4.2 Alkylbenzene sulfonation reaction.

The best sulfonation process is the falling-film sulfonation process. The sulfonation is done in a falling-film reactor (FFR). Alkylbenzene is fed from the top of the reactor and is blown into the tube by 3%–5% SO_3 in dry air mixture and is distributed on the inner walls of the reaction tubes as thin film (2 mm to 2 cm thick) and sulfonated by SO_3. The residence time is about 30 s. The outside of the tube is cooled with cool water. Since the falling film has a very high surface area, heat generated from the reaction is quickly dispersed, thus avoiding overheating and oxidation. The color of the resulting sulfonic acid is light, and there is less resulting sulfuric acid.

Even though the sulfonation reaction is instantaneous, the sulfonation reaction does not complete that fast, and an "aging" process is needed. For example, after C10–C13 LAB has gone through the FFR, the sulfonation yield ranges from 90% to 92%; after aging at 45°C–55°C for 30–40 min, the yield is improved to 97%–98%. The reason for this is that after SO_3 electrophilic addition to the benzene ring to form sulfonic acid, additional SO_3 molecules react with the sulfonic acid more easily to form pyrosulfonic acid, which sulfonates the remaining alkylbenzene at a slower rate [49]. After "aging," a small amount of water is added to quench the reaction, to decompose/hydrolyze the pyrosulfonic acid. A small amount (1%–2%) of sulfuric acid is inevitably formed. Sulfuric acid will also be neutralized in the subsequent overbasing process and maybe incorporated into the "inverse micelle," and does not contribute to the alkaline reserve, yet it negatively impacts the overbased detergent product properties. Sulfuric acid could be removed by washing with water, allowing for a longer settling time of the sulfonic acid. High-quality sulfonic acid for overbased detergent production typically contains less than 1% sulfuric acid or lower.

When monoalkylbenzene (MAB) is being sulfonated, the sulfonic acid group is mostly added on the *para*-position of the alkyl group. Since the sulfonic acid group is a strong electron-withdrawing group (EWG) and deactivates the benzene ring for further electrophilic addition of another SO_3 molecule, di-sulfonation on the same benzene ring is seldom observed under normal sulfonation conditions.

The sulfonation yield depends on the structure and molecular weight of the alkylbenzene. MAB has less steric hindrance on the benzene ring, and is more susceptible to sulfonation than DAB, and tri-alkylbenzene. Higher–molecular weight alkylbenzene sulfonation typically leaves 5%–15% alkylbenzene unsulfonated. It is possible that the sulfonation of alkylbenzene needs a longer "aging" time to reach a higher yield, or certain alkylbenzene isomers have more steric hindrance and resist sulfonation more. Some tri-substituted benzenes, such as alkylxylenes, could also be sulfonated [50] but with a lower yield.

Oleum typically contains 15–30 wt% SO_3 and is a commonly used sulfonating agent. It has lower reactivity than SO_3 in a dry air mixture and requires a simple reactor. It is relatively easy to operate, but the process generates large amounts of waste "spent" sulfuric acid. In comparison, the FFR process generates much less waste sulfuric acid.

4.2.1.5 Synthesis of Neutral and Overbased Sulfonates

The neutralization of sulfonic acid with a base, such as a metal oxide or a metal hydroxide, forms metallic salt of organic sulfonic acid detergent. For an oil-soluble alkylbenzene sulfonic acid with a longer alkyl chain (C14+), the reaction between the strong acid and strong base simply does not begin when they are mixed together because of poor contact. A number of compounds, called "promoters," are needed to improve the acid–base contact and facilitate the neutralization and subsequent overbasing reaction. Promoters have certain solubility both in the polar aqueous phase where the metal base is present and in the nonpolar oil phase where the oil-soluble sulfonic acid resides. The promoters function as phase transfer catalysts. Typical promoters include alcohols, carboxylic acids, lower–molecular weight alkylbenzene sulfonic acids, alkylphenols, and so on. Some promoters, such as methanol, ethanol, and isopropanol, can be removed easily during the process and will not stay in the finished product. Some promoters remain, such as carboxylic acids and low–molecular weight alkylbenzene sulfonic acid, and may affect the properties and performances of the final detergent product.

The degree of overbasing is often expressed as the metal ratio [51], also called the basicity index (BI) [52], which is the ratio of the total equivalent amount of metal to the acid. For neutral sulfonate detergents, the metal ratio is usually less than 1.5–2.0 and close to 1. Commercial neutral sulfonate typically has 40–50 wt% soap, 50–60 wt% diluent oil, and typically contains small amounts of excess alkaline reserve in $Ca(OH)_2$ or $CaCO_3$ form. The TBN is less than 50 mg KOH/g and typically less than 30 mg KOH/g, or less than 10 mg KOH/g. A highly overbased detergent can be diluted with oil to obtain a lower TBN product, but the metal ratio does not change during dilution. Thus, the metal ratio can more accurately reflect the degree of overbasing.

Overbased calcium sulfonate can be synthesized using the "one-step" process or the "two-step" process. In the two-step process, sulfonic acid is first neutralized with CaO or $Ca(OH)_2$ and is processed to produce neutral calcium sulfonate, and then through "overbasing" the neutral calcium sulfonate the overbased calcium sulfonate is obtained. The two-step process is redundant. A one-step process is generally preferred. In the one-step process, sulfonic acid and an excess amount of calcium oxide or calcium hydroxide, together with some water, a promoter (such as methanol), some base oil, and a hydrocarbon solvent, are mixed and CO_2 is blown through the reaction. The reaction mixture is subsequently filtered to remove the excess amount of solid and stripped to remove volatile alcohols and hydrocarbon solvent to obtain overbased calcium sulfonate (Scheme 4.3). The mechanism of the overbasing process is through micro emulsion [53]: alkylbenzene sulfonate, water, and promoter form "inverse micelle" with $Ca(OH)_2$ suspended inside, and CO_2 diffuses into the "inverse micelle" and reacts with $Ca(OH)_2$ to form hard $CaCO_3$ nanoparticles, the soap molecules anchor on the surface of the hard nanoparticle, with small amounts of unreacted $Ca(OH)_2$ entrapped [54].

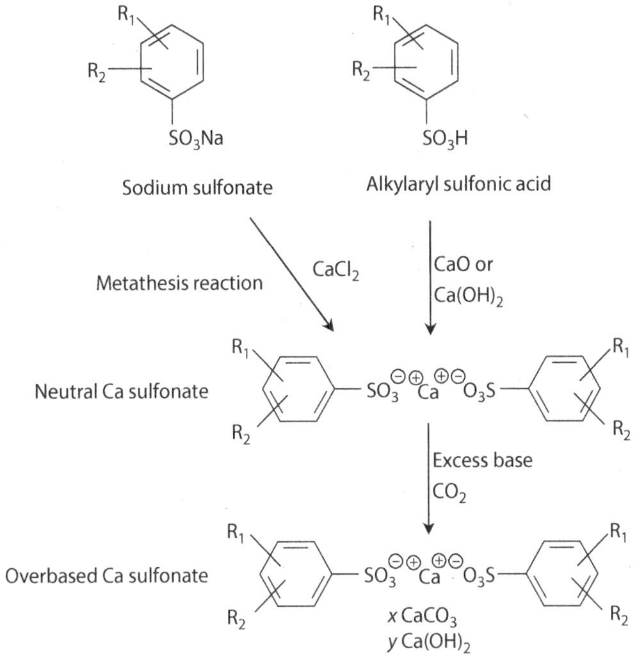

SCHEME 4.3 Synthesis of overbased calcium sulfonates.

The overbasing reaction relates to a complicated phase diagram involving alkylbenzene sulfonate, promoters, and water. Many parameters, such as reaction temperature, stirring rate, amount and quality of sulfonic acid (molecular weight and molecular weight distribution), promoter type and amount, acid/base ratio, and so on, can impact the outcome of the reaction. Incorrect reaction conditions will easily lead to a product with a high degree of turbidity, sediment, or even a failed reaction. When the amount of promoter used exceeds a certain level, the entire reaction product may solidify [55]. When the degree of carbonation is too high, the amorphous $CaCO_3$ can turn into a crystalline form, and the nanoparticle grows bigger, the product becomes hazy, contains more sediment, and even turns into a gel [56]. The crystalline form of $CaCO_3$ sulfonate has less usage as an overbased detergent for engine oil lubricant additive due to its high haze and sediment levels, but it can be used as a rheology modifier. Overbased products with different TBNs from the same sulfonic acids can be produced under different conditions [57].

The sulfonate group has high polarity. Sulfonates also have a higher–molecular weight oil-soluble tail than that of phenates and salicylates. Overbased sulfonate can pack up more alkaline reserves in the core and reach a high overbasing degree: highly overbased calcium sulfonates can reach a metal ratio of 25 (500 TBN calcium sulfonate with $CaCO_3$ core); highly overbased magnesium sulfonates can reach a metal ratio of 45 (600 TBN magnesium sulfonate with $MgCO_3$ core) [58] or 60 (with MgO core) [59]. Commercial barium sulfonates can only reach TBN at about 70 mg KOH/g with a metal ratio of less than 3, since $BaCO_3$ has higher density and higher molecular weight than that of $CaCO_3$ (density: 4.29 vs. 2.71 g/cm³; Mn: 197.34 vs. 100.08), therefore a stable inverse-micelle dispersion with packing of more TBN from $BaCO_3$ in it is difficult.

The number of sulfonate soap molecules attached to the hard alkaline reserve inorganic nanoparticle in the inverse micelle (aggregate number) depends on the degree of overbasing (metal ratio). The higher the metal ratio, the higher is the aggregate number, and the higher the molecular weight of the inverse micelle (the molecular weight of inorganic nanoparticle + the molecular weight of the soap molecules attached). Based on the nanoparticle size and structure, it is estimated that, for a neutral sulfonate ($CaCO_3$ core, 10 mg KOH/g TBN, metal ratio about 1.2), the average aggregate number ranges from 10 to 20, and the corresponding average molecular weight of the inverse micelle ranges from 8,000 to 10,000 Da; for an overbased sulfonate ($CaCO_3$ core, 400 mg KOH/g TBN, metal ratio about 20), the aggregate number can reach up to 1000–1500, and the corresponding molecular weight of the inverse micelle can reach up to 2,000,000 Da.

The overbasing process, reaction temperature, and stirring rate kinetically affect the inverse micelle size distribution; the alkyl chain length, and solvent/promoter/sulfonic acid/base ratio thermal-dynamically affect the inverse micelle size distribution (Table 4.1).

4.2.2 PHENATES

Phenates (alkylphenate sulfides) are the second most widely used detergents after sulfonates. They are mostly used in marine lubricants and diesel engine oils because of their fast acid neutralization ability and antioxidant properties. Phenates, together with salicylates, have antioxidant properties while sulfonates are pro-oxidants.

Tetrapropenylphenol (TPP) is the most common alkylphenol used in phenates. The proton in alkylphenol is weakly acidic ($pK_a \sim 10.0$) [60], giving alkylphenol a weaker amphiphilic structure than that of alkylaryl sulfonic acid. Alkylphenol could be overbased directly to form alkylphenate with alkaline earth metals under more vigorous overbasing conditions. However, free alkylphenol is known to be toxic, and alkylphenate salt could release alkylphenol in an acidic environment. TPP is acutely toxic to aquatic organisms and may cause long-term adverse effects in the environment. A reproductive toxicity study in rats shows that free or unreacted TPP may cause reduction in fertility, reduction in the number of offspring, and reduction in the size of reproductive organs [61]. It remains to be seen whether these effects observed in studies depend upon the concentration of residual TPP and its Ca salt in the material tested. Alkylphenol could be economically sulfurized with sulfur or SCl_2 to link the alkylphenol moieties with sulfide or a polysulfide bridge to form alkylphenol sulfide. It can also reduce the amount of free toxic alkylphenol, improving the antioxidant property. Studies conducted with alkylphenate sulfides diluted to the concentrations used by consumers did not result in any adverse effects on reproduction. Alkylphenate sulfide has a complex structure resulting from the free radical sulfurization process.

TABLE 4.1
Parameters of Typical Calcium, Magnesium, and Barium Sulfonates

Salt Form	Ca				Mg	Ba
Properties	Neutral	C300	C400	C500	M600	B70
Mn of alkylates	320–720	320–720	320–720	320–720	320–720	320–720
Mn of soap anion	400–800	400–800	400–800	400–800	400–800	400–800
TBN, mg KOH/g	0–30	280–320	380–420	480–520	500–640	65–70
Oil, wt%	50–60	40–50	40–50	30–40	35–45	40–50
Soap, wt%	40–50	15–30	15–25	15–25	10–15	40–45
Active, wt%[a]	40–50	40–60	50–60	60–70	55–65	50–60
Sulfur, wt%	2.5–3.5	1–2	1–1.5	1–1.5	0.5–1	2–3
Ca, wt%	1.8–2.2	10.5–12.5	14.5–15.5	18.0–19.5	—	—
Mg, wt%	—	—	—	—	12–15	—
Ba, wt%	—	—	—	—	—	13–14
Metal ratio	1–1.2	10–15	18–22	22–25	40–50	2–3

[a] Active% for sulfonate is defined as soap weight% combined with the alkaline reserve weight%.

4.2.2.1 Synthesis of Alkylphenols

Alkylphenate sulfides commonly use C10–C15 alkylphenols (predominantly C12) as starting raw material. Commercial manufacture of C10–C15 alkylphenols began in the mid-1940s. Complex olefin feedstocks with a range of alkyl chain lengths and branching patterns (propylene tetramers or butylenes trimers) react with phenol in a closed constant flow reactor using solid-phase Friedel-crafts acid catalysts. The reactants are pumped up through the catalyst bed in a continuous mode, and the crude alkylphenols are fractionally distilled to remove any unreacted olefin and phenol. The unreacted olefin and phenol are recycled and purified TPP is obtained. Commercial TPP is made up of >99% w/w single alkyl-substituted phenols, the majority of which (>95% w/w) are substituted at the *para*-position to the phenol group on the benzene ring. There is a very small amount of *ortho*- and *meta*-substitution. The DAP level is at <0.1% w/w. The C12-alkyl content is around 70%, but there are significant amounts of lower– and higher–molecular weight substances [62].

Phenol alkylation with linear α-olefin under normal alkylation conditions using a macroporous sulfonic acid resin catalyst or acidic clay affords higher amounts, which are substituted at the *ortho*-position to the phenol group on the benzene ring; the *para* isomer/*ortho* isomer ratio content is close to 50/50 [63].

4.2.2.2 Synthesis of Neutral and Overbased Phenates

Calcium phenate salt is the most common phenate salt. Magnesium is less basic and Mg salt is relatively more difficult to prepare than Ca salt. Other salts are less commonly used.

Alkylphenol does not overbase easily like alkylaryl sulfonic acid, since alkylphenol is weakly acidic. Sulfurized alkylphenates are commonly prepared by the "one-pot" vigorous high-temperature "glycol" process (Scheme 4.4): (1) neutralizing the alkyl phenol (TPP) with a base (e.g., calcium oxide, calcium hydroxide) in the presence of a glycol promoter to form a metal alkylphenate (e.g., alkyl calcium phenate) in base oil; (2) sulfurizing the metal alkylphenate with sulfur to cross-link the phenol rings of the metal alkylphenate to form metal alkylphenate sulfide (bridged by one to three sulfur atoms); and (3) overbasing the cross-linked sulfurized metal alkylphenate with carbon dioxide to incorporate calcium carbonate nanoparticles as alkaline reserve. During the process,

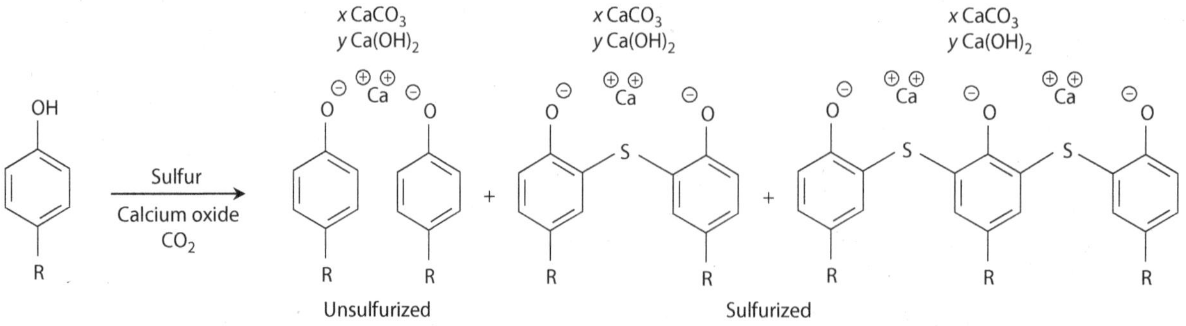

SCHEME 4.4 Synthesis of overbased calcium phenates.

hydrogen sulfide (H_2S) is formed first and may react with the calcium base, most likely with calcium alkoxide formed from the reaction of CaO with glycol, to form CaS, and CaS subsequently reacts with CO_2 to release the H_2S as a gas, since carbonic acid is a stronger acid than hydrogen sulfide (H_2S $pK_{a1} = 7.0$, $pK_{a2} = 19$; H_2CO_3 $pK_{a1} = 6.33$, $pK_{a2} = 10.33$ [64]). H_2S gas is usually scrubbed to avoid pollution. Ethylene glycol used in the "glycol" process has a relatively high boiling point (197.3°C) and is relatively difficult to remove completely through distillation; residue glycol may destabilize the "inverse micelle" system under certain conditions, such as in the presence of water, and other polar additives, such as ZDDP, friction modifiers, phosphate esters, and so on. Furthermore, residual ethylene glycol may contribute to sludge, varnish, and deposit formation in engine oil [65].

Other overbasing processes involving non-ethylene glycol solvents need similar polar promoters as in the sulfonate overbasing process.

Aminic catalysts, such as 2-mercaptobenzothiazole, primary and secondary alkylamines, polyalkyleneamines, amino acid, and so on, are usually employed to help initiate and facilitate the free-radical sulfurization process [66].

Sulfurized metal alkylphenates may still contain a certain level (5%–30%, w/w) of unreacted alkylphenol [67] (e.g., TPP). To reduce any potential health risks to customers and to avoid potential regulatory issues, there is still a need to reduce the amount of free alkylphenol in the sulfurized metal alkylphenates. Furthermore, there is a need for new lubricating oil detergents with a low free alkylphenol content. For example, the TPP level in phenate could be further reduced by reacting the TPP with an aldehyde to form a phenolic resin [68]. Phenates are not used in hydraulics, cutting and drilling fluids, or in oil-based or aqueous metalworking fluids due to their poor dispersancy and detergency, and environment and health concerns for TPP.

Alkylphenate soap molecules and alkylphenate sulfide soap molecules have poorer detergency than sulfonate and salicylate soap due to the relatively short oil-soluble alkyl side chain (average C12) in commercial phenate detergents, in comparison to sulfonates (usually average C18–C30). Phenoxy anions in phenates also have relatively weaker affinity to the counter salt cation in the alkaline reserve nanoparticle core due to the weak acidic phenol proton (pK_a for alkylphenol is ~10.0, and alkylphenol sulfide is less acidic with pK_a ~ 10.5 due to the electron-donating effect of sulfur atom linking the phenol ring [60]) in comparison to sulfonic acid (pK_a for dodecylbenzene sulfonic acid is 0.7 [69]). However, upon oxidation of the sulfide linkage to the polar sulfone group, the dispersancy and detergency of phenates will improve. Phenates usually have a lower degree of overbasing than sulfonates. The basicity index (metal ratio) for phenates is typically less than 4–6. Unlike neutral sulfonates, which have 0 TBN when the metal ratio is 1, commercial "neutral" phenates are still basic and have a relatively high TBN (80–150 mg KOH/g) since the "neutral" phenate soap molecules could be titrated as base in ASTM D2896 even though there are no entrapped alkaline hard-core nanoparticles. The "soap" compositions in phenates are more complicated than in sulfonates. They include the alkylphenol/alkylphenates, alkylphenol/alkylphenate sulfides, and some surfactants, which are used as promoters during overbasing, such as alkylsulfonates.

The shorter alkyl side chain and the weaker amphiphilic nature of the alkylphenate result in the relatively "thinner" inverse-micelle shell and relatively more unstable phenate inverse-micelle system. The alkaline reserves in phenates are more available to acids generated in the combustion process and phenates may neutralize acids faster than the relatively tightly packed, "thicker" overbased sulfonate inverse micelles.

Phenates are often used in combination with highly overbased sulfonates in marine cylinder oils and diesel engine oils to utilize the antioxidant property/fast acid neutralization property of phenates and the high alkaline reserve capacity/high-temperature detergency of the sulfonates. However, phenate–sulfonate is known in the industry to have compatibility problems [70]. Phenates tend to interact with highly overbased sulfonates to produce haze and sediment when blended into finished oil, which can happen during blending and storage. The haze and sediment most likely come from the aggregated alkaline reserve nanoparticles within the overbased sulfonates when they are destabilized by the phenates. This phenate–sulfonate incompatibility is influenced by the process producers employed to produce the phenates and sulfonates and their quality, such as molecular weight and distribution of the alkylates used, promoters used in the overbasing process, degree of overbasing, and so on. This phenomenon worsens in the presence of some polar components in the finished oils, such as zinc dialkyldithiophosphate (ZDDP), and water from combustion or seawater intrusion in marine cylinder oils. Polar species disrupt the inverse-micelle system and cause the suspended alkaline reserve nanoparticles to fall out of the inverse micelles, not only losing their TBN but also changing the filterability of the oil. High water tolerance is key for both sulfonates and phenates and the finished oils.

Chevron Oronite is the largest phenate producer in the world and offers several grades of phenate products. The two most common grades are 120 mg KOH/g TBN (Oloa® 216Q) and 250 mg KOH/g TBN (Oloa® 219). The lower TBN grade phenate is close to a "neutral" product. A typical overbased phenate contains around 40%–50% (w/w) base oil and 40%–50% (w/w) soap (Table 4.2).

4.2.3 SALICYLATES

Salicylates are another group of widely used detergents after sulfonates and phenates. They are mostly used in marine lubricants and diesel engine oils because of their fast acid neutralization ability, detergency, and antioxidant properties. Salicylates are unique to phenates and sulfonates in that they do not contain sulfur in the soap molecules. Salicylates are particularly suitable for lubricant formulations that do not want to introduce additional sulfur from the detergent system. Salicylates have better detergency than phenates.

Overbased salicylates are prepared by overbasing the corresponding alkylsalicylic acids. The alkyl group of the salicylate

TABLE 4.2
Parameters of Typical Commercial Calcium Phenates

Properties	P120	P250
Mn of soap mono anion[a]	260–900	260–900
TBN, mg KOH/g	110–130	230–270
Oil, wt%	40–60	40–60
Soap, wt%	45–55	25–35
Active, wt%[b]	40–60	40–60
Sulfur, wt%	5–6	3–4
Ca, wt%	3.9–4.6	8.2–9.6
Metal ratio[c]	1–2	4–6

[a] Sulfur commonly bridges 2–3 TPP molecules; 4 TPPs and above are less common.

[b] Active% for phenate is defined as total less diluent oil, including unsulfurized alkylphenol, sulfurized alkylphenol, some co-surfactant like sulfonate and their salt, and the dispersed $CaCO_3$.

[c] The metal ratio is calculated based on all phenols existing in salt form.

is typically a long-chain alkyl group of more than 14 carbon atoms used to provide adequate oil solubility.

4.2.3.1 Synthesis of Alkylsalicylic Acids

Commercial alkylsalicylic acids are conventionally prepared by the alkylation of phenol with linear α-olefin mixture containing certain molecular weight distribution to form alkylphenol, with a mixture of approximately 50:50 *ortho*- and *para*-alkylphenol isomers. This differs from phenol alkylation with branched C12-olefin used in phenate preparation, in which the *para*-isomer is more than 95% w/w. Alkylated salicylic acids are then obtained by carboxylation of the alkylphenols using the Kolbe–Schmitt reaction [71] (Scheme 4.5).

The Kolbe–Schmitt route requires high temperatures and pressure during the process. Also, when substantially linear alkylation feeds are employed, not all of the long-chain alkylphenol is readily carboxylated. The long-chain *para*-alkylphenol is more active and readily carboxylated, whereas the long-chain *ortho*-alkylphenol is less reactive and only about 70% of the total amount of the alkylphenol derived from a substantially linear alkylation feed is typically converted to alkylsalicylic acid during this reaction.

Alkylsalicylic acids can also be produced by direct alkylation of the salicylic acid with olefin mixtures using sulfuric acid [72] or methane sulfonic acid as catalysts [73].

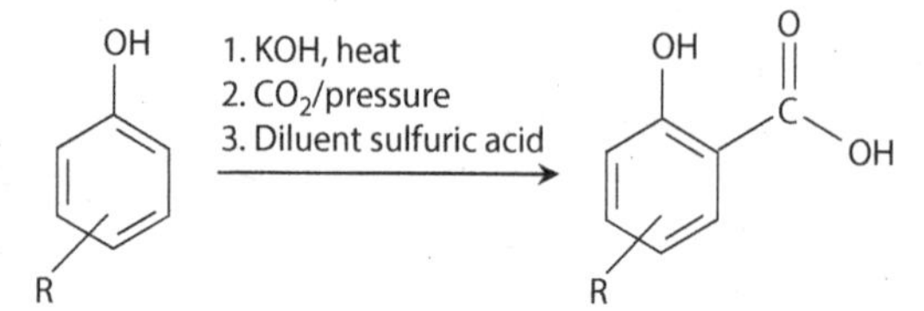

SCHEME 4.5 Synthesis of alkylsalicylic acids via Kolbe–Schmitt route.

Salicylic acid is a solid with a high melting point (159°C) and is not soluble in olefin mixtures. The alkylation mixture is therefore heterogeneous and does not offer good contact between the alkylating species and the substrate being alkylated. Some producers use liquid salicylic acid methyl ester to allow better mixing of the salicylic acid ester and olefin during alkylation using solid acidic resin catalyst [74]. The solid catalyst can be separated from the product easily through filtration.

Certain amounts of alkylphenols may still exist in the alkylsalicylic acid obtained from both synthesis routes. However, the alkylphenols in the commercial salicylates have longer carbon chains (C14 and above) than that of TPP, thus they may be less soluble in water and less toxic than TPP.

4.2.3.2 Synthesis of Neutral and Overbased Salcylates

Alkylsalicylic acid is a moderately strong acid (salicylic acid, $pK_{a1} = 2.97$, $pK_{a2} = 13.74$ at 25°C, alkylsalicylic acid is slightly less acidic than salicylic acid due to the electron-donating effect of the alkyl side chain) [75]. Thus, alkylsalicylic acid is easier to be overbased than alkylphenol, similar to sulfonic acid. In the overbasing process, self-assembly inverse-micelle structure needs to be formed to pick up alkaline reserve as TBN. Calcium and Mg cations behave differently from each other, more obviously for alkylsalicylates than for sulfonates, but both magnesium salicylate and calcium salicylate surfactants produce colloidally stable surfactant structures [76]. Neutral and overbased alkylsalicylates are synthesized through the "one-pot" process by mixing with base oil, solvent, calcium oxide, or calcium hydroxide, followed by carbonation with CO_2, either using the low-temperature process [77] (55°C–60°C) with low-boiling point alcohols as promoters similar to the sulfonate overbasing process, or the "high-temperature" "glycol" process [78] (~160°C) similar to the typical phenate overbasing process. Also, since the phenolic proton is much less acidic (H_2CO_3 $pK_{a1} = 6.33$, $pK_{a2} = 10.33$; alkylsalicylic acid $pK_{a2} > 13.7$ at 25°C) due to the neighboring carbonyl group, it is likely that the phenol will remain in the hydroxy form in the overbased product (for Ca or Mg salts) (Scheme 4.6).

Infineum is the largest salicylate producer in the world and offers several grades of salicylate products. Chemtura also offers a series of salicylates in addition to the wide range of sulfonate products. The two most common grades of salicylate in the market are 150 mg KOH/g TBN and 250 mg KOH/g TBN. Salicylate with 400 mg KOH/g TBN and above can also be produced. The lower TBN grade salicylate is close to a "neutral" product. A typical overbased salicylate contains around 40%–60% (w/w) base oil, and the amount of alkylsalicylate soap is about 40%–60% (w/w) (Table 4.3).

Salicylates are often used in combination with highly overbased sulfonates in marine cylinder oils and diesel engine oils. This utilizes the antioxidant property/fast acid neutralization, and the good detergency property of salicylates and the high

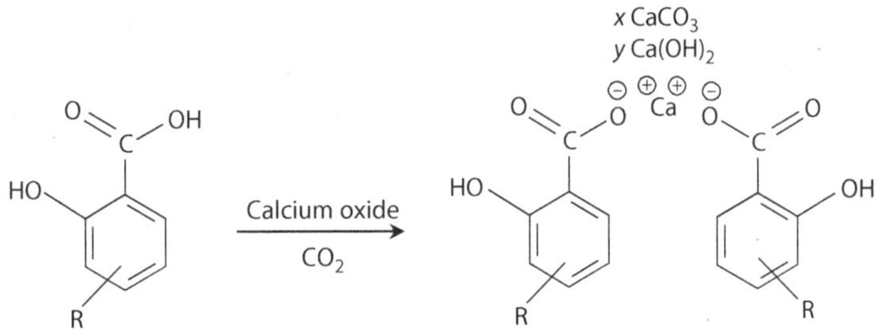

SCHEME 4.6 Synthesis of overbased calcium salicylates.

TABLE 4.3
Parameters of Typical Commercial Calcium Salicylates

Properties	S150	S250
Mn of soap mono anion	300–500	300–500
TBN, mg KOH/g	130–170	230–270
Oil, wt%	40–60	25–50
Soap, wt%	40–60	40–60
Active, wt%[a]	40–60	50–75
Sulfur, wt%	0	0
Ca, wt%	4.6–6.1	8.2–9.6
Metal ratio[b]	1–3	2–4

[a] Active% for salicylate is defined as total less diluent oil.

[b] The metal ratio for salicylate is calculated as free-phenol structure.

TABLE 4.4
Typical Overbased Detergent Property Comparison

Properties	General Ranking
Polarity of soap molecule polar group	Sulfonate > salicylate > phenate
Chain length of oil-soluble alkyl soap molecule	Sulfonate > salicylate > phenate
Detergency (high to low)	Sulfonate > salicylate > phenate
Dispersancy (high to low)	Sulfonate > salicylate > phenate
Antioxidation (good to poor)	Phenate > salicylate > sulfonate
Film formation capability on polar surface (good to poor)	Sulfonate > salicylate > phenate
Corrosion inhibition (good to poor)	Sulfonate > salicylate > phenate
Ease of overbasing (easy to difficult)	Sulfonate > salicylate > phenate
Alkaline reserve capability (high to low)	Sulfonate > salicylate > phenate
Acid neutralization rate (fast to slow)	Phenate > salicylate > sulfonate
Hydrolytical stability (high to low)	Sulfonate > salicylate > phenate

alkaline reserve capacity/high-temperature detergency of the sulfonates. Unlike phenates, salicylates are more compatible with sulfonates due to the stronger interaction of the salicylates with the overbased core and the relatively longer alkyl side chain in alkylsalicylates (average C16 and above) than in phenates. Salicylates are sulfur-free, and their dispersancy/detergency lies between sulfonates and phenates; their antioxidant property also lies between phenates and sulfonates. Salicylates have the combined benefit of sulfonates and phenates. However, salicylates are relatively more expensive to produce than sulfonates and phenates, thus hindering their application.

4.2.4 Summary

The physical properties of inverse-micelle self-assembly are fundamentally influenced by the geometric constraints of the soap molecules and their interactions with the hard nanoparticle cores, similar to but may be more complicated than the micelle system [79]. The physical property and performance of the detergents are closely related to the type of soap molecules (sulfonates, phenates, salicylates, and other types of detergents such as phosphonate, carboxylate, etc.), as well as raw material structure, purity, molecular weight, molecular weight distribution, type of metal salt, particle size, its crystallinity, degree of carbonation, etc. (Table 4.4).

4.3 DETERGENT PARAMETERS AND TESTING

Detergents can be described in terms of many parameters. Some of the parameters are related directly to each other: for example, the metal content and TBN and sulfated ash; soap% and S%; active% and soap% and TBN; active% and diluent oil%, and so on. Some parameters are more subtly related to each other. For example, detergent viscosity and particle size distribution; sediment% and particle size distribution; metal ratio and particle size distribution, and so on.

The most important parameters are TBN and soap%. They reflect the two most important functions of detergents: acid neutralization ability and detergency/dispersancy (Table 4.5).

4.3.1 TBN

There are two types of alkaline reserve in "overbased" detergents: total base number (TBN) and direct base number (DBN), and both are expressed as milligram KOH per gram of sample. DBN refers to the stronger, more reactive base, such as level of calcium alkoxides, CaO (both of which typically come from glycol process), and $Ca(OH)_2$ in the "inverse" micelle, and is closely related to the degree of carbonation [82]. TBN is the total acid neutralization capacity, which includes all types of alkaline reserve in the sample, including contribution from $CaCO_3$. Since CaO and $Ca(OH)_2$ are more water soluble,

TABLE 4.5

Typical Parameters of Detergents and Testing Methods

Properties	Method
Kinematic viscosity, 100°C, cSt	ASTM D445
Kinematic viscosity, 40°C, cSt	ASTM D445
Flash point, °C, open cup	ASTM D92
Flash point, °C, closed cup	ASTM D93
TBN, mg KOH/g	ASTM D2896, D4739
DBN, mg KOH/g	Titration [80]
Diluent oil content, wt%	ASTM D3712
Soap content, wt%	ASTM D3712
Active, wt%	Membrane dialysis [81]
Metal, wt%	ASTM D4951, D4927, D5185
S, wt%	ASTM D1552, D4294, D4951
Water, wt%	ASTM D95, D6304
Sulfated ash, wt%	ASTM D874
Sediment, vol%	ASTM D2273
Metal ratio (basicity index)	Calculation
Aggregate number	Calculation
Particle size (nm)	Light scattering, TEM[a]
Turbidity, NTU	ASTM D6181
Color	ASTM D1500

[a] Transmission electron microscopy.

their levels in the overbased detergent have a strong impact on the stability and performance of the overbased detergent; thus, it is important to control the level of DBN in the production process. DBN typically does not exceed more than 10% of the TBN in the product, especially for marine applications, such as marine cylinder lubricants (MCL) and trunk piston engine oils (TPEO). Where lubricants have close contact with water, DBN is typically less than 5% of the TBN [83].

The alkaline reserve in lubricants is usually measured by acid–base titration. The two methods are ASTM D2896 and ASTM D4739. Stronger perchloric acid is used as the titrant in ASTM D2896, while relatively weaker hydrogen chloride acid is used as the titrant in ASTM D4739. Thus, ASTM D4739 cannot titrate all the basic species in the finished lubricant, and the value of TBN obtained using ASTM D4739 is usually less than that obtained from ASTM D2896. However, for overbased detergents, both ASTM D2896 and ASTM D4739 give similar results [84].

TBN, metal content, and sulfated ash are closely related to each other. TBN and sulfated ash are only related to the metal content in detergents. For sulfonates, since sulfonic acid is a strong acid, metals bonded to the soap molecules do not contribute to TBN, and only metals from the overbased parts contribute to TBN; for phenates (ignore the contribution from small amounts of alkylsulfonate promoters if any) and salicylates, the alkylphenols, alkylphenol sulfides, and alkylsalicylic acids are all weak acids, thus all the metals contribute to TBN. Pure Ca metal's TBN is 2800 mg KOH/g, pure Mg metal's TBN is 4617 mg KOH/g, and pure Ba metal's TBN is 817 mg KOH/g.

For calcium phenates and calcium salicylates:

$$TBN = Ca\% \times 2800 \, mg \, KOH/g$$

$$Sulfated \, ash \, wt\% = Ca\% \div 40.08 \times 136.14 = Ca\% \times 3.40$$

$$Sulfated \, ash \, wt\% = TBN \div 2800 \div 40.08 \times 136.14$$
$$= TBN \div 824.30$$

For magnesium phenates and magnesium salicylates:

$$TBN = Mg\% \times 4617 \, mg \, KOH/g$$

$$Sulfated \, ash \, wt\% = Mg\% \div 24.31 \times 120.37 = Mg\% \times 4.95$$

$$Sulfated \, ash \, wt\% = TBN \div 4617 \div 24.31 \times 120.37$$
$$= TBN \div 932.32$$

For barium phenates and barium salicylates:

$$TBN = Ba\% \times 817 \, mg \, KOH/g$$

$$Sulfated \, ash \, wt\% = Ba\% \div 137.33 \times 233.43 = Ba\% \times 1.70$$

$$Sulfated \, ash \, wt\% = TBN \div 817 \div 137.33 \times 233.43$$
$$= TBN \div 480.74$$

For the same metal%, Mg salts have the highest TBN and the lowest sulfated ash among the three.

4.3.2 SOAP%

The amount of soap in a detergent is closely related to the ability of the detergent to disperse polar species and deterge varnish and deposit on a metal surface. Typically neutral detergent products contain higher soap%. The purpose of the neutral detergents is to provide soap for deterging and dispersing. High TBN overbased detergent products contain less soap% and more alkaline reserve. The purpose of the overbased detergents is to provide adequate alkaline reserve to neutralize acid. For sulfonates, ASTM D3712 measures not only the soap% and the diluent oil content but also the number average molecular weight of the soap molecules. Soap% measured by ASTM D3712 is expressed in the Na salt form, which is very close to the wt% in Ca form. For phenates and salicylates, other similar but more sophisticated methods, which are typically proprietary, are needed. Infrared (IR) spectrum, nuclear magnetic resonance (NMR), and mass spectrum analysis are other powerful ways to determine the soap molecule structure and to measure the molecular weight and distribution.

4.3.3 PARTICLE SIZE, TURBIDITY, AND VISCOSITY

Detergent nanoparticle size and distribution in overbased detergents are related to turbidity, viscosity, and sediment. Larger particles in overbased detergent scatter more light and contribute more to turbidity. For the same type of detergents with the same TBN, higher turbidity generally means the detergent contains bigger particles and thus is less stable and generates more sediment in ASTM D2273 test.

In 1900, water turbidity was measured by the light absorption method. By adding samples into an upright standing Jackson tube with a standard candle underneath until no clear candle image could be observed from above, the turbidity of the sample was obtained from the ratio of the thickness of the sample and the standard scale on the tube. The unit used to measure turbidity is JTU (Jackson turbidity unit) in this test. This method measures the incident light through the sample at a 180° angle, uses natural diatomite or kaolin suspension in water as standard, has poor consistency and reproducibility, and has now been eliminated. In 1970, turbidity was measured by the turbidity meter at 90° angle by measuring scattered light (ASTM D6181). It has a high sensitivity and good accuracy for a wide range of particle sizes and concentrations. ASTM D6181 is suitable for turbidity ranging from 0.1 to 500 turbidity unit (nephelometric turbidity units, NTU). NTU is calibrated using highly reproducible formazin polymer water suspension (average particle size $Dn = 1.5 \pm 0.6$ μm). Thus, NTU is also known as FTU (formazin turbidity unit), or FNU (formazin nephelometric unit), but NTU is the preferred turbidity unit. A formazin suspension turbidity of 40 NTU is roughly equal to 40 JTU, but NTU and JTU use different calibrating materials and different measuring instruments, hence they are not equal units [85]. The same method and principle are used to determine detergent turbidity.

Detergent turbidity, excluding foreign contaminants, such as filter aid and water, is influenced by detergent particle size and distribution. Particle size and distribution also influence the stability of detergents. Overbased detergent inverse-micelle particles contain $CaCO_3$ or other types of dense metal salt and thus have a higher density than the dispersing oil and tend to settle down. The higher the metal ratio of the detergents, the higher is the density of the micelles. Stokes' law may describe particle-settling velocity [86].

$$v_s = \frac{2}{9} \frac{\left(\rho_p - \rho_f\right)}{\mu} gR^2$$

where

v_s is the settling velocity
ρ_p is the density of the inverse micelle
ρ_f is the density of surrounding oil
g is the gravitational acceleration
R is the radius of the entire inverse micelle
μ is the dynamic viscosity of the surrounding oil

The settling velocity of the inverse-micelle particle is proportional to the square of the radius of the entire particle. Smaller particles contain smaller $CaCO_3$ cores, thus the density is smaller, and closer to the base oil; larger inverse-micelle particles contain larger $CaCO_3$ cores, and thus, the density is higher. ASTM D2273 quickly measures the volume of larger particles including impurities of the detergent in lower-viscosity solvents when subjected to high centrifugal acceleration force. High-quality products typically have less than 0.05% volume of settlements.

Overbased detergent production should minimize the size distribution of particles to reduce the adverse effect of the big particles (such as high turbidity and high settlement).

The viscosity of overbased detergents is also closely related to the size and distribution of the nanoparticular inverse micelles. Principally, the Krieger–Dougherty equation [87] can describe the volume of suspended $CaCO_3$ hard spheres depending directly on its rheological properties.

$$\frac{\eta}{\eta_0} = \left[1 - \frac{\phi}{\phi_m}\right]^{-[\eta]\phi_m}$$

where

η is the viscosity of the overbased detergent sample
η_0 is the viscosity of the carrier diluent oil
ϕ is the volume% of the solid in the overbased detergent
ϕ_m is the maximum theoretical volume% of the solid in the overbased detergent (random close packing is 63%)
$[\eta]$ is the intrinsic viscosity (equal to 2.5 for spherical particles)

For overbased detergent production, when the TBN increases, the volume of the suspended $CaCO_3$ solid increases, and the product viscosity increases. For overbased detergents with the same TBN, that is, the same volume of suspended $CaCO_3$ solid, if the number of inverse micelle increases (i.e., the particle size becomes smaller), the viscosity increases [88]; if the size distribution of the particle is wide, the viscosity decreases [89], because smaller particles may fill in the gaps among the larger particles. Thus, the maximum solid volume% (ϕ_m) increases from the random packing 63%–74%.

The shape of the nanoparticles also has an impact on the rheological properties. For particles of the same size, particles with a smooth surface have a lower viscosity; elongated particles have a higher viscosity at low shear rates in comparison to spherical particles, but have lower viscosity at higher shear rates from forced streamline arranging.

4.3.4 Performance Testing

Detergents are formulated into lubricant oils mostly for acid neutralization, detergency/dispersancy, corrosion inhibition, and antiwear. However, some types of detergents may negatively impact certain performances, and these need to be addressed during formulation.

Detergent type and detergent alkyl side chain length, structure, and distribution of overbased detergents directly impact their performance.

Sulfonates and salicylates may form thin films with tight packing on the metal surface, thus preventing the metal from rusting; phenates have no such effect. Polysulfide bridges linking the alkylphenol in phenates with more than two sulfur atoms may be more corrosive to yellow metals such as copper. Detergent molecular weight and distribution influence the water separation ability of the detergent-containing oil. Detergent alkyl chain structure and length may influence the foaming and air release properties of the detergent-containing oil.

Detergents also offer some EP/antiwear performance alone or together with other EP/antiwear additives [90].

Overbased detergents' inverse-micelle structure may be disrupted by interaction with certain polar species, such as water [91], additives, or other types of detergents, leading to gel formation and precipitation. For example, in marine lubricants, overbased sulfonates and overbased phenates are frequently used together to utilize the high alkaline reserve and high-temperature detergency of sulfonates and the antioxidant property of phenates against adverse effects from the use of high sulfur- and asphaltene-containing fuel. Marine lubricants are frequently exposed to seawater; the hydrolytical stability of the detergents and their compatibility are crucial. The hydrolytical stability of detergents is strongly influenced by the molecular weight and distribution of the soap molecules and the structure of the alkyl side chain of the soap molecules, such as branching. Longer side chains with certain degrees of branching can effectively increase the stability of the inverse micelle, thus improving the hydrolytical stability and detergency. Hydrolytical stability is also influenced by the degree of overbasing, which is related to the close packing of the soap molecule self-assembly on the hard nanoparticles. Poor hydrolytical stability of detergents also negatively affects oil filterability (Table 4.6).

TABLE 4.6
Typical Performance Tests Related to Detergents

Test Method	Objective
ASTM D665	Rust preventing
ASTM D130	Corrosion to yellow metal
ASTM D1401	Demulsibility
ASTM D2711	Demulsibility
ASTM D892	Foaming property
ASTM D3427	Air release property
Panel coker test	Determines the tendencies to form coke when in contact with surfaces at elevated temperatures
Hot tube test	Assesses high temperature thermal stability [92]
ASTM D2272	Assesses oxidative stability
ASTM D7097	Designed to predict the deposit-forming tendencies of engine oil in the piston ring belt and upper piston crown area
ASTM D2670	Assesses antiwear performance
ASTM D2882	Assesses wear characteristics of hydraulic fluids in constant volume Vane pump
ASTM D4172	Assesses antiwear performance
ASTM D2782	Assesses extreme pressure performance
ASTM D2783	Assesses extreme pressure performance
ASTM D2619	Evaluation of hydrolytic stability of hydraulic fluids
ISO 13357	Evaluation of the filterability of lubricants

4.4 DETERGENT FUTURE DEVELOPMENT

Lubricant technology has advanced significantly in recent decades. High-quality base fluids such as group II/III/IV base oils, base oils from gas to liquid (GTL) process, poly-α-olefin (PAO) with less unsaturation, less sulfur, better low-temperature properties, better thermal oxidative stability, and better viscosity index are replacing conventional less thermal-oxidatively stable solvent-refined group I–based lubricant oils. Renewable, biodegradable oils are finding ways into lubricant oil formulations. High-quality thickeners or viscosity modifiers, such as high-viscosity PAO, mPAO, OCP, PMA, and so on, are being developed continuously; additive chemistries are systematically studied and understood, and additive quality is improving; and new additive chemistry, new additive combinations are continuing to be developed. Detergents must ensure their performance and compatibility with the new base stocks and additives in finished lubricant formulations. With more understanding of the fundamentals among detergent performance, chemistry, raw material selection, production process, and interactions with other components in the lubricant, the traditional detergent technology is still evolving.

To ensure a high level of protection of human health and the environment from the risks that can be posed by the use of chemicals, the promotion of alternative test methods, the free circulation of substances on the internal market, and for enhancing competitiveness and innovation, the European Union (EU) started the Regulation on Registration, Evaluation, Authorisation and Restriction of Chemicals (REACH) on June 1, 2007 [93]. All chemicals entering the EU, including detergents, must be accompanied by a detailed chemical composition and health environment evaluation. This forces the chemical producers to study, characterize, understand, and disclose their products more carefully, but to a certain degree it also discourages producers from introducing new chemistries to avoid lengthy and costly registration.

More than 75% of detergents are used for the internal combustion engine oil market. New engine technology is developing toward greater output power, reduced engine oil sump size, improved energy efficiency, low fuel consumption, reduced emission (particulates, hydrocarbons, CO, NO_x, etc.), and long oil drain intervals. These developments result in higher stress of lubricant oils, thus requiring detergents with improved high-temperature detergency and thermal oxidative stability. With the wide use of the exhaust gas recirculation (EGR) system in internal combustion engines to reduce NO_x emission, low SAPS TBN [94] is desired to prolong catalyst life. With the development of electric cars, electric motors replace internal combustion engines and do not require engine lubricants. This may have a profound impact on the lubricant industry in the coming decades.

Detergents, together with lubricants, are evolving continuously as more advanced products are being developed to meet the rising demands of modern machinery for better productivity, performance, reliability, energy efficiency, and environmental responsibility.

ACKNOWLEDGMENT

I thank my daughter, Clara Ma, for her valuable comments and editing of the manuscript.

REFERENCES

1. van Dam, W., D.H. Broderick, R.L. Freerks, V.R. Small, W.W. Willis, TBN retention—Are we missing the point? SAE Paper, 972950, 1997, doi:10.4271/972950.

2. (a) Redmore, D., F.T. Welge, Magnesium carboxylated-sulfonate complexes. U.S. Patent 4,056,479, Petrolite Corporation, November 1, 1977. (b) Muir, R.J., T.I. Eliades, Overbased magnesium deposit control additive for residual fuel oils. U.S. Patent 6,197,075, Crompton Corporation, March 6, 2001.

3. Vollath, D., *Nanomaterials: An Introduction to Synthesis, Properties and Applications*, 2nd edn., Wiley-VCH, Weinheim, September 2013.

4. Schmid, G., *Nanoparticles: From Theory to Application*, Wiley-VCH, Weinheim, March 2006.

5. Moroi, Y., *Micelles: Theoretical and Applied Aspects*, Plenum, New York, 1992.

6. (a) Kandori, K., K. Konno, A. Kitahara, Formation of ionic water oil microemulsion and their application in the preparation of $CaCO_3$ particles. *J. Colloid Interface Sci.*, 122, 78–82, 1988. (b) Delfort, B., B. Daoudal, L. Barre, Particle size determination of (functionalized) colloidal calcium carbonate by small angle x-ray scattering—Relation with antiwear properties. *Tribol. Trans.*, 42, 296–302, 1999. (c) Glavati, O.L., S.M. Kurilo et al., Structure of micelles of verbased salicylate lube oil additives. *Chem. Technol. Fuel Oils*, 25, 273–275, 1989. (d) Griffiths, J.A., R. Bolton, D.M. Heyes, J.H. Clint, S.E. Taylor, Physico-chemical characterisation of oil-soluble overbased phenate detergents. *J. Chem. Soc. Faraday Trans.*, 91, 687–696, 1995.

7. (a) Marsh, J.F., Colloidal lubricant additives. *Chem. Ind.*, 20, 470–473, 1987. (b) Papke, B.L., L.S. Bartley, Process for preparing overbased metal sulfonates. U.S. Patent 4,995,993, Texaco Inc. February 26, 1991.

8. (a) Muir, R.J., T.I. Eliades, K. Niece, W.A. Mackwood, Oil soluble calcite overbased detergents and engine oils containing same. U.S. Patent 6,107,259, Witco Corporation, August 22, 2000. (b) Muir, R.J., Clarification method for oil dispersions comprising overbased detergents containing calcite. U.S. Patent 6,239,083, Crompton Corporation, May 29, 2000.

9. (a) Cizaire, L., J.M. Martin, T.L. Mogne, E. Gresser, Chemical analysis of overbased calcium sulfonate detergents by coupling XPS, TOFSIMS, XANES, and EFTEM. *Colloids Surf. A Physicochem. Eng. Aspects*, 238, 151–158, 2004. (b) Hudson, L.K., J. Eastoe, P.J. Dowding, Nanotechnology in action: Overbased nanodetergents as lubricant oil additives. *Adv. Colloid Interface Sci.*, 123–126, 425–431, 2006.

10. (a) O'Sullivan, T.P., M.E. Vickers, R.K. Heenan, The characterization of oil-soluble calcium carbonate dispersions using small-angle x-ray scattering (SAXS) and small-angle neutron scattering (SANS). *J. Appl. Crystallogr.*, 24, 732–739, 1991. (b) Ottewill, R.H., E. Sinagra, J.P. McDonald, J.F. Marsh, R.K. Heenan, Small-angle neutron scattering on non-aqueous dispersions. Part 5: Magnesium carbonate dispersions in hydrocarbon media. *Colloid Polym. Sci.*, 270, 602–608, 1992.

11. (a) Mcmillen, R.L., Basic metal-containing thickened oil compositions. U.S. Patent 3,242,079, Lubrizol Corporation, March 22, 1966. (b) Eliades, T.I., One-step process for preparation of overbased calcium sulfonate greases and thickened compositions. U.S. Patent 4,597,880, Witco Corporation, July 1, 1986. (c) Olsen, W.D., J.M. Muir, T.I. Eliades, T. Steib, Sulfonate greases. U.S. Patent 5,308,514, Witco Corporation, May 3, 1994.

12. (a) Kreuz, K.L., Gasoline engine chemistry as applied to lubricant problems. *Lubrication*, 55, 53–64, 1969. (b) Lachowicz, D.R., K.L. Kreuz, Peroxynitrates. The unstable products of olefin nitration with dinitrogen tetroxide in the presence of oxygen. A new route to α-nitroketones. *J. Org. Chem.*, 32, 3885–3888, 1967.

13. Pawlak, Z., *Tribochemistry of Lubricating Oils*, Elsevier, Amsterdam, the Netherlands, 2003.

14. (a) Ingold, K.U., Inhibition of the autoxidation of organic substances in the liquid phase. *Chem. Rev.*, 61, 563–589, 1961. (b) Rizvi, S.Q.A., Lubricant additives and their functions. In *American Society of Metals Handbook*, 10th edn., Henry, S.D., ed., ASM International, Materials Park, OH. pp. 98–112, 1992. (c) Bouman, C.A., *Properties of Lubricating Oils and Engine Deposits*, MacMillan and Co., London, U.K., pp. 69–92, 1950.

15. (a) Lamb, G., C. Loane, J. Gaynor, Indiana stirring oxidation test for lubricating oils, *Ind. Eng. Chem. Anal. Ed.*, 13(5), 317–321, 1941, DOI:10.1021/i560093a011. (b) Vipper, A.B. Antioxidant properties of engine oil detergent additives. *Lubr. Sci.*, 9, 1, 1996.

16. Erukhimovich, Z.S., M.A. Zubareva, V.M. Shkol'nikov, E.K. Ivanova, Sulfonate-type corrosion inhibitors based on mixed petroleum synthetic raw materials. *Chem. Technol. Fuels Oils*, 21, 508–510, 1985.

17. Mansot, J.L., M. Hallouis, J.M. Martin, Colloidal antiwear additives—Part two: Tribological behaviour of colloidal additives in mild wear regime. *Colloids Surf. A Physicochem. Eng. Aspects*, 75, 25–31, 1993.

18. Ma, Q.G., R.J. Muir, Overbased calcium sulfonate detergent technology overview. *Lubr. Oil*, 2, 41–45, 2009.

19. Hsu, S., R. Lin, Interactions of additives and lubricating base oils. *SAE Technical Paper*, 831683, 1983, doi:10.4271/831683.

20. Seeta, R.K., Art of purifying petroleum sulphonic acids derived from the treatment of mineral oils with sulphuric acid. U.S. Patent 1,930,488, Sonneborn Sons Inc., October 17, 1933.

21. Eckard, A., J.A. Weaver, I. Riff, Sodium sulfonate blends as emulsifiers for petroleum oils. U.S. Patent 6,225,267, CK Witco Corporation, May 1, 2001.

22. (a) Brown, A.B., J.O. Knobloch, *Symposium on Composition of Petroleum Oils, Determination and Evaluation*, ASTM Special Technical Publication, Vol. 224, 213, 1958. (b) Sandvik, E.I., W.W. Gale, M.O. Denekas, Characterization of petroleum sulfonates. *SPE J.*, 17(3), 184–192, 1977.

23. Rysek, J.J., J.W. Forsberg, Aqueous compositions containing overbased materials. U.S. Patent 4,468,339, The Lubrizol Corporation, August 28, 1984.

24. Roberts, D.W., D.L. Williams, The derivation of quantitative correlations between skin sensitisation and physio-chemical parameters for alkylating agents, and their application to experimental data for sultones. *J. Theor. Biol.*, 99, 807, 1982.

25. (a) Gard-Terech, A., J.C. Palla, Comparative kinetics study of the evolution of freshwater aquatic toxicity and biodegradability of linear and branched alkylbenzene sulfonates. *Ecotoxicol. Environ. Saf.*, 12, 127–140, 1986. (b) Tarring, R.C., The development of a biologically degradable alkylbenzene sulfonate. *Int. J. Air Water Pollut.*, 9, 545, 1965. (c) Visoottiviseth, P., S. Ekaehote, B.S. Upatham, Microbial degradation of anionic detergent in natural water. *J. Sci. Soc. Thailand*, 14, 209–217, 1988.

26. (a) Robeck, G.G., J.M. Cohen, W.T. Sayers, R.L. Woodward, Degradation of ABS and other organics in unsaturated soils. *J. Water Pollut. Control Fed.*, 35, 1225–1237, 1963. (b) Kosswig, K., Surfactants, in *Ullmann's Encyclopedia of Industrial Chemistry*, Wiley-VCH, 2000. DOI: 10.1002/14356007. a25_747.

27. (a) Pujado, P.R., Linear alkylbenzene (LAB) manufacture in *Handbook of Petroleum Refining Process*, Meyers, R.A., ed., 2nd edn., McGraw-Hill, NY, 1.53–1.66, 1996. (b) Moulden, H.N., Continuous process for alkylating an aromatic hydrocarbon. U.S. Patent 3,355,508, Chevron Research, November 28, 1967. (c) Huang, S.K., Production of alkylaryl sulfonates including the step of dehydrogenating normal paraffins with improved catalyst. U.S. Patent 3,585,253, Monsanto Corporation, June 15, 1971.

28. (a) Lenack, A.L., R. Tirtiaux, Calcium sulphonate process. U.S. Patent 4,387,033, Exxon Research & Engineering Corporation, June 7, 1983. (b) Tirtiaux, R., R.M. Laurent, Overbased sulphonates. U.S. Patent 4,259,193, Exxon Research & Engineering Corporation, March 31, 1981.

29. (a) Lew, H.Y., A.E. Straus, Process for the preparation of a low viscosity alkyl toluene or alkyl xylene sulfonate. U.S. Patent 4,608,204, Chevron Research Company, August 26, 1986. (b) Marie Le Coent, J., Superalkalinized isomerized linear alkylaryl sulfonates of alkaline earth metals, useful as detergent/dispersant additives for lubricating oils, and processes for their preparation and intermediate alkylaryl hydrocarbon. U.S. Patent 6,476,282, Chevron Chemical S.A., November 5, 2002. (c) Smith, G.A., P.R. Anantaneni et al., Alkyl toluene sulfonate detergent. U.S. Patent 6,995,127, Huntsman Petrochemical Corporation, February 7, 2006. (d) Duncan, C.B., D.W. Turner et al., Preparation of alkylaromatic hydrocarbons and alkylaryl sulfonates. U.S. Patent 7,622,621, Exxonmobil Chemical Patents Inc., November 24, 2009.

30. Ashjian, H., Q.N. Le et al., Naphthalene alkylation process. U.S. Patent 5,034,563, Mobil Oil Corporation, July 23, 1991.

31. Olah, G.A., *Friedel-Crafts and Related Reactions*, Vol. II, Interscience-Wiley Publishers, New York, 1964.

32. Meriadeau, P., B.Y. Taarit, A. Thangaraj, J.L. Almeida, Zeolite based catalysts for linear alkylbenzene production—Dehydrogenation of long-chain alkanes and benzene alkylation. *Catal. Today*, 38, 243–247, 1997.

33. Alul, H.R., Control of isomer distribution of straight-chain alkylbenzenes. *Ind. Eng. Chem. Prod. Res. Dev.*, 7, 7–11, 1968.

34. Brown, H.C., K.L. Nelson Aromatic substitution – Theory and mechanism in *Chemistry of Petroleum Hydrocarbons*, Vol. III, Brooks, B.T., ed., Reinhold Publishing Corp., New York, 1955, Chapter 56.

35. (a) Brown, H.C., C.R. Smoot, Disproportionation of the alkylbenzenes under the influence of hydrogen bromide and aluminum bromide; the nature of the transition state in disproportionation reactions. *J. Am. Chem. Soc.*, 78, 2176–2181, 1956. (b) Olah, G.A., O. Farooq, S.M.F. Farnia, J.A. Olah, Friedel-Crafts chemistry. 11. Boron, aluminum, and gallium tris(trifluoromethanesulfonate) (triflate): Effective new Friedel-Crafts catalysts. *J. Am. Chem. Soc.*, 110(8), 2560–2565, 1988.

36. Matheson, K.L., T.P. Matson, Effect of carbon chain and phenyl isomer distribution on use properties of linear alkylbenzene sulfonate: A comparison of "high" and "low" 2-phenyl LAS homologs. *J. Am. Oil Chem. Soc.*, 60, 1693, 1983.

37. Alul, H.R., Control of isomer distribution of straight-chain alkylbenzenes. *Ind. Eng. Chem. Prod. Res. Dev.*, 7(1), 7–11, 1968, doi:10.1021/i360025a002.

38. Lappin, G., *Alpha Olefins Applications Handbook*, CRC Press, New York, April 1989.

39. Reuben, B., H. Wittcoff, The SHOP process: An example of industrial creativity. *J. Chem. Educ.*, 65, 605, 1988.

40. Friedel, R.A., R.B. Anderson, Composition of synthetic liquid fuels. I. Product distribution and analysis of $C_5–C_8$ paraffin isomers from cobalt catalyst. *J. Am. Chem. Soc.*, 72, 2307, 1950.

41. Rogers, M., T. Long, *Synthetic Methods in Step-Growth Polymers*, John Wiley & Sons, Hoboken, NJ, August 2003.

42. (a) Kocal, J.A., Detergent alkylation process using a fluorided silica-alumina. U.S. Patent 5,196,574, UOP, March 23, 1993. (b) Tanabe, K., W.F. Hoelderich, Acid–base catalysis with metal oxides. *Appl. Catal. A*, 181, 399, 1999. (c) Kocal, J.A., B.V. Vora, T. Imai, Production of linear alkylbenzenes. *Appl. Catal. A*, 221, 295, 2001.

43. Beşergil, B., B.M. Baysal, Determination of the composition of post dodecyl benzene by IR spectroscopy. *J. Appl. Polym. Sci.*, 40, 1871–1879, 1990.

44. Gilbert, E.E., B. Veldhuis, Sulfonation with sulfur trioxide—High boiling alkylated benzene. *Ind. Eng. Chem.*, 50, 997–1000, 1958.

45. ASTM D1159-07(2012), Standard test method for bromine numbers of petroleum distillates and commercial aliphatic olefins by electrometric titration, ASTM International, West Conshohocken, PA, 2012.

46. ASTM D5768-02(2014), Standard test method for determination of iodine value of tall oil fatty acids, ASTM International, West Conshohocken, PA, 2014.

47. (a) Roberts, D.W., Optimisation of the linear alkyl benzene sulfonation process for surfactant manufacture. *Org. Process Res. Dev.*, 7, 172–184, 2003. (b) Roberts, D.W., Sulfonation technology for anionic surfactant manufacture. *Org. Process Res. Dev.*, 2, 194–202, 1998.

48. Herman de Groot, W., *Sulfonation Technology in the Detergent Industry*, Kluwer Academic Publishers, Dordrecht, the Netherlands, 1991.

49. Morley, J.O., D.W. Roberts, S.P. Watson, Experimental and molecular modeling studies on aromatic sulfonation. *J. Chem. Soc. Perkin Trans.*, 2, 538–544, 2002.

50. (a) Bolsman, T.A., Alkylxylene sulfonate compositions. U.S. Patent 4,873,025, Shell Oil Company October 10, 1989. (b) Campbell, C.B., G. Sinquin. Alkylxylene sulfonates for enhanced oil recovery processes, U.S. Patent 7,332,460, Chevron Oronite Company LLC, February 19, 2008. (c) Campbell, C.B., G. Sinquin. Method of making a synthetic alkylaryl sulfonate, U.S. Patent 7,598,414, Chevron Oronite Company LLC, October 6, 2009.

51. (a) Shiga, M., K. Hirano, M. Matsushita, Method of preparing overbased lubricating oil additives. U.S. Patent 4,057,504, Karonite Chemicals Corporation, Ltd., November 8, 1977. (b) Bakker, N., Overbased sulfonates. U.S. Patent 4,137,184, Chevron Research Company, January 30, 1979.

52. (a) Wijngaarden, G.D., H.M. Brons, Preparation of a basic salt. U.S. Patent 4,869,837, Shell Oil Company, September 26, 1989. (b) Zon, A.V., B. Coleman, Lubricating oil composition. U.S. Patent 4,876,020, Shell Oil Company, October 24, 1989. (c) Hammond, S., M.A. Price, P. Skinner, Overbased detergent additives. U.S. Patent 6,599,867, Infineum International Ltd., July 29, 2003.

53. Bandyopadhyaya, R., R. Kumar, K.S. Gandhi, Modeling of $CaCO_3$ nanoparticle formation during overbasing of lubricating oil additives. *Langmuir*, 17, 1015–1029, 2001.

54. Roman, J.P., P. Hoornaert, D. Faure, C. Biver, F. Jacquet, J.M. Martin, Formation and structure of carbonate particles in reverse microemulsions. *J. Colloid Interface Sci.*, 144, 324–339, 1991.

55. (a) Gragson, J.T., Overbasing calcium petroleum sulfonates in lubricating oils employing monoalkylbenzene. U.S. Patent 4,165,291, Phillips Petroleum Company, August 21, 1979. (b) Derbyshire, P.E., H.M. Silva, Overbasing chemical process. U.S. Patent 4,206,062, Edwin Cooper and Company Ltd., June 3, 1978.

56. Vinci, J.N., W.R. Sweet, Mixed carboxylate overbased gels. U.S. Patent 5,401,424, The Lubrizol Corporation, March 28, 1995.

57. (a) Beşergil, B., A. Akın, S. Çelik, Determination of synthesis conditions of medium, high, and overbased alkali calcium sulfonate. *Ind. Eng. Chem. Res.*, 46, 1867–1873, 2007. (b) Le Suer, W., Anion exchange process and composition. U.S. Patent 3,496,105, Lubrizol Corporation, February 17, 1967. (c) Clippeleir, G.D., A. Vanderlinden, Process for preparing overbased calcium sulfonates. U.S. Patent 4,086,170, Labofina S.A., April 25, 1977. (d) Whittle, J., Method of preparing overbased calcium sulfonates. U.S. Patent 4,427,559, Texaco Inc., January 24, 1984. (e) Muir, R.J., Process for the preparation of overbased magnesium sulfonates. U.S. Patent 4,617,135, Witco Corporation, October 14, 1986.

58. Muir, R.J., T.I. Eliades, Overbased magnesium deposit control additive for residual fuel oils. U.S. Patent 6,197,075, Crompton Corporation, March 6, 2001.

59. Ma, Q.G., C.A. Migdal, K.A. Schlup, J.L. Diflavio, R.J. Muir, Overbased magnesium oxide dispersions. U.S. Patent 8,580,716, Chemtura Corporation, November 12, 2013.

60. Carlisle, J., D. Chan et al., *Toxicological Profile for Nonylphenol*, Office of Environmental Health Hazard Assessment, California Environmental Protection Agency, http://www.opc.ca.gov/webmaster/ftp/project_pages/MarineDebris_OEHHA_ToxProfiles/Nonylphenol%20Final.pdf.

61. (a) Edwards, T.L. A Dietary Two-generation Productive Toxicity Study of Tetrapropenyl Phenol in Rats. WIL Research Laboratories, LLC, Study No. WIL186053, 2012. (b) Committee for Risk Assessment, Opinion on setting Specific Concentration Limits (SCLs) for Phenol, dodecyl-, branched; Tetrapropenylphenol (TPP) as proposed by Chevron Oronite SAS. https://echa.europa.eu/documents/10162/c5a96284-107c-4ad4-99a2-097ce3fe3434.

62. SIAR, SIDS Initial Assessment Report for SIAM 22, April 2005. Phenol, (tetrapropenyl) derivatives. HERTG Consortium and UK, 2006.

63. Campbell, C.B., J.L. Le Coent, Process for producing alkylated hydroxyl-containing aromatic compounds. U.S. Patent 6,670,513, Chevron Oronite Company, December 30, 2003.

64. Perrin, D.D., *Ionization Constants of Inorganic Acids and Bases in Aqueous Solution*, 2nd edn., Pergamon Press, Oxford, U.K., 1982.

65. Hanneman, W.W., Process for basic sulfurized metal phenates. U.S. Patent 3,178,368, California Research Corporation, April 13, 1965.

66. (a) Moore, C.G., R.W. Saville, The reactions of amines and sulphur with olefins. Part I. The reaction of diethylamine and sulphur with cyclohexene. *J. Chem. Soc.*, issue 0, 2082–2089, 1954. DOI: 10.1039/JR9540002089. (b) Bateman, L., C.G. Moore, M. Porter, B. Saville, Chemistry in vulcanization in *The Chemistry and Physics of Rubber-Like Substances*, Bateman, L., ed., Wiley, New York, 1963, Chapter 15.

67. RATG, Environmental life cycle analysis for the lubricant oil additive industry. Phenol, (tetrapropenyl) derivatives—CASRN 74499-35-7 etc., American Chemistry Council Risk Assessment Task Group, August 2005.

68. Harrison, J.J., J. McDonald, Sulfurized metal alkyl phenate compositions having a low alkyl phenol content. U.S. Patent 8,198,225, Chevron Oronite Company LLC, June 12, 2012.

69. Brooke, D., R. Mitchell, C. Watts, *Environmental Risk Evaluation Report: Para-C12-Alkylphenols (Dodecylphenol and Tetrapropenylphenol)*, Environment Agency, May 2007. https://www.gov.uk/government/uploads/system/uploads/attachment_data/file/290856/scho0607bmvn-e-e.pdf.

70. Chang, Y., Lubricant overbased phenate detergent with improved water tolerance. U.S. Patent 4,865,754, Amoco Corporation, September 12, 1989.

71. Lindsey, A.S., H. Jeskey, The Kolbe-Schmitt reaction. *Chem. Rev.*, 57(4), 583–620, 1957.

72. Feilden, A.D., D.J. Moreton, C.B. Thomas, Alkylation process. U.S. Patent 5,734,078, BP Chemicals Ltd., March 31, 1998.

73. Hobbs, S.J., Method for the alkylation of salicylic acid. U.S. Patent 7,045,654, Crompton Corporation, May 16, 2006.

74. Campbell, C.B., Alkylation of alkyl salicylate using a long chain carbon feed. U.S. Patent 5,434,293, Chevron Chemical Company, July 18, 1995.

75. Bruice, P., *Organic Chemistry*, 5th edn., Appendix II, Prentice Hall, Upper Saddle River, NJ, March 31, 2006.

76. Lee, C.L., P.J. Dowding, A.R. Doyle, K.M. Bakker, S.S. Lam, S.E. Rogers, A.F. Routh, The structures of salicylate surfactants with long alkyl chains in non-aqueous media. *Langmuir*, 29, 14763–14771, 2013.

77. Skinner, P., A.L.P. Lenack, Phenate overbased detergents with high TBN: % surfactant ratios are useful lubricant additives. U.S. Patent 6,417,148, Infineum USA L.P., July 9, 2002.

78. Miao, J.Q., Z.M. Shi, Detergent for lubricant oil and production process thereof. U.S. Patent 9,102,895, Wuxi South Petroleum Additive Corporation, Ltd., December 13, 2012.

79. Israelachvili, J.N., D.J. Mitchell, B.W. Ninham, Theory of self-assembly of hydrocarbon amphiphiles into micelles and bilayers. *J. Chem. Soc., Faraday Trans. II*, 72, 1525–1568, 1976.

80. (a) Rolfes, A.J., S.E. Jaynes, Process for making overbased calcium sulfonate detergents using calcium oxide and a less than stoichiometric amount of water. U.S. Patent 6,268,318, The Lubrizol Corporation, July 31, 2001. (b) Kocsis, J.A., J.P. Roski et al., Lubricating oil compositions containing saligenin derivatives. U.S. Patent 6,310,009, The Lubrizol Corporation, October 30, 2001.

81. Altgelt, K.W., T.H. Gouw, *Chromatography in Petroleum Analysis*, Marcel Dekker, New York, 1979.

82. Lallement, J., G. Parc, G.D. Gaudemaris, Improved process for preparing superbasic detergent additives. U.S. Patent 4,059,536, Institut Francais Du Petrole, November 22, 1977.

83. Bertram, R.D., P.J. Dowding, P.D. Watts, Detergent. U.S. Patent 8,012,918, Infineum International Ltd., September 6, 2011.

84. van Dam, W., D.H. Broderick, R.L. Freerks, V.R. Small, W.W. Willis, The impact of detergent chemistry on TBN retention. *Tribotest*, 6, 227–240, 2000.

85. (a) Duchrow, R.M., W.H. Everhart, Turbidity measurement. *Trans. Am. Fish. Soc.*, 100, 682–690, 1971. (b) Sadar, M.J., Understanding turbidity science, Hach Company Technical Information Series—Booklet No. 11, 1996. http://www.hach.com/asset-get.download-en.jsa?code=61792.

86. Mason, M., C.E. Mendenhall, Theory of the settling of fine particles, *Proc. Natl. Acad. Sci. USA*, 9(6), 202–207, 1923.

87. Krieger, I.M., T. Dougherty, A mechanism to non-Newtonian flow in suspensions of rigid spheres. *Trans. Soc. Rheol.*, 3, 137, 1959.

88. Willenbacher, N., L. Borger, D. Urban, Tailoring PSA-dispersion rheology for high speed coating, *Adhesives & Sealants Industry*, 10(9), 26, 2003.

89. (a) Einstein, A., Eine neue Bestimmung der molekueldimensionen. *Ann. Phys.*, 19, 289–306, 1906. (b) Farris, R., Prediction of the viscosity of multimodel suspensions from unimodel viscosity data. *J. Trans. Soc. Rheol.*, 12, 281–301, 1960. (c) Chong, J.S., E.B. Christiansen, A.D. Baer, Rheology of concentrated suspensions. *J. Appl. Polym. Sci.*, 15, 2007–2021, 1971. (d) Mueller, S., E.W. Llewellin, H.M. Mader, The rheology of suspensions of solid particles. *Proc. R. Soc. A*, 466, 1201–1228, 2010.

90. (a) Chinas-Castillo, F., H.A. Spikes, The behavior of colloidal solid particles in elastohydrodynamic contacts. *Tribol. Trans.*, 43(3), 357–366, 2000. (b) Chinas-Castillo, F., H.A. Spikes, Mechanism of action of colloidal solid dispersions. *J. Tribol.*, 125(3), 552–557, 2003. (c) Kasrai, M., M.S. Fuller, G.M. Bancroft, P.R. Ryason, X-ray absorption study of the effect of calcium sulfonate on antiwear film formation generated from neutral and basic ZDDPS. Part I. Phosphorus species. *Tribol. Trans.*, 46(4), 534–542, 2003. (d) Han, N., L. Shui, W.M. Liu, Q.J. Xue, Y.S. Sun, Study of the lubrication mechanism of overbased Ca sulfonate on additives containing S or P. *Tribol. Lett.*, 14(4), 269–274, 2003. (e) Costello, M.T., Study of surface films of amorphous and crystalline overbased calcium sulfonate by XPS and AES. *Tribol. Trans.*, 49(4), 592–597, 2006. (f) Costello, M.T., R.A. Urrego, Study of surface films of the ZDDP and the MoDTC with crystalline and amorphous overbased calcium sulfonates by XPS. *Tribol. Trans.*, 50(2), 217–226, 2007.

91. Tavacoli, J.W., P.J. Dowding, D.C. Steytler, D.J. Barnes, A.F. Routh, Effect of water on overbased sulfonate engine oil additives. *Langmuir*, 24, 3807–3813, 2008.

92. Ohkawa, S., K. Seto, T. Nakashima, K. Takase, Hot tube test-analysis of lubricant effect on diesel engine scuffing. SAE Technical Paper 840262, 1984, doi:10.4271/840262.

93. REACh, http://ec.europa.eu/environment/chemicals/reach/ reach_en.htm. Accessed on April 28, 2017.

94. Sasaki, S., D. Sawada et al. Effect of EGR on direct injection gasoline engine. *JSAE Rev.*, 19, 223–228, 1998.

Section II

Film-Forming Additives

5 Organic Friction Modifiers

Dick Kenbeck and Thomas F. Bunemann

CONTENTS

5.1 INTRODUCTION

Friction modifiers (FMs) or friction reducers have been applied for several years. Originally, the application was for limited slip gear oils, automatic transmission fluids, slideway lubricants, and multipurpose tractor fluids. Such products made use of friction modification to meet requirements for smooth transition from static to dynamic condition as well as reduced noise, frictional heat, and startup torque.

Since fuel economy became an international issue, initially to reduce crude oil consumption, FMs have been introduced into automotive crankcase lubricants, as well, to improve fuel efficiency through the lubricant. In the United States, additional pressure is imposed on original equipment manufacturers (OEMs) by the corporate average fuel economy (CAFE) regulation.

Following the introduction of vehicle exhaust emission regulations in various regions around the world, emphasis on friction reduction further increased. This can be well understood if it is realized that 20%–25% of the energy generated in an engine by burning fuel is lost through friction [1]. The biggest part is lost by friction on the piston liner/piston ring interface and a smaller part by bearing and valve train friction. It is predicted that in future engines the contribution of the piston group to engine friction will increase up to 50% [2].

Reduction of fuel consumption and emissions can be achieved through [3] engine design changes and modifications, such as

Application of roller followers
Use of coatings
Surface modifications
Material selection
Fuel quality
The engine lubricant

All these aspects are looked at and applied in the automotive industry. This chapter concentrates on the engine lubricant.

The need to measure fuel savings has led to the development of American Petroleum Institute (API) test sequences such as VI and VIA in the United States. Sequence VIB will be used for International Lubricant Standards Approval Committee GF-3. In Europe, a fuel economy test has been developed by Conseil Européen de Co-ordination pour le Développements des Essais de Performance des Lubrifiants et der Combustible pour Moteurs (CEC) (test number CEC L-54-T-96) for the Association des Constructeurs Européens d'automobiles A1 and B1 specifications using the DBM 111 engine. Both tests require that the candidate lubricant shows decreased fuel consumption relative to reference oil.

5.2 FRICTION AND LUBRICATION REGIMES

Friction is defined as *the resistance a body meets while moving over another body in respect of transmitting motion.* The friction coefficient is defined as

$$\frac{F_W}{F_n} \tag{5.1}$$

where

F_W is the frictional force
F_n is the normal force or load

For a lubricated surface, the coefficient of friction is determined by the lubrication regime. In simple terms, the following three lubricant regimes can be distinguished:

1. Elasto-hydrodynamic lubrication (EHL) regime characterized by a (relatively) thick lubricant film [4]. The mating surfaces are far enough from one another to prevent metal-to-metal contact. The load on the system is completely carried by the lubricant film, and the viscosity of the lubricant determines the friction coefficient. Viscosity depends on temperature and pressure/viscosity coefficient.
2. Boundary lubrication (BL) regime characterized by a thin lubricant film [5]. Under high loads, high temperature, or with low viscosity oils, most of the lubricating film is squeezed out between the metal surfaces, and metal-to-metal contact occurs. The load is entirely carried by the metal asperities. A thin layer of absorbed or otherwise deposited molecules is necessary to prevent the two surfaces and their asperities from plowing into one another.
3. Mixed lubrication (ML) regime characterized by a lubricant film of intermediate thickness [6]. The two metal surfaces have come closer compared to hydrodynamic lubrication, and metal-to-metal contact occasionally occurs. The load is carried by both the lubricant and the asperities.

These regimes are related to the friction coefficient f by a lubricant parameter defined as

$$\frac{su}{F} \quad \text{or} \quad \frac{su}{p} \tag{5.2}$$

where

s is the system speed
u is the lubricant dynamic viscosity
F is the load (F_n)
p is the contact pressure

The so-called Stribeck curve gives the relationship between f and these lubricant parameters. The shape of the Stribeck curve and the transitions from BL to ML and ML to EHL

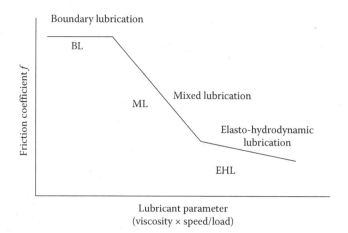

FIGURE 5.1 Stribeck curve at high contact pressure.

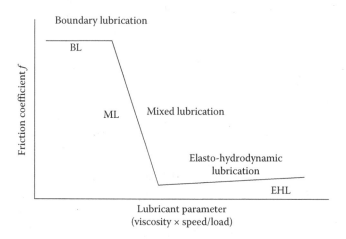

FIGURE 5.2 Stribeck curve at low contact pressure.

depend on a number of parameters such as material roughness (microgeometry), contact pressure, and lubricant viscosity. High contact pressure such as that present at point contacts leads to a different Stribeck curve as at line contact (lower contact pressure) (see Figures 5.1 and 5.2).

5.2.1 FRICTION REDUCTION THROUGH THE LUBRICANT

Engine friction originates from several components, that operate at different conditions of load, speed, and temperature. Hence, these components may experience various combinations of EHL, ML, and BL during engine operation. For each of these regimes, a number of factors govern engine friction.

Basically, two options to reduce friction and improve fuel efficiency come forward [7,8].

1. Use of low-viscosity engine oils (SAE 0W/5W-20/30) when fluid lubrication (EHL) is the governing factor [9–11]. Fluid lubrication is especially prevalent in the bearings. The gradual reduction of engine oil viscosity over the years has already brought significant fuel savings (see Figure 5.3).

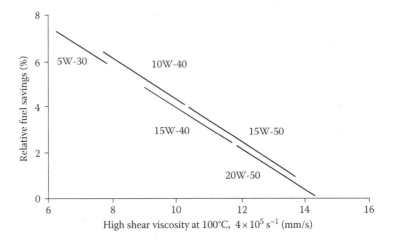

FIGURE 5.3 Relationship between SAE viscosity grades and fuel savings based on fleet car trials.

In the preceding case, oil selection is crucial. In terms of frictional characteristics, one must emphasize low kinematic viscosity, high viscosity index, low "high-temperature, high-shear" (HTHS) viscosity, and a low-pressure/viscosity coefficient [12,13]. However, it has to be realized that other base fluid properties, such as volatility and thermal/oxidation stability, must not be ignored.

2. Addition of friction-reducing agents when BL and ML are the governing factors [14]. These are prevalent in the valve train and the piston group.

In the preceding case, additive system design is the crucial factor. One must emphasize selecting proper FMs and controlling additive–additive and additive–base fluid interactions.

To assess possible fuel economy improvements in the engine sequences prescribed, an overview of the lubrication regimes existing in various test engines is provided. The data those used for current and previous ILSAC specifications are given in Table 5.1.

In Sequence VIA, which is prescribed for ILSAC GF-2, EHL is dominating, leading to a substantial effect of engine oil viscosity on fuel economy. Effects of FMs will be small due to the low presence of BL and ML conditions, which is due to the application of roller followers. Hence, the Sequence VIA test engine is often indicated as "a very expensive viscometer."

TABLE 5.2
Lubrication Regimes in the DB M111E Engine

	Frictional Loss (%)	Main Lubrication Regime
Valve train	25	Boundary lubrication
Piston assembly	40	Mixed lubrication
Bearings	35	Elasto-hydrodynamic lubrication

This characterization of the Sequence VIA engine will be addressed by Sequence VIB, to be used for ILSAC GF-3, for which an engine, a bucket tappet sliding valve train will be used, leading to an increase of the BL and ML regimes [3].

In Europe, the M111 engine is used for the CEC L-54-T-96 fuel economy test, which is prescribed in the ACEA A1 and B1 engine oil specifications. Similar data as aforementioned are not available to the authors, but data given in a Shell paper [15] indicate the frictional loss occurring in this engine, which can be translated to lubrication regimes (Table 5.2).

On the basis of the relatively high amount of frictional loss in the valve train and piston assembly, the M111 engine should be sensitive to FMs. This is due to the use of four valves per cylinder to improve combustion efficiency and so to obtain more power from a given amount of fuel.

However, compared to other engine designs, the frictional loss in the M111E valve train will be higher. Provided that the higher-power output obtained from the four-valve assembly is significantly higher than that is lost by higher valve train friction, this approach is favorable with regard to fuel economy.

5.3 FRICTION MODIFIERS VERSUS ANTIWEAR/EXTREME-PRESSURE ADDITIVES

A point of debate is often about the difference between FMs and antiwear/extreme-pressure (AW/EP) additives, especially when it is about FMs active at BL conditions. For a good understanding, this should be clarified; therefore, this section deals with the principal difference between these two additive categories [16].

TABLE 5.1
Lubrication Regimes in API Sequences VI and VIA

	API Sequence VI (%)	API Sequence VIA (%)
Boundary lubrication	37	24
Mixed lubrication	15	4
Elasto-hydrodynamic lubrication	48	72

TABLE 5.3
Lubrication Modes versus Friction Coefficient

Lubrication Mode	Friction Coefficient	Comparison
Nonlubricated surface	0.5–7	Dragging an irregular rock over rocky ground
AW/EP films	0.12–0.18	Dragging a flat stone over a flat rock
Friction-modified films	0.06–0.08	Ice skating
EHL	0.001–0.01	Hydroplaning

TABLE 5.4
FM Type and Mode of Action

Mode of Action/Type of FM	Products
Formation of reacted layers	Saturated fatty acids, phosphoric and thiophosphoric acids, sulfur-containing fatty acids
Formation of absorbed layers	Long-chain carboxylic acids, esters, ethers, amines, amides, imides
Formation of polymers	Partial complex esters, methacrylates, unsaturated fatty acids, sulfurized olefins
Mechanical types	Organic polymers

AW/EP additives are types of compounds that provide good BL. Such materials have the capacity to build strong BL layers under severe load conditions. Hence, AW/EP additives protect closely approaching metal surfaces from asperities damaging the opposite surface. On the contrary, most AW additives have little friction-modifying properties.

The crucial differences between AW/EP and FM films are their mechanical properties. AW/EP films are semiplastic deposits that are difficult to shear off. Thus, under shearing conditions, their coefficient of friction is generally moderate to high. Conversely, FM lubricant films are built up of orderly and closely packed arrays of multimolecular layers, loosely adhering to one another and with the polar head anchored on the metal surface. The outer layers of the film can be easily sheared off, allowing for a low coefficient of friction.

The difference between the two types of films and other lubrication modes is best illustrated by the data presented in Table 5.3.

5.4 CHEMISTRY OF ORGANIC FRICTION MODIFIERS

Organic FMs are generally long, slim molecules with a straight hydrocarbon chain consisting of at least 10 carbon atoms and a polar group at one end. The polar group is one of the governing factors in the effectiveness of the molecule as an FM. Chemically, organic FMs can be found within the following categories [16]:

Carboxylic acids or their derivatives, for example, stearic acid and partial esters
Amides, imides, amines, and their derivatives, for example, oleylamide
Phosphoric or phosphonic acid derivatives
Organic polymers, for example, methacrylates

Another classification can be given by mode of action and FM type (Table 5.4).

Owing to the different mode of actions, the mechanism of friction reduction varies for each category.

The next section further deals with details about their mode of action, and another section deals with the current chemistry used as well as specific products.

5.4.1 FRICTION MODIFIER MECHANISMS

5.4.1.1 Formation of Reacted Layers

Similar to AW additives, protective layers are formed by chemical reaction of the additive with the metal surface. However, the principal difference is that the reaction has to occur under the relatively mild conditions (temperature and load) of the ML regime. These conditions require a fairly high level of chemical activity as reflected by the phosphorus and sulfur chemistry applied.

An exception to this is stearic acid. Theoretically, the friction-reducing effect of stearic acid should decrease with increasing temperature due to desorption of the molecule from the metal surface. However, stearic acid experimentally shows a remarkable drop of friction with increasing temperature, which can only be explained by the formation of chemically reacted protective layers.

5.4.1.2 Formation of Absorbed Layers

The formation of absorbed layers occurs due to the polar nature of the molecules. FMs dissolved in oil are attracted to metal surfaces by strong absorption forces, which can be as high as 13 kcal/mol. The polar head is anchored to the metal surface, and the hydrocarbon tail is left solubilized in the oil, perpendicular to the metal surface (see Figure 5.4). Next the following steps occur:

1. Other FM molecules have their polar heads attracted to one another by hydrogen bonding and Debye orientation forces, resulting in dimer clusters. Forces are 15 kcal/mol.
2. Van der Waals forces cause the molecules to align themselves such that they form multimolecular clusters that are parallel to one another.
3. The orienting field of the absorbed layer induces further clusters to position themselves with their methyl groups stacking onto the methyl groups of the tails of the absorbed monolayer [17,18].

As a result, all molecules line up, straight, perpendicular to the metal surface, leading to a multilayer matrix of FM molecules (see Figure 5.5).

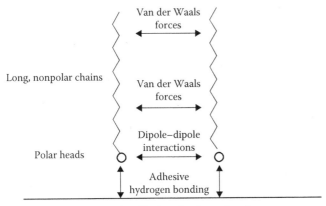

FIGURE 5.4 Organic friction modifiers—formation of adsorbed layers.

```
Metal surface
//////////////////////
H H H H H  = Polar head
T T T T T T  = Hydrocarbon tail
T T T T T T
H H H H H
H H H H H
T T T T T T
OIL OIL OIL
OIL OIL OIL
OIL OIL OIL
T T T T T T
H H H H H
H H H H H
T T T T T T
T T T T T T
H H H H H
//////////////////////
Metal surface
```

FIGURE 5.5 Multilayer matrix of friction modifier molecules.

The FM layers are difficult to compress but very easy to shear at the hydrocarbon tail interfaces, explaining the friction-reducing properties of FMs. Owing to the strong orienting forces, mentioned earlier, sheared-off layers are quite easily rebuilt to their original state.

The thickness and effectiveness of the absorbed FM films depend on several parameters, four of which are explained here.

1. *Polar group.* Polarity itself is not necessarily sufficient for adsorption; the polar group must also have hydrogen-bonding capability. Molecules with highly polar functional groups that are not capable of forming hydrogen bonds, such as nitroparaffins, do not adsorb. Hence, these do not function as friction-reducing additives. However, polarity plays a major role among the various lateral surface interactions through strong electrostatic dipole–dipole interactions. These may be either repulsive or attractive, depending on the orientation of the adsorbed dipoles with respect to the surface [19].

2. *Chain length.* Longer chains increase thickness of the absorbed film, and the interactions between the hydrocarbon chains increase as well [18].
3. *Molecular configuration.* Slim molecules allow for closer packing as well as increased interaction between adjacent chains, leading to stronger films. Therefore, straight chains may be preferred.
4. *Temperature.* Temperature influences FM film thickness and tenacity. Adsorption of friction-reducing compounds to the metal surface does occur at relatively low temperatures. AW additives form protective layers by chemical reactions for which higher temperatures are needed.

If the temperature is too high, enough energy might be provided to desorb the friction-reducing molecules from the metal surface.

5.4.1.3 Formation of *In Situ* Polymers

The formation of low-friction-type polymer films can be considered a special case. Instead of the usual solid films, fluid films are formed under influence of contact temperature (flash temperature) and load. Another difference is that the polymers are developed at the interface between metal asperities without reacting with the metal surface.

The requirements of such polymers are

1. Polymers must have relatively low reactivity. Polymerization must be generated by frictional energy.
2. The polymers formed must be mechanically and thermally stable and should not be soluble in the lubricant.
3. The polymers must develop a strong bond to the metal surface either by absorption or by chemical bonding.
4. The formation and regeneration of films must be fast to prevent competitive adsorption by other additives.

Examples of polymer-forming FMs are

Partial complex esters, for example, a sebacic acid/ ethylene glycol partial ester methacrylates

Oleic acid (olein), which may be explained through thermal polymerization (formation of dimers and higher oligomers)

5.5 CHEMISTRY OF OTHER FRICTION MODIFIERS

Within this group, the following categories can be distinguished by chemical type:

1. Metallo-organic compounds
2. Oil-insoluble materials

Classification by type appears in Table 5.5.

TABLE 5.5
Classification of Other FMs

Types of FMs	Products
Metallo-organic compounds	Molybdenum and copper compounds
Mechanical types	Molybdenum disulfide, graphite, teflon (PTFE)

5.5.1 METALLO-ORGANIC COMPOUNDS

Molybdenum dithiophosphate, molybdenum dithiocarbamate, and molybdenum dithiolate as well as copper-oleate, copper-salicylate, and copper-dialkyldithiophosphate are examples of friction-reducing metallo-organic compounds.

The mechanisms of operation of this class of products are not fully understood, but the following hypotheses are presented:

Diffusion of molybdenum into the asperities
Formation of polymer-type films
In situ formation of molybdenum disulfide (most accepted hypothesis)
Selective transfer of metal (copper) leading to the formation of thin, easy-to-shear metal films

5.5.2 MECHANICAL TYPES

In this group, the classical types such as graphite and molybdenum disulfide as well as some more recent FMs such as teflon (polytetrafluoroethylene, PTFE), polyamides, fluoridized graphite, and borates can be found. The friction-reducing mechanisms can be explained by

The stratified structure and formation of easy-to-shear layers
The formation of elastic or plastic layers on the metal surface

5.6 FACTORS INFLUENCING FRICTION-REDUCTION PROPERTIES

This section lists the main factors that impact friction-reducing properties.

1. *Competing additives.* Other polar additives with affinity to metal surfaces such as AW/EP and anticorrosion additives as well as detergents and dispersants may compete with FMs. This emphasizes that lubricant formulations have to be balanced carefully to achieve optimal performance.
2. *Contaminants.* Short-chain acids, which are formed by oxidative degradation of the lubricants, may compete at the metal surface, resulting in a loss of friction-modifying properties.
3. *Metallurgy.* The type of steel alloy used will affect the adsorption of FMs.
4. *Concentration.* Increase of FM concentration results in an increase of friction reduction up to a point above which improvements are marginal. Generally, the friction-reducing effect is most (cost-)effective at concentrations of 0.25%–1% for organic FMs and 0.05%–0.07% for molybdenum dithiocarbamates.

5.7 FRICTION MODIFIERS: CURRENT PRACTICE

The most frequently used organic FMs include

1. Long-chain fatty amides, specifically oleylamide (Figure 5.6). This is a reaction product of olein (main component oleic acid, a straight-chain unsaturated C18 carboxylic acid) and ammonia (NH_3).
2. Partial esters, specifically glycerol mono-oleate (GMO) (Figure 5.6). GMO is a reaction product of glycerin (natural alcohol with three hydroxyl groups) and olein (as mentioned previously). Investigations have shown that the alpha version (terminal hydroxyl groups esterified) is the active component rather than the beta one (middle hydroxyl group esterified). Special production techniques are required to manufacture high-alpha-containing products.

The mode of action of both product groups is based on the formation of adsorbed layers that can easily be sheared off, leading to reduced friction. It is expected that further research will result in new and improved types to cope with more severe requirements with regard to friction retention over time.

Within the group of metallo-organic compounds, molybdenum dithiocarbamate [$Mo(dtc)_2$] seems almost exclusively to be used to obtain friction reduction (Figure 5.7). Research [7,20] has shown that the friction-reducing activity of $Mo(dtc)_2$

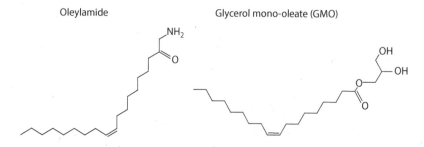

FIGURE 5.6 Organic friction modifiers—structural drawings.

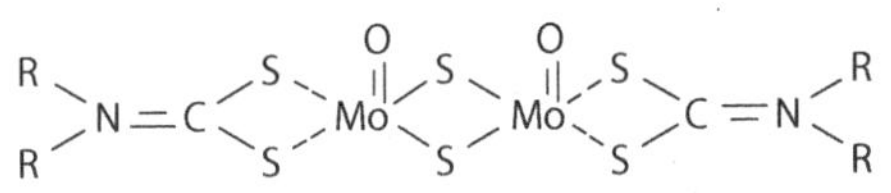

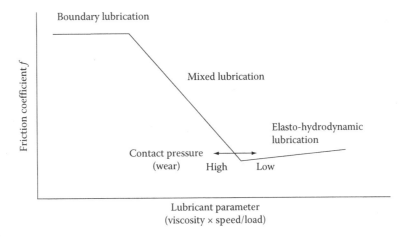

FIGURE 5.7 Molybdenum dithiocarbamate—structural drawing.

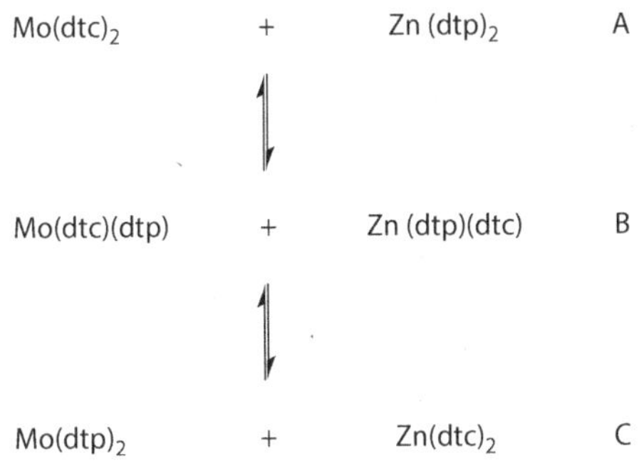

Mo(dtc)$_2$ +   Zn (dtp)$_2$ A

Mo(dtc)(dtp)   +   Zn (dtp)(dtc)   B

Mo(dtp)$_2$ +   Zn(dtc)$_2$ C

FIGURE 5.8 Molybdenum dithiocarbamate—exchange of functional groups.

is based on the exchange of functional groups with zinc dialkyldithiophosphates [Zn(dtp)$_2$] (Figure 5.8).

It was found that oxidation affects these exchange reactions significantly and that the most effective friction reduction is achieved at the later stages of oxidation when the concentrations of the single-exchange product [Mo(dtc)(dtp)] and the double-exchange product [Mo(dtp)$_2$] are high. When both products are nearly consumed by oxidation, friction reduction ceases.

5.8 FRICTION MODIFIER PERFORMANCE

Literature suggests that FMs act both in the BL and ML regimes [7,16,21]. Their mode of action should depend on FM chemistry and prevailing engine conditions. It is further suggested that organic FMs are most active in the mixed regime, whereas metallic types are predominantly active in the BL regime. Recent investigations by the authors, carried out with a pin-on-ring tribometer, showed that it is likely that organic FMs act predominantly in the BL regime as well.

Tests were carried out with CEC reference oil RL 179/2, which is applied in the CEC L-54-T-96 fuel economy test. RL 179/2 is a formulated 5W/30 engine oil that does not contain any FM and that has a proven fuel economy benefit CEC round-robin tests.

Frictional behavior was investigated by establishing stabilized Stribeck curves. By determining these, both boundary and mixed friction can be investigated. Stabilized Stribeck curves are obtained by measuring the coefficient of friction over a speed range from 0.0025 to 2 m/s at appropriate steps. A number of runs are carried out until two consecutive runs give a good match. Usually, after four runs, the curve has stabilized, meaning that process roughness has stabilized to a large extent.

Performance criteria in considering the results are frictional level in the BL and ML regimes in combination with specimen wear. The reason for looking at wear is that this parameter corresponds with contact pressure, which in turn influences the ML/EHL transition.

The relationship between wear and contact pressure is given by the expression

$$\frac{F_n}{A} = p \tag{5.3}$$

where

F_n is the normal force (load)
A is the wear scar
p is the contact pressure

Consequently, a high-wear scar leads to a lower contact pressure, and a lower contact pressure does shift the ML/EHL transition in the Stribeck curve to the left (see Figure 5.9).

5.8.1 Stribeck Curve Determinations

Stabilized Stribeck curves have been determined with a pin-on-ring tribometer at which the ring was a 100Cr6 stainless steel ring with a 730 mm diameter. These rings are high-quality materials used in standard bearings and therefore

FIGURE 5.9 Influence of wear/contact pressure on ML/EHL transition.

easily available. The pin used was a cylinder from the same material with an 8 mm diameter, also used in bearings. To get proper line contact, the cylinders have been provided with *flexible* ends to allow full alignment with the ring.

Ring roughness, R_a, was 0.15 μm, although the cylinder was very smooth. Hence, the roughness of the ring determined the shape of the Stribeck curve, specifically the BL/ML transition. The load (normal force F_n = 100 N) was chosen such that heat development in the contact zone was negligible, so that the viscosity was constant. Hence, it was possible to determine the Stribeck curve only as a function of speed. The temperature chosen was 40°C.

The following graph (Figure 5.10) shows comparative data for RL 179/2 as well as this oil with addition of 0.5% GMO and addition of 0.5% of organic FMs A and B. (A and B are products with both free and esterified hydroxyl groups.)

All the FMs studied here show a significant reduction of the friction coefficient in the BL regime. Organic FMs A and B show a reduction in the mixed regime as well. On first sight this looks favorable.

The next graph (Figure 5.11) shows the wear, taken at similar sliding distances.

The wear of the oil containing A and B is twice as high as those of the reference oil and that oil with addition of 0.5% GMO. Consequently, the contact pressure p is twice as low as those of the others and is what makes the ML/EHL transition shift to the left.

Thus it seems that organic FMs are predominantly active in the BL regime and that the shifts observed in the mixed regime are likely to be caused by other phenomena that must not be ignored.

5.8.2 FRICTION AS A FUNCTION OF TEMPERATURE

Another aspect of FM performance is friction as a function of temperature. Temperature plays an important role with regard to adsorption/desorption phenomena as for the formation of adsorbed layers as well as regarding those of reacted layers.

The graph in Figure 5.12 shows the frictional behavior of some organic FMs as a function of temperature, using the pin-on-ring tribometer as before with the same specimens and configuration. Again, CEC reference oil RL 179/2 was used, and the speed chosen (0.03 m/s) assured operation well within the BL regime.

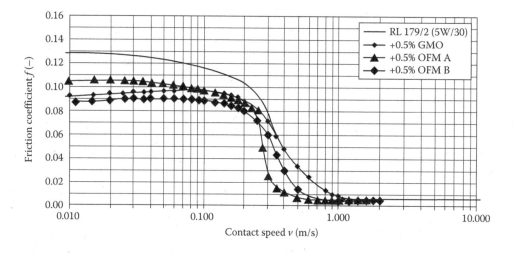

FIGURE 5.10 Stribeck curves of CEC RL 179/2 plus organic friction modifiers.

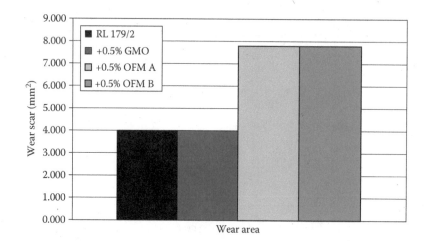

FIGURE 5.11 Wear scars of CEC RL 179/2 plus organic friction modifiers.

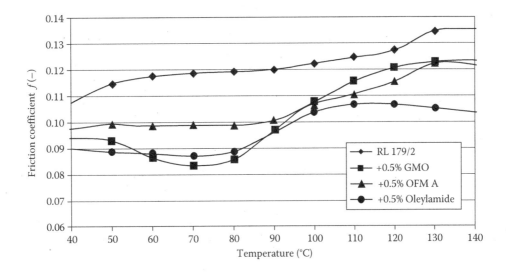

FIGURE 5.12 Friction coefficient versus temperature—CEC RL 179/2 plus organic friction modifiers.

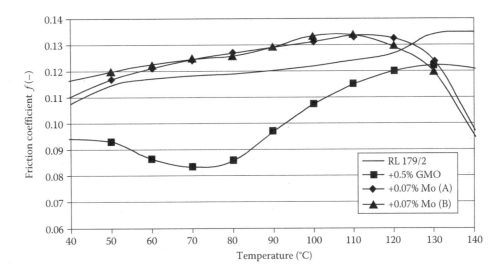

FIGURE 5.13 Friction coefficient versus temperature—CEC RL 179/2 plus molybdenum dithiocarbamates.

All the organic FMs studied show a significant friction reduction over the temperature range tested. GMO and oleylamide perform best, and the optimum adsorption seems to be obtained at 70°C. At higher temperatures, desorption may start to occur as well as some kind of competition with other surface-active additives, leading to a higher coefficient of friction. Oleylamide, however, continues to show high friction-reducing properties at elevated temperatures.

Figure 5.13 shows a comparison between organic FM GMO and metallic-type FM molybdenum dithiocarbamate. Two sources of the latter were used at a concentration equivalent to 0.07% molybdenum.

GMO and the molybdenum dithiocarbamates show a marked performance difference. Although GMO is active over a wide temperature range, the molybdenum dithiocarbamates start to reduce friction at temperatures of 120°C and above only. This has to be considered as an induction period that can be explained by the necessary exchange of ligands between

molybdenum dithiocarbamate and zinc dialkyldithiophosphate (see Section 5.7). Once molybdenum dithiocarbamate has "lighted off," a fast drop of friction is noticed. At the end of the test cycle 140°C, the system has not stabilized and the friction coefficient might decrease further.

The difference in friction-modifying characteristics between organic FMs and molybdenum dithiocarbamate suggests that it might be beneficial to use a combination of these materials.

5.9 CONSEQUENCES OF NEW ENGINE OIL SPECIFICATIONS AND OUTLOOK

Although initial fuel economy requirements were focused on fresh oil only, new engine oil specifications will address fuel economy longevity as well. A good example is Sequence VIB, which has been developed for the ILSAC GF-3 specification.

Sequence VIB includes aging stages of 16 and 80 h to determine fuel economy as well as fuel economy longevity.

These aging stages are equivalent to 4000–6000 mi of mileage accumulation required before the EPA metro/high-highway fuel economy test. That test is used in determining CAFE.

To obtain engine oil formations that are optimized with regard to fuel economy longevity, high requirements are demanded for base oil selection and additive system design [3,7,22]. These requirements are

To minimize the increase of viscosity thereby maintaining a low electrohydrodynamic friction coefficient
To maintain low boundary/mixed friction

A minimum increase of viscosity can be obtained by base fluid selection (in terms of volatility, oxidation stability, and antioxidant susceptibility) and selection of antioxidants and their treat level. The market is already anticipating requirements by increasing the production capacity of groups II (HIVI) and III (VHVI) base fluids and by increased interest in groups IV (PAOs) and V (a.o. esters) base fluids.

To achieve low friction under BL and ML conditions, the use of effective friction-reducing additives is needed. To maintain low boundary and mixed friction over time, it is necessary to prevent consumption of these additives by processes such as oxidation and thermal breakdown. Therefore, selecting suitable antioxidant systems for molybdenum compounds and organic FMs and developing organic FMs with highest thermal/oxidative stability will be key for high fuel economy longevity and a successful application in engine oil formulations.

Further studies on the mechanisms of FM action, for example, through molecular modeling techniques, could also speed up the development of optimized additives and additive systems. Apart from frictional properties, other important tribological parameters such as wear rate and surface-metal geometry should be investigated as well. In most papers studied, this seems to be ignored, although all three parameters should be considered in relation to one another.

5.10 BENCH TESTS TO INVESTIGATE FRICTION-REDUCING COMPOUNDS

Several bench tests can be thought of to investigate the frictional properties of base fluids and formulated products. In recent literature [8,23,24], the following test equipment has been used:

1. The high-frequency reciprocating rig (HFRR) to measure boundary friction. Although originally developed to measure diesel fuel lubricity, the equipment can be successfully applied to measure lubricant properties as well.

Frequency	10–200 Hz
Stroke length	20–2000 μm
Load	0–1000 g
Ambient temperature	200°C

A 6 mm diameter ball is the upper specimen and a 3 mm thick smooth disk with a 10 mm diameter is the lower specimen. HFRR specifications are the test conditions that the authors applied to screen FMs; these include a 40 Hz frequency, a stroke of 1000 μm, and a 400 g load.

2. A mini traction machine (MTM) to measure mixed and (E)HD friction, for example, by the determination of Stribeck curves. The MTM rig is capable of measuring at either constant or varying slide/roll ratios if required.

Speed range	Up to 5 m/s
Slide/roll ratio	0%–200% (full rolling to full sliding)
Load	0–75 N
Ambient temperature	150°C

Standard specimens are a 19.05 mm diameter ball as upper specimen and a 50 mm diameter disk as lower specimen. Both are manufactured from AISI 52100 bearing steel. The standard disk is smooth, which allows measurement of mixed-film and full-film friction. Alternatively, rough disks are available for measurements in the BL regime. MTM specifications are the test conditions that the authors applied to test FMs; these comprise a speed range of 0.001–4 m/s, a 30 N load, and a 200% slide/roll ratio.

3. An optical rig provided with a disk coated with a spacer layer to measure EHD film thickness. Such a rig enables film-thickness measurements down to <5 nm with a precision between 1 and 2 nm.

Some other literature refers to the low-velocity friction apparatus (LVFA). Alternative reciprocating rigs may be suitable as well.

REFERENCES

1. Wilk, M.A., W.D. Abraham, B.R. Dohner. An investigation into the effect of zinc dithiophoshpate on ASTM sequence VIA fuel economy. SAE Paper 961,914, 1996, Society of Automotive Engineers, West Conshohocken, PA.
2. Houben, M. Friction analysis of modern gasoline engines and new test methods to determine lubricant effects. *10th International Colloquium*, Esslingen, Germany, 1996.
3. Korcek, S. Fuel efficiency of engine oils—Current issues. *53rd Annual STLE Meeting*, Detroit, MI, 1998.
4. LaFountain, A., G.J. Johnston, H.A. Spikes. Elastohydrodynamic friction behavior of polyalphaolefin blends. *Tribology Series* 34:465–475, 1998.
5. Spikes, H.A. Boundary lubrication and boundary films. *Tribology Series* 25:331–346, 1993.
6. Spikes, H.A. Mixed lubrication—An overview. *Lubrication Science* 9(3):221–253, 1997.
7. Korcek, S. et al. Retention of fuel efficiency of engine oils. *11th International Colloquium*, Esslingen, Germany, 1998.

8. Sorab, J., S. Korcek, C. Bovington. Friction reduction in lubricated components through engine oil formulation. SAE Paper 982,640, 1998, Society of Automotive Engineers, West Conshohocken, PA.

9. Goodwin, M.C., M.L. Haviland. Fuel economy improvements in EPA and road tests with evine oil and rear axle lubricant viscosity reduction. SAE Paper 780,596, 1978, Society of Automotive Engineers, West Conshohocken, PA.

10. Waddey, W.E. et al. Improved fuel economy via engine oils. SAE Paper 780,599, 1978, Society of Automotive Engineers, West Conshohocken, PA.

11. Clevenger, J.E., D.C. Carlson, W.M. Keiser. The effects of engine oil viscosity and composition on fuel efficiency. SAE Paper 841,389, 1984, Society of Automotive Engineers, West Conshohocken, PA.

12. Dobson, G.R., W.C. Pike. Predicting viscosity related performance of engine oils. *Erd-1 und Kohle-Erdgas* 36(5):218–224, 1982.

13. Battersby, J., J.E. Hillier. The prediction of lubricant-related fuel economy characteristics of gasoline engines by laboratory bench tests. *Proceedings of International Colloquium*, Technische Akademie Esslingen, Ostfildern, Germany, 1986.

14. Griffiths, D.W., D.J. Smith. The importance of friction modifiers in the formulation of fuel efficient engine oils. SAE Paper 852,112, 1985.

15. Taylor, R.I. Engine friction lubricant sensitivities: A comparison of modern diesel and gasoline engines. *11th International Colloquim*, Esslingen, Germany, 1998.

16. Crawford, J., A. Psaila. Miscellaneous additives and vegetable oils. In R.M. Mortier and S.T. Orszulik, eds. *Chemistry and Technology of Lubricants, Miscellaneous Additives*. Blackie Academic and Professional, an imprint of Chapman & Hall, London, U.K., 1992, pp. 160–165.

17. Allen, C.M., E. Drauglis. Boundary lubrication: Monolayer or multilayer. *Wear* 14:363–384, 1969.

18. Akhmatov, A.S. Molecular physics of boundary lubrication. Gos. Izd. Frz.-Mat. Lit., Moscow, Russia, 1969, p. 297.

19. Beltzer, M., S. Jahanmir. Effect of additive molecular structure on friction. *Lubrication Science* 1(1):3–26, 1998.

20. Arai, K. et al. Lubricant technology to enhance the durability of low friction performance of gasoline engine oils. SAE Paper 952,533, 1995, Society of Automotive Engineers, West Conshohocken, PA.

21. Christakudis, D. Friction modifiers and their testing, additives for lubricants. *Kontakt Stud* 433:134–162, 1994.

22. Society of Automotive Engineers. Effects of aging on fuel efficient engine oils. *Automotive Engineering*, 102(2):99, February 1996.

23. Moore, A.J. Fuel efficiency screening tests of automotive engine oils. SAE Paper 932,689, 1993, Society of Automotive Engineers, West Conshohocken, PA.

24. Bovington, C., H.A. Spikes. Prediction of the influence of lubricant formulations on fuel economy from laboratory bench tests. *Proceedings of International Tribology Conference*, Yokahama, Japan, 1995.

6 Selection and Application of Solid Lubricants as Friction Modifiers

Gino Mariani

CONTENTS

6.1 INTRODUCTION

Solid lubricants are considered to be any solid material that reduces friction and mechanical interactions between surfaces in relative motion against the action of a load. Solid lubricants offer alternatives to the lubricant formulator for situations where traditional liquid additives fall short on performance. An example is a high-temperature lubrication condition in which oxidation and decomposition of the liquid lubricant will certainly occur, resulting in lubrication failure. Another example is for situations that generate high loads and contact stresses on bearing points of mating surfaces, producing a squeeze-out of the liquid lubricant and a resulting lubricant starvation (see Figure 6.1).

Solid lubricants, used as a dry film or as an additive in a liquid, provide enhanced lubrication for many different types of applications. Typical hot-temperature applications include oven chain lubrication and metal deformation processes such as hot forging. Solid lubricants are also helpful for ambient-temperature applications such as drawing and stamping of sheet metal or bar stock. Solid lubricants are effectively used in antiseize compounds and threading compounds, which provide a sealing function and a friction reduction effect for threaded pipe assembly [1]. Applications involving low sliding speeds and high contact loads, such as for gear lubrication, also benefit from solid lubricants. The solid lubricant effectively provides the required wear protection and load-bearing performance necessary from gear oil, especially capable when used with lower-viscosity base oils.

Solid lubricants also assist applications where the sliding surfaces are of a *rough* texture or surface topography. Under this circumstance, the solid lubricant is more capable than liquid lubricants for covering the surface asperity of the mating surfaces. A typical application is a reciprocating motion that requires lubrication to minimize wear. Another application for solid lubricants is for cases where chemically active lubricant additives have not been found for a particular surface, such as polymers or ceramics. In this case, a solid lubricant would function to provide the necessary protection to the mating surfaces, which would normally occur due to the reaction of a liquid component with the surface [2].

Graphite and molybdenum disulfide (MoS_2) are the predominant materials used as solid lubricants. These pigments are effective load-bearing lubricant additives due to their lamellar structure. Because of the solid and crystalline nature of these pigments, graphite and MoS_2 exhibit favorable tolerance to high-temperature and oxidizing atmosphere environments, whereas liquid lubricants typically will not survive. This characteristic makes graphite and molybdenum disulfide lubricants necessary for processes involving extreme temperatures or extreme contact pressures.

Other compounds that are useful solid lubricants include boron nitride, polytetrafluoroethylene (PTFE), talc, calcium fluoride, cerium fluoride, and tungsten disulfide. Any one of these compounds may be more suitable than graphite or MoS_2 for specific applications. Boron nitride and PTFE are discussed along with graphite and molybdenum disulfide in this chapter.

What are the basic requirements for an effective solid lubricant? Five properties must be met in a favorable way [3]:

1. *Yield strength*. This refers to the force required to break through the lubricant or deform its film. There should be high yield strength to forces applied

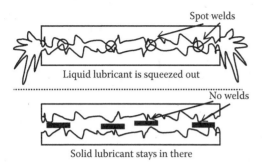

FIGURE 6.1 Contact stresses on bearing points of mating surfaces cause a squeeze-out.

perpendicular to the lubricant. This will provide the required boundary lubrication and protection to loads between the mating surfaces. Low yield strength of the film should be present in the direction of sliding to provide reduced coefficient of friction. This dependency on directional application of forces is considered an anisotropic property.

2. *Adhesion to substrate.* The lubricant must be formulated in a manner that maintains the lubricant film on the substrate for a sufficient period necessary for the lubrication requirements. The force of adhesion should exceed that of the sheer forces applied to the film. Any premature adhesion failure will result in a nonprotective condition between the two sliding surfaces that require lubrication.

3. *Cohesion.* Individual particles in the film of solid lubricant should be capable of building a layer thick enough to protect the high asperities of the surface and to provide a "reservoir" of lubricant for replenishment during consumption of the solid film (see Figures 6.2 and 6.3).

4. *Orientation.* The particles used must be oriented in a manner that parallels the flow of the stress forces and provide the maximum opportunity for a reduction in the coefficient of friction. For this to occur, it is necessary for the dimensions of the particles to be greatest in the direction of low shear.

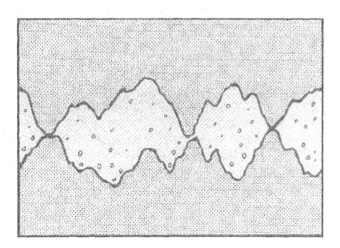

FIGURE 6.2 Surface asperities.

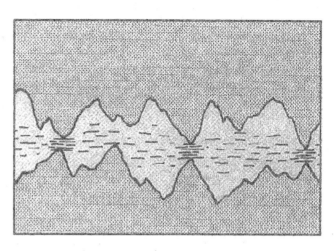

FIGURE 6.3 Burnished lubricant.

5. *Plastic flow.* The lubricant should not undergo plastic deformation when loads are applied directly perpendicular to the direction of motion. The solid should be able to withstand the intimate contact between the mating surfaces so that a continuous film of lubrication is maintained.

This chapter attempts to guide the formulator toward making successful choices in solid lubricants. It briefly summarizes the physical and chemical properties of the solid lubricant and discusses the merits of each type of major lubricant as well as the recommended application. The information will assist in understanding the chemistry of the lubricant and its general mechanism of lubrication.

6.2 SOLID LUBRICANT PROPERTIES

6.2.1 GRAPHITE

Graphite is most effective for applications involving high-temperature and high load-carrying situations. These capabilities make graphite the solid lubricant of choice for forging processes. Solid lubricants such as MoS_2 will oxidize too rapidly to be of any value at the typical hot forging temperature range of 760°C–1200°C, although MoS_2 has a greater lubrication capability than graphite.

Why is graphite such a good lubricant? The answer lies in the platelet, lamellar structure of the graphite crystallite (see Figure 6.4). Graphite is structurally composed of planes of polycyclic carbon atoms that are hexagonal in orientation. Short bond lengths between each carbon atom within the plane are the result of strong covalent bonds (see Figure 6.4).

Weaker van der Waals forces hold together a number of planes to create the lattice structure. The *d*-spacing bond distance of carbon atoms between planes is longer and, therefore, weaker than the bond distance between carbon atoms within the planes. As a force is applied perpendicular to the crystallite, a strong resistance is applied against the force. This high yield strength provides the load-carrying capacity for the lubricant. Concurrent with the force applied perpendicular to

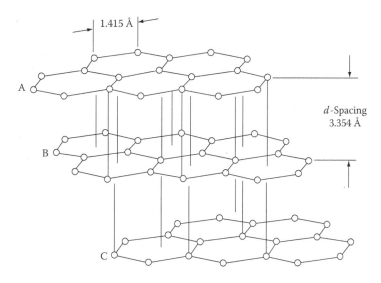

FIGURE 6.4   Structure of graphite.

the substrate is a sliding force applied parallel to the direction of sliding. The weak bond between the planes allows for easy shearing of the planes in the direction of the force. This creates a cleaving of the planes and results in friction reduction. The lamellar motion of graphite cleavage can be illustrated by the concept of a hand applying a force on a deck of playing cards as shown in Figure 6.5. Forces applied perpendicular to the deck are resisted by the stack's thickness and yield strength. Yet, a far easier force is required to rupture the stack when the force is applied parallel along the face of the deck, resulting in the shearing of the cards.

The effects of the lamellar structure of graphite can be observed when sliding conditions are applied onto metal surfaces. Coefficient of friction data can be generated by various bench test methods for measuring the lubricity of sliding conditions. In comparison to unlubricated or oil-lubricated metal surfaces, graphite provides excellent lubricity [4]. This is summarized in Table 6.1.

TABLE 6.1
Coefficients of Friction Provided by Graphite Films

Test Method	Graphite Film	Unlubricated Metal	Mineral Oil on Metal
Three-ball slider	0.09–0.12	0.16–0.18	0.15–0.17
Bowden–Leben machine	0.07–0.10	0.40	0.17–0.22

6.2.1.1   Sources of Graphite

There are many types and sources of graphite. These sources influence the properties of the graphite, which affects the performance of the end product that uses graphite. Graphite is characterized by two main groupings: natural and synthetic.

Natural graphite is derived from mining operations worldwide. The ore is processed to recover the usable graphite. Varying quality of the graphite will be evident from the ore quality and the postmining processing of the ore. High-purity natural graphite will normally be highly lubricating and resistant to oxidation. This is due to the high degree of crystal structure and graphitization usually associated with naturally derived graphite.

Natural graphite of lesser quality is also available. A lower total carbon content and a lower degree of graphitization characterize the lesser quality. The end product is graphite that is more amorphous in nature, with a higher content of ash components, which are mostly oxides of silicon and iron. Lubrication functionality decreases as crystallinity and graphitization decrease. Lubrication functionality also decreases as total ash content of the graphite increases.

Commercially available natural graphite is provided in a variety of grades. The suitability of the grades depends on the intended application and economic constraints. Table 6.2 characterizes examples of commercially available natural graphite.

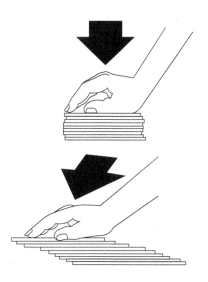

FIGURE 6.5   Representation of lamellar lubrication.

TABLE 6.2
Natural Graphite

	Amorphous	Crystalline Flake 1	Crystalline Flake 2
% Carbon	–85.0	90–95	96–98
% Sulfur	–0.30	0.15–0.20	0.10–0.70
% SiO$_2$	6.0–7.0	0.20–0.30	0.05–0.2
% Ash	10–15	7–10	2.0–3.0
Mesh	–325	–325	–325

Selecting the type of natural graphite to use is based on the degree of lubrication required for the application, the particle size of the graphite necessary for the application, and the economic constraint. For situations where the lubrication demand is severe, a high-carbon crystalline flake or crystalline vein graphite is desired. The high degree of crystallinity and graphitization provides superior lubrication. A more economical alternative is the lower-carbon-content flake graphite. For most situations, these types of graphite perform adequately in lubricating conditions that do not require the purity and lubricity of higher-quality crystalline graphite. For occasions where only minor lubricity is needed and perhaps a more thermally insulating coating is required, amorphous graphite would be chosen. Amorphous graphite is also the least expensive of the commercially available natural graphite grades. Combining amorphous and crystalline graphite can also be done to modify the amount of lubrication to suit the requirements of the application.

Synthetic graphite is an alternative source for lubricating graphite. Synthetic graphite is characterized as primary or secondary grade (see Table 6.3). Primary grade is derived synthetically from production within an electric furnace, utilizing calcined petroleum coke as well as very high temperatures and pressures to produce the graphite. The result is usually a product of high purity and can approach the quality of natural graphite flake in terms of percent graphitization and lubrication capability.

Secondary synthetic graphite is derived from primary graphite that has been used for the fabrication of electrodes. This type of graphite is usually less lubricating than natural or primary grades of graphite because of its lesser degree of crystallinity and graphitization and the presence of binding agents and surface oxides that do not contribute to lubrication. Secondary synthetic graphite is perfectly capable of

TABLE 6.3
Synthetic Graphite

	Typical Values	
	Primary	Secondary
% Carbon	99.9	99.9
% Sulfur	Trace	0.01
% SiO$_2$	0.02	0.05
% Ash	0.1	0.1

lubricating effectively for many applications that can afford a lesser degree of lubricity. The chief benefit in using secondary synthetic graphite is the cost, with the secondary graphite costing significantly less than primary-grade synthetic graphite or high-purity natural graphite.

6.2.1.2 Lubrication

Appropriate-quality graphite is able to meet the five criteria for an effective solid lubricant. Graphite possesses the necessary yield strength for successful lubrication. It is able to adhere sufficiently to metal surfaces due to its affinity to metal and its packing within and above the microstructure of the surface. Graphite has a burnishing capacity desirable for lubrication mechanisms that require a "memory" effect. Proper orientation of graphite particles is achieved by the natural tendency for the graphite crystal to orient itself parallel to the substrate and in the direction of lowest shear. The anisotropic characteristic of graphite lends itself well to its lubricating capability and friction reduction property. The planar orientation of the graphite particles on the substrate takes advantage of the anisotropic property. Proper orientation allows the lamellar functionality of graphite where easy shear is achieved along the crystal plane when sliding forces are put along the length of the particles. The high yield strength in graphite is maintained in the direction perpendicular to the direction of shear force, providing for the load-carrying capability.

Graphite is best suited for lubrication in a regular atmosphere. Water vapor is a necessary component for graphite lubrication. The role that adsorbed water vapor plays in the lubricating properties of graphite has been studied [5]. It is theorized that water vapor helps to reduce the surface energy of the graphite crystallite. The adsorption of a water monolayer onto the planar surface of the graphite likely reduces the bonding energy between the hexagonal planes of the graphite to a level that is lower than the adhesion energy between a substrate and the graphite crystal. This allows for lamellar displacement of the graphite crystals when shear forces are applied to the graphite film. The result is a reduction of friction and corresponding lubrication. Because water vapor is a requirement for lubrication, graphite is usually not effective as a lubricant in a vacuum atmosphere.

The lubricating ability of graphite as a function of temperature is very good. Graphite is able to withstand continuous temperatures of up to 450°C in an oxidizing atmosphere and still provide effective lubrication. The oxidation stability of graphite depends on the quality of the graphite, the particle size, and the presence of any contaminants that might accelerate the oxidation. Graphite will also function at much higher temperatures on an intermittent basis. Peak oxidation temperatures are typically near 675°C. For these instances, modifying the composition of the graphite mixture may be necessary as a way to control its rate of oxidation.

The thermal conductivity of graphite is generally low. For example, primary-grade synthetic graphite has a conductivity of ~1.3 W/mK at 40°C. Amorphous graphite is even less conducting and is sometimes considered for providing some degree of thermal insulation for specific applications.

6.2.2 Molybdenum Disulfide

Molybdenum disulfide is the second significant solid lubricant widely used in industry. It has been used since the early nineteenth century for lubrication applications. MoS_2, also known as molybdenite, is a mined material found in thin veins within granite. Lubricating-grade MoS_2 is highly refined by various methods to achieve a purity suitable for lubricants [6]. This purity usually exceeds 98%. MoS_2 is commercially available in a variety of particle size ranges. Table 6.4 lists basic properties for molybdenum disulfide. The low friction of MoS_2 is an intrinsic property related to its crystal structure, whereas graphite requires the adsorption of water to behave as an effective lubricant. Molybdenum disulfide achieves its lubricating ability with a mechanism similar to graphite. Just like graphite, MoS_2 has a hexagonal crystal lattice structure.

Sandwiches of planar hexagonal Mo atoms are interspersed between two layers of sulfur atoms. Similar to graphite, the bond strength between the hexagonal planes between the sulfur atoms is a weak van der Waal–type bond when compared to the strong covalent bond between molybdenum and sulfur atoms within the hexagonal crystal. Orientation of the MoS_2 crystallites is important if effective friction reduction is to be achieved. MoS_2 has anisotropic properties that are comparable to graphite. When a force is applied parallel along the hexagonal planes, the weak bond strengths between the planes allow for easy shearing of the crystal, resulting in a lamellar mechanism of lubrication. At the same time, the crystal structure and strong interplanar bond forces of MoS_2 allow for high load carrying against forces applied perpendicular to the plane of the crystal. This is necessary for the prevention of metal-on-metal contact for high-load applications such as gearbox lubrication.

MoS_2 scores well in the other criteria for an effective solid lubricant. It forms a strong cohesive film that is smoother than the surface of the substrate on which it is bonded. MoS_2 film has sufficiently high adhesion to most metal substrates, which it successfully burnishes onto the wearing surfaces, thus minimizing metal wear and prolonging friction reduction. This characteristic is an exception, however, with titanium and aluminum substrates due to the presence of an oxide layer on the metal surface, which tends to reduce the tenacity of the MoS_2 film.

The lubrication performance of MoS_2 often exceeds that of graphite. It is most effective for high load-carrying lubrication when temperatures are <400°C. Another advantage of MoS_2 is that it lubricates in dry, vacuum-type environments, whereas graphite does not. This is due to the intrinsic lubrication property of MoS_2. On the contrary, the lubricating ability of MoS_2 deteriorates in the presence of moisture because of oxidation of MoS_2 to MoO_3. The temperature limitation of MoS_2 is due to similar decomposition issues of the material as that experienced with moisture. As MoS_2 continues to oxidize, MoO_3 content increases, which induces abrasive behavior and increases coefficient of friction for the surfaces to be lubricated.

The effectiveness of MoS_2 improves as contact forces increase on the lubricated surface. Burnished surfaces exhibit coefficient of friction reduction as a function of increasing contact forces [7]. In contrast, graphite does not necessarily exhibit this behavior. The frictional property of MoS_2 systems has been reported to be generally better than graphite in many instances, up to the service temperature limitations for the lubricant.

The particle size and film thickness of MoS_2 will affect lubrication. Generally, the particle size should be matched to the surface roughness of the substrate and the type of lubrication process considered. Too large a particle distribution may result in excessive wear and film reduction as mechanical abrasion is experienced. Too fine a particle size may result in accelerated oxidation in normal atmospheres as the high surface area of the particles promotes the rate of oxidation.

6.2.3 Boron Nitride

Boron nitride is a ceramic lubricant with interesting and unique properties. Its use as a solid lubricant is typically for niche applications when performance expectations render graphite or molybdenum disulfide unacceptable. The most interesting lubricant feature of boron nitride is its high-temperature resistance. Boron nitride has a service temperature of 1200°C in an oxidizing atmosphere, which makes it desirable for applications that require lubrication at very high service temperatures. Graphite and molybdenum disulfide cannot approach such higher service temperatures and still remain intact. Boron nitride also has a high thermal conductivity property, making it an excellent choice for lubricant applications that require rapid heat removal.

A reaction process generates boron nitride. Boric oxide and urea are reacted at temperatures from 800°C to 2000°C to create the ceramic material. Two chemical structures are available: cubic and hexagonal boron nitride. As one might expect, the hexagonal boron nitride is the lubricating version. Cubic boron nitride is a very hard substance used as an abrasive and cutting tool component. Cubic boron nitride does not have any lubrication value. The hexagonal version of boron nitride is analogous to graphite and molybdenum disulfide. The structure consists of hexagonal rings of boron and

TABLE 6.4
Characteristics of Hexagonal Molybdenum Disulfide

Property	Value
Bulk hardness	1.0–1.5 Ʊ
Coefficient of friction	0.10–0.15
Color	Blue-gray to black
Electrical conductivity	Semiconductor
Luster	Metallic
Melting point	>1800°C
Molecular weight	160.08
Service temperature	Up to 700°F
Specific gravity	4.80–5.0
Thermal conductivity	0.13 W/mK at 40°C

TABLE 6.5
Hexagonal Boron Nitride

Property	Value
Coefficient of friction	0.2–0.7
Color	White
Crystal structure	Hexagonal
Density	2.2–2.3 g/cm^3
Dielectric constant	4.0–4.2
Dielectric strength	~35 kV/mm
Molecular weight	24.83
Service temperature	1200°C (oxidizing atmosphere)
Thermal conductivity	~55 W/mK
Size (grades)	1–10 μm

nitrogen, which are connected to each other, forming a stack of planar hexagonal rings. As with graphite, boron nitride exhibits a platelet structure.

The bond strength within the rings is strong. The planes are stacked and held together by weaker bond forces. Similar to graphite and molybdenum disulfide, this allows for easy shearing of the planes when a force is applied parallel to the plane. The ease of shear provides the expected friction reduction and resulting lubrication. Concurrently, the high bond strength between boron and nitrogen within the hexagonal rings provides the high load-carrying capability that is necessary to maintain metal–metal separation of the substrates. Similar to MoS_2, boron nitride has intrinsic lubrication properties. Boron nitride effectively lubricates in a dry as well as a wet atmosphere. It is very resistant to oxidation, more so than either graphite or MoS_2, and maintains its lubricating properties up to its service temperature limit.

Commercial grades are available in a variety of purities and particle sizes. These varieties influence the degree of lubrication provided by boron nitride since particle size affects the degree of adhesion to substrate, burnishing ability, and particle orientation within a substrate. Impurities such as boric oxide content need to be considered with respect to the lubrication capability of boron nitride powder since this will influence the ability of the powder to reduce the coefficient of friction for an application. The variation in grades will also influence the thermal conductivity properties and ease of suspension in a liquid carrier. Table 6.5 summarizes typical properties for hexagonal boron nitride.

6.2.4 POLYTETRAFLUOROETHYLENE

PTFE has been in use as a lubricant since the early 1940s. Structurally, the polymer is a repeating chain of substituted ethylene with four fluorine atoms on each ethylene unit:

$$-\left(CF_2-CF_2\right)_n^-$$ (6.1)

Contrary to the other lubricants discussed, PTFE does not have a layered lattice structure. The lubrication properties are at least partially the result of its high softening point. As frictional heat begins to increase from sliding contact, the polymer maintains its durability and is able to lubricate.

Various grades are produced and applied to specific applications as a result of the properties imparted by the grade. For example, molecular weight and particle size are two characteristics that can alter the performance of the polymer as a lubricant.

The critical characteristic of PTFE—the one it is widely known for [8]—is the outstandingly low coefficient of friction imparted by the molecule. PTFE has one of the smallest coefficients of static and dynamic friction than any other solid lubricant. Values as low as 0.04 for sliding conditions have been reported for various combinations of PTFE films on substrates [9]. The low-friction property is attributed to the smooth molecular profile of the polymer chains, which orient in a manner that facilitates easy sliding and slip. It is postulated that the PTFE polymer results in rod-shaped macromolecules, which can slip along each other, similar to lamellar structures. Its chemical inertness makes it useful in cryogenic to moderate operating temperatures and in a variety of atmospheres and environments. Operating temperatures are limited to ~260°C due to the decomposition of the polymer.

One consideration in using PTFE is the cold weld property of the material. This could eliminate its use for some applications where extreme pressure is encountered. Such pressure may result in the destruction of the polymer particle and in the lubrication failure, as the PTFE congeals and fails to remain intact on the rubbing surface.

PTFE finds many uses in bonded-film lubrication at ambient temperature. These applications include fasteners, threading compounds, and chain lubrication and engine oil treatments. PTFE is widely used as an additive in lubricating greases and oils for both industrial and consumer applications (see Table 6.6 for basic properties).

Although difficult to accomplish due to the low surface energy of PTFE, colloidal dispersions of PTFE in oil or water can be produced. This is useful for applications requiring the stable suspension of PTFE particles in the lubricating medium such as for crankcase oil or hydraulic oil. The nature and feedstock of the PTFE influence the ability to create a stable, unflocculated dispersion, which is necessary for effective lubrication.

TABLE 6.6
Typical Physical Properties of PTFE

Property	Value
Coefficient of friction (ASTM D1894)	0.04–0.1
Dielectric constant	2.1–2.4
Hardness	50–60 Shore D
Melting point	327°C
Service temperature	Up to 260°C
Specific gravity	2.15–2.20

6.3 PREPARATION FOR LUBRICANT APPLICATION

For a lubricant to be effective, the solid has to be applied in a manner that provides an effective interface between the mating substrates that require wear protection or lubrication. Dry powder lubrication can be used, but it is limited in its scope of application. In other words, the dry powder can be *sprinkled* onto the load-bearing substrate. By a combination of the rubbing action from sliding and the natural adhesion properties of the solid lubricant, some measure of attachment to the substrate will occur by burnishing to provide lubricating protection [10]. MoS$_2$ seems to function particularly well from this manner of application, as it has an effective burnishing capability.

The use of free powder has limitations. The films tend to have a short duration of service since adhesion is usually insufficient to provide any longevity for a continuous application. The use of dry powder also makes it difficult in many circumstances to accurately apply the lubricant to the place intended, with the possible exception of tumbling metal billets for achieving a coating over phosphated substrates.

This can be overcome by the use of bonded films. Bonded films will provide a strong adhesion to the substrate requiring protection. It also allows for a more controlled rate of film wear, which depends on the properties of the bonding agent and the film thickness of the bonded film. Bonded films can be achieved by a number of ways, all by the use of secondary additives that promote a durable and longer-lasting film. The intended application will dictate the appropriate type of bonding agent. For applications of continual service, resin and polymer bonding agents are typically used. These include phenolic resins, acrylics, celluloses, epoxies, polyimides, and silicones. Some of the binders such as epoxies are curable at room temperatures. Others such as the phenolic resins require elevated-temperature curing. Service temperature may be the limiting consideration for the chosen bonding agent.

To overcome service temperature limitations, alternative type bonding agents are also widely used. Most typical are inorganic salts such as alkali silicates, borates, and phosphates. These types of salts overcome temperature limitations of organic bonding agents, transferring the burden of temperature consideration to the solid lubricant. Conversely, the use of inorganic salts as bonding agents typically does not provide for a coating life that is as durable as an organic bonded coating. This usually limits the application to those requiring constant replenishment of the lubricant.

To facilitate the application of the solid lubricant, dispersion in a liquid is most commonly used. The liquid can be a solvent, oil, synthetic oil, or water. Suspension within a liquid allows for the easy and precise application of the solid lubricant to the intended areas that require protection. Compared to dry powder application, film control is easily achieved through spray, dip, or flow methods onto the substrate. Environmental cleanliness is also improved since the solid particles are entrapped within a liquid matrix, preventing the airborne dispersion of the particles. For applications in which the solid lubricant is a secondary additive in a liquid, proper suspension is critical for achieving effective lubrication.

A consideration for liquid suspensions is that the shelf life of the lubricant is limited. Because the particles require suspension within a liquid carrier, eventual sedimentation of the solid lubricant will occur. This necessitates proper mixing procedures for the handling of the suspension to provide for consistent lubricant performance within the stated shelf life of the material. Adjustment to formulations with respect to dispersion and viscosity controls will influence the time it takes for the suspension to destabilize. The quality of the suspension will also determine how easily the settled pigment is redispersed with mild agitation (see Figure 6.6).

To create the suspension, the solid lubricant particles require treatment of the particle surface to make it amenable to suspension within the carrier liquid. This is similar to paint, where the colorant is chemically treated to provide the required dispersion characteristics and form what is considered a colloidal suspension (see Figure 6.7). This treatment is necessary to maximize the available particles for lubrication

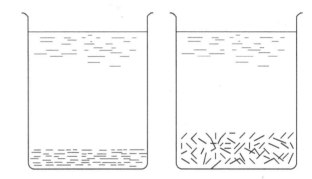

FIGURE 6.6 Particle sedimentation.

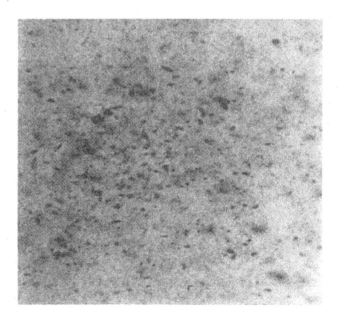

FIGURE 6.7 Colloidal dispersion.

and provide the degree of dispersion stability required for the job. Without such treatment, particle agglomeration and rapid sedimentation will occur. This would negatively influence the application of the lubricant onto the substrate in a manner that creates an inferior and ineffective film. Wetting agents and suspending agents such as polymeric salts, starches, and polyacrylics are used to treat the surface of the solid lubricant to render it capable of suspension within the liquid carrier.

When creating the dispersion, the particle size distribution of the solid lubricant has to be considered. Small, submicron particles are easier to suspend and retain physical stability than large, coarse particles. To this end, milling action on the solid lubricant is usually necessary to alter the size distribution to the desired range of sizes (see Figure 6.8).

Fine-sized particles are not necessarily the best distribution for a particular lubricating application (see Figure 6.9). Some consideration is required for the most beneficial particle size to match up with the surface roughness and nature of the application. This consideration could run contrary to what is the best particle size for dispersion stability. Therefore, some degree of compromise may be necessary to achieve a balance of dispersion stability and lubrication performance.

Some type of substrate preparation for the load-bearing surface may be required to facilitate the application of the solid lubricant. This is usually necessary for metal deformation processes so that the film thickness, film uniformity, and durability of the applied lubricant on a billet will be robust enough to lubricate. Typical treatments of the surface include

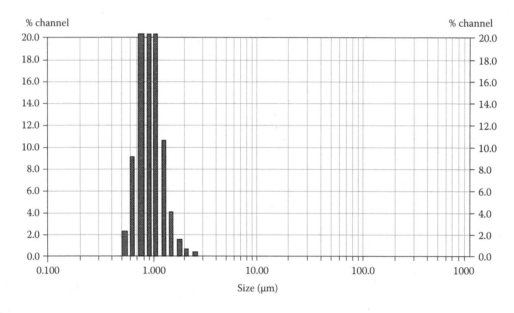

FIGURE 6.8 Particle size distribution of colloidal graphite suspension.

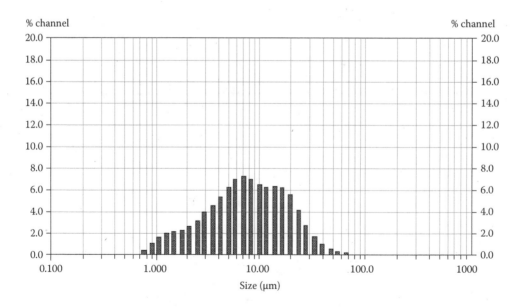

FIGURE 6.9 Coarse graphite particle size distribution.

phosphating, peening, and shot blasting, which are especially useful for powder tumbling applications. With water-based dispersions, heating the substrate to some elevated temperature is often necessary to activate the bonding agents. Substrate heating serves a dual purpose: it facilitates the evaporation of the water carrier, and it also initiates the physical/chemical bonding of the film onto the substrate.

6.4 APPLICATIONS

Two major lubrication applications are considered here: metal wear protection lubrication and lubrication for plastic deformation of metal. The former concerns applications such as constant sliding or reciprocating motion, for example, gear, chain, or journal lubrication. The latter concerns applications where metal is under plastic flow, such as metal-forming or metal-cutting applications.

6.4.1 WEAR PROTECTION AND GENERAL LUBRICATION

Wear protection and general lubrication applications are meant to include processes requiring hydrodynamic lubrication, elastohydrodynamic lubrication, and boundary lubrication. Examples of such applications include chain lubrication, gear lubrication, and engine oil treatments. In essence, any application where repetitive sliding or rolling contact occurs between two surfaces can be considered under the umbrella of wear protection lubrication. The intention is for the lubricant to reduce the coefficient of friction and protect against wear (see Figure 6.10). The benefits include savings in power consumption and service life of the component and efficiency gains due to the increased uptime resulting from proper lubrication.

Solid lubricants are useful and required for applications and conditions when conventional liquid lubricants are inadequate. These conditions include the following:

1. High operating temperatures that eliminate or reduce the functionality of the liquid lubricant
2. Contact pressure of sufficient magnitude that breaches the integrity of the liquid lubricant
3. Performance enhancement that extends the capability of the conventional liquid lubricant

4. Performance enhancement that extends the service life of the conventional liquid lubricant
5. Applications that undergo a "start/stop" routine
6. Applications that require low sliding speed but heavy bearing load
7. Applications that require "fool-proofing" for potential catastrophic lubrication failures that result from lubricant starvation

For successful incorporation of a solid lubricant as a secondary additive into liquid lubricants, a well-formulated colloidal dispersion is required. As an example, consider a case study where gear oil performance is enhanced above that of a conventional liquid lubricant by the use of colloidal solids. The addition of 1% colloidal molybdenum disulfide to AGMA No. 7 and AGMA No. 8 gear oils reduced the break-in times and steady-state operating temperatures of low-viscosity synthetic oils as compared to nonfortified gear oils [11]. Table 6.7 summarizes a comparison of the performance of various blended gear oils to the measured output criteria as tested on a worm gear dynamometer.

Another example concerns the potential lubrication improvement from solid lubricants for friction-modified engine oils. Because of the burnishing property that solid lubricants such as colloidal graphite or colloidal MoS_2 would have on metal surfaces, friction reduction in engine and axle components might be expected. Along with friction reduction, there should be a corresponding increase in fuel efficiency for motor vehicles. Various studies seem to support that conclusion. One report claims that in fleet trials conducted according to EPA 55/45 fuel economy testing with reference motor oils fortified with either MoS_2 or graphite, both in a colloidal dispersion, the fuel economy was improved by 4.5% [12]. In another fuel economy study using a fleet of taxicabs, the use of 2% colloidal graphite or colloidal MoS_2 in low-viscosity-formulated engine oils and rear axle lubricants improved the fuel economy by 2.5% [13].

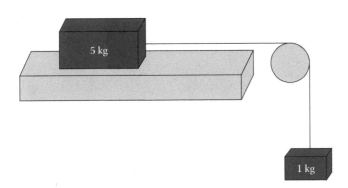

FIGURE 6.10 Lubrication of sliding surfaces—friction reduction.

TABLE 6.7
Worm Gear Dynamometer Tests

Description	Performance Parameters Output Torque = 113 N m		
	Mean Input Torque (N m)	Percent Efficiency	Mean Oil Sump Temperature (°C)
AGMA #8 gear oil	6.02	62.6	92.1
AGMA #8 gear oil + 1% colloidal MoS_2 dispersion	5.92	63.6	95.5
AGMA #7 gear oil	6.05	62.3	93.6
AGMA #7 gear oil + 1% colloidal MoS_2 dispersion	5.89	64.0	93.4
Synthetic PAG #2 oil	6.09	61.8	108.8
Synthetic PAG #2 oil + 1% MoS_2 dispersion	5.79	65.1	88.4

TABLE 6.8
Coefficient of Friction for Bonded Films

	Coefficient of Friction[a]
MoS_2	0.23
Graphite	0.15
PTFE	0.07

Source: Watari, K. et al., High-performance lubricant oil,
U.S. Patent 5,985,802, November 16, 1999.

[a] Evaluated at room temperature, ASTM D4918.

The friction-reducing influence of colloidal graphite in oil is illustrated in one study by a dynamometer evaluation conducted on a 2.3 L engine [14]. The study indicates that graphite properly dispersed in an appropriate liquid lubricant will considerably reduce friction with the subsequent benefit of fuel economy savings.

Solid lubricants are also applied as bonded films for certain applications. For example, applications requiring a permanent or semipermanent lubricating film would require a bonded film. Bonded coatings are commonly formulated with MoS_2 or PTFE. One example would be for self-lubricating composites that require high-temperature stability, such as for what may be needed for engine piston ring protection [15]. Other examples that benefit from a bonded lubricant include fasteners, chains, and reciprocating mechanisms that require a persistent lubricating film. For these applications, PTFE stands out due to its low coefficient of friction. This is summarized in Table 6.8 by comparative coefficient of friction data for PTFE, graphite, and MoS_2, which are bonded onto cold-rolled steel substrates.

In assessing the lubrication potential for dispersed solid lubricants, some type of bench testing is utilized to characterize the apparent lubrication performance of the material. The most typical lubrication tests are Shell 4-Ball Wear method, Shell 4-Ball EP method, Falex Pin–Vee method, Plint Reciprocating method, Incline Plane method, and FZG Gear Lubrication method. In many cases, custom lubrication tests are developed for the specific application to be considered. When conducting bench testing for lubricant performance, correlation is best achieved when the mode of contact and conditions of the application are closely replicated by the bench test. The configuration of the contact points for the application is matched with a similar mode of contact for the bench test.

For an illustration of laboratory lubrication assessments, see Table 6.9 [16] to compare the empirical performance of the four solid lubricants dispersed in an oil carrier. The lubricants were tested according to two common methods of lubrication evaluation.

In this example, the dispersion of MoS_2 and PTFE provides effective load bearing, wear resistance, and coefficient of friction reduction when evaluated by a point-to-point contact (4-ball) and line-to-point contact (Falex Pin–Vee). Interpretation of any bench test result must be done carefully to ensure the validity of extrapolating the test performance to the actual application.

What criteria should be considered for an application when selecting the preferred or optimal solid lubricant? First, consider the service temperature for the application. This dictates which solid lubricant can be used. For example, MoS_2 generally has a higher load-carrying capability than graphite. Yet, at service temperatures above 400°C, MoS_2 degrades and loses its lubricating capacity. MoS_2 is, therefore, eliminated from consideration if the service temperature is above 400°C.

The second consideration is environment. Atmospheric restrictions will eliminate the use of certain solid lubricants. For example, a vacuum environment will eliminate the use of graphite. As mentioned previously, graphite requires adsorption of water molecules to its surface to function as an effective lubricant. MoS_2, on the contrary, and PTFE and boron nitride have intrinsic lubrication properties and do not require water molecules on their surface to provide friction reduction value.

The third criterion is the nature of the lubricant: either a liquid fortified with solid lubricant additives or a bonded solid lubricant film. Some pigments are easier to disperse in liquid than others. For example, graphite and MoS_2 are comparatively easier to disperse in liquids than PTFE and boron nitride. This is mostly due to particle size-reducing capability, surface energy, and surface chemistry of the solid lubricant.

TABLE 6.9
Bench Lubrication Test Results

	Four-Ball Lubrication Test				Falex Lubrication Test		
	Wear ASTM D4172		Extreme Pressure ASTM D2783		Wear ASTM D2670	EP ASTM D3233	Coefficient of Friction
	20 kg mm	40 kg mm	Weld (kg)	Load Wear Index (kg)	Teeth	lb to Failure	Calculated
Base oil	0.678	1.060	126	17.20	Fail	875	0.159
With 1% colloidal graphite	0.695	0.855	160	18.7	78	1000	0.132
With 1% colloidal MoS_2	0.680	0.805	200	24.3	8	4375	0.077
With 1% colloidal PTFE	0.50	0.84	200	29.04	10	4500+	0.0568
With 1% colloidal BN	0.37	0.72	126	19.9	Fail	500	0.1602

Source: Acheson Colloids test data.

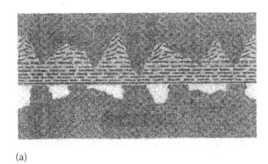 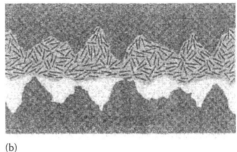

(a)   (b)

FIGURE 6.11 Orientation of solid lubricant particles in the direction of motion.

The particle size of the pigment has an influence on lubrication performance. The size of the particulate and the size distribution of the particles should be optimized for the application (see Figure 6.11). For example, larger particles tend to give better performance for applications that are slow in speed or oscillating in nature.

Large particles also tend to give better performance on substrates where the surface roughness is relatively coarse.

A finer particle size tends to provide superior results for applications with constant motion and high speeds. Finer particles tend to function better where the surface roughness is relatively fine. Although not always predictable, the influence of particle size needs to be considered not only for dispersion requirements but also for the intended use application.

The fourth criterion involves cost-effectiveness of the lubricant. When the application conditions are met with two or more solid lubricants, cost will dictate the choice. Generally, graphite will be the least expensive. High-purity graphite is more expensive than lower-purity natural graphite or secondary synthetic graphite, which are more expensive than low-quality graphite. Molybdenum disulfide will be next, followed by PTFE and boron nitride as the more expensive solid lubricants. Cost-effectiveness for any of the solid lubricants will be influenced by the quality of the lubricant and formulation that utilizes the lubricant. The effectiveness of the final formulation may prove that a costlier solid lubricant is more cost-effective in use. Table 6.10 attempts to rate the effectiveness of the solid lubricants for various criteria of application.

6.4.2 LUBRICATION FOR PLASTIC DEFORMATION OF METALS

Lubrication requirements for assisting metal deformation operations such as forging and metal drawing are far more demanding than those for wear lubrication. The metal movement process creates very fast metal flow and rapid new surface generation. This creates a demand for a lubricant to flow with the metal, remain adhered to the surface, maintain sufficient film cohesion to "meter" out the lubricant with the advancing metal, and interact rapidly with the newly formed metal surface. Metal-forming operations are inherently high-load and high-stress processes, which put a significant demand on protective lubrication.

TABLE 6.10
Solid Lubricant Selection Comparison and Rating

Criteria	Graphite	MoS$_2$	PTFE	Boron Nitride
Normal atmosphere	1	1	1	1
Vacuum atmosphere	3	1	1	1
Ambient temperature	1	1	1	1
Continuous service temperature to 260°C in air	1	1	1	1
Continuous service temperature to 400°C in air	1	1	1	1
Continuous service temperature to 450°C in air	2	3	N/A	1
Burnishing capability	1	1	3	2
Hydrolytic stability	1	2	1	1
Thermal conductivity	2	3	3	1
Load-carrying lubrication	2	1	1	2
Friction reduction	2	2	1	3
Dispersability	1	1	3	2
Color	Black	Gray	White	White
Relative cost	1	2	2	3

Note: 1 = best, 2 = good, 3 = ok.

Most applications are conducted at an elevated temperature region. Under this circumstance, conventional liquid lubricants fail to withstand the stresses for the application. Solid lubricants are most appropriate for such applications because of their ability to withstand the operating temperatures, orient and adhere to the substrate surface, provide the coefficient of friction reduction necessary to promote metal flow, and provide the required load-carrying properties to prevent metal-on-metal contact. Indeed, most applications that involve plastic deformation of metal will utilize solid lubricants as either the primary or the secondary lubricant within a formulation.

What application criteria are used for determining the necessity for a solid lubricant? Severity of metal movement is the most significant factor. In cases where it is judged that metal movement would be considered extreme, solid lubricants will most likely be required. Application examples include forward, backward, and extreme lateral extrusion of metals. For example, forging of spindles, constant velocity joints, crankshafts, and hubs would fall in this category. For these and similar cases, liquid lubricant technology falls short of providing the necessary lubrication, coefficient of friction reduction, and die wear protection.

Once it has been determined that a solid lubricant is necessary, the temperature criteria need to be determined. Metalworking applications done at ambient temperature can utilize MoS_2 as the solid lubricant. MoS_2 has the best lubrication properties among the four lubricants discussed. In fact, for applications such as cold forging, MoS_2 is the preferred lubricant because of its ability to handle the very high load and stress applied onto the part being deformed.

In some cases, application of the MoS_2 is by dry powder tumbling of the billets. Usually, the billets are phosphated before applying powder to anchor the MoS_2 onto the surface and within the structure of the phosphate coating. The phosphate coating acts as an anchor for the powder and allows the lubricant to advance with the metal deformation. Table 6.11 compares forging performance for bare versus coated steel. Lubrication is improved as press tonnage falls and spike height of the forged billet increases.

Dry powder tumbling is an effective application method for some cases. Other situations will require a more detailed and accurate depositing of MoS_2 film onto the substrate. This requires the use of a dispersed MoS_2 to provide a controlled coating thickness and particle size distribution considered appropriate for the job.

There may be instances where MoS_2 is not desirable—for example, environmental concerns or housekeeping issues. In these instances, PTFE or boron nitride would be appropriate. The white color of the pigments alleviates concerns regarding cleanliness of using graphite and molybdenum disulfide. Situations that require a reduction in emissions and material reactivity would favor boron nitride since PTFE will decompose at typical warm and hot forging temperatures. Both would effectively lubricate, with perhaps boron nitride faring better than PTFE for applications with significant metal flow.

PTFE can, however, stand out as a lubricant for cold metal-forming operations involving sheet stock and bar stock. The low coefficient of friction imparted by PTFE will provide the necessary lubrication to assist metal flow in a manner far better than boron nitride and much cleaner than graphite or molybdenum disulfide.

All the solid lubricants would be appropriate for bonded-film applications for metal deformation processes. Bonded films are desirable for sheet metal applications where coil or blank metal is prepared with a dry film lubricant. When developing bonded-film lubricants, consider the formulation of effective binders and bonding agents so that the solid lubricant can function as intended.

For metalworking applications at elevated temperatures, the operating temperature will determine which solid lubricant can be used. All the solid lubricants mentioned would be suitable for temperatures up to 260°C. Above that temperature, PTFE will be eliminated from consideration due to its decomposition. MoS_2 will be suitable for applications up to 400°C in an oxidizing environment. Above that temperature, decomposition of MoS_2 will occur. Both graphite and boron nitride will lubricate effectively above an operating temperature of 400°C. Graphite is the predominant lubricant used for plastic deformation at elevated temperatures.

The use of graphite is common and preferred for what is considered warm and hot forging situations. The forging process is considered warm forging when billet temperatures are up to 950°C. The process is considered hot forging when billet temperatures exceed 950°C. In both cases, oxidation of graphite will occur. But the rate of oxidation depends on temperature and is regulated by the formulation and characteristics of graphite. Graphite quality, contaminants, crystallite size, and particle size will influence the rate of oxidation. The components of the finished formulation also play a role in controlling the oxidation rate of graphite, allowing it to survive for an appropriate length of time necessary for lubricating the process.

The type and quality of graphite play an important role in performance. Its consideration is the first step in a selection process. The first choice is to choose between natural and synthetic graphites. Often, the choice is dictated by the degree of graphite quality suitable for the application. For instances where average lubrication is required, natural graphite of lesser quality can be used. More demanding lubrication will require the use of high-purity synthetic or natural graphite.

Selection of the particle size of graphite will vary depending on the intentions for the job. Particle size should be matched

TABLE 6.11
Cold Forging Lubrication

Sample	Press Tonnage	Spike Height (mm)
Bare steel	80.2	10.67
Bare steel + zinc phosphate	79.6	11.11
Bare steel + zinc phosphate + MoS_2	78.4	11.46

Source: Acheson Colloids test data.

to the type of metal movement expected from the process, the surface roughness of the die and part, and the degree of stability required for the formulated lubricant. If a large particle distribution is desired, then concern about physical stability of the lubricant must be addressed. Rapid settling and hard packing of graphite could occur due to the large particle size if countermeasures are not taken. This would create handling costs and product inconsistency for the end user.

For most circumstances, high-quality graphite should be used so as to minimize performance inconsistency. The quality and characteristic of graphite can affect the lubricating performance. Table 6.12 illustrates a lubricity comparison of standard formulations produced with different graphites. In this example, the application is warm forging of steel. Actual forging of a steel billet generates lubrication data where the spike height is determined using preset forging press parameters (see Figure 6.12). A greater spike height and lower coefficient of friction suggest better lubrication from the coating.

Once the type of graphite to be used is selected, then the cost of the powder needs to be considered versus the benefit derived from its use. In general, high-purity natural or primary synthetic graphite will be costlier than secondary synthetic graphite. However, the performance benefit of using the higher-cost material may justify its selection for the application. Benefits normally associated with the higher-cost materials are consistency, lubricating performance, and reduced oxidation rates of graphite.

The chosen graphite should be of a specific particle size distribution to derive certain benefits in performance. These benefits

TABLE 6.13
Lubrication Comparison of Forging Lubricants (800°C Forging Temperature)

Lubricant	Spike Height (mm)	Coefficient of Friction
Graphite A	1.5	0.05
Graphite B	1.3	0.08
Nongraphite lubricant	0.7	0.15

include the ease of dispersing graphite into a liquid carrier, the stability of graphite within the concentrated product, the application and film formation of the product onto the workpiece, and the optimized lubrication for the deformation process.

Forging processes normally require a temporary bond of the lubricant onto the workpiece and tool. This is achieved by the use of the type of bonding agents mentioned previously in this chapter. The use of dry powder or simple liquid–powder mixes will not perform adequately because of the poor adhesion onto the substrate.

To illustrate the value of graphite for hot-temperature metalworking applications, consider the example cited in Table 6.13. A comparison is made between two formulated graphite products and a nongraphite product tested under the same procedures of warm forging. In this example, the degree of spike height and coefficient of friction generated by the forging process are determined. The lower spike height and higher coefficient of friction for the nongraphite lubricant are indications of reduced lubrication capability in comparison to the graphite-containing materials.

In certain instances, graphite is not desirable due to either the operating temperature or concern about housekeeping and cleanliness. Hexagonal boron nitride is a capable alternative to graphite for these conditions. It is considered the "white graphite" due to its lamellar structure. It has a reasonably low coefficient of friction that approaches and sometime exceeds that of graphite. It is able to withstand operating temperatures up to 1200°C in oxidizing environments. This makes boron nitride an effective material for high-alloy isothermal forging, where extremely high temperatures and long contact times are encountered. A profile of oxidation characteristics provides a comparison of oxidation stability between boron nitride and graphite (see Figure 6.13). The ability for boron nitride to remain intact at a very high temperature makes it ideal for applications that require a long residency time for lubricant coating.

Another advantage of using boron nitride is the heat conductivity property of the material. For applications that would require rapid heat dissipation, boron nitride serves quite well and is superior to graphite in that regard. Thermal conductivity values of boron nitride powder will vary depending on its quality. But boron nitride in any of its grades is invariably more thermally conductive than graphite or MoS_2. Applications such as high-performance cutting oils are claimed to deliver benefits of enhanced lubrication and heat withdrawal when finely dispersed submicron particles of boron nitride are incorporated into the fluid [17–19].

TABLE 6.12
Graphite Influence on Forging Lubrication (800°C Forging Temperature)

Graphite	Spike Height (mm)	Coefficient of Friction
A	1.5	0.05
B	1.3	0.08
C	1.1	0.10

FIGURE 6.12 Deformed billet and spike.

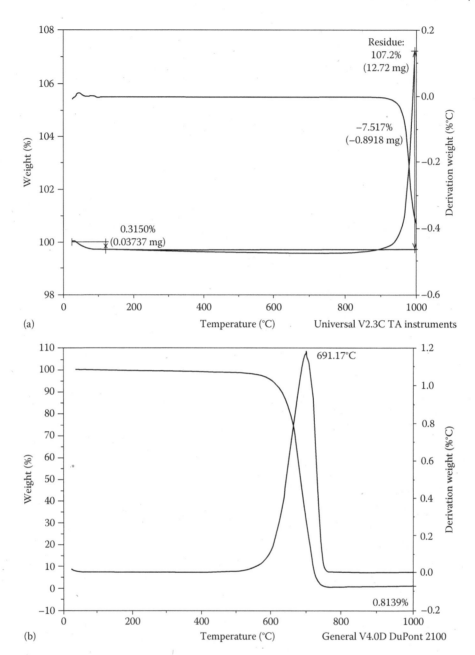

FIGURE 6.13 Comparison of peak oxidation temperatures of (a) boron nitride and (b) graphite. (From Acheson Colloids test data.)

REFERENCES

1. Jacobs, N.L. Metal-free lubricant composition containing graphite for use in threaded connections. U.S. Patent 5,180,509, January 19, 1993.
2. Ludema, K.C. *Friction, Wear, Lubrication, A Textbook in Tribology.* Boca Raton, FL: CRC Press, 1996, p. 123.
3. Acheson Colloids Company. J. Brian Peace Lecture.
4. Clauss, F.J. *Solid Lubricants and Self-Lubricating Solids.* New York: Academic Press, 1972, p. 45.
5. Savage, R.H. Graphite lubrication. *J Appl Phys* 19:1, 1948.
6. Barry, H.F. Factors relating to the performance of MoS$_2$ as a lubricant. *J Am Soc Lubr Eng* 33(9):475–480, 1977.
7. Kohli, A.K., B. Prakash. Contact pressure dependency in frictional behavior of burnished molybdenum disulphide coatings. *Tribol Trans* 44(1):147–151, 2001.
8. Du Pont Teflon®. Fluoroadditives brochure.
9. Bowden, F.P., D. Tabor. *The Friction and Lubrication of Solids.* New York: Oxford University Press, 1986, p. 165.
10. Kaur, R.G., C.F. Higgs, H. Hesmat. Pin-on-disc tests of pelletized molybdenum disulfide. *Tribol Trans* 44:79–87, 2001.
11. Pacholke, P.J., K.M. Marshek. Improved worm gear performance with colloidal molybdenum disulfide containing lubricants. ASLE paper presented at the *41st Annual Meeting,* Toronto, Ontario, Canada, May 12–15, 1986.

12. Haviland, M.L., M.C. Goodwin. Fuel economy improvements with friction-modified engine oils in Environmental Protection Agency and road tests. Society of Automotive Engineers Technical Paper 790,945, October 1979, Society of Automotive Engineers, Warrendale, PA.

13. Haviland, M.L., J.L. Linden. Taxicab fuel economy and engine and rear axle durability with low-viscosity and friction modified lubricants. Society of Automotive Engineers Technical Paper 821,227, October 1982, Society of Automotive Engineers, Warrendale, PA.

14. Broman, V.E. et al. Testing of friction modified crankcase oils for improved fuel economy. Society of Automotive Engineers Technical Paper 780,597, June 1978, Society of Automotive Engineers, Warrendale, PA.

15. Peters, J.A. Method for coating a substrate with a sliding abrasion-resistant layer utilizing graphite lubricant particles. U.S. Patent 5,702,769, December 30, 1997.

16. Acheson Colloids test data.

17. Watari, K., H.J. Huang, M. Turiyama, A. Osuka, O. Yamamoto. High-performance lubricant oil. U.S. Patent 5,985,802, November 16, 1999.

18. ZYP Coatings technical data sheet, Boron Nitride Powders for Research and Industry.

19. Booser, R.E. *Theory and Practice of Tribology, Vol. II: Theory and Design.* Boca Raton, FL: CRC Press, 1983, p. 276.

Section III

Antiwear and Extreme Pressure (EP) Additives

7 Ashless Antiwear and Antiscuffing (Extreme Pressure) Additives

Liehpao Oscar Farng and Tze-Chi Jao

CONTENTS

7.1   INTRODUCTION

To optimize the balance between low wear and low friction, machine designers specify a lubricant with a viscosity sufficient to generate hydrodynamic or elastohydrodynamic lubrication (EHL) films that separate the machine's interacting surfaces, but not too high to induce excessive viscous drag losses. In practice, the variety of contact types in a machine, the incidence of operating conditions beyond the design range, and the pressure to improve efficiency by reducing oil viscosity all conspire to reduce oil film thickness below the optimum. The high spots, or asperities, on the interacting surfaces then start to interact with each other, initially through micro-EHL films, and at the extreme through direct surface contact, resulting in increased friction and the likelihood of surface damage. Antiwear and antiscuffing ([AS] also known as extreme pressure [EP], an older terminology) additives are added to lubricating oils to decrease wear and prevent seizure under such conditions.

A common way to demonstrate the viscosity optimization is shown in Figure 7.1; this is known as a Stribeck curve. The curve is a composite of a boundary friction curve that decreases as viscosity increases and, therefore, film thickness increases and a viscous friction curve that increases as viscosity and speed increase. Slightly to the right of the minimum in the curve represents a good operation target. Improving the surface finish of contacting surfaces can move the minimum in the curve to a lower viscosity range, saving energy but increasing the cost of components. The hardening or coating of surfaces can increase their durability under increased

levels of contact with lower viscosity, but again at an increase in component cost. Notwithstanding these component manufacturing improvements, antiwear and AS additives will continue to be needed, but the nature of their chemistry is likely to change due to environmental constraints, component material developments, and the continuing increase in severity of machine operating conditions.

The distinction between antiwear and AS additives is not clear-cut. Some are classed as antiwear in one application and AS in another and some have both antiwear and AS properties. To add to the confusion, AS comes in mild and strong flavors and some AS additives are only effective in low-speed, high-load situations and others only in high-speed, high-temperature applications. Generally, antiwear additives are designed to deposit surface films under normal operating conditions and thereby reduce the rate of continuous, moderate wear, whereas AS additives are expected to react rapidly with a surface under severe distress and prevent more catastrophic modes of failure such as scuffing (scoring), galling, and seizure.

Antiwear and AS additives are designed to provide protection over a broad spectrum of operating conditions and both act to protect against adhesive wear that ranges from mild, moderate, to severe. Mild adhesive wear is confined to the oxide layers of gear tooth surfaces. During initial operation, the asperities are smoothed. This usually subsides with time and is considered normal. At the other extreme, scuffing is severe adhesive wear, and it can cause catastrophic damage. Since antiwear and AS additives both act in similar ways, it is

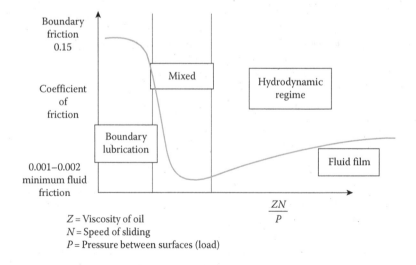

$$\frac{ZN}{P}$$

Z = Viscosity of oil
N = Speed of sliding
P = Pressure between surfaces (load)

FIGURE 7.1   Regions of Lubrication–Stribeck curve.

best to classify them in terms of their activation temperature. Antiwear additives become effective at relatively low contact temperatures and become ineffective at moderate contact temperatures, whereas AS additives remain on gear tooth surfaces until they are rubbed off or melt, as intended, at relatively high contact temperatures.

The choice between antiwear and AS additives depends on the gear application. Antiwear additives such as tricresyl phosphate (TCP) and zinc dialkyl dithiophosphate (ZnDDP) might be adequate for high-speed, lightly loaded gears that are not subjected to shock loads, whereas slow-speed, highly loaded gears that are subjected to shock loads might require AS additives such as those containing sulfur and phosphorus, alone or in combination. In many applications, lubricants with both antiwear and AS additives are required to protect against the full range of adhesive wear, but care should be taken to avoid aggressive chemistry that can result in polishing wear, micropitting, or degradation of other components.

Therefore, it has recently been suggested that EP additives be renamed as AS additives, since there is no "pressure" distinction between them and antiwear additives, only an expectation of a performance boost under severe conditions. AS additives tend to be very reactive and some can have adverse effects on oxidative stability of oils, can be corrosive to nonferrous materials, and can reduce the fatigue life of bearing and gear surfaces. They should only be used when severe distress is a distinct possibility.

Antiwear additives function in a variety of ways. Some deposit multilayer films thick enough to supplement marginal hydrodynamic films and prevent asperity contact altogether. Some develop easily replenishable monolayer films that reduce the local shear stress between contacting asperities and are preferentially removed in place of surface material. Others bond chemically with the surface and slowly modify surface asperity geometry by controlled surface material removal until conditions conducive to hydrodynamic film generation reappear.

AS additives are designed to prevent metal–metal adhesion or welding when the degree of surface contact is such that the natural protective oxide films are removed and other surface-active species in the oil are not reactive enough to deposit a protective film. This is most likely to occur under conditions of high-speed, high-load, and/or high-temperature operation. AS additives function by reacting with the metal surface to form a metal compound, such as iron sulfide. They act in a manner similar to that of antiwear additives, but their rate of reaction with the metal surface and therefore the rate of AS film formation are higher and the film itself is tougher. Some AS additives prevent scoring and seizure at high speed and under shock loads; others prevent ridging and rippling in high-torque, low-speed operations. In both cases, AS additives and surface metal are consumed, and a smoother surface is created with an improved chance of hydrodynamic action, resulting in less local distress and lower friction. In the absence of such additives, heavy wear and distress well beyond the scale of surface asperities would occur, accompanied by very high friction.

A wide variety of antiwear and AS additives are commercially available, and many other chemicals with antiwear and AS functionality have been reported in the literature and in patents. To be commercially feasible, additives must be adequately soluble in lubricant formulations and reasonable in cost and must not overly reduce the lubricant's oxidative stability nor increase the corrosivity of metals contacted by the lubricant [1–3]. Lead naphthenates were extensively used early in the industry's history, but environmental concerns have led to their virtual disappearance. Similarly, chlorine-containing additives are in decline. Zinc dialkyl dithiophosphates (ZDDP or ZnDTP) are the best known and most widely used antiwear additives in engine oils, transmission fluids, and hydraulic oils. However, the concern for phosphorus poisoning of automotive catalysts and for zinc as an environmental contaminant has resulted in a pressure to find metal- and phosphorus-free replacements for both automotive and industrial applications. This has resulted in a move toward ashless antiwear and AS additives, and this chapter covers these additives in terms of their chemistry, properties, and performance characteristics, applications, marketing, sales, and outlook.

7.2 CHEMISTRY, PROPERTIES, AND PERFORMANCE (CLASSIFIED BY ELEMENTS)

7.2.1 SULFUR ADDITIVES

Sulfur-containing additives are used to provide protection against high-pressure, metal-to-metal contacts in boundary lubrication. The magnitude of the EP activity is a function of the sulfur content of the additive; high-sulfur-content additives are usually more effective EP agents than are low-sulfur-content additives. The sulfur content of the additive must be balanced against requirements for thermal stability and noncorrosiveness toward copper-containing alloys. The additive's composition and structure represent a chemical compromise between conflicting performance requirements. In general, any compound that can break down under an energy-input stress, such as heat, and allow for a free sulfur valence to combine with iron would do well as an antiwear and AS additive. Sulfur additives are probably the earliest known, widely used AS compounds in lubricants.

Sulfurization by addition of sulfur compounds (elemental sulfur [ES], hydrogen sulfide, and/or mercaptans) to unsaturated compounds has been known to the chemical industry for years [4–8]. The two most common classes of additives are called sulfurized olefins [9,10] and sulfurized fatty acid esters [11], because they are produced from reactions of olefins and naturally occurring or synthetic fatty acid esters with sulfur compounds.

In the absence of initiators, the addition to simple olefins is by an electrophilic mechanism, and Markovnikov's rule is followed. However, this reaction is usually very slow and often cannot be done or requires very severe conditions unless an acid catalyst is used. In the presence of free radical initiators, H_2S and mercaptans add to double and triple bonds by a free radical mechanism, and the orientation is anti-Markovnikov.

By any mechanism, the initial product of addition of H_2S to a double bond is a mercaptan, which is capable of adding to a second molecule of olefin, so that sulfides are often produced (reaction 7.1):

$$\tag{7.1}$$

7.2.1.1 Sulfurized Olefins

Sulfurized olefins are prepared by treating an olefin with a sulfur source, under proper reaction conditions. The more sulfur used, the higher is the sulfur content. Suitable olefins preferably include terminal olefins and internal olefins, monoolefins, and polyolefins. However, in order to provide adequate oil solubility, the olefin should provide a carbon chain of at least four carbon atoms. Accordingly, suitable alpha olefins are the butenes, pentenes, hexenes, and preferably higher alpha olefins such as the octenes, nonenes, and decenes. Isobutylene is a very unique olefin that not only exhibits very high reactivity toward sulfur reagents (high conversion rate) but also can produce sulfurized products having very good stability and lubricant compatibility. Therefore, sulfurized isobutylene (SIB) has been by far the most cost-effective, widely used EP additive in lubricants.

7.2.1.1.1 Chemistry and Manufacture

Initially, sulfurized olefins were synthesized through a two-step chloride process, and often, the products were referred to as "conventional sulfurized olefins." Sulfur monochloride and sulfur dichloride were used in the first step to produce chlorinated adducts, and then the adducts were treated with an alkali metal sulfide in the presence of free sulfur in an alcohol–water solvent, followed by further treatment with an inorganic base (reactions 7.2 and 7.3) [12]. The final product is a light yellow–colored fluid with oligomeric monosulfides and disulfides as the main compositions, as typified by

$$\tag{7.2}$$

$$\tag{7.3}$$

The manufacture of conventional sulfurized olefins involves sulfur monochloride, and the final product contains some residual chlorine. The process also generates aqueous waste with halogen- and sulfur-containing by-products that must

be disposed of. Chlorine in lubricants and other materials is increasingly becoming an environmental concern because chlorinated dioxins can be formed when chlorine-containing materials are incinerated. Chlorinated waxes have been eliminated from many lubricants for this reason. Residual chlorine content is also becoming a major concern in many areas of the world. Germany currently has a 50 ppm maximum limit on the chlorine content of automotive gear oils. This requirement is a problem for automotive gear oil suppliers as well as additive suppliers if their technology is based on conventional sulfurized olefins, since the residual chlorine content is a consequence of the chemistry required to manufacture conventional sulfurized olefins. By fine-tuning the manufacturing process, the chlorine content of conventional sulfurized olefins may be reduced from a typical 1500 ppm to less than 500 ppm. However, manufacturing changes to reduce the residual chlorine content will probably slow the production process, require additional capital investments, and possibly generate more aqueous waste.

In the late 1970s, the high-pressure sulfurized isobutylene (HPSIB) process was developed to replace the conventional, low pressure, chlorine process. HPSIBs are usually mixtures of di-tert-butyl trisulfides, tetrasulfides, and higher-order polysulfides [13–16]. Some HPSIB contains oligomeric polysulfides of poorly defined composition or other materials such as 4-methyl-1,2-dithiole-3-thione (Structure A [8,17] and reaction 7.4). The higher-order polysulfides generally favor EP activity at the expense of oxidative stability and copper corrosivity compared to the monosulfides and disulfides of conventional sulfurized olefins. In the absence of other reagents, the straight reaction of ES and isobutylene results in a dark-colored liquid that contains a significant amount of dithiolethiones (thiocarbonates). 4-Methyl-1,2-dithiole-3-thione is a pseudoaromatic heterocyclic compound. Due to its rigid ring structure, dithiolethione can be easily precipitated as yellowish solids that cause severe staining problems. Therefore, dithiolethione is often not a desirable side product in SIB.

$$\tag{7.4}$$

In the presence of various catalysts (or basic materials), such as aqueous ammonia, alkali metal sulfides, or metal dithiocarbamates, amounts of dithiolethiones (Structure A) and oligomeric polysulfides can be reduced and low-molecular-weight polysulfides (X = 2–6 in Structure B) are the predominant products [18].

Structure A   Structure B

The use of hydrogen sulfide in the high-pressure sulfurized olefin process can ease the reaction complexity and also yield high-quality, low-molecular-weight polysulfides. The compositions of products prepared from this process usually have good clarity, low odor, light color, and high AS activity. Hydrogen sulfide is a very foul smelling and very toxic gas. Collapse, coma, and death from respiratory failure may come within a few seconds after one or two inhalations. Liquefied hydrogen sulfide has a high vapor pressure that requires additional, adequate protective equipment. There are considerable risks associated with its routine use on an industrial scale, but hydrogen sulfide is a low-cost, commodity chemical, which can often offset the additional costs for safe use. High-pressure sulfurized olefins can also be prepared with reagents that generate hydrogen sulfide within the reactor during the course of the reaction. Direct handling of hydrogen sulfide is thus avoided, but there can be processing penalties, usually in the area of aqueous waste handling. Performance-wise, high-pressure sulfurized olefins could replace conventional sulfurized olefins in suitable applications. A decision to manufacture high-pressure sulfurized olefins by one process or another will require a careful assessment of acceptable risks versus economic requirements.

Other olefins or mixed olefins are also used in preparation of various sulfurized olefins. Among these, di-tert-nonyl and di-dodecyl trisulfides and pentasulfides are very popular additives. Diisobutylene (2,4,4-trimethyl-1-pentene) is also used extensively to make higher-viscosity sulfurized products. In addition, other sulfurized hydrocarbons, such as sulfurized terpene, sulfurized dicyclopentadiene, or sulfurized dipentene olefin, and sulfurized waxes are also widely used due to low raw material costs.

7.2.1.1.2 Applications and Performance Characteristics

Sulfurized olefins played a key role in establishing superior ashless sulfur/phosphorus (S/P) additive systems for lubricating automotive and industrial AS gear oils in the late 1960s [19–21]. The early AS gear oil additives were clearly dominated by chlorine, zinc, and lead, which had difficulty in adequately protecting heavy-duty equipment. On the other hand, the S/P gear oil additive technology, based on ashless and chlorine-free components, possesses very good thermal/oxidative stability and rust inhibition (CRC L-33 and ASTM D665B); therefore, this is a significant performance improvement over the metal- and chlorine-based technology.

Sulfurized olefins function mainly through thermal decomposition mechanisms. Sulfur prevents contact between interacting ferrous metal surfaces through the formation of an intermediate film of iron sulfide. By doing this, sulfur usually decreases the wear rate but accelerates the smoothing of the surfaces. This smoothing actually helps reduce the wear rate. Furthermore, a higher percent of active sulfur in a molecule increases the chances of reaction with the metal surface and favors AS (antiseizure) more than antiwear properties. Thus, SIB is mainly a strong AS additive, with outstanding scuffing protection properties (e.g., CRC L-42 performance). Table 7.1 shows the coefficients of friction and dimensions with respect to metal surface, oil molecules, and sulfide layers. It can be seen that the friction coefficients of the sulfide layers are about half of those for metal-to-metal surfaces. The sulfide

TABLE 7.1

Typical Surface Characteristics

Surface	Coefficients of Friction
Steel:Steel	0.78
FeS:FeS	0.39
Copper:Copper	1.21
CuS:CuS	0.74
Material	**Dimension (Angstrom)**
Size of oil molecules	50
Size of sulfide layers	3000
Surfaces with "superfinish"	1000

layers retard the welding of the moving metal surfaces, but do not prevent wearing. Particles of iron sulfide are constantly sloughed off from the metal surface. This wear can be determined by the analysis of the lubricating oils (residual iron content), and subsequent sludge formation can be controlled by the use of dispersants.

Besides heavy-duty gear oil applications [22], sulfurized olefins have also found usefulness in other lubricant areas, such as metal processing oils, greases, marine oils, and tractor-transmission oils.

7.2.1.2 Sulfurized Esters and Sulfurized Oils

The oldest widely used sulfur-based additive that is still found in commercial lubricants is sulfurized lard oil (SLO), a sulfurized animal triglyceride. In 1939, H.G. Smith made one of the most important discoveries in the history of lubricant additives. He found that sulfurized sperm whale oil (SSWO) was more soluble in paraffinic base oils, even at low temperatures, and had a much higher thermal stability than SLO. Thus, over 60 years ago, he recognized that the improved stability of sulfurized sperm oil resulted from its monoester structure compared with the triester structure of SLO. With long-chain monoesters, the sulfur has little potential to form bridges between the molecules, as it does when triglycerides are being sulfurized. SSWO is an excellent boundary lubricant and is highly resistant to gumming, resin formation, or viscosity increase, when subjected to high temperature and high pressure [23]. Unfortunately, from a lubricant cost-performance viewpoint, this additive is no longer available due to restrictions on the use of sperm whale oil. Available but an expensive alternative is sulfurized jojoba oil, which is also a mixture of long-chain alcohol-fatty acid compounds. All these sulfurized fats or esters usually are manufactured to contain 10%–15% sulfur and are often good antiwear and mild AS agents.

7.2.1.2.1 Chemistry and Manufacture

Sperm oil is a waxy mixture of esters of fatty alcohols and fatty acids with a small amount of triglycerides. After the separation of the solid waxes by filtration or centrifuge, a liquid wax remains, consisting mainly of an ester of oleic alcohol and oleic acid. Such a structure could not be better for sulfurizing purposes.

Similar to sulfurized olefins, sulfurized esters can be made by either direct sulfurization with ES or sulfurization with hydrogen sulfide under superatmospheric pressures. Nowadays, they are mainly made from vegetable oils having one or more double bonds. Sulfurized esters are made from unsaturated fatty acids, like oleic acid, and esterified with an alcohol such as methanol. Frequently, the sulfurization of fats is made in the presence of an olefin, preferably of long chain, and the resulting commercial product is a mixture of the two types. Equation 7.5 shows a typical example of sulfurization of methyl oleate with ES. When sulfurized with hydrogen sulfide, products usually possess lighter color and lower odor.

$$(7.5)$$

7.2.1.2.2 Properties, Performance Characteristics, and Applications

The load-carrying property of sulfurized oils is directly linked to the amount of active sulfur in the additive. Percent active sulfur (which is believed to provide AS activity) and total sulfur can be determined by proper analytical methods, and the difference is the percent inactive sulfur. The more active sulfur present, the higher the load-carrying property. However, there is also a direct correlation between active sulfur and copper corrosiveness—the more active sulfur, the poorer the copper corrosion protection. The more active sulfur can also lead to cleanliness and stability challenges. Therefore, the ultimate product properties for a specific lubricant product will dictate which sulfurized products to use.

Although the sulfur content may not be as high as in many sulfurized olefins, sulfurized esters are attractive for their exceptionally good frictional properties in many applications. This is because combining sulfur with fat in a lubricant additive provides a synergistic effect. In this instance, the fat provides reduced friction, and sulfur provides wear and scuffing protection. Of all the elements, sulfur probably gives the best synergistic results in combination with other components and organic compounds. As to AS characteristics, sulfurized esters have a surface activity conferred by a small amount of their normal free fatty acids. These are polar species that tend to be adsorbed in layers of molecular dimensions at the metal interface. The interposition of such films is effective in preventing metal seizure under conditions of EP

or under conditions tending to displace the lubricating film between the bearing surfaces. Here, film strength and EP phenomena are often used synonymously. Film strength implies that metal-to-metal contact and welding are prevented as a result of the film formation (or replenishment) by the chemical reaction of the metal and an AS additive. Also, fatty oils and sulfurized fatty oils because of their affinity for metal surfaces are less easily displaced from metal surfaces by water than are mineral oils.

Ferrophilic ester groups improve AS properties. Depending upon the molecular structure and its polarity, the surface activities vary. Since the surface activity or polarity of the substances used for sulfurization plays an equally decisive role in lubricating action, it should be taken into serious consideration when one formulates a product for a specific application. Comparing sulfurized triglycerides (e.g., SLO) with sulfurized monoesters (e.g., SSWO), the AS properties of the triglycerides are better. Two factors may be responsible for this phenomenon: (1) as the triester structure is more ferrophilic, hydrogen bridging may occur; (2) as triglycerides decompose at high temperatures to form acrolein moieties during the lubrication process, the polymerized acrolein film can add strength to the sulfide film and improve the AS characteristics. However, this AS activity of triglycerides has limited value due to their poor stability and oil solubility. Stability tests at elevated temperatures show faster and heavier sludging for SLO than for SSWO. Therefore, a proper balance of all properties is an essential part of product formulations.

Sulfurized fats or esters are used extensively in lubricants such as metalworking fluids, tractor-transmission fluids, and greases.

7.2.1.3 Other Sulfur Additives

ES provides good AS properties; however, it leads to corrosion. It dissolves in mineral oils up to certain levels depending upon the type of base oils. Low-polarity paraffinic/naphthenic type Group II and III base oils usually have very limited solubility of ES. Sulfurized aromatics such as dibenzyl disulfide (DBDS), butylphenol disulfide, diphenyl disulfide, or tetramethyldibenzyl disulfide generally containing less active sulfur improve only moderately the AS characteristics of lubricants; they are therefore used predominately in combination with other sulfur- or phosphorus-containing AS additives [24,25]. Other sulfur carriers such as sulfurized nonylphenol, dialkyl thiodipropionates ($S[CH_2CH_2C(=O)OR]_2$) derivatives of thioglycolic acid esters ($HS–CH_2C(=O)OR$), derivatives of thiosalicylic acid, and trithians are also available [26]. However, materials with low sulfur content are usually less active as antiwear/AS additives, but more effective as antioxidants.

7.2.2 PHOSPHORUS ADDITIVES

Phosphorus-containing additives are used to provide protection against moderate- to high-pressure, metal-to-metal contacts in boundary lubrication and EHL. Unlike sulfur additives, where their AS activity must be balanced against performance requirements for thermal stability and mild corrosivity toward copper-containing alloys, phosphorus additives usually possess very good corrosivity control. Due to totally different mechanisms involved in surface film formation rates and film strengths, phosphorus additives cannot replace sulfur additives in many applications and vice versa. Typically, phosphorus additives are extremely effective in applications with slow sliding speeds and high surface roughness.

7.2.2.1 Phosphate Esters

Phosphate esters have been produced commercially since the 1920s and have gained importance as lubricant additives, plasticizers, and synthetic base fluids for compressor and hydraulic oils. They are esters of alcohols and phenols with a general formula $O=P(OR)_3$, where R represents alkyl, aryl, alkylaryl, or, very often, a mixture of alkyl and aryl components. The physical and chemical properties of phosphate esters can be varied considerably depending on the choice of substituents, and these can be selected to give optimum performance for a given application. Phosphate esters are particularly used in applications that benefit from their high-temperature stability and excellent fire-resistance properties in addition to their adequate antiwear properties [27].

7.2.2.1.1 Chemistry and Manufacture

Phosphate esters are produced by the reaction of phosphoryl chloride with alcohols or phenols as shown in the following:

$$3ROH + POCl_3 \quad \Rightarrow \quad O=P(OR)_3 + HCl \qquad (7.6)$$

Early production of phosphate esters was based on the so-called "crude cresylic acid" fraction or "tar acid" derived by distillation of coal tar residues. This feedstock is a complex mixture of cresols, xylenols, and other heavy materials and includes significant quantities of ortho-cresol. The presence of high concentrations of ortho-cresol results in an ester that has been associated with neurotoxic effects, and this has led to the use of controlled coal tar fractions, in which the content of ortho-cresol and other ortho-n-alkylphenols is greatly reduced. Phosphate esters using coal tar fractions are generally referred to as "natural," as opposed to "synthetic" where high-purity raw materials are used.

The vast majority of modern phosphate esters are "synthetic," using materials derived from petrochemical sources. For example, t-butylated phenols are produced from phenols by reaction with butylene. The reaction of alcohol or phenol with phosphoryl chloride yields the crude product, which is generally washed, distilled, dried, and decolorized to yield the finished product. Low-molecular-weight trialkyl esters are water soluble, requiring the use of nonaqueous techniques. When mixed alkylaryl esters are produced, the reactant phenol and alcohol are added separately. The reaction is conducted in a stepwise process and the reaction temperature kept as low as possible to avoid transesterification reactions from taking place.

The most commonly used phosphate esters for antiwear performance features are TCP, trixylenyl phosphates, and tributylphenyl phosphates.

7.2.2.1.2 Physical and Chemical Properties

The physical properties of phosphate esters vary considerably according to the mix and type of organic substituents, molecular weights, and structural symmetry, all proving to be particularly significant. Consequently, phosphate esters range from low-viscosity, water-soluble liquids to insoluble high-melting solids.

As mentioned previously, the use of phosphate esters as synthetic base fluids arises mostly from their excellent fire resistance and superior lubricity but is limited due to their hydrolytic and thermal stability, low-temperature properties, and viscosity index. While phosphate esters are widely used as antiwear additives for lubricants, the concerns about hydrolytic stability, thermal stability, and, of course, satisfactory antiwear properties are equally important. In that sense, triaryl phosphates are dominant over trialkyl phosphates, because their hydrolytic–thermal stability is much better.

The thermal stability of triaryl phosphates is considerably superior to that of the trialkyl esters, which degrade thermally by a mechanism analogous to that of the carboxylic esters (reaction 7.7).

$$(7.7)$$

With respect to hydrolytic stability, aryl phosphate esters are superior to the alkyl esters. Increasing chain length and degree of branching of the alkyl group leads to considerable improvement in hydrolytic stability. However, the more sterically hindered the substituent, the more difficult it is to prepare the ester. Alkylaryl phosphates tend to be more susceptible to hydrolysis than the triaryl or trialkyl esters.

The low-temperature properties of phosphate esters containing one or more alkyl substituents tend to be reasonably good. Many triaryl phosphates are fairly high-melting-point solids, but an acceptable pour point can be achieved by using a mixture of aryl components. Coal tar fractions, used to make "natural" phosphate esters, are already complex mixtures and give esters with satisfactory pour points.

Phosphate esters are very good solvents and are extremely aggressive toward paints and a wide range of plastics and rubbers. When selecting suitable gasket and seal materials for use with these esters, careful consideration is required. The solvency power of phosphate esters can be advantageous in that it makes them compatible with most other common additives and enables them to be used as carriers for other less soluble additives to generate additive slurry.

7.2.2.2 Phosphites

Phosphites are the main organophosphorus compounds used to control oxidative degradation of lubricants. They eliminate hydroperoxides and peroxy and alkoxy radicals, retard the darkening of lubricants over time, and also limit photodegradation. In addition to their important role as antioxidants, phosphites are also found to be useful antiwear additives. Dialkyl hydrogen phosphites and diaryl hydrogen phosphites are neutral esters of phosphorous acid. These materials have two rapid equilibrating forms: the keto form, $(RO)_2P(=O)H$, and the acid form, $(RO)_2P-O-H$. Physical measurements indicate that they exist substantially in the keto form, associated in dimeric or trimeric groupings by hydrogen bonding. Trialkyl phosphites and triaryl phosphites are neutral trivalent phosphorus esters. These materials are clear, mobile liquids with characteristic odors.

7.2.2.2.1 Chemistry and Manufacture
Phosphites are produced by reaction of phosphorus trichloride with alcohols or phenols as shown in the following:

$$3ROH + PCl_3 + 3NH_3 \Rightarrow P(OR)_3 + 3NH_4Cl \quad (7.8)$$

When mixed alkylaryl phosphites are produced, the reactant phenol and alcohol are added separately with the reaction temperature being controlled carefully. High-molecular-weight phosphites can be produced from transesterification reaction of either alcohols or phenols with trimethyl phosphite under catalytic (acidic) conditions.

$$P(OCH_3)_3 + 3ROH \Rightarrow P(OR)_3 + 3CH_3OH \quad (7.9)$$

With acid-catalyzed hydrolysis, dialkyl or diaryl hydrogen phosphites can be produced from trialkyl or triaryl phosphites as shown in reactions 7.10 through 7.12.

$$P(OR)_3 + H_2O \Rightarrow (RO)_2P(=O)H + ROH \quad (7.10)$$

$$P(OR)_3 + HCl \Rightarrow (RO)_2P(=O)H + RCl \quad (7.11)$$

$$2P(OR)_3 + HP(=O)(OH)_2 \Rightarrow 3(RO)_2P(=O)H \quad (7.12)$$

By carrying out these reactions in the presence of hydrogen chloride acceptors such as pyridine, the isolation of mono-, di-, and trialkyl phosphites is feasible. However, with alcohols of normal reactivity, the product is often mainly dialkyl hydrogen phosphite. This can be made in up to 85% yield, by adding PCl₃ to a mixture of methanol and a higher alcohol at low temperature. The methyl and hydrogen chlorides are then removed by heating under reduced pressure on a steambath.

$$PCl_3 + 2ROH + CH_3OH \Rightarrow (RO)_2P(=O)H + CH_3Cl + 2HCl \quad (7.13)$$

The commonly used phosphites available in the marketplace are dimethyl hydrogen phosphite, diethyl hydrogen phosphite, diisopropyl hydrogen phosphite, dibutyl hydrogen phosphite, bis(2-ethylhexyl) hydrogen phosphite, dilauryl hydrogen phosphite, bis(tridecyl) hydrogen phosphite, dioleyl hydrogen phosphite, trisnonylphenyl phosphite, triphenyl phosphite, triisopropyl phosphite, tributyl phosphite, triisooctyl phosphite, tris(2-ethylhexyl) phosphite, trilauryl phosphite, triisodecyl phosphite, diphenylisodecyl phosphite, diphenylisooctyl phosphite, phenyldiisodecyl phosphite, ethylhexyl diphenyl phosphite, and diisodecyl pentaerythritol diphosphite.

7.2.2.2.2 Chemical and Physical Properties
Phosphites tend to hydrolyze when exposed to humidity in the air or moisture in the lubricant. The extent of hydrolysis depends upon the moisture content of the ambient atmosphere, the temperature, and the duration of exposure. Generally, liquid phosphites are more stable than solids because of the reduced surface area available for moisture pickup. But hydrolysis can be minimized if proper precautions, such as a dry nitrogen atmosphere, cool storage, and use of tight seals, are observed. The lower dialkyl hydrogen phosphites hydrolyze in both acidic and alkaline solutions to monoalkyl esters and phosphorous acid. Rates of hydrolysis normally decrease with increasing molecular weight. The lower esters of trialkyl phosphites are rapidly hydrolyzed by acids; however, they are relatively stable in neutral or alkaline solutions. In general, the hydrolytic stability of the trialkyl phosphites increases with molecular weight.

Since the dialkyl hydrogen phosphites are predominately in the keto form, they are somewhat resistant to oxidation and do not complex with cuprous halides. Both of these reactions are characteristic of trivalent organic phosphorus compounds [23–25]. These esters are relatively resistant to reaction with oxygen and sulfur but react quite readily with chlorine and bromine giving the corresponding dialkyl phosphorohalidates ($(RO)_2P(=O)X$, where $X = Cl$ or Br) [27].

The hydrogen atom of the dialkyl hydrogen phosphites is replaceable by alkali but is not acidic in the usual sense. The alkali salts are readily obtainable by the reaction of ester with metals. In contrast with the parent compound, these salts readily add sulfur to form the corresponding phosphorothioates. Sodium salts of phosphites can be reacted with alkyl chlorides to produce alkyl phosphonates. These salts react with halophosphites to produce pyrophosphites and with chlorine or bromine to yield the corresponding hypophosphates. Dialkyl hydrogen phosphites add readily to ketones, aldehydes, olefins, and anhydrides, and these reactions are catalyzed by bases and free radicals. This type of reaction provides an excellent method for preparing phosphonates.

Sulfur reacts readily with trialkyl or triaryl phosphites to form corresponding trialkyl or triaryl phosphorothioates, which are also very useful antiwear additives. The reaction of trialkyl phosphites with halogens is an excellent method for preparing dialkyl phosphorohalidates. Acyl halides and most polyfunctional primary aliphatic halides can be used. Triisopropyl phosphite provides a unique means for the preparation of unsymmetrical phosphonates and diphosphonates because the by-product isopropyl halide reacts very slowly and thereby does not compete with the primary reaction.

7.2.2.2.3 Applications and Performance Characteristics

Dialkyl (or diaryl) hydrogen phosphites, besides being excellent antiwear agents, are considered the most potent form of phosphorus, suited to high-torque, low-speed operations. This is the area where antiwear processes are taken to the extreme and is one of the most important sections of the AS performance spectrum. Sulfur can be quite incapable of giving protection under such conditions. Only a phosphorus source, if active enough and in sufficient concentration, can help here. Conversely, phosphorus components are of little use in high-speed and shock operations where sulfur components can be excellent. Dialkyl or diaryl phosphites are also potent antioxidants.

With dialkyl phosphites, it has been reported that oxidation produces a phosphate anion that tends to act as a bridging ligand to form an oligomeric iron (III) complex, that is, an iron oxide complex resembling Structure C.

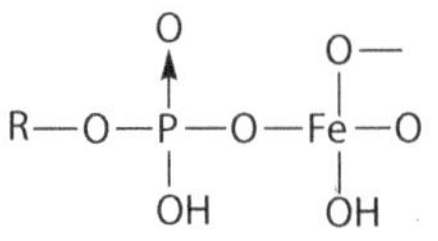

Structure C

However, there is also a weaker, high-viscosity, nonsolid film that increases the overall thickness of the total film at high speeds [24,31].

Dialkyl phosphites are widely used in gear oils, automatic transmission fluids (ATFs), and many other applications. Spiro bicyclodiphosphites are also reported to be used in continuously variable transmission fluids [32] (Structure D).

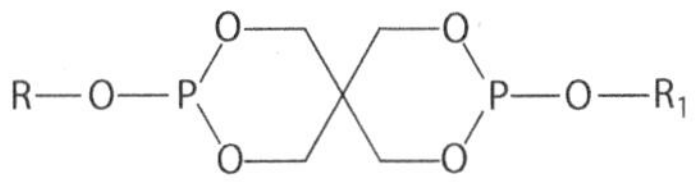

Structure D

7.2.2.3 Dialkyl Alkyl Phosphonates

Dialkyl alkyl phosphonates $[R–P(=O)(OR)_2]$ are stable organic phosphorus compounds that are miscible with ether, alcohol, and most organic solvents. Besides being used as additives in solvents and low-temperature hydraulic fluids, they can also be used in heavy metal extraction, solvent separation, and as preignition additives to gasoline, antifoam agents, plasticizers, and stabilizers. Dialkyl alkyl phosphonates are prepared from either dialkyl hydrogen phosphites or trialkyl phosphites as described in the following (Michaelis–Arbuzov reaction):

$$(RO)_3P + R'X \quad \Rightarrow \quad (RO)_2P(=O)R' + RX \qquad (7.14)$$

$$(RO)_2P(=O)H + R'OH + CCl_4 \quad \Rightarrow \quad (RO)_2P(=O)R' + H_2O \qquad (7.15)$$

$$(RO)_2P(=O)H + NaOH \Rightarrow (RO)2P \cdot O \cdot Na + R'X \Rightarrow (RO)_2P(=O)R' + NaX \qquad (7.16)$$

In principle, the thermal isomerization of all phosphites to phosphonates can be carried out. The stability of these

compounds varies greatly; however, depending upon the nature of the R group, other products may be formed during heating. For R = methyl, complete conversion occurs at 200°C in 18 h, but for R = butyl, the compound is stable at 223°C. It is thought by some that isomerization of phosphites may be possible only if traces of phosphonate are already present as an impurity [33].

$$(RO)_3P + Heat \quad \Rightarrow \quad (RO)_2P(=O)R \qquad (7.17)$$

7.2.2.4 Acid Phosphates

Acid phosphates are also potent additives, useful in similar areas of antiwear and AS to the dialkyl phosphites. Orthophosphoric (monophosphoric) acid $[H_3PO_4]$, the simplest oxyacid of phosphorus, can be made by reacting phosphorus pentoxide with water. It is widely used in fertilizer manufacture. Orthophosphoric acid has only one strongly ionizing hydrogen atom and dissociates according to the following reaction:

$$H_3PO_4 \Leftrightarrow H^+ + H_2PO_4^- \Leftrightarrow H^+ + HPO_4^{2-} \Leftrightarrow H^+ + PO_4^{3-} \qquad (7.18)$$

Since the first dissociation constant, K_1 (7.1×10^{-3}), is much larger than the second ($K_2 = 6.3 \times 10^{-8}$), very little of the H_2PO_4 produced in the first equilibrium goes on to dissociate according to the second equilibrium. Even less dissociates according to the third equilibrium since the third constant, K_3, is very small ($K_3 = 4.4 \times 10^{-13}$). The acid gives rise to three series of salts containing these ions, for example, NaH_2PO_4, Na_2HPO_4, and Na_3PO_4.

7.2.2.4.1 Chemistry and Manufacture

Alkyl (aryl) acid phosphates are made from alcohol (phenol) and phosphorus pentoxide. In general, a mixture of monoalkyl (aryl) and dialkyl (aryl) phosphates is produced:

$$3ROH + P_2O_5 \quad \Rightarrow \quad (RO)_2P(=O)OH + (RO)P(=O)(OH)_2 \qquad (7.19)$$

Pure mono- or dialkyl (aryl) phosphates can be synthesized through different reaction routes as described in the following:

$$ROH + POCl_3 \Rightarrow ROP(=O)Cl_2 \Rightarrow (Hydrolysis) \Rightarrow (RO)P(=O)(OH)_2 \qquad (7.20)$$

$$(RO)_2P(=O)H + Cl_2 \Rightarrow (RO)_2P(=O)Cl \Rightarrow (Hydrolysis) \Rightarrow (RO)_2P(=O)(OH) \qquad (7.21)$$

7.2.2.4.2 Properties, Performance Characteristics, and Applications

Phosphoric acids tend to hydrolyze further when exposed to humidity. The extent of hydrolysis depends upon the moisture content of the ambient atmosphere and the duration of exposure.

Wherever possible, phosphoric acids should be handled in a dry nitrogen atmosphere to prevent hydrolysis. Therefore, for applications where incidental moisture contact is inevitable, acid phosphates are not recommended.

Acid phosphates are used as rust inhibitors and antiwear additives. However, they are not as widely used as their amine-neutralized derivatives, for example, amine phosphates.

7.2.3 Sulfur–Phosphorus Additives

Sulfur–phosphorus additives are used to provide protection against moderate- to high-pressure, metal-to-metal contacts in boundary lubrication and EHL. Metallic sulfur–phosphorus additives, such as zinc dithiophosphates (ZnDTP), are the most important antiwear/AS components used in engine oils. Ashless sulfur–phosphorus additives are used less extensively, and the most commonly available S/P additives in the marketplace are based on chemistries of dithiophosphates, thiophosphates, and phosphorothioates. Other important applications of S/P compounds are in matches, insecticides, flotation agents, and vulcanization accelerators.

7.2.3.1 Ashless Dithiophosphates

Numerous patents were issued on the use of phosphorodithioic acid esters in lubricating oils in the early days. U.S. Patent 2,528,732 describes alkyl esters of phosphorodithioic acid. U.S. Patent 2,665,295 describes the S-terpene ester, while U.S. Patent 2,976,308 describes an anti-Markovnikov addition of phosphorodithioic acid ester to various olefins, both aromatic and aliphatic. Amine dithiophosphates and other novel dithiophosphate esters are reported in the literature [34–38]. Coupling with vinyl pyrrolidinone, acrolein or alkylene oxides (to make hydroxyl derivatives) are also known [39–41].

7.2.3.1.1 Chemistry and Manufacture

Similar to metallic dithiophosphates, ashless dithiophosphates are also based on phosphorus pentasulfide (P_2S_5) chemistry. They can be prepared from the same precursor of ZnDTP, dithiophosphoric acid (Equation 7.22) through the reaction of alcohol (or alkylphenol) and P_2S_5.

$$4ROH + P_2S_5 \Rightarrow 2(RO)_2P(=S)SH + H_2S \qquad (7.22)$$

The dithiophosphoric acids are further reacted with an organic substrate to generate ashless derivatives. Typical organic substrates are compounds such as olefins, dienes, unsaturated esters (acrylates, methacrylates, vinyl esters, etc.), unsaturated acids, and ethers. The efficiency and stability of the ashless dithiophosphates very much depend on components used in their manufacture and the reaction conditions.

The most common ashless dithiophosphate used in the marketplace is a dithiophosphate ester made from ethyl acrylate and o,o-diisopropyl dithiophosphoric acid as described in the following:

$$\left[C_3H_7-O-\right]_2-P(=S)S-CH_2-CH_2-C(=O)O-C_2H_5$$

Treatment of terpenes, polyisobutylene (PIB), or polypropylene (PP) with phosphorus pentasulfide and hydrolysis give thiophosphonic acids (R–P(=S)(OH)₂, where R = PIB, terpenes, or PP). They can be further reacted with propylene oxide or amines to reduce acidity. However, this type of additive belongs to the same class of chemicals called ashless dispersants. Hence, they can be dual-functional dispersants with improved antiwear/EP properties.

7.2.3.1.2 Applications and Performance Characteristics

Unlike ZnDTP, ashless dithiophosphates are usually not as versatile and therefore cannot be considered as multifunctional additives. Although ashless dithiophosphates have fairly good antiwear and EP properties, their anticorrosion properties are not as good as ZnDTP. This is closely related to the stability and decomposition mechanisms of ashless dithiophosphates. Relatively weak corrosion protection also limits their application at high concentrations in engine oils as well as some industrial oils.

Ashless dithiophosphates can be useful in metalworking fluids, ATFs, gear oils, greases, and nonzinc hydraulic fluids [42,43].

7.2.3.2 Ashless Phosphorothioates and Thiophosphates

Numerous esters of the phosphorothioic acids are known. In salts and esters of these oxygen/sulfur acids, there may be a preferred location of the multiple bond, but in general this is not well known. Thus, in principle, there are two series of possible acids, each of which might give rise to salts and esters as described in the following:

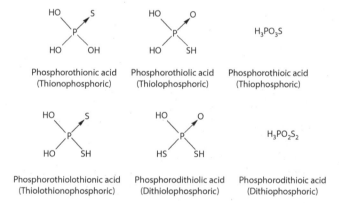

Phosphorothionic acid (Thionophosphoric) Phosphorothiolic acid (Thiolophosphoric) Phosphorothioic acid (Thiophosphoric) H_3PO_3S

Phosphorothiolothionic acid (Thiolothionophosphoric) Phosphorodithiolic acid (Dithiolophosphoric) Phosphorodithioic acid (Dithiophosphoric) $H_3PO_2S_2$

The "thionic" acids contain the group P=S, while the "thiolic" acids contain the group P–SH. The term "thioic" is often used when the molecular form is unknown or when specification is not desired. One form of these acids is usually more stable than the other, and it may not be possible to prepare both esters as, for example, the isomers of phosphorothioic acid:

In the case of some esters, the thiolo form is the most stable, but the phenyl ester exists 80% in thiono, $(PhO)_2P(=S)OH$, and 20% in thiolo, $(PhO)_2P(=O)SH$, forms. The equilibrium of these compounds is liable to be dependent on the nature of the R groups, the solvent used, and even the concentration. Intermolecular hydrogen bonding may be expected to play a part in such equilibrium [33].

7.2.3.2.1 Chemistry and Manufacture

The creation of a compound with a phosphorus–sulfur linkage can often be carried out simply by heating the appropriate phosphorus compound with sulfur [44]. Likewise, the replacement of oxygen by sulfur in compounds containing P–O linkages can also be achieved simply by heating them with P_2S_5. Inorganic phosphorothioates (thiophosphates) are usually prepared from sulfur-containing phosphorus compounds. They are produced during the hydrolytic breakdown of phosphorus sulfides and are often themselves unstable in water. They hydrolyze to the corresponding oxy compounds with the evolution of H_2S. Phosphorus–sulfur compounds are often thermally less stable than their oxy analogs. A few examples are listed as follows:

$$P_4S_{10} + 12NaOH \Rightarrow 2Na_3PO_2S_2 + 2Na_3PS_3O + 6H_2O \quad (7.23)$$

$$(BuO)_2P(=O)SH + RI \Rightarrow (BuO)_2P(=O)SR + HI \quad (7.24)$$

$$(PhO)_3P + S \Rightarrow (PhO)_3P=S \quad (7.25)$$

$$(PhO)_3P + PSCl_3 \Rightarrow (PhO)_3P=S + PCl_3 \quad (7.26)$$

Hydrolysis of phosphorothioate esters results in a progressive loss of sulfur as hydrogen sulfide (H_2S) and its replacement by oxygen.

$$(RO)_3P=S + H_2O \Rightarrow (RO)_3P=O + H_2S \quad (7.27)$$

7.2.3.2.2 Applications and Performance Characteristics

It has been known for many years that sulfur compounds form a film of iron sulfide, and phosphorus compounds form iron phosphate, on the mating metal surfaces. Generally, the films formed from sulfur sources such as SIB are expected to contain FeS, $FeSO_4$, as well as organic fragments from the additive decomposition. With phosphorus sources, such as dialkyl phosphites, films containing $FePO_4$, $FePO_3$, as well as organic fragments are expected. When both sulfur and phosphorus are present, both elements contribute to the nature of the film, and which one predominates depends on the S/P ratio, the decomposition mechanisms, and the operating conditions, for example, high speed and shock or high torque/low speed.

Ashless phosphorothioates are widely used as replacements for metallic dithiophosphates in many lubricant applications where metal is less desirable [43,44]. Phosphorothioates are often present (generated in situ) in lubricant formulations when both sulfur and phosphorus additives are used. Aryl

phosphorothioates provide good thermal stability and good antiwear/AS properties as evidenced by their strong FZG performance.

7.2.4 SULFUR–NITROGEN ADDITIVES

Sulfur- and nitrogen-containing additives are used to provide protection against moderate- to high-pressure, metal-to-metal contacts in boundary lubrication and EHL. Both open chain and heterocyclic compounds have attracted a considerable amount of research effort to explore their potential as antiwear and AS additives. Among open-chain additives, dithiocarbamates are the most widely used. Other additives, such as organic sulfonic acid ammonium salts [45] and alkyl amine salts of thiocyanic acid [46], are reported in the literature but are of relatively low commercial value. Nitrogen- and sulfur-containing heterocyclic compounds, such as 2,5-dimercapto-1,3,4-thiadiazole (DMTD, Structure E), 2-mercapto-1,3-benzothiazole (MBT, Structure F), and their derivatives, have been used for many years as antioxidants, corrosion inhibitors, and metal passivators, generally at relatively low concentrations.

Structure H   Structure I

7.2.4.1 Dithiocarbamates

The dithiocarbamates, the half amides of dithiocarbonic acid, were discovered as a class of chemical compounds early in the history of organosulfur chemistry [47,48]. The strong metal-binding properties of the dithiocarbamates were recognized early by virtue of the insolubility of the metal salts and the capacity of the molecules to form chelate complexes. Other than applications in lubricant areas, dithiocarbamates have been used in the field of rubber chemistry as vulcanization accelerators and antiozonants.

7.2.4.1.1 Chemistry and Manufacture

Organic dithiocarbamates can be made by a one-step reaction of dialkylamine, carbon disulfide, and an organic substrate. The organic substrate is preferably an olefin, diene, epoxide, or any other unsaturated compounds as exemplified in the literature [49,50]. Organic dithiocarbamates can also be made through a two-step reaction involving ammonium or metal dithiocarbamate salts and organic halides [51]. In the case of their ammonium salts, N-substituted dithiocarbamic acids, $RNHC(=S)SH$ or $R_2NC(=S)SH$, are formed by the reaction of carbon disulfide with a primary or secondary amine in alcoholic or aqueous solution before they are further reacted with ammonia. In order to conserve the more valuable amine, it is common practice to use an alkali metal hydroxide to form the salt:

$$RNH_2 + CS_2 + NaOH \Rightarrow RNHC(=S)S-Na + H_2O \quad (7.28)$$

The dithiocarbamic acid can be precipitated from an aqueous solution of dithiocarbamate by adding strong mineral acid.

The acids are quite unstable but can be held below 5°C for a short time. The most common additive, methylene bis-dibutyl dithiocarbamate, is prepared from sodium dibutyl dithiocarbamate and methylene chloride:

$$2(C_4H_9)_2NC(=S)\text{–}S\text{–}Na + CH_2Cl_2 \quad \Rightarrow \quad [(C_4H_9)_2NC(=S)S]_2CH_2 + 2NaCl$$

(7.29)

7.2.4.1.2 Applications and Performance Characteristics

Unlike metallic dithiocarbamates that have been widely used in lubricants, ashless dithiocarbamates have only been gaining more attention recently. Relatively high cost certainly is a major factor in limiting wider use. The success of metallic dithiocarbamates also overshadows their ashless counterpart. Certain metallic dithiocarbamates, such as molybdenum dithiocarbamates, offer exceptionally good frictional properties that also cannot be matched by their ashless analogs. However, ashless dithiocarbamates have been found to be versatile, multifunctional additives in a few areas. They can be effective antiwear/AS additives as well as good antioxidants and metal deactivators [52–55] (Structure Ga and Gb). They tend to generate less sludge or deposits than mostly metallic additives and they are very compatible with various base oils.

Structure Ja

Structure Jb

7.2.4.2 Dimercaptothiadiazole and Mercaptobenzothiazole Additives

Additives derived from 2,5-dimercapto-1,3,4-thiadiazole (DMTD) and 2-mercaptobenzo-thiazole (MBT) are well documented in the literature. Due to strong ring stability (partial aromaticity and resonance delocalization), balanced sulfur–nitrogen distributions, and reactive mercaptan groups, both heterocyclic compounds can be versatile core molecules to make many useful additives with many beneficial characteristics, such as improved thermal/oxidative stability and reduced corrosivity. Unfortunately, some potentially good reactions are hampered by the limited solubility of DMTD and MBT in common petrochemical solvents. Therefore, a suitable sample preparation procedure is very critical to help achieve desirable antiwear/AS additives.

7.2.4.2.1 Chemistry and Manufacture

Many differing organic reactions can be applied to functionalize the mercaptan groups of DMTD and MBT. Oxidative coupling reactions involving other alkyl mercaptans can

bring in additional sulfur for EP performance and additional alkyl chains for improved solubility [51]. Addition reactions with organic compounds containing activated double bonds can link DMTD or MBT heterocyclic core molecules with long-chain esters, ketones, ethers, amides, and acids together [52–55]. Likewise, ring opening with epoxides to generate alcohol derivatives is also known [56]. Direct amine salt formation and linking alkyl amines through Mannich base condensation are also extensively studied [57–59]. A number of examples are listed in the following where TD is the abbreviation for the thiadiazole moiety and BT is for the benzothiazole moiety:

Oxidative coupling
$$DMTD + 2RSH + 2H_2O_2 \quad \Rightarrow \quad RS\text{–}S\text{–}(TD)\text{–}S\text{–}SR + 4H_2O \quad (7.30)$$

Mercapto alkylation and Mannich alkylation
$$DMTD + 2CH_2{=}O + 2RSH \quad \Rightarrow \quad RS\text{–}CH_2\text{–}S\text{–}(TD)\text{–}S\text{–}CH_2\text{–}SR + 2H_2O$$

(7.31)

$$MBT + CH_2{=}O + RNH_2 \quad \Rightarrow \quad (BT)\text{–}S\text{–}CH_2\text{–}NHR + H_2O \quad (7.32)$$

Amine salt formation
$$DMTD + 2RNH_2 \quad \Rightarrow \quad RNH_3\text{–}S\text{–}(TD)\text{–}S\text{–}NH_3R \quad (7.33)$$

7.2.4.2.2 Applications and Performance Characteristics

MBT is a light yellow powder with limited solubility in hydrocarbons. It is more soluble in aromatic solvent (~1.5% in toluene), polar solvents, and highly aromatic oils. MBT is used as a copper corrosion inhibitor in fuels as well as a corrosion inhibitor/deactivator in numerous industrial lubricants such as heavy-duty cutting and metalworking fluids, hydraulic oils, and lubricating greases. DMTD is also a light yellow powder with very limited solubility in hydrocarbons. It is considered a versatile chemical intermediate suitable for making various oil-soluble derivatives.

Both MBT and DMTD derivatives are widely used as copper passivators and nonferrous metal corrosion inhibitors. Some proprietary load-carrying additives are substituted MBT and DMTD compounds that are used in various applications either as a component or a part of additive packages with a specific purpose [60,61]. In the absence of any phosphorus moiety in MBT and DMTD, their oil-soluble derivatives are suitable for replacing zinc dithiophosphates in some lubricant applications. For example, a commercial, high-density, powderlike DMTD derivative is used as a dual-functional antioxidant/EP agent in greases.

7.2.4.3 Other Sulfur–Nitrogen Additives

In addition to DMTD, MBT, and dithiocarbamate additives, there are other sulfur–nitrogen-containing additives available in the marketplace or reported in the literature. Among these, phenothiazine derivatives (Structure H, PTZ), substituted thiourea additives (Structure I, TU), thionoimidazolidine derivatives (Structure J, TIDZ), thiadiazolidine and oxadiazole derivatives (Structure K, L), thiuram monosulfides, thiuram disulfides, and benzoxazoles are of particular interest because they are all sulfur- and nitrogen-rich molecules [62–68].

Thiuram disulfides, chemically similar to dithiocarbamates, can be used in the rubber industry as vulcanizers. 2-Alkyldithiobenzoxazoles also offer good frictional properties in addition to strong antiwear/EP properties [69].

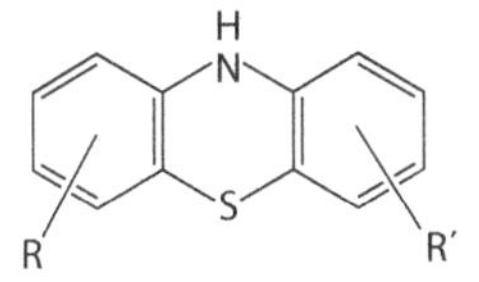

Structure L, PTZ

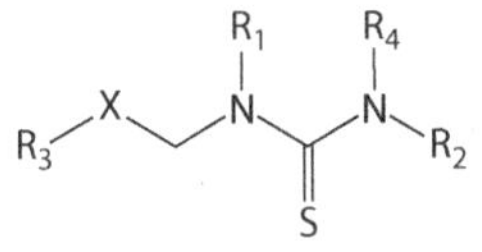

Structure M, TU

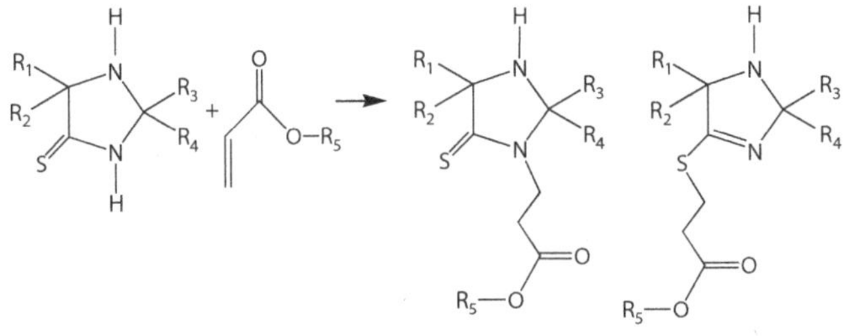

Structure N, TIDZ

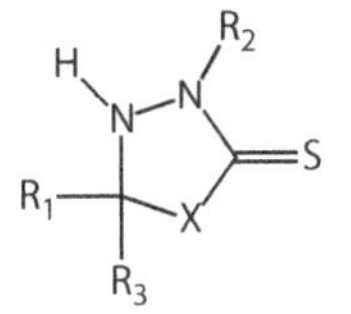

Structure O, TDZL

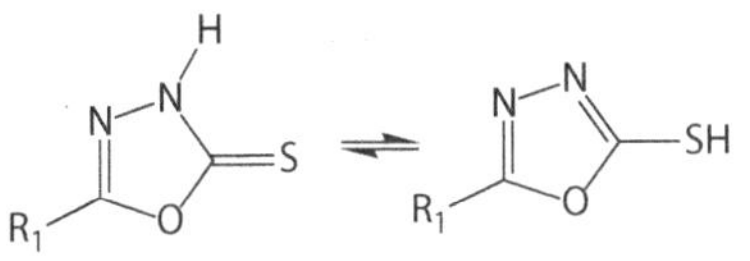

Structure P, ODZ

7.2.5 Phosphorus–Nitrogen Additives

Phosphorus–nitrogen additives are used to provide protection against moderate- to high-pressure, metal-to-metal contacts in boundary lubrication and EHL. Ashless phosphorus–nitrogen additives are used as dual-functional antiwear/antirust additives extensively, and the most commonly available in the marketplace are based on chemistries of amine dithiophosphates, amine thiophosphates, amine phosphates, and phosphoramides.

7.2.5.1 Amine Phosphates

Amine phosphates are by far the most important phosphorus–nitrogen-containing additives used in lubricants. In fact, they are multifunctional additives possessing very good antirust properties as well as antiwear/AS properties.

7.2.5.1.1 Chemistry and Manufacture

Amine phosphates are produced by treating acid phosphates with alkyl or aryl amines. Under various conditions, neutral, overbased, and underbased amine phosphates can be synthesized. If mixed mono- and dialkyl acid phosphates are used as starting materials, mixed mono- and dialkyl amine phosphates are produced. The final additives usually possess high total acid number (TAN) and high total base number (TBN), although reaction adducts are considered fairly neutral. It is known that a complete neutralization of both phosphoric acid groups in monoalkyl acid phosphates with amines cannot be easily achieved, and therefore, under normal conditions, a partially neutralized amine phosphate is formed.

$$(RO)_2P(=O)(OH) + R'NH_2 \Rightarrow (RO)_2P(=O)O \cdot NH_3R' \qquad (7.34)$$

$$(RO)P(=O)(OH)_2 + R'NH_2 \Rightarrow (RO)P(=O)(OH)O \cdot NH_3R' \qquad (7.35)$$

7.2.5.1.2 Applications and Performance Characteristics

Amine phosphates are extensively used in industrial oils, greases, and automotive gear oils. They offer very good rust protection as demonstrated in various bench rust tests (ASTM D665 and CRC L-33). They also show very good antiwear/AS characteristics (four-ball wear and four-ball EP, FZG, Timken, and CRC L-37). Since amine phosphates are very polar species, they interact strongly with other additive components, making their performance very dependent on the formulation. Hence, extra attention is needed when amine phosphates are used.

7.2.5.2 Amine Thiophosphates and Dithiophosphates

Amine thiophosphates and amine dithiophosphates can be found in engine oils and industrial oils where zinc dithiophosphates and other nitrogen-containing additives are used, either as decomposition products or in situ produced products. They are critical to the lubricant performance because of their high activity toward metal surfaces.

7.2.5.2.1 Chemistry and Manufacture

Amine thiophosphates are produced by reacting thiophosphoric acid with alkyl or aryl amines [75]. Likewise, amine dithiophosphates are synthesized from dithiophosphoric acid and amines.

$$(RO)_2P(=S)SH + H_2NR' \Rightarrow (RO)_2P(=S)S \cdot H_3NR' \qquad (7.36)$$

$$(RO)_2P(=O)SH + H_2NR' \Rightarrow (RO)_2P(=O)S \cdot H_3NR' + (RO)_2P(=S)O \cdot H_3NR' \qquad (7.37)$$

7.2.5.2.2 Applications and Performance Characteristics

Amine thiophosphates and dithiophosphates are also multifunctional additives providing good rust inhibition and antiwear properties. Due to their high activity and low stability, amine thiophosphates and dithiophosphates are not as extensively used as either amine phosphates or metallic dithiophosphates. A detailed study of their antiwear mechanisms suggested that a tribofragmentation process is involved [76,77]. Relatively poor corrosion control is one area of concern that needs attention. With proper formulation adjustments, it is quite feasible to overcome certain intrinsic weaknesses and apply both chemistries to various lubricant products.

7.2.5.3 Other Phosphorus–Nitrogen Additives

There are many other phosphorus–nitrogen-containing ashless antiwear additives reported in the literature. Some are proprietary technologies and their commercial status is unknown. Organophosphorus derivatives of benzotriazole are a group of additives based on triazole and dialkyl or dialkylphenyl phosphorochloridate chemistry [78]. Arylamines and dialkyl phosphites can be coupled through a Mannich condensation reaction to form unique phosphonates that are used as multifunctional antioxidant and antiwear additives [79]. Bisphosphoramides are also reported [80].

7.2.6 Nitrogen Additives

Nitrogen-containing additives are used to provide rust inhibition and cleanliness features in various lubricant applications. For example, nitrogen-containing ashless dispersants are a key component for engine oils, and alkoxylated amine compounds are used in lubricating greases to provide corrosion inhibition [81]. Furthermore, arylamines are widely used as antioxidants due to their abilities to terminate radical chain propagation and decompose peroxides. Very few nitrogen additives alone are considered effective antiwear/AS additives, and their performance is either very specific to industrial applications or fairly dependent upon product formulations. However, when used in combination with other sulfur, phosphorus, or boron additives, nitrogen-containing additives can be very effective supplements to enhance antiwear/AS performance.

7.2.6.1 Chemistry, Manufacture, and Performance

Several novel chemistries are available in the literature for nitrogen-only antiwear additives. Among these, dicyano compounds were tested and exhibited very good four-ball wear activities [82]. Polyimide-amine salts of styrene–maleic anhydride copolymers are also reported as antiwear additives; however, high additive concentrations (5%–10%) are needed [83]. Alkoxylated amines (Structure M) and mixtures of fatty acid, fatty acid amide, imide, or ester derived from substituted succinic acid or anhydride have been identified to be good fuel lubricity additives [84] (Structure M). Alkyl hydrazide additives possessing two adjacent nitrogen atoms have also been claimed to exhibit good antiwear properties [85] (Structure N). Products of nitrogen heterocycles, such as oxadiazole ODZ (Tables 7.2 and 7.3 for performance evaluations), benzotriazole (BZT), tolyltriazole (TTZ), alkyl succinhydrazide (SHDZ), and borated hydroxypyridine (BHPD) (Structures O, P, Q, R, and S, respectively), with pendant alkylates, amines, or carboxylic acids have been found to be effective antiwear additives in both lubricants and fuels [86–92]. Although triazoles are costly chemicals, they have unique geometric structures that contribute high surface film–forming efficiency.

TABLE 7.2
SAE 5W-20 Prototype Motor Oil Formulation

Component	Formulation A (wt%)
Solvent neutral 100	22.8
Solvent neutral 150	60
Succinimide dispersant	7.5
Overbased calcium phenate detergent	2
Neutral calcium sulfonate detergent	0.5
Rust inhibitor	0.1
Antioxidant	0.5
Pour point depressant	0.1
OCP VI improver	5.5
Antiwear additive[a]	1

[a] In the case of no antiwear additive present in the formulation, solvent neutral 100 is put in its place at 1.0 wt%.

TABLE 7.3
Four-Ball Wear Results

Compound	Formulation	Wear Scar Diameter (mm)
No antiwear additive	A	0.73 (0.74)[a]
1.0 wt% ZDDP	A	0.50 (0.51)
0.5 wt% ZDDP	A	0.70 (0.67)
5-Heptadecenyl-1,3,4-oxadiazole	A	0.38 (0.38)
5-Heptyl-1,3,4-oxadiazole	A	0.54 (0.56)
5-Heptadecenyl-2,2-dimethyl-1,3,4-oxadiazole	A	0.7
5-Heptadecenyl-2-furfuryl-1,3,4-oxadiazole	A	0.38 (0.39)

[a] Repeat data points in parentheses.

$$RO(C_4H_8O)_nCH_2CH_2CH_2NH_2$$

Structure Q

Structure R, AHDZ

Structure S, ODZ

Structure Ta, BZT   Structure Tb, TZ

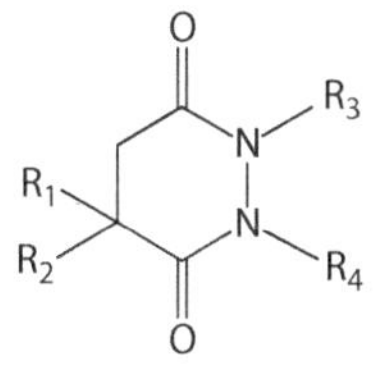

Structure U, SHDZ

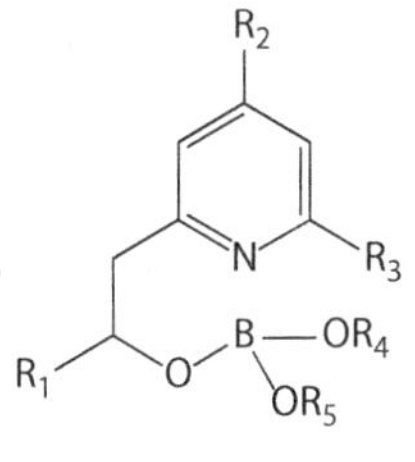

Structure V, BHPD

Both BZT and TTZ derivatives are also effective copper deactivators at low concentrations. Therefore, these types of additives are indeed dual functional. They find applications in industrial oils, greases, and fuels.

Table 7.2 listed a prototype engine oil formulation used for the evaluation of oxadiazole additives where various oxadiazoles can be blended at 1 wt% in place of the same amount of light base oil. Table 7.3 listed the four-ball wear performance data where a series of oxadiazoles were evaluated against 0.5 and 1 wt% ZDDP. As demonstrated, those oxadiazole additives exhibited fairly good antiwear properties in this bench test [86].

7.2.7 Additives with Multiple Elements

Complex additives with multiple elements can be derived from various S/P, sulfur/nitrogen, phosphorus/nitrogen, and many other traditional additive building blocks. As a result, molecules with more than four, five, six, or even more elements are created (S/P/N/B in addition to C/H/O). Derivatization frequently adds a degree of complexity yet provides a chance of achieving better synergisms among all critical elements that not only can satisfy the performance needs but also could help to neutralize any potentially added costs associated with the new chemistry under development.

Many examples are available in the literature as well as in the marketplace, such as amine salts of dithiophosphates and thiophosphates (Section 7.2.5.2); borated derivatives of dithiophosphates [41], dithiocarbamates [50], and dimercaptothiadiazole [93]; urethane derivatives of dithiophosphates [94] (Structure T); and reaction adducts of dialkyl phosphites, sulfur, and acylated amines [95].

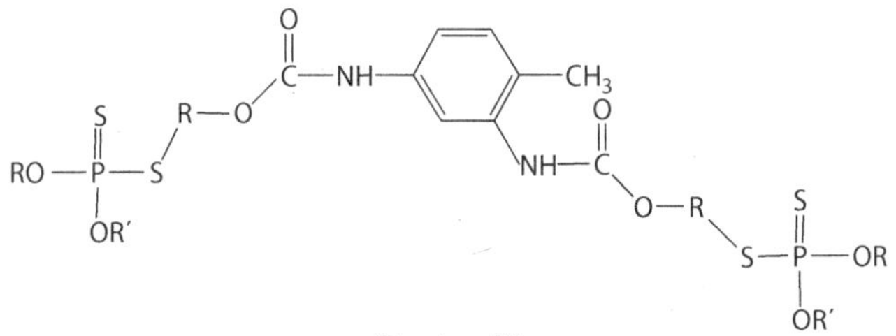

Structure W

Several complex additives have different chemistries involved with the same element in a single molecule in order to attain strong synergisms. As exemplified in the following case, where both phosphite chemistry and phosphate chemistry are incorporated into the same molecule, greater antiwear performance can be achieved ([96]; reaction 7.38):

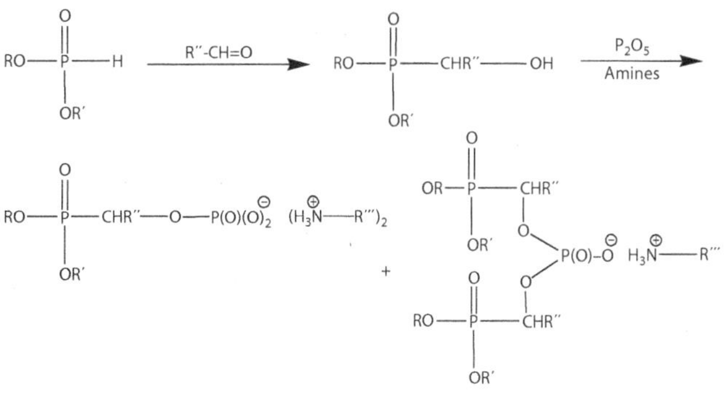

$$(7.38)$$

The synergistic antiwear performance of the aforementioned complex phosphorus additives is illustrated in Chart 7.1. Nine different analogs were synthesized and tested at 1 wt% in base oils using three different conditions in the four-ball wear test. As demonstrated, they all exhibit exceptionally good antiwear properties.

7.2.8 Halogen Additives

Chlorine was one of the earliest antiwear and AS elements used in the lubricant industry. Chlorine-containing additives are still used in cutting oils and related metalworking lubricants, in combination with sulfur additives. Iodine was mentioned in aluminum-processing lubricants for wear control. Fluorine, in perfluorinated compounds, is well known to reduce wear and especially friction.

The chlorine compounds act and function in that they coat the metal surface with a metal chloride film under the influence of the high pressure at the point of lubrication and in the presence of traces of moisture. $FeCl_2$ melts at 672°C and has low shear strength in comparison with steel.

The effect of the chlorine compounds depends on the reactivity of the chlorine atom, the temperature, and the concentration. Hydrogen chloride formed in the presence of larger quantities of moisture can cause severe corrosion of the metal surfaces. As the corrosion hazards increase along with the AS properties with increasing reactivity of the chlorine atoms, a compromise must be found in the development of chlorine-containing additives.

Chlorinated paraffins (CPs) such as trichlorocetane represent a group of important AS additives used in the past. They can significantly increase the load stages in the FZG test with increasing concentration. The chain length has practically very little influence on the AS effect; on the other hand, the load-carrying capacity increases with increasing degree of chlorination. In practice, CPs with ~40 to 70 wt% chlorine are used; however, they are sensitive to moisture and light and can easily evolve hydrogen chloride [97]. Compounds such as

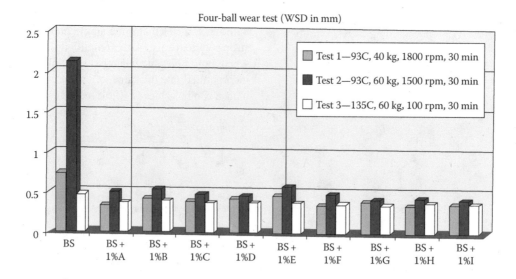

CHART 7.1 Complex phosphorus additives. A and B: Dibutyl phosphite-butyraldehyde-primene JMT with different JMT ratios; C and D: Dibutyl phosphite-butyraldehyde-Adogen 183 with different Adogen amine ratios; E, G, and I: Dibutyl phosphite-butyraldehyde-Duomeen O with different Duomeen ratios; F and H: Dibutyl phosphite-butyraldehyde-bis 2-EH amine with different bis-amine ratios.

phenoxy-propylene oxide, amines, or basic sulfonates neutralize the hydrogen chloride and thus act as stabilizers.

Good results are also obtained with chlorinated fatty acids and their derivatives, particularly those with trichloromethyl groups in the end position, since the additives with –CCl$_3$ groups are particularly effective.

Due to their high stability, chlorinated aromatics have less favorable AS properties than the chlorinated aliphatics. Alkylaromatics with chlorinated side chains improve the load-carrying capacity much more than those chlorinated in the ring; the efficiency increases with the number of carbon atoms in the side chain. Chlorinated fatty oils and esters as well as chlorinated terpenes and amines have also been patented as AS additives.

The U.S. Environmental Protection Agency (EPA) aims to eliminate only medium- and long-chain CPs by May 2016. The EPA determined that very long-chain versions (C21 and longer) are not a risk to the environment, so these may still be sold. Suppliers still must file Remanufacture Notifications and obtain CAS numbers for very long-chain CPs [98].

Sulfur–chlorine additives were found to be satisfactory for gear lubrication in passenger cars in the mid-1930s. Apparently, this type of additive could satisfy the high-speed and moderate-load operation of passenger cars used in that time period. When sulfur and chlorine are combined in the organic molecule, the sulfur somewhat reduces the corrosive tendency of chlorine; on the other hand, the AS properties of the combined moieties are improved in comparison with the individual compounds. Chlorinated alkyl sulfides, sulfurized chloronaphthalenes, chlorinated alkyl thiocarbonates, bis-(p-chlorobenzyl) disulfide, tetrachlorodiphenyl sulfide, and trichloroacrolein mercaptals [Cl$_2$C=CCl–CH(SR′)–SR″, where R′ and R″ are alkyl or aryl] must be mentioned in this class. Reaction products of olefins and unsaturated fatty acid esters with sulfur chlorides contain highly reactive β-chlorosulfides,

which due to their reactive chlorine and sulfur atoms give very good AS agents yet show more or less strong corrosive tendencies. However, severe wear was frequently encountered in truck axles where performance under high-torque, low-speed conditions is of greater importance. Later on, the presence of chlorine, although a good AS agent, was found to be detrimental to lubricant thermal stability. Hence, for the last 30 years, chlorine has not been used in gear oils.

Chlorinated trioleyl phosphate, condensation products of chlorinated fatty oils with alkali salts of dithiophosphoric acid diesters, and reaction products of glycols with PCl$_3$ are examples of chlorine–phosphorus additives used in earlier years.

The most serious drawback for chlorine antiwear and AS additives is in the environmental area. Legislation around the industrial world limits the chlorine content of many lubricants to parts per million. Therefore, except for the cutting oil industry, which is also under pressure to change, chlorine additives are not considered a viable option for modern lubricants.

7.2.9 Nontraditional Antiwear/ Antiscuffing Additives

Traditional sulfur, phosphorus, and halogen-related compounds are considered to be the dominant antiwear/AS additives in the marketplace. However, as environmental concerns escalate, the future trends will favor products that diminish potential hazard and disposal problems. Recent clean fuel activities are driving sulfur levels toward 10–50 weight parts per million ranges. Subsequently, the petroleum industry is favoring lower sulfur lubricants since sulfur is also known to poison the catalytic system used for NO$_x$ reduction. Therefore, the use and development of nontraditional antiwear additives are becoming more valuable.

A number of nonsulfur, nonphosphorus ashless antiwear additive technologies have been reported in the literature

[99–103]. Among these, high hydroxyl esters, dimer acids, hydroxyamine esters, acid anhydrides, cyclic and acyclic amides, and boron derivatives are recognized as leading technologies. Graphite, graphene, and polytetrafluoroethylene (PTFE) possess excellent friction reduction properties and indirectly contribute some antiwear/AS characteristics. However, these materials need to be dispersed in the oil as they have very limited lubricant solubility, which hampers their usefulness. Organic borates are considered effective friction modifiers, antioxidants, and cleanliness agents. Recent studies indicate that some borates can be good antiwear additives. Potassium borates have been used in gear oils for years, but these types of metallic borates are outside the scope of this chapter. Esters are known to possess good lubricity properties. The properties can be further improved to offer antiwear characteristics through proper functionalization. Several companies have marketable products in this area. Among all new technologies, ionic liquids and nanoparticles are most widely studied in recent decade; therefore, the authors would like to place them into two special topics as described in the following.

7.2.10 Ionic Liquids as Ashless Antiwear and Antiscuffing Lubricant Additives

As mentioned in the previous section, ionic liquids are an important class of chemicals, which show very good antiwear and AS performance in lubricants. They are nontraditional sulfur–phosphorus- and halogen-containing compounds because these elements present not only in sufficient amount but also linked through quite different functional compositions and synthesis schemes in chemistries. For example, phosphorus-containing ionic liquid is usually made from anionic phosphate ligands.

7.2.10.1 Definition of Ionic Liquids

Ionic liquids are molten salts that melt below 100°C. As a subset of ionic liquids, room temperature ionic liquids are those that melt lower than ambient temperature. They are made of ion pairs consisting of bulky asymmetric cations and weakly coordinating, more highly symmetric anions. Due to the large size and asymmetry of the cations and diffusive nature of the charge distribution of the ion pairs, it is difficult for the ionic liquid molecules to pack and undergo crystallization, so they prefer to stay as liquids at room temperature [104,105]. Literature has shown that there are a large number of cations and anions, from which ionic liquids can be selectively combined [104,106]; typical ionic liquids that are being investigated for potential applications in lubrications are shown in Figure 7.2.

Since ionic liquids can be obtained from any combination of a large number of currently available cations and anions and new ones that can be readily synthesized, one can virtually claim that there would be close to infinite number of ionic liquids at anyone's disposal [106]. Ionic liquids possess many useful properties. In particular, they are low volatility, nonflammability, high thermal stability, wide liquid range, high solvent power (miscible with water and organic solvents), and good conductivity [106]. Because of these outstanding properties, they have been used to improve chemical process efficiency and increase product yield, replace volatile and flammable solvents in battery electrolytes, maximize safety of delivery of gases to electronics market, facilitate the production of valuable pharmaceutical intermediates, and enhance petroleum refinery processes and other applications in chemical industry [104].

7.2.10.2 Ionic Liquids as Neat Lubricants

The potential of ionic liquids as neat lubricants was first reported by Ye et al. in 2001 [107]. In their study,

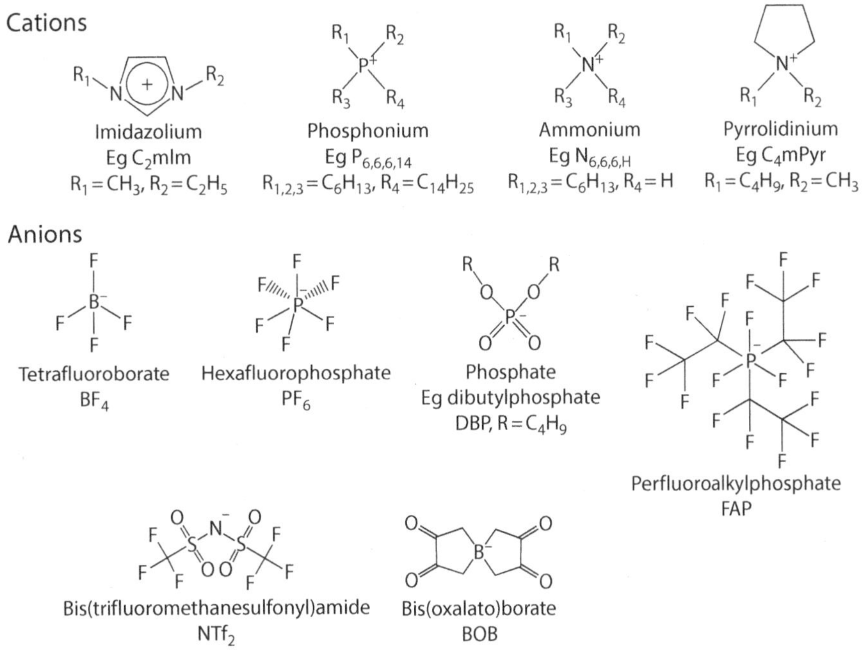

FIGURE 7.2 Typical cations and anions of ionic liquid molecules. (From Somers, A.E. et al., *Lubricants*, 1, 3, 2013.)

1-ethyl-3-hexylimidazolium tetrafluoroborate ionic liquid, $[C_6C_2Im][BF_4]$, as a neat lubricant was found to lower friction coefficient and wear volume in steel–steel contact compared with traditional lubricants, such as phosphazene (X-1P) and perfluoropolyether. Since then, more ionic liquids as neat lubricants were found to lower friction and wear in steel–steel and steel–aluminum contacts [108]. For steel–steel contact, the most studied cation was imidazolium ion. The variants of this type of cation include (1) changes in the chain length of the two alkyl groups from 1 to 18 carbon atoms, (2) one or two imidazolium cations (Yao) [109], and (3) one of the alkyl chains contain an ester moiety. Phosphonium cation was also studied. The number of counter anions investigated was more. The list includes tetrafluoroborate (BF_4), hexafluorophosphate (PF_6), perfluoroalkylphosphate (FAP), bis(trifluoromethanesulfonyl)amide (NTf_2), and various dialkylphosphates. For steel–aluminum contact, in addition to simple dialkyl imidazolium cations, phosphonylimidazolium cations, phosphonium cations, and pyrrolidinium cations were studied. The counter anions included BF_4, PF_6, FAP, NTf_2, tosylate, triflate, phosphates, phosphite, and various borates. Several reviews of neat ionic liquid lubricants are available in the literature [105,106,108,109].

The excellent properties of low volatility, nonflammability, and high thermal stability attracted many studies to explore for applications of ionic liquids as lubricants in high vacuum machines used in space, in equipment operated in high temperatures to prevent fire, and in low pressure and clean environment as in microelectrochemical machines and nanoelectrochemical machines [108,109]. Ionic liquids as neat lubricants can be improved by some additives developed for mineral oils or synthetic lubricants. For example, 1% of TCP in an ionic liquid rapidly establishes a tribofilm and reduces the wear volume by 64% compared to the same test for the neat ionic liquid or neat TCP [108,110].

7.2.10.3 Ionic Liquids as Friction Modifier and Antiwear Lubricant Additives

Since the work of Ye et al. in 2001 showed the potential of imidazolium-based ionic liquids as effective lubricants, numerous studies started investigating ionic liquids as lubricant additives. One of the incentives of using ionic liquids as lubricant additives instead of neat lubricants is the cost of materials. The other is to replace the traditional lubricant systems with potentially more efficient and/or more environmentally friendly alternatives. An earlier attempt to explore imidazolium-based ionic liquids as lubricant additives was carried out by Jimenez et al., who reported the results in 2006 [108,111]. They studied a series of ionic liquids by combining imidazolium cation with anion of BF_4, triflate (CF_3SO_3), or tosylate ($MeC_6H_4SO_3$), as 1 wt% lubricant additives in mineral oil for steel–aluminum contact. They found that as lubricant additives in mineral oil, the more effective aluminum wear-reducing ionic liquids at room temperature using the same anion of BF_4 were the cations with shorter chains, that is, $C_2mIm > C_6mIm > C_8mIm$. For the same cation, C_2C_1Im, the order of wear-reducing capability

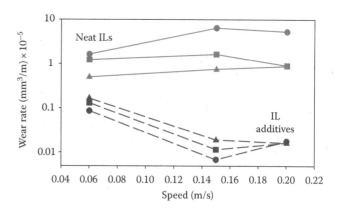

FIGURE 7.3 Comparative antiwear performance of $[C_2C_1Im^+]$ ionic liquids as neat lubricants and 1 wt% additives [108]; counter anions: ● BF_4^-, ■ triflate, ▲ tosylate.

for various anions is $BF_4 > CF_3SO_3 > MeC_6H_4SO_3$. The imidazolium cation alkyl chain length effect seen is opposite to the trend observed when ionic liquids were tested as neat lubricants. More interestingly, ionic liquids as lubricant additives made with 1-ethyl-3-methylimidazolium cation ($C_2C_1Im^+$) not only reduce the friction and wear with respect to paraffinic–naphthenic base oils, but they show better performance than neat ionic liquid lubricants for some ionic liquids, as shown in Figure 7.3 for steel–aluminum at room temperature, particularly for a determined sliding velocity that improves ionic liquid miscibility with the base oil [108,111]. The excellent tribological properties of ionic liquid additives are attributed to the formation of physically adsorbed films, formation of tribochemical products during friction without tribocorrosion, and good miscibility with the base oil.

The effort to explore other ionic liquids as lubricant additives was extended to other cations by Qu et al. [112]. They found in 2006 that adding 10 vol% of $[(C_8H_{17})_3NH][NTf_2]$ ionic liquid in mineral oil decreased slightly friction coefficient in steel–aluminum contact compared with the mineral oil alone while it had a higher friction coefficient than the neat ionic liquid but produced lower wear on steel–aluminum contact than either the mineral oil or the ionic liquid alone. Later, the same research group discovered that in a standardized reciprocating sliding test with 5% $[(C_8H_{17})_3NH][NTf_2]$ ionic liquid blended into the oil, the specific ionic liquid can reduce friction by 9% and wear by 14% compared with a 15W40 fully formulated diesel engine oil [113]. The test involved a segment of a Cr-plated diesel engine piston ring against a gray cast iron flat specimen. In the same study, they found that blending 5% $[C_{10}mIm][NTf_2]$, 1-n-decyl-3-methyl imidazolium bis(trifluoromethanesulfonyl)amide, and ionic liquid into engine oil, it could reduce friction by 9% but decreased wear by 34% compared with the same 15W40 diesel engine oil [113]. However, while these ionic liquids as lubricant additives showed reduction in friction and wear, their solubility in the mineral oil was low.

As a lubricant additive, adding 1 wt% of 1-n-hexyl-3-methyl imidazolium tetrafluoroborate ionic liquid in Group III base oil, which contained 2% succinimide dispersant and 2% sulfonate detergent, was found by Mistry et al. [114] in 2008 to be very effective in reducing friction coefficient from 0.078 for mineral oil to 0.028 for Cr-coated steel on Ni–SiC-coated aluminum contact though it increased wear. However, mixing 1 wt% of this ionic liquid with 1 wt% TCP further reduced the friction coefficient to 0.015 while improving wear performance. The succinimide dispersant and sulfonate detergent helped solubilizing ionic liquid in the oil.

In a special base oil like polyethylene glycol (PEG), antioxidant ionic liquids consisting of an imidazolium incorporated with hindered phenol such as 1-(3,5-di-tert-butyl-4-hydroxybenzyl)-3-methyl-imidazolium, or [BTHmlm], with different anions were found to be very potent antiwear additives [115]. The anions used were hexafluorophosphates, PF_6, bis(trifluoromethylsulfonyl)-imide, NTf_2, and tetrafluoroborates, BF_4. In particular, at 1 wt% in PEG, the ionic liquid combined with PF_6 reduced wear greater than 100 times compared with PEG alone in a SRV wear test involving steel–steel contact [115]. All three ionic liquids exhibited good antioxidancy in the rotary pressure vessel oxidation test, previously known as rotary bomb oxidation test [115].

Anions containing fluorocarbon ionic liquids have propensity to produce corrosive HF by-product. To overcome corrosivity and oil-immiscibility issues, Qu's research group developed an ionic liquid consisting of a phosphonium cation and phosphate anion as a lubricant additive. This ionic liquid is named "trihexyltetradecylphosphonium bis(2-ethylhexyl) phosphate," or $[P_{6,6,6,14}]$[DEHP]. In a simple blend of mixing 5% of this ionic liquid in a PAO base oil, it performed nearly as well as a fully formulated Mobil 1™ 5W-30 engine oil in a Plint TE77 reciprocating sliding tribological test with engine piston ring on cast iron cylinder liner. More interestingly, top treating 5% of this ionic liquid into the 5W-30 engine oil, it decreased the wear rates by ~70% for both sliding surfaces while maintaining a similar friction behavior [116]. At 1 wt% of this ionic liquid in PAO, it provides wear reduction almost equal to secondary ZDDP in the same tribotest at room temperature [117]. Later, Qu's research group showed that at equal phosphorus concentration of 800 ppm, tetraoctylphosphonium bis(2-ethylhexyl)phosphate or $[P_{8,8,8,8}]$[DEHP] ionic liquid in a gas-to-liquid (GTL) base oil performs better than secondary ZDDP in reducing wear [118]. In particular, keeping the total phosphorus concentration the same at 800 ppm, a 1:1 mixture of $[P_{8,8,8,8}]$[DEHP] or $[P_{6,6,6,14}]$[DEHP] ionic liquid and a secondary ZDDP produced a synergistic effect that showed an approximately 70% reduction in wear and an approximately 30% reduction in friction coefficient as compared with the same base oil additized with the secondary ZDDP alone [118]. Such synergistic effect was also observed with the combination of $[P_{8,8,8,8}]$[DEHP] ionic liquid and a secondary ZDDP in a fully formulated experimental SAE 0W-16 engine oil blended with a GTL base oil [119].

Toward the same goal of developing ionic liquids as engine oil additives, two ionic liquids, choline bis(2-ethylhexyl)phosphate (P-IL) and choline dibutyldithiophosphate (TP-IL), were evaluated in a SRV tribometer at a phosphorous concentration of 1000 ppm. In both cases, moderate reductions in friction and wear compared with the base oil were observed. However, keeping the total phosphorous concentration at 1000 ppm, a mixture of either P-IL or TP-IL with the octadecylphosphorofluoridothiolate produced a very significant synergistic effect on reduction of both friction and wear compared to the base oil and ZDDP [120].

When dicationic bis(imidazolium) ionic liquids with the same long side-chain substituted cation and different anions (DIL $C_{10}1$, DIL $C_{10}2$, DIL $C_{10}3$ in Figure 7.4) were evaluated as 2% additives in PEG at room temperature, reduction in friction and wear of steel–steel sliding pairs was observed for all three additives compared with PEG as base oil without additives [121]. Among the three additives, DIL $C_{10}1$ gave the best result in wear reduction, while DIL $C_{10}3$ produced the lowest friction coefficient. More recently, different dicationic bis(imidazolium) ionic liquids functionalized with PEG and with different anions (DIL PEG1 and DIL PEG2) were evaluated in mineral base oil [120]. At 1% in base oil, DIL PEG1 reduced friction and wear by approximately 37% and 70%, respectively. The mixture of two at 1% reduced wear by 82%. Some of the excellent tribological properties of ionic liquids as additives can be explained by the mechanism of forming adsorbed layers and protective tribofilms as proposed by Liu et al. [122].

7.2.10.4 Ionic Liquids as Lubricant Additives for Greases

Fox and Priest [110] showed that in a four-ball EP test addition of 1 wt% of 1-methyl-3-hexyl imidazolium hexafluorophosphate or $[C_6mlm][PF_6]$ ionic liquid to a base grease, which contained 1 wt% each of a detergent and a dispersant, increased the weld load by 60% for the same wear scar. In a similar experiment, top treating 1 wt% of the same ionic liquid to fully formulated commercial high-temperature grease increased the weld load by 32% from 290 kg to 380 kg or 32% increase. However, further increasing the ionic liquid concentration only produced a moderate increase in weld load for both cases. In a tribological test using Plint TE77 reciprocating tribometer, addition of 1 wt% TCP and 1 wt% $[C_6mlm][PF_6]$ ionic liquid to a Group III XHVI base oil reduced wear volume by 54% and 52%, respectively, compared to the same base oil with 1 wt% TCP alone and with 1 wt% $[C_6mlm][PF_6]$ ionic liquid alone [110].

7.2.11 Carbon-Based Nanomaterials as Antiwear and Antiscuffing Additives

This section will cover potential applications of carbon-based nanomaterials as nonmetal-containing antiwear and AS additives. The list of carbon-based nanomaterials includes carbon nano-onions, carbon nanotubes, carbon nanohorns, fullerenes, nanodiamonds, graphenes, and nano-PTFEs. Before carbon-based nanomaterials were discovered, two carbon materials most known to exist in nature, graphite and diamond, had long been investigated for their tribological properties. Graphite has been used extensively

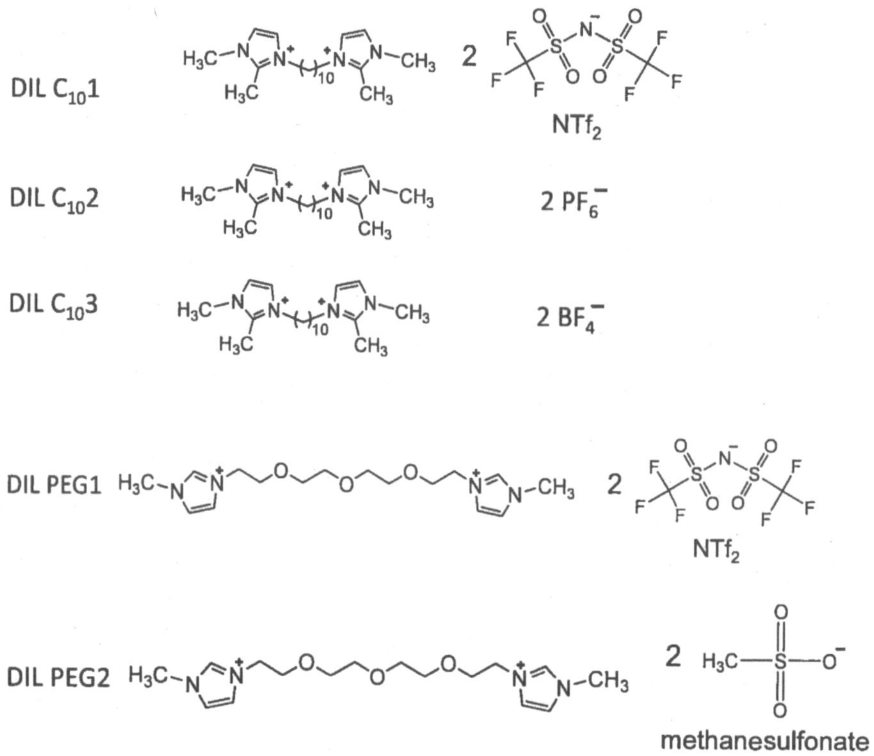

DIL C$_{10}$1

DIL C$_{10}$2   2 PF$_6^-$

DIL C$_{10}$3   2 BF$_4^-$

DIL PEG1   H$_3$C—N⁺...N—CH$_3$

DIL PEG2   H$_3$C—N⁺...N—CH$_3$ 2 H$_3$C—S—O⁻
                                         methanesulfonate

FIGURE 7.4 Various dicationic ionic liquids. (Listed from Aswath, P. et al., Synergistic mixtures of ionic liquids with other ionic liquids and/or with ashless thiophosphates for antiwear and/or friction reduction applications, U.S. Patent Application Publication No. 2013/0331305; Yao, M. et al., *ACS Appl. Mater. Interfaces*, 1, 467, 2009.)

as a solid lubricant for more than half of a century [123]. Its friction and wear reduction lubricating capacity was attributed to its low shear strength, layer lattice structure, and the ability to form a surface film on surfaces; it was an important solid lubricant because it works well in high-temperature and vacuum conditions. In addition to its friction and wear performance, it has also been reported to have moderate load-carrying capacity when it was added to grease formulation. It is interesting to note that a certain combination of graphite and molybdenum disulfide produces a synergistic effect to increase both AS and antiwear performances [124]. As for diamond, unlubricated diamond/diamond contact was known to have a low friction coefficient of 0.05 with excellent wear resistance [125,126].

7.2.11.1 Carbon Nano-Onions and Carbon Nanotubes

Carbon nano-onions, which are also known as multilayer fullerenes, were first synthesized by Ugarte from carbon soot under high-energy electron beam radiation [127]. Since then, several different synthetic methods have been developed. For example, He et al. developed a low-temperature synthesis of 5–50 nm carbon onions from methane by chemical vapor decomposition using aluminum-supported nickel oxide catalyst [128,129]. The product contains a mixture of onions with hallow core and a nickel particle and up to 1.5% nitrogen depending on the reaction time. Carbon onions synthesized from carbon soot particles by

high-energy electron beam radiation have quasispherical shape. Their structures are shown in Figure 7.5. Not all carbon onions have spherical shape; for example, some have polyhedral shape [130].

A very extensive tribological characterization of carbon onions and carbon nanotubes has been carried out by Joly-Pottuz and Ohmae [125]. A brief summary of their study is described here. They used carbon onions made by annealing carbon nanodiamonds in a graphite crucible. Two different

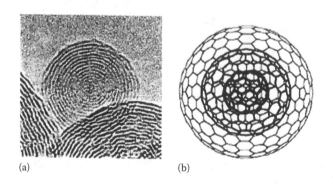

(a)   (b)

FIGURE 7.5 (a) TEM image carbon onions and (b) drawing of an onion-like graphitic particle formed by three concentric layers (C$_{60}$, C$_{240}$, C$_{540}$) (*Note:* only a half part of each shell is shown). (Reprinted from *Carbon*, 33(7), Ugarte, D., Onion-like graphitic particles, 989–993, Copyright 1995, with permission from Elsevier.)

carbon onion samples were made, one with a residual diamond core (CO1), which has an average diameter of 5–10 nm, and the other without (CO2), whose average size is slightly smaller.

Environmental pin-on-plat tribometer was used to characterize the tribological properties of both carbon onions and carbon nanotubes, which will be described later. The test conditions for boundary lubrication are described in details in the appendix of the *Nanolubricants* book [125]. The effect on friction was found to be practically independent of carbon onion concentration. At 0.1 wt% dispersion in PAO carbon onions (CO1 and CO2) gave friction coefficient of around 0.12 compared with approximately 0.25 for pure PAO at 0.83 GPa contact pressure. As contact pressure increases from 0.83 to 1.72 GPa, pure PAO decreases the friction coefficient to around 0.08–0.1 level, while both forms of carbon onions decrease the friction coefficient to approximately 0.06. After the friction tests, comparison of the wear scars measured on the pin between pure PAO and carbon onions indicated that both CO1 and CO2 reduce wear significantly. The range of wear reduction for CO1 is between 18% and 29% as contact pressure varies and for CO2 the reduction is greater ranging between 25% and 47%. To account for the friction and wear scar reductions, the authors proposed a crude model of lubrication mechanism. According to this model, a carbon onion–induced tribofilm is composed of a mosaic of graphitic sheets and intact carbon onions bound with slippery iron oxides and iron hydroxide nanoparticles.

Carbon nanotubes were first discovered by Iijima in 1991 as a by-product of the reaction of fullerenes [131]. There are two main categories of methods to prepare carbon nanotubes: high-temperature methods and low-temperature methods [125]. High-temperature methods involve a temperature of 3200°C for sublimation of carbon feedstock, and low-temperature methods use temperature range between 800°C and 1100°C for chemical vapor deposition of CO, CH_4, or C_2H_2 feedstock; the former is used to make single-walled carbon nanotubes (SWNTs) and the latter is for multiwalled carbon nanotubes (MWNTs). All carbon nanotube syntheses require catalysts for the processes.

Carbon nanotubes of both SWNTs and MWNTs with and without catalysts removed tend to gather in bundles [125,132]. SWNTs and MWNTs with the catalysts removed are shown in Figure 7.6.

The average diameter of SWNTs was found to be 1.3 nm, while in bundles the average diameter is 30–60 nm. The diameter of MWNTs ranges from 20 to 120 nm. The length of SWNTs and MWNTs can be found in the range of 1–100 μm but more common to be around 5–15 μm [132]. (Example of 5–15 μm length is shown in Reference 164.)

SWNTs without residual catalyst at 1 wt% dispersion in PAO have very little impact on friction and wear in the pin-on-flat tribometer test, but those with residual catalyst like nickel/yttrium (Ni/Y-SWNTs) at the same concentration reduce friction and wear significantly [125]. The effect of residual catalyst in the SWNT sample on tribological performance was also observed by another research group [133]. This SWNT sample dispersed in a SAE 20 mineral base oil

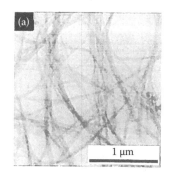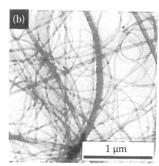

FIGURE 7.6 TEM images of SWNTs (a) and MWNTs (b). (From Joly-Pottuz, L. and Ohmae, N.: *Nanolubricants*. 2008. Copyright Wiley-VCH Verlag GmbH & Co. KGaA. Reproduced with permission.)

at 0.5 wt% appears to be more efficient than ZDDP at the equal treat level in friction and wear reduction and slightly better in EP performance [133]. On the contrary, MWNTs without residual catalyst were found to have a significant reduction in both friction and wear at 0.1 wt% in PAO compared with pure PAO [125]. For example, as contact pressure increases from 0.83 GPa through intermediate steps to 1.72 GPa, the friction coefficient reduces from 0.14–0.17 to around 0.05–0.06 compared with around 0.25 to around 0.08–0.1 for pure PAO. The ones with residual catalyst do not perform as well as those without it. The difference in tribological performance between SWNTs and MWNTs was explained based on a proposed lubrication mechanism, in which MWNTs can stand against a higher contact pressure to keep the bearing rolling ability before being flattened than SWNTs [125].

7.2.11.2 Carbon Nanohorns and Fullerenes

Single-walled carbon nanohorns (SWCNHs) can be synthesized by CO_2 laser ablation and arc discharge methods, both of which were developed by Iijima's research group [134,135]. Nanohorns tend to form aggregates. Because of its horn shape, the aggregates shown in Figure 7.7 look radically different from bundles of SWNTs.

Currently commercial samples of SWCNHs with dahlia-like shape of a mean diameter of about 80 nm are available from Carbonium S.r.l. in Selvazzano, Italy. Zin et al. studied the tribological properties behavior of the dahlia-like shape SWCNH nanoparticles in SAE 40 grade Mobil Pegasus 1005 synthetic base oil using a ball-on-disc Cetr UMT-2 tribotester at 25°C [137]. The study covered three regimes from boundary to EHL. The concentrations studied were 0.005, 0.01, and 0.02 vol%. The best results were obtained from 0.01 vol% dispersion in base oil. It reduced overall average friction coefficient from 0.085 for the base oil to 0.074 or 9% and wear rate (ν) from 109 ± 7 to 74 ± 5 (μm³/mm/min) or 30%. Sedimentation of the nanoparticles in the steel surface valleys coupled with rolling action was proposed as the lubrication mechanism for the tribological effect observed.

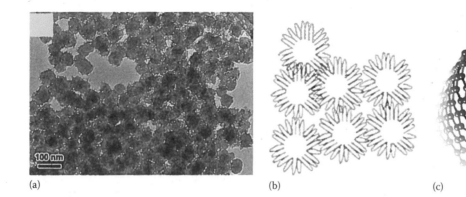

(a)   (b)   (c)

FIGURE 7.7 (a) TEM image of SWCNH aggregates, (b) schematic picture of SWCNH aggregates, and (c) structure of individual carbon nanohorn. (a: Reprinted from *Chem. Phys. Lett.*, 309, Iijima, S., Yudasaka, M., Yamada, R., Bandow, S., Suenaga, K., Kokai, F., Takahashi, K., Nano-aggregates of single-walled graphitic carbon nano-horns, 165–170, Copyright 1999, with permission from Elsevier; b and c: Reprinted from *Top. Appl. Phys.*, 111, Yudasaka, M., Iijima, S., Crespi, V.H., Single-wall carbon nanohorns and nanocones, 605–629, Copyright 2008 with permission from Springer-Verlag GmbH.)

C_{60} and C_{70} are fullerene molecules discovered in 1985 at Rice University from the evaporation of graphite by a laser method [138]. Since then, arc and combustion methods have been developed to produce these fullerenes [139,140]. There are a large variety of fullerenes; the number of carbon atoms in a fullerene is 2n, where n is an integer. The most stable ones are C_{60} and C_{70} [141]. Besides having 12 pentagons, C_{60} and C_{70} have 20 and 25 hexagons, respectively. C_{60} has a hollow carbon cage diameter of 0.71 nm. Fullerene is an antioxidant and has high elasticity, weak molecular interactions, and low surface energy [142]. The structures of C_{60} and C_{70} are shown in Figure 7.8.

Friction and wear performance of a heavy-duty SAE 10 base oil treated with 1 wt% C_{60} or a one-to-one mixture of C_{60} and C_{70} were evaluated with a specially made wear–friction test machine with a steel roller rubbing against a bronze shoe [142]. At the sliding speed of 0.61 m/s and a test length of 150 min, the best wear reduction from the base lubricant observed was 22% for C_{60} and 50% for C_{60} + C_{70} at a real contact pressure of about 1.2 GPa (estimated from the 480 N load run). However, friction reduction for either additive system was found to be insignificant [142].

In another study, 0.1 vol% C_{60} was added to compressor base mineral oils of different ISO grades (ISO 12 to

ISO 145) and evaluated their tribological properties against the respective unadditized base oils [143]. Friction performance was conducted in a disc-on-disc tester; load-carrying capacity and wear resistance were evaluated with ASTM D2783-03 and ASTM D4172-94, respectively. Biggest improvements over the raw oils in friction, load-carrying capacity, and wear resistance were found for the lowest viscosity grade oil of ISO 12 grade. In this ISO 12 grade, 0.1 vol% C_{60} oil increases the weld load from 900 N to about 1000 N while reducing the wear scar diameter from 0.92 to 0.84 mm.

At a higher additive concentration (5 wt%) of C_{60}-rich (85% C_{60} and 15% C_{70}) nanoparticles in a mineral oil, one research group found that the reduction in the width of the wear scar on the steel disc in a ball-on-disc tribotester was from 300–380 μm to 120–130 μm [144]. The diameter of the wear scar on the mating steel ball was reduced from approximately 200 to 60 μm. The friction coefficient was reduced about 20% from the base oil. Furthermore, the improvement in friction and wear at this treat level for fullerenes was found to be comparable or better than MoS_2.

7.2.11.3 Graphenes, Carbon Nanodiamonds, and Nano-Teflons

The most famous technique for the preparation of graphene is the use of a micromechanical method (or adhesive tape) to pull graphene layers from graphite and transferring them onto thin SiO_2 on a silicon wafer [145]. Since its discovery in 2004, many other methods have been developed. Among them, chemical exfoliation is a popular method because it possesses the potential of a bulk-scale production [146]. Graphene is a flat monolayer of carbon atoms tightly packed into a two-dimensional honeycomb lattice. Figure 7.9 shows a typical structure of grapheme [147].

Graphene has been evaluated as a solid lubricant under humid condition by a ball-on-disc tribometer. With 2–3-layer graphene on the steel surface, it was found to reduce friction coefficient by a factor of 6 and reduce wear by 4 orders of

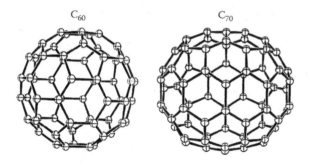

C_{60} C_{70}

FIGURE 7.8 The structures of C_{60} and C_{70} fullerene molecules. (Reprinted from *J. Mol. Struct.*, 202, Slanina, Z., Rudzinski, J.M., Togasi, M., Osawa, E., Quantum-chemically supported vibrational analysis of giant molecules: the C_{60} and C_{70} clusters, 169–176, Copyright 1989, with permission from Elsevier.)

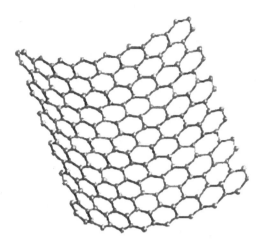

FIGURE 7.9 A segment of graphene structure. (Reprinted by permission from Macmillan Publishers Ltd. *Nat. Mater.* Geim, A.K. and Novoselov, K.S., The rise of grapheme, 6, 183–191, copyright 2007.)

magnitude [148]. Later, the same research group found that, unlike graphite, similar friction and wear performance was obtained under dry nitrogen.

Graphene platelets derived from graphite generally bear hydroxyl and carboxylic functional groups, which make them hydrophilic and easy to coagulate in oil [149]. To improve their dispersibility in base oil so they can be used as a lubricant additive, graphene platelets have been functionalized with steric acid and oleic acid. Using the 350SN base oil as the reference, the modified graphene platelets in base oil were evaluated using ASTM D4172-82 method for friction and wear performance and ASTM D2783 method to measure maximum nonseizure load. At an optimal concentration of 0.075 wt%, the modified graphene platelets reduce the friction from the base oil alone by approximately 28%–36% and the wear rate by approximately 50%–65%. The maximum nonseizure load was increased from 418.5 N (~42.7 kg) to 627.2 N (~64 kg) [149]. The tribological properties of the modified graphene platelets were shown to be significantly better than

graphite, which was modified by the same process as the graphene platelets.

Ultrathin graphene with reduced hydrophilicity has been prepared by solar exfoliation of graphite oxide to increase its dispersibility in engine oil [150]. The solar exfoliation process produced highly deoxygenated, less defective, and superhydrophobic graphene. The process enhances its dispersion stability in engine oil. A 0.075 wt% dispersion of this graphene sample in engine oil was prepared. The dispersion after standing for 7 days was evaluated for its friction coefficient, wear scar, and nonseizure load using a four-ball test machine. The load-carrying capacity was determined according to ASTM D2783 at a rotating speed of 1760 rpm. The friction and wear tests were performed using the same machine following the ASTM D5183 procedure at a rotating speed of 600 rpm and under a constant load of 392 N for test duration of 60 min at a temperature of 75°C. The improvement in friction characteristics, antiwear, and EP properties was found to be 80%, 33%, and 40%, respectively, compared with base oil [150].

Carbon nanodiamonds were first synthesized in 1962 by the detonation method in Russia [151]. Since then, various synthetic methods have been developed. They include ultrasonic cavitation of graphite, high-energy pulsed laser irradiation of graphite [152], and microplasma dissociation of ethanol vapor [153]. The majority of tribological studies on carbon nanodiamonds have used samples produced by the detonation method. The detonation nanodiamonds (DNDs) have a primary particle size of 3–5 nm. However, they form tightly bonded aggregates during detonation synthesis and purification processes. The size of aggregate particles ranges between 200 and 300 nm. The detonation reaction feedstock and conceptual structure of carbon nanodiamonds produced by detonation is shown in Figure 7.10. Various deagglomeration and fractionation procedures have been developed to reduce the aggregates to smaller sizes and even down to primary particle sizes for applications of carbon nanodiamonds as a lubricant additive [154].

DNDs with aggregate sizes of 120, 90, and 10 nm dispersed in PAO-6 were evaluated as a lubricant additive using

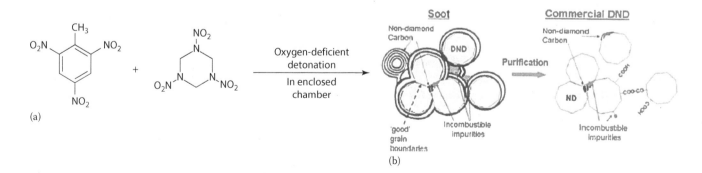

FIGURE 7.10 (a) Detonation reaction and (b) conceptual structure of nanodiamond product formed. (Information on reaction feedstock—a: Reprinted from *Nat. Nanotechnol.*, 7, Mochalin, V.N., Shenderova, O., Ho, D., Gogotsi, Y., The properties and applications of nanodiamonds, 11–23, Copyright 2012, with permission from Macmillan Publishers Ltd.; product soot and commercial DND—b: Reprinted from *Crit. Rev. Solid State Mater. Sci.*, 34(1–2), Schrand, A.M., Ciftan Hens, S.A., Shenderova, O.A., Nanodiamond particles: properties and perspectives for bioapplications, 18–74, Copyright 2009 of Taylor and Francis Group.)

a test apparatus of SMT-1 for ring-on-ring and the shaft/bushing test and four-ball EP test [157]. Close to 50% reduction in both friction and wear was observed for 0.05 wt% DND (120 nm) in PAO-6. No change in load-carrying capacity was observed. Dispersion of 0.03 wt% DND (10 nm) and 0.3 wt% PTFE in PAO-6 obtained a friction coefficient, EP failure load, and wear spot of 0.016, 750 kg, and 0.350 mm, respectively, compared with 0.106, 150 kg, and 0.567 mm for PAO-6 alone. The improvement in friction and wear was attributed to surface polishing. Aggregate size effect and synergistic effect with PTFE were particularly noted [157]. DND polishing effect in fresh oil was also observed by another research group in a reciprocating ring-on-liner test of a SAE 30 engine oil top treated with 200 ppm nanodiamonds [158]. However, the results seen in aged oil raised some concern that the abrasiveness of nanodiamonds may not go away after the initial running-in period. The fresh oil polishing results are somewhat in line with the results observed by another research group, who used a ball-on-disc test of UMT-2 tribometer, with 3–10 nm DNDs at 1 wt% in P100N paraffinic oil, 1.2–1.7 GPa load, and 100–500 mm/s sliding speed, friction was reduced, but in general wear increased [159]. Similar mixed results would be apparent from the data obtained from top treating various commercial oils with a certain amount of DNDs if the data had been examined more closely [160]. More extensive studies are needed to establish the true benefits of DNDs as lubricant additive to improve friction and/ or wear.

PTFE is well known for its excellent tribological properties in friction and wear reduction and load-carrying EP agents. Applications of PTFE in lubricants include greases, chain oils, dry-film lubricants, and engine oils [157]. The particle size of nanoparticles can influence the dispersibility of the nanomaterial in oil and impact significantly on its tribological properties. Fernandez Rico et al. investigated PTFE nanoparticle powder as an EP additive in two base oils of different viscosities using an ASTM D2783-88 four-ball EP test [161]. They found that at 3% and 10% PTFE in SN-350 base oil, the weld point increased from 160 to 315 and 630 kg, respectively, and the load-wear index increased from 25.7 by 81.3% to 222.6% for 3% and 10% PTFE, respectively. Dubey et al. showed that there is PTFE nanoparticle size effect on friction coefficient, wear reduction, and weld load; smaller particle size is better [162,163]. The optimal concentration for the PTFE nanoparticles of 70 nm by TEM was found to be 3 wt% in 150 N Group II base oil. At this concentration, the weld load was found to be 620 kg; the wear decreased up to 28% at 784 N as evaluated by ASTM D4172, while the friction coefficient was reduced up to 35% at 100 N as measured by Optimol SRV-III oscillating friction and wear tester. It was proposed that formation of the PTFE film on the wear surfaces was responsible for reducing shearing stress and thus improving the tribological properties. The high load-carrying capacity certainly qualifies this PTFE nanoparticle as a very efficient AS/EP additive [163]. It is noteworthy that all carbon-based nanomaterials discussed in this section do not possess any load-carrying capacity as high as sulfur-based EP additives except nano-PTFEs.

7.2.11.4 Methods to Improve Dispersibility of Carbon-Based Nanomaterials in Oils

In general, there is a concern about the dispersibility of carbon-based nanomaterials in oils and the associated long-term stability of the nanoparticle dispersion. If the particle size is too large, it would be very difficult to disperse it in oil and it tends to drop out of the oil phase easily. One way to improve the hydrophilicity is to incorporate sufficient hydrocarbon moieties. If the length of the nanomaterial is too long as in the case of some carbon nanotubes, shortening it and functionalizing it with alkyl or aryl alkyl group will improve the dispersibility. This approach has been demonstrated to be effective [164,165]. Alternately, other types of carbon-based nanomaterials, such as nanoparticles constructed from diblock polymers, which have built-in hydrocarbon moieties, need to be considered [166].

7.2.11.5 Environmental Impact of Nanoparticles

Materials with nanosize particles have always existed in nature and the atmosphere. More recent uses of nanotechnology mean that more and more man-made nanoparticles could in their lifetime enter our atmosphere, soil, or water environments. As nanotechnology is an emerging field, there is great debate regarding to what extent industrial and commercial use of nanomaterials will affect organisms and ecosystems.

Nanotechnology's environmental impact can be split into two aspects: the potential for nanotechnological innovations to help improve the environment and the possibly novel type of pollution that nanotechnological materials might cause if released into the environment.

Nanoremediation is the use of nanoparticles for environmental remediation. During nanoremediation, a nanoparticle agent must be brought into contact with the target contaminant under conditions that allow a detoxifying or immobilizing reaction. Nanofiltration is a membrane filtration based on method that uses nanometer-sized cylindrical through-pores that pass through the membrane at a perpendicular angle. Nanofiltration membranes have pore size from 1 to 10 Å, smaller than that used in microfiltration and ultrafiltration, but just larger than that in reverse osmosis. Some water treatment filtration devices incorporating nanotechnology are already on the market, with more in development.

Nanopollution is a generic name for waste generated by nanodevices or during the nanomaterials manufacturing process. Ecotoxicological impacts of nanoparticles and the potential for bioaccumulation in plants and microorganisms are a subject of great concern. There is a lack of information on nanoparticles entering the environment and what the health risks and consequences to the environment these particles could have. Once the nanoparticles enter one ecosystem, they could move to another, for example, the movement of nanoparticles between water and sediments, or the absorption of atmospheric particles into water. There are therefore many factors to be taken into account when aiming to detect and analyze the nanoparticles in an environment. The influence of the environment on the particles and follow-on effects must be considered alongside the toxicity of the nanoparticles themselves.

It is important to remember that nanotechnology can be used in a positive way in the environment, for example, the use of nanoparticles for groundwater and contaminated land remediation as mentioned earlier. However, the risks of the nanoparticles in their life cycle in the environment must be established.

7.2.12 MECHANISMS OF METAL SULFIDE FORMATION BY SULFUR-CONTAINING ANTISCUFFING ADDITIVES ON METAL SURFACES

Sulfur-containing AS additives are widely used as described in Section 7.2.1. Formation of iron sulfide on metal surfaces to prevent seizure between metal–metal sliding or rolling contact under heavy load is critical to smooth operation of machines under boundary lubrication. There has been a long history of study on the mechanism of metal sulfide formation from sulfur-containing AS agents, particularly the formation of iron sulfide from rubbing metal surfaces between steel and steel.

7.2.12.1 Early Model of Iron Sulfide Formation Process

The first proposed mechanism of iron sulfide formation by sulfur-containing AS agents based on the study of disulfide and monosulfide in mineral oils under load-carrying conditions was put forward by Davey and Edwards [167] and Forbes et al. [168,169]. The iron sulfide formation by disulfide was envisioned to proceed by the two-stage process as shown in Figure 7.11.

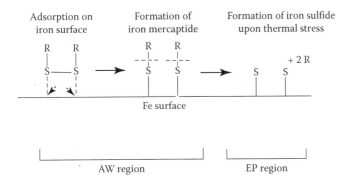

FIGURE 7.11 The iron sulfide formation mechanism scheme. (Reprinted from *Wear*, 15, Forbes, E.S., The load-carrying action of organo-sulphur compounds—A review, 87–96, Copyright 1970, with permission from Elsevier.)

This mechanism was proposed because it was the understanding at the time S–S bond was the weakest in a disulfide molecule [168]. In an attempt to confirm this mechanism, Forbes and his coworker carried out adsorption and reaction study of DBDS with iron powder and found the main reaction product at 150°C to be benzyl monosulfide [169]. To rationalize the observation, they postulated that the formation of thiyl radicals should be the first step of chemical reaction of DBDS with the metal surface; the dissociation into thiyl radicals is accompanied by cleavage of the C–S bond, and the dibenzyl monosulfide is then formed by a combination of the benzyl and benzylthiyl radicals. Analysis of the iron after 16 h at 150°C showed that the sulfur unaccounted for as soluble products was found on iron metal surface as inorganic sulfur–containing layer. Similar adsorption and reaction studies of alkyl disulfides with iron powder also produced measureable amounts of inorganic sulfur–containing layer on iron metal surface.

According to this proposed mechanism, in the antiwear stage, the disulfide is initially absorbed on the metal surface under mild loading condition and then cleavage of sulfur–sulfur bond occurs as the loading increases to boundary lubrication conditions and metal thiolate (mercaptide) is formed. The formation of iron thiolate protects the metal against wear. In the EP stage, under severe loading the carbon–sulfur bond in the thiolate would break down to form iron sulfide as a result of elevated temperatures. For monosulfide, it was rationalized that compared with disulfide the main reason for its inferior performance to prevent seizure under comparable loading conditions is that thiolate could not be readily formed. This mechanism was well received in the 1960s and the 1970s.

7.2.12.2 Modification of the First Model for Iron Sulfide Formation

Since the thiolate formation step in the two-stage action mechanism of disulfide had not been directly observed when the mechanism was proposed, more studies in this area were pursued by several research groups to see if alternate mechanisms could be found. Bovington and Dacre in the early 1980s carried out three thermal decomposition studies of DBDS and dibenzyl monosulfide with the rationale that thiolate formation may not be the necessary step for the formation of iron sulfide. They found that the major thermal decomposition products of DBDS in hexadecane obtained between 210°C and 270°C were free sulfur, toluene, trans-stilbene, and toluene-α-thiol [170]. To account for the major decomposition products observed, they attributed the thermal decomposition to a radical chain reaction initiated by C–S bond fission shown as Equation 7.39:

$$\langle\!\!\!\bigcirc\!\!\!\rangle\!-\!\overset{H_2}{C}\!-\!S_2\!-\!\overset{H_2}{C}\!-\!\langle\!\!\!\bigcirc\!\!\!\rangle \longrightarrow \langle\!\!\!\bigcirc\!\!\!\rangle\!-\!CH_2^\bullet + {}^\bullet S_2\!-\!\overset{H_2}{C}\!-\!\langle\!\!\!\bigcirc\!\!\!\rangle \tag{7.39}$$

Followed by Equation 7.40

$$\langle\!\!\!\bigcirc\!\!\!\rangle\!-\!\overset{H_2}{C}\!-\!S_2^\bullet \longrightarrow \langle\!\!\!\bigcirc\!\!\!\rangle\!-\!CH_2^\bullet + S_2 \tag{7.40}$$

S_2 generated can polymerize to S_8 or longer chains. Propagation reactions involve Equations 7.41 and 7.42:

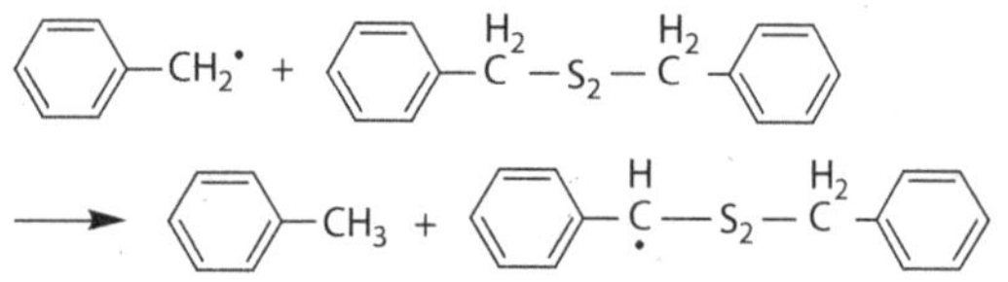

$$ \text{(7.41)} $$

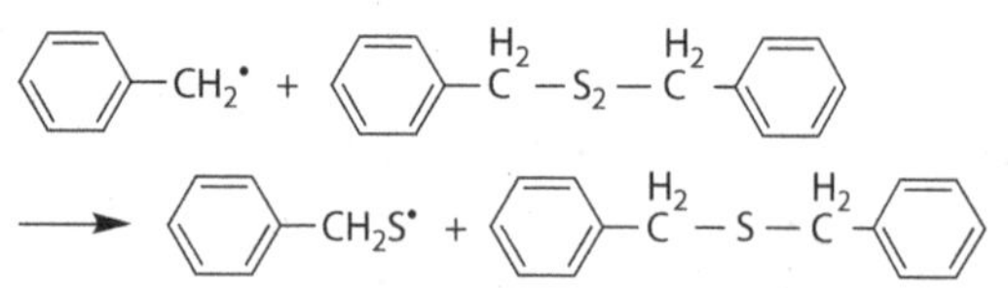

$$ \text{(7.42)} $$

Propagation reactions continue with hydrogen abstraction by free sulfur and benzyl and benzylthiyl radicals.

The propagation can be terminated by Equation 7.43:

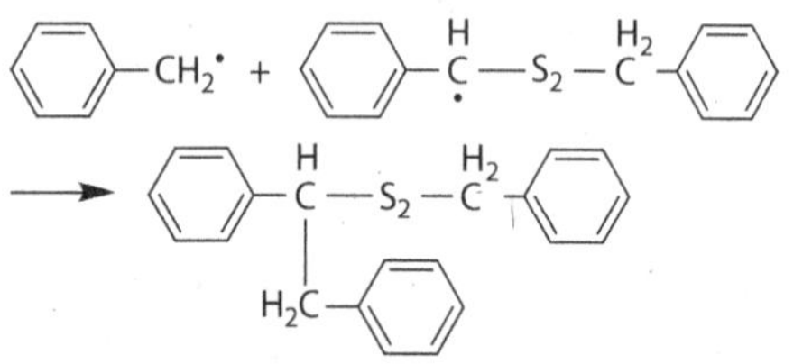

$$ \text{(7.43)} $$

The authors believed that the decomposition reactions could be more complex than the reactions described earlier. Nevertheless, they proposed the following overall reaction for the decomposition:

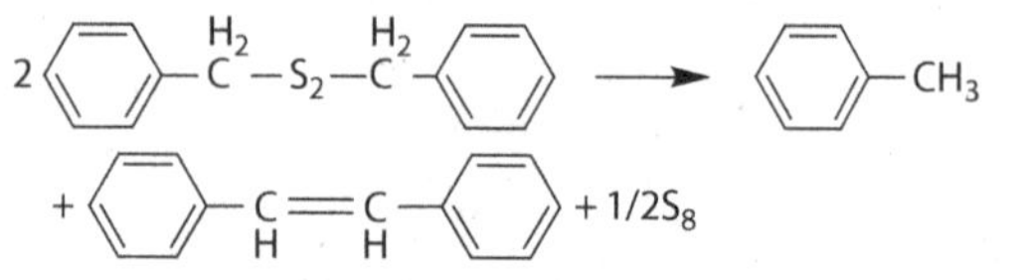

$$ \text{(7.44)} $$

The rate of decomposition of DBDS was shown to be relatively high from 210°C to 270°C. This temperature range was chosen to simulate the lower end of the thermal stress expected for EP lubrication conditions. Equation 7.44 implies that the more likely mechanism for the formation of iron sulfide is the reaction of iron with the ES produced by the thermal decomposition rather than directly with DBDS.

Thermal decomposition of DBDS and dibenzyl monosulfide was also carried out in the presence of iron powder to study the catalytic effect on thermal decomposition of these two compounds [171]. The decomposition products for DBDS are dibenzyl monosulfide, ES, toluene, and toluene-α-thiol. For dibenzyl monosulfide, the decomposition products are toluene and sulfur with trace amount of dibenzyl. Both the homogeneous decomposition and catalytic decomposition by iron powder apparently follow first-order kinetics. The catalytic effect of iron powder enables the thermal decomposition to proceed at lower temperatures. Nevertheless, both decompositions readily produce free sulfur as a major decomposition product.

Adsorption and desorption of DBDS and dibenzyl monosulfide on steel were studied at 100°C using both ^{35}S- and ^{14}C-labeled additives in hexadecane [172]. For each additive,

the physisorption reaction is reversible and follows Langmuir kinetics. DBDS physisorbs about 100–1000 times faster than dibenzyl monosulfide over the temperature range of 167°C–212°C. Each additive eventually produces a metal sulfide layer upon which a monolayer of physisorbed additive is present.

7.2.12.3 Observation of Elemental Sulfur in Tribological Process

The work of Bovington and Dacre was mainly based on immersion studies. Plaza did similar studies but carried out some of the studies under boundary conditions provided by a four-ball tester [173] and some immersion studies as well except the conditions were slightly different. He analyzed the decomposition products by a thiomercurimetric analytical method. He found that before initial seizure load (ISL, ~80 kg), the main by-products of tribochemical reactions from DBDS were ES and thiol. Formation of FeS commenced at loads higher than ISL, which also corresponded to rapid decrease of the ES content in oil suggesting that ES was converted by further reaction with steel surface to iron sulfide. Increasing the load caused the surface temperature to increase, which at some load became sufficient to promote chemical reaction of ES with iron, resulting in the abrupt increase of iron sulfide formation. Oils run at the most severe conditions showed strong corrosive tendencies. At the same time, there was a corresponding change in the rate of iron sulfide formation from mild to high. The author suspected that at the highest temperatures the sulfide formed by the corrosive attack of monoatomic ES. This is shown in figure 1 of that paper [173].

Plaza rationalized the relatively large amount of ES formation in the four-ball test by DBDS through a rearrangement of the molecule to a thiosulfoxide intermediate as shown in the following:

$$ RSSR \rightleftharpoons \begin{bmatrix} RSR \\ \parallel \\ S \end{bmatrix} \rightarrow RSR + S $$

In his 1987 paper where this path to form ES was proposed, the concept was taken from the review paper by Kutney and Turnbull [174]. The ES formed reacts with iron to form iron sulfide.

To account for the formation of thiol, he used the theory developed by Kajdas [175], who proposed the formation of three-electron-bonded radical anions with exoemitted electrons from frictional surfaces, which dissociates practically instantaneously to free thiol radical and thiolate as shown in the following:

$$
\begin{aligned}
RSSR + e &\rightarrow \left(RS \therefore SR \right)^{-} \rightleftharpoons RS^{\bullet} + RS^{-} \\
2RS^{\bullet} + 2Fe &\rightarrow 2FeS + RR \\
2RS^{-} + 2Fe^{+2} &\rightarrow 2FeS + RR \\
RS^{\bullet} &\xrightarrow{RH} RSH
\end{aligned}
\tag{7.45}
$$

Therefore, the RS$^\bullet$ radical can either abstract a hydrogen atom from oil to form thiol or react with iron to form iron sulfide while RS$^-$ anion can react with iron to form thiolate first and then iron sulfide. In another study, thiol has also been proposed to be formed from RSR in homogeneous decomposition of DBDS by Plaza et al. [177]. Incidentally, in this study, the homogeneous decomposition products under argon atmosphere obtained by Plaza et al. are S_8, RSR, H_2S, and RSH, while the major homogeneous decomposition products under air atmosphere formed from DBDS obtained by Bovington and Dacre [170] are ES, toluene, and trans-stilbene with minor amounts of dibenzyl, thiol, and RSR. A similar difference in the decomposition products was observed in the heterogeneous decomposition studies carried by the two research groups [171,176].

Nevertheless, in the last paper of a series of four studies on this subject by Plaza et al. [177], the authors admitted that since C–S bond is the weakest link in DBDS (implying S–S bond's dissociation energy is higher than that of C–S), in the four-ball testing condition DBDS is more likely to decompose to R$^\bullet$ and RS$_2$$^\bullet$, which releases the sulfur. Such conclusion agrees with the observation of Bovington et al. that DBDS decomposes mainly by breaking the C–S bond first and releases S_2 that way [170].

7.2.12.4 Direct Correlation between Elemental Sulfur and Load-Carrying Capacity

Recently, Pavelko conducted a study to correlate the thermal decomposition products of benzyl disulfide and other organic sulfur compounds with their AS characteristics [178]. The thermal decomposition products analyzed were hydrogen sulfide, ES, and thiols, while the AS characteristics measured by a four-ball ChMT-1 friction machine according to GOST 9490-75 were critical loads (P_c), welding loads (P_w), and initial seizure loads (I_s). The homogeneous thermal decomposition reactions were carried out at 400°C for organic sulfur compounds in a closed reaction system equipped with a reflux condenser. A 0.06 mol/kg solution of each organic sulfur compound was studied in three different base oils: Vaseline, oligodiethylsiloxane, and hexadecane. Different appropriate procedures were used to determine the respective amounts of hydrogen sulfide, ES, and thiols produced in the thermal decomposition reaction. In addition, the decomposition onset temperature (T_o) was determined for each additive. The correlation between the amounts of thermal decomposition products along with their onset decomposition temperatures and their AS characteristics (P_c, P_w, and I_s) was analyzed. The best correlation was obtained between the amounts of ES and AS characteristics, while among all sulfur species hydrogen sulfide and thiols have the worst correlation.

In an earlier study, Pavelko found that during organic sulfur compound EP additive thermal decomposition, the formation of hydrogen sulfide follows the radical-based mechanism and correlates reasonably well with AS characteristics [179]. Thus, he believed that organic sulfur compound EP additive AS action would also follow the radical-based mechanism. Following this reasoning, he proposed the use of cage effect to account for the AS activity of the organic sulfur compound additives. In the cage, there exists an equilibrium as shown in the following:

$$\overline{RS^{\bullet\bullet}SR} \rightleftharpoons R_2S_2 \rightleftharpoons \overline{R^{\bullet\bullet}S_2^{\bullet\bullet}R} \qquad (7.46)$$

Biradicals ($^\bullet$S$^\bullet$ or $^\bullet$S$_2$$^\bullet$), hydrogen sulfide, and other chemical species are expected through radical-based reactions to form in the cage. When leaving the cage, biradicals ($^\bullet$S$^\bullet$ or $^\bullet$S$_2$$^\bullet$) of sulfur react with iron surface to form iron sulfide or abstract a hydrogen atom to form additional hydrogen sulfide. Therefore, the yield of hydrogen sulfide is the total yield of hydrogen sulfide that is released from the cage and from the hydrogen sulfide formed by biradicals of ES abstracting hydrogen off the base oil (Vaseline oil). In one of his recent studies, he found that ES has a significantly higher load-carrying capacity than hydrogen and thiols [180]. From this, he reasons that the high AS effectiveness of organic sulfur compound AS additives is mainly governed by the amount of ES evolved from the cage and reaches the iron surface rather than the amount of the hydrogen sulfide, which arises under boundary condition.

The escape of ES from the cage assumes not only that it interacts with the solvent and iron surface but also that it accumulates on friction surfaces as follows:

$$\begin{aligned} Fe + {}^\bullet S_2{}^\bullet &\rightarrow FeS_2; \\ FeS_2{}^\bullet + {}^\bullet S_2{}^\bullet &\rightarrow FeS_4, etc.; \qquad (7.47) \\ FeS_{10}{}^\bullet &\rightarrow FeS_2{}^\bullet + S_8 \end{aligned}$$

ES at high temperatures (160°C–187°C) polymerizes. In the polymerized state, it is not dissolved in organic solvent. Under boundary friction, ES can be generated from hydrogen sulfide oxidation as follows:

$$H_2S + \frac{1}{2}O_2 \rightarrow S + H_2O \qquad (7.48)$$

or be regenerated by the scheme [181] as shown in the following:

$$FeS_n + \frac{1}{2}O_2 \rightarrow S_n + FeO \qquad (7.49)$$

The last reaction implies that ES while accumulating on the friction surface must improve the organic sulfur compound additive's AS properties.

The finding of Pavelko implies that the organic sulfur compound additive, which can generate more ES, will have a higher AS efficiency. This seems readily applicable to explain why t-dibutyl S_4 has a higher AS efficiency than t-dibutyl S_3 because the former can release more ES than the latter. It is also readily applicable to explain why t-dibutyl S_4 is more effective in AS than t-diadamantyl S_4. This is because upon the decomposition of t-dibutyl S_4, light and more movable low-molecular-weight t-butyl radicals arise in the cage and leave it rapidly while upon decomposition of t-diadamantyl S_4, less movable t-adamantyl radicals arise, which contain active hydrogen at the tertiary carbon atoms, which reacts with ES. As a result, lower amount of ES is evolved from the cage than on the decomposition of t-dibutyl S_4.

7.3 MANUFACTURE, MARKETING, AND ECONOMICS

All major additive suppliers produce ashless antiwear and AS additives that are available as components and as packages. A list of major producers (arranged in alphabetical order) is shown in the following.

Afton Corporation
Akzo Nobel
Atofina Chemicals (former Elf Atochem NA and Pennwalt Corporation)
BASF (Ciba Specialty Chemicals)
Chemtura (former Great Lakes Chemical's Durad Division)
Chevron Corporation (Oronite Division)
Clariant
Dover Chemical (former Keil Chemical Division, Ferro)
Dow Chemical (former Angus Division)
Elco Corporation (Detrex)
FMC
Hampshire Chemical Corporation (former Evans Chemetics, a subsidiary of Dow Chemical)
ICI America (Uniqema)
Infineum International Limited
Lubrizol Corporation
Rhein Chemie (integrated into Lanxess)
Solvay (former Albright & Wilson and Rhodia, which was acquired by Solvay Group in 2011)
Zeneca (became AstraZeneca since 1999)

Ashless antiwear and AS additives are supplied in a variety of chemistries, including single and multiple blends formulated to maximize performance and minimize adverse effects (e.g., dropout, corrosion). Product designations vary by chemical class and concentration. Many of them are formulated into additive packages according to applications, such as passenger car engine oils, heavy diesel engine oils, automotive transmission oils, automotive gear oils, and hydraulic fluids. Since the product offering information can be supplier specific, it is recommended to contact the suppliers directly or go to their corresponding website for further information.

There has been some consolidation in the additive business, but the market has not changed much as a result. The following are major changes by year:

1992 Ethyl acquired Amoco Petroleum Additives (U.S.) and Nippon Cooper (Japan).
1996 Ethyl acquired Texaco Additives Company.
1997 Lubrizol bought Gateway Additives (Spartanburg, SC).

(Continued)

1999 Infineum, the new petroleum additives enterprise, a joint venture between Exxon Chemical, Shell International Chemicals Ltd., and Shell Chemical Company, unveiled its new corporate identity and became fully operational on January 1, 1999 (the largest merge of additive companies in history)
1999 Crompton completed a merger with Witco.
2001 Texaco Oil merged into Chevron Oil while Chevron Chemical Oronite Division was kept intact.
2003 Dover Chemical acquired the Keil Chemical petroleum additives business from Ferro Corporation.
2004 Ethyl Corporation transformed into NewMarket Corporation, the parent company of Afton Chemical Corporation and Ethyl Corporation in order to maximize the potential of its operating divisions—petroleum additives and tetraethyl lead fuel additive business.
2005 Chemtura was formed by the merger of Crompton and Great Lakes Chemical Corporation.
2007 Chemtura bulked up its specialty lubricants business with the assets of Kaufman Holdings Corporation, parent company of Anderol and Hatco.
2008 BASF acquired the Swiss specialty chemicals maker Ciba to expand its specialty chemicals business.
2010 Afton Chemical Corporation acquired Polartech gaining strength in metalworking fluid additives business.
2011 Berkshire Hathaway announced an agreement to purchase Lubrizol.
2015 Dow and DuPont to combine in merger of equals and intend to subsequently spin into three independent, publicly traded companies.

Since most lubricant additives are produced through batch processes, consolidation can lead to improved operations and reduced costs (e.g., reducing plant idle time with better chemical manufacturing management systems). There are still many manufacturing facilities using equipment and procedures that are 30–40 years old. Hence, any investments in automation and continuous processing for a plant will be a competitive advantage. However, the business is so cost-competitive that most suppliers have difficulties in justifying major capital expenditures.

7.4 EVALUATION EQUIPMENT/SPECIFICATION

7.4.1 Lubricant Specifications

Lubricant components and formulated products are manufactured to tight specifications in petroleum refineries and lubricant blending plants and must also meet detailed commercial, industrial, and military specifications. As an example, the U.S. Military has rigid specifications for automotive lubricants, while the automotive manufacturers have similarly rigid but not necessarily the same specifications to assure quality and consistency of lubricant manufacture. In addition, there are performance specifications that must be met from such original equipment manufacturers (OEMs) as farm machinery and other off-highway automotive equipment. These specifications are designed to enable the user to select appropriate lubricants and to be assured of adequate performance over a specified service life.

The industry is, for the most part, adequately self-regulating with minimal government input concerning performance specifications. The most elaborate system for developing and upgrading lubricant and fuel specifications is for automotive lubricants. The American Society of Testing and Materials (ASTM), the Society of Automotive Engineers (SAE), and the American Petroleum Institute (API) all have defined roles in determining specifications for products such as passenger car motor oils and heavy-duty motor oils. These three organizations, working together in the United States, are known as the Tripartite. Extending internationally, the International Lubricant Specification Advisory Committee (ILSAC) (formerly known as International Lubricant Standardization and Approval Committee) is also active in all phases of engine lubricant category development. The International Organization for Standardization (ISO) is a worldwide federation of national standards bodies (ISO member bodies). The work of preparing International Standards is normally carried out through ISO technical committees. Each member body interested in a subject for which a technical committee has been established has the right to be represented on that committee. International organizations, governmental and nongovernmental, in liaison with the ISO, also take part in the work. The ISO collaborates closely with the International Electrotechnical Commission on all matters of electrotechnical standardization.

The main task of technical committees is to prepare International Standards. Draft International Standards adopted by the technical committees are circulated to the member bodies for voting. Publication as an International Standard requires approval by at least 75% of the member bodies casting a vote.

Attention is drawn to the possibility that some of the elements of this document may be the subject of patent rights. The ISO shall not be held responsible for identifying any or all such patent rights.

ISO 14635-3 was prepared by Technical Committee ISO/TC 60, *Gears*, Subcommittee SC 2, *Gear capacity calculation*.

In other product categories, lubricant and additive suppliers, OEMs, and industry trade associations work together to determine performance requirements and product specifications. In addition to the three industry organizations mentioned earlier, the National Lubricating Grease Institute, the National Marine Manufacturers Association, the American Gear Manufacturers Association, the Society of Tribologists and Lubrication Engineers, and other groups, associations, and key equipment builders can influence lubricant specifications.

In addition to meeting all military and industrial specifications, many leading lubricant marketers and finished lubricant suppliers develop their own internal specifications to be used for new product launching, competitive product analysis, and future product development. Proprietary field testing is an integral part of the overall new lubricant product development processes and often is the most critical step to assure technical success and customer satisfaction for new products.

7.4.2 Additive Specifications

Specifications for antiwear/AS additives focus primarily on application, base oil compatibility, and quantification of elemental constituents. In addition, specifications typically identify specific and critical performance standards for applications. Common specifications for antiwear/AS additives are shown in Table 7.4.

In addition to typical specifications as reported in the Certificate of Analysis from additive suppliers, individual lubricant marketers often prefer to conduct their own internal additive specifications, such as infrared analysis and key performance testing.

7.4.3 Test Methods and Equipment

In the United States, a number of bench and advanced tests were developed and approved by ASTM, and these tests have gained widespread reception throughout the industry. However, there are also a few selected lab-bench and advanced tests that were developed and approved only by specific OEMs but represent certain critical and desirable performance features (Figure 7.12, lubrication regimes, and Figure 7.13, ashless antiwear additives: availability, applicability, selection, and future needs). Please note that more details on test methods and designations are available in another chapter entitled "Testing Methods for Additives/Lubricant Performance" and Chapter 32 is not intended to cover all evaluation tests in detail, but rather to illustrate a few representative tests to highlight the key assessment criteria:

1. *Four-Ball Wear and EP Test*: This tester was developed to evaluate the antiwear, AS/EP, and antiweld properties of lubricants. It is a simple bench test

TABLE 7.4
Typical Specifications for Antiwear and Antiscuffing Additives

Chemical Class	Property	Performance Test
Amine phosphates	% nitrogen, phosphorus, and TAN/TBN	Four-ball wear, four-ball EP, FZG, rust/oxidation test
Methylene bis-dialkyl dithiocarbamate	% sulfur, nitrogen, and residual chlorine, amine	Four-ball EP, FZG, Falex EP, oxidation/corrosion test
Sulfurized lard, esters, fatty acids	% total sulfur	Four-ball wear, four-ball EP, stick-slip, Cu corrosion
	% active sulfur	
Triphenyl phosphorothioate	% sulfur, phosphorus, and melting point	Four-ball EP, FZG, Falex EP, oxidation/corrosion test
Chlorinated paraffins, fatty acids	% chlorine	Four-ball wear, Falex EP, Timken, Cu corrosion
	Acid value	

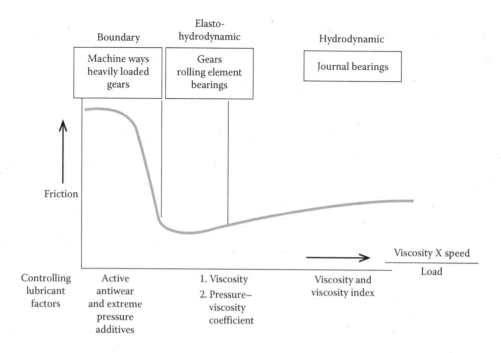

FIGURE 7.12 Lubrication regimes.

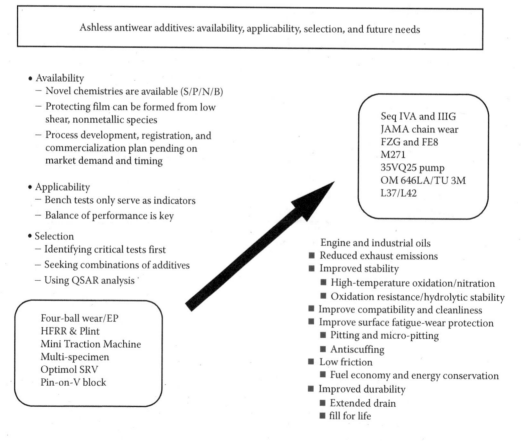

FIGURE 7.13   Ashless antiwear additives: availability, applicability, selection, and future needs.

machine designed to measure the protection a lubricant provides under conditions of high unit pressures and various sliding velocities. The four-ball wear tester consists of four one-half inch diameter steel balls arranged in the form of an equilateral tetrahedron. The three lower balls are held immovably in a clamping pot, while the fourth ball is caused to rotate against them. Test lubricant is added in the test pot, covering the contact area of the test balls. During a test, wear scars are formed on the surfaces of the three stationary balls. The diameter of the scars depends upon the load, speed, temperature, duration of run, and type of lubricant. The four-ball EP tester runs at a fixed speed of 1770 ± 60 rpm and has no provision for lubricant temperature control. A microscope is used to measure the wear scars. Two of the standard tests run on the four-ball machine are Mean-Hertz Load and load-wear index. ASTM D2596 covers the detailed calculation procedure of load-wear index for greases and D2783 for oils. These procedures involve the running of a series of ten second tests over a range of increasing loads until welding occurs. From the scar measurements, the mean load (load-wear index) is calculated and it serves as an indicator of the load-carrying properties of the oil being tested.

2. *FZG Four-Square Gear Test Rig (ISO 14635)*: The FZG test equipment consists of two gear sets, arranged in a four-square configuration, driven by an electric motor. The test gear set is run in the test fluid while increasing load stages (from 1 to 13) until failure. Each load stage is run for a 15 min period at a fixed speed. Two methods are used for determining the damage load stage. The visual rating method defines the damage load stage as the stage at which more than 20% of the load-carrying flank area of the pinion is damaged by scratches or scuffing. The weight loss method defines the damage load stage as the stage at which the combined weight loss of the drive wheel and pinion exceeds the average of the weight changes in the previous load stages by more than 10 mg. The test is used in developing industrial gear lubricants, ATFs, and hydraulic fluids to meet various manufacturers' specifications.

ISO 14635-3 was prepared by Technical Committee ISO/TC 60, *Gears*, Subcommittee SC 2, *Gear capacity calculation*.

ISO 14635 consists of the following parts, under the general title *Gears—FZG test procedures*:

- Part 1: FZG test method A/8,3/90 for relative scuffing load-carrying capacity of oils
- Part 2: FZG step load test A10/16, 6R/120 for relative scuffing load-carrying capacity of high EP oils
- Part 3: FZG test method A/2,8/50 for relative scuffing load-carrying capacity and wear characteristics of semifluid gear greases

3. *Falex EP/Wear Tester*: The Falex test machine provides a rapid method of measuring the load-carrying capacity and the wear properties of lubricants. The test consists of rotating a test pin between two loaded journals (V-Blocks) immersed in the lubricant sample. There are two common tests run in this machine: one is a scuffing test (subjecting a test lubricant to increasing loads until a failure occurs) and the other is a wear test (subjecting a lubricant to a constant load for a definite period of time while measuring the wear pattern).

4. *Timken EP Test*: The Timken EP test provides a rapid method of measuring abrasion resistance and the load-carrying capacity of lubricants. A number of lubricant specifications require Timken "OK" loads above certain minimum values. The mode of operation consists of rotating a Timken tapered roller bearing cup against a stationary, hardened steel block. Fixed weights force the block into contact with the rotating cup through a lever system. The OK load is the highest load the cup and block will carry without scoring during a 10 min run. Timken abrasion tests are run under fixed loads for extended time periods, and the weight loss of the cup and block is a measure of the abrasion resistance of the lubricant.

5. *L-37 High Torque Test*: The CRC L-37 test operates under low-speed, high-torque conditions. It evaluates the load-carrying ability, wear stability, and corrosion characteristics of gear lubricants. The test differential is a Dana Model unit driven by a Chevrolet truck engine and four-speed transmission. A complete, new axle assembly is used for each test after a careful examination of gear tooth and bearing tolerance. After break-in at reduced load and high speed, the test continues for 24 h under low-speed (80 axle rpm) and high-torque conditions.

6. *L-42 High Speed Shock Test*: The CRC L-42 test is established to evaluate the antiscore performance of EP additives in gear lubricants under high-speed, shock load conditions. The test axle is a Dana Model unit driven by a Chevrolet engine through a four-speed truck transmission. The procedure requires 5 accelerations in fourth gear with inertia loading and 10 accelerations in third gear with dynamometer loading. The lubricant evaluation is based on the amount of scoring, and test results are expressed as percent tooth contact area scored.

7. *FAG FE-8 Test (DIN 51819)*: FAG developed this test frame to be a flexible "tribological" system to conduct tests over a wide range of operating conditions with different test bearings. Short-duration standardized tests have been developed for different applications. FAG also uses longer-term testing (e.g., fatigue) for comprehensive evaluations. The FE-8 gear oil test was developed specifically to evaluate the effectiveness of antiwear additives. The test runs under heavy load and low speed that forces the bearing to operate under boundary lubrication conditions.

DIN stands for "Deutsches Institut für Normung," meaning "German institute for standardisation." DIN standards that begin with "DIN V" ("*Vornorm,*" meaning "pre-issue") are the result of standardization work, but because of certain reservations on the content or because of the divergent compared to a standard installation procedure of DIN, they are not yet published standards.

Bearing Test Conditions	
Bearings	Cylindrical roller/thrust loaded
Speed	7.5 rpm
Load	114 kN
Bearing temperature	Variable
Test duration	80 h

Other tests including Optimol SRV, Cameron-Plint, High Frequency Reciprocating Rig, Falex Multi-Specimen, Vickers Vane pump, Vickers 35-VQ-25 pump, and Denison high-pressure pump tests are also used widely in the evaluation of various lubricants and greases. Appropriate field tests are also arranged in proprietary test sites to ensure good product quality and equipment compatibility/friendliness prior to new product introduction to the marketplace.

On the engine oil side, the ILSAC GF-x specifications are the key passenger car engine oil specifications for North America and the Asia Pacific region. With the introduction of ILSAC GF-3 in 2001, the industry moved to a completely new set of engine tests for the validation of passenger car engine oil performance. Although some new tests replaced previous tests, which were running out of parts, ILSAC GF-4 superseded GF-3 in the summer of 2004.

Among the GF-4 tests, the most critical engine tests related to antiwear/AS performance are the Sequence IVA (ASTM D6891) and the Sequence IIIG (ASTM D7320). The Sequence IVA was previously referred to as the Nissan KA24E and was developed by the Japan Automotive Manufacturers Association. It is included to replace the wear component of the Sequence VE. The Sequence IVA is designed to evaluate an oil's ability to prevent cam lobe wear in slider valve train design engines operated at low-temperature, short trip, "stop and go" conditions (low-speed/low-temperature operation). The test conditions and specifications are listed in the following:

Engine	Nissan 2.4 L in-line 4 cylinder
Engine speed	800 and 1500 rpm cycles
Engine torque	25 Nm
Oil temperature	50°C–60°C
Cycle duration	50 min (low speed)/10 min (high speed)
Test length	100 h
7-Point cam lobe wear:	120/90/90 µm max (GF-3/GF-4/GF-5)

The Sequence IIIG is a replacement for the Sequence IIIF and uses a current production version of the GM 3800 Series II V-6

engine. Special camshaft and lifter metallurgy and surface finishing are used to induce wear. The Sequence IIIG procedure is designed to evaluate the oil resistance to oxidation and wear in high-speed and high-temperature vehicle operation. The test conditions and specifications are summarized in the following:

Engine	GM 3800 Series II V-6 (231 CID)
Engine speed	3600 rpm
Engine load	250 Nm
Valve spring load	205 lb
Oil temperature	150°C
Coolant temperature	115°C
Test length	100 h
Average cam and lifter wear	60 µm max

The Sequence IIIG and Sequence IVA tests were retained as the industry wear tests for ILSAC GF-5, which became licensable in October 2010.

The next ILSAC specification, GF-6, will likely be introduced in late 2018 or early 2019. This category will include two new wear tests. These are the Sequence IVB and the Ford Chain Wear Test. The Ford Chain Wear Test uses a downsized turbocharged gasoline direct injection engine. In a GDI engine, chain stretching can impact the timing of inlet and exhaust valve opening and closing, which impacts emission control and general engine performance. Overall operating conditions are summarized in Table 7.5. The test cycles between two different stages. The details for the two test stages are shown in Table 7.6. At this time, the ILSAC GF-6 limit for the Ford Chain Wear Test is not known.

TABLE 7.5
Ford Chain Wear Test Operating Conditions

Engine	2012 Ford EcoBoost
Displacement (L)	2.0
Cylinders (#)	4
Fuel delivery	Gasoline direct injection
Induction system	Turbocharged
Test duration	216 h
Test cycles per test	72
Soot at end of test	1%–2%

TABLE 7.6
Stage Details for the Ford Chain Wear Test

	Stage 1	Stage 2
Duration (min)	120	60
Engine speed (rpm)	1550	2500
Engine torque (Nm)	50	128
Oil gallery temperature (°C)	50	100
Air-to-fuel ratio	0.78	1.00

TABLE 7.7

Toyota Sequence IVB Test Operating Conditions

Engine	Toyota 2NRFE
Displacement (L)	1.5
Cylinders (#)	4
Fuel delivery	Port fuel injection
Test duration (h)	200
Test cycles per test	24,000
Valve train design	Bucket lifter
Fuel dilution at end of test	10%–15%

TABLE 7.8

Stage Details for the Toyota Sequence IVB Test

	Stage 1	Stage 2	Stage 3	Stage 4
Duration (s)	8	7	8	7
Engine speed (rpm)	4300—800	800	800–4300	4300
Engine torque (Nm)	25	25	25	25
Oil gallery temperature (°C)	55–53	53	53–55	55

The Toyota Sequence IVB test is designed to measure low-temperature valve train wear. Overall operating conditions are summarized in Table 7.7. The test cycles between four different stages. The details for the four test stages are shown in Table 7.8. Wear measurements include bucket lifter area loss, bucket lifter mass loss, and bucket lifter Keyence volume loss. The sequence IVB ILSAC GF-6 wear parameter that will be used is average Keyence volume loss (mm^3).

In the heavy-duty diesel engine oil area, there are a number of industry standard engine tests that measure the wear performance. These tests are required to meet both industry and engine manufacturer requirements such as API CJ-4 and various specifications from Caterpillar, Cummins, Detroit Diesel, Mack, and Volvo. The key wear tests assess the ability of an oil to control valve rain or ring and liner wear under severe operating conditions, which include high-load duty cycles, use of exhaust gas recirculation, and high levels of soot contamination.

The API CJ-4 category requires three tests that include valve train wear as a pass/fail parameter. The Roller Follower Wear Test (ASTM D5966) is run in a 6.5 L V-8 GM diesel engine; it was initially developed for the older API CG-4 category, which was developed for the introduction of low-sulfur (500 ppm maximum) fuel. However, this test remained as a requirement in all subsequent specifications. At the end of this 50 h test, the used oil soot level is typically 3.5%–4.0%. The level of wear on the stationary pin in the hydraulic cam followers is measured. The Cummins ISB test (ASTM D7484) was introduced as an industry requirement for API CJ-4. This 350 h test runs in a 5.9 L in-line 6-cylinder engine running of ultralow-sulfur (15 ppm maximum) diesel fuel. The first 100 h is run at steady-state conditions to generate 3.25% soot in the oil. The final 250 h is run under cyclic conditions to stress cam and tappet wear, which are the primary pass/fail criteria. The third diesel engine test that measures valve train wear is the Cummins ISM test (ASTM D7468). The Cummins ISM is the third in a series of Cummins heavy-duty wear tests developed for API and engine builder diesel oil specifications. Similar to previous Cummins M11 HST (ASTM D6838) and Cummins M11 EGR (ASTM D6975) tests, the Cummins ISM alternates between 50 h soot generation and 50 h wear stages. This test runs for 200 h using 500 ppm sulfur diesel fuel. The used oil typically contains 6%–7% soot, and the key pass/fail wear parameters are focused on the crossheads (bridges for the inlet and exhaust valves) and the adjusting screw for the fuel injectors.

The Mack T-12 test (ASTM D7422) measures ring and liner wear under severe operation using 15 ppm sulfur fuel. This 300 h test runs with a very high EGR rate for the first 100 h to generate 4.3% soot. During the final 200 h, the engine runs overfueled at peak torque conditions to create a very severe environment for top ring weight loss and liner wear at the point of top ring reversal, which are the key wear parameters for this test.

All of the wear tests mentioned earlier will continue to be part of the requirements to qualify oils against the latest API heavy-duty engine oil specifications—API CK-4, which represents an upgrade to the existing CJ-4 category, and FA-4, which is a new classification for oils having viscosity below the range covered by CK-4 and previous categories. The reduction in viscosity for FA-4 oils could lead to some formulation challenges since wear requirements for CK-4 and FA-4 are identical. Both new categories were introduced for first use in December 2016.

7.5 OUTLOOK

The additives business has experienced an economic upturn in recent years, primarily due to the imbalance between demand and supply as a result of tight feedstock availability and increased demand in the Far East region. The basic chemicals used to produce additives are subject to short supply as new and large capacity has not been effectively added to the manufacturing side for several years. A number of natural disasters, such as the Hurricane Katrina (Summer 2005), certainly made the situation even worse. The additive suppliers have successfully passed the raw material costs to their customers resulting in escalating unit pricing and improved profitability. The significant drop of crude oil price in 2015 certainly started to shift the basic chemicals toward the lower price range; however, the market is still not very sure whether this recent movement may impact the demand–supply balance as many additive components still belong to specialty chemicals business. The increased volume demand has been neutralized by several factors, such as longer-drain lubricants and the reduction of ash additives. Despite the push for new engine oils meeting more stringent requirements, a major rationalization is occurring because of the ability to use additives longer and the recycling of products in the industry. Consequently, the total additive volume demand is growing slowly. Ashless antiwear and AS additives should not be any different from other additives in terms of market demand [182].

Antiwear additives are a mature function class, and business opportunities in the next few years will be modest. The dominant position of zinc dithiophosphates in engine oils is gradually diminishing, but is not expected to be in jeopardy in the near term. Therefore, a total switch to ashless antiwear additives in engine oils is not likely to occur very soon, but minor changes are in progress.

The impetus for significantly improved lubricant additives is found on a number of fronts. Governmental and regulatory requirements continue to challenge the industry for improved products with lower toxicity. New engine developments, such as increased use of diamond-like carbon–coated engine parts and ceramic components for wear resistance and higher contact temperatures, are on the horizon and present opportunities for antiwear additives that can function at very high operating temperatures. Space technology and other advanced transportation needs present new challenges to the industry. And, of course, there will always be a need for low product costs and ease of production.

Five particular developments may have a major impact on the lubricant industry in the near term: (1) a move toward low-sulfur hydroprocessed (Group II and III), sulfur-free GTL and synthetic (Group IV and V) base stocks; (2) a move toward low-viscosity engine oils to improve fuel economy; (3) the imminent trend toward lower ash, sulfur, and phosphorus in engine oils; (4) a desire to reduce or eliminate chlorine in lubricants, particularly in metalworking fluids; and (5) a move to eliminate heavy metals and/or achieve low ash or even ash-free in both engine oils and industrial oils.

To meet the growing needs for better thermal/oxidative stability and better viscometrics, synthetic base stocks such as poly-α-olefins (PAOs), together with hydrotreated petroleum base stocks and GTL base stocks, are continuing to expand in all lubricant sectors. These types of materials have no aromatic hydrocarbons or greatly reduced amounts of aromatic hydrocarbons, which are potentially problematic for additive solvency; as a result of removing these solubilizing aromatics, the additives tend to precipitate out of the oil. This is particularly true for surface-active, polar components, such as antiwear additives. Therefore, greater compatibility with nonconventional base stocks (Groups II–IV) will be an essential requirement for all ashless antiwear and AS additives. Meanwhile, there are noticeable synergies identified among certain ashless antiwear and AS additives and nonconventional base stocks in a number of lubricant applications. Therefore, the choice of proper ashless additives will be vitally important.

To meet the growing needs for better fuel economy, low-viscosity engine oils are of high interest to automakers. The next passenger car specification GF-6 will introduce SAE 0W-16 oil, and several automakers have already started to market 0W-xx oils with xx being less than 20. For example, in Japan, Honda introduced 0W-8 oil recently for their N-box engine. As the oil becomes lighter in viscosity, wear protection is of great concern. The choice of proper antiwear additives including ashless antiwear components to work synergistically with surface-active friction modifiers as well as low level of ZDDP will be vitally important.

Because of the large number of automobiles equipped with catalytic converters that are sensitive to phosphorus derived from zinc dithiophosphates in the crankcase oil (possible reduction of catalytic efficiency), strong needs exist for engine oils with lower phosphorus content. Initially, ILSAC GF-4 aimed to reduce phosphorus levels to as low as 0.05% (about one-half of the former GF-3 level) but settled on a maximum phosphorus level of 0.08% instead (a 20% reduction). In addition to phosphorus limits, GF-4 oils also offered improved oxidation stability (including nitration control), high-temperature wear discrimination, high-temperature deposit control, and used oil pumpability [183]. The ILSAC GF-5 has kept the same maximum (0.08 wt%) and minimum (0.06 wt%) phosphorus levels as GF-4 and maintained the same performance requirements in valve train wear (sequence IVA); however, a new phosphorus volatility (ASTM D7320) requirement was added to maintain good phosphorus retention to address the critical phosphorus emission issue. The proposed ILSAC GF-6 specification will enable a new level of performance for passenger car engine oils in the era of efficiency. In moving beyond GF-5 requirements, GF-6 will incorporate increased fuel economy through the oil change interval, enhanced oil robustness for spark-ignited internal combustion engines, formulations to help minimize the occurrence of low-speed engine preignition, and wear protection for various engine components. The new GF-6B category defines the performance requirements for ultralow viscosity grades below SAE XW-20 and as such is distinct from GF-6A with no overlap at all. The challenge in moving to GF-6B is that oils are thinner at higher temperatures, which impacts the film thickness and can result in a faster transition from the EHL to a mixed lubrication regime. More effective friction modifiers and antiwear additives are likely to play an important role in providing the much needed performance level in GF-6B.

The new PC-11 requirements will be met by examining the critical areas of fuel economy in heavy-duty diesel engines and learning how to maintain engine durability in a low-viscosity regime. Therefore, the new additive platforms for PC-11 will offer the ability to cover multiple viscosity grades to suit the diverse needs of the market. The use of ashless antiwear additives to assist in boosting the performance of low phosphorus-containing HDEOs is critical for PC-11B. Likewise, universal oils that can operate in both gasoline and diesel engines will definitely need to operate at a lower phosphorus limit (800 vs. 1200 ppm) to meet the GF-6 specification.

As vehicle emission regulations become more challenging, increasing restrictions are likely to be placed on other lubricant elements besides phosphorus that can impact emission control systems. Sulfur and metals are also under scrutiny as sulfur is suspected as a poison of deNOx catalysts and ash (from metals) may plug after-treatment particulate traps. Modern engine oils rely heavily on zinc dialkyl dithiophosphates to provide antiwear, antioxidation, and anticorrosion protection. Since ZDDP is rich in phosphorus, sulfur, and zinc, it becomes an obvious target for emission control. In fact, at former use level, ZDDP was almost solely accountable for more than two-third of the sulfur and all the phosphorus and zinc present in engine oils, excluding sulfur from base oils. Oils with ZDDP at former levels could make it difficult for OEMs to optimize (for cost and life) their exhaust after-treatment systems. Therefore, the future trend will likely be toward further reduced ZDDP in engine oils providing that the performance integrity can be maintained through the use of alternate additives [184].

In order to satisfy performance requirements in terms of oxidation control and deposit levels, more antioxidants could be added to the engine oil formulation. These ashless antioxidants (hindered phenols and arylamines) may effectively compensate for the loss of oxidation protection due to the reduction in ZDDP concentrations. However, since ZDDP is such a cost-effective additive and is the sole antiwear component used in many engine oils, a reduction in ZDDP treat levels may not provide the needed wear protection. Recently, engine builders are requiring even greater antiwear protection and more demanding test protocols are being put in place to ensure that lubricants can meet these more stringent specifications. Therefore, there is a strong need for advanced ashless antiwear systems to replace or supplement ZDDP to satisfy emission regulations while ensuring high levels of wear protection [185,186].

Not all phosphorus-containing additives behave the same in engine oils. Furthermore, even within the same ZDDP family, not all ZDDP respond the same to after-treatment devices as evidenced by their relative volatility performance. Data indicated that volatilized phosphorus showed very low statistical dependence on either oil volatility or phosphorus concentration in the fresh oil. Rather, the data seemed to indicate that the chemistries of the phosphorus-containing additives and their formulation with other additives were the controlling cause of phosphorus volatility and, by extension, emission level. Selby's Phosphorus Emission Index and Sulfur Emission Index shed some insights into the volatility impact on emission issues and better S/P volatility control than the current ZDDP is highly desired for future ashless antiwear additives [187–189].

Chlorine in lubricants and other materials is becoming an increasing environmental concern. Legislation around the industrial world limits the chlorine content of many lubricant products to 50 parts per million or less. The Montreal Protocol mandated a gradual phaseout of the use of chlorine-containing refrigerants, such as hydrochlorofluorocarbon and chlorofluorocarbon (CFC) and replacement with alternative hydrofluorocarbons (HFC). Increased wear occurred in the refrigeration compressor when HFC refrigerants were substituted for CFC, and the cause of this increased wear was believed to be inferior antiwear capability of the alternative HFC refrigerant as the environmental gas, compared to that for CFC [190]. This offers some opportunities for the development of new ashless antiwear additives for refrigeration compressor oils.

Similar ecological pressures are facing the cutting oil industry, and future changes to reduce or eliminate chlorine are expected. The most significant opportunity is perhaps driven by human health and waste disposal issues concerning the use of CPs. CPs are used extensively as EP additives in metalworking fluids. The National Toxicology Program (NTP) listed CPs, derived from C12 feedstocks and chlorinated at 60%, as a suspect carcinogen. Although few metalworking fluids are formulated with this class of CP, the image of CPs in general has suffered due to uncertainties about future NTP reclassification of all such additives.

Gear additives are another area of concern. Because of the problems associated with chlorine additives, their use in gear oils has been greatly reduced. However, a number of processes for making gear additives utilize chlorine or chlorine-containing reagents at some point in the reaction sequence. Small amounts of chlorine still remain in the final product. The complete removal of the chlorine is therefore expected to become an important priority but will be difficult to attain in the near future.

Last, the use of metallic antiwear and AS additives is diminishing due to the influence of environmental concerns. Heavy metals are considered pollutants and their presence is no longer welcomed in the environment. Given equal performance and costs, ashless antiwear additives will be preferred for many future lubricants.

In the future, the lubricant additive business will continue to grow and will need more ashless antiwear and AS additives [191]. Possible new markets include biodegradable lubricants, biodiesel fuel–friendly lubricants, advanced transportation lubricants, robotics, ceramics, and space technology lubricants. Traditional markets in engine oils, ATFs, marine, aviation, gear, hydraulic, circulating oils, metalworking, and other industrial lubricants are also expanding. Healthy growth for nonconventional base oils (Groups II–V) is expected in many of these areas. Clearly advanced ashless antiwear additives with environmentally friendly features, excellent stability, and unique performance properties, especially for nonconventional base oils, will be the additives of choice for increasingly demanding lubricant applications.

ACKNOWLEDGMENT

We thank many of our colleagues, especially Dr. Douglas Deckman, Dr. Steven Kennedy, Dr. David Blain, Dr. Angeline Cardis, Dr. Angela Galiano-Roth, and Dr. Andy Jackson (currently a professor at the University of Pennsylvania) for their valuable comments. We also thank Mr. Larry Cunningham and Dr. Joe Roos of Afton Chemical Corporation for their support in providing the time needed by the second author (TCJ) to work on this chapter.

REFERENCES

1. Ranney, M.W., *Lubricant Additives*, Chemical Technology Review No. 2, 1973, Noyes Data Corporation.
2. Ranney, M.W., *Synthetic Oils and Additives for Lubricants—Advances since 1977*, Noyes Data Corporation.
3. Ranney, M.W., *Synthetic Oils and Additives for Lubricants—Advances since 1979*.
4. March, J., *Advanced Organic Chemistry: Reactions, Mechanisms, and Structure*, 2nd edn., McGraw-Hill Series in Advanced Chemistry, McGraw-Hill, Inc., p. 703, 1977.

5. Pozey, J.S. et al., in *Reactions of Sulfur with Organic Compounds*, J.S. Pizey, ed., Consultants Bureau, A Division of Plenum Publishing Corporation, New York, 1987.

6. Reid, E.E., *Organic Chemistry of Bivalent Sulfur*, Vols. I–V, 1958–1963, Chemical Publishing Co., Inc., New York, NY 1963.

7. Kharasch, N., *Organic Sulfur Compounds*, Vol. 1, Symposium Publications Division, Pergamon Press, Inc., New York, NY and Los Angeles, CA. Chapters 8, 10, and 20, 1961.

8. Landis, P.S., The chemistry of 1,2-dithiole-3-thiones, *Chem. Rev.* 65, 237, 1965.

9. Jones, S.O. and Reid, E.E., The addition of sulfur, hydrogen sulfide and mercaptans to unsaturated hydrocarbons, *J. Am. Chem. Soc.* 60, 2452, 1938.

10. Louthan, R.P., Preparation of mercaptans and thioether compounds, U.S. Patent 3,221,056; Other related patents—U.S. Patent 3,419,614, 4,194,980 and 4,240,958, 1965.

11. U.S. Patent 2,012,446, Method of sulphurizing pine oil and product thereof, 1935 and U.S. Patent 3,953,347, Novel sulfur-containing compositions, 1976.

12. Myers, H., Lubricating compositions containing polysulfurized olefin, U.S. Patent 3,471,404, 1969; U.S. Patent 3,703,504 and 3,703,505, 1972.

13. Davis, K.E., Sulfurized compositions, U.S. Patent 4,119,549, 1978 and 4,191,659, 1980.

14. Davis, K.E. and Holden, T.F., Sulfurized compositions, U.S. Patent 4,119,550, 1978 and 4,344,854, 1982.

15. Dibiase, S.A., Hydrogen sulfide stabilized oil-soluble sulfurized organic compositions, U.S. Patent 4,690,767, 1987.

16. Horodysky, A.G. and Law, D.A., Additive for lubricants and hydrocarbon fuels comprising reaction products of olefins, sulfur, hydrogen sulfide and nitrogen containing polymeric compounds, U.S. Patent 4,661,274, 1987.

17. Horodysky, A.G. and Law, D.A., Sulfurized olefins as antiwear additives and compositions thereof, U.S. Patent 4,654,156, 1987; U.S. Patent 2,995,569, 1961.

18. Johnson, D.E. et al., Sulfurized olefin extreme pressure/antiwear additives and compositions thereof, U.S. Patent 5,135,670, 1992 ; U.S. Patent 2,999,813, 1961, 2,947,695, 1960, and 2,394,536, 1946.

19. Papay, A.G., *Lubr. Eng.* 32(5), 229–234, 1975.

20. Korosec, P.S. et al., *NLGI Spokesman* 47(1), 1983.

21. Macpherson, I. et al., *NLGI Spokesman* 60(1), 1996.

22. Buitrago, J.A., Gear oil having low copper corrosion properties, EP Patent Application 1,471,133 A2, 2004.

23. Rohr, O., *NLGI Spokesman* 58(5), 1994.

24. Papay, A.G., *Lubrication Science 10-3*, 1998.

25. Mortier, R.M. and Orszulik, S.T., *Chemistry & Technology of Lubricants*, 1992.

26. Habeeb, J.J. and Haigh, H.M., Premium wear resistant lubricant containing non-ionic ashless anti-wear additives, U.S. Patent 7,754,663B2, 2010.

27. Samuel, D. and Silver, B.L., The mechanism of the reaction of nitrosyl chloride with dialkyl phosphonates (the Michaelski-Zwierzak reaction), *J. Chem. Soc.* 3582, 1963.

28. Smith, T.D., *J. Chem. Soc.* 1122, 1962.

29. Venezky, D.L. et al., *J. Am. Chem. Soc.* 78, 1664, 1956.

30. Orloff, H.D., *J. Am. Chem. Soc.* 80, 727–734, 1958.

31. Lacey, I.N., Macpherson, P.B., and Spikes, H.A., Thick antiwear films in EHD contacts, Part 2: Chemical nature of the deposited film, ASLE Preprint, 1985.

32. Ishikawa, M. and Watts, R.F., Continuously variable transmission fluid, U.S. Patent Application 2005/0250656 A1, 2005.

33. Corbridge, D.E.C., *Phosphorus: An Outline of Its Chemistry, Biochemistry and Technology*, pp. 213, 249, 401, 1985.

34. Lange, R.M., Norbornyl dimer ester and polyester additives for lubricants and fuels, U.S. Patent 4,707,301, 1987.

35. Pollak, K., Amine derivative of dithiophosphoric acid compounds, U.S. Patent 3,637,499, 1972.

36. Shaub, H., Amine salt of dialkyldithiophosphate, U.S. Patent 4,101,427, 1978.

37. Michaelis, K.P. and Wirth, H.O., Di- or trithiophosphoric acid diesters, U.S. Patent 4,244,827, 1981.

38. Horodysky, A.G. and Gemmill, R.M., Phosphosulfurized hydrocarbyl oxazoline compounds, U.S. Patent 4,255,271, 1981.

39. Farng, L.O. et al., Phosphorodithioate-derived pyrrolidinone adducts as multifunctional antiwear/antioxidant additives, U.S. Patent 5,437,694, 1995.

40. Ripple, D.E., Phosphorus acid compounds-acrolein-ketone reaction products, U.S. Patent 4,081,387, 1978.

41. Farng, L.O. et al., Lubricant additive comprising mixed hydroxyester or diol/phosphorodithioate-derived borates, U.S. Patent 4,784,780, 1988.

42. Farng, L.O. et al., Lubricant additives derived from alkoxylated diorgano phosphorodithioates and isocyanates to form urethane adducts, U.S. Patent 5,282,988, 1994.

43. Le Sausse, C. and Palotai, S., Ashless additives formulations suitable for hydraulic oil applications, U.S. Patent Application 2005/0096236 A1, 2005.

44. Cardis, A.B., Reaction products of dialkyl and trialkyl phosphites with elemental sulfur, organic compositions containing same, and their use in lubricant compositions, U.S. Patent 4,717,491, 1988.

45. Bosniack, D.S., Synthetic ester oil compositions containing organic sulfonic acid ammonium salts as load-carrying agents, U.S. Patent 4,079,012, 1978.

46. Nebzydoski, J.W. et al., Alkyl ammonium thiocyanate manufacture and lube containing same, U.S. Patent 3,952,059, 1976.

47. Debus, H., Uber die Verbindungen der Sulfocarbaminsaure, *Ann. Chem. (Liebigs)* 73, 26, 1850.

48. Thorn, G.D. and Ludwig, R.A., *The Dithiocarbamates and Related Compounds*, 1962.

49. Lam, W.Y., Lubricant composition, U.S. Patent 4,836,942, 1989.

50. Cardis, A.B. et al., Borated dihydrocarbyl dithiocarbamate lubricant additives and composition thereof, U.S. Patent 5,370,806, 1994.

51. Farng, L.O. et al., Dithiocarbamate-derived ethers as multifunctional additives, U.S. Patent 5,514,189, 1995.

52. Gatto, V.J., Dithiocarbamtes containing alkylthio and hydroxy substituents, U.S. Patent 6,852,680B2, 2005.

53. Cardis, A.B. and Ardito, S.A., Biodegradable non-toxic gear oil, U.S. Patent 6,649,574, 2003.

54. Daegling, S., Use of a noise-reducing grease composition, European Patent Application 1188814 A1, 2002.

55. Cartwright, S.J., Ashless lubricating oil composition with long life, Canadian Patent Application CA2465734 A1, 2004.

56. Little, R.Q., Process for preparing 2, 5-bis(hydrocarbondithio)-1, 3, 4-thiadiazole, U.S. Patent 3,087,932, 1963.

57. Gemmill, R.M. et al., Multifunctional lubricant additives and compositions there of, U.S. Patent 4,584,114, 1986.

58. Davis, R.H. et al., Antiwear/antioxidant additives based on dimercaptothiadiazole derivatives of acrylates and methacrylates polymers and amine reaction products thereof, U.S. Patent 5,188,746, 1993.

59. Karol, T.J., Maleic derivatives of 2,5-dimercapto-1,3,4-thiadiazoles and lubricating compositions containing same, U.S. Patent 5,055,584, 1991.

60. Fields, E.K., Corrosion resistant composition, U.S. Patent 2,799,652, 1957.

61. Davis, R.H. et al., Dimercaptothiadiazole-derived, organic esters, amides and amine salts as multifunctional antioxidant/ antiwear additives, U.S. Patent 4,908,144, 1990.

62. Vogel, P.W., Method of preparing reaction products of hydrazines, carbon disulfide and acylated nitrogen compositions, U.S. Patent 3,759,830, 1973.

63. Fields, E.K. et al., Corrosion inhibitors and compositions containing the same, U.S. Patent 2,703,784, 1955 and Soluble compositions containing a 2, 5-dimercapto-1, 3, 4-thiadiazole derivative, U.S. Patent 2,703,785, 1955.

64. Hsu, S.-Y. et al., Quaternary ammonium salt derived thiadiazoles as multifunctional antioxidant and antiwear additives, U.S. Patent 5,217,502 and 5,194,167, 1993.

65. Srinivasan S. et al., Automatic transmission fluid comprises major amount of base oil and minor amount of additives comprising dihydrocarbyl—Thiadiazole, sulfurized fat and ester and metal containing detergent, U.S. Patent Application 780998 20010209; EP Patent Application 1231256 20020814.

66. Srinivasan, S. et al., Automatic transmission fluid, for automatic transmission equipment platforms, comprises major amount of base oil and minor amount of additives comprising ashless dialkyl thiadiazole and amine antioxidants, U.S. Patent Application 800017 20010305, March 5, 2001; EP Patent Application 1239021 20020911, September 11, 2001.

67. Vann, W.D. et al., Lubricant containing a synergistic composition of rust inhibitors, antiwear agents, and a phenothiazine antioxidants, U.S. Patent 7,176,168 B2, 2007.

68. Esche, C.K., Gatto, V.J., and Lam, W.Y., Effective antioxidant combination for oxidation and deposit control in crankcase lubricants, U.S. Patent 6,599,865 B1, 2003.

69. Nalesnik, T.E. and Barrows, F.H., Substituted linear thiourea additives for lubricants, U.S. Patent 6,187,726 B2, 2001.

70. Mukkamala, R., Thioimidazolidine derivatives as oil-soluble additives for lubricating oils, European Patent Application EP 1,229,023 B1, 2003; EP 1,361,217 B1, 2005.

71. Nalesnik, T.E., Oxadiazole additives for lubricants, U.S. Patent 6,551,966 B2, 2003.

72. Nalesnik, T.E., Thiadiazolidine additives for lubricants, U.S. Patent 6,559,107 B2, 2003.

73. Camenzind, H. and Nesvadba, P., Lubricant composition comprising an allophanate extreme-pressure, anti-wear additive, U.S. Patent 5,084,195, 1992 and 5,300,243, 1994.

74. Zhang, J. et al., A study of 2-alkyldithio-benzoxazoles as novel additives, *Tribol. Lett.* 7, 173–177, 1999.

75. Polishuk, A.T. and Farmer, H.H., *NLGI Spokesman* 43, 200, 1979.

76. Schumacher, R. et al., Improvement of lubrication breakdown behavior of isogeometrical phosphorus compounds by antioxidants, *Wear* 146, 25–35, 1991.

77. Schumacher, R. et al., Tribofragmentation and antiwear behavior of isogeometric phosphorus compounds, *Tribol. Int.* 30(3), 199, 1997.

78. Okorodudu, A.O.M., Organophosphorus derivatives of benzotriazole, U.S. Patent 3,986,967, 1976.

79. Farng, L.O. and Horodysky, A.G., Phenylenediamine-derived phosphonates as multifunctional additives for lubricants, U.S. Patent 5,171,465, 1992.

80. Hotten, B.W., Bisphosphoramides, U.S. Patent 3,968,157, 1976.

81. Andrew, D.L. and Moore, G.G., Lubricating grease composition with increased corrosion inhibition, EP-903398, 1999.

82. Cier, R.J. and Bridger, R.F., Lubricant compositions containing nitrile antiwear additives, U.S. Patent 4,025,446, 1977.

83. Pratt, R.J., Product, U.S. Patent 3,941,808, 1976.

84. Daly, D.T., Adams, P.E., and Jackson, M.M., Additive composition, U.S. Patent 6,224,642 B1, 2001.

85. Nalesnik, T.E., Alkyl hydrazide additives for lubricants, U.S. Patent 6,667,282 B2, 2003.

86. Nalesnik, T.E., 1,3,4-Oxadiazole additives for lubricants, U.S. Patent 6,566,311 B1, 2003.

87. Avery, N.L. et al., Friction modifiers and antiwear additives for fuels and lubricants, U.S. Patent 5,538,653, 1996.

88. Farng, L.O. et al., Triazole-maleate adducts as metal passivators and antiwear additives, U.S. Patent 5,578,556, 1996.

89. Farng, L.O. et al., Fuel composition comprising triazole-derived acid-esters or ester-amide-amine salts as antiwear additives, U.S. Patent 5,516,341, 1996.

90. Nalesnik, T.E., Alkyl-succinhydrazide additives for lubricants, U.S. Patent 6,706,671 B2, 2004.

91. Nalesnik, T.E. and Barrows, F., Tri-glycerinate vegetable oil-succinihydrazide additives for lubricants, U.S. Patent 6,559,106 B1, 2003.

92. Levine, J.A. and Wu, S., Borate ester lubricant additives, U.S. Patent 7,291,581B2, 2007.

93. Farng, L.O. et al., Mixed alcohol/dimercaptothiadiazole-derived hydroxy borates as antioxidant/antiwear multifunctional additives, U.S. Patent 5,137,649, 1992.

94. Farng, L.O. et al., Lubricant additives, U.S. Patent 5,288,988, 1994.

95. Watts, R.F. et al., Power transmission fluids with improved extreme pressure lubrication characteristics and oxidation resistance, U.S. Patent 6,534,451 B1, 2003.

96. Farng, L.O. et al., Load-carrying additives based on organophosphites and amine phosphates, U.S. Patent 5,681,798, 1997.

97. Anon, The future of chlorine in metalworking fluids, *Lubr. Eng.* 35(5), 266–271, 1979.

98. (a) Gill, G., EPA nixes most chlorinated paraffins, *Lubes'N'Greases*, 21(5), pp. 45–48, May 2015. (b) Moon, M., Chlorinated paraffins: no quick solutions, *Lubes'N'Greases* 22(1), p38–42, January 2016.

99. Furey, M.J. and Kajdas, C., Wear reducing compositions and methods for their use, U.S. Patent 5,880072, 1999.

100. Baranski, J.R. and Migdal, C.A., Phenolic borates and lubricants containing same, U.S. Patent 5,698,499, 1997.

101. Roby, S.H. and Ruelas, S.G., Engine oil compositions, U.S. Patent Application 2005/0070450 A1, 2005; U.S. Patent 7,678,747B2, 2010; 7,926,453B2, 2011.

102. Williamson, W.F. and Rhodes, B., Non-phosphorus, non-metallic anti-wear compound and friction modifier, International Patent Application WO 00/42134, 2000.

103. Yoon, B.A. et al., Borated-epoxidized polybutenes as low ash anti-wear additives for lubricants, U.S. Patent 7,419,940B2, 2008.

104. Plechkova, N.V. and Seddon, K.R., Applications of ionic liquids in the chemical industry, *Chem. Soc. Rev.* 37, 123–150, 2008.

105. Somers, A.E., Howlett, P.C., MacFarlane, D.R., and Forsyth, M., A review of ionic liquid lubricants, *Lubricants* 1, 3–21, 2013.

106. Bart, J.C.J., Gucciardi, E., and Cavallaro, S., Advanced lubricant fluids, in *Biolubricants: Science and Technology*, Elsevier, Amsterdam, The Netherlands, pp. 824–846, 2013.

107. Ye, C.F., Liu, M.W., Chen, Y.X., and Yu, L.G., Room-temperature ionic liquids: A novel versatile lubricant, *Chem. Commun.* 21, 2244–2245, 2001.

108. Bermúdez, M.D., Jiménez, A.E., Sanes, J., and Carrión, F.J., Ionic liquids as advanced lubricant fluids, *Molecules* 14, 2888–2908, 2009.

109. Minami, I., Ionic liquids in tribology, *Molecules* 14, 2286–2305, 2009.

110. Fox, M.F. and Priest, M., Tribological properties of ionic liquids as lubricants and additives: Part 1: Synergistic tribofilm formation between ionic liquids and tricresylphosphate, *Proc. Inst. Mech. Eng. Part J: J. Eng. Tribol.* 222, 291–303, 2008.

111. (a) Jimenez, A.E., Bermudez, M.D., Iglesias, P., Carrion, F.J., and Martinez-Nicolas, G., 1-N-alkyl-3-methylimidazolium ionic liquids as neat lubricants and lubricant additives in steel-aluminium contacts, *Wear* 260, 766–782, 2006. (b) Jimenez, A.E., Bermudez, M.D., Carrion, F.J., and Martinez-Nicolas, G., Room temperature ionic liquids as lubricant additives in steel-aluminium contacts: Influence of sliding velocity, normal load and temperature, *Wear* 261, 347–359, 2006.

112. Qu, J., Truhan, J.J., Daic, S., Luo, H., and Blau, P.J., Ionic liquids with ammonium cations as lubricants or additives, *Tribol. Lett.* 22(3), 207–214, 2006.

113. Qu, J. et al., Ionic liquids as novel lubricants and additives for diesel engine applications, *Tribol. Lett.* 35, 181–189, 2009.

114. Mistry, K., Fox, M.F., and Priest, M., Lubrication of an electroplated nickel matrix silicon carbide coated eutectic aluminium-silicon alloy automotive cylinder bore with an ionic liquid as a lubricant additive, *Proc. Inst. Mech. Eng. Part J: J. Eng. Tribol.* 223, 563–569, 2009.

115. Cai, M., Liang, Y., Yao, M., Xia, Y., Zhou, F., and Liu, W., Imidazolium ionic liquids as antiwear and antioxidant additive in poly(ethylene glycol) for steel/steel contacts, *ACS Appl. Mater. Interfaces* 2, 870–876, 2010.

116. Qu, J. et al., Antiwear performance and mechanism of an oil-miscible ionic liquid as lubricant additive, *ACS Appl. Mater. Interfaces* 4, 997–1002, 2012.

117. Qu, J., Luo, H., Chi, M., Ma, C., Blau, P.J., Dai, S., and Viola, M.B., Comparison of an oil-miscible ionic liquid and ZDDP as a lubricant anti-wear additive, *Tribol. Int.* 71, 88–97, 2014.

118. Qu, J., Barnhill, W.C., Luo, H., Meyer III, H.M., Leonard, D.N., Landauer, A.K., Kheireddin, B., Gao, H., Papke, B.L., and Dai, S., Synergistic effects between phosphonium-alkylphosphate ionic liquids and zinc dialkyldithiophosphate (ZDDP) as lubricant additives, *Adv. Mater.* 27, 4767–4774, 2015.

119. Barnhill, W.C., Gao, H., Kheireddin, B., Papke, B.L., Luo, H., West, B.H., and Qu, J., Tribological bench and engine dynamometer tests of a low viscosity SAE 0W-16 engine oil using a combination of ionic liquid and ZDDP as anti-wear additives, *Front. Mech. Eng.* 1(Article 12), 1–8, 2015.

120. Aswath, P., Chen, X., Sharma, V., Igartua, M.A., Pagano, F., Binder, W., Zare, P., and Doerr, N., Synergistic mixtures of ionic liquids with other ionic liquids and/or with ashless thiophosphates for antiwear and/or friction reduction applications, U.S. Patent Application Publication 2013/0331305, 2013.

121. Yao, M., Liang, Y., Xia, Y., and Zhou, F., Bisimidazolium ionic liquids as the high-performance antiwear additives in poly(ethylene glycol) for steel–steel contacts, *ACS Appl. Mater. Interfaces* 1, 467–471, 2009.

122. Mu, Z., Zhou, F., Zhang, S., Liang, Y., and Liu, W., Effect of the functional groups in ionic liquid molecules on the friction and wear behavior of aluminum alloy in lubricated aluminum-steel contact, *Tribol. Int.* 38, 725–731, 2005.

123. Peterson, M.B. and Johnson, R.L., *Friction Studies of Graphite and Mixtures of Graphite with Several Metallic Oxides and Salts at Temperatures to 1000°F*, TN-3657, NACA, Cleveland, OH, 1956.

124. Antony, J.P., Mittal, B.D., Naithani, K.P., Misra, A.K., and Bhatgar, A.K., Antiwear/extreme pressure performance of graphite and molybdenum disulphide combinations in lubricating greases, *Wear* 174, 33–37, 1994.

125. Joly-Pottuz, L. and Ohmae, N., Carbon-based nanolubricants (Chapter 3), in *Nanolubricants*, Martin, J.M. and Ohmae, N., eds., Wiley, Chichester, West Sussex, U.K., pp. 93–148, 2008.

126. Bowden, F.P. and Tabor, D., *The Friction and Lubrication of Solids, Part 1*, Clarendon Press, Oxford, U.K., pp. 162–163, 1950.

127. Ugarte, D., Curing and closure of graphitic networks under electron-beam radiation, *Nature* 359, 707–709, 1992.

128. He, C., Zhao, N., Du, X., Shi, C., Ding, J., Li, J., and Li, Y., Low-temperature synthesis of carbon onions by chemical vapor deposition using a nickel catalyst supported on aluminum, *Scr. Mater.* 54, 689–693, 2006.

129. He, C., Zhao, N., Shi, C., Du, X., Li, J., Cui, L., and He, F., Carbon onion growth enhanced by nitrogen incorporation, *Scr. Mater.* 54, 1739–1743, 2006.

130. Ugarte, D., Onion-like graphitic particles, *Carbon* 33(7), 989–993, 1995.

131. Iijima, S., Helical microtubules of graphitic carbon, *Nature* 354, 56–58, 1991.

132. Dai, H., Carbon nanotubes: Synthesis, integration, and properties, *Acc. Chem. Res.* 35, 1035–1044, 2002.

133. Cursaru, D.-L., Andronescu, C., Pirvu, C., and Ripeanu, R., The efficiency of Co-based single-wall carbon nanotubes (SWNTs) as an AW/EP additive for mineral base oils, *Wear* 290–291, 133–139, 2012.

134. Iijima, S., Yudasaka, M., Yamada, R., Bandow, S., Suenaga, K., Kokai, F., and Takahashi, K., Nano-aggregates of single-walled graphitic carbon nano-horns, *Chem. Phys. Lett.* 309(3–4), 165–170, 1999.

135. Yamaguchi, T., Bandow, S., and Iijima, S., Synthesis of carbon nanohorn particles by simple pulsed arc discharge ignited between pre-heated carbon rods, *Chem. Phys. Lett.* 389, 181–185, 2004.

136. Yudasaka, M., Iijima, S., and Crespi, V.H., Single-walled carbon nanohorns and nanocones, *Top. Appl. Phys.* 111, 605–629, 2004.

137. Zin, V., Agresti, F., Barison, S., Colla, L., and Fabrizio, M., Influence of Cu, TiO_2 nanoparticles and carbon nano-horns on tribological properties of engine oil, *J. Nanosci. Nanotechnol.* 15, 3590–3598, 2015.

138. Kroto, H.W., Heath, J.R., O'Brian, S.C., Curl, R.F., and Smalley, R.E., C_{60}: Buckminsterfullerene, *Nature* 318, 162–163, 1985.

139. Loutfy, R.O., Lowe, T.P., Moravsky, A.P., and Katagiri, S., Commercial production of fullerenes and carbon nanotubes, in *Perspectives of Fullerene Nanotechnology*, Osawa, E., ed., Kluwer Academic Publishers, Dordrecht, The Netherlands, pp. 35–46, 2012.

140. Howard, J.B. and McKinnon, J.T., Combustion method for producing fullerenes, U.S. Patent 5,273,729, 1993.

141. Slanina, Z., Rudzinski, J.M., Togaso, M., and Osawa, E., Quantum-chemically supported vibrational analysis of giant molecules: the C_{60} and C_{70} clusters, *J. Mol. Struct.* 202, 169–176, 1989.

142. Titov, A., Effect of fullerene containing lubricants and wear resistance of machine components in boundary lubrication, PhD dissertation, The NJ Institute of Technology, Newark, NJ, 2004.

143. Ku, B.-C., Han, Y.-C., Lee, J.-E., Lee, J.-K., Park, S.-H., and Hwang, Y.-J., Tribological effects of fullerene (C_{60}) nanoparticles added in mineral lubricants according to its viscosity, *Int. J. Precis. Eng. Manuf.* 11(4), 607–611, 2010.

144. Gupta, B.K. and Bhushan, B., Fullerene particles as an additive to liquid lubricants and greases for low friction and wear, *Lubr. Eng.* 50(7), 524–528, 1994.

145. Novoselov, K.S., Geim, A.K., Morozov, S.V., Jiang, D., Zhang, Y., Dubonos, S.V., Grigorieva, I.V., and Firsov, A.A., Electric field effect in atomically thin carbon films, *Science* 306, 666–669, 2004.

146. Rao, C.N.R., Maitra, U., and Ramakrishina Matte, H.S.S., Synthesis, characterization, and selected properties of graphene (Chapter 1), in *Graphene: Synthesis, Properties, and Phenomena*, Rao, C.N.R. and Sood, A.K., eds., Wiley-VCH, Weinheim, Germany, pp. 1–47, 2013.

147. Geim, A.K. and Novoselov, K.S., The rise of grapheme, *Nat. Mater.* 6, 183–191, 2007.

148. Berman, D., Erdemir, A., and Sumant, A.V., Few layer graphene to reduce wear and friction on sliding steel surfaces, *Carbon* 54, 454–459, 2013.

149. Lin, J., Wang, L., and Chen, G., Modification of graphene platelets and their tribological properties as a lubricant additive, *Tribol. Lett.* 41, 209–215, 2011.

150. Eswaraiah, V., Sankaranarayanan, V., and Ramaprabhu, S., Graphene-based engine oil nanofluids for tribological applications, *ACS Appl. Mater. Interfaces* 3, 4221–4227, 2011.

151. Danilenko, V.V., On the history of the discovery of nanodiamond synthesis, *Phys. Solid State* 46(4), 395–399, 2004.

152. Kharisov, B.I., Kharissova, O.V., and Chavez-Guerrero, L., Synthesis techniques, properties, and applications of nanodiamonds, *Synth. React. Inorg. Metal-Org. Nano-Metal Chem.* 40(2), 84–101, 2010.

153. Kumar, A., Lin, P.A., Xue, A., Hao, B., Yap, Y.K., and Sankaran, R.K., Formation of nanodiamonds at near-ambient conditions via microplasma dissociation of ethanol vapour, *Nat. Commun.* 4, 2618, 2013.

154. Ivanov, M.G. and Ivanov, D.M., Nanodiamond nanoparticles as additives to lubricants (Chapter 14), in *Ultrananocrystalline Diamond: Synthesis, Properties, and Applications*, Shenderova, O.A. and Gruen, D.M., eds., 2nd edn., Elsevier, Amsterdam, The Netherlands, pp. 457–492, 2012.

155. Mochalin, V.N., Shenderova, O., Ho, D., and Gogotsi, Y., The properties and applications of nanodiamonds, *Nat. Nanotechnol.* 7, 11–23, 2012.

156. Schrand, A.M., Ciftan Hens, S.A., and Shenderova, O.A., Nanodiamond particles: Properties and perspectives for bioapplications, *Crit. Rev. Solid State Mater. Sci.* 34(1–2), 18–74, 2009.

157. Ivanov, M.G., Pavlyshko, S.V., Ivanov, D.M., Petrov, I., and Shenderova, O., Synergistic compositions of colloidal nanodiamond as lubricant-additive, *J. Vac. Sci. Technol. B* 28(4), 869–877, 2010.

158. Zhmud, B. and Pasalskiv, B., Nanomaterials in lubricants: An industrial perspective on current research, *Lubricants* 1, 95–101, 2013.

159. Novak, C., Kingman, D., Stern, K., Zou, Q., and Gara, L., Tribological properties of paraffinic oil with nanodiamond particles, *Tribol. Trans.* 57(5), 831–837, 2014.

160. Ivanov, M.G., Ivanov, M.D., Pavlyshko, S.V., Petrov, I., Vargasm, A., McGuire, G., and Shenderova, O., Nanodiamond-based nanolubricants, *Fuller. Nanotube Carbon Nanostruct.* 20, 606–610, 2012.

161. Fernandez Rico, E., Minondo, I., and Garcia Cuervo, D., The effectiveness of PTFE nanoparticle powder as an EP additive to mineral base oils, *Wear* 262, 1399–1406, 2007.

162. Dubey, M.K., Bijwe, J., and Ramakumar, S.S.V., PTFE based nano-lubricants, *Wear* 306, 80–88, 2013.

163. Dubey, M.K., Bijwe, J., and Ramakumar, S.S.V., Nano-PTFE: New entrant as a very promising EP additive, *Tribol. Int.* 87, 121–131, 2015.

164. Habeeb, J.J. and Bogovic, C.N., Reduced friction lubricating oils containing functionalized carbon nanomaterials, U.S. Patent 8,435,931, 2013.

165. Choudhary, S., Mungse, H.P., and Khatri, O.P., Dispersion of alkylated graphene in organic solvents and its potential for lubrication applications, *J. Mater. Chem.* 22, 21032–21039, 2012.

166. Jao, T.C. and Devlin, M.T., Diblock monopolymers as lubricant additives and lubricant formulations containing same, U.S. Patent 7,867,958, 2011.

167. Davey, W. and Edwards, E.D., The extreme-pressure lubricating properties of some sulfides and disulfides, in mineral oil, as assessed by four-ball machine, *Wear* 1(4), 291–304, 1958.

168. Forbes, E.S., The load-carrying action of organo-sulphur compounds—A review, *Wear* 15(2), 87–96, 1970.

169. Forbes, E.S. and Reid, A.J.D., Liquid phase adsorption/reaction studies of organo-sulfur compounds and their load-carrying mechanism, *ASLE Trans.* 16(1), 50–60, 1973.

170. Bovington, C.H. and Dacre, B., Thermal decomposition of dibenzyl disulfide in hexadecane, *ASLE Trans.* 25(2), 267–271, 1982.

171. Bovington, C.H. and Dacre, B., Catalytic decomposition of dibenzyl disulfide and dibenzyl sulfide on iron powder, *ASLE Trans.* 25(1), 44–48, 1982.

172. Dacre, B. and Bovington, C.H., The adsorption and desorption of dibenzyl disulfide and dibenzyl sulfide on steel, *ASLE Trans.* 25(2), 272–278, 1982.

173. Plaza, S., Some chemical reactions of organic disulfides in boundary lubrication, *ASLE Trans.* 30(4), 493–500, 1986.

174. Kutney, G.W. and Turnbull, K., Compounds containing the S=S bond, *Chem. Rev.* 82(4), 333–357, 1982.

175. Kajdas, Cz., On a negative-ion concept of EP action of organosulfur compounds, *ASLE Trans.*, 28(1), 21–30, 1985.

176. Plaza, S., Mazurkiewicz, B., and Gruzinski, R., Thermal decomposition of dibenzyl disulphide and its load-carrying mechanism, *Wear* 174, 209–216, 1994.

177. Plaza, S., Comellas, L.R., and Starczewski, L., Tribochemical reactions of dibenzyl and diphenyl disulphides in boundary lubrication, *Wear* 205, 71–76, 1997.

178. Pavelko, G.F. and Frict, J., Correlation between thermochemical and antiscuff characteristics of organosulfur compounds, *Wear* 33(6), 443–452, 2012.

179. Pavelko, G.F., Lubricating properties of organic sulfides as a function of hydrogen sulfide and sulfur quantity formed at friction, *Proceedings of Fourth Moscow Science Technology Conference 'Tribotechnics to Machine Engineering'*, Moscow, Russia, pp. 88–89, 1989.

180. Pavelko, G.F., Lubricating properties of chemically active products obtained from organic sulfides and halogen-hydrocarbons in the conditions of boundary friction, *Treie Iznos* 11(5), 926–928, 1990.

181. Matveevskii, R.M., Kaidas, Ch., Buyanovskii, I.A., and Dombrovski, Ya.R., Connection of lubricating properties of chemically active media with their reaction capacity, *Trenie Iznos* 7(6), 969–973, 1986.

182. Beeton, J., Will additives outpace the lubes market?, *Lubes'N'Greases* 22(1), 32–36, January 2016.

183. Tan, I., Lubricant additives—Treats and opportunities, *Lubricants World* 16–19, July/August 2003.

184. Farng, L.O. et al., Ashless anti-wear additives for future engine oils, *Fourteenth International Colloquium Tribology*, pp. 1547–1553, January 13–15, 2004.

185. Korcek, S., Jensen, R.K., and Johnson, M.D., Assessment of useful life of current long drain and future low phosphorus engine oils, *Proceedings of Second World Tribology Congress—Scientific Achievements—Industrial Applications—Future Challenges*, pp. 259–262, Vienna, Austria, 2001.

186. Korcek, S., Jensen, R.K., and Johnson, M.D., Engine oil performance requirements and reformulation for future engines and systems, SAE Paper #961146, 1996.

187. Selby, T.W., Development and significance of the phosphorus emission index of engine oils, *Thirteenth International Colloquium Tribology—Lubricants, Materials and Lubrication*, Technische Akademie Esslingen, Ostfildern, Germany, January 15–17, 2002.

188. Selby, T.W., Fee, D.C., and Bosch, R.J., Analysis of the volatiles generated during the Selby-Noack test by ^{31}P NMR spectroscopy, *Elemental Analysis Symposium, ASTM D02 Meeting*, Tampa, FL, December 2004.

189. Selby, T.W., Fee, D.C., and Bosch, R.J., Phosphorus additive chemistry and its effects on the phosphorus volatility of engine oils, *Elemental Analysis Symposium, ASTM D02 Meeting*, Tampa, FL, December 2004.

190. Mizuhara, K. et al., The friction and wear behavior in controlled alternative refrigerant atmosphere, *Tribol. Trans.* 37(1), 120, 1994.

191. Farng, L.O. and Deckman, D.E., Novel anti-wear additives for future lubricants, *Additives 2007 Conference: Applications for Future Transport*, London, U.K., April 17–19, 2007.

8 Ashless Phosphorus–Containing Lubricating Oil Additives

W. David Phillips and Neal Milne

CONTENTS

8.1 INTRODUCTION AND SCOPE

In any discussion of phosphorus-containing lubricating oil additives, the products that probably come most rapidly to mind are the zinc dialkyldithiophosphates (ZDDPs)—multifunctional additives that have been widely used in both automotive and industrial oils for many years. However, a wide variety of ashless phosphorus–containing additives are used in the lubricating oil industry. As with the metal-containing dithiophosphates, they have been in use over a long period and, despite considerable research into alternative chemistries, the basic structures introduced in the 1930s are still used today. In contrast, the technology of most other additive types, and that of the base stocks themselves, has steadily developed over this time.

Many different types of phosphorus-containing molecules have been examined as additives for lubricating oils, with most attention given to their potential as antiwear (AW) and extreme pressure (EP) additives. Consequently, the patent literature contains a host of references to different structures displaying this characteristic. However, regardless of composition, all the additives used in this application serve the same and specific function of bringing phosphorus into contact with the metal surface, where it can be adsorbed and, under certain conditions, react. The resulting surface film improves the lubrication properties of both mineral and synthetic oils.

This chapter discusses the use of chemicals that contain only phosphorus to improve the performance characteristics of oils, specifically neutral and acid phosphates, phosphites and phosphonates, and the amine salts of the acids (see Figures 8.1 and 8.2 for an outline of the main classes and their structures). These are the principal types of phosphorus compounds in current commercial use, but other types have also been examined and claimed in the patent literature, for example, phosphoramidates and, more recently, phosphorus-containing ionic liquids. There are also ashless compounds in which sulfur or chlorine has been incorporated into the molecule as, for example, in thiophosphates and chlorinated phosphates. They are, however, outside the scope of this discussion, but the performance of mixtures of compounds separately containing phosphorus and sulfur or chlorine will be discussed.

In addition to examining the impact of ashless phosphorus compounds on lubrication performance, this chapter also looks at their performance as antioxidants, rust inhibitors, and metal passivators. Additionally, their polar nature makes them good solvents and assists the solution of other additives in nonpolar base stocks. The versatility displayed by phosphorus-containing additives is such that usage of these products continues to grow nearly half a century after their introduction, and they find application in the latest technological developments.

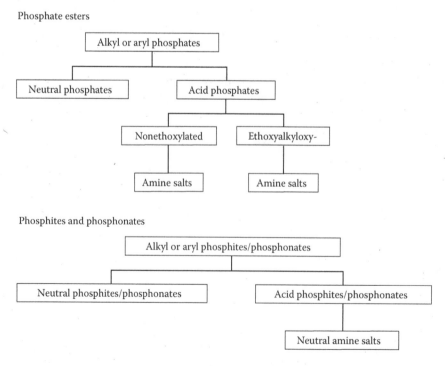

FIGURE 8.1 The main classes of ashless phosphorus-containing additives for lubricating oils.

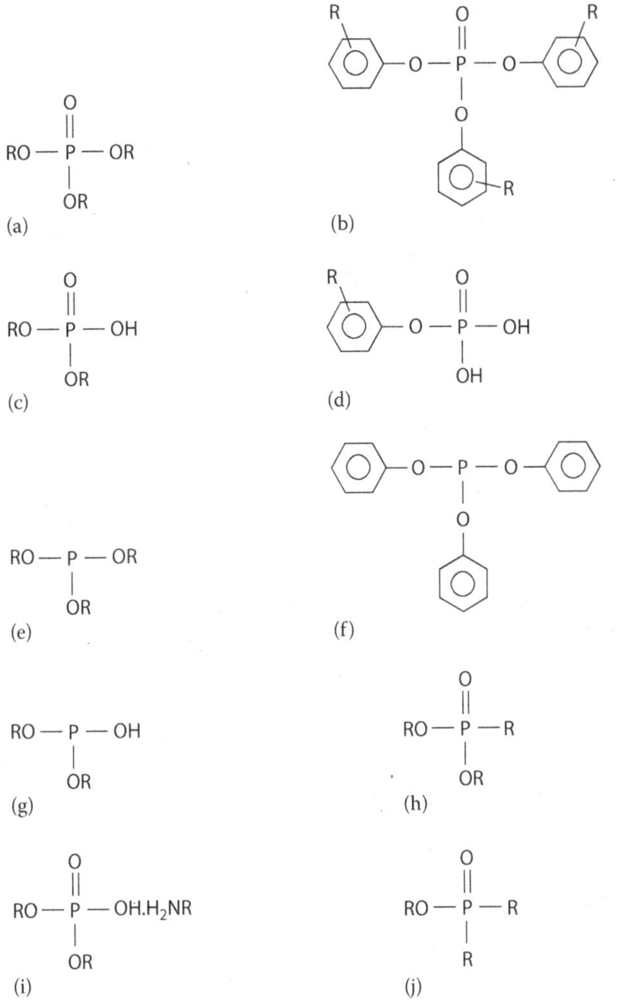

FIGURE 8.2 The structures of some common phosphorus-containing lubricating oil additives: (a) trialkyl phosphates, (b) triaryl phosphate, (c) alkyl monoacid phosphate, (d) aryl diacid phosphate, (e) trialkyl phosphite, (f) triaryl phosphite, (g) dialkyl phosphite, (h) dialkyl alkyl phosphonate, (i) amine phosphate, and (j) alkyl dialkyl phosphinate.

8.2 HISTORICAL BACKGROUND

Until the 1920s, additive-free mineral oils met the majority of industry's lubrication requirements. In the applications where their performance was unsatisfactory, an increase in viscosity or the sulfur content of the oils then available usually provided adequate lubrication. For very severe applications, the oil would be blended with animal or vegetable oils—for example, tallow or rapeseed oils were used for steam engine cylinder lubrication. Fish oils were used in the early locomotive axle boxes, while castor oil reduced friction in worm gear drives and flowers of sulfur were added to cutting oils. However, when hypoid gears were introduced, they quickly revealed the limited lubrication of oils then available. This resulted in the development of additives such as sulfurized lard oil and lead naphthenate. These were followed by sulfurized sperm oil, an additive that eventually became widely used in both industrial and automotive applications.

The earliest type of an organic phosphorus chemical to find use as a lubricating oil additive is thought to have been a neutral triaryl phosphate, specifically tricresyl phosphate (TCP). This material, which is sometimes called tritolyl phosphate, was originally synthesized in about 1854 [1] although trialkyl phosphates were synthesized slightly earlier, in about 1849 [2]. Commercial production of TCP began in about 1919, when this product was introduced as a plasticizer for cellulose nitrate, but it was not until the 1930s that patents began to appear claiming improved lubrication when TCP was blended with mineral oil. In 1936, this use was claimed in gear oils [3], but a detailed investigation into their behavior as AW additives was not published until 1940 [4,5], by which time TCP was already said to be in widespread use. During World War II, extensive research into phosphorus-containing additives took place in Germany [6,7]. This research was facilitated by the recent availability of test equipment for assessing wear and load-carrying behavior, for example, the four-ball machine [8]. The results of the research concluded that for high load-carrying (later known as EP) performance, the molecule must contain the following:

- A phosphorus atom
- Another active group, for example, Cl^- or OH^- (for attachment to the metal surface)
- At least one aryl or alkyl group (phosphoric acid was not thought to be active)

Subsequent to these studies, the market adopted chlorine-containing phosphates such as tris-(2-chloroethyl) phosphate, but they were later replaced in most applications by other EP additives as chlorine tended to produce corrosion.

The 1940s and 1950s saw significant development activity in the oil industry involving TCP, and patents appeared claiming the use of this AW/EP additive in general industrial oils [9], rolling oils [10], cutting oils [11], greases [12], rock drill lubricants [13], and aviation gas turbine lubricants [14,15]. Some military specifications, for example, on hydraulic oils (NATO codes H515/520/576), were published, which initially called for the use of this additive. However, in the late 1960s, the difficulty of obtaining good-quality feedstocks for the manufacture of *natural* phosphates based on cresol and xylenol, together with the concern regarding the neurotoxicity of TCP [16,17], led to the reformulation of many products with the less toxic *synthetic* triaryl phosphates based on alkylated phenols. TCP is still used today in aviation applications, but the quality of the phosphate in terms of its purity and freedom from the o-cresol isomers that were mainly responsible for its neurotoxicity behavior has significantly improved in the past 10–20 years.

In addition to its use as an oil additive, TCP was also used for a period in the 1960s as an ignition control additive for motor gasoline to avoid preignition arising from the deposition of lead salts. These were formed by the interaction of the lead tetraethyl antiknock additive and the alkyl halide scavenger [18–21]. Alkyl phosphates were claimed for this application in 1970 [22].

As a result of their polar nature, neutral triaryl phosphates have also been claimed as corrosion inhibitors for hydrocarbons [23,24], but they are unlikely to be promoted for this application today in view of the availability of more active species, such as the acid phosphates.

The use of trialkyl phosphates as AW and EP additives has been much less extensively evaluated. Although a flurry of patent activity took place in the late 1920s and 1930s covering methods for their manufacture [25–34], there was little interest in their use as lubricating oil additives for further 20 years. This was probably a result of the focus, in the interim period, on chlorinated derivatives. It was not until the late 1950s that tributyl phosphate (TBP) was disclosed in blends with isopropyl oleate [35] for use in gear oils and claimed in blends with chlorinated aromatics. In 1967, a patent appeared claiming the use of alkyl phosphates or the amine salts of alkyl acid phosphates in a water-based lubricating composition [36].

In addition to alkyl phosphates, various other types of phosphorus-containing compounds have been evaluated as AW/EP additives. Patents on acid phosphates claiming their use as EP additives for oil appeared in 1935 and 1936 [37–39], whereas the first publication with detailed information on the use of ethoxylated alkyl or aryl phosphate oil additives (in metalworking applications) appeared in 1964 [40]. Patents on the use of these products in mineral oils [41] and in synthetic esters [42] appeared later. Alkoxylated acid phosphates were also found to have good rust inhibition properties [43], a feature that was additionally observed for the alkyl (or aryl) acid phosphates [44,45].

Neutral amine salts of alkyl acid phosphates were claimed in 1964 [43] and, in 1969, in admixture with neutral phosphates [46]. These are, however, just a few examples of the patent estate covering these product groups.

The other main phosphorus-containing products to be discussed are the phosphites. The basic chemistry of alkyl and aryl phosphites, like that of the phosphate esters, was also uncovered in the nineteenth century. In a similar fashion, their utilization as oil additives was not exploited until much later. Patents appeared in 1940 on the use of mixed aryl phosphites as oil antioxidants [47] and on their activity as AW/EP additives at least as early as 1943 [48].

Isomeric with the acid phosphites are phosphonates (Figure 8.2). Dialkyl alkyl phosphonates were claimed as lubricants in 1952 and 1953 [49,50] but not until about 1971 as friction modifiers and EP additives [51,52].

Ionic liquids are low-melting-point salts usually defined as having melting points under 100°C. They were first developed in the 1970s and the 1980s and often based on imidazolium and pyridinium cations, with halide anions. Development has continued since then and phosphorus-containing ionic liquids are starting to appear in the literature as AW and EP additives in hydrocarbon lubricants and in other ionic liquids. The phosphorus can be found either in the cation, for example, as an organophosphate or in the anion as a phosphonium derivative

or in both charged species. So far, phosphorus-containing ionic liquids have not been commercialized and therefore are not mentioned in Section 8.3, but their performance is briefly discussed in Section 8.5.7.

The preceding summary focused on the use of phosphorus compounds alone. In reality, they are widely used in admixtures with sulfur-containing materials to provide good lubrication over a wider range of performance requirements. Examples of some of the combinations patented are given in Appendix 8.A.

8.3 MANUFACTURE OF PHOSPHORUS-CONTAINING LUBRICATING OIL ADDITIVES

8.3.1 NEUTRAL ALKYL- AND ARYL PHOSPHATE ESTERS

Although phosphate esters can be regarded as *salts* of *ortho*-phosphoric acid, they are currently not produced from this raw material because the yields are relatively low (~70% for triaryl phosphates). Instead, phosphorus oxychloride ($POCl_3$) is reacted with either an alcohol (ROH), a phenol (ArOH), or an alkoxide (RONa) as indicated in the following reactions:

$$3ROH + POCl_3 \rightarrow \underset{\text{Trialkyl phosphate}}{(RO)_3 PO} + 3HCl \qquad (8.1)$$

$$3ArOH + POCl_3 \rightarrow \underset{\text{Triaryl phosphate}}{(ArO)_3 PO} + 3HCl \qquad (8.2)$$

$$3RONa + POCl_3 \rightarrow \underset{\text{Trialkyl phosphate}}{(RO)_3 PO} + 3NaCl \qquad (8.3)$$

Reactions 8.1 through 8.3 pass through intermediate steps as shown in the following reaction (8.4):

$$ROH + POCl_3 \rightarrow ROPOCl_2 + HCl \xrightarrow{\ ROH\ } (RO)_2 POCl$$
$$+ HCl \xrightarrow{\ ROH\ } (RO)_3 PO + HCl \qquad (8.4)$$

The intermediate products are called *phosphorochloridates*, and, if desired, it is possible to obtain a mixture rich in a particular intermediate by changing the ratio of reactants.

In 2001, a two-stage process for the production of a tertiarybutylphenyl phosphate (TBPP) with low levels of triphenyl phosphate (TPP) was described [53]. In this process, the $POCl_3$ is first reacted with sufficient tertiarybutylphenol (C_4H_9ArOH) to mainly produce the phosphorodichloride. Phenol is then added to the reaction mixture to produce predominantly the mono-tertiarybutylphenyl diphenyl phosphate (reaction 8.5). This product was said to give unusually low air entrainment values and was therefore mainly of interest as a hydraulic fluid base stock.

$$C_4H_9ArOH + POCl_3 \rightarrow C_4H_9ArOPOCl_2$$
$$\xrightarrow{PhOH} C_4H_9ArOPO(OPh)_2 + 2HCl \qquad (8.5)$$

The production of mixed products can be achieved by using different alcohols or by an alcohol and alkoxide (reaction 8.6). These materials are not in significant use as AW/EP additives.

$$ROH + POCl_3 \rightarrow (RO)_2 POCl$$
$$\xrightarrow{ArONa} \underset{\text{Mixed alkyl-aryl phosphate}}{(RO)_2 PO(OAr)} + NaCl \qquad (8.6)$$

Trialkyl (or alkoxyalkyl) phosphates can be produced by either reaction 8.1 or 8.3, although in reaction 8.1, unless catalyzed, a considerable excess of alcohol is required to drive the reaction to completion. The hydrogen chloride (HCl) byproduct is removed as rapidly as possible—usually by vacuum or water washing while the reaction temperature is controlled to minimize the thermal degradation of the phosphate. In the alkoxide route (reaction 8.3), the chlorine precipitates as sodium chloride (NaCl), somewhat simplifying the purification treatment. After a water wash to remove the NaCl, purification consists of a distillation step to remove excess alcohol, an alkaline wash, and a final distillation to remove water [54]. With the alkoxide method, any residual chloride can be removed by water washing followed by a final distillation under vacuum.

Despite early research into the alkyl phosphates, a rigorous investigation into the preparation of the lower alkyl derivatives and their properties did not take place until 1930 [55]. In contrast to the large range of aryl phosphates available, the range of neutral alkyl phosphates is currently limited to tri-n-butyl phosphate and tri-iso-butyl phosphate, trioctyl phosphate, and tributoxyethyl phosphate. Other ether phosphates [56] have been claimed in the past but, as far as is known, are not currently manufactured.

Although the neutral trialkyl phosphates have been available for many years, they have not been widely used as additives for mineral oil. Those products in commercial production are used principally as components of aircraft hydraulic fluids, in turbine oils, rolling oils, or as solvents in industrial processes. However, interest in these materials as AW additives for applications where the release of phenols from the degradation of the phosphate is to be avoided currently exists. They also offer advantages as alternatives to the acid phosphates, alkoxylated acid phosphates, and their salts in metalworking applications, where there are concerns over instability in hard water and foam production in use (Canter, N., Private Communication, August 2001).

Triaryl phosphates, which are the most widely used of all ashless phosphorus–based AW additives, are currently manufactured almost exclusively by reaction 8.2. Phosphorus oxychloride is added to the reaction mass containing an excess of the phenolic feedstock in the presence of a small amount of catalyst, typically aluminum chloride or magnesium chloride, before heating slowly. The HCl is removed as it is formed under vacuum, followed by absorption in water. On completion of the reaction, the product is distilled to remove most of the excess phenols, the catalyst residue, and traces of polyphosphates. Finally, the product may be steam-stripped to remove volatiles including residual phenols and is dried under vacuum.

The raw material for the manufacture of triaryl phosphates was originally obtained from the destructive distillation of coal. This process yields coal tar, which is a complex mixture of phenol and alkyl phenols including cresols (methylphenols) and xylenols (dimethylphenols). Distillation of this mixture (sometimes known as cresylic acids) produces feedstocks rich in cresols and xylenols, which are then converted into the neutral phosphate. An early patent on the production of triaryl phosphates from tar acids was issued in 1932 [26].

Unfortunately, in the 1960s, as the number of coal tar distillers declined due to the move from coal to natural gas as a fuel, it became progressively more difficult to obtain cresols and xylenols from this source. As a consequence, the phosphate manufacturers turned their attention to the use of phenol, which was alkylated with propylene or butylene. The resultant mixtures of alkylated phenols were then converted into phosphates [57,58]. To distinguish phosphates from these two sources of raw materials, the cresol- and xylenol-based products became known as "natural" phosphates and the phosphates from alkylated phenols as "synthetic" phosphates. This distinction is no longer valid today as synthetic cresol and xylenol are now available and used in phosphate manufacture. However, the nomenclature remains a simple way of distinguishing between the cresol-/xylenol-based products and the newer products based on alkylated phenol. As the physical and chemical properties of each product type are slightly different, customer selection may depend on the application. For example, if the requirement is for a product that requires good oxidation stability, then the choice would be a TBPP, but a xylyl phosphate would be selected if the product required the best hydrolytic stability.

8.3.1.1 "Natural" Phosphates

The main products available in this category are TCP and trixylyl phosphate (TXP) (Figure 8.3). These products, based on cresols and xylenols, are complex mixtures of

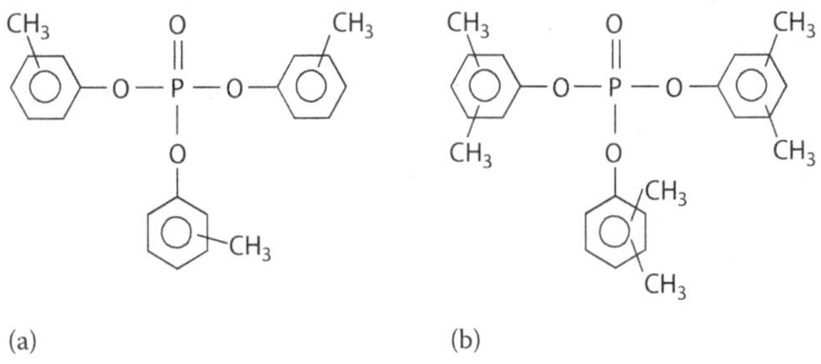

FIGURE 8.3 The structures of (a) tricresyl and (b) trixylyl phosphate.

isomeric materials [59]. However, the variation in phosphate isomer distribution, which arises from changes to the feedstock composition, has little impact on AW properties. Of greater importance are the actual phosphorus content and the level of impurities present, particularly those that are acidic.

In the past, the tri-*o*-cresyl phosphate content was a source of much concern in view of the high neurotoxicity of the material (see Section 8.7). However, the feedstock that is most widely used today in the production of TCP is predominantly a mixture of *m*- and *p*-cresol, and *o*-cresol levels are extremely low.

8.3.1.2 Synthetic Phosphates from Isopropylphenols

In this case, phenol is alkylated with propylene to produce a mixture of isomers of isopropylated phenol (Figure 8.4). Depending on the reaction conditions and the degree of alkylation, it is possible to produce a range of isopropylphenyl phosphates (IPPPs) with viscosities varying from ISO VG 22 to VG 100. In seeking an alternative product to TCP (an ISO 32 viscosity-grade fluid), the products with the closest phosphorus contents and viscosities (IPPP/22 and IPPP/32) are most widely used.

8.3.1.3 Synthetic Phosphates from Tertiarybutylphenols

In a similar fashion to the manufacture of IPPPs, it is possible to produce a range of phosphates from butylated phenols prepared by the reaction of isobutylene with phenol (Figure 8.4). The tertiarybutyl substituent is larger in size than the isopropyl substituent, and this reduces the overall level of alkylation in the molecule necessary to achieve the same viscosity, resulting in the presence of more unsubstituted phenyl groups. Again, the TBPPs from this range, which are used as AW additives in mineral oil, are those closest in phosphorus content and viscosity to TCP, that is, TBPP/22 and TBPP/32 (Table 8.1).

TABLE 8.1

Phosphorus Contents and Typical Viscosities for Neutral Trialkyl- and Triaryl Phosphate AW Additives

Phosphate Ester	Phosphorus Content (%)	Typical Viscosity at 40°C (cSt)
TiBP	11.7	2.9
TOP	7.8	7.9
TBEP	7.1	6.7
TCP	8.3	25
TXP	7.8	43
IPPP/22	8.3	22
IPPP/32	8.0	32
TBPP/22	8.5	24
TBPP/32	8.1	33
TBPP/100	7.1	95

8.3.2 Acid Phosphate Esters

8.3.2.1 Alkyl- and Aryl Acid Phosphates (Nonethoxylated)

The manufacture of acid phosphates, particularly alkyl acid phosphates, is also based on technology that has its roots in the nineteenth century but was commercialized only during the past 50 years. The process involves the reaction of phosphorus pentoxide with an alcohol in the absence of water (reaction 8.7).

$$P_2O_5 + 3ROH \rightarrow \underset{\substack{\text{Monoalkyl acid} \\ \text{phosphate}}}{ROP(OH)_2} + \underset{\substack{\text{Dialkyl acid} \\ \text{phosphate}}}{(RO)_2POH} \quad (8.7)$$

The ratio of monophosphate to diphosphate is usually 40%–50% monophosphate and 50%–60% diphosphate, with very small amounts of phosphoric acid ($\leq 1\%$). Small amounts of neutral ester may also be produced. Products commercially available are made from C_5, C_7–C_9 alcohols, mixtures of C_{10}–C_{12} alcohols, and C_{18} alcohols.

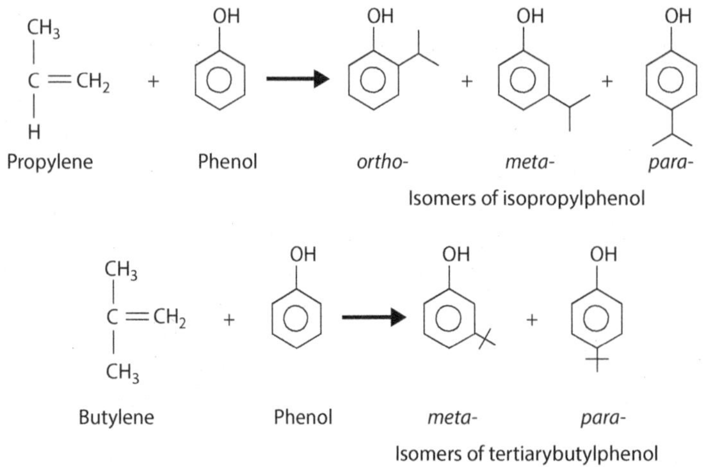

FIGURE 8.4 Process for the production of feedstocks used in the manufacture of synthetic phosphates.

Monoaryl and diaryl acid phosphates (also known as monoaryl and diaryl hydrogen phosphates) are by-products in the manufacture of triaryl phosphates and may be produced by stopping the reaction before completion and hydrolyzing the intermediate phosphorochloridates (reaction 8.8).

$$POCl_3 + ArOH \rightarrow ArOPOCl_2 + (ArO)_2 POCl$$

$$\xrightarrow{OH^-} \underset{\substack{\text{Monoaryl diacid} \\ \text{phosphate}}}{ArOPO(OH)_2} + \underset{\substack{\text{Diaryl monoacid} \\ \text{phosphate}}}{(ArO)_2 PO \cdot OH} \qquad (8.8)$$

Dialkylaryl monoacid phosphates, soluble in mineral oil, are reported to be produced by reacting phosphorus oxychloride with an alkylated phenol in the presence of base (reaction 8.9) or by the reaction of monoarylphosphorodichloridate with an alkylated phenol (reaction 8.10). The reactions are carried out at temperatures of about 60°C–90°C using less than equivalent amounts of phenol. The former process gives predominantly the monoacid phosphate with a small amount of neutral phosphate ester. The latter produces a somewhat greater amount of neutral mixed phosphate ester but mainly the mixed monoacid phosphate [60]. Examples of commercially available lubricating oil additives of this chemistry are amylphenyl- and octylphenyl acid phosphates.

$$RC_6H_4OH + POCl_3 \xrightarrow{OH^-} (RC_6H_4O)_3 PO$$

$$+ (RC_6H_4O)_2 PO \cdot OH \qquad (8.9)$$

$$RC_6H_4OPOCl_2 + ArOH \xrightarrow{OH^-} RC_6H_4O(ArO)_2 PO$$

$$+ RC_6H_4OArOPO \cdot OH \qquad (8.10)$$

Other processes reported for the production of mixtures of mono-2-ethylhexyl and di-2-ethylhexyl acid phosphates include the chlorination of bis-(2-ethylhexyl) hydrogen phosphite followed by hydrolysis (reaction 8.11) or the hydrolysis of the tris-(2-ethylhexyl) phosphate.

$$(C_8H_{17}O)_2 PO \cdot H \xrightarrow{Cl^-} (C_8H_{17}O)_2 PO \cdot Cl$$

$$\xrightarrow{OH^-} (C_8H_{17}O)_2 PO \cdot OH \qquad (8.11)$$

The alkyl acid phosphates are quite widely used as AW/EP additives in metalworking lubricant applications and as corrosion inhibitors for circulatory oils.

8.3.2.2 Alkyl- and Alkylarylpolyethyleneoxy Acid Phosphates

A range of polyethyleneoxy acid phosphate esters were introduced for metalworking lubricant applications in the early 1960s. These products, which consisted of both the free acids and their barium salts, were manufactured by reacting an ethoxylated alcohol with phosphorus pentoxide (reaction 8.12). The properties of the resulting acid phosphate mix can vary significantly depending on the chain length of the alcohol and the number of units of ethoxylation. For example, products that are only soluble in either oil or water can be produced as well as compounds that are soluble (or dispersible) in both media.

$$ROH + (C_2H_4O)_n H \rightarrow RO(C_2H_4O)_n H$$

$$\xrightarrow{P_2O_5} \left[RO(C_2H_4O)_n \right]_2 PO \cdot OH + RO(C_2H_4O)_n PO(OH)_2 \qquad (8.12)$$

8.3.3 Amine Salts of Acid Phosphates and of Polyethyleneoxy Acid Phosphates

Although the acidic products are very active AW/EP additives, their acidity can lead to precipitation problems in hard water and potential interaction with other additives. To minimize such adverse effects, acids are sometimes used as their neutral amine (or metal) salts. The salts are produced by reacting an equivalent weight of the base with that of the acid (reaction 8.13). The choice of base will depend on whether oil or water solubility is required. The use of short-chain amines will normally result in water-soluble additives, whereas using, for example, tertiaryalkyl primary amines with a chain length of C_{11}–C_{14} will tend to produce oil-soluble derivatives. The chain length of the acid phosphate also influences the solubility. The selection of the appropriate mixture of amine and phosphate for a given application is largely a compromise because the most active mixtures may also produce disadvantageous side effects, for example, on foaming and air release properties. The fact that a neutral salt is used also does not prevent the product from titrating as an acid and from forming a different salt in the presence of a stronger base.

$$RNH_2 + (R^1O)_{1-2} PO(OH)_{2-1} \rightarrow (R^1O)_{1-2} PO(OH \cdot H_2NR)_{2-1} \qquad (8.13)$$

where R is an alkyl group, typically C_8–C_{22}. It is also possible to use secondary and tertiary amines, R_2NH and R_3N, in the production of these salts.

8.3.4 Neutral Phosphite Esters

As with phosphate esters, it is possible to produce both neutral and acid phosphite esters. The neutral triaryl phosphites are produced by reacting phosphorus trichloride with a phenol or substituted phenol (reaction 8.14).

$$PCl_3 + 3ArOH \rightarrow \underset{\text{Triaryl phosphite}}{(ArO)_3 P} + 3HCl \qquad (8.14)$$

This is also a stepwise process occurring through the production of the monoaryl and diaryl hydrogen phosphite intermediates.

The production of neutral trialkyl phosphites using PCl_3 requires the addition of a tertiary amine base to neutralize the acid formed (reaction 8.15). Unless the HCl is removed quickly, it can cause the process to reverse with the production of an alkyl halide and the dialkyl acid phosphite (reaction 8.16).

$$PCl_3 + 3ROH + 3R_3N \rightarrow \underset{\text{Trialkyl phosphite}}{P(OR)_3} + 3R_3NHCl \qquad (8.15)$$

$$P(OR)_3 + HCl \rightarrow RCl + \underset{\text{Dialkyl acid phosphite}}{(RO)_2 POH} \qquad (8.16)$$

The use of mixtures of different alcohols or different phenols can result in the production of mixed alkyl or aryl phosphites. Mixed alkylaryl phosphites can be produced by reacting triaryl phosphite with alcohols to give a mixture of aryldialkyl- and alkyldiaryl phosphites (reaction 8.17). A commercial example of such a product promoted as an oil additive is decyldiphenyl phosphite.

$$(ArO)_3 P + ROH \rightarrow \underset{\text{Alkyldiaryl phosphite}}{(ArO)_2 POR} + ArOH$$

$$\rightarrow \underset{\text{Aryldialkyl phosphite}}{ArO(OR)_2} + 2ArOH \qquad (8.17)$$

Because of their widespread use in the plastics industry as stabilizers for polyvinyl chloride, polyolefinics, styrenics, and engineering plastics, many different neutral phosphites are commercially available. These range from C_2 to C_{18} alkyl (normally saturated) and from C_1 to C_9 alkaryl. In lubricant applications, the most common products are those with alkyl chains of C_8–C_{18} and C_{10}–C_{15} alkaryl, for example, trioctyl phosphite and tris-(2,4-ditertiarybutylphenyl) phosphite. The last type is increasingly important in view of its better hydrolytic stability.

8.3.5 ALKYL- AND ARYL ACID PHOSPHITES

Alkyl and aryl acid phosphites are manufactured by reacting together phosphorus trichloride, an alcohol (or phenol), and water (reaction 8.18):

$$PCl_3 + 2ROH + H_2O \rightarrow \underset{\text{Dialkyl acid phosphite}}{(RO)_2 POH} + 3HCl \qquad (8.18)$$

Mixtures of alcohols, as indicated earlier, may also be used to produce di-mixedalkyl phosphites. The commercially available dialkyl acid phosphites vary from C_1 to C_{18} with use as oil additives falling mainly in the range of C_8–C_{18}.

Little mention is made in the literature of the use of aryl acid phosphites, and there is no known oil industry use of ethoxylated neutral or acid phosphites. Phosphites are, however, generally unsuitable for applications where water contamination is likely in view of their hydrolytic instability, and ethoxylation, certainly in respect of water-soluble products, would not offer any obvious advantage.

8.3.6 DIALKYL ALKYL PHOSPHONATES

Although these products are isomeric with the dialkyl phosphites (Figure 8.2), they are a distinct class of materials with different properties. They are claimed as friction modifiers as well as AW/EP additives and are prepared by the Arbuzov rearrangement in which a trialkyl phosphite is heated with an alkyl halide, for example, an iodide (reaction 8.19):

$$P(OR)_3 + R'I \rightarrow \underset{\text{Dialkyl alkyl phosphonate}}{R'PO(OR)_2} + RI \qquad (8.19)$$

Commercially available materials range from the dimethyl methyl derivative to products based on dodecyl phosphite, although the higher-molecular-weight products are likely to be of greatest interest for oil applications. Polyethyleneoxy phosphonates, produced by the reaction of diphosphites with epoxides, have been claimed as friction modifiers [61], whereas diaryl hydrogen phosphonates, such as diphenyl phosphonate, are produced by hydrolysis of the corresponding phosphite with water.

8.4 FUNCTION OF LUBRICITY ADDITIVES

The earliest additives used for improving lubrication performance were known as oiliness additives and film strength additives. While these descriptions are no longer used, others are now employed. The current terminology together with typical examples of the chemistries employed is shown in Table 8.2.

TABLE 8.2

Different Types of Additives Used to Improve Lubrication Performance

Additive Description	Performance	Mechanism	Typical Chemistries
Friction modifier	Reduces friction under near-boundary lubrication conditions	Physical adsorption of polar materials on metal surfaces	Long-chain fatty acids and esters, sulfurized fatty acids, molybdenum compounds, long-chain phosphites, and phosphonates
Antiwear additive (usually with mild EP properties)	Reduces wear at low to medium loads	Reacts chemically with the metal surface to form a layer (normally a metal soap) that reduces frictional wear at low-medium temperature and loads	Neutral organic phosphates and phosphites, zinc di-alkyldithiophosphates
Extreme-pressure (EP) additive, also known as • film strength additive • load-carrying additive • antiscuffing additive • antiseize additive	Increases the load at which scuffing, scoring, or seizure occurs	Reacts chemically with the metal surface to form a layer, e.g. as a metal halide or sulfide that reduces frictional wear at high temperatures/loads	Sulfurized or chlorinated hydrocarbons, acidic phosphorus-containing materials, and mixtures thereof; some metal soaps, for example, of lead, antimony, and molybdenum

TABLE 8.3
A General Classification of Chemicals as Friction Modifiers, AW, and EP Additives

Additive	Friction Modifier	AW Additive	EP Additive
Natural oils and fats	1	4	5
Long-chain fatty acids, amines, and alcohols	1	4	5
Organo-molybdenum compounds	1	2	4
Synthetic esters	2	3	4
Organo-sulfur compounds	2	2	3
ZDDP	3	1	3
Phosphorus compounds	3	1	3
Sulfur compounds	4	3	1
Chlorine compounds	5	4	1

Source: Anonymous, *Lubrication*, 57, 1, 1971. With permission from the Energy Institute.

Note: The lower the number, the better the rating.

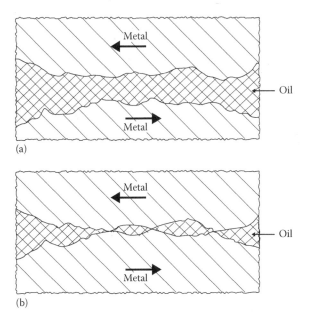

FIGURE 8.5 A diagrammatic representation of (a) full-film (hydrodynamic) and (b) mixed-film lubrication.

Table 8.3 [62] offers a generalized classification of the different chemical types of additives used to improve lubrication performance, but, depending on the structure of the additive, some variation in the performance can be expected. In reality, the distinction between AW and EP additives is not clear-cut. AW additives may have mild EP properties, whereas EP additives can have moderate AW performance, and both produce coatings on the metal surface. In fact, EP additives have been described as additives that reduce or prevent severe wear [63]. However, as seen from Table 8.3, EP additives are unlikely to function satisfactorily as friction modifiers and vice versa.

8.4.1 BASIC MECHANISM OF LUBRICATION AND WEAR AND THE INFLUENCE OF ADDITIVES

An understanding of the basic mechanism of lubrication is useful to appreciate the way in which additives behave and their relative performance. The following is therefore a somewhat simplified explanation of a complex process.

Lubrication can be described as the ability of oil (or another liquid) to minimize the wear and scuffing of surfaces in relative motion. It is a function of the properties of the lubricant (e.g., viscosity), the applied load, the relative movement of the surfaces (e.g., sliding speeds), temperature, surface roughness, and the nature of the surface film (hardness and reactivity, etc.).

All surfaces, even those that appear smooth to the naked eye, when examined microscopically, consist of a series of peaks and troughs. The simplest situation arises when the lubricating film is thick enough to completely separate the two surfaces so that metal-to-metal contact does not occur (Figure 8.5). Such a situation could arise at low loads or with highly viscous liquids, and the lubricating characteristics depend on the properties of the lubricant as the load is fully supported by the lubricant. This condition is known as *hydrodynamic* or *full-film lubrication.*

As the load increases, the lubricating film becomes thinner and eventually reaches a condition where the thickness is similar to the combined height of the asperities on the mating surfaces. At this stage, metal-to-metal contact commences, and as the asperities collide, they are thought to weld momentarily (causing friction) before shearing with loss of metal (wear) (Figure 8.5). The wear particles then abrade the surface and adversely affect friction, with the resulting damage depending on the hardness of the particle and the surface it contacts. This condition is known as *mixed-film lubrication* as it is a mixture of full-film lubrication and *boundary lubrication* with the trend toward the latter with increasing load.

As the film thins still further, the load is increasingly supported by the metal surface and friction rises rapidly. When eventually a film that is only a few molecules thick separates the surfaces, the roughness, composition, and melting point of the surfaces strongly influence the resulting friction. At this stage, viscosity plays little or no part in the frictional behavior. This stage is known as boundary lubrication and is characterized by high frictional values that now change little with further increases in load or sliding speed. The wear process that takes place under boundary conditions is perhaps the most complicated of those involved in lubrication in that it involves four different types of wear: corrosive, fatigue, ploughing, and adhesive.

Corrosive wear occurs when the metal surfaces react with their environment to form a boundary film, whereas fatigue wear is the process of the fracture of asperities from repeated high stress. Micropitting is an example of this form of wear, which is the subject of considerable investigation today. Micropitting is the result of plastic deformation of the surface that eventually causes the fracture of the asperity, leaving a

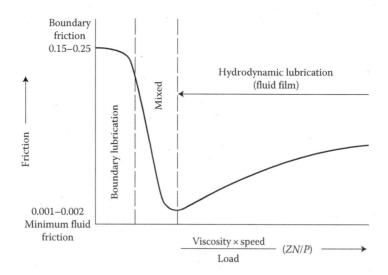

FIGURE 8.6 The relationship between coefficient of friction and *ZN/P*.

small *pit* in the surface. Ploughing wear arises when a sharp particle is forced along the surface, leaving a groove behind, whereas adhesive wear is the tendency of very clean surfaces to adhere to each other. However, this action requires the generation of fresh surfaces during the wear process, perhaps by plastic deformation. It is now thought that this mechanism is much less prevalent than was earlier believed [64].

The relationship between friction, viscosity, load, and sliding speeds can be represented graphically for a bearing by what is known as a *Stribeck curve*. This is shown in Figure 8.6 [65], where the frictional coefficient is plotted against the dimensionless expression *ZN/P*, where *Z* represents the fluid viscosity, *N* the sliding speed, and *P* the load. Friction is reduced as the value of *ZN/P* is lowered until a minimum is reached. For a bearing, this minimum value is ~0.002 for an ideal hydrodynamic condition. At this point, metal-to-metal contact begins; friction rises and continues to do so with increasing contact. In the mixed friction zone, the coefficient of friction lies in the region of 0.02–0.10. Eventually, when the film is very thin, friction becomes independent of viscosity, speed, and load and the coefficient of friction can then reach a value of 0.25.

By experiment, it was established that

- Continually increasing the load reduced the *ZN/P* value, assuming that speed and the viscosity remained constant. The same results can be obtained by reducing either the speed or viscosity, or both, provided the unit load remains constant or is increased.
- Friction varied directly with viscosity; it was proportional to velocity at lower speeds but varied inversely with velocity at higher speeds [65].

As the surfaces move closer together, the lubricant is squeezed out from between them. Some additives, when adsorbed onto the surface, display a molecular orientation perpendicular to the surface that reduces the level of contact and hence

lowers the friction. Such products are known as *friction modifiers*. Those additives effective in reducing wear and (usually) friction in the mixed friction zone are called *antiwear additives*, whereas products effective in reducing wear (and increasing seizure loads) in the boundary lubrication process are known as *extreme pressure additives*. However, due to the importance of temperature in the lubrication process, it has been pointed out in the past that the latter should, perhaps, be better described as *extreme temperature additives*.

The temperature at which an additive reacts physically or chemically with the metal or metal oxide surface significantly affects its activity. Each AW/EP additive type has a range of temperature over which it is active (Figure 8.7) [66]. The lowest temperature in the range would normally be the temperature at which physical adsorption takes place. This can occur at ambient or at higher temperatures depending on the polarity of the additive and the impact on surface energy. The greater the reduction in surface energy, the stronger will be the absorption of the surface film and the greater will be the likelihood that the additive remains in place for a chemical reaction with the surface. Additives that are only weakly bound to the surface may desorb as the temperature rises and cease to function further in the wear-reducing process.

As the temperature increases so does the surface reactivity. Fatty acids and esters react at fairly low temperatures to produce metal soaps followed by chlorine-containing compounds (to form chlorides), phosphorus (as phosphates, polyphosphates, and/or phosphides), and, finally, sulfur, which reacts at very high temperatures to form metal sulfides [66].

Chlorine-based additives can be film forming even at ambient temperatures, but as the temperature rises they become aggressive and, with the release of HCl, can cause significant corrosion. Although the $FeCl_2$ film has a fairly well-defined melting point at 670°C, the optimum operating temperature is much lower. Klamann [67] indicates that the efficiency of metal chlorides starts to drop above 300°C and that the friction coefficient at 400°C is already a multiple of the optimum value. However, the dry friction coefficient of the chloride film

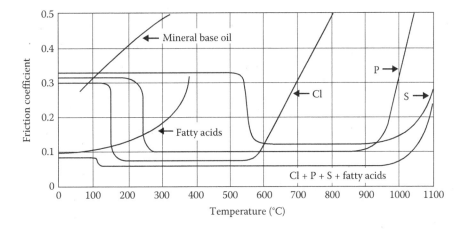

FIGURE 8.7 Effect of temperature on EP additive activity. (From Mandakovic, R., *J. Syn. Lubr.*, 16(1), 13, 1999. Copyright by John Wiley & Sons Inc. and used with permission.)

TABLE 8.4

Corrosion Films Formed on Sliding Iron Surfaces

Lubricant Type	Nature	Friction Coefficient (Dry)	Melting Point (°C)
Dry or hydrocarbon	Fe	1.0	1535
	FeO	0.3	1420
	Fe_3O_4	0.5	1538
	Fe_2O_3	0.6	1565
Chlorine	$FeCl_2$	0.1	670
Sulfur	FeS	0.5	1193

Source: Anonymous, *Lubrication*, 12(6), 61, 1957. Copyrighted by Chevron and used with permission.

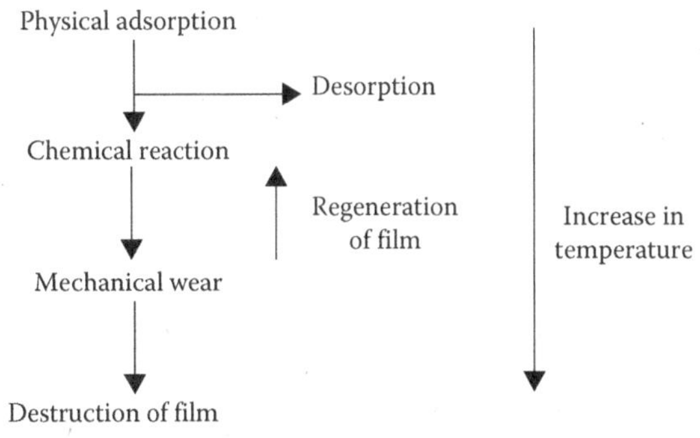

FIGURE 8.8 Basic processes involved in the mechanism of action of lubricity additives.

is substantially lower than that for iron sulfide (Table 8.4) [68]. The relatively low friction associated with this film is probably one reason why chlorinated products are so effective as EP additives. Phosphorus, by comparison, does not react until at higher temperatures and then at slower rates. However, the upper temperature limit of ~550°C in an air environment is thought to be a result of the oxidation of the carbon in the film rather than the degradation of a metal soap (Forster, N.H., Private Communication, July 2007).

The soaps, phosphates/phosphides, chlorides, and sulfides formed on the metal surface were originally considered to produce a lower melting and less-shear stable film than that of the metal/metal oxide. This film would cause a smoothing of the metal surface that was then able to support a higher unit loading. This is now thought to be an oversimplified explanation as research has found the EP films to be considerably different to those postulated and without the expected lower shear stability [69]. What it does not consider are additional "subprocesses" of removal of the film by mechanical wear and its possible regeneration *in situ* by further action of the AW/EP additive (Figure 8.8).

Since surface temperature is largely dependent on load, additives that might be effective at high loads may be

completely ineffective at low loads (and vice versa). Under such circumstances, therefore, significant wear could occur before the load-carrying properties of the EP additive come into play. To minimize this effect, additives are often used in combination, resulting in extending the temperature (and load) range over which they are active.

Although single AW/EP additives can be used to meet application and specification requirements, combinations of additives can produce both synergistic and antagonistic effects. The use of mixtures of phosphorus- and chlorine- or sulfur-containing compounds, to extend the temperature range over which a lubricating film is available, has already been mentioned. Another example of synergism was reported by Beeck et al. [5], who described the effect of combinations of TCP and long-chain fatty acids. It was suggested that the use of such mixtures in some way improved the *packing* of the film on the surface and therefore helped to reduce metal contact. Figure 8.9 [68] in fact shows that combinations of phosphate and fatty acid can result in lower wear rates than either component. Such synergy is useful in that it reduces additive costs and the possibility that the additives might have an adverse effect on product stability and surface active properties. An example of additive antagonism is given in Section 8.5.3.3.

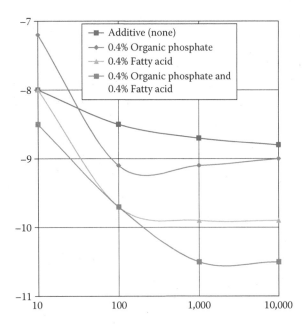

FIGURE 8.9 Effect of fatty acid and phosphate ester on wear rate. (From Anonymous, *Lubrication*, 12(6), 61, 1957. Copyrighted by Chevron and used with permission.)

8.5 INVESTIGATIONS INTO THE MECHANISM AND ACTIVITY OF PHOSPHORUS-CONTAINING ADDITIVES

Many papers have been written about the way in which TCP and other phosphorus-containing compounds work as AW/EP additives. As might be expected, researchers have had differences of opinion. These have probably arisen as a result of the different test conditions found in the wide variety of test equipment developed for measuring wear. For example, different test specimen geometries, surface finish, sliding speeds, and the use of additives with different levels of purity have meant that the data have not been strictly comparable.

After a brief review of the early development of AW/EP additives, a number of papers exploring the mechanism of action of different phosphorus-based additives are summarized in this section. It is not inclusive, and the results of many other workers could have been mentioned. An additional selection of papers on the topic is therefore given in Appendix 8.B. Some papers evaluate several classes of product (e.g., phosphates, phosphites, phosphonates); these may be located in sections other than that on neutral phosphates if information on these other structures is limited.

8.5.1 EARLY INVESTIGATIONS INTO ANTIWEAR AND EXTREME ADDITIVES

Some of the earliest experiments into the effects of different lubricants on friction were carried out by Hardy in 1919 [70], who noted the superior performance of castor oil and oleic acid. He found that good lubricating properties were closely related to the ability of substances to lower surface energy.

A series of papers from Hardy and Doubleday followed in 1922–1923 examining the activity of lubricants under boundary conditions.

In 1920, Wells and Southcombe [71] discovered that the addition of a small amount of a long-chain fatty acid significantly reduced the static coefficient of friction of mineral oil. Bragg postulated in 1925 [72] that long-chain molecules with a polar terminal group were attached to the surface by adsorption of the polar group and that the long hydrocarbon chains were orientated perpendicular to the surface. He also suggested that the formation of films on both the moving surfaces assisted lubrication by sliding over one another, with their long chains being "flattened" as the distance between the surfaces was reduced. However, in 1936, Clark and Sterrett [73] showed that the lubricating film could be up to 200 molecules in thickness but that only the first layer would have the strength to withstand the shearing stresses produced under sliding conditions. They also found that certain ring structures (e.g., trichlorophenol) that were active as "film strength" additives also showed molecular orientation, in this case, parallel to the metal surface, and attributed the good load-carrying performance to the ability of the layers to slide over one another. Orientation was not the only factor involved, as compounds with a similar orientation could show a wide difference in performance.

The mechanism and influence of additives on boundary lubrication were first investigated and reported by Beeck et al. [4]. They found that friction was reasonably constant with sliding velocity up to a critical velocity, beyond which there was a significant reduction. Additives were found to reduce the friction at low speeds relative to the base oil alone and also had a significant, but variable, effect on the reduction at different critical velocities. Low critical velocities were found for compounds that were strongly adsorbed and that showed orientation of the surface film. It was recognized that the adsorbed layer is thinner than the roughness of even the best machined surfaces and that high temperatures (or loads) at points of contact would cause decomposition of the molecules with the formation of a high-melting corrosion product and an increase in friction. If the surfaces were *highly polished*, then sliding could take place without destruction of the surface film. It was concluded that most of the friction-reducing compounds, principally, the long-chain fatty acids, were not able to produce a highly polished surface and therefore were not effective AW additives.

8.5.2 NEUTRAL ALKYL- AND ARYL PHOSPHATES

8.5.2.1 Historical Background

One additive examined by Beeck et al. [5] that was able to reduce both friction and wear was TCP, a product that was, at that time, beginning to find widespread commercial use as an AW additive. The authors proposed that TCP acted by a *corrosive* action, preferentially reacting with the high spot on the surface, where the surface temperatures are highest (from metal contact). It was thought that in the reaction, the phosphate ester formed a lower melting phosphide (or possibly an iron/iron phosphide eutectic) that flowed over the surface and caused a smoothing or chemical polishing effect. They also observed that there appeared to be an optimum level of addition of the TCP (1.5%), a conclusion later confirmed by other workers in the field.

Beeck et al. claimed in these papers that their research had produced a better understanding of the AW mechanism and enabled more precise distinctions to be drawn between the different types of additives—more specifically, that

> A wear prevention agent reduces pressure and temperature through better distribution of the load over the apparent surface. If the resulting minimum pressure is still too high for the maintenance of a stable film, metal to metal contact will take place in spite of the high polish. Since in this case the surface of actual contact is relatively very large, seizure and breakdown will follow very rapidly …

The intervention of World War II encouraged German researchers to prepare and evaluate a number of phosphorus compounds as EP/AW additives, principally phosphinic acid derivatives and also acid phosphates [6,7], while other workers [74] continued to investigate the behavior of TCP. The performance of the latter in white oil was examined, and it was suggested that the additive reacted with steel to form a thin, solid, nonconducting film that prevented seizure by shearing in preference to metal-to-metal contacts. The improved behavior of blends of TCP with fatty acids was explained as being due to the improved adsorption of the fatty acid onto the surface of the chemically formed film.

In 1950, an extensive evaluation of different neutral alkyl and aryl phosphates and phosphites, in some cases containing chlorine and sulfur, was undertaken [75]. The results of this investigation showed that the action of sulfur and chlorine on the surface is to form a sulfide and a chloride film, respectively. In the presence of phosphorus, mixed films of phosphide/sulfide or phosphide/chloride were formed. The presence of phosphide was established chemically by the liberation of phosphine in the presence of hydrochloric acid.

Although the concept of phosphide film formation was challenged at this time [76,77], it remained as the generally held theory until the mid-1960s when several papers appeared with contradictory data. Godfrey [78] pointed out that the experiments that had indicated the presence of phosphide had all been static, high-temperature investigations, and none had identified phosphide on a sliding surface lubricated with TCP. He experimented with the lubrication of steel-on-steel surfaces by TCP followed by an examination of the metal surface. This revealed the presence of white crystalline material, which was shown by electron diffraction measurements to be predominantly a mixture of ferric phosphate, $FePO_4$, and its dihydrate, $FePO_4 \cdot 2H_2O$. Phosphides, if present, were in extremely small quantities. Furthermore, a paste made from the dihydrate showed similar frictional characteristics to TCP, whereas a paste from iron phosphide showed no significant reduction in friction. Tests also suggested the importance of air to the performance of TCP as tests carried out under nitrogen revealed substantially increased wear. *Pure* TCP was

evaluated and, unlike *commercial* material, showed no significant friction-reducing properties.

The presence and role of *impurities* in the activity of commercial TCP was the subject of investigations using radioactive P^{32} [79]. Results suggested that the phosphorus-containing polar impurities—not the neutral TCP—were adsorbed onto the metal surface. The P^{32} found in the wear scar appeared to be the chemically bound—not physically adsorbed—but the latter process seemed to be the way that the phosphorus was initially made available on the surface. The authors indicated that the impurities resembled acid phosphates (rather than phosphoric acid, which Godfrey had assumed) and carried out wear tests comparing the neutral ester with both an acid phosphate (dilauryl acid phosphate) and hydrolyzed TCP. They found that these compounds generally gave equivalent performance to the neutral ester but at lower concentrations. Of interest was the observation that, although TCP showed no wear minimum in the reported tests (cf. the results given by Beeck et al. [5]), the data on acid phosphate, acid phosphite, and phosphoric acid did display such minima.

The work using radioactive P^{32} also allowed a study of the competition between TCP and different types of additives for the metal surface. This was determined by measuring the residual surface radioactivity after wear tests. Table 8.5 shows the effect of various types of additives on the adsorption of P^{32} from TCP. The lower the number of counts, the greater the interaction between the additive and TCP.

Radiochemical analysis was also the technique used to investigate the deposition of phosphorus on steel surfaces in engine tests [80]. In this study, the effect of different types of aryl phosphates (TPP, TCP, and TXP) on case-hardened tappets was examined. The results suggested that the efficacy of these additives correlated directly with their hydrolytic stability, that is, their ability to produce acid phosphates as degradation products. This was confirmed by tests on a series

TABLE 8.5

Effects of Various Additives on the Adsorption of P^{32}

Additive Concentration (wt%)	Activity (Counts/min)
0.5% TCP alone	280
+2% Barium sulfonate A	0
+2% Barium sulfonate B	80
+0.1% Rust inhibitor	16
+0.5% Di-isopropyl acid phosphite	25
+0.1% Dilauryl acid phosphate	24
+5.5% Acryloid dispersant	82
+7.9% Polymeric thickener	78
+0.7% Sulfur–chlorine EP additive	120
+0.5% Thiophosphate	150
+0.5% 2,2'-Methylene-bis(2-methyl,4-tertiarybutyl phenol)	250
+0.5% Sulfurized terpene	290

Source: Klaus, E.E. and Bieber, H.E., *ASLE Preprint* 64-LC-2, 1964. Reproduced with permission from STLE.

TABLE 8.6

Correlation between the Antiscuffing Performance and Ease of Hydrolysis (Acid Formation) of Organic Phosphates

Additive (0.08 wt% Added Phosphorus)	Relative Ease of Hydrolysis[a]	Time to Scuffing (min)[b] at a Spring Load	
		305 lb	340 lb
Benzyldiphenyl phosphate	100	>30	9
Allyldiphenyl phosphate	100	>30	Not tested
Ethyldiphenyl phosphate	80	28	Not tested
Octyldiphenyl phosphate	50	15	6
Triphenyl phosphate	50	15	5
Tritolyl phosphate	30	8	Not tested
2-Ethylhexyldiphenyl phosphate	5	2–3	Not tested
None	—	2–3	Not tested

Source: Barcroft, F.T. and Daniel, S.G., *ASME J. Basic Eng.*, 87, 761, 1965. Reproduced with permission of ASME.

Note: Camshaft, Ford Consul (cams phosphated); Tappet, Ford Consul (non-phosphated); Camshaft speed, 1500 rpm (equivalent engine speed 3000 rpm); Base oil, SAE 10W/30 oil without EP additive.

[a] Because of the wide range of hydrolytic stability of these compounds, it was not possible to compare the stabilities of all these compounds in the same acid medium. Consequently, an arbitrary scale was drawn up with benzyldiphenyl phosphate assigned a value of 100.

[b] Mean of several tests.

of other phosphates (largely alkyldiaryl phosphates), which showed good correlation between antiscuffing performance and hydrolytic stability (Table 8.6). Examination of the tappet surface revealed the presence of aryl acid phosphates on the surface and the absence of phosphides. Adsorption studies of the neutral aryl and acid phosphates on steel surfaces indicated that, although the film of neutral ester could be more easily removed, the adsorption of the acid phosphate was irreversible, suggesting salt formation. These studies led the authors to conclude that the mechanism involved initial adsorption of the phosphate on the metal surface followed by hydrolytic decomposition to give an acid phosphate. This reacted with the surface to give iron organophosphates, which then decomposed further to give iron phosphates.

The importance of impurities in determining the level of activity of TCP was confirmed in yet another paper [81]. The composition of impurities in commercial grades of TCP was determined using thin-layer chromatography and analysis by neutron activation. Acidic impurities, probably the monocresyl and dicresyl acid phosphate (and also small amounts [2 × 10^{-4}%] of phosphoric acid), were found at 0.1%–0.2%, that is, at levels that had previously been shown to produce a significant reduction in wear when added to mineral oil. Other impurities ranged from 0.2% to 0.8%. This latter category was assumed to contain chlorophosphates based on the amount of chloride ion produced. The authors commented that the TCP used for the investigation was the best grade available, but even this material contained up to 25% polar impurities. It was thought typical of the TCP used in the wear studies to date and reported in the literature.

Wear tests on TCP, acid phosphates, and phosphites in a super-refined mineral oil and a synthetic ester (di-3-methylbutyl adipate) indicated that relatively small amounts (0.01%) of additive can produce a significant wear reduction in mineral oils and that the acidic materials were more active. However, in the polar base stock, where there is competition for the surface, the amount of TCP required to provide a similar reduction in wear is substantially greater. The effectiveness of the alkyl acid phosphates is not significantly reduced in the synthetic ester, suggesting that their polarity (and hence adsorption) is greater than that of the neutral phosphate, the synthetic ester, and its impurities (Table 8.7). The authors concluded that the activity of TCP was due to the acidic impurities and that the neutral ester acted as a reservoir for the formation of these impurities during the life of the lubricant.

Until about 1969, the theory regarding the production of a phosphate film on the steel surface seemed to be widely accepted. Reports then appeared suggesting that the situation was more complicated. One paper [82] examined and compared the corrosivity toward steel, the load-carrying capacity, and the AW performance of several phosphorus compounds. Using the hot-wire technique at 500°C [83] followed by an x-ray analysis of the surface films that were produced, the reactivity (or corrosivity) was studied. Perhaps, not surprisingly, the neutral phosphate and phosphite evaluated showed relatively little reactivity with the steel, whereas the acid phosphate and phosphite produced substantially more corrosion. The anomaly was the behavior of a neutral alkyl trithiophosphite, which showed a very high reactivity but low load-carrying ability, suggesting a different mode of breakdown. Analysis of the films formed confirmed the major presence of basic iron phosphate (or principally iron sulfide in the case of the thiophosphite), but small amounts of iron phosphide were also found in all the x-ray analyses of the degradation products. Evaluation of

TABLE 8.7

Effect of Concentration on the AW Properties of Phosphorus-Containing Additives in a Synthetic Ester

Additive	Concentration (wt%)	Average Wear Scar Diameter (mm) at:		
		1 kg	10 kg	40 kg
None	—	0.39	0.71	0.91
TCP	1.0	0.38	0.71	0.97
	3.0	0.40	0.64	0.97
	0.5	0.23	0.25	0.78
Hydrolyzed TCP	0.1	0.57	0.74	—
	1.0	0.17	0.25	0.46
Dilauryl acid phosphate	0.01	0.21	0.41	0.84
	0.05	0.19	0.28	0.43
	1.0	0.17	0.28	0.42
Di-isopropyl acid phosphite	0.02	—	0.72	—
	0.05	0.16	0.25	—
	0.15	—	0.33	—
Phosphoric acid	0.001	0.41	0.69	0.90
	0.01	0.16	0.37	0.50
	1.0	0.38	0.60	0.78

Source: Bieber, H.E. et al., ASLE Preprint 67-LC-9, 1967. Reproduced with permission of STLE.

Note: ASTM D4172. Four-ball wear test conditions: test time, 1 h; test temperature, 167°C; test speed, 620 rpm.

the load-carrying capacity of the additives was found to vary directly with corrosivity except for the alkyl trithiophosphite. The authors surmised from this that the load-carrying capacity of phosphorus-containing additives was not only due to the reactivity of the films but also due to the properties of the film that was formed. The relationship between wear and reactivity also varied directly for several compounds, but in the case of the neutral phosphite and the alkyl trithiophosphite, there was no correlation. This was attributed to the different composition of the film in these cases. In fact, the authors proposed that the main reaction product of the phosphite could be iron phosphide. They suggested that the load-carrying capacity of the films formed by EP additives fell in the following order:

Phosphide > phosphate > sulfide > chloride

whereas the order of AW properties was

Sulfide > phosphate > phosphide

The first of these sequences is, of course, different to the order in EP activity predicted from the stability of the films formed on the metal surface and from the general perception that phosphorus is less active than either chlorine or sulfur. Similarly, for AW performance, phosphorus is normally regarded as more active than sulfur.

A paper by Goldblatt and Appeldoorn [84] cast doubt on the theory that the activity of TCP was due to the generation of acidic impurities. In this study, the activity of TCP in different atmospheres and in different hydrocarbon base stocks was examined. The resulting data showed that TCP was much more effective in a low-viscosity white (paraffinic) oil than

in an aromatic base stock. Aromatics are good AW agents and compete with the TCP for the surface. Under these conditions, either the iron phosphate reaction products are less stable or perhaps a thinner and less complete layer is produced and is worn away, leading to an increase in corrosive wear. Surprisingly, the AW performance of the mixed aliphatic/aromatic base stock was better than either of the components and was not improved by addition of TCP.

The behavior of TCP in different atmospheres focused on the effect of moisture in a wet-air atmosphere and also under dry argon, that is, in the absence of oxygen and moisture. No significant differences were found in the results indicated previously for the different hydrocarbon base stocks. However, in a further series of tests comparing the behavior under both wet and dry air and wet and dry argon in an ISO 32 grade white oil, TCP was shown to have a slight AW effect. The exception was in wet air, when it increased wear but also generally showed higher scuffing loads than when used in dry argon. In a naphthenic oil of similar viscosity, the use of wet air (or wet argon) again resulted in increased wear and exhibited higher scuffing loads. This behavior was also observed with other phosphates and phosphites. The authors suggested that in dry air the TCP film forms very rapidly and metallic contact quickly falls. In dry argon, the same thing happens, only at a slower rate. In wet air, the film is not as strong, and metallic contact remains high, whereas in the case of wet argon, it does not form at all. "Thus the formation of a protective film is enhanced by oxygen but hindered by the presence of moisture." The observation [78] that air was necessary for the action of TCP did not consider that moisture was present in the air and could have been responsible for the improvement in wear performance.

The previous theory indicating it was necessary for the TCP to hydrolyze to form acid phosphates before it became active was also challenged. Wear tests on standard and very low acid TCP in dry argon showed no significant difference in activity. It was concluded that TCP was reacting directly with the surface without first hydrolyzing to acid phosphate and without being preferentially adsorbed at the metal surface.

In 1972, Forbes et al. [85] summarized the current thinking on the action of TCP, which indicated that TCP was an effective AW additive at high concentrations independent of the base oil, but at low concentrations was adversely affected by the presence of aromatics. The acidic degradation products have similar properties but show better performance at low concentrations. It was felt that TCP adsorbed onto the metal surface decomposed to give acid phosphates that reacted with the surface to give metal organophosphates.

The results of further investigations into the effects of oxygen and temperature on the frictional performance of TCP on M50 steel were published in 1983 [86]. The critical temperature at which friction is reduced as surface temperature rises was measured under different conditions and was found to be 265°C in dry air (<100 ppm water) when full-flow lubrication is used, 225°C under conditions of limited lubrication, and 215°C under nitrogen, also with limited lubrication. Analysis of the surface indicated that TCP had reacted chemically at these temperatures, causing a substantial increase in the amount of phosphate deposited (phosphide was not observed). Oxygen was said to be necessary for this reaction, but the suggestion that prior hydrolysis of the phosphate was required could not be substantiated.

The debate regarding the formation of iron phosphate or phosphide as reaction products in the wear mechanism rumbled on into the late 1970s and the early 1980s. In 1978, Yamamoto and Hirano [87] carried out scuffing tests on several aryl and alkyl phosphates. The aryl phosphates showed better scuffing resistance, and it was suggested that the alkyl phosphates reacted with the steel surface, forming a film of iron phosphate under mild lubricating conditions, but that the aryl phosphates reacted only slightly until conditions became more severe with the formation of iron phosphide. The implication was that the phosphide (formed as a result of a reaction between the phosphate and the metal surface) acted as a good EP additive but that the iron phosphate had only AW activity. Surface roughness measurements showed a polishing action for the aryl phosphates (particularly for TCP) but not, under these conditions, for the alkyl phosphates.

The concept of corrosive wear and of phosphates as chemical polishing agents as expressed by Beeck et al. [5] was examined by Furey in 1963 [88]. In his work, surfaces of different roughness were prepared and friction measurements were made when in contact with a solvent-refined oil under different applied loads. In tests on an additive-free oil (unfortunately, no information was available on the sulfur or aromatic content), it was found that friction, in addition to being load dependent, was low for highly polished surfaces and rose with increasing roughness up to a roughness of ~10 μin. At about this roughness, the percentage metal contact was also found to be at its maximum but decreased thereafter. The explanation given for this was that with increasing roughness, the distances between the peaks and troughs increase but the peaks become flatter. The flatter the peak, the better the load-carrying capacity, whereas the deeper troughs allow for a greater reserve of oil available locally for lubrication and cooling. When several AW/EP additives were evaluated in the oil, it was found that, although there was a reduction in surface roughness, it was less than that found by the oil alone. Furthermore, at low loads, TCP was able to reduce metal contact significantly but had no effect on surface roughness. At moderate to high loads, although the metal contact was reduced, the surface roughness was increased. The author concluded that TCP was not acting through a polishing action.

In 1981, Gauthier et al. [89] looked again at the wear process and film formation. They categorized the process into three wear phases: an initial, very rapid phase, followed by a medium wear rate, and finally a slow wear phase. In the rapid wear phase, a brown film was formed that, on analysis, was found to be a mixture of ferrous oxide and phosphate. A blue film, which is formed as the wear rate slows (and the surface becomes smoother), contained no iron and was described as a polymeric acid phosphate. (No mention was made of the "white crystalline film" Godfrey reported.)

When both films were removed and the roughness of the underlying surface was measured, it was found that the surface below the brown film was very smooth. The surface under the blue film was much rougher and ~1000 Å thicker. The authors suggested that the smooth surface was the result of polishing arising from corrosive wear. They concluded that in the first phase of wear, a corrosive wear process is involved because of the presence of ferrous phosphate on the surface. When a "critical value" for the surface coverage by the phosphate was achieved, the organic phosphoric acids produced by the decomposition of TCP polymerized to form a polyphosphate. As a result, in the last two wear phases, "the wear of metal is almost completely replaced by the wear of the additive." In this way, the disparate observations of TCP behavior (polishing vs. increased surface roughness) could be related and combined.

The presence of polyphosphate was also noted by Placek and Shankwalkar [90] when investigating the films produced on bearing surfaces by pretreatment with phosphate esters. Tests were carried out on 100% phosphates and also on their 10% solutions in mineral oil, the latter condition because the combination had been reported to provide better wear protection than the individual components alone, apparently by the formation of a "friction polymer" [91,92]. Phosphates chosen for the work included both aryl and alkyl types. Analyses of the films formed by immersion in the phosphates at 250°C revealed the presence of a high level of carbon together with iron phosphate/polyphosphate and a small amount of phosphide. At 300°C, the hydrocarbon had all but disappeared and no phosphide was detected. The films formed by the mineral oil solutions were mainly hydrocarbon based, but the film formed by the alkyl phosphate was unique in that it contained needlelike fibers. The effect of pretreatment on wear found under four-ball test conditions is indicated in Table 8.8. The bearings treated with the mineral oil solutions displayed at least as good wear reduction as those treated with the 100% phosphate.

TABLE 8.8

Friction and Wear Reduction from Bearing Surface Pretreatment by Phosphate Esters

Bearing Preparation	Average Scar Diameter (mm)	Improvement (%)	Maximum Torque (gfm)	Improvement (%)
Untreated reference	1.00	—	46.1	—
TCP	0.72	28	18.4	60
IPPP	0.75	25	18.4	60
TOF	0.81	19	18.4	60
10% TCP in mineral oil	0.72	28	18.4	60
10% IPPP in mineral oil	0.72	28	15.0	68
10% TOF in mineral oil	0.64	36	19.6	58

Source: Reprinted from *Wear*, 173, Placek, D.G. and Shankwalkar, S.G., Phosphate ester surface treatment for reduced wear and corrosion protection, 207–217, Copyright 1994, with permission from Elsevier.

Note: ASTM D4172–88. Four-ball wear test conditions: test time, 60 min; test temperature, 75°C; test load, 40 kgf; test speed, 600 rpm. IPPP, isopropylphenyl phosphate; TCP, tricresyl phosphate; TOF, tris(2-ethylhexyl) phosphate. All wear tests performed in 100 solvent neutral paraffinic mineral oil.

8.5.2.2 Recent Technical Developments

In 1996, Yansheng et al. [93] reported on the effect of TCP on the wear performance of sulfurized, oxy-nitrided, and nitrided surfaces. A synergistic effect on nitrided and oxy-nitrided surfaces was found, resulting in significant increases in load-bearing capacity while reducing friction and wear, but no improvement was seen on sulfurized surfaces.

A recent application in this brief survey relates to the use of aryl phosphates as vapor-phase lubricants. Although not strictly an additive application, this development has been the focus of most recent analytical studies into the mode of action of these additives; therefore, the conclusions represent the current thinking. Aryl phosphates were chosen for this application because of their oxidation stability and good boundary lubrication performance at high temperatures. The initial studies took place with TCP [94] and involved examination of the films formed on tool steel balls and on iron, stainless steel, copper, nickel, tungsten, and quartz wire specimens. (TCP vapor had previously been shown to form tenacious films on graphite, tungsten, and aluminum at temperatures above its thermal decomposition point [95].) Wear tests on tool steel with vapor at 370°C showed low levels of wear even at 0.1 mol% concentration (Figure 8.10). An optimum concentration was reached at 0.5 mol%. Reaction with the metals indicated earlier is displayed in Figure 8.11, which shows that deposition on iron and copper is relatively fast but slow for quartz, nickel, and tungsten. Rates of formation are, of course, temperature dependent, but films are produced up to at least 800°C. Increases in temperature and TCP concentration caused an increase in deposit formation.

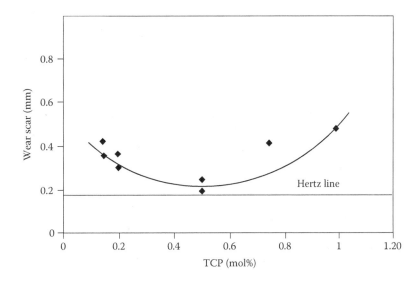

FIGURE 8.10 Four-ball wear values at 370°C with vapor lubrication as a function of tricresyl phosphate (TCP) vapor concentration. (From Klaus, E.E. et al., *Lubr. Eng.*, 45(11), 717, 1989. Reproduced with permission from STLE.)

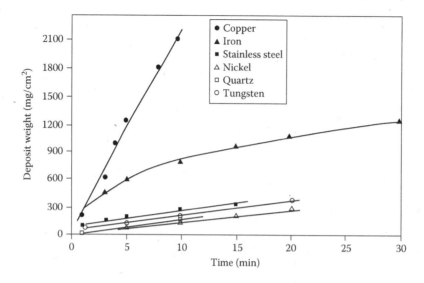

FIGURE 8.11 Deposition on various substrates with 1.55% TCP in a nitrogen stream at 700°C. (From Klaus, E.E. et al., *Lubr. Eng.*, 45(11), 717, 1989. Reproduced with permission from STLE.)

The use of TCP vapor to lubricate high-speed bearings made from M50 steel at 350°C was examined by Graham et al. in 1992 [96] with excellent results. In fact, the wear area was smoother than the unused surface. Surprisingly, similar results were found when lubricating silicon nitride surface without prior activation. Here, the results were clouded by the transfer of copper to the test specimens, and it was thought that activation could have occurred by the reaction of TCP with copper components of the vapor delivery system, which was then deposited onto the ceramic surface. Analysis of the film formed by TCP on a ceramic surface was also investigated by Hanyaloglu and Graham [97]. In this case, the ceramic was activated by a film ~20 atoms thick of iron oxide. The presence of TCP at 0.5% in nitrogen or air at 500°C gave a friction coefficient of 0.07 and produced a polymer containing mainly carbon, oxygen, and a small amount of phosphorus with a molecular weight range of 6,000–60,000 g/mol.

A combination of vapor and mist lubrication has also been evaluated in the lubrication of gas turbine bearings [98]. The data indicated that organophosphates worked well with ferrous metal due to the rapid formation of a predominantly iron phosphate film. This was followed by the development of a pyrophosphate-based film over the iron phosphate. As long as iron was present, the organophosphates worked well, but continued production of the phosphate/pyrophosphate film reduced access to iron and eventually led to surface failure. Morales and Handschuh [99] reported a solution to this problem in which the phosphate contained a small quantity of ferric acetylacetonate. Evaluation of this solution in comparison with the pure phosphate showed that the iron salt enabled a phosphate film to be successfully deposited onto an aluminum surface, which the pure phosphate is unable to do. (Neutral phosphates are known not to wet the surface of aluminum.) Vapor/mist lubrication of a gearbox using pure phosphate was compared with the performance of the phosphate containing the iron salt; a significant improvement in scuffing performance was noted. This was

enhanced when the mist was directed onto the gear teeth immediately before contact. Evaluation of the surface film on the gear teeth revealed no phosphorus when the pure phosphate was tested but showed the presence of "fair amounts" of both iron and phosphorus when using the soluble iron salt.

A recent study of the mechanism of film formation by aryl phosphates [100] involved examining the reaction of phosphates with metal in the form of foil or powder and also with various metal oxides in different oxidation states. The tests were carried out in both oxygen-rich and oxygen-depleted environments, and they revealed that the reactivity of both the commercial grade of TCP and the pure isomers increased with steel and other metals with increasing oxidation state of the metal/oxide. In comparison with little or no degradation in the absence of metal/metal oxides, limited degradation took place in the presence of metal, but almost complete breakdown of the phosphate occurred (at the same temperature—in the range of 440°C–475°C) in the presence of Fe_2O_3 and Fe_3O_4. The isomeric forms of TCP also displayed different levels of reactivity with tri-*ortho*cresyl phosphate (TOCP), more active than the *meta* and *para* isomers. The authors indicated that these relativities are consistent with the oxide's free energy of formation; those oxides with the highest free energy of formation show the lowest level of activity and vice versa. Different types of steel surface also displayed different levels of reactivity, with 316C stainless steel being the least active.

Surface analysis of the steel specimens used indicated that, depending on whether the metal surface was oxygen rich or poor, different mechanisms of degradation predominate. When excess oxygen was present, the film produced was a polyphosphate with good lubricating properties, whereas a surface with only a thin oxide coating produced iron phosphate, which has poor lubrication properties. No phosphide was found in the surface coating, but an iron/amorphous carbon layer, possibly rich in fused aromatics, arising from the degradation of the aromatic part of the phosphate was found when using the TBPP, but not when TCP was examined. Since these aromatics have a planar structure, they may assist with lubrication by allowing the surface to move more easily over one another. However, it is likely that the end result is a composite of the behavior of the polyphosphate and the carbonaceous film. Indeed, the author suggests that the polyphosphate may be acting as a "binder" for the carbon, and it is the latter that is providing the lubrication. The proposed mechanism for the formation of the polyphosphate film was thought to involve the cleavage of the C–O bond on one of the pendant groups as the phosphate attaches itself to the surface (presumably through the –P=O function), eliminating a cresyl radical. This is followed by the elimination of another cresyl radical as the second C–O bond breaks, and an Fe–O bond is formed. In this way, a "lattice of cross-linked PO_3 is formed with the Fe surface." Wear of the film is not a problem as it appears to be self-healing due to diffusion of Fe ions through the polyphosphate layer to the surface where reaction with phosphate continues. There was no suggestion that hydrolysis of the phosphate is involved.

Further evidence of the part played by carbon in the wear process has appeared in a recent patent application [101] claiming synergy between carbon nanoplatelets and ZDDPs or other phosphate AW additives. This suggests "the formation of a nano-composite wear-resistant and low friction tribofilm on both ferrous and non-ferrous surfaces."

8.5.2.3 Recent Commercial Developments

Although the majority of phosphates used as AW/EP additives are relatively low-viscosity products, interest has been expressed in materials of high molecular weight for aerospace applications, where low volatility is important, for example, high-temperature lubricants for aeroderivative gas turbines and greases for space vehicles. Three products have become commercially available and have been evaluated: an ISO 100 TBPP with low TPP content, resorcinol tetraphenyl bisphosphate (Figure 8.12a), and isopropylidene di-p-phenylene tetraphenyl bisphosphate (Figure 8.12b). The hydrolytic stability of the resorcinol diphenyl phosphate is relatively poor, but this would not be of major concern for aerospace applications, for example, in greases. However, this material has been claimed

(a)

(b)

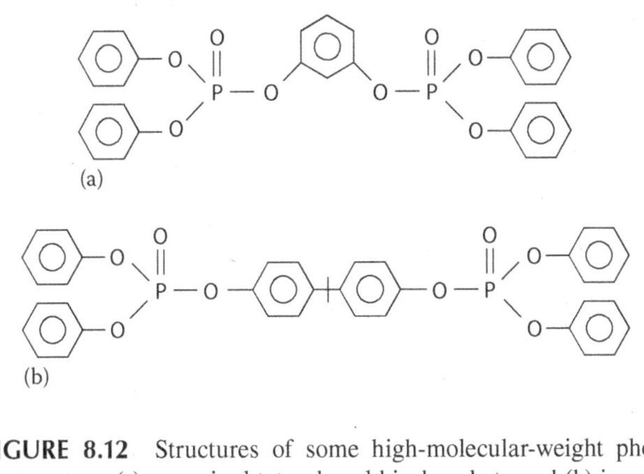

FIGURE 8.12 Structures of some high-molecular-weight phosphate esters: (a) resorcinol tetraphenyl bisphosphate and (b) isopropylidene di-p-phenylene tetraphenyl bisphosphate.

TABLE 8.9

The Effect of High-Molecular-Weight AW Additives on the Coking, Wear, and Magnesium Corrosivity of Ester-Based Gas Turbine Oil Formulations

AW Additive	Deposit Formation[a] (mg)	Wear[b] (mm)	Magnesium Corrosivity[c]
Blank—no additive	89	0.655	High
TCP	98	0.40	High
Tris-C_9–C_{10} alkylphenyl phosphorodithioate	103	0.505	Pass
TBPP	94	0.54	Pass
Resorcinol tetraphenyl bisphosphate	Not determined	0.425	Fail

Source: Gschwender, L. and Snyder, C., Private communication, August 2001. With permission.

Note: Additives used at 1% addition in the ester base.

[a] Fluid held at 300°C for 3 h: method described in paper by Gschwender et al., *Lubrication Engineering*, pp. 20–25, May 2000.

[b] ASTM D4172-88. Four-ball wear test for 1 h at 40 kg, 600 rpm, and 75°C.

[c] 20 mL sample held for 48 h at 232°C with 1 l/h, air flow.

as an AW additive for fuels and lubricants [102], whereas the TBPP has been incorporated into an aerospace grease formulation [103]. Other condensed bis- and polymeric phosphates have since been claimed as effective AW additives for industrial and automotive applications [104].

As part of an assessment of the high-molecular-weight additives for use in high-temperature aviation gas turbine oils, they were compared under coking, four-ball wear, and oxidation test conditions. The results are given in Table 8.9. Although the AW performance of the butylphenyl phosphate is not as good as that of TCP, the reduced impact on deposit formation and magnesium corrosion performance has made it the most promising candidate.

Although much of the recent focus of activity has been on aryl phosphates, there have also been developments with alkyl phosphates. TBP, for example, has been used as an EP additive for EP steam and gas turbine oils used when the turbine is driving a reduction gear (Ertelt, R., Private Communication, September 2001). About 1.5% of the additive is used to increase the FZG gear test performance (DIN 51354) from a load stage failure of about 6–8 to 10–11. Again, the neutral nature of the molecule is of advantage in minimizing interaction with other components of the formulation.

An additional application where interest has been expressed in alkyl phosphates is metalworking. Owing to a desire on environmental grounds to move away from chlorine, mixtures of neutral phosphates and sulfur-containing additives have been promoted as alternatives [105–108]. As concerns exist about the possible release of phenolic materials into the environment, the alkyl phosphates are, perhaps, best suited for this application and are able to provide similar or better performance to the chlorparaffins when used together with sulfur carriers. Table 8.10 summarizes the drill

TABLE 8.10
A Comparison of the AW/EP Performance of a Chlorinated Paraffin and an Alkyl or Aryl Phosphate/Active Sulfur Combination in a Simple Oil-Based Cutting Fluid Formulation

Formulation (w/w) and Test Data	A	B	C	D
ISO 22 paraffinic oil	92	95.7	96.9	96.5
Tri-isopropylphenyl phosphate	—	—	—	1.0
Trialkyl phosphate	—	—	0.6	—
Active sulfur compound (40% S)	—	4.3	2.5	2.5
Chlorinated paraffin (40% Cl)	8	—	—	—
Four-ball wear test (ASTM D4172) (mm)	0.65	—	—	0.43
Four-ball EP properties (ASTM D2783)				
Weld load (kgf)	400	—	—	620
Seizure load (kgf)	80	—	—	80
Load wear index	51	—	—	104
Pin and V-block wear (ASTM D3233)				
Failure load (lb)	>3100	—	—	2726
Drill Life test				
Holes drilled to failure (EN24T mild steel at 1200 rpm and 0.13 mm/min feed rate)	140	100	280	200

life and other AW/EP performance in a neat oil for both neutral IPPP and neutral alkyl phosphate in combination with a sulfurized olefin, when compared with a chlorparaffin.

In an extension to this work, drill life test data were obtained on tri-isobutyl and tributoxyethyl phosphate in comparison with a commercially available acid phosphate (oleyl acid phosphate). Each phosphate was evaluated at the same phosphorus level in the presence of a sulfur carrier (a 4:1 mixture of a sulfurized fatty acid ester with 26% total sulfur and a dialkyl polysulfide with 40% total sulfur content), and all additives were dissolved in a neat paraffinic mineral oil of ISO VG 22. The test was carried out on an automatic drilling machine, drilling holes of 18 mm depth in a 40 mm thick disk of stainless steel type 304 with a feed rate of 0.13 mm/rev at 1200 rpm. The test was concluded when either the drill broke or showed excessive wear. The results in Table 8.11 [109] show a significant improvement for the butoxyethyl phosphate over the isobutyl phosphate, whereas the oleyl acid phosphate showed little activity. The reason for the poor behavior of the acid phosphate is not known.

Also, in the field of metalworking, phosphates have been claimed as components of hot forging compositions [110,111].

8.5.3 ALKYL- AND ARYL ACID PHOSPHATES

8.5.3.1 Nonethoxylated

Although the range in commercial use is limited, acid phosphates are important components of metalworking oils—frequently in combination with chlorinated paraffins. However, because of environmental concerns associated with the use of specific types of chlorinated paraffins, their possible replacement by mixtures of phosphorus and sulfur compounds has been, and continues to be, investigated [66].

Mixtures of monophosphoric and diphosphoric acid esters were compared with a dithiophosphate acid amide in macroemulsions using a variety of EP tests. Performance in drilling and tapping tests (which are regarded as the conditions most closely simulating cutting performance) indicated that the dithiophosphate amide gave the best performance, whereas the mono- and diacid phosphates produced levels

TABLE 8.11
Results of Drill Life Tests on Alkyl and Alkoxyalkyl Phosphates in the Presence of Sulfur Carriers

Formulation (w/w) and Test Data	A	B	C	D
Sulfur carrier	5.2	5.2	5.2	5.2
Tri-isobutyl phosphate	—	4.17	—	—
Tributoxyethyl phosphate	—	—	6.4	—
Oleyl acid phosphate	—	—	—	10.0
Neat oil	94.8	90.63	88.4	84.8
Holes drilled to failure-(Type 304 stainless steel at 1200 rpm and 0.13 mm/min feed rate)	84	432	>500	18

of performance similar to or better than that of the chlorinated paraffin alone.

Traditionally, the acid phosphates in commercial use have high acid numbers (200–300 mg KOH/g). As a consequence, in addition to their use as AW/EP additives, they are used as corrosion inhibitors [112], and certain structures are promoted as copper passivators [112]. They are also used to minimize aluminum staining [113].

A recent development has been the availability of aryl phosphate–based products that have a relatively low level of acidity (typically 10–15 mg KOH/g) while offering a combination of good AW/EP performance with rust prevention and oxidation inhibition. The multifunctionality of this product type offers opportunities for the simplification of additive packages and use in a wide range of hydraulic and circulatory oils, metalworking, and gear applications, while the lower level of acidity reduces the potential for additive interaction and the promotion of foaming, etc.

Increased activity in alkyl acid phosphates has been reported in the patent literature. This arises from the use of long-chain alcohols (C_{16}–C_{18}) to produce an acid phosphate ester mix with a high monoacid content (preferably greater than 80%:20% monoacid–diacid ratio) [114]. With this acid distribution, it has been possible to achieve lower wear than for the conventional ethoxylated alkyl phosphates with a monoacid to diacid ratio of 60%:40%.

8.5.3.2 Alkyl- and Alkylarylpolyethyleneoxy Acid Phosphates

Polyethyleneoxy acid phosphates are a potentially very large class of compounds. Not only are variations possible in the type of alcohol or phenol chosen but also in the type and level of alkoxylation (although ethylene oxide [EO] is invariably used). Products of this process were originally claimed to be more active than the nonethoxylated variety, but recent developments with the latter types [114] suggest this may no longer be the case.

Depending on the choice of raw materials, the finished product may be oil or water soluble or water dispersible. Alkyl and (alk)arylpolyethyleneoxy acid phosphate esters containing <55% EO were found to be oil soluble; products with an EO content of more than 60% were water soluble as the free acids and their amine salts, whereas products with 40%–60% of EO were both oil and water soluble or water dispersible [40]. The free acids are used in oil applications, whereas amine (usually triethanolamine) or metal salts of the acids are used in aqueous applications. The alcohols and phenols selected for evaluation were lauryl and oleyl alcohols and nonyl, dinonyl, and dodecyl phenol. (Other raw materials in common use today include C_8–C_{10} alcohols, 2-ethyl hexanol, tridecanol, cetyl-oleyl mixed alcohols, and phenol). The products are nonionic surfactants with excellent wetting and emulsification properties, and certain types do not support bacterial growth. They are also good corrosion inhibitors—an important factor for their use in metalworking applications. The higher EO-content products tend to produce a large amount of stable foam, and

materials containing ~45% EO are therefore preferred for metalworking applications [40].

The effect of the alcohol or phenol and the impact of EO content on the wear behavior in a naphthenic oil can be seen in Figures 8.13 and 8.14, respectively [115].

The performance of the product based on oleyl alcohol is interesting in that it does not appear to change with EO content, yet is simultaneously capable of producing materials that vary from oil to water soluble. However, the four-ball or pin and v-block tests, although widely used as screening tests for the metalworking application, are not considered to be capable of predicting the performance under cutting conditions. This is confirmed in the paper given in 1995 by Werner et al. [116], which compared the performance of different ethoxylated acid phosphates under various test conditions. Of greater relevance than conventional four-ball or pin and v-block tests are actual cutting or tapping torque tests.

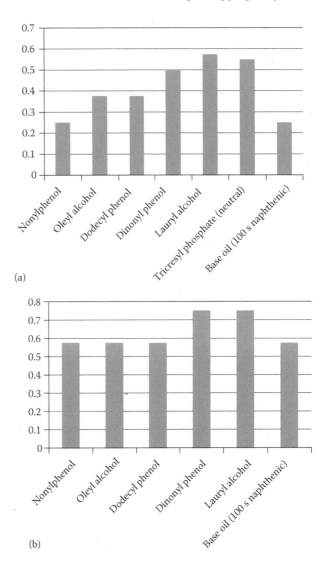

FIGURE 8.13 Effect of hydrophobe on wear properties-four-ball wear scar diameter. Acid phosphate esters based on non-ionics containing 23%–25% ethylene oxide. (a) Test conditions: 40 kg, 100 rpm, 60 min, 121°C; four-ball wear scar diameter (mm). (b) Test conditions: 100 kg, 100 rpm, 60 min, 121°C; four-ball wear scar diameter (mm).

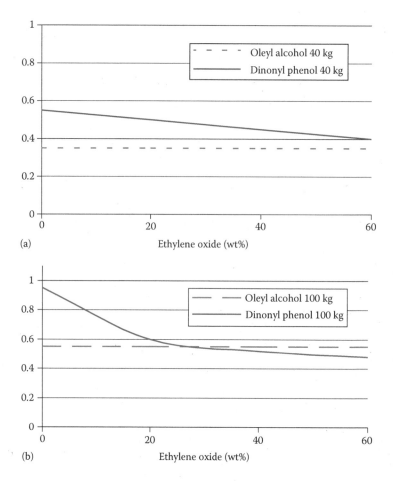

FIGURE 8.14 Effect of ethylene oxide content on wear properties: 1% of additive in 100 SUS [1000F] naphthenic base oil. (a) Test conditions: 40 kg, 100 rpm, 121°C, 60 min; four-ball wear scar diameter (mm). (b) Test conditions: 100 kg, 100 rpm, 121°C, 60 min; four-ball wear scar diameter (mm).

The results given in Table 8.12 show that (1) products with a hydrophilic–lipophilic balance value of 11–12 give the best results and (2), in general, the further the value deviates from this, the worse the results become. Unfortunately, no studies appear to have been made on the nature of surface film deposited on the metal, but the adsorption mechanism indicated previously is probably still valid.

8.5.3.3 Amine Salts of Acid Phosphates

One amine phosphate that appeared in the patent literature as early as 1934 as a corrosion inhibitor for aqueous systems (and still occasionally used) is triethanolamine phosphate [117]. Formed by the neutralization of phosphoric acid with triethanolamine, this product was widely used as a corrosion inhibitor in automotive antifreeze formulations for many years [118].

In 1970, Forbes and Silver [119] reported on their work investigating the effect of chemical structure on the load-carrying properties of different phosphorus compounds. In this case, the structures under review were di-*n*-butylphosphoramidates, amine salts of di-*n*-butyl phosphate, and derivatives of dialkylphosphinic and alkylphosphonic acids. The results indicated that the phosphoramidates

were more effective load-carrying additives than the neutral phosphates, TBP and TCP, but less active than the amine salts of di-*n*-butyl phosphate. The evaluation of the series of dialkylphosphinic and alkylphosphonic acid esters indicated that the AW performance related directly to the strength of the acid from which they had been produced (Figure 8.15), suggesting that adsorption through the polarity of the ester group was an important step in the process.

In addition to the work carried out in hydrocarbon base stocks, some testing was also performed in a synthetic ester. This fluid enabled a comparison to be made of tetra-alkyl-ammonium salts of dibutylphosphate (otherwise insoluble in mineral oil), which displayed the best AW/EP properties of all the amine phosphates tested (Table 8.13). The authors suggested that this was probably due to the stability of the ions.

A further study of the mechanism of amine phosphates by Forbes and Upsdell appeared in 1973 [120]. Adsorption/reaction studies of dibutyl and di-2-ethylhexyl phosphates with either *n*-octylamine or cyclohexylamine and iron powder showed that both the amine and the acid phosphate were adsorbed onto the metal surface and that the rate and extent of their adsorption/desorption varied with chemical structure. The higher the solubility of the iron–phosphate

TABLE 8.12

Phosphate Ester Surfactant Ranking on Steel in a Water-Based System as a Function of Composition

Alcohol	Phosphate EO Units	Sample Rankings				Overall Rating	HLB Value
		Pin-on-Vee Block	Four-Ball Wear	Tapping Torque	Total of Rankings		
C_9–C_{16}	5.5	4	1	4	9	1	13
C_{18}	4	6	4	2	12	2	12
Nonylphenol	4	10	2	1	13	3	11
C_{13}	10	2	9	5	16	4	14
C_{12}	6	1	12	7	20	5	12
C_8–C_{10}	6	8	7	6	21	6	11.5
C_{12}	12	5	8	12	25	7	15
C_{13}	6	9	6	10	25	8	11.5
C_{12}	9	7	5	14	26	9	14
Nonylphenol	6	14	3	9	26	10	8
Phenol	6	12	13	3	28	11	15.4
Dinonyl phenol	5	3	14	11	28	12	9
Nonylphenol	9.5	11	11	8	30	13	13
C_{13}	4	13	10	15	38	14	9.7
Butanediol	6	15	15	13	43	15	—

Source: Werner, J.J. et al., Relationship of structure to performance properties of phosphate-ester surfactants in metal working fluids, *STLE Annual Conference*, Chicago, IL, 1995. Reproduced with permission from STLE.

Note: EO = unit of ethylene oxide.

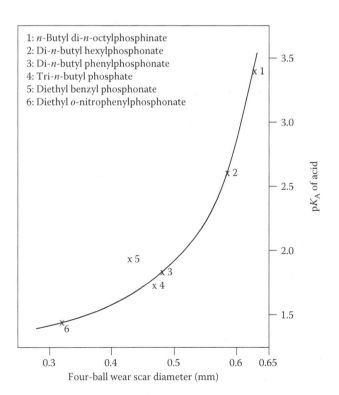

1: n-Butyl di-n-octylphosphinate
2: Di-n-butyl hexylphosphonate
3: Di-n-butyl phenylphosphonate
4: Tri-n-butyl phosphate
5: Diethyl benzyl phosphonate
6: Diethyl o-nitrophenylphosphonate

FIGURE 8.15 Effect of acid strength on AW performance. (From Forbes, E.S. and Silver, H.B., *J. Inst. Pet.*, 56(548), 90, 1970. Reproduced with permission from the Energy Institute.)

complex formed, the greater was the likelihood of desorption. Furthermore, good AW performance depended on high phosphate and amine adsorption and retention of the phosphate moiety on the surface.

The conversion of a dialkyl acid phosphite to an amine phosphate and the use of the mixed product as a multifunctional AW/EP additive, antioxidant, and corrosion inhibitor with improved metal passivation properties were claimed in 1997 [121].

An amine salt and TCP were studied as AW agents for different synthetic esters by Weller and Perez [122] and compared with a sulfurized hydrocarbon. The neutral ester (TCP) generally showed an increasing amount of wear up to 1% addition before reducing at higher levels. The amine salt, however, rapidly reduced wear to very low levels. Friction coefficients were also consistently lower with the amine salt.

Kristen [123] reported the effect of additive interaction between amine phosphates and a phosphorothionate. The additives were evaluated under FZG gear test conditions (DIN 51354). The results showed that the additives respond differently in nonpolar and polar base stocks, specifically a poly-α-olefin (PAO) and a synthetic ester (Figures 8.16 and 8.17). In the synthetic hydrocarbon base, a level of 0.75% amine phosphate and 0.25% phosphorothionate (or perhaps 0.5% of each) provided a borderline FZG 12 load stage pass/fail. In comparison, 0.75% of amine phosphate ester and 1% of phosphorothionate were required to achieve the same level of performance in the ester. Monitoring the response of additive combinations reveals not only the most cost-effective mixtures but also any antagonisms between additives,

TABLE 8.13

Four-Ball Test Results of Various Amine Dibutyl Phosphates in Di-Isooctyl Sebacate

	EP Test			AW Test		
				WSD (mm) After		
$(BuO)_2 PO_2 NR^1 R^2 R^3 R^4$	MHL (kg)	WL (kg)	ISL (kg)	30 min	45 min	60 min
$nC_4H_9NH_3$	40.5	130	125	0.38	0.38	0.36
$nC_6H_{11}NH_3$	40.0	130	115	0.37	0.38	0.39
$PhNH_3$	43.6	140	120	0.37	0.38	0.39
$[nC_4H_9]_2NH_2$	34.8	140	100	0.26	0.27	0.33
$[nC_4H_9]_3NH$	37.8	150	110	0.37	0.38	0.42
$[nC_4H_9]_4N$	54.4	150	150	0.25	0.25	0.26
$[CH_3]_2 \cdot [nC_8H_{17}]_2N$	56.1	165	165	0.30	0.35	0.40
None	18.6	120	55	0.57	0.61	0.64
TCP	19.8	110	60	0.31	0.33	0.35

Source: Forbes, E.S. and Silver, H.B., *J. Inst. Pet.,* 56(548), 90, 1970. Reproduced with permission from the Energy Institute.

Note: % additive = 4 milliatoms of P/100 g of fluid. (MHL, Mean Hertz Load; WL, Weld Load; ISL, Incipient Seizure Load; WSD, Wear Scar Diameter.)

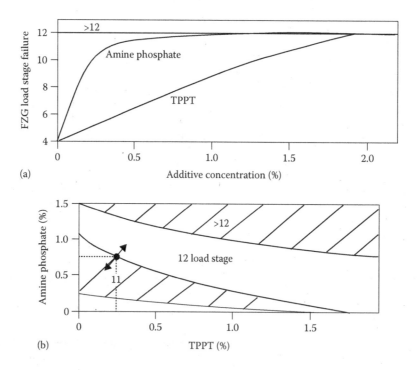

(a)

(b)

FIGURE 8.16 FZG performance of an amine phosphate and TPPT separately (a) and in mixtures (b) in an ISO VG 32 poly-α-olefin (PAO). (From Kristen, U., Aschefreie extreme-pressure-und verschleiss-schutz-additive, in: W.J. Bartz, ed., *Additive für Schmierstoffe*, Expert Verlag, Renningen, Germany, 1994. With permission.)

as were found here in the ester base at higher additive levels. Such information is invaluable to formulators when trying to meet specification requirements and ensuring that the performance level is consistently above the minimum limit.

Amine salts, for example, triethanolamine salts of alkyl and arylpolyethyleneoxy acid phosphates, are widely used in metalworking applications. Some of these products are not only commercially available but are also produced *in situ* when the pH of the product is adjusted by the addition of base

to ensure the product is alkaline in use. This is to avoid corrosion and minimize skin irritation.

8.5.4 NEUTRAL ALKYL- AND ARYL PHOSPHITES

8.5.4.1 Use as Antiwear/Extreme Pressure Additives

The earliest known reference to the evaluation of phosphites as AW/EP additives is in a 1950 paper by Davey [75]. As a result of these investigations, which also included a comparison

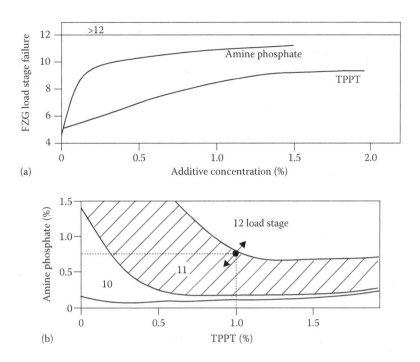

FIGURE 8.17 FZG performance of an amine phosphate and TPPT separately (a) and in mixtures (b) in an ISO VG 22 pentaerythritol ester. (From Kristen, U., Aschefreie extreme-pressure-und verschleiss-schutz-additive, in: W.J. Bartz, ed., *Additive für Schmierstoffe*, Expert Verlag, Renningen, Germany, 1994. With permission.)

with phosphates and the effect of incorporating chlorine into the phosphate/phosphite molecule, it was found that

- Phosphites have superior EP properties to the phosphates, and long alkyl chains are more effective than aryl groups.
- Evaluation of phosphates TBP and TXP revealed similar optimum concentrations of between 1% and 2% as were found in the previous study with TCP [5].
- Polar compounds such as acids or esters improve the lubricating (AW) properties of phosphites and phosphates by being strongly adsorbed onto the surface.
- The incorporation of chlorine or sulfur into the molecule (or the addition of small amounts of free sulfur) improves the EP properties. Chlorine is more effective when part of an alkyl residue, and when sulfur is added to a P/Cl compound (e.g., a chlorinated phosphite), the EP properties are further improved.

Following the study by Davey, a number of patents appeared claiming the use of phosphites in lubricant applications [124–127], but it was not until 1960 that a further detailed study of the behavior of phosphites, this time by Sanin et al. [128], was published. The study emphasized the correlation between structure and activity, and the short-chain derivatives were found to be the most active.

In 1993, Ohmuri and Kawamura [129] carried out fundamental studies into the mechanism of action of phosphite EP additives. They found that initial adsorption rates of phosphorus-containing esters depended largely on the existence of −OH and −P=O bonds in the structure. The extent of

adsorption was influenced by the hydrolytic stability of the esters, and this process was found to occur through reaction with water adsorbed onto the iron surface. Adsorption of the phosphites varied depending on the degree of esterification; triesters were adsorbed after being decomposed hydrolytically to monoesters, whereas diesters were adsorbed without hydrolysis. Phosphite esters eventually hydrolyzed to *inorganic acid* regardless of the degree of esterification, followed by its adsorption and conversion to the iron salt. It was suggested that the adsorbing and hydrolyzing properties of the esters depended on the arrangement of the molecules physisorbed onto the surface.

Evaluation of a range of alkyl phosphites as EP additives in gear oils was reported by Riga and Rock Pistillo [130]. The most effective products were those with short chains, particularly dibutyl phosphite, which resulted in a wear layer of >1000 Å and the formation of both iron phosphate and phosphide. Other phosphites formed only traces of phosphide, and as the chain length increased, the resulting film became thinner and contained less phosphorus, possibly due to steric hindrance. Long-chain (C_{12}) alkyl phosphites have also been claimed as AW additives for aluminum rolling oil [131] and, in fact, are still used in metalworking applications.

In view of the work carried out on the use of phosphates as vapor-phase lubricants, an investigation into the effect of phosphites on the frictional properties of ceramic-on-ceramic and ceramic-on-metal surfaces was carried out in 1997 [132]. The phosphites (and other additives evaluated) had no effect on ceramic-on-ceramic friction; in fact, short-chain phosphites significantly increased friction. When several types of metal were slid against oxide ceramics, the alkyl phosphites were found to lower the friction for each metal except copper. Apparently, the reaction products between copper and the phosphite had adhesive properties and increased friction.

The decomposition of trimethylphosphite on a nickel surface was also studied to obtain insight into the initial steps in the decomposition of phosphates when used as vapor-phase lubricants [133]. The main breakdown path is the cleavage of the –P–O– bond to yield the methoxy species, which then degrades to CO and H_2 or reacts with the nickel surface. Following heating to 700 K, the surface loses adsorbed species other than phosphorus, which is seen as a simple way for the controlled deposition of phosphorus onto a metal surface.

8.5.4.2　Use as Antioxidants for Lubricating Oils

In addition to their use as AW/EP additives, neutral (and acid) phosphite esters have long been used as antioxidants or stabilizers for hydrocarbons. They were originally introduced as stabilizers for rubber and thermoplastics. Trisnonylphenyl phosphite, for example, was first used to stabilize styrene-butadiene rubber in the early 1940s; this was shortly followed by patents claiming phosphites as antioxidants for lubricants [48,125,126,134,135].

Phosphites function as decomposers of hydroperoxide, peroxy, and alkoxy radicals (reactions 8.20 through 8.22) rather than eliminating the hydrocarbyl-free radicals formed in the chain initiation process. They also stabilize lubricants against photodegradation [136].

$$R^1OOH + (RO)_3 P \rightarrow (RO)_3 P = O + R^1OH \quad (8.20)$$
Hydroperoxide

$$R^1OO^\cdot + (RO)_3 P \rightarrow R^1O^\cdot + (RO)_3 P = O \quad (8.21)$$
Alkylperoxy radical

$$R^1O^\cdot + (RO)_3 P \rightarrow R^1OP(RO)_2 + RO^\cdot \quad (8.22)$$
Alkoxy radical

This behavior as *secondary* antioxidants by destroying the hydroperoxides, formed in the chain propagation process results in their use in synergistic combination with those antioxidant types that are active as radical scavengers in the initiation process, for example, the hindered phenols and aromatic amines [137–141].

Phosphites are useful additives because of their multifunctionality. However, although they are still used as antioxidants in hydrocarbon oils, their relatively poor hydrolytic stability and the formation of acidic compounds that could affect the surface-active properties of the oil have prompted the introduction of "hindered" phosphites with better hydrolytic stability: for example, tris-(2,4-ditertiarybutylphenyl) phosphite or tris-(3-hydroxy-4,6-ditertiarybutylphenyl) phosphite, and, where solubility permits, cyclic phosphites, for example, based on pentaerythritol such as bis-(2,4-ditertiarybutylphenyl) pentaerythritol diphosphite (Figure 8.18). These types are claimed as stabilizers or costabilizers for lubricating oils [142–145].

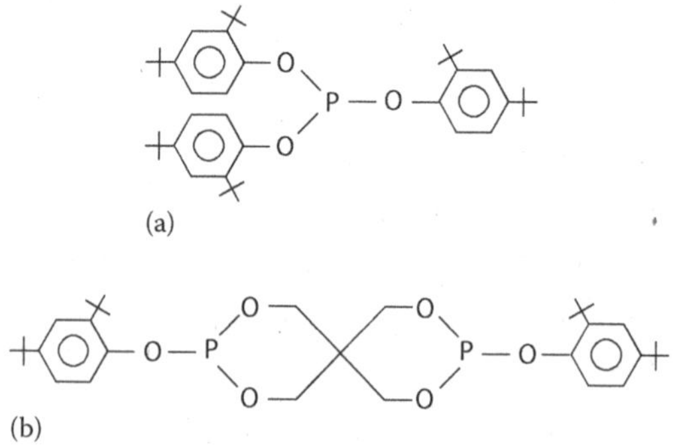

(a)

(b)

FIGURE 8.18 Structures of some commonly available hindered phosphites: (a) tris-(2,4-ditertbutylphenyl) phosphite and (b) bis-(2,4-ditertbutylphenyl) pentaerythritol diphosphite.

TABLE 8.14
Antioxidant Synergism between Hindered Aryl Phosphites and a Hindered Phenol

Base Stock	Antioxidant	Oxidation Stability	
		Viscosity Change (%)	Total Acid Number Increase (mg KOH/g)
1	Hindered phenol (0.5%)	357	11.5
	Hindered phosphite A (0.5%)	438	12.2
	Hindered phenol (0.1%) + phosphite A (0.4%)	8.7	0.01
	Hindered phenol (0.17%) + phosphite A (0.33%)	9.4	0.06
2	Hindered phenol (0.5%)	712	14.2
	Hindered phosphite B (0.5%)	452	10.6
	Hindered phenol (0.1%) + phosphite B (0.4%)	8.1	0.05
	Hindered phenol (0.17%) + phosphite B (0.33%)	8.7	0.03

Source:　U.S. Patent 4,652,385, Petro-Canada Inc., 1987.

Note:　Hindered phenol is tetrakis-(methylene-3,5-ditertbutyl-4-hydroxy-hydrocinnamate) methane; Phosphite A is tri-(2,4-ditertbutylphenyl) phosphite; Phosphite B is bis-(2,4-ditertbutylphenyl) pentaerythritol diphosphite. Test conditions: IP 48 (modified), 200°C for 24 h, air at 15 1/h in an ISO VG 32 mineral oil.

Table 8.14 [144] illustrates the significant improvement in oxidation stability shown by such blends.

In common with most other types of phosphorus-containing products, neutral (and acid) phosphites have also been claimed as corrosion inhibitors [146,147].

8.5.5　Alkyl- and Aryl Acid Phosphites

As might be predicted from the behavior of the other types of phosphorus-containing additives, the acid phosphites have good AW/EP properties; the nonylphenyl acid phosphite is particularly effective [148,149]. When used in aviation gas turbine lubricants, the acid phosphites were sometimes formulated in combination with neutral phosphates (TCP); blends of the two products showed synergy even when the amount of the phosphite was very low [150]. The acid phosphites are also claimed to be corrosion inhibitors [151] and antioxidants [47,152,153].

The effects of structure on the AW and load-carrying properties of dialkyl phosphites were studied by Forbes and Battersby [151] in a liquid paraffin. AW performance was best with long-chain compounds (Figure 8.19), whereas the short-chain (highest phosphorus content) derivatives displayed the best load-carrying performance.

Scuffing behavior, however, appeared to reach a minimum at about a C_8 carbon chain length (Figure 8.20). This parallels

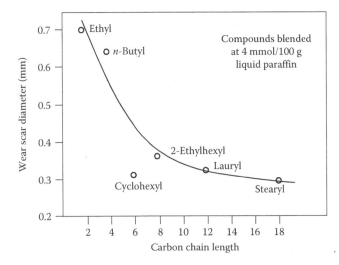

FIGURE 8.19 Effect of chain length on the four-ball AW performance of dialkyl phosphites. (From Forbes, E.S. and Battersby, J., *ASLE Trans.*, 17(4), 263, 1974. Reproduced with permission from STLE.)

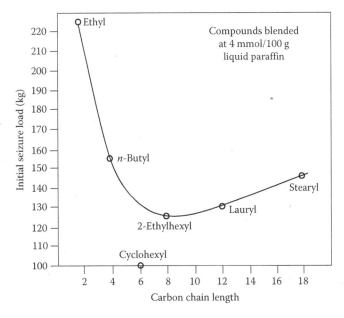

FIGURE 8.20 Effect of chain length on the initial seizure loads of dialkyl phosphites. (From Forbes, E.S. and Battersby, J., *ASLE Trans.*, 17(4), 263, 1974. Reproduced with permission from STLE.)

TABLE 8.15

Comparison of the Load-Carrying Properties of Dialkyl Phosphates and Dialkyl Phosphites at 4 mmol/100 g Base Oil

	Initial Seizure Load (kg)	Wear Scar Diameter—1 h (mm)
Diethyl phosphite	225	0.70
Diethyl phosphate	160	0.43
Dibutyl phosphite	155	0.64
Dibutyl phosphate	85	0.42
Di-2-ethylhexyl phosphite	125	0.36
Di-2-ethylhexyl phosphate	80	0.29
Dilauryl phosphite	130	0.32
Dilauryl phosphate	80	0.35

Source: Forbes, E.S. and Battersby, J., *ASLE Trans.*, 17(4), 263, 1974. Reproduced with permission from STLE.

the behavior of the neutral phosphites. Adsorption studies also showed that the phosphorus content of the solution was depleted in the same order as the load-carrying performance, namely, the most active products showed the highest loss from solution. The presence of water increased the uptake of phosphorus from solution. Comparison of the performance of the phosphites against the corresponding acid phosphate revealed that the phosphites had better load-carrying but inferior AW behavior (see Table 8.15). The authors suggest that the activity of the phosphites is due to an initial hydrolysis to produce an intermediate either in solution or on the metal surface (Figure 8.21). This reacts with the metal surface to give an iron salt that was thought to be responsible for the AW properties of the product. Under much more extreme conditions as are found with scuffing, the aforementioned salt was thought to decompose further to give the phosphorus-rich layer as shown in Figure 8.21.

8.5.5.1 Amine Salts of Acid Phosphites

In 1975, Barber [152] investigated the four-ball test performance of several long-chain amine salts of short-chain acid phosphites, which were found to be very active. Unfortunately, he did not investigate the effect of increasing the chain length of the phosphite while reducing the length of the amine. Most of the paper concerns the behavior of a wide range of phosphonate esters (see Section 8.5.6).

8.5.6 PHOSPHONATE- AND PHOSPHINATE ESTERS

A large group of phosphonate esters was prepared by Barber in 1975 [152] and evaluated using the four-ball machine. Although short-chain esters were more effective in preventing scuffing, the most effective products were those containing chlorine. However, even at high levels of chlorine, the performance was still inferior to the amine phosphite reaction products reported earlier. In comparison with TCP, incipient seizure loads were generally higher, but the weld

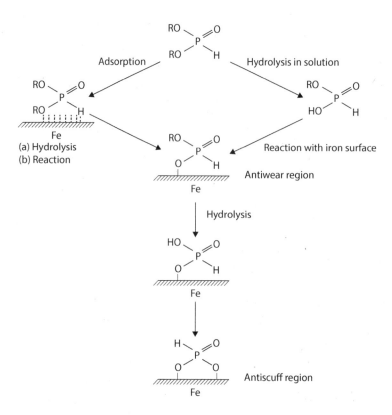

FIGURE 8.21 Mechanism of the load-carrying action of dialkyl phosphites. (From Forbes, E.S. and Battersby, J., *ASLE Trans.*, 17(4), 263, 1974. Reproduced with permission from STLE.)

loads were broadly similar. Unfortunately, there were no directly comparable data under wear test conditions. A limited number of phosphinate esters were evaluated and found (also by four-ball tests) to give similar performance to the phosphonate esters.

The activity of a range of phosphonates was studied by Sanin et al. [153], who concluded that their effectiveness depended on their structure and the friction regime, but esters containing no chlorine had "no effect at either low or high load." A further study, by the same authors, of the mechanism of activity of phosphonates again suggested the reaction of decomposition products with iron and the formation of a protective layer. Under severe conditions, this layer is removed, resulting in a sudden increase in friction followed by seizure or welding. Studies of the reaction of a dibutyltrichlorophosphonate ($Cl_3CPO(OBu)_2$) indicated that a reaction took place at 405–408 K to give chlorobutane and an iron-containing polymer. At 413 K, this polymer decomposed to give $FeCl_3$, which gave additional protection.

Phosphonate (and pyrophosphonate) esters, as their metal or amine salts, have appeared in the patent literature over many years as AW/EP additives. Amine salts of dinonylphosphonate are, for example, claimed in aircraft gas turbine lubricants [154], and dimethyltetradecyl phosphonate has been used in water-based formulations with good pump wear characteristics [155,156]. One of the most recent applications has been in refrigeration compressor oils (e.g., for automobiles) that are compatible with the more ecologically acceptable refrigerants. The reason for their selection in this application has probably been their good hydrolytic stability in view of the need for a long fluid life [157,158]. Other automotive industry applications for these products include use as friction modifiers, for example, in automatic transmission fluids [159], or possibly as detergents in engine oils [160–162] to keep insoluble combustion and oil oxidation products dispersed in the oil.

An alternative method for incorporating phosphorus into a dispersant is exemplified in Reference 160. This method involves reacting P_2S_5 with a sulfurized hydrocarbon, such as sulfurized polyisobutylene, at high temperatures to form a thiophosphorus acid (see reaction 8.23, Figure 8.22). This intermediate is then reacted with propylene oxide to form the hydroxypropyl esters of the phosphorus acid (see reaction 8.24, Figure 8.22).

Aminoethane phosphonate copolymers have also been claimed to provide dispersancy, corrosion protection, and pour point depression [163]. Among other applications mentioned in the literature for these products or their salts in lubricating oils are the extrusion, cold rolling, and cold forging of aluminum [164], offering improved rust inhibition [165] and antioxidant performance [166].

8.5.7 PHOSPHORUS-CONTAINING IONIC LIQUIDS

Ionic liquids are usually defined as salts with melting points under 100°C. They were first developed in the 1970s and the 1980s and based on imidazolium and pyridinium cations, with halide anions. Development has continued since

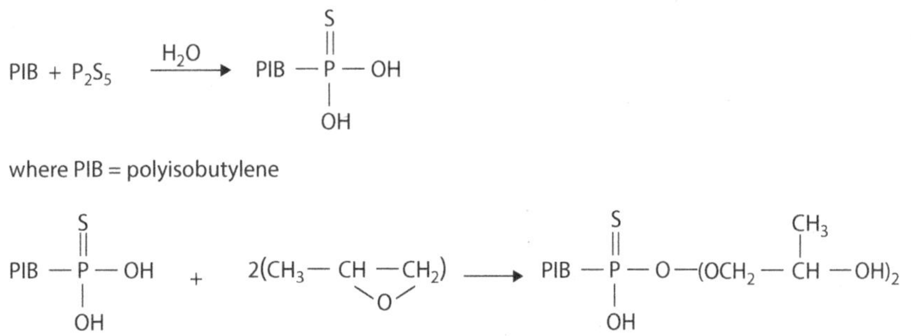

$$PIB + P_2S_5 \xrightarrow{H_2O} PIB-\overset{\overset{\displaystyle S}{\|}}{\underset{\underset{\displaystyle OH}{|}}{P}}-OH \tag{8.23}$$

where PIB = polyisobutylene

$$PIB-\overset{\overset{\displaystyle S}{\|}}{\underset{\underset{\displaystyle OH}{|}}{P}}-OH \ + \ 2(CH_3-CH-CH_2) \longrightarrow PIB-\overset{\overset{\displaystyle S}{\|}}{\underset{\underset{\displaystyle OH}{|}}{P}}-O-(OCH_2-\overset{\overset{\displaystyle CH_3}{|}}{CH}-OH)_2 \tag{8.24}$$

FIGURE 8.22 An example of the preparation of a phosphorus-based detergent. (From Colyer, C.C. and Gergel, W.C., *Chemistry and Technology of Lubricants*, VCH Publishers, New York, 1992. With permission.)

then and phosphorus-containing ionic liquids are starting to appear in the literature as AW and EP additives in hydrocarbon lubricants and in other ionic liquids. The phosphorus can be found either in the cation, for example, as an organophosphate or in the anion as a phosphonium derivative or in both charged species. So far, phosphorus-containing ionic liquids have not been commercialized.

In order to have sufficient solubility to dissolve in hydrocarbon solvents, either the phosphate or phosphonium moieties in the ionic liquid require the attachment of long hydrocarbon chains. Thus, trihexyltetradecylphosphonium bis(2-ethylhexyl)phosphate was found to be soluble up to 50% in PAO4 and when tested in a reciprocating contact between a piston ring and liner test piece exhibited reduced wear rates at 3% addition in both PAO and in fully formulated 5W-30 automobile engine oil [167]. In a more exotic application tri(2-ethylhexyl)tetradecylphosphonium bis(2-ethylhexyl)phosphate at a 1% treat in multialkylated cyclopentane, which is commonly used as a lubricant for space applications, was shown to reduce the friction coefficient by 31% and wear scar diameter by 71% when tested in a vacuum tribometer [168]. In addition, FePO$_4$ was detected in the wear scar indicating a tribochemical reaction.

By increasing the polarity of the lubricant base stock, less alkyl substitution is required, and when tributyl(methyl)phosphonium diphenylphosphate was formulated into synthetic ester base stocks, it demonstrated better AW performance than a conventional amine phosphate AW additive [169].

Phosphate ionic liquids have also been used as AW agents in other ionic liquids. Tributylmethylphosphonium dimethyl phosphate used as an additive in imidazolium borane ionic liquid exhibited reduced friction and wear in a ball on disk contact in an SRV tribometer [170]. In another alkylborane–imidazole ionic liquid, the phosphate ionic liquid tributylmethylphosphonium dimethyl phosphate showed improved friction and wear [171].

The work mentioned above demonstrates that in principle phosphate ionic liquids can act as AW and EP additives in a number of base stocks and tribological contacts although up to now there has been little published on their comparative performance with standard phosphorus-containing additives.

8.6 SUMMARY OF THE PROPOSED MECHANISM FOR ANTIWEAR AND EXTREME PRESSURE ACTIVITY OF PHOSPHORUS-BASED ADDITIVES

In attempting to produce an explanation for the activity of phosphorus-containing additives, it is not easy, as explained earlier, to compare the results of the preceding investigations because conditions vary from one investigation to another. No report evaluates all the different types of additives with the same (high) level of purity under identical test conditions. However, it is possible to draw together some of the more consistent "threads" running through the many papers. One parameter highlighted in past reports (and confirmed by recent observations) is that the presence of oxygen on the metal surface appears to be important for the activity of neutral aryl phosphates. This could perhaps be one of the major reasons why TCP is sometimes found to be inactive. The composition of the film formed on the surface is not yet completely defined, but current work points toward the formation of a self-regenerating polyphosphate layer in which amorphous carbon may be providing the lubrication benefits. The mechanism of formation of the polyphosphate layer and the role, for example, of moisture is not yet clear but appears to be a stepwise process as follows:

- The adsorption of the material onto the surface (occurring through the –P=O and –P–OH bonds in the molecule).
- Either the hydrolysis of a –P–OR bond to form –P–OH (probably arising from water on the surface but may also occur in solution) with the formation of acid phosphates/phosphites or, in the case of neutral phosphates, the cleavage of the C–O bond to release an aryl radical and a residual –P–O• radical.
- Either reaction of the –P–OH or –P–(OH)$_2$ with the metal surface to form an iron salt, possibly followed by further hydrolysis to release the remaining hydrocarbon moieties and reaction of the new –P–OH groups with the surface to form polyphosphate, or the reaction of the residual –P–O• radical with the iron surface to form a succession of Fe–O–P– bonds leading to the formation of polyphosphate.

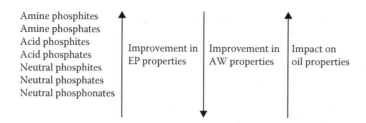

FIGURE 8.23 An approximate ranking of the effect of structure on the AW, EP, and stability properties of the base stock.

- Products that contain –P–C bonds (e.g., the phosphonates and particularly the phosphinates) are less likely to operate by a mechanism involving hydrolysis, and the stability of the P–C bond might be expected to prevent or delay the formation of the phosphorus-rich surface layer with an adverse effect on EP properties. However, the same stability could result in better friction-modification properties. The fact that phosphinates and phosphonates are active as AW/EP additives suggests that the –P=O bond is also involved in the surface adsorption process, but that either the nature of the surface film may be different or a polyphosphate film is produced as a result of the scission of a –P–C bond.
- The formation of amine salts results in an increase in activity, possibly as a result of the stability of the ion and improved adsorption on the metal surface.

The mechanism of formation of phosphide, which is reported in many instances, has not yet been clarified but might possibly involve the amorphous carbon that then acts as a reducing agent on the phosphate/phosphite layer as it forms on the surface. Carbon may also be involved in the formation of a composite wear-resistant film on the metal surface, the development of which is facilitated by the presence of phosphate esters. Further work may show whether this is a key factor in the function of phosphates as AW additives.

These conclusions lead, as a broad generalization, to the order in activity and impact on surface chemistry/stability as shown in Figure 8.23.

The preceding comments are, however, a simplification of the situation. Depending on the length of the alkyl or alkaryl chain, if the iron salts that are formed are soluble in the oil, they may desorb from the surface, leading to poor AW/EP performance. Interaction with other surface-active materials will inevitably influence the performance of AW/EP additives, whereas depletion in use due to oxidation, hydrolysis, and condensation, will also affect performance.

8.7 MARKET SIZE AND COMMERCIAL AVAILABILITY

Information on the market size for ashless phosphorus–containing AW/EP additives is limited. An approximate total market of 10,000 tpa is broken down in Table 8.16. The data exclude the use of phosphites as antioxidants in oil

TABLE 8.16
A Breakdown of the Market for Ashless Phosphorus–Containing AW/EP Additives

Product	Approx. Market Size (t)
Alkyl phosphates	100
Aryl phosphates	6000
Acid phosphates, ethoxylated alkyl and aryl phosphates, and amine salts of acid phosphates	3000
Phosphites, acid phosphites, dialkyl alkyl phosphonates, and amine salts of acidic products	1000

applications, which is separately estimated to be between 100 and 200 tpa.

The wide use of phosphorus-containing AW/EP additives is due, in addition to their good lubricity performance, to the following features of value to formulators:

- Ashless
- Low odor, color, and volatility
- Low acidity/noncorrosive (applies to the neutral esters only)
- Low toxicity
- Biodegradable (many but not all products)
- Compatible with most other types of additives (particularly the neutral esters)
- Soluble in a wide range of base stocks, both mineral oil and synthetic, and able to assist the solvency of other additives

Although the physical properties of phosphorus-containing additives are not critical, the values for the most widely used types of phosphate ester AW additives are given in Table 8.17, with TiBP as an example of an alkyl phosphate, TCP as a natural phosphate, and ISO 32 grades of both types of synthetic ester.

The major suppliers of phosphorus-containing lubricant additives are listed in Table 8.18, and their current oil industry applications are summarized in Table 8.19. Undoubtedly, the most important applications for the neutral aryl phosphates are hydraulic, turbine, and general circulatory oils, whereas almost the entire market for the ethoxylated alkyl and aryl

TABLE 8.17

Typical Physical Properties of the Most Widely Used Grades of Phosphorus-Based AW Additives

Property	Unit	TiBP	TCP	IPP/32	TBPP/32
Viscosity at 40°C	cSt	2.9	25.0	32.3	33
Viscosity at 0°C	cSt	10.0	1000	990	1500
Specific gravity	20/20°C	0.965	1.140	1.153	1.170
Pour point	°C	<−90	−28	−27	−26
Acid number	mg KOH/g	0.06	0.05	0.05	0.06
Water content	%	0.01	0.06	0.05	0.05
Phosphorus level	%	11.7	8.3	8.0	8.1
Flash point	°C	155	240	245	255

acid phosphates is to be found in metalworking. The acid phosphates, acid phosphites, and amine salts of these acidic materials are used in a mixture of metalworking, gear oils, hydraulic oils, etc., as indicated in Table 8.19.

The selection of an AW/EP additive depends on the specific requirements for the application, for example, whether both AW and EP performance is needed and what levels are required. When this has been ascertained, secondary considerations may be the level of stability required (oxidative or hydrolytic), potential interaction with other components of the formulation, and the effect on surface-active properties such as foaming. Table 8.20 attempts to identify the additives that should be given prime consideration when taking these secondary requirements into account. Products that demonstrate better AW than EP activity, and vice versa, are shown. However, the boundary between AW and EP performance is not clear-cut and much depends on the application requirements.

8.8 TOXICITY AND ECOTOXICITY

It was mentioned earlier in this chapter that concern had been expressed in the past regarding the toxicity of phosphorus-containing products, particularly TCP. Today, with increasing focus on the environmental behavior of chemicals, their ecotoxicity is also under scrutiny. As a result, detailed investigations into both the toxicity and the ecotoxicity have been carried out on alkyl and aryl phosphate esters. The results are summarized in recent publications [59,172], and most are available in the safety data sheets associated with different product types. The data demonstrate a relatively low (but variable) order of toxicity and ecotoxicity. No significant risks in handling are anticipated, provided the manufacturer's guidance, which is essentially the same as for mineral oils, is followed.

The concerns over TCP arose as a result of the o-cresol content in the feedstock as tri-orthocresyl phosphate (TOCP) was found to be a significant neurotoxin. Initially, the level of o-cresol in the feedstock was high (up to ~25%) and significant amounts of TOCP were present in the finished product. Although the initial focus was on TOCP, it was later acknowledged that any isomer containing the o-cresyl moiety was neurotoxic (e.g., mono-o-cresyldiphenyl phosphate

was said to be 10 times more neurotoxic than TOCP [173]). For these reasons, the o-cresol content of the feedstock used in the manufacture of TCP has been progressively reduced over time. In recent years, production has moved to the use of 99% minimum m- and p-cresol. Levels of o-cresol in the feedstock are now frequently <0.05%, and the TOCP content can be as low as parts per billion. Mackerer et al. [174] estimate that the toxicity of the TCP now available commercially is 400 times less than that of material available in the 1940s and 1950s, and a recent evaluation of the organophosphorus-induced delayed neurotoxicity (OPIDN) of a commercial aviation gas turbine oil containing TCP was negative [175]. However, in view of past concerns, the use of TCP is now largely restricted to aviation gas turbine oils. Most general industrial applications that require an aryl phosphate AW additive now use the isopropylphenyl- or, to a lesser extent, the tertiarybutylphenyl variants. In standard tests, neither of these types displays OPIDN from acute oral ingestion. There are, however, some differences in the toxicity and the ecotoxicity behavior between the different aryl phosphates. For example, the reproductive toxicity of the synthetic aryl phosphates, together with TXP, was recently studied in rats (according to Organisation for Economic Co-operation and Development [OECD] method 422). Both the IPPP and the TXP showed adverse effects at moderate to high dose levels, but these were reversible when exposure ceased. The TBPP (produced according to reaction 8.2) did not display any adverse effects.

The adverse reproductive toxicity findings for IPPP and TXP have resulted in REACH assigning the former Reproductive Toxicity-Category 2 with the risk phrase H361 and H373 and the latter being assigned Reproductive Toxicity-Category 1B with H373 and H360F risk phrases. Furthermore, TXP has now been placed on the REACH list of candidates of Substances of Very High Concern. This may result in TXP being placed on REACH Annex XIV. Once a substance has been placed on Annex XIV, the only way it will be possible to use substances will be to obtain an authorization from the EC and companies that register or use the substance will have to apply for authorization for specific uses. When a substance is placed on Annex XIV, a "sunset date" will be set after which its use will be prohibited unless an

TABLE 8.18
Principal Suppliers of Ashless Phosphorus-Containing Lubricating Oil Additives

Producer	Neutral Phosphates		Alkyl/Aryl Acid Phosphates		Amine Phosphates	Neutral Phosphites		Acid Phosphites	Alkyl Phosphonates
	Alkyl	Aryl	Non-Ethoxylated	Ethoxylated		Alkyl	Aryl		
Asahi Denka						×	×		
Chemtura	×	×				×	×	×	
BASF					×	×	×		
Croda	×		×	×					
Daihachi	×	×							
Dover Chemicals						×	×		
Johoku Chemicals			×			×	×	×	×
Krishna						×	×		
Lanxess	×	×							×
Libra Chemicals	×		×	×					
Rhein Chemie	×	×			×				
Rhodia	×	×	×	×	×	×	×	×	×
Sumitomo						×	×		
ICL-Industrial Products	×	×							
United Phosphorus			×			×			
Vanderbilt						×			×

TABLE 8.19

Principal Applications for Ashless Phosphorus–Containing AW/EP Additives

Application	Triaryl Phosphates	Trialkyl Phosphates	Amine Phosphates	Acid Phosphates	Alkyl/Aryl Phosphites
Automotive					
ATF				✓	✓
Gear oil			✓		
Power steering	✓				
Shock absorber	✓				
Electric motor	✓				
Industrial					
Hydraulic oil	✓		✓		
Gear oil			✓	✓	✓
Turbine oil	✓	✓			
Compressor oil	✓				
Gas oil	✓		✓		
Universal tractor oil	✓				
Metalworking fluids	✓	✓	✓	✓	✓
Grease	✓		✓	✓	
Way oil	✓				
Circulating oil	✓				
Vegetable oil		✓			
Aircraft					
Piston engine	✓				
Turbine engine	✓				
Grease	✓		✓	✓	

TABLE 8.20

The Selection of AW and EP Additives

Required Characteristic	Good AW Performance	Good EP Performance	Combination of AW and EP Performance
Non-phenolic additive	Neutral alkyl phosphates; dialkyl alkyl phosphonates	Acid alkyl phosphates; acid alkyl phosphonates and their salts; neutral and acid alkyl phosphites	Mixtures of neutral and acid phosphates, etc.
Good hydrolytic stability	TXP, dialkyl alkyl phosphonates	Acid alkyl phosphonates and their salts	Mixtures of TXP and acid alkyl phosphonates
Good oxidation stability	Neutral tertbutylphenyl phosphates	Hindered aryl phosphites	Mixtures of neutral TBPP phosphates and aryl phosphites
Low foaming/air release	Neutral phosphates, dialkyl alkyl phosphonates	Neutral phosphites	Mixtures of neutral phosphates and phosphites
Good toxicity performance	Neutral tertbutylphenyl-phosphates	Acid alkyl phosphates	Mixtures of neutral and acid phosphates, etc.
Good ecotoxicity performance	Neutral isopropylphenyl-phosphates	—	—
Multifunctionality, for example, rust inhibition, antioxidant	—	Neutral and acid phosphites, acid phosphates	Mixtures of neutral and acid phosphites and phosphates

authorization has been granted for that use. In anticipation of these changes some users have changed from IPPP and TXP to equivalent products based on TBPP, which do not have any reproductive toxicity issues and have no uncertainty regarding their continued supply.

Differences are also seen in ecotoxicity behavior. Owing to the high TPP content in the lower-viscosity grades of the synthetic phosphates (particularly ISO VG 22 and 32), these products have the worst ecotoxicity behavior. The TBPPs normally have a higher TPP content than the corresponding grade of IPPP and therefore, of the synthetic phosphates, possess relatively worse ecotoxicity. By comparison, the IPPPs generally show satisfactory behavior in these tests. One ISO VG 46 IPPP-based AW/EP additive has, for example, been approved by the German Environment Agency (Umweltbundesamt) for use in rapidly biodegradable hydraulic fluids, products that are eligible for the "Blue Angel" environmental award [176].

Another difference between the product types is displayed in biodegradability tests. The tests were carried out according to OECD method 301F (Manometric Respirometry). In this test, biodegradation is measured as the net oxygen uptake over that occurring in blank tests containing only inoculated medium. The extent of biodegradation is calculated from the mass of test material added to the test vessels and its theoretical oxygen demand for complete biodegradation. The test was carried out in triplicate on the ISO VG 46 grades of different types of aryl phosphates manufactured according to reaction 8.2. The results are summarized in Table 8.21.

The results are initially in the order of their hydrolytic stability, but it is interesting that TXP, after a slow start, eventually reaches the same level as the synthetic fluids and might have exceeded them had the test been extended.

In view of these data, the TBPP would be regarded as readily biodegradable (Pw1), whereas the TXP and IPPP would be classified as inherently biodegradable (Pw2).

Despite the relatively benign ecotoxicity of the *higher* viscosity grades of aryl phosphates, all these products are classified as marine pollutants because of the UN Marine Pollutant Classification. However, because they are used at low concentrations, they are unlikely to contribute significantly to the finished product's ecotoxicity.

The toxicity of other phosphorus-containing compounds is less well documented. Drake and Calamari [177] indicate that dialkyl alkyl phosphonates generally have a low level of acute toxicity, which decreases with increasing chain length, apparently a general observation for these classes of compounds. As with alkyl phosphates, certain short-chain products can be skin irritants. No clues were found to their environmental behavior, but in view of the absence of phenolics and improved hydrolytic stability, it might be surmised that fish toxicity could be good but biodegradability would be inferior to that of the phosphates. Neutral phosphites, particularly the alkyl phosphites, would be expected to have good toxicity and biodegradability behavior, but their ease of hydrolysis, which is the factor assisting the biodegradation, would probably result in poor aquatic toxicity. The future of the nonylphenyl phosphites is uncertain; the U.S. National Toxicology Program currently lists nonylphenol as an estrogen mimic and also as a thyroid disruptor. The acid phosphates, acid phosphites, and their salts, particularly amine salts, are likely to be classified as irritants and, due to their ease of hydrolysis, may again be toxic to aquatic organisms. In all cases, it is essential that reference be made to the health and safety information provided by the manufacturer.

8.9 FUTURE FOR ASHLESS PHOSPHORUS–BASED LUBRICATING OIL ADDITIVES

Although ashless phosphorus–containing additives are used in many industrial applications, there are certain market segments where they have not been acceptable. These are principally in automotive engine oils where the use of ZDDP dominates due to a combination of price and multifunctionality, and in gear oils where sulfur continues to be the preferred source of EP activity. However, the use of chlorine as an EP additive, particularly in metalworking applications, is in decline for environmental reasons and is expected to be slowly substituted by P/S combinations. The use of sulfur alone in applications requiring high EP performance may also move to P/S mixtures to reduce the total sulfur level and the ability to more readily "tailor" the balance of AW and EP performance to the application. The potential in other market segments including those traditionally using ZDDP is discussed in greater detail in the following.

8.9.1 LUBRICATING OIL FORMULATIONS (GENERAL)

The current trend toward the use of API Group II and III mineral oil base stocks for general industrial applications, with improved antioxidant response but inferior lubricity as a result of the removal of aromatics and sulfur compounds, could encourage the wider use of phosphorus. The lack of competition for the surface, which has previously been shown for TCP in stocks containing naphthenics and aromatics, should also result in the increased activity of phosphorus compounds. Their use may also be beneficial due to their ability to aid the dissolution of additives that might otherwise have limited solubility.

In Europe, legislation (Directive 2000/769/EC) implemented in 2004 requires a substantial reduction in sulfur dioxide (SO_2) emissions from the combustion of waste materials including waste oil. This may result in a move to lower

TABLE 8.21
OECD 301F Biodegradability Test Data on Different Types of Aryl Phosphates

Product (ISO VG 46 Base Stocks)	% Biodegradability After		
	10 Days	28 Days	68 Days
TXP	5	29	70
IPPP	18	47	65
TBPP	25	62	72

sulfur levels in lubricating oils (including metalworking oils) and a possible replacement by phosphorus to restore the level of AW/EP performance.

8.9.2 Hydraulic Oils

In recent years, there has been a move toward the use of ashless hydraulic oils. This is mainly for two reasons. First, as a result of the sensitivity of ZDDPs toward moisture and the resulting deposition of zinc oxide/sulfide. This deposit can adversely affect the filterability of the oil and reduce oxidation stability. Second, there is increasing concern regarding the environmental behavior of heavy metals. Regulatory controls, however, are likely to extend further to cover metals such as zinc, as in the Great Lakes Initiative between the United States and Canada. As the zinc cannot be easily removed from waste at the effluent plant, there has been a focus on the reduction in use levels.

Concern has also been expressed in certain countries regarding the smell of sulfur arising from the degradation of the ZDDP when the hydraulic oil is used, for example, in elevators (Dixon, R., Shell Global Solutions, Private Communication, November 2007).

8.9.3 Automotive Engine Oils

Vehicle emissions legislation (e.g., in the United States, Europe, and Japan) now exists to control and substantially reduce the levels of particulates, hydrocarbons, carbon monoxide, and oxides of nitrogen in the engine exhaust. The engine manufacturers have met these requirements by a variety of design changes but principally by the introduction of catalytic converters to oxidize the hydrocarbon and carbon monoxide components to carbon dioxide and water and to reduce the nitric oxide content to nitrogen. This development has been very successful in reducing emissions. When they operate at their normal operating temperature and optimum level of efficiency, they are almost 100% efficient and most of the remaining emissions occur in the time before the catalyst reaches "light-off" temperature. Many studies into reducing this period to achieve yet lower emissions have been conducted. Although much success has been achieved, further progress may be hindered by the formation of a deactivating film on the catalyst surface by the phosphorus from the ZDDP antioxidant and AW/EP additive. As a consequence, there is pressure to reduce the phosphorus content of engine oils to minimize catalyst fouling. Currently, oil specifications such as ILSAC GF-5 and ACEA Cx limit the phosphorus content of both diesel and gasoline engine oils to 0.05%–0.09% with the actual level being linked to the amount of catalyst used and the expected service interval. Further reductions below 0.05% are being considered, but there is a concern that such a low level could adversely affect the durability of certain engine parts, for example, the valve train and timing chain, as reducing the ZDDP content also reduces the wear protection. However, a study [178] suggested that the behavior of phosphorus compounds in

wear and catalyst tests varies according to the way in which phosphorus is incorporated into the molecule. Further work reports that it is possible to achieve improvements in catalyst protection (and fuel economy) by reducing the ZDDP content and then adding a metal-free phosphorus-containing AW additive [179].

While conventional alkyl and aryl phosphates are unlikely to displace ZDDP from engine oil formulations in view of their cost and effect on the catalyst, a recent study on the use of ionic liquids containing the phosphonium ion [180] has suggested that some of these compounds are much less poisonous to the catalyst while providing improved AW behavior and may enable the use of lower-viscosity oils in the search for more fuel-efficient engine lubricants.

8.10 CONCLUSIONS

Ashless phosphorus–containing additives are available in a wide range of structures and performance. Although most are used as AW and EP additives for industrial oils, they can also function as antioxidants, rust inhibitors, metal passivators, and detergents. In some cases, the multifunctionality can be found within the same molecule. Their advantageous physical properties—for example, low color and odor and good solubility for other additives—make them attractive components for additive packages. However, although the future looks promising in industrial oil applications in view of current pressure on sulfur and chlorine (mainly as a result of environmental concerns), it seems unlikely that conventional ashless phosphorus–containing additives will find any use in automotive engine oil due to cost and the continued downward pressure on phosphorus levels. Instead, there may be potential in this market for ionic liquids containing the phosphonium ion, and developments in this field will be watched with interest.

8.A APPENDIX: EARLY PATENT LITERATURE ON PHOSPHORUS-CONTAINING COMPOUNDS

8.A.1 Neutral Phosphates

U.S. Patent 2,723,237, Texas Oil Co.

8.A.1.1 Neutral Phosphites

As AW/EP Additives.
British Patent 1,052,751, British Petroleum (chlorethyl phosphite and a chlorparaffin).
British Patent 1,164,565, Mobil Oil Corp. (alkyl or alkenyl phosphite and a fatty acid ester).
British Patent 1,224,060, Esso Research and Engineering Co.
U.S. Patent 2,325,076, Atlantic Refining Co.
U.S. Patent 2,758,091, Shell Development Co. (haloalkyl or haloalkarylphosphites).
U.S. Patent 3,318,810, Gulf Research & Development Co. (phosphites and molybdenum compounds).

As Antioxidants
U.S. Patent 2,326,140, Atlantic Refining Co.
U.S. Patent 2,796,400, C.C. Wakefield & Co.

Acid Phosphates/Phosphites

British Patent 1,105,965, British Petroleum Co Ltd. (acid hydrocarbyl phosphite and phosphates or thiophosphates or phosphoramidates).

British Patent 1,153,161, Nippon Oil Co.

French Patent 797,449, E.I. du Pont de Nemours.

U.S. Patent 2,005,619, E. I. du Pont de Nemours.

U.S. Patent 2,642,722, Tide Water Oil Co.

Phosphonates

British Patent 823,008, Esso Research and Engineering Co. (dicarboxylic acid and either a haloalkane phosphonate, a haloalkyl phosphate or phosphite and optionally a neutral alkyl or aryl phosphate).

British Patent 884,697, Shell Research Ltd. (dialkenyl phosphonates).

British Patent 899,101, British Petroleum Co. (amino phosphonates).

British Patent 993,741, Rohm and Haas Co. (aminoalkane phosphonates).

British Patent 1,083,313, British Petroleum Co. (amino phosphonates).

British Patent 1,247,541, Mobil Oil Corp. (Dialkyl-n-alkylphosphonate or alkylammonium salts of dialkylphosphonates).

U.S. Patent 2,996,452, US Sec of Army (di-(2-ethylhexyl) lauroxyethyl phosphonate).

U.S. Patent 3,329,742, Mobil Oil Corp. (diaryl phosphonates).

U.S. Patent 3,600,470, Swift & Co. (hydroxy or alkoxy phosphonates and their amine salts).

U.S. Patent 3,696,036, Mobil Oil Corp. (tetraoctyl-(dimethylamino) methylene diphosphonate).

U.S. Patent 3,702,824, Texaco Inc. (hydroxyalkylalkane phosphonate).

Alkyl- and Arylpolyethyleneoxy-Phosphorus Compounds

U.S. Patent 2,372,244, Standard Oil Dev. Co.

Amine Salts

British Patent 705,308, Bataafsche Petroleum Maatschappij (substituted monobasic phosphonic acid and amine salts thereof).

British Patent 978,354, Shell International Research (alkali metal-amine salt of a halohydrocarbyl phosphonic acid).

British Patent 1,002,718, Shell International Research (alkylamine salt of diaryl acid phosphate).

British Patent 1,199,015, British Petroleum Co. Ltd. (quaternary ammonium salts of dialkyl phosphates).

British Patent 1,230,045, Esso Research and Engineering Co. (quaternary ammonium salts of alkyl phosphonic and phosphonic acids).

British Patent 1,266,214, Esso Research and Engineering Co. (neutral phosphate and a neutral alkylamine hydrocarbyl phosphate).

British Patent 1,302,894, Castrol Ltd. (tertiary amine phosphonates).

British Patent 1,331,647, Esso Research and Engineering Co. (quaternary ammonium phosphonates).

U.S. Patent 1,936,533, E. I. du Pont de Nemours (triethanolamine salts).

U.S. Patent 3,553,131, Mobil Oil Corp. (tertiary amine phosphonate salts).

U.S. Patent 3,668,237, Universal Oil Products Co. (tertiary amine salts of polycarboxylic acid esters of bis(hydroxyalkyl)-phosphinic acid).

Physical Mixtures of Phosphorus and Sulfur and Chlorine Compounds

British Patent 706,566, Bataafsche Petroleum Maatschappij (a phosphorus compound, e.g., a trialkyl phosphate, a glycidyl either and a disulfide).

British Patent 797,166, Esso Research and Engineering Co. (TCP and a metal soap of a sulfonic acid).

British Patent 841,788, C.C. Wakefield & Co. Ltd. (chlorinated hydrocarbon, a disulfide, and a dialkyl phosphite).

British Patent 967,760, The Distillers Co. Ltd. (disulfides, chlorinated wax, and a haloalkyl ester of an oxy-acid of phosphorus).

British Patent 872,899, Esso Research and Engineering Co. (trialkyl phosphates and chlorinated benzene).

British Patent 1,222,320, Mobil Oil Corp. (diorganophosphonate and a sulfurized hydrocarbon or sulfurized fat).

British Patent 1,287,647, Stauffer Chemical Co. (phosphonates or halogenated alkylphosphates, sulfurised oleic acid, and sebacic acid).

British Patent 1,133,692, Shell International (TCP and triphenylphosphorothionate).

British Patent 1,162,443, Mobil Oil Corp. (neutral or acid, alkyl or alkenyl phosphite, and a sulfurized polyisobutylene, triisobutylene, or a sulfurized dipentene).

U.S. Patent 2,494,332, Standard Oil Dev. Co. (thiophosphates and TCP).

U.S. Patent 2,498,628, Standard Oil Dev. Co. (sulfurized/phosphorized fatty material and TCP or tricresyl phosphite).

U.S. Patent 3,583,915, Mobil Oil Corp. (di(organo)phosphonate, and an organic sulfur compound selected from sulfurized oils and fats, a sulfurized monoolefin or an alkyl polysulfide).

Miscellaneous Phosphorus Compounds

British Patent 1,035,984, Shell Research Ltd. (diaryl chloralkyl phosphate or thiophosphate).

British Patent 1,193,631, Albright & Wilson Ltd. (hydroxyalkyl disphosphonic acid/alkylene oxide reaction products).

British Patent 1,252,790, Shell International Research (pyrophosphonic and pyrophosphinic acids and their amine salts).

U.S. Patent 3,243,370, Monsanto Co. (phosphinylhydrocarbyloxy phosphorus esters)

U.S. Patent 3,318,811, Shell Oil Co. (diacid diphosphate ester).

U.S. Patent 3,640,857, Dow Chemical Co. (tetrahaloethyl phosphates).

8.B APPENDIX: ADDITIONAL LITERATURE AND PATENT REFERENCES ON THE MECHANISM AND PERFORMANCE OF PHOSPHORUS-CONTAINING ADDITIVES

8.B.1 NEUTRAL PHOSPHATES

European Patent Appl. WO2012159828, Evonik Rohmax Additives GmbH, 2012.

Garaud, Y., M.D. Tran. Photoelectron spectroscopy investigation of tricresyl phosphate anti-wear action. *Analusis* 9(5): 231–235, 1981.

Ghose, H.M., J. Ferrante, F.C. Honecy. The effect of tricresyl phosphate as an additive on the wear of iron. NASA Technical Memo, NASA-TM-100103, E-2883, NASI. 15: 100103.

Han, D.H., M. Masuko. Comparison of antiwear additive response among several base oils of different polarities. *Tribol Trans* 42(4): 902–906, 1999.

Han, D.H., M. Masuko. Elucidation of the antiwear performance of several organic phosphates used with different polyol esters base oils from the aspect of interaction between the additive and the base oil. *Tribol Trans* 41(4): 600–604, 1998.

Kawamura, M., K. Fujito. Organic sulfur and phosphorus compounds as extreme pressure additives. *Wear* 72(1): 45–53, 1981.

Koch, B., E. Jantzen, V. Buck. Properties and mechanism of action of organism phosphoric esters as antiwear additives in aviation. *Proceedings of Fifth International Tribology Colloqium*, Esslingen, Germany, Vol. 1, 3/11/1–3/11/12, 1986.

Morimoto, T. Effect of phosphate on the wear of silicon nitride sliding against bearing steel. *Wear* 169(2): 127–134, 1993.

Perez, J.M. et al. Characterization of tricresyl phosphate lubricating films. *Tribol Trans* 33(1): 131–139, 1990.

Ren, D., A.J. Gelman. Reaction mechanisms in organophosphate vapor-phase lubrication of metal surfaces. *Tribol Int* 34(5): 353–365, 2001.

Riga, A., J. Cahoon, W.R. Pistillo. Organophosphorus chemistry structure and performance relationships in FZG gear tests. *Tribol Lett* 9(3,4): 219–225, 2001.

Riga, A., W.R. Pistillo. Surface and solution properties of organophosphorus chemicals and performance relationships in wear tests. *Proceedings of 27th NATAS Annual Conference on Thermal Analysis and Applications*, Savanna, GA, pp. 708–713, 1999.

U.S. Patent 8,822,712, US Secretary of Agriculture, 2014.

Weber, K., E. Eberhardt, G. Keil. Influence of the chemical structure of phosphoric EP-additives on its effectiveness. *Schmierungstechnik* 3(12): 372–377, 1972.

Wiegand, H., E. Broszeit. Mechanism of additive action. Model investigations with tricresyl phosphate. *Wear* 21(2): 289–302, 1972.

Yamamoto, Y., F. Hirano. Effect of different phosphate esters on frictional characteristics. *Tribol Int* 13(4): 165–169, 1980.

Yamamoto, Y., F. Hirano. The effect of the addition of phosphate esters to paraffinic base oils on their lubricating performance under sliding conditions. *Wear* 78(3): 285–296, 1982.

Yanshang, M. et al. The effect of oxy-nitrided steel surface on improving the lubricating performance of tricresyl phosphate. *Wear* 210(1–2): 287–290, 1997.

Neutral Phosphites

Barabanova, G.V. et al. Effect of phosphoric, thiophosphoric and phosphorus acid neutral ester-type additives on the lubricating capacity of C5–C9 synthetic fatty acid pentaerythritol ester. *Pererabotke Nefti* 17: 57–61, 1976.

Orudzheva, I.M. et al. Synthesis and study of some aryl phosphites. *3rd Tekh Konf Neftekhim* 3: 411–415, 1974.

Sanin, P.I. et al. *Tr Inst Nefti Akad Nauk SSSR*, 14: 98, 1960.

Sanin, P.I., A.V. Ul'yanova. Prisadki Maslam Toplivam, Trudy Naucha. *Tekhn Soveshch* 189, 1960.

Wan, Y., Q. Xue. Effect of phosphorus-containing additives on the wear of aluminum in the lubricated aluminum-on-steel contact. *Tribol Lett* 2(1): 37–45, 1996.

U.S. Patent 3,115,463/4, Ethyl Corp., NY, 1963.

U.S. Patent 3,652,411, Mobil Corp., NY, 1969.

U.S. Patent 4,374,219, Ciba-Geigy Corp., NY, 1981.

Alkoxylated Phosphates

Jia, X., X. Zhang. Antiwear property of water-soluble compound phosphate ester. *Runhua Yu Mifeng* 4(25–25): 67, 1999.

Wang, R. Study on water-soluble EP additives. *Runhua Yu Mifeng* 3(17–18): 51, 1994.

Zhang, X., X. Jia. Water soluble phosphate esters. *Hebei Ligong Xueyuan Xuebao* 21(2): 58–61, 1999.

Amine Salts

Forbes, E.S. et al. The effect of chemical structure on the load carrying properties of amine phosphates. *Wear* 18: 269, 1971.

Shi, H. Development trend of phosphorus-nitrogen-type extreme-pressure antiwear agents. *Gaoqiao Shi* 12(3): 37–41, 1997.

U.S. Patent Appl., 20080182770, Lubrizol Corp., 2008.

Dialkyl Alkyl Phosphonates

Cann, P.M.E., G.J. Johnston, H.A. Spikes. The formation of thick films by phosphorus-based antiwear additives. *Proceedings of the Institution of Mechanical Engineers Conference*, Savanna, GA, pp. 543–554, 1987.

Dickert, J.J., C.N. Rowe. Novel lubrication properties of gold O,O-dialkylphosphorodithioates and metal organophosphonates. *ASLE Trans* 20(2): 143–151, 1977.

Gadirov, A.A., A.K. Kyazin-Zade. Diphenyl esters of alpha-aminophosphonic acids as antioxidant and antiwear additives. *Khim Tekhnol Topl Masel* 3: 23–24, 1990.

Lashkhi, V.L. et al. Antiwear action of organophosphorus and chloroorganophosphorus compounds in lubricating oils. *Khim Tekhnol Topl Masel* 2: 47–50, 1975.

Lashkhi, V.L. et al. IR spectroscopic study of phosphonic acid ester-antiwear additives for oil. *Khim Tekhnol Topl Masel* 5: 59–61, 1977.

Lozovoi, Y., P.I. Sanin. Mechanism of action of phosphorus acid ester-type extreme pressure additives. *IZV Khim* 19(1): 49–57, 1986.

Tang, J., Q. Cang, X. Zong. Preparation of phosphorus friction modifiers used in automatic transmissions fluids. *Shiyou Lianzhi Yu Huagong* 31(3): 17–20, 2000.

Wan, G.T.Y. The performance of one organic phosphonate additive in rolling contact fatigue. *Wear* 155(2): 381–387.

Xiong, R.-G., W. Hong, J.-L. Zuo, C.-M. Liu, X.-Z. You, J.-X. Dong, Z. Pei. Antiwear and extreme pressure action of a copper (II) complex with alkyl phosphonic monoalkyl ester. *J Tribol* 118(3): 676–680, 1996.

Mixtures of Phosphorus and Sulfur Compounds

Kawamura, M. et al. Interaction between sulfur type and phosphorus type EP additives and its effect on lubricating performance. *Junkatsu* 30(9): 665–670.

Kubo, K., Y. Shimakawa, M. Kibukawa. Study on the load-carrying mechanism of sulfur-phosphorus type lubricants. *Proc JSLE Int Tribol Conf* 3: 661–666, 1985.

European Patent Appl. WO 2002053687A2, Shell Int Res.

Qiao, Y., B. Xu, S. Ma, X. Fang, Q. Xue. Study in synergistic effect mechanism of some extreme pressure and antiwear additives in lubricating oil. *Shiyou Xuebao, Shiyou Jiagong* 13(3): 33–39, 1997.

Qiao, Y., X. Fang, H. Dang. Synergistic effect mechanism of the combination system of two typical additives containing sulfur and phosphorus in lubricating oil. *Mocaxue Xuebao* 15(1): 29–38, 1995.

Xia, H. A study of the antiwear behaviour of S-P type gear oil additives in four-ball and Falex machines. *Wear* 112(3–4): 335–361, 1986.

General References

Hartley, R.J., A.G. Papay. Function of additives. Antiwear and extreme pressure additives. *Toraiborojisuto* 40(4): 326–331, 1995.

Palacios, J.M. The performance of some antiwear additives and interference with other additives. *Lubr Sci* 4(3): 201–209, 1992.

REFERENCES

1. Williamson, S. *Ann* 92: 316, 1854.
2. Vogeli, F. *Ann* 69: 190, 1849.
3. British Patent 446,547, The Atlantic Refining Co., 1936.
4. Beeck, O., J.W. Givens, A.E. Smith. On the mechanism of boundary lubrication—I. The action of long chain polar compounds. *Proc Roy Soc A* 177: 90–102, 1940.

5. Beeck, O., J.W. Givens, E.C. Williams. On the mechanism of boundary lubrication—II. Wear prevention by addition agents. *Proc Roy Soc A* 177: 103–118, 1940.

6. Tingle, E.D. Fundamental work on friction, lubrication and wear in Germany during the war years. *J Inst Pet* 34: 743–774, 1948.

7. West, H.L. Major developments in synthetic lubricants and additives in Germany. *J Inst Pet* 34: 774–820, 1948.

8. Boerlage, G.D., H. Blok. Four-ball top for testing the boundary lubricating properties of oils under high mean pressures. *Engineering* 144 (July): 1, 1937.

9. U.S. Patent 2,391,311, C.C. Wakefield, 1945.

10. U.S. Patent 2,391,631, E. I. du Pont de Nemours, 1945

11. U.S. Patent 2,470,405, Standard Oil Development Co., 1949.

12. U.S. Patent 2,663,691, The Texas Co., 1953.

13. U.S. Patent 2,734,868, The Texas Co., 1956.

14. U.S. Patent 2,612,515, Standard Oil Development Co., 1952.

15. British Patent 797,166, Esso Research and Engineering Co., 1958.

16. Morgan, J.P., T.C. Tullos. The jake walk blues. *Ann Intern Med* 85: 804–808, 1976.

17. Johnson, M.K. Organophosphorus esters causing delayed neurotoxic effect: Mechanism of action and structure/activity studies. *Arch Toxicol* 34: 259, 1975.

18. British Patent 683,405, Shell Refining and Marketing Co., 1952.

19. Greenshields, R.J. Oil industry finds fuel additives can help in controlling pre-ignition. *Oil Gas J* 52(8): 71–72, 1953.

20. Jeffrey, R.E. et al. Improved fuel with phosphorus additives. *Petrol Refiner* 33(8): 92–96, 1954.

21. Burnham, H.D. The role of tritolyl phosphate in gasoline for the control of ignition and combustion problems. *Am Chem Soc Petrol Div Symp* 36: 39–50, 1955.

22. U.S. Patent 3,510,281, Texaco Inc., 1970.

23. U.S. Patent 2,215,956, E. I. du Point de Nemours, 1940.

24. U.S. Patent 2,237,632, Sinclair Refining Co., 1941.

25. French Patent 681,1770, ICI, 1929.

26. U.S. Patent 1,869,312, Combustion Utilities Corp., 1932.

27. Russian Patent 47,690, R. L. Globus, S. F. Monakhov, 1936.

28. U.S. Patent 2,071,323, Dow Chemical Co., 1937.

29. British Patent 486,760, Celluloid Corp., 1938.

30. U.S. Patent 2,117,290, Dow Chemical Co., 1938.

31. Shlyakhtenko, A.I., P.P. Lebedev, R. Mandel. *Novosti Tekhniki* 20: 42, 1938.

32. Russian Patent 52,398, A. I. Shlyakhtenko, P. P. Lebedev, 1938.

33. Canadian Patent 379,529, Celluloid Corp., 1939.

34. U.S. Patent 2,358,133, Dow Chemical Co., 1944.

35. British Patent 872,899, Esso Research and Engineering Co., 1961.

36. French Patent Addn. 89,648, Establissements Kuhlmann, 1967.

37. British Patent 424,380, N. V. de Bataafsche Petroleum Maatschappij, 1935.

38. U.S. Patent 2,005,619, E. I. du Pont de Nemours, 1935.

39. French Patent 797,449, E. I. du Pont de Nemours, 1936.

40. Beiswanger, J.P.G., W. Katzenstein, F. Krupin. Phosphate ester acids as load-carrying additives and rust inhibitors for metalworking fluids. *ASLE Trans* 7(4): 398–405, 1964.

41. U.S. Patent 3,547,820, GAF Corp., 1970.

42. U.S. Patent 3,567,636, GAF Corp., 1971.

43. British Patent 1,002,718, Shell International Research Maatschappij, 1965.

44. U.S. Patent 2,381,127, Texas Co., 1945.

45. U.S. Patent 2,642,722, Tide Water Associated Oil Co., 1953.

46. British Patent 1,266,214, Esso Research and Engineering Co., 1972.

47. U.S. Patent 2,236,140, Atlantic Refining Co., 1944.

48. U.S. Patent 2,325,076, Atlantic Refining Co., 1944.

49. Dutch Patent 69,357, N. V. Bataafsche Petroleum Maatschappij, 1952.

50. U.S. Patent 2,653,161, Shell Development Co., 1953.

51. British Patent 1,247,541, Mobil Oil Corp., 1971.

52. U.S. Patent 3,600,470, Swift & Co., 1971.

53. U.S. Patent 6,242,631, Akzo Nobel, 2001.

54. Phillips, W.D., D.G. Placek, M.P. Marino. Neutral phosphate esters. In *Synthetics, Mineral Oils and Bio-Based Lubricants-Chemistry and Technology*, ed. L.R. Rudnick, 2nd edn., Boca Raton, FL: CRC Press, 2013.

55. Evans, D.P., W.C. Davies, W.J. Jones. The lower trialkyl phosphates. Part I. *J Chem Soc* 1310–1313, 1930.

56. U.S. Patent 2,723,237, Texas Co., 1955.

57. British Patent 1,165,700, Bush, Boake Allen Ltd., 1965.

58. British Patent 1,146,173, J. R. Geigy, 1966.

59. Phillips, W.D. Phosphate ester hydraulic fluids. In *Handbook of Hydraulic Fluid Technology*, 2nd edn., eds. G.E. Totten and V.J. De Negri. Boca Raton, FL: CRC Press, 2012.

60. U.S. Patent 5,779,774, K.J.L. Paciorek, S.R. Masuda, 1998.

61. Tang, J., Q. Cang, X. Zong. Preparation and evaluation of phosphorus friction modifiers used in automatic transmission fluids. *Shiyou Lianzhi Yu Huagong* 31(3): 17–20, 2000.

62. Anonymous. Product review—Lubricant additives. *Ind Lubr Tribol* 49(1): 15–30, 1997.

63. Lansdown, A.R. Extreme pressure additives. In *Chemistry and Technology of Lubricants*, eds. R.M. Mortier and S.T. Orszulik. New York: VCH Publishers Inc., 1992.

64. Anonymous. Boundary lubrication. *Lubrication* 57: 1, 1971.

65. Gunther, R.C. *Lubrication*. Philadelphia, PA: Chilton Books, 1971.

66. Mandakovic, R. Assessment of EP additives for water miscible metalworking fluids. *J Synth Lubr* 16(1): 13–26, 1999.

67. Klamann, D. *Lubricants and Related Products*. Weinheim, Germany: Verlag Chemie, 1984.

68. Anonymous. Fundamentals of wear. *Lubrication* 12(6): 61–72, 1957.

69. Anonymous. Lubrication fundamentals. *Lubrication* 59(October–December): 77–88, 1973.

70. Hardy, W.B. Note on the static friction and lubricating properties of certain chemical substances. *Philos Mag* 38: 32–49, 1919.

71. Wells, H.M., J.E. Southcombe. *Petrol World* 17: 460, 1920.

72. Bragg, W.H. The investigation of the properties of thin films by means of x-rays. *Nature* 115: 226, 1925.

73. Clark, G.L., R.R. Sterrett. X-ray diffraction studies of lubricants. *Ind Eng Chem* 28(11): 1318–1322, 1936.

74. Thorpe, R.E., R.G. Larsen. Antiseizure properties of boundary lubricants. *Ind Eng Chem* 41: 938–943, 1949.

75. Davey, W. Extreme pressure lubricants—Phosphorus compounds as additives. *Ind Eng Chem* 42(9): 1841–1847, 1950.

76. Larsen, R.G., G.L. Perry. Chemical aspects of wear and friction. In *Mechanical Wear*, ed. J.T. Burwell. Metals Park, OH: American Society for Metals, 73–94, 1950.

77. Bita, O., I. Dinca. Behaviour of phosphorus additives. *Rev Mecan Appl* 8: 441–442, 1963.

78. Godfrey, D. The lubrication mechanism of tricresyl phosphate on steel. ASLE Preprint 64-LC-1, 1964.

79. Klaus, E.E., H.E. Bieber. Effects of P impurities on the behaviour of tricresyl phosphate as an antiwear additive. ASLE Preprint 64-LC-2, 1964.

80. Barcroft, F.T., S.G. Daniel. The action of neutral organic phosphates as EP additives. *ASME J Basic Eng* 87: 761, 1965.

81. Bieber, H.E., E.E. Klaus, E.J. Tewkesbury. A study of tricresyl phosphate as an additive for boundary lubrication. ASLE Preprint 67-LC-9, 1967.

82. Sakurai, T., K. Sato. Chemical reactivity and load carrying capacity of lubricating oils containing organic phosphorus compounds. ASLE Preprint 69-LC-18, 1969.

83. Barcroft, F.T. A technique for investigating reactions between EP additives and metal surfaces at high temperature. *Wear* 3: 440–453, 1960.

84. Goldblatt, I.L., J.K. Appeldoorn. The antiwear behavior of TCP in different atmospheres and different base stocks. ASLE Preprint 69-LC-17, 1969.

85. Forbes, E.S., N.T. Upsdell, J. Battersby. Current thoughts on the mechanism of action of tricresyl phosphate as a load-carrying additive. *Proc Tribol Conv* 1: 7–13, 1972.

86. Faut, O.D., D.R. Wheeler. On the mechanism of lubrication by tricresylphosphate (TCP)—The coefficient of friction as a function of temperature for TCP on M-50 steel. *ASLE Trans* 26(3): 344–350, 1983.

87. Yamamoto, Y., F. Hirano. Scuffing resistance of phosphate esters. *Wear* 50: 343–348, 1978.

88. Furey, M.J. Surface roughness effects on metallic contact and friction. *ASLE Trans* 6: 49–59, 1963.

89. Gauthier, A., H. Montes, J.M. Georges. Boundary lubrication with tricresylphosphate (TCP). Importance of corrosive wear. *ASLE Preprint* 81-LC-6A-3, 1981.

90. Placek, D.G., S.G. Shankwalkar. Phosphate ester surface treatment for reduced wear and corrosion protection. *Wear* 173: 207–217, 1994.

91. Klaus, E.E., J.M. Perez. Comparative evaluation of several hydraulic fluids in operational equipment. SAE Paper No. 831680, Society of Automotive Engineers, 1983.

92. Klaus, E.E., J.L. Duda, K.K. Chao. A study of wear chemistry using a microsample four-ball wear test STLE. *Tribol Trans* 34(3): 426–432, 1991.

93. Yansheng, M., J. Liu, Y. Wu, Z. Gu. The synergistic effects of tricresyl phosphate oil additive with chemico-thermal treatment of steel surfaces. *Lubr Sci* 9(1): 85–95, 1996.

94. Klaus, E.E., G.S. Jeng, J.L. Duda. A study of tricresyl phosphate as a vapor-delivered lubricant. *Lubr Eng* 45(11): 717–723, 1989.

95. Cho, L., E.E. Klaus. Oxidative degradation of phosphate esters. *ASLE Trans* 24(1): 119–124, 1981.

96. Graham, E.E., A. Nesarikar, N.H. Forster. Vapor-phase lubrication of high-temperature bearings. *Lubr Eng* 49(9): 713–718, 1993.

97. Hanyaloglu, B., E.E. Graham. Vapor phase lubrication of ceramics. *Lubr Eng* 50(10): 814–820, 1994.

98. Van Treuren, K.W. et al. Investigation of vapor-phase lubrication in a gas turbine engine. *ASME J Eng Gas Turbines Power* 120(2): 257–262, 1998.

99. Morales, W., R.F. Handschuh. A preliminary study on the vapor/mist phase lubrication of a spur gearbox. *Lubr Eng* 56(9): 14–19, 2000.

100. Saba, C.S., N.H. Forster. Reactions of aromatic phosphate esters with metals and their oxides. *Tribol Lett* 12(2): 135–146, 2002.

101. U.S. Patent Appl. 20140038662, ExxonMobil Research and Engineering Co., 2014

102. European Patent 0521628, Ethyl Petroleum Additives, 1992.

103. Didziulis, S.R., R. Bauer. Volatility and performance studies of phosphate ester boundary additives with a synthetic hydrocarbon. Aerospace Report TR 95-(5935)-6, Aerospace Corp.

104. U.S. Patent Appl. 20120309656, Umehara, K. & Yamamoto, K., 2012.

105. Japanese Patent 05001837, Toyota Central Res. & Dev. Lab, 1988.

106. Japanese Patent 02018496, New Japan Chemical Co., 1990.

107. Japanese Patent 02300295, Toyota Jidosha & Yushiro Co., 1991.

108. U.S. Patent 6,204,277, PABU Services, 2001.

109. European Patent Appl., EP1618173, Great Lakes Chemical Corp., 2004.

110. U.S. Patent 5,584,201, Cleveland State University, 1986.

111. Soviet Union Patent 810,767, Berdyansk Experimental Petroleum Plant, 1981.

112. Metal Passivators, Newsletter No. 10, ADD APT AG, June 2000.

113. Canter, N. Metal deactivators: Inhibitors of metal interactions with lubricants. *Tribol Lubr Technol* 68(9): 10–22, 2012.

114. Werner, J.J., R.L. Reierson, J.-L. Joye. European Patent Application WO 00/37591, 2000.

115. Katzenstein, W. Phosphate ester acids as load-carrying additives and rust inhibitors for metalworking fluids. *Proceedings of the 11th International Tribology Colloqium*, Esslingen, Germany, pp. 1741–1754, 1998.

116. Werner, J.J., M. Dahanayake, D. Lukjantschenko. Relationship of structure to performance properties of phosphate-ester surfactants in metal working fluids. *STLE Annual Conference*, Chicago, IL, 1995.

117. U.S. Patent 1,936,533, E. I. Du Point de Nemours, 1933.

118. British Standard 3150, Corrosion-inhibited antifreeze for water-cooled engines, Type A. British Standard Institution, 1959.

119. Forbes, E.S., H.B. Silver. The effect of chemical structure on the load-carrying properties of organo-phosphorus compounds. *J Inst Pet* 56(548): 90–98, 1970.

120. Forbes, E.S., N.T. Upsdell. Phosphorus load-carrying additives: Adsorption/reaction studies of amine phosphates and their load-carrying mechanism. *First European Tribology Congress*, London, U.K., Paper C-293/73, pp. 277–298, 1973.

121. Farng, L.O., W.F. Olszewski. U.S. Patent 5,681,798, 1997.

122. Weller, D., J. Perez. A study of the effect of chemical structure on friction and wear, Part 1—Synthetic ester fluids. *Lubr Eng* 56(11): 39–44, 2000.

123. Kristen, U. Aschefreie extreme-pressure-und verschleiss-schutz-additive. In *Additive für Schmierstoffe*, ed. W.J. Bartz. Renningen, Germany: Expert Verlag, 1994.

124. U.S. Patent 2,722,517, Esso Research & Engineering Co., 1955.

125. U.S. Patent 2,763,617, Shell Development Co., 1957.

126. U.S. Patent 2,820,766, C. C. Wakefield & Co., 1958.

127. U.S. Patent 2,971,912, Castrol Ltd., 1961.

128. Sanin, P.I. et al. Effect of synthetic additives in lubricating oil on wear under friction. *Wear* 3: 200, 1960.

129. Ohmuri, T., M. Kawamura. Fundamental studies on lubricating oil additives. In *Adsorption and Reaction Mechanism of Phosphorus-Type Additive on Iron Surface*, eds. R. Toyota Chuo Kenkyusho and D. Rebyu, Vol. 28(1). Japan: Toyota, pp. 25–33, 1993.

130. Riga, A., W. Rock Pistillo. Surface and solution properties of organophosphorus chemical in wear tests. *NATAS Annual Conference of Thermal Analysis and Application*, Orlando, CA, Vol. 28, pp. 530–535, 2000.

131. French Patent 1,435,890, Albright & Wilson Ltd., 1996.

132. Ren, D., G. Zhou, A.J. Gellman. The decomposition mechanism of trimethylphosphite on Ni(III). *Surf Sci* 475(1–3): 61–72, 2001.

133. British Patent 682,441, Anglamol Ltd., 1952.

134. U.S. Patent 2,764,603, Socony Vacuum Oil Co., 1954.

135. Rasberger, M. Oxidative degradation and stabilisation of mineral oil-based lubricants. In *Chemistry and Technology of Lubricants*, eds. R.M. Mortier and S.I. Orszulik. New York: VCH Publishers, 1992.

136. European Patent 0049133, Sumitomo Chemicals, 1982.

137. Japanese Patent 63156899, Sumiko Junkatsu-Zai, 1988.

138. Japanese Patent 2888302, Tonen Corp., 1991.

139. Japanese Patent 05331476, Tonen Corp., 1994.

140. Japanese Patent 06200277, Tonen Corp., 1994.

141. U.S. Patent 4,656,302, Koppers Co. Inc., 1987.

142. European Patent Appl. 475560, Petro-Canada Inc., 1992.

143. U.S. Patent 4,652,385, Petro-Canada Inc., 1987.

144. U.S. Patent 5,124,057, Petro-Canada Inc., 1993.

145. U.S. Patent 3,329,742, Mobil Oil Corp., 1967.

146. U.S. Patent 3,351,554, Mobil Oil Corp., 1967.

147. U.S. Patent 3,321,401, British Petroleum Co., 1967.

148. U.S. Patent 3,201,348, Standard Oil Co., 1965.

149. Messina, V., D.R. Senior. South African Patent 07230, 1967.

150. U.S. Patent 3,115,463, Ethyl Corp., 1963.

151. Forbes, E.S., J. Battersby. The effect of chemical structure on the load-carrying and adsorption properties of dialkyl phosphites. *ASLE Trans* 17(4): 263–270, 1974.

152. Barber, R.I. The preparation of some phosphorus compounds and their comparison as load-carrying additives by the four-ball machine. ASLE Preprint 75-LC-2D-1, 1975.

153. Sanin, P.I. et al. Antiwear additives of the phosphonate type. *Neftekhimiya* 14(2): 317–322, 1974.

154. U.S. Patent 3,553,131, Mobil Oil Corp., 1971.

155. U.S. Patent 4,246,125, Ethyl Corp., 1981.

156. U.S. Patent 4,260,499, Ethyl Corp., 1981.

157. European Patent Appl. 510633, Sakai Chemicals, 1992.

158. Japanese Patent 05302093, Tonen Corp., 1993.

159. U.S. Patent 4,225,449, Ethyl Corp., 1980.

160. Colyer, C.C., W.C. Gergel. Detergents/dispersants. In *Chemistry and Technology of Lubricants*, eds. R.M. Mortier and S.T. Orszulik. New York: VCH Publishers, 1992.

161. Japanese Patent 62215697, Toyota Res. and Devt., 1987.

162. Czech Patent 246897, Kekenak, 1988.

163. U.S. Patent 3,268,450, Sims, Bauer and Preuss, 1966.

164. Japanese Patent 8126997, Showa Aluminium Co., 1981.

165. U.S. Patent 4,123,369, Continental Oil Co., 1978.

166. U.S. Patent 3,658,706, Ethyl Corp., 1972.

167. Qu, J. et al. Antiwear performance and mechanism of an oil-miscible ionic liquid as a lubricant additive. *Appl Mater Interfaces* 4: 997–1002, 2012.

168. Zhang, S., L. Hu, D. Qiao, D. Feng, H. Wang. Vacuum tribological performance of phosphonium-based ionic liquids as lubricants and lubricant additives of multialkylated cyclopentanes. *Tribol Int* 66: 289–295, 2013.

169. Khemchandani, B., A. Somers, P. Howlett, A.K. Jaiswal, E. Sayanna, M. Forsyth. A biocompatible ionic liquid as an antiwear additive for biodegradable lubricants. *Tribol Int* 77: 171–177, 2014.

170. Totolin, V., I. Minami, C. Gabler, N. Dorr. Halogen-free borate ionic liquids as novel lubricants for tribological applications. *Tribol Int* 67: 191–198, 2013.

171. Totolin, V., I. Minami, C. Gabler, N. Dorr. Lubrication mechanism of phosphonium phosphate ionic liquid additive in alkylborane–imidazole complexes. *Tribol Lett* 53: 421–432, 2014.

172. Goode, M.J., W.D. Phillips. Triaryl phosphate ester hydraulic fluids—A reassessment of their toxicity and environmental behaviour. SAE Paper 982004, Society of Automotive Engineers, 1998.

173. Henschler, D., H.H. Bayer. Toxicological studies of triphenyl phosphate, trixylenyl phosphate and triaryl phosphates from mixtures of homologous phenols. *Arch Exp Pathol Pharmakol* 233: 512–517, 1958.

174. Mackerer, C., M.L. Barth, A.J. Krueger. A comparison of neurotoxic effects and potential risks from oral administration or ingestion of tricresyl phosphate and jet engine oil containing tricresyl phosphate. *J Toxicol Environ Health A* 57(5): 293–328, 1999.

175. Daughtrey, W., R. Biles, B. Jortner, M. Ehrlich. Delayed neurotoxicity in chickens: 90 day study with Mobil Jet Oil 254. *The Toxicologist* 90: Abstract 1467, 2006.

176. Durad® 310M. Great Lakes Chemical Corp., Manchester, U.K., May 2002.

177. Drake, G.L., Jr., T.A. Calamari. Industrial uses of phosphonates. In *Role of Phosphonates in Living Systems*, ed. R. Hildebrand. Boca Raton, FL: CRC Press, 1983.

178. Devlin, M.T., R. Sheets, J. Loper, G. Guinther, K. Thompson, J. Guevremont, T.-C. Jao. Effect of ashless phosphorus antiwear compounds on passenger car emissions and fuel efficiency. *Additives 2007 Conference*, London, U.K., April 2007.

179. Trautwein, W.-P. AdBlue as a reducing agent for the decrease of NOx emissions from diesel engines of commercial vehicles, Research Report 616-1, DGMK, Hamburg, Germany, 2003.

180. Qu, J. Ionic liquids as next generation antiwear additives—From molecular design to engine dynamometer testing, Oak Ridge National Laboratory, April 2014, www.api.org/media/~/files/certification/engine-oil.

9 Sulfur Carriers

Thomas Rossrucker, Sandra Horstmann, and Achim Fessenbecker

CONTENTS

9.1 INTRODUCTION

In the lubricant industry, a great variety of sulfur-containing additives are known and in use today. We list only a few of the most common types:

- Sulfur carriers (sulfurized olefins, esters, and fatty oils)
- Sulfur/phosphorus derivatives (dithiophosphates, thiophosphonates, thiophosphites, etc.)
- Thiocarboxylic acid derivatives (dithiocarbamates, xanthogenates, etc.)
- Heterocyclic sulfur (mercaptobenzothiazoles, thiadiazoles, etc.)
- Sulfonates (Na-, Ca-salts of alkylbenzenesulfonic acids, nonylnaphthalenesulfonates, etc.)
- Others (sulfated fatty oils/Turkish red oils, sulfur-chlorinated fatty oils, sulfur-linked phenols and phenates)

This list gives a good impression of the versatility and importance of sulfur chemistry in lubricants. But the more versatile their chemistry is, the more versatile is the range of application of sulfur-containing lubricant additives. In this chapter, we attempt to review the major aspects of a group of additives commonly known as "sulfur carriers." This is a generic name that has been accepted in the marketplace and used to summarize a group of additives that provide extreme pressure (EP) and antiwear (AW) properties and are used in gear oils, metalworking fluids, greases, and engine oils. The vast majority of them are sulfurized fats, esters, and olefins. To distinguish them from other sulfur-containing products and avoid misunderstandings, a suitable definition of sulfur carriers is in order:

Sulfur carriers are a class of organic compounds that contain sulfur in its oxidation state, 0 or −1, where the sulfur atom is bound either to a hydrocarbon or to another sulfur atom

- That does not contain other hetero atoms except oxygen
- Produced by adding sulfur to all kinds of unsaturated, double-bond-containing compounds such as olefins, natural esters, and acrylates or by substitution reaction with reactive organic halides and alike

Lubricant additives fitting this definition are the main focus of this chapter.

Due to the overwhelming versatility of sulfur chemistry, other sulfur-containing product groups cannot be discussed in depth but are mentioned in the context where appropriate.

Despite the fact that this group of additives has been used in the lubricant industry for more than eight decades, sulfur carriers are not at all an endangered species. In fact, we see their increasing usage even today. This is partly due to continuous R&D work done in this area that brings about innovation and product improvement. Also, many chemical aspects and applications are waiting to be discovered. Furthermore, sulfur carriers are essential additives for the solution of upcoming lubricant market requirements such as "chlorinated paraffin substitution," "heavy metal replacement," and "health, safety, and environmental issues." Therefore, we expect to see substantial future growth of light-colored and low-odor sulfur carriers.

9.2 HISTORY

As we look back more than 100 years of sulfurized compounds, the authors had to rely on literature sources from before the 1950s. During the literature studies, it turned out that one of the most fruitful sources from before 1950 was the review articles of Helen Sellei [1,2] published in 1949. Much of what followed was based on her content, but we have tried to reinterpret the information with today's background knowledge.

9.2.1 First Synthesis and Application (1890–1918)

Sulfurized fatty oils have been commercially produced for more than 100 years. Long before they were used as additives in lubricants, they had become important additives for the rubber industry. The addition of 4%–8% sulfur to an unsaturated natural oil, such as rapeseed oil, at high temperatures (120°C–180°C) gives a flexible, gummy polymer called "factis." The sulfur undergoes an additional reaction due to the double bonds of the natural oil and builds up a three-dimensional structure of sulfur bridges between the triglyceride molecules. This is comparable to the vulcanization process of latex, which results in "rubber."

In the late nineteenth century, rubber was an expensive natural raw material, and with the rapid industrialization in general and the growing automobile industry in particular, rubber tires were needed in increasing amounts. It soon turned out that factis also provided special, positive properties to rubber goods during the vulcanization process. This was the starting point of smaller chemical factories producing additives for the rubber industry. In 1889, Carl Benz submitted the patent for the world's first automobile in Mannheim, Germany. In the same year and city, Rhein Chemie Rheinau GmbH was founded and started to produce sulfurized natural oils. Germany had seen a special national aspect to the industrial history of sulfurized fats and rubber before 1914. Because Germany had very few colonies, all rubber had to be imported. During the national tensions in the first decade of the twentieth century and subsequent trade boycotts, the search for alternatives had been strongly pushed, leading to the development of synthetic rubber ("Buna"). Subsequently, because it was cheap and based on locally available raw materials, factis had found increasing use as a rubber substitute and a rubber diluent.

9.2.2 First Application in Metalworking Oils (1920–1930)

In the very early days of modern lubrication, it had become known that sulfur is an important element to improve frictional properties and prevent seizure under high loads. Free sulfur as well as sulfur-containing heterocyclic molecules is known to be part of natural crude oil (thiophenes, thioethers, etc.). In early refining technologies, they were not removed effectively, especially from the higher-viscosity oils that were typically used for gear oils that had up to 3%–4% of sulfur. This natural sulfur contributed to mild EP performance

(antiwelding). After the positive effects of sulfur on lubricant oil formulations were recognized, the next step was to physically dissolve sulfur flower at elevated temperatures into the lubricant oil. This sulfur, however, is very reactive and corrosive against copper and its alloys. Also, sulfur flower has a limited solubility in mineral oil, which limits its maximum dosage and final EP performance achievable.

Sulfurized esters were first used in metalworking. For heavy-duty operations with a high degree of boundary lubrication conditions, it was realized that the addition of oil-soluble sulfur compounds had a tremendous effect on the performance. The first milestone literature that reports this effect, on cutting oils, was published in 1918 by the E.F. Houghton Corporation [3]. It is claimed that a mixture of lard oil, mineral oil, and wool fat treated with sulfur flower at elevated temperatures results in a sulfurized product that increases the performance of cutting oils enormously. In particular, the tool life is extended and smoking of the coolant is reduced greatly due to friction and temperature reduction. These observations are still valid today and may be considered the starting point of the application of sulfur carriers as additives for lubricants.

In comparison to the solid, rubberlike material factis, which has been commercialized for several decades, Houghton Corp's breakthrough was to produce a liquid fatty material that was soluble in mineral base oil at any ratio. It overcomes the solubility limits of sulfur flower and allows the adjustment of the EP performance level according to treat rate. They achieved this simply by using nonreactive mineral oil and wool wax as chain-breaking agents and diluent to control the polymerization reaction of lard oil to keep it liquid. From thereon, the use of sulfurized oils has become quite common in metalworking.

9.2.3 Sulfurized Compounds for Gear Oils and Other Lubricants (1930–1945)

Some years later, the idea of improving load-carrying capacity under high-pressure and high-temperature conditions had been picked up by automotive lubricant researchers and applied to oils for the newly constructed hypoid gear boxes. With the advent of hypoid gears in automotive applications in the 1920s and the 1930s, wear and seizure under high-load conditions became a major technical problem that lubricant companies needed to solve. Most of development work had been done within these lubricant companies, and the new technology had not been published in detail. However, the number of patents on sulfur compounds for lubricants developed very rapidly throughout the 1930s and the 1940s.

- 1936 First patent review on EP lubricants [4]
- 1940 Patent review 1938–1939 [5]
- 1941 General publication on lubricating additives including an extensive patent bibliography [6]
- 1946 Review article on sulfurization of unsaturated compounds [7]

This clearly indicates that the ideas that had been invented and first applied by the metalworking people also worked for gears. Combinations of sulfurized products with lead soaps and lubricity esters were the first high EP performance technology in gear oils. Many years later Musgrave stated in an article on hypoid gear oils [8] that it was just by chance that the synergistic effect of sulfur with lead soaps had been discovered in the early 1930s.

An interesting historical dimension was added to the EP gear oil development during World War II. Most of the EP gear oil development occurred in the United States due to the great importance of automotive industry in the 1920s and the 1930s. German gear oil technology was not as advanced. Eyewitnesses report of frequent gear box failures of German tanks and heavy equipment during the attack against Russia. The reason was that in autumn Russian roads were turning into mud and the heavy vehicles were operated most of the time at maximum power. The only mildly additized gear oils just were not good enough to prevent scoring and welding.

9.2.4 SCIENTIFIC RESEARCH ON CHEMISTRY AND APPLICATION (1930–1949)

Between 1930 and 1950, the important basics of sulfur carrier technology had been developed. Patents from this period include most of today's raw materials and reaction pathways. Raw materials used were animal oils [3], vegetable oils/organic acids [9], pine oils [10,11], whale oil (sperm oil), acrylates, olefins [12], alcohols [13], synthetic esters [14], and salicylates [15]. Even thiocarbonates [16] and xanthogenates [17] were synthesized and used as organic, oil-soluble sulfur-containing EP additives.

Reaction pathways mentioned in earlier patents include

- Sulfur flower reaction with and without H_2S, aminic, and other suitable catalysts
- Sulfur chlorination with S_2Cl_2 [18]
- Organic halides with alkali polysulfides
- Mercaptan route [19]

Important product properties that are still part of today's development work were also mentioned in that period:

- Stability of sulfurized products, for example, diisobutene [20]
- Active and inactive sulfur compounds
- Corrosive and noncorrosive compounds
- Light- and dark-colored derivatives
- High- and low-odor products

Parallel to new chemistry, the development of test machines for tribological research progressed quickly along with publications on mechanistic studies of additives. In 1931, at an API meeting Mougay and Almen [21] presented the first chemical interpretation for the load-carrying capacity of sulfur-containing EP additives and their synergy with lead soaps.

They attributed the performance to the formation of a separating film between the frictional partners—a theory generally accepted today in tribological science. In 1939, this film-forming theory of sulfur compounds was proven using the four-ball tester [22]. In 1938, Schallbock et al. [23] published standard-setting results on investigations in the field of metalworking. Empirical correlations were found between cutting speed, temperature, and tool life that are still valid. In 1946, synergistic effects of chlorinated additives with sulfur additives were explained based on a chemical reaction theory [24] under the aspect of newest generation hypoid gear formulations. Phosphorus additives (tricresyl phosphate [25], zinc dialkyldithiophosphates [ZnDTPs]), primary antioxidants (AOs) (phenyl-α-naphthylamine, butylated hydroxytoluene ["BHT"]), and detergent/dispersants [26] (salicylates) also joined the world of lubricant additives during this period and have been used since in combination with sulfur carriers.

Most of the development work at that time had been done in a deductive way in a trial-and-error approach. Theoretical explanations and tribological and chemical modeling always trailed behind (looking back from today's point of view, it is quite astonishing that not so much has changed in 80 years).

9.2.5 SUMMARY OF THE LAST 60 YEARS

With the fundamentals of sulfur carriers being explored so early in lubricant additive history, the literature from 1950 until today concentrates around improvements in production procedures, combinations and synergies with other additives, improvement in product qualities, and search for special applications. The use of sulfur carriers has been extended from metalworking and automotive engine to industrial oils and greases. Ashless hydraulic oils may contain sulfurized EP additives for special applications. Now this product group is used throughout the lubricant oil industry.

"Tribology" has been defined as a particular field of scientific research, and several basic models of additive response have been worked out. A review article [27] in 1970 summarizes the state of the art at that time, including many literature references.

Until the 1950s, the sulfur carriers were mostly made by the lubricant manufacturers. However, increasing environmental awareness, growing market, and the need for more specialized products brought about change. As the sulfurization process involves deep chemical knowledge and production know-how and includes extremely high safety risks, specialty chemical companies became active in this area. It is expected that the few lubricant companies that still produce a small quantity of black sulfurized fats in-house may discontinue sooner or later.

Since the 1950s, the sulfur carrier market was split into two segments: automotive and industrial. In automotive gear oil applications, sulfurized isobutene (SIB) soon became the standard EP product because it is high in sulfur content but low in corrosivity. The typical, rather strong smell of SIB is no real problem in this field of application because gear boxes are totally closed systems. In any open lube systems, this EP technology is not acceptable. The big oil companies had their petrochemical subsidiaries (Mobil Chemicals, BP Chemicals, Shell Chemicals, Exxon Chemicals, Chevron Chemicals, etc.) and added additive manufacturing as the market grew including SIB production units. So SIB production has always been the target for those companies with focus on automotive additives and lubricants. Over the decades, the SIBs have gone through changes in chlorine level due to environmental requirements [28]. Starting at 2%–3% in the early days, today high qualities no longer contain chlorine because of a chlorine-free, high-pressure H_2S production processes. Also, the amount of active sulfur in SIBs has been reduced to improve the long-term abrasive wear of gear oil formulations in bearings and to meet today's "fill-for-life" requirements. But in principle in automotive applications, the same sulfur chemistry is in use today as it was some 70 years ago.

The other big field of application of sulfurized EP additives is industrial lubrication. This area traditionally is less regulated and restricted by OEM approvals, general specifications, and standards—it is particularly true for the metalworking market. Here much more differentiated, problem-solving additives have been and still are in use. This environment has generated a greater variety of smaller volume sulfur carriers that address strongly differentiated technical requirements of metalworking processes and grease applications. Subsequently smaller, more specialized chemical companies entered the lube additive business. The first products were dark in color, but as early as 1962 Rhein Chemie commercialized its first light-colored and low-odor sulfurized synthetic ester based on chlorine-free production technology (see the following text).

A big milestone in the history of sulfur carriers has been the international banning of sperm oil (whale oil) in 1971. Up to this year, sperm oil and lard oil (pig fat) have been the dominant fatty raw materials for sulfur carriers. The sperm oil–based products in particular showed excellent solubility and lubricity in addition to their sulfur-related EP properties. The extensive research activities of this period resulted in a variety of patents [29,30]. The new raw materials turned out to be vegetable oils in combination with either synthetic esters or olefins.

Another aspect that strongly influenced the sulfur carrier market has been the change of refinery technology for base oils. The driving force behind was the necessity to improve environmental as well as health and safety aspects of the major refinery products: fuels. These requirements lead to drastic reduction of aromatic components and sulfur content in fuels and subsequently of base oils. From a lubricant point of view, the reduction of aromaticity of the base oils had strong negative impact on their solvency for additives and thus triggered intensive adjustment work on the additive producer's side including sulfur carrier manufacturers.

The reduction in sulfur content however has contributed substantially to the market growth of sulfurized additives. The reduction of naturally occurring sulfur in base oils through the desulfurization units now needs to be balanced for specific applications via addition of synthetic, oil-soluble sulfur components in order to keep EP/AW properties as well as AO performance. This trend started already in the 1970s but is getting stronger today with the increase in availability of XHVI base oils/Groups II and III as well as fully synthetic base stocks (poly-α-olefins [PAOs]).

In the late 1970s and the early 1980s, a new class of sulfur carrier has been introduced into the industrial lubrication market: dialkylpolysulfides. They are based on C8, C9, or C12 olefins and contain up 40% sulfur in a very reactive form. They can be looked at as liquid, oil-soluble sulfur flower. The starting point for the development of these additives was the requirement of many lubricant blending companies for an alternative to sulfurization of base oils with sulfur flower. It is a very time-consuming step and may generate toxic gas (hydrogen sulfide, H_2S) and sulfur dropout during application. Sulfur flower can be dissolved in mineral oil just above its melting point of 115°C in concentrations of typically 0.4%–0.6% and is used if appropriate in heavy-duty metalworking applications or running-in gear oils (see Section 9.5.1.3). The solution that has been offered from additive manufacturers has been the new class of organic polysulfides of light color and rather low odor. Diisobutenepentasulfide and tertiary nonyl- and dodecylpentasulfide have been introduced as easy-to-blend liquids as substitutes for sulfur flower. Today, these active-type pentasulfides have become the most important and widespread class of sulfur carriers on the industrial oil side.

In 1985, it was found that sulfur carriers, preferably polysulfide types, show a strong synergistic EP/AW behavior when combined with high TBN sodium and calcium sulfonates [31]. This has become known as the "PEP technology" (passive extreme pressure) in neat oil metalworking. In the beginning, it was hoped that this combination would be a general and simple solution to upcoming chlorinated paraffin replacement issue that started in Western Europe and Scandinavia in the mid- to late 1980s. But as it turns out today the PEP technology can only partially match the universal properties of chlorinated paraffin formulations especially under low-speed/high-pressure operation conditions (for more details, see Section 9.5.4.2).

In the late 1980s to the early 1990s, a totally new aspect of sulfurized esters and fats has gained substantial ground—the toxicology and ecotoxicology of these chemicals. Workers' safety, environmental compatibility, biodegradability, and similar requirements need to be addressed in industrial more than in automotive lubrication, for example, because workers in machine shops often cannot avoid constant direct contact with the lubricant. The fact that the use of natural, renewable raw materials and optimized production procedures may give low toxic and biodegradable sulfur carriers refreshed the interest of development chemists in these special, environmentally safe but classic additives.

The twenty-first century's central question of additive and lubricant R&D departments is how to further optimize energy efficiency and reduce friction. And again sulfur carriers play a role. Spanning from engine oils to wind turbine gear boxes, formulators take advantage of their multipurpose character.

9.3 CHEMISTRY

9.3.1 General Aspects of the Sulfurization Process

On the market are light- and black-colored sulfur carriers. Both types could be produced by a basic sulfurization process using just sulfur and unsaturated compounds. A more sophisticated process with additional educts like hydrogen sulfide or mercaptans is necessary for the synthesis of light-colored products.

The black and the light sulfurization processes follow different reaction mechanisms. The black sulfurization is a radical (see Section 9.3.3), whereas the light sulfurization (see Section 9.3.4) is an ionic mechanism. In both sulfurization processes, inter- as well as intramolecular polysulfide chains are formed depending on the reaction conditions.

For the sulfurization high temperatures above 120°C are used and different side reactions, like the polymerization of the unsaturated compounds, are observed. Furthermore, mercaptans, thioethers, thioketones, and other heterocyclic compounds like thiophene (chromophoric groups) can be formed.

Besides the previously mentioned sulfurization processes, sulfur carriers are also produced by other synthetic routes. One is the so-called mercaptan route and based on mercaptans and sulfur (see Section 9.3.5).

Another synthetic route is the sulfur chlorination (see Section 9.3.6). Raw materials of this process are disulfur dichloride, sodium hydrogen sulfide solution, and unsaturated compounds. The products contain discrete S_2 chains. This is the typical production route for SIB.

A close related process to the sulfur chlorination is the use of alkyl halides and alkali (poly)sulfide (see Section 9.3.7). Today's large-scale production technology in general avoids any halogen-containing reaction steps because there are low limits in the final lubricants that may not be exceeded, for example, in automotive gear oils. Also, expensive removal/workup steps, for example, by washing, can be avoided if halogens are not used. Therefore, sulfur carriers are normally produced by the light or black sulfurization.

9.3.2 Raw Materials

In principle, any single- or multi-double-bond-containing molecule may be sulfurized. Therefore, the list of olefinic raw materials is long. For unsaturated compounds, patent literature reports of the following:

- Vegetable oils (soybean, canola, rapeseed, cotton seed, rice peel, sunflower, palm, tall oil, terpenes, etc.)
- Animal fats and oils (fish oils, lard oil, tallow oil, sperm oil, etc.)
- Fatty acids
- Synthetic esters
- Olefins (isobutene, diisobutene, triisobutene, tripropylene, tetrapropylene, α-olefins, n-olefins, cyclohexene, styrene, polyisobutene, etc.)
- Acrylates and methacrylates
- Succinic acid derivatives

The list of sulfur-containing material is rather short. It is mainly sulfur flower (S_8), hydrogen sulfide gas (H_2S), little S_2Cl_2, and little alkali polysulfide (e.g., NaS_x). The choice of commercially applied raw material is certainly limited to those compounds that have a reasonable price level and give certain performance benefits. Sulfur carriers based on low-boiling olefins (e.g., C4 types) are limited to closed lubricating systems due to the volatility of the decomposition products and associated offensive smell. For water-based lubricant oil systems, sulfurized fatty acids that can be easily emulsified and active types of olefins that cannot be hydrolyzed are preferred. In oil application, one can find the full range of products.

9.3.3 Black Sulfurization

The black sulfurization is the simplest and oldest of all production technologies for sulfur carriers. The first patented sulfur carrier was done this way. The manufacturing equipment needs to withstand pressure above 1–2 bars (it may even be pressureless). Raw materials may be olefins as well as natural or synthetic esters with a certain degree of unsaturation. The other reactant introduced into the olefin-containing reaction vessel is sulfur flower. The mixture is heated above the melting point of sulfur. An uncatalyzed reaction starts to become exothermic above 150°C–160°C, with the evolution of substantial amounts of H_2S. Catalyzed reactions start already just above the melting point of sulfur in the range of 120°C–125°C. Typical catalysts are organic amines, metal oxides, and acids.

Mechanistic studies of this reaction have been reported [32–34] and are very complex. Under high temperatures between 120°C and 190°C, also vinylic mercaptans, vinylic thioethers, vinylic alkyl- and dialkylpolysulfanes, vinylic thioketones, and even sulfur-containing heterocycles like thiophenes can be formed [34].

For the black sulfurization, a radical mechanism is postulated. The first reaction step is the ring opening of the sulfur flower (S_8-ring structure) by forming a diradical (chain initiation, Equation 9.1).

$$S_x \; \rightleftharpoons \; \cdot S - S_{x-2} - S \cdot \qquad (9.1)$$

In the second step, the sulfur diradical abstracts an allylic hydrogen atom from the unsaturated organic compound (Equation 9.2). Thereby, a new radical is formed and the growth reaction is started.

$$(9.2)$$

The addition of a sulfur ring on the organic radical is another growth reaction step (Equation 9.3).

$$R_3 \sim\!\!\!\sim R_4 \;+\; S_8 \;\rightleftharpoons\; R_3 \sim\!\!\!\sim R_4 \quad (9.3)$$

The organic radical polysulfide can react with another organic radical polysulfide by forming a sulfur-bridged compound (Equation 9.4). This is one possible termination reaction.

$$\text{(structure)} \quad (9.4)$$

Besides the termination reaction mentioned earlier, also other known radical termination reactions like allylic hydrogen abstraction or disproportionation can happen. Due to the fact that these termination reactions can occur on both allylic carbon atoms, higher sulfur-bridged compounds could be formed.

The chain length of the organic polysulfanes could not be defined by this mechanism, because the polysulfide chains are also homolytically split under the influence of high temperature. Therefore, these polysulfide radicals can react in the previously mentioned way by forming further radicals or recombine with other radicals in a termination reaction. Furthermore, by abstraction of hydrogen atoms, hydrogen sulfide is produced, which is needed for the light sulfurization (see Section 9.3.4). Therefore, during the black sulfurization also the light sulfurization occurs and causes the high variety of compounds in the black sulfur carriers.

The final product consists of a full range of organic sulfur derivatives. Some of them are still unsaturated, with isomerized double bonds and conjugated, chromophoric (color-deepening) sulfur compounds such as thioketones and thiophenes, which cause the product to be dark black in color and rather smelly. From an application point of view, these products exhibit EP/AW performance, but because of their remaining double bonds, they have the following negative characteristics:

- They will continue to polymerize during use and even under normal storage conditions.
- They are easily oxidizable and form residues on fresh metal surfaces/discoloration.
- They will cause TAN increase within a short time in circulation systems and cause short oil drain intervals.
- They will even generate H_2S/mercaptan during high-temperature usage in lubricant systems (see Section 9.4.2.1.7).

So today's main use of these black sulfurized products is total loss lubricants where long-term stability and bad smell are not an issue. It is the cheapest way of making sulfurized additives.

9.3.4 LIGHT SULFURIZATION (HIGH-PRESSURE H_2S REACTION)

High-quality sulfur carriers, which have improved properties compared to the black materials, are produced today using high-pressure/high-temperature equipment. The handling of toxic H_2S under high-pressure conditions requires sophisticated handling techniques and safety measures. Furthermore, H_2S is an expensive gas. All these aspects contribute to significantly higher production costs compared to the simple black sulfurization.

In this process, the olefins, sulfur, and H_2S are added to a high-pressure-resistant reactor and heated to 120°C–170°C. The reaction is also catalyzed by amines, metal oxides, acids, etc. For low-boiling olefins such as isobutene, the pressure may go up as high as 50–60 bar. For higher-boiling olefins such as diisobutene, typical pressures are in the range of 2–15 bar.

The presence of H_2S makes a total difference to the black sulfurization mechanism. It changes from a radical to a nucleophilic mechanism [35]. The oxidative attack of sulfur on the vinylic carbon–hydrogen (C–H) bond is effectively suppressed. The side reaction of the black sulfurization process becomes the main reaction here.

Hydrogen sulfide is not a strong nucleophile. Due to this fact, the first step is the deprotonation of the hydrogen sulfide. This anion reacts under ring opening with sulfur to form a negatively charged polysulfide (Equation 9.5).

$$B + H_2S \rightleftharpoons BH^+ + HS^-$$
$$HS^- + S_8 \rightleftharpoons HS_x + S_{9-x} \quad (9.5)$$

The negatively charged polysulfide HS_x^- is a stronger nucleophile compared to hydrogen sulfide and can directly attack the double bond (Equation 9.6). This addition reaction forms an organic polysulfide $R–S_xH$.

$$R_1 \sim\!\!\!\sim R_2 + H^+ + HS_x^- \rightleftharpoons \text{(structure)} \quad (9.6)$$

In the next step, the organic polysulfide added on another unsaturated organic compound and a polysulfane is formed (Equation 9.7).

$$\text{(structure)} \quad (9.7)$$

The sulfur chain of the organic polysulfanes can split under the influence of mercaptans, sulfide, and polysulfide anions, and a sulfur chain distribution is observed (Equation 9.8). Furthermore, radicals can be formed due to the high temperature. Therefore, also the black sulfurization occurs also during the light sulfurization. However, the nucleophilic mechanism is extremely preferred due to the high hydrogen sulfide pressure.

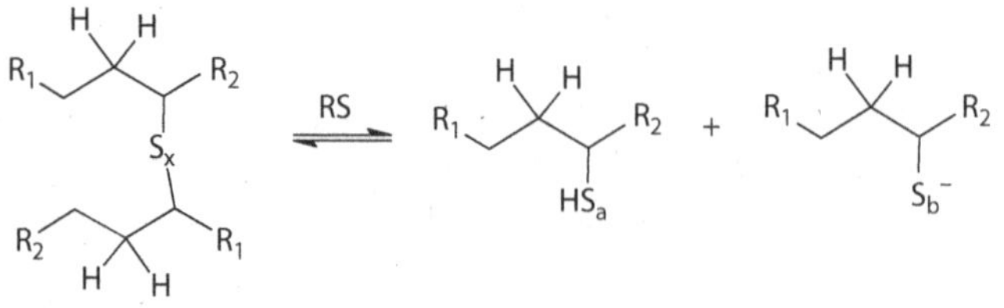

$$(9.8)$$

This procedure gives much more controlled reaction conditions and finally fewer side products. The most important effect of this reaction pathway is the fact that the double bonds are gone after the reaction and that no conjugated systems with chromophore (color-deepening) properties can be formed. The sulfur carriers produced this way are much more oxidatively stable and they are of light color. This one-step process is an advantage in terms of total production time and turnover.

9.3.5 Mercaptan Route

Few producers synthesize sulfur carriers in a two-step process:

1. In the first step only under catalytic action of Lewis acids, hydrogen sulfide is added to olefins. If such strong activators as boron trifluoride (BF_3) are used, the reaction takes place as low as $-20°C$. Another procedure works at $60°C–90°C$. The resulting alkylmercaptans are distilled from the reaction mixture and isolated as intermediates (Equation 9.9). The nonreacted olefins are circulated back to the reaction vessel.

$$R-HC=CH-R + H_2S \rightarrow R-H_2C-CH(SH)-R$$
Olefin   Alkylmercaptan

$$(9.9)$$

2. The mercaptans are oxidized with either hydrogen peroxide (H_2O_2, Equation 9.10) to the dialkyldisulfides:

$$2R-H_2C-CH(SH)-R + H_2O_2$$
Alkylmercaptan
$$\rightarrow R-H_2C-CHR-S-S-CHR-CH_2-R \quad (9.10)$$
Dialkyldisulfide

or by stoichiometric amounts of sulfur to trisulfides (Equation 9.11) and polysulfides (Equation 9.12):

$$2R-H_2C-CH(SH)-R + 2S$$
Alkylmercaptan
$$\rightarrow R-H_2C-CHR-S-S-S-CHR-CH_2-R + H_2S$$
Dialkyltrisulfide

$$(9.11)$$

$$2R-H_2C-CH(SH)-R + S_x$$
Alkylmercaptan
$$\rightarrow R-H_2C-CHR-S-(S_{x-1})-S-CHR-CH_2-R + H_2S$$
Dialkylpolysulfide

$$(9.12)$$

This may be summarized in the equation shown in Figure 9.1.

This process is mainly applied to olefin-based sulfur carriers based on tri- and tetrapropylene as starting material because the resulting tertiary dodecylmercaptan may be used in other applications such as rubber processing and as chemical intermediate.

9.3.6 Sulfur Chlorination Route

Sulfur carriers can be synthesized in a two-step process using disulfur dichloride and sodium sulfide solution. It had been widely used because it is a controlled way of adding discrete S_2-bridges to double bonds with little side reactions occurring.

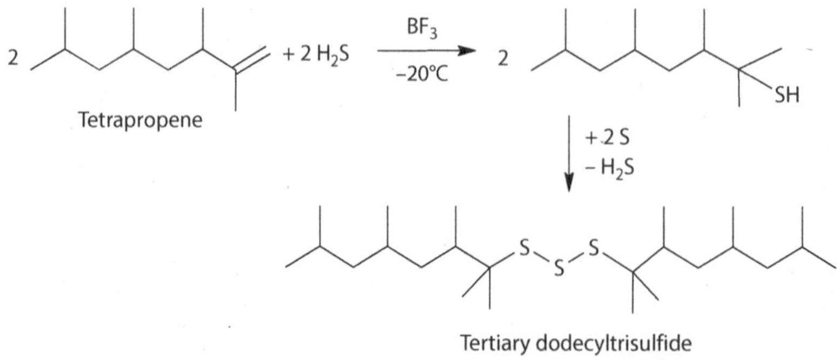

Tetrapropene

Tertiary dodecyltrisulfide

FIGURE 9.1 Summary of the mercaptan route two-step process.

In the first step, disulfur dichloride is added to the double bond (Equation 9.13)

$$(9.13)$$

In case of fatty oil as olefin source, the resulting product is a sulfur-chlorinated fatty oil useful as chlorine- and sulfur-containing EP additives in metalworking. From a technical point of view, their biggest problem is the split-off of chlorine and subsequent severe corrosion problems that are difficult to control. From today's point of view, the presence of chlorine is not favorable anymore because of environmental concerns.

In order to remove the chlorine, the sulfur-chlorinated products are subsequently treated with sodium sulfide (NaS_2) solution in water.

It is a substitution reaction of sulfur versus chlorine (Equation 9.14). Intermolecular linkage as well as ring closure may occur. The water-soluble sodium chloride is washed out.

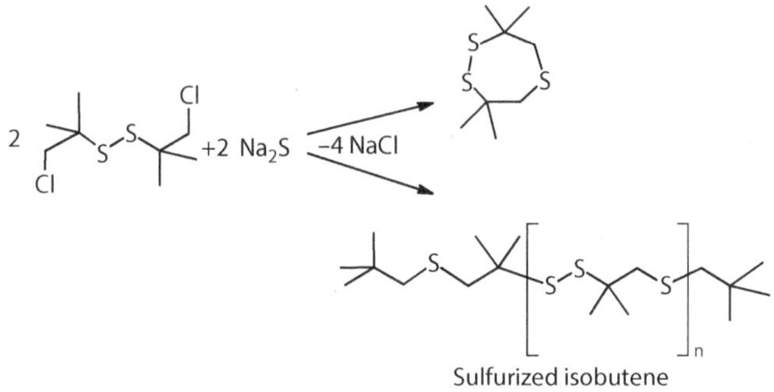

Sulfurized isobutene

$$(9.14)$$

9.3.7 Alkyl Halide/NaS$_x$

This process is closely related to the first step reaction discussed in Section 9.3.6. Starting materials may be alkyl or aryl halides. As shown in Figure 9.2, it is possible to substitute halogens with sulfur using alkali sulfides. If Na_2S is used, monosulfides are generated. In case alkali polysulfide is applied, alkyl or aryl polysulfanes are the resulting derivatives.

This route has not found commercial interest as raw material costs are too high compared to other synthetic methods.

9.3.8 Chemical Structure of Sulfur Carriers

9.3.8.1 General Considerations

For the majority of sulfur carriers, discrete structures are very hard to sketch for several reasons. One reason is the nature of the raw materials. They are very often mixtures of isomers. For example, olefins, like diisobutene, consist of five main isomers, and tetrapropylene shows some 35 different components in the gas chromatogram. Natural fatty oils have a distribution with mono-, double-, and triple-unsaturated acids plus unsaponifiable matter [36].

As already explained in Sections 9.3.1 through 9.3.7, depending on the reaction conditions and raw materials, different compounds are formed. The main compounds in the sulfur carriers are polysulfanes. However, also other functional groups could be observed. Furthermore, the catalyst directs the addition of sulfur in a certain way ("Markovnikov, Markovniko-Anti-Markaovnikov").

It is a fact that sulfur carriers are technical products based on technical raw materials. In the following, the most typical structures of sulfur carriers based on different, contemporary raw materials are shown. Taking the rather complex reaction pathways of a sulfurization reaction into account, they necessarily are simplified model structures.

9.3.8.2 Sulfurized Isobutene

SIB is the standard EP additive for gear oils with typical sulfur contents in the range of 40%–50%. It is produced typically by the sulfur chlorination route (Figure 9.3).

9.3.8.3 Active-Type Sulfurized Olefins

These are the polysulfide types of sulfur carriers that have been introduced as substitute for sulfurization of base oil and are widely used today in metalworking applications (Figure 9.4).

9.3.8.4 Inactive Sulfurized α-Olefins

Inactive sulfurized α-olefins are used in noncorrosive lubricant applications ranging from metalworking, greases, to even engine oil applications (Figure 9.5).

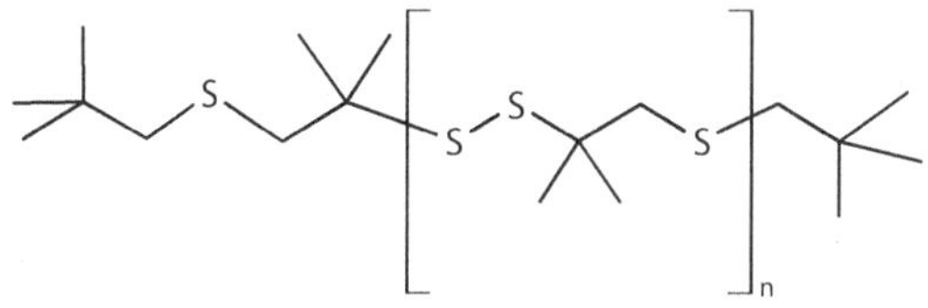

FIGURE 9.3 Sulfurized isobutene (SIB).

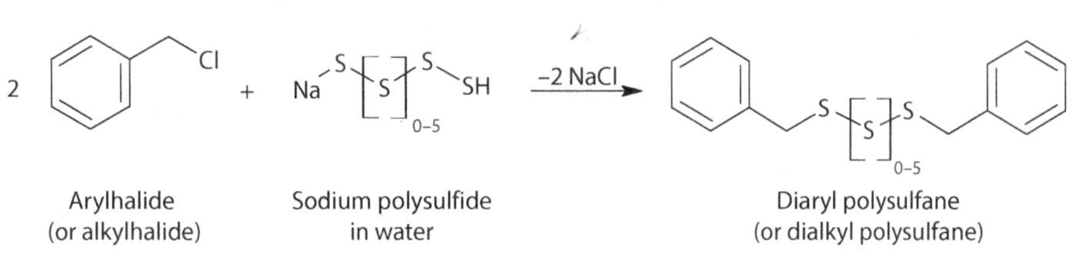

FIGURE 9.2 Dechlorination of arylhalogenides.

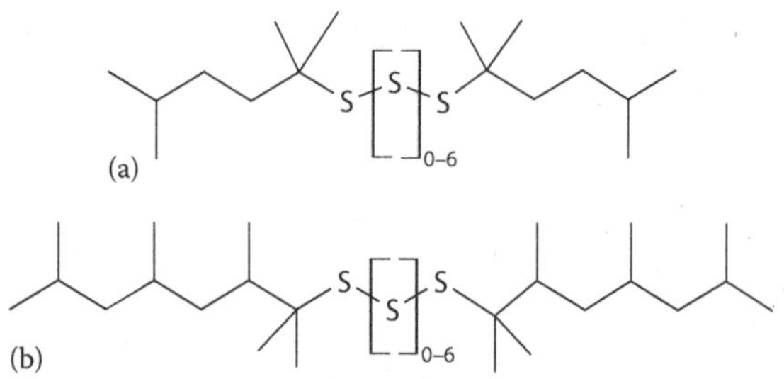

FIGURE 9.4 (a) Sulfurized diisobutene and (b) sulfurized tetrapropylene.

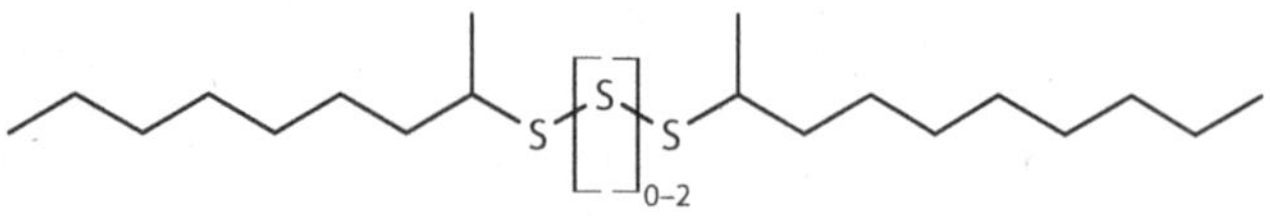

FIGURE 9.5 Inactive sulfurized α-olefins.

9.3.8.5 Sulfurized Synthetic Esters (Light Color)

Sulfurized synthetic esters are widely used in metalworking and grease applications. Depending on the type of synthetic ester chosen, special properties such as low-temperature stability/fluidity and low viscosity may be achieved (Figure 9.6).

9.3.8.6 Sulfurized Fatty Oil (Black Color)

See Section 9.3.3 (Figure 9.7).

9.3.8.7 Sulfurized Fatty Oil/Olefin Mixture (Light Color)

A special group of sulfur carriers are based on fatty oil and olefin mixtures. These sulfur carriers are outstanding in properties as they combine the positive effects of sulfurized olefins (e.g., hydrolytic stability, high sulfur content) with the excellent lubricity and film-forming properties of sulfurized fatty oils (Figure 9.8).

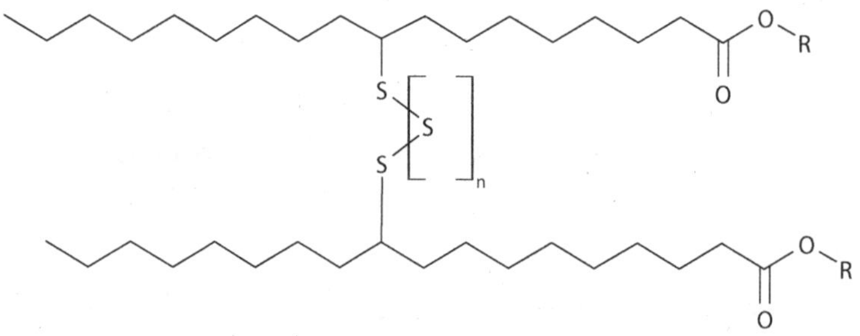

FIGURE 9.6 Sulfurized synthetic esters (light color).

Chromophoric group

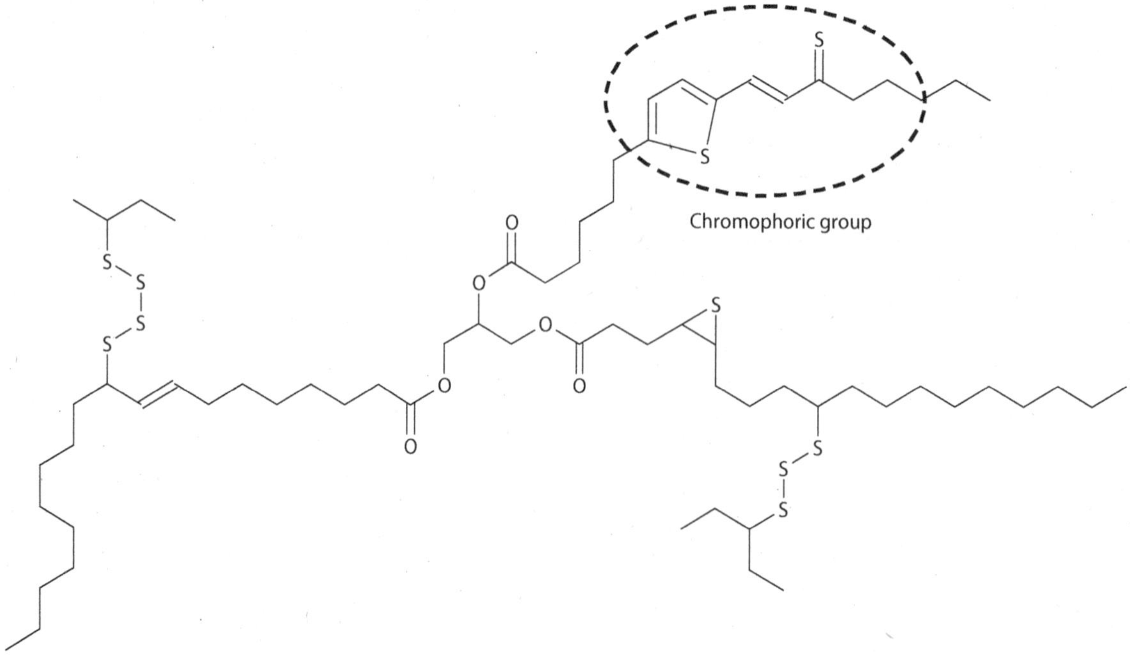

FIGURE 9.7 Sulfurized fatty oil (black color).

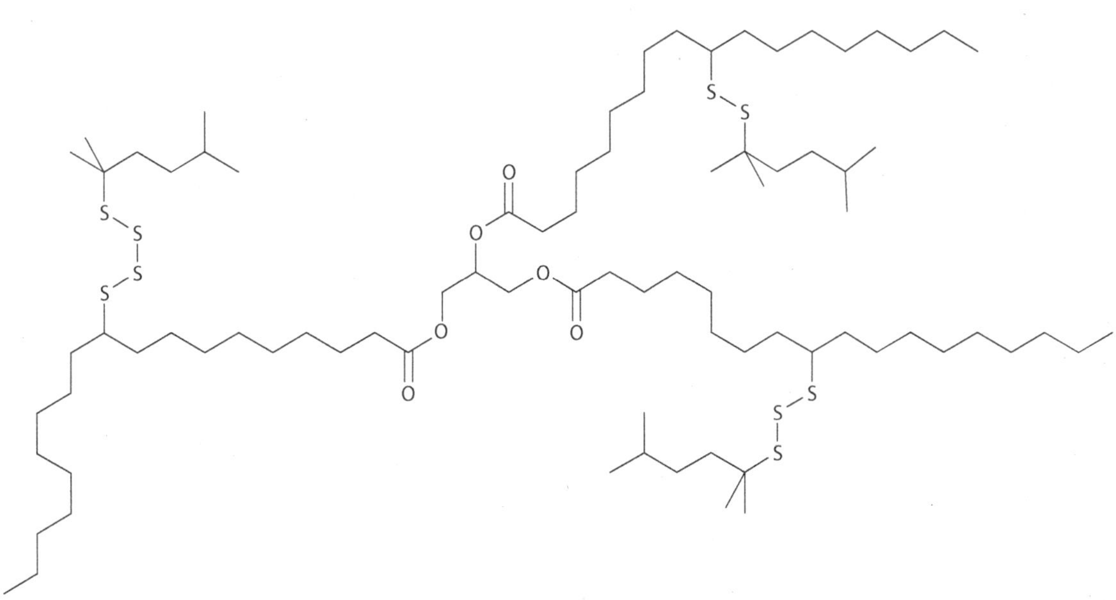

FIGURE 9.8 Sulfurized fatty oil/olefin mixture (light color).

9.4 PROPERTIES AND PERFORMANCE CHARACTERISTICS

9.4.1 CHEMICAL PROPERTIES

9.4.1.1 Effect of Additive Structure on Performance

9.4.1.1.1 Raw Materials

The additive structure is mainly influenced by the choice of the raw materials and the sulfurization method. A general overview of the performance properties of sulfurized products based on different raw materials is shown in Table 9.1.

9.4.1.1.2 Influence of Raw Materials on Extreme Pressure and Antiwear

The raw material determines the polarity and, therefore, the affinity of the product to a metal surface [37]. With increasing polarity, an increasing EP performance can be observed. Straight sulfurized olefins are nonpolar and show a relative poor affinity to metal surfaces (see Section 9.4.2.1.4). As the polarity increases from olefin < ester < triglyceride, the EP performance increases in the same order. This behavior is demonstrated in a simple four-ball EP test. Figure 9.9 shows

the four-ball weld load (DIN 51350 Part 2) of sulfurized additives over the sulfur level in oil. The products with a high polarity (C,D) show considerably higher EP loads than the nonpolar additives (A,B).

The content of active sulfur is only of minor importance on the EP performance, but the polarity and chemical structure play a major role.

9.4.1.1.3 Activity

Active sulfur is the amount of sulfur available for a reaction at a certain temperature. A common method for its determination is ASTM D1662 [38]. The amount of active sulfur is determined by reacting copper powder with the sulfurized product for 1 h at 149°C. Depending on the raw materials and on the sulfurization method, the active sulfur content can vary very much. The activity is a function of the temperature. Figure 9.10 shows typical active sulfur contents of sulfurized products based on different chemistry and sulfurization methods.

The activity depends mainly on the sulfur chain in the molecule. Mono- and disulfides are not aggressive against yellow metals. Pentasulfides are highly reactive and, therefore, suitable for heavy-duty machining of steel. The long-term

TABLE 9.1
Performance Properties of Sulfurized Products

	Ester		Triglyceride		Olefins	
	Inactive	Active	Inactive	Active	Inactive	Active
Extreme pressure	Fair	Good	Good	Very good	Low	Fair
Antiwear	Good	Low	Very good	Low	Good	Poor
Reactivity	Low	High	Low	High	Low	Very high
Cu corrosion	Low	High	Low	High	Low	High
Antioxidant	Good	Low	Good	Poor	Good	Poor
Lubricity	Fair	Fair	Very high	Very high	Poor	Poor

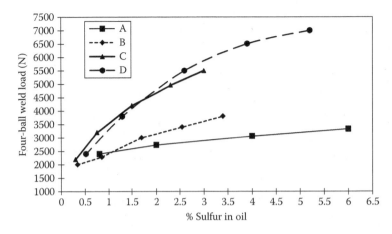

	Type	Total sulfur	Active sulfur	Activity (%)	Polarity
A	Olefin	40	36	90	
B	Ester	17	8	47	
C	Triglyceride	10	0.5	5	
D	Triglyceride	18	9	50	

FIGURE 9.9   Influence of raw materials on EP performance.

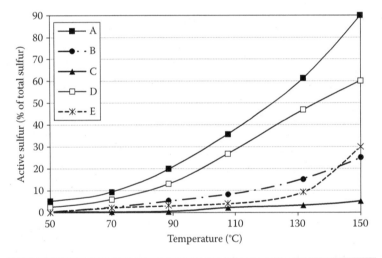

	Type	Total sulfur	Active sulfur absolute at 149°C	Activity (%)
A	Olefin	40	36	90
B	Olefin	20	5	25
C	Triglyceride	10	0.5	5
D	Triglyceride	18	10.5	58
E	Olefin/triglyceride	15	4.5	30

FIGURE 9.10   Active sulfur of various sulfurized products.

inhibition of these products against yellow metals is hardly possible. Long-chain sulfur bridges in polysulfides (A) are thermally less stable than short sulfur bridges, where the sulfur is linked to the carbon atom of the raw material. For this reason, the reaction with the metal surface is possible at relatively low temperatures. Mono- and disulfides show only a medium activity, because the sulfur will be released only at higher temperatures [39]. The active sulfur at a given temperature is an indication on the ability of the product to provide sufficient reactive sulfur to form metal sulfides.

Published work on the mechanism of the influence of organo-sulfur compounds on the load-carrying properties of lubricating oils indicates that this is due to their ability to form sulfide films that are more easily sheared than the metallic junctions under EP conditions [40].

Therefore, active sulfur has a significant influence on the AW performance. Higher sulfur activity results in faster formation of the metal sulfide and higher wear. This performance is visualized in Figure 9.11. The figure shows the four-ball wear scar (DIN 51350 Part 3) of sulfurized products with various activities.

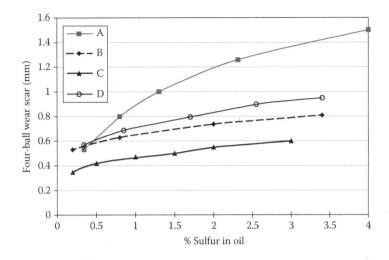

	Type	Total sulfur	Active sulfur (ASTM D1662)	Activity (%)
A	Hydrocarbon	40	36	90
B	Hydrocarbon	20	5	25
C	Triglyceride	10	0.5	5
D	Triglyceride	18	10.5	58

FIGURE 9.11 Influence on activity on AW performance.

9.4.1.1.4 Copper Corrosion

ASTM D130 [41] is a common method to determine the copper corrosion of additives. This copper corrosion does not necessarily reflect the activity of a sulfurized product, because very often yellow metal deactivators are used to mask the active sulfur.

The degree of copper corrosion depends on the amount of active sulfur and the presence of yellow metal deactivators. Inactive sulfurized products will show a long-term inactivity toward yellow metals, whereas active sulfur, masked with yellow metal deactivators, will react with the yellow metal as soon as the deactivator is consumed/reacted.

Therefore, the only statement that can be made is that a product will not stain copper under the given test parameters. This method is not suitable to determine the activity that is of major relevance for the performance of a sulfurized product (see Section 9.4.1.1.3).

9.4.1.1.5 Antioxidant

Sulfurized products with low active sulfur content are suitable to improve the AO behavior of lubricants. This is particularly important if hydrocracked, almost sulfur-free base fluids are used. During the synthesis of these oils, the natural sulfur (mainly heterocycles, inactive) is removed. The reintroduction of inactive sulfur carriers improves the oxidation stability, especially in combination with other secondary AOs.

9.4.1.1.6 Lubricity

Lubricity can be described as friction reduction under low-pressure conditions. Under these conditions, physical adsorbed lubricating films are effective (see Section 9.4.2.1.4). Inactive

sulfurized triglycerides are widely used to improve the lubricity of a lubricant. In general, the lubricity of sulfur carriers increases with the polarity. Sulfurized olefin (no lubricity) < sulfurized ester (medium lubricity) < sulfurized triglyceride (high lubricity). Special products with enhanced lubricity are based on synergistic raw material blends such as triglyceride/long-chain alcohol, triglyceride/fatty acid, and triglyceride/olefin.

9.4.1.1.7 Color

The color of sulfurized compounds is mainly influenced by the production method and by virtue of the raw materials. Light color is not only a matter of "cosmetics" but also a quality feature. Light-colored products manufactured with high-pressure hydrogen sulfide processes or by mercaptan oxidation do not have unsaturated double bonds left, and, therefore, they show better oxidation stability in general.

9.4.2 Physical Properties

9.4.2.1 Effect of Additive Structure on Properties

9.4.2.1.1 Raw Materials

The selection of the raw materials and the production process determine the chemical structure of the compound. The physical properties of a sulfurized product are dependent on the chemical structure. An overview is given in Table 9.2.

9.4.2.1.2 Polymerization

During the sulfurization process, the molecules of the raw materials are linked through sulfur. Depending on the structure of the raw material, two or more raw material molecules

TABLE 9.2
Physical Properties of Sulfurized Products

	Ester	Triglyceride	Olefin
Polymerization	Low	High	Very low
Solubility	Good	Fair—good	Very good
Polarity	Moderate	High	Low
Viscosity	Low	High	Very low
Biodegradability	Good	Excellent	Poor

will be linked. Triglycerides like lard oil and soybean oil do polymerize and form solid, rubberlike products, if the polymerization is not controlled through chain terminators like esters or olefins, containing only one double bond. The polymerization of olefins with only one double bond is limited. Two molecules are linked by a sulfur chain where length depends on the production process. Esters behave in a similar way but due to varying amounts of multiple unsaturated compounds in natural esters, some polymerization takes place. Dark sulfurized products do not only show less oxidation stability compared to light-colored, fully saturated compounds but will also resume polymerization after the production process is finished.

9.4.2.1.3 Solubility

The solubility is mainly a function of the polarity of the product. As the polarity increases from olefin < ester < triglyceride, the solubility in nonpolar mineral oils decreases. Polarity as well as the grade of polymerization determines the solubility. In general, sulfurized olefins have excellent solubility in solvents and all mineral oils. Depending on the sulfurization method, esters can exhibit good solubility even in Group II and Group III base oils if their polymerization grade is controlled during production. Sulfurized triglycerides are, in general, limited in their solubility due to their high polarity. But even more, the grade of polymerization plays a predominant role. A controlled reaction/polymerization can lead to light-colored products that will be soluble in paraffinic base oils, whereas uncontrolled polymerization will lead to dark-colored products, soluble only in oils with a higher polarity and aromatic content such as naphthenic base oils.

9.4.2.1.4 Polarity

Polarity determines the adhesion of a sulfurized product to the metal surface. The polarity depends on the raw materials used for the sulfurization. The organic portion of the molecule is responsible for the polarity and the affinity of the sulfurized product to the metal surface [39]. As the polarity increases from sulfurized hydrocarbon < sulfurized ester < sulfurized triglyceride, the affinity (physical adsorption) to metal surfaces increases also. Therefore, sulfurized products based on triglycerides, fatty acids, or alcohols provide superior lubricity compared to sulfur carriers based on less polar esters or nonpolar olefins.

9.4.2.1.5 Viscosity

Viscosity of a sulfurized product depends on the type of raw material used for the sulfurization and on the polymerization grade. A higher degree of polymerization (molecular weight) results in higher viscosity. The raw materials determine the viscosity index (VI) of a sulfur carrier. While short-chain sulfurized olefins show low VIs, sulfurized triglycerides can have VIs of above 200.

9.4.2.1.6 Biodegradability

Depending on the raw materials and on the sulfurization process, sulfurized products cover the whole range from not biodegradable to readily biodegradable. Besides the raw material, the production technology plays a predominant role. Catalysts used, impurities in the raw materials, and side components formed during the synthesis have a strong influence on the biodegradability. Therefore, biodegradability cannot be predicted but has to be tested for every single product. Biodegradable sulfur carriers are available for various applications [42].

9.4.2.1.7 Stability

Storage stability is obtained by the total reaction of the double bonds in the sulfur carrier and in eliminating hydrogen sulfides and mercaptans. Especially mercaptans, but also hydrogen sulfide, are left over from the sulfurization process. If H_2S or mercaptans are not removed completely, they will evaporate under severe conditions in the final application or even under unfavorable storage conditions. Mercaptans can react with the polysulfanes and thereby release H_2S. Depending on the raw materials and the type of sulfurization process, some sulfurized products continue to polymerize during storage. Especially triglycerides, sulfurized with flower of sulfur under atmospheric pressure, show a steady and sometimes very strong polymerization during storage.

9.5 COMPARATIVE PERFORMANCE DATA IN PERTINENT APPLICATION AREAS

9.5.1 METALWORKING

9.5.1.1 Cutting/Forming

In principle, we have to deal in all cutting processes with abrasive wear (i.e., cutting) and adhesive wear (i.e., buildup edges). Depending on the particular process, respectively the machining parameters, one of these wear types plays a dominant role. At low machining speeds, like in most of the forming operations, adhesive wear (cold welding), the formation of buildup edges, and wear on the flank of the cutting edge are very often the limiting factors for tool life. At high machine speeds and increasing contact temperatures, the abrasive wear determines the tool life. The reactivity of additives depends on temperature and pressure. Field and laboratory tests showed that different types of sulfur carriers (same sulfur content but varying raw materials and/or production processes) lead to significantly different results in a metalworking operation [43].

9.5.1.2 Contribution of Sulfur Carriers to Metalworking

Sulfurized products can be designed to meet technical and ecological requirements in metalworking processes. They are used successfully since almost 100 years to avoid abrasive and adhesive wear and enhance lubricity. In cutting operations, their main function is to support the cut and to prevent wear of the tool, whereas in forming processes, sulfurized products should form a pressure-stable lubricant film and prevent adhesive wear.

9.5.1.3 Replacement of Flowers of Sulfur

In the past, it was very common to dissolve flowers of sulfur in metalworking fluids to obtain a high reactivity and good EP properties. This procedure is very cost-intensive because it has to be done under controlled temperature conditions below the melting point of sulfur and has a lot of disadvantages like limited solubility (max 0.8% S), limited stability (sulfur dropout), and high corrosivity toward yellow metals. Also, there is a risk of H_2S generation, a highly toxic gas well known because of its rotten-egg odor. Today, this process is widely substituted by using sulfurized products. If just reactivity is required, sulfurized olefins with high total and active sulfur content are used, though it is possible to adjust almost any activity/lubricity ratio while using a combination of appropriate sulfurized products.

9.5.1.4 Copper Corrosion

Depending on the process and on the metals machined, corrosion control toward yellow metals can be a requirement. If inactivity (no staining) toward yellow metals is required, it is important to use either absolute inactive sulfurized products or medium-active sulfur carriers in combination with yellow metal deactivators. Active sulfurized products can be inhibited short term, but as soon as the inhibitor is used up, they will turn active again.

9.5.1.5 Substitutes for Chlorinated Paraffins (CLP)

The driving forces for the replacement of chlorinated paraffins are mainly ecological and toxicological reasons. Users and waste oil disposal facilities have additional concerns over the corrosivity of the chlorinated paraffin decomposition products, primarily hydrochloric acid.

Chlorinated paraffins work because of their ability to form a highly persistent lubricating film even at low temperatures or moderate pressure. At high-temperature/high-pressure conditions, they decompose and the formed hydrogen chloride forms iron chloride with the metals involved in the process [43].

Chlorinated paraffins can be substituted by sulfurized products. Depending on the main function of the chlorinated compound in the particular process used, lubricity, or activity, suitable sulfurized products are available that can function as alternatives. The lubricity performance is mainly covered by highly polar, inactive sulfur compounds (see lubricity), whereas the activity will be covered by reactive sulfurized olefins or mixed sulfurized olefins/triglycerides. Alternatives like sulfurized triglycerides and special esters do show a strong adsorption to ferrous metals, but do not create such corrosive layers while chemically reacting with the metal surface. If a high removal rate is desired, chemically active sulfurized compounds can be used as substitute. Like the CLP, the polysulfides can form shear instable chemical layers, which prevent welding and guarantee an efficient removal process.

9.5.1.6 Substitute for Heavy Metals

Organic heavy metal compounds based on antimony, molybdenum, bismuth, or zinc are used as EP and AW additives in severe metalworking processes and in grease applications. Sulfur carriers have proven to be suitable substitutes, particularly when used with synergistic compounds like polymer esters, phosphates, phosphites, dialkyldithiophosphates, and sulfonates.

9.5.1.7 Carbon Residue Reducing in Rolling Oils

Sulfur carriers are used in cold rolling of steel to prevent carbon residues built up on the surface of the metal sheets during the annealing process. Carbon residues are generated by oxidation/polymerization of additives used in rolling oils. Therefore, typical rolling oils contain sulfur-based antisnakey edge and carbon-reducing additives [44]. Clean burning of the lubricant is important to obtain a clean metal surface that can be evenly coated in subsequent process steps.

Typical products used for this application are sulfurized olefins with low to medium activity or inactive sulfurized triglycerides.

9.5.1.8 Water-Miscible Metalworking Products

Sulfurized products are used in water-miscible metalworking systems to provide EP performance and, depending on the type of sulfur carrier, lubricity. By far, the biggest applications are soluble oils or emulsions. Standard sulfurized products are not water soluble. Surfactants must be used to keep the sulfur carrier in the emulsion. Compared to applications in nonwater-based systems, the water-based systems require hydrolytically stable products that can react at relative low temperature. Therefore, active sulfurized olefins, preferably pentasulfides, are widely used for this application. Sulfurized esters and triglycerides are also used, especially if additional lubricity is required. Specialty sulfur carriers are reaction products of sulfurized fatty acids and sulfurized olefins. These sulfur carriers combine good emulsifying properties with relative high hydrolytic stability and activity. Straight sulfurized fatty acids, like sulfurized oleic acid, are used in semisynthetic metalworking fluids. The sulfurized fatty acid will be reacted with alkaline compounds like amines or potassium hydroxide to form a soap. This soap is water dispersible and needs much less emulsifiers than a sulfurized olefin. However, hard water stability can become a problem with this type of sulfur carrier.

9.5.2 Grease

High demands on load-carrying capacity of machine parts require the use of EP and AW additives to avoid material loss and the destruction of the surfaces of the friction partners.

Older technology still uses typical gear oil sulfur carriers based on short-chain olefins like isobutene. These sulfur carriers provide a high sulfur content, but their distinct, strong odor prohibits their use in open lubricating systems.

As it is almost impossible to mask the activity of sulfurized products in greases by using yellow metal deactivators or sulfur scavengers, inactive sulfurized products are widely used as EP additives in greases. Especially if the grease is designed for a wide application range, it is imperative that truly inactive sulfur carriers are used, because yellow metals are widely present as friction partners (e.g., brass cages in bearings). In addition, there are increasing demands on high-temperature stability for various grease types. This also calls for inactive, oxidation-stable sulfurized products. Typical sulfur carriers for greases are shown in Table 9.3.

Sulfurized products are also used to substitute heavy-metal-containing compounds, which are traditionally used as EP additives in greases. Besides their excellent performance, these heavy-metal-containing compounds show some weak points. Antimony and bismuth compounds are known to have some weakness regarding copper corrosion, and lead compounds are toxic. In the meantime, many of these products have been replaced by special sulfurized products either as a direct replacement or in combination with synergistic compounds such as ZnDTPs, phosphate esters, or overbased sulfonates [45]. Sulfurized products are also used in greases for constant velocity joints (CVJs) [46]. They are very efficient in combination with molybdenum compounds (e.g., molybdenum dithiocarbamate, molybdenum dithiophosphate, Mo-organic salts) as a sulfur source to support the formation of lubricating active molybdenum disulfide in the friction zone.

There is an increasing demand for EP greases for environmentally sensitive applications like railroad wheel flange lubrication, railroad switches, and agricultural equipment like tractors or cotton picker spindles. Some sulfurized products are biodegradable and show excellent ecological data [39]. Therefore, these products are used rather than heavy-metal-containing compounds to enhance EP and AW properties in such applications.

9.5.3 INDUSTRIAL OILS

An increasing variety of industrial fluids use sulfurized products as EP and AW additives.

9.5.3.1 Industrial Gear Oils

Typical sulfur carriers for this application are short-chain sulfurized olefins. Sulfurized isobutene (SIB) or diisobutene is widely used as an EP additive in industrial gear oils. SIB is used for some decades as the EP additive of almost all industrial gear oil packages. The high sulfur content, combined with a relative low active sulfur level, is ideally suited to match the requirements. Unfortunately, these products have a very distinct odor and, depending on the manufacturing process, can contain chlorine compounds. Newer developments are based on sulfurized olefins with a longer chain length. Specialty products with additional demands on lubricity are based on sulfurized triglycerides or mixtures of sulfurized olefins and triglycerides.

9.5.3.2 Slideway Oils

Slideway oils are a special type of gear oils with very good antistick-slip properties. Besides the austere requirements on coefficient of friction, there are also demands on compatibility and demulsibility with metalworking emulsions. Inactive sulfurized triglycerides are suitable to reduce the coefficient of friction. Unfortunately, most of these products are easy to emulsify and will, therefore, not meet the requirements on demulsibility without extensive formulation work. Modern slideway oils are based on demulsifying sulfurized olefin/triglyceride-based products that combine the advantage of a low coefficient of friction, good demulsibility, and high EP loads.

9.5.3.3 Hydraulic Fluids

It is possible to use inactive sulfur carriers in hydraulic systems with only moderate requirements on thermal stability. Typical products are sulfurized olefins and triglycerides or mixtures thereof.

9.5.3.4 Multifunctional Lubricants

Multifunctional lubricants cover more than just one lubrication application. There are increasing demands for this lubricant type, especially in metalworking shops. As one lubricant will be used for different applications with sometimes very different requirements, it is important that multifunctional additives are used. Depending on the overall performance requirements, sulfur carriers are used as EP, AW, or lubricity additive (see Tables 9.1 and 9.2). Multifunctional lubricants are almost always a compromise in their formulation. For example, metalworking machines with combined gear oil/process oil sump require a fine-tuned additive, especially on the EP side. Sulfur carriers with a medium activity, additionally passivated with sulfur scavengers, are widely used for this application [47].

TABLE 9.3
Typical Sulfurized Products for Greases

Type	Total Sulfur	Active Sulfur	Features
Triglyceride	8–12	0.5–3	Mainly inactive, limited EP performance
Triglyceride	13–15	4–7	Mainly active, hard to mask Cu corrosion long term; good EP
Olefin	45	10–15	High EP performance, very distinct odor, only for encapsulated systems
Triglyceride/olefin	15	4	Mainly inactive, high EP performance
Ester	9–11	1–3	Mainly inactive, limited EP performance, excellent low-temperature pumpability

9.5.3.5 Agricultural Applications

Lubricants in agricultural applications can be spilled on soil either because of the machine design or because of leaks in hydraulic and gear systems. Therefore, there are increasing requirements on environmentally compatible or less harmful lubricants. Sulfurized products are ideally suited for this type of applications. They can be designed to meet performance and ecological requirements (e.g., biodegradability). A wide range of lubricants exist for outdoor equipment based on vegetable oils (e.g., soybean, canola, rapeseed, sunflower oil).

Sulfur carriers for these applications are mainly based on vegetable oils and synthesized in strictly controlled manufacturing processes.

Typical sulfur carriers for agricultural applications are shown in Table 9.4.

9.5.3.6 Automotive Applications

It is disclosed in U.S. Patent Nos. 4,394,276 and 4,394,277 that various sulfur-containing alkane diols may be formulated with lubricating oils to effectively reduce fuel consumption in an internal combustion engine. Sulfurized products in general and inactive sulfurized, oxidatively stable olefins in particular are known to reduce friction efficiently in engines. They provide not only AW and antifriction but also antioxidation properties. However, they cannot substitute the multifunctional ZnDTPs in this application.

Besides crankcase applications, the use of sulfurized products in automotive gear lubricants is far more important. Since the middle of the twentieth century, almost every gear lubricant for automotive applications has been formulated with SIB.

The advantages of SIB are its high sulfur content, oxidation stability, and low corrosivity, but the very distinct odor and the low lubricity are disadvantageous.

Sulfurized esters and triglycerides are used in special transmission fluids to adjust the stick-slip properties. These sulfur carriers are also used in other lubricants in the automotive area such as wheel-bearing or CVJ grease.

9.5.4 Synergies/Compatibility with Other Additives

Sulfurized products are compatible with most of the additives used in lubricants. Only strong acids and bases must be avoided in combination with sulfur carriers.

9.5.4.1 ZnDTP

ZnDTPs are used in combination with sulfurized products in various applications. Besides their primary functions as AW and AOs, there is a well-known synergistic effect in regard of stabilization and improvement of copper corrosion of sulfur carriers. This behavior is demonstrated in the ASTM D130 copper corrosion test (Table 9.5).

In addition, ZnDTP can have a very positive effect on the odor of sulfur carriers.

9.5.4.2 Basic Alkali Metal Salts

Sulfurized products show a very strong synergistic effect in combination with basic alkali metal salts [48], often referred to as overbased sulfonates or carboxylates. Particularly active sulfur in combination with overbased calcium or sodium sulfonates exhibits advantageous performance with regard to improved load-carrying and AW properties. These additive combinations are used in lubricants for severe metalworking operations.

It is disclosed in International Patent WO 87/06256 [45] that the load-bearing characteristics of a grease composition and a gear lubricant may be unexpectedly improved by formulating these compositions with an additive mixture comprising overbased salts of alkaline earth metals or alkali metals and at least one sulfurized organic compound. From today's point of view, the overbased products/sulfur combination has its advantages in some stainless steel cutting and forming operations, but the high alkalinity of such formulations shows big compatibility problems when in contact with lubricity esters and other types of acidic additives. Alkaline washing baths get used up quickly and need much more frequent changes as calcium soaps built up. Welding

TABLE 9.4
Sulfur Carriers for Agricultural Applications

Sulfur Carrier Type	Application
Ester, inactive	Gear greases (NLGI class 000), cotton picker spindle lubricants
Triglyceride, inactive	Gear lubricants, hydraulic fluids, greases for bearing lubrication, chassis lubricants
Triglyceride, medium active	Gear lubricants, chain saw and bar saw lubricants
Olefin, inactive	Gear greases (NLGI class 000), cotton picker spindle lubricants
Triglyceride/olefin, inactive	Gear lubricants, hydraulic fluids, greases for bearing lubrication, chassis lubricants

TABLE 9.5
ZnDTP, Synergistic Effect on Copper Corrosion

Type	Total Sulfur	Active Sulfur	Treatment Level (%)	Cu Corrosion 3 h at 100°C	+1.5% ZnDTP[a] Cu Corrosion 3 h at 100°C
Triglyceride	18	10.5	5	4c	3b
Ester	17	8.5	5	3b	1b
Triglyceride	15	5	5	3a	1b

[a] Thermally stabilized ZnDTP based on 2-ethylhexyl alcohol.

without cleaning the metal surface is also impossible as the high TBN sulfonates are generating high amounts of oxide ash.

9.5.4.3 Antioxidants

Inactive sulfur carriers show a synergism with aminic AOs. This effect is very distinct in low or even sulfur-free base fluids. Active sulfurized products do not show this synergy. In the contrary, the active types deteriorate the oxidation stability. Table 9.6 demonstrates the oxidation stability (ASTM D2270, RPVOT Test) of an active and inactive sulfur carrier based on the same raw materials.

The inactive product improves the oxidation stability (250 minutes) twice as much as the active type (120 min). In combination with the aminic AOs, the synergistic effect is obvious. While the inactive sulfur carrier improves the AO properties of the aminic AO, the active type has a detrimental effect and reduces the oxidation stability.

9.5.4.4 Esters/Triglycerides

Esters are used either as base fluids or as additives. It is important to coordinate ester type and sulfur chemistry to achieve optimum performance. Unsaturated esters show strong synergistic effects with active sulfur, whereas inactive sulfur shows distinct synergies with saturated esters. The performance of sulfur carriers in saturated esters is similar to their performance in mineral oil. These synergies are widely used in the formulation of lubricants.

Combinations of active sulfur and unsaturated esters or triglycerides (mainly vegetable or animal oils like canola, rapeseed, tall, sunflower oil, and esters thereof) are very common in all types of metalworking fluids, in oils, as well as in water-based systems. The combination of these products shows better EP and AW properties than the single components. This performance is illustrated in the four-ball test (see Table 9.7).

Other applications for the combination of active sulfur and unsaturated ester are heavy-duty gear oils in agricultural applications and environmentally friendly chain saw and bar saw lubricants (see agricultural applications).

If Cu corrosion presents a problem, the use of medium-active sulfurized products in combination with unsaturated esters is of advantage. Some of these sulfur carriers are active enough to create the synergistic effects, but their Cu corrosion can be controlled with suitable yellow metal inhibitors.

Aside from the improvement of EP and AW performance, inactive sulfur carriers can boost the AO properties.

Saturated esters are used where good oxidation stability is required. The performance of sulfurized products in saturated esters is comparable to the performance in mineral oil. Inactive sulfurized olefins and sulfurized triglycerides, and mixtures thereof, are typically used as additives in lubricants based on saturated esters.

9.5.5 COST-EFFECTIVENESS

Sulfurized products cover the whole range from relative cheap commodity to high-price specialty. Depending on the type and treatment level, sulfurized products can be a major cost factor in lubricants. However, the use of sulfur carriers enables us to run processes and to overcome lubrication problems that cannot be solved in a cost-efficient way with other additives. For example, it is possible to increase machine speeds and thus productivity while using appropriate sulfur carriers instead of esters or chlorinated paraffins. Depending on the type of sulfur carrier, other commonly used additives in a formulation, for example, esters or yellow metal deactivators, can be saved. In comparison with heavy-metal- or chlorine-containing lubricants, the disposal costs for the used lubricant can be much lower.

Besides direct cost savings, respectively, cost efficiency due to higher productivity and lower disposal costs, there are secondary cost factors. In comparison with some traditionally used EP additives like chlorinated paraffins (HCl formation -> rust), overbased sulfonates (difficult degreasability, incompatibility with other additives), or heavy-metal-containing additives (residue formation), sulfurized products show, in general, less cost-effective side effects.

9.6 MANUFACTURE AND MARKETING ECONOMICS

9.6.1 MANUFACTURERS

Legislation and environmental conditions require substantial investment in safe and no emission production facilities globally. In recent years, there is a concentration of larger manufacturers who invested in new technology. About a handful of manufacturers are able to synthesize light-colored products starting with an olefin, hydrogen sulfide, and sulfur. Globally, there are about 10 manufacturers with significant capacity, many of them still making dark sulfurized products.

TABLE 9.6
AO Properties of Active and Inactive Sulfur

	Base Oil Hydrocracked, Dewaxed, No Sulfur (Minutes)	1.0% Inactive Olefin, 20% S, 5% Active S (Minutes)	1.0% Active Olefin, 39% S, 30% Active S (Minutes)
Base oil	40	250	120
0.2% aminic AO (alkylated diphenylamine)	400	540	135

TABLE 9.7
Synergy of Unsaturated Ester with Active Sulfur

Sulfurized Olefin 40% S, 36% Active	TMP Oleate	Four-Ball Weld Load DIN 51350 Part 2 (N)	Four-Ball Wear Scar DIN 51350 Part 3 (mm)
1.5%	—	2800	0.8
—	5.0%	800	0.6
1.5%	3.5%	3200	0.55

Additionally, there are some lubricant manufacturers who still sulfurize dark-colored products for their own use, and some local sulfurization plants also sulfurize commodity-type products.

9.6.2 MARKETERS

In general, the manufacturers are also marketing the products. Some local manufacturers will buy sulfurized products and blend them with esters, mineral oil, etc., and sell them under their own brand name. Sulfur carriers are intermediate and not consumer products.

9.6.3 ECONOMICS

Market prices vary depending on the raw materials, the production process, and the performance level of the products. Low-quality, dark sulfurized fats with distinct odor and limited stability sell for less than USD 1.5 per kg. High-performance, top-quality, low-odor, light-color products achieve prices of more than USD 5.0 per kg. Not only the raw materials but much more the production process determine the price for a sulfurized product. For example, the sulfurization of a typical gear oil sulfur carrier with disulfur dichloride and the necessary subsequent washing steps are more costly than the sulfurization of a fat with flower of sulfur.

9.6.4 GOVERNMENT REGULATIONS

9.6.4.1 Competitive Pressures

There are no government regulations concerning the use of sulfurized products, but depending on the location of the manufacturing plant, very stringent regulations and conditions concerning emission standards can apply. Therefore, there is a competitive distortion in production between more and less environmentally aware countries. Production technology and in particular low-emission production are key cost factors.

9.6.4.2 Product Differentiation

Apart from some commodities, there is a clear product differentiation mainly derived from quality and performance. A first criterion for differentiation is color, followed by sulfur content, raw materials, and odor. A classification is hardly possible because many of the products are tailor-made either to cover a specific performance profile or to meet specifications in various applications. A simple categorization by sulfur content or raw material bases would be to coarse and would not take performance into account. Even products based on the same raw materials but manufactured with a different process can be completely different in performance.

Modern, light-colored high-performance products with low odor are sulfurized using hydrogen sulfide or mercaptans. Even the appearance distinguishes these products from conventionally sulfurized dark-colored, smelly products.

9.7 OUTLOOK

9.7.1 CRANKCASE/AUTOMOTIVE APPLICATIONS

Steady demands on reduction of phosphorus levels in motor oils as well as requirements for increased fuel efficiency, that is, friction reduction, will open new opportunities for sulfur chemistry. Sulfurized products are already used in this type of application (see Section 9.5.3.6).

9.7.2 INDUSTRIAL APPLICATIONS

Multifunctional and multipurpose lubricants are on the wish list of many end users. The development for sulfurized products that can be used in these types of lubricants is in full progress, and products have already been commercialized. Mainly light-colored products based on mixed, well-balanced raw materials to ensure a broad performance range are used for multipurpose applications.

Replacement of heavy metals and chlorinated paraffins in almost all industrial lubricants is also an ongoing project that is widely found in the lubricants industry. Sulfur carriers are playing a predominant role as substitutes for these products.

Increasing demands for environmentally more acceptable lubricants have led many formulators into the development of lubricants based on natural triglycerides like canola oil, soybean oil, tall oil, or esters. Biodegradable sulfur carriers are used as EP and AW additives as well as secondary AOs in these applications.

9.8 TRENDS

9.8.1 CURRENT EQUIPMENT/SPECIFICATION

Sulfurized products are single components and not complete performance packages such as hydraulic or crankcase packages. Therefore, sulfur carriers are used in the whole variety of lubricants rather than in a specific equipment. Also, no national or international specification standards exist for these products. The manufacturer sets the specification in agreement with the user.

9.8.1.1 Types of Equipment

As already mentioned, the biggest use of sulfurized products (excluding SIB) is in industrial applications. Metalworking and grease applications followed by industrial gear oils are formulated with sulfur carriers. A lot of old equipment is still in use. Many midsize and small companies have not modernized their metalworking machines for more than three decades. This older, robust equipment is very often running at low machining speeds and nonoptimized machining parameters. Modern machining equipment requires thermally stable fluids, based on highly refined or synthetic base fluids. Improved solubility in nonpolar oils and thermal stability of sulfurized products gain importance.

9.8.1.2 Additives in Use

Today's additive usage depends very much on regional technological requirements and local legislation. In countries with low, old, or standard technology and little environmental concerns, additives such as chlorinated paraffins or heavy metals are used for the formulation of lubricants, often in combination with sulfurized products. In countries where legislation has put some pressure onto the formulators and users of lubricants (higher disposal costs for chlorine-containing lubricants, limits on heavy metals in wastewater, etc.), sulfurized compounds play an even more important role. They are the main EP additives, very often combined with sulfonates, salicylates, phosphoric acid esters, dialkyldithiophosphates, or carboxylic esters to complement AW and lubricity performance.

9.8.1.3 Deficiencies in Current Additives

All available sulfur carriers are limited in their thermal stability. This is a desired feature, because reactive sulfur will only be released while the molecule breaks down. However, there are applications running at high temperature where a fast decomposition of the EP product is not desired. Corrosion toward yellow metals is another deficiency of sulfurized compounds. In high-temperature applications, the active sulfur will react with copper to form copper sulfide.

9.9 MEDIUM-TERM TRENDS

In general, there is a trend toward higher economy and ecological and toxicological safe lubricants.

9.9.1 METALWORKING

Increased lubricant temperatures are a consequence of higher machine speeds, fully encapsulated machines, and reduced process steps. In the future, the thermal stability of metalworking lubricants and their toxicological safety will be on the focus. Due to integrated applications (e.g., one lubricant for process and machine lubrication), the additives need to cover wide temperature ranges. The trend to replace multiple cutting steps with forming operations exists. Therefore, the type of additives will also change.

Minimum amount lubrication requires new lubricant concepts in regard of performance and marketing. Maintenance of lubricants will further be reduced. Again this trend calls for increased stability of additives.

Sulfurized products for metalworking application will need improved thermal stability in combination with good solubility in high paraffinic or even synthetic base fluids. Ecological and toxicological safety will be basic requirements. Improved lubricity and excellent compatibility with process materials such as cleaners and paints will be essential for the formulation of modern lubricants for deformation processes (e.g., cold forging, deep drawing).

9.9.2 INDUSTRIAL OILS

Synthetic fluids such as PAOs, polyalkylene glycols, extra high VI mineral oils (XHVI), or synthetic esters are being used in increasing volumes for the formulation of high-performance industrial lubricants. Smaller lubricant sumps, reduced sizes of components, and increased performance will place high demands on the lubricants. Especially in mobile equipment (e.g., excavator, lawn mower), ecologically and toxicologically harmless lubricants will become a demand. Reduced maintenance and longer lubricant change intervals require high lubricant stability.

Improved thermal stability, low copper corrosion, and excellent solubility in synthetic fluids are demands on sulfurized products for the new generation of industrial lubricants.

REFERENCES

1. H. Sellei, Sulfurized extreme-pressure lubricants and cutting oils part 1, *Petroleum Processing*, 4, 1003–1008, 1949.
2. H. Sellei, Sulfurized extreme-pressure lubricants and cutting oils part 2, *Petroleum Processing*, 4, 1116–1120, 1949.
3. G.W. Pressel, Base for metal-cutting compounds and process of preparing the same, U.S. Patent 1,367,428, German Patent 129132, 1921.
4. J.T. Byers, Patents show trend in Extreme Pressure lube technology, *National Petroleum News*, 28, 79, 1936.
5. M.G. van Voorhis, 200 lubricant additive patents issued in 1938 and 1939, *National Petroleum News*, 32, R-66, 1940.
6. F.L. Miller, W.C. Winning, J.F. Kunc, Use of additives in automotive lubrication, *Refiner and National Gas Manuf*, 20(2), 53, 1941.
7. H.E. Westlake Jr., The sulfurization of unsaturated compounds, *Chemical Review*, 39(2), 219, 1946.
8. F.F. Musgrave, The development and lubrication of the automotive hypoid gear, *Journal of the Institute of Petroleum*, 32(265), 32, 1946.
9. L.R. Churchill, Lubricating Compound and process of making the same, U.S. Patent 1,974,299, 1932.
10. M.C. Edwards, J.V. Congdon, Method of sulfurizing pine oil and product thereof, U.S. Patent 2,012,466, 1934.
11. J.N. Borglin, Method of sulfurizing terpenes, abietyl compounds, etc., U.S. Patent 2,111,882, 1934.
12. A.E. Wade, G.M. McNulty, Pure compounds as extreme-pressure lubricants, U.S. Patent 2,110,281, 1934.
13. K. Baur, Production of mercaptanes, U.S. Patent 2,116,182, 1935.
14. B.H. Lincoln, W.L. Steiner, Sulphurized oils, U.S. Patent 2,113,810, 1936.
15. E.A. Evans, Lubricating oil, U.S. Patent 2,164,393, 1935.
16. B.B. Farrington, R.L. Humphreys, Extreme pressure lubricating composition, U.S. Patent 2,020,021, 1933.
17. E.W. Adams, G.M. McNulty, Lubricant composition, U.S. Patent 2,206,245, 1936.
18. J.B. Werder et al., Lubricating oil, U.S. Patent 1,971,243, 1932.
19. L.A. Mikeska, F.L. Miller, Lubricant containing organic sulphides, U.S. Patent 2,205,858, 1932.
20. C. Winning, D.T.R. Westfield, Compounded lubricating oil, U.S. Patent 2,422,275, 1942.
21. H.C. Mougay, G.O. Almen, Extreme pressure lubricants, *Ibid.*, 12, 76, 1931.

22. J.P. Baxter et al., Extreme pressure lubricant tests with pretreated test species, *Journal of the Institute of Petroleum*, 25(194), 761, 1939.

23. Schallbock et al., *Vorträge der Hauptversammlung der deutschen Gesellschaft für Metallkunde*, VDI Verlag, pp. 34–38, 1938.

24. C.F. Prutton et al., Mechanism of action of organic chlorine and Sulfur compounds in E.P. lubrication, *Journal of the Institute of Petroleum*, 32, 90–118, 1946.

25. H.G. Smith, Lubricant, U.S. Patent 2,179,067, 1938.

26. E.W. Cook et al., Lubricant, U.S. Patent 2,311,931, 1941.

27. E.S. Forbes, Load carrying capacity of organo-Sulfur compounds—A review, *Wear, Lausanne*, 15, 341, 1970.

28. A.G. Horodysky, Process for producing sulfurized olefins, U.S. Patent 3,703,504, 1972.

29. Rhein Chemie Rheinau GmbH, EP 1371949, 1972.

30. B.W. Hotten, Cross-sulfurized olefins and fatty acid monoesters in lubricating oils, U.S. Patent 4,053,427, German Publication 2235608, 1971.

31. J.N. Vinci, Metal working using lubricants containing basic alkali metal salts, U.S. Patent 4,505,830, 1985.

32. Hugo, *Rhein Chemie Rheinau GmbH Internal Studies*, Freie Universität Berlin, Berlin, Germany, 1980.

33. E.H. Farmer, F.W. Shipley, Modern views on the chemistry of vulcanization changes. I. Nature of the reaction between sulfur and olefins, *Journal of Polymer Science*, 1(4), 293–304, 1946.

34. M.G. Voronkov, N.S. Vyazankin, E.N. Derygina, A.S. Nakhmanovich, V.A. Usov, Reactions of sulfur with organic compounds, Chapter 3. In *The Action of Sulfur upon Hydrocarbons*, Consultants Bureau, New York, pp. 59–128, 1987, ISBN 0-306-10978-6.

35. D. Jungk, N. Schmidt, J. Hahn, P. Verlsoot, J.G. Haasnoot, J. Reedijk, *Tetrahedron*, 50(38), 11187–11196, 1994.

36. *Typical Composition of Natural Oils and Fats*, Flyer of the company Jacob Stern & Sons, Inc., Division Acme-Hardesty Co., Blue Bell, PA, 2016.

37. J. Korff, *Additive fuer Kuehlschmierstoffe, Additive fuer Schmierstoffe*, Expert Verlag, 3-8169-0916-7, Renningen Germany, 1994.

38. ASTM D 1662: May 2008 Standard Test Method for Active Sulfur in Cutting Oils.

39. A. Fessenbecker, Th. Rossrucker, E. Broser, Performance and ecology, two inseparable aspects of additives for modern metalworking fluids, Rhein Chemie Rheinau GmbH, International Colloquium, TAE Esslingen, 1992.

40. K.G. Allum, J.F. Ford, The influence of chemical structure on the load carrying properties of certain organo-sulfur compounds, *Journal of the Institute of Petroleum*, 51(497), 53–59, May 1965.

41. ASTM D 130: May 2004. Detection of copper corrosion from petroleum products by the copper strip tarnish test.

42. I. Roehrs, T. Rossrucker, Performance and ecology—Two aspects for modern greases, *NLGI Annual Meeting*, Rancho Mirage, CA, 1994.

43. T. Rossrucker, A. Fessenbecker, *Performance and Mechanism of Metalworking Additives: New Results from Practical Focused Studies*, Rhein Chemie Rheinau GmbH, *STLE Annual Meeting*, Las Vegas, NV, 1999.

44. J. Deodhar, Steel rolling oils—Cold rolling Lubrication, The metalworking fluid business, The College of Petroleum Studies, December 1989.

45. J.N. Vinci, Grease and gear lubricant compositions comprising at least one metal containing composition and at least one sulfurized organic compound, The Lubrizol Corporation, Wickliffe, OH, International Publication No. WO 87/06256, 1987.

46. G. Fish, Greases, GKN Technology Limited, WO 94/11470, EP 0668 900 B1, Wolverhampton West Midlands, Great Britain, 1994.

47. Hydraulic oils with detergent properties, according to Daimler specification DBL 6721, H. Pfander, Daimler Chrysler AG, Stuttgart, 2000.

48. J.N. Vinci, Metal working using lubricants containing basic alkali metal salts, The Lubrizol Corporation, Wickliffe, OH, US 4,505,830, 1985.

10 Chlorinated Paraffins

James MacNeil

CONTENT

It has been said that chlorinated paraffins were first used in the 1930s.[1] They are produced by the chlorination of linear hydrocarbons (Figure 10.1) and have been found to be useful in a variety of applications. Worldwide, the largest use is as a plasticizer for polyvinyl chloride. However, in the United States, that application is much smaller due to the ready availability of low-cost alternative plasticizers such as phthalates. The largest producer of chlorinated paraffins in the world is Inovyn, headquartered in the United Kingdom. There are several additional manufacturers in Europe and numerous local manufacturers in China and India. In North America, there are two producers: Qualice LLC and Dover Chemical.

Chlorinated paraffins are used in the metalworking industry as extreme pressure (EP) additives. This use dates back at least to the 1950s. They are used in the rubber industry as flame retardant plasticizers. In paints and coatings, they improve chemical and moisture resistance in applications such as swimming pool and traffic paint. They can provide flame retardancy as well in intumescent coatings. In adhesives, caulks, and sealants, they offer moisture resistance, promote adhesion, and can also increase flame retardancy. Flame retardancy is provided by thermal decomposition of the chlorinated paraffin at elevated temperatures. The decomposition produces hydrogen chloride, which in turn can disrupt the free-radical combustion chain reaction.

Chlorinated paraffins have a variety of physical properties that are determined by several factors. The choice of hydrocarbon feedstock is a major determinant of the physical properties. The longer the length of the carbon chain of the feedstock, the higher the viscosity of the chlorinated paraffin. There is also a direct correlation of the chlorine content and viscosity (Table 10.1).

A wide variety of physical properties allow chlorinated paraffins to be tailored to individual applications.

Viscosity is frequently the predominant physical property of interest to metalworking formulators. A wide variety of viscosities available in chlorinated paraffins provide numerous options for the formulator. Of secondary interest is chlorine content. Chlorine is the "active ingredient" in chlorinated paraffins, so the more chlorine on the molecule, the more cost-effective it can be. However, this principle has to be balanced by the viscosity, or the formulator runs the risk of producing a composition that has undesirable physical characteristics.

Chlorinated paraffins are prized for their EP performance. When two metal surfaces are moved against one another, microscopic surface asperities will come into contact and create friction. As the pressure applied to the two surfaces increases, so does the friction. The friction can then create heat and the asperities can thus become welded together. This effect occurs to a large extent on the areas where tools and workpieces come into contact. The welding activity creates tool wear and poor surface finishes. As the speed and severity of the operation increase, so does the tool wear. By chemically modifying the metal surface at the point of interface, the wear can be reduced and the finish of the newly created surface can be improved.

Chlorinated paraffins are used to perform the chemical modification. Under circumstances of high temperatures and pressures, also called EP, the chlorine on the chlorinated paraffin molecule will react with the metal producing metal chlorides.[2] Metal chlorides are typically planar, like graphite, in their structure or form complex polymers[3] that are very effective solid lubricants that slide past one another preventing welding. This effect occurs on the molecular level and can be quite dramatic enabling more severe operations to be performed at higher speeds. The reduction in friction results in reductions in operating temperatures and in tool wear. The process by which the metal chlorides are produced most likely proceeds via the thermal decomposition of the chlorinated paraffin at elevated temperatures. The decomposition produces minute quantities of hydrogen chloride, which then is the species that reacts with the interacting metal surfaces. The thermodynamics of iron–chlorine bond formation are favorable using an intermediate of hydrogen chloride as opposed to direct cleavage of the carbon–chlorine bond with subsequent metal chloride bond formation.

Other halogenated hydrocarbons can exhibit the same EP performance. Studies have been performed[4] using iodine compounds and they were found to perform very well. Unfortunately, they also caused a great deal of corrosion of the metal surface and so are considered to be too difficult to formulate into useful metalworking fluids. Bromine compounds

FIGURE 10.1   Preparation of chlorinated paraffins.

TABLE 10.1
Chlorinated Paraffin Physical Properties

Carbon Chain Length	% Cl by Weight	Viscosity (SUS at 210°F)
14–17	40	35
14–17	45	48
14–17	52	70
14–17	58	271
20+	40	83
20+	50	450
24–28	43	161
24–28	47	225

have also been studied[3] but have limitations due to cost and corrosion. The carbon chain length of the chlorinated paraffin does not seem to matter, and even carbon tetrachloride, with one carbon atom, is an effective EP additive.[5] However, it has associated environmental and toxicological issues and so is not used. The carbon chain length is more a function of available raw materials and solubility in common oils.

With nonferrous metals, particularly aluminum, the mechanism of action may be more related to viscosity than actual chemical reaction with the metal surface. Thermodynamics suggest that even with a hydrogen chloride intermediate the reaction is not very favorable[6]. Furthermore, there has been no observation of aluminum chloride documented.

Chlorinated paraffins are used predominantly in straight oil formulations and in emulsion metalworking fluids also known as soluble oils. They are almost never used in oil-free synthetic formulas. They are used in general-purpose metalworking fluids and with virtually all metals. They work synergistically with sulfurized fats and olefins and usually with phosphorus additives. Some straight oil formulations for general-purpose machining are shown in Tables 10.2 and 10.3.

An unusual application for chlorinated paraffins is fine blanking where very high pressures are used. For this process, it is not unusual for the metalworking fluid to be composed of 50%–100% chlorinated paraffin as the demands on the fluid to perform under EPs are very high.

TABLE 10.2
General-Purpose Machining Formula for Ferrous Metals

Additive	% by Weight
Base oil	85
Chlorinated paraffin	10
Sulfurized fat	5

TABLE 10.3
General-Purpose Machining Formula for Nonferrous Metals

Additive	% by Weight
Base oil	87
Chlorinated paraffin	8
Sulfurized fat or olefin	5

TABLE 10.4
General-Purpose Soluble Oil Concentrate Formula

Additive	% by Weight
Base oil	57
Surfactant package (soluble base)	20
Chlorinated paraffin	16
Additional surfactant	3
Biocide	3
KOH	1

TABLE 10.5
General-Purpose Semi-synthetic Concentrate Formula

Additive	% by Weight
Base oil	13
Surfactant package (soluble base)	16
Chlorinated paraffin	15
Water	45
Additional surfactant	3
Biocide	3
Tall oil fatty acid	5

Chlorinated paraffins are used extensively in emulsion metalworking fluids, commonly called soluble oils. Formulations are usually designed for a concentrate that is then diluted with water. Typical general-purpose soluble oil and semi-synthetic concentrate formulas are shown in Tables 10.4 and 10.5.

In drawing and stamping applications, chlorinated paraffins are used extensively. The amount that is used in each formulation is directly related to the severity of the operation. For heavy-duty applications such as deep drawing of stainless steel, a straight oil with 50% or more of a high-viscosity chlorinated paraffin will typically be used. The viscosity is important because it helps maintain the integrity of the lubricating film throughout the drawing process. Lighter-duty operations can use a lower-viscosity chlorinated paraffin at lower concentrations.

Chlorinated paraffins are very useful materials in metal-working fluids and provide outstanding cost-effective performance, enhancing the operation in a very wide variety of applications. They are useful in both straight and soluble oils and are compatible with all manner of additives. They are very important tools in the metalworking fluid formulator toolbox.

REFERENCES

1. Comments from Chlorinated Paraffins Industry Association to EPA, *Federal Register*, March 2016.
2. C. G. Williams, *Proceedings of the Royal Society of London. Series A, Mathematical and Physical Sciences*, 1952, *212*(1111), 512–515.
3. G. E. Barker, Effects of sulfur and chlorine content of water soluble cutting fluids on tool life–some popular misconceptions, Paper No. MR66-124, SME, Dearborn, 1966.
4. G. Barrow, Wear of Cutting Tools, *Tribology*, 1972, *5*, 22–30.
5. V. A. Tipnis and J. D. Christopher, *Machinability Testing and the Utilization of Machining Data*, ASM, Metals Park, OH, 1979, pp. 3–35.
6. G. W. Brady, M. B. Robin, and J. Varimbi, The structure of ferric chloride in neutral and acid solutions, *Inorganic Chemistry*, 1964, *3*(8), 1168–1173.

Section IV

Viscosity Control Additives

11 Olefin Copolymer Viscosity Modifiers

Michael J. Covitch

CONTENTS

11.1 INTRODUCTION

Olefin copolymer (OCP) viscosity modifiers are oil-soluble copolymers comprising ethylene and propylene and may contain a third monomer, a nonconjugated diene, as well. By virtue of their high thickening efficiency and relatively low cost, they enjoy a dominant share of the engine oil viscosity modifier market [1]. First introduced as a lubricant additive by Exxon in the late 1960s, the chemical and physical properties of OCPs continue to evolve to achieve improvements in low-temperature rheology, thickening efficiency, and bulk handling characteristics.

Several excellent reviews of OCP viscosity modifiers have been published [1–3,97]. This chapter serves as an update and current compilation of information relating to the chemistry, properties, and performance characteristics of this important class of lubricant additives.

11.2 CLASSES OF OLEFIN COPOLYMERS

There are many ways to classify OCP viscosity modifiers. From a user's perspective, OCPs are marketed as either solids, liquid concentrates, or viscous liquids. The physical state of the solids depends upon several factors, primarily the ethylene/propylene mass ratio (E/P) and molecular weight. When E/P is in the 45/55 to 55/45 range, the material is amorphous and cold-flows at room temperature. Thus, OCPs of this composition are most commonly sold as bales, packaged in rigid boxes to maintain bale shape. When E/P is higher than 60/40, the copolymer becomes semicrystalline in nature and does not cold-flow under ambient conditions. Thus, both bales and pellets can be produced.

OCP viscosity modifier compositions wherein propylene is the predominant monomer (60–90 mol%) have also been reported [99]. The remaining monomers may be ethylene or other alpha-olefins with at least four carbons, such as butene-1.

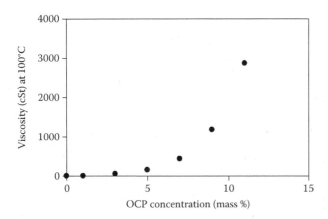

FIGURE 11.1 Kinematic viscosity of 50 PSSI amorphous OCP dissolved in 100N mineral oil.

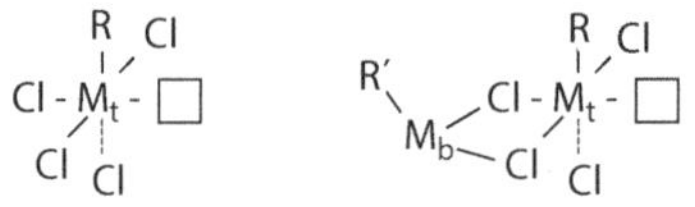

FIGURE 11.2 Active center [4] in Ziegler–Natta catalysts. (a) At crystal surface and (b) base metal complex (soluble specie). M_t, transition metal (such as Ti); M_b, base metal (such as Al).

Liquid concentrates of OCP in mineral oil contain enough rubber to raise the kinematic viscosity (KV) into the 500–1500 mm²/s (cSt) range at 100°C. A typical viscosity/concentration relationship is shown in Figure 11.1.

Very low-molecular-weight OCPs are viscous liquids at ambient temperature, with KV values ranging from 10 to 2000 mm²/s at 100°C. They are marketed by Lubrizol under the Lucant™ trade name and by Lion Elastomers as Trilene™.

From the preceding discussion, OCPs can also be classified according to crystallinity, which is measured by x-ray diffraction or differential scanning calorimetry (DSC). The influence of crystallinity on rheological performance will be discussed in Section 11.5.

Shear stability is another parameter by which OCP viscosity modifiers are categorized. The higher the molecular weight of a polymer, the more prone it is to mechanical degradation when elongational forces are imposed by the fluid flow field. This subject is dealt with in detail in Section 11.5.2.2.3.

Finally, chemical functional groups can be grafted to the OCP backbone, providing added dispersancy, antioxidant activity, wear protection, and/or low-temperature viscosity enhancement. A number of chemical routes for functionalizing OCPs will be described in Section 11.3.3.

11.3 CHEMISTRY

11.3.1 SYNTHESIS BY ZIEGLER–NATTA POLYMERIZATION

Although methods for synthesizing high-molecular-weight polymers of ethylene were commercialized in the 1930s (the ICI high-pressure process), the polymers contained a significant number of short- and long-chain branches that limited the ability to produce high-density polyethylene. The first commercially viable synthesis of linear polyethylene at low monomer pressure was pioneered by Ziegler in 1953, and the stereoregular polymerization of α-olefins was demonstrated by Natta the following year [4]. The secret to their success was the discovery of catalysts (called Ziegler or Ziegler–Natta catalysts), which are molecular complexes between

halides and other derivatives of Group IV–VIII transition metals (Ti, V, Co, Zr, Hf) and alkyls of Group I–III base metals. A typical catalyst of this type comprises an aluminum alkyl and a titanium or vanadium halide having the general structure shown in Figure 11.2. Electron donors, such as organic amines, esters, phosphines, and ketones, may be used to enhance reaction kinetics. Finally, molecular weight control is often aided by the use of chain transfer agents such as molecular hydrogen or zinc alkyls [5], which are effective in terminating chain growth without poisoning the active metal center.

Ziegler–Natta polymerization is probably the best-known example of insertion, coordination, and stereoregular or stereospecific polymerization. This nomenclature has been adopted to describe the mechanism(s) by which olefin monomers insert into the growing polymer chain, as directed by both steric and electronic features of the coordination catalyst. A commonly accepted chain propagation mechanism involves monomer insertion at the transition metal–carbon bond [4]. The main purpose of the base metal alkyl is to alkylate the transition metal salt, thus stabilizing it against decomposition. As pointed out by Boor [4], Ziegler–Natta catalysts may be modified to produce copolymers with varying degrees of randomness or, from a different perspective, blockiness of one or both comonomers. Due to the higher reactivity ratio of ethylene to that of propylene [4–7], the formation of long runs of ethylene is more favored than long sequences of propylene. This is substantiated by ¹³C NMR spectroscopy [8–13].

Ziegler–Natta catalysts come in two forms: heterogeneous and homogeneous. Heterogeneous catalysts are insoluble in the reaction medium and are suspended in a fluidized-bed configuration. Reaction takes place at the exposed faces of the metal complex surface. Since each crystal plane has a slightly different atomic arrangement, each will produce slightly different polymer chains in terms of statistical monomer insertion and molecular weight distribution. Thus, they are often called multisite catalysts. Homogeneous Ziegler–Natta catalysts are soluble in the reaction solvent and, therefore, function more efficiently since all molecules serve as potential reaction sites. Since the catalyst is not restrained in a crystalline matrix, it tends to be more "single site" in nature than heterogeneous catalysts. Polymers made by homogeneous polymerization generally are more uniform in microstructure and molecular weight distribution and, therefore, are favored for use as viscosity modifiers [1,4].

Nonconjugated dienes are often used in the manufacture of E/P copolymers, known as Ethylene–Propylene–Diene Monomer (EPDM) copolymers, to provide a site for cross-linking (in nonlubricant applications) or to reduce the tackiness of the rubber for ease of manufacture and handling. Certain dienes promote long-chain branching [2,5,14,15], which, in turn, increases the modulus in the rubber plateau region. The terpolymer is then easier to handle as it is dried and packaged [1]. A disadvantage of long-chain branching is that it reduces the lubricating oil thickening efficiency relative to a simple copolymer of similar molecular weight and copolymer composition, although low levels of vinyl norbornene or norbornadiene are claimed [15] to improve cold flow without loss in thickening efficiency or shear stability.

11.3.2 Synthesis by Metallocene Polymerization

The desire to achieve higher levels of control over stereoregularity, composition, and molecular weight distribution led to the development of activated metallocene catalysts. Although known to Ziegler and Natta, the technology was rediscovered by Kaminsky and Sinn in 1980 and further developed by workers such as Brintzinger, Chien, Jordan, and others [16–20]. Metallocene catalysts consist of compounds of transition metals (usually Group IVB: Ti, Zr, Hf) with one or two cyclopentadienyl rings attached to the metal. The most common activator is methylaluminoxane (MAO). A large number of variants have been reported, but the highest levels of stereospecificity have been achieved with bridged, substituted bis-cyclopentadienyl metallocenes (Figure 11.3). One of the major advantages of metallocenes over Ziegler–Natta catalysts is the ability to incorporate higher α-olefins and other monomers into the ethylene chain.

The first commercial use of metallocene single-site catalysts to manufacture EPDM elastomers was Dow Chemical Company's Plaquemine, LA, facility, which began operation in 1996 using Dow's Insite® constrained geometry catalyst [22,23]. The catalyst is described as "monocyclopentadienyl Group 4 complex with a covalently attached donor ligand...requiring activation by strong Lewis acid systems [such as] MAO...." Several advantages of this technology over conventional Ziegler–Natta processes were reported. Since the catalyst is highly efficient, less is needed; therefore, the process does not require a metal's removal or deashing step. In addition, the copolymers produced by this chemistry are reported to have narrow molecular weight distributions for good thickening efficiency and shear stability as well as good control over copolymer microstructure. Metallocene-catalyzed polyolefins also differ from Ziegler–Natta polymers in that the former contains a predominance of unsaturated ethylidene end groups [24].

Among various "post-metallocene" ethylene/α-olefin polymerization catalyst systems that have been reported in the literature, Dow Chemical's Olefin Block Copolymer (OBC) technology has been utilized to make multiblock ethylene–propylene copolymers for use as viscosity modifiers [94,95]. OBCs comprise at least one hard segment and at least one soft segment, but preferably many of each. The hard segments contain 60–95 wt.% ethylene, and the polymer overall contains 55–75 wt.% ethylene. The multiblock copolymer utilizes a three-component polymerization catalyst system: (1) one with a high comonomer incorporation index, (2) one with a lower comonomer incorporation index, and (3) a chain shuttling agent that transfers the active propagating center from one catalyst to the other, thereby affecting the number and distribution of blocks. Hydrogen can be used to control molecular weight by promoting chain termination. As the degree of blockiness increases, improvements in clarity, oil solubility, and compatibility with other polymers are attributed to this class of OCPs.

11.3.3 Functionalization Chemistry

Traditionally, OCPs are added to lubricating oil to reduce the degree to which viscosity decreases with temperature, that is, to function solely as a rheology control agent. Other lubricant additives—such as ashless succinimide dispersants, a variety of antioxidants, detergents, antiwear agents, foam inhibitors, friction modifiers, and anticorrosion chemicals—provide other important functions (dispersing contaminants, keeping engines clean, maintaining piston ring performance, preventing wear, etc.). It has been recognized for many years that it is possible to combine some of these performance and rheology control features on the same molecule. Some report that "both dispersant and antioxidant functionality may exhibit more potent activity when attached to the polymer backbone than in their monomeric form" [25]. Three hybrids have been commercialized, although many more have been disclosed in the patent literature. They include dispersant OCPs (DOCPs), dispersant antioxidant OCPs (DAOCPs), and polymethacrylate-grafted OCPs (gOCPs). The addition of antiwear functionality has also been reported [25,26].

Although many grafting reactions have been described in the literature, two general classes have received the most attention. Free radical grafting of nitrogen-containing monomers or alkylmethacrylates onto the OCP molecule is one class. Nitrogen-containing monomers such as vinyl pyridines, vinylpyrrolidones, and vinylimidazoles are often cited in the patent literature [27]. Free radical grafting with phenothiazine is claimed [27] to provide antioxidant functionality as well.

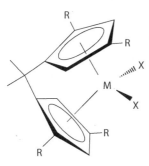

FIGURE 11.3 Chemical structure of generalized bridged bis-cyclopentadienyl metallocene catalyst. M is a Group IVB transition metal; X is a halogen or alkyl radical. (From Montagna, A.A. et al., *Chemtech*, 26, December 1997.)

The grafting reaction may be conducted with the OCP molecule dissolved in mineral oil or another suitable solvent; alternately, solvent-free processes have been disclosed [27] in which the reaction is conducted in an extruder.

Mixtures of alkylmethacrylate monomers, which are typical of those found in PMA viscosity modifiers, may be grafted to OCPs [28] to provide improved low-temperature properties. Adding nitrogen-containing monomers to the alkylmethacrylate mixture provides dispersancy characteristics as well. A common side reaction is homopolymerization, which can be minimized by process optimization. Homopolymers of N-containing monomers are usually not very soluble in mineral oils and often lead to hazy products and can attack fluoroelastomer seals. Homopolymers of alkylmethacrylates are fully soluble in oil, however. Thus, optimizing the grafting process is much more critical when working with N-containing monomers.

A second class of grafting reactions involves two steps [25,29]. In the first step, maleic anhydride or a similar diacyl compound is grafted onto the OCP chain, assisted by free radical initiators, oxygen [45], and/or heat. In the second step, amines and/or alcohols are contacted with the anhydride intermediate to create imide, amide, or ester bonds. In many respects, this second step chemistry is very similar to that used to create ashless succinimide dispersants. An advantage of this approach over free radical monomer grafting is that homopolymerization is avoided. The patent literature describes a related functionalization process in which free primary or secondary nitrogens of highly basic succinimide dispersants may be used to couple preformed dispersants to the maleic anhydride–grafted OCP molecule [31]. Amine derivatives of thiadiazole, phenolic [25], and amino-aromatic polyamine [32] compounds have been reacted with maleic anhydride–grafted OCP to provide enhanced antioxidant character to the additive [25]. Aromatic amine derivatives are claimed to perform well as soot dispersants in heavy-duty diesel lubricating oils [101].

Although maleic anhydride is the most common chemical "hook" for attaching functional groups to OCP polymers, a number of other approaches have been reported [30,33]. Further elaboration is beyond the scope of this review.

Another approach for attaching functionality to the OCP chain is via the nonconjugated diene in the terpolymer [26,34]. For example, 2-mercapto-1,3,4-thiadiazole is attached to the ethylidenenorbornene site on an EPDM polymer via addition of the thio group across the ethylidene double bond. The thiadiazole group is claimed to provide antiwear, antifatigue characteristics to lubricants containing the grafted OCP.

11.4 MANUFACTURING PROCESSES

Two polymerization processes have been used for the manufacture of ethylene–propylene copolymer viscosity modifiers: solution and slurry. In the solution process, the gaseous monomers are added under pressure to an organic solvent such as hexane, and the polymer stays in solution as it forms. By contrast, the slurry or suspension process utilizes a solvent such as liquid propylene in which the resultant copolymer is not soluble. It is reported [35] that removing the catalyst residue

from the polymer is more difficult in the slurry process, although some contend [36] that the levels of catalyst are so low that catalyst removal is not necessary.

Ethylene–propylene rubber was reported [37] to have been successfully manufactured in a fluidized-bed gas-phase reactor. However, the use of fluidization aids such as carbon black is necessary to process low-molecular-weight grades that are typical of lubricating oil viscosity modifiers. Thus, the gas-phase process is not appropriate for manufacturing OCP viscosity modifiers.

11.4.1 SOLUTION PROCESS

The most common method for manufacturing OCP viscosity modifiers is the solution process as described in Figure 11.4. It is made up of four sections: polymerization, polymer isolation, distillation, and packaging. In the polymerization section, monomers, an organic solvent such as hexane, and the soluble catalyst are introduced into a continuously stirred polymerization reactor. During polymerization, the polymer remains in solution and causes the bulk viscosity of the reaction medium to increase. To maintain good agitation, monomer diffusion, and thermal control, polymer concentration in the polymerization reactor is typically limited to 5–6 wt.% [36]. Up to five reactors arranged in series have been reported in the literature [39]. The effluent from the last reactor is contacted with an aqueous shortstop solution to terminate polymerization and wash away the catalyst, although this step is often omitted when using metallocene catalysts due to their high reactivity and, therefore, low concentration [22]. While the copolymer is still in solution, extender oils or antioxidants can be added. A laboratory method for producing low-molecular-weight OCPs (number-average molecular weight from 200 to 3000) has been reported [98].

In the isolation section, three techniques have been described in the literature. In the most common method (shown in Figure 11.4), the polymer is flocculated with steam, and the solvent and unreacted monomers are recovered, purified, and recycled. The aqueous polymer slurry is mechanically dewatered, granulated, and air-dried. A second nonaqueous method for isolating the polymer has been described in which the polymer is concentrated in a series of solvent removal steps [40,41]. The final step may be conducted in a devolatilizing extruder. A third technique does not isolate the polymer as a solid; rather, it mixes the polymer solution into mineral oil and distills off the solvent, producing a finished liquid OCP product [2].

Another type of solution polymerization process that has received a great deal of attention has been Exxon's tubular reactor technology. Its purpose is to generate a polymer with long blocks differing in monomer composition for improved performance as a viscosity-improving polymer [3,42]. A schematic of this process is shown in Figure 11.5. Monomers and solvent are premixed with a highly active Ziegler–Natta polymerization catalyst and metered into a plug flow reactor under conditions that minimize chain transfer and termination reactions. Ethylene and/or propylene is injected into the tube at different points to adjust the local monomer concentration and, thereby, the monomer composition along the growing polymer chain. In comparing the rheological properties

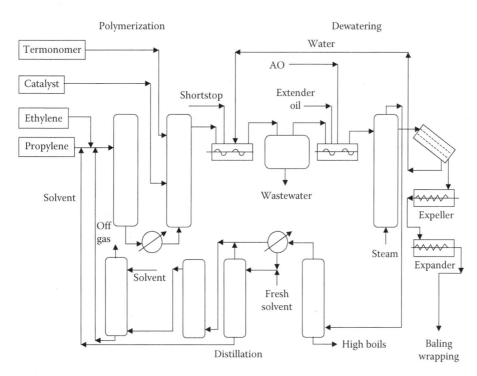

FIGURE 11.4 Solution process for the manufacture of EPDM. (From Ethylene–propylene rubber, Hydrocarbon Processing, p. 164, November 1981.)

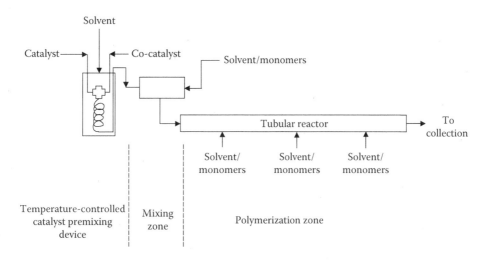

FIGURE 11.5 Tubular reactor process [42] for the preparation of multiblock ethylene/propylene copolymers.

of different A-B-A-type block compositions, Ver Strate and Struglinski reported [12] that "chains with high ethylene section in the center of the chain … associate at low temperature with little intermolecular connectivity." When the high-ethylene (crystallizable) segments are at the ends, polymer networks can form at low temperatures, which can impart a gelatinous texture to the solution.

11.4.2 SUSPENSION PROCESS

Ethylene and a nonconjugated diene, if desired, are contacted with liquid propylene, which acts as both a monomer and the reaction medium [36,43]. In the presence of a suitable catalyst,

polymerization takes place rapidly, producing a suspension of copolymer granules that are insoluble in the reaction medium. The heat liberated during the polymerization reaction is dissipated by propylene evaporation, thus providing a convenient mechanism for temperature control. In addition, since the polymer is not soluble in the reaction medium, viscosity remains low. Thus, relative to typical solution processes, the polymer concentration in a suspension reactor can be five to six times higher. Upon exiting the polymerization reactor, the polymer suspension is contacted with steam to strip off unreacted propylene that is then recycled. According to Corbelli and Milani [43], the Dutral® process does not include a catalyst washing step. The copolymer product, in aqueous suspension, is dewatered, dried, and packaged in a similar fashion to polymer made by the solution process.

11.4.3 POSTPOLYMERIZATION PROCESSES

There are two main types of packaging processes for medium- to high-molecular-weight OCPs in practice today [37]. In one, the isolated polymer is mechanically compressed into rectangular bales. The bales are often wrapped in a polyolefin packaging film to prevent the bales from adhering to one another during storage and to prevent foreign matter from sticking to the tacky rubber surface. Typical types of polyolefins films include poly(ethylene-co-vinyl acetate), low-density polyethylene, and ethylene/α-olefin copolymers. Another method for packaging solid OCP rubber is to extrude the polymer, pass the melt stream into a water-cooled pelletizer, and dry the final product. The pellets may be packaged in bags or boxes or may be compressed into rectangular bales.

The mechanical properties of the rubber often dictate what type of isolation and packaging processes are the most appropriate. Amorphous E/P copolymers are often too sticky to successfully traverse the conventional flocculation/drying/baling process. One way to modify these compositions to improve their handling characteristics is by introducing long-chain branching [5,44,45] through the use of low concentrations of nonconjugated dienes or other branching agents. For non-functionalized OCPs, this is the main reason that some commercial viscosity modifiers contain dienes [2]. Copolymer compositions higher in ethylene content (>60 wt.% [5]) are often semicrystalline and may be amenable to packaging in pellet form. In some cases, the pellets may contain an anticaking or antiblocking agent to prevent agglomeration.

Another type of manufacturing process has been used to manufacture low-molecular-weight OCP viscosity modifiers that are difficult to isolate and package in conventional equipment. A higher-molecular-weight feedstock of the appropriate composition may be fed into a masticating extruder or Banbury mixer to break down the polymer chain to lower-molecular-weight fragments using a combination of heat and mechanical energy [46]. Several patents describe the use of oxygen or free radical initiators [47,48] to enable this process.

11.4.4 MAKING THE OCP LIQUID CONCENTRATE

After the solid OCP viscosity modifier has been manufactured, it must be dissolved in oil before it can be efficiently blended with base stocks and other additives. The first stage entails feeding the rubber bale into a mechanical grinder [2] and then conveying the polymer crumb into a high-quality diluent oil that is heated to 100°C–130°C with good agitation. The rubber slowly dissolves, raising the viscosity of the oil as shown in Figure 11.1. Certain high-intensity homogenizers can also be used in which the entire rubber bale is fed directly into a highly turbulent diluent oil tank at high temperature; this bypasses the pregrinding step.

When the solid polymer is supplied in pellet form, the rubber can be fed directly into hot oil, or if it is slightly agglomerated, it first may be passed through a low-energy mechanical grinder.

11.5 PROPERTIES AND PERFORMANCE CHARACTERISTICS

11.5.1 EFFECT OF ETHYLENE/PROPYLENE RATIO ON PHYSICAL PROPERTIES OF THE SOLID

The comonomer composition of E/P copolymers has a profound influence on the physical properties of the rubber. These properties, in turn, dictate the type of containers in which the product can be stored and how it is handled during distribution and use.

^{13}C NMR has been used extensively to characterize the sequence distribution of E/P copolymers [49–54]. As the ratio of E/P increases, the fraction of ethylene–ethylene sequences (dyads) rises, as demonstrated by the data in Figure 11.6. Concurrently, the total fraction of ethylene–propylene dyads decreases (forward and reverse propylene insertion are designated p and p*, respectively). Thus, the average length of contiguous ethylenes increases with ethylene content. Above about 60 wt.% ethylene, these sequences become long enough to crystallize, as measured by DSC (Figure 11.7) or x-ray diffraction.

When the degree of crystallinity exceeds about 25%, E/P copolymers become unsuitable as viscosity modifiers due to limited solubility in most mineral oils. As the propylene content

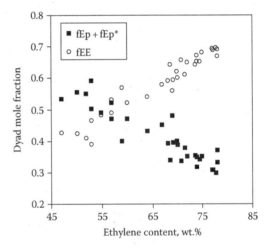

FIGURE 11.6 The effect of ethylene content (wt.% as measured by NMR) on ethylene–ethylene and ethylene–propylene dyad concentration.

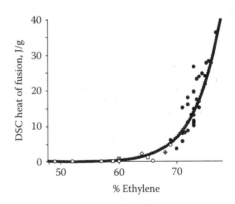

FIGURE 11.7 The effect of ethylene content (wt.%) on crystallinity as measured by DSC for a range of experimental EPDM copolymers.

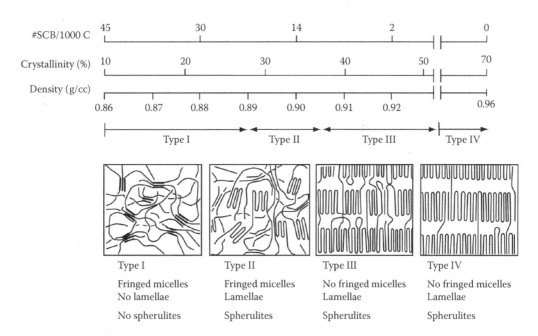

FIGURE 11.8 Schematic illustration of the solid-state morphologies of four types of poly(ethylene-co-octene) copolymers. # SCB/1000 C is defined as the number of side chain branches per 1000 backbone carbon atoms. (From Minick, J. et al., *J. Appl. Polym. Sci.*, 58, 1371, 1995.)

approaches zero, the copolymer takes on the physical characteristics of high-density polyethylene, which, due to its inertness to oil, is used as the packaging material of choice for engine oils and other automotive fluids. Microstructural investigations of metallocene ethylene/α-olefin copolymers by Minick et al. [55] concluded that the relatively short ethylene sequences of low-crystallinity (<25%) samples are capable of crystallizing into fringed micelle or short bundled structures (Figure 11.8). Higher-order morphologies such as lamellae or spherulites are not observed. Therefore, the physical properties of semicrystalline OCPs fall in between those of polyethylene and amorphous E/P rubber.

Polyethylene is a rigid, high-modulus solid at room temperature. Amorphous E/P rubber is a relatively soft material under ambient conditions, which cold-flows and exhibits a tacky feel. The degree of tack is inversely proportional to molecular weight and can be reduced by the incorporation of long-chain branching. Solid bales of this type of rubber are easily compressed and further densify during storage. Semicrystalline OCPs hold their shape during storage but are slightly tacky to the touch. Higher compression pressures, longer compression times, and higher finishing temperatures are required to successfully produce dense bales. Typical physical properties of E/P copolymers are summarized in Table 11.1.

11.5.2 Effect of Copolymer Composition on Rheological Properties in Solution

11.5.2.1 Low-Temperature Rheology

Rubin et al. [56–62] and others [63,100] measured the intrinsic viscosity [η] of E/P copolymers as a function of temperature in various solvents (Figure 11.9). Intrinsic viscosity, a measure of polymer coil size in dilute solution, is fairly insensitive to temperature for noncrystalline OCPs. Semicrystalline copolymers undergo a precipitous drop in [η] as temperature

TABLE 11.1

Typical Physical Properties of Ethylene–Propylene Copolymers

Property	Typical Value
Density, kg/m^3	860
Heat capacity, cal/g °C	0.52
Thermal conductivity, cal/cm·s °C	8.5×10^{-4}
Thermal diffusivity, cm^2/s	9.2×10^{-4}
Thermal coefficient of linear expansion, /°C	2.2×10^{-4}

Source: Corbelli, L., *Dev Rubber Technology*, 2, 87, 1981.

drops below about 10°C. In this region, the polymer begins to crystallize, forming intramolecular associations that effectively cause the molecule to shrink in on itself yet remain sufficiently solvated to remain suspended in oil solution.

The viscosity of a dilute polymer solution often follows the Huggins equation [64]

$$\frac{\eta_{sp}}{c} = [\eta] + k'[\eta]^2 c$$

where

c is the polymer concentration (g/dL)
η_{sp} is the specific viscosity ($(\eta - \eta_o)/\eta_o$)
η is the solution viscosity
η_o is the solvent viscosity
k′ is a constant

Thus, for a specific lubricant composition, c and η_o are fixed, and the temperature dependence of the solution viscosity is directly related to that of the intrinsic viscosity.

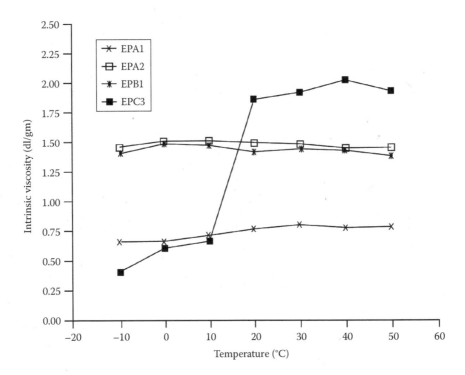

FIGURE 11.9 Intrinsic viscosities of EP copolymers in SNO-100 base oil at various temperatures [61]. EPC3, EPB1, EPA2, and EPA1 have 73, 61, 50, and 50 wt.% ethylene, respectively. Differences in intrinsic viscosity above 20°C are attributable to differences in molecular weight.

Low-temperature viscosity is an important rheological feature of automotive lubricants. For the vehicle to start in cold weather, the lubricant viscosity in the bearings should be below a critical value as determined by low-temperature engine startability experiments [7] and defined within SAE J300 [65] for all "W" grades. The cold cranking simulator (CCS), ASTM D5293, is a high-shear-rate rheometer operating at fixed subambient temperatures, designed to simulate oil flow in automotive bearings during start-up. After the engine starts, the lubricant must also be able to freely flow into the oil pump and throughout the internal oil distribution channels of the engine. This is the other half of the low-temperature viscosity specification for motor oils [65]. The mini-rotary viscometer (MRV), ASTM D4684, is a low-shear-rate rheometer designed to simulate pumpability characteristics of a multigrade oil in a vehicle that was sitting idle for about two days in cold weather. SAE J300 also contains upper viscosity limits for MRV viscosity and yield stress for all "W" grades.

Thus, the mechanism of intramolecular crystallization, which leads to molecular size contraction in solution, affords high-ethylene OCP viscosity modifiers the opportunity to contribute less to viscosity at low temperatures than noncrystalline or amorphous copolymers of similar molecular weight. For this reason, the class of E/P copolymers having ethylene content greater than about 60 wt.% is often called low-temperature OCPs, or LTOCPs. A rheological comparison of two LTOCPs and one conventional OCP viscosity modifier in several SAE 5W-30 oil formulations may be found in Figures 11.10 and 11.11. These data illustrate the low-temperature rheological benefits of LTOCPs.

A number of workers have cautioned, however, that the long ethylene sequences of LTOCPs are similar in structure to paraffin wax and can interact with waxy base oil components at low temperatures. In many cases, they require specially designed pour point depressants to function properly in certain base stocks. Thus, LTOCP viscosity modifiers have been implicated in problems such as MRV failures in comingled fresh oils [66] and used passenger car lubricants [67,68]. Attempts to modify the microstructure of LTOCPs by process optimization to improve MRV performance have been reported [96].

11.5.2.2 High-Temperature Rheology

Copolymer composition has less influence on high-temperature rheological behavior than it has at low temperatures, partly because lightly crystalline OCPs have melting points well below 100°C [58]. Since both high-temperature KV and high-temperature high-shear-rate (HTHS) viscosity are used to classify multigrade engine oils [65], it is important to understand how copolymer composition and molecular weight influence these key parameters.

11.5.2.2.1 Kinematic Viscosity

For both economic and performance reasons, it is desirable to limit the amount of polymer needed to achieve a given set of rheological targets. Therefore, it is important to quantify the effects of molecular weight, molecular weight distribution, and branching on thickening efficiency. Thickening efficiency has been defined in many ways, but the most common definitions are (1) the amount of polymer necessary to increase the KV of a reference oil to a certain value or (2) the KV or

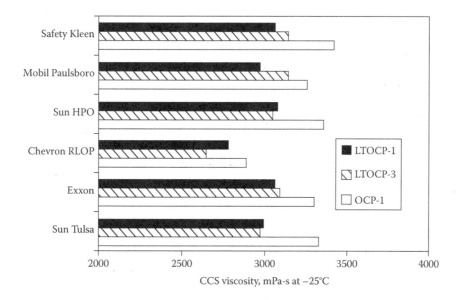

FIGURE 11.10 Cold cranking simulator viscosity for six SAE 5W-30 lubricant formulations, each blended with one of three viscosity modifiers: LTOCP-1, LTOCP-3, or OCP-1. Within each base oil type, the ratio of high- and low-viscosity base oil was kept constant.

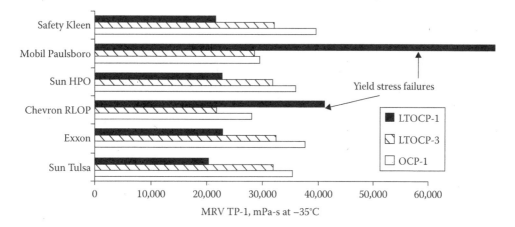

FIGURE 11.11 Mini-rotary viscosity results for six SAE 5W-30 lubricant formulations, each blended with one of three viscosity modifiers: LTOCP-1, LTOCP-3, or OCP-1. Within each base oil type, the ratio of high- and low-viscosity base oil and the type and concentration of pour point depressant was held constant.

specific viscosity (see Section 11.5.2.1) of a given polymer concentration in a reference oil. For polymers of equal molecular weight, thickening efficiency increases with ethylene content and is highest for copolymers with narrow molecular weight distributions [1,2]. A plot of intrinsic viscosity versus weight average molecular weight (Figure 11.12) demonstrates the familiar Mark–Houwink power law relationship:

$$[\eta] = K'M^a$$

Assuming a single value for the power law constant a = 0.74, Crespi and coworkers [5] published a table of K′ values as a function of ethylene content, which is reproduced in Table 11.2. This clearly shows that thickening efficiency can be improved by increasing ethylene concentration.

This is further illustrated by plotting data from Kapuscinski et al. [59] in Figure 11.13. The 80 mol% ethylene copolymer requires less polymer to achieve a target viscosity than a 60 mol% copolymer; therefore, the former has a higher thickening efficiency than the latter.

Long-chain branching has a directionally detrimental effect on thickening efficiency for polymers of equal molecular weight. This is not surprising, since the average chain end-to-end distance of a random coil in solution is controlled, in large part, by the length of the main chain. Branching essentially shortens the chain and lowers its hydrodynamic radius. For example, Table 11.3 contains thickening efficiency data for two sets of noncrystalline OCP viscosity modifiers, one linear and the other containing 2% branching agent, each set differing only in molecular weight.

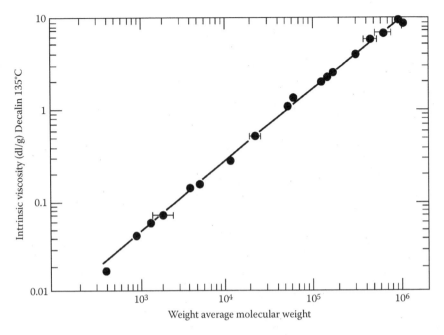

FIGURE 11.12 Intrinsic viscosity as a function of weight average molecular weight [1] for ethylene/propylene copolymers of narrow polydispersity ($M_w/M_n \sim 2$).

TABLE 11.2
Mark–Houwink K′ Constants [5] for Ethylene–Propylene Copolymers Containing Different Mole Percentages of Ethylene (a = 0.74)

Mol% Ethylene	K′ × 10⁴	Mol% Ethylene	K′ × 10⁴
5	2.020	55	3.020
10	2.115	60	3.140
15	2.205	65	3.260
20	2.295	70	3.385
25	2.390	75	3.515
30	2.485	80	3.645
35	2.585	85	3.790
40	2.690	90	3.940
45	2.795	95	4.240
50	2.910	—	—

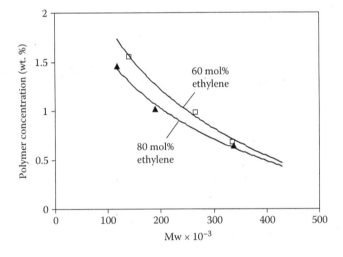

FIGURE 11.13 Polymer concentration needed to raise the kinematic viscosity of a 130N base oil to 11.5 cSt. (Data from Kapuscinski M.M. et al., Solution viscosity studies on OCP VI improvers in oils, *Soc. Auto. Eng. Technol.*, Paper Ser. 892152, 1989.)

11.5.2.2.2 High-Temperature High-Shear-Rate (HTHS) Viscosity

Since concentric journal bearings operate in the hydrodynamic or elastohydrodynamic lubrication regimes, oil film thickness is a critical factor influencing wear [69,70]. For this reason, SAE J300 specifies a minimum HTHS viscosity for each viscosity grade [65]. HTHS viscosity is measured at very high shear rates and temperatures (10^6 s⁻¹ and 150°C, respectively), which is similar to the flow environment in an operating crankshaft bearing at steady state. At these rates of deformation, most high-molecular-weight polymers will align with the flow field [71], and a temporary reduction in

TABLE 11.3
Thickening Efficiency of Linear and Branched OCP Viscosity Modifiers

Mw	Linear	Branched
230,000	13.50	12.03
180,000	11.17	10.87

Thickening efficiency is defined as the kinematic viscosity (at 100°C) of a 1.0 wt.% polymer solution in 6.05 cSt mineral oil.

TABLE 11.4

Rheological Comparison of Lubricants Containing OCP Viscosity Modifiers Differing in Molecular Weight

Viscosity Modifier	Wt. Average Molecular Weight	PSSI	Capillary Viscosity, cP at 150°C	HTHS, cP at 150°C	% TVL
OCP1	160,000	45	5.33	3.43	36
OCP2	80,000	30	5.33	3.77	29
OCP3	50,000	22	5.33	3.88	27

Source: Spiess, G.T. et al., Ethylene propylene copolymers as lube oil viscosity modifiers, in Bartz, W.J., ed., Addit. Schmierst. Arbeitsfluessigkeiten., 5th Int. Kolloq., 2,8., 10–1, Tech. Akad. Esslingen, Germany, 1986.

viscosity is measured. The difference between low- and high-shear-rate viscosities at 150°C is termed temporary viscosity loss (TVL) or percent temporary viscosity loss (relative to the low-shear-rate KV). As is true for most polymers [71], TVL is proportional to molecular weight [1] as shown in Table 11.4. For polymers of equal weight average molecular weight, those with narrow molecular weight distributions undergo less TVL than those with broad M_w/M_n values [1].

HTHS viscosity can be adjusted by increasing the viscosity of the base oil or by increasing the viscosity modifier concentration, as shown in Figure 11.14. Since the formulation also has to meet KV and CCS viscosity limits, there is often only limited flexibility to adjust HTHS viscosity within the bounds of a given set of base oils and additives.

11.5.2.2.3 Permanent Shear Stability

The tendency of an OCP molecule to undergo chain scission when subjected to mechanical forces is dictated by its molecular weight, molecular weight distribution, ethylene content,

and degree of long-chain branching. Mechanical forces that break polymer chains into lower-molecular-weight fragments are elongational in nature, causing the molecule to stretch until it can no longer bear the load. This loss in polymer chain length leads to a permanent degradation of lubricant viscosity at all temperatures. In contrast to temporary shear loss, permanent viscosity loss (PVL) represents an irreversible degradation of the lubricant and must be taken into account when designing an engine oil for commercial use.

PVL is similar to TVL, except that the viscosity loss is measured by KV before and after shear. Permanent shear stability is more commonly defined by the Permanent Shear Stability Index (PSSI) or simply SSI, according to ASTM D6022:

$$PSSI = SSI = 100 \times \frac{\left(V_o - V_s\right)}{\left(V_o - V_b\right)}$$

where

V_o is the viscosity of unsheared oil
V_s is the viscosity of sheared oil
V_b is the viscosity of the base fluid (without polymer)

SSI represents the fraction of viscosity contributed by the viscosity modifier that is lost during shear. SSI is proportional to $\log_{10}$ molecular weight (M_w), as shown in Figure 11.15. Commercial engine oil OCP viscosity modifiers have SSI values in the 23–55 range.

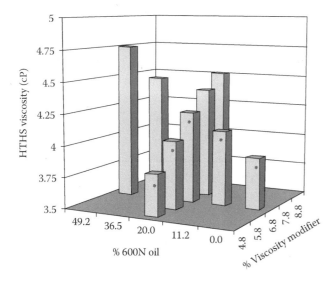

FIGURE 11.14 The effects of base oil composition and viscosity modifier concentration on high-temperature high-shear-rate viscosity. SAE 15W-40 engine oil consisting of European API Group I base oils (150N + 600N), an ACEA A3-98/B3-98 quality performance additive, an oil diluted amorphous OCP VM, and a pour point depressant. Bars marked with an asterisk comply with ASTM D445 and D5293 limits for SAE 15W-40 oils.

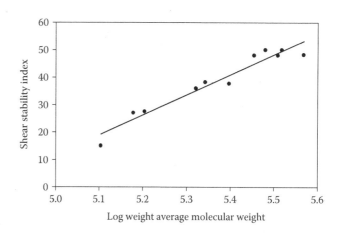

FIGURE 11.15 Relationship between weight average molecular weight and SSI (ASTM D3945) for a series of OCP viscosity modifiers.

Although ASTM D6022 provides a definition for SSI, it is important to recognize that the only component that is responsible for viscosity loss during shear is the high-molecular-weight polymer. If the additive for which SSI is to calculated happens to be a concentrated polymer solution in oil, according to the strict definition of ASTM D6022, the composition of the "base fluid" does not include the VM diluent oil. Since the diluent oil viscosity is usually lower than the base blend viscosity for most viscosity grades, V_b is higher than it would be if the VM diluent oil viscosity were factored into V_b. For example, take an SAE 15W-40 engine oil formulated with a liquid OCP concentrate containing 10 wt.% polymer in a 5.1 cSt mineral oil. V_o and V_s are 15.2 and 12.8 cSt, respectively. The base blend viscosity (when the VM component is a liquid) is 9.4 cSt. When the VM component is defined as the solid polymer, V_b is 9.15 cSt. The calculated shear stability index values are 41.4 and 39.7, respectively. Thus, the numerical value of SSI is dependent upon the definition of the polymeric additive in question.

The concept of "stay-in-grade" is generally used to refer to a lubricating oil, when tested in vehicles or laboratory shearing devices, which maintains its KV within the limits of its original SAE viscosity grade. The problem with viscosity measurements of engine drain oils is that many factors other than permanent polymer shear influence viscosity—such as fuel dilution, oxidation, and soot accumulation. Therefore, it is customary to measure PVL after shear in a laboratory rig, the most common being the Kurt Orbahn test, ASTM D6278. Several reviews of methods for determining the shear stability of polymer-containing lubricating oils have been published [72–74].

Selby devised a pictorial scheme for mapping the effects of shear rate and PVL on high-temperature viscosity, the viscosity loss trapezoid (VLT) [75], shown in Figure 11.16. The corners of the "trapezoid" are defined by viscosity data, and the points are connected by straight lines. Note that the straight lines do *not* imply that there is a linear relationship between viscosity and shear rate. The VLT is a convenient graphical representation of the temporary and permanent shear loss characteristics of polymer-containing oils. Molecular weight degradation causes a permanent loss in both KV and HTHS, but the magnitude of

the former is always larger than the latter. The shape of the VLT is characteristic of polymer chemistry and molecular weight.

It is experimentally observed [76] that the Kurt Orbahn shear test breaks molecules above a threshold molecular size; molecules smaller than the threshold value are resistant to degradation. Selby [75] uses this observation to derive certain qualitative conclusions of the polymer molecular weight distribution from the shape of the VLT.

11.5.3 EFFECT OF DIENE ON THERMAL/OXIDATIVE STABILITY

There has been little solid scientific data published in the literature to compare the relative thermal and oxidative stability of oil solutions containing E/P copolymers versus EPDM terpolymers. Marsden [2] states that "introduction of a termonomer… can…detrimentally affect shear and oxidation stability, dependent on the monomer," but he offers no data. Others [5,36] cite high-temperature aging experiments on solid rubber specimens, which demonstrate that EP copolymers are more stable (in terms of tensile properties) than EPDM terpolymers of similar E/P ratio. Copolymers containing higher levels of ethylene are claimed to have better thermal/oxidative stability than more propylene-rich copolymers, presumably due to the lower concentration of oxidatively labile tertiary protons contributed by the propylene monomer. High thermal stresses are sufficient to promote hydrogen abstraction by a free radical mechanism. The relative susceptibility of protons to hydrogen abstraction follows the classical order tertiary > secondary > primary. In the presence of oxygen, peroxy radicals are formed, which can accelerate the degradation process.

Despite these suggestions that diene-containing E/P copolymers may be less thermally stabile than EP copolymers, the author is not aware of any definitive studies that have shown that EPDM viscosity modifiers are more likely to degrade in service than EP copolymers. Indeed, engine oils formulated with both types have been on the market for years, although EP copolymers dominate.

11.5.4 COMPARATIVE RHEOLOGICAL PERFORMANCE IN ENGINE OILS

The most influential factors governing the rheological performance of OCPs in engine oils are molecular weight and monomer composition. The effects of molecular weight and molecular weight distribution were discussed in Section 11.5.2.2.3, and the influence of E/P ratio on low-temperature rheology was covered in Section 11.5.2.1. In this section, two comparative rheological studies are presented to further illustrate the links between OCP structure and rheological performance.

11.5.4.1 Comparative Study of OCP Viscosity Modifiers in a Fixed SAE 5W-30 Engine Oil Formulation

There are two ways to compare the relative performance of several viscosity modifiers. One is to choose a fixed engine oil formulation where the base oil composition and additive

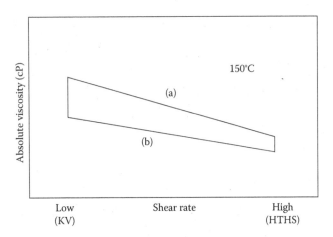

FIGURE 11.16 Viscosity loss trapezoid, per Selby [75]. (a) Fresh oil viscosities and (b) oil viscosities after permanent shear.

TABLE 11.5
Properties of OCP Viscosity Modifiers Used in Table 11.6

Viscosity Modifier Code	Shear Stability Index (ASTM D6278)	Copolymer Type
OCP1	55	EP, LTOCP
OCP2	50	EPDM, amorphous
OCP3	37	EP, amorphous
OCP4	25	EP, amorphous

concentrations are held constant, and the VM level is adjusted to achieve a certain 100°C KV target. The other is to adjust both base oil composition and VM concentration to achieve predetermined KV *and* CCS viscosity targets. Section 11.5.4.1 offers an example of the first and Section 11.5.4.2 illustrates the second approach.

Four OCP viscosity modifiers were blended into an SAE 5W-30 engine oil composition consisting of a 95/5 w/w blend of Canadian 100N/250N mineral base stocks, an ILSAC GF-2 quality performance additive, and a polyalkylmethacrylate pour point depressant. The viscosity modifiers are described in Table 11.5. OCP1 and OCP2 are high SSI polymers differing in both E/P ratio and diene content. OCP3 and OCP4 are progressively more shear stable and have essentially 0% crystallinity.

Rheological data are summarized in Table 11.6. Comparing OCP1 and OCP2, the former is a more efficient thickener because it contains no long-chain branching (no diene monomer) and it has a higher ethylene content (see Figure 11.13). The latter property also manifests itself in lower CCS viscosity. The MRV viscosity is also lower for OCP1 relative to the other amorphous OCPs, but this is highly dependent upon the particular pour point depressant that was chosen for this study. The fact that most of the oils displayed yield stress failures at −35°C shows that the PPD was not optimized for this particular set of components.

As SSI decreases from OCP1 to OCP4, the polymer concentration needed to reach a KV target of 10 cSt increases. In other words, polymer thickening efficiency is proportional to shear stability index. Among the noncrystalline OCPs, increasing polymer level causes the CCS viscosity to increase.

Since all oils were formulated with the same base oil composition, HTHS viscosity is relatively constant, independent of OCP type. OCP4, the lowest-molecular-weight polymer, should have the lowest degree of TVL, and it indeed has the highest HTHS viscosity of the group.

11.5.4.2 Comparative Study of 37 SSI OCP Viscosity Modifiers in an SAE 15W-40 Engine Oil Formulation

In this example, the base oil ratio and polymer concentration were adjusted to achieve the following targets: KV = 15.0 cSt and CCS = 3000 cP at −15°C. The base stocks were API Group I North American mineral oils, the additive package was of API CH-4 quality level, and the pour point depressant was a styrene ester type that was optimized for these base oils. All viscosity modifiers (see Table 11.7) were nominally 37 SSI according to ASTM D6278. Rheological results are summarized in Table 11.8.

OCP3 is the same polymer as in Table 11.5. Although OCP8 and OCP9 are semicrystalline LTOCPs, they represent different manufacturing technologies, broadly described in Figures 11.4 and 11.5, respectively. Incidentally, OCP1 in Table 11.5 was also manufactured by the tubular reactor technology described in Figure 11.5.

The rheological data in Table 11.8 further illustrate several features of LTOCPs mentioned earlier. Their inherently lower CCS viscosity contributions permit the greater use of higher-viscosity base oils, which can be beneficial in meeting volatility requirements. The low-temperature MRV performance of OCP9 was far inferior to that of the other copolymers, indicating that the pour point depressant chosen for this particular study was not optimized for OCP9 in these base stocks.

Another polyalkylmethacrylate pour point depressant was found to bring the MRV viscosity of the OCP9 formulation down to 7,900 and 18,000 cP at −20°C and −25°C, respectively. More will be said about interaction with pour point depressants in Section 11.5.7.

Again, the higher thickening efficiency of E/P copolymers versus EPDMs of similar molecular weight (shear stability) is clearly demonstrated in Table 11.8. Another feature worth noting is that increasing base oil viscosity can nudge HTHS viscosity upward (compare OCP7 with OCP8 or OCP3 with OCP9).

TABLE 11.6
Rheological Properties of SAE 5W-30 Engine Oils Containing Different OCP Viscosity Modifiers

	OCP1	OCP2	OCP3	OCP4
Polymer content, wt.%	0.58	0.71	0.73	1.05
Kinematic viscosity, cSt 100°C	10.17	10.09	9.99	10.10
Viscosity index	156	160	160	158
CCS viscosity, cP −25°C	3,080	3,280	3,510	3,760
MRV viscosity, cP −30°C	13,900	26,500	26,700	28,100
MRV viscosity, cP −35°C	40,100	Yield stress	Yield stress	Yield stress
HTHS, cP	2.95	2.88	2.96	3.07

TABLE 11.7

OCP Viscosity Modifiers Used in Rheological Study in Table 11.8

Viscosity Modifier Code	Copolymer Type
OCP7	EPDM, amorphous
OCP3	EP, amorphous
OCP8	EPDM, LTOCP
OCP9	EP, LTOCP

11.5.5 INTERACTION WITH POUR POINT DEPRESSANTS

Although base oil and VM play a role in determining low-temperature oil pumpability, the pour point depressant provides the primary control in this area. SAE J300 [65] specifies the MRV test (ASTM D4684) as the sole guardian of pumpability protection, although it acknowledges that other tests may also be useful in the development of lubricants from new components. The Scanning Brookfield test (ASTM D5133) and Pour Point (ASTM D5873), although not required within SAE J300, are often contained in other standards established by original equipment manufacturers (OEMs), oil marketers, and governmental agencies and, therefore, must also be considered in the development of modern engine oils.

Advances in base oil technology have led, in recent years, to a wide range of mineral and synthetic lubricant base stocks [77], classified as API Group I, II, III, IV, and V stocks. The API system classifies oils according to viscosity index (VI), saturates content, and sulfur level. Group I mineral oils are defined as having less than 90% saturates and VI greater than 80 but less than 119 and more than 0.03% sulfur. Group II and III oils have less than 0.03% sulfur and greater than 90% saturates, but they differ mainly in VI. Group II oils have VI values from 80 to 119, whereas Group III stocks have VI values of 120 or more.

Formulating these conventional and highly refined oils to meet all of the rheological requirements of SAE J300 is not always straightforward. An important aspect of base oil technology that is not embodied within the API Group numbering scheme is the type of dewaxing process or processes

employed. It is well known [78–84] that the low-temperature oil pumpability performance of engine oils is often impeded by the nucleation and growth of wax crystals, which can coalesce and restrict the flow of oil at low temperatures. The type and amount of wax that forms dictates the type and concentration of pour point depressant that will be effective in keeping wax crystals small so that they do not form network structures and lead to high viscosity and yield stress. Both the feedstocks and dewaxing steps used in the manufacture of a given base oil determine wax composition and, in turn, PPD response.

Certain types of viscosity modifiers can interact with base oils and pour point depressants at low temperatures and can lead to excessively high MRV viscosities in some situations (see, e.g., OCP9 in Table 11.8). Formulating with amorphous OCPs has not, in the authors' experience, posed many difficulties. Conversely, LTOCPs possess longer ethylene sequences that have the potential to interact with wax crystals at low temperatures. Several reports in the literature [66,76] suggest that high-ethylene OCPs can, under certain circumstances, have a negative impact on MRV viscosity and yield stress and may be more sensitive to the type of pour point depressants that will be effective in some formulating systems.

A study of low-temperature interactions among base oils, OCP viscosity modifiers, and pour point depressants was carried out in the author's laboratory using components listed in Tables 11.9 through 11.11.

Fully formulated SAE 5W-30 and 15W-40 engine oils were blended using performance additive packages DI-1 (at 11 wt.%) and DI-2 (at 13 wt.%), API SJ quality, and CH-4 quality, respectively, and all combinations of base oil, VM, and PPD.

Figures 11.17 through 11.20 summarize VM/PPD effects on MRV viscosity for each base oil type. In these graphs, the letter Y adjacent to a vertical bar denotes a yield stress failure. For the API Group I base oil B1 (SAE 5W-30, Figure 11.17), only PPD-3 is effective with all three viscosity modifiers. Both VM-1 and VM-3 suffer yield stress failures with at least one PPD. In the 15W-40 formulation, VM-3 exhibits yield stress behavior with all four PPDs, even in one case in which the MRV viscosity is quite low (PPD-1). In the authors' experience, it is quite unusual to observe yield stress failures in the MRV test when the viscosity is below about 40,000 cP.

TABLE 11.8

Rheological Properties of SAE 15W-40 Engine Oils Containing Different OCP Viscosity Modifiers

	OCP7	OCP3	OCP8	OCP9
Polymer content, wt.%	0.95	0.85	0.85	0.64
150N base oil percentage	76	76	70	70
600N base oil percentage	24	24	30	30
Kinematic viscosity, cSt 100°C	15.04	14.97	15.25	15.12
Viscosity index	141	140	140	135
CCS viscosity, cP −15°C	3,080	3,040	3,070	3,010
MRV viscosity, cP −20°C	10,000	9,900	8,800	Solid
MRV viscosity, cP −25°C	20,500	18,600	18,300	Solid
HTHS, cP	4.17	4.38	4.25	4.42

TABLE 11.9

Base Oils Used in Low-Temperature Viscosity Modifier/Pour Point Depressant Interaction Study

Base Oil Code (API Group)	Saturates, wt.%	Kinematic Viscosity, cSt at 100°C	Sulfur, wt.%	Viscosity Index	CCS Viscosity, cP at −25°C
B1-L (I)	73.6	3.88	0.276	104	1170
B1-M (I)	71.8	5.15	0.553	102	4060
B1-H (I)	61.8	12.10	0.381	97	—
B2-L (I)	75.2	4.18	0.193	105	1510
B2-M (I)	75.0	4.91	0.544	106	3060
B2-H (I)	72.3	12.73	0.412	99	—
B3-L (II)	100	4.20	0.006	100	1570
B3-M (II)	100	5.49	0.011	117	2430
B3-H (II)	100	10.72	0.016	98	—
B4-L (III)	100	4.50	0.007	123	1120
B4-H (III)	100	6.49	0.006	131	2710

TABLE 11.10

Viscosity Modifiers Used in Low-Temperature Viscosity Modifier/Pour Point Depressant Interaction Study

VM Code	OCP Type	SSI (ASTM D6278)
VM-1	Amorphous	37
VM-2	LTOCP	35
VM-3	LTOCP	37

TABLE 11.11

Pour Point Depressants Used in Low-Temperature Viscosity Modifier/Pour Point Depressant Interaction Study

Pour Point Depressant Code	Chemistry
PPD-1	Poly(styrene-maleate ester)
PPD-2	Poly(alkylmethacrylate)
PPD-3	Poly(styrene-maleate ester)
PPD-4	Poly(alkylmethacrylate)

The other API Group I blended oils, B2 (Figure 11.18), respond to PPDs in a similar manner to B1, but only in the SAE 5W-30 formulation. In the 15W-40 case, VM-3 is the only VM to experience yield stress failure, but only in the presence of PPD-2; all other combinations demonstrate acceptable pumpability performance.

Figure 11.19 describes the MRV map of B3, the API Group II oil. Overall low-temperature viscosities are all quite low for all VM/PPD combinations, although VM-3 again experiences one yield stress failure in each viscosity grade. VM-1 fails the yield stress criterion once.

Finally, the API Group III SAE 5W-30 formulation (Figure 11.20) was the most difficult to pour depress. All four oils blended with VM-3 were yield stress failures, and each of the other two VMs showed significant yield stresses for one PPD each.

In summary, the number of MRV failures due to yield stress may be found in Table 11.12. Clearly, one of the LTOCP viscosity modifiers, VM-3, is substantially more sensitive to pour point depressants than the other two polymers. It is especially incompatible with PPD-2. The other LTOCP in this study, VM-2, and the amorphous VM-1 were found to be far more tolerant of PPD type. Similar to the discussion in Section 11.5.4.2, one of the major differences between VM-3 and VM-2 is that they were manufactured by different technologies, broadly described in Figures 11.5 and 11.4, respectively.

11.5.6 FIELD PERFORMANCE DATA

Multigrade lubricants containing E/P viscosity modifiers have been tested in passenger car and heavy-duty truck engines for over three decades, but very few studies devoted to VM effects on engine cleanliness and/or wear have been published. It is generally believed that adding a polymer to the engine lubricant will have a detrimental effect on engine varnish, sludge, and piston ring-pack deposits [1,2], but the performance additive can be formulated to compensate for these effects. Kleiser and coworkers [85] ran a taxicab fleet test designed to compare the performance of a nonfunctionalized OCP viscosity modifier with a highly dispersant-functionalized OCP (HDOCP) as well as to test other oil formulation effects. They were surprised to find that an SAE 5W-30 oil containing a higher concentration of nondispersant OCP showed statistically better engine deposit control when compared with a similar SAE 15W-40 oil with lower polymer content. They also observed significant improvements in sludge and varnish ratings attributed to the use of HDOCP. Others have also reported that DOCPs can be beneficial in preventing buildup of deposits in laboratory engines such as the Sequence VE [86], VD [1], and Caterpillar 1H2/1G2 [1] tests. These authors found that engine oils containing certain dispersant-functionalized OCPs need less ashless dispersant to achieve an acceptable level of engine cleanliness than nonfunctionalized OCPs. The actual level of deposit prevention is highly influenced by the functionalization chemistry as well as the number of substituents per 1000 backbone carbon atoms, as shown in Table 11.13.

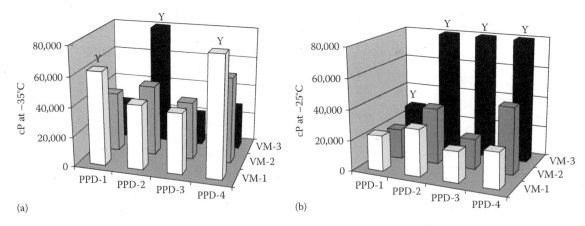

FIGURE 11.17 Rheological results for oil B1. (a) SAE 5W-30 and (b) SAE 15W-40.

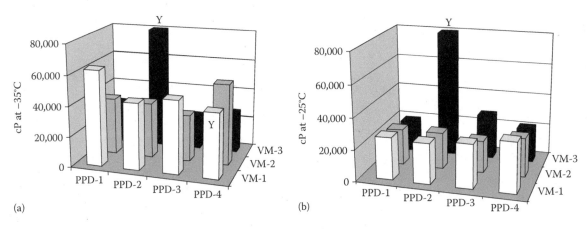

FIGURE 11.18 Rheological results for oil B2. (a) SAE 5W-30 and (b) SAE 15W-40.

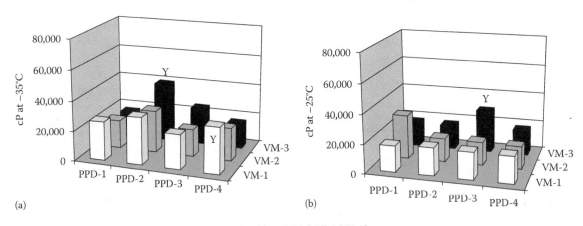

FIGURE 11.19 Rheological results for oil B3. (a) SAE 5W-30 and (b) SAE 15W-40.

11.5.7 Manufacturers, Marketers, and Other Issues

11.5.7.1 EP/EPDM Manufacturers

E/P copolymers and EPDM terpolymers are manufactured by a number of companies around the globe. Table 11.14 contains a listing of those with production capacity greater than 80,000 MT/year (176 million pounds per annum). Not all are necessarily supplying rubber into the viscosity modifier market. The vast majority of the capacity goes into other applications such as automotive (sealing systems, radiator hoses, injection molded parts), construction (window gaskets, roofing/sheeting, cable insulation,

cable filler), and plastics modification. Less than 2% of global production makes its way into the oil additives market [87].

Various grades are often classified by melt viscosity, E/P ratio, diene type and content, physical form, and filler type and level (carbon black, pigments, or extender oils). Melt viscosity is measured by two main techniques, Mooney viscosity (ASTM D1646 or ISO 289) or melt index (ASTM D1238 or ISO 1133-1991, also called melt-mass flow rate). Mooney viscosity is proportional to molecular weight, whereas melt index is inversely proportional to molecular weight.

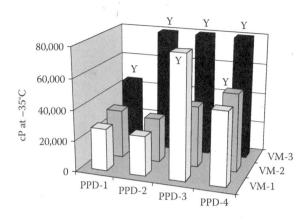

FIGURE 11.20 Rheological results for oil B4. SAE 5W-30.

TABLE 11.12

Number of MRV Failures due to Yield Stress in Low-Temperature Viscosity Modifier/Pour Point Depressant Interaction Study

VM	PPD-1	PPD-2	PPD-3	PPD-4	Total
VM-1	1		1	3	5
VM-2				1	1
VM-3	2	6	3	2	13
Total	3	6	4	6	19

TABLE 11.13

120-Hour Caterpillar 1H2 Piston Deposit Ratings of SAE 10W-40 Oils Formulated with N-Vinylpyrrolidone Grafted OCPs

Sample	Nitrogen, wt.%	TGF	WTD
OCP	0	—	>800
MFOCP 5	0.3	46	244
MFOCP 6	0.5	39	173
MFOCP 7	0.7	47	149
MFOCP 6[a]	0.26	28	156
MFOCP 2[a]	0.28	11	139

Source: Spiess, G.T. et al., Ethylene propylene copolymers as lube oil viscosity modifiers, in Bartz, W.J., ed., Addit. Schmierst. Arbeitsfluessigkeiten., 5th Int. Kolloq., 2,8., 10–1, Tech. Akad. Esslingen, Germany, 1986.

[a] OCP grafted with maleic anhydride and subsequently reacted with amines.

11.5.7.2 Olefin Copolymer VM Marketers

Companies that provide E/P copolymers and terpolymers to the viscosity modifier market are listed in Table 11.15. A wide variety of products, varying in shear stability and level of crystallinity, are available in both solid and liquid forms. Functionalized polymers that provide added dispersancy or other performance characteristics are available from several suppliers. The reader is advised to update this information periodically, since each company's product lines change over time.

Mergers and acquisitions have also contributed to significant flux in the OCP market. For example, the Paratone® product line was originally developed and marketed by the Paramins Division of Exxon Chemical Company. When Exxon and Shell combined their lubricant additives businesses to form Infineum in 1998, the Paratone business was sold to Oronite, the lubricant additives division of Chevron Chemical Company. Ethyl's purchase of Amoco and Texaco OCP product technology in the 1990s resulted in rebranding of Texaco's TLA-XXXX products to Ethyl's Hitec® product line. Ethyl Additives changed its name to Afton Chemical Company in 2004. DuPont originally marketed EPDM—manufactured at its Freeport, TX, facility—into the viscosity modifier market under the Ortholeum® trademark until it was sold to Octel in 1995. Thereafter, DuPont adopted the NDR brand name. DuPont and Dow Chemical Company formed a 50/50 Joint Venture in 1995, merging their elastomers businesses. Shortly following the successful start-up of their metallocene plant in Plaquemine, LA, 2 years later, the Freeport facility was closed. Dow Chemical Company acquired control of the EPDM product line, marketed under the Nordel® IP trade name, in 2005. Bayer transferred its EPDM business to a new company named LANXESS in 2004, which purchased the DSM Elastomers business in 2011. Although primarily a poly(alkylmethacrylate) company, Rohm GmbH of Darmstadt, Germany, developed several OCP-based viscosity modifiers under the Viscoplex® trade name, currently owned and marketed by Evonik Oil Additives. In mid-2007, Crompton sold its EPDM business to Lion Copolymer, which changed its name to Lion Elastomers in 2014.

11.5.7.3 Read Across Guidelines

Various lubrication industry associations have published highly detailed guidelines [88–90] for defining conditions under which certain additive and base oil changes to a fully or partially qualified engine oil formulation may be permitted without requiring complete engine testing data to support the changes. The purpose of these standards is to minimize test costs while ensuring that commercial engine oils meet the performance requirements established by industry standards, certification systems, and OEMs. From a viscosity modifier perspective, changes are often driven by one or more of the following three needs:

1. To optimize viscometrics within a given viscosity grade
2. To improve the shear stability of the formulation
3. To interchange one polymer for another (cost, security of supply, customer choice)

There are similarities and differences among codes of practice adopted by the American Chemistry Council (ACC) and the two European agencies, ATC (Technical Committee of Petroleum Additive Manufacturers of Europe) and ATIEL (Technical Association of the European Lubricants Industry). All permit minor changes in VM concentration (no more than 15% relative on a mass basis) to accomplish (1). The European codes explicitly allow the interchange of one VM for another (if both are from the same supplier) if the VM supplier deems them to be "equivalent and interchangeable." VMs from different suppliers, or those from the same supplier that are not

TABLE 11.14

Manufacturers with Production Capacity Greater than 80,000 MT/Y

Company	Manufacturing Location(s)	Capacity (Metric Ton/Year)	Technology	Trade Name	Comments
Dow Chemical	Plaquemine, LA	151,000	Metallocene, solution process	Nordel IP	EPDM
LANXESS Corporation	Geleen, the Netherlands Triunfo, Brazil Orange, TX Marl, Germany	>332,000	Ziegler–Natta, solution and suspension processes	Keltan	EP and EPDM
ExxonMobil Chemical	Baton Rouge, LA; Notre Dame de Gravenchon, France	265,000	Ziegler–Natta and metallocene, solution processes	Vistalon	EP and EPDM
Kumho Polychem	Yeosu, Republic of Korea	160,000 [93]	Ziegler–Natta, solution process	KEP	EP and EPDM
Lion Elastomers	Geismar, LA	130,000 [92]	Ziegler–Natta, solution process	Royalene Trilene[b]	EP and EPDM
Mitsui	Chiba, Japan	95,000[a] 11,000 [91][b]	Ziegler–Natta and metallocene, solution processes	Mitsui EPT LUCANT[b]	EP and EPDM
Versalis S.p.A.	Ferrara, Italy	85,000	Ziegler–Natta, suspension process	Dutral	EP and EPDM

Source: Ormonde, E. and Yoneyama, M., *Chemical Economics Handbook—IHS Chemical*, CEH Marketing Research Report, Ethylene-Propylene Elastomers, June 2012.

[a] Mitsui EPT.

[b] Liquid rubber.

TABLE 11.15

Marketers of Ethylene/Propylene Copolymers and EPDM Terpolymers as Engine Lubricating Oil Viscosity Modifiers

Company	Headquarters	Trade Name	Product Classes	Product Form
Afton	Richmond, VA	Hitec 5700 series	NDOCP, DOCP, DAOCP	Liquid concentrates
Chevron Oronite	Richmond, CA	Paratone	NDOCP, DOCP	Pellets, bales, and liquid concentrates
Dow Chemical	Midland, MI	Nordel IP	NDOCP	Bales and pellets
Infineum	Linden, NJ	Infineum V8000 series	NDOCP	Bales, pellets, and liquid concentrates
Lubrizol	Wickliffe, OH	Lubrizol 7000 series LUCANT	NDOCP, DOCP	Bales, pellets, viscous liquids, and liquid concentrates
Evonik	Darmstadt, Germany	Viscoplex	NDOCP, mixed PMA/OCP	Liquid concentrates

NDOCP, nonfunctionalized OCP; DOCP, dispersant-functionalized OCP; DAOCP, OCP with dispersant and antioxidant functionality; PMA, poly(alkylmethacrylate).

judged to be "equivalent," must undergo a rigorous engine testing program such as that outlined in Table 11.16.

The ACC guidelines impose two levels of data needed to support viscosity modifier interchange. Level 1 support is defined as analytical and rheological test data. Level 2 support includes both Level 1 data and full-length valid ASTM engine tests, intended to demonstrate that the proposed VM interchange presents no harm in terms of lubricant performance. There are three categories of engine tests that can be used to satisfy the ACC Level 2 criterion: (a) statistically designed engine test matrices, (b) complete programs, or (c) partial data sets from the same technology family. This broad definition of additive interchange testing is more open for interpretation than the ATIEL guidelines as represented in Table 11.16. Minor formulation modifications needing only Level 1 data do not permit changes in VM type, defined as polymers of a "specific molecular structure with a specific shear stability characterized by a specific trade name, stock or code number." When a change in shear stability is required, Level 2 support is sufficient for polymers of the same chemical family (e.g., OCPs) and from the same manufacturer. Otherwise, a full engine testing program is needed. The ACC guidelines also specify that if a DVM is used in a core multigrade formulation, the additional dispersant needed to read across to a monograde or other multigrades with lower VM concentration requires a Sequence VG test and Level 2 support in other tests.

11.5.7.4 Safety and Health

Copolymers of ethylene and propylene as well as EPDMs under normal conditions of use do not pose a risk to human health or

TABLE 11.16

Engine Test Requirements for Interchanging Viscosity Modifiers within the ATIEL Code of Practice

Performance Category	NDVM to NDVM (1, 2, 3)	DVM to DVM or NDVM to DVM (1, 2, 3)
Gasoline/light-duty diesel	TU572	TU572
	M111 and VG	M111 and VG
	VW DI	OM602A or OM646LA
	M111FE	DV4
		VWDI
		M111FE
Gasoline/light-duty diesel with after-treatment devices	TU572	TU572
	M111 and VG	M111 and VG
	VW DI	OM602A or OM646LA
	M111FE	DV4
		VW DI
		M111FE
Heavy-duty diesel	Mack T8E (5)	OM602A or OM646LA
	OM 441 LA or OM501LA	Mack T8E
	ISM (4, 6)	OM 441 LA or OM501LA
		ISM (4, 6)

Source: ATIEL, Technical Association of the European Lubricants Industry, Code of Practice for Developing Engine Oils Meeting the Requirements of the ACEA Oil Sequences, Issue No. 17, Appendix C, August 1, 2011.

Notes: 1. Full testing is required for VMI not listed earlier. 2. Physical mixes of NDVM and DVM are treated as DVM. 3. Only the tests included in the ACEA sequence for which read across is required have to be run. 4. ISM not required if the new oil formulation has the same or greater HTHS value compared with the original tested formulation. 5. T8E requirement is waived, if the replacement NDVM is within the same chemical type as the tested NDVM ("chemical type" means chemical family such as, but not limited to, styrene ester, polymethacrylate, styrene butadiene, styrene isoprene, polyisoprene, olefin copolymer, and polyisobutylene). 6. The M11 or M11EGR test may be used in place of the ISM test.

the environment and are typically not classified as hazardous according to OSHA HCS 2012 and the European Economic Community. They are generally considered to be not acutely toxic, similar to other high-molecular-weight polymers. Material that is heated to the molten state can emit fumes that can be harmful and irritating to the eyes, skin, mucous membranes, and respiratory tract, especially copolymers containing nonconjugated diene termonomers. Proper ventilation and respiratory protection are recommended when handling EP and EPDM copolymers under these conditions. Appropriate personal protective equipment is also advised to guard against thermal burns.

EP/EPDM grades are indexed by the Chemical Abstract Service as summarized in Table 11.17.

TABLE 11.17

CAS Index of EP and EPDM Copolymers

EP/EPDM	CAS Index
EP	9010-79-1
EPDM (ENB termonomer)	25038-36-2
EPDM (DCPD termonomer)	25034-71-3
EPDM (1,4-hexadiene termonomer)	25038-37-3

REFERENCES

1. Spiess, G.T., J.E. Johnston, and G. VerStrate, Ethylene propylene copolymers as lube oil viscosity modifiers, in W.J. Bartz, ed., *Addit. Schmierst. Arbeitsfluessigkeiten., 5th Int. Colloq. Addit. Lubricant. Operat. Fluids*, 2, paper 8.10, p. 1–8, Technical Academy Esslingen, Ostfildern, Federal Republic of Germany (1986).

2. Marsden, K., Literature review of OCP viscosity modifiers, *Lubr. Sci.*, **1**(3), 265 (1989).

3. Ver Strate, G. and M.J. Struglinski, Polymers as lubricating-oil viscosity modifiers, in D.N. Schulz and J.E. Glass, eds., *Polymers as Rheology Modifiers*, ACS Symposium Series, Miami, FL, Vol. **462**, 256 (1991).

4. Boor, J. Jr., *Ziegler-Natta Catalysts and Polymerizations*, Academic Press, New York (1979).

5. Crespi, G., A. Valvassori, and U. Flisi, Olefin copolymers, in *Stereo Rubbers*, John Wiley & Sons, New York, pp. 365–431 (1977).

6. Brandrup, J. and E.H. Immergut, eds., *Polymer Handbook*, 2nd ed., John Wiley & Sons, New York, p. II-193 (1975).

7. ASTM Res. Report RR-DO2-1442, Cold starting and pumpability studies in modern engines (1999).

8. Wilkes, C.E., C.J. Carmen, and R.A. Harrington, Monomer sequence distribution in ethylene-propylene terpolymers measured by ^{13}C nuclear magnetic resonance, *J. Polym. Sci.: Symp.*, **43**, 237 (1973).

9. Carmen, C.J. and K.C. Baranwal, Molecular structure of elastomers determined with Carbon-13 NMR, *Rubber Chem. Technol.*, **48**, 705 (1975).

10. Carmen, C.J., R.A. Harrington, and C.E. Wilkes, Monomer sequence distribution in ethylene-propylene rubber measured by ^{13}C NMR. 3. Use of reaction probability model, *Macromolecules*, **10**(3), 536 (1977).

11. Randall, J.C., *Polymer Sequence Determinations: Carbon-13 NMR Method*, Academic Press, New York, 53 (1977).

12. Carmen, C.J., Carbon-13 NMR high-resolution characterization of elastomer systems, in *Carbon-13 NMR in Polymer Science*, ACS Symposium Series, Vol. **103**, 97 (1978).

13. Kapur, G.S., A.S. Sarpal, S.K. Mazumdar, S.K. Jain, S.P. Srivastava, and A.K. Bhatnager, Structure-performance relationships of viscosity index improvers: I. Microstructural determination of olefin copolymers by NMR spectroscopy, *Lubr. Sci.*, **8**(1), 49 (1995).

14. Kresge, E.N. and G.W. Ver Strate, Ethylene polymer useful as a lubricating oil viscosity modifier, US Patent 4666619 (1987).

15. Hall, J.R., Ethylene, C3-16 monoolefin polymer containing 0.02%-0.6% by weight vinyl norbornene bound in the polymer having improved cold flow, US Patent 4156767 (1979).

16. Olabisi, O. and M. Atiqullah, Group 4 Metallocenes: Supported and unsupported, *J. Macromol. Sci. Rev. Macromol. Chem. Phys.*, **C37**(3), 519 (1997).

17. Soares, J.B.P. and A.E. Hamielec, Metallocene/aluminoxane catalysts for olefin polymerization. A review, *Polymer React. Eng.*, **3**(2), 131 (1995).

18. Gupta, V.K., S. Satish, and I.S. Bhardwaj, Metallocene complexes of group 4 elements in the polymerization of monoolefins, *J. Macromol. Sci. Rev. Macromol. Chem. Phys.*, **C34**(3), 439 (1994).

19. Hackmann, M. and B. Rieger, Metallocene Catalysis, *Cattech*, 1(2), 79–92, 1997.

20. Chien, J.C.W. and D. He, Olefin copolymerization with metallocene catalysts. I. Comparison of catalysts, *J. Polym. Sci. A Polym. Chem.*, **29**, 1585 (1991).

21. Montagna, A.A., A.H. Dekmezian, and R.M. Burkhart, The evolution of single-site catalysts, *ChemTech*, 27(12), 26–31, 1997.

22. McGirk, R.H., M.M. Hughes, and L.C. Salazar, Evaluation of polyolefin elastomers produced by constrained geometry catalyst chemistry as viscosity modifiers for engine oil, SAE Technical Paper 971696 (1997).

23. Rotman, D., DuPont Dow debuts metallocene EPDM, Synthetic rubber, *Chemical Week*, p. 15, (May 14, 1997).

24. Struglinski, M.J., Oleaginous compositions containing novel ethylene alpha-olefin polymer viscosity index improver additive, US Patent 5151204 (1992).

25. Mishra, M.K. and R.G. Saxton, Polymer additives for engine oils, *ChemTech*, **25**(4), 35–41, 1995.

26. Baranski, J.R. and C.A. Migdal, Lubricants containing ashless antiwear-dispersant additive having viscosity index improver credit. US Patent 5698500 (1997).

27. Kapuscinski, M.M. and L.D. Grina, Hydrocarbon compositions containing polyolefin graft polymers, European Patent 199453 (1986); Chapelet, G., H. Knoche, and G. Marie, Novel lubricating compositions containing nitrogen containing hydrocarbon backbone polymeric additives, US Patent 4092255 (1978); Kapuscinski, M.M. and R.E. Jones, Dispersant-antioxidant multifunction viscosity index improver, US Patent 4699723 (1987); Kapuscinski, M.M., L.D. Grina, and L.A. Brugger, Clear high-performance multifunction VI improvers, US Patent 4816172 (1989); Boden, F.J., R.P. Sauer, I.L. Goldblatt, and M.E. McHenry, Polar grafted polyolefins, methods for their manufacture, and lubricating oil compositions containing them, US Patent 5814586 (1998); Boden, F.J., R.P. Sauer, I.L. Goldblatt, and M.E. McHenry, Polar grafted polyolefins, methods for their manufacture, and lubricating oil compositions containing them, US Patent 5874389 (1999).

28. Pennewib, H. and C. Auschra, The contribution of new dispersant mixed polymers to the economy of engine oils, *Lubr. Sci.*, **8**(2), 179–197 (1996).

29. Engel, L.J. and J.B. Gardiner, Polymeric additives for fuels and lubricants, US Patent 4089794 (1978); Stambaugh, R.L. and R.A. Galluccio, Polyolefinic copolymer additives for lubricants and fuels, US Patent 4160739 (1979); Waldbillig, J.O. and I.D. Rubin, Fatty alkyl succinate ester and succinimide modified copolymers of ethylene and an alpha olefin, US Patent 4171273 (1979); Girgenti, S.J. and J.B. Gardiner, Stabilized amide-imide graft of ethylene copolymeric additives for lubricants, US Patent 4219432 (1980); Hayashi, K., Multi-purpose additive compositions and concentrates containing same, US Patent 4320019 (1982); Hayashi, K., Carboxylic acylating agents substituted with olefin polymers of high/low molecular weight mono-olefins, derivatives thereof, and fuels and lubricants containing same, US Patent 4489194 (1985); Chung, D.Y. and J.E. Johnston, Olefin polymer viscosity index improver additive useful in oil compositions, US Patent 4749505 (1988); Gutierrez, A. and D.Y. Chung, Multifunctional viscosity index improver derived from polyamine containing one primary amino group and at least one secondary amino group exhibiting improved low temperature viscometric properties, US Patent 5210146 (1992); Chung, D.Y., A. Gutierrez, J.E. Johnston, M.J. Struglinski, and R.D. Lundberg, Multifunctional viscosity index improver derived from amido-amine exhibiting improved low temperature viscometric properties, US patent 5252238 (1993); Chung, D.Y., A. Gutierrez, and M.J. Struglinski, Multifunctional viscosity index improver derived from amido-amine and degraded ethylene copolymer exhibiting improved low temperature viscometric properties, US Patent 5290461 (1994); Chung, D.Y. and M.J. Struglinski, Viscosity index improver, US Patent 5401427 (1995); Chung, D.Y., P. Brice, S.J. Searis, M.J. Struglinski, and J.B. Gardiner, Mixed ethylene alpha olefin copolymer multifunctional viscosity modifiers useful in lube oil compositions, US Patent 5427702 (1995); DeRosa, T.F., N. Benfaremo, M.M. Kapuscinski, B.J. Kaufman, and R.J. Jennejahn, Process for making a polymeric lubricant additive designed to enhance anti-wear, anti-oxidancy, and dispersancy thereof, US Patent 5534171 (1996).

30. Gordon, C.D. and G.L. Fagan, Method for oxidatively degrading an olefinic polymer, US Patent 4743391 (1988); Chung, D.Y. and J.E. Johnston, Olefin polymer viscosity index improver additive useful in oil compositions, US Patent 4749505 (1988); Kiovsky, T.E., EPR dispersant VI improver, US Patent 4169063 (1979); Waldbillig, J.O. and C.M. Cusano, Lubricating oil additives and composition containing same, US Patent 4132661 (1979).

31. Bloch, R., T.J. McCrary, Jr., and D.W. Brownawell, Ethylene copolymer viscosity index improver-dispersant additive useful in oil compositions, US Patent 4517104 (1985); Gutierrez, A., D.W. Brownawell, R. Bloch, and J.E. Johnston, Ethylene copolymer viscosity index improver-dispersant additive useful in oil compositions, US Patent 4632769 (1986); Lange, R.M., Dispersant-viscosity improvers for lubricating oil compositions, US Patent 5540851 (1998); Lange, R.M., Metal containing dispersant-viscosity improvers for lubricating oils, US Patent 5811378 (2000).

32. Nalesnik, T.E., Viscosity index improver, dispersant and antioxidant additive and lubricating oil composition containing same, European Patent 338672 (1989); Migdal, C.A., T.E. Nalesnik, N. Benfaremo, and C.S. Liu, Dispersant and antioxidant additive and lubricating oil composition containing same, US Patent 5075383 (1991).

33. Kaufman, B.J., C.S. Liu, and R.L. Sung, Reaction products of glycidyl derivatized polyolefins and 5-aminotetrazoles and lubricating oils containing same, US Patent 4500440 (1985); DeRosa, T.F.,

B.J. Kaufman, and R.J. Jennejahn, Dispersant, VI improver, additive and lubricating oil composition containing same, US Patent 5035819 (1991); Nalesnik, T.E., Oil dispersant and antioxidant additive, European Patent 470698 (1992); Kapuscinski, M.M. and T.E. Nalesnik, Dispersant, antioxidant and VI improver and lubricating oil composition containing same, European Patent 461774 (1991); Kapuscinski, M.M., T.F. Derosa, R.T. Biggs, and T.E. Nalesnik, Dispersant-antioxidant multifunctional viscosity index improver, US Patent 5021177 (1991); Liu, C.S., W.P. Hart, and M.M. Kapuscinski, Lubricating oil containing dispersant viscosity index improver, US Patent 4790948 (1988); Liu, C.S. and M.M. Kapuscinski, Multifunctional viscosity index improver, European Patent 284234 (1988).

34. Adams, P.E., R.M. Lange, R. Yodice, M.R. Baker, and J.G. Dietz, Intermediates useful for preparing dispersant-viscosity improvers for lubricating oils, US Patent 6117941 (2000).

35. Synthetic rubber is resilient, *Chem. Eng.*, **105**(12), 30 (November 1998).

36. Corbelli, L., Ethylene-propylene rubbers, *Dev. Rubber Tech.*, **2**, 87–129 (1981).

37. Italiaander, E.T., The gas-phase process—A new era in epr polymerization and processing technology, *NGK Kautschuk Gummi Kunststoffe*, **48**, 742–748 (October 1995).

38. Bunawerke Huls GmbH, Ethylene-propylene rubber, *Hydrocarbon Processing*, Vol. **60**, p. 164 (November 1981).

39. EPM/EPDM, *Gosei Gomu*, **85**, 1–9 (1980).

40. Darribère, C., F.A. Streiff, and J.E. Juvet, Static devolatilization plants, *DECHEMA Monogr.*, **134**, 689–704 (1998).

41. Anolick, C. and E.W. Slocum, Process for isolating EPDM elastomers from their solvent solutions, US Patent 3,726,843 (1973).

42. Cozewith, C., S. Ju, and G.W. VerStrate, Copolymer compositions containing a narrow MWD component and process of making same, US Patent 4874820 (1989); VerStrate, G., R. Bloch, M.J. Struglinski, J.E. Johnston, and R.K. West, Viscosity modifier polymers, US Patent 4804794 (1989); Cozewith, C., G.W. VerStrate, R.K. West, and G.A. Capone, Ethylene-alpha olefin block copolymers and methods for production thereof, US Patent 5798420 (1998).

43. Corbelli, L. and F. Milani, Recenti Sviluppi Nella Produzione Delle Gomme Sintetiche Etilene-Proilene, *L-Industria Della Gomma*, **29**(5), 28–31, 53–55 (1985).

44. Young, H.W. and S.D. Brignac, The effect of long chain branching on EPDM properties, *Proceedings of Special Symp. Advanced Polyolefin Tech., 54th Southwest Regional Meeting American Chemical Society*, Baton Rouge, LA (1998).

45. Garbassi, F., Long chain branching: An open question for polymer characterization? *Polym. News*, **19**, 340–346 (1994).

46. Elliott, J.S., I.R.H. Crail, and P.J. Hattersley, Degraded polymers, British Patent 1372381 (1974); Johnston, T.E., Continuous process for the manufacture of oil soluble ethylene-propylene copolymers for use in petroleum products, Canadian Patent 991792 (1976); Joffrion, R.K., Process for dissolving EPM and EPDM polymers in oil, US Patent 4464493 (1984); Chung, D.Y. and J.E. Johnston, Olefin polymer viscosity index improver additive useful in oil compositions, US Patent 4749505 (1988); Chung, D.Y., A. Gutierrez, and M.J. Struglinski, Multifunctional viscosity index improver derived from amido-amine and degraded ethylene copolymer exhibiting improved low temperature viscometric properties, US Patent 5290461 (1994); Olivier, E.J., R.T. Patterson, and P.N. Nugara, Solid sheared polymer blends and process for their preparation, US Patent 5391617 (1995); Chung, D.Y. and M.J. Struglinski, Viscosity index improver, US Patent 5401427 (1995); Olivier, E.J., R.T. Patterson, and P.N. Nugara, Sheared polymer blends and process for their preparation, US Patent 5837773 (1998).

47. Dorer, C.J. Jr., Functional fluid containing a sludge inhibiting detergent comprising the polyamine salt of the reaction product of maleic anhydride and an oxidized interpolymer of propylene and ethylene, US Patent 3316177 (1967); Hu, S-en., Oleaginous compositions containing sludge dispersants, US Patent 3326804 (1967).

48. Gordon, C.D. and G.L. Fagan, Method for oxidatively degrading an olefinic polymer, US Patent 4743391 (1988); Gardiner, J.B., A.A. Loffredo, R. Bloch, N.C. Nahas, K.U. Ingold, and T.V. Kowalchyn, Catalytic process for oxidative, shear accelerated polymer degradation, US Patent 5006608 (1991).

49. Carman, C.J. and K.C. Baranwal, Molecular structure of elastomers determined with carbon-13 NMR, *Rubber Chem. Technol.*, **48**, 705–718 (1975).

50. Carman, C.J., R.A. Harrington, and C.E. Wilkes, Monomer sequence distribution in ethylene-propylene rubber measured by ^{13}C NMR. 3. Use of reaction probability model, *Macromolecules*, **10**(3), 536–544 (1977).

51. Carman, C.J., Carbon-13 NMR high-resolution characterization of elastomer systems, *Carbon-13 NMR in Polymer Science*, ACS Symposium Series, Vol. **103**, 97–121 (1978).

52. Wilkes, C.E., C.J. Carman, and R.A. Harrington, Monomer sequence distribution in ethylene-propylene terpolymers measured by ^{13}C nuclear magnetic resonance, *J. Polym. Sci.*, **43**, 237–250 (1973).

53. Di Martino, S. and M. Kelchtermans, Determination of the composition of ethylene-propylene-rubbers using ^{13}C-NMR spectroscopy, *J. Appl. Polym. Sci.*, **56**, 1781–1787 (1995).

54. Randall, J.C., *Polymer Sequence Determinations: Carbon-13 NMR Method, 53-138*, Academic Press, New York (1977).

55. Minick, J., A. Moet, A. Hiltner, E. Baer, and S.P. Chum, Crystallization of very low density copolymers of ethylene with α-Olefins, *J. Appl. Polym. Sci.*, **58**, 1371–1384 (1995).

56. Rubin, I.D. and M.M. Kapuscinski, Viscosities of ethylene-propylene-diene terpolymer blends in oil, *J. Appl. Polym. Sci.*, **49**, 111 (1993).

57. Rubin, I.D., Polymers as lubricant viscosity modifiers, *Polymer Preprints, Am. Chem. Soc. Div. Polym. Chem.*, **32**(2), 84 (1991).

58. Rubin, I.D., A.J. Stipanovic, and A. Sen, Effect of OCP structure on viscosity in oil, SAE Technical, Paper 902092 (1990).

59. Kapuscinski, M.M., A. Sen, and I.D. Rubin, Solution viscosity studies on OCP VI improvers in oils, SAE Technical, Paper 892152 (1989).

60. Kucks, M.J., H.D. Ou-Yang, and I.D. Rubin, Ethylene-propylene copolymer aggregation in selective hydrocarbon solvents, *Macromolecules*, **26**, 3846 (1993).

61. Rubin, I.D. and A. Sen, Solution viscosities of ethylene-propylene copolymers in oils, *J. Appl. Polym. Sci.*, **40**, 523–530 (1990).

62. Sen, A. and I.D. Rubin, Molecular structures and solution viscosities of ethylene-propylene copolymers, *Macromolecules*, **23**, 2519 (1990).

63. LaRiviere, D., A.A. Asfour, A. Hage, and J.Z. Gao, Viscometric properties of viscosity index improvers in lubricant base oil over a wide temperature range. Part I: Group II base oil, *Lubr. Sci.*, **12**(2), 133–143 (2000).

64. Huggins, M.L., *J. Am. Chem. Soc.*, **64**, 2716 (1942).

65. Engine oil viscosity classification, SAE International, Surface Vehicle Standard No. J300 (Rev. January 2015).

66. Rhodes, R.B., Low-temperature compatibility of engine lubricants and the risk of engine pumpability failure, SAE Technical, Paper 932831 (1993).

67. Papke, B.L., M.A. Dahlstrom, C.T. Mansfield, J.C. Dinklage, and D.J. Rao, Deterioration in used oil low temperature pumpability properites, SAE Technical Paper 2000-01-2942 (2000).

68. Matko, M.A. and D.W. Florkowski, Low temperature rheological properties of aged crankcase oils, SAE Technical Paper 200-01-2943 (2000).

69. J.A. Spearot, ed., High-temperature, high-shear (HTHS) oil viscosity: Measurement and relationship to engine operation, ASTM Special Tech. Pub., STP 1068 (1989).

70. Bates, T.W., Oil rheology and journal bearing performance: A review, *Lubr. Sci.*, **2**(2), 157–176 (1990).

71. Wardle, R.W.M., R.C. Coy, P.M. Cann, and H.A. Spikes, An 'In Lubro' study of viscosity index improvers in end contacts, *Lubr. Sci.*, **3**(1), 45–62 (1990).

72. Alexander, D.L. and S.W. Rein, Relationship between engine oil bench shear stability tests, SAE Technical Paper 872047 (1987).

73. Bartz, W.J., Influence of viscosity index improver, molecular weight, and base oil on thickening, shear stabilit, and evaporation losses of multigrade oils, *Lubr. Sci.*, **12**(3), 215–237 (2000).

74. Laukotka, E.M., Shear stability test for polymer containing lubricating fluids—Comparison of test methods and test results, *Third International Symposium—The Performance Evolution of Automotive Fuels and Lubricants, Co-ordinating European Council Paper No. 3LT*, Paris, France (April 19–21, 1987).

75. Selby, T.W., The viscsosity loss trapezoid—Part 2: Determining general features of VI improver molecular weight distribution by parameters of the VLT, SAE Technical Paper 932836 (1993).

76. Covitch, M.J., How polymer architecture affects permanent viscosity loss of multigrade lubricants, SAE Technical Paper 982638 (1998).

77. Kramer, D.C., J.N. Ziemer, M.T. Cheng, C.E. Fry, R.N. Reynolds, B.K. Lok, M.L. Sztenderowicz, and R.R. Krug, Influence of group II & III base oil composition on VI and oxidation stability, *National Lubricating Grease Inst. Publ. No. 9907, Proceedings from 66th NLGI Annual Meeting*, Tucson, AZ, October 25, 1999.

78. Rossi, A., Lube basestock manufacturing technology and engine oil pumpability, SAE Technical Paper 940098 (1994).

79. Rossi, A., Refinery/additive technologies and low temperature pumpability, SAE Technical Paper 881665 (1988).

80. Mac Alpine, G.A. and C.J. May, Compositional effects on the low temperature pumpability of engine oils, SAE Technical Paper 870404 (1987).

81. Reddy, S.R. and M.L. McMillan, Understanding the effectiveness of diesel fuel flow improvers, SAE Technical Paper 811181 (1981).

82. Xiong, C.-X., The structure and activity of polyalphaolefins as pour-point depressants, *Lubr. Eng.*, **49**(3), pp. 196–200 (1993).

83. Rubin, I.D., M.K. Mishra, and R.D. Pugliese, Pouir point and flow improvement in lubes: The interaction of waxes and methacrylate polymers, SAE Technical Paper 912409 (1991).

84. Webber, R.M., Low temperature rheology of lubricating mineral oils: Effects of cooling rate and wax crystallization on flow properties of base oils, *J. Rheol.*, **43**(4), 911–931 (1999).

85. Kleiser, W.M., H.M. Walker, and J.A. Rutherford, Determination of lubricating oil additive effects in taxicab service, SAE Technical Paper 912386 (1991).

86. Carroll, D.R. and R. Robson, Engine dynamometer evaluation of oil formulation factors for improved field sludge protection, SAE Technical Paper 872124 (1987).

87. CEH Marketing Research Report, Ethylene-Propylene Elastomers, E. Ormonde and M. Yoneyama, Chemical Economics Handbook—IHS Chemical (June 2012).

88. *Petroleum Additives Product Approval Code of Practice*, American Chemistry Council, Appendices H and I (December 2010).

89. *ATC Code of Practice*, Technical Committee of Petroleum Additive Manufacturers in Europe, Section H (June 2011).

90. Code of Practice for Developing Engine Oils Meeting the Requirements of the ACEA Oil Sequences, ATIEL, Technical Association of the European Lubricants Industry, Issue No. 17, Appendix C (August 1, 2011).

91. Mitsui Chemicals Co., Mitsui Chemicals to Expand Production Capacity and Start Commercial Operation of Ethylene/α Olefin Olygomer (LUCANT™), www.mitsuichem.com, news release (April 10, 2008).

92. *Lion Copolymer to idle its Baton Rouge facility and complete expansion of its EPDM facility*, www.lioncopolymer.com, news release (December 3, 2013).

93. Kumho Polychem, Kumho Polychem leaps to global top 3 EPDM manufacturer with completion of 2nd plant construction, www.polychem.co.kr, news (September 12, 2013).

94. Li Pi Shan, Colin, Kuhlman, Roger L., Rath, Gary L., Kenny, Pamela J., Hughes, Morgan M., Cong, Rongjuan, Viscosity index improver for lubricant compositions, US Patent 8486878B2 (2013) and US Patent 8492322B2 (2013).

95. Arriola, D.J., E.M. Carnahan, P.D. Hustak, R.L. Kuhlman, and T.T. Wenzel, Catalytic production of olefin block copolymers via chain shuttling polymerization, *Science*, **312**(5774), 714–719 (May 5, 2006), doi: 10.1126/science.1125268.

96. Kolb, R., P.T. Matsunaga, P.S. Ravishankar, L.B. Stefaniak, and Q.P.W. Costin, Process for the production of polymeric compositions useful as oil modifiers, US Patent Application 20130203641A1 (2013).

97. Chum, P.S. and K.W. Swogger, Olefin polymer technologies—History and recent progress at the dow chemical company, *Prog. Polym. Sci.*, **33**(8), 797–819 (August 2008). doi:10.1016/j.progpolymsci.2008.05.003.

98. Oda, H., T. Kinoshita, and A. Shimizu, Process for producing ethylene/alpha-olefin copolymer, US Patent 6153807 (2000).

99. Ravishankar, P.S. and K.A. Nass, Olefinic Copolymer Compositions for Viscosity Modification of Motor Oil, US Patent 20120015854A1 (2012); Ikeda, S., N. Kamiya, H. Hoya, J. Tanaka, and C. Huang, Viscosity modifier for lubricating oils, additive composition for lubricating oils, and lubricating oil composition, US Patent 20120190601A1 (2012).

100. Covitch, M.J. and K.J. Trickett, How polymers behave as viscosity index improvers in lubricating oils, *Adv. Chem. Eng. Sci.*, **5**, 134–151 (2015).

101. Mishra, M.K. and I.D. Rubin, Multifunctional copolymer and lubricating oil composition, US Patent 5429757 (1995); Valcho, J.J., M. Rees, P. Growcott, M.T. Devlin, and E.J. Olivier, Highly grafted, multi-functional olefin copolymer VI modifiers, US Patent 6107257 (2000); Goldblatt, I.L., S.-J. Chen, and R.P. Sauer, Multiple-Function Dispersant Graft Polymer, US Patent 20080293600A1 (2008); Goldblatt, I.L., S.-J. Chen, M.R. Patel, and S.T. McKenna, Preparation of monomers for grafting to polyolefins, and lubricating oil compositions containing grafted copolymer, US Patent 7371713B2 (2008); Covitch, M.J., J.K. Pudelski, C. Friend, M.D. Gieselman, R.A. Eveland, M.G. Raguz, and B.J. Schober, Dispersant viscosity modifiers containing aromatic amines, US Patent 7790661B2 (2010); Gieselman, M.D. and A.J. Preston, Lubricating composition containing a functionalized carboxylic polymer, US Patent 8557753B2 (2013); Gieselman, M.D., C. Friend, and A.J. Preston, Lubricating composition containing a polymer, US Patent 8637437B2 (2014).

12 Polymethacrylate Viscosity Modifiers and Pour Point Depressants

Boris M. Eisenberg, Alan Flamberg, and Bernard G. Kinker

CONTENTS

12.1 HISTORICAL DEVELOPMENT

12.1.1 FIRST SYNTHESIS

The first synthesis of a polymethacrylate (PMA) intended for potential use in the field of lubricant additives took place in the mid-1930s. The original work was conducted under the supervision of Herman Bruson, who was in the employ of the Rohm and Haas Company (a parent of Evonik Oil Additives), and it was conducted in Rohm and Haas' Philadelphia Research Laboratories. Bruson was exploring the synthesis and possible applications of longer alkyl side chain methacrylates [1]. He had proposed poly(lauryl methacrylate) as a product that might serve as a potential thickener or viscosity index improver (VII) for mineral oils. The result of the work was the 1937 issuance of two U.S. patents, for "Composition of matter and process" [2] and for "Process for preparation of esters and products" [3].

12.1.2 FIRST APPLICATION

Bruson's invention did indeed thicken mineral oils, and it was effective in increasing viscosity at higher temperatures more so than at lower, colder temperatures. Since this behavior influences the viscosity–temperature properties or viscosity index (VI) of a fluid, these materials eventually became known as VIIs. Although PMAs successfully thickened oils, there were other competitive thickeners at that time that increased the viscosity of mineral oils; these were based on polyisobutylene and alkylated polystyrene. The commercial success of PMA was not at all assured. The driving force behind PMA eventually eclipsing the other commercial thickeners of the era was PMA's value as a VII rather than as a simple thickener of oils. In other words, PMAs have the ability to contribute relatively little viscosity at colder temperatures, such as those that might be encountered at equipment start-up, but have a much higher contribution to viscosity at hotter temperatures at which equipment tends to operate. This desirable behavior enabled oil formulators to prepare multigrade oils that could meet a broader range of operating temperature requirements. The positive enhancement of VI ensured the future success of PMAs.

12.1.3 First Manufacture and Large-Scale Application

The commercial development of PMAs as VIIs lagged until the beginning of World War II, when a U.S. government board *rediscovered* Bruson's VII invention. The board was charged with searching the scientific literature for useful inventions that might aid the war effort. When considering potential utility, they hypothesized about a PMA VII providing more uniform viscosity properties over a very broad range of temperatures, particularly in aircraft hydraulic fluids. The fluids of that era were judged to be deficient, particularly in fighter aircraft, because of the exaggerated temperature/time cycles experienced. On the ground, the fluids could experience high-temperature ambient conditions and engine waste heat, and then after a rapid climb to high, very cold altitudes, the fluid might experience temperatures below −40°C. After successful trials of the multigrade aircraft hydraulic fluid concept, Rohm and Haas, in cooperation with the National Research Defense Committee, rapidly proceeded to commercialize PMA VIIs and delivered the first product, ACRYLOID® HF, in 1942. These multigrade hydraulic fluids were quickly adopted by the U.S. Army Air Corps and were followed by other multigrade hydraulic fluids and lubricants in ground vehicles that incorporated VIIs.

After the war, Rohm and Haas introduced PMA VIIs to general industrial and automotive applications. Early passenger car engine oil VIIs were first introduced to the market in 1946. The adoption of *all-season* oils in the commercial market was greatly influenced by two events. First, the automotive manufacturers' viscosity specification introduction of the new designation "W" (for winter grades), and then by Van Horne's publication [4] pointing to the possibility of making and marketing cross-graded oils such as the now well-known *10W-30* as well as other cross grades. By the early 1950s, the use of multigrade passenger car oils was becoming widespread in the consumer market. Methacrylates played a major role in enabling the formulation of that era's multigrade engine oils. The use of PMA VIIs has since been extended to gear oils, transmission fluids, and a broad array of industrial and mobile hydraulic fluids, in addition to the early usage in aircraft hydraulic fluids.

12.1.4 Development of Other Applications

Another important application area for PMA chemistry is in the field of pour point depressants (PPDs). When a methacrylate polymer includes at least some longer alkyl side chains,

relatively similar to the chain length of waxes normally present in mineral oil, it can interact with growing wax crystals at sufficiently low temperatures. Wax-like side chains can be incorporated into a growing wax crystal and disrupt its growth. The net effect is to prevent congealing of wax in the oil at the temperature where it would have occurred in the absence of a PPD. Early PMA PPDs were used first by the military and later by civilian industry when Rohm and Haas offered such products to the industrial and automotive markets in 1946. Although PMAs were not the first materials used as PPDs (alkylated naphthalenes were), PMAs are probably the predominant products in this particular application now.

Another use of wax-interactive PMAs is as refinery *dewaxing aids*. The process of dewaxing is carried out primarily to remove wax from paraffinic raffinates in order to lower the pour point of the resulting lube oil base stocks. PMA dewaxing aids are extremely interactive with waxes found in raffinates and thus function as nucleation agents to seed wax crystallization and to promote the growth of relatively large crystals. The larger crystals are more easily filtered from the remaining liquid so that lube oil throughputs and yields are improved while pour points are lowered by virtue of lower wax concentrations.

Incorporating monomers more polar than alkyl methacrylates into a PMA provides products useful as ashless *dispersants* or *dispersant VIIs*. The polar monomer typically contains nitrogen and/or oxygen (other than the oxygen present in the ester group), and its inclusion in sufficient concentration creates hydrophilic zones along the otherwise oleophilic polymer chain. The resulting dispersant PMAs (d-PMAs) are useful in lubricants since they can suspend in solution that might otherwise be harmful materials ranging from highly oxidized small molecules to soot particles.

PMAs have also been used in a number of other petroleum-based applications, albeit in relatively minor volumes. An abbreviated list would include asphalt modifiers, grease thickeners, demulsifiers, emulsifiers, antifoamants, and crude oil flow improvers and paraffin inhibitors. PMAs have been present in lubricants for about 70 years now, and their longevity stems from the flexibility of PMA chemistry in terms of composition and process. Evolution of the original lauryl methacrylate composition to include a variety of alkyl methacrylates and/or non-methacrylates has brought additional functionality and an expanded list of applications. Process chemistry also has evolved such that it can produce polymers of almost any desired molecular weights (shear stability) or allow the synthesis of complex polymer architectures. The evolution of efficient processes for controlled radical polymerization (CRP) in the 1990s has led to the development of taper and block copolymers and has permitted the development of products with narrower molecular weight distributions.

12.2 CHEMISTRY

12.2.1 General Product Structure

Typically, a methacrylate VII is a linear polymer constructed from three classes (three distinct lengths) of hydrocarbon side chains. These would be short, intermediate, and long chain lengths.

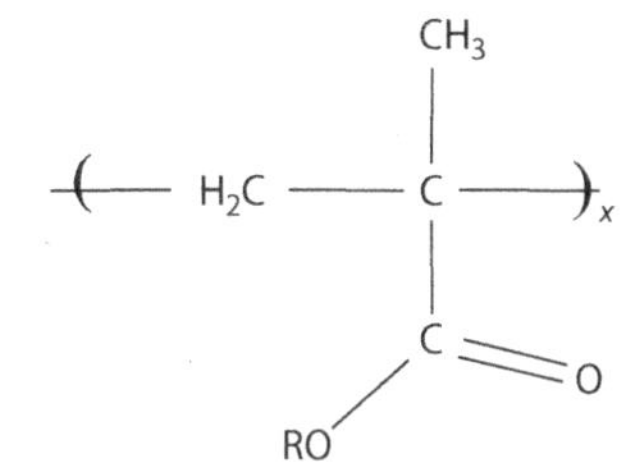

FIGURE 12.1 Polyalkylmethacrylate generalized structure.

A more extensive discussion is given in Section 12.3, but an abbreviated description is given here in order to better understand the synthesis and chemistry of methacrylate monomers and polymers.

The first class consists of short-chain alkyl methacrylates of one to seven carbons in length. The inclusion of such short-chain materials influences polymer coil size, particularly at colder temperatures, and thus influences the VI of the polymer in oil solutions. The intermediate class contains 8–13 carbons, and these serve to give the polymer its solubility in hydrocarbon solutions. The long-chain class contains 14 or more carbons and is included to interact with wax during its crystallization and thus provides PPD properties.

The structure of polymethacrylates used as PPDs differs from that of a VI improver by virtue of normally containing only two of these sets of components. These are the long-chain, wax crystallization interactive materials and intermediate chain lengths.

The selected monomers are mixed together in a specific ratio in order to provide an overall balance of the earlier mentioned properties. This mixture is then polymerized to provide a copolymer structure where R represents different alkyl groups and x indicates various degrees of polymerization. The simplified structure is shown in Figure 12.1.

12.2.2 Monomer Chemistry

Before discussing lubricant additives based on PMAs, it is necessary to give an introduction to the chemistry of their parent monomers. The basic structure of a methacrylate monomer is shown in Figure 12.2.

There are four salient features of this vinyl compound:

1. The carbon–carbon double bond that is the reactive site in addition to polymerization reactions.
2. The ester functionality adjacent to the double bond that polarizes and thus activates the double bond in polymerization reactions.

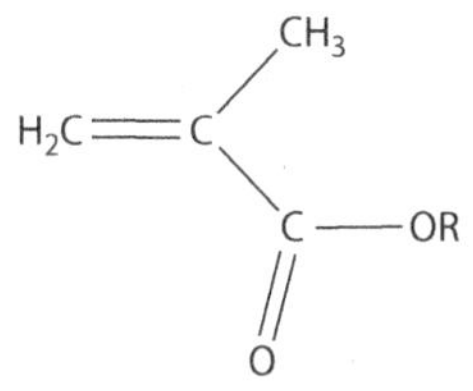

FIGURE 12.2 Alkyl methacrylate monomer.

3. The pendant side chain attached to the ester (designated as R). These chains may range from an all-hydrocarbon chain to a more complex structure containing heteroatoms. A significant portion of the beneficial properties of PMAs is derived from the pendant side chain.
4. The pendant methyl group adjacent to the double bond, which serves to shield the ester group from chemical attack, particularly as it relates to hydrolytic stability.

As mentioned earlier, a variety of methacrylate monomers, differing by length of the pendant side chains, is normally used to construct PMA additives. The synthesis chemistry of these monomers falls into two categories: shorter chains with four or fewer carbons and longer chains with five or more carbons. The commercial processes used to prepare each type are quite different.

The short-chain monomers are often mass produced because of their usefulness in applications other than lubricant additives. For instance, methyl methacrylate is produced in large volumes and used primarily in production of PLEXIGLAS® acrylic plastic sheet and as a component in emulsion paints and adhesives. It is also used in PMA lubricant additives, but the volumes in this application pale in comparison to its use in other product areas. Methyl methacrylate is generally produced by either of two synthetic routes. The more prevalent starts with acetone, then proceeds through its conversion to acetone cyanohydrin, followed by its hydrolysis and esterification. The other route is oxidation of butylenes followed by subsequent hydrolysis and esterification. Recently, additional processes for methyl methacrylate have been introduced based on propylene and propyne chemistry, but their use remains comparatively small.

The long-chain monomers are typically but not exclusively used in lubricant additives and can be produced by either of two commercial processes. The first is direct esterification of an appropriate alcohol with methacrylic acid. This well-known reaction is often used as a laboratory model of chemical strategies used to efficiently drive a reaction to high yield. These strategies involve a catalyst, usually an acid, an excess of one reagent to shift the equilibrium to product, and removal of at least one of the products, typically water of esterification, again to shift the equilibrium. The relevant chemical equation is given in Figure 12.3.

A second commercial route to longer side chain methacrylate monomers is transesterification of methyl methacrylate with an appropriate alcohol. The reaction employs a basic compound or a Lewis base as a catalyst. The equilibrium is shifted to product by the use of an excess of methyl methacrylate and by the removal of a reaction product, that is, methanol (if methyl methacrylate is used as a reactant). Figure 12.4 shows the reaction equation.

12.2.3 Traditional Polymer Chemistry

A combination of alkyl methacrylate monomers chosen for a given product is mixed together in specific ratios and then polymerized by a solution, free-radical initiated, addition

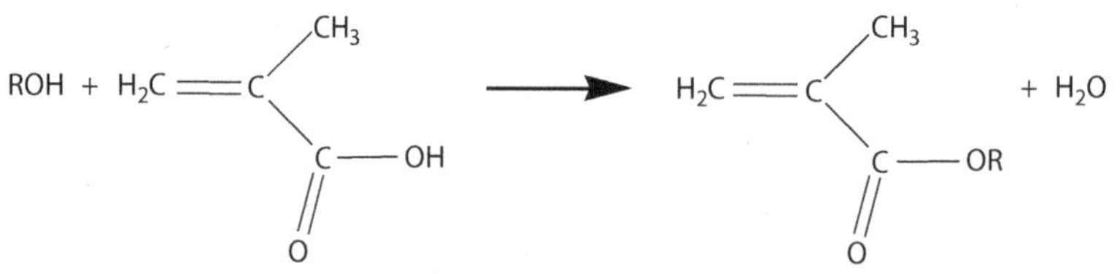

FIGURE 12.3   Direct esterification of methacrylic acid and alcohol.

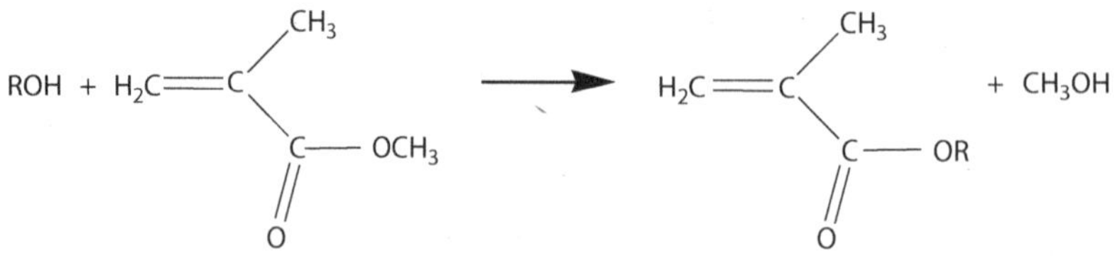

FIGURE 12.4   Transesterification of methylmethacrylate and alcohol.

polymerization process that produces a random copolymer. The reaction follows the classic pathways and techniques of addition polymerization to produce commercial materials [5].

Commercial polymers are currently synthesized through the use of free-radical initiators. The initiator may be from either the oxygen-based or nitrogen-based families of thermally unstable compounds that decompose to yield two free radicals. The oxygen-based initiators, that is, peroxides, hydroperoxides, peresters, or other compounds containing an oxygen–oxygen covalent bond, thermally decomposed via homolytic cleavage to form two oxygen–centered free radicals. Nitrogen-based initiators also thermally decompose to form two free radicals, but these materials quickly evolve a mole of nitrogen gas and thus form carbon-centered radicals. In any event, the free radicals attack the less-hindered, relatively positive side of the

methacrylate vinyl double bond. These two reactions are the classic initiation and propagation steps of free-radical addition polymerization and are shown in Figures 12.5 and 12.6.

The reaction temperature is chosen in concert with the initiator's half-life and may range from 60°C to 140°C. Generally, a temperature–initiator combination would be selected to provide an economic, facile conversion of monomer to polymer and to avoid potential side reactions. Other temperature-dependent factors are taken into consideration. Chief among these might be a need to maintain a reasonable viscosity of the polymer in the reactor as it is being synthesized. Obviously, the temperature can be utilized (as well as solvent) to maintain viscosity at a level appropriate for the mechanical agitation and pumping systems within a production unit. Excessive temperatures must be avoided in order to avoid the ceiling temperature of the

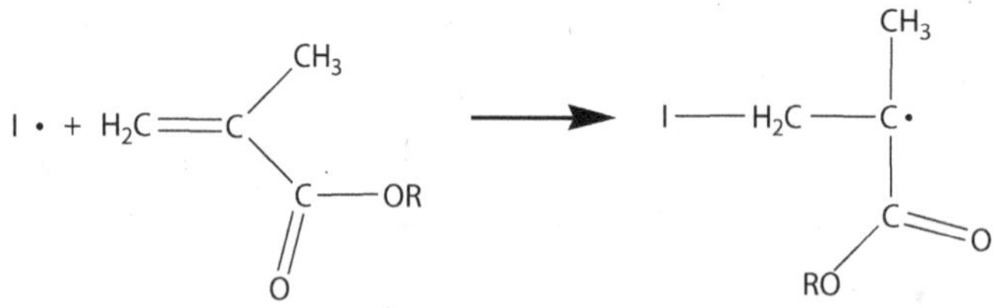

FIGURE 12.5   Free-radical initiation of methacrylate polymerization.

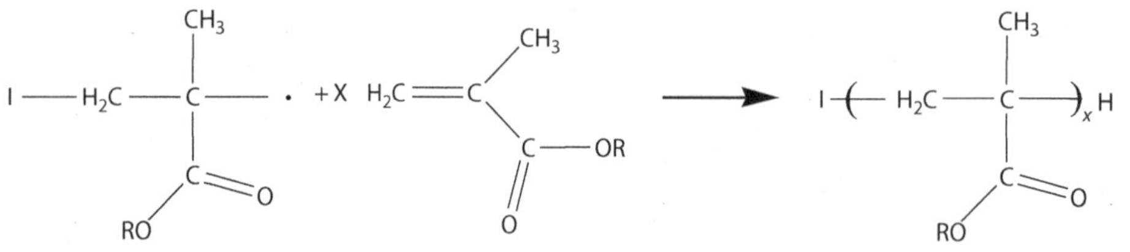

FIGURE 12.6   Monomer addition—propagation step.

polymerization, which is the temperature where the depolymerization reaction commences (see Section 12.3.1.2).

Normally, a mixture of alkyl methacrylate monomers is used to produce a random copolymer. No special reaction techniques are needed to avoid composition drift over the course of the reaction since reactivity ratios of alkyl methacrylates are quite similar [6].

The most important concern during a synthesis reaction is to provide polymer at a given molecular weight so as to produce commercial product of suitable shear stability. As normal for vinyl addition polymerizations, methacrylates can undergo the usual termination reactions: combination, disproportionation, and chain transfer. Chain transfer agents (CTAs), often mercaptans, are the most commonly chosen strategy to control molecular weight. Selection of the type and amount of CTA must be done carefully and with an understanding that many other factors influence molecular weight. Numerous factors can impact the degree of polymerization: initiator concentration, radical flux, solvent concentration, and opportunistic chain transfer with compounds other than the CTA. An undesirable opportunistic chain transfer possibility is hydrogen abstraction at random sites along the polymer chain leading to branched polymers that are less-efficient thickeners than strictly linear chains. The mercaptan chain transfer reaction is shown in Figure 12.7. In addition to chain transfer, other usual termination reactions of chain combination and/or disproportionation can occur with methacrylates.

Commercial products cover a broad range of polymer molecular weights ranging from approximately 20,000 to about 750,000 Da. Molecular weight is carefully controlled and targeted to produce products that achieve suitable shear stability for a given application.

Higher-molecular-weight PMAs are rather difficult to handle as neat polymers, so it is necessary in almost all commercial cases to use a solvent to reduce viscosity to levels consistent with reasonable handling properties. Additionally, it is important to maintain reasonable viscosity during the polymerization reaction (even though it is always increasing as monomer-to-polymer conversion increases) so that sufficient agitation can be maintained. Thus, solvents are almost invariably employed. An appropriate solvent would (1) be nonreactive, (2) be nonvolatile (at least at the reaction temperature), (3) avoid chain transfer reactions, and (4) be consistent with the intended application of the resulting product. It turns out that mineral oil meets the earlier mentioned criteria reasonably well so that a solvent choice can be made

from higher-quality, lower-viscosity-grade mineral oils. Nonreactivity demands relatively higher saturated contents, so better quality API Group I (or higher API Group) mineral oil can be used. The choice of solvent viscosity primarily depends on the end application; choices range from very low-viscosity oils of 35 SUS to light neutrals typically up to 150N. Alternatively, one can use a nonreactive but volatile solvent when mineral oil might interfere with a sensitive polymerization and then do a solvent exchange into a more suitable carrier oil. The amount of solvent added to commercial PMA VIIs is sufficient to reduce viscosity to levels consistent with reasonable handling or container pump out properties. This amount is dictated by polymer molecular weight, as this also heavily influences product viscosity. Generally, a higher-molecular-weight polymer requires more solvent. Commercial products may thus contain polymer concentrations over a very broad range of approximately 30–80 wt.%.

Dispersant PMAs were first described by Catlin in a 1956 patent [7]. The patent claims the incorporation of diethylaminoethyl methacrylate as a way of enhancing the dispersancy of VIIs and thus providing improved deposit performance in engine tests of that era. The original dispersant methacrylate polymers utilized monomers that copolymerized readily with alkyl methacrylates and did not require different polymerization chemistry. Beyond these original random polymerizations, grafting is also an important synthetic route to incorporate desirable polar monomers onto methacrylate polymers. Stambaugh [8] identified grafting of N-vinylpyrrolidone onto a PMA substrate as a route to improved dispersancy of VIIs. Another approach is to graft both N-vinylpyrrolidone and N-vinyl imidazole [9]. An obvious benefit of grafting is an ability to incorporate polar monomers that do not readily copolymerize with methacrylates due to significant differences in reactivity ratios. Grafting reactions are carried out after achieving high conversion of the alkyl methacrylates to polymer. Bauer [10] identified an alternate synthetic route to incorporate dispersant functionality by providing reactive sites in the base polymer and then carry out a postpolymerization reaction. For example, maleic anhydride copolymerized into or grafted onto the polymer backbone can be reacted with compounds containing desirable chemical functionality such as amines. This strategy is a route to incorporate compounds that are otherwise not susceptible to addition polymerization because they lack a reactive double bond.

The earlier discussion characterizes most of the chemistry used to prepare the great majority of commercial PMA products.

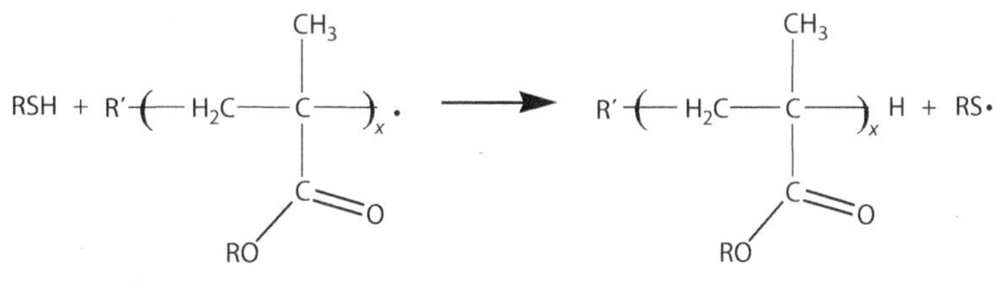

FIGURE 12.7 Termination by chain transfer.

Additional chemical strategies and some novel processes and polymer blend strategies are reviewed in the following literature and patent section.

12.2.4 PATENT REVIEW

A review of pertinent literature and patents shows PMAs to have been the subject of numerous investigations over the course of years. A huge body of PMA patent literature exists, and a large subset of it is related to lubricant additives. A summary from the additive-related patents suggests five major areas of investigation that can be categorized as variation of polymer composition; incorporation of functionality to enhance properties other than rheology, that is, dispersancy; improved processes to improve economics or to enhance a performance property; polymer blends to provide unique properties; and finally polymer architecture. A short discussion of only a few of the more important patents within the five categories ensues.

Since the first PMA patents [2,3], there has been a continuing search for *composition modifications* to methacrylate polymers to improve some aspect of rheological performance. As expected, much of the earlier work explored uses of various alkyl methacrylate monomers and examined the ratios of one to another, this work is part of the well-established art. But even more modern patent literature includes teachings about PMA compositions. For instance, highly polar PMA compositions made with high concentrations of short-chain alkyl methacrylates are useful in polar synthetic fluids such as phosphate esters to impart rheological advantages [11]. There are numerous examples of incorporating non-methacrylates into polymers; a good example is the use of styrene [12] as a comonomer to impart improved shear stability. However, styrenic monomers have different reactivity ratios than methacrylates, and the usual processes lead to relatively low conversions of styrene. This can be overcome by a process utilizing additional amounts of methacrylate monomers near the end of the process to drive the styrene to high conversion [13].

Incorporation of *functional monomers* to make dispersant versions of PMA has been discussed in the preceding section on chemistry. Despite the well-known nature of dispersant PMA, it remains an area of active research as exemplified by [14] that describes a dispersant for modern diesel engine oil soot. Although nitrogen-based dispersants are the focus of much research, oxygen-based dispersants such as hydroxyethyl methacrylate [15] and ether-containing methacrylates [16] have also been claimed. In addition to incorporating dispersant functionality, significant efforts to incorporate other types of chemical functionality, such as antioxidant moieties [17,49], have been made.

Novel processes have been developed to improve either economics or product properties. Tight control of molecular-weight distribution and degree of polymerization can be achieved through constant feedback of conversion information to a computer control system that adjusts monomer and initiator feeds, as well as temperature [18]. Coordinated polymerizations are useful in preparing alternating copolymers of

methacrylates with other vinyl monomers [19]. A process has been described to prepare continuously variable compositions that can obviate the need to physically blend polymers [20].

Polymer blends of PMA and olefin copolymer (OCP) VIIs provide properties intermediate to the individual products with OCP imparting efficient thickening (and economics) and PMA imparting high VI and good low-temperature rheology. However, a physical mixture of the two VIIs in concentrated form is incompatible. This problem is overcome by the use of a compatibilizer, actually a graft polymer of PMA to OCP, to make a ~70% PMA and ~30% OCP mixture compatible [21]. Very high-polymer-content products can be prepared by emulsifying the mixture so that the PMA phase is continuous in a slightly polar solvent while the normally very viscous OCP phase is in micelles [22,23]. Blends of PMAs can provide synergistic thickening and PPD properties [24].

PMA polymer architecture is being very actively investigated today. Preparation of PMA blocks, stars, combs, and narrow molecular weight distribution polymers is all the subject of relatively recent patents or patent applications. For instance, the newer polymerization technique of CRP, specifically atom transfer radical polymerization (ATRP), has been used to prepare PMAs of very narrow molecular weight distribution in order to improve the thickening efficiency/shear stability balance of the resulting product [25]. Similarly, a CRP nitroxide-mediated polymerization (NMP) has been described [26] as providing products with similar improved properties. The ATRP technique has been used to prepare PMAs with functional (polar) monomers in blocks, in order to enhance physical attraction to metal surfaces and thus improve frictional properties under low speed conditions [27,28]. Star-shaped PMAs made via CRP processes, including reversible addition–fragmentation chain transfer (RAFT) polymerization, NMP, and ATRP, have been described as providing improved solution properties [29]. Star shapes and other polymer architectures via various CRP processes are described [30,31] as having enhanced thickening efficiency/shear stability and VI contribution relative to the more traditional linear polymers. Another new polymer architecture of interest are comb polymers with polyolefin and PMA elements as described by [32,46–48]; the same enhancement of properties as mentioned earlier applies to these structures.

12.3 PROPERTIES AND PERFORMANCE CHARACTERISTICS

12.3.1 CHEMICAL PROPERTIES

PMAs are rather stable materials and do not normally undergo chemical reactions under moderate to even relatively severe conditions. The chemical design of any VII or PPD clearly entails avoiding reactive sites in their structure in order to provide as high a degree of stability as possible in the harsh environments to which lubricants are exposed. It is expected that these PMA additives are not chemically active, as they are added to alter only physical properties (i.e., viscosity and wax crystallization phenomena). When considering dispersant

PMAs, which include chemistry other than alkyl methacrylate, even these have essentially the same fundamental stability as the nondispersant polyalkylmethacrylates. The most notable reaction of any VII, including PMA, is molecular weight degradation by either mechanical shear or thermal cracking in hot spots.

12.3.1.1 Hydrolysis

PMAs are not very susceptible to hydrolysis reactions; however, the question of hydrolytic stability is often posed because of the presence of the ester group. In the polymeric form, methacrylate ester groups are quite stable since they are well shielded by the surrounding polymer as well as the pendant side chains. The immediate chemical environment surrounding the ester is definitely hydrophobic and not compatible with compounds that participate in or catalyze hydrolytic reactions. Extraordinary measures can be used to induce hydrolysis; for instance, lithium aluminum hydride hydrolyzes PMAs to yield the alcohols from the side chain. Nevertheless, there is no evidence that PMA hydrolysis is of any significant consequence in lubricant applications.

12.3.1.2 Thermal Reactions: Unzipping and Ester Pyrolysis

These reactions are known to occur with PMAs, but only under severe conditions. Thus, there appear to be no important consequences in lubricant applications since bulk oil temperatures are usually lower than the onset temperatures of these reactions [33].

A purely thermal reaction of PMA is simple depolymerization (unzipping of polymer chains). PMA chains at sufficiently high temperatures unzip to produce high yields of the original monomers, the unzipping reaction is merely the reverse of the polymerization reaction. One consequence of unzipping is that care must be taken to avoid depolymerization during polymer synthesis by simply avoiding excessive temperatures. Thus, synthesis temperatures are designed to be well below the *ceiling temperature* of the polymerization/depolymerization equilibrium. Onset temperatures for unzipping of PMAs are on the order of 235°C, the temperature listed in the literature for polymethylmethacrylate depolymerization [34]. There may be a minor dependence on detailed side chain structure, but relatively little investigation has been done on longer side chain alkyl methacrylate polymers useful in lubricants. It is also thought that terminal double bonds on the polymer chain are the point where unzipping most readily starts. When terminal double bonds are present in the structure, they are most likely the product of termination by disproportionation.

Another potential reaction is the thermal decomposition of individual ester units within the polymer chain. Usually termed *beta ester pyrolysis*, this reaction degrades polymer side chains to yield an alpha olefin from the pendant side chain. The olefin is of the same length as the carbon skeleton of the side chain (as long as the original alcohol used to make the monomer). The other reaction product is an acid

TABLE 12.1
Pyrolysis of PMA

	Pyrolysis Temperature (°C)	
% Weight loss after:	290	315
2 minutes	—	18.7
3 minutes	93.2	
5 minutes	93.1	94.6
10 minutes	94.9	96.1
15 minutes	96.3	97.7

that presumably remains in the polymer chain. The acid may react with an adjacent ester to yield alcohol and a cyclic anhydride. Another possibility is the reaction of two adjacent acid groups to form a cyclic anhydride with the elimination of a molecule of water. The pathway for the pyrolysis reaction proceeds through the formation of a six-membered intermediate ring formed from a hydrogen bond of side chain beta carbon hydrogen to the ester carbonyl oxygen. The intermediate ring decomposes to alpha olefin and acid products [35]. Temperatures in the range of 250°C and even higher are actually needed to initiate this reaction.

Ester pyrolysis produces volatile alpha olefins and acids or anhydrides along the polymer backbone, or perhaps monomeric acids or anhydrides should the reaction occur after depolymerization. The fate of acid or anhydride in the polymeric backbone would still be depolymerization under the severe thermal conditions present. The ultimate products are volatile, given the conditions needed to initiate the reaction.

Consequences of either thermal reaction are loss of activity as a VII or PPD and the generation of volatile small molecules. The products distill from the high-temperature reaction zone and thus offer no further opportunity for chemical reaction. The data in Table 12.1 indicate the very high volatility of a PMA VII that has been exposed to extreme temperatures in air. There is no evidence that unzipping or ester pyrolysis is important in normal lubricant applications. Most applications generate temperatures less than the reaction onset temperatures; thus, these reactions do not appear to be an issue. A limited potential might exist in a microenvironment, such as if a VII molecule was trapped in a piston groove, for example, confined in a carbonaceous deposit, where temperatures near the combustion chamber exceed onset temperature [6].

12.3.1.3 Oxidative Scissioning

Like any hydrocarbon when exposed to severe oxidative conditions, PMAs can be subject to classic oxidation reactions resulting in polymer scissioning [33]. The scissioning reaction yields two fragments of various lengths each of which is obviously of lower molecular weight than the parent chain, and consequently, there is some loss of viscosity contribution. The reaction takes place at random sites along the backbone since oxidative or free-radical attack may occur anywhere along the polymer chain. Allylic, benzylic, or tertiary hydrogen are most susceptible to oxidative or free-radical attack.

Methacrylates do not normally contain those structural elements; thus, the reaction is not normally an important consideration. The pyrolysis data in Table 12.1 would seem to support this conclusion, as scission fragments would, by and large, not be volatile under the conditions of the experiment.

Proof of PMA oxidative stability and continued effectiveness has been demonstrated by comparing viscosities of used oils exposed to the very severe Sequence IIIG oxidative engine procedure. One oil contained a PMA PPD. The other had all the same components but had no PPD; after the engine aging procedure, this oil was treated with exactly the same PPD in the same concentration as the PPD-containing oil. The resulting viscosities were essentially the same, and of particular note, the low-temperature, low-shear rate viscosities did not differ in any significant way indicating that the PMA PPD was not degraded in the severe environment of a Sequence IIIG [36].

On the whole, PMAs are not prone to thermal or oxidation reactions under normal conditions of use, and there is little evidence that these reactions are important in the vast majority of PMA-based lubricant applications.

12.3.1.4 Mechanical Shearing and Free-Radical Generation

A well-known, very important degradation reaction of any VII, including PMAs, is mechanical shearing. Although polymer shearing begins as a physical process, it does generate free radicals. For each polymer chain rupture, two transitory carbon-centered free radicals are generated. In lubricants, the free radicals are apparently quickly quenched, presumably by abstracting hydrogen from the surrounding hydrocarbon solvent, perhaps by the antioxidants in formulated lubricants, or most likely by dissolved oxygen molecules (from air) in the lubricant. Overall, there appear to be few, if any, further chemical consequences. On the other hand, there are important viscometric consequences since the rupture leads to two lower-molecular-weight fragments that provide a reduced viscosity contribution. The shearing process is initiated through the concentration of sufficient energy within the polymer chain to induce homolytic cleavage of a carbon–carbon bond in the backbone of the polymer. The susceptibility of the polymer to mechanical shearing is not related to its chemical structure; rather, it is very clearly a function of polymer molecular weight or even more appropriately to the end-to-end distance of the polymer chain [37]. Overall VII shear stability, while an important physical process, does not appear to carry any appreciable chemical consequences. Further discussion of shearing can be found in the section on the effect of structure on physical properties.

12.3.2 Physical Properties

The paramount properties of PMAs are those associated with their use in solution as PPDs, VIIs or, dispersants. The dispersants may also be utilized for their VII or thickening properties, but in some cases, thickening properties are not needed and the molecules are used solely as dispersants. The useful properties of PMAs are related to both their physical nature (primarily molecular weight) and to their chemical nature (primarily side chain structure).

12.3.2.1 Pour Point Depressants

PPDs are used to modify and control wax crystallization phenomena in paraffinic mineral oils. As temperature decreases, waxy components begin to form small, plate-like crystals. The plates eventually grow together to form an interlocking network that effectively traps the remaining liquid. Flow ceases unless a force strong enough to break the relatively weak wax gel matrix structure is applied. Control of wax crystallization in lubricants is often described as pour point depressancy, since one of the quantifiable effects is to reduce the ASTM D97 pour point. The pour point test is fairly archaic, as it utilizes a very rapid cooldown to measure only flow versus no-flow conditions. PPDs also control wax crystallization during a variety of slower, more realistic cooling conditions that better favor crystal growth. PPDs are used to maintain fluidity of lubricants under a variety of cooling conditions in order to expand the operating temperature window into colder regimes. How much the operating window can be expanded is a complex function of wax chemistry, its concentration in base oil, the presence or absence of other waxy additives, the cooling conditions, the final cold temperature, and, of course, PPD chemistry and concentration [38].

PPDs do not affect either the temperature at which wax crystallizes from solution or the amount of wax precipitate. PPDs co-crystallize on the edges of the growing wax plates by virtue of their longer alkyl side chains. Thus, the growing wax crystal is attached to the polymer, and then, because of the presence of the molecularly large polymer backbone, crystal growth is sterically hindered in-plane. Further growth is redirected in a perpendicular direction, resulting in the formation of more needle-like crystals. Thus, the usual tendency to form a three-dimensional structure based on plate-like crystals is disrupted, and wax gel matrices are prevented at least temporarily. At exceedingly low temperatures, oils may eventually become so viscous as to appear to cease flowing, but this is irrespective of wax issues [39].

PMAs were the first polymeric PPDs and were commercialized in the 1940s by Rohm and Haas Company. Today, they are the predominant chemistry in this application, enjoying a majority of the worldwide market. The reason for this success is related to the molecular structure, as shown in Figure 12.8, and its inherent chemical flexibility in terms of polymer chain length, but more importantly, its ability to include various side chain lengths, and at appropriate concentrations.

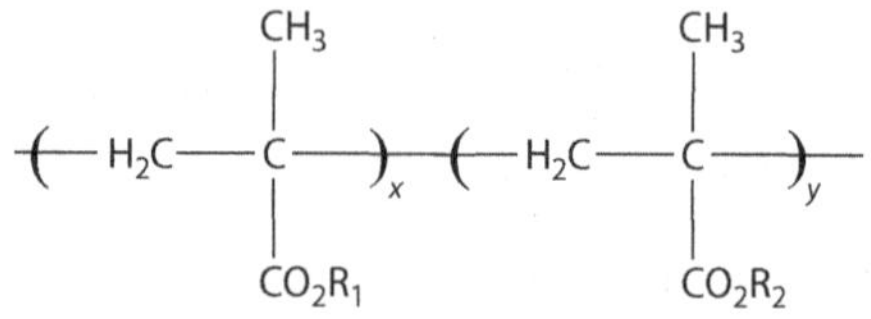

FIGURE 12.8 PMA pour point depressant.

In Figure 12.8, R1 and R2 represent two different lengths of alkyl side chains; one is wax interactive and the other is *neutral*, or *noninteractive* with wax. But the side chains are actually complex mixtures of alkyl groups that may be anywhere from 1 to over 20 carbons. The longer carbon side chains are intended to interact with wax; in order to do so, they should be linear and typically be at least 14 carbon atoms in length. The interaction of a waxy alkyl side chain with wax intensifies as its length increases. However, shorter chains are added to serve as inert diluents, thereby ensuring a controlled degree of wax interaction or to act as *spacers* between the longer side chains so as to better fit into crystal lattice structures.

Pour point depressancy is largely independent of molecular weight over a broad range and degree of polymerization that may vary from about 200 to near 3000. But it is important to achieve a minimum degree of polymer backbone size to provide enough steric hindrance to crystal growth as described previously.

The optimum positive interaction with wax requires a careful balance of the waxy alkyl groups in terms of both type and concentration. This thought is sometimes expressed as a function known as the wax interaction factor (WIF) that takes into account the amount of each alkyl group that interacts with wax and the strength of the interaction. Since mineral oils contain a distribution of wax chain lengths, PPDs often contain a distribution of waxy side chains to best interact with wax in a specific situation. As the wax structure and content change because of different base stocks and/or additives, then a different PPD with a different WIF may be the optimum. This effect is shown in Table 12.2 where different base stocks with different wax chemistries and concentrations respond differently to PPDs with different WIFs.

Finished lubricant formulations often respond differently to PPDs than do the base oils used to make the fully formulated oils. For example, in the MRV TP-1 measurement (ASTM D4684) two different 150N base stocks respond to PPD 3: one (A) with some yield stress and the other (B) with no yield stress; however, when these same base stocks are additized each with the same DI and VII, they respond in an opposite way. The SAE 10W-40 formulated oil based on oil A has passing rheology, while the SAE 10W-40 formulated oil based on oil B is a clear failure. PPD 4 with different WIF does provide passing results with the formulation based on oil B. These data are shown in Table 12.3.

Previously (Section 12.3.1.3), the oxidative and thermal stability of PMA was discussed, and a conclusion reached that PMA PPDs remain effective even after exposure to a severe oxidative environment. The background for the experimentation is a low-temperature pumpability requirement for used gasoline engine oils that first appeared in the ILSAC GF-4 Standard for Passenger Car Engine Oils [40]. Pumpability is measured by ASTM D4684 on an oil after undergoing a Sequence IIIG engine aging procedure (ASTM D7320), which is a severe oxidative and volatilization environment. One normally associates low-temperature pumpability with wax-related phenomena and ultimately with PPD activity; however, the used oil pumpability requirement was added to address OEM concerns about oil degradation in the field that could be identified by low-temperature viscosity measurements, and not necessarily by the usual higher-temperature kinematic viscosity determinations [41]. The need for a low-temperature measurement is because of severe oxidation that may lead to the formation of polar molecular species that associates to form gel-like structures at relatively cold temperatures, but not necessarily at warmer temperatures.

12.3.2.2 Viscosity Index Improvers

VIIs are used to achieve the advantages of multigrade lubricating oils in numerous applications, including crankcase engine oils, automatic transmission fluids (ATFs), high VI hydraulic fluids, gear oils, and other lubricants used primarily (but not necessarily) outdoors. VIIs, also known as viscosity modifiers, are high-molecular-weight, oil-soluble polymers that ideally provide increased viscosity at higher temperatures and minimal viscosity contribution at lower temperatures [42]. Current commercial chemistries are based on either of two chemical families: hydrocarbons such as ethylene–propylene copolymers or ester-containing materials such as PMAs.

TABLE 12.2
PPD Response[a] in Different Base Oils

| | Base Stock 1 | | Base Stock 2 | |
| | PPD 1 | PPD 2 | PPD 1 | PPD 2 |
Wt.%	Low WIF	High WIF	Low WIF	High WIF
None	−18		−12	
0.05	−24	−21	−18	−18
0.10	−33	−30	−21	−27
0.20	−36	−30	−27	−33

[a] ASTM D97 pour point (°C).

TABLE 12.3
Comparison of PPD Response[a] in Base Stocks versus Fully Formulated Fluids

PPD Treat Rate	Base Stock A	Base Stock A + DI and VII	Base Stock B	Base Stock B + DI and VII
0.1 wt.%	150N	SAE 10W-40	150N	SAE 10W-40
PPD 3	31,100/35	39,700/<35	26,500/<35	62,000/70
PPD 4	—	—	28,100/<35	35,800/<35

[a] ASTM D4684 TP-1 MRV at −30°C, viscosity (mPa · s)/yield stress (Pa).

There are other examples of each chemical family. PMA chemistry dominates applications where high VI and/or superior low-temperature properties are required [36]. These benefits can usually be observed in typical lubricant industry low-temperature, low-shear-rate rheology tests such as MRV TP-1, Scanning Brookfield, and ASTM D2983 Brookfield, as well as numerous others. Polymer molecular size influences thickening at all temperatures, and the larger the coil size, the higher the thickening power. Conversely, with smaller coil size, less thickening occurs. PMA VIIs thicken oils well at higher temperatures but contribute relatively little low-temperature viscosity. This desirable viscosity–temperature behavior stems from PMA ester functionality imparting polarity to the polymer in the nonpolar hydrocarbon solvent, mineral oil, leading to a relatively small molecular coil size at low temperature. The ester polarity can be accentuated by the use of short pendant side chain monomers.

Thickening properties for any chemical class of VII or viscosity modifier are related to their immensely greater molecular size compared to that of the solvent in which they are dissolved. The long polymeric strand, the backbone of the polymer, is configured in a random coil shape. The size of the coil, or more appropriately its hydrodynamic volume, is proportional to polymer molecular weight as a first approximation; but more precisely it is proportional to the cube of the root mean-square end-to-end distance of the polymer. In the lattice theory of viscous flow, segments of polymer molecules fill holes in the lattice (constructed of all surrounding molecules) and thereby limit the ability of smaller molecules to participate in movement through the lattice [38]. The degree of viscosity increase depends on coil size; thus, higher-molecular-weight polymers provide more thickening. The overall viscosity of a polymer-thickened solution is related to polymer concentration and molecular weight through the following equation developed by Stambaugh [6].

$$\text{Ln}\,\eta = KM_v^a c - k''\left(M_v^a\right)^2 c^2 + \ln \eta_0 \qquad (12.1)$$

where

M_v is the VII viscosity average molecular weight
c is the VII (or thickener) concentration
η_0 is the solvent viscosity
Exponent a relates to solubility of the specific polymer
    chemistry, solvent, and temperature

For a PMA VII in solution, the coil size expands and contracts with temperature [43]. At lower temperatures, PMA is, on a relative basis, poorly soluble in oil. This is *not* meant to say that PMA precipitates from solution, but the relatively poor solubility results in a contracted, smaller-volume polymer coil, which has a relatively low-viscosity contribution. As temperature increases, solubility improves and polymer coils eventually expand to some maximum size, and, in so doing, donate more and more viscosity. The process of coil expansion is entirely reversible. The polymer coil will continue to expand or contract with temperature changes irrespective of

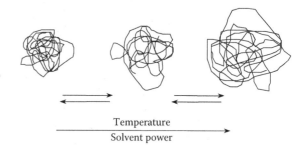

FIGURE 12.9 PMA coil expansion.

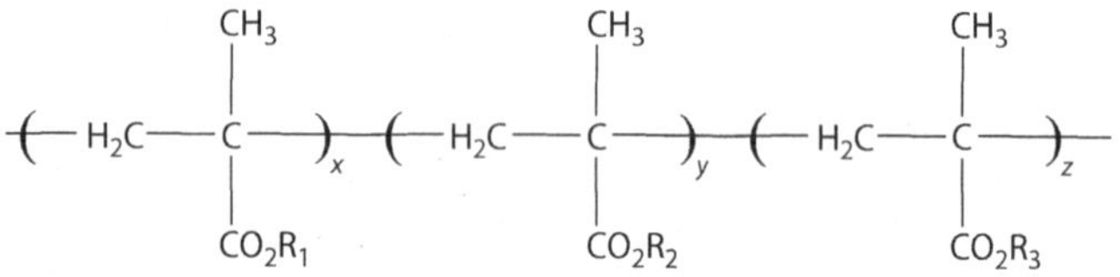

FIGURE 12.10 PMA VII structure.

the aging history of the solution (see Figure 12.9). In contrast, nonpolar polymeric thickeners are well solvated by oils at all temperatures and experience far less change to coil size with temperature [44]. For any VII or viscosity modifier chemistry, these solubility factors relate to the value of the exponent α in Equation 12.1 and, specifically for PMA, are a function of the average length side chain chemistry: short, intermediate, and long alkyl chains. This structural concept of three distinct chain lengths is represented in Figure 12.10.

A typical average side chain length of about eight carbons will provide PMA solubility in oil, even down to extremely low temperatures. So an intermediate chain length monomer is used to provide overall oil solubility and is normally selected from linear or branched chains composed of 8–14 carbons. The longer chains, consisting of more than 14 carbons, may be incorporated to provide wax crystallization interactions as described earlier in the section on PPDs. Very short chains, usually C1 or C4, are used to balance the composition and make the average chain length at least eight carbons for paraffinic mineral oil solubility. To a very good first approximation, the average side chain length will determine viscosity–temperature properties rather than the detailed nature of the side chain structures.

Building pour point depressancy into PMA VIIs by including longer side chain monomers may involve compromises. An optimized wax interaction provided by the VII may not be possible because of the many different base stocks (with different wax types and contents) in which the VII might be used. On the other hand, VII treat rates are relatively high compared to PPDs, and the high polymer concentration may simply overwhelm wax crystallization as it occurs and thereby prevent wax gel matrix consequences (see Section 12.3.2.1).

Commercial PMA VIIs are available in various chemical compositions and molecular weights ranging from about 20,000 to 740,000 Da. The higher-molecular-weight materials are the most efficient thickeners and provide the greatest VI lift but are also the most susceptible to shearing effects.

Selection of a suitable VII for a given application should focus on shear stability, thickening efficiency, and VI lift.

Shearing effects, either temporary or permanent, are related to polymer backbone molecular weight. High-molecular-weight polymers are subject to both temporary viscosity loss via shear thinning and to permanent viscosity loss when polymer chains are broken in mechanical degradation processes, both of which result in loss of thickening.

Temporary viscosity loss occurs when polymer molecules become oriented along the axis of flow at sufficiently high shear rates. This phenomenon, known as shear thinning, occurs at a minimum, nominal value on the order of 10^4/s. Shapes of individual polymer coils change from a spherical to an elongated configuration that occupies a smaller hydrodynamic volume and thus contributes less viscosity. With further increases in shear rate, molecules increasingly deform, leading to a corresponding greater loss of viscosity contribution until maximum distortion is reached. Within the lubricant community, non-Newtonian, shear-thinning behavior is better known as the temporary loss of viscosity, since the process is reversible upon removal of high shear rates. Distorted polymer molecules resume random coil shapes, reoccupy original hydrodynamic volumes, and contribute viscosity just as before the application of a higher shear rate. The degree of temporary viscosity loss depends on the level of shear stress and the molecular weight (size) of the polymer. Because of their small coil sizes, low-molecular-weight polymers are less susceptible to shear thinning.

The temporary viscosity loss of PMAs is directly related to molecular weight or, even more appropriately, to backbone molecular weight. PMAs are not associative thickeners and do not experience viscosity losses through loss of molecular associations in high shear stress fields. Any loss of viscosity is related merely to at-rest molecular size and subsequent molecular shape distortion, leading to lower hydrodynamic volume. Figure 12.11 is compiled with data taken from a set of SAE 10W-40 oils blended to constant kinematic viscosity (14.5 mm²/s) and constant cold-cranking simulator viscosity with the same compounding materials, except for the VII. Three chemically equivalent, dispersant PMA VIIs, differing only in molecular weight, were used in the formulations.

The resulting high-temperature, high-shear-rate viscosities are clearly a function of polymer MW, showing essentially an inverse linear relationship with polymer MW.

Permanent viscosity loss via mechanical degradation occurs when very high shear stresses, perhaps coupled with turbulent flow, lead to extreme polymer coil distortion and concentrate enough vibrational energy to cause polymer chain rupture. Cavitation probably also plays an important role by producing intense velocity gradients. Polymer chain rupture occurs through homolytic cleavage of a carbon–carbon bond, statistically near the middle of the polymer chain. The cleavage produces two molecules, each having on average approximately half the molecular weight of the original molecule; Figure 12.12 represents molecular elongation and rupture concepts. The total hydrodynamic volume of the two smaller molecules is less than that of the single parent molecule, resulting in lower viscosity contribution. Since the bond scission is not reversible, the viscosity loss is permanent. Higher-molecular-weight polymers are more susceptible to distortion and mechanical degradation, while polymers of sufficiently low molecular weight may not even undergo permanent shearing. Since the sheared polymer molecules are of lower molecular weight compared to unsheared ones, they are less susceptible to further mechanical degradation. Thus, the degradation process is self-limiting under any given intensity of shearing.

As with shear thinning phenomena, PMAs of sufficient molecular weight are subject to mechanically induced permanent loss of viscosity. Again, the amount of viscosity loss for linear PMA is a reasonably straightforward function of molecular weight. Molecular weight distribution plays a secondary role. If the MW distribution were skewed to larger fractions of high MW polymer, then the loss of viscosity would be greater than that of a polymer of similar average MW but with a Gaussian distribution. It should also be noted that different applications can create very different shear stresses, and because of this, viscosity losses of polymers of a given MW vary by application. The net result is that viscosity loss to a first approximation is directly related to molecular weight and to the amount of shearing force in a given application or laboratory bench test. The relationship of shear stability index to polymer MW and shearing severity by application is shown in Figure 12.13 for a set of PMA VIIs [45].

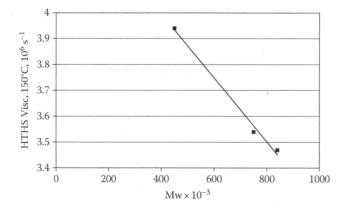

FIGURE 12.11 Relationship of PMA VII molecular weight to temporary shear stability (SAE 10W-40 oils blended to 14.5 mm²/s at 100°C).

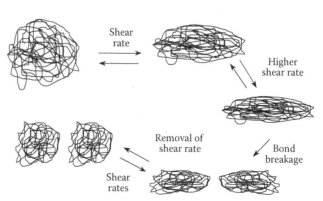

FIGURE 12.12 Temporary and permanent shearing of polymer.

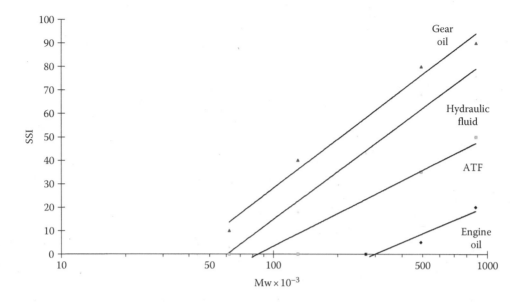

FIGURE 12.13   Relationship of shear stability index to PMA VII molecular weight and shear severity by application.

Dispersant functionality may be incorporated into PMA chemistry by utilizing a monomer that contains a heteroatom that will create polar zones along the otherwise oleophilic polymer chain. Typically, the heteroatom is nitrogen incorporated in an amine, amide, or lactam, but oxygen-containing monomers are also sometimes used. As previously discussed, incorporation can be achieved through monomer copolymerization, grafting a monomer onto the preformed polymer chain, or incorporating a reactive chemical site on the polymer followed by post-reacting it with an appropriate chemical species. For instance, incorporating an anhydride, such as maleic anhydride, as the reactive chemical site, then after the polymerization reacting with an amine, and thus creating succinimide sites along the polymer chain. The resulting polar region(s) serves to attract and peptize small polar molecules or particles, that is, oxidized oil, oxidized fuel, or even soot. These undesirable materials are often unintentional by-products of combustion or oxidation of the lubricant. Left undispersed, the small molecules may undergo further reactions, resulting in the formation of harmful deposits, while particulate matter may become a source of abrasive wear.

Dispersant PMAs are used as stand-alone dispersant products in a few applications but more typically as dispersant VIIs, which augment the dispersancy provided by classic ashless dispersant molecules found in detergent/inhibitor (DI) packages. An example of a stand-alone dispersant PMA is its use in SAE J1899 piston aircraft engine oils. PMA dispersants can also be employed to boost the overall dispersancy of a formulation. These materials have been used in engine oils either to supplant some of the traditional ashless dispersants or simply to enhance dispersancy as well as imparting the usual rheological properties. Table 12.4 contains data from Sequence VE engine testing of a SAE 5W-30 API SG oil made from *dispersant* PMA VII (45 diesel injector SSI) and another SAE 5W-30 oil containing all the same compounding materials but with a *nondispersant* PMA VII (also 45 diesel

TABLE 12.4

Engine Performance[a] Dispersant PMA versus Nondispersant PMA in SAE 5W-30 API SG Oils

PMA V1Is (45 SSI)	Ave. Sludge	Ave. Varnish	Ave. Cam Wear
Dispersant	9.23	6.25	1.5
Nondispersant	4.55	4.56	5.8

[a] Sequence VE test—summary data (average from six tests of each formulation).

injector SSI). The results in each case are the average of six engine tests. Clearly, the dispersant VII provides enhanced dispersancy relative to its nondispersant analog.

12.4   MANUFACTURERS, MARKETERS, AND ECONOMICS

12.4.1   MANUFACTURERS AND MARKETERS

The following companies offer commercial quantities of PMA additives:

Afton Chemical Corporation
330 South Fourth Street
P.O. Box 2189
Richmond, VA 23219

BASF SE
Fuel and Lubricant Solutions
67056 Ludwigshafen, Germany

Chevron Oronite Company LLC
6001 Bollinger Canyon Road
San Ramon, CA 94583

Evonik Resource Efficiency GmbH
Oil Additives Business
Kirschenallee D-64293
Darmstadt, Germany

Infineum UK Limited
P.O. Box 1 Milton Hill
Abingdon Oxfordshire
OX 13 6BB, U.K.

Kusa Chemicals Pvt. Ltd.
101, Varun Apartment
Dattatray Road
Santa Cruz (West)
Mumbai 400054, India

The Lubrizol Corporation
29400 Lakeland Boulevard
Wickliffe, OH 44092-2298

Sanyo Chemical Industries, Ltd.
11-1, lkkyo Nomoto-cho. Higashiyama-ku
Kyoto 605-0995, Japan

Toho Chemical Industry Co.
Ltd. 6-4. Akashi-Cho, Chuo-Ku
Tokyo 104-0044, Japan

Within the realm of lubricant additives, NSC and Kusa primarily offer PMA VIIs and PPDs. Evonik Oil Additives, Sanyo Chemical Industries, and Toho, although focused on PMAs, do offer other product types to the lubricant and oil-refining industries. BASF offers other oil additives besides methacrylates, most notably antioxidants and ashless antiwear components. Afton, Chevron Oronite, Infineum, and Lubrizol offer PMAs as well as numerous other additive types and additive packages for lubricants.

12.4.2 Economics and Cost-Effectiveness

In order to do a meaningful economic study, one must know an additive's treat rate and its unit selling price. While this is true for any additive, it is particularly true for VIIs as their treat rates may vary widely, depending on their chemical nature and base oil viscosities and by the desired viscosity grade of the treated oil. The cost-effectiveness of PMA VIIs varies widely by application. When enhanced low-temperature performance, high VI, and/or excellent shear stability are requirements, then PMAs often enjoy excellent cost competitiveness. Examples of these applications are ATFs, multigrade hydraulics, and the lower SAE W grade gear oil grades. In these situations, it may often be impractical to employ other VII chemistries because of some technical deficiency related to the earlier mentioned properties.

In engine oils where these technical criteria are often not as stringent, PMA VIIs are often at a cost disadvantage. Fundamentally, PMAs are less-efficient thickeners than many

competitive hydrocarbon VIIs such as ethylene–propylene copolymers or isoprene-based polymers as described earlier. VII thickening is a function of polymer backbone molecular weight (actually the unperturbed root mean-square end-to-end distance), but only a minor portion (about 15%) of a typical PMA's molecular weight is in the backbone, compared to an olefin polymer with about 80%–90% backbone MW. Recall that a large majority of the molecular weight of a PMA is found in the pendant side chain simply to provide oil solubility. So, when comparing a PMA to a polyolefin of similar MW, the PMA has a shorter end-to-end distance and is clearly a less-efficient thickener. In situations where a PMA may bring some unique value to the formulation such as additional dispersancy (i.e., soot control in heavy-duty engine oils) or superior low-temperature properties, then these technical advantages may make PMA more economic.

For PPDs, it is again difficult to assign a typical treat rate. PPD concentrations may vary widely due to numerous factors including the types of base stocks employed, the impact of other additives on desired properties, and even the types of cold-temperature tests used for a given application.

12.4.3 Other Incentives

In addition, if the VII brings value-added features such as pour point depressancy or dispersancy, then the economic credit for these features must be accounted for in the net treat cost.

PPD economics also involve the commercial reality that, as few PPD products as possible, more likely a single PPD would be favored at a blend plant. Thus, performance robustness in a variety of formulations is highly leveraged. Methacrylates with their inherent flexible chemistry can be well tuned to meet this objective and thus often provide a further degree of cost-effectiveness. Overall, it is difficult to quantify cost-effectiveness in a simple way; however, PMA PPDs are believed to be the most widely used chemistry in this application and command over 50% of the world market.

12.5 OUTLOOK AND TRENDS

12.5.1 Current and Near-Term Outlook

The commercial presence of PMAs over so many decades is a compliment to adaptability and the innovative effort to evolve the chemistry to meet more and more demanding rheological requirements. Competitive pressures from other chemistries will always remain high in the modern, global economy, so the continued use of PMAs depends on the market's continuing and evolving needs for higher rheological performance. Given this, PMA compositions will almost assuredly evolve as well. At the same time, raw materials and processes will be under constant investigation in order to create more cost-effective materials. Finally, advanced polymer processes are being developed to provide unique polymer architectures that provide enhanced rheological properties.

The outlook for PMA VIIs is, to a large degree, tied to transportation and industrial multigrade fluid viscosity requirements, which are driven by equipment and operating conditions needs. Both equipment and operating conditions are in a time of flux. The evolution of automatic transmission equipment offers a pronounced example of such change with a concomitant change to ATF requirements. This has created a need in ATFs for more stringent low-temperature viscometrics (cold-temperature operation), lower high-temperature viscosity and higher VIs (improved fuel economy contribution), and better shear stability performance (long fluid life) while maintaining the traditional properties of fluid film strength and thermal/oxidative stability. PMAs offer the best commercial route to low-temperature viscometric performance and high VI contribution at the same time as being entirely adaptable relative to shear stability needs. Obviously, the evolution of ATFs has favored the continuing use of PMA VIIs.

Fluid requirements may often appear to be contradictory. For instance, modern ATFs frequently require low overall viscosity but high VI for fuel economy reasons. Another example is the constant push to better low-temperature viscometric performance at the same time as improved shear stability; typically, better shear stability implies higher polymer treat rates, which imply higher low-temperature viscosity. In so much as these contrasting requirements emphasize one or more of the desirable properties of PMAs, these factors further reinforce PMA.

Transmission oils satisfying lubricant needs for more complex step transmissions, continuously variable transmissions, and double clutch transmissions are not the only multigrade fluids undergoing change. Transportation gear oils will evolve to better meet fuel economy expectations and more thermally demanding environments. In industrial and tractor hydraulics, higher-pressure hydraulic pumps and the ubiquitous desire for fuel economy improvement will drive fluid change. In so much as these changes involve higher VI, better low-temperature viscometrics, and enhanced shear stability, PMA VII will be involved.

It seems quite probable that such equipment trends and usually more severe operating environments will continue to evolve over the near and midterm future and thus require further enhancement of fluid viscometric properties. In those applications where PMA VIIs provide valued advantages, this should not only reinforce their use but also spur further innovation. Even in the most cost-conscious applications, such as engine oils, the constant demand to further improve fuel economy has reinstated the use of PMA VIIs. For the last three decades, hydrocarbon-based VIIs, such as ethylene–propylene copolymers or isoprene-based polymers, have been dominating in this application due to their cost-effectiveness and their ability to meet current rheological requirements for the finished lubricant. As in other applications, the need to improve fuel economy performance has led to increased utilization of advanced VIIs, such as the comb polymers with polyolefin and PMA elements.

The outlook for PMA *PPDs* is obviously tied to low-temperature rheology specifications as well as to the wax characteristics of lubricant fluids. Certainly, low-temperature specifications have trended to become more restrictive for many lubricants in addition to the ATF situation discussed previously.

North American engine oil specifications have now added used oil low-temperature requirements to deal with oxidative thickening (light duty engines) and soot-related thickening (heavy duty engines). The existence of these requirements adds a new dimension to PPD evaluations since used oils must now be evaluated. The latter point on wax characteristics of lubricant fluids deals with wax chemistry and wax concentration in base stocks, as these factors will continue to be of paramount importance to the future use of PPDs. As base oil processing moves forward to provide more advanced oils, these will likely contain different combinations of wax structures than past base oils. PPDs will need to be developed to control the resulting different wax crystallization phenomena. In addition, the interaction of base stock wax with other additives that contribute wax-like character (i.e., detergents) provides a complex overlay that often impacts the choice of PPD chemistry. PMA PPD structure can be modified to provide appropriate crystallization interaction with the various types and levels of wax in any fluid. Given the ability of PMA to meet both current and future wax-related needs, their future as PPDs seems assured.

For PMA *dispersant VIIs* (disregarding thickening effects), the outlook is not as clear. In those applications where these materials are currently employed, their use will probably continue for the foreseeable future with little change in terms of chemistry or commercial volumes. There are potentially important commercial possibilities, if industry needs for dispersancy either reach higher levels or needs to achieve a certain dispersancy at minimized low-temperature thickening. While incumbent dispersants, such as polyisobutylene succinimides, provide sufficient dispersancy, their detrimental effect on low-temperature thickening is well known. In order to further improve fuel economy by minimizing low-temperature viscosities at equipment start-up, without lowering the high-temperature viscosities because of durability reasons, it may be necessary to formulate with dispersant VIIs, in order to minimize the amount of *classical*, PIB-based dispersants in the formulation, and thus their contribution to low-temperature viscosities.

12.5.2 Long-Term Outlook

Just as for many other additives and base stocks, the future of PMAs depends on the rheological appetites of future equipment. In the absence of revolutionary equipment development, longer-term lubricant trends will continue to be driven by the concerns mentioned earlier: emissions control, fuel efficiency, and longer fluid life times. These three concerns certainly apply in varying degrees to all multigrades and will cause fluid requirements to continue their migration to enhanced rheological properties. This should solidify the case for PMA in those applications where it holds prominent technical advantages. At the same time, PMAs will continue to evolve to provide improvements in shear stability, low-temperature viscometrics, wax interaction, VI, and dispersancy. In addition to these properties, some other less typical attributes and functionalities of VII polymers, such as antiwear performance and improved, but viscosity independent, fluid film strength under load, will gain higher importance, especially if the trend to lower-viscous lubricants continues. Current polymer architecture research noted in the patent review section certainly points in this direction.

Another important factor for future use of PMAs depends intimately on future base stocks. For instance, should entirely synthetic, poly-α-olefin (PAO) base stocks be the choice of the future to provide enhanced properties, then this would obviate the use of any PPDs, as these base stocks are wax-free. On the other hand, additives like PMA PPDs may be an enabling factor in the use of newer base stocks, such as the so-called API Group III+ base stocks, in meeting performance goals that match or are quite similar to those of API Group IV PAO base stocks.

For the case of revolutionary equipment changes such as ceramic engine parts or vehicles based on batteries, fuel cells, or even clean burn systems (say, hydrogen or natural gas), lubricant requirements will surely change and lubricants themselves will undergo a similar revolution. Most of these changes would radically alter the chemical and physical character and/or volumes of lubricants, since the interesting but hostile environment of an internal combustion engine would no longer dictate the complex chemistry of engine lubricants. If engine lubricant volumes declined because of radical new materials or power sources, then surely all PPDs and VIIs would experience a proportional decline. However, in other applications, such as hydraulics, where PMA VIIs are primarily employed, such radical equipment changes do not seem to be on the horizon.

REFERENCES

1. S Hochheiser. *Rohm and Haas: History of a Chemical Company*. Philadelphia, PA: University of Pennsylvania Press (1986), p. 145.
2. Rohm and Haas Co. Composition of Matter and process, U.S. Patent 2,091,627 (1937).
3. Rohm and Haas Co. Process for Preparing Esters and Products, U.S. Patent 2,100,993 (1937).
4. WL Van Horne. Polymethacrylates as viscosity index improvers and pour point depressants. *Industrial and Engineering Chemistry*, 41(5): 952–959 (1949).
5. FW Billmeyer. *Textbook of Polymer Science*. New York: Wiley-Interscience (1971), pp. 280–310.
6. RL Stambaugh, BG Kinker. Viscosity index improvers and thickeners. In: RM Mortimer, MF Fox, ST Orszulik, eds. *Chemistry and Technology of Lubricants*. 3rd edn. The Netherlands: Springer Science+Business Media B.V. (2010), pp. 153–187.
7. E.1. Du Pont de Nernours Co. Lubricating Oil Compositions Containing Polymeric Additives, U.S. Patent 2,737,496 (1956).
8. Rohm and Haas Co. Lubricating Oils and Fuels Containing Graft Copolymers, U.S. Patent 3.506,574 (1970).
9. Rohm GmbH. Graft Copolymeric Lubricating Oil Additives, U.S. Patent 3,732,334 (1973).
10. Rohm and Haas Co. Preparation of copolymers useful as dispersants in oil, Canadian Patent 683,352 (1964).
11. Rohm and Haas Co. Viscosity Index Improving Additives for Phosphate Ester-containing Hydraulic Fluids, U.S. Patent 5,863,999 (1998).
12. Institute Français du Petrol. Copolymer compositions usable as additives for lubricating oils, U.S. Patent 4,756,843 (1988).
13. Rohm and Haas Co. Process for making a viscosity index improving copolymer, U.S. Patent 6,228,819 (2001).
14. RohMax Additives. Oil composition for lubricating an EGR equipped diesel engine and an EGR equipped diesel engine comprising same, U.S. Patent 7,560,420 (2009).
15. Shell Oil. Lubricating composition containing non-ash forming additives, U.S. Patent 3,249,545 (1966).
16. Rohm and Haas Co. Preparation of copolymers useful as dispersants in oils, U.S. Patent 3,052,648 (1962).
17. Texaco Inc. Lubricating oil containing dispersant viscosity index improver, U.S. Patent 4,790,948 (1988).
18. Rohm and Haas Co. Dispersant polymethacrylate viscosity index improvers, EP. Patent 569,639 (1993).
19. Sumitomo Corp. Japanese Patent 4,704,952 (1972).
20. Rohm and Haas Co. Process for preparing continuously variable-composition copolymers, U.S. Patent 6,140,431 (2000).
21. Rohm GmbH. Lubricating Oil Additives, U.S. Patent 4.149,984 (1979).
22. Rohm GmbH. Lubricating Oil Additives, U.S. Patent 4,290,925 (1981).
23. Rohm GmbH. Concentrated Emulsion of Olefin Copolymers, U.S. Patent 4,622,358 (1986).
24. Shell, Intl. Lubricating Oil Compositions, Great Britain Patent 1,559,952 (1977).
25. RohMax Additives GmbH. Method for preparation of a composition that contains polymer ester compounds with long-chain alkyl residues and use of this composition, U.S. Patent 6, 403, 746 (1999).
26. Arkema. Controlled radical acrylic copolymer thickeners, U.S. Patent 7,691,797 (2010).
27. Evonik RohMax Additives GmbH. Lubricating oil composition with good frictional properties, U.S. Patent 8,288,327 (2012).
28. Evonik RohMax Additives GmbH. Polyalkyl (meth)acrylate copolymers having outstanding properties, European Patent EP 1866351 (2007).
29. The Lubrizol Corporation. Star polymers and compositions thereof, European Patent EP 1833868 (2007).
30. The Lubrizol Corporation. Process for preparing polymers and compositions thereof, European Patent EP 1833852 (2007).
31. RohMax Additives GmbH. Oil soluble polymers, European Patent EP 1919961 (2014).
32. RohMax Additives GmbH. Oil soluble comb polymers, European Patent EP 1899393B1 (2012).
33. JP Arlie, J Dennis, G Parc. *Viscosity Index Improvers 1. Mechanical and Thermal Stabilities of Polymethacrylates and Polyolefins*. IP Paper 75-005. London, U.K.: Institute of Petroleum (1975).
34. J Branderup, EH Irmergul, eds. *Polymer Handbook*. New York: Interscience (1966), pp. V-5–V-11.
35. J March. *Advanced Organic Chemistry: Reactions, Mechanisms, and Structures*. New York: McGraw-Hill (1968), p. 747.
36. B Kinker, R Romaszewski, J Souchik. Pour point depressant robustness after severe use in passenger car engines in the field and in the sequence IIIGA engine. SAE Paper 2005-01-2174 (2005).
37. BG Kinker. Automatic transmission fluid shear stability trends and viscosity index improver trends. *Proceedings of 98th International Symposium of Tribology of Vehicle Transmissions*, Yokohama, Japan (1998), pp. 5–10.
38. BG Kinker. Fluid viscosity and viscosity classification. In: G Totten, ed. *Handbook of Hydraulic Fluid Technology*. New York: Marcel Dekker, Inc. (2000), pp. 318–329.
39. Evonik publication. Pour point depressants, a treatise on performance and selection. http://oil-additives.evonik.com/product/oil-additives/en/Pages/default.aspx (2015).

40. International Lubricant Standardization and Approval Committee. ILSAC GF-4 standard for passenger car engine oils (January 14, 2004).

41. M Bartko, D Florkowski, V Ebeling, R Geilbach, L Williams. Lubricant requirements of an advanced designed high performance, fuel efficient low emissions V-6 engine. SAE Paper 2001-01-1899 (2001).

42. P Neudoerfl. State of the art in the use of polymethacrylate VI improvers in lubricating oils. In: WJ Bartz, ed. *Fifth International Colloquium: Additives for Lubricants and Operations Fluids*, vol. 11. Technische Akademic Esslingen, Ostfildern, Germany (1986), pp. 8.2-1–8.2-15.

43. TW Selby. The non-Newtonian characteristics of lubricating oil. *ASLE Transactions* 1: 68–81 (1958).

44. HG Mueller, G Leidigkeit. Mechanism of action of viscosity index improvers. *Tribology International* 11(3): 189–192 (1978).

45. RJ Kopko, RL Stambaugh. Effect of VI improver on the in-service viscosity of hydraulic fluids. SAE Paper 750693 (1975).

46. Evonik Oil Additives. Use of comb polymers for reducing fuel consumption, European Patent EP 2164885 (2010).

47. Evonik Oil Additives. Use of comb polymers as antifatigue additives, International patent application WO 2010/102903 (2010).

48. Evonik Oil Additives. Transmission oil formulation for reducing fuel consumption, International patent application WO 2014/170169 (2014).

49. Evonik Oil Additives. Polyalkyl (meth) acrylate for improving lubricating oil properties, International patent application WO 2012/013432 (2012).

13 Hydrogenated Styrene–Diene Copolymer Viscosity Modifiers

Isabella Goldmints and Sonia Oberoi

CONTENTS

13.1 HISTORICAL OVERVIEW

With the advent of multigrade lubricants in the early 1930s and with ever-evolving needs for better fuel economy and durability, engine oil formulators rely on the use of viscosity modifiers (VMs), also called viscosity index improvers, to meet the lubricant flow properties for engine oil as per SAE J300. VMs are oil-soluble polymers that modify the rheological properties of the motor oil. There are different classes of VMs, namely, (1) olefin copolymers (OCPs), (2) polyalklymethacrylates (PMAs), and (3) hydrogenated styrene–diene (HSD) copolymers. The focus of this chapter is on HSD copolymers.

The first commercial HSD polymers were introduced in the early 1970s. Shell [1–3] and Phillips Petroleum [4,5] were the pioneers in the field of HSD copolymers. In 1971, Phillips Petroleum [6] introduced the first hydrogenated random styrene–butadiene copolymer having 30%–44% butadiene. These random copolymers were prepared by conventional anionic polymerization as described by Zelinski [4,5], where a mixture of butadiene and styrene monomers can be polymerized using butyllithium as catalyst and tetrahydrofuran as a randomizing agent. Hydrogenation was carried out over a reduced nickel–kieselguhr catalyst or nickel octoate–triethylaluminum catalyst system [6–9].

During this period, Shell was also developing processes (and catalyst systems) for making block copolymers. In 1964, Lee Porter of Shell introduced a new process for preparing block copolymers using a special organometallic catalyst [1]. In 1966, Holden et al. (Shell) found that the use of secondary or tertiary alkyl lithium reduces the induction period during the polymerization so that the block molecular weight can be better controlled, thereby improving the physical properties of the polymers. They further discussed the process of making triblock copolymers of polystyrene–polyisoprene–polystyrene using secondary butyllithium. They continued to define processes for making tapered copolymers [3].

The properties of the HSD copolymer depend on the polymer architecture, molecular weight, and monomer chemistry, and literature is full of different copolymer designs and compositions that can find application as viscosity index improvers.

In 1973, Anderson [10] filed the first patent on the use of diblock copolymers of isoprene and styrene with improved properties over 800,000 Da molecular weight PMA. St. Clair and Evans [11] introduced the diblock copolymers for use in lubricating oils with enhanced performance. This preceded Shell's commercializing Shellvis 40 and Shellvis 50. The polystyrene block in HSD copolymers not only thickens the oil at higher temperature but also provides important nonviscometric performance benefits such as reduced soot-induced viscosity increase and improved piston cleanliness and deposit control. In 2001, Wedlock and Jong from Infineum proposed a mini diblock copolymer of styrene and hydrogenated isoprene as dispersant additive in lubricant oils [12]. In 2006, Lubrizol described a smaller diblock of hydrogenated styrene and butadiene with average molecular weight between 10,000 and 50,000 and a polystyrene content between 5% and 45%, which is also useful for reducing soot-induced viscosity increase in a compression-ignited internal combustion engine equipped with an exhaust gas recirculation [13]. Besides diblock copolymers, triblock copolymers of hydrogenated polyisoprene–polystyrene–polyisoprene described by Olson and Handlin [14] exhibit good balance of thickening efficiency (TE) and mechanical shear stability combined with relatively higher contribution to high-temperature high shear (HTHS) viscosity [15].

In addition to linear HSD block copolymers, star polymer also finds application in lubricating crankcase oils. Fetter and Eckert describe star polymer wherein the arms are hydrogenated homopolymers and copolymers of conjugated diene, hydrogenated copolymers of conjugated diene, and monoalkenyl arenes [16,17]. Eckert [17] exemplified certain star polymers comprising hydrogenated poly(butadiene/isoprene) tapered block copolymer. However, the long ethylene sequences of these hydrogenated polybutadiene blocks were found to cause the same low-temperature performance problems associated with high ethylene content OCPs. To provide an improvement in TE, while maintaining low-temperature performance, Rhodes et al. developed with star polymers comprising triblock copolymer arms of hydrogenated polyisoprene–polystyrene–polyisoprene [18] and polyisoprene/polybutadiene/polyisoprene [19]. The hydrogenated polybutadiene block provided an increased ethylene content, which improves TE. The patent suggested that by placing the hydrogenated polybutadiene block more proximal to the nucleus, the adverse effect on low-temperature properties could be minimized. Such polymers were found to provide improved low-temperature properties relative to the tapered arm polymers of Eckert [16]. However, when such polymers were provided with a hydrogenated polybutadiene block of a size sufficient to provide a credit in TE, a debit in low-temperature performance remained relative to the pure polyisoprene polymers. In 2007, Briggs and Chu provided a class of hydrogenated linear, radial, and star-shaped random copolymers of isoprene and butadiene with TE comparable to high ethylene content OCP without any harm in cold-temperature performance [20]. Duggal describes the use of hydrogenated styrene–butadiene copolymer for lubricating oil with improved viscometric properties [21]. With the advent of more stringent fuel economy regulations, VMs have been defined to reduce the viscometric contribution of the oil under engine operating conditions. Oberoi and Briggs describe a triblock polymer composition of hydrogenated polydiene–polystyrene–polydiene with better shear stability and higher viscosity index [22] versus hydrogenated polydiene star polymer. The polymer composition of this triblock also allows for better engine durability [23].

13.2 CLASSES OF HYDROGENATED STYRENE–DIENE COPOLYMER

As mentioned in the previous section, HSD copolymers are available to consumers in two different architectures: (1) linear block copolymers and (2) star copolymers.

Selective architectures offer discrete rheology advantages. The diblock copolymer can consist of carefully designed segments of hydrogenated polyolefin (isoprene and/or butadiene) and polystyrene. Such diblock polymer chains in oil and the associated micelles provide the necessary thickening at higher temperature (100°C–150°C). Butadiene monomer when used in the olefin segment either on its own or in combination with isoprene must be distributed in a way that there are no crystalline polyethylene segments from 1,4-addition of butadiene, and the resulting polymer is completely amorphous. The amorphous nature of HSD copolymer is important as it provides excellent low-temperature pumpability (as measured by mini-rotary viscometer, MRV) and low contribution to cold-cranking simulator (CCS) viscosity. The polystyrene segment in HSD copolymer plays a critical role and can be used to define copolymer architecture with added performance dimensions for improved soot dispersancy, low-temperature pumpability, and improved viscometrics (Figure 13.1).

The star-shaped HSD polymers with varying arm architecture can be defined using the same set of monomers as mentioned earlier to suit a particular application. Star copolymers are characterized as having a dense center or nucleus of cross-linked poly (polyalkenyl coupling agent) and a number of polymer arms extending outwardly therefrom.

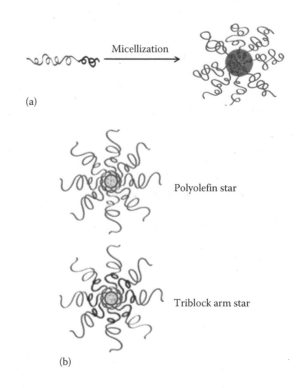

FIGURE 13.1 (a) Schematic of diblock copolymer (black = Polystyrene, gray = Polyolefin). (b) Schematic of different star architectures (black = Polystyrene, gray = Polyolefin, light gray = core).

Molecular weight, polymer chemistry, and architecture of the polymer govern its TE and shear stability. Multigrade oils containing a VM are non–Newtonian fluids, which means their viscosity depends on the shear rate. There can be either a temporary viscosity loss under high shear conditions where the polymers align with the flow, resulting in reversible oil thinning, or a permanent shear loss where the polymers break. Cleavage of polymer backbone can lead to drop in viscosity that can compromise engine durability. Therefore, it is very important that the viscosity of the lubricant oil is retained throughout the entire drain interval. Molecular weight and shear stability of the VM are inversely correlated to each other, for example, the higher the molecular weight of the polymer, the less shear stable the polymer. ASTM D7109/ASTM D6278 test method covers the evaluation of the shear stability of polymer-containing fluids. The test method measures the viscosity loss at 100°C of polymer-containing fluids when evaluated procedure uses the European diesel injector test equipment. The shear stability index (SSI) measures the loss of polymer-added kinematic viscosity (KV) after 30 cycles. The higher the SSI, the more viscosity loss the polymer will undergo upon oil shearing. SSI of commercially available HSD copolymers varies from 4 to 60 SSI. The choice of the polymer depends on the application, for example, North American market is mostly 25–35 SSI for both passenger car motor oil (PCMO) and heavy-duty diesel application, and European PCMO application mostly desires for a very shear VM (5–25 SSI).

$$SSI = 100 \times \frac{\begin{array}{c} KV100 \text{ of polymer solution after } 30 \text{ cycles} \\ - \text{ Fresh } KV100 \text{ of polymer solution} \end{array}}{\begin{array}{c} KV100 \text{ of polymer solution} \\ - KV100 \text{ of the base oil} \end{array}}$$

These polymers are used by lubricant blenders as liquid concentrates typically at a 5%–20% concentration in mineral oil. They are supplied as ready-made concentrates or solids (bales/pallets) for dissolution at a blending plant.

13.3 CHEMISTRY

13.3.1 DIBLOCK COPOLYMERS

As mentioned previously, HSD copolymers are typically made using conjugated diene monomers (e.g., isoprene, butadiene) and alkenyl aromatics (e.g., styrene). Anionic polymerization offers the best route to make different architectures (block, stars, telechelic, random copolymer, etc.) [3,6,10–12,14,16,17,24–27]. Nucleophilic (basic) initiators are used to initiate anionic polymerization, for example, ionic metal amides, $Na\,NH_2$, alkoxides, amines, cyanides, and organometallic compounds.

Alkyl lithium compounds are the most widely used organometallic initiators. The organolithium initiators, RLi, are those that are hydrocarbon soluble, wherein the R represents an aliphatic, cycloaliphatic, or aromatic radical. Such compounds are well-known initiators, and selection of a suitable one, for example, sec butyllithium, may be made merely on the basis of availability, and the choice of the initiator type may have an impact on the polydispersity of the polymer. The amount of initiator employed will depend upon the desired molecular weight of the polymer units. Suitable reaction conditions for the first step require that the polymerizations be conducted in a solvent free from ionizable protons, such as cyclohexane, or toluene alone, or with a polar solvent including cyclic or linear ethers, tertiary amines, or phosphines. The reaction temperatures can range from about 20°C to 50°C, for a period of time ranging from about 1 to 2 hours [1,2,4,5,28–31].

$$(i) \quad nBuLi + CH_2 = CHY \rightarrow nBu - CH_2 - \overset{H}{\underset{Y}{C}} \colon^- (Li^+)$$

$$(ii) \quad nBu - CH_2 - \overset{H}{\underset{Y}{C}} \colon^- (Li^+) + CH_2 = CHY$$
$$\rightarrow Bu(CH_2CHY)n - CH_2 - \overset{H}{\underset{Y}{C}} \colon^- (Li^+)$$

In a typical anionic polymerization, AB block copolymer is made by initiating polymerization of monomer A (isoprene/butadiene) in the presence of butyl lithium and propagation of the polymerization to completion. Monomer B is then added to the polycarbanion of monomer A as shown in the subsequent equation. Since anionic polymerization is a living polymerization, a terminating agent is added when monomer B has completely reacted. If needed, the unsaturated polymer can be isolated by addition of either methanol or water that breaks the carbon–lithium bond and causes coagulation of the polymer. AB block copolymer is then separated from the alcohol by filtration or any other suitable method. This step of isolating the polymer is not usually done until the hydrogenation step is complete.

1. $nBu^+Li^- + A \rightarrow nBuA^-L^+$ (*initiation*)
2. $nBuA^-L^+ + A \rightarrow nBu(A)_n^-L^+$ (*propagation to polycarbanion A*)
3. $nBu(A)_n^-L^+ + B \rightarrow nBu(A)_nB^-L^+$
4. $nBu(A)_nB^-L^+ + B \rightarrow nBu(A)_n(B)_m^-L^+$
5. $nBu(A)_n(B)_m^-L^+ + MeOH \rightarrow$ Block copolymer AB (*termination and isolation to final product*)

Similar to this, random, tapered, or triblock copolymers can be produced by properly sequencing the addition of monomers, provided that each carbanion can initiate polymerization of the next monomer [3,32]. Crossover from one carbanion to another occurs only when the new carbanion is comparable or higher in stability than the original carbanion. Block copolymers of styrene with isoprene or 1,3-butadiene require no specific sequencing since crossover occurs either way. The length of each segment of the copolymer block is controlled by the ratio of each monomer to initiator [28]. Structure–performance maps of block copolymers can be designed by varying the block sizes of individual segment [33–39]. Copolymers prepared via anionic polymerization have narrow molecular weight distribution (see Section 13.5.1).

The unsaturated living block copolymer prepared in step (4) is then hydrogenated to make it useful as VMs for engine oil application [11,14–16,18–21,27]. Moczygemba described a process where the hydrogenation can be carried out after the polymerization has been terminated and the polymer has been isolated [35]. Removal of unsaturation is important for improved oxidative stability of the polymer. This is done with caution to not hydrogenate the aromatic segment. Examples of suitable heterogeneous catalyst systems are nickel on kieselguhr, Raney nickel, copper–chromium oxide, molybdenum sulfide, and finely divided platinum or noble metals on suitable carriers. Homogeneous catalysts are preferred. Such homogeneous hydrogenation catalysts can be prepared by reducing a cobalt, nickel, or iron carboxylate or alkoxide with an alkyl aluminum compound. An example of a preferred homogeneous catalyst is that formed through the reduction of nickel octoate by triethylaluminum.

Typical hydrogenation conditions are described in the subsequent equation. The polymer is diluted with an inert solvent, and the polymer solution and hydrogenation catalyst are added to a high-pressure autoclave. The autoclave is pressured with hydrogen to about 100–7000 kPa and then heated to 10°C–220°C for about 10 minutes to 24 hours. When treating the polymer in solution, the pressure is sufficiently high to maintain the reaction mixture substantially in the liquid phase. The reactor is then depressurized, the catalyst removed by filtering or, where the catalyst components can be converted to water-soluble salts, washed from the polymer solution. The hydrogenated polymer is recovered from the solvent simply by evaporating the solvent [7,26,27,34,35]. An antioxidant can be added if desired followed by coagulation of the polymer and, finally, removal of solvent traces under reduced pressure [35].

more monoalkenyl aromatic hydrocarbon and copolymers of one or more conjugated dienes (e.g., isoprene, butadiene), where diene is the major component. These star-block copolymers are prepared by first forming linear block polymers having active lithium atom on one end of the polymer chain. These active, linear polymer chains are then coupled by the addition of a polyfunctional compound having at least two reactive sites capable of reacting with the carbon to lithium bond on the polymer chains to add the polymer chain onto the functional groups of the compound. Preferred linking compounds that may be employed are either divinyl or trivinyl aromatic compounds or aliphatic or aromatic diisocyanates and polyisocyanates. Other difunctional linking compounds may be employed to form the nucleus, such as various diepoxides, diketones, and dialdehydes, either aliphatic or aromatic.

Linking of the polymeric arms to the molecules of linking compound generally occurs rapidly with the amount and type of coupling agent controlling the number of arms per star-block molecule; see Scheme 13.1. During the linking reaction, a lithium active chain end of a polymeric arm combines with one vinyl carbon of the divinylbenzene while the lithium ion is added to the remaining vinyl carbon. The resultant product may combine with another molecule of the linking compound by the attack of the lithium ion on the vinyl group of the second molecule. Simultaneously, the lithium active chain ends of other available polymeric arms attach the available vinyl group of the linking compound as they combine to form a nucleus. The ratio of linking compound to lithium active chain ends must range at least 2.4:1 and upward. During the linking reaction, molecules of the linking compound are joined together forming a nucleus, while the polymer arms are linked together forming the star product. The number of arms may vary but is typically in the 10–20 range [16–22]. The star polymer is finally hydrogenated as explained in Section 13.3.1.

$$\begin{array}{c} \text{Polymer} \\ \text{solution} \\ \text{(inert solvent)} \end{array} \xrightarrow[\substack{\text{H}_2, \text{ Pressure: 100–7000 kPa} \\ \text{Time: 10 minutes to 10 hours}}]{\substack{\text{Hydrogenation catalyst} \\ \text{Temperature: 10°C–220°C}}} \begin{array}{c} \text{Hydrogenated} \\ \text{polymer} \end{array}$$

13.3.2 STAR COPOLYMERS

Highly branched block copolymers, sometimes called star-block copolymers, are well-established old art of anionic polymerization. Star copolymers are as effective as block copolymers as viscosity index improvers [16,20,40–45]. The arms can also be either random or block copolymers of one or

13.4 CONCENTRATE MANUFACTURING

Hydrogenated star and block copolymers of styrene–diene are finished as solid crumbs, pellets, or bales. They are dissolved in oil at typically 5%–20% to make a liquid concentrate that can be used for blending the finished motor oil. In the case of bales, the polymer is first shredded into small pieces for effective dissolving. The use of a polymer shredder, grinder, or pulverizer enhances the dissolving rate by breaking up the polymer into small particles. Dissolution is conducted in a conventional dissolving vessel. The vessel size is based on

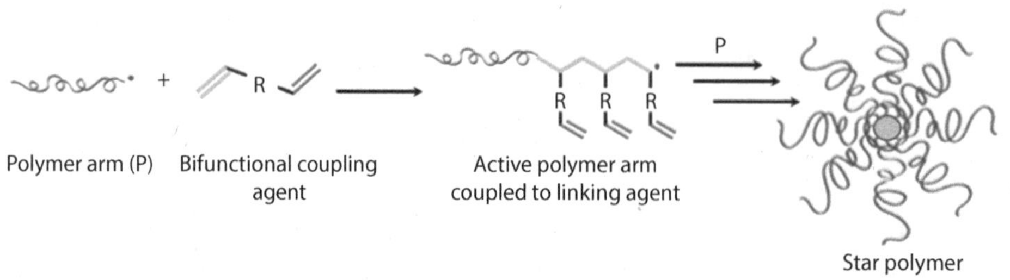

Polymer arm (P)   Bifunctional coupling agent   Active polymer arm coupled to linking agent   Star polymer

SCHEME 13.1   Formation of a star polymer using difunctional coupling agent.

annual throughput and resultant desired capacity. Each vessel must be equipped with agitation and heating systems. Base oil is charged into a tank and heated, and polymer is then added. Dissolving may be done in the presence of nitrogen (if available) in order to minimize the oxidation of base oil. The mixer must produce a good vortex, which will result in a fast turnover of vessel contents and efficient mixing, high-speed mixers providing high shear are preferred for efficient dissolution of these polymers in high-quality diluent oils. Diluent oil can be API Group I to Group IV base stock. The concentration of polymer in diluent oil is generally dictated by the maximum viscosity of the resulting mixture at the pump-out temperature and the corresponding pump requirement. The molecular weight, architecture, and chemistry of the VM affect the viscosity of the concentrate. The optimum concentration of solid polymer in the concentrate is determined for each polymer for suitable handling in standard manufacturing equipment.

13.5 PROPERTIES

13.5.1 POLYDISPERSITY

Linear hydrogenated diene copolymers are produced by anionic polymerization with subsequent hydrogenation that results in a saturated carbon–hydrogen backbone with pendant alkyl groups (methyl or ethyl) similar to OCP produced from ethylene/propylene or ethylene/1-butene [11,25,46]. The anionic living polymerization ensures narrow molecular weight distribution of the polymer. The gel permeation chromatograph (GPC) traces of the hydrogenated diene block copolymer and the OCPs produced by metallocene process are shown in Figure 13.2a.

The narrow molecular weight distribution is not only characteristic of the random and block linear copolymer structures created in living polymerization process. As described in Section 13.3.2, the hyperbranched "star" polymers are made by first forming linear block polymers with narrow polydispersity and with active lithium atom on one end of the polymer chain. These linear polymer chains are then coupled by the addition of a polyfunctional cross-linking compound. The reaction kinetics result in a narrow distribution of the number of these linear chains per

molecule and thus in a narrow molecular weight distribution of the final coupled molecule (Figure 13.2b) [16,44,47,48].

The low polydispersity of the HSD molecules plays a significant role in their properties in the lubricating oils.

13.5.2 THICKENING EFFICIENCY

There are a number of important parameters that characterize the behavior of polymeric VMs, in a particular solvent. Some of them are defined as follows:

$$\text{Thickening efficiency:} \frac{2}{\ln 2}\frac{1}{c}\ln\left(\frac{KV_{\text{solution}}}{KV_{\text{solvent}}}\right)$$

$$\text{Inherent viscosity:} \frac{1}{c}\ln\left(\frac{\eta_{\text{solution}}}{\eta_{\text{solvent}}}\right)$$

$$\text{Intrinsic viscosity:} \lim_{c\to 0}\left(\frac{1}{c}\ln\left(\frac{\eta_{\text{solution}}}{\eta_{\text{solvent}}}\right)\right)$$

Thickening efficiency is usually measured at 1% and is proportional to inherent viscosity. It characterizes the behavior of a particular polymer in a particular solvent.

Starting in the 1960s, the solution behavior of HSD copolymers in model organic solvents and base oils has been investigated in great detail, resulting in a number of academic publications [47,49–65] and lubricant composition patents [11,16,25,44,46]. While hydrogenation of the dienes is required for the polymer to qualify as an effective lubricant component with high oxidation stability, the fundamental studies of the solution behavior of the polymers often use nonhydrogenated diene copolymers. In these cases, the solution behavior is similar to hydrogenated polymers and the conclusions apply to both hydrogenated and nonhydrogenated versions. Three types of structures are generally used as a viscosity index improver in lubricating oils: linear hydrogenated polydiene, linear HSD copolymer, and star-branched or hyperbranched HSD copolymer. Linear polydienes have similar chemistry and architecture to OCP VMs but have lower polydispersity. Thus, there is little technical and economic advantage for using these versus mainline OCPs, and they are rarely used in practice. However,

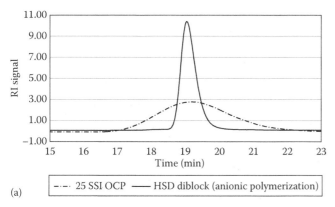

(a)

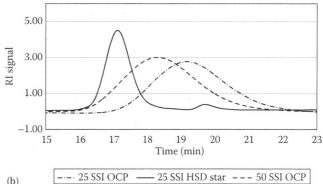

(b)

FIGURE 13.2 (a) GPC of 25 SSI OCP made by metallocene process and HSD diblock made by anionic polymerization. (b) GPC of 25 and 50 SSI OCP and 25 SSI HSD star polymer.

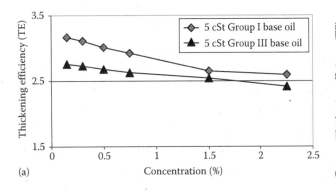

(a)

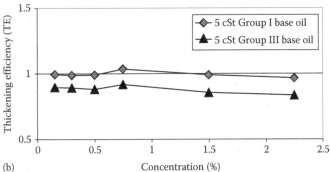

(b)

FIGURE 13.3 (a) Thickening efficiency of linear hydrogenated poly(isoprene), MW 170k, in two different base stocks. (b) Thickening efficiency of linear hydrogenated poly(isoprene), MW 39k, in two different base stocks.

studying their behavior in the solvent (base oil) is useful for understanding the behavior of more complex diblock copolymers and star polymers. Base oil is generally a good solvent for the hydrogenated diene block, while the polystyrene block is not readily soluble in the base oil with its solubility rapidly decreasing with lower temperature and increasing block size. Thus, it is believed that the fluid-thickening property of HSD polymers comes from the hydrogenated diene part of the molecule.

The TE of hydrogenated diene polymer depends on the solvent (base oil properties) and the molecular weight of the polymer similar to the other OCP. The thickening efficiencies of hydrogenated dienes of different molecular weight in various base oils are shown in Figure 13.3.

In dilute regime, when the higher-order concentration terms can be ignored, the intrinsic viscosity or TE can be approximated by Mark–Houwink power law dependence on the weight average molecular weight of the polymer.

$$[\eta] = K'M^a, \tag{13.1}$$

where K′ and a are constants.

Thus, the higher the molecular weight of the polymer, the higher the TE, provided that the structure and chemistry of the polymer, as well as the solvent, are the same. This is demonstrated in Figure 13.4 by plotting TE at low concentration for linear HSDs of various molecular weights. These polymers also exhibit 15%–20% lower TE in Group III base oil than in Group I oil. Lower TE does not necessarily mean a higher treat rate in the finished oil since Group III oils have higher viscosity index.

Thickening efficiency in the same base stocks for a HSD star is shown in Figure 13.5. The number of arms per star is between 15 and 20, making it a much larger molecule than the linear hydrogenated polymer shown in Figure 13.3b. Generally, the higher the molecular weight of the polymer, the higher the TE, as shown in Figure 13.4. However, this is only true for the same architecture of the polymer. Different architectures, such as branching versus linear polymer, result in different polymer density distribution in solution, and thus different thickening efficiencies. The star polymer forms a more compact, dense structure in the base oil than

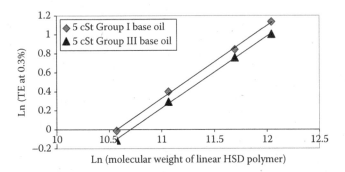

FIGURE 13.4 Thickening efficiency depends on molecular weight of the HSD polymer, as well as on solvent (base oil).

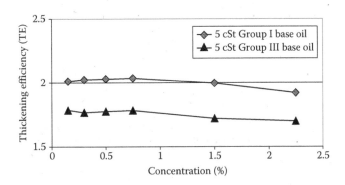

FIGURE 13.5 Thickening efficiency of star hydrogenated poly(isoprene), MW of the arm 37k, in two different base oils.

the linear one. This results in the TE of the star (Figure 13.5) being lower than that of the linear polymer of lower molecular weight shown in Figure 13.3a. The fact that the star polymer is more compact with higher polymer density in solution has been confirmed by light scattering and viscometric studies [53], as well as small angle neutron scattering [56], showing that in the semidilute solution the stars resemble linear polymers with slightly denser star centers. It is interesting to note that when analyzing the dependence of the intrinsic viscosity on the molecular weight of linear versus star polymers in a good solvent, it has been shown that the exponent a in Equation 13.1 remains roughly the same for linear and star polymers with various number of arms, while coefficient K′

decreases with the number of arms in the star confirming the effect of the arm crowding in the star architecture [53].

The fact that the linear polymer is more spread out in the solution is also manifested in the decrease in TE with concentration for 170k linear polymer. This can be explained by the fact that the polymer chains start "seeing" each other at higher concentrations and the chains become more compact. The chains however are not involved in any cooperative behavior. Because they are self-avoiding, the volume available to each chain becomes smaller as the concentration of chains increases, the TE decreases with concentration. This is not the case for the star polymer within the same range of weight concentrations (which corresponds to a much lower molar concentrations).

The thickening behavior of the styrene–diene diblock copolymers depends on their self-assembly. The diblocks themselves have relatively low molecular weight and thus contribute little to viscosity increase. They are designed to form micelles in base oils due to poor solubility of the styrene block forming the core of a micelle, and the good solubility of the hydrogenated diene block forming a highly solvated corona around a polystyrene core. Once diblocks assemble into a micelle, the total micellar weight is high and the TE, which is proportional to the molecular weight, is also high. The aggregation behavior and the type of the aggregates formed as a function of block sizes, concentration, temperature, and solvent quality was extensively studied [58–65]. In a given solvent, the aggregation behavior, that is, aggregation temperature and the aggregation number of the micelle, depends mostly on the polystyrene block size and to a much lesser degree on the diene block size.

The TE of diblocks shows similar behavior to linear and star polymers with respect to base stocks, that is, solvent quality. The TE decreases in Group III compared to Group I. While there are a number of factors that affect such behavior (polystyrene block solubility in the base stock being one of them), hydrogenated diene block TE contributes to the difference in overall TE. With respect to the concentration, the TE behavior is quite different from linear diene. The increase in TE with concentration indicates cooperative behavior of diblock aggregates. Thus, there could be a significant difference in inherent viscosity within the diblock lubricant formulation window even in the same base stock (Figure 13.6).

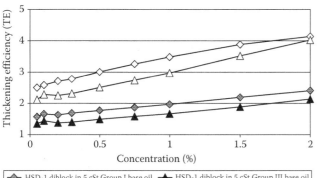

FIGURE 13.6 Thickening efficiency of two HSD diblock copolymers of different molecular weight in Groups I and III base oils.

13.5.3 Shear Stability Index

The SSI of the VM is a predictor of the stability of the polymer in the high shear environment that can be seen in the typical field service. SSI is determined by the percentage loss of polymer-derived viscosity after a 30-cycle Kurt Orbahn test as per ASTM D6278. The lower the SSI is, the more shear stable the VM polymer is. To compare VM polymers, SSI is measured in the same base oil and at the same or similar concentrations of polymer. However, SSI depends on base oil viscosity [66–68], base oil solvent quality, polymer architecture [67–69], and to a lesser extent on polymer concentration. As demonstrated in the previous section, the relative viscosity or TE of the same polymer depends on the quality of the solvent. Figure 13.5 demonstrates the TE of the same HSD star copolymer in Group I and Group III base oils of about the same viscosity. The polymer TE in Group I base oil is higher than in Group III indicating that the polymer is more extended due to better solubility compared to Group III base oil. The polymer conformation in the base oil plays a role in its stability under high shear conditions: the more extended the polymer is, and the more volume it occupies, the more time it takes for the chain to align in the shear flow. And when the polymer is not in the aligned conformation to minimize the system energy, breaking the covalent bond to create smaller molecules becomes more energetically favorable. The effect of the base oil solubility is demonstrated in Figure 13.7. Here, the same HSD star copolymer is dissolved in Group I and Group III base oils of different viscosities. Thickening efficiency in Group III base oil is lower than in Group I, and the polymer exhibits more shear stability under the same conditions and in the same viscosity base oil as Group I.

Base oil viscosity plays a role in shear stability: the higher the base oil viscosity is, the more mechanical degradation under shear will the polymer experience. This phenomenon is demonstrated in Figure 13.8. While the magnitude of the effect can be different in different base oils, the trend for better shear stability in lighter base oils holds for all relevant solvents as demonstrated by a number of studies [68]. This phenomenon has an effect on the measured SSI, which depends on the base oil viscosity and group or viscosity grade and base oil of the fully formulated oil when measured this way. The effective SSI decreases (the polymer becomes more shear stable) in the lower viscosity oils and in the higher group base stocks.

13.5.4 TE/SSI Balance

The TE/SSI balance is an important characteristic of VMs. The TE increases with the molecular weight (see Section 13.5.2), while the shear stability decreases (increasing SSI indicates the loss of polymer-added viscosity under shear). The TE/SSI balance for different classes of polymers is shown in Figure 13.9. The chemistry of the polymer and its solubility in the base oil are the major factors in defining TE for a particular VM.

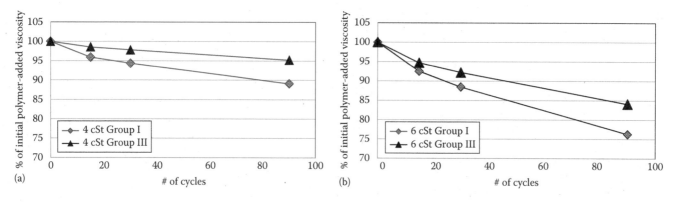

FIGURE 13.7 (a) Polymer-derived viscosity loss in 4 cSt base oils with different solubility for VM polymer (API Groups I and III). (b) Polymer-derived viscosity loss in 6 cSt base oils with different solubility for VM polymer (API Groups I and III).

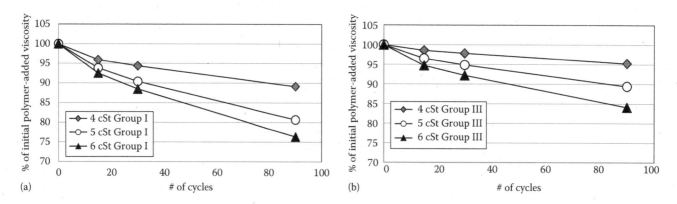

FIGURE 13.8 (a) Polymer-derived viscosity loss in API Group I base oils of various viscosities. (b) Polymer-derived viscosity loss in API Group III base oils of various viscosities.

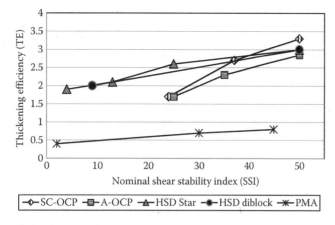

FIGURE 13.9 TE/SSI balance for different classes of VMs.

This can be easily demonstrated by the difference in TE of PMAs and OCPs. PMAs' solubility in the petroleum-derived base oils is much lower than that of the OCPs and HSDs so the TE is also much lower. However, the chemical nature of HSDs and OCPs is very similar, so their thickening efficiencies are much closer to each other. Here, the differences in

molecular architecture (stars vs. linear), polydispersity, and other features such as self-assembly play a role. For more shear stable molecules, the hyperbranched, or a star, architecture is a big credit. While the molecular weight of star polymer is high to provide efficient thickening, the more compact star architecture allows it to have faster dynamics under shear and to realign in the shear flow fast enough not to be mechanically degraded [67,68]. This is also true for self-assembled diblock molecules, which not only align in shear flow but can also partially disassemble to assume the more favorable configuration [70,71]. However, as molecular weight of HSD star and diblock molecules increases, they can no longer react fast enough to external shear and behave similar to OCP molecules that provide similar thickening.

13.5.5 VISCOMETRIC PROPERTIES OF HSD VISCOSITY MODIFIERS IN MULTIGRADE OILS

A wide variety of multigrade oils are formulated with HSD VMs including oils for heavy-duty diesel and passenger car applications. Due to the advantageous TE/SSI balance, they are typically characterized by lower polymer treat rate than

TABLE 13.1

SAE 15W-40 Heavy-Duty Diesel Oil Formulated with 25 SSI HSD VM and 25 SSI A-OCP and SC-OCP VMs

VM Chemistry	HSD Star	A-OCP	SC-OCP
VM SSI	25	25	25
SAE viscosity grade	15W-40	15W-40	15W-40
Additive technology	API CI-4	API CI-4	API CI-4
Base oil	API Group II	API Group II	API Group II
Solid VM treat (%)	0.7	0.94	0.77
KV100 (cSt)	15.17	15.14	15.14
CCS at −20°C (cP)	5320	6297	5845
HTHS at 150°C (cP)	3.92	4.18	4.03
Noack volatility (%)	9.8	9.7	9.6

TABLE 13.2

SAE 5W-30 Passenger Car Oil Formulated with 50 SSI HSD VM and 50 SSI A-OCP and SC-OCP VMs

VM Chemistry	HSD Star	A-OCP	SC-OCP
VM SSI	50	50	50
SAE viscosity grade	5W-30	5W-30	5W-30
Additive technology	API SN	API SN	API SN
Base oil	API Group II/ III	API Group II/ III	API Group II/ III
Solid VM treat (%)	0.59	0.59	0.52
KV100, cSt	10.42	10.4	10.38
CCS at −30°C (cP)	5854	5898	5823
HTHS at 150°C (cP)	2.94	3.01	3.01

contributes to a more pronounced non–Newtonian behavior. The low CCS viscosity contribution by a VM is widely viewed as beneficial as it allows formulators to use heavier, less volatile cuts of the base stock, minimize the use of higher group base stocks, and formulate the lubricant with better low-temperature start properties. HTHS viscosity has been widely correlated to wear protection property of the lubricant, and thus until recently, low HTHS viscosity contribution was viewed as detrimental. Recently however, lower HTHS viscosity has been correlated to better fuel efficiency. With the global lubricant industry seeking to provide more fuel economy through the use of advanced lubricants as well as with OEMs accepting low viscosity oils and designing the engines to work with these oils, low HTHS is seen as a tool to provide fuel efficiency.

The advantages of TE/SSI balance of HSD VMs at low SSI numbers (more shear stable polymers) disappear for higher SSI polymers (see Section 13.5.4). This is demonstrated in passenger car oil formulated with various 50 SSI VMs: HSD star and OCPs (Table 13.2).

These examples demonstrate that the greatest value for HSD polymers in terms of TE/SSI balance and cost can be realized at lower SSI, generally below 30. The higher SSI HSD polymers are successfully used to formulate multigrade oils but do not possess the feature of superior TE/SSI balance characteristic of low SSI HSD polymers.

Typical heavy-duty diesel oils formulated with HSD diblock VMs and HSD star VM for comparison are shown in Table 13.3. All four oils are formulated to the same HTHS at 150°C viscosity. Although all three diblock VMs have the same nominal SSI, their thickening efficiencies are significantly different mostly due to the chemistry of the hydrogenated diene block. Diblock 3 is produced by 1,2-butadiene polymerization that results in ethyl side groups, while Diblock 2 is made by 1,4-isoprene polymerization, which yields methyl side groups. These side groups, or short branches, contribute very little to the TE of the polymer, thus the polymer with the larger side group is less thickening and requires higher solid polymer treat rate. Despite the higher treat rate, the use of hydrogenated styrene–butadiene VMs is often economically justified by cheaper raw material cost for polymer production.

OCPs at the same shear stability. Examples of such formulations can be seen in Tables 13.1 and 13.2. Another characteristic of HSD polymers is their relatively lower contribution high shear viscosity, such as CCS viscosity and HTHS viscosity. This phenomenon is explained by the HSD ability to align fast in the high shear flow. The mechanism is similar to that described in Section 13.5.4, although the magnitude of shear is lower. In this case, the fast alignment in the shear flow

TABLE 13.3

SAE 10W-40 Heavy-Duty Diesel Oil Formulated with 5 SSI HSD Star VM and Three 10 SSI HSD Diblock VMs

VM Chemistry	HSD Star	HSD Diblock 1	HSD Diblock 2	HSD Diblock 3
VM SSI	5	10	10	10
SAE viscosity grade	10W-40	10W-40	10W-40	10W-40
Additive technology	ACEA E4	ACEA E4	ACEA E4	ACEA E4
Base oil	API Group III	API Group III	API Group III	API Group III
Solid VM treat (%)	0.66	0.7	0.76	0.92
KV100 (cSt)	13.4	13.9	13.9	13.2
CCS at −25°C (cP)	5550	5760	6550	5980
HTHS at 150°C (cP)	3.9	3.9	3.9	3.9

13.5.6 Performance Characteristics

13.5.6.1 Base Stock and PPD Sensitivity

While the most important properties of the VMs are TE and shear stability, some other performance characteristics play a role. Low-temperature pumpability as measured by ASTM D4684 [72] can be significantly affected by VM in both fresh and aged oils [73–79]. The ability of a VM to interact with other fresh oil components, such as base stock waxes, pour point depressants (PPDs), and other additive components, can lead to formation of large-scale structures in the lubricant at low temperature, thus leading to high viscosity and sometimes yield stress. This feature of VM is usually attributed to polyethylene segments within the molecule that can interact with base oil waxes and other components. However, HSD polymers as well as A-OCP VMs are amorphous, that is, lack the long ethylene stretches that can co-crystallize or align with waxes and other components. This is achieved by regular side groups: methyl in polymers made by 1,4-isoprene addition and ethyl in 1,2-butadiene polymers. These side groups provide steric barrier for co-crystallization of base oil waxes with polymer chains. In this respect, HSD polymers sensitivity to base stocks and PPDs are similar to A-OCPs, that is, very low. This phenomenon is demonstrated in Figure 13.10 showing an oil formulated with several base stocks and the same PPD. The SC-OCP that has polyethylene stretches is sensitive to the base stock choice and would require careful selection of PPD to prevent interactions with base oil waxes as well as between wax molecules in the base oil itself. HSD star VM is insensitive to the choice of base stock because it does not interact with the wax components. This feature of the HSD polymers allows formulators to treat it as an inert component when formulating for low-temperature pumpability performance in fresh oils as well as in oils aged in bench, engine, and field service. The robustness of HSD VMs in low-temperature pumpability in various aging conditions in bench and engine tests as well as in the field service of varying severity has been well documented [76–79].

13.5.6.2 Soot Handling Credits

The incorporation of styrene monomers into the VM polymer gives it distinct advantage in soot handling capability. The soot surface is largely aromatic allowing the interaction and some adsorption of polystyrene stretches of VM molecule onto the soot particle surface. The rest of the polymer molecule creates steric hindrance and prevents soot agglomeration [12,80–82]. While these molecules cannot be considered dispersant VMs because of the lack of truly dispersant functionality, that is, they do not contain nitrogen or oxygen, they can bring some credit to soot dispersancy. This is especially true for diblock copolymers with longer styrene blocks and thus better affinity to the soot surface [12,80–81].

This phenomenon is best demonstrated in borderline additive packages with low dispersant activity. When the dispersant concentration is below the minimum that is needed to cover the surface of the soot generated by an engine, the HSD VM adsorption onto soot particle can help prevent soot agglomeration and corresponding viscosity growth. The effect of the diblock HSD polymer in an oil with low dispersant concentration in the Mack T-11 test is shown in Figure 13.11. The addition of the HSD diblock helps to control soot agglomeration and viscosity increase. The credit in soot handling depends on the styrene content and distribution in the molecule and is a unique property of the HSD non-dispersant VMs.

Another demonstration of the boost in soot control provided by an HSD diblock VM in the PSA DV4E3 test is shown in Figure 13.12. A baseline SAE 5W-40 passenger car diesel oil formulated with an amorphous VM is compared to a formulation with a mixed VM system in which a portion of the amorphous VM is replaced by the HSD diblock. The oils were blended to a constant HTHS viscosity of 3.5 cP.

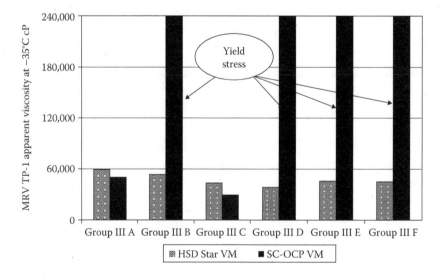

FIGURE 13.10 HSD VM does not interact with base oil waxes at low temperature. SC-OCP VM that has polyethylene stretches that can co-crystallize with base oil waxes is shown for comparison.

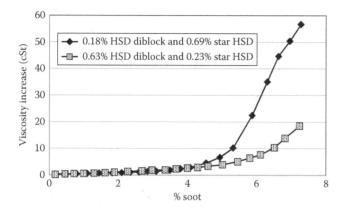

FIGURE 13.11 Mack T-11: addition of HSD diblock VM helps to control soot agglomeration and viscosity increase in SAE 5W-40 oil.

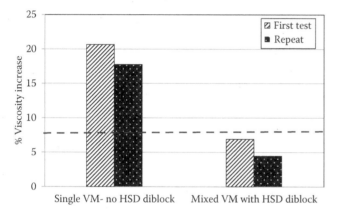

FIGURE 13.12 HSD diblock polymer addition reduces soot-induced viscosity increase in DV4E3 engine test.

13.5.7 FIELD PERFORMANCE

Field performance is the ultimate test for a lubricant. No engine or bench test performance can anticipate every single situation that can happen in actual service. HSD polymers, both diblock and star, are used to formulate multigrade oils for both passenger car and heavy-duty diesel engines and widely used in the field since the 1970s. In addition to over 40 years in a wide variety of field applications, HSD VMs were tested in a range of controlled field trials including ones required for top-tier formulation approvals such as Scania LDF claims.

VMs can impact several important properties of the lubricating engine oil including low-temperature pumpability [83–85], viscosity retention (shear stability/oxidation), durability (wear), and cleanliness of the engine. The performance of a VM depends on the engine hardware and the lubrication regime and temperature, as well as fuel quality. Thus, field trials are generally conducted not only for formal approvals but also to verify no harms performance of the VM in radically new types of engine hardware.

Below is an example of a heavy-duty field trial where the low-temperature pumpability of several oils in a modern 2010 emission complaint heavy-duty diesel engine was evaluated using different viscosity index modifier chemistries. The field

test was conducted using a fleet equipped with emissions-compliant Mack MP8 engines using ultralow sulfur diesel fuel. The MP8 features 415 HP to 485 HP and has torque, with ratings from 1540 to 1700 lb-ft. It has a single overhead cam with ultrahigh-pressure fuel injection. All trucks had similar medium severity service type. Oil drain for this study was at 35,000 miles. The drain data from the field are the average of drains from four trucks. Four SAE 15W-40 oils were used in this study. The oils were formulated in the same Group II base oil and using the same additive package designed to perform at API CJ-4 level. The oils differed only in the choice of VM. All oils were treated with PPD at the same concentration resulting in excellent fresh oil MRV TP-1 viscosity and no measurable yield stress (<35 Pa). The oils had similar kinematic viscosity (KV100), CCS viscosity, and MRV TP-1 viscosity.

The drained oils from at least four different trucks using the same fresh oil were analyzed. Kinematic viscosity (KV100), CCS viscosity, MRV TP-1 viscosity, oxidation, nitration, concentration of wear metals, fuel dilution, and soot concentration were measured. All these parameters were similar for the same oil drained from different trucks. The results for the same oil drained from four different trucks were then averaged and reported in Table 13.4. It is interesting to note that all of the oils had moderate levels of oxidation and nitration and low levels of soot (1%–2%). Oils formulated with HSD copolymers retained their low-temperature pumpability at the end of 35,000 miles [77] emphasizing the importance of the amorphous nature of the polymer in this type of hardware and service.

In addition to low-temperature pumpability, wear metals, soot levels, and TBN retention were analyzed for the oils at the drain. The content of wear metals, including copper, iron, and lead, is all quite low, indicating the formulation used in the field trial was able to protect the engine well during the trial.

As part of the same field trial, the change in viscosity over the drain was measured [86]. The viscosity change in the field trial mainly includes the contributions from oil shearing and

TABLE 13.4

SAE 15W-40 Field Trial Oils (Data for Drained Oils Are an Average from Four Different Trucks)

Name	Oil A	Oil B	Oil C	Oil D
VM	A-OCP	Styrene–diene 1	Styrene–diene 2	SC-OCP
Fresh oil viscometrics				
KV100 (cSt)	15.5	15.5	15.7	15.7
CCS at −20°C (cP)	6650	6180	6500	6730
MRV TP-1 at −25°C (cP)	24,300	25,600	22,300	21,200
Yield stress, Pa	No	No	No	No
Drained oil at 35 k miles				
CCS at −20°C (cP)	6395	6560	5770	6370
MRV TP-1 at −25°C (cP)	21,100	23,930	21,130	199,750
Yield stress, Pa	No	No	No	250[a]

[a] Every drained oil exhibited yield stress in the range between 210 and 315 Pa.

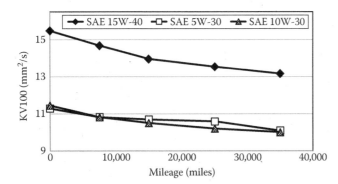

FIGURE 13.13 Viscosity of the oils of various SAE viscosity grades formulated with HSD VM in the Mack MP8 field trial.

fuel dilution. The Mack MP8 engines represent a challenging shear environment, and thus, it is important to verify that the VM-containing oils do not shear out of grade in these engines. Because the HSD VMs are used in top-tier lubricating formulations that include lower viscosity grades for fuel economy benefit, it is also important to demonstrate that these thinner oils can maintain their viscosity in the field. The results of the field test for various viscosity grades using HSD VM are shown in Figure 13.13. They confirm the ability of HSD VM to keep in-grade viscosity throughout the drain interval in a range of viscosity grades in the challenging shear environment.

13.5.8 SPECIALIZED VISCOSITY MODIFIERS

HSD copolymers are produced by anionic polymerization (a living polymerization), a process that not only inherently allows narrow molecular weight distribution but also gives an opportunity for precise control of block sizes and monomer distribution within the polymer chain. This in turn allows one to design and manufacture a polymer structure with targeted composition and properties. These kinds of HSD VMs usually target specific properties, and their structures are optimized to deliver maximum performance in the targeted area. The example of such targeted performance can be higher viscosity index and better fuel economy (FE) contribution at a fixed TE and shear stability. Table 13.5 shows viscometrics of the formulations with the

TABLE 13.5

Viscosities of SAE 5W-30 Oils Formulated with Conventional HSD and Specialized Fuel Economy HSD VMs to HTHS Viscosity at 150°C of 3.5 cP

	Conventional HSD	New FE HSD
Solid VM (%)	1.1	1.1
KV at 100°C (cSt)	11.9	11.7
KV at 40°C (cSt)	67.1	58.1
HTHS at 150°C (cP)	3.5	3.5
HTHS at 100°C (cP)	7.7	6.9
HTHS at 70°C (cP)	14.2	13.1
HTHS at 45°C (cP)	30.1	27.8

conventional HSD and the FE HSD designed to minimize HTHS viscosity at engine operating temperatures at the fixed HTHS viscosity at 150°C. While treat rate and SSI are the same, the lubricant viscometric profile of FE HSD is different. Consequently, the fuel economy performance of the oil with the special VM as measured by NEDC, FTP-75, and JC08 is improved [87–89].

13.6 OUTLOOK AND TRENDS

Globally, there are diverging trends affecting the demand for VMs. Monograde lubricants are mostly replaced by multigrade even in the developing markets, requiring the use of VMs. The worldwide vehicle population is expected to double during the period of 2015–2035, with the significant growth coming from the developing markets, thus leading to the increased demand for VM. Longer drain intervals linked with advancements in hardware and in lubricants quality will partially offset the volume growth of lubricants and VMs. At the same time, lubricants are shifting to lower viscosity grades for better fuel economy with corresponding reduction in VM treat in the finished oils. Thus, the overall global volumes for VM are expected to grow slightly faster than lubricants in the short to medium term, but then plateau in the long term. Quality upgrades will determine the choice of VM chemistry between segments, with faster growth in high-quality lubricants typically requiring HSD VMs.

The higher-tier markets are experiencing continued tightening and fragmentation of technical requirements, requiring more performance features from the VM than just viscosity index improvement. Therefore, specialized VMs providing better fuel economy or better cleanliness, or soot and sludge performance, are becoming more popular in top-tier market. Due to their chemistry, HSD polymers will continue to find application in top-tier niches, and it can be easily designed to have these specialized features. However, the emergence of new chemistries affording additional performance features is also possible.

ACKNOWLEDGMENTS

The authors are grateful to Dr. Stuart Briggs for reviewing the manuscript and for many helpful suggestions as well as to Andrew Anastasio for providing experimental data.

REFERENCES

1. Porter L.M., Process for preparing block copolymers utilizing organolithium catalysts, US 3149182, September 15, 1964.
2. Holden G., Milkovich R., Process for the preparation of block copolymers, US 3231635, January 25, 1966.
3. Holden G., Milkovich R., Block polymers of monovinyl aromatic hydrocarbons and conjugated dienes, US 3265765, August 9, 1966.
4. Zelluski R.P., Preparation of copolymers in the presence of an organo-lithium catalyst and a solvent mixture comprising a hydrocarbon and an ether, thioether or amine, US 2975160, March 14, 1961.
5. Hsieh H.L., Zelinski R.P., High molecular polymers and method for their preparation, US 3078254, February 19, 1963.

6. Johnson M.M., Schiff S., Streets W.L., Viscosity index improvers, US 3554911, January 12, 1971.

7. Jones R.V., Moberly C., Hydrogenated polybutadiene and process for producing same, US 2864809, December 16, 1958.

8. Breslow D.S., Matlack A.S., Hydrogenation of unsaturated hydrocarbons, US 3113986, December 10, 1983.

9. Lapporte S.J., Preparation of complex organic metallic hydrogenation catalysts and their use, US 3205278, September 7, 1965.

10. Anderson W., Block copolymers as viscosity index improvers for lubricating oils, US 3763044, October 2, 1973.

11. Claire D., Evans D., Lubricating compositions, US 3772196, November 13, 1973.

12. Wedlock D.J., Jong F.D., Lubricating oil composition, US 6303550B1, October 16, 2001.

13. Abraham W., Covitch M., Galic M., Aromatic diblock copolymers for lubricant and concentrate compositions and methods thereof, US 20060052255 A1, March 9, 2005.

14. Olson D.H., Handlin D.L., Polymeric viscosity index improver and oil composition comprising the same, US 4788361, November 29, 1988.

15. Rhodes R.B., Stevens C.A., Asymmetric triblock copolymer viscosity index improver for oil compositions, CA2155686 C, March 21, 2006.

16. Eckert R.J.A., Hydrogenated star-shaped polymer, US 4116917, September 26, 1978.

17. Fetters J.J., Star polymers and process for the preparation of the same, US Patent 3985830, October 12, 1976.

18. Rhodes R.B., Stevens C.A., Star polymer viscosity index improver for oil lubricating compositions, CA2152992 C, May 2, 2006.

19. Rhodes R.B., Handlin D.L., Stevens C.A., Star polymer viscosity index improver for oil compositions, US 5460739, October 24, 1995.

20. Briggs S., Chu C., Viscosity index improvers for lubricating oil compositions, US7163913, January 16, 2007.

21. Duggal A., Lubricant additive, US8999905 B2, April 7, 2015.

22. Oberoi S., Briggs S., Viscosity index improvers for lubricating oil compositions, US9133413 B2, September 15, 2015.

23. Cui J., Oberoi S., Briggs S., Goldmints I., A viscosity modifier solution to reconcile fuel economy and durability in diesel engines, *Tribology International* 101 (2016): 43–48.

24. Charleux B., Faust R., Synthesis of adavnaced polymers by catioinc polymerization, *Advances in Polymer Science*, 142, 1, Springer, New York, 1998.

25. Hadjichristidis N., Pispas S., Floudas G., *Block Copolymers: Synthetic Strategies, Physical Properties and Applications.* Wiley, New York, 2002.

26. Anderson W.S., Block copolymers as viscosity index improvers for lubricating oils, US Patent No. 3,668,125, June 6, 1972.

27. Pappas J.J., Makowski H.S., Rossi A., Hydrogenated copolymer butadiene with another conjugated diene are useful oil additives, US Patent No. 3795615, March 5, 1974.

28. Odian G., *Principles of Polymerization*, 4th edn. Wiley Interscience, New York, 2004.

29. Young R.J., Lovell P.A., *Introduction to Polymers*, 3rd edn. CRC Press, Taylor & Francis Group, New York, 2011.

30. Auguste S., Edwards H.G.M., Johnson A.F., Meszena Z.G., Nicol P., Anionic polymerization of styrene and butadiene initiated by n-butylilithium in ethylbenzene: Determination of the propagation constants using Raman spectroscopy and gel permeation chromatography, *Polymer*, 37(16) (1996): 3665–3673.

31. Pispas S., Pitsikalis, hadjichristidis, anionic polymerization of isoprene, butadiene and styrene with 3-dimethylaminiopropyllithium, *Polymer*, 36(15) (1995): 3005–3011.

32. Ayano S., Yabe S., Anionic polymerization of isoprene II: Synthesis and characterization of A-B-A type block copolymers by oligomeric dilithium initiator, *Polymer Journal*, 1(6): 706–715.

33. De La Mare H.E., Shaw A.W., Block copolymers having reduced solvent sensitivity, US Patent No. 3,670,054, June 13, 1972.

34. Jones R.C., Hydrogenated block copolymers of butadiene and a monovinyl aryl hydrocarbon, US Patent 3,431,323, March 4, 1969.

35. Moczygemba G.A., Tetrablock polymers and their hydrogenated analogs, US Patent 4,168,286, September 18, 1979.

36. Himes G.R., Multiblock hydrogenated polymers for adhesives, US Patent 5,627,235, May 6, 1997.

37. Joly G., Maris C.A.L., Ossterbosch S.M., Blockcopolymer compositions, having improved mechanical properties and processability and styrenic blockcopolymer to be used in them, US Patent 7,268,184 B2, September 11, 2007.

38. Vermunicht G.E.A., Van De Vliet B.M.L., Southwick J.G., Poly(styrene-butadiene-styrene) polymers having a high vinyl content in the butadiene block and hot melt adhesive composition comprising said polymers, US Patent 7,517,932, April 14, 2009.

39. Southwick J.G., Ewins E.E., St. Clair D.J., 100% triblock hydrogenated styrene-isoprene-styrene block copolymer adhesive composition, US Patent 5,210,147, May 11, 1993.

40. Wald M.M., Selectively hydrogenated block copolymer, US Patent 3,700,633, October 24, 1972.

41. Rhodes B.R., Polymeric viscosity index additive and oil composition comprising the same, US Patent 4,900,875, February 13, 1990.

42. Kiovsky T.E., Star-shaped dispersant viscosity index improver, US Patent No. 4077893A, March 7, 1978.

43. Thomas E.K., Star-shaped polymer reacted with dicarboxylic acid and amine as dispersant viscosity index improver, US Patent 4,141,847, February 27, 1979.

44. Eckert R.J.A., Hydrogenated star-shaped polymer, US, Patent 4,156,673, May 29, 1979.

45. Sutherland R.J., Oil compositions containing functionalised polymers, US Patent 4,427,834, November 25, 1991.

46. St Clair D.J., Ronald K., Crossland: Lubricating compositions containing hydrogenated block copolymers as viscosity index improvers, US Patent No. 3,965,019, June 22, 1976.

47. Hadjichristidis N., Roovers J.E.L., Synthesis and solution properties of linear, four-branched, and six-branched star polyisoprenes, *Journal of Polymer Science: Polymer Physics Edition*, 12(12) (1974): 2521–2533.

48. Nguyen A.B., Hadjichristidis N., Fetters L.J., Static light scattering study of high-molecular weight 18-arm star block copolymers, *Macromolecules*, 19(3) (1986): 768–773.

49. Bi L.-K., Fetters L.J., Synthesis and properties of block copolymers. 3. Polystyrene-polydiene star block copolymers, *Macromolecules*, 9(5) (1976): 732–742.

50. Bauer B.J., Fetters L.J., Synthesis and dilute-solution behavior of model star-branched polymers, *Rubber Chemistry and Technology*, 51(3) (1978): 406–436.

51. Raju V.R. et al., Concentration and molecular weight dependence of viscoelastic properties in linear and star polymers, *Macromolecules*, 14(6) (1981): 1668–1676.

52. Bauer, B.J. et al. Chain dimensions in dilute polymer solutions: a light-scattering and viscometric study of multiarmed polyisoprene stars in good and THETA solvents. *Macromolecules* 22(5) (1989): 2337–2347.

53. Bauer B.J. et al., Chain dimensions in dilute polymer solutions: A light-scattering and viscometric study of multiarmed polyisoprene stars in good and THETA solvents, *Macromolecules*, 22(5) (1989): 2337–2347.

54. Fetters L.J. et al., Rheological behavior of star-shaped polymers, *Macromolecules*, 26(4) (1993): 647–654.

55. Willner L. et al., Ordering phenomena of star polymers in solution by SANS, *EPL (Europhysics Letters)*, 19(4) (1992): 297.

56. Willner L. et al., Structural investigation of star polymers in solution by small-angle neutron scattering, *Macromolecules*, 27(14) (1994): 3821–3829.

57. Grest G.S. et al., Star polymers: Experiment, theory, and simulation, *Advances in Chemical Physics*, 94(1996): 67.

58. Mandema W., Zeldenrust H., Emeis C.A., Association of block copolymers in selective solvents, 1. Measurements on hydrogenated poly (styrene-isoprene) in decane and in trans-decalin, *Die Makromolekulare Chemie*, 180(6) (1979): 1521–1538.

59. Mandema W., Emeis C.A., Zeldenrust H., Association of block copolymers in selective solvents 2. Viscosity, diffusion and light-scattering measurements on hydrogenated poly (styrene-isoprene) in trans-decalin/decane mixtures, *Die Makromolekulare Chemie*, 180(9) (1979): 2163–2174.

60. Higgins J.S. et al., Study of micelle formation by the diblock copolymer polystyrene—b-(ethylene-co-propylene) in dodecane by small-angle neutron scattering, *Polymer*, 27(6) (1986): 931–936.

61. Higgins J.S. et al., Comparison of the structural and rheological consequences of micelle formation in solutions of a model di-block copolymer, *Polymer*, 29(11 (1988): 1968–1978.

62. Velichkova R. et al., Styrene–isoprene block copolymers. II. Hydrogenation and solution properties, *Journal of Applied Polymer Science*, 42(12) (1991): 3083–3090.

63. Bang J. et al., Sphere, cylinder, and vesicle nanoaggregates in poly (styrene-b-isoprene) diblock copolymer solutions, *Macromolecules*, 39(3 (2006): 1199–1208.

64. Choi S.-H., Bates F.S., Lodge T.P., Structure of poly(styrene-b-ethylene-alt-propylene) diblock copolymer micelles in squalane, *The Journal of Physical Chemistry B*, 113(42) (2009): 13840–13848.

65. Choi S.-H., Bates F.S., Lodge T.P., Small-angle x-ray scattering of concentration dependent structures in block copolymer solutions, *Macromolecules*, 47(22) (2014): 7978–7986.

66. Bartz W.J., Influence of viscosity index improver, molecular weight, and base oil on thickening, shear stability, and evaporation losses of multigrade oils, *Lubrication Science*, 12(3) (2000): 215–237.

67. Covitch M., How polymer architecture affects permanent viscosity loss of multigrade lubricants, SAE Technical Paper 982638, 1998.

68. Cui J., Oberoi S., Goldmints I., Briggs S., Field and bench study of shear stability of heavy duty diesel lubricants, *SAE International Journal of Fuels and Lubricants*, 7(3) (2014): 882–889.

69. Covitch M.J., Weiss J., Kreutzer I.M., Low-temperature rheology of engine lubricants subjected to mechanical shear: Viscosity modifier effects, *Lubrication Science*, 11(4) (1999): 337–364.

70. Choi S.-H., Lodge T.P., Bates F.S., Mechanism of molecular exchange in diblock copolymer micelles: Hypersensitivity to core chain length, *Physical Review Letters*, 104(4) (2010): 047802.

71. Choi S.-H., Bates F.S., Lodge T.P., Molecular exchange in ordered diblock copolymer micelles, *Macromolecules*, 44(9) (2011): 3594–3604.

72. ASTM D 4684, Standard test method for determination of yield stress and apparent viscosity of engine oils at low temperature, ASTM International, West Conshohocken, PA, 2014.

73. Rhodes R.B., Low temperature compatibility of engine lubricants and the risk of engine pumpability failure, SAE International, SAE Technical Paper 932831, 1993, doi:10.4271/932831.

74. May C.J., Habeeb J.J., Factors affecting the low temperature pumpability of used engine oils, SAE International, SAE Technical Paper 872049, 1987, doi:10.4271/872049.

75. Batko M.A., Florkowski D.W., Devlin M.T., Shoutain L., Eggerding D.W., Lam W.Y., McDonnell T.F., Jao T.C., Low temperature rheological properties of aged crankcase oils, SAE International, SAE Technical Paper 2000-01-2943, 2000, doi:10.4271/2000-01-2943.

76. Goberdhan D., Goldmints I., Assessment of ageing mechanisms in lubricants and their effects on retained low temperature pumpability of top tier oils, SAE International, SAE Technical Paper 2010-01-2177, 2010, doi:10.4271/2010-01-2177.

77. Oberoi S., Goldmints I., In-service low temperature pumpability: Field performance vs. bench tests, SAE International, SAE Technical Paper 2012-01-1708, 2012, doi:10.4271/2012-01-1708.

78. Bansal J.G., Chu C., Outten E., In-service low temperature pumpability of crankcase lubricants- effect of viscosity modifiers, SAE International, SAE Technical Paper 2004-01-1932, 2004, doi:10.4271/2004-01-1932.

79. Goberdhan D., Goldmints I., In-service low temperature pumpability issues of crankcase lubricants, *Proceedings of 16th International Colloquium Tribology Automotive and Industrial Lubrication*, 2008.

80. Ritchie Andrew J.D. et al., Lubricating oil compositions, US Patent No. 6,869,919, March 22, 2005.

81. Ritchie Andrew J.D. et al., EGR equipped diesel engines and lubricating oil compositions, US Patent No. 6,715,473, April 6, 2004.

82. Growney D.J. et al., Star diblock copolymer concentration dictates the degree of dispersion of carbon black particles in nonpolar media: Bridging flocculation versus steric stabilization, *Macromolecules*, 48(11) (2015): 3691–3704.

83. Mc Geehan J., Eiden K., Low temperature oil pumpability in emission controlled diesel engines, SAE Technical Paper 2000-01-1989, 2000, doi:10.4271/2000-01-1989.

84. Stehouwer D. et al., Sooted diesel engine oil pumpability studies as the basis of a new heavy duty diesel engine oil performance specification, SAE International, SAE Technical Paper 2002-01-1671, 2002, doi:10.4271/2002-01-1671.

85. Galbraith R., May C., A cold start and pumpability study of fresh and highly sooted engine oils in 1999 heavy duty diesel emission engines, SAE International, SAE Technical Paper 2003-01-3224, 2003, doi:10.4271/2003-01-3224.

86. Cui J., Goldmints I., Briggs S., Oberoi S., Field and bench study of shear stability of heavy duty diesel lubricants, *SAE International Journal of Fuels and Lubricants*, 7(3) (2014): 882–889, doi:10.4271/2014-01-2791.

87. Oberoi S., Goldmints I., Briggs S., Baxter D., Designing lubricants for superior fuel economy and wear protection, *19th International Colloquium Tribology-Industrial and Automotive Lubrication*, ISBN 978-3-943563-10-8 for the proceedings.

88. Carvalho M., Richard K., Goldmints I., Tomanik E., Impact of lubricant viscosity and additives on engine fuel economy, SAE International, SAE Technical Paper 2014-36-0507, 2014, doi:10.4271/2014-36-0507.

89. Cui J., Oberoi S., Briggs S., Goldmints I., Baxter D., Measuring fuel efficiency in various driving cycles: How to get maximum fuel economy improvement from the lubricant, SAE Technical Paper 2015-01-2042, 2015, doi:10.4271/2015-01-2042.

14 Pour Point Depressants

Joan Souchik

CONTENTS

14.1 INTRODUCTION

Pour point depressants (PPDs), also known as low-temperature flow improvers and wax crystal modifiers, are polymeric molecules that are added to mineral oil–based lubricants to improve the cold-flow properties of the lubricants. Without the addition of a PPD, many lubricants at cold temperatures would be too viscous to flow easily, or might even be gelled, and the result would be little or no lubricant moving through the system or machine requiring lubrication. For various lubricant applications such as automatic transmission fluids (ATFs), engine oils, gear oils, and hydraulic fluids, paraffinic base stocks are the preferred lubricant. These paraffinic base stocks are typically derived from crude petroleum and are composed of non-aromatic saturated hydrocarbons. Paraffinic stocks, also called base oils, make excellent lubricants because they are chemically stable and resistant to oxidation and have good viscosity index values. However, paraffinic stocks, by their very nature, contain molecular species that have linear carbon chains of 14 carbons or more. These species, recognized as waxy materials, can cause oil pumpability failures at low temperatures. In addition to the inherent waxy base stock components, other sources of waxy material are added to the lubricant as part of its product-specific formulation. These additional components include some viscosity index improvers (VII) and components of the detergent-inhibitor (DI) package.

14.2 WAX-RELATED FLUID MECHANICS

Mineral oils are commonly understood to be Newtonian fluids, meaning that they behave according to the following equation:

$$\text{Shear stress} = \text{Shear rate} \times \text{Viscosity}$$

Experiments show that this equation holds for mineral oils as long as the temperature is above the cloud point of that oil. The cloud point is the temperature at which some of the waxy components of a mineral oil start to crystallize and precipitate from solution, leading to a hazy appearance. This visual evidence of the onset of wax crystals can be tested using American Society for Testing and Materials (ASTM) D2500 [1]. A plot of viscosity versus temperature is shown in Figure 14.1. Above the cloud point, denoted by the data point in Figure 14.1, the viscosity decreases proportionally with temperature. At temperatures below the cloud point, the viscosity increases steeply as the temperature decreases. Proper selection of a PPD will improve the viscosity of a fluid below the cloud point allowing fluid viscosities that approach that

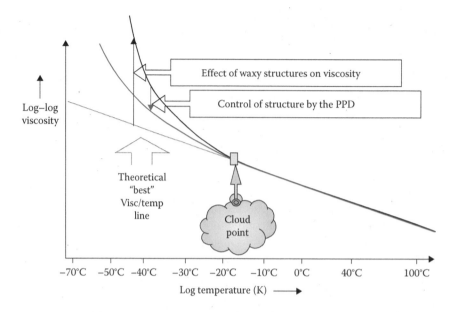

FIGURE 14.1   Relationship between viscosity and temperature for a mineral oil.

fluid's theoretical best viscosity temperature line, the dotted line in Figure 14.1. Below the cloud point, it is not uncommon to observe one of the two non-Newtonian behaviors in these otherwise Newtonian fluids: Bingham fluid behavior or unpredictably high viscosity.

14.2.1   BINGHAM FLUID BEHAVIOR

Bingham fluid behavior is the failure of a fluid to move under low-shear conditions, that is, unless some energy is added to the system. The following equation describes Bingham fluid behavior:

$$\text{Shear stress} = (\text{Shear rate} - \text{Yield stress}) \times \text{Viscosity}$$

The relationship between shear stress and shear rate for a Bingham fluid is shown in Figure 14.2. This failure to flow is similar to the initial flow observed on opening a bottle of ketchup. When a bottle of ketchup is turned over, the ketchup does not flow out of the bottle due to weak associations among some molecular components of the ketchup. However, adding energy to the system by tapping the bottom of the bottle triggers the ketchup to move. Likewise, mineral oils at cold temperatures often do not flow because they contain molecules that have a crystalline nature at low temperatures. First, two-dimensional crystals (platelets) form and then ultimately three-dimensional needlelike structures form. The needlelike structure containing platelets is shown in Figure 14.3. The needlelike structures layer on each other, forming a network of crystals that trap the noncrystalline oil molecules within the gel network, which impedes the flow of the oil (see Figure 14.4). This process is known as gelation, and the source of the yield stress (YS) in the preceding equation is the wax–gel matrix in which the noncrystalline oil molecules are immobilized.

Bingham fluid behavior can be observed by running two widely recognized industry tests: ASTM D4684 [2] and

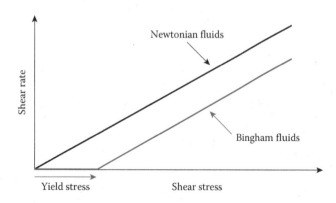

FIGURE 14.2   Bingham fluid behavior.

FIGURE 14.3   Three-dimensional needlelike structure of crystal platelets.

FIGURE 14.4 Gel network of needlelike structures.

ASTM D5133 [3]. A YS observation in the ASTM D4684 test or a high gelation index (>12) measurement in the ASTM D5133 test indicates the formation of a crystal network and quantifies the extent of it. Air binding is an example of problematic Bingham fluid behavior that can occur in an automobile engine. When the engine is at rest, the oil drains and collects in the oil pan. At cold temperatures, the waxy materials crystallize in the oil creating a gel network. When the engine is started, there is enough energy created by the oil pump to break apart the fragile gel structure just at the oil filter. A small bit of oil is pumped into the engine leaving an air gap at the filter. It is possible that the oil in the pan is of a viscosity that it could pump but is locked in place by the wax crystal network that has formed. At this point, air is being pumped into the engine and air is a very poor lubricant.

14.2.2 High-Viscosity Behavior

The second non-Newtonian behavior, often observed in a mineral oil at temperatures below its cloud point, is unpredictably high viscosity. As waxy molecules in the oil crystallize, they cocrystallize but possibly do not form a strongly organized network. However, these cocrystals have enough hydrodynamic volume that they impede the flow of the noncrystalline oil molecules, greatly increasing the viscosity of the oil. In this situation, the oil is so viscous that it cannot be pumped at all into the engine, creating a serious situation where the engine runs with limited, if any, lubrication.

It is important to note that flow-limited behavior can occur in an oil for another reason, unrelated to wax crystal precipitation and growth. All fluids will become more viscous as the temperature decreases, eventually reaching a viscosity where the fluid cannot be pumped. The temperature at which the fluid no longer flows under gravity, not due to wax crystal growth, but simply due to the viscosity contribution of the other molecular species present is called the viscous pour point. The addition of a PPD to a fluid will not change its viscous pour point.

A PPD can resolve both gel structure and high-viscosity modes of failure. PPDs work by controlling the wax crystallization phenomenon in two principal ways: delaying the formation of wax–gel matrix to significantly lower temperatures than would normally occur or reducing the viscosity contribution of the crystal wax particles. PPDs act by interrupting the three-dimensional growth of wax crystals.

14.3 POUR POINT DEPRESSANT CHEMISTRY AND MECHANISM OF ACTION

Before the use of PPDs in the early 1930s, options were few for controlling wax crystallization. One method was to use heat. For example, fires were built under the oil reservoir of a vehicle. Another technique was to increase the solvency of the lubricant fluid portion by the addition of kerosene. The kerosene would then evaporate during use. A third alternative was the use of naturally occurring materials such as microcrystalline waxes or asphalt resin. Although these natural additives were somewhat effective, they were often specific to an application. This motivated the investigation of synthetic PPDs with structures based on the hydrocarbon structures of the naturally occurring materials. In 1931, small molecule chemistry efforts identified alkylated naphthalenes as one type of PPD, and in 1937, polymer chemistry studies by Rohm and Haas established polyalkylmethacrylates as the first polymeric PPDs [4]. Since the early 1930s, other synthetic PPDs have been introduced. However, polymeric PPDs remain the most commercially viable option and include, but are not limited to, acrylates, alkylated styrenes, alpha olefins, ethylene/vinyl acetates, alkylmethacrylates, olefin/maleic anhydrides, styrene/acrylates, styrene/maleic anhydrides, and vinyl acetate/fumarates. All of these polymers work according to the same principles of operation.

The chemical structure of a PPD resembles a "comb," as shown in Figure 14.5. Long waxy side chains are added into the polymer side chains and interspaced with short neutral (nonwax interacting) side chains. The long waxy side chains interact with the wax in the oil in a manner that is proportional to the length of the side chain. These side chains can be linear or branched and should contain at least 14 carbon atoms for the PPD to interact with the wax in the oil. The short neutral side chains act as inert diluents and help to control the extent of wax interaction. Also, a distribution of hydrocarbon side chain lengths can be utilized for best interaction with the wax in the oil because the wax contains a range of molecule chain lengths. The backbone molecular weight of a PPD, generally, can vary with little effect on performance; however, there is a minimum backbone size below which a PPD will become largely ineffective. Figure 14.6 depicts the three-dimensional structure of a PPD. The long coil represents the backbone of the PPD, and the shorter zigzag pieces attached to the backbone represent the side chains. Figure 14.7 shows the side chains of the PPD crystallizing with the oil crystals on the edges of the platelets. This cocrystallization sterically inhibits the formation of the three-dimensional network (see Figure 14.8),

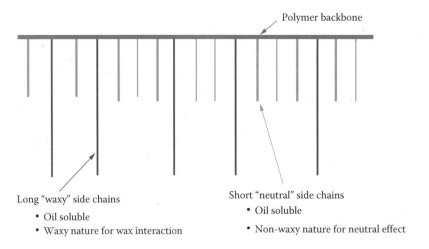

FIGURE 14.5   PPD comb polymer diagram.

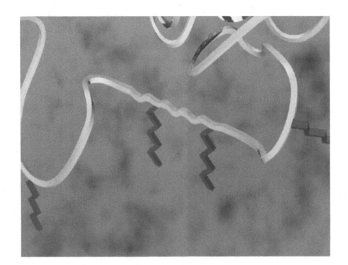

FIGURE 14.6   Three-dimensional representation of a PPD.

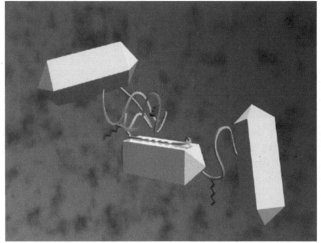

FIGURE 14.8   PPD prevention of gel network formation.

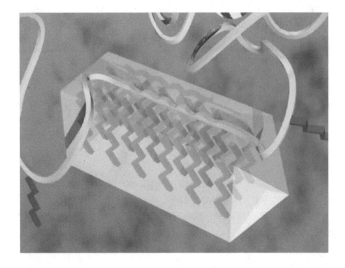

FIGURE 14.7   Crystallization of a PPD with a needlelike structure.

thus preserving the wax as a distribution of tiny crystals and ensuring the complete fluidity of the oil.

The intensity of interaction between the wax and the alkyl side chains is a function of the types and amounts of side chains. This interaction can be expressed by a wax interaction factor (WIF). WIF is a method of ranking PPDs by their wax nature, accounting for the amount of waxy side chains in a PPD and their strength of interaction. A low-wax fluid usually responds best to a low-WIF PPD, and likewise, optimum performance per unit will be achieved in a more waxy fluid with a higher-WIF PPD. Because of the need to treat a vast variety of wax-related situations, a multitude of PPD products have been developed [5]. As described, the mode of operation for a PPD is physical interference. In selecting an optimum PPD, consideration needs to be given to the nature and amount of waxy molecules that may be present in a fluid. The waxy side chains present in the PPD need to be in balance with those waxy molecules found in the fluid. This balance is critical to blending oils with good low-temperature flow properties.

14.4 POUR POINT DEPRESSANT PERFORMANCE TESTING

Mineral oil is typically non-Newtonian at low temperatures; therefore, no single test can ensure that a lubricant blended with mineral oil will remain fluid over a wide range of conditions. Thermal history, including temperature cycling and cooling rates, can also affect the flow behavior of the oil. Various test methods have been developed over the years to evaluate oils under conditions that are experienced during operation. Each test method has specifications, such as viscosity at a certain temperature or YS less than 35 Pa, that the results must pass for the oil to be considered acceptable. A test, to be meaningful in predicting wax crystal growth, must incorporate three critical factors: a low-temperature end point, a defined cooling profile, and a low shear rate. Obviously, wax crystal growth is a low-temperature phenomenon, so a low-temperature end point is required. The low temperature finally achieved in a test needs to reflect the temperature requirements of the application. Not all waxy molecules in a fluid are crystalline at the same temperature, and changing the temperature point for a measurement will greatly impact the quantity of wax present as a crystal; hence, making an appropriate choice of a final test temperature is critical. The rate of cooling affects the competing factors of crystal growth and nucleation of new crystals, thus affecting the number of crystals formed and their relative sizes. The existence of temperature cycling within the cooling profile can further affect the number or size of wax crystals. Clearly, a defined cooling rate is critical in understanding wax crystal growth. Last, low shear rate is also a critical part of characterizing low-temperature behavior. Wax networks are fragile and easily disrupted under higher-shear conditions. To understand whether wax crystal networks have formed, it is important to study the fluid under low-shear conditions, allowing any networks that formed to be preserved.

A comparison of the cooling and shear rates for the different low-shear test methods is given in Table 14.1. The ASTM D97 [6] test for pour point employs a rapid and unrealistic cooling rate for wax crystal growth. Quick cooling does not allow the wax crystals that form to associate to their full extent. However, this test has been used by the industry for many years, and its quickness certainly has benefit in quality control situations. The stable pour point [7] is more realistic than the ASTM D97 pour point because of its longer cooling time, but its seven-day cycle time has made this test largely obsolete. ASTM D3829 [8] and ASTM D4684 have comparable shear rates, but ASTM D3829 employs a programmed cooling rate over a shorter time period, on average 2°C/h over 16 h, versus 0.33°C/h over approximately 48 h for ASTM D4684. Figure 14.9 illustrates the cooling rates for pumpability bench tests. Because of the differences in the tests, analysis of several test method results is critical to selecting the best PPD for a lubricant formulation. A PPD that gives satisfactory performance at both rapid and slow cooling rates under low-shear conditions is presumed to be better able to deliver performance over all foreseeable conditions than is a PPD that already shows performance deficits under one of the limiting test conditions. In the end, it is impossible to test all conditions that might possibly arise.

The cold-cranking simulator (CCS) test, ASTM D5293 [9], is a common oil testing method that is inappropriate for wax–gel testing because it uses a rapid cooling rate and high shear rate. One set of results from ASTM D5293 testing to determine whether PPD addition can affect the CCS viscosity of a Society of Automotive Engineers (SAE) 5W-30 oil is depicted in Table 14.2. As the PPD content was increased from 0% to 0.3%, virtually no change in the viscosity was observed. This result is contrary to what many hope for with the addition of a

TABLE 14.1
Comparison of Low-Temperature Low-Shear Rate Tests

Test	Cooling Rate (°C)	Shear Rate (s^{-1})	Test Duration
ASTM D97 (Pour Point)	0.6/min	0.1–0.2	2 h
Stable Pour Point	0 to –40 over 7 days	0.1–0.2	7 days
ASTM D3829 (MRV)	2/h	17.5	16 h
ASTM D4684 (MRV TP-1)	0.33/h	17.5	48 h
ASTM D2983 (Brookfield Viscosity)	Shock	0.1–12	16 h
ASTM D5133 (Scanning Brookfield Viscosity and Gelation Index)	1/h	0.25	35 h

MRV, mini-rotary viscometer; MRV TP-1, mini-rotary viscometer temperature profile 1.

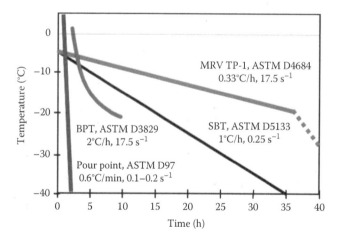

FIGURE 14.9 Cooling rates of pumpability bench tests.

TABLE 14.2
ASTM D5293 CCS Viscosity of SAE 5W-30 Oil with PPD

% PPD	As is	0.15%	0.3%
Viscosity at –30°C (mPa · s)	4750	4750	4790

PPD and is explained by the high-shear nature of the CCS test. PPDs are added to a fluid to control wax crystal growth. Wax crystal networks are fragile. Any wax network that formed would be broken up under high-shear conditions of the CCS test, and as a result, PPDs cannot improve the CCS viscosity of an oil.

14.5 PRINCIPLES OF POUR POINT DEPRESSANT SELECTION

14.5.1 TREAT RATE AND POUR POINT REVERSION

Using a PPD to improve the pour point means the addition of a waxy material to the lubricant. The treat rate, or concentration, of a PPD should be optimized so that the PPD itself does not cause crystallization. The effect of the PPD treat rate on the pour point of Group I 100N base oil is shown in Figure 14.10. The addition of a PPD at 0.2% by weight reduces the pour point from −15°C to −33°C. Doubling the concentration to 0.4% produces only an incremental decrease of about 3°C. An additional increase in the PPD content does not reduce the pour point any further. For this particular oil–PPD system, the optimum treat rate is approximately 0.3%. Continued increases in the PPD content result in a phenomenon known as pour point reversion. The addition of PPD beyond the optimum treat rate is in effect adding wax to the system and consequently reversing the benefit of the PPD by contributing to the formation of crystals or PPD polymer networks as demonstrated in Figure 14.10. The pour point begins to increase at a treat rate of 0.7% from its minimum value at the treat rate of 0.3%. At the highest concentration tested, 1.2%, the pour point was similar to the untreated oil. Pour point reversion can also occur with improper selection of a PPD with a WIF that is too high for the oil being treated.

Figure 14.11 is a typical response curve demonstrating the effect of a PPD as a function of the WIF in one specific base oil. The lowest treat rate of 0.05% (top curve of Figure 14.11) indicates that a specific WIF is responsible for

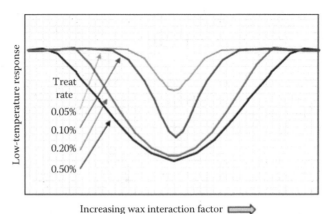

Increasing wax interaction factor ⟹

FIGURE 14.11 Low-temperature response of PPD as a function of the WIF.

optimum performance. As the treat rate increases, the size of the response window increases. If a less-specific response is desired, then a higher treat rate can be used. Also, WIF and treat rate can, to some degree, be traded off, meaning that a nonoptimum WIF can be offset by a higher treat rate and vice versa. This flexibility can allow for one PPD to function well in multiple oils or across plant systems.

14.5.2 EFFECT OF PERFORMANCE ADDITIVES

14.5.2.1 Base Oil Wax Chemistry and Content

The wax content of base oils from which lubricants are formulated varies with the source of the crude oil, the refining process used for the crude oil, the dewaxing process used for the refined oil, and the final viscosity grade of the base oil. The wax removal process is certainly a key determinant for the low-temperature behavior of an oil. Different types of hydrocarbons remain in the oil after refining, depending on the process. Dewaxing of the refined crude oil by solvent extraction removes the long-chain hydrocarbons. In contrast, catalytic dewaxing changes the structure of the hydrocarbons either by cracking them into smaller pieces or by isomerizing them rather than removing them. Last, the viscosity grade of the base oil is directly dependent on the wax content of the oil given that the higher-viscosity oils are produced by including longer-chain hydrocarbons.

The effect of two different PPDs on the viscosity of solvent-extracted and catalytically dewaxed SAE 10W-30 formulated oils is shown in Figure 14.12. ASTM D4684 was used to determine the viscosity of the four PPD–oil formulations. Both oils had the same DI and VII, and PPD 1 had a lower WIF than PPD 2. For the solvent-extracted oil, the viscosity remained fairly low for both PPDs. However, the catalytically dewaxed oils had a higher viscosity with PPD 2 than with PPD 1 and extremely high YS of 105 Pa with PPD 2. To pass current SAE J300 engine oil specifications, YS must be less than 35 Pa. The higher WIF of PPD 2 most likely caused the increase in viscosity and YS in the catalytically dewaxed oil. One explanation for this is that the catalytically dewaxed stock contains

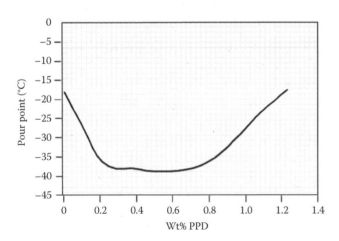

FIGURE 14.10 Effect of PPD treat rate on the pour point of Group I 100N base oil.

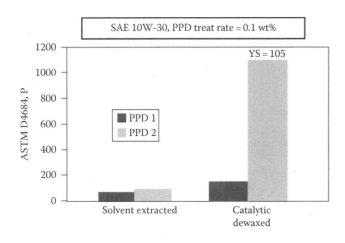

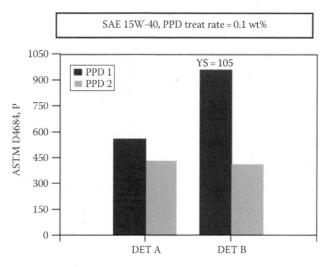

FIGURE 14.12   Solvent-extracted versus catalytically dewaxed oil.

FIGURE 14.14   Detergent effects. *Note:* P = Poise, 100 P = 10,000 mPa · s.

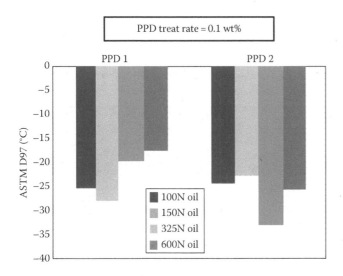

FIGURE 14.13   Viscosity grade effects—Group I base oils. *Note:* P = Poise, 100 P = 10,000 mPa · s.

fewer linear molecules of C14 or greater. The waxy side chains of the PPD, having less natural waxes to interact with, found themselves and built viscosity and structure through polymer-to-polymer interactions.

The effect of the WIF of a PPD on different viscosity grades of Group I base oils is shown in Figure 14.13. For the lower-viscosity grades of 100 and 150N, increasing the WIF from 1 to 2 had a small effect on decreasing the pour point temperature. For the higher-viscosity grades of 325 and 600N, the result was reversed, and the higher WIF decreases the pour point temperature significantly. This result can be explained by the interaction of the PPD with the longer-chain hydrocarbons present in the higher-viscosity grades.

14.5.2.2   Other Wax Sources

Other wax sources include DI packages and friction modifiers that may themselves contain waxy hydrocarbon chains. Also, VIIs, which can typically contain long ethylene sequences,

should be considered. These waxy components can affect the low-temperature low-shear performance of the fully formulated oil, both positively and negatively, and therefore, it is important to understand their contribution when selecting a PPD for a lubricant.

Detergent effects using two different PPDs in SAE 15W-40 are shown in Figure 14.14. The viscosity of two oils was tested by ASTM D4684. PPD A had a lower wax content than PPD B. In the oil with detergent A, PPD A was able to control the viscosity to 13,000 mPa · s (100 mPa · s = 100 centipoise = 1 poise), which was below the test specification of 30,000 mPa · s. However, PPD A could not control the YS, which, at 105 Pa, exceeded the test specification of less than 35 Pa. The excessive YS indicated that a gel network was forming. One explanation for this is that detergent A was waxy in nature, and the waxy side chains PPD A interacted with detergent A leaving fewer PPD side chains available for interaction with the wax in the oil. The results for both PPDs were comparable in detergent B testing, passing both viscosity and YS specifications.

Chemistry effects of VIIs and a PPD in SAE 5W-30 oil are shown in Figure 14.15. The oils containing both VII 1 and VII 2 had comparable performance, and the oils met the test specification of less than 30,000 mPa · s at −30°C. As the temperature was lowered to −35°C, a much larger change in viscosity was observed in the oil containing VII 2. The oil containing VII 1 doubled in viscosity, whereas the oil containing VII 2 tripled in viscosity. Although both VIIs met the test specification of less than 60,000 mPa · s at −35°C, this PPD was not able to control viscosity growth in the oil with VII 2. With proper PPD selection, the fluid formulated with VII 2 certainly can show equal performance to one with VII 1. This demonstrates the effect that different cold-temperature conditions can have on PPD selection and how seemingly small changes in an oil formulation, VII 1 or VII 2, can change the optimum PPD for a lubricant. Olefin copolymers (OCPs) are used commonly as VIIs. The ethylene content of the OCP can greatly affect the

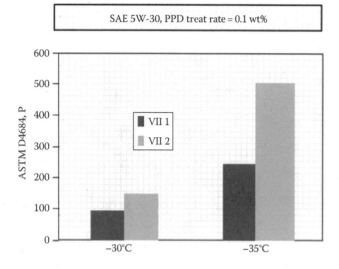

FIGURE 14.15   VII chemistry effects.

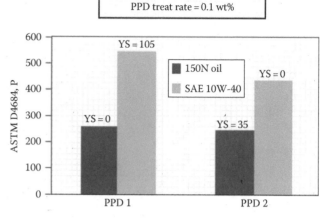

FIGURE 14.16   PPD selection in base oil versus fully formulated oil. *Note:* P = Poise, 100 P = 10,000 mPa · s.

TABLE 14.3
Same VII Chemistry Effects

VII	Traditional	HE OCP	HE OCP
PPD properties	1	1	2
ASTM D97 (°C)	−30	−27	−30
ASTM D4684, viscosity (mPa · s)	13,700	13,900	9,700
ASTM D5133°C at 30,000 mPa · s	−28.8	−25.9	−30.2
Gelation index	4.6	9.8	4.8

Note: Same base oils, SAE 10W-30, PPD treat rate = 0.1% weight.

choice of PPD for optimum viscosity and structure control of a lubricant. The importance of testing PPDs and considering all pertinent tests is summarized in Table 14.3, which presents the results of testing two different PPDs in SAE 10W-30 oil containing either a traditional OCP or a higher ethylene (HE) OCP. Using PPD 1, a lower-WIF PPD, the oil containing the HE OCP had results similar to the oil contsssaining traditional OCP for all the tests except the Gelation Index, which approached the spec limit of 12, indicating that a structure is starting to develop. When PPD 2, a higher-WIF PPD, was used with the HE OCP, the results were similar to or better than the oil containing traditional OCP and PPD 1. Assuming that the longer ethylene sequences in the HE OCP can contribute to wax interactions, these data demonstrate that high-WIF PPD is needed to disrupt those waxy VII crystals as well as to interact with the waxy species in the base oil.

14.5.2.3   Fully Formulated Oil

It should be obvious from the previous discussion that another important aspect of PPD selection is the viscosity behavior of fully formulated oil versus base oil. The fully formulated oil almost always contains additives and wax sources not present in the base oil, and these waxes may affect the viscosity behavior. The effect of two different PPDs on 150N base oil and a SAE 10W-40, a fully formulated oil, is shown in Figure 14.16. PPD 2 has a higher WIF than PPD 1. PPD 1 and PPD 2 gave similar viscosity results when tested in the base oil alone; however, YS was detected with PPD 2. These results indicate that PPD 1 would be a better candidate for the 150N base oil. With the fully formulated oil, however, PPD 1 resulted in a higher viscosity and a YS of 105 Pa, which exceeded the test specifications of less than 35 Pa. In this case, PPD 2 is the better choice because the additional waxy components from the DI or VII in the SAE 10W-40 formulation required more waxy sites on the PPD polymer to effectively control the low-temperature low-shear properties of the fully formulated oil.

14.6   POUR POINT DEPRESSANT ROBUSTNESS

The usefulness of PPDs has been clearly established through laboratory testing. However, the question of whether a PPD loses its effectiveness during use must be considered. The robustness of the PPD can be determined from field and laboratory tests. Experiments were performed to determine whether the PPD response was deteriorated by severe field and laboratory engine tests. The severe field test was a New York City 10,000 miles taxi test; the laboratory test was the Sequence IIIG test (ASTM D7320) [10], a 100-h, high-temperature engine test. The pumping viscosity (ASTM D4684 MRV TP-1) was measured to evaluate the PPD response. Figures 14.17 and 14.18 depict the comparison of responses for fresh oil, used oil without the PPD, used oil with the PPD, and posttest addition of the PPD to used oil in the taxi test and the Sequence IIIG test, respectively. All fresh oil measurements were made at −35°C and used oil measurements at −30°C. Both tests demonstrate that, without the PPD, the used oil had a significantly higher viscosity than the fresh oil, becoming too viscous to measure during the testing at −30°C. On the contrary, the used oil containing PPD

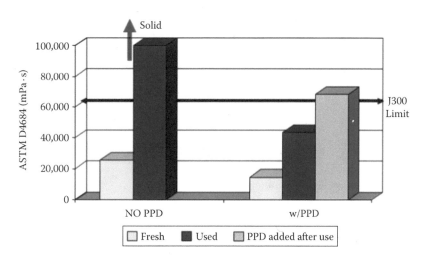

FIGURE 14.17 PPD stability after taxi field test. *Note:* SAE 5W-30 as fresh oil (−35°C), SAE 10W-40 as used oil after 10,000 miles taxi service (−30°C), PPD at 0.1% wt—when present.

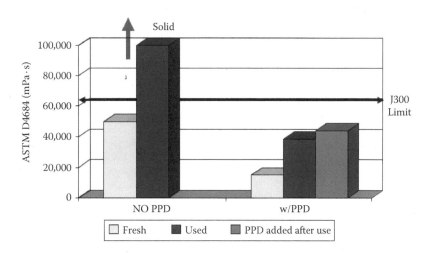

FIGURE 14.18 PPD stability after Sequence IIIGA testing. *Note:* SAE 5W-30 as fresh oil (−35°C), SAE 10W-40 as used oil after Sequence IIIGA (−30°C), PPD 0.3%—when present.

from both tests easily passed the SAE J300 limit of 60,000 mPa · s with no YS. Additionally, when PPD was added to the used oil without PPD from both taxi and Sequence IIIG tests, the used oil viscosity was greatly improved, showing slightly higher but similar results to the used oil with PPD results. Clearly, the PPD had not deteriorated during use since the PPD-pretreated used oil provides good MRV TP-1 viscosity, the untreated used oil has failing MRV TP-1 viscosity, and the post-added PPD used oil results are similar to the oil pretreated with PPD [11].

14.7 LUBRICANT APPLICATIONS

PPDs are selected and engineered for different lubricant applications ranging from automotive engine and transmission oils to industrial oils and biodegradable fluids. Engine oils tend to drive the PPD selection process for most blenders as engine oils are required to meet several low-temperature low-shear specifications with severe limits, and they are often the high-volume products for a blend plant. Additionally, other lubricants more often are less selective for PPD.

14.7.1 AUTOMOTIVE ENGINE OILS

A PPD is always needed for automotive engine oils formulated with mineral oil base stocks, and given the complexity of these oils, the optimum PPD can usually only be identified by undertaking a PPD study on the fully formulated oils. The low-temperature low-shear tests that are most often used today to access the pumpability of fresh engine oils are ASTM D4684, ASTM D5133, and ASTM D97. Various national, international, and OEM–specific engine oil specifications exist, each varying in the low-temperature test recommended or in the acceptable limits for a particular test.

More recently, it has been recognized that oils that pass pumpability requirements as fresh oil can still give pumpability failures when the oil has been aged in an engine.

As a result, both ILSAC and ACEA have introduced requirements for aged oil low-temperature viscosity. For ILSAC GF-5, MRV TP-1 pumpability after aging by Sequence IIIGA or by ASTM D7528 [12], also known as the ROBO test, is a requirement. ILSAC GF-6 will certainly include an aged oil low-temperature viscosity requirement. In addition to this ILSAC requirement for gasoline engine oils, a number of ACEA grades now require MRV TP-1 pumpability after aging in the presence of biodiesel by CEC L-105 [13]. The addition of aged oil specifications do make the selection of a proper PPD more complex but certainly not insurmountable. With careful PPD selection, it is possible to provide for good pumpability in both fresh and aged oils.

14.7.2 AUTOMOTIVE TRANSMISSION OILS

Selection of PPDs for automotive transmission oils, including gear oils and ATFs, normally entails testing with ASTM D2983 [14] (Brookfield viscosity) at −12°C, −26°C, and −40°C. A PPD is always needed for mineral oil–based fluids, but it may be a constituent of the VII, if one is used.

14.7.3 INDUSTRIAL OILS

In contrast to the automotive oils, which typically require specialized low-temperature viscosity testing, industrial oils often have just a simple pour point (ASTM D97) requirement. Therefore, the targets are often not difficult, and a low treat level of a traditional PPD may be suitable. For these oils, base stock choice is the primary driver for PPD selection. Some fluids, such as multigraded hydraulic or tractor fluids, have additional low-temperature low-shear requirements, namely, Brookfield viscosity. These fluids generally need a low dose of PPD to meet requirements, and a PPD lab study will identify the optimum product and treat rate. Specific industrial oils, such as refrigeration oils, may be blended with synthetic or naphthenic stocks. These oils have excellent low-temperature low-shear properties, and the addition of a PPD will not generally give further improvement.

14.7.4 BIODEGRADABLE FLUIDS

Another application of PPDs is in biodegradable fluids such as canola and soy oil. The PPDs for this application have to be specially designed and engineered because the mechanism of interaction for these biodegradable fluids is different from those of traditional automotive lubricants. For these biodegradable fluids, which consist almost exclusively of long-chain fatty acid triglycerides, the high wax content is responsible for the observed low-temperature viscosity problems, and therefore, the normal interference mechanism of small quantities of PPD containing long waxy side chains is not sufficient to address the observed problems. This leads to the need for different PPD structures and alternative approaches [15,16].

TABLE 14.4
Major Pour Point Depressant Manufacturers and Marketers

Manufacturer	Trade Name
Afton Petroleum Additives, Inc.	HiTEC®
BASF SE	Irgaflo™
Chevron Oronite LLC	OLOA®
Evonik Oil Additives	Viscoplex®
Infineum	Infineum V300®
Kusa Chemicals Pvt. Ltd.	KUSAPOUR®
Sanyo Chemical Industries, Ltd.	Aclube™
The Lubrizol Corporation	LUBRIZOL®

14.8 POUR POINT DEPRESSANT MANUFACTURERS AND MARKETERS

Many companies manufacture and market PPDs and some of those are listed in Table 14.4. Certainly, this is not a complete list of companies offering commercial quantities of PPDs; it does however attempt to reflect the current major suppliers and their trade names. The chemical nature of commercial PPDs varies, and the most common chemistries are listed in Section 14.3.

REFERENCES

1. ASTM Standard D2500-11, *Test Method for Cloud Point of Petroleum Products*, Annual Book of ASTM Standards, West Conshohocken, PA, Sec. 5, Vol. 05-01, 2016.
2. ASTM Standard D4684-14, *Test Method for Determination of Yield Stress and Apparent Viscosity of Engine Oils at Low Temperature*, Annual Book of ASTM Standards, West Conshohocken, PA, Sec. 5, Vol. 05-02, 2016.
3. ASTM Standard D5133-15, *Standard Test Method for Low Temperature, Low Shear Rate, Viscosity/Temperature Dependence of Lubricating Oils Using a Temperature-Scanning Technique*, Annual Book of ASTM Standards, West Conshohocken, PA, Sec. 5, Vol. 05-02, 2016.
4. Hochheiser, S., *Rohm and Haas: History of a Chemical Company*. University of Pennsylvania Press, Philadelphia, PA, 1986.
5. Evonik publication. *Pour Point Depressants, A Treatise on Performance and Selection*. Darmstadt, Germany. www.evonik.com/oil-additives, 2015.
6. ASTM Standard D97-15, *Test Method for Pour Point of Petroleum Products*, Annual Book of ASTM Standards, West Conshohocken, PA, Sec. 5, Vol. 05-01, 2016.
7. FED-STD-791D, *Pour Stability of Lubricating Oil*, Federal Standard Testing Methods of Lubricants, Liquid Fuels and Related Products, Government Printing Office, Washington, D.C., November 2007.
8. ASTM Standard D3829-14, *Test Method for Predicting the Borderline Pumping Temperature of Engine Oil*, Annual Book of ASTM Standards, West Conshohocken, PA, Sec. 5, Vol. 05-02, 2016.
9. ASTM Standard D5293-15, *Test Method for Apparent Viscosity of Engine Oils between −5 and −35°C Using the Cold-Cranking Simulator*, Annual Book of ASTM Standards, West Conshohocken, PA, Sec. 5, Vol. 05-02, 2016.

10. ASTM Standard D7320-15a, *Test Method for Evaluation of Automotive Engine Oils in the Sequence IIIG, Spark-Ignition Engine*, Annual Book of ASTM Standards, West Conshohocken, PA, Sec. 5, Vol. 05-04, 2016.

11. Kinker, B.G., R. Romaszewski, J. Souchik. *Pour Point Depressant Robustness After Severe Use in Passenger Car Engines in the Field and in the Sequence IIIGA Engine*, SAE Paper 2005-01-2174, 2005.

12. ASTM Standard D7528-13, *Test Method for Bench Oxidation of Engine Oils by ROBO Apparatus*, Annual Book of ASTM Standards, West Conshohocken, PA, Sec. 5, Vol. 05-04, 2016.

13. CEC L-105-12(S), *Low Temperature Pumpability*, The Coordinating European Council, Brussels, Belgium, www.cectests.org, 2012.

14. ASTM Standard D2983-09, *Test Method for Low-Temperature Viscosity of Lubricants Measured by Brookfield Viscometer*, Annual Book of ASTM Standards, West Conshohocken, PA, Sec. 5, Vol. 05-01, 2016.

15. Webb, M., Biolubes: Cradle to grave. *ILMA Annual Meeting October 2011*.

16. Ethyl Corporation, US Patent US5658864A, *Biodegradable pour point depressants for industrial fluids from biodegradable base oils*, 1995.

Section V

Miscellaneous Additives

15 Evaluating Tackiness of Polymer-Containing Lubricants by Open-Siphon Method
Experiments, Theory, and Observations

Brian M. Lipowski and Daniel M. Vargo

CONTENTS

15.1 INTRODUCTION

A tackifier is a lubricant additive that imparts a tack or stringiness to a substance and is typically used to provide adherence in fluid lubricants and stringiness in greases. Tackifiers are used to inhibit dripping, removal, and flinging of lubricating oils and to impart texture in greases. In lubricant applications, most tackifiers are dissolved polymers in either mineral oil or vegetable oil–based diluents. The polymers are often polyisobutylene (PIB) with a molecular weight of 1000–4000 kDa or rarely an ethylene–propylene copolymer (OCP) with a molecular weight of approximately 200 kDa. The tackiness of a solution generally increases with polymer molecular weight. The operational environment of the lubricant dictates polymer selection; for example, in high mechanical shear and high temperature applications, OCPs are generally preferred over PIBs. The base oils may be paraffinic, naphthenic, or vegetable, depending upon application. Lubricant dripping, splashing, and aerosol formation (misting) are major problems for many lubricants. Tackifier additives in lubricating oils are useful in minimizing these problems. Application areas where lubricant retention and enhanced film thickness is benefited by tackifiers are open gear oils, slideway lubricants, wire ropes, textile operations, chain saws, chains, and rock drill oils. Tackifier additives in greases enhance the greases ability to stay in place and to resist removal by water in operation. Specialty applications include food grade, aerospace, and biobased lubricants.

Tackifier treat rates range from 0.02% in antimist applications to 3% or more in grease applications. A key challenge among manufacturers and users of tackifiers is to evaluate and compare tackifier formulations in a systematic and repeatable manner. This chapter provides a theoretical foundation and practical test method to address this challenge.

In many industrial applications, the lubricating oil must not drip or mist from the surfaces to be lubricated. This can be practically achieved by increasing the cohesive energy of the lubricating fluid in order to prevent atomization. The oil viscosity should remain as low as possible to prevent excess wasted energy. In this case, the properties of the optimal lubricating oil should be stringy to prevent oil loss and increase lubrication lifetime for machinery in which oil waste is a problem. To satisfy the earlier industrial need, the lubricating industry has formulated and utilized lubricating oils containing tackifiers. Typical formulations of tacky lubricants are limited to high-molecular-weight PIB dissolved in petroleum oil.

For polymer solutions to be tacky, the polymer chains should have the capacity to extend and thus behave as viscoelastic liquids. To be less viscous and reduce cost, tackifiers should be very dilute solutions of high-molecular-weight polymer in base fluid. It is well known that the viscoelastic properties of polymer solutions depend on polymer concentration and several molecular parameters of the polymer determined by the chemical structure [1–3]. The most important parameters are the polymer molecular weight and, to a lesser degree, the molecular weight distribution, and the flexibility of the polymer chains. The chain flexibility is responsible for uncoiling of chains with smaller Kuhn segments, and, for more rigid chains, orientation of chains in the direction of extension. Another important parameter of a tacky polymer solution is the viscosity of the base fluid. Preparation and testing of various lubricants with enhanced tackiness are in high demand for lubricant applications. There have been few studies of tackiness phenomena in polymer solutions designed for use in lubricating oils.

This chapter investigates the tackiness of very dilute solutions of polymer-containing lubricants using the open-siphon method introduced and described in paper [4] (see also the monographs [5,6]). In this method, viscoelastic liquids are vertically withdrawn from a container through a capillary tube connected to a vacuum source. The vacuum pulls a tacky liquid out of the container forming a free jet (string). Tackier fluids result in a longer jet in air than less tacky ones, while nontacky fluids are not drawn upward from the free surface.

Basic experiments and theory analyzing the open-siphon phenomena on water solutions of polyethylene oxide (PEO) were initiated in paper [7], using a viscoelastic approach. Further studies [8] utilized the idea of relaxation liquid–solid transition [6] and developed the theory further, considering the free jet withdrawn from the viscoelastic solution as a purely elastic gel. Good agreement was found between the calculations and the experimental data of paper [7]. These results were later reviewed in a monograph [6]. The experimental and theoretical results in papers [7,8] described only the stationary processes of withdrawal of free viscoelastic jets by a rotating drum, where the constant rate of withdrawal and the flow rate of the fluid were controlled by the rotational speed of the drum. Additionally, the experiments [7,8] were conducted on relatively concentrated (0.5% by volume) water solutions of very high-molecular-weight PEO. Thus, to accurately analyze and describe the withdrawal of viscoelastic jets of very dilute polymer solutions in the nonstationary open-siphon process, the previously developed stationary theory [6,8] must be modified.

It should also be mentioned that the problem of withdrawal of viscoelastic liquids may seem similar to the withdrawal of viscous liquids by a vertically moving flat (or cylindrical) plate. In this problem, a viscous liquid forms a thin layer near the rigid plate under the action of viscosity, gravity, and surface tension. The solution of the problem [9,10] uses a matching condition between the viscous flow and static meniscus. However, despite the seeming similarity of these two problems, the withdrawal of polymer solutions from a free surface is more complicated. This is because the radius of the extendable free jet varies with height and is unknown *a priori*.

The chapter is organized as follows: Section 15.2 describes the experimental setup, procedures, and the fluids used in the experiments; Section 15.3 introduces some basic facts of viscoelasticity for polymeric liquids; Section 15.4, using a pseudo-steady-state approach, modifies the theory [6,8] in the non-steady-state case of the open-siphon experiment and applies it to very dilute polymer solutions;

Section 15.5 discusses the quantitative experimental findings and describes the data using the theory of Section 15.4; Section 15.6 applies the open-siphon method to evaluate the tackiness in two lubricating oils; and Section 15.7 consists of concluding remarks.

15.2 EXPERIMENTAL SETUP, PROCEDURES, AND FLUIDS

The experimental device used for evaluating the tackiness of lubricating fluids is similar to those described in previous papers [4,5]. The setup is shown in Figure 15.1. The glass capillary tubing has an inner diameter of 1.58 mm and a length of 120 mm and is connected to a vacuum pump. Three vacuum pressures, p_v, of 68, 77.3, and 84 kPa were examined. The graduated glass cylinder filled with test fluid was of inner diameter 28 mm and height 190 mm. To quantify the jet profiles, we used enlarged photographs of the jet.

To perform the test, the capillary was lowered into the container filled with test liquid until the open end of the capillary was below the liquid surface. The capillary was stationary during the remainder of the experiment. Vacuum was used to pull the liquid into the capillary that decreased the level of liquid in the container. The experiment began at the moment when the dropping liquid surface in the container reached the lower end of the capillary. From this point, a free jet of tacky liquid was formed. The siphon draws down the level of liquid in the container, increasing the length and decreasing the radius of the free jet. Flow rate, q, as measured using the graduated cylinder, was dependent on the applied vacuum; the higher the vacuum, the higher the flow rate. At low flow rates, the jet was broken; while at higher flow rates, the jet lost its axial symmetry and the flow rate oscillated with time. Vacuum pressures were selected for this test that would prevent breakage and high oscillation. Sporadic oscillation of the jet did occur during testing. Each measurement was run in triplicate and the mean was used for calculations. The photos of the jets in Figure 15.2a and b show a relatively large viscoelastic meniscus near the free surface, which is characteristic of the free jet. Some traces of instability are also shown in these figures.

The size and shape of the free jet were examined by taking photographs every 25 seconds during the experiment. The maximum length of the free jet supported by the vacuum is recorded as a "string length" and is representative of the tackiness. Photos were enlarged using Adobe Photoshop CS2. Using the printed pictures, the jet radius, r, at several distances, z, from the liquid surface and at several flow rates, q, was measured. Special attention was given to measurements of parameters of the meniscus that appears at the moment when the capillary disconnects from the fluid. These are the radius of the meniscus, R, and the radius of the jet, r_0, measured at different flow rates, q.

A 0.025% by weight PIB solution with viscosity average molecular weight of $M_\eta = 2.1 \times 10^6$ [11] in ISO 68 viscosity grade oil, which has viscosities of $\eta_s \approx 0.138, 0.0585,$

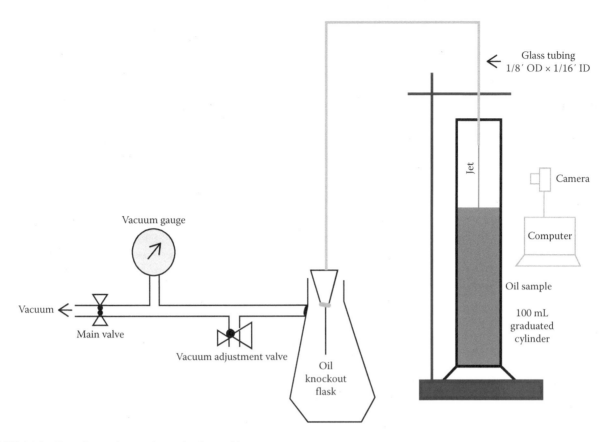

FIGURE 15.1   Experimental setup for evaluating tackiness.

(a)   (b)

FIGURE 15.2   (a, b) Photographs of free jets of 0.025% PIB tacky solutions in lubricating oil.

and 0.0073 Pa · s at 20°C, 40°C, and 100°C, respectively, was used. Viscosities were measured using capillary viscometers by ASTM D445. The density, ρ_s, of this oil at 25°C is 0.86 g/cm^3. Surface tension, γ_s, at 20°C is 2.7 Pa · cm [12].

In previous experiments, whose results are briefly discussed in Section 15.6, various PIBs with $M_\eta = 0.9 \times 10^6$, 1.6×10^6, 4.0×10^6 Da, with weight concentration varied

from 0.005% to 0.12% in mineral oil with a viscosity η_s at 40°C of 0.068 and 0.022 Pa · s (ISO 68 and ISO 22) were also used. A solution of an OCP with molecular weight of about 200 kDa was also used. Polymers were granulated and then dissolved in the oil in a glass container on a hot plate with low-shear agitation. The time of dissolution was approximately 48 hours.

15.3 VISCOELASTIC EFFECTS DURING WITHDRAWAL OF TACKY LUBRICANTS

Dilute polymer solutions can be characterized using three basic parameters: solvent viscosity, η_s; polymer volume concentration, c; and relaxation time, θ. It is well known that at very low concentrations of polymer additives in mineral oil, the viscosity, η, of the polymer solution essentially equals η_s. Adding low levels of PIB to mineral oil dramatically increases the relaxation time, θ, of the solutions.

Materials that are intermediate in behavior between viscous liquids and elastic solids in their response to applied forces are referred to as viscoelastic liquids. They behave as viscous liquids at low rates of external force (deformation) and as elastic solids when these rates are high. This type of behavior is commonly estimated by the dimensionless Weissenberg number, We. In extensional flows, including the problem of withdrawal, it is presented as [6]

$$We = \theta \cdot \dot{\varepsilon} \qquad (15.1)$$

Here, $\dot{\varepsilon}$ is the elongation rate, that is, the velocity gradient in the direction of extension (withdrawal). When extensional rate is low and $We \ll 1$, a viscoelastic liquid behaves as a viscous one. When $We \gg 1$, the solid-like properties of viscoelastic liquids dominate and they behave as elastic solids.

Many viscoelastic liquids display a rapid transition from liquid-like to solid-like behavior when passing through a critical point in the Weissenberg number, We_c. This phenomenon is referred to as *fluidity loss* and has been well documented in polymers with narrow-molecular-weight distributions and is treated as a *relaxation transition* (e.g., see the monograph [6] and references therein). The underlying physics of this transition is that highly oriented polymer molecules that arise under certain flow conditions create physical cross-links that effectively cause gelation of the polymer chains preventing the molecules from flowing freely. In the case of withdrawal of dilute polymer solutions, the fluidity loss effect assumed in References 6 and 8 could also be caused by an effective increase in polymer concentration in strong extensional flows near the axis of extension. This may occur because the fluid trajectories in extensional flows cause the polymer macromolecules to align with each other and closely pack together.

15.4 THEORETICAL MODEL

In order to make clear the basic physics of processes in the withdrawal of tacky lubricants, we now discuss modification of the theoretical model developed in previous studies [6–8]. This modification is based on the following assumptions: (1) the effect of *fluidity loss* plays a dominant role in jet withdrawal, (2) the *inertia phenomena* are negligible in the dynamics of polymer jet withdrawal, and (3) *exudation of solvent* out of the withdrawn jet is important in case of dilute polymer solutions.

The first and second assumptions have been employed and proved valid in paper [8] for an example of a concentrated aqueous solution of polymer. A solution of 0.5% very-high-molecular-weight PEO was used. In the case of dilute polymer solutions, the validity of these assumptions is unknown *a priori*. Neglecting inertial effects allows one to extend the stationary theory to the nonstationary case, using a pseudo-steady-state approach. The third assumption comes from observations of withdrawn lubricant jets and plays an important role in the following modeling.

To develop a formal model, the vertical coordinate, z, which coincides with the jet centerline and is determined from the moving free surface as shown in Figure 15.3, is used. The origin, $z = 0$, is located at the free surface, and the upper coordinate, $z = l(t)$, located at the capillary entrance, indicates the length of the visible jet at time, t. It is convenient for theoretical treatment to roughly separate the domain of the liquid flow into three regions: *Region 3*, $\{z < 0\}$, located under the free surface, the meniscus; *Region 2*, $\{0 \leq z \leq R(t)\}$, located from the free surface upwards to the end of the meniscus; and *Region 1*, $\{R(t) \leq z \leq l(t)\}$, where motion of the free jet occurs. In this region, the functions $R(t)$ and $l(t)$ are unknown and must be determined. Basic flow effects that occur in the three regions can be qualitatively described as follows [6–8].

In Region 3, the vacuum from the capillary causes a specific extensional flow. This flow, changing slowly in time, looks like an effective undersurface jet that narrows from the bottom to the surface. Therefore, the vertical velocity of the jet and the characteristic extensional velocity gradient, $\dot{\varepsilon} \approx dV/dz$, are increased when approaching the surface from below. Substituting this value of $\dot{\varepsilon}$ into Equation 15.1 explains the increase in the Weissenberg number, which may cause the relaxation driven fluid–solid transition. It was speculated in previous papers [6,8] that the complete relaxation transition occurs in Region 2, where the polymer solution behaves as a viscoelastic liquid and forms a free jet that is compressed under the additional action of surface tension. In Region 1, the free jet can be treated as an elastic gel swollen with solvent and is under the action of extensional force, gravity, and surface tension.

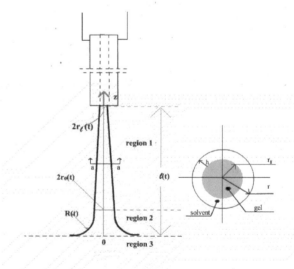

FIGURE 15.3 Schematics of jet withdrawal.

To describe the free jet behavior in Region 2, a semiempirical approach is used [6–8] rather than analyzing the complicated viscoelastic flow in Regions 2 and 3. This approach roughly approximates the shape of a static meniscus by the expression:

$$r(z,t) \approx r_0 + R - \sqrt{R^2 - (R-z)^2} \quad (0 \le z \le R) \quad (15.2)$$

where

$R(t)$ is the maximum height of the meniscus
$r_0(t)$ is the initial (maximal) radius of the free jet

The geometric picture of this approximation is shown in Figure 15.3, although the circle of radius R with a horizontal tangent at $z = 0$ surely cannot smoothly touch the meniscus surface at $r = r_0$. In spite of its very approximate character, Equation 15.2 allows for the description of some critical phenomena at low flow rates and for interpretation of the measurements of the jet radius near the free surface at $z = 0$.

Despite the very dilute character of the polymer solutions, it is assumed that the analysis of jet behavior in Region 2 is identical to that described in previous studies [6,8]. Utilizing dimensional and geometric arguments, this approach yields the equations:

$$R = \alpha(q\theta)^{1/3}; \quad r_0 = \beta(q\theta)^{1/3} \left(S_0 = \pi r_0^2 = \pi\beta^2(q\theta)^{2/3}\right), \quad (15.3)$$

$$\sigma_0 \approx \rho g m R + 2\nu\gamma/R = \rho g m \nu r_0 + 2\gamma/r_0. \quad (15.4)$$

where

σ_0 is the initial stress at jet radius r_0
$q(t)$ is the flow rate
θ is a characteristic relaxation time
ρ and γ are the density and surface tension of the solution, respectively, whose values will be determined using the corresponding solvent values ρ_s and γ_s

In Equations 15.3 and 15.4, α and β are numerical parameters, whose values are estimated by fitting, and

$$\nu = \frac{\alpha}{\beta}, \quad m = m(\nu) = \nu^2 \left[\frac{2}{3} + \left(1 + \frac{1}{\nu}\right)^2 - \left(\frac{\pi}{2}\right)\left(1 + \frac{1}{\nu}\right)\right]. \quad (15.5)$$

The first and second terms in the identical expressions of Equation 15.4 for σ_0 describe, respectively, the contributions of weight and surface tension of the liquid column in the meniscus (Region 2). The stress, σ_0, given by Equation 15.4, is a function only of R (or r_0) and has a minimum at $R = R_c$ (or r_{0c}). The minimum value of initial stress σ_{0min} and corresponding values of R_c (or r_{0c}) and q_c are given by

$$\sigma_{0min} = (c_1 \rho g \gamma)^{1/2}, \quad R_c = \left[\frac{c_2\gamma}{\rho g}\right]^{1/2}, \quad r_{0c} = \left[\frac{\hat{c}_2\gamma}{\rho g}\right]^{1/2} \quad \theta q_c = \left[\frac{c_3\gamma}{\rho g}\right]^{3/2}$$

$$c_1 = 8m\nu, c_2 = 2\nu/m, \quad \hat{c}_2 = 2/(\nu m), \quad c_3 = c_2/\alpha^2$$

(15.6)

Here, numerical parameters c_1, c_2, $\hat{c}_2$, and c_3 are calculated using Equation 15.5. The values of R_c (or r_{0c}) and q_c are critical physical parameters, below which the jet structure does not exist [6,8]. Remarkably, these values depend only on the equilibrium physical parameters of the fluid, its density, ρ, and the surface tension, γ, and are independent of the withdrawal conditions and the viscoelastic constants of the solution.

In order to analyze the jet behavior in Region 1, we mention the first unusual flow phenomenon in the capillary, recorded in experiments and shown in Figure 15.4.

Here, the jet entered the capillary with a considerably smaller diameter than the inner capillary diameter and, at higher vacuum, seemingly continued extending to the upper capillary end. Thus, instead of visible jet length $l(t)$ introduced earlier, the total jet length $L(t)$ should be considered as the real dynamic variable at high vacuum, where

$$L(t) = l(t) + l_T \quad (15.7)$$

and l_T is the capillary length.

In case of withdrawal of dilute polymer solutions, an additional effect of *strain-induced exudation of solvent* should also be taken into account. Although the kinetics of this process is unknown, the flow of a thin film of solvent covering the gelled jet is expected to be much the same as in the case of a thin film withdrawn from a vessel by a two-dimensional surface moving upward, that is, controlled by the vertical drag speed, viscosity, gravity, and surface tension [9].

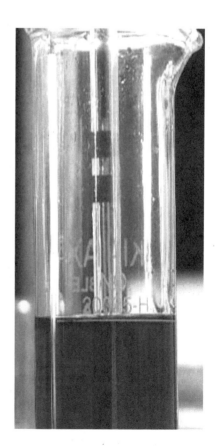

FIGURE 15.4 Photographs of two-phase motion of jet in capillary.

To take solvent exudation into account, a *two-phase model* of the jet where the actual radius, r, of the jet is represented as the sum of actual radius r_g of the gelled jet and the precipitated film thickness h, that is, $r = r_g + h$ is necessary. This is shown in the inset of Figure 15.3. At any value of r, the core of the swollen jet with radius r_g can be treated as an elastic solid with large deformations, whereas the peripheral thin film of solvent with thickness h is treated as a viscous liquid. The contribution of the solvent film to the axial stretching stress is neglected. In this two-phase model, the local surface tension effect is considered to act on the total radius r.

The hypothesis that the film thickness h depends only on time is employed. Using the scaling argument, it is assumed that $h \approx \xi r_0(t)$ where $\xi = \xi(q_c/q) \leq 1$ is a positive increasing function and $r_0(t)$ is the maximum radius of the jet at time t (see Figure 15.3). This two-phase approach yields the kinematic relationship:

$$r = r_g + h(t) = r_g + \xi\left(\frac{q_c}{q(t)}\right)r_0(t). \qquad (15.8)$$

Neglecting the effect of the solvent film in Region 2, the dependence $r_0(t)$ will be calculated at the end of this section, using the kinetics of jet withdrawal and analyzing the effects in Region 2. The function $\xi(q_c/q(t))$ will be proposed in the next section.

Using the earlier assumption that in Region 1 the core of the gelled jet behaves as a weakly cross-linked purely elastic solid with a very low elastic modulus, μ, and very large elastic strain, λ ($\gg 1$), a slightly inhomogeneous, quasi-one-dimensional approach similar to the case of homogeneous extension (e.g., see [6,8]) is used in Region 1. Using Equation 15.8 yields

$$\lambda = \lambda_0 \left(\frac{r_{0g}}{r_g}\right)^2 = \lambda_0 \left[\frac{(1-\xi)r_0}{(r-\xi r_0)^2}\right] \qquad (15.9)$$

$$\sigma \approx \left(\frac{\mu}{n}\right)\lambda^n - \frac{\gamma}{r} \quad (n > 1) \qquad (15.10)$$

where,

λ and σ are stretch ratio and stress, respectively

n is a numerical parameter characterizing a specific elastic potential [8]

λ_0 is an extensional stretch ratio attributed to the liquid–solid transition in Region 2

The noninertial momentum and mass balance equations, averaged over the jet cross section, can be written in a form similar to those used in previous studies [6,8]:

$$\frac{d}{dz}\left(\sigma S_g - 2\gamma\sqrt{\pi S}\right) = \rho g S \quad \left(S = \pi r^2, S_g = \pi r_g^2\right) \qquad (15.11)$$

$$uS = u_g S_g + u_f S_f = q(t) \quad \left(S_f = S - S_g\right) \qquad (15.12)$$

where

S, S_g and S_f are the total cross-sectional area and the cross-sectional area occupied by the gel and the solvent film, respectively

r is the radius of the jet

u, u_g and u_f are the total gel and solvent film vertical velocities, respectively

Note that if $u_g \approx u$, then $u_f \approx u$. In this case, the solvent film is drawn upward with the same speed as the gel.

Substituting Equations 15.9 and 15.10 into Equation 15.11 yields the following solution of the stress–strain problem (15.4) through (15.6) described by the two-phase model:

$$\sigma = G_0\left(\frac{r_0(1-\xi)}{r-\xi r_0}\right)^{2n} - \frac{\gamma}{r},$$

$$z - R = \frac{n-1}{n}\cdot\frac{G_0}{\rho g}\left[\left(\frac{r_0(1-\xi)}{r-\xi r_0}\right)^{2n} - 1\right] - \frac{\gamma}{\rho g}\left(\frac{1}{r} - \frac{1}{r_0}\right)$$

$$\left(G_0 = \sigma_0 + \gamma/r_0\right) \qquad (15.13)$$

Equations 15.13 in the limit $\xi \to 0$ have the same form as in References 6 and 8. The parameters r_0, S_0, σ_0, G_0 in Equation 15.8 are slowly changing functions of time. They represent the boundary values of the respective variables at $z = R$. These boundary values should be determined by matching the behavior of the liquid in Regions 1 and 2. Also the function $\xi(q_c/q_0)$ is a slowly changing function of time. As soon as these values are determined, the jet profile and the stress distribution along the jet in the region ($R < z < L(t)$) are determined for any time from Equation 15.13.

In the case of long jets with such large values of z that the surface tension effects on the stress are negligible as compared to the gravity, the solution of Equation 15.13 of the withdrawal problem has the asymptotic form found in previous papers [6,8]:

$$\sigma \approx \rho g\left(\frac{n}{n-1}z + \frac{m(c-1)}{c}R\right)\left(c = \frac{m(n-1)}{n}\right) \qquad (15.14)$$

Here, the numerical parameter c is again expressed via α and β using Equation 15.12. When the first term in Equation 15.14 dominates, the simplified expression will be used:

$$\sigma \approx \frac{\rho g z n}{n-1} \quad (z \gg R) \qquad (15.15)$$

The earlier Equations 15.9 through 15.15 have been obtained in Reference 8 for the problem of stationary withdrawal in the limit $\xi \to 0$ using the noninertial approach. In Reference 8, the characteristic sizes, R and r_0, of the meniscus, as well as the flow rate, q, have constant values. In the non-steady-state case of jet withdrawal under study, these equations remain valid because of a slow, noninertial approach, although the basic kinematic variables of the process, the length of withdrawn jet $L(t)$, and the flow rate, q, are now functions of time t. To determine the functions $L(t)$ and $q(t)$, two additional physical conditions are required.

One kinematic condition evident from Figures 15.1 and 15.3 is

$$q(t) = A \cdot \frac{dL}{dt} \qquad (15.16)$$

Here, $A = \pi r_j^2$ is the cross-sectional area of the container. Equation 15.16 shows that the change in length of the withdrawn jet is caused by the decrease of the liquid level in the container.

The second kinematic condition describes the dependence of flow rate on the pressure drop for the liquid flow in the capillary. If there was a gel–fluid transition in the capillary, this dependence would be described by the well-known linear Poiseuille equation. However, this gel–fluid transition was never observed in the capillary. Instead, a very complicated two-phase flow, as shown in Figure 15.4, occurs. It seems that at the highest vacuum level, the jet continues to be extended to the end of the capillary. Above the capillary, the jet breaks down to a two-phase liquid–air mixture in the adjacent tube. At lower levels of the vacuum, this breaking process occurs in the capillary. Because of the relative slowness of the withdrawal process, this complicated viscous flow can still be described by a linear hydraulic relationship between the pressure drop and the flow rate:

$$q \approx k(\sigma_v - \sigma_L)(k = const), \quad \sigma_v = p_a - p_v. \qquad (15.17)$$

Here,

σ_v is the stress due to vacuum

p_v and p_a are the absolute vacuum and atmospheric pressures, respectively

σ_L is the acting elastic stress in the jet at $z = L(t)$

The constant k of dimensions $cm^3/(Pa \cdot s)$ describes the hydraulic resistance of the jet at the end of the capillary. It should be evaluated by comparison of the theory with experimental data.

Determining σ_L from the asymptotic Equation 15.15 at $z = L(t)$ and substituting it along with Equation 15.17 into Equation 15.16 yield the kinetic equation describing the time evolution of $L(t)$:

$$\frac{dL}{dt} + sL = sL_u \qquad (15.18)$$

Here, s is a parameter of dimensionality of $1/s$, L_u is the ultimate length of the whole jet achievable with a given vacuum; these parameters being described as follows:

$$s = \frac{n}{n-1} \cdot \frac{\rho g k}{A}, \quad L_u = \frac{n-1}{n} \cdot \frac{\sigma_v}{\rho g}. \qquad (15.19)$$

Beginning with a time t_* where the asymptotic Equation 15.15 is valid, the solution of Equation 15.18 is presented as follows:

$$L(t) = L_u \cdot \{1 - \exp[-s(t-t_*)]\} \quad (t \geq t_*) \qquad (15.20)$$

Finally, Equations 15.16 and 15.20 yield the asymptotic expression for the flow rate:

$$q = s \cdot L_u \cdot \exp[-s(t-t_*)] \quad (t \geq t_*) \qquad (15.21)$$

Equations 15.19 and 15.21 will be used in the next section for the evaluation of parameters s and elastic constant of the gel, n. These relationships are reliable only after a certain time t_* from the beginning of the withdrawal process when Equation 15.15 is valid.

To describe the process from its beginning, one should employ numerical calculations using Equations 15.3 through 15.5, 15.13, 15.16, and 15.17 where $z = L(t)$ and $S = S(l(t)) \equiv S_l(t)$. Although these calculations can be performed relatively easily, the problem is that the beginning of withdrawal is accompanied by a poorly understood two-phase motion of the jet in the capillary.

15.5 EXPERIMENTAL RESULTS WITH A 0.025% PIB SOLUTION: COMPARISON WITH THE THEORY

The photographs in Figure 15.2a and b show horizontal ripples on the jet that may be explained by secondary instability of the solvent film exuded out of the solution. This instability is well known in the case of liquid films flowing down inclined surfaces [10] and indirectly confirms the exudation of the solvent from the withdrawn jet.

With increasing withdrawal time, the radius of the jet dramatically decreases. The amount of liquid removed from the container was measured every 5 seconds. Using these data, the time dependence of flow rate q (cm^3/s) was determined for three applied vacuum levels as shown in Figure 15.5.

At the beginning of withdrawal, the flow rate q substantially increases with time and reaches a maximum at about

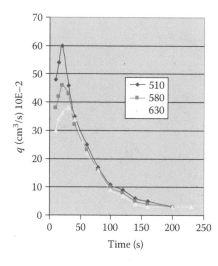

FIGURE 15.5 Time dependence of flow rate for three vacuum levels in the case of a 0.025% PIB solution in lubricating oil.

25 seconds. The higher the vacuum, the higher the q and the earlier the maximum is achieved. A possible explanation of this effect is that the initially short jet is under the action of surface tension and extension from the applied vacuum in the capillary. The action of the surface tension compresses the jet causing an increase in the flow rate. With increasing jet length, gravity becomes more important and eventually overcomes the surface tension effect, causing a decrease in flow rate.

Figure 15.6 demonstrates that the flow rate is proportional to the rate of change in the jet length given by $dl/dt(= dL/dt)$. This is direct confirmation of the evident kinematic relation given by Equation 15.16 in the case of the highest vacuum level.

After passing through the maximum, the flow rate exponentially decreases. This effect, predicted by Equation 15.21, is illustrated in Figure 15.7, where the time dependence of $\ln q$ is presented by a straight line with a slope of about -0.02 (1/s). As shown in Figure 15.7, a decrease in applied (constant)

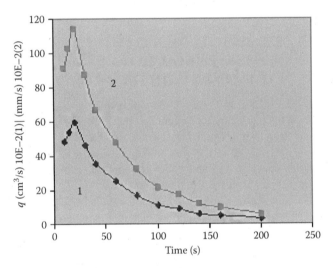

FIGURE 15.6 Comparison of the time dependence of withdrawal rate $i(t)$ (curve 1) and flow rate (curve 2) with vacuum value $\sigma_v = 32$ kPa for a 0.025% PIB solution in lubricating oil.

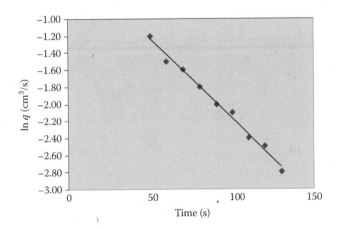

FIGURE 15.7 The decreasing time dependence of flow rate shown in Figure 15.5, represented in semilogarithmic coordinates.

vacuum level also causes a decrease in flow rate; after passing through the maxima, the differences between curves with different vacuum levels are negligible. Thus, the data presented in Figure 15.7 illustrate the change in average flow rate with time for three vacuum levels.

Another important observation is that, independent of vacuum level, all jets break at a flow rate of approximately 0.03 cm^3/s. The very existence of a lower critical value of flow rate is predicted by the fourth formula in Equation 15.6. This fact will be utilized in the modeling that follows.

Using the data presented in Figure 15.5, four data points from the decreasing portion of the flow rate function, $q(t)$, $q_1 = 0.35$, $q_2 = 0.25$, $q_3 = 0.20$, and $q_4 = 0.14$ cm^3/s, corresponding to various times t_k. Using photographs taken at these t_k, the jet profiles corresponding to the flow rates q_k related to these instances t_k were measured. In this region of flow rates, the measured jet profiles were well reproduced. Figure 15.8 presents these jet profiles as decreasing dependences of the jet radius r versus distance z from the liquid surface, up to the maximum visible jet length at those time instances t_k. In order to describe the jet profiles shown in Figure 15.8, the function $\xi(q_c/q(t))$ describing the thickness of exuded solvent in Equation 15.13, three fitting parameters, numerical parameters ν related to the meniscus Equations 15.3 through 15.5, the numerical parameter n describing the gel elastic potential given by Equation 15.10, and parameter κ describing the stress–flow rate hydraulic relation of Equation 15.17 were determined.

The parameters of the meniscus were evaluated starting with the highest value of flow rate, $q_1 = 0.35$ cm^3/s, and following the meniscus approximation procedure previously outlined, it was found that $r_{01} \approx 0.49$ mm corresponds to the maximum flow rate q_1. All other values r_{0k} corresponding to different values of q_k have been calculated using the scaling Equation 15.3 as $r_{0k} = r_{01}(q_k/q_1)^{1/3}$. As shown in Figure 15.8, the calculated data presented in Table 15.1 provide a good approximation of the experimental values. Then, utilizing the fact that the lower critical value of flow rate is $q_{0c} \approx 0.03$ cm^3/s, the lower critical value of r_{0c} as $r_{0c} = r_{01}(q_1/q_c)^{1/3} = 0.216$ mm was calculated. Using this value in Equation 15.6 for r_{0c} and Equation 15.5 for function $m(\nu)$ yields the equation: $\nu m(\nu) \equiv \nu(0.0959\nu^2 + 0.429\nu + 1) \approx 137$. The solution of this equation is $\nu \approx 9.7$, and therefore, $m \approx 14$. Using these data, the values for large meniscus radii $R = \nu r_0$, meniscus stress σ_0 given by Equation 15.4, and the values of parameter G_0 in Equation 15.13 were calculated. These variables for the earlier four profiles are presented in Table 15.1.

The elastic potential for the highly swollen gel is unknown. It is characterized by the parameter n that is determined by a fitting procedure as follows: Consider Equation 15.19 for the total ultimate length L_u of the jet. Using the values of vacuum pressure reported in Section 15.2, the corresponding values of vacuum stress σ_v are 32, 22.7, and 16 kPa, respectively. Consider the maximum value of $\sigma_v = 32$ kPa. Experimental results show that for this vacuum level, the visible maximum length of withdrawn

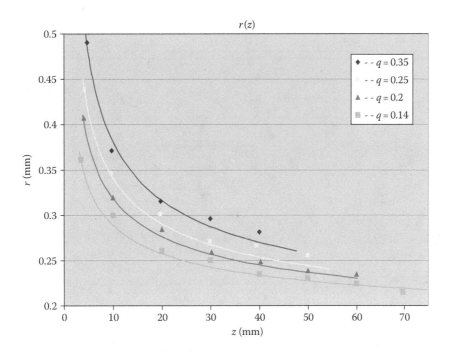

FIGURE 15.8 Jet profiles for several values of flow rate established by photographing a 0.025% PIB solution in lubricating oil. Symbols represent experimental data and solid lines indicate model calculations.

TABLE 15.1
The Values of Meniscus Variables in the Experimental Profiles (Figure 15.8)

q cm³/s	q_c 0.03	0.14	0.20	0.25	0.35
r_0 mm	0.216	0.361	0.407	0.438	0.490
R mm	2.10	3.50	3.94	4.25	4.75
$\sigma_0 \times 10^{-2}$ Pa	4.97	5.63	5.98	6.25	6.71
$G_0 \times 10^{-2}$ Pa	6.22	6.38	6.65	6.87	7.26

jet $l_u \approx 10$ cm, so the ultimate total jet value $L_u \approx 22$ cm. Thus, from Equation 15.19 for L_u, $n \approx 1.062$.

As the strain-induced exudation kinetics of solvent out of gel are unknown, it is necessary to simply parameterize the function $\xi(q_c/q(t))$ as a power relation $\xi = (q_c/q)^a$. Fitting the jet profiles in Figure 15.8 with Equation 15.13 results in a value of $a \approx 1/2$, that is, the thickness of the solvent film exuded out of the gel under extension is described by

$$h = \xi \cdot r_0 = r_0 \sqrt{\frac{q_c}{q}} \qquad (15.22)$$

The simplicity of this result could indicate a fundamental physical relationship of strain-induced solvent exudation out of cross-linked swollen gel; however, this is beyond the scope of this chapter.

Using the previous values for the meniscus parameters and parameter n along with Equation 15.22, jet profiles were calculated according to Equation 15.13. Figure 15.8

shows good agreement between the calculations and experimental data.

Modeling of the kinetics of long jet withdrawal, asymptotically described by Equations 15.16 through 15.20, is considered. Figure 15.6, as discussed earlier, verifies the kinematic Equation 15.16 for the case of higher vacuum stresses. Finally, the parameter, k, introduced in the hydraulic relationship given by Equation 15.17, is evaluated using Equation 15.18 and value $s = 0.02$ s⁻¹ as $k \approx 2.71 \times 10^{-5}$ cm³/(Pa · s). Figure 15.5 and the use of Equation 15.20 and obtained values for s and L_u allow calculation of the time $t_* \approx 32$ s, after which the flow rate decays exponentially with time.

The relaxation time, θ, could not be directly determined from the earlier experimental data. Using these data, the value can be calculated as follows:

$$\beta\theta^{1/3} \approx 6.95 \times 10^{-2} (s)^{1/3} \qquad (15.23)$$

Direct measurement of θ in very dilute polymer solutions is complex. One possible method is to use the formula roughly

TABLE 15.2

The Values of Basic Constants in Equations 15.21 Obtained Using a Fitting Procedure

β	ν	$\alpha = \beta\nu$	$m(\nu)$	n	k cm³/(Pa · s)	θ s
0.215	9.7	2.08	14	1062	2.71×10^{-5}	3.38×10^{-2}

describing the longest relaxation time in the Rouse model (e.g., see [3], p. 222):

$$\theta_R = \frac{[\eta]_0 \, \eta_s M}{N_A k_B T} \qquad (15.24)$$

where

[η]₀ is the intrinsic viscosity with a value of ~4 known for lubricating liquids [11]
η_s is the solvent viscosity
M is the polymer molecular weight
N_A is the Avogadro number
k_B is the Boltzmann constant
T is the absolute temperature

Calculations preformed for conditions at room temperature for a 0.025% PIB solution in lubricating oil yield the value $\theta_R \approx 4.54 \times 10^{-4}$ s.

Another method of evaluating the relaxation time is hypothesizing that near the sol–gel transition, the polymer concentration in the jet increases strongly resulting in cooperative relaxation effects. Thus, the transition may happen under the condition $We_c = const$ with a universal value of We_c. Roughly estimating the critical value $\dot{\varepsilon}_c$ of strain rate in the transition as $\dot{\varepsilon}_c = q / \left(\pi r_0^2 R\right)$ results in the relationship:

$$We_c = \theta_c \cdot \dot{\varepsilon}_c \approx \frac{\theta_c q}{\pi r_0^2 R} = \frac{1}{\pi \nu \beta^3} \qquad (15.25)$$

Assuming a universal value for We_c, the data in papers [7,8] can be utilized, where, in the case of a 0.5% PEO water solution, $\nu_{PEO} = 2.6$, $\beta_{PEO} = 0.334$, and $We_c \approx 3.3$. Then the value of β, in this case where $\nu = 9.7$, may be calculated as $\beta = (\pi\nu We_c)^{-1/3} \approx 0.215$. Substituting this value of β in Equation 15.22 yields $\theta_c \approx 3.38 \times 10^{-2}$ s. The calculated value of θ_c is almost two orders of magnitude higher than θ_R. In this case, the value θ_c and related value of parameter β seem preferable. The values of basic constants in Equation 15.21, obtained using a fitting procedure, are presented in Table 15.2.

15.6 USING THE OPEN-SIPHON METHOD FOR EVALUATION OF TACKINESS OF LUBRICATING FLUIDS

The ultimate jet length (or tackiness) strongly depends on the viscosity of oil, molecular weight of dissolved polymer and its concentration in solution. Figure 15.9 shows the dependence of jet length, l, on the concentration of PIB with $M_\eta \approx 2,000,000$ in two paraffinic oils with respective viscosities 0.068 and 0.022 Pa · s at 40°C.

One can see that the dependence of tackiness on polymer concentration in different oils is linear. These dependences are in fact linear for any molecular weight polymer in any oil. It is clear from Figure 15.9 that decreasing oil viscosity is accompanied by a large decrease in tackiness. For example, at the concentration 0.025% of PIB, the jet length in oil with viscosity 0.068 Pa · s is equal to 100 mm, whereas in oil with viscosity 0.022 Pa · s, it is equal to 20 mm, that is, five times less than in the first case.

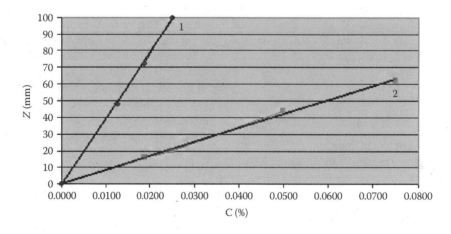

FIGURE 15.9 Ultimate jet length, l, versus the concentration of PIB with $M_\eta \approx 2,100,000$ in two paraffin oils with respective viscosities 0.068 and 0.022 Pa · s at 40°C.

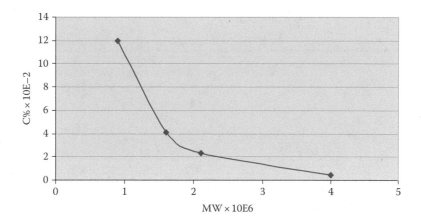

FIGURE 15.10 Concentration C% of PIB in oil with viscosity 0.068 Pa · s corresponding to the jet length $l = 100$ mm versus viscosity average molecular weight M_η of polymer.

Data presented in Figure 15.10 demonstrate what the concentration of PIB with different molecular weights in oil with viscosity 0.068 Pa · s should be used to reach a jet length of 100 mm. It is possible to substantially increase the jet length or tackiness simply by increasing the polymer molecular weight. However, increasing the molecular weight of a polymer causes a decrease in both thermal/oxidative stability and shear stability.

Combining the data presented in Figures 15.9 and 15.10, it is possible to evaluate the jet length or tackiness of PIB solutions with different molecular weights and at different polymer concentrations.

Unlike PIBs, which are commonly used as tackifiers, OCPs usually do not display tackiness. Nevertheless, we obtained some unusual data for a blend of PIB with $M_\eta \approx 2,000,000$ and very small amount of added OCP. Figure 15.11 demonstrates that adding 0.01% of the OCP to the PIB solution increases the jet length by about 30%. It should also be mentioned that addition of 0.01% of copolymer to the solution of 0.025% of PIB does not significantly change the viscosity of the solution. As shown in Figure 15.11, a higher OCP concentration in the PIB solution results in a decrease in tackiness.

This effect may be explained as follows. Solutions that contain 0.01% OCP could be considered as very dilute, that is, the macromolecules are well separated. During flow-induced orientation of long flexible PIB chains, much shorter and more rigid OCP molecules are involved in the process of orientation and support orientation of the PIB macromolecules, causing an increase in tackiness. However, with an increase in the concentration of OCP, the rigid macromolecules will form ensembles that cannot be involved in the orientation process and moreover may restrict the orientation of the PIB macromolecules and decrease tackiness.

15.7 CONCLUSION

This chapter applies the open-siphon method for the evaluation of tackiness of several lubricating oils. In these experiments, an evacuated capillary tube withdraws a vertical free jet of liquid from a container with a free surface. Dilute solutions of PIB of different molecular weights and polymer concentrations in lubricating fluids, as well as the blends of PIB with OCP, were used in these experiments. Time dependence of the length and shape of the free jet and the flow rate in the process were recorded in the experiments for several values of vacuum pressure in a capillary tube and for several lubricating fluids. The tackiness of lubricating fluids was quantified by the ultimate length of the free jet. Several specific phenomena were observed in the experiments such as solvent exudation out of the extended jet, a maximum in the time dependence of the flow rate during the process, and two-phase flow in the capillary under certain conditions.

The theoretical model developed in this chapter is similar to that used in paper [8], where the withdrawn jet is treated as a slightly cross-linked elastic gel. The previous model [8] was modified by including two new features. First, the exudation of the swollen gel under extension, a common effect for dilute polymer solutions and observed in our experiments, was accounted for. Second, a non-steady-state extension of an earlier stationary theory [8] was used. Fitting the theory with experimental data allowed for interpretation and validation of the theory for the case of lubricating oils.

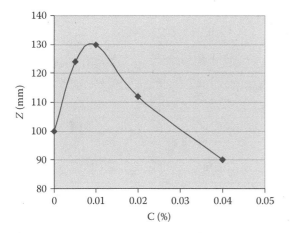

FIGURE 15.11 Tackiness effect versus concentration of OCP added to the 0.025% PIB solution in lubricating oil.

The results of the chapter clearly demonstrate that the evaluation of tackiness using the open-siphon technique presents a simple and reliable method useful in lubricant industry applications.

ACKNOWLEDGMENT

The authors thank Functional Products Inc. for initiating and supporting this work.

REFERENCES

1. H.G. Elias, ed., *Macromolecules 1 & 2*, Plenum Press, New York (1984).
2. L.R.G. Treloar. *Physics of Rubber Elasticity*, 3rd edn., Clarendon Press, Oxford, U.K. (1975).
3. R.G. Larson, *Constitutive Equations for Polymer Melts and Solutions*, Butterworth, Boston, MA (1988).
4. G. Astarita and L. Nicodemo, Extensional flow behavior of polymer solutions, *Chem. Eng. J.*, **1**, 57–61 (1970).
5. D.V. Boger and K. Walters, *Rheological Phenomena in Focus*, Volume 4, Elsevier, New York (1993).
6. A.I. Leonov and A.N. Prokunin, *Nonlinear Phenomena in Flows of Viscoelastic Polymer Fluids*, Chapman and Hall, New York (1994), pp. 106–108, 338–342.
7. A.I. Leonov and A.N. Prokunin, On spinnability in viscoelastic liquids, *Trans. Acad. Sci. USSR, Fluid Gas Mech.*, **4**, 24–33 (1973).
8. A.N. Prokunin, A model of elastic deformation for the description of withdrawal of polymer solutions, *Rheol. Acta*, **22**, 374–379 (1984).
9. L.D. Landau and V.G. Levich, Entrainment of fluid by the driven plate, *Acta Phys. Chem. URSS* (Russian) **17**, 42 (1942).
10. V.G. Levich, *Physico-Chemical Hydrodynamics* (Russian), Fizmatgiz, Moscow, Russia (1959).
11. Exxon Corporation, *Vistanex, Properties and Applications*, Exxon Corporation, Irving, TX (1993).
12. R.J. Irvin, ed., *Environmental Contaminate Encyclopedia, Mineral Oils, General Entry*, National Park Service, Water Resources Divisions, Water Operations Branch, Fort Collins, CO (July 1, 1997).

16 Seal Swell Additives

Ronald E. Zielinski and Christa M.A. Chilson

CONTENTS

16.1 INTRODUCTION

The use of additives in lubricants, both military and civilian, began around 1930. Prior to that time, additives in lubricants were not required since most fluid systems operated at low-pressure and low cycling rates. Seal swell agents were not required since seals were generally made from leather or similar materials not subject to swelling by the petroleum-based fluids in use at that time. The use of additives to improve fluid performance began with the military fluids. Industrial fluid development paralleled or followed the military fluid development.

With the enhanced development of military aircraft and reciprocating engines spurred by World War II starting in approximately 1935, advanced fighters and bombers required new, high-performance lubricating fluids and seals. The new military fluids were based largely on petroleum base stocks or mineral oils for hydraulic systems. These consisted of paraffinic, aromatic, and alicyclic (naphthenic) components. Various additives including rust inhibitors, oxidation inhibitors, detergents/dispersants, viscosity index improvers (VI), pour point depressants, antiwear components, and antifoam materials were used to develop additive packages for these hydraulic fluids depending on aircraft type and intended missions. These development efforts resulted in the creation of a U.S. military specification, MIL-O-5606, for these fluids. The specification allowed for some flexibility in formulation for competitive purposes and guaranteed a performance band for the fluid and its compatibility with elastomer seals.

Similarly, new reciprocating engine lubricants, based primarily on diesters of both aromatic and aliphatic acids plus additive packages for fluids intended for each aircraft engine type, that is, high-flying bombers (B-17s) and fighters (P-47s), were developed. These were the early versions of what has now become MIL-PRF-7808 (formerly MIL-L-7808).

High, acrylonitrile content acrylonitrile–butadiene rubber (NBR) seal materials were used for these aircraft hydraulics and worked fine except when strategic aircraft flew for long time periods at very high altitudes (i.e., >30,000 ft), then the seals leaked, and thus, the base polymer had to be modified with a lower acrylonitrile content to improve low-temperature performance. Seals made from NBR with reduced acrylonitrile content have higher butadiene content and because of this have increased volume swell in petroleum-based fluids. This caused many problems. Programs driven by the military had material and chemical engineers working together and adjusting both fluid and seal material formulations to solve the problems. The results of these coordinated efforts were the development of fluids and seal materials that were complementary.

It is very difficult to speak in specific terms with regard to seal swell additives, since commercial fluids have additive packages that are considered very proprietary and highly confidential. In order to address seal swell additives, we will follow the development of U.S. military fluids and seal materials in four basic time frames: 1935–1960, 1961–1980, 1981–2000, and 2001–2015. In general, specific information of compositions of most aircraft fluids, including seal swell additives, is extremely limited and is usually, again, company confidential. However, because these developments were at least in part supported by the military, more information is available. Here, we describe what is known about swelling agents yet recognize that generalities will have to suffice due to the confidential/proprietary nature of formulated fluids. Industrial fluid development, including seal swell additives, followed the military fluid development. The additives used in military fluids were used in the industrial fluids, both used essentially the same seal swell additives. Therefore, by following the U.S. military fluids through time frames from 1935 to 2015, we are also following the industrial fluids.

16.2 FLUID/SEAL SWELL ADDITIVES DURING 1935–1960

During this period, petroleum-based fluids predominated as the fluid of choice for hydraulic fluids. The earliest versions of the military fluid, designated MIL-O-5606, which became MIL-H-5606 in 1957 (mineral oil), had a higher aromatic content and thus provided sufficient volume swell for the high-acrylonitrile (ACN) NBR seal materials used in the low-pressure (approximately 1500 psi) hydraulic systems in piston engine aircraft. As aircraft development proceeded, more powerful piston engines were developed for the very long-range bomber missions, primarily the B-17, to be used by the U.S. Army and U.S. Air Force for attacks on enemies in Europe and the Far East theaters. The long-range bombers flying at very high altitudes, 35,000 to 45,000 ft, caused oil thickening and seal leakage that resulted in sluggish performance. The seals used were of nitrile material designated by the military as MIL-P-25732. Fighter aircraft, both land-based and aircraft carrier-based, did not experience problems because they flew at lower altitudes and for shorter missions. To solve the bomber fluid problem, the mineral oil base fluid was further refined to replace some aromatic content with aliphatic content to improve low-temperature viscosity at −65°F. This resulted in a revised version of MIL-H-5606 hydraulic fluid. Low-temperature sealing performance of the high-ACN NBR content MIL-P-25732 seal material was improved by replacing some of the high-ACN NBR with medium or low ACN NBR as the base polymer and/or replacing part of di-2-ethylhexylphthalate (DOP) processing oil used in seal compounding formulations with di-2-ethylhexylsebacate (DOS) oil. Fluid formulators added sufficient diester fluid, such as DOS, to MIL-H-5606 to maintain 18%–30% seal volume swell. Although very early piston engines used petroleum-based oils as lubricants, this gave way to pentaerythritol ester-based fluids to improve low-temperature performance to −65°F and maintain high-temperature performance at 275°F.

Seals made out of MIL-P-25732 materials were used for engine lube sealing. For naval aircraft, antirust additives were added to fluid formulations for obvious reasons. Commercial aircraft versions of military aircraft during this era were developed, that is, Douglas DC-3 (C-47 military transport), Douglas DC-6, and Lockheed Constellation. These aircraft initially used military aircraft hydraulic oils/seals and engine lubrication oils/seals. However, because of their relatively short flight duration and many takeoffs and landings, a new, more fire-resistant fluid was introduced. This new fluid had a phosphate ester base stock, which created the need for seals made with a polymer compatible with this new base stock.

16.3 FLUID/SEAL SWELL ADDITIVES DURING 1961–1980

As advanced military aircraft with enhanced performance (jet aircraft) and missions were developed, both hydraulic system and engine operating temperatures increased. Oil companies and seal manufacturers responded by enhanced synthetic oil development and corresponding seal material development to meet aircraft and gas turbine engine manufacturers needs for −65°F to 350°F operational performance. Oil companies synthesized thermally stable poly-α-olefin (PAO) fluids to meet the desired 350°F high-temperature fluid requirement; these were formulated with various additives such as antioxidants, lubricity additives, antifoams, metal deactivators, rust inhibitors, viscosity index improvers, and seal swell materials to meet 18%–30% seal volume swell. The first synthetic military hydraulic fluid developed was less flammable than MIL-H-5606 and was designated as MIL-H-83282. MIL-H-83282 had an operational range of −40°F to 350°F. Seal materials based on sulfur-donor or peroxide-cured NBRs were developed for this fluid and to meet the −65°F to 350°F operational requirements. The specification for the new seal material was designated as MIL-P-83461.

MIL-H-83282 fluid contained a significant amount of diesters to achieve desired seal volume swell. This fluid found almost immediate use in most military aircraft with the exception of the U.S. Air Force strategic aircraft (i.e., B-52, B-1B, KC-135, and KC-10). The U.S. Air Force strategic aircraft, which required good −65°F viscosity, did not use this fluid because it became very viscous below −40°F. This low-temperature shortcoming was later resolved by the development of a PAO-based fluid designated MIL-H-87257. This low-temperature PAO fluid was achieved by the synthesis of lower-molecular-weight PAOs and replacing part of the higher-molecular-weight fractions present in MIL-H-83282. This approach caused a small loss in nonflammability capability but met the desired capability of a universal less-flammable hydraulic fluid for all U.S. military aircraft and with a −65°F to 350°F operational range capability.

During this time frame, engine lubrication fluid requirements were also changed to −65°F to 350°F, which led to the development of diesters of both adipic and sebacic acids, especially dioctyladipate (DOA) and DOS, with the latter predominating. Neopentyl polyol esters (POEs) were also being investigated for elevated thermal stability. POEs lack hydrogen on their beta carbon, the first site of thermal attack on diesters, which gives POE increased thermal properties over DOA and DOS and allows POE to be used at higher temperatures. Eventually, POE was formulated to later versions of MIL-L-7808 for both military and commercial aircraft. During this time, seal materials were also formulated with fluorocarbon (FKM) polymers to meet a −40°F to 350°F operational requirement for these fluids. The specification for this new seal material was designated by the military as MIL-R-83485. Seal swell additives were not needed for the MIL-L-7808 lubrication fluids, since both NBR and FKM seals exhibited swelling in these fluids.

During the 1961–1980 time frame, commercial aircraft were now using the nonflammable phosphate ester fluids. Ethylene–propylene (EPM or EP)- or ethylene propylene diene (EPDM)–based elastomeric seals were developed for these fluids. The specification for the new seal material was designated as NAS1613. Since the phosphate ester base stock acts as a plasticizer, there was no need for seal swell additives in these fluids.

16.4 FLUID/SEAL SWELL ADDITIVES DURING 1981–2000

As mentioned previously, MIL-H-83282, which became MIL-PRF-83282 in 1997, a PAO-based hydraulic fluid, was developed during the 1970s and during the 1980s, with implementations in most nonstrategic military (i.e., F-16, F-14, and F-18 fighters) aircraft beginning in 1982 after extensive flight testing. The U.S. Air Force's strategic aircraft, long-range/high-altitude bombers, tankers, and reconnaissance aircraft, continued to use MIL-H-5606 to accommodate its need for a hydraulic fluid with a −65°F to 275°F capability.

The desire for a less-flammable fluid with better low-temperature viscosity led to the development of MIL-H-87257, which became MIL-PRF-87257 in 1997. This fluid balanced the PAO content with the proper molecular weight range of PAO components. With the proper additive package, including sufficient quantities of diesters (i.e., DOS) to achieve seal swell of 18% to 30%, dynamic cycling testing indicated that this new fluid had the desired capability of −65°F to 350°F as well as compatibility with both MIL-R-83485/AMS-R-83485 FKM and MIL-P-83461 NBR seal materials. Flight testing was conducted on all military aircraft, including strategic aircraft, and it appeared likely that MIL-PRF-87257 (formerly MIL-H-87257) should eventually be the universal hydraulic fluid (−65°F to 350°F) for military aircraft. In March 1996, the U.S. Department of Defense issued a notice for MIL-H-5606G (the current version of this specification at the time) stating that the specification was inactive for new design and is no longer used (for the U.S. military), except for replacement purposes.

Engine lubrication fluids were still primarily MIL-L-7808, which became MIL-PRF-7808 in 1997, and other DOS and POE type fluids for both military and commercial jet aircraft. Seal swell additives were not required since the ester fluids accomplish this purpose.

During the 1981–2000 time frame, commercial aircraft continued to use phosphate ester fluids for hydraulics with the previously mentioned NAS1613 EPM-/EPDM-based seal materials.

16.5 FLUID/SEAL SWELL ADDITIVES DURING 2001–2015

MIL-PRF-83282 (formerly MIL-H-83282) and MIL-PRF-87257 (formerly MIL-H-87257), PAO-based hydraulic fluids, have replaced the mineral oil–based MIL-H-5606, which became MIL-PRF-5606 in 2002, as the principal hydraulic fluids used in military aircraft. Older aircraft that had been using MIL-PRF-5606 have been gradually changing over to MIL-PRF-83282 and/or MIL-PRF-87257 by attrition. The newer advanced military fighter aircraft, namely, the F-22, F-35, and the latest model F-18, require either MIL-PRF-83282 or MIL-PFR-87257. Most U.S. military liquid-cooling systems now use MIL-PRF-87252. The seal materials for these systems are AMS-P-83461 (formerly MIL-P-83461) NBR and/or the higher-temperature capability AMS-R-83485 (formerly MIL-R-83485) FKM. During this time frame, AMS-R-83485 has been canceled and superseded by AMS3384 (published 2013) and AMS7278 (published 2012).

While MIL-PRF-83282 and/or MIL-PRF-87257 are currently required in the main systems of all fleet, aircraft previously using MIL-PRF-5606 are still used in some areas, mainly in landing gear shock strut applications, due to its better low-temperature (−65°F) characteristics; however, even in these applications, there is an ongoing trend to change to MIL-PRF-87257 fluid. The seals for MIL-PRF-5606 are molded out of MIL-P-25732 NBR.

MIL-PRF-7808-type fluids are still primarily used for military jet aircraft as they are designed to operate at lower-temperature environments. MIL-PRF-23699 (formerly MIL-L-23699) engine oils are also used by the military; however, the biggest users are commercial aircraft. During this time period, the trend has been for commercial engines to operate at higher temperatures and pressures then in past periods. The second-generation MIL-PRF-23699 oils were stressed in these operating conditions, and new third-generation engine oils with improved thermal stability, known as "HTS" oils, have been developed. These meet the high thermal stability (HTS) classification of MIL-PRF-23699. A new FKM seal material, AMS 7379, was developed for the HTS oils to work at low temperatures (−40°F). AMS7287 materials can also be used in the HTS fluids but do not have the lower-temperature capability as the AMS 7379 materials. AMS 7379 FKM can also be used with MIL-PRF-7808 fluids in addition to AMS3384. Like MIL-PRF-7808, seal swell materials are not needed for these new HTS oils; no new seal swell additives have been introduced for engine lubrication fluids during this period.

Even with the trend of newer commercial aircraft, namely, Airbus A380 and Boeing 787, to have higher hydraulic system pressures up to 5000 psi, commercial aircraft continue to use phosphate ester fluids with the previously mentioned EPM-/EPDM-based seal materials (NAS1613). Improvements in fire properties and other areas of performance continue to be made with these phosphate ester fluids; however, seal swell materials are still not required because the diester fluids accomplish this purpose.

16.6 TRENDS AND THE FUTURE

It is very likely that advanced military fighter aircraft will continue to have the requirement of hydraulic fluid with either a −40°F to 340°F operational range or −65°F to 392°F operational range with pressures ranging from 3000 to 4000 psi. As MIL-PRF-83282 meets the higher-temperature requirement and MIL-PRF-87257 meets the lower-temperature requirement, these should continue to be the primary hydraulic fluids for these aircraft. MIL-PRF-5606 will still be used in some areas requiring better low-temperature (−65°F) performance, but its use in military aircraft will continue to decline. MIL-PRF-87257 will continue to replace MIL-PRF-5606 in landing gear shock strut applications.

If future performance demands cause changes in operational temperature ranges (below −65°F and at/or above 400°F), new base hydraulic fluids and seal materials will be needed.

Newer polymers will continue to be investigated as potential seal materials for the current military systems since NBR elastomers oxidize when exposed to temperatures above 250°F for extended periods of time and FKM elastomers cannot achieve the low-temperature requirement of −65°F.

Strategic aircraft needs over the next 20 years will likely be met by the C-17 transport, the KC-10, or tanker versions of the A330 (A330MRTT) and the B767 (KC767); strategic bomber needs are likely to be met by the B-1B, B-2, and an advanced version of the B-2 stealth aircraft. While there could be some changing needs in both hydraulic and engine lubrication fluids, it is currently not possible to determine specific fluid/seal needs or seal swell agent needs.

Commercial aircraft requirements over the next 20–30 years are not likely to involve major changes to hydraulic fluids. Airframe designs and engine requirements will not change substantially but will likely feature incremental improvements in performance and, as always, be dependent on customer requirements. Therefore, a dramatic change is not expected in hydraulic or lubrication fluid/seal needs for commercial aircraft.

16.7 INDUSTRIAL FLUIDS

As stated earlier, industrial fluid development has closely followed the development of military fluids. Natural base oils including mineral oils have been used in industrial fluids. These fluids are not as controlled as military fluids, which have developed specifications to meet and essentially perform the same within a narrow band of additive variability, allowed by the specification for competitive bid requirements. Industrial fluids vary widely as a result of their crude source, for example, whether paraffinic, naphthenic, mixed, and paraffinic–naphthenic; as well as their formation, for example, distillation range, straight run or cracked, hydrorefined, and solvent extracted. Because these fluids vary widely, the additive packages are very proprietary, but again, the seal swell additives parallel those used in military fluids.

DOS, DOP, and POE are used in industrial fluids. Other seal swell additives include mineral oils with aliphatic alcohols such as tridecyl alcohol. Oil-soluble, saturated, aliphatic, or aromatic hydrocarbon esters such as dihexylphthalate are very popular. Trisphosphite ester in combination with a hydrocarbonyl-substituted phenol is also used as a seal swell additive. As more highly refined mineral oils and synthetics based upon PAO enter the market, it is not likely that there will be significant differences in seal swell additives.

16.8 SEAL SWELL DILEMMA

When a fluid is formulated, the usual concentration of seal swell additive is 0.6%–1.0% by volume. During development, the fluid manufacturer uses standard test elastomers to determine fluid–seal swell, which is normally targeted at 18%–30%. These tests are normally 70–168 hours in duration and do not allow for a full measure of fluid/seal compatibility evaluation. The fluid is then marketed as compatible with the elastomer seal. The dilemma is that the seal used in the actual system is much different than the elastomer used for seal swell determination, and swell beyond 20% is normally not allowed by elastomer specification or desired from a performance standpoint. Thus, we have fluid developed to produce high seal swell and an elastomer seal that is not allowed to swell beyond 20%. This dilemma occurs because fluid and seal development are no longer coordinated. In the past with the military driving parallel development programs, fluids and seals were simultaneously developed; this is not always true in today's commercial arena. The fluid manufacturer has been taught that high seal swell is good. It allows for more seal force and less compression set. The problem is that seals that swell excessively soften and completely fill the seal gland. These conditions for dynamic seals lead to nibbling of the seal as it overflows the gland and to excessive seal wear. This then leads to premature seal failure and leakage, which results in costly downtime while seals are replaced.

To put fluid developers on the same track as seal developers, ASTM Committee D02 developed ASTM D6546. This specification details a fluid–elastomer seal test program, which involves the testing of commercially available elastomers in fluid at actual service temperatures for long periods of time (1000 hours). These test conditions with recommended limits allow for a more complete evaluation of fluid–seal compatibility and should allow for optimized fluid formulations with a minimum amount of swell additive.

Changes to increase test times are also gradually being made in new fluid specifications as more importance is placed on reliability and long-term performance. The commercial aerospace industry is becoming the driving force behind newer and updated specifications, such as AS5780, with increased test times using commercially available fluids tested with commercially available seal materials. The military is slowly following.

16.9 COST-EFFECTIVENESS

It is important to minimize the amount of seal swell additive in fluid formulations. Seal swell additives increase the cost of mineral oil– and PAO-based hydraulic fluids. This is because the rather large amounts of DOS are expensive when compared to the cost of the base stocks, but it accomplishes the purpose of seal swell. It must be remembered that qualification testing of fluids and seals, not to mention quality control testing of both, is inherently expensive. This is why the implementation of ASTM D6546 should be a cost-effective, integral part of the fluid development process. The fluid developer can also use the results of his or her tests to instill confidence in the fluid users that they are provided with a fluid and recommendations for seals substantiated by long-term testing.

The long-term testing can also provide useful information to the fluid users, which can help them plan preventative maintenance programs to avoid or minimize costly downtimes.

Fluid stability tests are already performed for 1000 hours; these additional tests would be extremely cost-effective in optimizing the fluid formulation and addressing total system needs—fluid and seals.

16.10 MANUFACTURING

Manufacturing of both fluids and seals is an extremely competitive and very proprietary business. This has not changed over the years and is not likely to change in the foreseeable future. In general, the following is true:

1. The competitive nature of the two business types has controlled product costs.
2. The fluids with the reputation for long life with a minimum of problems are the most successful.

Fluid formulation plans that include testing to ASTM D6546 or similar specifications can be very cost-effective over the long run since fluids would be developed with optimum concentrations of expensive additives, and long service life with compatible seal materials would be demonstrated.

16.11 SUMMARY

Seal swell additives are necessary in many fluid formulations and can be beneficial in slowing compression set. However, excessive seal swell can be deleterious to seal life and system performance. Seal swell additives add cost to final fluid formulation, so minimizing the concentration of these additives is important from a financial standpoint. For both the fluid and seal developer, ASTM D6546 represents a method to ensure that the seal and the fluid with its additive package are fully compatible. Whatever the proprietary formulation is, the fluid developer can demonstrate seal compatibility with confidence over a long period of time. He or she can also tailor the amount of seal swell additive for maximum seal performance and minimum cost.

REFERENCES

The following are publications of the U.S. Department of Defense; 700 Robbins Ave., Philadelphia, PA 19111-5094; http://assist.dla.mil:

MIL-O-5606, Oil; hydraulic, aircraft, petroleum base, January 31, 1950.

MIL-H-5606A, Hydraulic fluid, petroleum base, aircraft and ordinance, February 21, 1957.

MIL-H-5606G, Hydraulic fluid, petroleum base; aircraft; Missile and ordinance—Notice 1, March 29, 1996.

MIL-PRF-5606H, Hydraulic fluid, petroleum base; Aircraft missile, and ordinance, June 7, 2002.

MIL-PRF-7808L, Lubricating oil, aircraft turbine engine, synthetic base, May 2, 1997.

MIL-L-23699E, Lubricating oil, aircraft turbine engine, synthetic base, NATO code number O-156, August 25, 1994.

MIL-PRF-23699F, Lubricating oil, aircraft turbine engine, synthetic base, NATO code number O-156, May 21, 1997.

MIL-PRF-23699G, Lubricating oil, aircraft turbine engine, synthetic base, NATO code numbers: O-152, O-154, O-156, and O-167, March 13, 2014

MIL-P-25732C Packing, preformed, petroleum hydraulic fluid resistant, limited service at 275°F (135°C), February 25, 1980.

MIL-P-25732C, Packing, preformed, petroleum hydraulic fluid resistant, limited service at 275°F (135°C)—Notice 1, November 15, 1989.

MIL-H-83282B, Hydraulic fluid, fire resistant, synthetic hydrocarbon base, metric, NATO code number H-537, February 10, 1982.

MIL-PRF-83282D, Hydraulic fluid, fire resistant, synthetic hydrocarbon base, metric, NATO code number H-537, September 30, 1997.

MIL-P-83461B, Packing, preformed, petroleum hydraulic fluid resistant, improved performance at 275°F (135°C), February 25, 1980.

MIL-R-83485 (USAF), Rubber, fluorocarbon elastomer, improved performance at low temperatures, September 8, 1976.

MIL-H-87257, Hydraulic fluid, fire resistant; Low temperature, synthetic hydrocarbon base, aircraft and missile, March 2, 1992.

MIL-PRF-87257 B, Hydraulic fluid, fire resistant; low temperature, synthetic hydrocarbon base, aircraft and missile, April 22, 2004.

The following are publications by SAE International, 400 Commonwealth Drive, Warrendale, PA 15096-0001 USA; www.sae.org:

AS5780, Core requirement specification for aircraft gas turbine engine lubricants, September 2000.

AS5780B, Specification for aero and aero-derived gas turbine engine lubricants, February 2013.

AMS3384, Rubber, fluorocarbon elastomer (FKM), 70 to 80 hardness, low temperature sealing Tg −22°F (−30°C), for elastomeric shapes or parts in gas turbine engine oil, fuel and hydraulic systems, August 2013.

AMS7287, Fluorocarbon elastomer (FKM) high temperature/HTS oil resistant/fuel resistant low compression set/70 to 80 hardness, low temperature Tg −22°F (−30°C) for seals in oil/fuel/specific hydraulic systems, August 2012.

AMS 7379, Rubber: Fluorocarbon elastomer (FKM) 70 to 80 hardness, low temperature sealing Tg −40°F (−40°C) for elastomeric seals in aircraft engine oil, fuel and hydraulics systems, July 2008.

AMS-P-83461, Packing, preformed, petroleum hydraulic fluid resistant, improved performance at 275°F (135°C), April 1998.

AMS-R-83485 Rev. A, Rubber, fluorocarbon elastomer, improved performance at low temperatures, May 1998.

The following are publications by ASTM International, West Conshohocken, PA, 2000, www.astm.org:

ASTM D6546-00, Standard test methods for and suggested limits for determining compatibility of elastomer seals for industrial hydraulic fluid applications, 2000.

ASTM D6546-15, Standard test methods for and suggested limits for determining compatibility of elastomer seals for industrial hydraulic fluid applications, 2015.

The following are publications by Aerospace Industries Association of America Inc., 1000 Wilson Boulevard, Suite 1700, Arlington, VA, 22209; www.aia-aerospace.org:

NAS1613, Packing, O-ring, phosphate ester resistant, July 31, 1969.

NAS 1613 Rev. 6, Packing, preformed, ethylene propylene rubber, November 2012.

17 Antimicrobial Additives for Metalworking Lubricants

Alan C. Eachus and William R. Schwingel

CONTENTS

17.1 INTRODUCTION

Metalworking fluids, the class of lubricants (along with water-containing hydraulic fluids) most susceptible to microbial attack, provide a number of vital functions in metal removal and forming operations. Designed to cool (to remove heat), to lubricate (to reduce heat generation), and to provide corrosion resistance, electrochemical resistance, extended resistance to microbial degradation, and eventual biodegradability in the environment, a metalworking fluid must also be safe for human use and exposure. The failure of a fluid to perform any one of these has the potential to result in operational problems, process shutdowns, decreased tool life, and product quality issues, all of which will result in increased costs.

Perhaps one of the most common and controllable complications is microbial degradation. Metalworking fluids contain the nutrients that can permit unchecked microbial growth, including mineral oil base stocks, glycols, fatty acid soaps and esters, amines, sulfonates, and other organic components. Use-diluted emulsions, synthetic chemical solution fluids, and aqueous fluid concentrates are all vulnerable because they contain water, an essential microbial growth requirement and the major source of microbiological contamination. Conversely, straight oils, while not immune to contamination problems, show greater overall resistance to microbial degradation simply because they do not intentionally contain water. To ensure acceptable long-term performance of any fluid, protection strategies for minimizing microbial degradation should be an integral part of the metalworking fluid system.

17.2 MICROBIOLOGICAL GROWTH

The two types of microorganisms that contaminate metalworking fluids, bacteria and fungi, often proliferate independently, but they can also coexist in a fluid. The types of organisms present, and their ability to coexist in the same fluid, depend on fluid composition and environmental conditions.

17.2.1 BACTERIA

Bacteria, single-cell organisms that lack the internal cell organelles found in higher forms of life, are classified as either Gram-positive or Gram-negative depending on their cell wall structure, and as aerobic or anaerobic according to their requirement for oxygen. Anaerobic bacteria cannot grow in the presence of oxygen; however, many strict anaerobes can tolerate very short exposure to oxygen before dying. Another group of bacteria, facultative anaerobes, is able to use metabolic pathways like those of aerobes when oxygen is present and then switch to anaerobic metabolism when oxygen becomes limited. The compounds produced from anaerobic pathways create the unpleasant odors associated with the degradation of fluid quality and performance. The compounds produced through anaerobic pathways may include organic acids, as well as poisonous and explosive gases such as H_2S, NH_3, and H_2.

Not all bacteria are able to survive in the environment of metalworking fluids, and their occurrence is often specific to certain fluid formulations. However, *Pseudomonas aeruginosa*, *Enterobacter* spp., *Escherichia coli*, *Klebsiella pneumoniae*, and *Desulfovibrio* spp. are commonly found in all types of fluids. Because these bacteria are the species most predominantly found in rivers, streams, lakes, and soil, it follows that makeup water is viewed as a significant source of bacterial contamination. Other contamination sources include air, raw materials, and animals, including humans. When bacteria are present, they are generally distributed throughout the metalworking fluid system and multiply rapidly when conditions are favorable.

17.2.2 FUNGI

Fungi may occur as either yeasts or molds. Yeasts, like bacteria, are unicellular and usually spherical in shape. Conversely, molds are composed of more than one cell and form complex mazes of filamentous hyphae with spore-bearing structures that give them their powdery, mat-like appearance. They are widespread in the environment and are commonly seen on decaying foods such as bread, fruit, and cheese. Fungi are often found in synthetic metalworking fluids but are usually found in lower concentrations and distributed less evenly than bacteria. Fungi may be found in the fluid itself (planktonic), but they also tend to grow on solid surfaces (sessile) such as splash areas and in sluices, troughs, filters, and pipes, providing an important constituent of biofilm. Although many different species of fungi

have been isolated from metalworking fluids, *Fusarium* sp., *Candida* sp., and *Cephalosporium* sp. occur most frequently.

17.2.3 BIOFILMS

Although microorganisms can exist in the free-flowing metalworking fluids in central systems, a large proportion of the microbial population exists within biofilms. Often composed of diverse populations of microorganisms and complex in structure, biofilms adhere to system surfaces. Both chemical and biological material can become trapped in the secretions of a biofilm's living cells. Biofilms, or "slime," can vary in thickness but usually range from a few millimeters to several centimeters. The environment within the biofilm is highly influenced by the microorganisms present and may be very different from the conditions in the fluid itself. Within a biofilm community, the constituent cells communicate through a variety of signaling chemicals, thus driving metabolic processes that cannot be achieved by its individual members. Because the microorganisms growing within a biofilm are protected from the conditions that may be affecting the bulk fluid, the ability of antimicrobial agents to effectively control their growth is severely limited. This explains why chemical control treatments shown to be effective in laboratory testing sometimes fail in field situations [1].

17.3 A MICROBE-FRIENDLY ENVIRONMENT

The primary factors affecting microbial growth are the availability of water and nutrients, system pH and temperature, and the presence of inhibitory chemicals [2]. Like all living organisms, bacteria and fungi require water for survival, which is present in all use-diluted emulsifiable, semisynthetic, and synthetic metalworking fluids, as well as in some concentrates. In regard to nutrients, metalworking fluids are capable of providing all the essential nutrients needed to support microbial growth [3]. These include carbon, nitrogen, phosphorus, sulfur, and other trace elements. Small quantities of inorganic salts are also crucial for microbiological growth, which means variances in water hardness and quality can influence microbial survival. And, while at least some level of inorganic salts must be present, very high salt concentrations may actually inhibit the growth of microbes commonly found in metalworking fluids.

In regard to pH, the ideal level for optimum bacterial growth is a neutral or slightly acidic pH of 6.5–7.5. Fungi prefer a slightly lower pH of about 4.5–5.0. However, both bacteria and fungi can survive in fluids with pH levels outside these ranges. A large contingent of microorganisms thrives in typical metalworking fluid pH ranges; however, species diversity decreases in fluids with pH levels of 8.5 and higher.

The optimum temperature range for mesophiles is 30°C–35°C, the bulk temperature range of most metalworking fluid systems. However, many organisms can exist at extreme temperatures; for example, bacteria have been found growing in both Arctic climates and at the mouths of geysers and hot springs.

By manipulating these various growth factors, microbial survival and proliferation can be controlled and even prevented.

Water quality and pH are the two factors most easily managed. However, the availability of nutrients can be regulated through the strategic selection of raw materials that are inherently more resistant to microbial degradation. Components such as borates, certain amines, and biocides can all inhibit microbial proliferation, as detailed later. While little can be done economically to control bulk fluid temperature, factors such as oxygen availability can, if not slow microbial growth, at least alter microbial populations. For instance, a shift from aerobic bacteria to facultative or obligate anaerobe populations is likely to occur if metalworking fluids are not aerated or circulated.

17.4   DESIGN, MAINTAIN, AND MONITOR

In the ongoing quest for microbial control in metalworking fluids, system design, proper housekeeping, fluid maintenance, and microbiological monitoring programs all play an essential role. Microorganisms are somewhat less likely to establish healthy film populations in areas of high uniform flow, so a well-designed system, one that eliminates or minimizes dead legs and stagnant zones, is useful. For example, anaerobic bacteria are particularly likely to establish populations where flow is minimal, since this often limits the diffusion of O_2 into the fluid.

Good, consistent housekeeping practices at metalworking fluid end-use sites can significantly reduce the occurrence and proliferation of microbes. By preventing contaminants from entering the system, good housekeeping practices reduce the amount of chemical and human effort needed to maintain a system. An all-too-common scenario occurs when metalworking fluid splashes around machines. These small puddles of fluid can harbor population densities of 10^8–10^9 microbes/mL. When debris or built-up material is washed off or falls into central sumps, an extremely high microbiological burden is placed on the metalworking fluid's preservation package. Therefore, machine surfaces and work areas should be kept clean and free of debris or residue buildup. Operators should not rinse or sweep debris into return sluices, as this can inoculate central systems with contaminated material. Tramp oil is another significant cause of contamination that should be removed, since it introduces its own microbial population into the fluid and helps prevent oxygen contact with the fluid surface. High levels of tramp oil are associated with microbiological preservation problems in metalworking fluids and are known to reduce the efficacy of chemical biological control agents [4].

Good housekeeping practices also require that workers wear clothing that is cleaned and kept on site. This reduces the risk of workers introducing contaminants such as dirt into the plant and transporting contaminants and other possible chemical and biological irritants into their homes. Food and human waste should also be precluded from entering fluid systems, as they are both likely sources of contamination.

17.5   COMPLICATIONS AND CHALLENGES

Uncontrolled microbiological growth can significantly impact both the performance of metalworking fluids and the operation of the central system. The poor fluid quality and performance that microbial degradation causes can result in corrosion of machines, tools, and workpieces, loss of lubricity and fluid stability, and the formation of slime, which can plug filters and delivery lines and cause unacceptable odors. Microbial growth may also contribute to skin irritation and respiratory problems.

Fluid deterioration can be defined as any change that adversely affects a fluid's utility. Direct deterioration of metalworking fluids can be caused by a variety of microorganisms with the capability to degrade a number of the organic compounds that make up the intricate formulation of a metalworking fluid [5,6]. While components such as petroleum sulfonates, fatty acids and fatty acid esters, organophosphorus products, and nitrogen compounds may or may not remain unaffected by any one singular group of organisms, they are all susceptible to attack by combinations of microbes (microbial consortia). The degradation of these fluid components is responsible for the loss of fluid performance. Resourceful groups of microorganisms may even develop the ability to attack and degrade biocides, especially if they are used at sub-optimum levels.

Microbial contamination is also an indirect source of problems in fluid applications, including the generation of odors. The odors so produced are associated with volatile organic acids (primarily acetate, propionate, and butyrate) and hydrogen sulfide, by-products of microbial metabolism. These compounds are formed when oxygen levels drop, allowing anaerobic bacteria to proliferate. A complication of weekend shutdowns occurs when fluids are not circulated, causing fluid oxygen levels to drop; odors generated by this means are often referred to as "Monday morning odor," since fluid agitation at start-up volatilizes them. The odor can become so offensive that employees refuse to work until it is resolved.

Another complication of microbial contamination is a decrease in fluid pH. The organic acids generated from microbial metabolism have the potential to decrease the pH of the fluid, and a fluid with a lower pH can become corrosive to ferrous materials. A decrease in pH also can result in emulsion splitting and decreased efficacy of other fluid components, including antimicrobial agents or biocides. To maintain proper performance, the alkalinity of the fluid must be buffered at the desired pH. Microorganisms also can have a direct effect on corrosion through a process known as microbiologically influenced corrosion. Here, certain bacterial populations either establish an electrolytic cell or stimulate anodic or cathodic reactions on metal surfaces or contribute to both processes. This can cause pitting and corrosion on both machining tools and workpieces, although it is more often associated with the pipeline, chemical process and water treatment industries [7]. Microorganisms can also degrade corrosion inhibitors, leaving metals more vulnerable to corrosion.

The formation of biofilm is a source of significant problems within a central system. Biofilm growth on machine surfaces can affect the heat-exchange characteristics of the system. Also, if it sloughs off, it can interfere with mechanical operations such as fluid flow and filtration. Fungal contamination in particular plays a major role in the clogging of lines and filters. Considerable time and effort are required to remove established biofilm from the system and resume efficient operations.

Microbial contamination is sometimes blamed for worker skin irritation and dermatitis, but such issues may turn out to be chemically caused, instead. In recent years, respiratory problems among machine shop workers have been thought to be linked to metalworking fluid microbial contamination. The most severe condition, hypersensitivity pneumonitis (HP), was once attributed to the presence of *Mycobacterium immunogenum* in contaminated fluid mists [8]. However, it must be noted that more than a dozen other commonly detected metalworking fluids (MWF) microbes are also known to cause HP. Other possible causes of respiratory impairment include endotoxins, which are cell wall fragments of Gram-negative bacteria that may be found in metalworking fluids, and mycotoxins, produced by fungi in contaminated fluids. These situations continue to be explored; Passman has provided a detailed discussion of worker health issues [9]. The American Society for Testing and Materials (ASTM) Subcommittee E34.50 on Health and Safety Standards for Metalworking Fluids has produced a standard test method for enumeration of mycobacteria (E2564) [10].

17.6　TESTING FOR CONTAMINATION

Test procedures are vital in the fight to control and prevent microbiological contamination in fluid concentrates and coolant systems, through the timely and accurate diagnosis and treatment of the problem. Many test methods exist, and each has its own set of advantages and limitations. The best method to use will vary by situation, but ideally, it should balance ease of use with accurate, practical, and timely results. It is important to remember that different testing methods measure different aspects of microbial contamination.

Changes in fluid or system properties such as odor or slime formation, corrosion, emulsion splitting, and color changes are all good indicators of microbiological problems. However, by the time these symptoms are noted, microorganism populations may be very high and fluid properties significantly impaired. Other methods estimate the amount of microbial growth present through either direct or indirect measure of the biomass growing in the fluid.

The most common direct procedure includes taking microscopic counts. Requiring observation by a trained professional, this method is tedious and time-consuming but yields results within a matter of minutes. Another direct method that requires a skilled professional is staining. This identifies both living and dead cells among specific groups of organisms. Direct methods can also help account for organisms that will not grow on the various nutrient media specifically designed for enumerating microorganisms. However, because of their level of sophistication, direct test methods are not typically implemented in plants using metalworking fluids.

Indirect measures include obtaining estimates of the number of organisms present through the replication of these organisms and obtaining measurements of microbiological activity that translate into the amount of microbial biomass present in the system. Standard plate counts and dip slide counts are the most common indirect methods for enumerating numbers of bacteria or fungi. Here, counts are estimated based on the development of bacterial or fungal colonies on the surface of solid nutrient media. In a standard plate count, a serial dilution is made of the metalworking fluid onto sterile petri dishes. The dishes are then filled with nutrient agar. Agar is a liquid above 45°C but solidifies at room temperature. After several days of incubation, bacteria and fungi form visible colonies on the agar and can be visually counted. Dilutions are taken into account in determining the number of organisms in the original sample.

Because plate count methodology can be expensive and time-consuming, a similar method has been developed that uses paddles coated with the agar media to form a dip slide. The dip slide is immersed into a fluid sample for a few seconds, returned to its sterile holder, and incubated for a few days at ambient temperature. Dip slides generally contain a color indicator that changes color as bacteria or fungi begin to grow, providing a simple visual way to determine colony counts. An additional benefit of many dip slide paddles is their ability to test for bacteria and fungi at the same time by offering a different medium on each side of the paddle. Dip slide technology is available from several commercial sources.

In both standard plate count and dip slide methodology, the assumption is made that each colony comes from a single cell; however, cells often aggregate together, invalidating this conclusion. In addition, a given nutrient medium will only support the growth of certain microbes and may actually inhibit the growth of others, thereby limiting the ability to gain an accurate count of the total organisms present. It has been estimated that direct plate count methods may recover as few as 10% of the organisms present. Another major drawback to plate count methods is the time required for the organisms to grow. Typically, results are available in 24–72 h, but they may take longer. This provides ample time for microorganism levels to reach detrimental levels in plant environments. The long lag times associated with both plate count and dip slide methods have prompted a search for more rapid techniques Many rapid test methods focus on the determination of the concentration of cell constituents such as proteins, enzymes, lipids, or adenosine triphosphate (ATP, the energy unit of living cells) (see ASTM E2694 [11]).

One of the more common methods for rapid detection of microorganisms involves the use of devices to measure ATP [12,13]. ATP is a molecule found only in and around living cells, and as such, it can provide a direct measure of microbiological concentration and health. ATP is quantified by measuring the light produced through its reaction with the naturally occurring firefly enzyme luciferase, using a luminometer. The amount of light produced is directly proportional to the amount of biological energy present in a sample. Using ATP bioluminescence does not provide a direct measurement of the number or types of microorganisms present, but does give an overall estimate of the microbial load in a particular sample. These systems generally take only about five minutes to produce results. Several commercial systems are available for measuring ATP and thus provide estimates of microbial growth; however, differentiation among microorganism types is not possible. Kits have been designed to optimize the measurement of ATP for a particular sample matrix being measured.

Microrespirometry technology has also been developed as a possible rapid detection method for microorganisms. Such systems utilize the production of carbon dioxide and the consumption of oxygen that in turn cause pressure changes to occur in a closed container. These pressure changes can be measured and correlated to the number of microbes present. Results are generally provided in about a 30-min time period, but differentiation between bacteria and fungi is not possible with this technology.

Another technology developed for rapid assessment of microbial growth integrates electrical flow impedance and fluorescence to determine the number, size, and fluorescent characteristics of individual cells in a conductive fluid [14]. The instrumentation is optimized to detect and enumerate viable and nonviable cells in fluid samples with varied particulate content and can discriminate between individual cell statuses. In addition, differences determined in cell size can be utilized to differentiate among bacteria, yeasts, and other microorganisms. This technology permits measurement of each microbe in a sample as it passes through the instrument. The process takes a matter of minutes, allowing results to be obtained in real time. Currently, this technology is relatively expensive.

Recently, Saha and Donofrio [15] have discussed several ways of using molecular techniques to rapidly determine microbiological contamination in metalworking fluids. Both real-time polymerase chain reaction (PCR) (qPCR) and fluorescent *in situ* hybridization are discussed in detail in this review. Currently, these test methods are relatively expensive to utilize on a large-scale commercial basis and require a high level of skill and sophistication to perform.

17.7 TREATMENT OPTIONS

Because metalworking fluids provide an environment conducive to microorganism growth, prevention techniques are vital to minimize the threat of contamination and to facilitate trouble-free, consistent operations. Both chemical and physical methods are available, and often a combination of various methods is warranted when trying to establish and maintain microbiological control.

17.7.1 CHEMICAL

The use of biocides (also known as "antimicrobial pesticides") is the most common method for controlling microbiological growth in metalworking fluids. Used to reduce or maintain bacteria and fungi at acceptable levels in water-based concentrates and final use–diluted fluids, the use of biocides ultimately maintains the integrity of the final product. While they are capable of totally eradicating microbial growth at high concentrations, biocides are more often used at levels that control growth and manage contamination, due to economic and toxicological considerations.

Governmental agencies in most countries regulate the use of biocides. In the United States, biocides are regulated by the Environmental Protection Agency (EPA), under the Federal Insecticide, Fungicide, and Rodenticide Act (FIFRA), and must meet certain toxicological and environmental impact standards. In theory, biocides are approved based on the relative risk resulting from their use. In the European Union, the Biocidal Products Regulation (BPR 528/2012) performs a similar function; approval here is based on hazard. The EPA requires that all biocide labels include specific end-use applications, and it is up to the user to determine whether a registered biocide is appropriate for a specific application. Metalworking fluids are a specific EPA end-use application, and of the approximately 60 chemicals registered as biocides for this end use, fewer than two dozen are routinely used by fluid formulators and end users. In addition to the Federal government, state and local governmental agencies can also regulate hazardous materials.

Two main biocide categories exist: bactericides and fungicides. Though some biocide chemistries possess both capabilities, it is common to use both antibacterial and antifungal functionality to preserve a fluid.

A brief summary of the most common chemistries follows. For a more complete listing of biocides and their properties, several excellent references on the subject exist, including Rossmoore [16], Ash and Ash [17], and Paulus [18]. ASTM Standard Practice E2169 provides a complete list of EPA-registered MWF biocides, which is updated periodically [19]. Biocide suppliers are another valuable source of information and can supply product information and safety data sheets (SDS) that provide detailed information on biocide chemistry, efficacy, toxicological properties, and handling. Additionally, appropriate biocide product labels must be consulted in determining dosages for a particular metalworking fluid application, since Federal law requires that biocide use levels be within label limits. Dosage rates also will vary depending on fluid formulation, housekeeping practices, time of year, operating temperature, pH, and type of contamination likely to be encountered. Testing should also be performed to determine the optimal biocide dosage.

17.7.2 POPULAR CHEMISTRIES

This discussion of antimicrobial chemistries involves the active biocidal molecules themselves, identified by their Chemical Abstracts Service (CAS) numbers. EPA-registered forms of them are offered commercially by various suppliers, often in a formulation that will contain solvents, stabilizers, diluents, carriers, or other active ingredients. Each commercially offered version will have a specific EPA Registration Number associated with it.

17.7.2.1 Formaldehyde Condensates

Formaldehyde condensates remain the most popular and proven biocide chemistry for metalworking fluid applications. They control microbiological growth through their ability to generate formaldehyde *in situ*. Like other aldehydes, the antimicrobial activity of formaldehyde is derived from an electrophilic active group that reacts with nucleophilic cell sites. These targets include amino acids or proteins, which are often important components of enzymes or other functional proteins critical to cell function. Because of this more-generalized mode of action, formaldehyde condensate biocides exhibit a broad spectrum of antimicrobial activity but are viewed as more effective against bacteria than against fungi. A general disadvantage of formaldehyde is the tendency of microorganisms to develop resistance to it [20,21].

Hexahydro-1,3,5-tris[2-hydroxyethyl]-s-triazine (HHT, CAS# 4719-04-4) is the most commonly used and most cost-effective formaldehyde condensate biocide for metalworking fluids. Water soluble and stable at moderately alkaline pH levels, it is the cyclic trimer made from formaldehyde and monoethanolamine (MEA). It is usually supplied as a 78% active aqueous solution, and its use can add significant alkalinity to a system. HHT can be added tankside to an in-use fluid or incorporated into an aqueous concentrate. Viewed primarily as an antibacterial agent, its customary dose rate is about 1500 ppm (0.15%) in a use-diluted fluid.

N,N-Methylenebismorpholine (MBM, CAS# 5625-90-1) is a formaldehyde condensate biocide, meant to be incorporated into MWF concentrates at levels that will provide 1000–1500 ppm in the final use dilution. Its solubility in both oil and water causes it to exhibit good emulsion stability, and its alkaline pH can aid in maintaining desired fluid alkalinity. MBM is said to release its formaldehyde more slowly than do other formaldehyde condensate biocides, thereby lessening the exposure of end users to airborne formaldehyde.

Dimethyloldimethylhydantoin (DMDMH, CAS# 6440-58-0) is a formaldehyde condensate antimicrobial well known as a preservative in personal care products. It is totally water soluble and has a broad spectrum of antibacterial activity. DMDMH is best suited for addition to use-diluted MWF to achieve levels of 1000 ppm. Hydantoin chemistry has been shown to aid in protecting carbon alloy steel against corrosion [22].

Another formaldehyde condensate biocide that is well established as a preservative for personal care products is 1-(3-chloroallyl)-3,5,7-triaza-1-azoniaadamantane chloride (CTAC, CAS# 4080-31-3). It is a quaternary ammonium salt, a cationic material, and is therefore not compatible with anionics such as sodium petroleum sulfonate. Completely water soluble, CTAC is designed to be utilized as a tankside antibacterial additive, at levels between 1000 and 2500 ppm in the use-diluted fluid.

17.7.2.2 Isothiazolinones

Several isothiazolinone chemistries have also seen extensive use in metalworking fluid applications. A blend of the active ingredients 5-chloro-2-methyl-4-isothiazolin-3-one (CMIT, CAS# 26172-55-4) and 2-methyl-4-isothiazolin-3-one (MIT, CAS# 2682-20-4) in a ratio of about 2.7:1 (CMIT/MIT) is the most commonly used isothiazolinone chemistry. Its typical dosage range for the total active ingredients is 10–21 ppm. Offered in the form of a stabilized aqueous solution, this

biocide is effective against bacteria and fungi, including both yeasts and molds, and can be used in all fluid types [23]. Like formaldehyde condensates, isothiazolinones are strong electrophilic agents that react with nucleophilic cell entities to exert their antimicrobial effect. Although resistance of microbial populations to isothiazolinones has been reported [24], its development was strongly affected by the dosing pattern of the biocide. When a regimen of fewer high-dose biocide treatments is compared to more frequent low-dose treatments, with the total biocide addition being equal, the low-dose treatment regimen is more likely to result in selection of resistant microbial populations. Stable over a pH range of about 3–9, this chemistry shows greater stability at the lower end of the pH range. While isothiazolinone chemistry is compatible with a wide variety of metalworking fluid additives, it is incompatible with reducing agents, various amines, mercaptobenzothiazole corrosion inhibitors, and the antifungal agent sodium 2-pyridinethiol-1-oxide.

Due to the incompatibility of isothiazolinones with certain amines (especially at higher concentrations) and the issues of sensitization that could result from using high levels of isothiazolinones in concentrates, the CMIT/MIT blend is used in tankside applications only and is not added to metalworking fluid concentrates. Combining CMIT/MIT with formaldehyde condensates or copper [25] has been shown to stabilize the isothiazolinone molecules, indicating that these combinations may be better suited for fluid concentrate incorporation. The MIT component by itself has been offered as a fluid concentrate biocide. While requiring a higher effective dose level than the CMIT/MIT blend (20–150 ppm active ingredient), MIT demonstrates significantly greater stability in aqueous solution and is less aggressive in human contact.

An aromatic isothiazolinone derivative, 1,2-benzisothiazolin-3-one (BIT, CAS# 2634-33-5), is well established as a metalworking fluid biocide. It can be incorporated into fluid concentrates or added tankside to use-diluted fluids. Its customary dosage level is about 40–360 ppm active ingredient in the final use–diluted fluid, but its antimicrobial spectrum demonstrates a deficiency against *Pseudomonas aeruginosa*.

Another isothiazolinone chemistry, 2-n-octyl-4-isothiazolin-3-one (OIT, CAS# 26530-20-1), is also registered for use in metalworking fluids. Unlike CMIT/MIT, which demonstrates broad-spectrum activity, OIT functions only as a fungicide. Stable in the pH range of 2–10 and compatible with a wide variety of metalworking fluid additives, OIT is often used in combination with other metalworking fluid biocides including CMIT/MIT, triazine, oxazolidines, and sodium pyrithione. Its use with sulfides, mercaptans, bisulfites and metabisulfites, and strong oxidizing agents, however, should be avoided. It may be added either to concentrates or to use-diluted fluids as a tankside addition. Dosage rates range from 25 to 75 ppm active OIT based on the final use–diluted fluid.

17.7.2.3 Other Halogenated Biocides

A reaction product of nitromethane and 2 mol of formaldehyde, plus a mole of bromine, is known as bronopol (BNPD, 2-bromo-2-nitro-1,3-propanediol, CAS# 52-51-7).

An antibacterial product, bronopol, does not release free form-aldehyde in its mechanism of antimicrobial action, and it is considered especially effective against resistant strains of *Pseudomonas aeruginosa*. This water-soluble product is not chemically stable, long term, at pH levels above about 8.0. Its usual dosage rate is about 50–400 ppm active material. Bronopol is well known as a preservative in personal care products, but it is not approved in Europe for use in MWF under the BPR.

The biocide 2,2-dibromo-3-nitrilopropionamide (DBNPA, CAS# 10222-01-2) has recently seen increased use in metal-working fluid applications. This biocide chemistry is viewed as a bactericide with quick killing properties that are achieved at very low dosage rates and is meant for use only as a tank-side treatment. Because of the short half-life of DBNPA at a pH of greater than 8, this chemistry is not persistent enough to provide prolonged preservation when added to metalwork-ing fluid concentrates. An effective dosage rate for DBNPA in tankside applications is 1–2 ppm active ingredient. More recently, a timed-release tablet [26] has been developed that provides for a slow, continuous dose of DBNPA of 0.5–1 ppm active ingredient to central systems. The tablet eliminates the need for pumping liquid material tankside and is especially useful in smaller sumps.

Another biocide chemistry used as a fungicide in metal-working fluid applications is 3-iodo-2-propynylbutylcarba-mate (IPBC, CAS# 55406-53-6). Available in both powder and liquid forms, it has very limited water solubility but is miscible in both alcohols and aromatics. Effective dosage rates for IPBC range from 100 to 300 ppm active ingredient, and it can be formulated into metalworking fluid concentrates or used as a tankside additive.

17.7.2.4 Phenolic Compounds

Phenolic compounds, including the commonly used phenolic active ingredient 2-phenylphenol (OPP, CAS# 90-43-7) and the sodium salt of OPP (CAS# 6152-33-6), have long been used as biocides in metalworking fluids. While environmen-tal concerns have at times lessened their popularity, these compounds have seen a resurgence due to their proven favor-able human toxicity profile and to restrictions prohibiting the release of metalworking fluids into the environment. As a biocide, OPP can be used either tankside or in metalworking fluid concentrates. The sodium salt of OPP has greater solu-bility in aqueous systems. OPP is more easily formulated into concentrates with low water contents. It is a broad-spectrum compound, having efficacy against bacteria and fungi includ-ing both yeast and molds. Both OPP and sodium OPP can be used over a broad pH range; however, their performance and solubility are best at a pH above 9. Dosage rates range from about 500 to 1500 ppm active ingredient for OPP and from 500 to 1000 ppm active ingredient for the sodium salt of OPP. These active ingredients are available for various formulations in both solid and liquid forms.

Another phenolic compound is *para*-chloro-*meta*-cresol or 4-chloro-3-methylphenol (PCMC, CAS# 59-50-7). A very broad-spectrum biocide, PCMC is effective against bacteria,

yeasts, and molds at dosages of 500–2000 ppm active ingredi-ent. The active ingredient PCMC is effective in a pH range of 4–8 but is less stable at an alkaline pH than either OPP or sodium OPP. It can be added to metalworking fluid concen-trates or used as a tankside additive. In tankside addition, it is generally preferable to make a solution of it rather than add-ing the dry powder directly. PCMC has also received renewed interest recently as a result of its efficacy against mycobacte-ria. Additionally, PCMC is one of only two industrial micro-biocides to be approved for use in food-grade lubricants.

17.7.2.5 Other Biocide Chemistries

A biocide consisting of 4-(2-nitrobutyl)morpholine (CAS# 2224-44-4) and 4,4′-(2-ethyl-2-nitrotrimethylene)dimorpho-line (CAS# 1854-23-5) has proven effective as an antibacte-rial and antifungal metalworking fluid biocide. Although it is produced from formaldehyde, morpholine, and nitropropane, this blend does not exert its antimicrobial effect through the release of formaldehyde. With limited water solubility and high oil solubility, the active ingredients formulate easily into many metalworking fluid concentrates, especially soluble oils. In fact, it is one of the few oil-soluble fungicides available. To optimize both its fungal and bacterial efficacy, it should be dosed at 500–1000 ppm active ingredient. This product is not approved under the BPR for use in MWF in Europe.

Sodium 2-pyridinethiol-l-oxide (sodium pyrithione, CAS# 15922-78-8), a broad-spectrum fungicide, can be added to either water-based metalworking fluid concentrates or tank-side into use-diluted fluids in metalworking fluid sumps. It has an effective pH range of 4.5–9.5 and is compatible with most metalworking fluid formulations; however, it is not rec-ommended for use in combination with CMIT-/MIT-based biocides. Strong oxidizing or reducing agents also impair the efficacy of sodium pyrithione. The appearance of a blue color or black specks is common with this chemistry, as it reacts with ferric ions to form insoluble ferric pyrithione, which, incidentally, also has antifungal properties. Its zinc salt is used as an active antifungal agent in personal care products. Dosage rates in MWF range from 46 to 64 ppm active sodium 2-pyridinethiol-1-oxide.

A dialdehyde, known as glutaraldehyde (1,5-pentanedial, CAS# 111-30-8), also offers quick killing properties and can be used in the cleanup of contaminated systems or as a tank-side additive in fluids that do not contain amines, since its activity is impeded by the presence of amines. Its customary dose level in use-diluted fluids is 60–300 ppm active material.

A listing of major producers of MWF biocides who offer several different chemistries includes Dow Microbial Control (Midland, MI); Lanxess (Pittsburgh, PA); Troy Corporation (Florham Park, NJ); Thor Specialties, Inc. (Shelton, CT); Lonza Microbial Control (Atlanta, GA); and The Lubrizol Corporation (Wickliffe, OH).

17.7.3 Biocide Combinations

The use of more than one biocidal active ingredient is a well-known practice and works to broaden or complement the antimicrobial spectrum of a single product. Here, an antifungal product may be combined with an antibacterial product to provide a dual mode of action, or in cases where an antibacterial biocide may have limited efficacy against certain bacteria, another antibacterial agent may be added to expand the system's overall antibacterial spectrum. For example, sodium 2-pyridinethiol-l-oxide, a proven fungicide, is often blended with triazine or oxazolidines, both proven bactericides, to form a product effective against both types of organisms. An added benefit of commercial blends is that they reduce the number of products that an end user must hold in inventory. Several suppliers offer preblended antimicrobial products.

Biocides may also be used together to improve the overall performance of either biocidal components. This concept, known as synergy, can boost efficacy in several different ways. One biocide molecule may stabilize another while broadening the spectrum of antimicrobial coverage or providing additional modes of action for killing or inhibiting microorganisms. The use of isothiazolinones with certain formaldehyde condensate biocides is an example of this type of synergistic blend [27]. Some isothiazolinone molecules when used alone have limited persistence in metalworking fluids. However, these formaldehyde condensates stabilize the isothiazolinones and increase the persistence of the blend. It is common also to use an antibacterial compound together with an antifungal product for broad-spectrum protection.

It is important to remember that in the United States, if a blend of registered active ingredients is sold, the blend itself must be separately registered by the EPA as a pesticide.

17.7.4 Methods of Biocide Application

Biocides can be either formulated into a metalworking fluid concentrate or added to a system tankside. Each strategy has advantages and limitations, and the biocide chemistry may dictate which application method is best. A biocide added into a concentrate will not only protect the concentrate but will provide antimicrobial protection in the use-diluted fluid as well. This method of biocide addition reduces handling of the concentrated biocide at plant locations and end-use sites and ensures that the biocide is added proportionally to the systems in the correct ratio to the fluid. The limitation to this type of addition is that different metalworking fluid systems may exert different demands on the biocide. For instance, there may be fluid ingredients in a central system that are not compatible with a certain biocide chemistry. Thus, the biocide in the newly diluted fluid may be deactivated and consequently fail. In systems that are already heavily contaminated, the organisms may create a heavy demand on the biocide and rapidly deplete it, leaving the remaining formulation components unprotected. Furthermore, adding a biocide to the concentrate for preservation of the use-diluted fluid assumes that the fluid will be diluted properly at the end-use site. If for any reason the fluid is overdiluted, the biocide level may not be sufficient to control microbiological growth.

Alternatively, tankside addition of a biocide at the plant site allows a more direct and specific way to address individual plant situations. With this method, biocide selection can be based on unique plant-operating parameters such as water quality (microbiological and chemical), housekeeping, fluid residence time, degree of expertise among personnel, and waste treatment processes used in the plant. Tankside treatment does require close, if not constant, monitoring, but the added benefit of this is the ability to respond immediately to problems as they arise. Routine tankside addition of a biocide, with specific targeted levels of biocide in the fluid, is more likely to provide effective control if these levels are maintained at all times. It is much better to integrate biocide use into a data-driven dosing program than to merely add it on an apparent as-needed basis. Waiting until a microbiological problem is noted before providing treatment may result in large amounts of dead microbial biomass, which can plug pipes, blind filters, and release odors associated with decaying microbial cells. Limitations associated with tankside treatment include user concerns about maintaining and handling undiluted (concentrated) biocide products due to the health and safety issues associated with these products and the labor-intensive aspect of having to constantly monitor and maintain biocide levels. Tankside addition also requires in-depth training in the proper handling of biocides, as well as a deeper understanding of how to properly calculate and add a given dose of biocide. When using tankside treatment methods, it is very important that all personnel in the plant be informed of the necessary precautions to take when neat biocides are present.

Many metalworking fluid operations use both methods of biocide application to provide a comprehensive treatment program. While biocides added to fluid concentrates help preserve the concentrate and help reduce the volume of tankside preservative that must be kept on site, they may become depleted in the working fluid over time. The routine addition of a tankside biocide, once the biocide provided in the concentrate has been depleted, prevents uncontrolled microbial growth and ultimately helps extend fluid life. However, compatibility between the concentrate biocide and the tankside biocide must be established.

17.7.5 Biocide Selection

Though a seemingly straightforward task, biocide selection can prove challenging because of the number of options available and the fact that no single biocide will provide optimum performance in all situations. Of the various factors to consider when choosing a biocide, the first should be whether the product has an EPA end-use registration for metalworking fluid applications. In addition, the compatibility of the biocide with the fluid must be considered. The biocide must not affect the functional properties of the fluid, including lubricity, corrosion inhibition, and emulsion stability. Other factors such as cost, known chemical incompatibilities, and mode of application also need to be examined before laboratory or field evaluation of the biocide can begin. The type of metal on which the fluid is being used should also be considered, as ferrous and nonferrous metals may have different compatibilities with various biocides. The ASTM document E2169, entitled "Standard Practice for Selecting Antimicrobial Pesticides for Use in Water-Miscible Metalworking Fluids," [19] provides a detailed discussion of the issues to be considered in matching the preservative with coolant chemistry and performance requirements. As indicated earlier, this document also lists the chemicals registered by the EPA for use in preservation of metalworking fluids.

Biocides must also be cost-effective and protect against the spectrum of microorganisms affecting the fluid. While a variety of laboratory test methods is available for measuring biocide performance, simulating real-use conditions in metalworking fluid systems is difficult. The most rigorously documented laboratory test procedure is that published by ASTM. It is designated as ASTM E2275, "Standard Practice for Evaluating Water-Miscible Metalworking Fluid Bioresistance and Antimicrobial Pesticide Performance" [28]. It replaces two previous ASTM methods, D3946 and E686, and can be used to evaluate initial fluid bioresistance, biocide speed of kill, and fluid resistance to repeated microbial challenges. In this test procedure, use-diluted fluid samples of less than 1 L in volume are inoculated with microbes and then aerated to simulate recirculation conditions. The samples may also contain metal chips. The test duration may range between 24 h and 3 months; for longer test periods, a portion of the test fluid is periodically replaced with fresh fluid to simulate fluid turnover in use. Fluid performance is determined by measuring properties such as increase in viable cell recovery or biomass, and changes in physical and chemical fluid properties. This test procedure is meant to compare the relative performance of fluids and/or biocides, not as an absolute predictor of performance in the field. Despite the best efforts and assumptions, laboratory testing is not a perfect indicator of real-use conditions. Consequently, when a biocide has been chosen for a system based upon laboratory trials, it is imperative to confirm the laboratory results with a field trial.

17.7.6 Biocide Handling

It is important to remember that biocide products are designed to control and arrest microbiological growth and, therefore, have toxic properties. Special precautions should always be taken when handling biocides. A product's safety data sheet is the best resource to consult, as it will present comprehensive information on the proper handling methods for that particular chemical hazard. The product label should also be consulted, since Federal law requires that registered pesticide labels provide detailed safety procedures and use instructions. Although the recommended types of protection will vary depending on biocide chemistry, in the interest of good chemical hygiene, protective gloves, aprons, and eye and face protection, at a minimum, are recommended when handling any biocide.

17.7.7 Physical and Nonchemical Control Methods

Although chemical methods are by far the most common means of controlling microbiological growth in metalworking fluids,

several nonchemical methods have received some degree of attention. These include heat treatment or pasteurization, ultraviolet (UV) treatment or irradiation, and filtration. Compared to chemical treatments, these technologies share some common disadvantages. Because these methods treat metalworking fluids at a single point in a system, microbiological growth that may not be circulating in the fluid, or is held up in dead areas of the system or in biofilms, is not treated.

Pasteurization of metalworking fluids heats the fluid to a temperature high enough to destroy the microorganisms responsible for fluid deterioration but temperate enough not to damage the functional properties of the fluid. While many microorganisms are particularly susceptible to heat treatment, others are more thermotolerant and can survive this treatment, leaving the remaining microorganisms to degrade the fluid. This process is also energy-intensive, requiring fluids to remain at about 142°F (63°C) for periods of greater than 30 min. Consequently, the success of this method depends on the availability of an affordable energy source and the feasibility of heating the fluid in an entire system. While Elsmore and Hill [29] obtained good results with heat treatment, they found that heat-resistant populations developed upon intermittent pasteurization. However, the use of fluid heating together with biocidal treatment is more effective in microbial control than either method alone [30].

Several types of radiation technology might appear to prove useful in controlling microbial growth in metalworking fluid systems, including UV irradiation, high-energy electron irradiation, and gamma irradiation. With all these methods, fluids flow through an irradiation bank, contained in a thin film layer. The radiation causes lethal mutations or kills organisms by direct ionizing effects. All three methods have at least some degree of difficulty penetrating opaque materials like metalworking fluids. UV technology is probably the least effective due to this challenge, followed by high-energy electron radiation, which is also energy-intensive, making it less cost-effective. Gamma irradiation offers the most potential in penetrating fluids and killing organisms; however, handling of this radiation source is costly and requires well-trained personnel [31].

Filtration systems are designed to remove particulate material from fluid streams; however, they do not remove microbial cells directly. Microbial populations tend to grow on particulate surfaces rather than in free-flowing fluids. By removing these particles, filtration systems can keep metalworking fluids cleaner and, therefore, less vulnerable to microbial growth. Naturally, cleaner systems often require less biocide. It is important to remember, however, that filter media themselves can also harbor microorganisms and may actually add to contamination problems.

Currently, nonchemical treatment methods remain largely unpopular for metalworking fluid applications because of their significant capital expenditures and energy costs and the scarcity of well-documented technical success stories from the field. However, as pressure to remove toxic chemicals from fluid formulations increases, there may be a greater impetus for the industry to investigate more of these nonchemical treatment technologies. No single-point treatment is 100%

effective, and surviving microbes that settle on downstream-system surfaces will never again be exposed to the treatment. This is an inherent disadvantage of such methods.

17.7.8 Enhancement of Fluid Protection

Today's metalworking fluid operations emphasize extending fluid life as long as possible. Not only does this reduce waste disposal costs, but customers also wish to eliminate or at least reduce the concentration of fluid components viewed as toxic, such as biocides. In an attempt to reach these goals, considerable effort has been spent on trying to formulate fluids with greater inherent biological stability.

One way to attack the problem is to incorporate nonbiocidal materials that will enhance biocide activity. The best-known example of such technology is the use of chelating agents such as ethylenediaminetetraacetic acid (EDTA) salts.

Chelants are ordinarily used as conditioners in water treatment to soften hard water by complexing calcium and magnesium cations out of the solution. However, EDTA has also been known for a long time as a potentiator of antimicrobial agent activity [32]. This is due to its effect in disrupting the cell wall of Gram-negative bacteria (e.g., *Pseudomonas aeruginosa*), rendering the cells more susceptible to damage from biocidal chemistries. Thus, the presence of EDTA in a use-diluted metalworking fluid could lower the level of actual biocide required for microbial control and/or lengthen the useful life of a working fluid. However, as is often the case in the development of chemical formulations, a balance must be maintained between desired properties and detrimental ones imparted by an ingredient. Too much EDTA or other chelating agent can lead to excessive fluid foaming and increased potential for corrosion of metal parts and machine tools.

Another approach is the use of bioresistant functional ingredients in the fluid. A bioresistant material is one that, while not actually killing microorganisms, is not readily destroyed by microbial attack. Bioresistant materials do not provide a readily available food source for microorganisms, but at the same time, these materials cannot stand up to repeated insults of heavy microbial contamination without degradation. Bioresistant products are not subject to the requirements for U.S. EPA registration under FIFRA as long as no claims of antimicrobial efficacy are made. A metalworking fluid's overall potential for microbial degradation depends on the susceptibility of each of its raw materials to microbial attack. A bioresistant component remains unchanged in structure and functional properties in the presence of a microbial population. Fluids with greater bioresistance do not necessarily show lower counts of microorganisms, and inhibition of microbial activity or growth is not ordinarily a property of a bioresistant raw material.

One metalworking fluid component chemistry that has received considerable attention in this regard is the alkanolamines, employed to maintain an alkaline pH and assist in corrosion protection. While all alkanolamines eventually become susceptible to biodegradation, their specific bioresistance characteristics can vary [33,34]. In general, 2-amino-2-methyl-1-propanol (CAS# 124-68-5) appears more bioresistant than

do 2-(2-aminoethoxy)ethanol (CAS# 929-06-6), MEA (CAS# 141-43-5), or triethanolamine (CAS# 102-71-6). However, differences exist in the bioresistance of these alkanolamines, depending on the fluid formulation and the situation in which the fluid is being used. At least some of the bioresistance attributed to alkanolamines appears to be related to pH.

Rossmoore [35] showed that high pH attributed to alkanolamines provides a degree of bioresistance. However, in some fluids, bioresistance was maintained when pH was lowered, while in others, bioresistance was lost. Sandin et al. [36] reported that bioresistance of metalworking fluids was directly proportional to pH and alkyl chain length of the particular alkanolamine tested. Other alkanolamines providing bioresistance include N-butylethanolamine (CAS# 111-75-1) and N-butyldiethanolamine (CAS# 102-79-4).

Amine reaction products with boric acid can be used as water-soluble inhibitors for corrosion of ferrous metals. A major additional benefit of these materials is that they resist microbial attack very effectively [37]. However, upon drying, they may tend to form hard and/or sticky residues that are difficult to remove from machines and parts [38]. In Europe, boric acid in MWF has come under regulatory pressure as a reproductive toxin, Category 1B. There is concern in the industry that boric acid derivatives may suffer similar fates, resulting in major increases in the severity of warning statements on product label and SDS, and the required generation of massive amounts of analytical data [39].

17.8 FUTURE OF MICROBIOLOGICAL CONTROL

Although the metalworking fluid industry is not viewed as an aggressively expanding market, it is filled with impending changes and challenges. Metalworking operations and fluid technologies will continue to become more sophisticated. As many experts predict, this will drive the use of synthetic and semisynthetic fluids, increasing the demand for innovative and effective microbial control strategies. Environmental restrictions regarding fluid disposal are escalating worldwide, thus increasing disposal costs and creating even more pressure to extend fluid life. The once-popular option of simply adding more biocide to increase fluid utility will likely see counterpressure as safety-conscious workers demand the use of lower-toxicity fluids. As environmental and exposure concerns increase, the industry's biocide options will decrease, since more biocidal substances will come under regulatory pressure [40,41].

17.8.1 Formaldehyde Regulation

In 2014, the National Research Council of the U.S. National Academies issued a review that concluded that formaldehyde causes nasal and sinus cavity tumors, as well as myeloid leukemia. This upheld a contested 2011 Department of Health and Human Services opinion, based on a National Toxicology Program assessment that classified formaldehyde as a known human carcinogen in its 12th Report on Carcinogens (http://www.nap.edu/catalog.php?record id=18948). In an attempt to eliminate risk to machine operators from airborne formaldehyde released by formaldehyde concentrate biocides, the EPA proposed lowering the maximum permissible metalworking fluid concentration of HHT, the most widely used formaldehyde condensate metalworking fluid biocide, from 1500 ppm down to 500 ppm. The industry contends that this level would be antimicrobially ineffective, and the EPA has placed this proposal on indefinite hold, pending further data generation and discussion.

In 2012, the European Risk Assessment Committee, acting on behalf of the European Chemicals Agency, announced its decision to reclassify formaldehyde as a Category 1B carcinogen (presumed human carcinogen) and a Category 2 mutagen (suspected of causing genetic defects). This classification will be implemented in 2016, under an amendment to the Classification, Labeling, and Packaging (CLP) of chemicals and mixtures regulation (1272/2008/EC). Such action will result in significantly more severe cautionary wording on the labels and SDS of affected chemicals. Additionally, in Europe, the former Biocide Products Directive, or BPD (98/8/EC), has been superseded by the Biocidal Products Regulation, or BPR (528/2012/EC). It covers the EU countries, plus Iceland, Switzerland, and Norway. Under this new law, there are only 25 active biocidal substances that can be used for MWF preservation. They are classified as Product-type 13 in the Article 95 list and can only be purchased legally from approved suppliers. Furthermore, each active ingredient must be evaluated by a BPR member state for toxicity and environmental hazard prior to its ultimate EU registration. Eleven of these 25 provisionally approved active ingredients are based on formaldehyde condensate chemistry, and there is concern that they may be treated in the same fashion as formaldehyde *per se* with regard to the labeling and SDS wording of the active material and even of the MWF concentrates containing them [42].

The flow of new biocide chemistries coming to the market is also anticipated to slow due to the high cost of registering new active ingredients [43]. Further, regulatory schemes such as the European law known as Registration, Evaluation and Authorization of CHemicals (REACH, Regulation 1907-2006-EC), and the aforementioned CLP, may make it economically unattractive to either develop or continue to market some specialty chemical materials, including biocide enhancers and bioresistant ingredients. At the same time, a recent trend in "green" chemistry is to replace at least a part of a hydrocarbon lubricant material with a vegetable oil or a vegetable oil–derived ester or triglyceride. While such materials offer greater lubricity *per se* than do mineral oil, their readily biodegradable nature can provide new opportunities for microbiological attack in metalworking fluids. Moreover, in the United States and Europe, the recently mandated elimination of many chlorinated paraffin compounds from MWF formulations will remove a bioresistant ingredient. Their replacement with more biologically soft chemistries, such as those based on phosphorus, will add further difficulty to antimicrobial preservation.

These are the industry's challenges, fighting more battles with fewer and more expensive weapons, but they may also serve as the impetus for the advancement of the industry and evolution of the next generation of microbial control strategies.

APPENDIX 17A: SUPPORTIVE TECHNICAL/TRADE ORGANIZATIONS

Society of Tribologists and Lubrication Engineers (STLE, Park Ridge, IL)

Independent Lubricant Manufacturers Association (ILMA, Alexandria, VA)

Biocides Panel, American Chemistry Council (ACC, Washington, DC)

Independent Union of the European Lubricants Industry (UEIL, Brussels, Belgium)

Verband Schmierstoff-Industrie e.V. (VSI, Association of the German Lubricant Industry, Hamburg, Germany)

United Kingdom Lubricants Association (UKLA, Herts, UK)

UKLA Metalworking Fluid Product Stewardship Group

European Chemical Industry Council (CEFIC, Brussels, Belgium)

Formaldehyde Biocide Interest Group (FABI, a CEFIC registration group)

APPENDIX 17B: OTHER MWF-BIOCIDE CHEMISTRIES PERMITTED UNDER BPR BUT NOT EPA-REGISTERED FOR SUCH USE IN THE UNITED STATES

2-Phenoxyethanol (CAS# 122-99-6)

N-(3-aminopropyl)-*N*-dodecylpropane-1,3-diamine (diamine, CAS# 2372-82-9)

(Ethylenedioxy)dimethanol (EGForm, CAS# 3586-55-8)

Tetrahydro-1,3,4,6-tetrakis(hydroxymethyl)imidazo[4,5d]imidazole-2,5(1H,3H)-dione (TMAD, CAS# 5395-50-6)

7a-ethyldihydro-1H,3H,5H-oxazolo[3,4c]oxazole (EDHO, CAS# 7747-35-5)

(Benzyloxy)methanol (CAS# 14548-60-8)

α,α′,α″-trimethyl-1,3,5-triazine-1,3,5(2H,4H,6H)-triethanol (HPT, CAS# 25254-50-6)

3,3′-methylenebis[5-methyloxazolidine] (MBC, CAS# 66204-44-2)

2-Butyl-benzo[d]isothiazol-3-one (BBIT, CAS# 4299-07-4)

2-Methyl-1,2-benzisothiazol-3(2H)-one (MBIT, CAS# 2527-66-4)

REFERENCES

1. EM White. The low down on the slime—Biofilm control in metalworking fluid distribution systems. *Compoundings*, **54**(9):22–23, 2004.
2. FJ Passman. Microbial problems in metalworking fluids. *Tribol Lubr Technol*, **60**:24–27, April 2004.
3. KL Buers, EL Prince, CJ Knowles. The ability of selected bacterial isolates to utilize components of synthetic metalworking fluids as sole sources of carbon and nitrogen for growth. *Biotechnol Lett*, **19**:791–794, 1997.

4. M Abanto, J Byers, H Noble. The effect of tramp oil on biocide performance in standard metalworking fluids. *Lubr Eng*, **50**(9):732–737, 1994.

5. RG Almen et al. Application of high-performance liquid chromatography to the study of the effect of microorganisms in emulsifiable oils. *Lubr Eng*, **38**(2):99–103, 1982.

6. HW Rossmoore, LA Rossmoore. Effect of microbial growth products on biocide activity in metalworking fluids. *Int Biodeterior*, **27**:145–156, 1991.

7. PBJ Scott et al. Expert consensus on MIC—Prevention and monitoring, Part 1. *Mater Performance*, 2–6, March 2004.

8. HW Rossmoore, L Rossmoore, D Bassett. Life and death of mycobacteria in the metalworking environment. *Lubes'N'Greases*, **10**:21–27, April 2004.

9. FJ Passman. Metalworking fluid microbes: What we need to know to successfully understand cause-and-effect relationships. *Tribol Trans*, **51**:1–11, January 2008.

10. ASTM E2564-13. Standard practice for enumeration of mycobacteria in metalworking fluids by direct microscopic counting (DMC) method. ASTM International, West Conshohocken, PA, 2013, www.astm.org.

11. ASTM E2694-11. Standard test method for measurement of adenosine triphosphate in water-miscible metalworking fluids. ASTM International, West Conshohocken, PA, 2011, www.astm.org.

12. JM Baker et al. Bacterial bioluminescence: Applications in food microbiology. *J Food Protect*, **55**:62–70, 1992.

13. A Lundin. Use of firefly luciferase in ATP-related assays of biomass, enzymes and metabolites. *Methods Enzymol*, **305**:346–370, 2000.

14. P Schwartzentruber. CellFacts II—Single cell analysis in real time. *Macromol Symp*, **187**:543–552, 2002.

15. R Saha, R Donofrio. The microbiology of metalworking fluids. *Appl Microbiol Biotechnol*, **94**:1119–1130, 2012.

16. HW Rossmoore, ed. *Handbook of Biocide and Preservative Use*. Blackie Academic & Professional, Glasgow, U.K., 1995.

17. M Ash, I Ash. *The Index of Antimicrobials*. Gower, Aldershot, England, 1996.

18. W Paulus, ed. *Directory of Microbicides for the Protection of Materials—A Handbook*. Springer, Berlin, Germany, 2005.

19. ASTM E2169-12. Standard practice for selecting antimicrobial pesticides for use in water-miscible metalworking fluids. ASTM International, West Conshohocken, PA, 2012, www.astm.org.

20. M Sondossi, HW Rossmoore, JW Wireman. Observations of resistance and cross-resistance to formaldehyde and a formaldehyde condensate biocide in *Pseudomonas aeruginosa*. *Int Biodeterior*, **21**:105–106, 1985.

21. M Sondossi, HW Rossmoore, R Williams. Relative formaldehyde resistance among bacterial survivors of biocide-treated metalworking fluids. *Int Biodeterior*, **25**:423–437, 1989.

22. HZM Al-Sawaad et al. The inhibition effects of dimethylol-5-methyl hydantoin and its derivatives on carbon steel alloy. *Mater Environ Sci*, **1**(4):227–238, 2010.

23. DF Heenan et al. Isothiazolinone microbiocide-mediated steel corrosion and its control in aluminum hot rolling emulsions. *Lubr Eng*, **47**(7):545–548, 1991.

24. M Sondossi, HW Rossmoore, ES Lashen. Influence of biocide treatment regimen on resistance development to methylchloro-/methylisothiazolinone in *Pseudomonas aeruginosa*. *Int Biodeterior Biodegrad*, **43**:85–92, 1999.

25. AB Law, ES Lashen. Microbiocidal efficacy of a methylchloro-/methylisothiazolinone/copper preservative in metalworking fluids. *Lubr Eng*, **47**(1):25–30, 1991.

26. FJ Passman, J Summerfield, J Sweeney. Field evaluation of a newly registered metalworking fluid biocide. *Lubr Eng* **56**(10):26–32, 2000.

27. M Sondossi et al. Factors involved in bacterial activities of formaldehyde and formaldehyde condensate/isothiazolone mixtures. *Int Biodeterior*, **32**:243–261, 1993.

28. ASTM E2275-14. Standard practice for evaluating water-miscible metalworking fluid bioresistance and antimicrobial pesticide performance. ASTM International, West Conshohocken, PA, 2014, www.astm.org.

29. R Elsmore, EC Hill. The ecology of pasteurized metalworking fluids. *Int Biodeterior*, **22**:101–120, 1986.

30. LA Rossmoore. Magnets & magic wands—Exploring the myths of metalworking fluid microbiology. *Compoundings*, **55**(9):19–21, 2005.

31. TF Heinrichs, HW Rossmoore. Effect of heat, chemicals and radiation on cutting fluids. *Dev Ind Microbiol*, **12**:341–345, 1971.

32. MRW Brown, RME Richards. Effect of ethylenediamine tetraacetate on the resistance of *Pseudomonas aeruginosa* to antibacterial agents. *Nature*, **20**:1391–1393, 1965.

33. U Aumann, P Brutto, A Eachus. Boost your resistance: Metalworking fluid life can depend on alkanolamine choices. *Lubes'N'Greases*, **6**:22–26, June 2000.

34. U Aumann. Comparative study of the role of alkanolamines with regard to metalworking fluid longevity-laboratory and field evaluation. *Proceedings of 12th International Colloquium Tribology*, Vol. 10.8, pp. 787–794. Technische Akademie Esslingen, Ostfildern, Germany, 2000.

35. HW Rossmoore. Biostatic fluids, friendly bacteria, and other myths in metalworking microbiology. *Lubr Eng*, **49**(4):253–260, 1993.

36. M Sandin, I Mattsby-Baltzer, L Edebo. Control of microbial growth in water-based metalworking fluids. *Int Biodeterior*, **27**:61–74, 1991.

37. S Watanabe et al. Antimicrobial properties of the products from the reaction of various aminoalcohols and boric anhydride. *Mater Chem Phys*, **19**:191–195, 1988.

38. JP Byers, ed. *Metalworking Fluids*, 2nd edn. CRC Press, New York, 2006.

39. Boric acid and disodium tetraborates, White Paper, UKLA Metalworking Fluid Product Stewardship Group of UKLA, July 2015.

40. S Baumgaertel. New European Legislation on chemicals and its estimated influence on raw material availability: Lubricant industry and users. *Fifth International Conference on Metal Removal Fluids*, Chicago, IL, September 2015.

41. M Hentz. How to preserve water-miscible coolants in the context of changing EU-legislation—What does the future look like?. *Fifth International Conference on Metal Removal Fluids*, Chicago, IL, September 2015.

42. R Beercheck. Replacing formaldehyde—Regulatory threats loom for key biocides. *Lubes'N'Greases*, Europe-Middle East-Africa, 12–18, July 2012.

43. FJ Passman, AC Eachus. Impact of biocide products regulation on microbial contamination control in metalworking fluids, *TAE 20th International Colloquium Tribology—Industrial and Automotive Lubrication*, Technische Akademie Esslingen, Ostfildern, Germany, January 2016.

18 Surfactants in Lubrication

Girma Biresaw

CONTENTS

18.1 INTRODUCTION

Surfactants are one of the most widely applied materials [1–9]. The application of surfactants varies from everyday mundane tasks such as cleaning to highly complex processes involving the formulation of pharmaceuticals, foods, pesticides, lubricants, etc. Although surfactants have been known and applied for centuries, they continue to be the subject of vigorous research and development effort. This is because of the continued development of new and complex applications (e.g., nanotechnology) that require new and improved surfactants and a deeper understanding of their performance.

The objective of this chapter is to provide an overview of the applications of surfactants in lubrication. However, before delving into the lubrication aspects of surfactants, the chapter provides basic information about surfactants. This basic information is critical to understanding the role of surfactants in lubrication discussed in the subsequent sections.

The chapter begins with a review of the basic information about the chemical structure of surfactants. Although there are thousands of surfactants being used in thousands of applications, they all have similar basic structures.

Another important property of surfactants is their ability to dissolve in both polar and nonpolar solvents. A brief description of this phenomenon is provided followed by a discussion of surfactant-organized assemblies. The ability of surfactants to spontaneously form various types of microstructures or organized assemblies is responsible for many of their applications [1–9].

Surfactants are used in lubrication to provide a number of desirable outcomes (e.g., lower friction, lower wear, better filtration, lower emission, and better cooling). The list of surfactant functions in lubrication is numerous, and a brief description of some examples is given. In addition, an in-depth review of two of the most important functions of surfactants in lubrication is highlighted. These two functions are solubilization and boundary lubrication.

Solubilization deals with the ability of surfactants to solubilize various polar and nonpolar ingredients [1–13]. This property of surfactants makes it possible to formulate the various water- and oil-based lubricants currently in use for various applications.

In boundary lubrication, friction surfaces are pressed together at a very high load and rubbed against each other at low speeds. Such a process leads to the generation of high friction and wear. Surfactants adsorb on friction surfaces and prevent direct contact between the friction surfaces, thereby lowering both friction and wear. This property of surfactants is illustrated from a review of an extensive study into the boundary lubrication properties of agriculture-based surfactants [14–32].

18.2 STRUCTURE OF SURFACTANTS

The term *surfactant* is used to describe organic molecules that comprise two dissimilar groups in the same molecule. One group is a hydrocarbon chain of varying structures and is soluble only in nonpolar solvents. Examples of nonpolar solvents include benzene, toluene, and hexanes. Because it is soluble only in nonpolar solvents, the group is also referred to as the lipophilic segment or the lipophilic tail of the surfactant.

The second group in the surfactant is composed of polar groups that readily associate with or dissolve in polar solvents such as water, glycerol, and formamide. Because of its strong affinity for polar solvents, the group is also referred to as the hydrophilic segment or hydrophilic head of the surfactant.

Figure 18.1 displays schematics of surfactant structure of varying complexity. The simplest surfactants comprise a linear or branched lipophilic chain directly attached to the hydrophilic component (Figure 18.1a and b). More complex structures include more than one lipophilic chain attached to one or more hydrophilic component (Figure 18.1c through e) or polymeric surfactants with various repeating hydrophilic and lipophilic components (Figure 18.1f).

Depending on the nature of the hydrophilic component, surfactants are classified into four categories: anionic, cationic, nonionic, and zwitterionic. Anionic surfactants are negatively charged and contain one or more positively charged organic or inorganic counterion. Similarly, cationic surfactants are positively charged with negatively charged organic or inorganic counterions. Nonionic surfactants have no charge, and zwitterionic surfactants have both positive and negative charges with no counterions. Examples of ionic and nonionic surfactants of varying hydrophilic and lipophilic structures are given in Table 18.1, along with their common names.

18.3 SOLUBILITY OF SURFACTANTS

Almost all ionic (anionic, cationic, and zwitterionic) surfactants display good solubility in water but poor solubility in nonpolar solvents. The reverse is true for most nonionic surfactants, except for ethoxylated surfactants ($C_{12}EO_6$ in Table 18.1). In the latter case, the solubility of the surfactant in hydrophilic (e.g., water) versus lipophilic (e.g., hexanes) solvents depends on the relative *strength* of the hydrophilic versus the lipophilic segments of the surfactant. Becher [9] developed the following empirical equation to quantify the net *strengths*, or the hydrophilic–lipophilic balance (HLB) of such nonionic surfactants.

$$HLB = \frac{20 \times (MW - EO)}{MW} \qquad (18.1)$$

where

MW − EO is the molecular weight of the ethoxylated segments of the surfactant

MW is the molecular weight of the surfactant

To illustrate the application of Equation 18.1, let us examine the HLB of a hypothetical ethoxylated nonionic surfactant

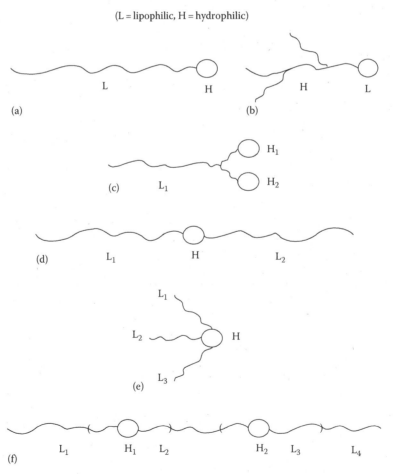

(L = lipophilic, H = hydrophilic)

FIGURE 18.1 Schematics of surfactant structure of varying complexity. Hydrophilic and lipophilic segments are indicated by H and L, respectively. The presented hypothetical examples represent surfactants with simple (a), branched lipophile (b), multiple hydrophile (c), multiple lipophile (d,e), and polymeric structures (f).

TABLE 18.1

Examples of Surfactants of Varying Hydrophilic and Lipophilic Structures

General Structure: RXY

Type	R	X	Y	Name
Anionic	$C_{11}H_{23}^-$	COO^-	K^+	Potassium laurate (soap)
	$C_{12}H_{25}^-$	SO_4^-	Na^+	Sodium dodecyl sulfate (SDS)
Cationic	$C_{16}H_{33}^-$	$(CH_3)_3N^+$	Br^-	Cetyltrimethyl ammonium bromide
	$C_8H_{17}^-$	$(C_8H_{17})_2CH_3N^+$	Cl^-	Tri(n-octyl)-methyl ammonium chloride
Zwitterionic	$C_{12}H_{25}CH^-$	$(CH_3)_2N^+$	$-CH_2CO_2^-$	Laurylbetaine
Nonionic	$C_{12}H_{25}^-$	OH		Lauryl alcohol
	$C_{17}H_{33}^-$	$COOH$		Oleic acid
	$C_{12}H_{25}O^-$	$(CH_2CH_2O)_6H$		$C_{12}EO_6$, ethoxylated lauryl alcohol

$$H-(C_xH_{2x}O)-(CH_2CH_2O)_y-H$$

FIGURE 18.2 General structure of a polyethoxylated nonionic surfactant.

shown in Figure 18.2. According to Equation 18.1, the maximum value of HLB is 20, and it is for a nonionic surfactant without a lipophilic segment, that is, $x = 0$ in Figure 18.2. Using a similar argument, the minimum value of HLB is zero and is for a nonionic surfactant with no hydrophilic segments, that is, $y = 0$ in Figure 18.2. The general relationship between HLB and solubility of ethoxylated nonionic surfactants is as follows [9,10]: nonionic surfactants with high HLB (>11) are soluble in water and, hence, are good candidates for formulating oil-in-water (o/w) emulsions. On the contrary, nonionic surfactants with low HLB (<9) are soluble in oil and are good candidates for formulating water-in-oil (w/o) emulsions.

18.4 ORGANIZED SURFACTANT ASSEMBLIES

Surfactants form various types of structures when dissolved in hydrophilic and lipophilic solvents. The types of structures formed depend on various factors including the nature of the surfactant, type of solvent, surfactant concentration, temperature, and nature and concentration of cosolutes (e.g., salts and other surfactants).

At low concentrations, surfactants adsorb on the surface of the solvent, with one segment dissolved in the solvent and the other segment in contact with air. Which segment contacts the solvent depends on whether the solvent is hydrophilic or lipophilic. In hydrophilic solvents, the polar group is in contact with the solvent, whereas the lipophilic chain is in contact with air. The structure is reversed in a lipophilic solvent. These structures are depicted in Figure 18.3a and b.

When the concentration of the surfactant reaches a certain critical value, the surfactant molecules spontaneously form micelles or aggregate. The concentration at which such microstructures spontaneously form is called critical micelle concentration or critical aggregation concentration. In hydrophilic solvents, normal micelles or aggregates are formed. In such systems, the lipophilic chains of the surfactants form the

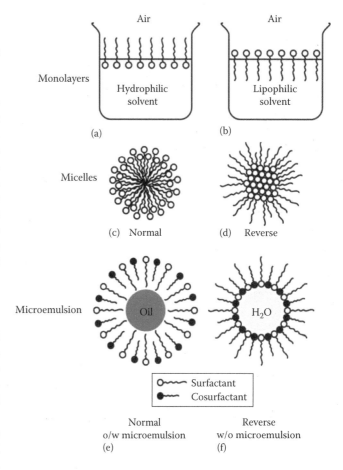

FIGURE 18.3 Microstructures of surfactant assemblies in solution: (a, b) monolayers, (c) normal micelles, (d) reverse micelles, (e) o/w microemulsions, and (f) w/o microemulsions.

core of the spherical structures, away from the hydrophilic solvent, whereas the hydrophilic heads are in contact with the hydrophilic solvent. In lipophilic solvents, reverse micelles and aggregates are formed where the hydrophilic heads are buried in the core and away from the lipophilic solvent, whereas the lipophilic tails are in contact with the solvent. The structures of normal and reverse micelles are depicted in Figure 18.3c and d.

Micelles differ from aggregates in that they have a uniform size and a constant aggregation number, that is, they comprise the same number of surfactant molecules per micelle. Aggregates, on the contrary, have no fixed size or aggregation number but a range of sizes and aggregation numbers [11].

Incorporation of a cosurfactant (e.g., short-chain alcohol) allows normal micelles and reverse micelles to solubilize lipophiles (e.g., oils) and hydrophiles (e.g., water) in their cores, respectively. The resulting micelles have bigger diameters and are referred to as swollen micelles. Normal micelles with large volumes of solubilized oil are called o/w microemulsions, whereas reverse micelles with large volumes of solubilized water are called w/o microemulsions. The structures of o/w and w/o microemulsions are illustrated in Figure 18.3e and f.

The diameters of micelles and microemulsions are smaller than the wavelength of light. As a result, solutions with micelles and microemulsions are clear and transparent and remain so indefinitely since they are thermodynamically stable.

In the absence of cosurfactants, normal micelles dissolve little or no oil. As a result, mixing oil in micelle solutions will result in the formation of emulsions. Emulsions differ from micelles and o/w microemulsions in that they comprise oil droplets with diameters larger than the wavelength of light. As a result, they reflect visible light and have a milky appearance. Emulsions are also thermodynamically unstable; as a result, they separate into their respective oil and water phases over time.

This distinction between micelles, o/w microemulsions, and o/w emulsion is very important in lubrication. In water-based lubricants, these three fluids are referred to by different names by lubricant suppliers and users [12]. Generally, emulsions are referred to as soluble oils because they comprise relatively large quantities of solubilized oils. Microemulsions are referred to as semisynthetic lubricants because they comprise relatively smaller quantity of solubilized oils. Micelles are called synthetic lubricants because they comprise no oil, and lubrication is achieved by synthetic ingredients dissolved in the water phase of the micelle. More details about water-based lubricants are given in Section 18.6.

Increasing the concentration of the surfactant after the formation of micelles results in the formation of lamellar liquid crystals. These are three-dimensional surfactant multilayer structures. The exact orientation of the surfactants in these structures depends on whether the solvent is polar or nonpolar, as illustrated in Figure 18.4a and b.

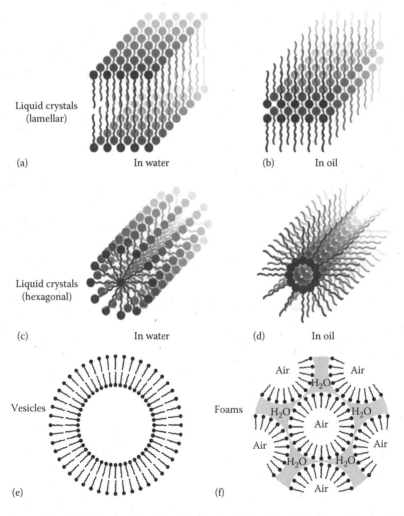

FIGURE 18.4 Microstructures of surfactant assemblies in solution: (a, b) lamellar liquid crystals, (c, d) hexagonal liquid crystals, (e) vesicles, and (f) foams.

Further increase in the concentration of surfactant leads to the formation of normal and reverse hexagonal liquid crystal structures, which are depicted in Figure 18.4c and d. Other surfactant-based microstructures include vesicles (Figure 18.4e) and foams (Figure 18.4f). Control of foam is a major problem in lubrication [12].

18.5 FUNCTIONS OF SURFACTANTS IN LUBRICATION

Surfactants play a number of critical roles in lubrication and, thus, are important ingredients in lubricant formulations. Different types of surfactants are used to attain different lubrication objectives. Examples of lubrication objectives attained by surfactants include solubilization, antifriction, demulsification, dispersion of fines and debris, anticorrosion, defoaming, antiwear, and antimicrobes.

Sometimes, the performance objectives that surfactants are supposed to provide the lubricant are contradictory. Examples of contradictory objectives include emulsification versus demulsification, biodegradability versus bioresistance, and agglomeration versus dispersion of debris and fines. These contradictions occur because of the changing requirements of lubricant performance over time or process. For example, the performance requirements of the lubricant during use are different from that during waste treatment, disposal, or recycling. Another example is the difference in performance requirements of the lubricant during the lubrication process (e.g., hot rolling) versus lubricant handling process (e.g., filtration).

In this section, a brief description of some of the surfactant functions is given. Detailed description of two of the functions, solubilization and boundary friction, is provided in subsequent sections. It is clear that surfactants have numerous functions in lubrication, and it is beyond the scope of this chapter to provide detailed discussions of each. Readers are encouraged to go through the various references cited in this chapter for more details.

Emulsification is one of the most important lubrication functions of surfactants and is discussed in detail in Section 18.6. Emulsification allows for preparing a stable dispersion of the lubricant formulation in water so that the lubricant satisfies both lubrication and cooling functions of the process. The surfactants used in these functions are selected so as to provide the appropriate emulsion properties without interfering with the lubrication function of the formulation.

Demulsification is one of the requirements for waste treatment of used lubricants. In this process, a surfactant is used to help the oil and water in the lubricant to cleanly separate into their respective phases so that they can be properly processed and disposed.

Dispersion of fines and debris is an important function of surfactants in lubrication processes that generate a lot of metal fines and debris. The function of the surfactant is to prevent fine metal particles and other debris from aggregating into large mass. If not dispersed, the fines and debris will agglomerate and accumulate on tool and workpiece surface. This can cause undesirable outcomes such as poor product quality (e.g., debris rolled into the workpiece), accelerated tool wear, and unsafe working conditions due to accumulation of debris on equipment and work area.

Agglomeration of fines and debris may be important in some processes where removing the fines and debris by filtration is important. Dispersed fines and debris are hard to filter out and might require the use of expensive filtration equipment and media to attain the degree of lubricant cleanliness required for the process. The function of the surfactant here is to allow the fines and debris to attract to each other, aggregate, and grow into large particles. Thus, the application of the proper surfactant to help the debris and fines agglomerate for easy filtration will save cost and result in cleaner lubricant.

Bioresistance is the property of a lubricant to resist attack by bacteria and fungi (mold and yeast) microbes. Attack by microbes will result in a number of undesirable outcomes (e.g., changes in pH, depletion of critical lubricant ingredients, bad odor, tool, and workpiece corrosion) that compromise the proper functioning of the lubricant. The function of the surfactant here is to protect the lubricant from attack by microbes. Surfactants used in such application are called biocides.

Biodegradability is the property of a lubricant to be easily and naturally degraded when disposed. This requires that the lubricant ingredients are easy to digest by microbes employed in composting and similar processes. Lubricants that are not biodegradable require expensive waste treatment processing before disposal to comply with tightening environmental laws. Surfactants that are easily biodegradable will improve the biodegradability of the lubricant in which they form a part.

18.6 SOLUBILIZATION WITH SURFACTANTS

Liquid lubricants are broadly classified into two groups: oil based and water based. Oil-based lubricants comprise a base oil of appropriate viscosity with various solubilized additives. The function of the additives is to reduce friction and wear and to impart the lubricant other important characteristics such as oxidation stability, low pour point, and bioresistance. Oil-based lubricants are used in processes where the heat generated during lubrication does not cause concern about tool wear, product quality, productivity, safety, and other issues. Another reason for the use of oil-based lubricants is when the presence of water causes undesirable outcomes such as poor product quality (e.g., water stain), corrosion of tools and machine elements, and rust.

Water-based lubricants comprise 0%–20% of all or portions of the oil-based lubricant ingredients mentioned previously, solubilized in 80%–99% water. Water-based lubricants are employed when the lubrication process generates excessive heat capable of damaging tools, negatively affecting product quality, causing fire, and other undesirable results.

Examples of such lubrication processes include metal forming (e.g., hot rolling of aluminum and steel) and metal removal (e.g., grinding and machining).

Various types of surfactants (e.g., cosurfactants and coupling agents) are employed to achieve proper dissolution of the oil phase in water. Water-based lubricants are classified into the following categories depending on their composition and physical properties [12,13].

1. *Soluble oils*: These are o/w emulsions with up to 20% of the oil phase in water. They are produced by vigorously mixing the oil and water phases in the presence of surfactants. These emulsions comprise large droplets of oil stabilized by surfactants in water. Unlike o/w microemulsions, the o/w emulsions have a large diameter of oil droplets. As a result, o/w emulsions reflect visible light and hence appear milky to the naked eye. Another major difference between o/w emulsions and microemulsions has to do with their stability. O/w emulsions are thermodynamically unstable and, hence, separate into the respective oil and water phases over time. A schematic comparing o/w emulsions and o/w microemulsions is given in Figure 18.5.

2. *Semisynthetic oils*: These are water-based lubricants with 1%–5% (w/w) of solubilized oil phase. These are the o/w microemulsions that have been discussed previously (Section 18.4) and depicted in Figure 18.3e. Semisynthetic oils are clear and transparent solutions. This is attributed to the fact that the size of the oil droplets is smaller than the wavelength of light. As mentioned earlier, semisynthetic oils are thermodynamically stable (see Figure 18.5 for comparison of o/w emulsions versus o/w microemulsions).

3. *Synthetic oils*: These are normal micellar solutions discussed previously (Section 18.4) and depicted in Figure 18.3c. Synthetic oils contain no solubilized oil. However, they contain additives solubilized in the continuous water phase and the lipophilic core of the normal micelle, to provide the necessary friction, wear, and other properties to the synthetic oils. Because of the presence of solubilized additives, synthetic oils can be considered as swollen micelles. Again, because of the small diameter of the swollen micelles relative to the wavelength of light, synthetic oils are transparent and clear solutions. They are also thermodynamically stable, that is, various additives will not appear as a separate phase over time.

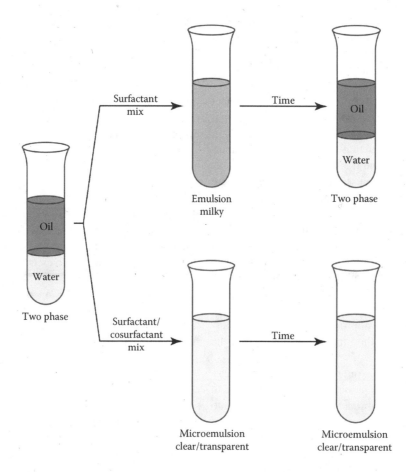

FIGURE 18.5 Emulsions versus microemulsions.

18.7 BOUNDARY LUBRICATION PROPERTIES OF SURFACTANTS

Boundary lubrication occurs in processes where the friction surfaces are rubbing at relatively low speeds and under very high load [13]. Under such conditions, no lubricant film is formed to separate the rubbing surfaces from each other. The rubbing surfaces contact each other, but direct contact is prevented by surfactants adsorbing on the friction surfaces. This is another important function of surfactants in tribology. The effectiveness of surfactants at reducing friction and wear under boundary conditions is highly dependent on the chemistries of their polar and nonpolar segments, concentrations, properties of the friction surfaces, and process conditions (e.g., temperature). This will be illustrated by a brief review of an extensive study into the boundary properties of agriculture-based surfactants.

Proteins [14,15], starches [16,17], and vegetable oils [18–22] obtained from plant-based agricultural products, with or without further modification (through enzymatic, chemical, or other methods), have been found to possess surface-active properties. This has dramatically increased the potential application areas where these materials can be used to develop new uses from surplus agricultural crops. Surface-active materials are used in a wide range of consumer and industrial products including food, cosmetics, pharmaceuticals, paints, coatings, adhesives, polymers, and lubricants [1–9].

Among the various ag-based products, vegetable oils are the most widely investigated raw materials for lubricant applications [18–23]. This is because most vegetable oils are liquid at room temperature. This, along with the fact that they are surface active, allows vegetable oils to be used in lubrication processes that occur in all the lubrication regimes, that is, boundary, hydrodynamic, and mixed film.

Vegetable oils have two types of chemical structures: triglycerides and waxy esters (Figure 18.6). However, most vegetable oils are triglycerides. In both triglycerides and waxy esters, the structure of the polar segment of the molecules comprises one or more ester linkages of fatty acid with glycerol or long-chain fatty alcohols. In all cases, various structures of fatty acids are involved. The structural variations of the fatty acids include chain length; degree, position, and stereochemistry of unsaturation; and the presence of additional functional groups (such as hydroxy or epoxy groups), on the fatty acid chain. As a result, triglycerides and waxy esters can have a multitude of structures. In addition, the triglycerides and waxy esters from specific crops (e.g., soybean oil) are mixtures of various structures. Despite these enormous possible structural variations, the structural composition of triglycerides from a specific crop (e.g., soybean oil) is fixed and well known [23]. Thus, for example, triglycerides of regular soybean oil comprise 23% oleic acid and 54% linoleic acid, whereas regular canola oil comprises 61% oleic acid and 23% linoleic acid [23].

The boundary lubrication properties of vegetable oils were investigated in steel/steel and steel/starch contact. In addition to varying the properties of the contacting surfaces (steel/steel versus steel/starch), vegetable oils of varying chemical structures were employed. Also, simple methyl esters of fatty acids were included in these studies as model compounds to the more complex triglycerides and waxy esters. A summary of these studies along with subsequent data analysis is reviewed next.

18.7.1 BOUNDARY FRICTION OF VEGETABLE OILS IN METAL/METAL CONTACT

The boundary lubrication properties of vegetable oils, in metal/metal contact, were investigated using a ball-on-disk

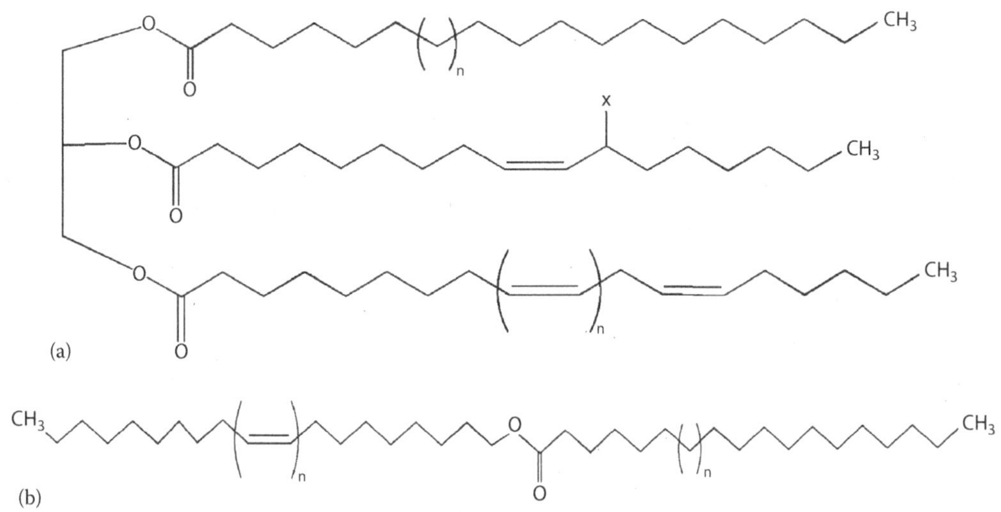

FIGURE 18.6 Chemical structures of vegetable oils: (a) triglycerides and (b) waxy esters.

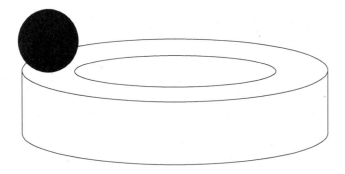

FIGURE 18.7 Schematic of ball-on-disk tribometer. (From Biresaw, G. et al., *J. Am. Oil Chem. Soc.*, 79(1), 53, 2002. With permission.)

tribometer. A schematic depiction of the ball-on-disk tribometer is shown in Figure 18.7. In these studies, the steel ball and steel disk were immersed in a lubricant of hexadecane, with varying concentrations of vegetable oils. Friction between the stationary steel disk and the rotating steel ball was measured under the following conditions: speed, 5 rpm or 6.22 m/s; load, 1788 N or 400 lb; temperature, room temperature; and test duration, 15 min. The concentration of the vegetable oils in hexadecane varied in the range 0.0–0.6 M. A typical data from a ball-on-disk tribometer are shown in Figure 18.8. The data show that the coefficient of friction (COF) increases with time and levels off after 6 min. The COF for the specific lubricant is obtained by averaging the COF values in the steady-state region.

The effect of vegetable oil concentration on boundary COF from a ball-on-flat tribometer is shown in Figure 18.9. The data in Figure 18.9 have four important features that need to be pointed out: (1) the COF without vegetable oil is very high

(>0.5), (2) addition of vegetable oil results in a sharp reduction of COF, (3) the COF decreases with increasing concentration of vegetable oil and reaches a minimum value, and (4) once the minimum value is attained, the COF remains constant and independent of any further increase in vegetable oil concentration. These observations are important for understanding the mechanism of boundary lubrication by vegetable oils. It also provides the basis for proper analysis of the friction data to elucidate the effect of vegetable oil and substrate properties on boundary friction.

18.7.2 BOUNDARY FRICTION OF VEGETABLE OILS IN STARCH/METAL CONTACT

Vegetable oils can be encapsulated into a starch matrix by a steam-jet cooking process called Fantesk™ [24]. The resulting starch–oil composite can be redispersed in water and applied on sheet metal surfaces as a dry film lubricant [25–28]. Dry film lubricants are those that are applied on sheet metals to help protect the surface from damage (dents, rust, and corrosion) during transportation and storage or as lubricants in subsequent forming processes [13]. Studies were conducted to investigate the friction properties of vegetable oils in starch–oil composites. In this investigation, composites with varying concentrations of vegetable oil (0–40 parts per hundred) were redispersed in water and applied (using various methods) onto steel surfaces, and its friction properties evaluated. Boundary friction of the dry film was measured using a ball-on-flat tribometer (Figure 18.10) at room temperature and the following conditions: speed, 2.54 mm/s; load, 14.7 N (1500 gf); temperature, room temperature; and test duration, 24 s. A typical data from a ball-on-flat tribometer are

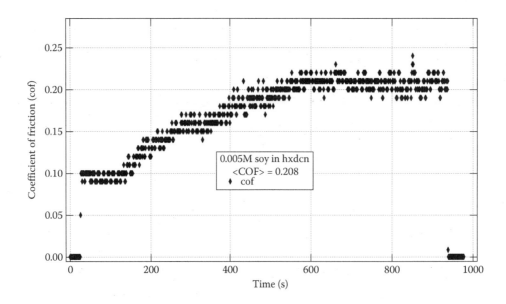

FIGURE 18.8 Typical data from a ball-on-disk tribometer. SBO, soybean oil. (From Biresaw, G. et al., *J. Am. Oil Chem. Soc.*, 79(1), 53, 2002. With permission.)

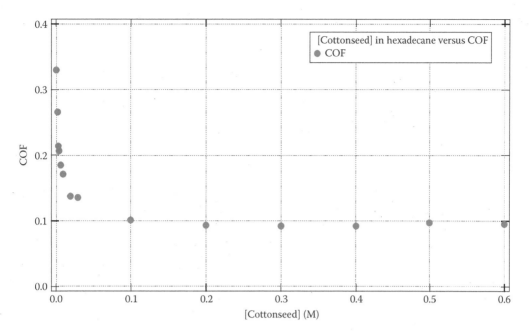

FIGURE 18.9 Effect of vegetable oil concentration in hexadecane on steel/steel boundary friction. (From Adhvaryu, A. et al., *Ind. Eng. Chem. Res.*, 45(10), 3735, 2006. With permission.)

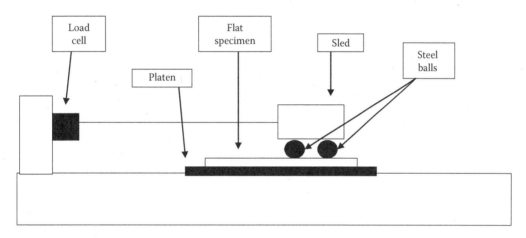

FIGURE 18.10 Schematic of a ball-on-flat tribometer. (From Biresaw, G. and Erhan, S.M., *J. Am. Oil Chem. Soc.*, 79, 291, 2002. With permission.)

shown in Figure 18.11. As can be seen in Figure 18.11, initially, the friction force rose sharply to a maximum and immediately went down to a steady-state value for the remainder of the test period. The maximum value corresponds to the static friction of the steel ball against the starch, whereas the steady-state value corresponds to the kinetic friction. The COF for the dry film lubricant is obtained by dividing the average of the steady-state friction force data by the load.

The effect of vegetable oil concentration in starch matrix on starch–steel boundary friction is illustrated in Figure 18.12. The data show that, in the absence of vegetable oil, the COF between starch and steel is very high (0.8). However, incorporation of vegetable oil into the starch matrix results

in a sharp decrease in COF, which eventually levels off to a constant value. Further addition of vegetable oil will cause no change in COF. The vegetable oil concentration versus COF profile in starch/metal contact shown in Figure 18.12 is similar to that discussed earlier for metal/metal contact (Figure 18.9). Thus, it is reasonable to assume that a similar mechanism may be responsible for the boundary friction properties of vegetable oils in both systems.

18.7.3 Mechanism of Boundary Lubrication of Vegetable Oils

The effect of vegetable oil concentration on metal/metal and starch/metal boundary COF can be explained using an

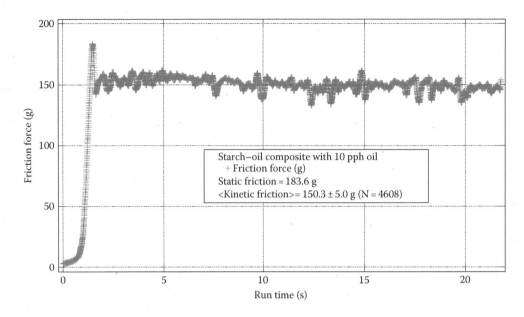

FIGURE 18.11 Typical time vs friction force data from a ball-on-flat tribometer. (From Biresaw, G. and Erhan, S.M., *J. Am. Oil Chem. Soc.*, 79, 291, 2002. With permission.)

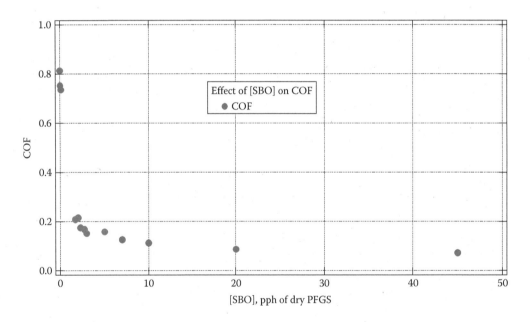

FIGURE 18.12 Effect of vegetable oil concentration in starch matrix on starch/steel boundary friction. PFGS, purified food grade starch. (From Biresaw, G. and Erhan, S.M., *J. Am. Oil Chem. Soc.*, 79, 291, 2002. With permission.)

adsorption model [29,30]. According to this model, in the absence of vegetable oil, high COF is observed because of rubbing between two high-energy surfaces (metal/metal or starch/metal), which are in direct contact (boundary conditions). When vegetable oil is introduced into the lubricant, the COF is reduced because the vegetable oil adsorbs on the rubbing surfaces (metal or starch) and prevents direct contact between the two high-energy surfaces.

Adsorption occurs due to a strong polar–polar interaction between the polar groups of the vegetable oils and

the polar adsorption sites on the starch or metal surfaces. According to the adsorption model, there are a finite number of adsorption sites (S_t) on the surfaces, and once all adsorption sites are occupied, full surface coverage is attained, and no further adsorption occurs and friction remains unchanged. The adsorption sites could be oxides or hydroxides in the metal surfaces and hydroxyl groups on starch. Depending on the concentration of the vegetable oil in hexadecane or starch matrix, the adsorption sites on the steel or starch surface may be fully or partially occupied by the vegetable oils. The surface concentration of vegetable oils is expressed in terms of fractional surface coverage, θ, which is defined as follows:

$$\theta = \frac{[S_o]}{[S_t]} \qquad (18.2)$$

where $[S_t] = [S_o] + [S_u]$ and S_o and S_u are occupied and unoccupied adsorption sites, respectively.

The fractional surface coverage is obtained from boundary friction data as follows [17–21,28–30]:

$$\theta = \frac{f_s - f_o}{f_s - f_m} \qquad (18.3)$$

where

f_s is COF without vegetable oil
f_m is the minimum COF or COF at full surface coverage
f_o is COF at various vegetable oil concentrations

Note that in Equation 18.3, without added vegetable oil, $f_o = f_s$ and $\theta = 0$ and, at full surface coverage, $f_o = f_m$ and $\theta = 1$. Thus, the value of θ varies in the range of 0–1.

The adsorption model also stipulated that adsorption of vegetable oil is an equilibrium process between vegetable oil in solution (O_b) and that on the surface as follows:

$$O_b + S_u \xrightleftharpoons{K_o} S_o \qquad (18.4)$$

where K_o is the equilibrium constant in moles per liter, defined as follows:

$$K_o = \frac{[S_o]}{[O_b][S_u]} \qquad (18.5)$$

K_o in Equation 18.5 can be used to calculate the free energy of adsorption (ΔG_{ads}) of vegetable oils using an appropriate adsorption model. However, to be able to do that, two things have to be accomplished: First, an adsorption isotherm must be constructed. An adsorption isotherm shows the relationship between the concentration of vegetable oil in solution (moles per liter) and that on the surface, expressed as fractional surface coverage, θ. Second, the adsorption isotherm must be analyzed using an appropriate adsorption model. An adsorption model is a mathematical expression describing how surface concentration is related to solution concentration. The equilibrium constant K_o is obtained by analyzing the adsorption isotherm data with the adsorption model.

Typical adsorption isotherms (vegetable oil concentration in hexadecane or starch versus fractional surface coverage, θ), calculated from boundary COF data using Equation 18.3, are given in Figures 18.13 and 18.14. θ represents the concentration of the vegetable oil on the metal or starch surface, and the COF data are used in the calculation of θ (Figures 18.13 and 18.14). As can be seen in Figures 18.13 and 18.14, the concentration versus θ profile (or the adsorption isotherm) is a mirror image of the concentration versus COF profile.

Various models are available for analyzing adsorption isotherms [5,6]. The simplest such model is called the Langmuir adsorption model [31], which assumes that adsorption is only due to primary interaction between the vegetable oil and the surface. The model assumes no

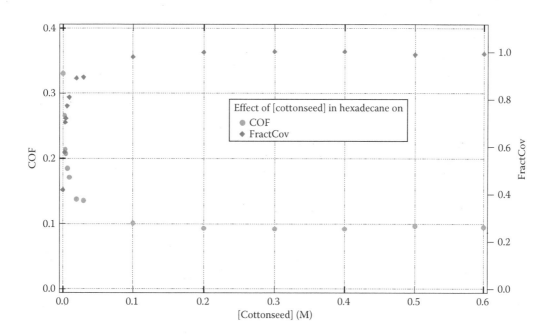

FIGURE 18.13 Concentration in hexadecane versus fractional surface coverage, θ, of vegetable oil from boundary friction data on a ball-on-disk tribometer.

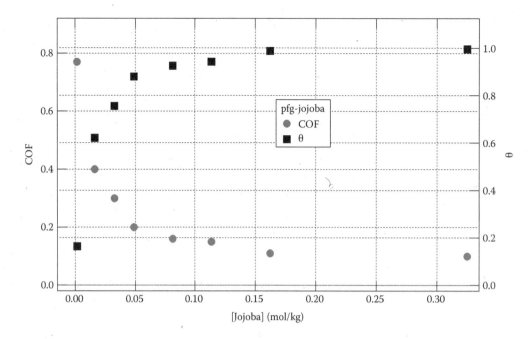

FIGURE 18.14 Concentration in starch versus fractional surface coverage, θ, of vegetable oil from boundary friction data on a ball-on-flat tribometer.

contribution to adsorption from lateral interaction between the adsorbed molecules. The Langmuir model predicts the following relationships between surface and solution concentrations of vegetable oils:

$$\frac{1}{\theta} = 1 + \left(\frac{1}{\{K_o [O_b]\}} \right) \quad (18.6)$$

According to Equation 18.6, a plot of $[O_b]^{-1}$ versus θ^{-1} should give a straight line with a slope of K_o^{-1} and an intercept of 1. The free energy of adsorption, ΔG_{ads}, is then obtained by substituting the value of K_o obtained from Equation 18.6 in the following equation:

$$\Delta G_{ads} = \Delta G_o = -RT \left[\ln (K_o) \right] \quad (kcal/mol) \quad (18.7)$$

where
  ΔG_o is the free energy of adsorption due to primary interaction in kilocalories per mole
  R = 1.987 cal/K mol is the universal gas constant
  T is the absolute temperature in Kelvin

Figures 18.15 and 18.16 show analyses of boundary friction data of vegetable oil in steel–steel and starch–steel contact using the Langmuir model. In all cases, plots of $[O_b]^{-1}$ versus θ^{-1} gave straight lines, with intercepts close to 1. From the slopes, K_o values were obtained and used to calculate ΔG_{ads} using Equation 18.7.

Another model that has been used to analyze friction-derived adsorption isotherms is the Temkin model [32]. Unlike the Langmuir model, the Temkin model assumes a net repulsive lateral interaction, and the free energy of adsorption is given as follows:

$$\Delta G_{ads} = G_o + \alpha \theta \quad (kcal/mol) \quad (18.8)$$

where
  ΔG_o is the free energy of adsorption due to primary interaction calculated using Equation 18.7
  α the free energy of adsorption due to lateral interaction ($\alpha > 0$ for repulsive lateral interactions)

Estimation of ΔG_{ads} using the Temkin model (Equations 18.7 and 18.8) requires values of K_o and α. For $0.2 \leq \theta \leq 0.8$, the Temkin model predicts the following relationship between solution concentration, $[O_b]$, and surface concentration, θ, of vegetable oil [32]:

$$\theta = \left(\frac{RT}{a} \right) \ln \left(\frac{[O_b]}{K_o} \right) \quad (18.9)$$

According to Equation 18.9, a plot of $\ln([O_b])$ versus θ in the specified θ range will result in a straight line. K_o and α are then obtained from the analysis of the slope and intercept of the line using Equation 18.9. The values of K_o and α are then used to calculate ΔG_{ads} using Equations 18.7 and 18.8.

Figure 18.17 shows the results of Temkin analysis of adsorption isotherms data derived from boundary friction measurements of vegetable oils in steel–steel contact. As can be seen in Figure 18.17, the analysis gave a straight line in the range $0.2 \leq \theta \leq 0.8$, as predicted by the Temkin model. From the slope and intercept of such analysis, K_o and α are obtained and used to calculate ΔG_{ads} using Equations 18.7 and 18.8.

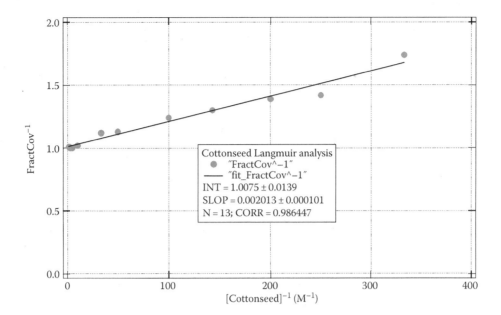

FIGURE 18.15 Langmuir analysis of adsorption isotherms derived from steel/steel boundary friction of vegetable oils in hexadecane. (From Adhvaryu, A. et al., *Ind. Eng. Chem. Res.*, 45(10), 3735, 2006. With permission.)

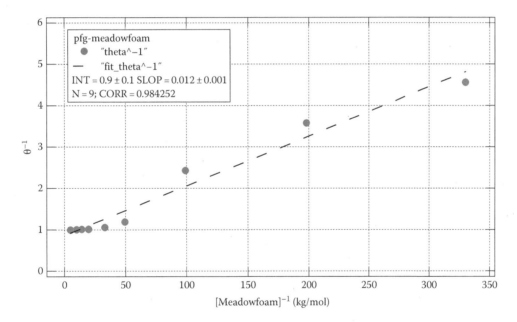

FIGURE 18.16 Langmuir analysis of adsorption isotherms derived from starch/steel boundary friction of vegetable oils in starch–oil composites.

Table 18.2 summarizes ΔG_{ads} values for various vegetable oils obtained from Langmuir and Temkin analyses of steel–steel boundary friction data obtained using the ball-on-disk tribometer. The data in Table 18.2 show that the Temkin model predicts a weaker adsorption (a smaller negative number) than the Langmuir model. This observation is consistent with the fact that the Temkin models assume a net repulsive lateral interaction ($\alpha > 0$), which will counter the strong primary interaction. Table 18.2 also shows that triglycerides generally adsorb stronger than the waxy ester jojoba, or the simple fatty acid methyl esters. This has been attributed to the ability of the triglycerides to engage in multiple binding with the surface, whereas the waxy esters and the methyl esters cannot do so since they have only one functional group per molecule. The data in Table 18.2 also indicate that the adsorption of methyl esters weakens with decreasing fatty acid chain length. This may be an indication of decreasing van der Waals interaction due to the decreasing molecular weight of the oil molecules.

Table 18.3 compares ΔG_{ads} data obtained from the Langmuir analysis of boundary friction data from steel/steel and starch/steel contact. The data in Table 18.3 show that, in general, ΔG_{ads}

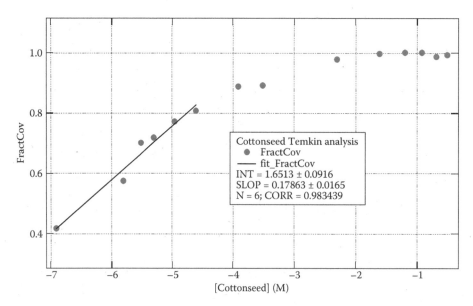

FIGURE 18.17 Temkin analysis of adsorption isotherms derived from steel/steel boundary friction of vegetable oils in hexadecane.

TABLE 18.2

ΔG_{ads} Values Obtained from Langmuir and Temkin Analyses of Steel/Steel Boundary Friction Data Obtained on a Ball-on-Disk Tribometer

Oil	Friction Pairs	Test Geometry	ΔG_{ads} (kcal/mol)	
			Langmuir	Temkin
Cottonseed	Steel/steel	Ball-on-disk	−3.68	−2.16
Canola	Steel/steel	Ball-on-disk	−3.81	−2.04
Meadowfoam	Steel/steel	Ball-on-disk	−3.66	−2.28
Jojoba	Steel/steel	Ball-on-disk	−3.27	−1.31
Methyl oleate	Steel/steel	Ball-on-disk	−2.91	−1.02
Methyl palmitate	Steel/steel	Ball-on-disk	−2.7	−1.63
Methyl laurate	Steel/steel	Ball-on-disk	−1.9	−0.6

TABLE 18.3

ΔG_{ads} Values of Vegetable Oils Obtained from Langmuir Analysis of Steel/Steel and Starch/Steel Boundary Friction Data

	Friction Pairs			
	Starch/Steel (Starch–Oil Composite Dry Film Lube)		Steel/Steel (Liquid Lube)	
Vegetable Oil	Starch Matrix	ΔG_{ads} (kcal/mol)	Base Oil	ΔG_{ads} (kcal/mol)
Meadowfoam	PFGS	−2.41[a]	Hexadecane	−3.57[b]
Jojoba	PFGS	−2.63[a]	Hexadecane	−3.27[c]
Soybean	PFGS	−2.96[a]	Hexadecane	−3.6[d]
Soybean	Waxy	−2.91[a]		

[a] Biresaw, G., Kenar, J.A., Kurth. T.L., Felker, F.C., Erhan, S.M., Investigation of the mechanism of lubrication in starch-oil composite dry film lubricants, *Lubr. Sci.*, 19(1), 41–55, 2007.

[b] Adhvaryu, A., Biresaw, G., Sharma, B.K., Erhan, S.Z., Friction behavior of some seed oils: Bio-based lubricant applications, *Ind. Eng. Chem. Res.*, 45(10), 3735–3740, 2006.

[c] Biresaw, G., Adharyu, A., Erhan, S.Z., Friction properties of vegetable oils, *J. Am. Oil Chem. Soc.*, 80(2), 697–704, 2003.

[d] Biresaw, G., Adharyu, A., Erhan, S.Z., Carriere, C.J., Friction and adsorption properties of normal and high oleic soybean oils, *J. Am. Oil Chem. Soc.*, 79(1), 53–58, 2002.

is smaller for adsorption on steel than on starch. This means that vegetable oils will adsorb stronger on metal than on starch. This observation is consistent with the fact that steel has much higher surface energy than starch and will result in a stronger interaction with the polar ester groups of vegetable oils.

18.8 CONCLUSION

Surfactants are widely used in various consumer and industrial applications. They are invaluable in various products ranging from simple cleaning to complex formulations of pharmaceuticals, foods, lubricants, etc.

Surfactants owe their success to their unique chemical structure, which involves the presence of two dissimilar segments in the same molecule. One segment is hydrophilic and is soluble only in hydrophilic or polar solvents such as water, whereas the other is lipophilic and is soluble only in lipophilic or nonpolar solvents such as hexanes. As a result, surfactants display various degrees of solubility in both hydrophilic and lipophilic solvents. By varying the structures of the hydrophilic and lipophilic segments, thousands of surfactants have been developed and used in thousands of applications.

Depending on the nature of the hydrophilic component, surfactants can be broadly categorized into anionic, cationic, zwitterionic, and nonionic. In general, ionic surfactants are more soluble in hydrophilic solvents such as water than in lipophilic solvents. The reverse is true with nonionic surfactants except for polyethoxylated alcohols. In the latter case, solubility depends on the relative *strength* of the hydrophilic and lipophilic segments, which can be systematically varied and quantified using the HLB system. The HLB of polyethoxylated alcohols varies from 0 to 20, those with low HLB (<7) are oil soluble, whereas those with high HLB (>11) are water soluble, and those in between are soluble in both oil and water.

An important property of surfactants is their ability to spontaneously organize into various assemblies when in solution. The structures of these assemblies depend on the nature of the surfactant, the nature of the solvent, the concentration of surfactant, the presence of cosolutes (e.g., salts and cosurfactants), temperature, etc. At low concentrations, surfactants form monolayers. Progressive increase in surfactant concentration leads to the formation of aggregates, micelles, lamellar, and hexagonal liquid crystals. Other organized assemblies of surfactants in solution include vesicles and foams.

With proper selection of surfactant combinations, micelles can be used to solubilize large quantities of oil in water, or large quantities of water in oil. The resulting single-phase solutions, which are transparent due to the small size of the droplets, are called o/w or w/o microemulsions.

Surfactants perform a number of functions in lubrication so that the lubricant can meet its lubricity, cooling, handling, waste treatment, disposal, and other requirements. Some requirements of lubricants are contradictory and require proper selection of surfactant combinations to achieve the desired objectives. An example of a contradictory requirement is that the lubricant possesses both biostability and biodegradability. During use, the lubricant should be biostable, and the purpose of the surfactant is to protect it from attack by microbes (bacteria and fungi). On the contrary, used surfactants that are subjected to waste treatment and disposal processes should be biodegradable, that is, easy to be broken down by composting microorganisms, and the purpose of the surfactant is to enhance biodegradation.

Lubricant formulations can be broadly categorized into oil based (straight oils) and water based. Formulation of water-based lubricants is one of the most important functions of surfactants in lubrication. Water-based lubricants are used in processes that generate a lot of heat and must be cooled to prevent tool wear, workpiece damage, fire, and other undesirable outcomes. Surfactants allow dispersion of the oil-phase lubricant formulation in water, without interfering with the lubrication function of the oil phase and the cooling function of the water phase. Depending on the concentration of solubilized oil phase, water-based lubricants are classified into three groups: soluble oils, with up to 20% oil; semisynthetic, with up to 5% oil; and synthetic, with no solubilized oil. Soluble oils are o/w emulsions, which, unlike o/w microemulsions, are milky and thermodynamically unstable. Semisynthetic oils are o/w microemulsions, thermodynamically stable single-phase clear solutions. Synthetic oils are micellar solutions with solubilized additives that provide lubricity and other functions.

Another important function of surfactants in lubrication is adsorbing on solid and liquid surfaces, and interfaces, and modifying surface and interfacial properties. Agriculture-based surfactants derived from farm products such as proteins, starches, and vegetable oils are effective at reducing surface tension, interfacial tension, surface energy, and boundary friction. Boundary friction investigations in metal/metal and starch/metal contacts revealed that vegetable oils are effective boundary additives. Free energy of adsorption values obtained from the analysis of friction-derived adsorption isotherms was consistent with the expected effect of vegetable oil chemical structures and friction surface properties on adsorption.

Surfactants have been known, investigated, and applied for centuries. Despite that the current rapid development of new and complex products, processes, and equipment has produced a great deal of demand for new and improved surfactants and a deeper understanding of their performance. As a result, surfactants will continue to be the subject of vigorous research and development effort well into the future.

ACKNOWLEDGMENT

We are grateful to Danielle Wood for his help in preparing this manuscript.

REFERENCES

1. Milton, J.R. *Surfactants and Interfacial Phenomena*, 3rd edn., Wiley, New York, 2004.
2. Lange, K.R. *Surfactants: A Practical Handbook*, Hanser-Gardner, Cincinnati, OH, 1999, pp. 1–237.

3. Myers, D. *Surfactant Science and Technology*, 3rd edn., VCH, Berlin, Germany, 1992, pp. 1–380.
4. Mittal, K.L., ed. *Contact Angle, Wettability, and Adhesion*, VSP, Utrecht, the Netherlands, 1993.
5. Hiemenz, P.C., Rajagopalan, R. *Principles of Colloid and Surface Chemistry*, 3rd edn., Marcel Dekker, New York, 1997, p. 327.
6. Adamson, A.P., Gast, A.W. *Physical Chemistry of Surfaces*, Wiley, New York, 1997.
7. Israelachvili, J. *Intermolecular and Surface Forces*, 2nd edn., Academic Press, London, U.K., 1992, p. 150.
8. Holmberg, K. *Surfactants and Polymers in Aqueous Media*, 2nd edn., Wiley-Interscience, New York, 2002, pp. 1–562.
9. Becher, P. *Emulsions, Theory and Practice*, 3rd edn., Oxford University Press, New York, 2001.
10. Anonymous. *The HLB System—A Time-Saving Guide to Emulsifier Selection*, ICI United States Inc., Wilmington, DE, 1976.
11. Biresaw, G., Bunton, C.A. Application of stepwise self-association models for the analysis of reaction rates in aggregates of tri-n-octylalkylammonium salts, *J. Phys. Chem.*, 90(22): 5854–5858, 1986.
12. Byers, J.P. *Metalworking Fluids*, Marcel Dekker, New York, 1994.
13. Schey, J.A. *Tribology in Metalworking: Friction, Lubrication and Wear*, American Society of Metals, Metals Park, OH, 1983.
14. Mohamed, A.A., Peterson, S.C., Hojilla-Evangelista, M.P., Sessa, D.J., Rayas, D., Biresaw, G. Effect of heat treatment and pH on the thermal, surface, and rheological properties of *Lupinus albus* protein, *J. Am. Oil Chem. Soc.*, 82(2): 135–140, 2005.
15. Mohamed, A., Hojilla-Evangelista, M.P., Peterson, S.C., Biresaw, G. 2007 Barley protein isolate: Thermal, functional, rheological and surface properties, *J. Am. Oil Chem. Soc.*, 84: 281–288, 2007.
16. Shogren, R., Biresaw, G. 2007 Surface properties of water soluble starch, starch acetates and starch acetates/alkenyl-succinates, *Colloids Surf. A Physicochem. Eng. Asp.*, 298(3): 170–176, 2007.
17. Biresaw, G., Shogren, R. Friction properties of chemically modified starch, *J. Synth. Lubr.*, 25: 17–30, 2008.
18. Biresaw, G., Adharyu, A., Erhan, S.Z., Carriere, C.J. Friction and adsorption properties of normal and high oleic soybean oils, *J. Am. Oil Chem. Soc.*, 79(1): 53–58, 2002.
19. Kurth, T.L., Byars, J.A., Cermak, S.C., Biresaw, G. Nonlinear adsorption modeling of fatty esters and oleic estolides via boundary lubrication coefficient of friction measurements, *Wear*, 262(5–6): 536–544, 2007.
20. Biresaw, G., Adharyu, A., Erhan, S.Z. Friction properties of vegetable oils, *J. Am. Oil Chem. Soc.*, 80(2): 697–704, 2003.
21. Adhvaryu, A., Biresaw, G., Sharma, B.K., Erhan, S.Z. Friction behavior of some seed oils: Bio-based lubricant applications, *Ind. Eng. Chem. Res.*, 45(10): 3735–3740, 2006.
22. Birsaw, G., Liu, Z.S., Erhan, S.Z. Investigation of the surface properties of polymeric soaps obtained by ring opening polymerization of epoxidized soybean oil, *J. Appl. Polym. Sci.*, 108: 1976–1985, 2008.
23. Lawate, S.S., Lal, K., Huang, C. Vegetable oils—Structure and performance, in *Tribology Data Handbook*, Booser, E.R. ed., CRC Press, New York, 1997, pp. 103–116.
24. Fanta, G.F., Eskins, K. Stable starch-lipid compositions prepared by steam jet cooking, *Carbohydr. Polym.*, 28: 171–175, 1995.
25. Biresaw, G., Erhan, S.M. Solid lubricant formulations containing starch-soybean oil composites, *J. Am. Oil Chem. Soc.*, 79: 291–296, 2002.
26. Erhan, S.M., Biresaw, G. Use of starch-oil composites in solid lubricant formulations, in *Biobased Industrial Fluids and Lubricants*, Erhan, S.Z., Perez, J.M., eds., AOCS Press, Champaign, IL, 2002, pp. 110–119.
27. Biresaw, G. Biobased dry film metalworking lubricants, *J. Synth. Lubr.*, 21(1): 43–57, 2004.
28. Biresaw, G., Kenar, J.A., Kurth, T.L., Felker, F.C., Erhan, S.M. Investigation of the mechanism of lubrication in starch-oil composite dry film lubricants, *Lubr. Sci.*, 19(1): 41–55, 2007.
29. Jahanmir, S., Beltzer, M. An adsorption model for friction in boundary lubrication, *ASLE Trans.*, 29: 423–430, 1986.
30. Jahanmir, S., Beltzer, M. Effect of additive molecular structure on friction coefficient and adsorption, *J. Tribol.*, 108: 109–116, 1986.
31. Langmuir, I. The adsorption of gases on plane surfaces of glass, mica and platinum, *J. Am. Chem. Soc.*, 40: 1361–1402, 1918.
32. Temkin, M.I. Adsorption equilibrium and the kinetics of processes on non-homogeneous surfaces and in the interaction between adsorbed molecules, *J. Phys. Chem. (USSR)*, 15: 296–332, 1941.

19 Foaming Chemistry and Physics

Kalman Koczo, Mark D. Leatherman, Kevin Hughes, and Don Knobloch

CONTENTS

19.1 INTRODUCTION

Foam control agents (antifoams, defoamers) are important, although often neglected, components of lubricating fluids. In this chapter, we will attempt a systematic discussion of the chemistry, properties, mechanism of operation, and use of these components. During use, air can be incorporated into a lubricating fluid by mechanical action, causing air entrainment and then foam. In the context of lubricating fluids, foams are undesirable. The reasons for this include possible overflow of fluids (through breather tubes, vents, etc.) leading to equipment failure, impaired power transfer, accelerated oxidation, erratic shifting (transmission fluids), reduced hydrodynamic lubrication, and reduced heat transfer capabilities. Foam problems can sometimes be mitigated with proper equipment design, but in the other cases, the addition of foam control agents is necessary. In lubricating oils, these foam control agents take the form of small (~micron scale), insoluble liquid droplets that are dispersed in the oil. In spite of the potential troubles, the formulators often neglect to select an optimal antifoam for a particular lubricant. Although antifoams are dosed at very low concentrations, typically 10–100 ppm, their importance requires more than 0.01%–0.001% of the formulator's attention!

In this chapter, first, the basic properties of foams and antifoams will be reviewed, emphasizing the significant differences between water-based and oil-based foams (Section 19.2). In Section 19.3, the various root causes of foam in lubricating fluids will be discussed, as well as the methods available to test for foaming tendency and stability.

Section 19.4 will describe the chemistry, manufacturing, and properties of the most important types of silicone- and nonsilicone-based foam control agents, for both aqueous and nonaqueous lubricating fluids. This section will include discussion of organomodified silicone foam control agents, which are less known to formulators. Among the foam control agents, one can distinguish between the terms *antifoams* and *defoamers*, in that antifoams are added to the foaming liquid to prevent foam *before* its generation, whereas defoamers are added *after* foam formation to eliminate it. In this chapter, we will mostly use the former term, since this is the typical way these chemicals are used with lubricants. However, these terms are often mixed in practice, and the same agents can often act both to prevent and to eliminate foam. Other terms are also used, such as *foam suppressors*, *foam breakers*, and *foam inhibitors*.

In the recent decades, there has been significant progress in the understanding of the mechanisms of antifoaming action, including the deactivation of antifoams, especially for water-based foams. The main theories will be discussed in Section 19.5.

Finally, we will give some practical information on the use of antifoams in lubricating oils, including information on the incorporation of antifoams into lubricating oils. One important aspect of antifoams that sets them apart from many other additives is that their performance and stability are strongly dependent on the method used to incorporate them into the finished fluid or concentrated additive package. This is because the insoluble droplets that make up the antifoam are formed during the blending process, which will be described in detail in Section 19.6. Since antifoams are insoluble droplets, a challenging aspect to their use is that they can settle out of the product during shipment and storage, which is a particular concern in lubricant applications. Therefore, smaller antifoam droplet size is preferred, since this will slow the eventual settling of the antifoam droplets. These discussions will be illustrated with data not only from the literature but often with original data generated by the authors and their companies.

Throughout this entire chapter, it is important to keep in mind the significant differences between oil-based (nonpolar liquid based) and water-based systems, such as the high surface tension of water compared to oils. Among lubricating fluids, only metalworking fluids (MWFs) and some hydraulic fluids are used in water dilution, the rest are generally oil based. However, water-based liquids are much more commonly used in other industries (paper, textile, agriculture, coatings, etc.), and these solutions are generally easier to study. Therefore, the literature and knowledge related to surfactants, micellization, thin liquid films, foams, and antifoaming mechanisms are much more detailed with aqueous systems than in oil-based systems. In fact, several aspects of foaming (e.g., antifoaming mechanisms) are barely investigated, so one must speculate based on analogies with water-based systems to explain phenomena in oil-based liquids.

19.2 FOAM BASICS

In this section, we will summarize the main phenomena that control the behavior of liquid films and foams. In the next section, we will discuss how these phenomena relate to foaming in lubricating oil specifically.

Gas can be incorporated into a liquid in a number of ways. In lubricating oil applications, a common route for air incorporation is by mechanical mixing, though other routes exist, for example, the precipitation of dissolved air into bubbles upon sudden pressure change. When gas is dispersed into a liquid, it generally exists in the form of subsurface bubbles, which then rise due to buoyancy, unless the liquid is under intensive mixing. When the bubbles reach the surface of the liquid, they form layers and then foam [1,2].

As a rising bubble approaches the liquid surface, it forms first a thick lamella (see Figure 19.1).

Initially, the foam contains only roughly spherical bubbles separated by relatively thick liquid films (also known as a *wet foam*). After some drainage has taken place, a polyhedral foam

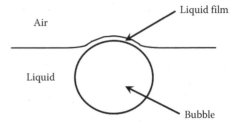

FIGURE 19.1 Formation of liquid film as a bubble approaches the liquid surface.

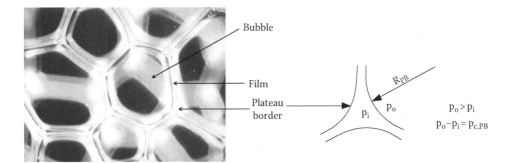

FIGURE 19.2 Picture of a polyhedral foam.

is formed, which is essentially composed of air chambers separated by thin liquid films called *lamellae*. These are clearly seen in the picture of a real foam shown in Figure 19.2. Lamellae always intersect in groups of three, meeting in channels called *Plateau borders*. These channels form a drainage system inside the foam, where most of the liquid is located in a polyhedral foam. It will be discussed in Section 19.5 that the Plateau borders play an important role in the antifoaming mechanism.

In pure liquids with low bulk viscosity (e.g., pure water or low-viscosity solvents), the film generally ruptures very fast and no stable foam can form. However, if the fluid has appreciable bulk viscosity [3], or contains foam-stabilizing components [4], then more or less stable foam films, and thus a stable foam layer, can form. Both are possible in the case of lubricating oils.

Foam-stabilizing materials generally include

- Surface-active molecules, including
  - Smaller molecules, such as traditional surfactants
  - Polymers
- Fine solid particles

Foam stabilizers, especially in water-based systems, are often amphiphilic surfactants, containing hydrophilic (ionic or nonionic) and hydrophobic (lipophilic) parts, typically one of each. Examples of polymeric (macromolecular) foam stabilizers are proteins in water, which also must have both hydrophobic and hydrophobic sections, typically many of each. In *aqueous systems*, these surface-active molecules spontaneously adsorb at the water surface due to their hydrophobic groups because the attraction between the water molecules is stronger than between the water molecules and the hydrophobic section of the surfactant molecules (or ions), resulting in decreased surface tension. The amount of surfactant adsorbed at the surface increases, and thus the surface tension decreases, with increasing surfactant concentration, until saturation is reached. At surfactant concentrations above the critical micelle concentration, the surface tension levels off and the surfactant molecules aggregate into micelles due to intermolecular interactions. In aqueous systems, the hydrophobic groups orient toward the inner core of the micelle, while the hydrophilic groups orient on the outside. With nonpolar solvents, analogous processes happen, but the role

of hydrophobic and hydrophobic groups switch and reverse micelles can form in which the hydrophilic groups are inside the reverse micelles and the hydrophobic groups are outside.

A number of additives present in lubricating fluids may fit into the category of *smaller molecule* surfactants, including detergents, friction modifiers (FMs), or any other component with a polar group attached to a long-chain hydrocarbon. Some viscosity modifiers, including so-called dispersant viscosity modifiers, fall into the category of surface-active polymers.

Solid particles can also stabilize liquid films and foams very well, if the contact angle of the particles is intermediate (partial wetting). A special case is the so-called Janus particles, which have a hydrophobic and a hydrophilic section, making them adsorb especially well at the surface. Solid stabilized foams are not typical in lubricating fluids and we will not discuss them further.

As stated previously, foams are not thermodynamically stable structures. The amount of foam in a lubricating oil application should instead be considered to be the sum of a number of kinetic effects. At the most basic level, the degree of foaming is the overall sum of two competing processes, which take place simultaneously: (1) the generation of foam originating through air entrainment (by mechanical action or some other mechanism) and (2) the destruction of the foam by drainage and rupture of foam films. Since the rate of foam generation is not generally under the formulator's control, we will focus our discussion on factors that affect the drainage and rupture of foam films.

The processes of drainage and rupture are themselves controlled by a number of competing effects. Driving forces for foam drainage and thinning include gravity, which draws fluid downward into the bulk phase, and capillary pressure that forces liquid from the lamellae into the Plateau borders.

Because of surface tension, the pressure inside a curved liquid surface is higher than outside of it, and the difference is the *capillary pressure* (Laplace pressure). According to the Young–Laplace equation, the capillary pressure (p_c) is

$$p_c = \sigma\left(\frac{1}{R_1} + \frac{1}{R_2}\right), \tag{19.1}$$

where

R_1 and R_2 are the principal radii of curvature
σ is the surface tension

For a bubble (sphere), the radii of curvature are equal, and the equation simplifies to

$$p_c = \frac{2\sigma}{R}, \qquad (19.2)$$

where R is the bubble radius. The capillary pressure in the Plateau borders is

$$p_{c,PB} = \frac{\sigma}{R_{PB}} \qquad (19.3)$$

Note that $R_2 = \infty$ in Equation 19.1, since the Plateau border is practically straight perpendicular to the drawing of Figure 19.2. Since the capillary pressure is close to zero in the adjoining liquid films (which are flat or have high radii of curvature), the pressure inside the Plateau borders is lower, thereby forcing liquid from the films into the Plateau borders.

Another consequence of Equation 19.2 is that, for a foam in which the bubbles have a size distribution (shown in Figure 19.2), the pressure inside the smaller bubbles is higher than in the larger ones. Therefore, the gas will diffuse from the smaller bubbles to the larger ones through the films, and the small bubbles eventually disappear. However, the rate of this interbubble gas diffusion is generally slow relative to other processes and has a minor effect on foam stability.

Drainage of fluid from the films by gravity and capillary pressure is opposed by the bulk viscosity of the liquid, the Gibbs–Marangoni effect, and other related surfactant effects. Considering viscosity first, a fluid with higher viscosity will drain slower in the presence of gravity and capillary pressure. Therefore, a highly viscous fluid can form a foam with considerable kinetic stability, even if other surfactant-driven stabilizing forces are not present. However, the effect of bulk viscosity on foaming of lubricants is not straightforward. While lamellae composed of viscous fluids are more kinetically stable, it is more difficult to incorporate small bubbles into a more viscous fluid, and it is these small bubbles that ultimately form a foam layer. In other words, it is more difficult to generate the foam structure in a more viscous fluid. For this reason, the relationship between bulk viscosity and foam can be non-monotonic [3].

The *Gibbs–Marangoni effect* is another important film-stabilizing phenomenon that requires the presence of surfactants. As the foam film between two bubbles drains, by either gravity or capillary pressure, the viscous drag by the outflow of liquid sweeps adsorbed surfactants toward the adjoining Plateau borders and creates a surface tension gradient. This gradient opposes film drainage, as shown in Figure 19.3.

If the effect is low (due to low or very high surfactant concentration, or because the change in surface tension as a function of surfactant surface concentration is small), then the film surfaces will be very mobile and the drainage will be fast. If, however, the surface tension gradients are large, the film surface will be rigid and the drainage rate will be slow. Wasan and coworkers [5,6] found that the difference in the drainage rates of rigid and mobile films can be several fold.

A closely related effect that also plays a role in determining foam stability is surface viscoelasticity. Analogous to bulk viscoelasticity, surface viscoelasticity originates from intermolecular forces present at the air/liquid phase boundary. Surface shear viscoelasticity, for example, refers to the viscoelastic response when the surface is sheared laterally. Surface dilational viscoelasticity refers to changes in surface tension when the surface is dilated. The purely elastic response to a change in surface area can be expressed as the Gibbs elasticity [7,8], which gives the change in surface tension σ, by the increase of area, A:

$$E_0 = \frac{2d\sigma}{d \ln A} \qquad (19.4)$$

In other words, when a liquid surface containing adsorbed surfactants is expanded, the local surfactant concentration decreases, and therefore the surface tension increases. If the surface expansion is fast relative to the rate of surfactant diffusion and adsorption at the air/fluid surface, then there is not enough time for the surfactant molecules to reestablish

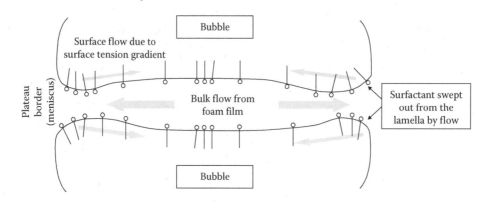

FIGURE 19.3 Gibbs–Marangoni effect in a draining liquid film containing surfactants. The viscous drag created by the outflow of liquid from the films toward the meniscus creates a surface tension gradient that opposes film drainage. (Reprinted from *Progr. Surf. Sci.*, 33, Wasan, D.T., Nikolov, A.D., Lobo, L.A., Koczo, K., and Edwards, D.A., Foams, thin films and surface rheological properties, 119–154, Copyright 1992, with permission from Elsevier.)

equilibrium at the surface, and the increased (dynamic) surface tension will act as a restoring force on the film. Fast diffusion and adsorption of surfactant from the bulk to the surface have the effect of decreasing the change in surface tension to below that expressed by the Gibbs elasticity, making perturbations, that is, drainage and deformation, of the film easier. The viscous response to dilation or compression of a surface is driven by the exchange (adsorption/desorption) of surfactant between the surface and the bulk fluid, or by the rearrangement of surfactant molecules at the surface [8]. High surface viscosity and elasticity, imparted by surfactants, act to resist deformation of the lamellae and are therefore associated with more stable foams [9].

It is important to note that the magnitude of surface tension change due to the presence of adsorbed surfactant is necessarily less in oil-based systems compared to water based. This is due to the lower surface tension of the *clean* (surfactant free) oil surface (generally 25–35 mN/m), compared to the clean water surface (~72 mN/m). The consequence of this is that the phenomena described here—Gibbs–Marangoni effect and surface viscoelasticity—are generally much weaker in oil-based systems.

Another mechanism of film drainage was studied by Wasan and coworkers [5,10–14] with solutions containing micelles or particles of similar size. In this case, the micelles or particles can spontaneously form ordered layers in the thinning film, and the film drains step by step, layer by layer.

As drainage takes place, the two opposite surfaces of the thinning films become closer to one another. Below about 100 nm film thickness, the two surfaces, which may have layers of adsorbed surfactant, can interact via Van der Waals attraction, ion–ion repulsion (with ionic surfactants) and steric repulsion. The net interaction is expressed as disjoining pressure, Π [5,15–17].

With a stable foam, an equilibrium can be reached when the capillary pressure is equal to the disjoining pressure (so that the drainage stops from the films to the Plateau borders), and at the same time, the capillary pressure in the Plateau borders is equal to the (negative) hydrostatic head measured from the bottom of the foam, that is, the capillary pressure is the highest and the films are the thinnest at the top of the foam and the bottom part of the foam is spherical with thick films [18]. Therefore, foams generally start rupturing from the top. In oil-based liquid films, the ion–ion repulsive interactions are mostly absent, one key reason why oil-based films and foams are less stable than with aqueous systems.

As the thin liquid film drains further, it can reach a critical thickness (often around 30 nm in aqueous liquids) where it either ruptures or becomes a stable, common or Newton black film (about 10 nm thickness) [5,15–17,19–21]. It is this spontaneous rupture at the critical thickness or at very low film thickness that destroys foams in the absence of antifoams. The role of an antifoam, to be discussed in detail in Section 19.5, is to *short-circuit* and thus highly accelerate the process of thinning and spontaneous rupture by introducing artificial instability in the foam structure.

19.3 FOAMING IN LUBRICATING FLUIDS

In this section, we will discuss the connection between the fundamental mechanisms driving the formation of stable foams as discussed in Section 19.2, as they relate to the properties and composition of typical lubricating fluids. First, the testing methods of lubricating fluids will be reviewed, followed by a discussion of the relationships between fluid properties and foaming tendency/stability. This latter discussion will be further divided into the effects of bulk properties of the fluid (e.g., viscosity and solvency), and the surfactant effects of components in the additive package.

It is important to note that there are a number of key differences between the aqueous systems discussed in much of the literature on the topic of foams, and the oil-based systems that are relevant to the lubricant formulator (see Section 19.1). One of the most significant among these is the bulk viscosity of typical lubricating oils, which tends to be higher than water. This means that the bulk viscosity will play a very significant role in foam stability. The second key difference is that the surface tension of a typical base oil (roughly 30 mN/m) is significantly lower than water (~72 mN/m), which means that the magnitude of the surfactant-related forces will be less significant than in water-based systems [22].

19.3.1 LUBRICANT FOAM TESTING METHODS

A variety of empirical methods have been developed to test the foaming tendency of fluids and to compare the effectiveness of various antifoams. These methods will be reviewed in this section. Foam test methods generally consist of two parts: (1) a mechanism to introduce air bubbles into the fluid, which rise and produce a surface foam and (2) a scheme to measure the height of foam produced, and/or the stability of that foam.

An ideal foam testing method for lubricants should meet several requirements, the most important of which are the following:

- The test should cover typical temperatures during lubrication and the temperature control should be accurate in the procedure.
- The test should be relatively easy to perform and standardize, making it accessible to many users.
- Good reproducibility and repeatability.
- Cleaning of the reusable equipment between tests should be relatively easy; any consumables should be inexpensive.
- The test should mimic the real-use conditions (shear, turbulence, temperature, etc.).

In practice, all existing foam test methods are some sort of compromise of all these requirements and have advantages and disadvantages. It is the last of these—mimicking real-use conditions—which is the most difficult and most commonly overlooked. This is because the real-use conditions can be difficult to reproduce on a lab scale, and because these conditions can vary significantly between different applications. Therefore, most test methods in common use attempt to optimize the remaining criteria. However, it should be noted that some OEMs have developed foam test methods using relevant equipment.

The methods can be classified by the mechanism by which the air bubbles are introduced [23,24]:

- *Shaking tests*, in which the fluid is shaken by hand or by a machine. Machine shaking, using a wrist action shaker, for example, can lead to better automation and reproducibility. The main advantage to these types of tests is that that they are fast and simple and that generally it is not necessary to clean the equipment between tests. The simple versions also do not require specialized equipment. However, care must be taken to ensure repeatability, especially in the case of hand shaking by different operators [25]. Furthermore, shaking is not a close representation of the type of fluid flow experienced by lubricants during use in a device; therefore, these tests may introduce unwanted bias in foam test results that is not representative of the fluid in use. Accurate heating to the typical use temperature of lubricating fluids is also problematic.
- *Liquid cascade or Ross–Miles tests* [26], in which the fluid falls in a stream from a fixed height onto the surface of an existing volume of fluid.
- *Aeration tests*, in which air is introduced into the fluid through a subsurface delivery tube or similar device. This type of test delivers a fixed volume of air to the fluid and provides better ability to control temperature precisely. The ASTM tests D892 and D6082, which were developed specifically for lubricating fluids and are widely used, fall into this category [27,28]. These tests will be discussed in more detail in Section 19.3.1.1. A disadvantage of the aeration tests is that the foam generation conditions are far from the operating conditions (shear and mixing rates, etc.) of lubricating fluids.
- *Mechanical agitation tests*, in which the fluid is agitated, for example, by spinning mixer blades, gears, moving a perforated disc, or by full devices like gearboxes, running engines, or circulation systems. These tests can provide the best reproduction of how air is incorporated into the fluid during use. However, they require the use of specialized, often expensive equipment. With these types of tests, one should keep in mind that the test may be probing other aspects of the fluid besides its tendency to support a stable foam. Examples include the tendency of a foam inhibitor to break down over time, due to temperature or shear, or the tendency of the foam inhibitor to be caught in an oil filter. An example of this type of test is the so-called Flender foam test, or ISO 12152 [29]. In this test, a set of two horizontally oriented spur gears rotate in the test fluid, incorporating air by shear.

In many cases, standard test methods using these approaches have been published for general use. Table 19.1 lists a number of these tests that are relevant to lubricating fluids. As shown in the "Scope" column, many of these methods were not developed for lubricating fluids specifically. However, the general strategies used in these methods may be applicable across aqueous and nonaqueous fluids, provided appropriate modifications are made (in consideration of fluid viscosities, the need for alternate cleaning solvents, measurement time scales, and other factors).

Included in these test methods are detailed recommendations and requirements for performing the tests, including instructions for calibrating and cleaning test equipment and handling liquid samples and other materials used in the test. In addition, there are a number of other practical considerations unique to foam testing that are not detailed in these specifications. The most important of these is the *history* of the sample to be tested. Because the foam inhibitor consists of insoluble liquid droplets, these droplets can fall to the bottom of the storage vessel, or otherwise migrate or degrade over time. Therefore, the length of shelf storage since blending, as well as the details of how the sample was collected (i.e., whether the sample was drawn from the bottom of a holding tank, or off the top, or poured from a larger sample container, whether that container was shaken or stirred before sampling), can have a significant effect on the results of the test. There is no set of general rules that can be given for the handling of a sample before testing, since all storage and sample movement situations are different. The best recommendation that can be given is to ensure that the storage and handling of samples for foam testing are done in a consistent way, and in a way that is representative of how the sample will be stored and handled in practice by the end user. Needless to say, sample history can also include any prestressing of the sample (e.g., heating, shearing, filtration, or centrifugation), or use of the fluid in a field, bench, or engine test. Again, in these cases, care must be taken to treat all samples consistently to ensure the validity of sample-to-sample comparisons.

19.3.1.1 D892 and D6082 Tests

The ASTM D892 and D6082 methods are of particular relevance, because they are widely used in the area of lubricating fluids and are commonly found industry-wide (i.e., ILSAC GF and PC specifications) and in OEM-specific lubricant specifications [27,28].

These two methods both describe air-blowing tests, meaning that air is introduced at the bottom of a graduated cylinder containing the test fluid, through a perforated *diffuser*, at a controlled rate (see Figure 19.4). After 5 min of aeration, the air flow is stopped and the height of foam is recorded. The foam is then allowed to settle for a fixed period of time, commonly 10 min, and then the foam height is recorded a second time. Taken together, the two methods involve repeating this procedure four times:

D892 Sequence I: Fresh aliquot of fluid tested at 24°C
D892 Sequence II: Fresh aliquot of fluid tested at 94.5°C

TABLE 19.1
Summary of Standard Methods for Foam Testing Relevant to Lubricating Fluids

Test Type	Designation	Data Reported	Scope (If Specified in Method)	Last Updated or Reapproved
Shaking	ASTM D3601	Total foam height after shaking; time to reach 10 mm of foam height; foam remaining after 5 min	Low-viscosity aqueous liquids	2007 (withdrawn 2013)
	ASTM D4921	Total foam height after shaking	Solutions of aqueous engine coolant in water	2012
Mechanical agitation	ASTM D3519	Foam height after switch off; time to 10 mm height (or height at 5 min if >10 mm)	Low-viscosity aqueous liquids	2007 (withdrawn 2013)
	DIN EN 12728	Beating method using a perforated disc	Surface-active agents	2000
	E2407	Reduction in foam volume 7 min after adding defoamer	Aqueous surfactant solutions	2009
	ISO12152 (Flender foam test)	Level of entrained air and foam height over 90 min	Oils used for the lubrication of gears	2012
Liquid cascade	D655212 (CNOMO)	Foam height at 5 h circulation, or time to reach 2000 mL foam		
	ASTM D1173	Foam height after 0, 1, 3, and 5 min		2015
	DIN 53902-2	Foam height after 0.5, 3, and 5 min after flow. Modified Ross–Miles method	Aqueous surfactant solutions	
Air blowing	ASTM D892, D892_A IP 146, DIN 51566, JIS K-2518, ISO 6247	Foam height at 0 and 10 min	Lubricating oils	2013 (ASTM D892)
	ASTM D6082 ASTM D6082_A	Foam height at times from 0 to 10 min (static). Entrained air at 0 min (kinetic)	Lubricating oils	2012
	ASTM D7840	Initial foam height; time for foam to collapse	Nonaqueous engine coolants	2012
	ASTM D1881	Initial foam height; time for foam to collapse	Solutions of engine coolants in water	2009
	ASTM D3427, DIN 51381	Time for fluid to release 99.8% of entrained air	Turbine, hydraulic, and gear oils	2014 (ASTM D3427)

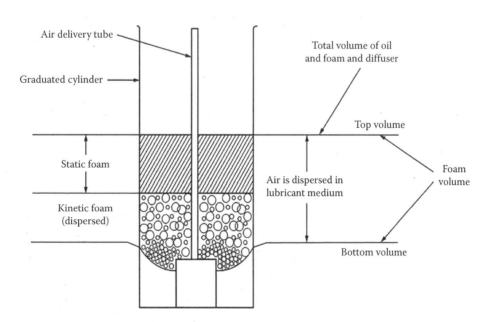

FIGURE 19.4 Schematic of the test cylinder used in the D892 and D6082 tests. (Reprinted with permission from ASTM D6082-12, Standard test method for high temperature foaming characteristics of lubricating oils, Copyright ASTM International, West Conshohocken, PA. A copy of the complete standard may be obtained from ASTM, www.astm.org.)

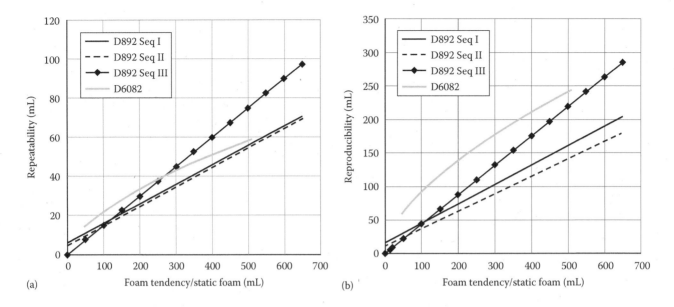

FIGURE 19.5 (a) Repeatability and (b) reproducibility of the ASTM D892 and D6082 foam test methods, according to information given in the respective methods [27,28]. Data for D6082 refer to the static foam result.

D892 Sequence III: Same aliquot of fluid used in Seq II tested, at 24°C (after foam from Seq II collapses to zero height)

D6082 (sometimes referred to as "Sequence IV"): Fresh aliquot of fluid tested at 150°C

There are two notable differences between the D892 and the D6082 procedures. First, the air flow rate for the D892 test is specified to be 94 mL/min and 200 mL/min in the D6082. Also, in the D892 procedure, only the height of the foam above the fluid surface is reported. In the D6082, this is reported as the *static foam*, and, in addition, the height of the layer of aerated oil is reported as the *kinetic foam* (also shown in Figure 19.4).

A key part of the test apparatus in the D892 and D6082 tests is the diffuser that introduces the air bubbles into the fluid. In general, there are two types of porous diffusers commonly in use: sintered steel canisters and *stone*-type diffusers made from fused alumina particles. As of the latest revision/reapproval of the two methods, either type may be used in the D892 test, while only the sintered steel type is approved for D6082. Both ASTM methods specify the permeability and maximum pore size of the diffuser. Testing of both types of diffusers has shown that the sintered metal type is generally more reliable in terms of meeting required specifications of the test procedure [30]. However, studies into whether the type of diffuser has an effect on the results of the test have produced conflicting results.

The repeatability and reproducibility of the D892 and D6082 tests are shown graphically in Figure 19.5, based on information provided in the test specifications. One should keep the fairly poor precision of these tests in mind when evaluating antifoams. For example, the reproducibilities of Seq I, II, and III for a foam tendency of 100 mL are all about 40 mL in Figure 19.5b; that is, results, for example, between 100 and 140 mL are not statistically different, nor are repeat results of 10 and 30 mL (see Figure 19.5a).

A useful way to conceptualize the D892 and D6082 tests is as a competition between two rates. The first is the rate of foam *generation*, which is generally held constant by the constraints of the test. This rate is in competition with the rate of foam *destruction*, which is determined by the processes described in Section 19.2, and by the foam inhibitor. If the foam inhibitor system is performing well, the rate of destruction is higher than the rate of generation, and the result of the test will be a very low or zero foam height. If the rate of destruction does not exceed the rate of generation throughout the test, then the result will be significantly nonzero. In the D892 test sequences, the fluid charge volume combined with the total volume of air blown through the fluid during the course of the test is ~660 mL, so this represents the theoretical maximum for the test. In the D6082 test, the fluid charge + total volume of air blown into the fluid is ~1190 mL, which exceeds the volume of the cylinder specified in the test method, so it is possible that the test fluid can foam out of the cylinder during the test.

19.3.2 Effect of Base Oil Properties on Foaming

As discussed elsewhere in this volume, base oils used in lubricating fluids can span a wide range of viscosity, polarity, and solvency. These properties can affect foaming in finished lubricating oils aside from the impact of additives.

The effect of *bulk viscosity* on foam tendency is related to the ease of air incorporation into the fluid (which requires less energy in low-viscosity fluids) and the speed of drainage of fluid from foam once it is formed (which is slower in high-viscosity fluids). We can isolate the effect of bulk viscosity alone by measuring the foaming tendency of base oils with no additives present. The foam tendency in an ASTM D892–like test for a variety of mineral and synthetic base oils, over a wide range of temperatures, has been measured by Hubman

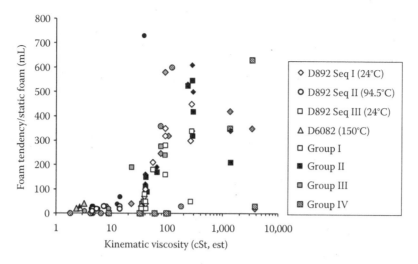

FIGURE 19.6 ASTM D892 and D6082 foam tendency/static foam results for a variety of pure base oils, plotted against the estimated bulk viscosity of the base oil at the temperature of the measurement.

and Lanik [3]. These measurements showed that, at low viscosity (i.e., high temperature), all the oils tested exhibited low foam tendencies. The foam tendency (foam volume) increased with increasing viscosity, reaching a maximum at 300–3000 cSt, while the foam lifetime always increased with the viscosity. Similar behavior of increasing foam lifetime with increasing viscosity has also been shown for other lubricating fluids [31] and crude oils [32].

Figure 19.6 shows the D892 or D6082 foam tendency as a function of kinematic viscosity, at the temperature of the foam test, for a wide range of API Groups I, II, III, and IV base oils. Similar to the results in [3], we see low foam at high temperature/low viscosity (mainly in D6082 or Seq IV testing), with increasing levels of foam up to a maximum in

foam tendency at ~300 cSt. Significant scatter in the data prevents any conclusions as to whether foam tendency decreases at high viscosity, as was observed in [3].

Furthermore, it has been shown that, even in a case where additives such as viscosity modifiers, dispersants, detergents, and FMs are present, one of the strongest predictors of foam lifetime is bulk viscosity [33]. Figure 19.7 shows foam half-life versus viscosity of synthetic oil–based fluids containing single additives at varying concentration. The viscosity of the fluids varied within this set because of variations in temperature and concentration of the additives. However, we see a clear correlation between foam half-life and viscosity that transcends the effects of any single additive type [33].

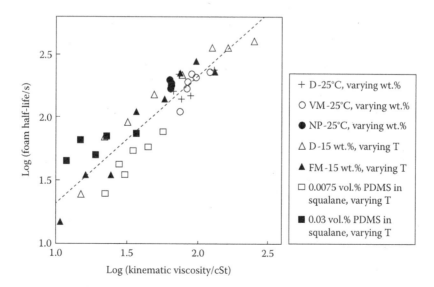

FIGURE 19.7 Correlation between foam half-life and the kinematic viscosity of the foaming solution for a range of additives, concentrations, and temperatures. D, dispersant; VM, viscosity modifier; NP, overbased detergent particles. (Reprinted from *Colloids Surf. A: Physicochem. Eng. Asp.*, 360, Binks, B.P., Davies, C.A., Fletcher, P.D.I., and Sharp, E.L., Non-aqueous foams in lubricating oil systems, 198–204, Copyright 2010, with permission from Elsevier.)

A second property of base oils that can affect foam is the ability of the base oil to *solubilize any antifoaming agents* that are present. As will be discussed in Section 19.5, a key requirement for an antifoam to function is its insolubility in the base fluid. If the base oil is able to solubilize the antifoaming additive, performance will be lost.

This phenomenon is demonstrated in Figure 19.8. Here, two finished lubricant fluids were made by diluting the same additive package in either a mixed Group I/II base oil or a pure Group IV base oil. Two versions of each fluid were made, one with no foam inhibitor and another with an acrylate-based foam inhibitor. The foam inhibitor was the same in both cases. In the Group IV–based fluid, the foam inhibitor was effective, reducing the foam tendency to zero in all test sequences (Figure 19.8, *right*). However, in the Group I/II–based fluid, the foam inhibitor was effective only in D892 Seq I and Seq III, the test sequences in which the fluid is held at 25°C (Figure 19.8, *left*). These results are related to the *solubility* of the foam inhibitor at the temperature of the foam test. In the Group IV fluid, the foam inhibitor is insoluble, and therefore effective, at the temperatures of all foam test sequences (25°C, 94.5°C, and 150°C). However, in the Group I/II–based fluid, the foam inhibitor appears to be only insoluble at room temperature (effective at Seq I and III temperatures) but is soluble in the fluid at the elevated temperatures of D892 Seq II and D6082 and therefore loses effectiveness at these temperatures.

Another experiment using the same set of fluids (also shown in Figure 19.8) demonstrates the insolubility of the antifoam using a different approach. In this experiment, both fluids were centrifuged, after which the top layer of the fluid was carefully removed. The foam tendency of the top layer was then measured with selected sequences (D6082 for the mixed Group I/II base oil and D892 Sequences II and III for the Group IV oil). In the case of the Group IV–based fluid, centrifugation significantly increases the level of foam in

D892 Seq II and III (500/200 mL), confirming the removal of insoluble antifoam droplets, which settle to the bottom during centrifugation because they are heavier than the bulk oil. For the mixed Group I/II fluid, centrifugation has no effect on D6082 foam tendency, consistent with the antifoam being soluble. In fact, the increase in the D6082 result for this fluid when antifoam is added suggests that the antifoam is actually acting like a soluble profoamer in this case. Note that in this chapter the term *profoamer* refers to materials, usually surfactants, which increase the tendency of a liquid to form a foam. We will return to the subject of profoaming components later in this section. For further discussion on the performance of acrylate foam inhibitors, see Section 19.4.2.

19.3.3 EFFECT OF THE ADDITIVE PACKAGE ON FOAMING

As discussed earlier, one of the strongest predictors of foam tendency in lubricating fluids is the bulk viscosity of the fluid. In previous sections, the effect of surfactants on the stability of foams has also been discussed. Therefore, the effect of the additive package on foam in a finished fluid can be considered to be divided into its effects on the bulk properties of the fluid and on its surface properties.

Regarding the effect of additives on the bulk properties of the fluid, some are straightforward, such as the effect of viscosity modifiers and high molecular weight (MW) dispersants to increase the bulk viscosity of the fluid, in turn influencing the tendency of the fluid to form foam [33]. This effect is shown in Figure 19.7. Less obvious are components that affect the fluid flow in a foam on the microscopic level; for example, overbased detergent particles. In one study, it was found that the presence of an overbased detergent contributed to foam lifetime, even though the components of the detergent were not strongly surface active [34]. Here, the effect was attributed to the change in rheological properties of the fluid due to the presence of the overbased detergent particles.

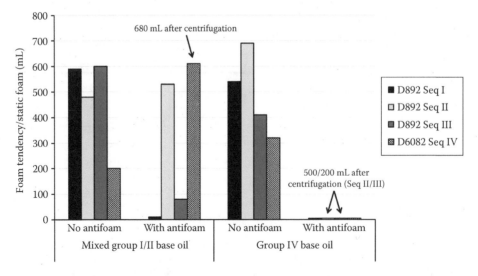

FIGURE 19.8 ASTM D892 and D6082 foam tendency/static foam for a Group I/II–based fluid (*left*) and a Group IV–based fluid (*right*). Both fluids used the same additive package.

The second way that the additive package can affect foam is through the surfactant-driven phenomena described in Section 19.2. Any impact of the additive package on foam due to surfactant effects should be considered to overlay the effect of bulk viscosity. By way of example, we consider an engine oil that shows high foam tendency in D892 Seq I, II, and III and in D6082. If we plot the foam tendency of the fluid as a function of viscosity of the fluid at the foam test temperature (filled symbols in Figure 19.9), we see that for Seq I and III (high-viscosity, low-temperature sequences), the result is comparable to what would be expected for the pure base oil alone (open symbols). Therefore, foam in these sequences is not significantly affected by the additive package. However, at lower viscosity/higher temperature (Seq II and D6082), we see that foam tendency is significantly higher than would be expected for a pure base oil; therefore, the additive package must be playing a role in this case.

Many lubricant additive molecules have a structure that includes small polar head groups connected to oily tail sections and are therefore potentially surface active. This makes the issue of surfactant-driven foam stability in lubricating fluids potentially very complicated, since there are many components in the additive package that could potentially cause foam. One technique to determine which specific additives are contributing to foam is to make test formulations without selected additives and then measure the foaming behavior of the test formulations. In the example illustrated in Figure 19.9, the additive package consists of components that are typically found in an engine oil formulation, including overbased detergents, dispersants, antiwear chemistry, antioxidants, and an FM. When the FM was removed from the formulation (leaving all other components in), the foam versus viscosity behavior was approximately what would be expected for the pure base oil, shown as the asterisks in Figure 19.9. This leads us to conclude that, while there are several potentially

surface-active components in the blend, the FM has the most significant effect on foam tendency.

Direct surface tension measurements explain this result, as shown in Figure 19.10. At room temperature, the pure base oil and fully formulated fluids with and without FM have equivalent surface tension. At 80°C, however, the formulation with FM shows lower surface tension relative to both the pure base oil and the fully formulated fluid with no FM. This indicates that (1) the FM is the only component in the formulation that significantly affects the fluid's surface tension and (2) the FM is only surface active at an elevated temperature. This is fully consistent with the FM being the only component that affects foam via a surface mechanism and that it only affects foam at elevated temperature (see Figure 19.9).

A number of other studies have been conducted to identify the foam-stabilizing agents in lubricants and other oil-based systems. Another strategy besides the *component deletion* presented earlier is to perform foam tests on oils containing single components. To that end, McBain et al. [35,36] have screened several hundred pure components and mixtures of components for their effect on lubricant oil foaming. Binks et al. [33] have also investigated the foam-stabilizing effect of several lubricant additives as a function of concentration by making solutions of individual additives in base oil.

In a similar way, Ross and Suzin [37] conducted detailed studies on the foaming properties of trimethylolpropane heptanoate (TMP heptanoate; heptanoic acid, ester with 2-ethyl-2-(hydroxymethyl)-1,3-propanediol), a lubricant ester, called *Base Stock 704*, alone and in the presence of various additives, such as *N*-phenyl-1-naphthylamine, phenothiazine, quinizarin (antioxidants), and tricresyl phosphate (antiwear additives). Here, they found that certain additive *combinations* could result in high foam tendency; any of these additives alone caused no foaminess, but several combinations of two or three of these additives resulted in copious amount of foam if their total concentration was about 2.5%.

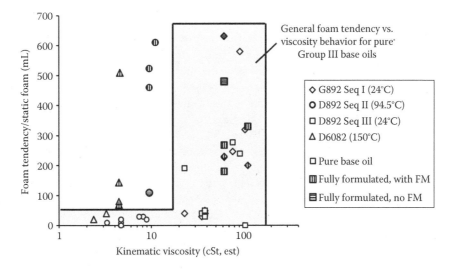

FIGURE 19.9 Foam tendency/static foam in D892 and D6082 as a function of fluid bulk viscosity, for a selection of Group III base oils, and for two versions of a fully formulated engine oil in a Group III base oil. The shaded area is the expected trend for pure base oils. FM, friction modifier.

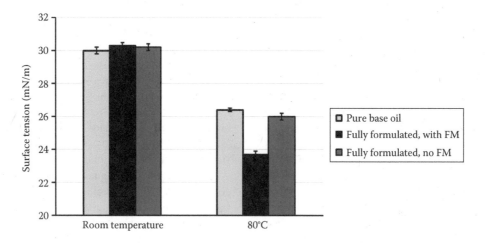

FIGURE 19.10 Surface tension of a pure Group III base oil and the two fully formulated fluids shown in Figure 19.9 (with and without FM).

A third strategy is to chemically isolate and analyze the foam-causing components from a complex mixture, a technique employed by Callaghan et al. [38] to show that low-MW, polar compounds were responsible for foaming in a crude oil.

19.3.4 Effect of Contamination on Foaming

An additional source of foaming that should be considered is the contamination of the lubricating fluid by foreign substances. There are relatively few systematic studies of the effect of contamination on lubricant foaming available in the literature, but there are some, including the products of root cause investigations or evaluation of sets of drain oils from field or bench tests, that will be discussed here.

A common contaminant that is arguably the most troublesome profoamer in hydrocarbon-based lubricants is low-molecular-weight silicone. Below a certain molecular weight, which is base oil dependent, polydimethylsiloxane (PDMS) are appreciably soluble in the base oil. When soluble, due to its low surface tension, PDMS is surface active and stabilizes foam by the same mechanisms discussed earlier in this section for surface-active additive components. Side chain substitutions on PDMS, such as aryl groups or fluorination, change the solubility profile and therefore influence this behavior (see Section 19.4.1.3). Though it may seem counterintuitive, the ability of silicones to promote foam or destroy foam, depending on molecular weight, is fully consistent with the mechanisms proposed in this chapter (see Section 19.5.2.2). The key property in the behavior of silicones is their low surface tension, allowing them to readily partition to the interface when soluble, and to disrupt foam films as insoluble droplets when insoluble.

Centers [39] studied the effect of PDMS MW on its antifoaming/profoaming action. PDMS grades with various viscosities, ranging from 0.65 to 60,000 cP were added to a very-low-foaming synthetic ester turbine lubricant at 10 ppm, and the foaming rate was measured with the FTM 3214 test. He found that very-low-viscosity PDMS

(below 10 cP) did not cause foaming and that PDMS in this viscosity range was soluble in the lubricant. At medium viscosities, 10–5000 cSt, PDMS was a strong profoamer, but at higher viscosities, it was a strong antifoam. The behavior in the 0.65 to 5000 cSt range could be explained by the decreasing bulk solubility and associated increase in the surface activity of PDMS with increasing MW, also known as the Ross–Nishioka effect [40], followed by insolubility above 5000 cSt. The overall effect is that PDMS can have no effect on foam at very low MW, stabilize foam at intermediate MW, and inhibit foam at high MW. The MW ranges corresponding to these regions of behavior depend on the solubility of the silicone in the particular lubricant, which in turn depends on the base oil group, temperature, and nature of the additive package. The effect of PDMS MW distribution and side chain modification will be further discussed in Sections 19.4 and 19.5.

There are several potential sources of silicone contamination. One common source is silicone-based greases, which can be in contact with the lubricant. Centers has shown that this situation can lead to leaching of low-MW silicone into the lubricant fluid, increasing foam tendency [41]. Another potential source of low-MW silicone contamination is silicone-based room temperature vulcanization gaskets, sometimes referred to as form in place gaskets, and any other silicone-based sealants/caulks. These materials, which are applied as liquids and cure to solids under ambient conditions, can contain low-MW silicone as a plasticizer. If the lubricant comes into contact with these materials, the low-MW silicone can leach into the fluid, causing an increase in foam. Silicone contamination as low as 10–100 ppm can lead to significant foaming issues, so care must be taken to avoid or mitigate the issue, through equipment design, careful application of the sealant, and robust fluid formulation. A final source of silicone contamination is through splashing or misting of heating baths or heat transfer fluids/greases used to heat oils during bench testing procedures.

Another common contaminant in lubricants is water. There are reports in the literature of water having a dramatic effect on foam and of water having no effect. These seemingly contradictory results are likely due to the fact that water, by itself, can have no direct effect on foam stability because of its high surface tension (~72 mN/m). Instead, it can only affect foaming through interactions with other components present in the oil, such as additive package components. Therefore, its effect on foam will be highly dependent on the additive package. For example, it was found that water, up to 1000 ppm, increased the foaming tendency of single-component blends of both overbased calcium phenate and polyamine derivatives of isostearic acid [42]. However, water had very different effects on directly measured surface properties in the two fluids, so the authors concluded that water was leading to foam tendency via different mechanisms in the two fluids. Another study, involving the measurement of foam tendency in 119 used turbine oils with water content from ~0 to ~1300 ppm, found no correlation between water concentration and foam tendency [43]. This inconsistent effect of water is in contrast to soluble, surface-active silicones, which are generally profoamers regardless of what additive components are present.

One additional consideration is the impact of water contamination on the foam control agent. Organomodified siloxane foam control agents with high degrees of ethylene oxide substitution and thus relatively high water solubility (see also Section 19.4.1.3) for use in *diesel and other liquid hydrocarbon fuels* have shown a reduction in foam control performance upon addition of 250 ppm or more water. The effect of water can be minimized by appropriate design and modification of these structures with additional substituents [44,45].

A final type of contamination includes unintentional mixing of lubricants in close proximity. For example, it has been observed that highly basic lubricating fluids, engine oils, for example, can react with acidic components of other lubricants to produce insoluble, highly surface-active compounds [46]. It is also possible that these compounds can also form if a lubricant with acidic compounds is contaminated with very hard water (i.e., water with a high mineral content).

19.3.5 Metalworking Fluids

MWFs represent a separate and special class of lubrication fluids, and therefore their foaming behavior can be different than the behavior discussed previously in this section. These fluids are injected onto the work pieces and tools during use so that they lubricate and cool the metals, extending the life of the tool. They also aid in providing a good finish and remove metal particles in a wide range of metal forming operations (cutting, grinding, drilling, etc.) [24]. The mechanism of their operation is complex and still not well understood [47]. In this section, the foaming problems and the use of antifoams with MWFs will be discussed.

The chemistry of these foam control agents will be further explained in Section 19.4.3 and their mechanisms in Section 19.5.1.

A traditional classification of MWFs is based on their oil content [48,49]:

1. Straight oils, which contain no water and are not diluted with water
2. Soluble oils, which contain 30%–85% oil
3. Semisynthetic fluids (semichemical solutions), which contain less than 20%–30% oil
4. Synthetic fluids that generally do not contain oils; sometimes they do not contain *synthetic* components either, and therefore it is better to call them chemical solutions

Soluble oils, semisynthetics, and synthetic fluids are diluted with water, typically 10–30 times, and therefore their volume during use is much higher than straight oils [48].

The oil phase primarily provides lubrication and can be natural (crude oil, distillate, or vegetable oil based) or synthetic. Synthetic oils can be hydrocarbons, for example, poly-α-olefins (PAOs), esters (polyesters, diesters), or renewable base fluids, such as esters based on vegetable oils or animal oil/fat. The synthetic (solution)-type MWFs contain polymers and glycols, both synthetic (polyalkylglycols) and renewable (cellulose, glycerol) types [47], and these additives perform the lubrication function of the oils.

The water-based (water dilutable) formulations contain a number of additives:

- Extreme pressure additives
- Antiwear
- FMs
- Emulsifiers
- Corrosion inhibitors
- Biocides
- Dispersants
- Antifogging additives
- Antifoams
- Buffers
- Antioxidants

and may contain 15–30 different chemicals [47]. Some are analogous to the additives found in lubricating oils, but some are not. One class of component that is found in MWFs, and not in other types of lubricating fluids, is the emulsifier.

19.3.5.1 Foaming Problems with Metalworking Fluids

MWFs that comprise oil/water mixtures, after dilution—and often even before that—with water, are typically oil-in-water emulsions, with drop sizes in the range of 0.1 μm to several microns [47]. To stabilize these emulsions, emulsifiers have to be used, which are typically anionic (such as sulfonates, fatty acid soaps, and derivatives of polyisobutylene succinic anhydride) or nonionic (such as alcohol ethoxylates and fatty alkanolamines) surfactants [24]. These surfactants can stabilize not only emulsions but also foams, and in fact they are the primary source of foaming problems with mixed water/oil MWFs.

Several other factors affect how much foaming occurs during metalworking operations, including water hardness, age of the fluid, filtration system design, fluid pressure and flow rate, fluid temperature, and contamination in the MWF [24]. Generally, hard water is better to use for diluting the MWF, since it foams less than soft water, and addition of calcium salts and/or calcium soaps also reduces foaming, although these salts can have adverse effects on the process and equipment [50]. Oil contamination from sources other than the MWF (also known as *tramp oil*) can also lead to enhanced foam stability.

Foaming becomes particularly severe if the effect of foam-stabilizing additives is coupled with mechanical problems, such as low fluid level in the reservoir, air leaks in the pump or piping on the intake side, too short residence time in the central circulation system, and incorrectly selected fluid nozzles (too high fluid velocity).

The presence of foam during metalworking operations can have several negative effects [50,51]:

- Insufficient lubrication, since foam is a poor lubricant.
- Heat exchange and heat capacity of foams are much worse than those of emulsions, so the cooling efficiency will be reduced and the metals will heat up, possibly changing shape.
- Faster corrosion process.
- Foam can float chips and fines and reduce filtration efficiency.
- The entrapped air can cause pump cavitation and damage.
- Foam can obstruct the worker's view of the workpiece.
- In the worst case, the foam overflows the equipment on the shop floor and becomes a safety hazard.

In a recirculating system, the foaminess of the MWF and the process conditions (flow rates, temperature, etc.) define the size of the reservoir that is necessary for sufficient retention time to burst the larger bubbles (and foam) and the release of the smaller bubbles [50].

19.3.5.2 Foam Control of Metalworking Fluids

With all types of water-diluted MWFs, foam control and air release have to be taken care of, and with the increasing cutting speeds and pressures, foam control is becoming more critical than before [52]. More oil in the formulation generally results in more tendencies to foam [50].

An approach to control foam problems with water-based MWFs is to use low-foaming ingredients, such as low-foaming surfactants as emulsifiers [53]. It was shown recently that the use of ether carboxylates as emulsifiers can provide much less foaming and less stable foam [52].

In many cases, however, foam control agents have to be added to solve foaming problems. The antifoams are either included in the MWF formulation and/or added into the

circulation system during use. The latter is needed in cases where the foam inhibitors are deactivated over time (see also in Section 19.5.1.7).

Foam control agents with several types of chemistries exist, but the most efficient with the widest range of applications is the *mixed type*, containing combinations of nonpolar oils and hydrophobic solid particles. The chemistry of these antifoams and their use for MWFs is discussed in Section 19.4.3, with a detailed analysis of their operating mechanisms in Section 19.5.1.

19.4 CHEMISTRY AND USE OF ANTIFOAMS FOR LUBRICANTS

As discussed in the previous sections, there are significant differences between the surface chemistries of water-based versus nonaqueous or oil-based systems. Therefore, different materials are used as antifoams in the two types of liquids.

In any application, an antifoam material generally has to meet the following criteria:

1. Lower surface tension than the continuous phase
2. Insoluble in the continuous phase
3. Chemically inert with respect to other components in the formulation
4. Thermally stable
5. Dispersible into small droplets

In the case of oil-based foams, materials that meet these criteria and are therefore used commonly as antifoams include various types of silicones (PDMSs, fluorosilicones, and organomodified silicones) and organic (non-silicone) polyacrylates. Aqueous foams can be controlled best by mixtures of nonpolar oils, including silicones or organic oils, and hydrophobic solid particles (see Section 19.4.3). Although many other types of chemicals have foam-breaking effects, and may meet these criteria, in most cases only these are utilized because they have much higher efficiency than other materials and therefore can be used in very low concentrations.

In the next sections, the chemical structure, manufacturing, properties, and use of these materials will be reviewed.

19.4.1 Silicones (Siloxanes)

For both water- and oil-based systems, silicone-based antifoam chemistries are very important and widely used. There are a vast number of silicon-based materials (mostly polymers or copolymers) that have been synthesized and used in many industries [54–57].

Silicones (or siloxanes) are a subset of silicon-containing molecules that contain combinations of inorganic (based on the silicon atom*) and organic groups [58,59]. The backbone of the polymer is a siloxane chain, containing $-Si-O-Si-O-$

---

* Note the slightly different spelling of the Si atom, *silicon* and polymers based on repeating Si—O units, *silicone*.

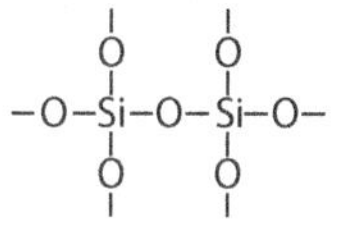

FIGURE 19.11 Structure of silicate network.

TABLE 19.2
Siloxane Shorthand Notations

Shorthand	Formula	Structure
M	$(CH_3)_3SiO_{1/2}$	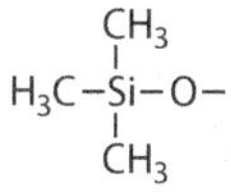
D	$(CH_3)_2SiO_{2/2}$	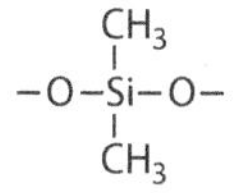
T	$(CH_3)SiO_{3/2}$	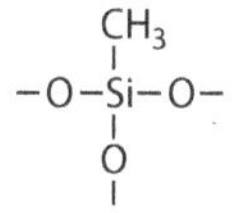
Q	$SiO_{4/2}$	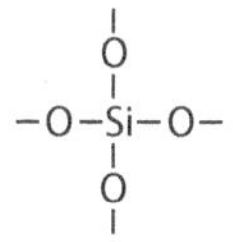

repeating units. The remaining valences of the Si form bonds to further siloxane or organic groups, for example, methyl groups. One of the simplest, and most common, siloxanes is PDMS, with a linear backbone and methyl groups in all the remaining valences of the Si atoms. The siloxane chain can also be branched, and in the extreme it can lack any organic groups, forming a silicate (silica) network (Figure 19.11).

To describe simply all the possible branched siloxane structures with methyl groups, a shorthand notation is generally used, as shown in Table 19.2, with monofunctional (M), difunctional (D), trifunctional (T), and tetrafunctional (Q) groups. "$O_{1/2}$" means that only another "$O_{1/2}$" can be connected to it, to form a siloxane bond.

Using this common notation, for example, MM is hexamethyldisiloxane and MDM is octamethyltrisiloxane (see Figure 19.12), PDMS is MD_nM, and many of the possible methylsiloxane polymers can be briefly described as $M_wD_xT_yQ_z$ (where the w, x, y, and z indices can be positive numbers or zero).

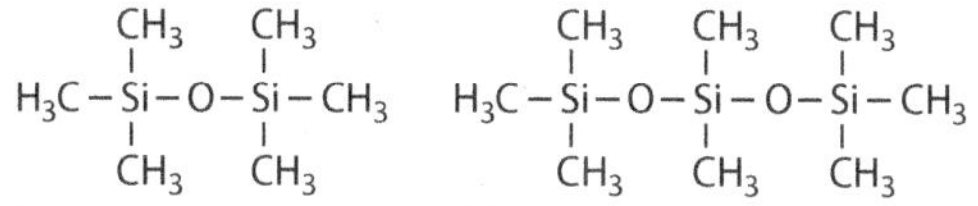

Silanes, containing Si—H, Si—X (X: halogen), or Si—OR (R: methyl, ethyl, or other alkyl) bonds, are generally reactive with water or similar reagents. Siloxanes, such as PDMS, lack such reactive functionality and instead act by physical mechanisms.

The unique physical properties of silicones are often attributed to the high flexibility of the siloxane chain, which is the result of the very low barrier of rotation of the Si—O bonds, which is much lower that for C—C bonds, and allows almost completely free rotation along the siloxane bonds. The Si—O—Si angle is also much wider than in a tetrahedron (140°–180°) [60].

Many other materials can be made by replacing one or more of the methyl groups in polymethylsiloxanes with other organic groups. The preparation and properties of these *organomodified silicones* as antifoams for lubrication fluids will be discussed in Sections 19.4.1.2 and 19.4.1.3.

19.4.1.1 Polydimethylsiloxanes

The silicones manufactured in the highest quantities are PDMSs (see Figure 19.13). These materials have applications beyond antifoams, including use as lubricating oils and greases, insulation, heat transfer media and release agents, in personal care, and for a wide range of other industries and applications.

19.4.1.1.1 Manufacture

PDMS is manufactured via a multistep process [58,60]. First, Si metal is generated from silica, then chlorosilanes are synthesized using the Si metal via the so-called Direct Process (or Rochow Process), invented in 1940 [61]. In this process, finely powdered Si is reacted with methyl chloride gas, in the presence of copper catalyst and other metal promoters to yield a mixture of methylchlorosilanes:

$$Si + 2CH_3Cl \rightarrow (CH_3)_2SiCl_2 + CH_3SiCl_3 + (CH_3)_3SiCl + SiCl_4 \ldots$$

$$(19.5)$$

From the reaction mixture, dimethyldichlorosilane is obtained by fractional distillation and is then converted into PDMS in two steps. First, it is hydrolyzed to yield linear and cyclic dimethylsiloxanes:

$$(m+n)(CH_3)_2SiCl_2 + (m+n+1)H_2O \rightarrow HO\underbrace{\left((CH_3)_2SiO\right)_m}_{\text{linear}} H + \underbrace{\left[(CH_3)_2SiO\right]_n}_{\text{cyclic}} + 2(m+n)HCl$$

$$(19.6)$$

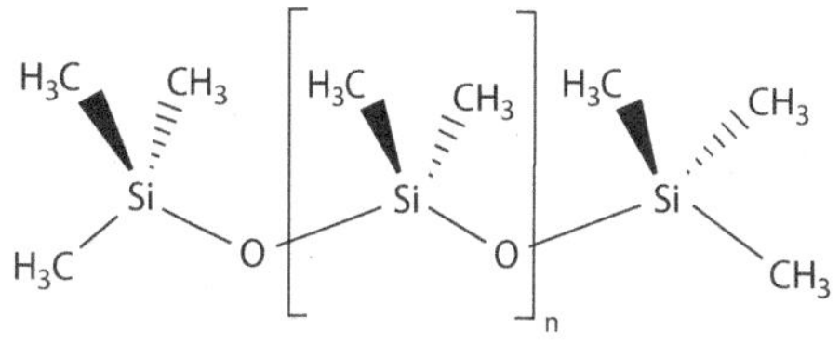

FIGURE 19.12 Structures of hexamethyldisiloxane (a) MM and octamethyltrisiloxane (b) MDM.

FIGURE 19.13 Structure of PDMS.

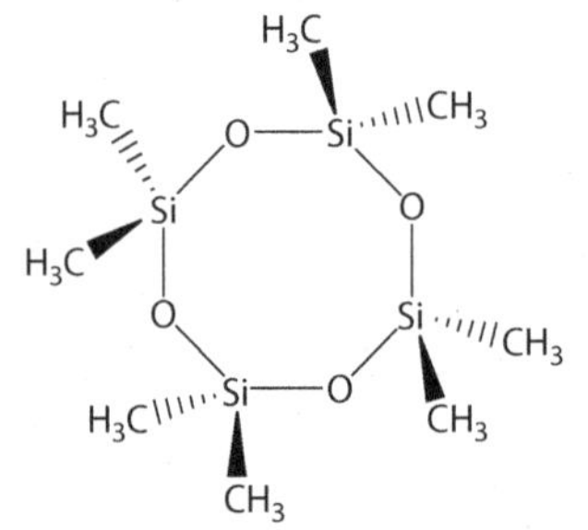

FIGURE 19.14 Structure of octamethylcyclotetrasiloxane, D_4.

Figure 19.14 shows the structure of D4, a typical cyclic methylsiloxane. The linear, noncapped siloxanes with Si—OH end groups, called silanols, were at one time used instead of PDMS [62]. Currently, methyl-terminated PDMS is used for antifoaming in most cases. To obtain the end-capping groups, trimethylchlorosilane is converted into MM (see Figure 19.12a):

$$2(CH_3)_3 SiCl + H_2O \rightarrow (CH_3)_3 SiOSi(CH_3)_3 + 2HCl \quad (19.7)$$

Then the hydrolysis products from both reactions (19.6) and (19.7) are reacted, in the presence of acid or base catalyst, to obtain PDMS via an equilibration polymerization process to yield a mixture of linear and cyclic species:

$$HO\big((CH_3)_2 SiO\big)_m H + \big[(CH_3)_2 SiO\big]_n + (CH_3)_3 SiOSi(CH_3)_3$$
$$\rightarrow (CH_3)_3 SiO\big[(CH_3)_2 SiO\big]_n Si(CH_3)_3 + H_2O$$
$$(19.8)$$

After removing the volatile cyclic siloxanes, water, and other contaminants, the linear PDMS polymer is obtained. The final polymer contains a distribution of molecular weights. The main variables of the PDMS structure are the degree of polymerization and its distribution. For industrial applications, the degree of polymerization of PDMS is normally characterized by viscosity. A wide range of grades, from 0.65 cSt up to millions of cSt, are manufactured commercially. The typical polydispersity of commercial PDMS grades is about two. For antifoaming of lubricating oils, generally PDMS viscosities in the 1,000–100,000 cSt (from 28,000 to 139,000 Da MW) range are used. A number of empirical formulas can be used to describe the bulk viscosity of PDMS based on its MW [63,64].

19.4.1.1.2 Chemical Properties

PDMS is highly stable and resistant to oxidation and has low reactivity under normal conditions. Their main reaction is hydrolysis, since the siloxane bond can be attacked by acids or bases due to the wide angle of the Si—O—Si bonds. For the hydrolysis reaction to occur, however, strong base, acid, high temperature, and/or extended reaction times are necessary due to the incompatibility of PDMS with water and other liquids.

19.4.1.1.3 Physical Properties

PDMS is a clear liquid, with low pour point (roughly −70°C to −40°C) and glass transition temperature (−120°C to −150°C). At 10,000 Da MW and above, the PDMS chains start to entangle, and therefore their viscosity increases steeply with the molecular weight above this critical value. Low-MW PDMS grades are Newtonian fluids up to high shear rates, but for the higher MW grades (>10,000 Da), the viscosity remains constant with the shear rate only up to a critical point, above which the polymers show shear thinning behavior.

PDMS has negligible vapor pressure (except for the smallest molecules and the cyclic siloxanes), and thus cannot be distilled, but decomposes slowly at high temperatures (200°C), and faster above 260°C. Based on TGA experiments, PDMS decomposes in air at temperatures >275°C and in an inert nitrogen environment at >400°C [65].

The specific gravity of PDMS at ambient temperature is about 0.97, which is higher than most lubricating oils. As such, one major issue with the use of PDMS as an antifoam in lubricating oils is that it can settle from finished fluids during storage. The severity of this problem can be reduced with efficient incorporation (see Section 19.6.1). The permeability of PDMS to gases is very high [66].

19.4.1.1.4 Solubility and Surface Tension

Solubility and surface tension are key properties of PDMS that affect its foaming/antifoaming action (see also Section 19.3.4). In spite of the relatively polar nature of the Si—O bond, PDMS is a relatively nonpolar liquid with low dielectric constant (2.75), due to the presence of the high number of methyl groups that shield the siloxanes. Therefore, it is soluble in small molecule, nonpolar solvents, such as C_5 to about C_{10}, aliphatic and aromatic hydrocarbons (hexane, toluene, and xylene, as well as kerosene and ketones). PDMS is generally insoluble or sparsely soluble both in polar solvents, such as water, alcohols, glycols, and acetone, and in long-chain hydrocarbons, such as paraffins and lubricating oils. In this sense, PDMS can be considered both a hydrophobic and an oleophobic liquid. All surfactants are insoluble in PDMS.

The solubility of PDMS in lubricating oils is very important for determining antifoaming effectiveness, particularly at elevated temperatures, as well as in the area of air entrainment, and this will be discussed in more detail in Section 19.5.2.1.

The surface tension of PDMS at ambient temperature is about 21 mN/m (smaller for the lower MW polymers, increasing slightly with MW for the larger molecules), which is lower than typical lubricating oils, most surfactants, and other liquids. The PDMS/water interfacial tension is high (42.7 mN/m) [67]. PDMS itself is not a surfactant in the traditional sense, since it does not contain hydrophilic groups but is surface active and readily spreads on the surface of water or oils in which it is insoluble. The PDMS surface layer is oriented with the methyl groups toward the air, shielding the siloxane groups and minimizing the surface tension due to the flexibility of the siloxane chains. Zisman [68] studied the wetting properties of low energy solid surfaces and found that they can be characterized by their critical surface tension, based on the functional group on the surface. Table 19.3 shows that fluorocarbons (—CF3 and —CF2 groups) provide the lowest surface tension and that methyl groups represent much lower

TABLE 19.3
Critical Surface Tension (γ_c) of Low-Energy Surfaces

Surface Group	γ_c (mN/m) at 20°C
—CF$_3$	6
—CF$_3$ and —CF$_2$—	17
—CH$_3$ (crystal)	22
—CH$_3$ (monolayer)	24
—CH$_2$—	31

Source: Based on Zisman, W.A., Relation of the equilibrium contact angle to liquid and solid constitution, in: Fowkes, F. (ed.), *Contact Angle, Wettability and Adhesion,* Advances in Chemistry, American Chemical Society, Washington, DC, 1964, pp. 1–51.

energy (22–24 mN/m) than methylene groups (31 mN/m), consistent with the low surface tension of PDMS relative to organic polymers.

19.4.1.1.5 Use of PDMS as Antifoam

PDMS is highly efficient at controlling foam in *oil-based liquids*, including lubricating oils, and thus is a commonly used antifoam in these systems. However, it has poor efficiency to break foams in *water-based liquids* by itself and therefore is normally mixed with hydrophobic solid particles when used in water-based systems (see Section 19.4.3).

The largest industrial use of antifoams for nonpolar liquids is in *crude oil* processing where foam control is often crucial. In particular, in an early step of processing, the produced crude oil flows into gas/oil separator(s) where the oil and gas are separated, generally with pressure reduction. As a result, a large amount of gas is liberated, which can quickly form a high volume of foam decreasing the efficiency of the separation and oil production if antifoams are not used. Silicones are used almost exclusively in this application, with PDMS (typically 10,000 to 100,000 MW), at 0.5–20 ppm use concentrations being the most common [69,70]. PDMS can be also combined with cross-linked silicone resins [71]. Another large-scale usage of PDMS antifoam in nonpolar liquids is in delayed cokers, where the temperature is extremely high (450°C–500°C) and high-viscosity (10^5–10^6 cSt) PDMS is generally the only option.

PDMS is the most common foam inhibitor material used in the area of lubricating oils and is found in engine oils, driveline/transmission fluids, and nonautomotive applications. Patents describing the high antifoaming efficiency of PDMS or silanols in lubricating oils appeared as early as the 1940s [62,69,72,73]. Historically, low-MW PDMS (<1000 cSt) have been used in lubricants [69], but the high-viscosity grades have become more common, similar to crude oils [74].

Generally, high-temperature performance improves with increasing PDMS MW/viscosity; however, incorporation and formation of small PDMS droplets become more difficult with increasing PDMS viscosity.

The complicated relationship between PDMS antifoam and *air entrainment/aeration* (microfoam) has been widely published [46,75,76], and this will be further discussed in Section 19.5.2.1.

19.4.1.2 Organomodified Silicones

Organomodified siloxanes (OMS; organopolysiloxanes) represent a class of antifoams less known among lubrication oil formulators. In this type of silicones, one or more methyl groups of PDMS are substituted by alternate organic groups. Figure 19.15 shows a general structure of the polymers that can be derived from PDMS with monofunctional organic groups (R$_1$ and R$_2$) [77]. The organic groups can be on the side of the chain (R$_1$ groups: pendant, comb, or rake type) or at the ends of the siloxane chain (R$_2$ groups: ABA type, linear) or in both positions.

The R$_1$ and R$_2$ groups most typically are

- Substantially nonreactive groups:
  - Polyethers (PEG, PPG)
  - Alcohols, diols, polyols
  - Alkyl, aryl groups
  - Fluorinated alkyl groups
- Substantially reactive groups:
  - Amines (including quaternized amines)
  - Epoxides (cyclic and linear aliphatic)
  - Acrylates
  - Vinyl
  - Various ionic groups
  - Isocyanates

For antifoaming applications, the nonreactive derivatives, especially those with polyether and fluorinated alkyl side groups (see Section 19.4.1.3), are more widely used.

The functional group substitutions on the siloxane polymer can be the same or a combination of the groups earlier mentioned. More complex structures, such as ABABABA, or [AB]$_n$ (where A is the organic, B is the siloxane chain), and their combinations, are also possible. While in PDMS, the main variable is the number of D units (and possibly the end groups, see Section 19.4.1.1), Figure 19.15 shows that with OMS many more structures are possible and a wide variety of new materials can be made. The organic substitution can fundamentally change the physical properties of the silicone molecule, especially its solubility, compatibility, surface activity, and reactivity.

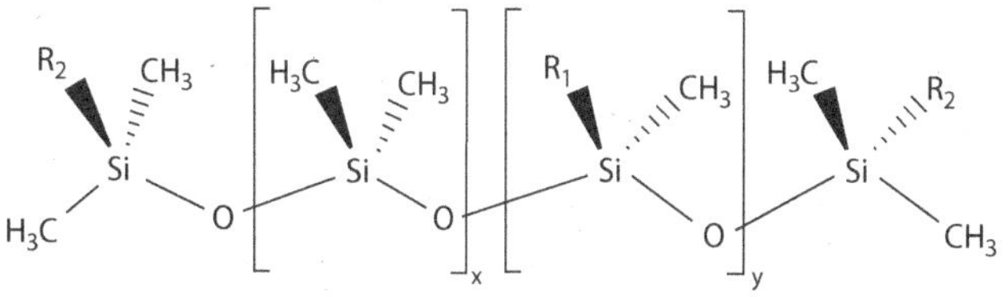

FIGURE 19.15 General formula of organomodified siloxanes with monofunctional organic groups (R$_1$ and R$_2$).

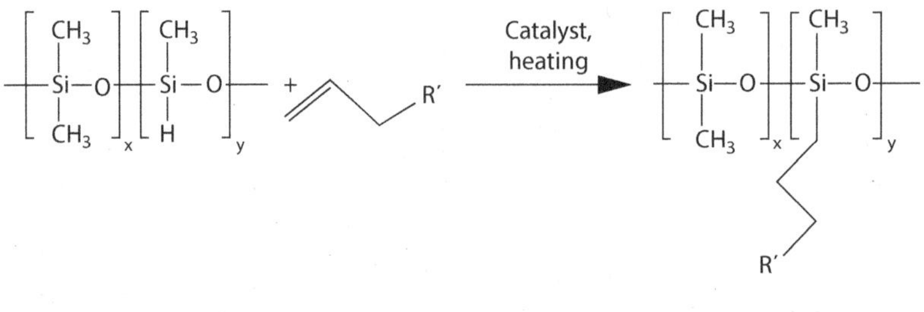

FIGURE 19.16 Scheme of hydrosilylation reaction with allyl-started reagent.

The siloxane polymers and the organic groups are generally connected via Si—O—C or Si—C linkages. A disadvantage of the Si—O—C bond is that it can hydrolyze easily, and therefore the Si—C linkage is more typical. The most common process to make these copolymers is by *hydrosilylation* (hydrosilation), an addition reaction involving hydride functional silicone polymers (SiH fluids), which can be described by the general structure in Figure 19.15 with R_1 or R_2 = H. These fluids are reacted with olefins, which are typically vinyl- or allyl-started derivatives of the organic groups, in the presence of a catalyst (see Figure 19.16 for making side chain–substituted OMS).

Since PDMS products, like most other polymers, contain a distribution of molecule sizes, the OMS also contain a distribution of "x" and "y" values (except the smallest molecules). Moreover, these polymers are often *not* block copolymers (as Figure 19.15 or 19.16 might indicate!); rather, generally, the two (or more) kinds of siloxane groups are randomly distributed in the molecule, which leads to further variation among OMS molecules.

The SiH fluid can be made by various processes, for example, by modifying the Direct Process (see Equation 19.5) and then re-equilibrating with cyclics (similarly to Equation 19.8) to obtain the required number of repeat units in the polymer backbone (x and y). Many types of catalysts can be used for hydrosilylation, but Pt-based ones are most typical, such as chloroplatinic acid (H_2PtCl_6) and Karstedt's catalyst ($Pt((CH_2\text{=}CH)(CH_3)_2SiOSi(CH_3)_2(CH\text{=}CH_2))_3$). The reaction is exothermic and temperature control is important to minimize side reactions. A significant side reaction in the hydrosilylation of allyl-functional species is the isomerization of the allyl group to propenyl:

$$H_2C\text{=}CH\text{–}CH_2\text{–}R' \rightarrow H_3C\text{–}HC\text{=}CH\text{–}R' \qquad (19.9)$$

The propenyl group typically does not react with SiH, and therefore, a molar excess of allyl reagent (10%–40% with polyethers) may be used to completely consume all of the SiH groups. The hydride fluids react not only with allyl groups, but with active hydrogens as well (such as those in water or alcohols), forming hydrogen gas via dehydrocondensation. This is not only a side reaction, but may also lead to increased flammability and pressure in closed systems, and therefore the purity of the starting reaction mixture is critical, as is complete reaction of the SiH groups. Additionally, if the allyl species being hydrosilylated contains reactive hydrogen groups, cross-linking and gelation may occur due

to dehydrocondensation. This can be minimized with careful temperature and pH control during the hydrosilylation reaction or can be avoided by protecting the reactive end groups (e.g., by converting a C—OH end group into a nonreactive group, such as C—OMe).

19.4.1.2.1 Silicone Polyether Copolymers

Silicone polyether (polyglycol, copolyol) copolymers (SPE) are important OMS antifoams. The polyether can be polyethyleneoxide (EO, polyethyleneglycol), polypropyleneoxide (PO, polypropylene glycol), or sometimes polybutyleneoxide (BO, polybutylene glycol), connected to the ends of the siloxane chain or grafted as side chains, in many combinations. Figure 19.17 shows the structure of typical pendant (graft) polyethers, with EO and PO, illustrating that there are five variables (x, y, m, n, Z) in these structures giving several degrees of freedom to the designers of such polymers. In the polyether copolymers with both EO and PO, the polyethers can be in blocks, but most often they are partially blocked or random in distribution. The Z end group is typically —H, —CH_3, or —$C(O)CH_3$ [59].

The main effect of the polyether substitutions is that they change the polarity, solubility, compatibility, and possibly the reactivity of PDMS. By functionalizing with EO, the completely water-insoluble PDMS can become dispersible or completely soluble in water, depending on the values of m and y, relative to x and n, that is, the ratio of EO to dimethylsiloxane.

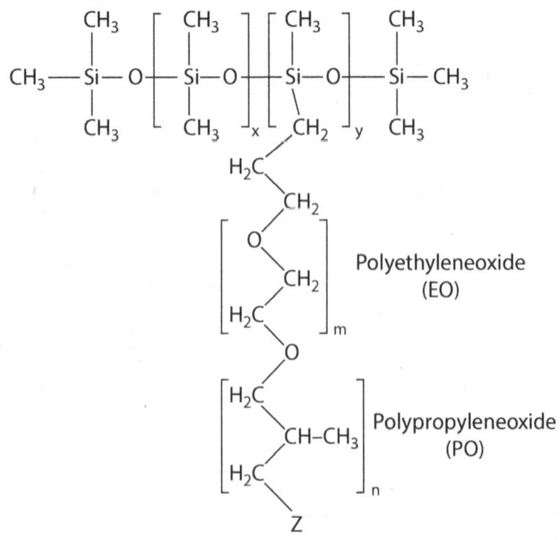

FIGURE 19.17 Structure of typical pendant silicone polyethers with EO and PO substitutions.

Unlike PDMS, these silicone polyethers can be considered surfactants, since they contain both hydrophobic and hydrophilic sections, and indeed they behave as nonionic surfactants since (depending on their composition) they can

- Reduce surface tension of water or other liquids
- Form micelles and/or liquid crystals
- Have a cloud point in aqueous solutions
- Adsorb on surfaces
- Act as profoamers or emulsifiers
- Act as antifoams or demulsifiers

Due to the low surface tension of PDMS, SPE tend to have lower surface tension than traditional surfactants, often not much higher than that of PDMS: 22–26 versus 21 mN/m for regular PDMS at 0.1 wt.%. The reason for this is the flexibility of the siloxane chain and the low energy of the methyl groups in the siloxane [78].

Some of the most surface-active versions are the trisiloxanes, which contain no D groups and only one polyether chain (x = 0, y = 1 in Figure 19.17). These surfactants (especially the ones with about 8 EO and no PO groups) are also called *superspreaders* because they greatly enhance the spreading of water drops on hydrophobic solid surfaces with a mechanism that has been studied and debated significantly [79–82].

For antifoaming, only some of the possible SPE and other OMS structures are usable, namely, those with limited solubility in the foaming liquid, since that is generally a requirement for antifoaming action. Therefore, in water-based systems, SPE should not be too hydrophilic (i.e., high EO content), and in oil-based systems, such as lubricating oils, the polyether should not contain too much oleophilic functionality (i.e., PO or BO). Most of the nonfluorinated, OMS-type antifoams for oils contain polyethers.

19.4.1.2.2 Use of Organomodified Silicones as Antifoams

Although epoxy or cyclohexane-modified siloxanes were patented earlier [83], OMS have been used as antifoams for *nonaqueous liquids* since the late 1980s. Callaghan et al. patented water-insoluble polyethers for crude oil separators [75], suggesting that they cause less air entrainment than PDMS, though experimental evidence was not presented.

Many OMS structures have been patented for antifoaming of hydrocarbons [84], especially for fuels.

In *diesel fuels* and similar fuels almost exclusively, OMS antifoams are used (at 0.5–20 ppm concentration) to avoid air entrainment issues associated with PDMS. Due to the immense number of possible structures, many patents have been filed, including SPE [75,85–87], which can be crosslinked [88], and OMS that contains not only polyether groups but also polyhydric, aliphatic, aromatic, unsaturated, alkylphenolic, and similar groups [44,71,89–96]. These patents specify diesel fuel or nonaqueous liquids in general for the applications. Many of these OMS antifoams may work with lubricants as well, and some of these publications explicitly specify that [94]. SPE-type antifoams were described recently

for low-viscosity lubricants, especially for transmission fluids [97]. Because of their partially organic nature, OMS-type antifoams can be more prone to be highly soluble in base oils compared to PDMS or fluorinated silicones. Because of this, OMS should be carefully tested in any formulation in which they are used.

OMS that work as antifoams have little solubility in lube oils, diesel fuel, crude oil, and similar hydrocarbons, but they are soluble in many polar and aromatic solvents. With fuels, compatibility of the antifoam with the fuel additive package is an important consideration in antifoam formulation. Dispersions and emulsions can be also formulated from them [98], and the drop size of the antifoam can be varied by changing the solvent, according to Fey and Combs [88].

19.4.1.3 Fluorosilicones

Siloxanes with fluorine-containing groups (fluorosilicones, FSs) are also organomodified PDMSs and generally can be described by the formula in Figure 19.15, where the R_1 and R_2 groups generally contain fluorocarbons, most often only side chain modified ($R_2 = -CH_3$ or $-OH$). The fluorocarbons contain $-CF_3$ and $-CF_2$ groups (perfluorinated sections) only. The simplest such groups on organomodified PDMSs would be $-CF_3$; however, such polymers are unstable due to the high electronegativity of the fluorine atoms. Therefore, generally only polymers with fluoro substitution only in the γ-position or farther to the Si atoms ($-CH_2-CH_2-CF_3$, etc.) are usable.

Several such structures are described in the patent literature, often with antifoaming of nonaqueous systems in mind. Borner et al. [99] described FS with $FC(CF_3)_2$ groups, while others added even more F-containing groups [100,101]. Callaghan and Taylor described FS with $C_nF_{2n+1}(CH_2)_2O$ groups for crude oil [102].

From the large number of possible polymers, only a few structures are commonly used. The most important ones are homopolymers with no D unit, only with a trifluoropropyl and a methyl group on all the siloxanes [103], as shown in Figure 19.18.

Wu et al. [104] prepared siloxanes grafted with both polyethers and fluorocarbons and tested them as antifoams in a fuel blend.

19.4.1.3.1 Manufacture

There are several ways to synthesize FS, but the most common method is from the cyclic, fluoro trimer, shown in

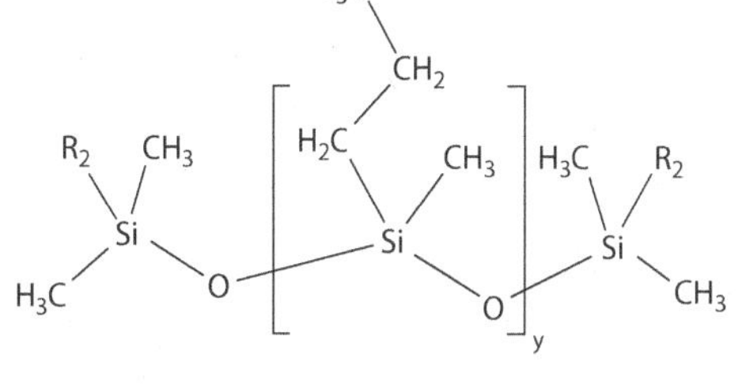

FIGURE 19.18 Structure of poly[methyl(3,3,3-trifluoropropyl) siloxane]. R_2, methyl or $-OH$.

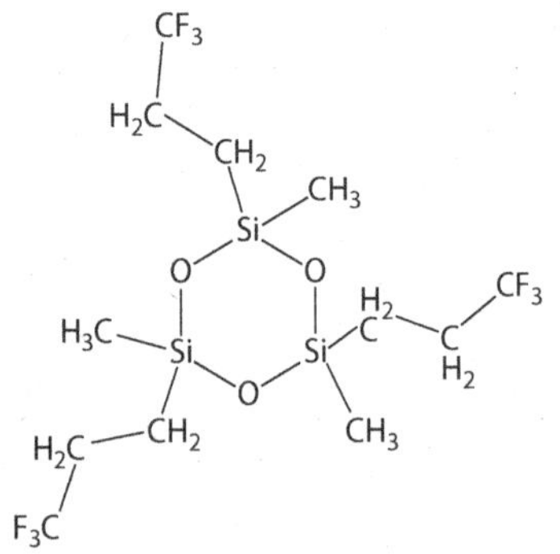

FIGURE 19.19 Structure of fluoro trimer.

Figure 19.19 [103]. First the dichloro monomer, methyl(3,3,3-trifluoropropyl)dichlorosilane of the cyclic FS is made from methyldichlorosilane, CH_3HSiCl_2, by hydrosilylating it with 3,3,3-trifluoropropene, using a Pt catalyst. Then, the fluoro-propyldichlorosilane is hydrolyzed to obtain the fluoro trimer, followed by ring opening with acid or base catalysis, analogously to PDMS production (see Equations 19.6 and 19.8). The fluoropropyldichlorosilane hydrolysis yields high amounts of cyclics, which makes their manufacture more complicated and more costly than PDMS.

The fluorinated siloxanes can be also copolymerized with D_x units, as in Figure 19.15 (where x and y > 0, R_1 = 3,3,3-trifluoropropyl and R_2 = Me or —OH). Such copolymers can be made by copolymerizing two silanol (Si—OH)-terminated diorganosiloxanes, one with fluoro-containing groups and the other with only methyl groups [105]. Another method of copolymerization is ring opening the fluoro trimer (Figure 19.19) as shown earlier and at the same time reacting either with triorganosilanol [106] or with cyclics, such as D_4 (see Figure 19.14) [107] or, alternatively, with MM (see Figure 19.12a) [108] in the presence of acidic or basic catalysts.

19.4.1.3.2 Properties

The fluorinated groups strongly increase the viscosity of siloxanes; although for antifoaming purposes, only viscosity grades no more than 10,000 cSt are generally available. The fluorocarbon groups significantly change the solubility, compatibility, and surface energy of siloxanes. FS are generally *not soluble in hydrocarbons* (even in small molecules), only in somewhat more polar solvents, especially in ketones and esters, and they are also insoluble in PDMS or in SPE. A possibility to avoid solvents is to emulsify the FS, for example, into mineral oil [109] or even into aqueous solution [110]. However, an additional concern in this case is the stability of the resulting emulsion.

As Table 19.3 shows, the —CF_3 groups can potentially provide a much lower surface energy to FS. This is true for solid surfaces, but surprisingly the surface tension of liquid poly[methyl(3,3,3-trifluoropropyl)siloxane] is somewhat *higher* (23–24 mN/m) than that of PDMS (21–22 mN/m) [103]. Similar to PDMS, FSs are also not surfactants in the traditional sense, since they generally do not contain hydrophilic groups (see Figure 19.18). Addition of polyether groups to the

FS polymers can make them amphiphilic [111], although this is not common.

The main application of FSs is in elastomers, which are very high-MW cross-linked materials, but the lower MW liquid polymers are primarily used as antifoams for oils.

19.4.1.3.3 Use of Fluorosilicones as Antifoams

FSs often show much higher antifoaming efficiency than PDMS in oil-based systems. Similar to PDMS, they are not commonly used as antifoams in aqueous systems, not even in mixtures with hydrophobic solid particles.

FSs tend to have much higher efficiency than PDMS, especially in *crude oil separators*, where the typical use level of PDMS is in the 0.5–20 (sometimes as high as 100) ppm range, while FS can be used at 0.1–5 ppm levels, which can compensate, especially in difficult cases, for their significantly higher prices. Evans first patented the use of FS (containing trifluoropropyl substitutions) for foam control of crude oils [112], and Berger et al. described fluorinated norbornylsiloxanes for the same application [113].

The advantage of FS antifoams is particularly strong with light crude oils. These oils contain smaller hydrocarbon molecules, in which PDMS is soluble and therefore tends to promote foam rather than reducing it [66,114]. In such oils, FS is the only choice and works efficiently due to its incompatibility with hydrocarbons (see 19.4.1.3.2).

As with PDMS, FS antifoams are widely used in automotive and other lubricant fluids, for example, in automatic transmission fluids (ATFs) [115].

19.4.2 POLYACRYLATES

19.4.2.1 Chemistry and Applications

Another class of commercially available antifoams widely used in lubricating oils is polyacrylates (and polymethacrylates). The chemistry of polyacrylates and polymethacrylates is very well studied and documented in the literature [116–118], and the application of acrylate polymers as foam inhibitors in nonaqueous systems was first developed and patented in the late 1950s to early 1960s by Monsanto Co. [119–121]. Since then, various process and formulation improvements have been developed and patented [122].

Discussion of the relationship between antifoaming performance and acrylate properties, most importantly surface tension and solubility, can be found in Section 19.4.2.3, and the mechanism of antifoaming action in lubrication oils will be further explained in Section 19.5.2.2. First, in this section, we will discuss the suppliers of acrylate antifoams and their manufacture.

Selected commercial sources for polyacrylate antifoams are given in Table 19.4. Manufacturers of other antifoams will be shown in Section 19.4.5.

A general polyacrylate structure is given in Figure 19.20.

A notable development in the past two decades has been the use of functionalized acrylic monomers such as fluoroalkyl acrylates/methacrylates, which have been reported as

TABLE 19.4

Commercial Polyacrylate Antifoam Suppliers for Nonaqueous Applications

Supplier	Component Trade Name	Description/Application	Additional Information
Afton Chemical	HiTEC® 2030 [123]	Acrylate-based foam inhibitor for driveline applications	Viscosity (40°C, mm²/s): 40 Active polymer content: 40% Rec. treat rate (ppm): 100–200
Lubrizol Corporation	LZ® 889D [124]	Acrylate-based foam inhibitor	Viscosity (40°C, mm²/s): 475 Active polymer content: 40% Rec. treat rate (ppm): 50–500
Munzing	Foam Ban® 152 [125]	Blend of polyacrylate technology and synthetic hydrocarbon for industrial and marine lubricants	Consistency:200 mPa·s Rec. treat rate (ppm): 100–1000
Munzing	Foam Ban® 155 [125]	Blend of silicone–polyacrylate and synthetic hydrocarbon for industrial and marine lubricants	Actives (%): 100 Consistency: 30–200 mPa·s/cP Rec. treat rate (ppm): 500–2000
Munzing	Foam Ban® 3633E [125]	Blend of polyacrylate technology and synthetic hydrocarbon for Industrial lubricants	Viscosity: 20–200 mPa·s/cP Rec. treat rate (ppm): 500–1000
ALLNEX	PC Defoamers [126]	Acrylic defoamers for nonaqueous systems such as lubricants and oils	

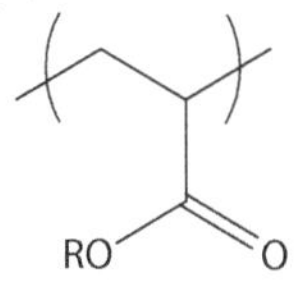

FIGURE 19.20 Polyacrylate structure.

a way to improve the effectiveness of antifoams in certain applications [127–129]. In addition to their use in lubricating oils, polyacrylate and functionalized polyacrylate antifoams also find use in papermaking, coatings and paints, HVAC, and crude oil–refining industries [130–134].

Polyacrylate antifoams are a viable alternative to silicones in cases where silicones are incompatible or cause undesirable level of haze and/or sedimentation in the lubricant, or where the presence of silicones causes issues related to surface wetting (Section 19.3.4), or the presence of silicone is either not preferred or prohibited from the fluid due to regulatory restrictions. At least one case also reports an improvement in the lubricant NOACK volatility upon substituting a silicone antifoam with a polyacrylate [135].

Additionally, polyacrylates may contribute less to promotion of air entrainment compared to silicones, due to their lower densities and generally higher surface tensions [69] (see Section 19.5.2.1 for more information on the relationship between foam inhibitors and air entrainment). There are several examples of acrylic-type polymers used for deaeration/air release in both aqueous and nonaqueous environments reported in the literature [136], although the use of acrylate antifoams with aqueous surfactant solutions is uncommon. In other cases, acrylates are used in combination with silicones to deliver excellent foam control while maintaining good air release [74], see also in Section 19.4.4.

Acrylates can be used alone or in combination with silicone-based antifoams (see also Section 19.4.4). In some

examples, combinations of acrylate and silicone antifoams provide a wider temperature range of performance compared to single antifoams [74,137].

19.4.2.2 Synthesis and Manufacture

Polyacrylate antifoams are typically synthesized via chain-growth polymerization of a mixture of alkyl acrylate monomers in hydrocarbon solvent or oil [118,138] (see Figure 19.21). Commonly, the polymerization is initiated by a peroxide or other free radical source, although examples of cationic and anionic polymerization are also known. The selection of monomer mixture is based on the solubility properties of the intended lubricating fluid and typically includes linear and branched short-chain alkyl acrylates ranging from C_2 to C_{12}. A common example is a copolymer of ethyl acrylate and 2-ethylhexyl acrylate [120]. There are also examples of incorporating functionalized acrylate monomers, for example, 2,2,2-trifluoroethyl acrylate [128].

19.4.2.3 Structure–Property–Performance Relationships

Acrylates offer the benefit of highly tailorable molecular weight, composition, surface tension, and solubility, which are key determinants in antifoam performance. However, as will be shown later, for acrylates, these properties are close to the boundaries of effectiveness. By way of background,

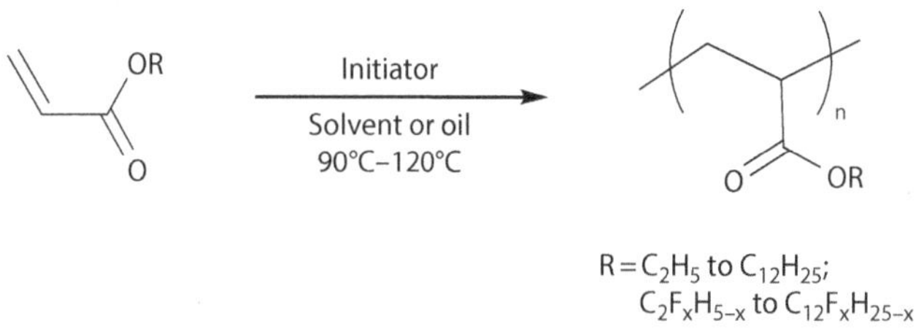

$R = C_2H_5$ to $C_{12}H_{25}$;
 $C_2F_xH_{5-x}$ to $C_{12}F_xH_{25-x}$

FIGURE 19.21 General polyacrylate synthesis.

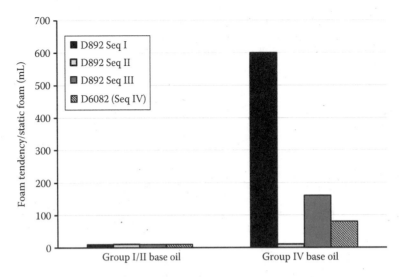

FIGURE 19.22   D892/D6082 foam performance of the same polyacrylate antifoam in a Group I/II–based finished lubricant (*left*) and a Group IV–based lubricant with identical additive package (*right*).

polyacrylate foam inhibitors were originally designed in the 1950s and 1960s, predating the existence of more highly refined (i.e., API Group II–IV) base oils. Thus, the composition of the polymers developed at that time was tailored to be effective specifically in Group I–based fluids. As Groups II to IV base oils have gained popularity, acrylates developed for Group I fluids may exhibit only limited effectiveness in these newer fluids. For example, Figure 19.22 shows that, an acrylate antifoam inhibits foam completely in a Group I/II–based finished oil in ASTM D892 and D6082 testing (Figure 19.22, *left*). However, when the same additive is blended into Group IV base oil, with the same additive package, the same acrylate antifoam is ineffective (except in Seq II), shown by the high foam tendency in this case (Figure 19.22, *right*). The loss in antifoaming performance is related to changes in the solubility properties and surface tension of the base oil. As discussed elsewhere in this chapter, for effective foam inhibition, the antifoam selected must be at least partially insoluble in the finished fluid and must exhibit a sufficiently low surface tension.

The composition of the acrylate polymer has a direct influence on its surface tension and its solubility. In theory, a range of surface tensions are accessible by using different monomers. For example, Table 19.5 gives a selection of acrylate and methacrylate homopolymers and their corresponding surface tensions [139]. In practice, copolymers consisting of at least two monomer types are used, in order to adjust the properties of the acrylate into an optimized range. We note that calculating the surface tension of a copolymer based on surface tensions of corresponding homopolymers of the constituent monomers is not straightforward, and it is perhaps best to measure surface tension of copolymers experimentally [140–142].

In one study, a series of polyacrylate antifoams were prepared targeting a surface tension range comparable to the range observed among the Group I–IV classes of base oil (Table 19.6). For reference, the static surface tensions of

typical formulations range roughly from 28 to 32 mN/m at room temperature, based on base oil group and additive package. In this case, the surface tension was varied by adjusting the comonomer ratio. Adjusting the comonomer ratio also necessarily affects the solubility of the polymer.

Static surface tension measurements were conducted on this series of acrylate polymers at 80°C using a Wilhelmy plate technique, and the results are shown in Table 19.6 and plotted as a function of composition in Figure 19.23. These suggest that the surface tension decreases with increasing nonpolarity of the bulk polymer, as expected.

Polyacrylate solubility in a Group III base oil was also assessed by a combination of pyrolysis and gas chromatography (GC) (Figure 19.23). In these experiments, a 50:50 wt.% mixture of neat polymer and the Group III oil was heated to 150°C for 24 h. The top oil layer with a portion of solubilized polymer was then removed, cooled, and diluted to a 50:50 wt.% ratio with toluene to fully solubilize the material. The samples were pyrolyzed and analyzed by GC to quantify the constituent monomer concentrations, which were then used to calculate the concentration of polymer solubilized in Group III base oil at 150°C. These solubilities, given in Figure 19.23, are expressed as a mole percentage based on a 50:50 wt.% mixture of polymer and oil. For example, a solubility of 100% means that the polymer and oil are fully miscible at 50:50 wt.% at 150°C. A solubility of 0% means the polymer and oil are completely immiscible. The point labels indicate visual appearance of a 5 wt.% blend of each polymer in the Group III base oil at room temperature. Overall, these results illustrate the inverse relationship between surface tension and solubility for this series of copolymers.

The ability of a polyacrylate to effectively control foams, based on its physical properties, can be understood using Figure 19.24. The two axes represent the surface tension of the pure acrylate and its solubility in the foaming liquid. A polyacrylate material would be represented by a point on the diagram.

TABLE 19.5
Selected Surface Tensions of Acrylate and Methacrylate Homopolymers

Polymer	γ_s (Surface Tension, Dynes/cm)	$d\gamma/dt$ (Surface Tension Change in Dynes/cm per °C)
Poly(*n*-butyl acrylate)	32.8	−0.070
Poly(*iso*-butyl methacrylate)	30.9	−0.060
Poly(*n*-butyl methacrylate)	33.1	−0.059
Poly(*t*-butyl methacrylate)	30.5	−0.059
Poly(ethyl acrylate)	35.1	−0.070
Poly(ethyl methacrylate)	34.0	−0.070
Poly(*n*-propyl methacrylate)	33.2	−0.065
Poly(2-ethylhexyl acrylate)	29.7	−0.070
Poly(2-ethylhexyl methacrylate)	28.8	−0.062
Poly(heptadecafluorooctyl methacrylate)	15.3	—
Poly(heptafluoroisopropyl acrylate)	14.0	—
Poly(heptafluoroisopropyl methacrylate)	15.0	—
Poly(nonafluoroisobutyl acrylate)	14.0	—
Poly(hexylmethacrylate)	30.0	−0.062
Poly(laurylmethacrylate)	32.8	—
Poly(stearylmethacrylate)	36.3	—
Poly(methylacrylate)	39.8	−0.070
Poly(methylmethacrylate)	41.8	−0.076

Source: Accu Dyne Test, Tables of polymer surface characteristics, www.accudynetest.com/polymer_tables. html, Diversified Enterprises, 2016. Accessed December 14, 2015.

TABLE 19.6
Series of Polyacrylates and Experimental Surface Tensions (SFT) at 80°C

Entry	Name	Monomer Composition (wt.%)[a] EHA	EA	GPC Results M_n	M_w	PDI	SFT at 80°C (mN/m)
1	AF A	50	50	13,653	93,069	6.8	26
2	AF B	72	28	11,741	68,586	5.8	23.5
3	AF C	85	15	10,281	52,313	5.1	21.7
4	AF D	100	—	7,352	26,226	3.6	22

[a] All reactions run at 50% actives in toluene at 110°C, initiated by *tert*-butylperoxyoctanoate (0.1 wt.% of total monomer). EHA, 2-ethylhexylacrylate; EA, ethyl acrylate.

If the acrylate has both a sufficiently low surface tension and low level of solubility in the foaming liquid (i.e., falling in the region labeled *potentially effective antifoam* in Figure 19.24), it can act as antifoam based on the mechanisms presented in Section 19.5.2.2. These limits are shown as the *critical surface tension* and *solubility limit* in the diagram. The absolute value of the critical surface tension is related to the surface tension of the foaming liquid, as well as the interfacial tension of the antifoam and the foaming liquid. Above this critical surface tension, the antifoam droplet does not have a driving force to emerge or spread on the fluid/air surface (see Section 19.5.2.2 for more detail on the antifoaming mechanism). The solubility limit is a function of the solvency of the foaming liquid for the antifoam material; above this limit, the antifoam is soluble in the base fluid and therefore does not form droplets. We can also understand changes in antifoaming performance associated with changes in base fluid and temperature, which will cause shifts in the positions of the critical limits shown in the diagram. For example, an increase in temperature will generally shift the solubility limit to the left, causing an antifoam that is in the *potentially effective* region at low temperature to be in the *soluble in base fluid* region at high temperature.

This diagram is valid for all antifoams, including silicones but, in general, is more relevant for acrylates, since the properties of acrylate antifoams are closer to the critical limits.

The relationship between surface tension and solubility for polyacrylates with only hydrocarbon side chains suggests a

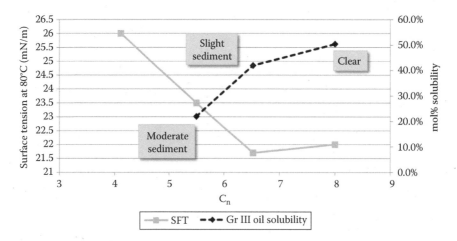

FIGURE 19.23 Solubility in a Group III base oil at 150°C and surface tension (SFT) at 80°C of selected polyacrylates as a function of composition. C_n, the average number of carbons in the monomer side chains. Point labels refer to visual appearance for a blend of 5 wt.% polymer in the Group III base oil at room temperature.

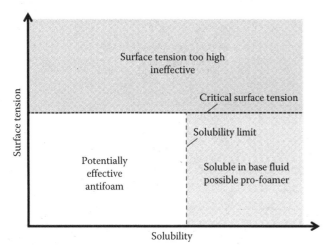

FIGURE 19.24 Conceptual diagram illustrating the performance of acrylate antifoams based on surface tension of the acrylate and solubility of the acrylate in the base fluid.

constrained range of surface tension–solubility combinations and has important implications for antifoam performance in lubricating fluids with respect to proper antifoam selection. This is illustrated in the following examples.

In the Group I/II–based gear oil, shown in Figure 19.25, AF A imparts only marginal improvement in foam control, while AF B (having higher C_n, the average number of carbons in the side chains) provides optimal foam performance. Increasing the C_n content in AF C and AF D gives rise to improved solubility in the base oil and thus to foam stabilization or profoaming (i.e., foam tendency greater than that of the no-AF control) at elevated temperature. An interpretation of the results is shown on the left side. At low C_n, the surface tension of the acrylate is too high to act as an antifoam. At high C_n, the polymer is soluble in the base fluid. Its low surface tension causes it to act as a foam-stabilizing surfactant.

When a second finished fluid is made by diluting the same additive into a Group IV base oil (Figure 19.26), the critical

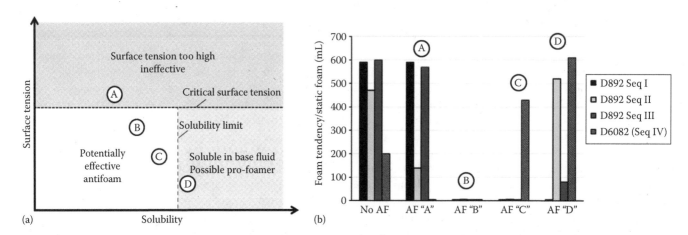

FIGURE 19.25 (a) AF A through AF D (see Table 19.6) mapped to a conceptual surface tension-solubility diagram for a Group I/II-based gear oil and (b) Sequence I–IV foam tendency (ASTM D892/D6082) results for acrylate copolymers in Group I/II–based gear oil.

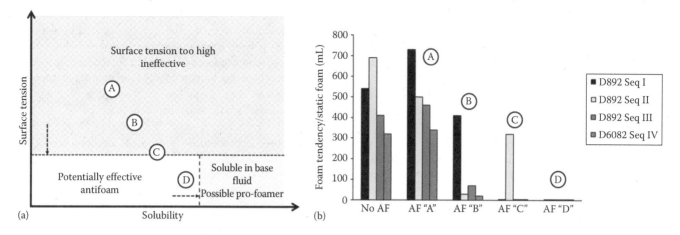

FIGURE 19.26 (a) AF A through AF D (see Table 19.6) mapped to a conceptual surface tension-solubility diagram for a Group IV-based gear oil, and (b) Sequence I–IV foam tendency (ASTM D892/D6082) results for acrylate copolymers in Group IV–based gear oil.

surface tension is lower and the effective solubility boundaries shift to the right in the conceptual diagram due to the decreased polarity of the base oil (Figure 19.25a). In this new chemical environment, AF B is less effective, while AF D eliminates foam completely.

The results described earlier highlight two important points. First, blending the same or a similar additive package in different base oils can result in very different polyacrylate antifoam requirements. Second, small compositional changes in the polyacrylate structure can yield dramatically different antifoam performances for a given finished fluid, based on surface tension and solubility effects.

To summarize in practical terms, the selection rules for acrylates are the same as those for silicones at a general level (i.e., lower surface tension than the lubricant, insoluble, and dispersible into small droplets); however, relative to silicones, the properties of acrylates are often closer to the critical surface tension and solubility properties of the fluid. There is therefore no universal acrylate foam inhibitor, as small chemical adjustments may make the difference between a soluble profoamer and an excellent foam inhibitor. Ultimately, the preference to use an acrylate foam inhibitor may arise when all silicone options have been exhausted or using silicones imparts an undesirable side effect (e.g., poor air release, sedimentation, or regulatory restriction).

19.4.3 MIXED (OIL + SOLID) ANTIFOAMS

To efficiently control foaming of *aqueous liquids*, generally, combinations of nonpolar oils and hydrophobic solid particles are used [143]. Mixed-type antifoams have been known since the 1950s, and since 1970, many patents have been issued [23,144]. Among lubricating fluids, the mixed-type antifoams are important only for MWFs in which water is the continuous phase.

As discussed in Section 19.4.1.1, PDMS alone (as well as FS alone) is a poor antifoam in water-based liquids. Organomodified silicones can break foams of aqueous systems; although in surfactant solutions, they often have poor efficiency. Nevertheless,

OMS antifoams, such as SPE, are commonly used for coatings (both water based and oil based) and latexes.

In the mixed-type antifoams, the nonpolar oil component is typically either of the following:

- Silicone-based oil, often PDMS
- Organic oil, such as mineral oil

In fact, a lubrication oil or a base oil could be also used as the nonpolar oil component of the mixed-type antifoam, but this is not common.

The hydrophobic solid particles are typically

- Hydrophobized silica
- Sometimes other hydrophobic solids, such as ethyl-*bis*-stearamide

The mixed-type antifoam actives are often called *antifoam compounds*. In most commercial compounds, the particles are homogenously mixed and they do not separate or separate only very slowly from the nonpolar oil.

The silica used in antifoams is amorphous and of high purity, typically either fumed or precipitated [145], and has to consist of fine particles, typically less than a micrometer in size. Fumed silica is typically made by burning silicon tetrachloride ($SiCl_4$) with oxygen and hydrogen, with HCl as a by-product. Very fine, round, nonporous particles are formed with primary particle sizes as low as a few nm [146], but they agglomerate to larger aggregates. Precipitated silica is made from alkaline silicates, by precipitating them with mineral acids, such as sulfuric acid, and then can be dried and ground. The precipitated silica particles are porous and their primary size is also very small, but can also aggregate into large units.

The hydrophilic silica then can be hydrophobized by heating it with PDMS or with silanes, such as hexamethyldisilazane (($CH_3)_3SiNHSi(CH_3)_3$) or dimethyldichlorosilane. A method of reacting with PDMS is dry roasting at about 250°C in a fluidized bed reactor to promote good contact between the silica and silicone oil [147]. The hydrophobization is either performed by the silica manufacturer [148] or by the manu-facturer of the antifoam. In the latter case, the oil and the par-ticles can be a simple mixture, or heated, in the presence of acid or base catalyst, to facilitate the *in situ* hydrophobization of silica. This reaction mixture can also contain silicone resins (*three-dimensional*), that is, branched siloxanes with T and Q groups (see Table 19.1). The concentration of the solids in the mixture is typically in the 3%–10% range, and the viscosity of the antifoam compound is generally at least 1000 cP, often more than 10,000 cP, and can be as high as 1,000,000 cP.

The combination of the oil and the hydrophobic particles results in a *strong synergy* between the two components, since either of the two components is a poor antifoam. The role of the components in the antifoaming mechanisms will be dis-cussed in Section 19.5.1.

The efficiency of this type of antifoam is very high, espe-cially with a wide range of surfactant-stabilized foams, and therefore, it is much more commonly used for aqueous systems than any other kinds of antifoams. Silicone oil–based products generally work better than organic mineral oil–based ones. An important phenomenon with antifoams, and especially with the mixed type, is that they have a finite durability: during pro-longed foam generation, their efficiency gradually diminishes, and after some time, their antifoaming efficiency disappears (although during storage, generally, no change happens in their efficiency). The mechanism of this antifoam deactivation phe-nomenon will be discussed in Section 19.5.1.7. Silicone-based antifoams have very often much longer durability than mineral oil–based ones. The reason for this is the high incompatibility of PDMS with water-based systems because of which PDMS does not get solubilized into surfactant micelles or get emulsi-fied as fine (and ineffective) drops.

Fluorosilicones are not commonly used as the oil component in mixed-type antifoams for aqueous systems, although Owen and Groh [149] prepared effective antifoams for solutions of commonly used surfactants. Their main target was, however, *fluorocarbon* (*FC*) surfactant solutions, which, because of their very low surface tension, can create foams that are extremely difficult to break. The FSs, containing $CF_3(CF_2)_nCH_2CH_2$ pen-dant groups (n = 3, 5, or 7; similar to Figure 19.18, but with more fluorocarbons) had, however, higher surface tensions than the FC surfactants, and probably due to this, the mixed-type antifoams prepared from them did not break the FC-stabilized foams. Kobayashi and Masatomi [150] patented an antifoam containing two types of FS oils, a fluoroalkyl containing alk-oxysilane, and silica and concluded that the antifoam can be used also for foam control of fluorocarbon surfactants.

19.4.3.1 Antifoam Formulations for Metalworking Fluids

Finding the optimum defoamer formulation for MWFs is often an empirical process [52], and there are several require-ments for antifoams:

- High efficiency at the lowest possible concentration.
- High durability (persistence, longevity) so that no or only minimum tankside post addition is necessary.

- It is important that the antifoam emulsion or concen-trate should be easily and evenly dispersible in the MWF, without forming clumps or large particles, and it must be stable as a concentrated component, as well as after blending into the MWF. The latter requirement is difficult if the diluted MWF is stored or otherwise stationary for a prolonged amount of time.

Foam control agents with several types of chemistries exist for MWF, but the most efficient and with the widest range of applications is the mixed type (see Section 19.4.3). The most efficient silicone-based antifoams may contain silicone resins, three-dimensional siloxanes, and may be used for all the three types of water-diluted MWFs: soluble oils, semisynthetic, and synthetic [50,52]. It is important to emphasize that there are great variations (sometimes orders of magnitude!) in the effi-ciencies of the various antifoam emulsions. The active ingre-dients of these types of antifoams are insoluble in water, and therefore, they are generally either emulsified as oil-in-water emulsions or blended into water-*soluble* (actually dispers-ible) concentrates, which contain no or only a small amount of water. The drop size in these emulsions has to be controlled for optimum efficiency, somewhere in the 5–50 µm range, and therefore, the emulsion has to be thickened to avoid the separa-tion of the active drops. The relation between the drop size and antifoaming action will be further discussed in Section 19.5.1.5. If the emulsion or dispersible concentrate is diluted and then stored, the drops will generally separate (cream to the top or settle to the bottom depending on their density relative to the density of the medium) with time (sometimes within hours) unless it is continuously agitated or restabilized by adding a thickener (to raise its viscosity) and biocide.

A potential disadvantage of silicone-based antifoams is that silicone oil (PDMS) can act as a release agent and there-fore can cause coating defects. This happens particularly if the efficiency of the formulation is low, that is, high concentration of silicones is dosed, especially after multiple additions to the circulation system, and/or if the dispersibility of the emulsion is poor and it forms large particles in the MWF. Therefore, MWF antifoam suppliers generally carry both silicone- and nonsilicone-containing products, and if the metal part will be painted, the use of silicone antifoams is sometimes avoided [151]. A list of antifoam suppliers is given in Section 19.4.5.

Alternatives to silicone fluids as actives in the foam control agents can be mineral oils, as the oil phase of the mixed-type antifoam [47,152], moreover, tributylphosphate, waxes, fatty acid esters (stearates, etc.), vegetable oils, and synthetic poly-mers. The use level of these antifoam formulations tends to be significantly higher, and their efficiency, especially their durability, is much lower than silicone-based antifoams.

A problem sometimes encountered in central metalwork-ing circulation systems is that the filtration system removes the antifoam actives. The defoamer drops can initially pass through larger pore size filters, but after some time, a filter cake can form that will then remove them. This is another rea-son to use low-foaming ingredients in the MWF formulation, especially in high-pressure applications [50].

19.4.4 COMBINATION OF VARIOUS TYPES OF ANTIFOAMS

Simultaneous use of more than one type of antifoam is also possible for controlling foaming of nonaqueous liquids so that all the requirements (e.g., requirements based on all the sequences of ASTM D892 test and the D6082 test) can be satisfied. Lomas described antifoam systems for nonaqueous liquids containing PDMS, silicone resins, SPE, and nonionic surfactant as dispersants [98,153]. Using more than one antifoam is not typical with *crude oils or diesel fuel*, although Gallagher et al. [154] reported synergy between PDMS and FS antifoams for crude oils.

Combination of various antifoam chemistries is also possible with lubricating oils. For example, Ward et al. patented combinations of a PDMS and a FS antifoam for lubricants [74,155] and found that combination of FS and polyacrylate antifoams can provide sufficient foam control of lubricants at both high and low temperatures without compromising the air release time.

Muchmore et al. [156] described an antifoam package containing PDMS, FS, and acrylate antifoams for ATFs that can fulfill the increased requirements of modern transmission with high-pressure pumps, where the traditional packages with 3–10 ppm silicone are not sufficient anymore. If the antifoam is not performing well, surface foam and entrained air can cause pressure ripples in hydraulic pumps, causing *pump whine* in some transmissions.

Loop et al. described recently an antifoaming system for diesel engine lubricants that comprised two types of PDMS with different viscosities and FS [157].

19.4.5 ANTIFOAM SUPPLIERS

Table 19.7 shows a list of selected antifoam suppliers, mostly from a North American and European perspective. It includes silicone and organic antifoam manufacturers, moreover, selected chemical companies and antifoam formulators, who manufacture antifoams primarily for MWFs. Polyacrylate antifoam suppliers are listed in Table 19.4 (see Section 19.4.2.1).

19.5 MECHANISMS OF ANTIFOAMING ACTION

Scientists and technologists dealing with antifoams have been asking for at least five decades: How do foam control agents work? Why can as little as 1–10 ppm antifoam effectively control foam? Why do the mixed-type antifoams (nonpolar oil + hydrophobic particles) have so much higher efficiency than their individual components? Why does the defoaming performance diminish after prolonged foam generation? What is the optimum drop size of antifoams? Research in the past three decades has given reasonable explanations to many of these questions. However, as discussed in Section 19.1, almost all of this work was done with *water-based foams*, and relatively, little work has been conducted with oil (nonpolar liquid)-based foams [23,144,158]. In this section, we will first review the antifoaming mechanisms of aqueous foams. We note that, among lubricants, this discussion will be directly relevant only to (aqueous) MWFs, and some hydraulic

TABLE 19.7

Selected Silicone and Organic Antifoam Suppliers and Their Trade Names

Company Name	Typical Trade Name
Silicone manufacturers	
BLUESTAR Silicones	Silcolapse®
Dow Corning Corp.	Dow Corning®, Xiameter®
EVONIK Industries	TEGO®
Momentive Performance Materials	Sag™, Silbreak™
Siltech	Siltech®
Siovation	TA
Wacker	Silfoam®
Shin-Etsu	FA, KF, KS
General chemical manufacturers	
Air Products	Surfynol®
Formulators (mostly MWFs)	
Afton Chemical	Antimus™, HiTec®
Emerald Performance Materials	Foam Blast®
Enterprise Specialty Products Inc. (ESP)	Foam-A-Tac™
Munzing	Foam Ban®
PMC Crystal	Surtech®

fluids. In the subsequent section, we will discuss the information available on antifoaming in oil-based foams, such as lubricating oils, with the goal of describing their antifoaming mechanisms, and any key differences in these mechanisms between aqueous and oil-based foams.

19.5.1 ANTIFOAMING THEORIES FOR AQUEOUS FOAMS (METALWORKING FLUIDS)

A great deal of work has been published on foam control mechanisms of water-based systems, and Garrett's recent book [23] gives an excellent account of them, with great details and analysis. While these mechanisms have mainly been proposed for aqueous systems, several key aspects may be generalized to nonaqueous fluids also.

Most (if not all) antifoams work by a heterogeneous mechanism: the foam control liquid, an apolar fluid (oil) in general, is not soluble or is only partially soluble in the foaming liquid, and therefore it forms droplets. The first theories on antifoaming mechanisms were based on the surface tension of the foaming liquid and the antifoam liquid drops in it, respectively, and the interfacial tension between them. Furthermore, it has been observed that antifoam materials should have lower surface tension than the foaming liquid. The phenomenon explaining this is that a lower surface tension material (pure fluid, surfactant, or surfactant solution) generally spreads on a liquid of higher surface tension with remarkable speed.

19.5.1.1 Role of Surface and Interfacial Tensions

To further refine the surface tension condition described earlier, one has to consider the configurations of the antifoam oil

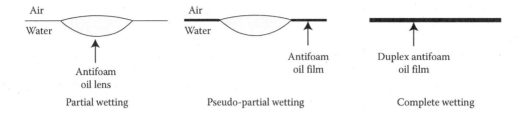

FIGURE 19.27 Configurations of oil after entering the liquid surface. (Adapted with permission from Garrett, P.R., ed., *The Science of Defoaming: Theory, Experiment and Applications,* Surface Science Series, Vol. 155, CRC Press, Boca Raton, FL, 2014, p. 83. Taylor & Francis Group LLC.)

in a foaming liquid. The antifoam oil drop inside the foaming liquid can enter the surface and then can be either of the following:

- Form a lens (partial wetting).
- It can spread on the surface and then form either of the following:
  - A thick (so-called duplex) film (complete wetting)
  - Lens(es) and a thin (mono- or multimolecular) film around the lens(es) (pseudo-partial wetting) [23]

These possibilities are illustrated in Figure 19.27. Entering and then spreading of the antifoam oil can spontaneously occur only if the total surface energy of the system decreases during the process. In order to predict these processes, various coefficients have been defined using the surface and interfacial tensions present in the system.

Since during drop entry a foaming liquid/air surface and a foaming liquid/antifoam oil interface are replaced by an antifoam oil/air surface, the *entry (entering) coefficient* was defined as

$$E = \sigma_{\text{Air/Foaming liquid}} + \sigma_{\text{Antifoam/Foaming liquid}} - \sigma_{\text{Air/Antifoam}}, \quad (19.10)$$

where

$\sigma_{\text{Air/Foaming liquid}}$ and $\sigma_{\text{Air/Antifoam}}$ are the surface tensions of the foaming liquid and the antifoam oil, respectively

$\sigma_{\text{Antifoam/Foaming liquid}}$ is the interfacial tension between the two liquids

The entry of the drop is energetically favorable if $E > 0$.

It is important to realize that the two liquid phases (antifoam and foaming liquid) can be in phase equilibrium (saturated) with each other, or in a nonequilibrated state before the entering occurs, and thus different entry coefficients can be calculated. The equilibrium value is smaller than the nonequilibrium one, and if it is larger than zero, then the oil will probably permanently spread on the liquid as a duplex film.

A second parameter, the spreading coefficient, S, determines how the drop will spread on the surface after entry:

$$S = \sigma_{\text{Air/Foaming liquid}} - \sigma_{\text{Antifoam/Foaming liquid}} - \sigma_{\text{Air/Antifoam}} \quad (19.11)$$

For the entered drop to spread, $S > 0$ is required, and S can be also defined with nonequilibrium (initial) or equilibrium

values. Since from the equations earlier presented, $E - S = 2\sigma_{\text{Antifoam/Foaming liquid}}$, the entry coefficient is always larger than the spreading coefficient, assuming that the two coefficients are calculated from the same type of surface and interfacial tensions (equilibrium or nonequilibrium). If $E > 0$ and $S < 0$, then the antifoam oil will form only lenses. In some cases, after equilibration S can become negative, at that point the film retreats to a lens or lenses (e.g., benzene on water) [23].

Due to its low surface tension (roughly 21 mN/m at room temperature), PDMS leads to strongly positive E and initial S values, and zero or near zero equilibrium S values on water, most surfactant solutions, and on most oils. As it spreads, PDMS readily enters the air/liquid interface, and can then form a duplex film (complete wetting), lenses, or a combination of both (pseudo-partial wetting) [23,159,160].

The classical antifoaming theories state that the foambreaking action correlates with the magnitude of E and S. However, more recently, several authors found no such correlation, although it is still generally accepted that an effective antifoam oil must have a lower surface tension than the foaming liquid [23,158,161,162]. Another significant problem of the classical theories using the E and S coefficients is that they cannot explain the high efficiency of the mixed-type antifoams, that is, the mixtures of apolar liquids (oils) + hydrophobic solid particles, since these coefficients do not take into account the major effect of particles. Research in the past 30 years has made major progress in the understanding of antifoaming mechanisms and will be reviewed in the next sections.

19.5.1.2 The Pseudoemulsion Film

Most of the discrepancies between the classical theories with entry and spreading coefficients and the observed antifoaming action of antifoam oils and mixed-type antifoams can be explained by the role of the so-called pseudoemulsion film. It was discovered by Nikolov and Wasan [163] that when an antifoam oil drop approaches the water surface, first, an antifoam oil/water/air film forms between the drop and the air. The drop can then enter the surface only if this film breaks (see Figure 19.28).

They called this asymmetrical film a *pseudoemulsion film*, since it is neither an emulsion (oil/water/oil) nor a foam (air/water/air) film. Due to its asymmetry, this film is practically never flat, but curved. For antifoaming action, the pseudoemulsion film has to break first, and this process

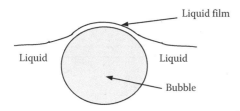

FIGURE 19.28 Formation of an (aqueous) pseudoemulsion film.

is often the rate controlling step with aqueous foams. Therefore, the concept of pseudoemulsion film became a cornerstone of the current antifoaming mechanism theories [23,164,165]. Since the foaming liquid can contain surface-active foam stabilizers, these surfactants can also adsorb at the antifoam oil/water interface and stabilize the pseudoemulsion film from this side as well, thereby preventing the entering of the drop, with similar mechanisms as the stabilization of foam films (see Section 19.2) regardless of the values of E or S.

Apolar liquids (including PDMS, hydrocarbon oils, etc.) alone, without solid particles in them, are generally ineffective in breaking aqueous foams especially those that contain surfactants because the surfactants make the pseudoemulsion film too stable for quick antifoaming action. Using oil as a *defoamer* (see Section 19.1), for example, by spraying it onto a foam, can be quite effective, because the oil is added directly to the surface from the air phase, so there is no need for entering. However, when the same oil is pre-emulsified, it generally becomes ineffective as an antifoam. In certain cases, the pseudoemulsion film can be broken by pure fluid antifoams, as shown by Arnaudov et al. [162]. Here it was shown that some alkanols and esters showed significant antifoaming action in aqueous surfactant solutions.

19.5.1.3 Role of Solid Particles

One of the main roles of the hydrophobic solid particles in mixed-type antifoams is that they destabilize the pseudoemulsion film. Koczo et al. [166] showed this effect through experiments involving the formation of a single pseudoemulsion film. Bergeron et al. [159] also observed that the pseudoemulsion film breaks at much lower capillary pressure if hydrophobic particles are present in the system. The experimental technique of Denkov et al. [158,167] was used to show that, for example, for PDMS in a sodium

dodecylbenzene sulfonates solution with no particles, the entry barrier was >3000 Pa, while with mixed-type antifoams, it was less than 10 Pa.

The hydrophobic particles in an antifoam oil drop are not completely covered with the antifoam oil, but partially penetrate into the water phase (see Figure 19.29c), and their sharp edges are able to pierce the pseudoemulsion film. Using fluorescent labeling, Wang et al. [168] have shown where the particles are located at the surface of the antifoam oil drops. Therefore, it is important that the particles are not smooth but have a high degree of surface roughness or sharp edges. For best effect, the particles should not have complete hydrophobicity (less than 180° contact angle) because their penetration into the water film would be zero or too small [169].

As will be shown in the next sections, the antifoaming process is related to the formation of antifoam oil bridges in the foam. Frye and Berg [170] suggested first that a function of the particles in the mixed-type antifoams is that they increase the *penetration depth* of the oil lens into the aqueous phase, as illustrated in Figure 19.29. With deeper penetration, the lens can reach the other surface of the film more easily, and thus the bridge formation is faster. It can be also seen that hydrophobic particles and the oil lenses alone (Figure 19.29a and b) have low penetration depth, which is a possible explanation why solid particles alone are generally poor antifoams [166].

19.5.1.4 Location of Antifoam Oil Drops Inside the Foam

In a freshly generated foam with antifoam oil drops in it, the spherical bubbles first approach each other and then get compressed together forming foam films, due to gravity (buoyancy). The foam films then start to drain. Initially, the antifoam drops are uniformly distributed in the foam. During film drainage, however, a powerful flow inside the films is moving the liquid out of the films into the adjoining Plateau borders, and it is also sweeping out the antifoam oil drops. Figure 19.2 shows the picture of a real foam with antifoam oil drops (small white circles) in it. It can be observed that all the antifoam oil drops are located inside the Plateau borders, and no drop is inside the foam films [23,158,161]. It is important to note that it takes time, at least a few seconds, until the antifoam drops get out of the draining films. Therefore, the location of the antifoaming action depends on the rate of foam rupture (see in Section 19.5.1.6).

It was demonstrated that massive amounts of emulsified oil can be incorporated into foams of aqueous surfactant

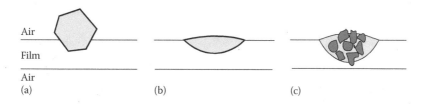

FIGURE 19.29 Illustration of penetration depth of (a) hydrophobic solid particle, (b) antifoam oil lens, and (c) mixed-type antifoams into aqueous film.

solutions without destabilizing the foam, if the system contained stable pseudoemulsion films, even with highly positive entry or spreading coefficients [5,161]. This phenomenon may play a role in water-based MWFs that contain a high amount of lubrication oil drops. Since these oil drops do not contain solid particles, the pseudoemulsion film is probably stable and the lubrication oil drops can stabilize the foam.

19.5.1.5 Bridging

If an antifoam oil drop enters both sides of a foam film, then a liquid bridge forms, as shown in Figure 19.30 for two possible configurations. Garrett [171] who gave first an analysis of the stability of liquid bridges found that the top configuration shown in Figure 19.30 ($\Theta_{OW} < 90°$) cannot be mechanically stable and that the configuration shown on the bottom ($\Theta_{OW} > 90°$) can be stable only if the film has a particular thickness.

Garrett also defined a bridging coefficient to characterize the instability:

$$B = \sigma^2_{\text{Air/Foaming liquid}} + \sigma^2_{\text{Antifoam/Foaming liquid}} - \sigma^2_{\text{Air/Antifoam}}$$

$$(19.12)$$

and concluded that the bridge will be unstable if $B > 0$.

The bridge is generally very unstable and ruptures quickly [168]. Most of the current antifoaming theories are based on bridging.

There are various theories as to how the antifoam bridge actually breaks (see Figure 19.31). A commonly cited theory is that the aqueous foaming liquid film dewets the bridge, and it pinches off [170,172,173] (see Figure 19.31a).

Denkov suggested a different mechanism, based on experiments with a horizontal film, where the oil bridge ruptures by stretching (see Figure 19.31b) [158]. A reason for this is that the antifoam oil surface is not stabilized against stretching by the Gibbs–Marangoni effect (see Section 19.2). This mechanism, however, requires that the antifoam drop is deformable, and therefore it is less likely to work with mixed-type antifoams with highly viscous liquid phase (which generally have the highest efficiency) than with low viscosity antifoams.

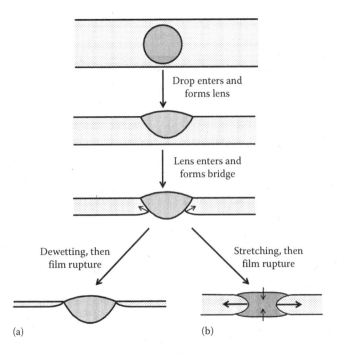

FIGURE 19.31 Rupture mechanisms of antifoam liquid bridge: (a) Dewetting, (b) stretching. (Reprinted with permission from Denkov, N.D., Mechanisms of foam destruction by oil-based antifoams, *Langmuir*, 20, 9463–9505. Copyright 2004, American Chemical Society.)

19.5.1.6 Antifoaming Mechanisms

Early theories on antifoaming action were based on the spreading of the highly surface-active antifoam oil on the aqueous surface, according to the entry and spreading coefficients (see Section 19.5.1.1). Ross [174] suggested that the viscous drag caused by the spreading of oil on a foam film thins and breaks it. This theory is less accepted today, although it may have merit in oil-based foams (see Section 19.5.2.2).

Direct observation, in other words monitoring of the quick movement of microscopic antifoam drops inside macroscopic foams, involves significant experimental difficulties. Therefore, the only experiments reported in the literature that we are aware of are done using simplified systems, with one or a few foam films, and often large antifoam droplets.

The current theories are based on bridging, and they try to explain the observations described in the previous sections, such as the role of the pseudoemulsion film, the role of the particles in the mixed antifoams, the location of the antifoams, and thus the foam-breaking action in the foam [23,164,168,173].

Koczo et al. [166] suggested that mixed-type antifoams work in the following steps:

- The oil drops + hydrophobic particles collect in the Plateau borders and as the foam drains are trapped in them.
- The antifoam drops, due to the presence of particles partially penetrating into the water phase, break the pseudoemulsion film and enter the foam surface, forming a lens.

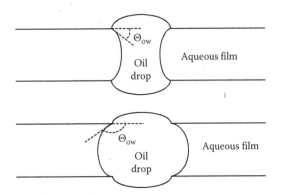

FIGURE 19.30 Possible configurations of antifoam oil bridge in a foam film. (Republished with permission of P.R. Garrett, ed., *The Science of Defoaming: Theory, Experiment and Applications*, Surf. Sci. Series Vol. 155. Boca Raton, FL, CRC Press, 2014.)

- On further foam drainage, the lens will enter the opposite side of the film, in the Plateau border region, and form a bridge.
- The bridge then quickly breaks and eliminates a section of the foam.
- The antifoam drops then fall to another section of the foam and break more foam.

In this way, an antifoam drop can eliminate multiple foam films, so only a small amount of antifoam is sufficient. The mechanism of antifoaming in the Plateau borders could explain that for effective foam control, the antifoam drop sizes have to be larger than about 3–5 μm so that they can get trapped inside Plateau borders. It is well known that antifoams with small sizes, less than 1–2 μm, have poor efficiency in aqueous systems.

The video frames in Figure 19.32 by Wang et al. [168] show an interesting way of bridge formation. It shows a layer of bubble (between two glass slides under microscope) inside a water phase containing a commercial antifoam emulsion. Frames A–C show that some of the antifoam drops entered the bubbles and formed lenses. In frame C, the lenses of a smaller bubble and a large bubble on the left meet. At this point, the two lenses coalesced, forming a bridge that then ruptured immediately causing the small bubble to disappear in frame D. Thus, here the last step before bridge formation is the rupture of an *emulsion film* between two lenses.

Denkov [158] conducted exhaustive research on most aspects of antifoaming phenomena, which not only confirmed some of the theories above but also suggested several changes. Although it was noted earlier that in many cases antifoams act so fast (sometimes immediately) that there is no time for the drops to collect in the Plateau borders [158,166] Denkov showed systematically that with highly effective antifoams (typically with mixed type), there is no time for the drops to leave the draining films or for the films to get thinner than micrometers. Therefore, he concluded that the antifoams, which act fast, must bridge the foam films directly, when their thickness is still in the few μm range. The extremely low entry barrier of the pseudoemulsion film (10 Pa or less) with the antifoam oil + particle antifoams is a reason for this fast action. *Slow* antifoams, which are typically liquids, such as PDMS or long-chained, branched alcohols or esters [162] without particles, form pseudoemulsion films with high entry barriers and therefore act slowly enough, after the foam has drained substantially, that they collect and then act in the Plateau borders, also by bridging. The antifoam bridge breaks then in both mechanisms either by the dewetting or by the stretching mechanism (see Figure 19.31). They also emphasized the importance of prespread antifoam oil (PDMS) layer on the film surfaces.

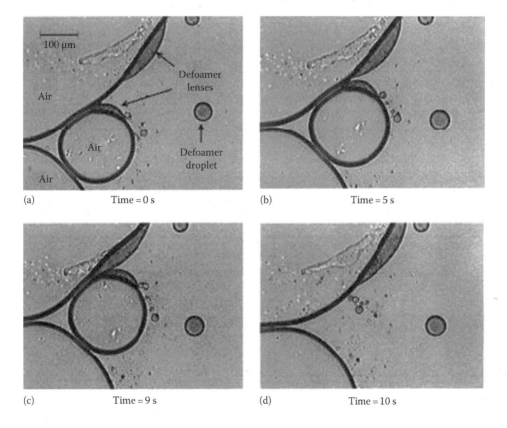

(a) Time = 0 s

(b) Time = 5 s

(c) Time = 9 s

(d) Time = 10 s

FIGURE 19.32 Sequence of video frames showing the migration of a defoamer lens into the film between two air bubbles, causing coalescence. (Reprinted with permission from Wang, G. et al., On the role of hydrophobic particles and surfactants in defoaming, *Langmuir*, 15, 2202–2208. Copyright 1999, American Chemical Society.)

In the presence of the antifoam oil film, the entry barrier of the pseudoemulsion was significantly lower than without it. Several details of the bridge formation and rupture are still not well understood [23].

The *optimum drop size* of the antifoam is of great practical importance. In aqueous systems, small drops (i.e., <1–3 µm) are ineffective, which is explained in various ways, either by their inability of getting trapped in the Plateau borders [166] or that it takes longer time for the films to drain to their size [159].

On the other hand, if the drop size is too high, the number of drops becomes drastically smaller, which can slow the foam-breaking action. Based on these considerations and experience, the optimum drop size range for aqueous systems is around 5–30 µm [158]. It is well known that antifoam emulsions are shear sensitive: if subjected to high shear, their efficiency can be severely reduced, which can be explained by the earlier arguments on the optimum drop size. Interestingly, antifoam compounds (100% actives) can also be shear sensitive, which is harder to explain.

19.5.1.7 Antifoam Deactivation (Durability)

It was observed decades ago [175] that after prolonged foaming (bubbling, shaking, etc.) of *water-based liquids*, the antifoaming action diminishes and antifoams become inactive. During storage, no change in the foam control potency of antifoams occurs (assuming good chemical stability, and stability against separation); hence, the diminishing efficiency should be caused only by processes that occur during the antifoaming action. The phenomenon is also called the durability (persistence) of antifoams and is an important parameter of antifoam performance since high durability is one of the main requirements for these products. There can be great differences between the durabilities of various mixed-type antifoams. Another key antifoam parameter is the initial antifoaming action, also called *knockdown*. The knockdown performance of an antifoam is generally *not* related to its durability; either one can be good while the other could be poor or good. It is not clear whether or not antifoam deactivation occurs with *oil-based liquids* (see Section 19.5.2.2.5).

It was first Pouchelon and Araud [176] followed by Koczo et al. [166] and Racz et al. [177] who observed that during deactivation an emulsification process takes place, in which the antifoam drops become smaller (the solution becomes hazy). Moreover, the smaller drops that form during deactivation are different from the original ones: they do not contain particles, only antifoam oil, and the particles get concentrated in a few large clusters. This dispersion is an ineffective antifoam since the oil drops contain no particles (the pseudoemulsion films will not break easily) and the large drops are too few and solid-like.

The phenomenon was explained by the spreading of the antifoam [176,177]: when the antifoam drop or lens enters the water/air surface, the surface-active antifoam oil (only the oil and not the particles) spreads on the surface, and after rupture, this oil film rolls up to small oil drop(s). This way at every rupture, the antifoam drop that caused it loses a little oil and gets more concentrated in particles, until it is so concentrated

that it is solid-like and can stick to other such particles on the surface of the solution.

Denkov et al. [178,179] offered a different explanation to the silica disproportionation phenomenon. They found that the size of the antifoam drops does not decrease (or decreases only a small amount) and is therefore not relevant to deactivation. Further, they found that the deactivation point coincides with the disappearance of the PDMS oil film on the surface of the aqueous solution, and the critical capillary pressure to break the pseudoemulsion film is significantly higher without the oil film. They suggested that the particle-free oil drops form during the stretching of the bridge (see Figure 19.31b), causing an oil emulsification. However, the stretching and bridge breakup process should yield several smaller drops, and it is not clear why they did not observe decreasing drop size, as the other previously mentioned authors did.

It was also observed by several authors that the deactivated antifoam could be (at least partially) reactivated by mixing only antifoam oil (no particles) to it [159,177,178]. The explanation for this is probably that the oil remixes and dilutes the concentrated hydrophobic particles and thus regenerates the antifoam.

The *effect of antifoam viscosity* on antifoaming efficiency and deactivation (durability) is of great practical importance. Although it is claimed in the literature, without experimental evidence, that there is an optimum oil viscosity [178], our experience shows that increasing the antifoam oil viscosity improves the efficiency and significantly slows the deactivation of antifoams (in water-based applications), although the incorporation/emulsification of more viscous compounds is a higher challenge to the formulator. Koczo et al. [166] studied the efficiency and deactivation (durability) of PDMS + hydrophobic silica blends with PDMS viscosities ranging from 5 to 60,000 mPa·s and found that both improve with increasing viscosity. A logical explanation would be that the rate of antifoam oil spreading decreases with the molecular size. However, Figure 19.33 shows how the rate of spreading varies with the PDMS viscosity (or MW) [159]: it is constant up to about 1000 mPa·s and then starts to decrease with the PDMS viscosity. Hence, the improved durability cannot be explained with the spreading rate for oils under 1000 mPa·s. Koczo et al. [166], however, found that the pseudoemulsion film also becomes less stable with increasing oil viscosity, and therefore, the antifoam action is faster and there is less time for spreading and losing oil from the antifoam drops. We note that the optimal viscosity range for PDMS antifoams in oil-based systems may be significantly different than for water-based systems due to the different solubility of PDMS in the two environments.

19.5.1.8 "Cloud Point Antifoams"

With increasing temperatures, micelles of nonionic surfactant solutions grow, and at the cloud point temperature, the micelles become large enough that they become visible, making the solution turbid. At this point, the *micelles* become so large that they also separate, leaving behind a solution with

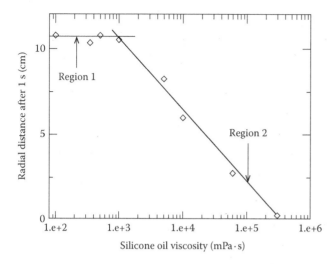

FIGURE 19.33 Average spreading rates of PDMS oils of various viscosities on the surface of 14.8 mM aqueous $C_{14}TAB$ solution. (Reprinted from *Colloids Surf. A*, 122, Bergeron, V. et al., Polydimethylsiloxane (PDMS)-based antifoams, 103-120, Copyright 1997, with permission from Elsevier.)

low surfactant concentration. The foaminess of the surfactant solution also drops above the cloud point temperature. Bonfillon-Colin and Langevin [180] studied the mechanism of this foam destabilization and found that small drops from the surfactant-rich phase separating out at the cloud point bridge the foam films.

As we discussed in Section 19.4.1.2, silicone polyethers are nonionic surfactants, and they can also act as antifoams. If the silicone surfactant is water soluble, it generally acts as an antifoam only above its cloud point, and for this reason, these materials are called *cloud point antifoams*. Organic polyethers, such as EO–PO–EO block polymers, also behave similarly. There are only a few studies published on the mechanism of *cloud point antifoams*. Nemeth et al. [181] studied the foaming of nonionic, Triton™ X-100 surfactant solutions, in the presence of water-soluble silicone polyethers as a function of temperature. In these solutions, the micelles of the antifoam and profoaming surfactants combine, and mixed micelles form having only one cloud point that is between the cloud points of the two nonionic surfactants. They suggested that above the cloud point, small drops form and they get trapped in the thinning foam films and break them by bridging. Interestingly, foam control started a few degrees under the cloud point when the solution was clear, so technically, this could be considered *homogeneous* antifoaming—a rare phenomenon.

There are also silicone surfactants or organomodified silicones, which are insoluble in water under all practical conditions, and therefore cannot have a cloud point, that can be even better antifoams than the *cloud point antifoams* described earlier. These water-insoluble silicones always form drops, and their behavior resembles that of PDMS as an antifoam (especially in oil-based systems, see later in Section 19.5.2).

19.5.2 NONAQUEOUS FOAMS (LUBRICATION OILS)

19.5.2.1 Foaming versus Air Entrainment

While both involve the incorporation of air into the lubricating fluid, air entrainment and foaming are distinct phenomena, and when an issue arises with either, they must generally be addressed in different ways. There is some overlap in the two, however. The air contained in foams generally starts as the incorporation of air bubbles into the liquid (entrainment/aeration). The bubbles then ascend to the surface and may become foam if not immediately destroyed. Therefore, foam control agents can play different roles in an aeration versus foam context. This is mainly because foam inhibitors are designed to act by breaking the thin film–Plateau border structure of the foam, while subsurface entrained air bubbles do not have such a structure and are separated by a significant amount of bulk liquid. Therefore, in this section only, the phenomena related to air entrainment will be discussed, and the possible mechanism of foam control in oil-based systems will be reviewed in Section 19.5.2.2.

Air entrainment in the form of small bubbles (microbubbles, microfoam) can be a serious problem for lubricating oils [46,76]. The route for eliminating entrained air is for it to rise to the top surface of the fluid, so the only way that additives can affect the problem is to accelerate or decelerate the rise of the bubbles. The rate of rise of the bubbles depends on

- Viscosity and density of the liquid
- Size of the bubbles
- Presence of surface-active components
- Presence of insoluble droplets (i.e., antifoam) that can stick to the surface of a bubble and affect its buoyancy

If we approximate a rising bubble as a single rigid sphere (larger than a few microns so that Brownian motion is negligible), in a large container, and we assume that the Reynolds number is less than one, then v, its rate of rise, according to Stokes' law is

$$v = \frac{2R^2(\rho_1 - \rho_2)g}{9\eta}, \qquad (19.13)$$

where
R is the radius
ρ_1 is the density of the bubble
ρ_2 is the density of the liquid
g is the gravitational acceleration
η is the viscosity of the liquid

Equation 19.13 shows that smaller bubbles rise much slower and thus can cause more problems for lubrication oils, especially with viscous oils. For example, based on this equation, in a lubrication oil with 0.8 g/mL density and 50 cP viscosity, a bubble with 1000 µm (1 mm) diameter would rise 10 cm distance in 11 s, while it will take 19 min for a 100 µm bubble, or 132 days for a 1 µm bubble.

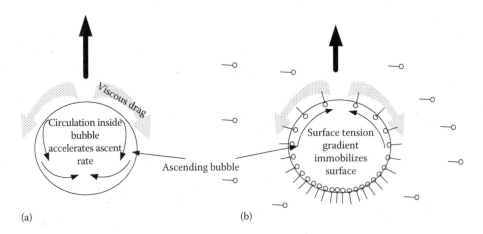

FIGURE 19.34   Effect of surface-active molecules on bubble ascent rate. (a) In pure liquid (no surface-active molecules) viscous drag creates circulation inside the bubble, which accelerate ascent rate. (b) With surface-active molecules, Gibbs-Marangoni effect immobilizes the surface of rising bubble, eliminates inside circulation, and reduces the ascent rate of the bubble.

However, a gas bubble is not solid, and its movement can create circulation inside the bubble and flow at the bubble–oil interface. This circulation reduces the velocity difference at the interface between the rising bubble and the oil, reducing drag. Theoretical calculations predict that a bubble (or drop) with mobile surface will rise 50% faster than Stokes' law predicts due to circulation (see Figure 19.34a) [182]. This is the case for a liquid containing no surface-active materials.

However, the presence of a very small amount of surface-active material at the bubble surface can drastically change the mobility due to the Gibbs-Marangoni effect (see Figure 19.3). Figure 19.34b illustrates the effect of surface-active molecules. Here, the viscous drag of the liquid flowing around the bubble sweeps the adsorbed surfactants from the top toward the bottom, creating a surface tension gradient that opposes the movement of the surface, thereby increasing drag and reducing the rate of rise.

This was found by Ross et al. [183,184], who studied the rate of bubbles ascending in nonpolar liquids, with and without PDMS and oil-soluble surfactants. It has to be emphasized that the effect of PDMS on bubble rise considered here is due to the surface-active character of the *soluble portion of the silicone* at the air/oil interface, and not the insoluble PDMS droplets, which may also be present.

Figure 19.35 shows the relative rate of rise for bubbles ($K = v/v_s$), compared to the calculated rate for bubbles with

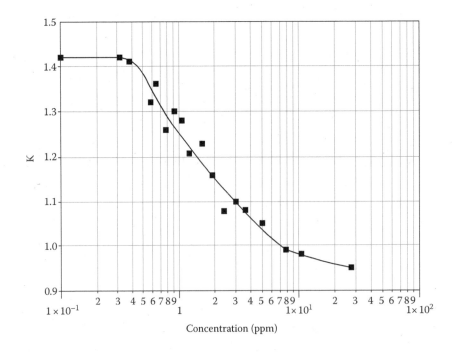

FIGURE 19.35   Comparative rates of ascent (corrected for the effect of the wall) of air bubbles in TMP heptanoate containing various concentrations of PDMS (1000 cSt) at 22°C. (Reprinted from *J. Colloid Interface Sci.*, 103, Suzin, Y. and Ross, S., Retardation of the ascent of gas bubbles by surface-active solutes in nonaqueous solutions, 578–585, Copyright 1985, with permission from Elsevier.)

a rigid surface (v_s from Stokes' law) in trimethylolpropane heptanoate (TMP heptanoate, 2,2-diheptanoyloxymethyl-*n*-butyl heptanoate), containing PDMS (1000 cSt). Below about 0.3–0.4 ppm PDMS concentration, the bubbles moved faster, with the ratio close to 1.5, the theoretical value for a bubble with completely mobile interface.

Starting at a few ppm PDMS concentration, the rate of rise of the bubbles slowed down as the surface tension gradient formed and finally made the bubble surface rigid (K = 1 within experimental error) above about 20 ppm PDMS. The results were similar with solutions of the same PDMS in mineral oil. Sorbitan monooleate showed similar behavior; however, about 500 times higher concentration was necessary to immobilize the bubble surface relative to PDMS.

Mannheimer [185] found similar results with PDMS, 12,500 cSt (which is closer to the PDMS viscosities used as lubrication oil antifoams), at 1 and 10 ppm concentrations, in mineral oils of various viscosities. Interestingly, the relative mobility of the bubble surfaces, with the same amount of PDMS, increased with the bubble size.

Ross et al. [183,184] also estimated the magnitude of surface tension gradient (the difference between the surface tensions of the top and the bottom of the bubble), based on the theory of Griffith [182] and found that a relatively small surface tension gradient (less than 1 mN/m) is enough to make the bubble surfaces rigid and slow down the rise of the bubbles by about 50%. Although the earlier examples involve PDMS, this same phenomenon applies to other soluble, surface-active materials. Therefore, the presence of highly surface-active lubricant additives, which are known to increase foam stability, can also slow the ascent of subsurface air bubbles, leading to increased air entrainment.

As previously stated, these experiments and calculations are related to the surface-active character of the soluble portion of the silicone at the air/oil interface. As was discussed in Section 19.4.1.2, any commercial PDMS contains a range of molecular weights, and their solubility decreases with increasing MW and decreasing temperature. Therefore, this soluble silicone may be delivered to lubricating oils by antifoams through a small fraction of low-MW silicone present in the antifoam [37]. It is not trivial to determine the solubility of PDMS in lube oils, because the various MW cuts have different saturation levels. Furthermore, measuring the MW distribution of the soluble PDMS is experimentally challenging. Ross and Suzin [37] and Mannheimer [185] measured the total solubility of PDMS in lubricating oils, by measuring turbidity. Ross and Suzin [37] determined that about 25 ppm of 1000 cSt PDMS and 8 ppm of 50,000 cSt PDMS are soluble in TMP heptanoate at 20°C. These values show that the solubility indeed decreases strongly with increasing MW but appear to be too high, since these figures are within the typical treat rate ranges of PDMS antifoams, which are known to form insoluble droplets. Presumably, the dissolved PDMS is mainly the low-MW fraction.

The insoluble fraction of PDMS, responsible for its antifoaming action, can affect air entrainment in two different ways, depending on the situation. This is demonstrated in a study by Fowle [46], who reviewed the ASTM D3427 (and the similar Grosse-Oetringhaus) deaeration tests and observations with turbine oils. These are two-stage tests. In the first stage, a large amount of fine bubbles are injected into the oil, then the air flow is stopped. In the second stage, the amount of air remaining entrained is measured as a function of time, as the air bubbles rise to the surface and are eliminated. The figure normally reported from these tests is the air release value (arv), which is the time until the total air content falls to 0.2% by volume. In this study, it was noted that fluids containing a silicone antifoam had less air entrained at the end of the first stage (just after the injection of air is stopped) compared to fluids without silicone antifoams. During the air injection stage, entrained air is also being released from the fluid, so this suggests that the presence of the silicone antifoam is increasing the rate of air release during this stage of the test. However, in the second (quiescent) part of the test, the presence of silicone antifoam *reduced* the rate of air release. These results are shown conceptually in Figure 19.36.

Fowle interpreted these results to mean that one of two phenomena could dominate the overall effect of antifoams on the air release rate, depending on the level of turbulence in the fluid. During the air injection phase, collisions, both between air bubbles and between the bubbles and silicone droplets, are frequent. When an air bubble and a silicone drop collide, a thin silicone oil film (*oil sheath*) forms around the bubble, and when another bubble approaches, a bridge can be formed between the two bubbles, leading to coalescence, rather than the air bubbles simply moving apart after colliding. The *oil sheath* around the bubble is probably an oil lens, and the bridging mechanism shown in Figure 19.32 could be taking place [168]. In other words, the presence of PDMS droplets can increase the probability that a bubble–bubble collision will lead to coalescence.

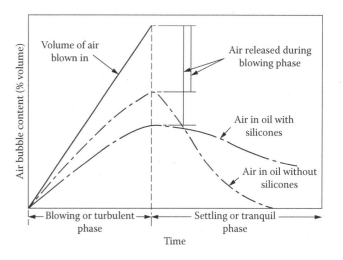

FIGURE 19.36 Effect of silicones on the two phases of air release tests. (Reprinted from *Tribology Intl.*, 14, Fowle, T.I., Aeration in lubricating oils, 151–157, Copyright 1981, with permission from Elsevier.)

If coalescence occurs, the combined bubble is larger and therefore rises more quickly according to Stokes' law, so bubble coalescence speeds up the process of air release. Extending this idea, any material that enhances the stability of the thin oil layer between bubbles (surface-active additive components, in particular) will decrease the likelihood of a collision leading to coalescence. This is another reason why additives that lead to increased foaming can also lead to increased air entrainment.

With little turbulence (infrequent bubble collisions), the silicones only had detrimental effect on air release [46]. As discussed in the previous section, silicones can enter and spread at the air/oil interface. This property also causes silicone droplets to stick onto submerged air bubbles. With relatively few bubble collisions, this is the effect that dominates air release. Because of the higher density of the silicone compared to the surrounding base oil, attached silicone droplets can cause a decrease in the rate of rise of the air bubble, leading to an increase in entrainment. With polyacrylate antifoams, this effect can be lessened by the lower density of acrylates, and their relatively lower surface activity compared to silicones.

19.5.2.2 Possible Mechanism of Antifoaming in Oils

As was emphasized in the previous sections, there are several differences in the foaming and foam control of water-based and oil-based systems. The behavior of oil-based liquids with respect to foams has been much less studied. In this section, foaming and antifoaming in oil-based systems will be discussed, based on comparison with the mechanisms discussed for aqueous systems in Section 19.5.1. Using new data, we will propose the possible mechanism of foam control in oil-based foams, including lubricating oils.

19.5.2.2.1 Physical Processes Underlying Foam Stability in Oils

A fundamental difference between water- and oil-based systems is the much lower surface tension of oil-based systems, and therefore, the magnitude of surface tension reduction that can be achieved with surfactants is much lower (see Section 19.3). Most traditional surfactants have surface tension close to, or higher than typical oils, so they cannot function as surface-active agents in nonaqueous media due to the lack of a thermodynamic driving force. This has a number of important physical effects. The notable effects here are that the Gibbs-Marangoni effect, an important foam-stabilizing mechanism in water-based liquids that operates based on surface tension gradients (see Section 19.2, Figure 19.3), and that surfactant-derived surface elasticity and surface viscosity are significantly lower in magnitude in oil-based systems.

In addition, in aqueous systems, thin liquid films can be stabilized by the adsorbed layers of surfactants on both sides of the film, producing a positive disjoining pressure that results from the combination of (negative) van der Waals attraction forces and the (positive) ion–ion repulsion and steric repulsion forces (see Section 19.2). However, in oil-based liquids, charged species are generally not single, surface-active molecules, but rather larger aggregates like inverse micelles, which are generally not surface active [186]. The reasons for this are beyond the scope of this chapter but are due in large part to the low dielectric constant of oils, which make charge separation and stabilization more difficult. Because the surface-active molecules adsorbed on foam lamellae are not charged, it can be expected that the electrostatic component of the disjoining pressure cannot stabilize the oil-based foam lamella. Without the presence of the strong, repulsive electrostatic forces, we expect that thin oil films will drain and become progressively thinner, eventually rupturing at a critical thickness, and that the rate of this process will be driven in large part by the viscosity of the fluid, as shown in Figure 19.7.

Nevertheless, this is not always the case, and surface-active components can play a role. Fully formulated lubricant oils can contain surface-active and foam-stabilizing components, which can increase foam lifetime/tendency by slowing the drainage process through additional surface tension–driven effects (shown in Figure 19.9). However, because these effects, which will be detailed later in this section, are much weaker in nonaqueous systems, foams composed of nonaqueous fluids tend to be much less stable and shorter-lived than aqueous foams, except in very rare cases not normally encountered in lubricant applications.

A significant amount of work has been done previously to determine which mechanisms are the main drivers of foam stability in nonaqueous fluids. *Crude oils* represent one set of nonaqueous fluids in which foaming has been studied extensively. In general, profoaming components in crude oils have been shown to contribute to foam stability by altering the air/oil surface rheology [32,38,187,188].

Measurements of surface viscosity and surface tension in lubricating oils have been used to show that additives can influence foam tendency just by altering the surface viscosity, or just through the Gibbs-Marangoni effect [42], depending on which additives are present. This suggests that a combination of these effects is at work in typical multicomponent additive packages. Direct observation of foams formed in oils containing typical lubricant additives (VM, dispersant, overbased detergent, friction modifier, and antiwear) show that the volume fraction of oil in the foam was between 20% and 50%, even in the later stages of foam collapse [33]. This indicates that the foams formed in lubricant fluids are generally the *wet* type, or spherical foams, in which surfactants act to stabilize foams through surface tension gradient effects that counter the flow of liquid from the foam lamella (i.e., Gibbs-Marangoni effect). This is in contrast to *dry* aqueous foams, which are stabilized by electrostatic and steric repulsion between opposing sides of very thin liquid films.

This was further investigated through thickness measurements of single foam lamellae composed of an overbased calcium sulfonate detergent in ethylbenzene. For thick films (>300 nm), drainage was driven by viscous effects, while for thinner films, weak attractive forces between the opposing walls of the foam lamellae begin to play a role, accelerating the rate of thinning [34].

The effect of additive package on foaming of lubrication oils was discussed in Section 19.3.3, and Figures 19.9 and 19.10 demonstrated a system where the friction modifier component acted as foam stabilizer.

There are some exceptions to nonaqueous foams being stabilized mainly by kinetic effects, for example, in cases where rigid or liquid crystalline structures are formed at the air–oil interface, or in foams stabilized by solid particles, where very stable foams can be formed [22,189], or in cases involving surfactants that have significantly lower surface tension than typical base oils (e.g., silicone or FS surfactants), as will be shown in the next section.

19.5.2.2.2   Surface Tension Effects

As discussed in Section 19.4, antifoams must be insoluble in and have a lower surface tension than the foaming liquid. For these reasons, silicone-based materials are most often used for foam control in lubrication oils. Table 19.8 shows the surface tension of various silicone antifoams, and of a fully formulated engine oil with these silicones at various concentrations. It can be seen that the surface tension of the oil with silicone at the typical use levels (10–60 ppm) was far from that of the neat silicones, even after storage for several months, except at much higher concentration (1000 ppm). Others [23,37,185] made similar observations. In reference to crude oils, Garrett [23] suggested that the reason for this is that the solubility of silicones in crude oil is so low that its surface concentration is not enough to form a close-packed monolayer.

As was discussed in Section 19.5.1.1, the entry and spreading coefficients (E and S, respectively) indicate whether entry and spreading of the antifoam drop are thermodynamically favorable and the bridging coefficient (B) shows if the antifoam bridge is unstable. Table 19.9 shows these coefficients calculated for PDMS and a FS antifoam in representative Group II, III, and IV base oils, using the appropriate surface and interfacial tensions in Equations 19.10 through 19.12.

It can be seen, that—as expected—all the three coefficients are highly positive for all base oil/antifoam combinations. This means that PDMS has surface and interfacial properties to function well as an antifoam in these fluids. However, it should be noted that in extreme cases—for example, high levels of low-MW silicone contamination—the surface tension of the base fluid approaches the surface tension of the PDMS antifoam, reducing its effectiveness.

TABLE 19.8
Surface Tension (ST) of a Fully Formulated Engine Oil with Various Silicone Antifoams, Measured at Ambient Temperature with the Wilhelmy Plate Method, Fresh and after Prolonged Aging (with Selected Samples). PDMS-100 k: PDMS-100,000 cSt; OMS-1, OMS-2, OMS-3, and OMS-4: Organomodified Silicones; FlSil-1: FS Antifoam. ΔST, TS of the Engine Oil Minus the ST of the Blend

Antifoam	Antifoam Concentration	ST (mN/m)	ΔST (mN/m)	ST After Aging	Age
None	0	30.18	0		
PDMS-100k	10 ppm	29.24	0.94		
PDMS-100k	60 ppm	29.64	0.54		
OMS-1	100%	27.47	2.71		
OMS-1	10 ppm	30.19	−0.01	30.07	8 months
OMS-1	60 ppm	30.36	−0.18	30.1	9 months
OMS-1	1000 ppm	27.74	2.44		10 months
OMS-2	100%	26.97	3.21		
OMS-2	10 ppm	30.19	−0.01	30.05	9 months
OMS-2	60 ppm	30.02	0.16		
OMS-2	1000 ppm	27.74	2.44	27.74	11 months
OMS-3	100%	27.02	3.16		
OMS-3	10 ppm	30.21	−0.03		
OMS-3	60 ppm	29.02	1.16		
OMS-4	100%	24.30	5.88		
OMS-4	10 ppm	30.00	0.18		
OMS-4	60 ppm	29.07	1.11		
FlSil-1	100%	22.99	7.19		
FlSil-1	3.75 ppm	29.92	0.26	30.0	8 months
FlSil-1	10 ppm	29.70	0.48		
FlSil-1	60 ppm	27.74	2.44		

Note: ST of 100,000 cSt PDMS Is about 20 mN/m.

TABLE 19.9

Entry (E), Spreading (S), and Bridging (B) Coefficients of Groups II, III, and IV Oils, with PDMS, 30,000 cSt (PDMS 30k), and a FS Antifoam (FlSil), Respectively, at Ambient Temperature

	E mN/m	S mN/m	B mN/m²
Gr. II/PDMS 30k	9.8	7.8	437
Gr. III/PDMS 30k	10.6	8.2	473
Gr. IV/PDMS 30k	9.2	7.6	414
Gr. II/FlSil	6.8	5.8	374
Gr. III/FlSil	7.4	6.2	414
Gr. IV/FlSil	6.4	5.6	353

19.5.2.2.3 Solubility of Antifoam in the Oil

One challenging aspect to using PDMS is that, if it has sufficiently low molecular weight, it can be soluble in lubricant fluids. Because of its low surface tension, soluble PDMS can act as a foam-stabilizing surfactant. Figure 19.37 shows the foaming tendency of a fully formulated engine oil with 50 ppm PDMS of various viscosities. Foam was formed on the liquid by bubbling nitrogen gas through oil with a stainless steel gas sparger, similar to the ASTM D892 procedure. It can be seen that the most viscous, that is, the highest MW, PDMS grades provided good foam control, but the foam volume increased with decreasing PDMS MW. With 50 cSt PDMS, the foam control disappeared or even became slightly worse than with no antifoam. This trend can be explained by the competition of the foam-stabilizing effect of the increasing amount of dissolved PDMS with decreasing MW versus the

foam-breaking effect of the insoluble part of the PDMS, also taking into account that the smaller fractions of the siloxane polymers have higher solubility (see Section 19.4.1.2).

In a similar experiment, a transmission fluid was spiked with 500 ppm PDMS of various viscosities and tested with ASTM D892. As shown in Figure 19.38, this oil did not produce foam by itself, but the low MW, 50 cSt PDMS acted as a strong profoamer. The 100 cSt PDMS had a profoaming effect, but much less, while the highest viscosity grade did not stabilize the foam. This further shows that low-MW silicones can be profoamers if they are soluble in the oil. An obvious conclusion of these experiments is that low viscosity PDMS should not be used as an antifoam in lubrication oils. The precise MW cutoff between pro- and antifoaming PDMS is a strong function of the temperature and to a lesser extent the base oil type. Care should be taken to avoid selecting a partially soluble antifoam because, according to the Ross–Nishioka effect [40], materials are highly surface active under conditions where they are near their solubility limit.

Ross and Suzin [37] studied the effect of 1000 cSt PDMS on the foaming of TMP heptanoate (2,2-diheptanoyloxy-methyl-*n*-butyl heptanoate) lubricant and also found that the foaminess increases with temperature, due to increasing solubility of the PDMS. They pointed out that due to the MW distribution of the silicone, it is not straightforward to measure the solubility of PDMS in a lubricant, because at a given concentration, the high-MW species may have exceeded their solubility limit, but the low-MW species may be still undersaturated. Moreover, the observed effect of PDMS on the foam is a combination of the foam-stabilizing effect of the dissolved (low MW) part of the silicone and the foam-inhibiting effect of the insoluble part. Therefore, it is possible that the foaminess *increases* with the PDMS

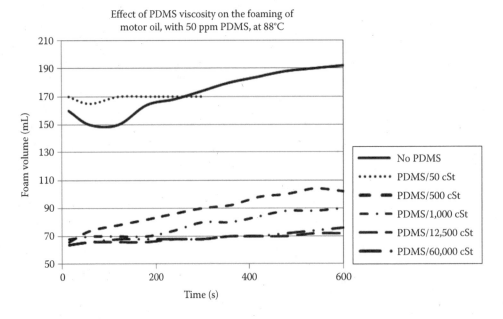

FIGURE 19.37 Foam volume as a function of time generated from 50 mL fully formulated engine oil at 88°C, in the presence of 50 ppm PDMS of various viscosities. The liquid was foamed in a 250 mL graduated glass cylinder bubbling nitrogen gas at 1.0 L/min flow rate, with a stainless steel gas sparger of 10 μm pore size.

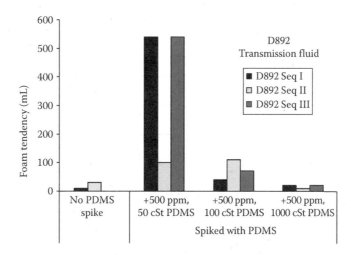

FIGURE 19.38 ASTM D892 foam tendency of a transmission fluid as-blended, and after spiking with 500 ppm PDMS with 50, 100, or 1000 cSt viscosity, respectively.

concentration, and they indeed observed this trend with TMP heptanoate. This explains why some foam problems originate from the overtreatment of the foam inhibitor itself.

19.5.2.2.4 Antifoaming Mechanism

As was shown in the previous sections, antifoams are understood to work by a *heterogeneous* mechanism, that is, the foam control agent forms a separate droplet phase in the foaming liquid, which breaks the foam film/Plateau border structure.

Figure 19.39 illustrates this for an engine oil with an organomodified silicone (OMS)-type antifoam at 30 and 60 ppm concentrations. The foaming tendency was also measured after centrifuging the oil + OMS-1 mixture at high speed (21,500 g, 1 h) and testing only the supernatant. The silicones, including

the organomodified PDMS types, are heavier than fully formulated lubrication oils; thus in the supernatant, there were significantly less OMS droplets than in the original mixture. It can be seen that foam control diminished after centrifuging, which shows that also with the OMS-type antifoams, as with PDMS, silicone droplets are responsible for foam control.

Due to the profoaming nature of the dissolved silicones, high-viscosity PDMS grades are often used. However, it is more difficult to disperse highly viscous silicones to small drops (see more on this in Section 19.6.1). Figure 19.40 shows the results of ASTM D6082 (Seq IV) tests with an engine oil, containing 10 ppm of 1000 (1k) or 100,000 (100k) cSt PDMS, respectively, both with and without *option A* (vigorous mixing before the test). In this case, the baseline level

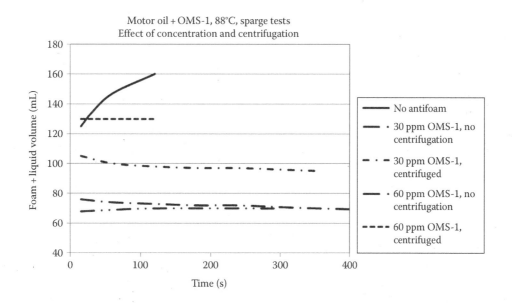

FIGURE 19.39 Foam volume with N_2 sparging of a fully formulated engine oil with 30 and 60 ppm organomodified silicone antifoam OMS-1. The same experimental setup was used in Figure 19.37, but in some of the tests, the engine oil + antifoam mixture was centrifuged and only the top phase was tested.

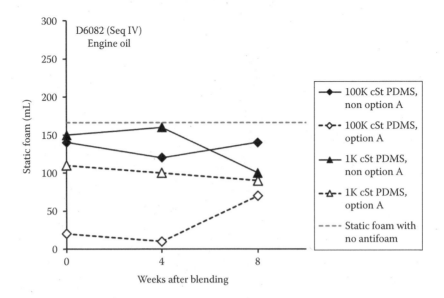

FIGURE 19.40 Static foam in ASTM D6082 (Seq IV, 150°C) with an engine oil containing 10 ppm of 1000 (1k) and 100,000 (100k) cSt PDMS, respectively, with and without *option A* (high shear mixing prior to testing) as a function of storage time after blending.

of mixing used to incorporate the antifoam into the fluid was not sufficient, and neither antifoam provided significant benefit compared to the baseline (no antifoam) case. With option A, antifoam droplets were more efficiently incorporated, and the 100,000 cSt antifoam performed well. However, even when antifoam was adequately dispersed with option A, the 1000 cSt PDMS showed poor efficiency. This is due to its increased solubility at 150°C. At this temperature, the antifoam droplets are solubilized and are therefore inactive.

Based on the results of the experiments described here, we can conclude that, similar to the mechanism of antifoaming in water-based systems, oil-based foams are also broken by insoluble droplets of the antifoam.

The same mechanistic principles that explain the behavior of silicones with respect to foam rupture also apply to

polyacrylates (i.e., destabilization of the foam lamellae by insoluble droplets). The insoluble nature of acrylate antifoams can also be shown using a centrifugation experiment. Here, three versions of a finished automotive gear oil were made. One contained a silicone-type antifoam, another one contained an acrylate-type antifoam, and the third contained no antifoam. As shown in Figure 19.41, the foam tendency of all three fluids was measured in the D892 test (Sequence II and III only). The two foam inhibitor–containing fluids were then centrifuged. D892 foam results after centrifugation were much higher compared to the precentrifuged values, approaching the value for the finished fluid with no antifoam. This indicates that, in both cases, the foam inhibitor exists as insoluble droplets that can be separated from the base fluid by centrifugation. For more information on polyacrylate-type antifoams, see Section 19.4.2.

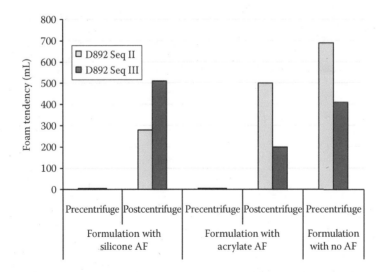

FIGURE 19.41 D892 foam tendency before and after centrifuging three versions of a finished gear oil. One fluid had a silicone-type antifoam, one had an acrylate-type antifoam, and one had no antifoam.

However, there are also differences how foam is controlled in water- and oil-based liquids. As was shown in Section 19.5.1.2, an important and often rate controlling step in the antifoaming mechanisms of aqueous foams is the destabilization of the *pseudoemulsion film*, because this film can act as a barrier to drop entry and thus slow or prevent foam rupture, even if the entry coefficient is positive. It was also shown that the role of the hydrophobic particles in the most effective, mixed-type antifoams is that they destabilize the pseudoemulsion film. In oil-based foams, including lubrication oil, diesel fuel, or crude oil, it is not necessary to blend particles with the antifoam to make it effective. This suggests that the pseudoemulsion film in the case of oil-based fluids must be less stable relative to aqueous liquids. Unfortunately, experimental evidence for this could not be found in the literature. Water-based pseudoemulsion films can be stabilized by surfactants, in similar mechanisms as aqueous foam films (or emulsion films), as was shown in Figure 19.2. Since surfactant stabilization is generally lacking in oil-based foams (see Section 19.5.2.2.1), it is highly likely that the oil-based pseudoemulsion film is not stabilized either, although experimental proof would certainly be more convincing.

Similarly, there are no experimental observations in the literature regarding how antifoam drops break oil-based films/foams. Based on analogy, and the data shown here, the best hypothesis appears to be that a similar bridging mechanism operates in oil foams as in the water-based systems, as Garrett also suggested [23] (see Sections 19.5.1.5 and 19.5.1.6).

19.5.2.2.5 Antifoam Deactivation

As was discussed in Section 19.5.1.7, mixed-type antifoams have a finite durability in aqueous systems, meaning that they eventually lose their activity in the process of repeatedly breaking foams over time. This is due to a deactivation process that results in a disproportionation of the particles when foam films are broken. This, however, cannot happen during the foam control of oil-based foams, due to the lack of particles in typical antifoams used with oils.

There are some observations that antifoams in oils may not deactivate. Ross and Suzin [37] found that a FS inhibited the foaming of TMP heptanoate synthetic ester lubricant for 24 h with no sign of deactivation. Similarly, Figure 19.42 shows

that foam control of an engine oil by 30 ppm OMS-1 did not fade even after nearly 5 h.

Further work is necessary to conclusively understand if and when antifoams deactivate in oil-based liquids simply through performing their foam-breaking function. However, in the often harsh conditions of lubricant use, the antifoam can lose activity by other routes, for example, chemical breakdown through reaction with other additive components or contaminants, very high temperature, or shear. It is also possible for the antifoam droplets to be filtered out, for example, by oil filters in circulation lines, or by filters at the point of lubricant filling. This removal can be modulated not just by the pore size of the filter, but by its thickness/tortuosity and material properties. Fiberglass filters, for example, have been shown to remove silicone antifoam droplets due to interaction between PDMS and the fiberglass surface [190]. Because use conditions are widely variable, no specific instruction can be given for how to prevent these issues. However, issues of antifoam loss or degradation can be diagnosed and understood through experiments similar to those presented in this chapter, including the following:

- Measurements of Si concentration during fluid use, by ICP, for example, to track loss of silicone antifoam (level too low), or ingress of low-MW silicone contamination (level too high). Silicone contamination, or contamination from other profoamers, will cause a decrease in surface tension, so this can also be used as a diagnostic technique.
- D892 and D6082 testing of fluid drains during or after use in the field.
- Microscopic analysis of fluid to confirm the presence of antifoam droplets.
- Isolation of possible contributors to antifoam breakdown/loss, for example, by heating the fluid to high temperature for an extended period of time, followed by foam testing, or foam testing after subjecting the fluid to filtration or shear. If possible, these results should be compared to foam test results on a foam inhibitor–free version of the blend. This will indicate whether the antifoam is partially lost or completely lost or if a profoamer has entered the fluid.

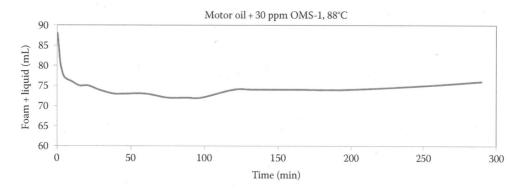

FIGURE 19.42 Foam volume of a fully formulated engine oil with N_2 sparging, in the presence of 30 ppm organomodified silicone antifoam OMS-1 versus time. The same experimental setup was used as in Figure 19.37.

19.6 APPLICATION NOTES

19.6.1 ANTIFOAM INCORPORATION

Antifoam additives can be included as embedded components in the concentrated additive package, or added as *top treats* to finished lubricants. Unlike many other lubricant additive components, the conditions under which the antifoam is incorporated into the additive package, or into the finished lubricant (amount of shear used, temperature, viscosity, solvency of the fluid environment, rate of addition), are critical to determining the final performance and stability of the antifoam. This is because the active antifoam droplets are actually formed during the blending process Figure 19.43).

For the purposes of this discussion, we will consider the lubricant fluid or additive package to be a homogeneous solution. In cases where liquid droplets are used as the foam inhibitor, as is the case with lubricating oils, we also consider the foam inhibitor to be a pure solution of the active molecule in a carrier solvent. During the blending process, these two fluids are combined, with the final state being a dispersion of droplets of the active molecule in the lubricant fluid (see Sections 19.4.1.2.2, 19.4.1.3.3, and 19.4.1.4.1). The carrier solvent has two important roles in the process. First, it enables the use of high-viscosity foam inhibitors by reducing their viscosity to the point where they can be easily handled. Second, the carrier solvent can act as a cosolvent in the initial stages of antifoam incorporation [191].

The complicated process of antifoam droplet formation can be broken down into two sequential stages. In the first stage of incorporation, just after addition of the antifoam, the base fluid, active antifoam molecule, and the carrier solvent are miscible, and no phase boundaries exist. Here, the carrier fluid acts as a cosolvent, solubilizing the active antifoam in the mixture. Instead of phase boundaries, there are regions of low and high concentration of antifoam polymer. Over time, the carrier solvent diffuses out of these regions, further enriching the local concentration of the antifoam, which is insoluble in the base fluid. This leads to precipitation of antifoam-rich droplets. From a thermodynamic point of view, the process of droplet precipitation can be described by a three-component phase diagram (Figure 19.44). At the point of introduction of the antifoam, the mixture is at point 1 on the diagram. The final state, in the area of the antifoam droplet, is shown as point 2. Between point 1 (injection) and point 2 (final state), the system passes through a phase boundary, where antifoam precipitation takes place. After precipitation of the antifoam,

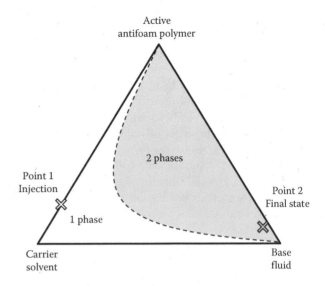

FIGURE 19.44 Conceptual three-component phase diagram for the active antifoam, the antifoam carrier solvent, and the base fluid (either concentrated additive package or fully formulated lubricant).

the second stage of incorporation begins. In this stage, the insoluble droplets of antifoam are broken down by shear. It is worth noting that the same conditions that break down antifoam droplets in the second stage, like turbulence and high shear, are also effective at efficiently dispersing and diluting the antifoam in the pre-precipitation stage. Other authors have previously described how high shear, in combination with elevated temperature, can be an effective method to reduce the size of antifoam droplets [191,192].

The property that connects incorporation and antifoam effectiveness/stability over time is *droplet size*. By forming smaller antifoam droplets, the final concentration of droplets (i.e., the number of droplets per mL) is higher at equal treat rates. In other words, breaking the antifoam down into smaller antifoam droplets generally increases its efficiency, as long as the drop size is still in the effective range. Furthermore, smaller droplets are more stable against gravity-driven settling over time. From Stokes' law (Equation 19.13), the settling rate (v_s) of a droplet is proportional to the square of its radius. Settling rate is also affected by bulk viscosity of the lubricant fluid, and the difference in density between the antifoam droplet and the lubricant (see Equation 19.13, Section 19.5.2.1). However, in practice, droplet size is the only parameter that can be easily varied without changing the composition of the formulation.

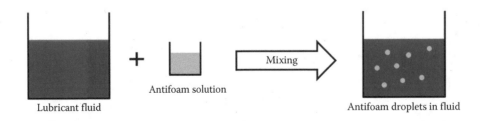

FIGURE 19.43 Conceptual diagram of antifoam incorporation. The process involves mixing two solutions to form a dispersion of insoluble droplets.

Situations in which the antifoam is not properly incorporated are indicated by poor antifoaming performance in the freshly blended fluid, the appearance of a hazy or separated antifoam-rich layer at the bottom of the lubricant storage container over time, inconsistent foam control efficiency in samples from the same lubricant batch, or gradual degradation of foam performance over time. Note that these observations can also be the result of other issues, including incorrect selection of antifoam materials, or chemical reaction between the antifoam and other components in the additive package. When diagnosing foam issues, it is advisable to include the variable of shear level when testing. If foaming performance improves significantly after high shear mixing (e.g., in ASTM D892 with option A versus D892 without option A), then suboptimal incorporation of the foam inhibitor should

be suspected. If high shear mixing does not improve results, then some other issue is indicated.

Antifoam incorporation methods are normally selected to achieve sufficiently small droplet sizes. Experiments performed on a number of lubricating fluids showed that silicone droplets of <100 μm were required to effectively control foams [193]. For lubricating fluids, which must have shelf stability of many months, up to years, the particle size requirements based on stability are generally stricter, on the order of <10 μm [192]. Particle size requirements for stability will vary in individual situations, in ways that can be roughly predicted by the Stokes' equation mentioned earlier.

An example of improved antifoam efficiency with smaller droplet size is shown in Figure 19.45. Here, three identical, concentrated additive packages were blended, using three

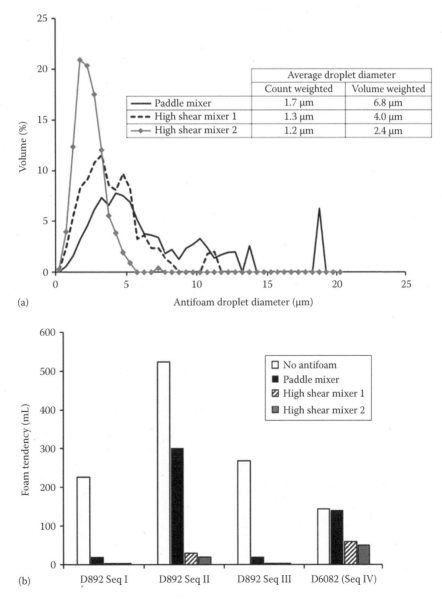

(a)

(b)

FIGURE 19.45 (a) Antifoam droplet size distributions for three identical additive packages blended using three mixing techniques and (b) D892 and D6082 foam test results for finished fluids made with these additive packages and a control fluid made using the same additive package with no antifoam.

different mixing methods, including a low-shear paddle mixer, and two high-shear mixers. The antifoam in this case was a high-viscosity PDMS and was treated at the same concentration in all three packages. Droplet size measurements indicated that smaller antifoam droplet size was achieved with the higher shear mixers (Figure 19.45a). These three additive packages were then diluted into base oil to make a finished lubricant (in this case, an engine oil). ASTM D892 and D6082 foam tests showed a correlation between antifoam effectiveness and droplet size/shear level (Figure 19.45b). One factor that improves the efficiency of the antifoam in this case is the dramatic increase in droplet concentration that comes as a result of breaking the droplets down to smaller sizes. In this experiment, for example, the droplet concentration by number is roughly 4–6 times higher for the packages where high shear mixing was used, compared to paddle mixing. The high efficiency of the sample with the highest shear indicates the drops with about 1–5 μm diameter work very efficiently with this lubricant.

The question of whether there is a minimum effective size for antifoam droplets in oil-based fluids has not been resolved to date. In water-based systems, there does appear to be a minimum effective size that is on the order of ~5 microns. For example, measurements of droplet size and antifoaming effectiveness over the course of repeat foam testing [159] indicated a lower size for effectiveness of 6 μm for PDMS in water-based solutions. However, droplet sizes below this limit are clearly effective foam inhibitors in lubricant oils. In theory, there must be some size below which antifoam droplets are ineffective in oil-based systems, related to the length scales of the foam lamellae and Plateau borders, but this size is apparently lower than the minimum effective droplet size in water-based systems and does not seem to be routinely reached using conventional lubricant blending techniques.

REFERENCES

1. Szekrényesy, T., K. Liktor, and N. Sándor. Characterization of foam stability by the use of foam models 1. Models and derived lifetimes. *Colloids Surf.*, 68 (4), 267–273, 1992.
2. Szekrényesy, T., K. Liktor, and N. Sándor. Characterization of foam stability by the use of foam models 2. Results and discussion. *Colloids Surf.*, 68 (4), 275–282, 1992.
3. Hubman, A. and A. Lanik. Air entrainment and foaming properties of synthetic lubricants at extreme temperatures. *J. Synth. Lubr.*, 2, 121–142, 1985.
4. Bikerman, J.J. *Foams*, Springer-Verlag, New York, 1973.
5. Wasan, D.T., A.D. Nikolov, L.A. Lobo, K. Koczo, and D.A. Edwards. Foams, thin films and surface rheological properties. *Prog. Surf. Sci.*, 33, 119–154, 1992.
6. Malhotra, A.K. and D.T. Wasan. Effects of surfactant adsorption-desorption kinetics and interfacial rheological properties on the rate of drainage of foam and emulsion films. *Chem. Eng. Commun.*, 55, 95–128, 1987.
7. Lucassen-Reynders, E.H., A. Cagna, and J. Lucassen. Gibbs elasticity, surface dilational modulus and diffusional relaxation in nonionic surfactant monolayers. *Colloids Surf. A Physicochem. Eng. Asp.*, 186 (1–2), 63–72, 2001.
8. Miller, R. and L. Liggieri. *Interfacial Rheology*, Koninklijke Brill NV, Leiden, the Netherlands, 2009.
9. Islam, M. and A.I. Bailey. Surface rheology of a nonaqueous system and its relationship with foam stability, in K.L. Mittal and P. Kumar, eds., *Emulsions, Foams and Thin Films*, Marcel Dekker, New York, pp. 131–150, 2000.
10. Nikolov, A.D. and D.T. Wasan. Ordered micelle structuring in thin films formed from anionic surfactant solutions: Part I-Experimental. *J. Colloid Interface Sci.*, 133, 1–12, 1989.
11. Nikolov, A.D., P.A. Kralchevsky, I.B. Ivanov, and D.T. Wasan. Ordered micelle structuring in thin films formed from anionic surfactant solutions: Part II-Model development. *J. Colloid Interface Sci.*, 133, 13, 1989.
12. Wasan, D.T., A.D. Nikolov, D.D. Huang, and D.A. Edwards. Foam stability: Effects of oil and film stratification, in D.H. Smith, ed., *Surfactant-Based Mobility Control*, ACS Symposium Series, ACS Publications, Vol. 373, pp. 136–162, 1988.
13. Vesaratchanon, J.S., A.D. Nikolov, and D.T. Wasan. The importance of oscillatory structural forces in the sedimentation of a binary hard-sphere colloidal suspension. *Ind. Eng. Chem. Res.*, 48 (14), 6641–6651, 2009.
14. Wasan, D.T. and A.D. Nikolov. Thin liquid films containing micelles or nanoparticles. *Curr. Opin. Colloid Interface Sci.*, 13, 128–133, 2008.
15. Sheludko, A. Thin liquid films. *Adv. Colloid Interface Sci.*, 1 (4), 391–464, 1967.
16. Ivanov, I. and D. Dimitrov. *Thin Liquid Films*, Surfactant Science Series, Vol. 29, Marcel Dekker, New York, 1988.
17. Langevin, D. Influence of interfacial rheology on foam and emulsion properties. *Adv. Colloid Interface Sci.*, 88, 209–222, 2000.
18. Racz, G., K. Koczo, and D.T. Wasan. Measurement of pressure distribution in the Plateau borders of fresh and aged foams. *Colloid Polym. Sci.*, 260, 720–725, 1982.
19. Vrij, A.J. and J. Th.G. Overbeek. Rupture of thin liquid films due to spontaneous fluctuations in thickness. *J. Am. Chem. Soc.*, 90, 3074–3078, 1968.
20. Mysels, K.J., K. Shinoda, and S. Frankel. *Soap Films: Studies of Their Thinning and a Bibliography*, Pergamon Press, New York, 1959.
21. Manev, E.D. and A.V. Nguyen. Critical thickness of microscopic thin liquid films. *Adv. Colloid Interface Sci.*, 114–115, 133, 2005.
22. Friberg, S.E. Foams from non-aqueous systems. *Curr. Opin. Colloid Interface Sci.*, 15, 359–364, 2010.
23. Garrett, P.R., ed. *The Science of Defoaming: Theory, Experiment and Applications*, Surfactant Science Series, Vol. 155, CRC Press, Boca Raton, FL, 2014.
24. Childers, J.C. The chemistry of metalworking fluids, in J.P. Byers, ed., *Metalworking Fluids*, 2nd edn., Chapter 6, CRC Press, STLE, Boca Raton, FL, pp. 127–146, 2006.
25. Garrett, P.R., J. Davis, and H.M. Rendall. An experimental study of the antifoam behaviour of mixtures of a hydrocarbon oil and hydrophobic particles. *Colloids Surf. A Physicochem. Eng. Asp.*, 85, 159–197, 1994.
26. Ross, J. and G.D. Miles. An apparatus for comparison of foaming properties of soaps and detergents. *Oil Soap*, 32, 99–102, 1940.
27. ASTM Standard D892-13. Standard test method for foaming characteristics of lubricating oils, ASTM International, West Conshohocken, PA, 2013.
28. ASTM Standard D6082-12. Standard test method for high temperature foaming characteristics of lubricating oils, ASTM International, West Conshohocken, PA, 2012.
29. ISO International Standard 12152:2012. Lubricants, industrial oils and related products—Determination of the foaming and air release properties of industrial gear oils using a spur gear test rig—Flender foam test procedure, ISO, Geneva, Switzerland, 2012.
30. Kishore Nadkarni, R.A. Foam tests for lubricating oils: Limitations of reliability and reproducibility. *J. ASTM Int.*, 6, 1–15, 2009.

31. Brady, A.P. and S.J. Ross. The measurement of foam stability. *J. Am. Chem. Soc.*, 66, 1348–1356, 1944.

32. Callaghan, I.C. and E.L. Neustadter. Foaming of crude oils: A study of non-aqueous foam stability. *Chem. Ind.*, 17, 53–57, 1981.

33. Binks, B.P., C.A. Davies, P.D.I. Fletcher, and E.L. Sharp. Non-aqueous foams in lubricating oil systems. *Colloids Surf. A: Physicochem. Eng. Asp.*, 360, 198–204, 2010.

34. Ottewill, R.H., D.L. Segal, and R.C. Watkins. Studies on the properties of foams formed from non-aqueous dispersions. *Chem. Ind.*, 17, 57–60, 1981.

35. McBain, J.W., S. Ross, A.P. Brady, J.V. Robinson, I.M. Abrams, R.C. Thorburn, and C.G. Lindquist. National advisory committee for aeronautics advanced restricted report 4105, 1944.

36. McBain, J.W. and J.W. Woods. Effect of various compounds in use with airplane engines upon foaming of aircraft lubricating oil. National Advisory Committee for Technical Note 1843, 1949.

37. Ross, S. and Y. Suzin. Lubricant foaming and aeration study. Report AFWAL-TR-82-2090, Renssealer Polytechnic Institute, Troy, NY, October 1982.

38. Callaghan, I.C., A.L. McKechnie, J.E. Ray, and J.C. Wainwright. Identification of crude oil components responsible for foaming. *Soc. Petrol Eng. J.*, 171–175, April 1985.

39. Centers, P.W. Behavior of silicone antifoam additives in synthetic ester lubricants. *Tribol. Trans.*, 36 (3), 381–386, 1993.

40. Ross, S. Profoams and antifoams. *Colloids Surf. A Physicochem. Eng. Asp.*, 118, 187–192, 1996.

41. Centers, P.W. Turbine engine lubricant foaming due to silicone basestock used in non-specification spline lubricant. *Lubr. Eng.*, 51, 368–371, 1995.

42. Tamai, Y., S. Koyama, and N. Takano. Relation between foaming and surface properties of detergent-containing lubricating oil. *Tribol. Trans.*, 21, 351–355, 1978.

43. Duncanson, M. Effects of physical and chemical properties on foam in lubricating oils. *J. Soc. Tribol. Lubr. Eng.*, May, 9–13, 2003.

44. Grabowski, W. Efficient diesel fuel antifoams of low silicon content. U.S. Patent 5,542,960, OSi Specialties, Inc., August 6, 1996.

45. Hansel, R., A. Vetter, S. Herrwerth, A. Lohse, J. Venzmer, and P. Seidensticker. Organofunctionallly modified polysiloxanes and use thereof for defoaming liquid fuels with biofuel additions. EP 2,011,813, Evonik Degussa GmbH, July 9, 2014.

46. Fowle, T.I. Aeration in lubricating oils. *Tribol. Int.*, 14 (3), 151–157, 1981.

47. Brinksmeier, E., D. Meyer, A.G. Huesmann-Cordes, and C. Herrmann. Metalworking fluids—Mechanisms and performance. *CIRP Annals Manuf. Tech.*, 64, 605–628, 2015.

48. Brown, W.L. and R.G. Butler. Metalworking fluids, in L.R. Rudnick, ed., *Synthetics, Mineral Oils, and Bio-Based Lubricants*, 2nd edn., Chapter 38, CRC Press, Boca Raton, FL, pp. 623–650, 2013.

49. Tucker, K.H. Metalforming applications, in J.P. Byers, ed., *Metalworking Fluids*, 2nd edn., Chapter 5, CRC Press, STLE, Boca Raton, FL, pp. 103–126, 2006.

50. Richter, A. Bursting bubbles. *Cutting Tool Eng. Plus*, 63 (8), 62–68, August 2011.

51. CTE Staff. Coolant bursts bubble on foam problem. *Cutting Tool Engr. Plus*, 66 (9), 89–91, September 2014.

52. Beercheck, R. New tools for foam control in metalworking fluids. *Lubes'n'Greases*, 18–25, March 2014.

53. Ineman, J. Advanced emulsifier technology. *Tribol. Lubr. Technol.*, 46–48, November 2011.

54. Noll, W. *Chemistry and Technology of Silicones*, Academic Press, New York, 1968.

55. R.G. Jones, W. Ando, and J. Chojnowski, eds., *Silicon-Containing, Polymers*, Kluwer Academic Publishers, Dordrecht, the Netherlands, 2000.

56. Brook, M.A. *Silicon in Organic, Organometallic, and Polymer Chemistry*, Wiley-Interscience, New York, 2000.

57. Clarson, S.J., M.J. Owen, S.D. Smith, and M.E. Van Dyke, eds., *Advances in Silicones and Silicone-Modified Materials*, ACS Symposium Series, ACS Publications, Vol. 1051, Washington, DC, 2010.

58. Perry, R. et al. Silicones, in L.R. Rudnick, ed., *Synthetics, Mineral Oils, and Bio-Based Lubricants*, 2nd edn., Chapter 12, CRC Press, Boca Raton, FL, pp. 213–226, 2013.

59. Hill, M.H., ed. *Silicone Surfactants*, Surfactant Science Series, Vol. 86, Marcel Dekker, New York, 1999.

60. Chojnowski, J. and M. Cypryk. Synthesis of linear polysiloxanes, in R.G. Jones, W. Ando, and J. Chojnowski, eds., *Silicon-Containing Polymers*, Kluwer Academic Publishers, Dordrecht, the Netherlands, pp. 3–41, 2000.

61. Rochow, E.G. Preparation of organosilicon halides. U.S. Patent 2,380,995, General Electric Company, August 7, 1945.

62. Trautman, C.E. and H.A. Ambrose. Prevention of foaming of hydrocarbon oils. U.S. Patent 2,416,503 and 2,416,504, Gulf Research & Development Co., February 25, 1947.

63. Warrick, E.L., W.A. Piccoli, and O.F. Stark. Melt viscosities of dimethylpolysiloxanes. *J. Am. Chem. Soc.*, 77, 5017–5018, 1955.

64. Barry, A.J. Viscometric investigation of dimethylsiloxane polymers. *J. Appl. Phys.*, 17, 1020, 1946.

65. Camino, G., S.M. Lomakin, and M. Lazzari. Polydimethylsiloxane thermal degradation. Part 1. Kinetic aspects. *Polymer*, 42, 2395–2402, 2001.

66. Pape, P.G. Silicones: Unique chemicals for petroleum processing. *J. Petrol Technol.*, 35, 1197–1204, June 1983.

67. Owen, M.J. Surface properties and applications, in R.G. Jones, W. Ando, and J. Chojnowski, eds., *Silicon-Containing Polymers*, Chapter 8, Kluwer Academic Publishers, Dordrecht, the Netherlands, pp. 213–231, 2000.

68. Zisman, W.A. Relation of the equilibrium contact angle to liquid and solid constitution, in F. Fowkes, ed., *Contact Angle, Wettability and Adhesion*, Advances in Chemistry, American Chemical Society, Washington, DC, pp. 1–51, 1964.

69. Callaghan, I.C. Antifoams for nonaqueous systems in the oil industry, in P.R. Garrett, ed., *Defoaming: Theory and Industrial Applications*, Chapter 2, Surfactant Science Series, Vol. 45, Marcel Dekker, New York, pp. 119–150, 1993.

70. Callaghan, I.C., F.H. Ferdi, C.M. Gould, K. Gotz, H.J. Patzke, and C. Weitemeyer. Oil gas separation. U.S. Patent 4,557,737, The British Petroleum Company, December 10, 1985.

71. Farminger, K.W. Antifoam process for non-aqueous systems. U.S. Patent 4,082,690, Dow Corning Corp., April 4, 1978.

72. Larsen, R.G. Antifoaming compositions. U.S. Patent 2,375,007, Shell Development Co., May 1, 1945.

73. Anzenberger, Sr., J.F. and M.P. Silvon. Method and composition for controlling foam in non-aqueous fluid systems. U.S. Patent 4,151,101, Stauffer Chemical Co., April 24, 1979.

74. Pillon, L.Z. and A.E. Asselin. Antifoam agent for lubricating oils (LAW455). U.S. Patent 5,766,513, Exxon research and Engineering Co., June 16, 1998.

75. Callaghan, I.C., C.M. Would, and W. Grabowski. Method for the separation of gas from oil. U.S. Patent 4,711,714, The British Petroleum Company, December 8, 1987.

76. Carey, J.T., A.S. Galiano-Roth, M. Hill, and G.K. Dudley. Alkylated naphthylene base stock lubricant formulations. U.S. Patent 8,716,201, ExxonMobil Research and Engineering Co., May 6, 2014.

77. Legrow, G.E. and L.J. Petroff. Silicone polyether copolymers: Synthetic methods and chemical compositions, in M.H. Hill, ed., *Silicone Surfactants*, Chapter 2, Surfactant Science Series, Vol. 86, Marcel Dekker, New York, pp. 49–64, 1999.

78. Czajka, A., G. Hazell, and J. Eastoe. Surfactants at the design limit. *Langmuir Article ASAP*, DOI: 10.1021/acs.langmuir.5b00336, 31 (30), 8205–8217, 2015.

79. Nikolov, A.D. and D.T. Wasan. Superspreading mechanisms: Overview, *Eur. J. Phys.*, The European Physical Journal-Special Topics 197.1, 325–341, 2011.

80. Hill, R.M. Silicone surfactants—New developments. *Curr. Opin. Colloid Interface Sci.*, 7 (5–6), 255–261, 2002.

81. Nikolov, A.D., A. Chengara, K. Koczo, G.A. Policello, and I. Kolossvary. Superspreading driven by Marangoni flow. *Adv. Colloid Interface Sci.*, 96 (1–3), 325–338, 2002.

82. Churaev, N.V., N.E. Esipova, R.M. Hill, V.D. Sobolev, V.M. Starov, and Z.M. Zorin. The superspreading effect of trisiloxane surfactant solutions, *Langmuir*, 17 (5), 1338–1348, 2001.

83. Camp, M. and P. Rostaing. Process for removing bubbles of gas from liquids. U.S. Patent 3,887,487, Rhone-Poulenc, SA. June 3, 1975.

84. Haensel, R., M. Fiedel, M. Ferenz, and J. Venzmer. Use of self-crosslinked siloxanes for the defoaming of liquid hydrocarbons. U.S. Patent Appl. 2013/0217930, Evonik Industries AG, August 22, 2013.

85. Adams, G. and M.A. Jones. Foam control. U.S. Patent 4,690,688, Dow Corning Ltd, September 1, 1987.

86. Burger, W., C. Herzig, M. Bloechl, P. Huber, and E. Innertsberger. Verfahren zum Enschaumen und/oder Entgasen von organischen Systemen. German Patent Application DE 40 32 006, Wacker-Chemie GmbH, April 16, 1992.

87. Herzig, C., W. Burger, E. Innertsberger, P. Huber, and M. Blochl. Process for defoaming and/or degasing organic systems. U.S. Patent 5,474,709, Wacker-Chemie GmbH, December 12, 1995.

88. Fey, K.C. and C.S. Combs. Middle distillate hydrocarbon foam control agents from cross-linked organopolysiloxane-polyoxyalkenes. U.S. Patent 5,397,367, Dow Corning Corp., March 14, 1995.

89. Grabowski, W. and R. Haubrichs. Diesel fuel antifoam compositions. U.S. Patent 6,093,222, CK Witco Corp., July 25, 2000.

90. Spiegler, R., M. Keup, K. Kugel, P. Lersch, and S. Silber. Verwendung von organofunktionell modifizierten Polysiloxanen zum Enschaumen von Dieselkraftstoff. German Patent Application DE 43 43 235, TH Goldschmidt AG, December 22, 1994.

91. Busch, S., M. Keup, K. Kugel, P. Lersch, and R. Spiegler. Verwendung von organofunktionell modifizierten Polysiloxanen zum Enschaumen von Dieselkraftstoff. German Patent Application DE 195 16 360, TH Goldschmidt AG, May 15, 1996.

92. Boinowitz, T., K. Kugel, R.-D. Langenhagen, I. Schlachter, and A.K. Weier. Use of silicone polyether copolymers for defoaming diesel fuel. European Patent 0,849,352, Goldschmidt AG, December 5, 1997.

93. Grabowski, W. and R. Haubrichs. Diesel fuel and lubricating oil antifoam and methods of use. U.S. Patent 6,221,815, Crompton Corp., April 24, 2001.

94. Grabowski, W. and R. Haubrichs. Diesel fuel and lubricating oil antifoams and methods of use. U.S. Patent 6,001,140, Witco Corp., December 14, 1999.

95. Herzig, C., W. Burger, B. Deubzer, and M. Bloechl. Organopolysiloxanes containing ester groups. U.S. Patent 5,446,119, Wacker-Chemie GmbH, August 29, 1995.

96. Battice, D.R., K.C. Fey, L.J. Petroff, and M.A. Stanga. Silicone foam control agents for hydrocarbon liquids. U.S. Patent 5,767,192, Dow Corning Corp., June 16, 1998.

97. Gauthier, D. and J.B. Carroll. Antifoam additives for use in low viscosity applications. U.S. Patent Application 2014/0073542, Afton Chemical Corp., March 13, 2014.

98. Lomas, A.W. Process to control foaming in non-aqueous systems. U.S. Patent 4,460,493, Dow Corning Corp., July 17, 1984.

99. Borner, D., H.-F. Fink, G. Koerner, and G. Rossmy. Antischaummittel, German Patent 24 44 073, TH Goldschmidt AG, March 25, 1976.

100. Okawa, T., H. Masayuki, and P. Hupfield. Fluorosilicones. U.K. Patent Application 2,443,626, Dow Corning Corp., May 14, 2008.

101. Hupfield, P., M. Hayaski, and T. Okawa. Fluorosilicone materials. PCT International Application WO 2008/057128, Dow Corning Corp., May 15, 2008.

102. Callaghan, I.C. and A.S. Taylor. Fluorosilicone antifoam additive composition for use in crude oil separation. U.K. Patent Application 2,234,978, The British Petroleum Company Plc, February 20, 1991.

103. Owen, M. Poly[methyl(3,3,3-trifluoropropyl)siloxane], in D.W. Smith, S.T. Iacono, and S.S. Iyer, eds., *Handbook of Fluoropolymer Science and Technology*, 1st edn., Chapter 9, John Wiley & Sons Inc., Hoboken, NJ, pp. 183–200, 2014.

104. Wu, F., C. Cai, W.-B. Yi, Z.-P. Cao, and Y. Wang. Antifoaming performance of polysiloxanes modified with fluoroalkyls and polyethers. *J. Appl. Polym. Sci.*, 109, 1950–1954, 2008.

105. Razzano, J.S. Process for the preparation of fluorosilicone polymers and copolymers. U.S. Patent 3,997,496, General Electric Company, December 14, 1976.

106. Razzano, J.S. and N.E. Gosh. Process for producing fluorosilicone polymers. U.S. Patent 6,492,479, General Electric Company, December 10, 2002.

107. Razzano, J.S. and V.G. Simpson. Fluorosilicone copolymers and process for the preparation thereof. U.S. Patent 3,974,120, General Electric Company, August 10, 1976.

108. Razzano, J.S. and N.E. Gosh. Polymerization process for fluorosilicone polymers. U.S. Patent 6,232,425, General Electric Company, May 15, 2001.

109. Keil, J.W. Oil emulsions of fluorosilicone fluids. U.S. Patent 4,537,677, Dow Corning Corp., August 27, 1985.

110. Taylor, A.S. Fluorosilicone antifoam additive. U.K. Patent 2,244,279, The British Petroleum Company Plc, November 27, 1991.

111. Abe, A. and N. Terae. Composition containing organopolysiloxane having polyoxyalkylene and perfluoroalkyl units. U.S. Patent 4,597,894, Shin-Etsu Chemical Co., Ltd., July 1, 1986.

112. Evans, E.R. Method of defoaming crude hydrocarbon stocks with fluorosilicone compounds. U.S. Patent 4,329,528, General Electric Company, May 11, 1982.

113. Berger, R., H.-F. Fink, G. Koerner, J. Langner, and C. Weitemeyer. Use of fluorinated norbornylsiloxanes for defoaming freshly extracted degassing crude oil. U.S. Patent 4,626,378, TH Goldschmidt AG, December 2, 1986.

114. Hera, J. and R. Gingrich. Development of an antifoam for subsea application for the Parquet Das Conchas project BC-10 block, offshore Brazil. OTC 20372, Paper prepared for presentation at the *Offshore Technology Conference*, Houston, TX, May 3–6, 2010.

115. Bloch, R.A., M. Devine, and J. Ryer. Power transmitting fluids with improved resistance to foaming. PCT International Application WO 97/04051, Exxon Chemical Patents Inc., February 6, 1997.

116. Odian, G. *Principles of Polymer Chemistry*, 4th edn., John Wiley & Sons, Inc., Hoboken, NJ, 2004.

117. Carraher, C.E. *Carraher's Polymer Chemistry*, 9th edn., CRC Press, Taylor & Francis, Baton Raton, FL, 2014.

118. Kent, J.A. *Riegel's Handbook of Industrial Chemistry*, 9th edn., Springer Science & Media, New York, 1992.

119. Monsanto Chemical Co. Polycarboxylates as foam inhibitors for hydrocarbon oils. GB Patent 757707, Monsanto Chemical Co., September 26, 1956.

120. Fields, J.E. Hydrocarbon oils of reduced foaming properties. U.S. Patent 3,166,508, Monsanto Chemical Co., January 19, 1965.

121. Fields, J.E. and E.H. Mottus. Mineral oil containing alkyl polymethacrylate antifoamant. U.S. Patent 3,340,193, Monsanto Chemical Co., September 5, 1967.

122. Lange, R.M. Novel foam inhibiting polymers and lubricants containing them. JP Patent 55,036,299, Lubrizol Corp., March 13, 1980.

123. Product Data Sheet. HiTEC® 2030, Afton® Chemical, Richmond, VA, 2016.

124. Product Data Sheet. LZ 889D, Lubrizol Corporation, Wickliffe, OH, 2016.

125. Product Data Sheet. Foam Ban® 152/155/3633E, Munzing, Abstatt, Germany, 2016.

126. Product Data Sheet. PC (R) Defoamer 2544/1644/3144/1344/1244/1844. Allnex, Brussels, Belgium, 2016.

127. Grolitzer, M. and M. Zhao. Foaming resistant hydrocarbon oil compositions. WO Patent 199920721 A1, Solutia Inc., April 29, 1999.

128. Fang, J.Z. and M. Zhao. Foaming-resistant hydrocarbon oil compositions. U.S. Patent Application 20070254819 A, Cytec Surface Specialties, November 1, 2007.

129. Fang, J.Z. and M. Zhao. Foaming resistant hydrocarbon oil compositions. U.S. Patent 7,700,527, Cytec Surface Specialties, April. 20, 2010.

130. Oppenlaender, K., R. Fikentscher, N. Greif, and F. Poschmann. Antifoaming agents for paper coatings. German Patent 2161772, Badische Anilin & Soda Fabrik, June 20, 1973.

131. Yoshihide, I. Antifoaming agent for emulsion coating use. JP Patent 2005279565, San Nopco Ltd., October 13, 2005.

132. Hu, W., F.O.H. Pirrung, P.J. Harbers, C. Zhu, and M. Zhu. Silicone-free defoamer for solvent-based coatings. WO Patent Application 2009138343, BASF SE, November 19, 2009.

133. Kujak, S.A., J.A. Majurin, and D.D. Steinke. Lubricant defoaming additives and compositions. WO Patent 2014144558, Trane International Inc., September 18, 2014.

134. Cevada, M.E. et al. Formulations of homopolymers based on alkyl acrylates used as antifoaming agents in heavy and superheavy crude oils. CA Patent Application 2872382, Instituto Mexicano del Petroleo, June 6, 2015.

135. Dixon, R.T. and G.E. Hunt. Lubricating composition with non-silicone antifoam additive. WO Patent Application 2014135549, Shell Oil Company, September 12, 2014.

136. Dietz, T. et al. Polyacrylate esters and their use as deaerating agents for lacquers and paints. German Patent 19841559, Th. Goldschmidt A.G., March 23, 2000.

137. Xu, W. et al. Composite antifoaming agent for lubricating oil. CN Patent 104232246 A, China petroleum & Chemical Corp., December 24, 2014.

138. Kricheldorf, H.R. Handbook of Polymer Synthesis, Marcel Dekker, Inc., New York, 1991.

139. Accu Dyne Test. Tables of polymer surface characteristics. www.accudynetest.com/polymer_tables.html, Diversified Enterprises, 2016. Accessed December 4, 2015.

140. Roe, R.J. Parachor and surface tension of amorphous polymers. J. Phys. Chem., 69 (8), 2809–2810, 1965.

141. Slow, K.S. and D. Patterson. The prediction of surface tensions of liquid polymers, Macromolecules, 4 (1), 26–30, 1971.

142. Krevelen, D.W. and K. Nijenhuis. Properties of Polymers—Their Correlation with Chemical Structure; Their Numerical Estimation and Prediction from Additive Group Contributions, Elsevier, Amsterdam, The Netherlands, 2009.

143. Sawicki, G.C. Silicone polymers as foam control agents. J. Am. Oil. Chem. Soc., 65 (6), 1013–1016, 1988.

144. Garrett, P.R., ed., Defoaming: Theory and Industrial Applications, Surfactant Science Series, Vol. 45, Marcel Dekker, New York, 1993.

145. EVONIK Industries. SIPERNAT® and AEROSIL® for defoamers. Technical information 1313. https://www.aerosil.com/product/aerosil/Documents/TI-1313-SIPERNAT-and-AEROSIL-for-Defoamer-EN.pdf, January 2017.

146. EVONIK Industries. AEROSIL® fumed silica and SIPERNAT® in sealants. Technical bulletin fine particles, TB 63, 2009.

147. Huber Engineered Materials, Huber silicas and silicates for defoamers will burst your bubble, Jan. 2017. http://www.hubermaterials.com/products/silica-and-silicates/defoamers.aspx.

148. CABOT. CAB-O-SIL fumed silicas. CABOT product bulletin, #TD-117, February 1999.

149. Owen, M.J. and J.L. Groh. Fluorosilicone antifoams. J. Appl. Polym. Sci., 40, 789–797, 1990.

150. Kobayashi, H. and T. Masatomi. Fluorosilicone antifoam. U.S. Patent 5,454,979, Dow Corning Toray Silicon Co., October 3, 1995.

151. Griffiths, R.A. Practical problems in using water mix cutting fluids when machining cast iron. Tribol. Int., 337–339, December 1978.

152. Przybylinski, J.L. Diethanol disulfide as an extreme pressure and anti-wear additive in water soluble metalworking fluids. U.S. Patent 4,250,046, Pennwalt Corp., February 10, 1981.

153. Lomas, A.W. Compositions used to control foam. U.K. Patent Application 2,119,394, Dow Corning Corp., June 4, 1982.

154. Gallagher, C.T., P.J. Breen, B. Price Brian, and A.F. Clemmit. Method and composition for suppressing oil-based foams. U.S. Patent 5,853,617, Baker Performance Chemicals, December 29, 1998.

155. Ward, W.C., C. Tipton, K.A. Murray. Lubrication fluids for reduced air entrainment and improved gear protection. European Patent 0,761,805, The Lubrizol Corporation, September 10, 1996.

156. Muchmore, R.A., R. Sarkar, A.J. Rollin, S.H. Tersigni, M. Deweerdt, and S. Grangeretl. Antifoam agent and method for use in automatic transmission fluid applications involving high pressure pumps. U.S. Patent 7,060,662, Afton Chemical Corporation, June 13, 2006.

157. Loop, J.G. and W.D. Abraham. Lubricant compositions comprising anti-foam agents. U.S. Patent Application US 2014/0018267, The Lubrizol Corporation, January 16, 2014.

158. Denkov, N.D. Mechanisms of foam destruction by oil-based antifoams, Langmuir, 20, 9463–9505, 2004.

159. Bergeron, V., P. Cooper, C. Fischer, J. Giermanska-Kahn, D. Langevin, and A. Pouchelon. Polydimethylsiloxane (PDMS)-based antifoams, Colloids Surf. A Physicochem. Eng. Asp., 122, 103–120, 1997.

160. Mann, E.K., L.T. Lee, S. Henon, D. Langevin, and J. Meunier. Polymer-surfactant films at the air-water interface. 1. Surface pressure, ellipsometry, and microscopic studies, Macromolecules, 26 (25), 7037–7045, 1993.

161. Koczo, K., L. Lobo, and D.T. Wasan. Effect of oil on foam stability: Aqueous foams stabilized by emulsions. J. Colloid Interface Sci., 150 (2), 492–506, 1992.

162. Arnaudov, L., N.D. Denkov, I. Surcheva, P. Durbut, G. Broze, and A. Mehreteab. Effect of oily additives on formability and foam stability. 1. Role of interfacial properties. Langmuir, 17, 6999–7010, 2001.

163. Nikolov, A.D., D.T. Wasan, D.D. Huang, and D. Edwards. The effect of oil on foam stability: Mechanisms and Implications for oil, displacement by foam in porous media. *Proceedings of 61st Annual Technical Conference of the SPE of AIME*, New Orleans, Preprint SPE 15443, 1986.

164. Wasan, D.T., K. Koczo, and A.D. Nikolov. Mechanisms of aqueous foam stability and antifoaming action with and without oil, in L.L. Schramm, ed., *Foams: Fundamentals and Applications in the Petroleum Industry*, ACS Symposium Series No. 242, Chapter 2, pp. 47–114, 1994.

165. Lobo, L. and D.T. Wasan. Mechanisms of aqueous foam stability in the presence of emulsified non-aqueous-phase liquids: Structure and stability of the pseudoemulsion film. *Langmuir*, 9, 1668–1677, 1993.

166. Koczo, K., J.K. Koczone, and D.T. Wasan. Mechanisms for antifoaming action in aqueous systems by hydrophobic particles and insoluble liquids. *J. Colloid Interface Sci.*, 166, 225–238, 1994.

167. Hadjiiski, A., S. Tcholakova, N.D. Denkov, P. Durbut, G. Broze, and A. Mehreteab. Effect of oily additives on foamability and foam stability. 2. Entry barriers. *Langmuir*, 17, 7011–7021, 2001.

168. Wang, G., R. Pelton, A. Hrymak, N. Shawafaty, and Y.M. Heng. On the role of hydrophobic particles and surfactants in defoaming. *Langmuir*, 15, 2202–2208, 1999.

169. Marinova, K.G., N.D. Denkov, S. Tcholakova, and M. Deruelle. Model studies of the effect of silica hydrophobicity on the efficiency of mixed oil-silica antifoams. *Langmuir*, 18, 8761–8769, 2002.

170. Frye, G.C. and J.C. Berg. Mechanisms of the synergistic antifoam action by hydrophobic solid particles and insoluble oils. *J. Colloid Interface Sci.*, 120 (1), 54–59, 1989.

171. Garrett, P.R. Preliminary considerations concerning the stability of a liquid heterogeneity in a plane-parallel liquid film. *J. Colloid Interface Sci.*, 76 (2), 587–590, 1980.

172. Aveyard, R., P. Cooper, P.D.I. Fletcher, and C.E. Rutherford. Foam breakdown by hydrophobic particles and nonpolar oil. *Langmuir*, 9 (2), 604–613, 1993.

173. Aveyard, R., B.P. Binks, P.D.I. Fletcher, T.G. Peck, and C.E. Rutherford. Aspects of aqueous foam stability in the presence of hydrocarbon oils and solid particles. *Adv. Colloid Interface Sci.*, 48, 93–120, 1994.

174. Ross, S. The Inhibition of foaming. II. A mechanism for the rupture of liquid films by anti-foaming agents. *J. Phys. Chem.*, 54 (3), 429–436, 1950.

175. Kulkarni, R.D., E.D. Goddard, and B. Kanner. Mechanism of antifoaming: Role of filler particle. *Ind. Eng. Chem. Fundam.*, 16 (4), 472–474, 1977.

176. Pouchelon, A. and C. Araud, Silicone defoamers: The performance, but how do they act. *J. Disp. Sci. Technol.*, 14 (4), 447–463, 1993.

177. Racz, G., K. Koczo, and D.T. Wasan. Mechanisms of antifoam deactivation. *J. Colloid Interface Sci.*, 181, 124–135, 1996.

178. Denkov, N.D., K.G. Marinova, C. Christova, A. Hadjiiski, and P. Cooper. Mechanisms of action of mixed solid–liquid antifoams: 3. Exhaustion and reactivation. *Langmuir*, 16 (6), 2515–2528, 2000.

179. Denkov, N.D., P. Cooper, and J.-Y. Martin. Mechanisms of action of mixed solid-liquid antifoams. 1. Dynamics of foam film rupture. *Langmuir*, 15, 8514–8529, 1999.

180. Bonfillon-Colin, A. and D. Langevin. Why do ethoxylated nonionic surfactants not foam at high temperature? *Langmuir*, 13 (4), 599–601, 1997.

181. Nemeth, Z., G. Racz, and K. Koczo. Foam control by silicone polyethers—Mechanisms of "cloud point antifoaming". *J. Colloid Interface Sci.*, 207, 386–394, 1998.

182. Griffith, R.M. The effect of surfactants on the terminal velocity of drops and bubbles. *Chem. Eng. Sci.*, 17, 1057–1070, 1962.

183. Ross, S. Foaminess and capillarity in apolar solutions, in H.-F. Eicke and G.D. Parfitt, eds., *Interfacial Phenomena in Apolar Media*, Surfactant Science Series, Vol. 21, Marcel Dekker, New York, pp. 1–39, 1987.

184. Suzin, Y. and S. Ross. Retardation of the ascent of gas bubbles by surface-active solutes in nonaqueous solutions. *J. Colloid Interface Sci.*, 103, 578–585, 1985.

185. Mannheimer, R.J. Factors that influence the coalescence of bubbles in oils that contain silicone antifoamants. *Chem. Eng. Commun.*, 113, 183–196, 1992.

186. Morrison, I.D. Electrical charges in nonaqueous media. *Colloids Surf A: Physicochem. Eng. Aspects*, 71, 1–37, 1993.

187. Callaghan, I.C. Non-aqueous foams: A study of crude oil foam stability, in A. Wilson, ed., *Foams: Physics, Chemistry, and Structure*, Springer-Verlag, New York, pp. 89–104, 1989.

188. Callaghan, I.C., C.M. Gould, R.J. Hamilton, and E.L. Neustadter. The relationship between the dilatational theology and crude oil foam stability I. Preliminary studies. *Colloids Surf.*, 8, 17–28, 1983.

189. Friberg, S.E. and W.-M. Sun. Foam stability in non-aqueous multi-phase systems, in D.M. Bloor and E. Wyn-Jones, eds., *The Structure, Dynamics and Equilibrium Properties of Colloidal Systems*, Kluwer Academic Publishers, Dordrecht, the Netherlands, pp. 529–539, 1990.

190. Friesen, T.V. Transmission-hydraulic fluid foaming. SAE Technical Paper Series #871624, SEA International, 1983.

191. Awe, R.W. Silicone antifoams for lubricating oils. SAE Technical Paper 630437, 1963.

192. Beerbower, A. and R.E. Barnum. Studies on the dispersion of silicone defoamant in non-aqueous fluids. *Lubr. Eng.*, 17, 282–285, 1961.

193. Shearer, L.T. and W.W. Akers. Foaming in lube. *J. Phys. Chem.*, 62, 1269–1270, 1958.

20 Antifoams for Nonaqueous Lubricants

Ernest C. Galgoci

CONTENTS

20.1 INTRODUCTION

Foam is defined as the concentrated dispersion of a gas in a liquid. This gaseous dispersion resides at the surface of the liquid, but in the extreme case, the liquid system can be converted entirely to foam. As shown in Figure 20.1, the foam exhibits a distinct phase boundary with the liquid, which itself can contain a large fraction of entrained air. The entrained air buoyantly moves to the surface to create and replenish or increase the foam. For both aqueous and nonaqueous industrial fluids (e.g., metalworking fluids) and lubricants (e.g., gear oils), minimizing the formation of foam during use is required in order to maintain the effectiveness of the fluid to provide maximum lubrication and, in the case of metalworking fluids, heat removal. For nonaqueous systems, it is also important to minimize the air content (foam and entrained air) to reduce fluid oxidation that will degrade the fluid and eventually lead to potentially costly and premature fluid replacement. Other negative effects of foam and entrained air in nonaqueous systems exist such as cavitation, pressure fluctuations, and, in the extreme, loss of lubricant through vents.

In order to mitigate these issues, an antifoam is a necessary and critical component of the fluid formulation. The term *antifoam* refers to a substance that suppresses foam as it forms. The antifoam is added directly to a formulation before use and is designed to disrupt the interfacial forces that stabilize the foam bubbles. In contrast, a defoamer is a material added during the use of the fluid to break any undesired foam buildup and to provide a period of foam suppression before another

addition may be necessary. In practice, the terms *antifoam* and *defoamer* are often used synonymously, since their compositions are essentially the same or very similar and they perform the same ultimate function. For nonaqueous systems, the term *antifoam* is generally more precise since, for practical reasons, defoamer additions are almost never done. Although the criteria for choosing an antifoam will vary depending on the specific fluid and requirements, the antifoam must generally exhibit strong initial defoaming, persistence (longevity) of the defoaming, and compatibility (no separation) with the fluid.

20.1.1 FOAM STABILIZATION MECHANISMS FOR AQUEOUS SYSTEMS

Foam is not thermodynamically stable in pure liquids such as water or a base oil. In aqueous systems, foam is stabilized through a combination of mechanisms that involve surface-active agents (e.g., surfactants) that stabilize the interfacial regions between foam bubbles and their surroundings (i.e., other bubbles and the external atmosphere). For example, in aqueous metalworking fluids, combinations of surfactants are used to produce dispersions of oil droplets when the fluids are diluted in water. These surfactants, in turn, stabilize foam bubbles that result from the incorporation of air into the fluid as a result of the vigorous mechanical machining operations that the fluids are designed to enable. In order to understand how antifoams work, it is helpful to understand the primary mechanisms of foam stabilization.

FIGURE 20.1 Foam layer (top) and entrained air (bottom) in a liquid.

A foam bubble has two interfaces where the bulk liquid contacts air. One of those interfaces is along the outside wall of the bubble itself, and the other is in adjacent contact with either another bubble or the environment (atmosphere). The two interfaces along with the intervening liquid layer are referred to as the foam lamella. In order for the bubble to break, the liquid from the lamella must drain so that the lamella becomes thin enough for the internal pressure of the bubble to exceed the strength of the lamella [1,2]. In aqueous systems, surfactants adsorbed at the lamella interfaces retard drainage through a number of interfacial effects that are described in the following:

- *Electrostatic repulsion*: Ionic surfactants adsorbed on the bubble lamella walls can lead to a strong stabilization of lamella thickness (and thus foam) through mutual charge repulsion. This mechanism is comparatively strong and is responsible for the relatively stable foams produced by soaps.
- *Steric hindrance*: Nonionic surfactants with sufficiently long hydrophilic chains (e.g., ethoxylates) retard the thinning of lamella due to the steric repulsion (entropic effects) of those chains.
- *Marangoni effect/surface transport* [3]: As a lamella thins during drainage, a surface tension gradient ($d\gamma/dA$; γ = surface tension, A = surface area) is induced by the local change in the surface area of the thinned region. This gradient occurs because the surfactant concentration on both the lamella walls and the liquid layer is reduced relative to adjacent regions. In order to eliminate the surface tension gradient, liquid will flow from areas of low surface tension to high surface tension, and this will replenish the liquid in the lamella. Additionally, transport (due to the surface tension gradient) of surfactant at the interfaces will drag a significant amount of liquid back to the thinned region. The net effect is to thicken and, thus, reinforce the thinned region.
- *Gibbs elasticity* [3]: In addition to the Marangoni effect, the earlier mentioned surface tension gradient

imparts an "elasticity" to a foam lamella. The elasticity is directly proportional to the surface tension gradient induced by the thinning lamella as given by $E = 2A \times d\gamma/dA$.

- *Surface viscosity*: At sufficient concentrations, surfactant molecules adsorbed at the liquid/gas interface orient to allow maximum packing. The resulting structure leads to an increase in the surface viscosity that slows drainage.
- *Bulk liquid viscosity*: Drainage rates are inversely proportional to the viscosity of the bulk liquid.

20.1.2 FOAM STABILIZATION MECHANISMS FOR NONAQUEOUS SYSTEMS

The mechanisms for foam stabilization in nonaqueous systems such as lubricating oils are not as clear as that for aqueous systems, which have been extensively studied. For nonaqueous systems, several of the stabilizing mechanisms mentioned earlier do not apply, since there is no equivalent to an aqueous surfactant added to the oil. Therefore, we can rule out electrostatic (also not important due to the low dielectric constants of base oils) and steric stabilizations and other effects due to surface tension gradients induced by surfactants. Instead, foam stabilization in oils appears to be more related to bulk and surface viscosities, as an increase in either or both of these leads to retardation of lamella drainage. Callaghan [4] described several studies [5,6] that showed these viscosities to be important for foam in several oil-based systems, but several other studies [7,8] that were cited reportedly did not show a correlation. Friberg [9] presented a review on the topic of the stability of nonaqueous foams and concluded that surface forces were not important and that the stability was dependent on the drainage rate. Further, he concluded that liquid crystal and/or solid phases in the liquid provide the foam stabilization. In another review, Blazquez [10] came to similar conclusions as Friberg. It is well known that the presence of additives (excluding the antifoam) in an oil will generally lead to increased foaminess, so at least some of these additives must influence the surface or intralamellar viscosities or affect the lamella surface. In a study by Ross and Suzin [11], it was shown that certain additives and especially their combinations led to increased foam stability. Binks et al. [12] studied additives in a poly-α-olefin (PAO) base oil and concluded that foam correlated with proximity to a phase separation boundary when the soluble additive concentrations were relatively high (above ~1 wt.%) and that the tendency to adsorb at the gas bubble surfaces was thereby optimized.

20.1.3 THERMODYNAMICS OF DEFOAMING

As shown in Figure 20.2, an antifoam droplet must enter the foam lamella and then spread at the interface. Another requirement is that the antifoam droplet must eventually bridge the gap between the two interfaces of a lamella. The properties

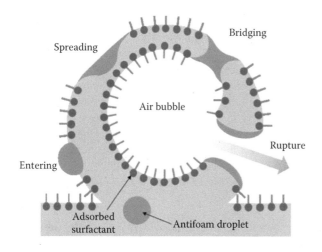

FIGURE 20.2 Schematic of the defoaming process.

of entering, spreading, and bridging are dependent on the surface tensions of the liquid film (γ_L) and the antifoam droplet (γ_D) and the interfacial tension (γ_{LD}) between the two. Entering (E), spreading (S), and bridging (B) coefficients can be defined such that, when maximized (i.e., greater than zero), optimal defoaming is achieved. The respective equations are shown as follows [13–16].

$$E = \gamma_L + \gamma_{LD} - \gamma_D > 0$$

$$S = \gamma_L - \gamma_D - \gamma_{LD} > 0$$

$$B = \left(\gamma_L\right)^2 + \left(\gamma_{LD}\right)^2 - \left(\gamma_D\right)^2 > 0$$

As can be deduced from these equations, a necessary requirement for defoaming is that the active component has a surface tension (γ_D) that is lower than the surface tension (γ_L) of the liquid medium. Because they possess the lowest surface tensions/energies (Table 20.1) of commonly available liquids, mineral oils, silicones (e.g., polydimethylsiloxane [PDMS]), and solids such as waxes and treated silicas are utilized in

TABLE 20.1

Surface Tensions of Liquids Used to Make Antifoams (γ_D) and Liquid Medium (γ_L) to Defoam

Material	Surface Tension (mN/m)
Water	72
Mineral oils	30–35
Silicone (PDMS) oil	20–22
Water + surfactant	26

Sources: Booser, E.R., ed., *CRC Handbook of Lubrication: Theory and Practice of Tribology Volume II: Theory and Design*, CRC Press LLC, Boca Raton, FL, 246, 1983; Triton GR-5M Surfactant (anionic sulfosuccinate) at 1 wt% actives in water at 25°C, Dow Chemical Company, Form No. 119-01907-1107.

antifoam formulations for aqueous systems. In the case of nonaqueous lubricating oils, silicone is the only common fluid with a sufficiently low surface tension, and this explains its general effectiveness.

20.1.4 KINETICS AND OTHER FACTORS OF DEFOAMING

Although thermodynamic considerations are important for considering why certain materials are useful as antifoams, a foaming system is not in equilibrium. Kinetic and other factors are equally if not more important for defoaming effectiveness. For an antifoam to function, it must have a low solubility in the liquid to be defoamed, such that it exists as dispersed droplets in the medium. The reason for this requirement is that if the antifoam is soluble, there is no thermodynamic driving force for it to act at the interface. With that requirement, it is apparent that the number of droplets and their size distribution are important. The number of droplets affects the kinetics, while the size distribution affects the effectiveness of spreading and bridging. Of course, for a given amount of antifoam, the number of droplets is directly related to the size distribution. With regard to the size distribution, Shearer and Akers [19] determined that the diameter of droplets of the silicone should be less than 100 µm to be effective antifoams and that if the silicone was completely soluble in the oil, it would be a "profoamant." A simple thought experiment regarding two extreme examples can further elucidate the point of droplet size distribution. In one case where the antifoam exists as a single droplet in the system, it will defoam effectively in its vicinity, but there will be far too few droplets to break the foam bubbles in the rest of the system. On the other hand, if the droplets are exceeding small, they will have a low mass and be less effective at spreading and bridging. For a given system, there will be an optimal droplet size distribution for defoaming. In practice, the droplet size distribution is rarely determined.

Another practical aspect of particle size is storage stability of the antifoam dispersion. Beerbower and Barnum [20] pointed out that a particle size less than 10 µm is required for adequate storage stability. Awe [21] proposed that a particle size range of 2–4 µm was needed for the stability of a PDMS dispersion in an SAE 10 oil; for a trifluoropropylmethylsiloxane (TFPMS) dispersion, the value was 1–2 µm due to the relatively higher density of the TFPMS.

Unfortunately, due to the complexity of factors (e.g., chemistry, additives, end-use conditions) for a given system, the choice of a specific antifoam (and its consequent optimal droplet size) is often only determined by experiential factors and empirical testing.

20.1.5 AIR ENTRAINMENT AND RELEASE

The rate at which entrained air from a lubricating oil moves to the surface is controlled by the viscosity of the fluid and the size of the air bubbles. Hadamard and Rybczynski [22,23] derived the following equation to describe the rise velocity of an air bubble in a liquid for the case where the internal gas in the rising air bubble can circulate ("soft bubble").

$$V_s = \frac{g\rho r^2}{3\eta}$$

In this equation,
   g is the gravitational constant
   ρ is the density of the liquid (assuming the density of air to be negligible)
   r is the bubble radius
   η is the viscosity of the liquid

In contrast, Stokes' law predicts the following for the case of a rigid sphere.

$$V_r = \frac{2g\rho r^2}{9\eta}$$

As can be seen, the ratio of V_s/V_r is 1.5. Levich [24] explains the ratio of 1.5 as arising because the velocity of the liquid is not zero at the air bubble boundary for a soft bubble but, for the case of a rigid sphere, it is zero. While air bubbles are not ordinarily considered rigid spheres, their rise velocities in formulated lubricating oils are often found to match more closely to that predicted by Stokes' law. This behavior appears to arise from the presence of surface-active agents that impart an elasticity (Gibbs) that immobilizes the surface of the air bubbles. For example, it was observed that as little as 10 ppm of silicone oil in a mineral oil could reduce the velocity of an air bubble (radius of 0.11 cm) from that predicted by Hadamard and Rybczynski to nearly that of Stokes' law [25]. Even sorbitan monolaurate, which is not a surfactant in the traditional sense for nonaqueous systems, was shown to have an effect on the rise velocity, but it required over 400 ppm to decrease the rise velocity to that of Stokes' law. This result with sorbitan monolaurate suggests that materials that are not surfactants in the traditional sense can play a role in slowing the speed of bubble rise.

Mannheimer [26] reported the effects of a number of variables on the rate of bubble rise in three base oils of different viscosities. In the case of the pure oils, the rate of bubble rise was close to the prediction of Hadamard and Rybczynski. However, when silicone oil was introduced, small bubbles behaved as rigid spheres, while larger bubbles deviated toward the behavior of a soft bubble. The effect was least pronounced for the lowest viscosity oil. Silicone at even 1 ppm had an effect, but except for the smallest bubbles, the rise velocity was always lowest at the level of 10 ppm. In the same study, it was shown that the coalescence time for bubbles below a radius of about 0.5 mm was increased by the presence of silicone but, above that radius, the coalescence time dropped significantly below that of the case of no silicone. The presence of 1 ppm silicone had a more dramatic effect than 10 ppm in the high- and medium-viscosity oils. However, there was no effect observed at 1 ppm of silicone in the lowest-viscosity oil

up to the largest bubble radius (~0.75 mm) studied. Bubble coalescence is important for speeding the elimination of entrained air, since the rise velocity is proportional to the square of the bubble radius.

As indicated previously, there are examples of silicones inhibiting bubble rise, and those observations are in agreement with general industry observations. Polyacrylates/polymethacrylates, on the other hand, are generally known to have less effect upon air entrainment. This is most likely due to their polar ester component that imparts a higher surface tension and solubility parameter relative to PDMS silicones.

20.2   ANTIFOAM FORMULATION AND CHEMISTRY FOR LUBRICATING OILS

For nonaqueous systems, antifoams are typically formulated only with liquid actives and carriers. Emulsifiers, which are commonly used for antifoams in aqueous systems, are not functionally applicable for nonaqueous systems. The most common actives used are silicones (most common chemistry employed) and polyacrylates/polymethacrylates. Silicones are most effective at defoaming, but they can negatively affect air entrainment. Polyacrylates, on the other hand, are not as effective at defoaming, but they tend to have little effect on air entrainment. Appropriate carriers will reduce viscosity to aid incorporation, improve compatibility, and sometimes enhance defoaming. In some cases, combinations of silicones and polyacrylates provide an effective balance of defoaming and air entrainment to lubricating oil systems. Further references to antifoams in oil-based systems can be found in References 4 and 27.

20.2.1   SILICONES

The most commonly used silicone antifoam is based on the linear polymer PDMS, which has the chemical formula shown in Figure 20.3. PDMS grades are designated by their kinematic viscosities in centistokes (cSt = mm²/s). For lubricating oils, PDMS grades vary from <1,000 cSt up to about 60,000 cSt [4,28], which corresponds to weight average molecular weights (M_w) of about 28,000 and 120,000 g/mol, respectively [29]. Generally, the 12,500 cSt grade is one of the most common, but there is a high degree of variability due to historical precedence. The PDMS used as antifoams for lubricating oils are usually terminated with trimethylsilyl groups, which serve to control molecular weight by blocking further polymerization.

Key advantages of PDMS are its relatively low surface tension and significantly different solubility parameters compared to oils. Both combine to make PDMS the most important antifoam for lubricating oils, since it fulfills the requirements

$$(CH_3)_3-Si-O-[-Si(CH_3)_2-O-]_n-Si(CH_3)_3$$

FIGURE 20.3   Chemical structure of PDMS.

of being insoluble in the oil and surface active. The solubility of PDMS in oils decreases as its molecular weight increases. Generally, higher-molecular-weight PDMS is more effective for defoaming, which is presumably due to its lower solubility in the oil and thus a greater tendency to form droplets. Sometimes, several grades of PDMS are combined to enhance defoaming.

The lower-viscosity grades of PDMS can be incorporated directly into lubricating oils; however, all PDMS grades are usually added to the oil as a component of an additive package or as a mixture (solution or fine dispersion) in hydrocarbon solvents. This approach has the advantage of easier handling and incorporation (mixing) of the antifoam in the lubricating oil. Kerosene is a common solvent, but other, more refined hydrocarbons are often used. The refined hydrocarbons offer the advantages of improved consistency and, in some cases, appear to increase antifoaming performance by, presumably, facilitating the formation of droplets with a more desirable size distribution.

In order to choose the most effective silicone antifoam for a given application, multiple grades of PDMS at several dosage levels should be tested. If a carrier solvent is needed, that should also be included in the study. It must be noted that using more antifoam does not always lead to better defoaming. Often, the reverse is found to be true, and that is why it is important to establish the proper use level. Use levels (PDMS actives basis) for PDMS antifoams in lubricating oils range from a few ppm to several hundred ppm. Typically, the range is 50–100 ppm.

In addition to PDMS, other types of silicone-based chemistries are available, whereby the methyl groups attached to the silicon atom are partially replaced by other organic groups. One of the most common chemistries is the polyglycol. The polyglycols are typically ethylene oxide and propylene oxide homopolymers or copolymers. These materials do not have a significant use as antifoams in lubricating oils, but they are used as antifoams in diesel and biodiesel fuels [30,31]. Another type of modification is to replace the methyl group(s) with aryl or alkyl groups. Those modifications tend to increase the solubility of silicone in oils and to increase the surface tension of the silicone. Additionally, fluorosilicones are available as antifoams [21]. In this case, fluoroalkyl groups (e.g., trifluoropropyl) are the modification. These products find limited applicability in lubricating oils due in part to their high cost, but they are utilized as defoamers in crude oil–gas separators (especially on offshore locations).

20.2.2 Polyacrylates

The second most common class of antifoams used in lubricating oils is the polyacrylates [4,32]. The class encompasses copolymers made from acrylic and methacrylic monomers. Usually, acrylic monomers are used exclusively to prepare the polymers, since they are relatively inexpensive compared to the corresponding methacrylic monomers. Two of the most common monomers are 2-ethylhexyl acrylate and ethyl acrylate. Other monomers may be used, but the basic structure of these polymers is that there is at least one monomer with a relatively long alkyl side chain (e.g., 2-ethylhexyl). The molecular weights of the polymers range from about 5,000 g/mol (oligomeric) to >100,000 g/mol. The viscosities of the neat polymers are relatively high, so they are almost always supplied as solutions in hydrocarbon solvents. Unlike the case for PDMS, the carrier solvent does not appear to have an effect on defoaming performance. Generally, use levels (polyacrylate actives basis) range from 50 to 500 ppm.

Polyacrylates do not have low surface tensions compared to PDMS, and so they do not usually defoam nearly as well as PDMS. Generally, they are used in applications where air entrainment is a concern or where silicone is not allowed or undesired (e.g., engine oils). Since air entrainment does not seem to be impacted by polyacrylates, they must not be surface active in oils or only weakly so. This poses a dilemma with regard to their mode of action as antifoams in lubricating oils, but it is most likely the cause of their relatively poor defoaming performance [4]. To compensate, polyacrylates are sometimes used in conjunction with PDMS to obtain better defoaming than the polyacrylate and better air release than the PDMS.

20.2.3 Combinations and Copolymers

As mentioned previously, polyacrylates and silicones are occasionally utilized together as the antifoam package, where each are added separately to the oil. However, there are products available that combine both silicone and polyacrylate [33]. The advantages of a combination product are that there is less inventory to stock and the product has generally wide applicability across many applications. Besides the combination products, there are silicone–polyacrylate copolymers [34–36]. The copolymers were claimed to provide improved defoaming performance relative to combinations or single antifoams in the lubricating oils in which they were tested.

20.2.4 Antifoam Selection

The ultimate choice of an antifoam for a given lubricating oil is complicated by a number of factors such as the oil composition, the additives, and the end-use environment. The complexity ultimately makes the empirical approach the most viable way to select the antifoam. As mentioned earlier for the silicones, screening should include several PDMS viscosity grades and solvents. In addition, it is also advised to include different chemistries and their variations. Unlike the case of antifoams for aqueous systems, the choice of antifoam chemistry for lubricating oils is less complicated, as there are currently only two significant options.

The empirical nature of antifoam selection is illustrated by the results reported by Galgoci and Brüning [35,36], who determined the best antifoams for three lubricating oils (wind turbine gear oil, automatic transmission fluid (ATF), and a slideway lubricant). Each oil required a different antifoam chemistry to meet the defoaming requirements (air release was not determined). For the wind turbine gear oil (PAO based),

TABLE 20.2
Recommended Antifoam Chemistries by Application

Application	Antifoam Chemistry: Primary	Secondary
Engine oil	Polyacrylate	Silicone
Transmission fluid	Silicone; polyacrylate–silicone combination	
Gear oil	Silicone; polyacrylate–silicone combination	
Turbine oil	Silicone; polyacrylate–silicone combination	Polyacrylate
Hydraulic fluid	Silicone; polyacrylate–silicone combination	Polyacrylate
Compressor oil	Silicone; polyacrylate–silicone combination	

Source: Technical Brochure MUE_2014_009_US-03/14, Industrial fluids: Defoamers, Münzing NA LP, 2014, p. 7.

TABLE 20.3
Recommended Antifoam Chemistries by Base Oil Type

Base Oil Type	Antifoam Chemistry
Synthetic	Silicone; polyacrylate–silicone combination
Group I	Silicone
Group III	Silicone; polyacrylate–silicone combination
PAG	Silicone
PAO	Silicone; polyacrylate–silicone combination
Ester	Silicone; polyacrylate–silicone combination

Source: Galgoci, E.C., Unpublished results.

a silicone (PDMS) was found to defoam better and at a much lower dosage than either a polyacrylate or a combination of polyacrylate–silicone (PA-S). In the ATF, the combination of PA-S performed significantly better than the silicone control and at <10% of use level of the control. With the slideway lubricant, the PA-S copolymer performed better than the other chemistries mentioned in this paragraph.

Although the empirical approach to selecting the appropriate antifoam is necessary, experiential knowledge is helpful to determine starting points for the antifoam selection for an application. This can often be done by examining product recommendations provided by suppliers. Table 20.2 provides recommended chemistries for various applications and Table 20.3 by base oil type.

20.2.5 ANTIFOAM INCORPORATION

As stated earlier, the antifoam can be incorporated into the lubricating oil either as a component in the additive package or as a separate addition. When additive packages containing antifoams are used, the recommendations of the supplier should be followed. For the case of separate addition,

the antifoam should be added very slowly to the oil with strong enough agitation to produce a vortex in the liquid. This will ensure that the antifoam is adequately dispersed and that undesired large droplets are not formed. For similar reasons, it is generally recommended that the addition take place below 50°C, although this may vary depending on the specific oil formulation and practical considerations of the blending process. Predilution of the antifoam may be necessary if the viscosity is too high for easy handling. Antifoams that are supplied as dispersions in solvents usually do not require predilution. Silicones that are prediluted are more readily incorporated into the oil without forming large droplets. Awe [21] suggested the use of a rotor stator to achieve the desired small droplet size, but this is usually not an option for most blending operations.

20.3 TEST METHODS

The most common, standard test methods for lubricating oils are ASTM D892-13 [38] and ASTM D3427-15 [39]. ASTM D892-13 is an air sparging method that involves conditioning the test sample under three different conditions (referred to as "Sequences" in the method). The intent is to simulate potential use conditions for the lubricant. The sparging is accomplished by the introduction of a porous spherical ceramic "stone" or a cylindrical, metallic "diffuser" into a graduated cylinder that contains the test oil. The metallic diffuser is typically the more difficult condition, as it produces smaller bubbles. Often, Sequence II is the most difficult to meet, since it is run at 95°C compared to 25°C for the other two sequences. However, depending on the fluid, other sequences can also be challenging. Generally, the target foam volume is a maximum of 50 mL at the end of the air sparging and 0 mL at 10 min thereafter. A related method used often for motor oils and transmission fluids and referred to as Sequence IV of ASTM D892-13 is ASTM D6082-12 [40]. This method is for testing defoaming at a high temperature (150°C), which can occur in the end uses of these fluids. Another variant of ASTM D892 is to test the effect of small amounts of water intentionally introduced into a transmission fluid [41]. ASTM D3427-15 is the standard test used for air release (entrainment) in nonaqueous systems. The design of the test is to measure the density of the oil over time after air is bubbled into it. Data of density over time after the air flow is stopped are collected until the density reaches 99.8% of the original value. Figure 20.4 shows a comparison of the air release (ASTM D3427) of a PAO-based lubricating oil that contains no antifoam and the same oil containing 100 ppm PDMS [29]. Interestingly, the 99.8% density time for the silicone antifoam is 1.5 times longer than the case of no antifoam and in line with that predicted by Stokes' law. For wind turbine gear oils, a standard test is the "Flender" test [42]. This test involves the use of a standard 2-gear set that is enclosed in a box that has a window from which the foam height is measured. A filter through which the lubricant is recirculated can be included to further simulate field conditions. There are other customized methods that are used. For example, in the cases where high shear occurs

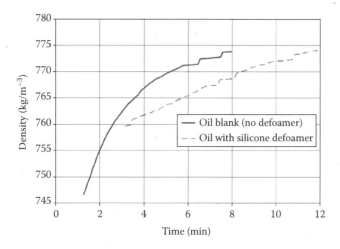

FIGURE 20.4 Air release profile of a PAO-based lubricant with and without PDMS.

(e.g., transmissions), tests with high shear (e.g., rotor stator) and varying temperatures are sometimes used.

In the selection of an antifoam for a given application, it is important to match the test method with the application conditions that are responsible for the foam. Even though ASTM D892 is the standard test employed for lubricating oils, it may not accurately predict performance in a given application. Tests such as the Flender test for wind turbine gear oils and the rotor stator for engine oils and transmission fluids may provide more realistic simulations of the end uses.

20.4   OTHER CONSIDERATIONS

Factors that can affect defoaming performance are contamination (e.g., dirt, water, and oxidation) of the oil and filtration. Many types of contaminants will adversely affect the foaming behavior of lubricating oils. The antifoam should be able to perform adequately when some degree of contamination is present, so it is advisable to anticipate some contamination and to test appropriately such as specified in Reference 41. Certain applications require filtration to ensure oil cleanliness so that contaminants do not adversely affect lubrication and thus gear wear. Because silicone antifoams are present as dispersed droplets and have a tendency to adsorb onto surfaces, they can be negatively affected by filtration. As filter pore sizes decrease (especially for sensitive applications such as wind turbine gear oils), it becomes more difficult to maintain defoaming performance, since some of the antifoam is inevitably removed from the oil. Pore sizes below about 10 μm seem generally to pose more problems for silicone antifoams. Ultimately, there will need to be a balance between the desire for extreme cleanliness of the lubricating oil and its defoaming properties.

Related to cleanliness are particle count requirements [43]. Various applications require that the oil contain less than a certain amount of particles within various size fractions. Unfortunately, the common methods employed to determine the particle counts do not discriminate between antifoam

droplets and undesired contaminants. In particular, particle counts can be significantly increased when silicone antifoams are employed.

In certain applications such as lubricants for food processing equipment, it is required that the antifoam ingredient be registered as HX-1 by NSF International [44]. Certified antifoam products from several suppliers can be found by a search on the NSF website. Silicone antifoams based on PDMS are generally acceptable for HX-1. Polyacrylates are generally not acceptable, since there is only a specific composition in FDA regulation 21 CFR 175.300 that is listed for incidental food contact.

20.5   FUTURE DIRECTIONS

One of the needs moving forward is to reconcile the necessities of adequate defoaming and cleanliness through filtration. Resolution of this situation may be through the development of new antifoam technologies that are not as easily affected by filtration or new filtration methods (e.g., different materials of construction). Improved particle count methods that can distinguish between antifoam particles and contaminants may also be required.

For many decades, silicones and polyacrylates have been the main antifoam chemistries available. As requirements for lubricants are becoming more stringent, there may be a need for other chemistry options, combination products of current and new chemistries, or copolymers thereof. It is not clear what those new chemistries might be, but they may include new molecular architectures.

20.6   SUMMARY AND CONCLUSION

Foam stabilization in lubricating oils is primarily controlled by the bulk and surface viscosities, which can be affected by various additives in the oil. Silicones are the most effective and most commonly used antifoams in nonaqueous systems, but they negatively affect air release properties. Polyacrylates do not affect air release, but they are generally relatively poor antifoams. Antifoam products, which combine polyacrylates and silicones and copolymers thereof, offer the potential of performance that offers the best of the individual components. The specific choice of antifoam for a given lubricating oil is best empirically determined, due to the large number of compositional and application variables. Several standardized foam test methods exist, but the choice of the method used should be driven by its applicability to the end-use conditions. Increasingly stringent filtration and particle count requirements are challenges facing current and future antifoam development.

REFERENCES

1. Wasan, D. T., Koczo, K., and Nikolov, A. D., Mechanisms of aqueous foam stability and antifoaming action with and without oil: A thin film approach, in Schramm, L. L. (ed.), *Foams: Fundamentals and Applications in the Petroleum Industry*, Chapter 2, Advances in Chemistry Series, Vol. 242, American Chemical Society, Washington, DC, p. 47, 1994.

2. Kruglyakov, P. M., Karakashev, S. I., Nguyen, A. V., and Vilkova, N. G., Foam drainage, *Current Opinion in Colloid & Interface Science*, 13, 163–170, 2008.

3. Rosen, M. J., *Surfactants and Interfacial Phenomenon*, 3rd edn., John Wiley & Sons, Hoboken, NJ, pp. 277–297, 2004.

4. Callaghan, I. C., Antifoams for nonaqueous systems in the oil industry, in Garrett, P. R. (ed.), *Defoaming: Theory and Industrial Application*, Surfactant Science Series, Vol. 21, Marcel Dekker, New York, 119–150, 1993.

5. Brady, A. P. and Ross, S., The measurement of foam stability, *Journal of American Chemical Society*, 66, 1348–1356, 1944.

6. McBain, J. W. and Robinson, J. V., Surface properties of oils, *National Advisory Committee for Aeronautics*, Washington, DC, Technical Note No. 1844, 1949.

7. Mannheimer, R. J. and Schecter, R. S., Shear-dependent surface rheological measurements of foam stabilizers in nonaqueous liquids, *Journal of Colloid Interface Science*, 32, 212–224, 1970.

8. Scheludko, A. and Manev, E., Critical thickness of rupture of chlorbenzene and aniline films, *Transactions of the Faraday Society*, 64, 1123–1134, 1968.

9. Friberg, S. E., Foams from non-aqueous systems, *Current Opinion in Colloid and Interface Science*, 15, 359–364, 2010.

10. Blazquez, C., Emond, E., Schneider, S., Dalmazzone, C., and Bergeron, V., Non-aqueous and crude oil foams, *Oil & Gas Science and Technology—Review of IFP Energies nouvelles*, 69(3), 467–479, 2014.

11. Ross, S. and Suzin, Y., Lubricant foaming and aeration study, Air Force Wright Aeronautical Laboratories, AFWAL-TR-82-2090, Wright-Patterson Air Force Base, OH, October 1982.

12. Binks, B. P., Davies, C. A., Fletcher, P. D. I., and Sharp, E. L., Non-aqueous foams in lubricating oil systems, *Colloids and Surfaces A: Physicochemical and Engineering Aspects*, 360, 198–204, 2010.

13. Kulkarni, R. D., Goddard, E. D., and Kanner, B., Mechanism of antifoam action, *Journal of Colloid and Interface Science*, 59(3), 468–476, 1977.

14. Hill, R. M. and Fey, K. C., Silicone polymers for foam control and demulsification, in Hill, R. M. (ed.), *Silicone Surfactants*, Surfactant Science Series, Vol. 86, Marcel Dekker, New York, p. 166, 1999.

15. Princen, H. M., The equilibrium shape of interfaces, drops, and bubbles. rigid and deformable particles at interfaces, in Matijevic, E. (ed.), *Surface and Colloid Science*, Vol. 2, Wiley-Interscience, New York, p. 56, 1969.

16. Garrett, P. R., Preliminary considerations concerning the stability of a liquid heterogeneity in a plane-parallel liquid film, *Journal of Colloid and Interfacial Science*, 76(2), 587–590, 1980.

17. Booser, E. R. (ed.), *CRC Handbook of Lubrication: Theory and Practice of Tribology Volume II: Theory and Design*, CRC Press LLC, Boca Raton, FL, p. 246, 1983.

18. Triton GR-5M, Surfactant (anionic sulfosuccinate) at 1 wt% actives in water at 25°C. Dow Chemical Company. Form No. 119-01907-1107.

19. Shearer, L. T. and Akers, W. W., Foaming in lube oils, *Journal of Physical Chemistry*, 62, 1269–1270, 1958.

20. Beerbower, A. and Barnum, R. E., Studies on the dispersion of silicone defoamant in non-aqueous fluids, *Lubrication Engineering*, 17, 282–285, 1961.

21. Awe, R. W., Silicone antifoams for lubricating oils, *Society of Automotive Engineers, National Fuels and Lubricants Meeting*, Tulsa, OK, October 30–31, 1963.

22. Rybczynski, W., Uber die fortschreitende Bewegung einer flüssigen Kugel in einem zähen Medium, *Bulletin International de Academie des Sciences de Cracovie, A*, 40–46, 1911.

23. Hadamard, J. S., Mouvement permanent lent d'une sphere liquide et visqueuse dans un liquide visqueux, *Comptes Rendus de l'Académie des Sciences*, 152, 1735–1738, 1911.

24. Levich, V. G., *Physiochemical Hydrodynamics*, 2nd edn., Chapter VIII, Prentice-Hall, Inc., Englewood Cliffs, NJ, p. 435, 1962.

25. Eicke, H. F. and Parfitt, G. D. (eds.), *Interfacial Phenomena in Apolar Media*, Surfactant Science Series, Vol. 21, Marcel Dekker, New York, 1987.

26. Mannheimer, R. J., Factors that influence the coalescence of bubbles in oils that contain silicone antifoamants, *Chemical Engineering Communication*, 113, 183–196, 1992.

27. Rudnick, L. R. (ed.), *Lubricant Additives: Chemistry and Applications*, 2nd edn., CRC Press, Boca Raton, FL, 2009.

28. Kichkin, G. I., Foam formation in lubricating oils, *Chemistry and Technology of Fuels and Oils*, 2(4), 272–275, 1966.

29. Galgoci, E. C., Unpublished results.

30. Gaham, A. and Jones, M. A., Foam Control, U.S. Patent 4,690,688, Dow Corning, Ltd., 1987.

31. Spiegler, R., Kugel, K., Lersch, P., and Silber, S., Use of Organofunctionally Modified Polysiloxanes for Defoaming Diesel, U.S. Patent 5,613,988, Th. Goldschmidt AG., 1997.

32. Fields, J. E., Hydrocarbon Olis of Reduced Foaming Properties, U.S. Patent 3,166,508, Monsanto Co., 1965.

33. Technical Datasheet for FOAM BAN® 155, Revision: 06/14, Münzing NA LP, Bloomfield, NJ, 2014.

34. Zhao, M. and Fang, J., Polymer Compositions, U.S. Patent 8,227,086 B2, Cytec Surface Specialties, S.A., 2012.

35. Galgoci, E. C. and Brüning, W., Antifoams for aqueous and non-aqueous industrial fluids and lubricants, *Proceedings of the 19th International Colloquium Tribology*, Stuttgart/Ostfildern, Germany, January 21–23, 2014.

36. Galgoci, E. C. and Brüning, W., Antifoams for aqueous and non-aqueous industrial fluids and lubricants, *Tribologie + Schmierungstechnik*, 61, 47–54, 2014.

37. Technical Brochure MUE_2014_009_US-03/14, Industrial fluids: Defoamers, Münzing NA LP, Bloomfield, NJ, p. 7, 2014.

38. ASTM D892-13, Standard test method for foaming characteristics of lubricating oils, ASTM International, West Conshohocken, PA, 2013.

39. ASTM D3427-15, Standard test method for air release properties of hydrocarbon based oils, ASTM International, West Conshohocken, PA, 2015.

40. ASTM D6082-12, Standard test method for air release properties of hydrocarbon based oils, ASTM International, West Conshohocken, PA, 2012.

41. Specification TO-4, Section 1, p. 3, Caterpillar Inc., Peoria, IL, 2005.

42. ISO 12152:2012, Lubricants, industrial oils and related products—Determination of the foaming and air release properties of industrial gear oils using a spur gear test rig—Flender foam test procedure, International Organization for Standardization, Geneva, Switzerland, 2012.

43. ISO 4406:1999, Hydraulic fluid power—Fluids—Method for coding the level of contamination by solid particles, International Organization for Standardization, Geneva, Switzerland, 1999.

44. Lubricant Ingredient (H1): Lubricants with incidental food contact, NSF International, Ann Arbor, IL, 2016.

21 Corrosion Inhibitors and Rust Preventatives

Michael T. Costello

CONTENTS

21.1 INTRODUCTION

Since the observation by Plato that "things in the sea … are corroded by brine" in the fourth century BC, the prevention of corrosion of metal parts has been a concern to manufacturers and is the cause of serious economic loss estimated in billions of dollars per annum. The exact mechanism of corrosion remained unclear for centuries after Plato. Then, in 1675, Robert Boyle wrote the first scientific description of the mechanism of corrosion: "The Mechanical Origine or Production of Corrosiveness and Corrosibility." Subsequently, another century passed before Luigi Galvani, in 1780, described the mechanism of bimetallic or Galvanic corrosion. After this advancement, the study of corrosion progressed more rapidly, and W.H. Wollaston introduced the "electrochemical theory of acid corrosion" around 1800 to explain the corrosion resistance of platinum boilers used to make concentrated sulfuric acid [1]. Finally, in 1903, W.R. Whitney [2] published a complete chemical description of corrosion in which he stated that oxygen and carbonic acid were necessary for the formation of rust when iron was dissolved into solution and that the addition of base inhibited corrosion. Whitney's paper firmly placed corrosion as a chemical and electrochemical phenomenon.

In the 1920s, methods to reduce corrosion involved the coating of a metal part with various natural products such as petrolatum, wax, oxidized wax, fatty acid soaps, and rosin [3–5]. By the 1930s, petroleum sulfonates derived as by-products from oil-refining processes were commonly used and during World War II (WWII) the military demand expanded so rapidly that synthetic sulfonates were produced to supplement the supply [6]. After the war, it was found that barium and overbased barium sulfonates were very effective as rust preventatives (RPs) in slushing oils, and this became the *de facto* standard [7,8]. In the 1940s and 1950s, it was discovered that corrosion could be prevented in gas pipelines by injection of $NaNO_2$, whereas in aqueous systems by using chromates and polyphosphates [9,10]. Subsequently, after WWII, the chemistry and various corrosion inhibitors that are in current use have grown exponentially and expanded to a multitude of applications.

Corrosion inhibitors can be divided into several categories depending on their duration of use, chemical composition, and end-use application. The common chemistries used for corrosion inhibitors include nitrites, chromates, hydrazines, carboxylates, silicates, oxidates, sulfonates, amines, amine carboxylates, borates, amine borates, phosphates, amine phosphates, imidazoles, imidazolines, thiazoles, triazoles, and benzotriazoles, and depending on the application different product families are desired (Table 21.1). Primarily the performance property that determines the type of chemistry desired for a corrosion inhibitor is the intended duration of use of the RP coating. In particular, some coatings require long service periods due to outdoor exposure in extreme climates where reapplication is difficult, whereas other corrosion inhibitors are only intended for short-term storage before a part is manufactured or painted. There are essentially three broad classes of corrosion inhibitors, which can be divided into the following areas:

Oil coatings. These are temporary liquid coatings used to prevent corrosion during transport of the metal part or for temporary indoor (months) or outdoor (weeks) storage. The RP oil (or slushing oil) can be applied during the metal-forming operation or after the part is completed.

Soft coatings. These are temporary soft solid coatings typically made of wax, petrolatum, or grease. These coatings are used to coat structures exposed to the elements such as bridges, cars, or trucks and last from a few months up to many years.

Hard coatings. These coatings are typically applied as an alkyd resin or as an inorganic coating (or galvanized coating) and are used to form a hard permanent barrier from corrosion. These products are typically used as a pure barrier to prevent corrosion and do not possess the RP additives used in lubricants.

Although the hard coatings are generally applied using aqueous solvents, the oil and soft coatings are applied to the metal surface by various methods. Many oil and soft coatings are typically applied using oil or solvent (naphtha)-based systems, but they can also be emulsified or solublized in an aqueous system where the volatile solvent is allowed to dry and cast a film of the RP oil. The coatings can be classified into three different categories based on the method of application to the metal surface as follows:

Water soluble. The additives are mainly inorganic materials (nitrite or arsenate) used for aqueous systems such as water treatment or drilling muds.

Soluble oils. The additives used are mainly oil-soluble sulfonates and organic amines that can be used in emulsions used for metal deformation and metal removal processes.

Oil soluble. The additives used are mainly oil-soluble additives that can be used in lubricant oils for machinery or slushing oils.

In this chapter, the study of corrosion inhibitors will be restricted to the use of additives that are used to prevent corrosion in lubricating oils or additives used to prevent corrosion in aqueous systems. In particular, although water treatment, refinery, drilling, and vapor-phase corrosion inhibitors (VCIs) are not strictly lubricating oil applications, many of the additives used in these aqueous systems have been adapted to oil-based systems and have been used as additives in lubricants.

21.2 INHIBITOR TYPES

Corrosion is a chemical or electrochemical reaction between a material, usually a metal, and its environment, which produces a deterioration of the material and its properties. For our purposes, it is instructive to envision corrosion as an electrochemical process in which the metal is oxidized at the anode, and a reduction takes place at the cathode (Figure 21.1). Depending on the pH of the system, the cathodic reaction can produce either H_2 in acidic media or OH^- in neutral/alkaline media.

TABLE 21.1

Common Rust-Preventative Chemistries and Applications

	Water Soluble	Soluble Oil	Oil Soluble
Oil coatings			Carboxylates
			Sulfonates
	Carboxylates		Alkyl amines
	Sulfonates	Carboxylates	Alkyl amine
	Alkyl amines	Sulfonates	Phosphates
	Alkyl amine	Alkyl amines	Alkyl amines
	Phosphates	Alkyl amines	Borates
	Alkyl amines	Borates	Imidazolines
	Borates		Thiazole
			Benzotriazoles
			Triazole
Soft coatings		Oxidates	Oxidates
	—	Sulfonates	Sulfonates
Hard coatings	Nitrites	Alkyl amines	Alkyl amines
	Chromates		
	Hydrazines		
	Phosphates	—	—

Acidic media

Anode:	$Fe \rightarrow Fe^{2+} + 2e^-$		$E_{ox} = -(-0.44)\,V$
Cathode:	$2H^+ + 2e^- \rightarrow H_2$		$E_{red} = +0.00\,V$
	$Fe + 2H^+ \rightarrow Fe^{2+} + H_2$		$E_{cell} = +0.44\,V$

Neutral or alkaline media

Anode:	$Fe \rightarrow Fe^{2+} + 2e^-$		$E_{ox} = -(-0.44)\,V$
Cathode:	$O_2 + 2H_2O + 4e^- \rightarrow 4OH^-$		$E_{red} = +0.40\,V$
	$2Fe + O_2 + 2H_2O \rightarrow 2Fe^{2+} + 4OH^-$		$E_{cell} = +0.48\,V$

FIGURE 21.1 Corrosion reactions in acidic and neutral/alkaline media.

The oxidized metal ions (Fe^{2+} or Fe^{3+} in the case of iron) formed at the anode then diffuse through the system to react with either an inhibitor or the products of the reduction process at the cathode. The end result is the formation of corrosion products (scale or rust) of poorly defined stoichiometry (Figure 21.2). In this standard model, inhibitors are used to prevent either the oxidation of metal or the reduction of oxygen (or H^+).

The mechanism of corrosion inhibition has been extensively investigated, and three basic mechanisms are prevalent. The inhibitor disrupts the anodic process, cathodic process, or a combination of both processes (Figure 21.3). The mechanism of inhibition of the process is significantly affected by the pH of the system and ultimately determines the type of protective film that is formed.

Anodic inhibitors (passivating inhibitors or dangerous inhibitors). Anodic inhibitors produce a large positive shift in the corrosion potential of a metal through the production of a protective oxide or hydroxide film (Figure 21.3). These can be quite dangerous since they can be corrosive at low concentration and need to be carefully monitored. There are two main modes of action that serve to disrupt the anodic processes.

Oxidizing anions. Oxidizing anions passivate a metal in the absence of oxygen. Typical examples include chromates and nitrites, which they function by shifting the potential into a region where insoluble oxides or hydroxides are formed. For example, the commonly used chromates will be reduced from Cr^{6+} to Cr^{3+}, which in turn oxidizes the Fe^{2+} on the surface to Fe^{3+}. The Fe^{3+}, which is less soluble in aqueous solutions than the Fe^{2+}, then forms a protective oxide coating and passivates the metal surface [10,11].

Nonoxidizing anions. Nonoxidizing anions contain species that need oxygen to passivate a metal. Typical chemistries include silicate, carbonate, phosphate,

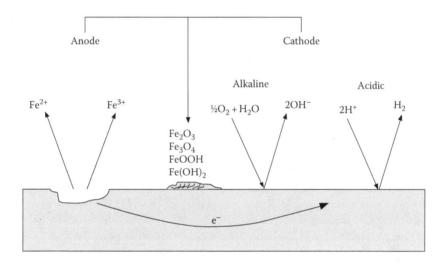

FIGURE 21.2 Formation of corrosion products in acidic or alkaline/neutral media.

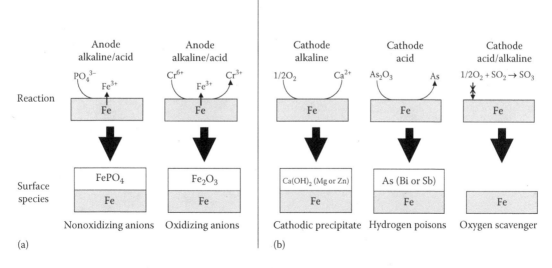

FIGURE 21.3 Schematic mechanism of (a) anodic and (b) cathodic corrosion inhibitors.

tungstate, and molybdate. The mode of action appears to promote the formation of a passivating oxide film on the anodic sites of the metal surface [11].

Cathodic inhibitors. Cathodic inhibitors act to retard the reduction of O_2 or H^+ or selectively precipitate onto cathodic areas (Figure 21.3). Although the cathodic inhibitors are not as effective at low concentration as their anodic counterparts, the cathodic inhibitors are not corrosive at low concentrations. There are three main modes of action that serve to disrupt the cathodic reaction.

Hydrogen poisons. Hydrogen poisons such as As (as As_2O_3 or Na_3AsO_4), Bi, or Sb primarily act in acidic media (Figure 21.3) to retard the hydrogen reduction reaction by reducing at the cathode and precipitating a layer of the poisoning metal. Unfortunately, they also promote hydrogen absorption in steel and can cause hydrogen embrittlement if not carefully monitored.

Cathodic precipitates. Cathodic precipitates are used in neutral or alkaline solutions (Figure 21.3) and act to form insoluble hydroxides (such as Ca, Mg, or Zn) that will reduce the corrosion rate where the metal is exposed. Typically when the hydroxyl ion (OH^-) concentration increases in the cathodic areas, cathodic precipitates such as the calcium or magnesium carbonates will react to absorb the excess hydroxide and precipitate $Ca(OH)_2$ or $Mg(OH)_2$ on the surface of the cathode which in turn inhibits the reduction of oxygen.

Oxygen scavengers. Oxygen scavengers reduced corrosion by capturing excess oxygen in the system. Typical aqueous oxygen scavengers used for water treatment include hydrazine, SO_2, $NaNO_2$, and Na_2SO_3, but there are also many organic antioxidants based on alkylated diphenylamine or alkylated phenols that are used in lubricating oils to scavenge oxygen, which could also be considered in this category.

Mixed (or organic) inhibitors. Mixed inhibitors are organic materials (not inorganic ions such as anodic and cathodic inhibitors) that absorb on a metal surface and prevent both anodic and cathodic reactions. These materials are the typical corrosion inhibitors used in lubricating oils and are more difficult to remove by chemical or mechanical action than the monolayer films formed by the anodic and cathodic inhibitors.

21.3 MECHANISM

The basic description of the mechanism for corrosion inhibition by additives was first proposed by Baker et al. as the adsorption of a monolayer of the inhibitor on the metal surface to form a protective barrier. This barrier is impervious to water and prevents contact with the outside environment. They found that the amount of rust prevention depended on a complex interaction of several variables including the lifetime of the adsorbed film, presence of oxygen, thermal and mechanical desorption, solubility (in water), and surface wettability [7,8]. Subsequently, Kennedy [12] added that there is a complex equilibrium between the water on the metal surface and the micellar or solublized water; he found that the concentration of water in the system effects the corrosion inhibition of a sulfonate. Later, Anand et al. [13] found that lowering the interfacial surface tension also improved the corrosion resistance.

Subsequently, it was found that the monolayer of adsorbed additive was oriented with the polar head adsorbed on the surface and the nonpolar tail closely packed and vertically aligned [14]. The matched chain lengths of the inhibitor and the basestock (e.g., C_{16}-fatty acid matched with C_{16}-alkane) prevent rust due to tight surface packing, which makes the inhibitor more difficult to remove through chemical action (Figure 21.4) [15]. Further investigation revealed that the absorption was accomplished by either an electrostatic (physisorption) or an electron transfer to a coordinate type of bonding (chemisorption) [13,15]. The film was found not only to restrict the access to the surface of aggressive species but also to prevent the passage of metal ions into solution, which restricts the cathodic process to prevent the evolution of H_2 or reduction of O_2 (Figure 21.4) [14].

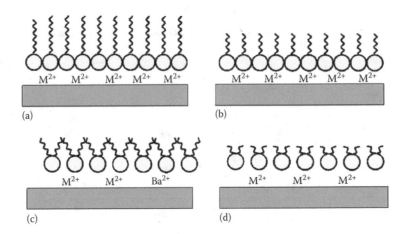

FIGURE 21.4 Schematic diagram of adsorbed films: (a) high MW linear, (b) low MW linear, (c) high MW branched, and (d) low MW branched.

Early studies on rust inhibitors focused on the widely used dinonylnaphthalene sulfonates. It was found that in water-saturated solutions the cation does not effect the adsorption of the sulfonate, whereas in anhydrous solutions the cation possesses a significant effect, where it is proposed that the cation coordinates directly with the oxide film (not solublized in the aqueous phase) [16]. In the general trends for sulfonate, it was observed that the corrosion inhibition of the cations increased in the order: Na < Mg < Ca < Ba [13,14,16].

Overall, it can be stated that a corrosion inhibitor is a substance that forms a protective barrier and creates a mechanical separation between the metal and environment. This barrier cannot be easily removed and prevents transport of aggressive agents from contact with the surface, as well as corrosion products from leaving the surface.

21.4 CHEMISTRY

21.4.1 NITRITES

Wachter and Smith first described the use of sodium nitrite ($NaNO_2$) as an anodic inhibitor to prevent internal corrosion by water and air in steel pipelines for petroleum products [9,17]. It was first speculated that the mechanism involved the formation of a tight oxide layer of oxygen and nitrite to prevent corrosion by forming a passivating layer [18], but subsequently it was proposed that the nitrite accelerated the production of Fe^{3+} on the metal surface and formed a less-soluble protective barrier than the Fe^{2+} species [19].

Historically, nitrites were primarily used in aqueous systems such as water treatment or concrete emulsions as an anodic inhibitors [19], but subsequently they have also been used in soluble oils for metalworking applications. Since the 1950s, the reaction of amines commonly used in the soluble oils with nitrite (Figure 21.5) was found to form the carcinogenic nitrosamines (R_2NNO) under acidic conditions [20,21]. As a result, the use of nitrites as RPs in metalworking fluids containing ethanol amine carboxylate salts was banned by the U.S. Environmental Protection Agency (EPA) in 1984, and the industry quickly modified their formulations to replace the nitrite with other inhibitors [22]. Typical replacements used were the borate, carboxylate, or phosphate salts of triethanolamine (TEA) [23,24]. Subsequently, it was found that of the three common amine ethoxylates (monoethanolamine [MEA], diethanolamine [DEA], and TEA), DEA was the most prone to the reaction with nitrite, whereas there was no evidence of nitrosamine formation in the TEA-containing soluble oils [25]. Despite low nitrosamine formation with TEA, the use of nitrite has been widely discontinued in metalworking applications.

21.4.2 CHROMATES

Chromates are another inhibitor that is part of the class of anodic passivators, which appear to inhibit corrosion through the formation of an oxide coating [18,26]. Historically, chromates have been used in steel plating and finishing, aqueous corrosion inhibition, and leather tanning, and although there have been attempts to make oil-soluble chromate derivatives [27], they have not been widely used in lubricants due to their instability in the presence of organics. In water treatment, these products were applied as chromic acid, sodium chromate (Na_2CrO_4), or sodium dichromate ($Na_2Cr_2O_7$), all of which contain hexavalent chromium. Subsequently, the hexavalent chromium (Cr^{6+}) was found to be a powerful respiratory carcinogen and possessed high aquatic toxicity. As a result, the use of chromates in aqueous water treatment was banned in 1990, and interest in using them in lubricants has disappeared [28].

21.4.3 HYDRAZINES

Hydrazine (N_2H_4) has been historically used for water treatment as a corrosion inhibitor and oxygen scavenger. The hydrazine prevents corrosion in a boiler by (1) reacting with O to form N_2 and H_2O; (2) reacting with Fe_2O_3 (hematite) to form the harder (passive) Fe_3O_4 (magnetite), which makes a protective skin over the iron surface; and (3) forming NH_3 at high temperatures and pressures to maintain a high alkalinity (Figure 21.6) [29]. Although it has been demonstrated that derivatives of hydrazine can be used to inhibit copper corrosion in sulfuric acid [30], due to health concerns there relatively few new oil-soluble derivatives have been developed, and the main focus has been to use alternative organic treatment chemicals [31].

21.4.4 SILICATES

These materials are mainly used as inhibitors in water treatment for potable water due to their low cost. Owing to their poor oil solubility, silicates have limited use as inhibitors in lubricant formulations.

21.4.5 OXIDATES

Oxidates are one of the oldest classes of RP additives and were historically made from oxidized oils, waxes, and petrolatums obtained from the refining process [3–5]. The production of oxidates can be accomplished by either heating the petroleum product air in a closed vessel in the presence of a catalyst or chemical treatment using nitric acid, sulfuric acid, or $KMnO_4$. Both these methods typically make a combination of polar compounds (including carboxylic acids, esters, alcohols, aldehydes, and ketones), where the total acid number (TAN)

$$2HNO_2 \rightarrow N_2O_3 + H_2O$$

$$R_2NH + N_2O_3 \rightarrow R_2N-N=O + HNO_2$$

FIGURE 21.5 Nitrosamine formation.

$$N_2H_4 \rightarrow N_2 + 4H^+ + 4e^-$$

$$4H^+ + O_2 + 4e^- \rightarrow 2H_2O$$

FIGURE 21.6 Hydrazine decomposition.

of the resulting mixture is between 10 and 200 mgKOH/g and the saponification number (SAP#) is between 10 and 200 mgKOH/g. Typically, commercial grade products possess a TAN of 50 and 100 mgKOH/g and a SAP# of 10 and 50 mgKOH/g. Depending on the application, oxidates can be applied to a surface in various methods as follows:

1. *Aqueous dispersions.* The oxidate can be dissolved or suspended in a water-based formulation that is then applied to the surface. Either the water is removed with washing or the film is allowed to dry to make an RP coating.
2. *Solvent based.* The oxidate is dissolved in a low-boiling solvent (like naphtha), and the solvent is allowed to quickly evaporate to form the RP coating.
3. *Solid film (typically wax).* Either the oxidate is heated to apply the coating as a liquid and then allowed to cool to solidify or the solid oxidate is applied through mechanical methods as a solid at ambient temperatures.

Although these materials can be used as effective RP additives in their acid form, they are typically neutralized to form more effective coatings. It has been found that oxidized wax neutralized with $Ca(OH)_2$ or $Ba(OH)_2$ are effective RPs [32], and that the use of $Ca(OH)_2$ as a neutralizing agent is effective at preventing gelation [33]. Basic materials that contain $CaCO_3$ have also been found to be effective when combined with petroleum oxidate. The combination of petroleum oxidates and overbased calcium sulfonates has demonstrated both improved lubricity and corrosion protection in forming and engine oils [34,35]. Additionally, the petroleum oxidates can be neutralized with amines to form the carboxylate amine salts. Although there are various amines listed in the patent literature (piperadines [36] and polyamines [37]), the most common amines used are the simple alkanolamines [3–5].

21.4.6 SULFONATES

The sulfonates comprise a class of compounds that can be derived from petroleum (natural) or synthetic feedstocks. The sulfonic acids are formed in the reaction of SO_3 with a feedstock and can be neutralized with various bases to form the Na, Ca, Mg, or Ba salts, all of which have demonstrated activity as RPs in various applications. Additionally, the neutral salts can be *overbased* by the addition of excess base and carbon dioxide. For example, in the case of a calcium petroleum sulfonate, an excess of $Ca(OH)_2$ and CO_2 can be added to the sulfonic acid to form a colloidal suspension of $CaCO_3$ in oil, where calcium sulfonate serves to disperse the $CaCO_3$ in the oil carrier (Figure 21.7). The overbased sulfonates serve a dual role in rust prevention, where the sulfonate acts to form a protective layer, and the calcium carbonate acts to absorb any acidic by-products of corrosion. As a result, a combination of neutral and overbased sulfonates can be a quite effective RP.

The two types of sulfonates that are commonly manufactured are as follows:

1. *Petroleum (natural) sulfonates.* The petroleum sulfonates were originally made from the by-products of the acid treating of petroleum oil but have also been intentionally made as coproducts from the acid-treating process to manufacture technical and medicinal white oil [6,38]. In this process, the oil-soluble petroleum sulfonic acids formed are typically isolated from the oil layer as the sodium salts by treatment with sodium carbonate followed by extraction with alcohol, where the typical activity of the extracted product is 60%, which significantly reduces the viscosity and facilitates easier storage and handling.
2. *Synthetic sulfonates.* The synthetic sulfonates are made from the sulfonation of long-chained alkyl aromatics. Depending on the aromatic structure used, the alkyl chain length varies, but it is important for

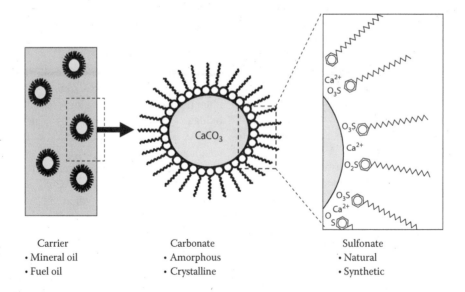

Carrier
• Mineral oil
• Fuel oil

Carbonate
• Amorphous
• Crystalline

Sulfonate
• Natural
• Synthetic

FIGURE 21.7 Overbased sulfonate structure.

the overall alkylation to be sufficient to render the compound oil-soluble. The alkylbenzene sulfonate derivatives are typically monoalkylated with a long-chained (C_{16}–C_{40}) moiety, whereas alkylnaphthalene derivatives are typically dialkylated with short-chain (C_9–C_{10}) lengths, where barium dinonyl naphthalene sulfonate (BaDNNS) is a common structure [6,39,40].

A natural or synthetic sulfonate can be neutralized and over-based with various cations depending on the application. For example,

Sodium sulfonate. In general, the high equivalent weight (500–550 EW) petroleum or (390–700 EW) synthetic alkylbenzene sulfonic acids are neutralized with NaOH to form sodium sulfonates, which are commonly used in soluble oils for metalworking applications, where the divalent cations (Mg, Ca, and Ba) are detrimental to the stability of the soluble oil. The sodium salts of the synthetic dialkyl naphthalene sulfonates have been used, but their high cost has limited their use in this application.

Calcium sulfonate. Both the synthetic and natural sulfonic acids have been neutralized with CaO or Ca(OH)₂ to form the neutral calcium sulfonates, but generally these products are more effective when they have been overbased. The overbased calcium sulfonates can contain either amorphous or crystalline form of calcium carbonate. The amorphous calcium carbonates are easily soluble in oil and are commonly used as a detergent in engine oils [41,42], whereas the crystalline form of calcium carbonate, called calcite, typically contains colloidal particles that are too large to be held in suspension in the fluid and precipitate to form a gel or gelled solid and are commonly used in RP coatings and greases [43–46].

Magnesium sulfonate. Both the natural and synthetic sulfonic acids can be neutralized with MgO or Mg(OH)₂ to form the magnesium sulfonates. Although magnesium sulfonates have not been generally used in RP oils, the overbased magnesium sulfonates are extensively used as fuel oil additives [47,48]. In particular, the undesirable contaminants in fuel oil such as V and Na can form low-melting corrosive slags on the fire-side of a commercial boiler used for power generation. The molten V_2O_5 can act as an oxygen carrier and can accelerate corrosion. The addition of the overbased magnesium sulfonate to the fuel oil serves to react with low-melting sodium vanadates to form high-melting magnesium vanadates that increase the viscosity, reduce the oxygen uptake, and counteract the destruction of the protective oxide film. Additionally, magnesium sulfonate reacts with sulfur oxides (SO_3/SO_2) forming high-melting friable $MgSO_4$, which can be easily removed by washing, whereas the carbonate neutralizes any free acids to reduce the pH and lower the acid deposition rate (ADR) [49,50].

TABLE 21.2
Ionic Radii of Cations

Type	Solubility Product (Ksp)	Ionic Radius (pm)	ΔH_{soln} (kJ/mol)	$\Delta H_{lattice}$ (kJ/mol)
MgCO₃	6.82×10^{-6}	65	−25.3	3180
CaCO₃	4.96×10^{-9}	99	−12.3	2804
SrCO₃	1.1×10^{-10}	113	−3.4	2720
BaCO₃	2.58×10^{-9}	135	4.2	2615

Barium sulfonate. Both the natural and synthetic sulfonic acids can be neutralized with BaO or Ba(OH)₂ to form the barium sulfonates. The barium sulfonates have been found to be effective when both neutral and overbased barium sulfonates are combined in a formulation and applied to a metal surface, where these products are generally used as RPs in mill and slushing oils [8,13,38–40,51,52]. The BaDNNS was found to be very effective at low concentrations [6,38–40] compared to Ca and Na derivatives, and, in general, the RP effectiveness of the sulfonate increases with ionic radius (Table 21.2) where Ba > Ca > Mg > Na.

The effectiveness of alkaline earth metals can be correlated to their ionic radius, which is inversely proportional to the solubility of the metal carbonate (and metal sulfonate) of the additive. For example, the large ionic radius of Ba results in a small release of energy (enthalpy of solution) due to the small amount of solvation necessary for this large polarizable cation. As a result, the large cations require more energy to solubilize and are more difficult to remove from a metal surface by a humid atmosphere or aqueous washing.

21.4.7 CARBOXYLATES

Carboxylates are some of the oldest known corrosion inhibitors and can be made from animal (lanolin, lard, or tallow), vegetable (tall oil fatty acids [TOFAs]), or mineral (naphthenic or aromatic acids) oils and were commonly used in early slushing oil formulations [3–5]. The carboxylic acids can be neutralized with many exotic cations (such as Bi [53]), but they are commonly reacted with NaOH to form the sodium carboxylate salts.

Owing to their corrosivity in aqueous systems [54], carboxylates are more typically combined with alkanolamines to form the alkanolamine salts *in situ* in soluble oil applications for metalworking. In metalworking, the use of the alkanolamine carboxylates provides good coupling with other lubricity additives, enhanced lubricity, and the formation of soft (noncalcium containing) residues, but the alkanolamine carboxylates do suffer from hard water sensitivity, and many short-chained derivatives can cause odor and excessive foaming. The dicarboxylates, which possess good corrosion protection and low foam, can remedy this problem, but they are not popular due to their high cost and poor coupling with other lubricity additives.

In other applications, carboxylates have been used with varying amounts of success, and it had been found that a small amount of the C_6–C_{18} carboxylic acids prevented corrosion in turbines, whereas acids with chain length $<C_6$ promoted corrosion [55]. Additionally, in stamping applications, it was found that a simple fatty acid ester provides good lubricity as well as possesses inherent RP properties [56].

21.4.8 ALKYL AMINES

The alkyl amines represent the largest and most diverse chemistry of all the RP types and are used in various metalworking and industrial oil applications. The most common alkylated amines that are used include MEA, DEA, and TEA, fatty amines, diamines, phenylene diamine, cyclohexylamine, morpholine, and ethylenediamine, triethylene tetramine (TETA), and tetraethylene pentamine (TEPA). It had initially been recognized that the amine was preferentially adsorbed onto the metal surface and inhibited the reaction of the corrosive species and the metal [57], but now it is believed that the alkyl amines first displace the water on the metal surface to form a bond between the lone pair on the nitrogen and the unoccupied orbitals on the metal through defects present in the oxide coatings. As a result, the amines provide cathodic protection by creating a barrier and inhibiting H_2 formation in acidic environments [58–61]. In general, it has been found that longer-chained amines are more effective than the shorter-chained amines [59] and that the nucleophilicity of the nitrogen strongly correlates with rust inhibition effectiveness where the tertiary is more effective than the secondary or primary amine [60].

Owing to their high volatility and low ash formation, products such as cyclohexyl amine, morpholine, and piperadine can be used as VCIs. In systems that are unsuitable for oil, grease, or hard coatings, they expand in the vapor phase to fill the void space in a metal enclosure to form an extremely thin film over the entire metal surface (including intricate interior parts in a fired engine) [58,62]. Owing to the absence of inorganic salts (such as the phosphates, borates, or carboxylates), they do not have the tendency to leave dry residues on the surface [63]. They are also commonly found in sweet (CO_2) and sour (H_2S) gas applications where they volatilize to fill the entire void space of the pipeline [64].

Unfortunately, it has also been found that the amines tend to cause occupational skin diseases (including irritant and allergic contact dermatitis) in workers exposed to metalworking fluids [65–67]. In particular, the commonly used mono- and diethanol amines were found to elicit a significant amount of positive reactions [68], and as a result, they have been largely phased out of this application in favor of the tertiary amines such as TEA. The tertiary amines possess the added benefit of low ecotoxicity and are used environmentally sensitive applications. For example, they are particularly effective in oil field acidizing operations as corrosion inhibitors where they are directly introduced to the outdoor environment [69].

21.4.9 ALKYL AMINE SALTS

Although the alkylamines are commonly used in the gas phase, the amine salts of the carboxylates, borates, and phosphates are most commonly used in metalworking fluids where they are formed *in situ* by reaction with the corresponding acid. For example, the combination TOFA neutralized with triethanol amine to form the amine carboxylate salt is commonly used as a RP. Each amine salt formed for metalworking and RP applications possesses its own unique chemistry and performance issues, each of which will be discussed.

21.4.10 AMINE CARBOXYLATE SALTS

In metalworking formulations, typically tertiary amines such as TEA are neutralized with organic fatty acids to form the amine carboxylates *in situ* to not only provide corrosion inhibition but also improve lubricity and emulsification. Although there have been many combinations of carboxylic acid and amines proposed, the most effective combinations include the long-chained carboxylic acids from C_{18} to C_{22} [70–72].

21.4.11 AMINE BORATE SALTS

The borates represent the most chemically diverse and least understood of the RP classes. Although the structure of the borates and borate esters can be written empirically as $B(OR)_3$, where R is a H or alkyl group, the actual three-dimensional structure of the borates are complex chains and rings that contain both sp^2 (3-coordinate)- and sp^3 (4-coordinate)-hybridized boron species. For example, the species $Na_2(B_4O_5(OH)_4)$ contains both sp^3 and sp^2-hybridized species, where two of the sp^2-hybridized B have empty p-orbitals that can be used to bond to an amine or any other lone pair of electrons as well as use the terminal oxygen to form borate ester linkages (Figure 21.8).

As a result, common ethoxylated amines, such as the triethanol amine, can bond to the borate using four different modes of connectivity (Figure 21.9), which can be described as follows:

1. The TEA can datively bond to the unoccupied sp^2-hybridized orbital of the boron to make the borate salt that is nominatively of the form $R_3N{:}B(OR)_3$.
2. The TEA can bond through one oxygen on an sp^2-hybridized boron to form the monodentate $-(N–CH_2–CH_2–O–B)-$ bond.
3. The TEA can datively bond to the unoccupied sp^2-hybridized orbital of the boron to make the borate salt and can bond through one oxygen on an sp^2-hybridized boron to form the bidentate $-(N–CH_2–CH_2–O–B)-$ bond.
4. The TEA can bond through two oxygen on an sp^2-hybridized boron to form the bidentate $-(N–(CH_2–CH_2–O)_2–B)-$ bond.

The advantages of the borate salts are their low cost, low toxicity, hard water stability, excellent antiwear, and reserve alkalinity. The large disadvantage is the possible formation of a tacky residue that remains after the part is machined due to the partially dehydrated products such as *meta*-boric acid and its esters depositing on the surface [73].

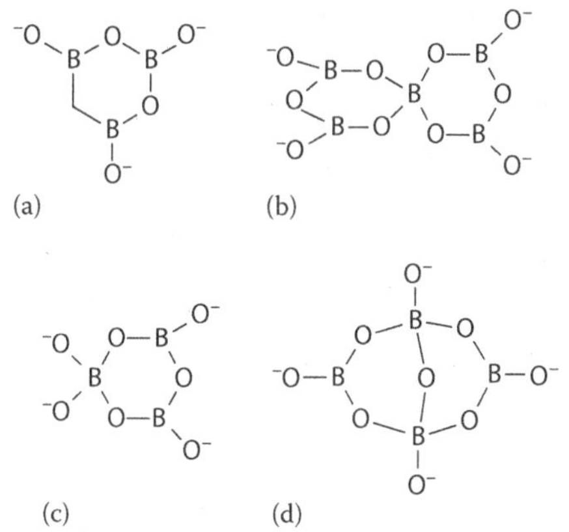

FIGURE 21.8 Proposed structures of borates. (a) Boroxyl ring, (b) pentaborate, (c) triborate, and (d) diborate.

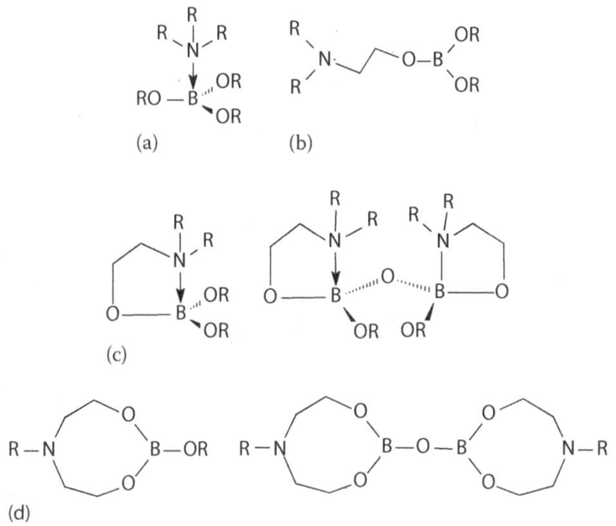

FIGURE 21.9 Proposed structures of borated amines: (a) Nitrogen dative bond, (b) oxygen single bridge, (c) nitrogen dative bond–oxygen single bridge, and (d) oxygen double bridge.

The borates not only serve to inhibit corrosion but have also been found to be bacteriostatic (biostatic) agents, and a large synergistic inhibitory effect has been found when combined with amines. Although the mechanism of the effect is still unclear, it is believed that it may be due to the release of boric acid at low pH that may react to form *cis*-diols with the ribonucleotides to promote the antimicrobial activity [74–76].

Although the borated amines have been cited for use in engine oils [77,78], hydraulic fluids [79], and slushing oils [80,81], they are most commonly used in metalworking fluids [82] for both their rust inhibition and biostatic properties.

21.4.12 PHOSPHATES

The phosphating of metals is a well-known technique to improve both wear and corrosion resistance and is primarily used as a surface preparation before painting. The process of phosphate coating is the treatment of iron, steel, or a steel-based substrate with a solution of phosphoric acid or as K, Na, or Ca phosphate salts, in the presence of heavy metal accelerators (Zn or Mg) whereby the surface of the metal is converted to a mildly protective layer of insoluble crystalline phosphate. Phosphating is considered the heart of pretreatment operations in a steel mill and where the top surface of the metal is converted into a highly insoluble, corrosion-resistant coating [83]. The mechanism of this cathodic inhibitor is believed to be the precipitation of a phosphate film on the surface of the steel that prevents the corrosive action of acids and water. Additionally, phosphates have been extensively used in water treatment where the phosphate combines with calcium to form calcium phosphate precipitates [10].

Although the inorganic phosphates (mono-, ortho-, or polyphosphates) are not oil soluble, the phosphate esters can be synthesized to provide oil-soluble derivatives for lubricants. In particular, the tertiary and secondary phosphate esters of the form $(RO)_3P{=}O$ or $(RO)_2OHP{=}O$, where R is typically an alkylaryl group have been used in slushing oils alone or in combination with other additives (Figure 21.10) [83–89]. The most common alkylaryl derivatives include the tricresyl phosphate (TCP) and trixylyl phosphate (TXP). Although the trialkyl phosphates (such as tributyl or trioctyl) could be used for this application, the instability of these species due to facile thermal, oxidative, and hydrolytic degradation makes them less desirable as RPs [90,91].

The use of phosphates in soluble oils is less extensive, but there are a few examples of trialkyl phosphates being used in drawing and ironing operations [92,93]. Although the phosphates possess low toxicity and excellent EP/AW and RP properties, their use in soluble oils for metalworking additives has been limited due to their stimulation of microbial growth that can lead to contact dermatitis or worse after prolonged worker exposure [10,68,94,95].

Phosphates have also been used in combination with amines, and this provides not only additional EP/AW protection but also both anionic (phosphate) and cationic (ammonium) RPs in a single salt [84].

Another area where phosphates have been extensively used is in aviation turbines in which low ash properties of TCP are desirable [96]. Although the TCP produced in the 1940s and 1950s did possess a considerable amount of tri*ortho*cresyl phosphate (TOCP), which increased concerns of neurotoxicity, the TCP now commercially available can have TOCP levels as low as parts per billion (ppb) and are not a significant concern [97].

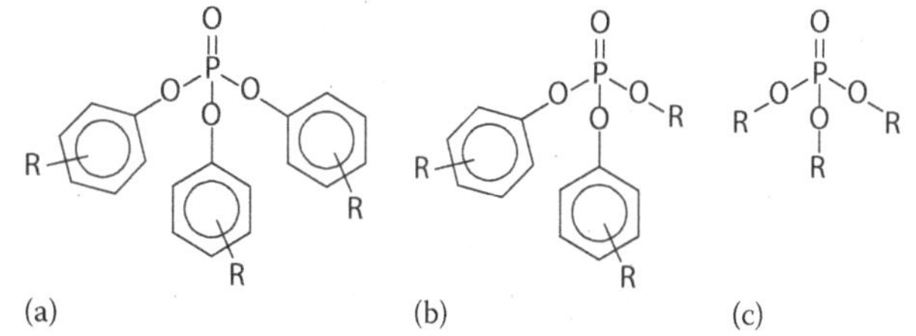

FIGURE 21.10 Phosphate ester structures: (a) Triaryl phosphate, (b) aryl-alkyl phosphate, and (c) trialkyl phosphate.

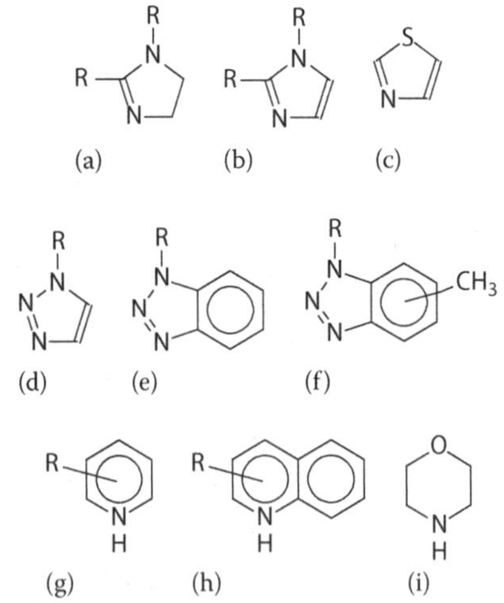

FIGURE 21.11 Common nitrogen ring structures: (a) Imidazoline, (b) imidazole, (c) thiazole, (d) triazole, (e) benzotriazole (f) tolyltriazole, (g) pyradine, (h) quinoline, and (i) morpholine.

21.4.13 NITROGEN RING STRUCTURES

Although tertiary amines have gained wide acceptance in various RP formulations for their ability to neutralize acids, the nitrogen heterocycles have also found use in many common RP applications. The most common type of nitrogen ring structures are imidazolines, benzotriazoles, thiazoles, triazoles, imidazoles, and the mercaptobenzothiazoles (Figure 21.7), as well as any of their common derivatives that include a basic nitrogen in their structure. Owing to the volatile nature, these materials can be used as VCIs and are particularly effective in metal equipment such as engine blocks made from cast iron or aluminum [98]. Additionally, the absence of an inorganic salt minimizes the possibility of dry residue formation.

It has also been found that some imidazoline derivatives have low toxicity in the marine environment and are useful in offshore oil and gas production [63,99]. Although the mechanism is still unclear, it appears as if the highly basic nitrogen in the ring structure is converted to a less basic salt (by reaction with organic acids), which renders the molecule less toxic in the marine environment [63].

Although many amine heterocycles have been proposed, due to their relative low cost and high RP effectiveness, primarily imidazoline and its derivatives have found use in grease, soluble oils, rolling, cutting, drilling, turning, grinding, wire drawing, stamping, and sheathing with varying levels of success. Additionally, the imidazolines can be used alone or combined with a number of organic acids to make the ashless amine salt, which have been demonstrated to be very effective in RP (Figure 21.11).

21.5 CORROSION TESTING

RPs can be used in various end-use lubricant applications, and as a result, there are a wide variety of industry standard rust and corrosion tests available to the formulator. Although industrial organizations such as the American

Society for Testing and Materials (ASTM) list many commonly performed test methods, there are a host of international (Deutsche Norm Test Methods [DIN], International Standards Organization [ISO], Association Française de Normalisation [AFNOR], or Institute of Petroleum [IP]) and Original Equipment Manufacturer [OEM] methods that are commonly used to screen corrosion and RPs. The most common ASTM corrosion testing used for RPs and corrosion inhibitors are listed in Table 21.3 along with cross-references to various industry and international methods.

- *ASTM D1401—Water separability characteristics of petroleum and synthetic fluids.* This test determines the ability of lubricating fluids, such as rust and oxidation-inhibited industrial gear lubricant, hydraulic oils, and turbine oils that have viscosities in the range of 28.8–90 cSt at 40°C to separate from water. Although not a standard corrosion test, it is widely used in the lubricant industry to screen potential RP lubricant formulations where in some applications water shedding is desired (slushing and rolling oils), whereas in other applications, emulsification is required (soluble oils for metalworking). The demulsibility characteristics of lubricating oils (ASTM D2711) measures an industrial gear lubricant's ability to separate from water, similar to the ASTM D1401 method, whereas the demulsibility characteristics of lubricating oils (ASTM D2711) test is primarily used to test medium and heavy-viscosity (ISO 220–ISO 1500) industrial gear lubricants' ability to separate from water.
- *ASTM D130—Copper strip corrosion test.* This test is used to evaluate the corrosive tendencies of oils to copper-containing materials, while ASTM D4048 is used for grease. A result of 1b slight tarnish or 2a moderate tarnish is considered a passing result in most applications. This test is particularly sensitive to active sulfur (ASTM D1662) and can be easily passed if an inactive sulfur additive is used in the RP formulation.
- *ASTM D665—Turbine oil rust test.* This test is designed to measure the ability of industrial oils to prevent rusting under conditions of water contamination. To pass the test, the specimen must be completely free from visible rust when examined without magnification under normal light. This test was developed for steam-turbine oils and is one of the oldest standard corrosion tests and has found utility in various applications (aviation oils, RP oils, gear oils, and hydraulic fluids).
- *ASTM D1748—Standard test method for rust protection by metal preservatives in the humidity cabinet.* This test method is used for evaluating the rust-preventive properties of metal preservatives under conditions of high humidity at 49°C. The polished steel panels coated with RP oil are hanged vertically in a humidity cabinet where the water vapors from the bottom of the cabinet contact the panel. At different time intervals, the panels are removed from the cabinet and observed visually for any rust or stains. The time

TABLE 21.3
Corrosion Test Cross-Reference

Description	Test Method				
	ASTM	DIN	ISO	IP	FTM
Oil					
Detection of copper corrosion from petroleum products by copper strip tarnish test	D130	51 759, 51 811	2160	154	791 5225
Standard method for rust-preventing characteristics of inhibited mineral oil in the presence of water	D665, D3603	51 355, 51 585	7120	135	791 4011, 791 5315
Standard test method for water separability of petroleum oils and synthetic fluids	D1401	—	—	—	—
Standard test method for rust protection by metal preservatives in the humidity cabinet	D1748	50 017[a]	—	366	791 5310
Standard test method for iron chip corrosion for water-dilutable metalworking fluids	D4627	51 360P2[a], 51 360P1[a]	—	287[a], 125[a]	—
Standard test method for corrosiveness of lubricating fluid to bimetallic couple	D6547	—	—	—	791 5322.2
Standard practice for operating salt spray (Fog) apparatus	B117	50 021-SS[a]	9227-SS[a]		791 4001.2
Method of acetic acid-salt spray (Fog) testing	B287	51 021-ESS[a]	9227-AASS[a]	—	—
Standard method for copper-accelerated acetic acid–salt spray (Fog) testing (CASS Test)	B368	50 021-CASS[a]	9227-CASS[a]	—	—
Standard practice for conducting moist SO_2 tests	G87	50 018-SWF[a]	3231	—	—
Grease					
Standard test method for determining the corrosion preventative properties of lubricating grease	D1743	—	—	—	791 4012
Standard test method for detection of copper corrosion from lubricating grease	D4048	—	—	—	791 5309
Standard test method for corrosion-preventive properties of lubricating greases in presence of dilute synthetic sea water	D5969	—	—	—	—
Standard test method to determine the corrosion preventative properties of lubricating greases under dynamic wet conditions (Emcor test)	D6138	51 802	—	220	—
Transport					
Standard test method for the evaluation of corrosiveness of diesel engine oil at 121°C	D5968				791 5308.7
Standard test method for evaluation of rust preventative characteristics of automotive engine oil	D6557				

[a] Similar test conditions to ASTM method.

is recorded when the rust or stains are observed or alternatively the area percent rust can be noted at a specified time interval. Depending on the application 1000, 2000, or 4000 h are considered passing results.

ASTM B117—Standard practice for operating salt spray (fog) apparatus. Although not a specific method, this practice covers the apparatus, procedure, and conditions required to create and maintain the salt spray (fog) test environment at 35°C. This practice prescribes neither the type of test specimen or exposure periods to be used for a specific product nor the interpretation to be given to the results. As a result, there are many OEM, military, and international variations to this method, and specifications must be agreed by the buyer and the seller. The most common form of this test is the FTM-791C 4001.2; it describes the most common time, temperature, and salt concentrations used for this testing. It is considered

as an industry standard test procedure and is used to compare different RP formulations on a relative basis. A typical passing rating can range from 100 to 2000 h, but an absolute number is generally not specified as the test formulation is most commonly related to a standard reference fluid for performance evaluation.

A list of the standard test conditions for the most common RP tests is provided in Table 21.4, where the order of severity increases as follows:

ASTM D665A < ASTM 665B < ASTM D1748 < FTM 4001.2 < CASS

ASTM D4627—Test method for iron chip corrosion for water-dilutable metalworking fluids. This method was designed to determine the relative corrosion rate of a

TABLE 21.4

Typical Corrosion Test Conditions

Method	ASTM D665A&B	ASTM D1748	FTM 4001.2	CASS
Type	Rust test	Humidity cabinet	Salt spray	Acetic acid–salt spray
Time (h)	4	100	24	5
Temperature (°C)	60	49	35	35
Conditions	Distilled water or synthetic sea water	Distilled water	5% NaCl	5% NaCl + 0.2% acetic acid
Test panel	Steel rod	Sandblasted cold rolled steel	Cold finished steel	Cold rolled steel
Grade	AISI/SAE 1010 (BS 970)	AISI/SAE 1009C	AISI/SAE 1010	AISI/SAE 4130
Dimensions	5″ × 2 × 11/16″	4″ × 2″ × 1/3″	3″ × 2″ × 1/8″	6″ × 4″ × 1/8″

water-dilutable metalworking fluid by using typical cast iron chips found during machining as the test specimen. The procedure uses ~2 g of clean cast iron chips that are spread onto a filter paper in a petri dish, the fluid mixture is then pipetted on to the chips, and the dish is covered. After a certain period of time (typically 24 h), the chips are removed, and the paper is examined for staining. The test is typically evaluated as a pass or fail criterion using successive dilutions of a water-dilutable metalworking fluid. There are many slight variations of this basic procedure (IP-287 or DIN 51 360P1 and IP-125 or DIN 51 360P2), and typically, the method is modified to accommodate prevailing conditions of the end user. Results are evaluated by determining the lowest concentration of soluble oil necessary to prevent rust appearing on the filter paper or on the iron chips where typically a concentration of 2%–5% is considered acceptable. This procedure remains an excellent low-cost screening tool to evaluate relative effectiveness of an RP in water-dilutable metalworking fluids.

ASTM D6547—Standard test method for corrosiveness of lubricating fluid to bimetallic couple. This test method covers the corrosiveness of hydraulic and lubricating fluids to a bimetallic galvanic couple and replicates Fed-Std No. 791, Method 5322.2. It uses the same apparatus, test conditions, and evaluation criteria, but it describes test procedures more explicitly.

ASTM D1743—Standard test method for determining corrosion preventive properties of lubricating greases. This test method covers the determination of the corrosionpreventive properties of greases using grease-lubricated tapered roller bearings stored under wet conditions. This method distributes a lubricating grease sample in a roller bearing by running a bearing under a light thrust load. This test method is based on CRC Technique L 41[2] that shows correlations between laboratory results and service for grease-lubricated aircraft wheel bearings.

ASTM D5969—Standard test method for corrosion-preventive properties of lubricating greases in presence of dilute synthetic sea water environments. This test method covers the determination of the corrosion-preventive properties of greases using grease-lubricated tapered roller bearings exposed to various concentrations of dilute synthetic sea water stored under wet conditions. It is based on test method D 1743, which is practiced using a similar procedure and distilled water. The reported result is a pass or fail rating as determined by at least two of the three bearings.

ASTM D6138—Standard test method for determination of corrosion-preventive properties of lubricating greases under dynamic wet conditions (Emcor test). This test measures the ability of a grease to protect a bearing against corrosion in the presence of water. This test method covers the determination of corrosion-preventive properties of greases using grease-lubricated ball bearings under dynamic wet conditions. It is a dynamic test where two sets of grease-coated bearings are immersed in water and a series of running and resting periods are in rotation. At the end of the test, the raceways of the bearing outer rings are inspected for rust and evaluated. This method is equivalent to IP-220, DIN 51 802, and the Emcor/SKF water wash test.

ASTM B368—Standard test method for copper-accelerated acetic acid salt spray (fog) testing (CASS test). Although the ASTM B368 method is typically used for hard coatings (galvanized or painted surfaces), extended outdoor exposure oil and soft coatings can be tested as an alternative to the milder conditions of the salt spray (fog) method. In ASTM B368, the corrosiveness of the standard salt spray test is increased by lowering the pH or increasing the operating temperature. Several domestic automotive OEMs have their own versions of this test (Ford BG105-01 and GM 4476P), which address their own specific materials requirements.

ISO 6988—Metallic and other nonorganic coatings, sulfur dioxide test with general condensation of moisture. The Kesternich Cabinet (with SO_2 added) is used for metallic and organic hard coatings for long-term outdoor exposure. The Kesternich test simulates the detrimental effects of acid rain and the test calls for dissolving sulfur dioxide in distilled water. The chamber is heated for 8 h at 100% relative humidity. After 8 h, the chamber vents the excess sulfuric dioxide and returns to room temperature. The cycle is repeated every day, and the results are reported as pass/fail after specified duration after a visual inspection for rust (Table 21.5).

TABLE 21.5
Commercial Rust Preventatives/Corrosion Inhibitors

Manufacturer	Chemistry	Trade Name
Air Products	Alkylamine	
Akzo Nobel	Fatty imidazoline	Armohib
Akzo Nobel	Amine sulfonate	Armohib, Armeen, Duomeen
Akzo Nobel	Alkylamine	Triameen
Akzo Nobel	Ammonium thiocyanate	
Akzo Nobel	Quaternary ammonium compounds	Arquad
Akzo Nobel	Ethoxylated alkyl amine	Ethomeen, Ethoduomeen
Arch Chemical	Hydrazine	Scav-Ox
Arizona Chemical	Dimerized fatty acid	Century, Unidyme
BASF (formerly PPG/Mazer)	Alkylamine	Mazon
Chemtura	Calcium sulfonate	Calcinate
Chemtura	Barium sulfonate	Surchem, Petronate
Chemtura	Oxidized petrolatum	Oxpet
Chemtura	Sodium sulfonate	Petronate
Ciba	Succinic acid ester	Irgacor
Ciba	Polycarboxylic acid	Irgacor
Ciba	Sodium sebecate	Irgacor
Ciba	Nonyl phenoxy acetic acid	Irgacor
Ciba	Amine phosphate	Irgalube
Ciba	N-acyl sarcosine	Sarkosyl
Ciba	Imidazoline derivatives	Amine
Cognis	Dimer acids	
Cognis	Fatty acid ethoxylate	Eumulgin
Cognis	Alkylamine	Texamin
Dover (formerly Keil)	Carboxylic acid salt	Synkad
Dover (formerly Keil)	Boramide	Synkad
Dover (formerly Mayco)	Carboxylic acid salt	Mayco
Dover (formerly Mayco)	Barium sulfonate	Mayco
Dow (Angus Chemical)	Amino alcohol	AMP AEPD, Corrguard
Dow (Angus Chemical)	Oxazoline	Alkaterge
Dupont	Dilauryl acid phosphate	Ortholeum
Dupont	Dibasic acids	Corfree
Grace Construction	Calcium nitrite	DCI Corrosion Inhibitor
Georgia Pacific	Ethanolamine borate	Actracor
Georgia Pacific	Ethanolamine carboxylate	Actracor
Georgia Pacific	Alkanolamide	Actramide
Halox	Alkyl ammonium salt	Halox
Halox	Alkanol amine/boric acid/phosphoric acid	Flash-X
Huntsman	Amine borates	Diglycol amine
Huntsman	Ethanol amines	MEA, DEA, TEA
Huntsman	Polyether amines	Jeffamines
Huntsman	Castor oil ethoxylates	
Huntsman	Aminomethyl propanol	AMP

(Continued)

TABLE 21.5 (*Continued*)
Commercial Rust Preventatives/Corrosion Inhibitors

Manufacturer	Chemistry	Trade Name
King Industries	Dinonylnaphthalene sulfonate	Na-Sul
King Industries	Alkyl amines	K-Corr
King Industries	Acid amine salts	K-Corr
Lonza	Quaternary ammonium carbonate	Carboshield
Lonza	Quarternary ammonium chloride	Uniquat
Lonza	Imidazoline derivatives	Unamine
Lubrizol	Calium sulfonate	Alox, Lubrizol
Lubrizol	Barium sulfonate	Alox, Lubrizol
Lubrizol (Alox)	Oxidized petrolatum	Alox
Oxy-Wax	Oxidized wax	Hypax
Pilot Chemical	Sodium sulfonate	Aristonate
Pilot Chemical	Ethanolamine sulfonate	Aristonate
Pilot Chemical	Calcium sulfonate	Aristonate
PMC Specialties	Benzotriazole	Cobratec
PMC Specialties	Tolyltriazole	Cobratec
PMC Specialties	Carboxybenzotriazole	Cobratec
PMC Specialties	Sodium tolyltriazole	Cobratec
PQ Corporation	Sodium silicate	PQ
R.T. Vanderbilt	Succinic acid ester	Vanlube
R.T. Vanderbilt	Barium sulfonate	Vanlube
R.T. Vanderbilt	Imidazole	Vanlube
R.T. Vanderbilt	Amine phosphate	Vanlube
Rohm & Haas	Alkylamines	Primene
Uniqema	Proprietary	Perfad
Uniqema	Fatty acid derivatives	Prifac, Priolene, Pripol
Uniqema (formerly Mona)	Borate esters	Monacor
Uniqema (formerly Mona)	Amine carboxylate	Monacor
Uniqema (formerly Mona)	Amine phosphate	Monacor
Uniqema (formerly Mona)	Phosphate ester	Monalube
Uniqema (formerly Mona)	Imidazone	Monazoline

21.6 FUTURE REQUIREMENTS

The demand for RPs in lubricants will continue for many years, in part due to higher demand in the developing countries such as Brazil, Russia, China, and India (BRIC) and the aging of existing OEM equipment that is currently used in the developed world. The current market drivers for new RPs will continue to be environmental and health concerns.

New environmental requirements for RPs address various shortcomings of the current commercial chemistries. For amine-based products, there is a desire for lower marine toxicity (ecotoxicity), which is driving the industry toward alternative eco-friendly products, whereas for the classic barium-based products, there is a drive to reduce the heavy metal content and replace them with less-hazardous metals (i.e., Na, Ca, or Mg). Additionally, the waste disposal of the final end product has been a great concern in both lubricant and metalworking applications and has driven the market to pursue more biodegradable products that could be based on new chemistries or natural fats and esters.

Although the majority of the severe health effects such as the carcinogenicity of the hexavalent chromium products have, for the most part, been addressed, there are still concerns for sporadic outbreaks of acute and chronic symptoms. In particular, the public health concerns for the reduction of contact dermatitis for machine shop workers has lead to increase demand for *amine-free* lubricant and metalworking fluids, using alternative technologies such as sulfonate or low-toxicity amines. Additionally, although contact dermatitis is generally viewed as an acute symptom, the occurrence of hypersensitivity pneumonitis (HP or machine operators lung) can lead to a chronic condition that may require hospitalization in its most extreme cases. As a result, regular worker health checkups and air quality monitoring are standard practices now in many metalworking shops.

Generally, workers employed in industrial manufacturing are safer now due to tougher regulations from industrial and governmental bodies, which has led to the search for alternative chemistries. As a result of both environmental and health concerns, the RP suppliers will continue to develop alternative technologies that incorporate a cradle-to-grave approach to product stewardship.

REFERENCES

1. McDonald, D., The beginnings of chemical engineering—The design of platinum boilers for sulphuric acid concentration, *Platin. Met. Rev.*, 1(2), 51–54, 1957.
2. Whitney, W.R., The corrosion of iron, *J. Chem. Soc.*, 25, 394–406, 1903.
3. Styri, H., Rust prevention by slushing, *Trans. Am. Electrochem. Soc.*, 40, 81–98, 1921.
4. Walker, P.H., L.L. Steele, Slushing oils, Technical Paper, U.S. Bureau of Standards, 176, 1–23, October 14, 1920.
5. Bishkin, S.L., Slushing type rust preventatives, *Nat. Pet. News*, 35, R225–R233, 1943.
6. Baker, H.R., C.R. Singleterry, E.M. Solomon, Neutral and basic sulfonates—Corrosion-inhibiting and acid-deactivating properties, *J. Ind. Eng. Chem.*, 46, 1035–1042, 1954.
7. Baker, H.R., W.A. Zisman, Polar-type rust inhibitors: Theory and properties, *J. Ind. Eng. Chem.*, 40, 2338–2347, 1948.
8. Baker, H.R., D.T. Jones, W.A. Zisman, Polar-type rust inhibitors. Methods of testing the rust-inhibition properties of polar compounds in oils, *J. Ind. Eng. Chem.*, 41, 137–140, 1949.
9. Wachter, A., S.S. Smith, Preventing internal corrosion—Sodium nitrite treatment for gasoline lines, *Ind. Eng. Chem.*, 35, 358–367, 1943.
10. Kalman, E., Routes to the development of low toxicity corrosion inhibitors for use in neutral solutions, *European Federation of Corrosion*, Publication No. 11, pp. 12–38, 1994.
11. Nathan, C.C., *Kirk-Othmer Encyclopedia of Chemical Technology*, 2nd edn., vol. 6, pp. 317–346, 1965.
12. Kennedy, P.J., The mechanism of corrosion inhibition by dinonylnaphthalenesulfonates, *ASLE Trans.*, 23(4), 370–374, 1980.
13. Anand, O.N., V.P. Malik, K.D. Neemla, Sulfonates as rust inhibitors, *Anti-Corros. Methods Mater.*, 33(7), 12–16, 1986.
14. Mertwoy, A., H. Gisser, J. Messina, Rust inhibition in petroleum and synthetic lubricant, *Corrosion*, 74, Paper No. 69, 1974.
15. Reis, H.E., J. Gabor, Chain length compatibility in rust prevention, *Chem. Ind.*, 1561, 1967.
16. Bascom, W.D., C.R. Singleterry, The adsorption of oil-soluble sulfonates at the metal/oil interface, *J. Phys. Chem.*, 65, 1683–1689, 1961.
17. Wachter, A., Internal corrosion prevention in conduits, U.S. Patent 2,297,666, 1942.
18. Cohen, M., Inhibition of steel corrosion by sodium nitrite in water, *J. Electrochem. Soc.*, 93(1), 26–39, 1948.
19. Hoar, T.P., Nitrite inhibition of corrosion: Some practical cases, *Corrosion*, 14, 103t–104t, 1958.
20. Zingman, P.A., T.Y. Fan, W. Miles, N.P. Sen, Nitrosamines and the potential cancer threat in metalworking lubricants, *Metalwork. Interface*, 3(1), 9–16, 1978.
21. Archer, M.C., J.S. Wishnok, Nitrosamine formation in corrosion-inhibiting compositions containing nitrite salts of secondary amines, *J. Environ. Sci. Health A Environ. Sci. Eng.*, A11(10–11), 583–590, 1976.
22. Bennett, E.O., D.L. Bennett, Metalworking fluids and nitrosamines, *Tribol. Int.*, 17(6), 341–346, 1984.
23. Vukasovich, M.S., Rust protection of synthetic metalworking fluids with nitrite alternatives, *Lubr. Eng.*, 40(8), 456–462, 1984.
24. Fette, C.J., Sodium replacement—The state of the art, *Lubr. Eng.*, 35(11), 625–627, 1979.
25. Lucke, W.E., J.M. Ernst, Formation and precursors of nitrosamines in metalworking fluids, *Lubr. Eng.*, 49(4), 217–275, 1993.
26. Hackerman, N., E.S. Snavelyin, *Corrosion Basics—An Introduction*, L.S.V. Delinder, ed., NACE, Houston, TX, 1984.
27. Cerveny, L., Anticorrosive lubricants, GB Patent 1,003,789, 1962.
28. Lake, D.L., Approaching environmental acceptability in cooling water corrosion, *Corros. Prev. Control*, 35(4), 113–115, 1988.
29. Collis, D.E., Hydrazine for boiler protection, *Kent Tech. Rev.*, 12, 19–21, 1974.
30. Siddagangappa, S., S.M. Mayanna, F. Pushpanadan, 2,4-Dinitrophenylhydrazine as corrosion inhibitor for copper in sulphuric acid, *Anti-Corros. Method. Mater.*, 23(8), 11–13, 1976.
31. Banweg, A., Organic treatment chemicals in steam generating systems—Using the right tool in the right application, *PowerPlant Chem.*, 8(3), 137–140, 2006.
32. Nagarkoti, B.S., A.K. Jain, M.F. Sait, Rust prevention characteristics of barium and calcium oxides in rust preventing oils, *Additives in Petroleum Refinery and Petroleum Product Formulation Practice, Proceedings*, Sopron, Hungary, pp. 125–127, 1997.
33. Rigdon, O.W., A. Macaluso, Petroleum oxidate and calcium derivatives thereof, U.S. Patent 4,089,689, 1978.
34. Eckard, A.E., I. Riff, Overbased sulfonates combined with petroleum oxidates for metal forming, U.S. Patent 5,439,602, 1995.
35. Slama, F.J., Process for overbased petroleum oxidate, U.S. Patent 5,013,463, 1991.
36. Hayner, R.E., Metal-sufonate/piperadine derivative combination protective coatings, U.S. Patent 4,650,692, 1987.

37. Carlos, D.D., B.K. Friley, Hydrocarbon oxidate composition, U.S. Patent 4,491,535, 1985.
38. Anand, O.N., V.P. Malik, P.K. Goel, Oil-soluble metallic petroleum sulfonates, *Petrol. Hydrocarbons*, 7(2), 59–67, 1972.
39. Baker, R.F., P.F. Vaast, Sulfonates as corrosion inhibitors in greases, *NLGI Spokesman*, 54(11), 465–469, 1991.
40. Baker, R.F., Barium sulfonates in lubricant applications, with a regulatory overview, *NLGI Spokesman*, 57(11), 463–466, 1994.
41. Hone, D.C., B.H. Robinson, J.R. Galsworthy, R.W. Glyde, Colloidal chemistry of lubricating oils, *Reactions and Synthesis in Surfactant Systems*, Surfactant Science Series, vol. 100, pp. 385–394, 2001.
42. Liston, T.V., Engine lubricant additives what they are and how they function, *Lubr. Eng.*, 48, 389–397, 1992.
43. Han, N., L. Shui, W. Lui, Q. Xue, Y. Sun, Study of the lubrication mechanism of overbased Ca sulfonates on additives containing S or P, *Tribol. Lett.*, 14(4), 269–274, 2003.
44. Eckard, A.E., H. Ridderikhoff, High performance calcium sulfonates for metalworking, *12th International Tribology Colloquium*, Esslingen, Germany, 2000.
45. Marsh, J.F., Colloidal lubricant additives, *Chem. Ind.*, 20, 470–473, 1987.
46. Riga, A.T., H. Hong, R.E. Kornbrekke, J.M. Calhoon, J.N. Vinci, Reactions of overbased sulfonates and sulfurized compounds with ferric oxide, *Lubr. Eng.*, 49(1), 65–71, 1992.
47. Muir, R.J., T.I. Eliades, Overbased magnesium deposit control additive for residual fuel oils, PCT International Application WO 99/51707, 1999.
48. Diehl, R.C., J.L. Walker, Composition and method of conditioning fuel oils, European Patent Application EP 0 0007 853, 1980.
49. Tiwari, S.N., S. Prakassh, Magnesium oxide as a corrosion inhibitor of hot oil ash corrosion, *Mater. Sci. Technol.*, 14(5), 467–472, 1998.
50. Pantony, D.A., K.I. Vasu, Corrosion of metals under melts I. Theoretical survey of fire-side corrosion of boilers and gas turbines in the presence of vanadium pentoxide, *J. Inorg. Nucl. Chem.*, 30(2), 423–432, 1968.
51. Sabol, A.R., Method of preparing overbased barium sulfonates, U.S. Patent 3,959,164, 1976.
52. Tebbe, J.M., L.A. Villahermosa, R.B. Mowery, Corrosion preventing characteristics of military hydraulic fluids, *2005 SAE World Congress*, Detroit, MI, SAE Technical Paper # 2005-01-1812, 2005.
53. Graichen, S., Fatty acid bismuth salts as bactericides and anticorrosion inhibitor suitable for lubricating or cooling fluids in metalworking, European Patent Application 1035192, 2000.
54. Watanabe, S., T. Fujita, New additives derived from fatty acids for water-based cutting fluids, *J. Am. Oil Chem. Soc.*, 62(1), 125–127, 1985.
55. Larson, R.G., G.L. Perry, Investigation of friction and wear under quasi-hydrodynamic conditions, *Trans. ASME*, 65(1), 45–49, 1945.
56. Carcel, A.C., D. Palomares, E. Rodilla, M.A. Perez-Puig, Evaluation of vegetable oils as pre-lube oils for stamping, *Second International Conference, Tribology in Environmental Design 2003: The Characteristics of Interacting Surfaces—A Key Factor in Sustainable and Economic Products*, Bournemouth, U.K., pp. 205–215, September 8–10, 2003.
57. Annand, R.R., R.M. Hurd, N. Hackerman, Adsorption of monomeric and polymeric amino corrosion inhibitors on steel, *J. Electrochem. Soc.*, 112, 138–144, 1965.
58. Subramanian, A., M. Natesan, V.S. Muralidharan, K. Balakrishnan, T. Vasudevan, An overview: Vapor phase corrosion inhibitors, *Corrosion*, 56(2), 144–155, 2000.
59. Ch'iao, S.J., C.A. Mann, Nitrogen-containing organic inhibitors of corrosion, *Ind. Eng. Chem.*, 39(7), 910–919, 1947.
60. Desai, M.N., Corrosion inhibitors for aluminium alloys, *Werstoffe Korrosion*, 23(6), 475–482, 1972.
61. Hausler, R.H., L.A. Goeller, R.P. Zimmerman, R.H. Rosenwald, Contribution to the "filming amine" theory: An interpretation of experimental results, *Corrosion*, 28(1), 7–16, 1972.
62. Murray, W.J., Vapor phase corrosion inhibition, U.S. Patent 2,914,424, 1959.
63. Naraghi, A., H. Montgomery, N. Obeyesekere, Corrosion inhibitors with low environmental toxicity, European Patent Application EP 1 043 423 A2, 2000.
64. Wu, Y., K.E. McSperritt, G.D. Harris, Corrosion inhibition and monitoring in sea gas pipeline systems, *Mater. Perf.*, 27(12), 29–34, 1988.
65. de Boer, E.M., W.G. van Ketel, D.P. Bruynzeel, Dermatoses in metal workers. (I) Irritant contact dermatitis, *Contact Dermatitis*, 20, 212–218, 1989.
66. de Boer, E.M., W.G. van Ketel, D.P. Bruynzeel, Dermatoses in metal workers. (I). Allergic contact dermatitis, *Contact Dermatitis*, 20, 280–286, 1989.
67. Grattan, C.E., J.S.C. English, I.S. Foulds, R.J.G. Rycroft, Cutting fluid dermatitis, *Contact Dermatitis*, 20, 372–376, 1989.
68. Geier, J., H. Lessman, H. Dickel, P.J. Frosch, P. Koch, D. Becker, U. Jappe, W. Aberer, A. Schnuch, W. Uter, Patch test results with the metalworking fluid series of the German Contact Dermatitis Research Group (DKG), *Contact Dermatitis*, 51, 118–130, 2004.
69. Williams, D.A., D.S. Sullivan, B.I. Bourland, J.A. Haslegrave, P.J. Clewlow, N. Carruthers, T.M. O'Brien, Amine derivatives as corrosion inhibitors, European Patent Application EP 0 593 294 A1, 1992.
70. Quitmeyer, J.A., Amine carboxylates: Additives in metalworking fluids, *Lubr. Eng.*, 52(11), 835–839, 1996.
71. Kyle, G.H. Maleated tall-oil fatty acid reacted with alkanolamines for a corrosion inhibitor suitable for lubricating or cooling fluids in metalworking, PCT International Application WO 00/52230, 2000.
72. Vuorinen, E., W. Skinner, Amine carboxylates as vapour phase corrosion inhibitors, *Br. Corros. J.*, 37(2), 159–160, 2002.
73. Yao, J., J. Dong, Improvement of hydrolytic stability of borate esters used as lubricant additives, *Lubr. Eng.*, 51(6), 475–479, 1995.
74. Sherburn, R.E., P.J. Large, Amine borate catabolism by bacteria isolated from contaminated metal-working fluids, *J. Appl. Microbiol.*, 87, 668–675, 1999.
75. Loomis, W.D., R.W. Durst, Chemistry and biology of boron, *Biofactors*, 3, 229–239, 1992.
76. Buers, K.L.M., E.L. Prince, C.J. Knowles, The ability of selected bacterial isolates to utilize components of synthetic metal-working fluids as sole sources of carbon and nitrogen growth, *Biotech. Lett.*, 19(8), 791–794, 1997.
77. Hellmuth, W.W., E.F. Miller, Boron amide lubricating oil additive, U.S. Patent 4,025,445, 1977.
78. Small, V.R., T.V. Liston, A. Onopchenko, Diethylamine complexes of borated alkyl catechols and lubricating oil compositions containing the same, U.S. Patent 4,975,211, 1990.
79. Sato, T., K. Kawakatsu, Hydraulic fluid comprising a borate ester and corrosion inhibiting amounts of an oxyalkylated alicyclic amine, U.S. Patent 4,173,542, 1979.

80. Herd, R.S., A.G. Horodysky, Products of reaction of organic diamines, boron compounds and acyl sarcosines and lubricant containing same, U.S. Patent 4,474,671, 1984.

81. Hunt, M.W., Amine boranes, U.S. Patent 3,136,820, 1964.

82. Jahnke, R.W., W.C. Woerner, Corrosion-inhibiting compositions, and oil compositions containing said corrosion-inhibiting compositions, PCT International Patent Application WO 86/03513, 1985.

83. Placek, D.G., G. Shankwalker, Phosphate ester surface treatment for reduced wear and corrosion protection, *Wear*, 173(1–2), 207–217, 1994.

84. Palmer, R.C., Lubricant, U.S. Patent 2,418,422, 1947.

85. Paciorek, K.J.L., S.R. Masuda, Rust inhibitor phosphate ester formulations, U.S. Patent 5,779,774, 1998.

86. Anzenberger, J.F., Rust-preventative compositions, U.S. Patent 4,263,062, 1981.

87. Bretz, J., Metal-containing organic phosphate compositions, U.S. Patent 3,411,923, 1968.

88. Cantrell, T.L., J.G. Peters, Rust inhibited lubricating oil compositions, U.S. Patent 2,772,032, 1956.

89. Pragnell, J.W.A., A.J. Markson, M.A. Edwards, Corrosion inhibiting lubricant composition, GB Patent 2,272,000 A, 1994.

90. Shankwalker, S.G., C. Cruz, Thermal degradation and weight loss characteristics of commercial phosphate esters, *Ind. Eng. Chem. Res.*, 33, 740–743, 1994.

91. Cho, L., E.E. Klaus, Oxidative degradation of phosphate esters, *ASLE Trans.*, 24(1), 119–124, 1981.

92. Malito, J.T., R.D. Wintermute, S.F. Ross, J.M. Ferrara, Polycarboxylic acid ester drawing and ironing lubricant emulsions and concentrates, U.S. Patent 4,767,554, 1988.

93. Tury, B., Ammonium organo-phosphorus acid salts, PCT International Patent Application WO 94/03462, 1994.

94. Isaksson, M., M. Frick, B. Gruvberger, A. Ponten, M. Bruze, Occupational allergic dermatitis from the extreme pressure (EP) additive zinc, bis ((O,O′-di-2-ethylhexyl)dithiophosphate) in neat oils, *Contact Dermatitis*, 46(4), 248–249, 2002.

95. The effects of water impurities on water-based metalworking fluids, Technical Report No. J/N 96/47, Milacron Marketing Co.

96. Phillips, W.D., Phosphate ester hydraulic fluids, *Handbook of Hydraulic Fluid Technology*, G.E. Totten, ed., New York: Marcel Dekker, 2000.

97. Mackerer, C., M.L. Barth, A.J. Krueger, A comparison of neurotoxic effects and potential risks from oral administration or ingestion of tricresyl phosphate or jet engine oil containing tricresyl phosphate, *J. Toxicol. Environ. Health, Part A*, 57(5), 293–328, 1999.

98. DeCordt, F.L.M., N. Svensson, J. Mihelic, M. Blackowski, Corrosion inhibiting composition, U.S. Patent Application 2005/0017220 A1, 2005.

99. Braga, T.G., R.L. Martin, J.A. McMahon, B. Oude Alink, B.T. Outlaw, Combinations of imidazolines and wetting agents as environmentally acceptable corrosion inhibitors, PCT International Patent Application WO 00/49204, 2000.

22 Alkylated Naphthalenes

Maureen E. Hunter

CONTENTS

22.1 INTRODUCTION

Alkylated naphthalenes are classified by the American Petroleum Institute as part of the Group V base oil category. However, alkylated naphthalenes are rarely used as the sole base fluid. They are typically incorporated into lubricant formulations as a base stock–type additive replacing a portion of a Group II, Group III, or PAO base oil. This is done to extend the lifetime of high-performance lubricants by improving their thermal and thermo-oxidative stability, enhancing the solubility and response of other additives, imparting seal swell, decreasing volatility, and providing varnish control for system cleanliness.

The technology to manufacture alkylated naphthalenes was first developed during WWII for use as synthetic fluids in engine oils because of the scarcity of available mineral oil, but further development stopped after the war [1]. In the last 20 years, with advanced processing technology and raw material availability, alkylated naphthalenes have emerged as cost-competitive, high-performance base stock–type additives that are used to help meet current lubricant trends in many types of automotive and industrial oils and greases, including

- Automotive and stationary engine oils
- Automotive and industrial gear oils
- High-temperature chain lubricants
- Paper machine oils
- Hydraulic oils
- Circulating oils/turbine oils/rust and oxidation (R&O) oils
- Screw compressor oils
- Heat transfer oils
- Automotive and industrial greases

Lubricant trends common to industrial fluids, automotive engine oils, driveline fluids, and greases include

- Higher operating speeds and temperatures
- Longer life lubricants for extended drain intervals

- Lower viscosity and reduced volatility for improved fuel efficiency
- Reduced lubricant volumes
- Cleanliness of moving parts
- Sealed-for-life systems

These trends lead to lubricant requirements for improved thermo-oxidative stability, improved thermal stability, improved additive response, improved compatibility with seals and housings, and system cleanliness. To meet these demanding trends, lubricant formulators can optimize their formulations by adding alkylated naphthalene.

22.2 CHEMISTRY OF ALKYLATED NAPHTHALENES

The basic structure of an alkylated naphthalene, as shown in Figure 22.1, consists of two fused six-membered rings with the alkyl groups attached. R_1–R_8 are independently a linear or branched alkyl group or hydrogen.

22.2.1 SYNTHESIS

In the lubricant industry, alkylated naphthalenes are prepared by a Friedel–Crafts reaction where naphthalene is reacted with an alkylating agent in the presence of an acid catalyst. This produces a mixture of alkylated naphthalenes having different numbers of alkyl groups attached to the naphthalene ring. The reaction is shown in Figure 22.2.

The alkylating agents may be linear or branched olefins, alcohols, and alkyl halides. Often, *alpha*-olefins are used because the double bond is located at one terminus of the alkyl chain, which enhances the reactivity of the compound. Chain lengths of 6–16 carbons are commonly used because

these produce high-molecular-weight products with low volatility and good viscometric properties.

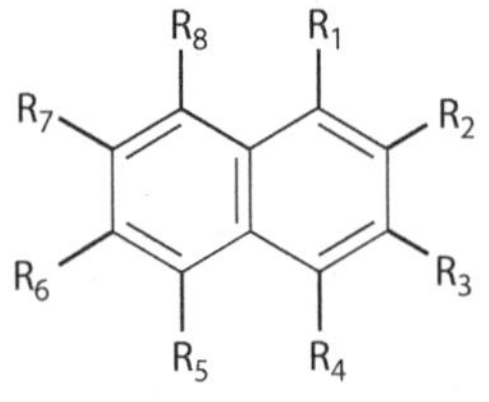

Naphthalene is highly activated for electrophilic alkylation, and once the alkylation reaction starts, the second and third alkyl groups are easily added onto the naphthalene ring. Thus, Friedel–Crafts reactions can produce complex compositions of alkylated naphthalenes with many possible isomers of different molecular weights and structures. This is because both the α and β carbons on the naphthalene ring can be alkylated.

Specific compositions for various applications can be custom designed by tailoring the alkylating agent chemistry, catalyst type, reaction temperature, ratio of the alkylating agent to naphthalene, and the manner in which the reactants are combined. Many different types of catalysts are suitable for the reaction, including Lewis acids, strong protic acids, and heterogeneous catalysts, such as zeolites or acid-treated clays [2].

22.2.2 NAPHTHALENE RING

Naphthalene has a cyclic, conjugated π electron system with *p* orbital overlap both around the 10-carbon periphery of the molecule and across the central bond. It is the ability of this electron-rich naphthalene ring to absorb energy, resonate, and then disperse that energy, much like antioxidants do, that gives alkylated naphthalenes their inherent excellent thermal and thermo-oxidative stability.

22.2.3 ALKYL CHAIN

By carefully controlling the type of alkylating agent used and the degree of alkylation, the chemical and physical properties of alkylated naphthalenes can be significantly influenced. In particular, the alkyl groups control most of the physical characteristics of the compound, including viscosity, viscosity index (VI), pour point, and aniline point.

The physical properties of alkylated naphthalenes primarily depend on three major factors:

- The number of carbons in the alkyl group
- The degree of branching of the alkyl group
- The number of alkyl groups on the naphthalene ring

The ability to choose and control the earlier mentioned variables allows for a great deal of flexibility to custom design

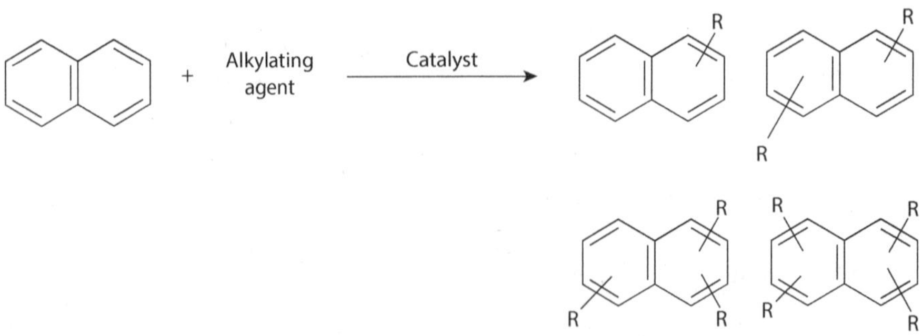

FIGURE 22.1 Alkylated naphthalene structure. R_1–R_8 are independently a linear or branched alkyl group or hydrogen.

FIGURE 22.2 Alkylated naphthalene synthesis.

alkylated naphthalenes, thereby providing the physical properties to meet the needs of a broad range of applications.

22.3 EFFECT OF ALKYL CHAINS ON PHYSICAL PROPERTIES

Table 22.1 describes the physical properties of alkylated naphthalenes made with linear and branched alkylating agents of equivalent carbon number greater than 10 carbon atoms per alkyl group [3]. The pure monoalkylated naphthalenes and polyalkylated naphthalenes were isolated by distillation. The various chemistries are designated as follows:

- MLAN = Monolinear alkylated naphthalene, so it has one linear alkyl group on the naphthalene ring
- MBAN = Monobranched alkylated naphthalene, so it has one branched alkyl group on the naphthalene ring
- PLAN = Polylinear alkylated naphthalene, so it consists of two or more linear alkyl groups on the naphthalene ring
- PBAN = Polybranched alkylated naphthalene, so it consists of two or more branched alkyl groups on the naphthalene ring

General trends can be observed in the physical properties of alkylated naphthalenes based on alkyl chain length, alkyl chain branching, and the number of alkyl groups on the naphthalene ring.

22.3.1 VISCOSITY

Increased viscosity is observed with an increased number of alkyl groups and with chain branching. The monoalkylated naphthalenes have significantly lower viscosities than the polyalkylated naphthalenes. The viscosity of the branched alkylated naphthalenes is also higher than the corresponding linear alkylated naphthalene of the same alkyl carbon

number. Consequently, the monolinear alkylated naphthalene has the lowest viscosity. Alkyl chain length will also increase the viscosity.

22.3.2 VISCOSITY INDEX

Highly branched alkylated naphthalenes typically have no appreciable VI, while long-chain linear alkylated naphthalenes have VI values, which increase with the number of alkyl groups. Linear alkyl chains have more degrees of freedom, which makes it easier for them to coil up at low temperatures and expand at high temperatures compared to branched alkyl chains. Branched alkylated naphthalenes are structurally more rigid, which decreases the ability of the alkyl chains to contract and expand with changes in temperature.

22.3.3 POUR POINT

Pour point increases with increasing number of alkyl groups and branching. Therefore, monolinear alkylated naphthalenes typically have the best (lowest) pour points.

22.3.4 ANILINE POINT, POLARITY, AND ADDITIVE SOLUBILITY

Aniline point, which is a measure of the polarity of a substance and its ability to solubilize polar material, increases with increasing number of alkyl groups but decreases with chain branching. The aniline point is defined as the lowest temperature at which equal volumes of aniline and test fluid exist as a single phase. The lower the number, the better the ability to solubilize polar material. Monobranched alkylated naphthalenes have the lowest aniline points.

22.4 COMMERCIAL PRODUCTS OF ALKYLATED NAPHTHALENES

22.4.1 PHYSICAL PROPERTIES

Currently, there are two suppliers of alkylated naphthalenes in the marketplace. Under the trade name NA-LUBE®, King Industries, Inc. manufactures a series of alkylated naphthalene products with a diverse ISO viscosity range from 22 to 193 cSt at 40°C. Properties of selected King Industries products are shown in Table 22.2. ExxonMobil Chemical Company manufactures two products under the trade name Synesstic™. Properties of these products are shown in Table 22.3 [4].

In general, as the viscosity of alkylated naphthalenes increases:

- Viscosity index increases
- Aniline point increases, which means its polarity decreases
- Volatility decreases significantly
- Pour point increases
- Flash point increases
- Thermo-oxidative stability remains excellent
- Thermal stability remains excellent

TABLE 22.1
Physical Properties of Alkylated Naphthalenes

Designation	Alkyl Group Structure	Number of Alkyl Groups		
MLAN	Linear	1		
MBAN	Branched	1		
PLAN	Linear	2–4		
PBAN	Branched	2–4		
	MLAN	**MBAN**	**PLAN**	**PBAN**
Kinematic viscosity at 40°C (cSt)	18.2	24.2	110.5	1050
Kinematic viscosity at 100°C (cSt)	3.4	3.3	12.8	23
Viscosity index	20	—	110	—
Pour point (°C)	−54	−36	−30	3
Aniline point (°C)	1.8	−9.2	94	42
Flash point (°C)	>200	>200	274	214

Linear and branched alkyl groups are of the same carbon number.

TABLE 22.2

Properties of Alkylated Naphthalene Products from King Industries, Inc.

Property	AN-7A	AN-8	AN-9	AN-15	AN-19	AN-23
Viscosity at 40°C (cSt)	21.8	36	37	114	177	193
Viscosity at 100°C (cSt)	3.8	5.6	5.7	13.5	18.7	19.8
Viscosity index	26	90	90	115	119	118
Aniline point (°C)	40	42	50	94	103	N/A
NOACK volatility (%)	39	12	8	2.2	1.4	<1.0
Evaporation loss, 6.5 h at 205°C	55	11	—	3.8	1.4	0.4
Pour point (°C)	<−48	−33	−36	−39	−26	−21
Flash point (°C)	206	236	>240	260	285	310

The generic product numbers roughly correspond to the viscosity at 100°C.

TABLE 22.3

Properties of Alkylated Naphthalene Products from ExxonMobil Chemical Co.

Property	ASTM Method	AN-5	AN-12
Kinematic viscosity at 40°C (cSt)	D445	29	109
Kinematic viscosity at 100°C (cSt)	D445	4.7	12.4
Viscosity index	D227	74	105
Pour point (°C)	D97 or D5950	−39	−36
Flash point, open cup (°C)	D92	222	258
Flash point, closed cup (°C)	D93	192	240
Kinematic viscosity at −40°C (cSt)	D445	43,600	392,500
Brookfield viscosity at −16°C (cP)	D2983	3,950	22,000
TAN (mg KOH/g)	D664	<0.05	<0.05
NOACK volatility, weight % loss	D5800	12.7	4.5
Specific gravity at 15.6°C/15.6°C	D4052	0.908	0.887

22.4.2 FOOD GRADE

Several of these alkylated naphthalene products are NSF HX-1 registered products that meet the requirements for incidental food contact as prescribed by FDA 21 CFR 178.3570, where they can be used up to 100% if needed to achieve the desired properties.

22.4.3 CHEMICAL PROPERTIES

Alkylated naphthalenes can improve the deficiencies of high-performance lubricants. They can impart several important chemical properties when incorporated into Group II and III base stocks, PAOs, and other synthetic fluids, including

- Exceptional thermo-oxidative stability
- Exceptional thermal stability
- Improved additive solubility
- Improved additive response (less surface competition than seen with esters)
- Improved seal swell, countering the shrinking effects of nonpolar fluids
- Good film thickness, film strength, and lubricity to reduce friction and wear
- Varnish control and system cleanliness

The required treat levels depend on the desired performance.

22.4.3.1 Thermo-Oxidative and Thermal Stability

22.4.3.1.1 Federal Test Method 3411

In Federal Test Method 3411, samples are held at 274°C for 96 h in the presence of a steel coupon in a sealed glass tube. The test evaluates the change in viscosity, increase in acid number, steel coupon weight loss, steel coupon appearance, and appearance of the oil. Table 22.4 shows the performance of a nonpolar 7 cSt Group III oil where 20% of AN-15, a trimethylolpropane (TMP) ester, and a diester were added to incorporate polarity. The Group III oil containing 20% alkylated naphthalene exhibited excellent performance, while the oils containing the esters resulted in thick, dark deposits and, in one case, extensive degradation of the metal.

TABLE 22.4
Thermal Stability (FTM 3411)

	7 cSt Group III	20% AN-15 80% 7 cSt Group III	20% TMP Ester 80% 7 cSt Group III	20% Diester 80% 7 cSt Group III
Percent change in viscosity	0	0	−10.0	−15.8
Change in acid number	0.03	0	6.0	0.52
Change in metal wt (mg/cm^2)	0	0	−3.0	0.05
Metal appearance	Gold shiny	Shiny	Etched	Blue–black shiny
Oil appearance	Clean	Light amber	Black	Very dark amber
Test cell appearance	Clean	Clean	Heavy black stains	Clean

Note: 274°C for 96 h with steel coupon in sealed glass tube.

22.4.3.1.2 Panel Coker: FTM 791-3462

In Panel Coker testing, oil is splashed against a test panel at elevated temperatures and the amount of coke deposited on the panel is determined by weight. As shown in Table 22.5, adding 10% of AN-8, AN-15, and AN-19 to a PAO significantly reduced the amount of coke formed. The neat PAO resulted in 9 mg of coke, while the PAO modified with alkylated naphthalene resulted in much less. This is also evident by the pictures showing the 100% PAO result on the top and the PAO containing 10% AN-15 on the bottom.

22.4.3.1.3 Rotating Pressure Vessel Oxidation Test: ASTM D2272

Rotating Pressure Vessel Oxidation Test (RPVOT) uses 50 g of test fluid and 10 mL of water with a copper coil catalyst. The system is pressurized to 90 psi with oxygen and then held at 150°C. The time required for the pressure to drop to 25 psi is defined as the lifetime of the sample.

Table 22.6 shows the AN-8 being used to boost the RPVOT performance of a Group III oil containing 0.7% of an R&O package. Replacing 15% of the Group III oil with alkylated

TABLE 22.5
Thermal Stability: Panel Coker FTM 791-3462

	100% PAO ISO VG 220	10% AN-8 90% PAO	10% AN-15 90% PAO	10% AN-19 90% PAO
Coking value (mg)	9.0	1.0	3.0	2.0

Temperature conditions: Test panel, 200°C and Oil sample, 140°C

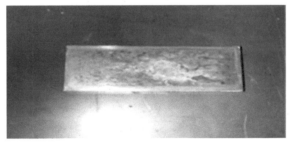

Panel with 100% PAO

Panel with PAO containing 10% AN-15

TABLE 22.6
Group III (ISO VG 46) versus Group III Modified with Alkylated Naphthalene

	0.7% R&O Package 99.3% Group III	0.7% R&O Package 15% AN-8 84.3% Group III
RPVOT (ASTM D2272)		
Lifetime (minutes)	1340	1930
CM thermal stability (ASTM D2070)		
Viscosity change (%)	0.19	0.93
Acid number change (mg KOH/g)	0.00	0.00
Condition of steel rod		
Color	2	2
Deposit (mg)	0.40	1.20
Metal loss (mg)	0.30	0.20
Condition of copper rod		
Color	5	5
Deposit (mg)	0.50	0.60
Metal loss (mg)	0.40	0.10
Total sludge (mg/100 mL)	10.75	5.30

naphthalene increased the oxidation lifetime from 1340 to 1930 min. The excellent thermo-oxidative stability imparted by the alkylated naphthalene is the result of the electron-rich naphthalene ring, which can scavenge radicals, resonate, and disperse the energy.

22.4.3.1.4 Cincinnati Milacron: ASTM D2070
Cincinnati Milacron testing was also conducted with the R&O package. In this test, 200 mL of the test fluid is held at 135°C for 7 days in the presence of copper and steel rods. Measurements include change in viscosity, change in acid number, steel rod rating, copper rod rating, and sludge formation. Table 22.6 shows that replacing 15% of

the Group III oil with alkylated naphthalene significantly reduced the amount of sludge formed in the Cincinnati Milacron test.

22.4.3.1.5 Pressurized Differential Scanning Calorimetry: ASTM D6186
Pressurized differential scanning calorimetry (PDSC) measures the oxidation induction time to an onset of an exotherm under specific conditions. Testing was conducted using 2 mg samples at 500 psi and 160°C. An ISO VG 46 Group II oil was compared to the Group II oil containing 20% of AN-8, AN-15, and AN-19, as shown in Figure 22.3. The addition of the alkylated naphthalene increased the induction time and reduced the rate of oxidation. The degree of improvement depended upon which alkylated naphthalene was used.

Also tested was a 4 cSt PAO modified with 20% AN-8 with and without antioxidant. Figure 22.4 shows that the addition of the alkylated naphthalene exhibited good additive response, increasing the oxidation induction time with 0.2% alkylated diphenylamine and 0.2% alkylated phenyl-alpha-naphthylamine.

22.4.3.2 Hydrolytic Stability
22.4.3.2.1 Beverage Bottle Test: ASTM D2619
In the Hydrolytic Stability Beverage Bottle test, a 75 g sample, 25 g of distilled water, and a copper test strip are sealed in a Coca-Cola bottle. The bottle is rotated for 48 h in an oven at 93°C, and measurements include the total acidity of the water layer.

Figure 22.5 shows the beneficial effect of adding 10% AN-15 to an ISO VG 46 Group II base oil compared to adding a diester and a polyol ester. The esters are hydrolytically unstable and this resulted in high acid values of the water layer. Alkylated naphthalenes do not impart hydrolytic instability to lubricants because they do not contain any functional groups that can hydrolyze.

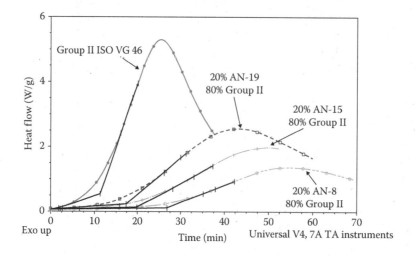

FIGURE 22.3 Thermo-oxidative stability—PDSC ASTM D6186. Isothermal at 160°C, 500 psi oxygen. Group II versus 20% alkylated naphthalene in Group II.

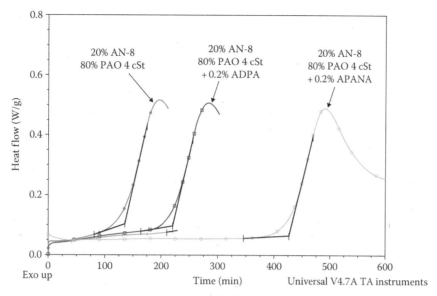

Isothermal @ 160°C, 500 psi oxygen
20% AN-8 in PAO 4 cSt with 0.2% antioxidant

FIGURE 22.4 Thermo-oxidative stability—PDSC ASTM D6186. Isothermal at 160°C, 500 psi oxygen. 20% AN-8 in PAO 4 cSt with 0.2% antioxidant.

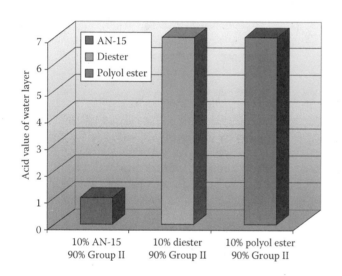

FIGURE 22.5 Hydrolytic stability—Beverage Bottle Test ASTM D2619.

22.4.3.3 Volatility

Thin film testing was conducted using 2 g of fluid in an aluminum pan for 24 h at temperatures of 200°C, 225°C, and 250°C. Table 22.7 shows that when 20% of AN-19 is added to a 40 cSt PAO with higher volatility, the volatility was brought down to the level of the alkylated naphthalene alone. For comparison, an ester at 20% in the PAO resulted in significantly higher volatility.

22.4.3.4 Low-Temperature Performance

Viscosity versus temperature profiles were run as shown in Figure 22.6. The upper line is for an 8 cSt PAO containing

TABLE 22.7
Thin Film Volatility

	Weight Loss %		
	200°C	**225°C**	**250°C**
AN-19	8.5	19.7	41.6
PAO 40	17.9	29.8	45.4
20% AN-19/80% PAO 40	9.4	20.2	39.6
20% Ester/80% PAO 40	28.5	43.1	56.7

2 g in aluminum pan for 24 h.

20% VI improver. When 20% AN-8 is added to the PAO formulation, the viscosity temperature profile is improved, and further improvement is realized with the addition of 0.3% of a pour point depressant.

22.4.3.5 Seal Swell Performance

22.4.3.5.1 Rubber Property Test: ASTM D471

Using a nitrile rubber seal material, a 20% addition of AN-15 to a 7 cSt Group III oil was compared to the Group III alone using ASTM D471. Figure 22.7 shows that the alkylated naphthalene was able to impart seal swell properties to the nonpolar Group III base oil.

22.4.4 Chemical Properties: Greases

Three NLGI #2 greases made using lithium 12-hydroxystearate, polyurea, and aluminum complex thickeners with 10 cSt PAO or white oil containing various amounts of alkylated naphthalene were evaluated.

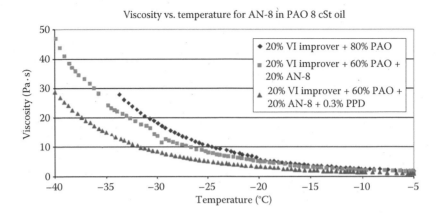

FIGURE 22.6 Low-temperature viscosity profile.

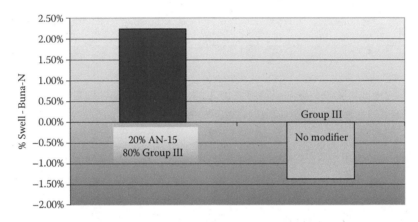

FIGURE 22.7 Seal swell—ASTM D471. Alkylated naphthalene can impart seal swell properties to nonpolar base oils.

22.4.4.1 Lithium 12-Hydroxystearate

PDSC (ASTM D5483) testing was conducted using greases made with a lithium 12-hydroxystearate thickener and PAO containing 10% and 50% AN-15. In this test, 2 mg of grease was placed in an aluminum test pan and the temperature was ramped at 100°C/min to the test temperature. Then the sample was allowed to equilibrate at the test temperature for 2 min. The oxygen valve was opened, and the system was pressurized to 500 psi within 2 min. When equilibrated, the oxygen was adjusted to a flow rate of 100 mL/min. The oxidation induction time is measured from the time when the oxygen valve is opened. Figure 22.8 shows the

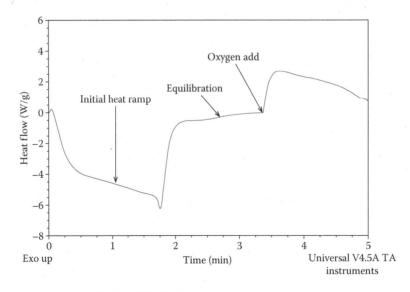

FIGURE 22.8 Thermo-oxidative stability—PDSC ASTM D5483 explanation.

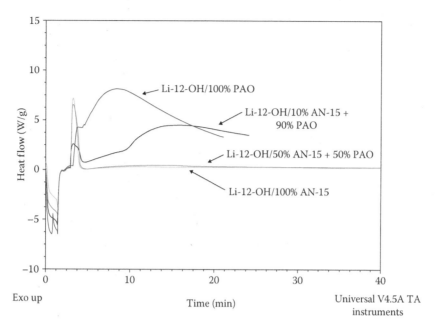

FIGURE 22.9 Thermo-oxidative stability—PDSC ASTM D5483 at 180°C.

temperature ramp to the test temperature, the 2 min equilibration, and the opening of the oxygen valve, which always causes a spike.

Figure 22.9 shows the PDSC test results at 180°C. The temperature ramp, the 2 min equilibration, the oxygen valve opening, and the spike can clearly be seen followed by whatever happens afterward. The grease made with 100% PAO oxidized immediately and quickly. Adding 10% AN-15 to the PAO imparted oxidation resistance to the grease. And adding 50% alkylated naphthalene to the PAO imparted significant oxidation resistance making the grease completely stable and equivalent to the grease made with 100% alkylated naphthalene.

Norma Hoffman (ASTM D942) oxidation testing was also conducted using the lithium 12-hydroxystearate greases. In this test, five glass dishes are filled with 4 g of test grease and placed in a pressure vessel. The vessel is sealed and pressurized to 110 psi with oxygen and then placed in a bath held at 99°C. The pressure in the vessel is recorded at various times throughout the test.

As shown in Figure 22.10, the Norma Hoffman oxidation testing showed the same results as the PDSC testing. The grease made with 100% PAO was the least stable and oxidized immediately and quickly. Adding 10% AN-15 to the PAO imparted oxidation resistance to the grease. And adding 50% alkylated naphthalene to the PAO imparted significant oxidation resistance making the grease very stable.

22.4.4.2 Polyurea

A polyurea grease made with 50% AN-15 and 50% PAO also had improved properties over the grease made with 100% PAO, as shown in Table 22.8. The grease made with the alkylated naphthalene showed superior oil separation properties and superior PDSC performance.

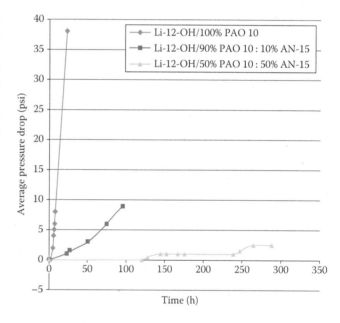

FIGURE 22.10 Thermo-oxidative stability—Norma Hoffman ASTM D942.

TABLE 22.8

AN-15/PAO versus PAO 10 in Polyurea Grease

	AN-15/PAO (1:1) Polyurea	PAO Polyurea
Oil separation		
18 h	2%	5.6%
168 h	6.2%	12.6%
Dropping point	>200°C	>200°C
PDSC (500 psi O$_2$, 200°C)	3 W/g at 60 min	22 W/g at 20 min

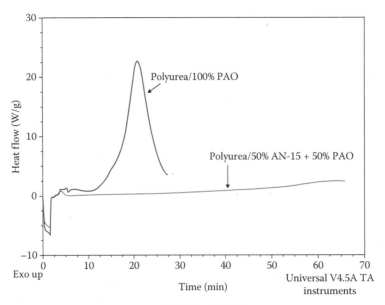

FIGURE 22.11 Thermo-oxidative stability—PDSC ASTM D5483 at 200°C.

TABLE 22.9

AN-15 and AN-19 versus White Oil in Aluminum Complex Grease

	White Oil Al Complex	White Oil/AN-15 (1:1) Al Complex	AN-15 Al Complex
PDSC—500 psi O$_2$, 170°C	8 W/g at 13 min	No exotherm	No exotherm
PDSC—500 psi O$_2$, 180°C	14 W/g at 10 min	1 W/g at 30 min	No exotherm
	White Oil Al Complex	**White Oil/AN-19 (1:1) Al Complex**	**AN-19 Al Complex**
PDSC—500 psi O$_2$, 170°C	8 W/g at 13 min	1 W/g at 40 min	No exotherm
PDSC—500 psi O$_2$, 180°C	14 W/g at 10 min	2 W/g at 25 min	No exotherm

Figure 22.11 shows the PDSC test results at 200°C. The grease made with 100% PAO showed significant oxidation degradation, while modifying the PAO with 50% AN-15 imparted oxidation resistance to the grease.

22.4.4.3 Aluminum Complex

PDSC testing was also conducted at 170°C and 180°C comparing aluminum complex greases made with white oil to greases modified with AN-15 and AN-19, as shown in Table 22.9. The PDSC curves are shown in Figures 22.12 and 22.13.

At both temperatures, the grease made with 100% white oil oxidized immediately and quickly. Modifying the white oil with 50% of either AN-15 or AN-19 imparted significant oxidation resistance to the aluminum complex grease. At 170°C, the grease containing AN-15 was completely stable, while the grease containing AN-19 liberated just a very small amount

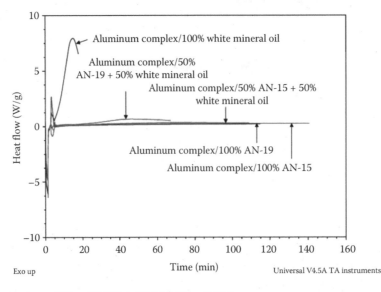

FIGURE 22.12 Thermo-oxidative stability—PDSC ASTM D5483 at 170°C.

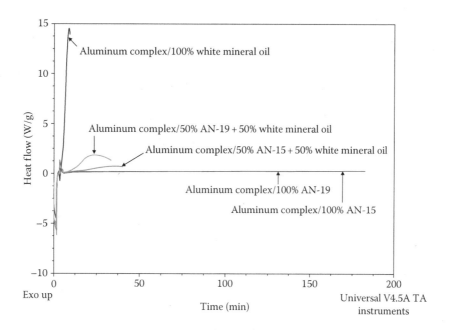

FIGURE 22.13 Thermo-oxidative stability—PDSC ASTM D5483 at 180°C.

of heat. At 180°C, the same trend was observed where the addition of either AN-15 or AN-19 to the white oil imparted significant oxidation resistance with the greases liberating just very small amounts of heat. The AN-15 showed slightly better oxidation protection than the AN-19. At both temperatures, the greases made with 100% of either AN-15 or AN-19 were completely stable.

22.5 SUMMARY

Alkylated naphthalenes are multifunctional, high-performance base stock–type additives available in a diverse ISO viscosity range from 22 to 193 cSt. They are typically incorporated into lubricant formulations replacing a portion of a Group II, Group III, or PAO base oil to extend the lifetime and performance of high-performance lubricants.

When incorporated into oil and grease formulations, alkylated naphthalenes can impart thermal and thermo-oxidative stability because the electron-rich naphthalene ring has the ability to absorb energy, resonate, and then disperse that energy. Alkylated naphthalenes are hydrolytically stable because they do not contain functional groups that can hydrolyze. Because of their intermediate polarities, alkylated naphthalenes are

the optimal chemistry to impart seal swell, reduce volatility, provide excellent dispersancy and varnish control for system cleanliness, and enhance the solubility and response of other additives without competing for the surface. They also have good low pour points and good VI and provide film thickness and film strength, which can reduce friction and wear. Most alkylated naphthalenes have HX-1 approvals for incidental food contact. All of these attributes make alkylated naphthalenes a superior base stock–type additive that provides flexibility in designing lubricants for specific applications.

REFERENCES

1. Koelbel, H. 1948. Synthesis of lubricants via the alkylation of naphthalene. *Erdoel Kohle* 1:308–318.
2. Hourani, M.J., Hessell, E.T., Abramshe, R.A., and Liang, J.G. 2007. Alkylated naphthalenes as high performance synthetic fluids. *Tribology Transactions* 50:82–87.
3. Hessell, E.T. and Abramshe, R.A. 2003. Alkylated naphthalenes as high performance synthetic fluids. *Journal of Synthetic Lubrication* 20(2):110–122.
4. Wu, M.M. and Ho, S.C. 2013. Alkylated aromatics. In: *Synthetics, Mineral Oils, and Bio-Based Lubricants*, ed. L.R. Rudnick, pp. 149–168. Boca Raton, FL: CRC Press/Taylor & Francis.

23 Additives for Bioderived and Biodegradable Lubricants

Mark Miller

CONTENTS

23.1 INTRODUCTION

There is increasing interest in biobased and biodegradable lubricants as alternatives to petroleum-based products. According to the National Oceanic and Atmospheric Administration (NOAA), 700 million gallons of petroleum is released into the ocean each year. Over half of that, 360 million gallons, is because of irresponsible maintenance practices as well as routine leaks and spills.

As environmental enforcement agencies increase pressures and costs for petroleum lubricant spills, many equipment operators are using or considering environmentally safer products. These types of fluids can protect the users against fines, cleanup costs, and downtime, but care must be given in selecting the right product for each specific application.

The benefits of biodegradable hydraulic fluids are well known. Their biodegradable properties allow them to break down in the environment reducing the negative impact from leaks and spills. They can be nontoxic, meaning they will not hurt operators, animals, or plants that come into contact with the fluid. They are renewable and reduce dependence on foreign petroleum oil.

Conventional knowledge has focused on the limitations of vegetable oils as base stocks for lubricants. The weakness of the oxidative stability, the cold temperature performance, and incompatibility with elastomers are well documented. Early generation biobased lubricant formulators utilized performance chemistry similar to those used in petroleum-based fluids creating lubricant products that did not meet industrial performance requirements. Over the past decade, however, improvements in vegetable base stocks, performance chemistry, and formulation expertise have allowed for the development of biodegradable products with performance similar to or better than conventional petroleum fluids.

Although there is a wide range of *biobased products*, most are readily biodegradable, meaning that they are biodegradable above 60% in 28 days as measured by the ASTM D5864 biodegradability test. Petroleum hydraulic fluids are either not biodegradable (<30% in 28 days) or inherently biodegradable (30% < X < 59% in 28 days, where X is biodegradability).

Most of these biobased products are free of toxic performance chemicals such as sulfated ash, zinc, calcium, and other heavy metals. Biobased products are also usually formulated to be *virtually* nontoxic and frequently measure 5–10 times less toxic compared with the ASTM *nontoxicity* specification. A conventional petroleum lubricant typically contains a zinc and calcium additive performance system and will generally be considered toxic. These fluids are toxic and persistent as the additive system kills the microbes responsible for biodegradation, and the petroleum fluid itself is not readily consumed by the microbes. There are, however, some *inherently biodegradable* petroleum hydraulic fluids containing ash-free additive systems and as such are less toxic than standard petroleum hydraulic fluids. However, these products are still petroleum-based, and the microbes cannot readily degrade (digest) these types of fluids. They will persist in the environment for many months or years and will dramatically reduce water quality, harming local wildlife and ecosystems.

Furthermore, standard petroleum products contain aromatic, cyclic (ring structure) hydrocarbons. These *aromatics* cause the familiar rainbow sheen on a water surface. Biobased products do not contain aromatics and as such do not produce a *rainbow* sheen on a water surface when spilled. Severely hydrotreating petroleum oil during the refining process will remove most aromatics. These fluids might not produce a sheen but will persist on the water surface harming the aquatic wildlife and ecosystem.

There are no universal definitions of biobased, environmental, or biodegradable for lubricants. The focus of this chapter is on vegetable or biobased lubricants with the vegetable oil content maximized at >60%. The Farm Security and Rural Investment Act (FSRIA or Farm Bill) was signed into law in 2002. A goal of that legislation is to increase the government's purchases and use of biobased products. Under this legislation, the United States Department of Agriculture (USDA) has selected and prioritized items for designation as *preferred* biobased products and set minimum biobased content levels. Each level of prioritization is issued in a *round* of designated products. In the first round, lubricant-related products were specified (Table 23.1) [1].

Despite the levels required by *federal minimum biobased content*, high-performance vegetable-based products can sometimes contain >90% biobased content. As a result, they will be readily (rapidly) biodegradable [2].

Also, this chapter looks primarily at environmentally compatible performance chemistry, which is defined as no heavy metals, no ash, low treatment volume, low toxicity, and low environmental persistence. Use of these types of chemistries will not adversely affect the environmental performance of biobased products and are very less toxic. One standard of measuring toxicity of a substance is the LC 50, that is, the concentration, in parts per million (ppm), of a substance that is lethal to 50% of the laboratory animals exposed to it in a 96 h test [3]. Therefore, the higher the LC 50, the lesser the toxicity. As most commercially available biobased lubricants can be formulated at LC 50s ranging from 5,000 to 10,000 ppm, only formulations less toxic than 5,000 ppm will be considered. This chapter, however, touches lightly on chemistry that seems to have a positive effect on vegetable-based products yet might be <5,000 ppm LC 50 level. The additives will include both traditional petroleum oil-type additives and other novel additives.

Significantly, more work has been done in hydraulic and tractor transmission fluids as well as total loss lubricants, that is, lubricants that are consumed through their use. Engine oils, although a large market, typically do not present an environmental threat due to leaks or spills. They require extremely high operating temperatures that are not suitable for vegetable oils.

Although some work has been done with engine oils to date, the Chemical Manufacturers Association (CMA) or International Lubricant Standardization and Approval Committee (ILSAC) has certified no vegetable-based engine oil. Some of the development work on engine oil performance has shown the performance of several additive types and will be addressed.

The use of refined, bleached, and deodorized (RBD) *food-grade vegetable* oils in industrial lubricant applications has been limited due to their inherent performance deficiencies regarding oxidative and thermal stabilities and their limited cold temperature flow properties. The use of genetically modified vegetable base stocks such as high oleic, reaction modification such as esterification, and chemical modification combined with improved vegetable oil-specific performance additives can fully address the concerns surrounding the use of vegetable oils in hydraulic applications. Commercial applications and previous research have demonstrated that mixtures of modified vegetable oils provide performance levels required for hydraulic original equipment manufacturers (OEM). Much work has been done evaluating the performance benefits of genetically modified vegetable oils and mixtures of base fluids including ester and poly-α-olefins (PAOs), but it is outside the scope of this chapter. For the most part, the benefits that an additive provides to a conventional vegetable oil will also apply to genetically modified oils and mixtures.

Before modern additives are explored, the history of vegetable oils will be reviewed.

23.2 HISTORY

Early generation biobased lubricant formulators utilized performance chemistry similar to those used in petroleum-based fluids creating lubricant products that frequently did not meet industrial performance requirements.

Historically, vegetable-based lubricants have not exhibited sufficient performance for industrial applications. There were several reasons for this inability. The first reason is that vegetable-based lubricants were misformulated. Traditionally, a lubricant is compounded from base oil and various performance chemistries. Early formulators in the vegetable-based lubricant market used the same chemistry that was used for petroleum lubricants for vegetable base oils. This approach was not effective as the characteristics of vegetable oils are vastly different than those of petroleum oils. Typical characteristics of vegetable oils as compared with petroleum are summarized in Table 23.2.

Vegetable oils have to be formulated for their individual characteristics. Some modern vegetable-based products currently in the market offer good performance and a fair price. Figure 23.1 shows relative costs for a wide range of lubricants.

Although all triglyceride vegetable-based lubricants have temperature limitations, there are some that are better than others. Although most vegetable-based lubricants have a maximum operating temperature of 140°F, there are some that offer protection as high as 220°F. Similarly, most vegetable-based lubricants offer good performance down to 30°F, yet there are some that flow below −30°F.

TABLE 23.1
Government-Specified Minimum Biobased Content

Product	Minimum Biobased Content (%)
Diesel fuel additives	90
Hydraulic fluids (mobile equipment)	44
Penetrating lubricants	68

Source: Extracted from United States Department of Agriculture Bio-preferred web site, http://www.biopreferred.gov.

TABLE 23.2
Typical Characteristics of Various Base Fluids

Characteristic	Petroleum Oil	Vegetable Oil	Saturated Ester	PAO
Lubricity	Low	High	High	Low
Oxidative stability RPVOT	300	50	180	300
Viscosity Index (VI)	100	200	165	150
Hydrolytic stability	High	High	Low	High
Polarity	Low polar	Highly polar	Polar	Low polar
Saturation	Saturated	Unsaturated	Saturated	Saturated
Flash point (°F)	200	450	400	350
Pour point (°F)	−35	−35	−40	−50

Source: Derived from Terresolve Technologies In-house Knowledge Base.

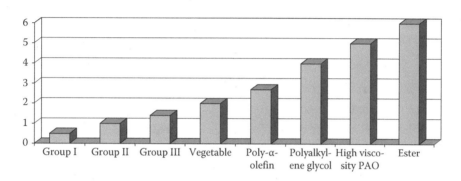

FIGURE 23.1 Relative cost of various base fluids. (Derived from Terresolve Technologies In-house Knowledge Base.)

These temperature limitations led to another major reason for early technical failures and that was the wrong fluid choice. Even the highest-performing biobased fluids have operating limitations in terms of temperature and life expectancy. Using a biobased fluid in an application over 220°F (and as low as 160°F for some fluids) will cause premature and possibly catastrophic equipment failure. This, combined with sensitivity to moisture, which is another characteristic of vegetable-based lubricants, has caused numerous cases in which using a vegetable-based fluid was a major contributor to the failure. In extreme high-temperature, environmentally sensitive applications, readily biodegradable synthetic fluids should be utilized.

Finally, biobased fluids require special care to maximize its useful life. Water in any lubricant system is bad for many reasons, which include causing many additives to fall out, increasing the onset of acid formation, deteriorating seals, creating rust, and accelerating wear. Most biobased fluids are more susceptible to hydrolytic breakdown, the result of which can be acid formation, more susceptible to additive precipitation, and as previously mentioned are prone to oxidative instability.

Over the past decade, however, improvements in vegetable base stocks including low–euricic acid rapeseed oils, high oleic soy and rapeseed oils, and improved low-temperature castor oil; improvements in additive chemistry; and improvements in formulation expertise have allowed the development of biodegradable products with performance similar to or better than conventional petroleum fluids.

23.3 VEGETABLE BASE FLUIDS

Vegetable oils and other fats are triglycerides that are essentially triesters of fatty acids and glycerol. They are soluble in most organic solvents and are insoluble in polar substrates such as water. They can have a broad range of fatty acid profiles, but most commonly used vegetable oils will be C18—oleic, linoleic, or lineleic acids. The proportion of each of these acids depends on the vegetable type, the growing season, and the geography. These factors can dramatically affect the performance of the vegetable oil in terms of oxidative stability, cold flow, hydrolytic stability, and other features of the final product. For example, the higher the oleic acid content, the better the oxidative stability but the higher (worse) the pour point. Further discussion of acid content and oil performance is reviewed in other sections.

23.4 OIL TYPES AND PERFORMANCE

Figure 23.2 shows the effect of oleic acid level on oxidative stability as measured by the pressure differential scanning calorimeter (PDSC). The PDSC is a test frequently used to evaluate thin-film oxidation of lubricants and base fluids. It is also known as the CEC L-85-T-99. The test measures the *minutes* to *induction time*, and the longer the time, the better the oxidative stability.

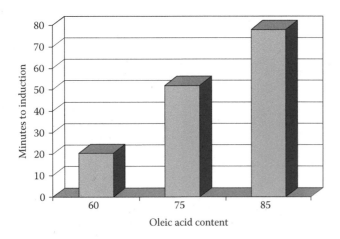

FIGURE 23.2 PDSC oxidation stability. (Based on Bergstra, R., Emerging opportunities for natural oil based chemicals, *Plant Bio-Industrial Workshop*, Saskatoon, Saskatchewan, Canada, MTN Consulting Associates, Edmonton, Alberta, Canada, February 27, 2007.)

Figure 23.2 [4] shows that the there is a direct correlation between the oleic acid content of a vegetable oil and its oxidative stability and that the higher the oleic content the higher the oxidative stability.

23.5 ADDITIVE PACKAGES

As previously mentioned, early vegetable oil formulators used performance additives similar to the performance additives used for petroleum base fluids. Petroleum oils are, for the most part, nonpolar, whereas triglycerides are highly polar. As such, conventional petroleum additives sometimes have solubility problems with vegetable oils. Frequently, to utilize typical additive packages, a solubilizing agent must be used. Additive packages without metals, such as *ashless* additives, are more soluble in vegetable oils.

Even today, several additive manufacturers try to promote additive packages targeted to vegetable oils. Although major additive companies are highly skilled at performance chemistry for petroleum base oils, the expertise needed for formulating vegetable oils is highly specialized. Adding to it is the complication of maintaining environmental integrity, which makes it a daunting task.

Most additive packages supplied by the major additive companies are specifically designed for petroleum base oils. Even those targeted toward vegetable oils utilize core chemistry tailored for petroleum. Additionally, the carrier fluid from the various components is usually petroleum, and when a multicomponent package is created, several percent of petroleum is included. This will reduce biodegradability and increase toxicity of the final fluid, that is, counter the objective of the product.

To make high-performance fluids from vegetable oils, a *clean sheet approach* must be utilized, and the fluid must be designed with the specific characteristics of the base fluids and the specific requirements of the end products. Some typical petroleum additives can be utilized and are effective in vegetable oils. Frequently food-grade additives are superior and have lower toxicity. When formulating vegetable-based lubricants, it is useful to expand the search into food additive and other nonlubricant industries. Table 23.3 [5] summarizes the performance of some conventional additive packages designed for both petroleum and vegetable base oils as compared with a custom-designed performance formulation using vegetable oil and specific additives. The rotary pressure vessel oxidation test (RPVOT) known as the ASTM D2272 is a common test procedure that compared the oxidative stability of lubricants and base fluids. The ASTM D665B rust test is also a typical indicator of the rust protecting characteristics of a finished fluid.

For the past several years, Valvoline has been working toward the development of a bio-containing engine oil. In this work, they compared results of a mid-oleic soy oil and petroleum oil mixture with a conventional additive package with the results from a *clean sheet* design. Taking the characteristic of biofluid under consideration, they were able to achieve vastly superior results in the Sequence VII bearing corrosion test (Table 23.4) [6].

TABLE 23.3
Performance of Various Additive Packages in Vegetable Oil

	Custom Formulation	Vegetable Additive Package 1	Vegetable Additive Package 2	Petroleum Additive Package 1	Petroleum Additive Package 2
Petroleum additive package 1				X	
Petroleum additive package 2					X
Vegetable additive package 1		X			
Vegetable additive package 2			X		
RPVOT D2272	125	76	16	41	16
Rust D665B	Pass	Pass	Fail	Pass	Fail

Source: Derived from In-house Terresolve Technologies Proprietary Development Research, October 11, 2005 to January 8, 2007.

TABLE 23.4
Sequence VII Performance

Engine Test Result	Weight Loss (mg)	Stripped Viscosity (cSt) at 10 h
GF 4 Spec limit	26	Stay in grade
Traditional package	108.6	11.51
Unique package	4.0	11.09

Source: Based on McCoy, S., United Soybean Council Technical Advisory Panel, The Valvoline Company, September 20, 2005.

23.6 DETERGENTS

Vegetable oils have a high level of solvency and by themselves act as a detergent for many applications. Vegetable oils and methyl esters of vegetable oils have been successfully used as solvents and cleaners for many years.

When attempting to add detergent performance, phenates and sulfonates should be avoided, as they are frequently toxic for *environmentally safe* fluids. The phenate and sulfonate functional groups are not the problem, but conventional petroleum detergent additives that are typically attached to heavy metals, such as calcium, sodium, and barium, are highly toxic. Specialty *metal-free* phenates and sulfonates have been used as a component to protect seal materials. These products have shown collateral benefits. For example, some classes of phenates have shown improvements in thermal stability to natural oils, and some sulfonates show an additional side benefit of corrosion inhibition.

Low amounts of phosphate and phosphate esters have been used successfully in vegetable products showing good performance without adversely affecting the environmental performance.

Sulferized phenates and salicylate chemistries are typically not used in bio oils as they adversely affect the environmental benefits of a vegetable oil. There has been some work, although very sparse, on the detergent and antiwear benefits of thiophosphates, phosphonates, and thiophosphonates. These additives and their benefits, however, are of limited value in formulating vegetable-based lubricants because vegetable oils on their own have sufficient detergent and antiwear capabilities and do not require supplementation.

23.7 DISPERSANTS

Almost all dispersants found are designed for petroleum oils. They typically contain a long *tail* that will be soluble in the oil and *polar head* to keep certain contaminants in the oil. These contaminants can be filtered out later or are swept away during an oil change. The vegetable oil molecular size is too small to act as a dispersant, and while the *polar head* is highly polar, the molecular chain length is too small to accommodate dirt and sludge.

Typical vegetable-based lubricant applications such as hydraulic fluids and total loss lubricants (bar-chain oil and 2-cycle oil) have little need for dispersants. There has, however, been some work done evaluating dispersants for vegetable oil-containing engine oils. This work has shown that the most effective way to add dispersancy without significantly reducing other performance characteristics is through a dispersant polymer. As it is one additive that can be used for multiple purposes, a side benefit to a dispersant polymer is that it keeps the amount of nonbiodegradable, nonbiobased products to a minimum. Conventional high-molecular-weight dispersants such as polyisobutylene (PIB), PIB succinimide, and PIB succinate ester are not used as they adversely affect the environmental performance.

23.8 CORROSION INHIBITORS/ANTIRUST

Rust is a chemical reaction between water and ferrous metals, whereas corrosion is a chemical reaction between chemicals (usually acids) and metals. As previously mentioned, water in any lubricant system is bad. It can cause rust on metal surfaces and reacts with chemicals in the system to produce acids that cause corrosion. Vegetable oils can utilize more novel (expensive) acidic rust inhibitors helping to protect against corrosion and rust. Typically, acid inhibitors in conventional additive packages form precipitates that cause filter plugging. Sulfonates and acid rust inhibitors are more surface active and perform better in vegetable-based oils.

23.9 ANTIOXIDANTS

Vegetable oil's primary weakness for use in industrial applications is its oxidative instability. Oxidation occurs with the interaction of oil and oxygen. Vegetable oils (triglycerides) have a number of double bonds depending on the type of the oil. Through use, these fluids develop free radical ions. The double bonds in the oil are *searching* for ions (free elections). The double bond reacts with the free electrons and will change the characteristics of the vegetable oil. This will open a double bond and allows for degradation of by-products. There is a very good correlation between the degree of unsaturation of a vegetable oil and its oxidative stability. Table 23.5 [7] summarizes several oil types and their oxidation characteristics

TABLE 23.5
Degree of Saturation and PDSC Onset Temperature in Soybean Oils

Base Fluid	Name	Unsaturation	PDSC Onset Temperature (°C)
Soybean oil	SO	1.5	173
High linoleic SO	HLSO	1.4	179
Mid oleic SO	MOSO	1.14	190
High oleic SO	HOSO	0.94	198

Source: Extracted from Erhan, S.Z., Oxidative stability of mid-oleic soybean oil: Synergistic effect of antioxidant–antiwear additives, National Center for Agricultural Utilization Research, USDA/ARS, Peoria, IL, 2006.

as measured by the PDSC (see Section 23.4 for a detailed description). The onset oxidation temperature and the oxidative stability can be predicted using fatty acid composition rather than individual fatty acid percentage.

Vegetable oils have significantly lower oxidative stability than petroleum fluids. The stability can be improved through the proper choice of vegetable type, chemical modification, and antioxidant additives.

The study on oxidative stability (Section 23.5, Table 23.3) mentioned earlier was continued to include antioxidant additives to both the vegetable additive package and the petroleum additive package. One can see from Table 23.6 that although both additive packs can have oxidative stability improvements, neither was able to reach the performance level of a custom formulation hydraulic fluid.

Some unique antioxidants such as zinc diamyl dithiocarbamate (ZDDC), alkylated diphenyl amines (ADPA), and butylated hydroxyl toluene (BHT) improve in both PDSC and RPVOT in soy oils. These also have shown a synergistic effect with some antiwear agents such as antimony dialkyl dithiocarbamate (ADDC) as shown in Figure 23.3.

These chemistries can have a positive effect on the oxidative performance of vegetable oils. However, they can be toxic and persistent and reduce the environmental benefits of the fluids.

There are two other widely used methods for oxidation control. The first is free radical scavenging and the other is hydroperoxide decomposition. Radical scavenging will *lock up* free radicals and prevent oxidation. Commonly used free radical scavengers for vegetable oils are arylamines and hindered phenol. Zinc dialkyldithiophosphate (ZDDP) is a widely used free radical scavenger and an excellent antiwear agent but is not often used in vegetable oil formulations due to its high level of toxicity.

The other widely used oxidation inhibitor mechanism is *hydroperoxide decomposition*. Hydroperoxide is another precursor to oxidation, and *breaking it down* can significantly improve oxidation stability. The main additive components for hydroperoxide decomposition are phosphorus-based additives as they are mostly environmentally benign and very effective. Other types of hydroperoxide decomposers such as sulfides, phosphates, and olefins are not used due to environmental incompatibility.

TABLE 23.6
Effect of Various Antioxidants on Vegetable Oil Formulations

	1	2	3	5	6
	Custom Formulation	Vegetable Additive Package	Vegetable Additive Package—Supplement 1	Petroleum Additive Package—Supplement 3	Petroleum Additive Package—Supplement 2
Petroleum additive package 1				X	X
Vegetable additive package 1		X	X		
AO supplement 1			X		
AO supplement 2					X
AO supplement 3			X	X	
RPVOT D2272	125	76	100	41	52
Rust D665B	Pass	Pass	Pass	Pass	Not tested

Source: Derived from In-house Terresolve Technologies Proprietary Development Research, October 11, 2005 to January 8, 2007.

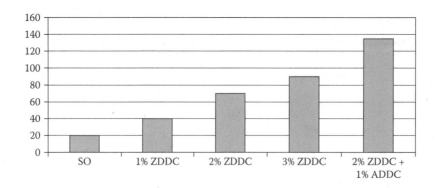

FIGURE 23.3 Effects of antioxidant and antiwear on soy oil RPVOT. (Extracted from Erhan, S.Z., Oxidative stability of mid-oleic soybean oil: Synergistic effect of antioxidant–antiwear additives, National Center for Agricultural Utilization Research, USDA/ARS, Peoria, IL, 2006.)

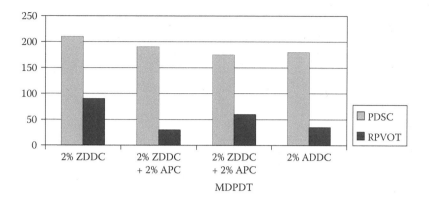

FIGURE 23.4 Oxidation effects of various antiwear additives in vegetable oil formulations. (Extracted from Erhan, S.Z., Oxidative stability of mid-oleic soybean oil: Synergistic effect of antioxidant–antiwear additives, National Center for Agricultural Utilization Research, USDA/ARS, Peoria, IL, 2006.)

23.10 ANTIWEAR

Vegetable oils can have excellent lubricity, far superior than that of mineral oil. The polarity of vegetable oil improves antiwear characteristics as they have an affinity to metal and protect the surface. Some RBD vegetable oils have passed hydraulic pump wear tests, such as ASTM D2882 and ASTM D2271, in their natural state with no additives. These unadditized oils thermally break down due to their oxidative instability but perform well in the wear tests.

Amine phosphate compounds (APCs) are used in readily and inherently biodegradable products. They do not work well in refined petroleum as they do in vegetable oil.

Some commonly used antiwear additives have been utilized in vegetable oils and found to reduce oxidative stability when combined with certain antioxidants. These include, for example, APC and molybdenum dialkyl phosophordithioate (MDPDT) with ZDDC (Figure 23.4).

These additives as well as other molybdenum, zinc, and boron compounds are excellent chemicals for antiwear and friction reduction, but are typically designed for petroleum fluids. They will work well in vegetable oils and highly refined petroleum oils, but will increase overall toxicity of these blends.

23.11 ANTIFOAM

Vegetable oils need additional antifoam agents to prevent air entrainment. Conventional antifoam agents have only a minor effect on vegetable oils. Silicon foam inhibitors are very effective and widely used. Silicon is very nonpolar and is an excellent foam inhibitor. The silicon chemicals are usually diluted in either petroleum oil base stock or vegetable oils and may need an additional solubility agent such as an ester to maintain solubility. Silicon acts as a surfactant and prevents air from reacting with the lubricant surface. Other foam inhibitors used with vegetable oils are dimethylsiloxane, alkylmethacrylate, and other alkylacrylates.

23.12 VISCOSITY MODIFIERS/POUR POINT DEPRESSANTS

Viscosity is commonly known as resistance to flow. Vegetable oils normally have a good natural viscosity for industrial lubricants. Some formulators will utilize this natural characteristic and the very high viscosity index (VI) and not use any additional viscosity modifiers.

VI is the change in viscosity compared to the change in temperature. A high VI indicates small viscosity changes with temperature changes. Vegetable oils have very high VI usually ·200 when compared to petroleum oil, which has a VI of ·100. This means that vegetable oils maintain their design viscosity over a broader temperature range.

Finding an environmentally safe viscosity modifier is very difficult. Typically, the long-chain polymers do not break down in the environment and therefore reduce biodegradability. Most viscosity modifiers are nonpolar and hard to solubilize in vegetable oils and will create hazy mixtures. Ethylene–propylene–diene monomer (EPDM) polymers are especially difficult to solubilize due to their *diene* double bond. There are some viscosity modifiers, however, that can be dispersed in vegetable oils including olefin copolymer (OCP) and polymethacrylate (PMA).

The pour point is the lowest temperature at which oil will flow. Most pour point depressants (PPDs) reduce the size and cohesiveness of the crystal structures and will thereby reduce pour point and improve the flow at reduced temperatures. PPDs are base oil specific. While the steric effect of the carbon side chains of vegetable oils have more of an effect on the pour point than any other factor, some PPDs can reduce the pour sufficiently for industrial usage.

Styrene esters, methacrylates, and alkylated naphthalenes work well for vegetable oils as well as petroleum fluids. Methacrylates typically are more effective for group II petroleum oils.

REFERENCES

1. United States Department of Agriculture Biopreferred web site, http://www.biopreferred.gov.
2. Terresolve Technologies In-house Knowledge Base.
3. Canadian Centre for Occupational Health and Safety web site, http://www.ccohs.ca/oshanswers/chemicals/ld50.html.
4. Bergstra, R., Emerging opportunities for natural oil based chemicals, *Plant Bio-Industrial Workshop*, Saskatoon, Saskatchewan, Canada, February 27, 2007, MTN Consulting Associates, Edmonton, Alberta, Canada.
5. In-house Terresolve Technologies Proprietary Development Research, October 11, 2005 to January 8, 2007.
6. McCoy, S., United Soybean Council Technical Advisory Panel, The Valvoline Company, September 20, 2005.
7. Erhan, S. Z., Oxidative stability of mid-oleic soybean oil: Synergistic effect of antioxidant–antiwear additives, National Center for Agricultural Utilization Research, USDA/ARS, Peoria, IL, 2006.

Section VI

Applications

24 Additives for Grease Applications

Robert Silverstein and Leslie R. Rudnick

CONTENTS

24.1   INTRODUCTION

Greases are one of the oldest forms of lubricating materials, and in fact, the first greases can be considered environmentally friendly and biodegradable by our modern conventions. Ancient Egyptians about 1400 BC made crude greases to lubricate the wheels of their chariots. These early greases consisted of both mutton fat and beef fat, which were sometimes mixed with lime. In fact, the word "grease" is derived from *crass us*, the Latin word for fat. As time progressed, there was only a minimal improvement in grease performance until about the eighteenth century. During the eighteenth and nineteenth centuries and the Industrial Revolution, society created a greater variety of new machines. This modern technology that developed during and after the Industrial Revolution required lubricants to handle greater loads and to operate under more severe conditions. New grease technology was needed to improve the performance of equipment and to improve the lifetime of machine components.

Grease is a semisolid or solid lubricant composed of a thickening agent dispersed in a liquid lubricant, or in combination of lubricants. The principal advantage of choosing grease over conventional lubricating oil in certain applications is because of grease's ability to remain in contact with the desired moving surfaces. Greases will generally not leak away from the point of application. Thickeners in the grease are designed to reduce any migration caused by gravity, pressure, or centrifugal action. Grease is best thought of as thickened oil.

There are at least three reasons where grease is chosen over oil as a lubricant:

1. In applications where leakage of oil would occur
2. Where the natural sealing action of grease is needed
3. When greater film thickness is necessary for the application

The liquid lubricant or base oil in grease usually accounts for 70%–95% of the grease. The base oil may be mineral oil, synthetic oil, or natural oil, such as a vegetable oil. Oils from animal sources are generally not used today. Generally, between 5% and 25% of the grease composition is thickening agents. The most common thickeners in use today are metallic soaps, particularly lithium soaps, which account for over 60% of the grease market. Silica, expanded graphite [1], or clays (bentonite or hectorite) are sometimes used as thickeners. Clay thickeners represent about 5%–10% of the grease market.

Additives, required to improve certain properties of the grease such as oxidative stability, wear protection, and corrosion inhibition, are used in 0.5%–10% of the grease depending on the grease type and application. When grease is to be used under more severe conditions, it generally contains a more enhanced additive package. These additives are often the same additives used in liquid lubricants. Exceptions are solid additives such as graphite and molybdenum disulfide, which are typically used as antifriction and extreme-pressure (EP) additives.

Further research will continue to provide newer and more effective components that will serve as thickeners and/or additives in the field of lubrications. These future materials are expected to provide greases and liquid lubricants with properties and performance capabilities exceeding those possible today.

In addition to the resistance to flow that maintains grease at the point of application, the primary performance features are to reduce friction and wear. Some of the main factors that affect grease properties and performance are

- Type and amount of thickener
- Consistency or hardness
- Dropping point
- Antioxidant additives
- EP additives
- Pumpability
- Volatility
- Environmental considerations as toxicity and biodegradability

Soaps, used as thickeners, can be considered additive in greases, even though they generally comprise a large percentage of the grease composition. Thickeners are, in fact, what make grease grease and what differentiate grease from conventional lubricating fluid. This chapter does not consider soaps, because they have been previously described by Klamann [2] and Boner [3]. Functionalized soaps have also been used to impart EP and antiwear (AW) performance to grease [2].

The purpose of this chapter is to serve as a review of some of the important additives used in greases and to present a variety of additive testing results that demonstrate their performance benefits. Because final grease formulations are a function of particular performance needs, application data described herein are focused on base oil/thickener/additive effects and not on fully formulated greases, many of which are proprietary.

24.2 WORLDWIDE USE OF GREASES

On a global basis, the regions that consume the largest percentage of the world's grease are North America (545 MM pounds), the People's Republic of China (510 MM pounds), and Europe (414 MM pounds). India and the Indian Subcontinent (182 MM pounds), Japan (184 MM pounds), Pacific and Southeast Asia (113 MM pounds), the Caribbean and Central and South America (99 MM), and Africa and the Middle East (71 MM) represent the remaining bulk of the global grease production [4]. Each of the world's different regions requires different grease products.

The National Lubricating Grease Institute (NLGI) Grease Production Survey Report for the year 2006 reports a grand total of 2.1 billion pounds of lubricating grease manufactured worldwide. Of this, conventional lithium soap and lithium complex were approximately 59% and 15% of the total, respectively. In the United States and Canada, conventional lithium and lithium complex greases represent 32% and 36%, respectively, of the 545 million lb of grease produced in 2006 [4].

Highly refined, chemically modified mineral oils or synthetic base oils are used in high-performance complex soap greases in more demanding applications that cannot be met with commodity greases [5]. Industrial applications compare the majority of grease used worldwide; the largest application areas are found in railroad lubrication, steel production, mining, and general manufacturing. Automotive applications include trucks, buses, agricultural and off-road construction equipment, and passenger cars [5].

24.3 PRODUCTION OF GREASES

Soap-thickened greases are manufactured using three basic sequential process steps. First, the soap is formed by the saponification of various fats. Then the soap is dehydrated and further optimized in properties. Finally, milling to provide a consistent dispersion of all components disperses the soap, base fluid, and additives. Milling is a process of shearing the grease to disperse all components. Brown et al. have demonstrated how the process of shearing during milling orients the soap fibers [6]. Dispersion and orientation can harden greases by 100 penetration units [7].

Grease is produced using a kettle process, a contactor process, or a continuous process. These processes differ in terms of the rate of grease production, the quantity of grease to be produced, and the initial investment in equipment. The kettle and contactor processes are both batch processes. The contactor process utilizes pressure to reduce reaction time.

When grease is prepared using a batch process, the base oil is heated together with the additives and thickening agents. If saponification of the fat is required to form soap, the reaction mixture is brought to a higher temperature to complete the reaction and to fully disperse all components. A typical batch kettle reaction will take 8–24 h and can range in scale from 2 to 50,000 lb. They are typically cylindrical in shape with a height-to-diameter ratio of 1–1.5. They can have dished or conical bottoms and either flat or dished tops. Because the principal objective in making grease is the dispersion of components, kettles may have either one set of rotating paddles or an arrangement of stationary and rotating paddles so as to more efficiently mix the components. Scrapers are sometimes used to remove material adhering to the kettle walls.

The advantages of the contactor process are reduced heating and reaction time. For example, a typical 20,000 lb batch of lithium grease may take 20 h in a kettle process, but only 5–7 h in a contactor process [8,9]. In the contactor process, fatty acids or glycerides such as hydrogenated castor oil, tallow, or vegetable oils and lithium hydroxide, for example, are added at the top. Base fluid is added and water may also be added. The contactor is pressurized, which accelerates the process of saponification. Other process steps include cooling after reaction, adding additional finishing oil, and pumping the grease to a finishing kettle.

There are three main sections in a continuous grease reactor: the soap reactor section, a dehydration section, and a finishing section. Saponification occurs in the reactor section, where the fatty component, alkali, and base fluids are added. The grease components are transferred to the dehydration section, which operates under a vacuum to remove volatiles. The product at this stage is a dehydrated soap base. The product of the dehydration section is moved to the finishing section, where additional base fluid and additives are combined and dispersed [10]. A comparison of the continuous and kettle processes is shown in Table 24.1 [8].

Many chemical and manufacturing factors can affect the properties and performance of greases. For example, several factors that affect the manufacture of lithium complex greases include the base fluid, fatty acid concentration, temperature,

moisture, and additive chemistry [11]. Several features of the various manufacturing processes need to be considered in preparing greases. These include the yield of the grease product, any energy requirements to prepare the grease, production time, and capital investment along with associated maintenance costs.

For applications in extreme environments, such as aerospace and military, specialized lubricating greases are necessary. These greases must be able to perform reliably under conditions that most other greases never experience. These materials must have excellent thermal stability, have low volatility, and be able to function under very-low- and very-high-temperature regimes. The base fluid and all additive components must meet these criteria. For example, high-performance greases have been made that incorporate multiply alkylated cyclopentanes as the base fluid [12]. Other specialized base fluids have been studied, including those containing fluorine and other heteroatoms.

24.4 GREASE COMPOSITION AND PROPERTIES

24.4.1 COMPOSITION

Base fluids used in conventional grease are of three types: mineral (petroleum derived), synthetic (synthesized from discrete chemical components), and natural (plant and animal derived). Mineral oils can be paraffinic or naphthenic. Synthetic base fluids commonly used include poly-α-olefins (PAOs), esters, polyglycols, and silicones. Natural oils can be derived from soybean, rapeseed, and other plant oils and high-oleic acid versions of these oils.

Base fluid viscosity, in general, directly affects the viscosity characteristics of the resultant grease. Base fluids used in grease formulation generally range from 150 to 600 N oils and may also include bright stock. High-viscosity base fluids are typically used for high-temperature, heavy-duty applications. Alternatively, high-speed applications generally require low-viscosity oils. Lubricating greases are also classified according to thickener type [13].

There are many different types of grease thickeners; however, a few major types are generally employed throughout the industry, depending on the application. Alkali components include lithium hydroxide, calcium hydroxide, sodium hydroxide, and aluminum hydroxide. These alkalis are reacted with materials from animal, marine, and vegetable sources. Nonsoap thickeners can also be used. These include silica, clays, urea, polyurea, and Teflon®.

Grease stability, oxidation resistance, effects of water, maximum operating temperature, and other properties have been summarized by Wills [13].

24.4.2 PROPERTIES

Grease is generally classified and evaluated for properties and performance using tests in the following categories:

- Consistency
- Stability

TABLE 24.1
Comparison of Continuous and Kettle Grease Processing

	Continuous Process	Kettle Process
Investment cost factor	1.0	1.2
Manufacturing cost factor	1.0	4.1
Heat load (Btu/h)	1,500,000	3,000,000
Cooling water (gal/min)	75	200
Electrical use (kW/h)	120	380

Note: Values are based on a 10 million lb/year output of lithium, calcium, and sodium greases.

- EP/AW
- Pumpability
- Corrosion
- Water tolerance
- Bench performance tests

24.4.3 Consistency

One of the most important properties of grease is its consistency. Just as lubricating oils are available in different viscosity grades, greases are specified in terms of their consistency, as measured by their ASTM *worked penetration*. *Worked penetration* means that the grease is *worked* for 60 strokes in a standard manner before measurement. Penetration is measured by distance, in tenths of millimeters, that a standard cone will penetrate a grease sample when a standard force is applied. The standard method of measurement is ASTM D217 [7]. However, a special procedure, ASTM D1403, has been designed to utilize one-quarter and one-half scale cones for smaller samples [14]. Equations have been developed to convert the results to compare to ASTM D217 results.

In these tests, which are performed at a constant temperature of 25°C (77°F), a harder grease will exhibit a lower penetration number than a softer grease. The NLGI classification system (see Table 24.2) ranks grease hardness based on this test.

Other tests include apparent viscosity (ASTM D1092) and evaporation loss (ASTM D972 and D2595).

24.4.4 Stability

24.4.4.1 Dropping Point

The hardness of grease is a function of temperature. Greases act as thickened lubricants only to a point, and then at some temperature they become fluid. A standard measure of the resistance of a grease to flow as temperature is increased is the dropping point (ASTM D2265) [15]. In essence, the dropping point is the measure of the heat resistance of the grease. In the dropping point test, a grease sample is packed

TABLE 24.3
General Temperature Properties of Various Thickened Greases

Thickener	Dropping Point (°F)	Maximum Usable Service Temperature (°F)
Aluminum soap	230	175
Calcium soap	270–290	250
Sodium soap	340–350	250
Lithium soap	390	275
Calcium complex	>500	350
Lithium complex	>500	350
Aluminum complex	>500	325
Polyurea (nonsoap)	>450	350
Organoclay (nonsoap)	>500	350

into a standard test cup with a small opening. The sample is heated by introducing the sample into a preheated aluminum block. The sample temperature plus one-third of the difference between that temperature and the block temperature when the first drop of fluid leaves the cup is defined as the dropping point. Guidelines for the maximum usable service temperature and ranges for dropping points of greases made with various thickeners are summarized in Table 24.3 [16,17]. The self-diffusion of oil in lubricating greases has been studied using nuclear magnetic resonance spectroscopy at temperatures between 23°C and 90°C. Greases based on naphthenic mineral oils and PAO synthetic oils were measured. It was shown in this temperature range (40°C–90°C) that, using the same base fluid, the concentration of the thickener affected diffusion [18].

Other measures of high-temperature stability include high-temperature bleed (ASTM D1742), trident probe (ASTM D3232), cone bleed (FTM 791B), evaporation loss (ASTM D972 and D2595), rolling stability (ASTM D1831), oxidative stability (ASTM D942), and a high-temperature bearing test (ASTM D3336).

24.4.5 Extreme Pressure/Antiwear

24.4.5.1 Four-Ball Wear

ASTM D2266 describes a test method using three hard steel balls in locked position and coated with lubricating grease. A fourth ball is rotated against the three stationary balls, producing a wear scar on each of the three balls, which is reported as the average scar diameter. This test is run at light loads; thus, seizure or welding does not occur [19].

24.4.5.2 Four-Ball EP

ASTM D2296 describes a test method similar to ASTM D2266 except that the loads are much higher. At a certain load, the four balls will weld together; this is referred to as the weld load. The weld load can be used to assess whether lubricating grease has a low, medium, or high level

TABLE 24.2
NLGI Classification System for Penetration of Greases

NLG Grade	Worked Penetration
000	445–475
00	400–430
0	355–385
1	310–340
2	265–295
3	220–250
4	175–205
5	130–160
6	85–115

of load-carrying ability. The ASTM defines the load wear index (or the load-carrying property of the lubricant) as an index of the ability of a lubricant to prevent wear at applied loads. Under the test conditions, specific loadings in newtons (kg-force) having intervals of approximately 0.1 logarithmic units are applied to three stationary balls for 10 runs prior to welding. The load wear index is the average of the corrected loads determined or the 10 applied loads immediately preceding the weld point [20].

Fretting wear is measured using ASTM D3704, and load-carrying capacity can be evaluated using Timken O.K. Load (ASTM D2509).

24.4.6 PUMPABILITY

Pumpability can be evaluated using a USS Low-Temperature Mobility Test. It is largely a function of thickener type and concentration and is not significantly affected by additives.

24.4.7 CORROSION

Three common tests for corrosion are a Copper Corrosion Test (ASTM D130) modified for grease, a Bearing Rust Test (ASTM D1743), and the Emcor Rust Test.

24.4.8 WATER TOLERANCE

Water washout is measured using ASTM D1264. Water spray-off is measured using ASTM D4049. The wet roll stability of grease is evaluated using a modified ASTM D1831 test.

24.4.9 BENCH PERFORMANCE TESTS

Several bench performance tests are used in the grease industry to evaluate greases prior to field evaluation. Some of these include a High-Temperature Wheel Bearing Test (ASTM D3527), an SKF R2F Test that simulates paper mill applications, the FE-8 Test, CEM Electric Motor Test, and a GE Electric Motor Test. Low-temperature torque is evaluated using ASTM D1478 and ASTM D4693 at a temperature of $-54°C$.

24.5 GREASE THICKENERS

Most greases are made from metal–salt soaps that serve as the thickeners for the base fluid. Metal–salt soap thickeners are prepared from alkali base and a fat or fatty acids. The fatty materials may be derived from animal, marine, or vegetable fatty acids or fats. Fatty acids from these sources are generally even-numbered, straight-chain carboxylic acids containing zero or one double bond. A common fat used for making grease is hydrogenated castor oil, which yields lithium 12-hydroxy stearate upon saponification. Lithium hydroxide, calcium hydroxide, aluminum hydroxide, and sodium hydroxide are frequently used as the alkali materials. Sample reactions for the formation of some of these soaps are shown in Figure 24.1 [2].

In modern greases, simple soaps and complex soaps are used. A simple soap is prepared from one fatty acid and one metal hydroxide. The NLGI defines complex grease as soap wherein the soap crystal or fiber is formed by cocrystallization of two or more compounds: the normal soap and the

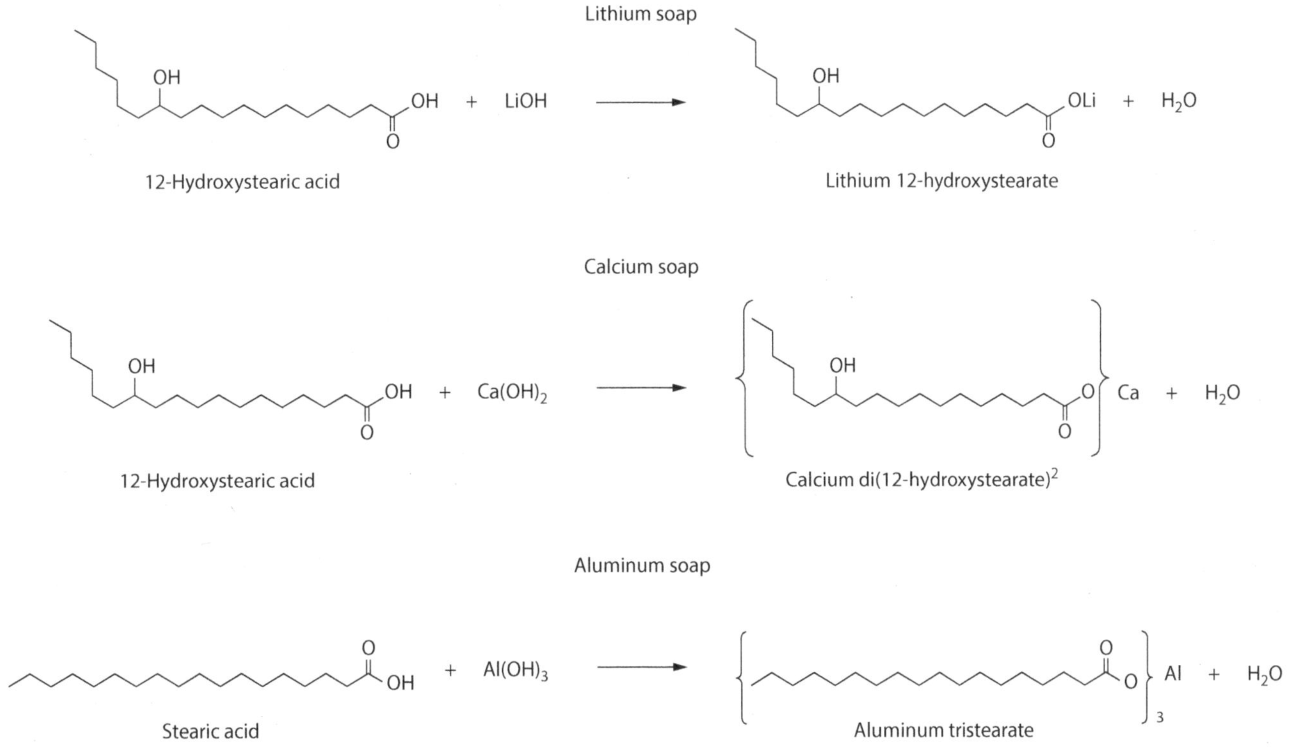

FIGURE 24.1 Chemical reactions for preparation of grease thickeners.

complexing agent [21]. The normal soap is the metallic salt of a long-chain fatty acid similar to a regular soap thickener, for example, calcium stearate and lithium 12-hydroxy stearate. The complexing agent can be the metallic salt of a short-chain organic acid, for example, acetic acid.

In summary, grease thickeners are generally one of the following types:

- Aluminum
- Aluminum complex
- Calcium
- Calcium complex
- Lithium
- Lithium complex
- Polyurea
- Clay

The properties of complex soap greases depend on [22]

- Metal type (lithium, calcium, etc.)
- Normal soap type (metallic stearate, metallic 12-hydroxy stearate, etc.)
- Complexing agent type (metallic acetate, benzoate, carbonate, chloride, etc.)
- Normal soap to complexing agent ratio
- Manufacturing conditions

Complex greases can be used over a greater temperature range because they have higher dropping points than normal greases.

The structure of grease thickeners directly affects the properties and performance characteristics of fully formulated grease. Thickeners range in structure from linear soaplike structures to more complex circular structures. Physically, they may be needles, platelets, or spherical and can vary significantly in dimension. Soaps form microscopic fibers, which form a matrix to hold the base fluid. Fibrous sodium soap structures can be 1 × 100 μm, whereas short-fibered lithium soap might be on the order of 0.2 × 2 μm. Aluminum soaps might have a diameter of only 0.1 μm [23]. Boner has reported that the rheological properties of lithium-based grease depend on the different fiber dimensions of the soap molecules [24]. Fiber structure and the surface-area-to-volume ratio of the fibers tend to vary with different soaps. It has been reported that a thin strip is the most effective shape for a thickener molecule. With an increase in the surface-area-to-volume ratio of the thickener molecules, the grease structure is strengthened as indicated by lower penetration [25]. Wilson has reported that lithium soap fibers are long, flat strips [26]. In general, sodium soap fibers range in size from 1.5 to 100 μm and lithium soap fibers from 2 to 25 μm. Aluminum soap fibers are essentially spherical on the order of 0.1 μm. Calcium soap fibers are short, generally about 1 μm long. Organophilic bentonite is about 0.1 × 0.5 μm.

In general, it is safe to conclude that greases made with different grease thickeners should not be mixed in the same application. This could result in poorer performance of the mixed greases relative to that expected from either individual grease. Incompatibility and poorer performance could also result in equipment malfunction. For example, it is generally cautioned that lithium and sodium greases are incompatible. There are always two sides to every story, and Meade, in an extensive study, tested more than 1200 grease combinations and found 75% of them to be compatible [27].

The type and structure of the grease thickener affect the stability of the grease during and after shear forces are applied. In bench tests, lithium-thickened greases have been shown to have the least reduction in shear stress and in film thickness. Less structural resistance was found for calcium-, sodium-, and bentonite (clay)-thickened greases [28]. Performance features and application of greases as a function of the types of thickener are summarized in Table 24.4 [2].

TABLE 24.4
Performance Features and Applications of Greases as a Function of Thickener

Thickener	Performance Features	Applications
Aluminum soap	Low dropping point; excellent water resistance	Low-speed bearings; wet applications
Calcium soap	Low dropping point; excellent water resistance	Bearings in wet applications; railroad rail lubricants
Sodium soap	Poor water resistance; good adhesive properties	Older industrial equipment requiring frequent relubrication
Lithium soap	Higher dropping point; resistance to softening and leakage; moderate water resistance	Automotive chassis; automotive wheel bearings; general industrial grease
Calcium complex	Excellent water resistance; inherent EP/load-carrying capability	High-temperature industrial and automotive bearing applications
Lithium complex	Resistance to softening and leakage; moderate water resistance	Automotive wheel bearings; high-temperature industrial rolling-element applications
Aluminum complex	Excellent water resistance; resistance to softening; good pumpability reversibility	Steel mill roll neck bearings; rolling-element and plain bearings; high-temperature industrial applications; food processing machinery
Polyurea (nonsoap)	Good water resistance; oxidation resistant; less resistance to softening and leakage	Industrial rolling-element bearings; automotive constant velocity joints
Organoclay (nonsoap)	Resistance to leakage; good water resistance; thickener has no melting point	High-temperature bearing with frequent relubrication

Greases are like other lubricants in that they exhibit limited useful lifetimes. The length of time that grease maintains the property and performance criteria for which it was designed depends on all the factors affecting other lubricants. In addition, grease structure is affected by the thermal, mechanical, and oxidative stresses related to the thickener structure. As grease ages, it may become dry and brittle under the conditions of the application and, therefore, will not exhibit the lubricating properties and performance for which the grease was originally designed.

The general performance and properties of grease as a function of thickener are as follows:

1. An aluminum soap–thickened grease generally exhibits excellent water resistance, poor mechanical stability, excellent oxidative stability, good oil separation, and poor pumpability and, in general, can be used to a maximum application temperature of 175°F (79.5°C).
2. A calcium soap–thickened grease generally exhibits excellent water resistance, fair mechanical stability, poor oxidative stability, excellent antirust performance, and fair pumpability and, in general, can be used to a maximum application temperature of 250°F (121°C).
3. A lithium soap–thickened grease generally exhibits good water resistance, excellent mechanical stability, good-to-excellent oxidative stability, poor-to-excellent antirust performance depending on the formulation, and fair-to-excellent pumpability and, in general, can be used to a maximum application temperature of 275°F (135°C).
4. An aluminum complex soap–thickened grease generally exhibits excellent water resistance, good-to-excellent mechanical stability, and good pumpability and, in general, can be used to a maximum application temperature of 350°F (177°C).
5. A calcium complex soap–thickened grease generally exhibits excellent oil separation and good mechanical stability, the thickener in this case provides a degree of AW and EP protection, and in general, it can be used to a maximum application temperature of 350°F (177°C).
6. A lithium complex soap–thickened grease generally exhibits excellent oil separation and moderate water resistance and, in general, can be used to a maximum application temperature of 350°F (177°C).
7. Clay-thickened grease has good-to-excellent water resistance, good pumpability, and excellent oil separation and, in general, can be used at maximum application temperatures greater than 350°F (177°C).
8. Polyurea-thickened grease has excellent oxidative stability, excellent pumpability, and excellent oil separation but has poor worked stability and fair-to-modest antirust performance. These greases can be used at a maximum application temperature of 350°F (177°C). Polyurea greases soften easily but are reversible.

Chemical and physical processes caused by thermal and shear stresses degrade greases [29]. These authors demonstrated that thermally aged lithium 12-hydroxy stearate greases were affected in terms of oil film thickness and oil release in a rolling contact under starved conditions.

24.6 OXIDATION INHIBITORS

The mechanism of hydrocarbon oxidation, because of its importance to lubricant chemistry and performance, has been well studied and is reviewed in Chapter 1 of this book.

Hydrocarbons react with oxygen to initially produce peroxides and hydroperoxides that further react to give alcohols, aldehydes, ketones, and carboxylic acids. These oxidation reactions proceed via free-radical chain processes. Grease, which is basically soap-thickened hydrocarbon, is also susceptible to oxidation. In addition, the metals of the metal soaps can catalyze oxidation.

Examples of classes of antioxidants used in grease are

- *Hindered phenols*: for example, 2,6-di-*t*-butyl phenol and 2,6-di-*t*-butyl-*p*-cresol
- *Aromatic amines*: for example, diarylamines, di-octyldiphenylamine
- *Metal dialkyldithiophosphates*: for example, zinc dithiophosphate
- *Metal dialkyldithiocarbamates*: for example, zinc and molybdenum dithiocarbamates
- *Ashless dialkyldithiocarbamates*
- *Sulfurized phenols*: for example, phenolic thioesters and phenolic thioethers
- *Phenothiazine*
- *Disulfides*: for example, diaryldisulfides
- *Trialkyl and triaryl phosphates and phosphites*: for example, tris(2,4-di-tert-butylphenyl) phosphite

Alkylated phenol antioxidants are most effective at low temperatures. Secondary aromatic amines such as phenyl alpha-naphthylamine, phenyl beta-naphthylamine, di-octyldiphenylamine, and phenothiazines are most useful at high temperatures [5]. In practice, grease is generally formulated to include a combination of alkylated or secondary amine-type and phenol-type antioxidants to provide performance over as wide a temperature range as possible.

In some cases, the combination of antioxidants (or other additives) provides an additive effect, while in other cases, synergy is observed when both a hindered phenolic and an aryl amine antioxidant are used together in the same formulation.

24.7 FRICTION AND WEAR

Friction is the force required to cause the motion of two surfaces or bodies in contact with each other. Lubricants are used to reduce the frictional forces. High friction results in heat, and because more force or power is necessary to move the parts relative to one another, this friction reduces operating efficiency. When the lubricant film is insufficient to protect the metal surfaces, there is wear on one or both components. Wear is material loss directly caused by the interaction of asperities on the two surfaces while in relative motion to each other. Since wear results in the loss of material and the scarring changes the size and shape of the machined components, wear reduces the useful life of the components. Extreme wear can result in failure of the equipment and in safety issues. There are three general types of wear: abrasive, adhesive, and corrosive. These are generally addressed by formulating a grease using additives designed to protect against these phenomena. When a lubricant is applied between the rubbing surfaces, the friction and wear can be minimized. Three lubrication regimes are defined depending on the amount of lubricant film separating the surfaces. These are

1. Boundary lubrication
2. Elastohydrodynamic
3. Hydrodynamic lubrication

These three lubrication regimes are indicated on a Stribeck curve shown in Figure 24.2 [30]. Approximate thickness of films in each regime and how they are related to the size of asperities and sliding wear debris in the boundary regime are shown in Table 24.5 [31].

Hydrodynamic lubrication is a regime where the moving surfaces are essentially separated from each other. In this regime, the viscosity of the oil in combination with the motion of the mechanical components can produce a fluid pressure high enough to completely separate the two surfaces.

Elastohydrodynamic lubrication is a regime where the film thickness is insufficient to completely separate the surfaces. In this regime, the surface asperities make contact, which leads to wear. Lubricant in the contact area is continually replenished at the front of the contact [32]. The film thickness in the elastohydrodynamic regime is larger than in boundary lubrication but smaller than film thickness in the hydrodynamic regime.

Boundary lubrication is a regime where film thickness between the moving surfaces is only a few molecules thick. In this regime, because of the closeness of the moving surfaces,

TABLE 24.5
Dimensions of Films, Asperities, and Debris Related to Boundary Lubrication

	Approximate Size Range (µm)
Monomolecular layer	0.002–0.2
Sliding wear debris	0.002–0.1
Boundary film	0.002–3
Elastohydrodynamic film	0.01–5
Asperity height	0.01–5
Rolling wear debris	0.07–10
Hydrodynamic film	2–100
Asperity tip radius	10–1000
Concentrated contact width	30–500

friction and wear are determined by the properties of both the surfaces and the lubricant. Boundary films form because they reduce the surface energy and, therefore, are thermodynamically favored [33]. These films form by molecules that contain polar functional groups. Because of this, they orient onto the surface by either chemical or physical adsorption. Even oxidation products derived from the breakdown of the lubricating fluid can adsorb onto metal parts and into contact areas that are being lubricated. Boundary lubrication can range from mild to severe conditions.

Physical adsorption is a reversible process where molecules adsorb and desorb from a surface without chemical change. Additives that provide protection by physical adsorption are polar structures. This is because at least two phenomena must occur: the molecule must have a preferential affinity for the surface and it should have a preferred orientation on the surface so that a more closely packed arrangement can be achieved. Alcohols, acids, and amines are examples of long-chain molecules with functional groups at the end. Molecules that can pack tightly and orient in a close-packed arrangement relative to the surface provide improved film strength. Because the forces involved in physical adsorption are relatively weak, these films are effective at low to moderate temperatures. New molecules from the bulk lubricant are constantly available to replace those that physically desorb or are mechanically removed from the surface.

Chemical adsorption, however, is an irreversible process where a lubricant fluid molecule or additive component reacts with the surface to form a low-shear-strength protective layer. As this new low-shear-strength material is worn away, additional additive reacts to form a new protective layer. Protection from chemical adsorption occurs at higher temperatures because chemical reactions are required to generate the actual species that form the surface films. EP additives can protect lubricated surfaces at temperatures as high as 400°C.

Wear protection and friction reduction over a wide temperature range can be achieved by combining additives that function by physical adsorption and chemical adsorption. Between the low-temperature physically adsorbed layer and the high-temperature chemically adsorbed layer can be a temperature

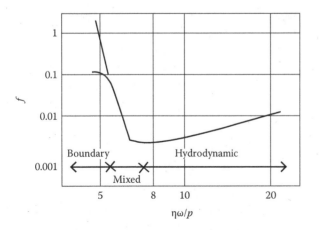

FIGURE 24.2 Stribeck curve for a journal bearing.

range over which there is poorer protection. This has been experimentally demonstrated where oleic acid was used as the normal wear additive and a chlorinated additive provided EP protection at higher temperatures [34].

24.8 EXTREME-PRESSURE AND ANTIWEAR AGENTS

EP and AW additives are used to reduce friction and prevent wear under moderate to more severe boundary lubrication conditions. Reactive compounds containing sulfur, phosphorus, or chlorine, metals, or combinations are known to provide EP protection.

Under high loads, opposing metal surfaces contact each other, and as a result, high local temperatures develop, enabling an EP agent to react with the metal surfaces, forming a surface film preventing the welding of opposing asperities [35].

Some of the major group materials that have been used as EP and AW additives are as follows:

- Sulfurized olefins, fats, and esters
- Chlorinated paraffins
- *Metal dialkyldithiophosphates*: including antimony and zinc
- *Phosphate and thiophosphate esters*: for example, tricresyl phosphate, di-*n*-octyl phosphite, isodecyl diphenyl phosphite
- Ammonium salts of phosphate esters
- Borate esters
- *Metal dithiocarbamates*: including antimony
- *Metal naphthenates*: including bismuth and lead
- *Metal soaps*: including lead
- *Sulfides and disulfides*: for example, diaryldisulfides
- High-molecular-weight complex esters

24.8.1 SOLID ADDITIVES

Solid additives are organic or polymeric solid materials or inorganic compounds used to impart EP and friction reduction properties to the grease and protection in case of lubricant loss. A more detailed description of solid additives can be found in Chapter 6 of this book. Examples include the following:

- Bismuth has recently been reported as being more environmentally friendly than lead for application as an EP additive [36,37].
- Boron-containing additives: boric acid, borax, and metal borates.
- Boron nitride.
- Molybdenum disulfide.
- Inorganic sulfur–phosphorus additive: patented blend of phosphate and thiosulfate [38].
- Fluorinated polymers, for example, perfluorinated polyolefins.
- Graphite—in various forms. The merits of expanded graphite have been reported [39].

- Calcium acetate, carbonate, phosphate, and cerium fluoride.
- Zinc stearate and zinc oxide.
- Copper powder.
- Nickel powder.
- Phosphate glasses.

EP properties and the mechanism of phosphate glasses in lubricating greases have been studied where the authors compared phosphate glasses to molybdenum disulfide, graphite, molybdenum dithiocarbamate, polytrifluoroethylene, and boron nitride [40]. An improvement in the load-carrying capacity of greases using phosphate glass—a white, relatively inexpensive powder compared with other solid additives—has been reported to provide very effective wear protection under severe conditions [41,42]. Under light loads, the finely divided phosphate glass particles were found to maintain their original round shape. The particles were performing as micro ball bearings under these low-load conditions. At very high loads, the phosphate glass particles compressed and formed a thick protective film on the wear surface.

Another area where greases find application is in space applications, where these lubricants need to demonstrate long-term use in situations involving vacuum. Additives used in these greases must have low volatility and excellent lubricity. Greases for these applications have included perfluoropolyalkylether (PFPAE) fluids for many years. Studies using deep-groove ball bearings filled with PFPAE-based grease have been reported with long run periods in vacuum of 10^{-4}–10^{-4} Pa at 2000 rpm [43].

24.9 RUST AND CORROSION INHIBITORS

Rust is a form of corrosion formed by electrochemical interaction between iron and atmospheric oxygen and is accelerated in the presence of moisture due to the catalytic action of water [44]. Rusting of iron and steel surfaces can reduce operating efficiencies and cause part and equipment damage.

The electrochemical oxidation of the surfaces of iron or steel can be prevented by the addition of specific water-blocking additives to lubricating grease that inhibit the formation of rust (or iron oxides). Rust inhibitors are typically highly polar surface active oil-soluble compounds that attach to metal surfaces by physical adsorption.

Rust inhibitors incorporated into lubricating grease provide a protective film against the effects of moisture, water, and air. Corrosion inhibitors work by neutralizing corrosive acids formed by the degradation of base fluids and lubricant additives.

Examples of various chemical classes of rust and corrosion inhibitors used in grease are

- *Carboxylic acids, including fatty acids*: for example, alkyl succinic acid half ester and nonylphenoxy acetic acid
- *Salts of fatty acids and amines*: for example, disodium sebacate

- *Succinates*: for example, alkyl succinic acid half ester
- *Fatty amines and amides*
- *Metal sulfonates*: including ammonium, barium, and sodium
- *Metal naphthenates*: including bismuth, lead, and zinc
- *Metal phenolates*
- *Nitrogen-containing heterocyclic compounds*: for example, substituted imidazolines
- *Amine phosphates*
- *Salts of phosphates esters*

24.9.1 METAL DEACTIVATORS

These materials reduce the catalytic effect of metal on the rate of oxidation. They act by forming an inactive film on metal surfaces by complexing with metallic ions. Several classes of materials have been reported to be effective:

- Organic complexes containing nitrogen or sulfur, amines, sulfides, and phosphates
- Derivates of 2,5-dimercapto-1,3,4-thiadiazole
- Triazoles, benzotriazoles, and tolyltriazoles
- Disalicylidene-propanediamine

24.9.2 TACKINESS ADDITIVES

Grease may be formulated to withstand the heavy impact common in heavy equipment applications. The adhesive and cohesive properties of grease can be improved to resist throw-off from bearings and fittings, while providing extra cushioning to reduce shock and noise through the use of tackiness agents. The water resistance of such grease can also be significantly improved through the use of tackiness additives.

High-molecular-weight polymers such as polyisobutylene, polybutene, ethylene–propylene Olefin copolymer (OCP) copolymer, and latex compounds are typical examples of tackiness additives. Like all long-chain polymers, tackiness additives are susceptible to breakdown when exposed to high rates of shear. A further discussion of these interesting materials is described in Chapter 13 of this book.

24.10 ENVIRONMENTALLY FRIENDLY GREASE

Fully formulated grease that combines needed high-performance features with environmental safety and compliance has been developed not only for civilian industrial and automotive applications but for the military as well [45].

The goal of environmentally friendly lubricants is to minimize or eliminate any potential harm or damage to humans, wildlife, soil, or water. Depending on the nature of the application, the use of environmentally friendly (or *green*) lubricants will enable industry to reduce some of, and perhaps even eliminate, the costs associated with the remediation and disposal of nonbiodegradable and/or toxic lubricants.

One key raw material used in formulating biodegradable grease is vegetable oil. Vegetable oils are obtained from renewable sources and are biodegradable and, as such, are more environmentally friendly than conventional mineral oil–based lubricants [46].

A special class of vegetable oils, containing a high oleic content (greater than or equal to 75% oleic) and low polyunsaturated fatty acid content (linoleic and/or linolenic), displays good oxidative stability with acceptable low-temperature properties. This makes them well suited for use in greases compared to conventional vegetable oils [46].

In addition to the lubricating fluids, the toxicity and biodegradability of additive components are important. Over the years, many of the additives that were originally based on fractions cut from petroleum- or coal-derived liquids are now synthetic and, therefore, are of much higher purity. Linear side-chain hydrocarbon groups have in many instances replaced branched and aromatic functional groups. This results in greater potential biodegradability. Toxicity of additives is related to metals in many cases. Current trends are to replace metal-containing additives with ashless varieties having similar or greater performance features. This will result in a lower pollutant load on the environment. Even military formulations are moving toward more environmentally friendly versions.

Environmentally friendly grease is used in applications such as agriculture, construction, forestry, marine, mining, and railroad. Specific applications include tramway tracks and railway switches, wheel flange lubricant for railways, and farm tractors.

24.11 SUMMARY

In summary, grease finds applications where fluid lubricants may drip or leak from the point of application. Grease reduces the need to frequently lubricate a particular site, because the grease structure serves as a lubricant reservoir. Grease is very effective where lubricant is needed on a vertical machine component or at positions that are difficult to reach. Grease acts as a physical barrier that is effective in sealing out external contaminants and provides better protection when contaminated than does a liquid lubricant. Grease can also provide noise reduction in certain applications and is effective in protecting equipment at high temperatures and pressures and under conditions where there is shock loading in the lubricated components.

The formulation of grease can be considered an art and/or a science depending on the researcher questioned, but certain guidelines are generally useful.

For low-temperature applications, grease can use low-viscosity, high-viscosity index lubricating fluids with a relatively low thickener content. The fluid lubricant should have a low pour point and good pumpability and rust protection. Grease for high-temperature applications will typically have a higher-viscosity lubricating fluid and high-temperature complex thickener. This will result in grease with a higher dropping point. This type of grease will also find application where low flammability and, therefore, low volatility are design requirements.

Grease for applications where water contamination is an issue should be designed with a water-resistant thickener and generally a high-viscosity lubricating fluid. The grease should have low water washout and low water spray-off, which can be improved by incorporating a tackiness additive. Any application in a wet environment requires that the grease have excellent rust and corrosion protection. Careful formulation to balance the chemistries of the needed additives is critical in achieving the described performance where surface protection is so important to machine life. Where shock or heavy loads are applied, grease is formulated using high-viscosity lubricating fluids and should include higher concentrations of AW and EP additives. This combination will provide both thicker lubricant film and appropriate additive chemistry to protect the metal surfaces.

Grease used in a centralized system needs to have a low-to-moderate thickener concentration and a relatively low-viscosity lubricating fluid so that the grease exhibits a lower apparent viscosity and can be readily pumped throughout the system.

24.12 EFFECTS OF INDIVIDUAL ADDITIVES AND GREASE SOAPS

This section includes a sample of application data where various grease additives were employed to modify the performance of various greases. The first section describes the use of individual additives with a particular type of grease. The second section demonstrates some synergistic effects of combining two or more additives that individually may not provide all the performance benefits needed.

An excellent general review of the fundamental characteristics of synthetic lubricating greases has been published [47]. This review describes the very wide range of potential synthetic greases in terms of properties and performance capabilities but also cautions that some of these higher-performance capabilities are not always needed, and thus, the higher cost of synthetics may not always be justified. Previously published chapters on greases can be found in earlier books [48,49].

These applications data are meant to serve as a guideline, not as a recipe for greases. The actual combination of all the components of grease, including the base oil, thickener, each additive, and the order of addition, will affect the properties and performance of a final grease formulation, and thus these issues are left for the formulator. A list of additive components has also been included in Table 24.6.

The data in Table 24.7 show the results of wear experiments on aluminum complex grease using four organomolybdenum-containing additives at the same concentration. Molyvan A is a powder, and any heterogeneity in the lubricant might result in the observed higher wear scars in grease made with this additive.

Comparison of aluminum complex and lithium complex grease with one level of Vanlube 829 showed an improvement in both four-ball EP weld load and four-ball wear scar in the lithium complex grease (Table 24.8). It should be noted that the weld load difference in these results is within experimental error, but the combined improvement in weld load and wear scar is directionally desirable. With Na-Lube 5665, there is an improvement in both weld load and scar diameter in these tests.

The data in Table 24.9 show the improved performance of an aluminum complex grease containing Vanlube 622 over the grease containing a similar amount of Vanlube 73 or Vanlube 7723. There should be concern over the high sulfur content of Vanlube 622 and its effect on copper corrosion for

TABLE 24.6
Grease Additives—Chemical Components

Molyvan L: Molybdenum di (2-ethylhexyl) phosphorodithioate
Molyvan 822: Molybdenum dialkyldithiocarbamate in oil
Molyvan 855: Organomolybdenum complex
Molyvan A: Molybdenum di-*n*-butyldithiocarbamate
Vanlube 829: 5,5-Dithiobis-(1,3,4-thiadiazole-2(3H)-thione)
Vanlube 73: Antimony tris(dialkyldithiocarbamate) in oil
Vanlube 7723: Methylene bis(dibutyldithiocarbamate)
Vanlube 622: Antimony o,o-dialkylphosphorodithioate in oil
Vanlube 8610: Antimony dithiocarbamate/sulfurized olefin blend
Vanlube NA: Alkylated diphenylamines
Desilube 88: Blend of phosphate and thiosulfate
Irgalube 63: Ashless dithiophosphate
Irgalube TPPT: Triphenyl phosphorothionate
Lubrizol 1395: Zinc dialkyldithiophosphate (ZDDP)
Lubrizol 5235: Sulfur–phosphorus–zinc additive package
Lubrizol 5034A: Sulfur–phosphorus industrial gear oil additive package
Amine O: Substituted imidazoline
Sarkosyl O: N-oleoyl sarcosine
Na-Sul ZS-HT: Zinc dinonylnaphthalene sulfonate/carboxylate complex
MoS$_2$: Molybdenum disulfide

TABLE 24.7
Organomolybdenum Compounds in Aluminum Complex Grease

Additive	Wt.%	Four-Ball EP Weld Load (kg)	Four-Ball Wear Scar Diameter (mm)
Base grease (880 SUS paraffinic base oil blend)	—	126	0.68
Molyvan L	3.0	200	0.48
Molyvan 822	3.0	250	0.60
Molyvan 855	3.0	250	0.58
Molyvan A	3.0	250	0.76

applications where the grease will be in contact with copper-containing components.

Comparison of two antimony-containing EP/AW additives in aluminum complex grease showed improved EP performance and lower wear scar diameter for the sulfurized antimony dithiocarbamate (Vanlube 8610) (Table 24.10). Comparison of two sulfur–phosphorus EP additives in aluminum complex and lithium complex grease shows extremely high four-ball EP performance using Desilube 88 compared to Irgalube 63 (Table 24.11).

24.12.1 FOOD-GRADE GREASES

Food-grade greases are a special class of grease that requires nontoxic components approved for this particular application. Additional discussion and details can be found in Chapter 22 of this book.

TABLE 24.8
Thiadiazole Compound in Grease

Additive	Wt.%	Four-Ball EP Weld Load (kg)	Four-Ball Wear Scar Diameter (mm)
Aluminum complex base grease (900 SUS paraffinic base oil blend)	—	140	0.59
Lithium complex base grease (600 SUS paraffinic base oil)	—	180	0.61
Vanlube 829 in aluminum complex	3.0	620	0.66
Vanlube 829 in lithium complex	3.0	800 kg pass	0.50
Aluminum complex base grease (700 SUS paraffinic base oil)		126	0.68
Na-Lube EP-5665	1.0	300	0.68
Na-Lube EP-5665	2.0	500	0.58

Na-Lube EP-5665: 5,5-dithiobis-(1,3,4-thiadiazole-2(3H)-thione).

TABLE 24.9
Effect of Antimony Compounds in Aluminum Complex Grease

Additive	Wt.%	Four-Ball EP Weld Load (kg)	Four-Ball Wear Scar Diameter (mm)
Aluminum complex base grease (980 SUS paraffinic base oil blend)	—	126	0.68
Vanlube 73	3.0	315	0.84
Vanlube 7723	3.0	315	0.81
Vanlube 622	3.0	500	0.46

TABLE 24.10
Antimony Dithiocarbamate (DTC) and Antimony DTC/Sulfurized Olefin in Aluminum Complex

Additive	Wt.%	Four-Ball EP Weld Load (kg)	Four-Ball Wear Scar Diameter (mm)
Aluminum complex base grease (880 SUS paraffinic base oil blend)	—	126	0.68
Vanlube 73	3.0	315	0.84
Vanlube 8610	3.0	400	0.74

TABLE 24.11
Sulfur–Phosphorus Compounds in Grease

Additive	Wt.%	Four-Ball EP Weld Load (kg)	Four-Ball Wear Scar Diameter (mm)
Aluminum complex base grease (900 SUS paraffinic base oil blend)	—	140	0.59
Lithium complex base grease (600 SUS paraffinic base oil blend)	—	180	0.61
Desilube 88 in aluminum complex	3.0	500	0.76
Desilube 88 in lithium complex	3.0	620	0.88
Irgalube 63 in aluminum complex	3.0	250	0.60
Irgalube 63 in lithium complex	3.0	315	0.55

TABLE 24.12
Food-Grade Aluminum Complex Grease

Additive	Wt.%	Four-Ball EP Weld Load (kg)	Four-Ball Wear Scar Diameter (mm)	Rust Test ASTM D1743
Aluminum complex base grease (500 SUS Technical White mineral oil)	—	140	0.62	Fail
Calcium carbonate	4.0	315	0.49	—
Calcium carbonate	4.0	315	0.55	Pass
Amine O	0.5			
Sarkosyl O	0.5			
Amine O	0.5	160	0.80	Pass
Desilube 88	3.0	500	1.04	—

The data in Table 24.12 demonstrate the better balance of EP and wear protection by the addition of calcium carbonate when compared to Desilube 88 alone. Rust protection is demonstrated by the synergistic effect of the two corrosion inhibitors using ASTIM D 1743. The individual additives are not effective alone in providing sufficient EP and wear protection in this formulation.

24.12.2 SYNERGISTIC ADDITIVE EFFECTS

Frequently, combinations of two or more additives show enhanced performance over that of the individual components. The occurrence and magnitude of synergistic behavior

TABLE 24.13
Synergistic Compounds in Aluminum Complex Grease

Additive	Wt.%	Four-Ball EP Weld Load (kg)	Four-Ball Wear Scar Diameter (mm)
Aluminum complex base grease (880 SUS paraffinic base oil blend)	—	126	0.68
Vanlube 829	1.0	400	0.88
Irgalube TPPT	3.0		
Vanlube 829	1.0	400	0.60
Irgalube 63	3.0		
Irgalube 63	3.0	250	0.50
Irgalube TPPT	3.0	200	0.55

involving specific compounds very likely depend on the nature of the base grease, including the base oil and the thickener.

The use of Vanlube 829 in combination with Irgalube 63 showed an improvement in the four-ball wear scar compared with Vanlube 829 in combination with Irgalube TIPPT (Table 24.13). These combinations were each better in EP performance than formulations containing only Irgalube 63 or Irgalube TIPPT in the absence of Vanlube 829. Desilube 88 and MoS_2 showed a significant improvement in EP and wear performance for aluminum complex grease when compared to lithium complex grease (Table 24.14).

A sulfonate rust inhibitor can enhance the four-ball wear performance of a lithium 12-hydroxy stearate grease in combination with an EP additive that by itself does not show reduced wear at low loads [50] (Table 24.15). A widely used AW additive, zinc dialkyldithiophosphate (ZDDP), exhibits moderate EP performance and low wear scars in both aluminum and lithium complex greases compared to the base greases alone (Table 24.16).

TABLE 24.14
Addition of MoS_2 in Grease

Additive	Wt.%	Four-Ball EP Weld Load (kg)	Four-Ball Wear Scar Diameter (mm)
MoS_2 in lithium complex (600 SUS paraffinic base oil)	3.0	500	0.72
MoS_2 in aluminum complex (900 SUS paraffinic base oil blend)	3.0	400	0.90
MoS_2	1.3	500	1.0
Desilube 88 in lithium complex	3.0		
MoS_2	1.3	800	0.8
Desilube 88 in aluminum complex	3.0		

TABLE 24.15
Effect of a Sulfonate Additive in Lithium 12-hydroxystearate Grease

Additive	Wt.%	Four-Ball EP Weld Load (kg)	Four-Ball Wear Scar Diameter (mm)
Vanlube 73	3.0	250	0.73
Vanlube 73	3.0	250	0.47
Na-Sul ZS-HT	1.0		

TABLE 24.16
ZDDP Component in Grease

Additive	Wt.%	Four-Ball EP Weld Load (kg)	Four-Ball Wear Scar Diameter (mm)
Aluminum complex base grease (900 SUS paraffinic base oil blend)	—	140	0.59
Lithium complex base grease (600 SUS paraffinic base oil)	—	180	0.61
Lubrizol 1395 in lithium complex	3.0	315	0.47
Lubrizol 1395 in aluminum complex	3.0	250	0.55

TABLE 24.17
S-P-Zn-N Additive Package in Aluminum Complex Grease

Additive	Wt.%	Four-Ball EP Weld Load (kg)	Four-Ball Wear Scar Diameter (mm)
Lubrizol 5235 in aluminum complex (900 SUS paraffinic base oil blend)	3.0	250	0.57
Lubrizol 5235 in lithium complex (600 SUS paraffinic base oil)	3.0	315	0.50

A multifunctional additive package specifically designed for greases provides good EP and AW performance. In general, this additive package also provides antioxidant and anticorrosion properties and copper metal deactivation (Table 24.17). EP additives can sometimes be detrimental to the oxidative stability of the grease. The addition of an amine antioxidant can improve the oxidation stability of EP-containing organoclay grease, as shown in Table 24.18.

TABLE 24.18
Oxidation Stability in Organoclay Grease

Additive	Wt.%	Four-Ball EP Weld Load (kg)	Four-Ball Wear Scar Diameter (mm)	Oxidation Stability psi Loss at 500 h, ASTM D942
Organoclay base grease (600 SUS paraffinic base oil)	—	140	0.61	13.0
Molyvan A in organoclay	1.4	200	0.52	17.0
Molyvan A	1.4	200	—	12.0
Vanlube NA in organoclay	0.5			

TABLE 24.19
S-P Gear Oil Package in Grease

Additive	Wt.%	Four-Ball EP Weld Load (kg)	Four-Ball Wear Scar Diameter (mm)
Lubrizol 5034A in aluminum complex (900 SUS paraffinic base oil blend)	3.0	250	0.45
Lubrizol 5034A in lithium complex (600 SUS paraffinic base oil)	3.0	315	0.50

Note: Additional additives such as rust and corrosion inhibitors and antioxidants are usually added to the examples mentioned earlier to provide fully formulated grease. The four-ball EP test has a repeatability and reproducibility of one loading. The four-ball wear test has a repeatability of 0.2 mm wear scar diameter and a reproducibility of 0.37 mm.

Because gear oils can require good EP and AW properties, they are sometimes used in grease formulations. A sulfur–phosphorus industrial gear oil additive package exhibited good EP and wear performance in both aluminum complex and lithium complex greases (Table 24.19).

ACKNOWLEDGMENTS

The authors would like to thank Henry Kruschwitz of FedChem LLC for his help and the use of his laboratory.

REFERENCES

1. A Polishuk. Expanded graphite as a dispersed phase for greases. *NLGI Spokesman* 61(2):38, May 1997.
2. D Klamann. *Lubricants and Related Products.* Weinheim, Germany: Verlag Chemie, 1984, pp. 177–217.
3. CJ Boner. *Modern Lubricating Greases.* New York: Scientific Publication Ltd., 1976.
4. The NLGI Grease Production Survey Report, 2006.
5. Lubrizol. *Ready Reference for Grease.* Wickliffe, OH: The Lubrizol Corp., October 1999.
6. JA Brown, CN Hudson, LD Loring. Lithium grease-detailed study by electron microscope. *The Institute Spokesman* 15(11):8–17, 1952.
7. ASTM D-217. Standard test methods for cone penetration of lubricating grease. In: *Annual Book of ASTM Standards.* Philadelphia, PA, Section 5, Vol. 5.01, 1996.
8. Texaco Inc. (New York). Technology of modern greases. *Lubrication* 77(1), 1991.
9. AC White, DA Turner. Grease making in the 90's. *NLGI Spokesman* 59(3):97–107, June 1994.
10. Texaco Inc. (New York). Continuous grease manufacture. *Lubrication* 55(1), 1969.
11. T Okaniwa, H Kimura. *NLGI Spokesman* 61(3):18–24, 1997.
12. PA Bassette. Presented at the *1997 Annual Meeting on the NLGI,* Carlsbad, CA, October 26–29, 1997.
13. JG Wills. *Lubrication Fundamentals.* New York: Marcel Dekker, 1980, pp. 59–84.
14. ASTM D 1403. Standard test methods for cone penetration of lubricating grease using one-quarter and one-half scale cone equipment. In: *Annual Book of ASTM Standards.* Philadelphia, PA, Section 5.01, 1996.
15. ASTM D 2265. Standard test method for dropping point of a lubricating grease over wide temperature range. In: *Annual Book of ASTM Standards.* Philadelphia, PA, Section 5, Vol. 5.01, 1996.
16. Grease—The world's first lubricant. *NLGI Spokesman* 59(5):1–8, insert, August 1995.
17. AT Polishuk. Properties and performance characteristics of some modern lubrication greases. *Lubr. Eng.* 33(3):133–138, March 1997.
18. M Hermansson, E Johansson, M Jansson. *J. Synthetic Lubric.* 13(3):279–288, 1996.
19. ASTM D 2266. Standard test method for wear preventative characteristics of lubricating grease (four-ball method). In: *Annual Book of ASTM Standards.* Philadelphia, PA, Section 5, Vol. 5.01, 1996.
20. ASTM D 2596. Standard test method for measurement of extreme-pressure properties of lubricating grease (four-ball method). In: *Annual Book of ASTM Standards.* Philadelphia, PA, Section 5, Vol. 5.01, 1996.
21. *NLGI Lubricating Grease Guide,* 4th edn. 1996.
22. G Fagan. Complex soap greases. Presented at *NGLI 64th Annual Meeting,* Carlsbad, CA, 1997.
23. RL Frye. An introduction to lubricating greases. Presented at *NLGI 64th Annual Meeting,* Carlsbad, CA, 1997.
24. CJ Boner. *Manufacture and Application of Lubricating Greases.* New York: Reinhold Publishing Corp., 1954.
25. BW Hotten. *Advances in Petroleum Chemistry and Refining,* Vol. 9. New York: Interscience Publishers, 1964, Chapter 3.
26. JW Wilson. Three-dimensional structure of grease thickener particles. *NLGI Spokesman* 3:372–379, 1994.
27. G Arbocus. When greases don't perform. *Lubr. World* 3(10):7–8, October 1993.
28. R Czarny. Effect of changes in grease structure on sliding friction. *Ind. Lubr. Tribol.* 47(1):3–7, 1995.
29. S Hurley, PM Cann, HA Spikes. *Tribol. Trans.* 43(1):9–14, 2000.
30. JA Williams. *Engineering Tribology.* New York: Oxford University Press, 1994.
31. ER Booser, ed. *CRC Handbook of Lubrication,* Vol. 2. Boca Raton, FL: CRC Press, 1984.

32. J Pemberton, A Cameron. A mechanism of fluid replenishment in elastohydrodynamic contacts. *Wear* 37:185–190, 1976.

33. Texaco Inc. Boundary lubrication. 57(1), 1971.

34. DD Fuller. *Theory and Practice of Lubrication for Engineers.* New York: John Wiley, 1984.

35. *Encyclopedia for the User of Petroleum Products.* Lubetext DG-400, Exxon Corp., 1993, p. 10.

36. O Rohr. *NLGI Spokesman* 61(6):10–17, 1997.

37. O Rohr. Presented at *NLGI 63rd Annual Meeting*, Scottsdale, AZ, October 27–30, 1996.

38. JP King. Presented at *1998 NLGI Annual Grease Meeting*, Dublin, Ireland.

39. YL Ischuk et al. *NLGI Spokesman* 61(6):24–30, 1997.

40. T Ogawa et al. Presented at *NLGI 64th Annual Meeting*, Carlsbad, CA, October 21–29, 1997.

41. L Tocci. Transparent grease bears a heavy load. *Lubes-n-Greases* 3:52–54, April 1997.

42. T Ogawa, H Kimura, M Hayama. *NLGI Spokesman*, 62(6):28–36, 1998.

43. K Seki, M. Nishimura. *J. Synthetic Lubr.* 9(1):17–27, 1992.

44. RJ Lewis Sr. *Hawley's Condensed Chemical Dictionary*, 13th edn. New York: Van Nostrand Reinhold, 1997, p. 977.

45. I-S Rhee. *NLGI Spokesman* 63(1):8–18, 1999.

46. S Lawate, K Lal, C Huang. In: ER Booser, ed. *Tribology Data Handbook.* Boca Raton, FL: CRC Press, 1977, pp. 103–106.

47. PS Coffin. Characteristics of synthetic lubricating greases. *J. Synthetic Lubric.* 1(1):34–60, 1984.

48. P Bassette, RS Stone. Synthetic Grease. In: LR Rudnick, RL Shubkin, eds. *Synthetic Lubricants and High-Performance Functional Fluids.* New York: Marcel Dekker, 1999, pp. 519–538.

49. RC Schrama. Greases. In: ER Booser, ed. *Tribology Data Handbook.* Boca Raton, FL: CRC Press, 1977, pp. 138–155.

50. RF Baker, RF Vaast, Jr. *NLGI Spokesman* 55(12):23-487–25-489, March 1992.

25 Additives for Crankcase Lubricant Applications

Ewa A. Bardasz and Gordon D. Lamb

CONTENTS

25.1   INTRODUCTION

Engine oil lubricants make up nearly one-half of the lubricant market and therefore attract a lot of technical and fundamental scientific interest. The principal function of engine oil lubricants is to protect and extend the life of moving parts operating under many different conditions of speed, temperature, and pressure. At low temperature, the lubricant is expected to flow sufficiently in order that moving parts are not starved of oil. At higher temperatures, they are expected to keep the moving parts apart to minimize potential for abrasive wear. The lubricant performs its function by reducing metal-to-metal friction and removing heat from the moving parts. Contaminants pose an additional problem, as they accumulate in the engine during operation. The contaminants may be wear metal debris, sludge, soot (carbon) particles, organic/inorganic acids, or peroxides. An important function of the lubricant is to prevent any of these contaminants from causing significant damage to engine hardware.

To function effectively, the lubricant composition contains a variety of chemical additives as well as a spectrum of petroleum or synthetic base oils. Depending on the specific engine size and application, various combinations of additives are used to meet the industry and/or original equipment manufacturer (OEM) required performance levels, the most important of which are listed as follows:

- Detergents
- Dispersants
- Antiwear
- Antioxidants
- Viscosity modifiers
- Pour point depressants
- Foam inhibitors

In addition to these key families of additives, some specific applications require usage of other additives, specifically designed to control corrosion, rust, seal swelling, that can act as biocides and promote demulsability.

25.2   DETERGENTS

25.2.1   INTRODUCTION

Detergents play an essential role in protecting critical metallic components of internal combustion engines by neutralizing acidic compounds formed during gasoline or diesel fuel combustion processes [1–3]. In comparison to other applications, such as industrial oils, gasoline and diesel engine oils account for over 75% of total detergent consumption. Detergent treatment in engine lubricants can reach 6–10 wt.%, with marine diesel engine lubricants containing the highest concentration levels due to combustion of high sulfur fuel, which leads to the formation of strong inorganic acidic combustion products such as sulfuric acid.

The purpose of detergents in crankcase oils is

1. To suspend/disperse oil-insoluble combustion products, such as sludge or soot (carbon) and oil oxidation products
2. To neutralize combustion products (inorganic acids)
3. To neutralize organic acid products of oil degradation processes
4. To control rust, corrosion, and deposit-forming resinous species [4]

Why are these specific functions critical to engine durability? Coke and varnish-like deposits can restrict the free movement of the upper piston rings, allowing a portion of the combustion gases to pass into the crankcase or combustion chamber, leading to heavy contamination of the oil, impacting engine exhaust emissions, and even causing piston seizure if the engine operates at high loads [5]. Heavy sludge can plug oil filters, leading to oil starvation and thus to catastrophic wear, especially during cold temperature start-ups [6]. Acidic fuel combustion products can cause corrosion.

Modus operandi of detergents can be described in the following away. Molecules react with the hydroxyacids, deposit precursors, formed during the oxidation of the oil. Deposit precursors are attracted to detergent micelles, eventually trapped within them, and, thus, cannot settle out onto metal surfaces and form resinous deposits. The cleaning action of detergent additives is attributed to metal surface chemisorption processes potentially leading to the formation of metal salts.

In order to satisfy these requirements, practically all detergent additive molecules contain the following:

- *Polar head*: hydrophilic, acidic groups (e.g., sulfonate, hydroxyl, mercapton, carboxylic, or carbonamide groups) which react with metal oxides or hydroxides
- *Hydrocarbon tail*: oleophilic aliphatic, cycloaliphatic, or alkylaromatic hydrocarbon radicals which provide oil solubility
- *One or several metal ions*: calcium, magnesium, sodium

An idealized representation of the detergent structures is shown in Figure 25.1.

Although several metals have been incorporated into detergents, only three metal cations are now commonly used: calcium, magnesium, and sodium. Heavy metals such as barium are no longer used.

Detergents are described chemically in terms of their metal ratio, soap content, percent sulfate ash, degree of overbasing or

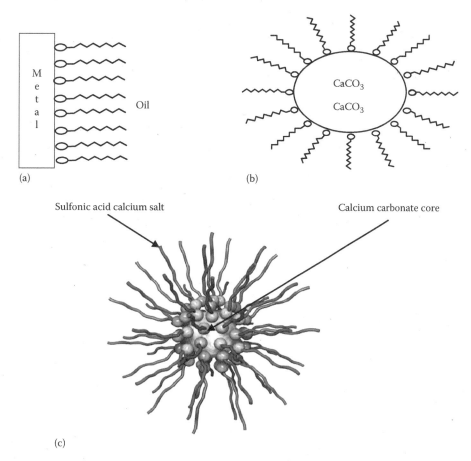

FIGURE 25.1 Idealized representations of neutral and overbased detergents (stabilized micelles): (a) Schematic representation of metal surface/neutral detergent interactions, (b) schematic representation of overbased detergent structure, and (c) computer generated representation of overbased sulfonate detergent structure.

conversion, and total base number (TBN) [2]. The *metal ratio* is defined as total equivalents of metal per equivalent of sulfonate acid. *Soap content* refers to the amount of neutral salt and reflects the detergent's cleansing ability, or detergency. The *percent sulfate ash* is the ash obtained after treating the detergent with sulfuric acid and complete combustion. The *degree of overbasing (conversion)* describes the ratio of equivalents of the metal base to equivalents of the acid substrate and is usually expressed as conversion. Conversion provides the amount of inorganic material relative to that of organic material and is expressed as the number of equivalents of base per equivalent of acid times 100. The overbased part of the detergent is needed to neutralize acid by-products. The *TBN* indicates the detergent's acid-neutralizing ability and is expressed as milligram KOH per gram of additive. It is measured using a potentiometric method (e.g., ASTM D2897).

The alkaline reserve of all modern detergents may vary considerably. Neutral detergents contain the stoichiometric amounts of metals, corresponding to the basicity of acids. Basic (or overbased) detergents contain a significant excess of metal oxides, hydroxides, carbonates, and so on, in colloidally dispersed form. The structure of detergents can be envisioned as a reverse micelle, with an amorphous carbonate molecule encapsulated by metal soap molecules with their nonpolar ends extended into the oil (Figure 25.1).

In practice, virtually all commercial detergents are overbased to some extent. For example, commercial "neutral" sulfonates have a TBN of 30 or less. "Basic" detergents have a TBN of 200–500.

Some detergents can act as oxidation inhibitors, depending upon the nature of their functional group. Most modern motor oils contain combinations of several detergent types, which are selected to give optimum performance.

Preparation of calcium detergents can be represented schematically as follows:

$$2RSO_3H + CaO \rightarrow (RSO_3)_2 Ca + H_2O$$

$$(RSO_3)_2 Ca + Ca(OH)_2 \xrightarrow[\substack{xCO_2 \\ promoters}]{xCa(OH)_2} (RSO_3)_2$$

$$Ca(CaCO_3)_x + H_2O \uparrow + Promoter \uparrow$$

25.2.2 Sulfonates

The salts of long-chain alkylarylsulfonic acids are being widely used as detergents. Basic calcium sulfonates make up 65% of the total detergent market.

As the demand for sulfonic acids has rapidly increased, synthetic products with the general structure $(RSO_3)_x Me_w (CO_3)_x (OH)_y$ are also used besides the sulfonated

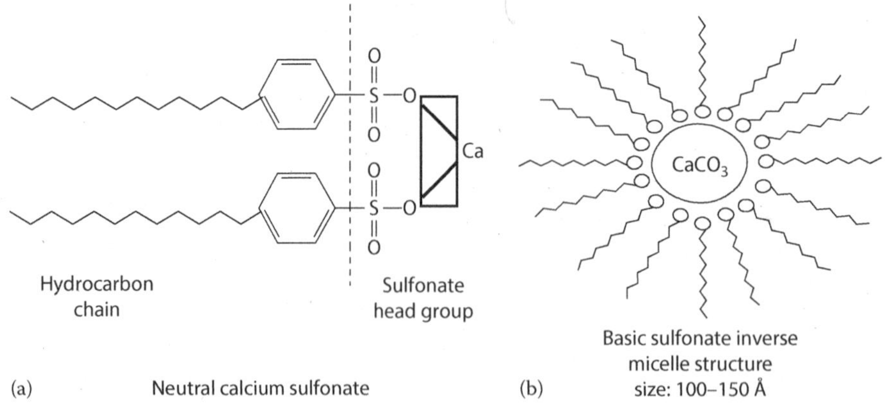

FIGURE 25.2 (a) Neutral and (b) overbased calcium sulfonate micelle structure.

alkylaromatics from petroleum refining known as "natural" sulfonates. Synthetic products are produced by the sulfonation of suitable alkylaromatics, for example, the dialkylation and polyalkylation products from dodecylbenzene production; their alkyl radicals should contain together at least 20 C atoms. Other starting materials are alpha-olefin polymers with mean molecular masses around 1000.

Neutral sulfonates (schematically shown in Figure 25.2) contain stoichiometric amounts of metal ion and acid. Besides Na, Ca, and Mg, patents have been issued describing detergents containing tin, chromium, zinc, nickel, and aluminum; however, the performance of these metals is found to be inferior to that of alkaline earth metals.

Neutral oil-soluble metal petroleum sulfonates can be converted to *basic sulfonates* by mixing and heating with metal oxides or hydroxides, followed by filtration. In these products, metal oxides and hydroxides are present in colloidally dispersed form (Figure 25.2). Such basic sulfonates have a considerably increased alkaline reserve and thus a higher acid-neutralizing power.

Treatment with carbon dioxide converts basic sulfonates into metal sulfonate–carbonate complexes that have the same alkaline reserve, yet a lower basicity. Efforts to produce additives with even higher neutralizing power have led to the development of *overbased sulfonates*. Besides high neutralizing power, the additives also possess a high dispersing capacity, due to the large amount of polar inorganic bases present.

Overbased sulfonates are produced, for instance, by heating an oil-soluble sulfonate with metal oxides in the presence of substances that act as catalysts, such as phenols, phosphoric acid derivatives, and so on.

25.2.3 PHENATES, SULFURIZED PHENATES, AND SALICYLATES

Basic phenates make up 31% of the total detergent market. Schematic structures of phenates and sulfurized phenates are shown in Figure 25.3.

Phenate detergents are available as calcium and magnesium salts. Metal salts of alkylphenols and alkylphenol sulfides (R) (OH) C_6H_3–Sx–C_6H_3 (OH) (R), where x = 1 or 2 and R is ~12 C, can be prepared at elevated temperatures by the reaction of alcoholates such as Mg ethylate with alkylphenols or by the reaction

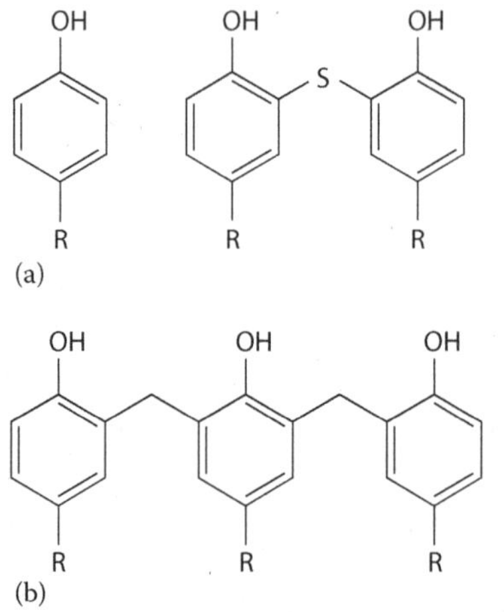

FIGURE 25.3 (a) Phenates and sulfurized phenates structures and (b) sulfur free phenates structures.

of phenols or phenol sulfides with an excess of metal oxide or hydroxide (particularly of Ca) sulfides with an excess of metal oxides or hydroxides in neutral phenates. Besides their neutralization power, phenates also possess good dispersant properties.

As in the case of sulfonates, phenates can be overbased. Overbased phenates are often used as components of marine diesel cylinder lubricants.

In many commercial lubricant applications, sulfonate and phenate detergents are used in combination and often contain various metals in order to obtain an optimum detergent action and neutralizing power. Besides better neutralizing power, the main incentive for the use of basic phenates is lower manufacturing cost compared to normal phenates.

Salicylates are less commonly used as detergents in crankcase lubrication. The typical structure of salicylate detergents is given in Figure 25.4.

Besides their detergent properties, metal alkylsalicylates also possess oxidation-inhibiting and anticorrosion properties. Their solubility in mineral oils can be improved in the case of esters by extending the chain length of the alcohol radicals or generally by alkylation of the aromatic ring. The alkaline earth salicylates are usually overbased by the incorporation of alkaline earth carbonates, stabilized in the form of micelles.

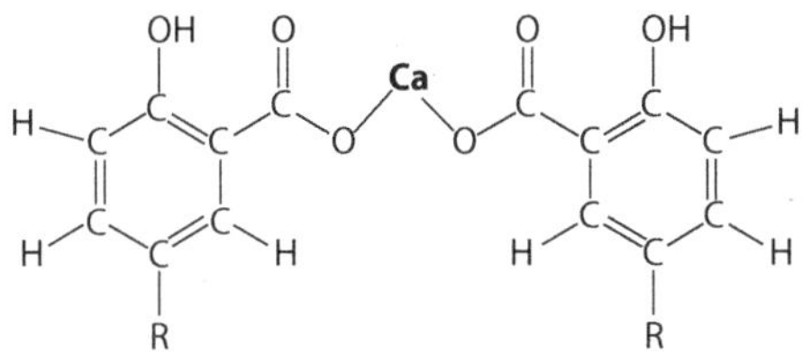

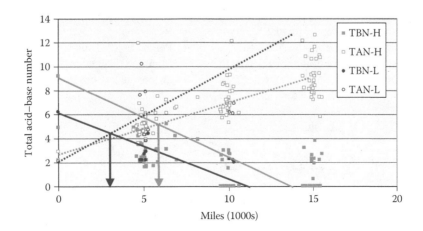

FIGURE 25.4 Typical calcium salicylate structure.

25.2.4 Other Detergents: Phosphates, Thiophosphates, Phosphonates, and Thiophosphonates

Besides their use as oxidation inhibitors, phosphates and thio-phosphates also serve in several variations and combinations as detergents. Thiophosphonates are obtained by the reaction of phosphorus pentasulfide with polyisobutenes (PIBs), olefins, fatty alcohols, and esters, which are neutralized after hydrolysis with metal hydroxide.

25.2.5 Performance in Lubricants

For crankcase engine oils, which include passenger car, heavy-duty diesel, marine diesel, and stationary gas applications, detergents provide several key performance functions. One of the primary functions of overbased detergents is to neutralize acidic combustion by-products [2–4]. In all reciprocating piston internal combustion engines, gases from the combustion chamber are forced around and through the piston rings and into the crankcase where they do interact with the lubricant. These combustion gases and by-products contain such components as oxides of sulfur, derived from the sulfur content of petroleum-based fuels. Particularly in diesel engines, these sulfur oxide compounds interact with oxidized components from the fuel and base oil to produce sulfuric acid and a variety of strong organic acids [6].

Another form of combustion by-products comes from oxides of nitrogen, derived from the high temperature combination of nitrogen and oxygen from the intake air. These by-products are predominant in gasoline engines in which the oxides of nitrogen materials can further react with water (from the combustion process), oxidized oil and fuel, and soot (if present) to produce engine sludge and piston varnish [5]. Obviously, these acidic combustion gases and by-products are detrimental to the extended life of both engine hardware components and the lubricant itself. They can give rise to increased rusting of steel parts and corrosive wear of bearings.

The use of high-TBN, overbased detergents can combat these problems. One, however, must be careful in formulating with an appropriate mix of detergents for acid control and corrosion performance. The use of several appropriate detergents in a lubricant for excellent engine rust and bearing corrosion performance may not necessarily be favorable to maintaining good valve train wear performance such as in the well-known gasoline engine specification tests: Sequence VE and Sequence IVA.

Figure 25.5 plots data from a passenger car field test showing the decrease in TBN and the increase in total acid number (TAN) with use. OEM oil drain recommendations are often determined by this type of used oil testing. Typically, it is considered desirable to change the oil before lines representing the TBN and TAN behavior cross. In this "severe service" example, the oil that starts at a TBN value of 6 would need to be changed twice as often as the oil that has an initial TBN value of about 9.

Evidence such as this results in oils with higher TBN being recommended for longer drain intervals.

A second function of a detergent is to retard deposit formation on engine parts, especially parts that are operating at high temperature such as pistons and piston rings (Figure 25.6). Detergency of North American diesel engine oils is evaluated using both single-cylinder engine tests (Caterpillar 1N, 1K, 1P, and 1R) and multicylinder engine tests (Caterpillar C13, Mack T-12). The selection of detergents to give the best piston and ring cleanliness is highly dependent on the temperature of the piston-ring area, the metallurgy of the piston, the ring pack design, and the base stock of the lubricant being tested. Metallurgy variances in engine designs such as aluminum versus articulated steel diesel pistons complicate proper

FIGURE 25.5 Passenger car field testing: drain oil TBN/TAN behavior relationship.

(a)   (b)

FIGURE 25.6 Detergents keep upper pistons clean throughout oil drain interval: (a) acceptable and (b) unacceptable.

detergent selection. A particular mixture of detergents may be excellent with aluminum hardware but may only perform marginally with steel hardware.

Some types of detergents perform additional functions in an engine oil formulation. For example, coupled-coupled alkyl phenols enhance high-temperature oxidation inhibition. Due to their specific structure and thermal stability, these detergents help prevent oxidation of the lubricant under high speed and load engine conditions; the result is a lower viscosity increase of the oil. Of course, high-temperature oxidation inhibition synergizes with the detergent's ability to enhance "cleanliness."

In summary, the best overall cost and performance compromise when using a selected combination of detergents in a lubricant depends on many factors. Several of these factors entail complete engine performance, customer desires, and regulations including the maximum total amount of metal or metal ash allowed in a lubricant as set by specification requirements.

25.3 DISPERSANTS

25.3.1 INTRODUCTION

Dispersants are typically the highest treated additive in an engine oil formulation. They are similar to detergents in that they have a polar head group with an oil-soluble hydrocarbon tail. While detergents are used to clean engine surfaces and neutralize acidic by-products, their effectiveness is limited when it comes to dispersing oil-insoluble products resulting from the by-products of combustion. The principal function of a dispersant is to minimize the deleterious effects of these contaminants. The most obvious contaminants related to engine lubricants are black sludge and soot particles. Sludges range from thick oil-like deposits to a harder deposit; soot is

composed primarily of carbon particles and is typically found in diesel engines. Dispersants are used to disperse these contaminants within the engine, thereby ensuring that the oil flows freely. The dispersing ability of dispersants helps keep the engine clean and, in some cases, will maintain piston cleanliness. Some formulators will actually refer to certain dispersants as ashless detergents.

Over many years, dispersants have played a major part in keeping the engine clean, and they continue to play an important part in oil formulation. In modern top-quality engine oil formulations, dispersants will range from 3% to 6% by weight. Typically, this would account for around 50% of the total oil additive in the lubricant.

25.3.2 DISPERSANT STRUCTURE

Dispersants consist of an oil-soluble portion plus a polar head group. The polar group is attached to the oil-soluble group by means of a "hook." The schematic in Figure 25.7 gives a simplistic representation of a dispersant structure based on the reaction of polyisobutenyl succinic anhydride (PIBSA) with either a polyamine or a polyol.

Many different types of dispersants have been used in lubricant additive packages. Over the years, these have evolved as the lubricant requirements for both OEMs and the testing organizations have become more demanding. The following list, while not exhaustive, covers the most popular types of dispersant chemistries in use today:

- Polyisobutenyl succinimide
- Polyisobutenyl succinate ester
- Mannich dispersants
- Dispersant viscosity modifiers (dVMs; e.g., dispersant olefin copolymer, dispersant polymethacrylate)

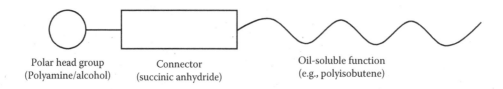

Polar head group   Connector   Oil-soluble function
(Polyamine/alcohol)   (succinic anhydride)   (e.g., polyisobutene)

FIGURE 25.7 Schematic of a dispersant molecule.

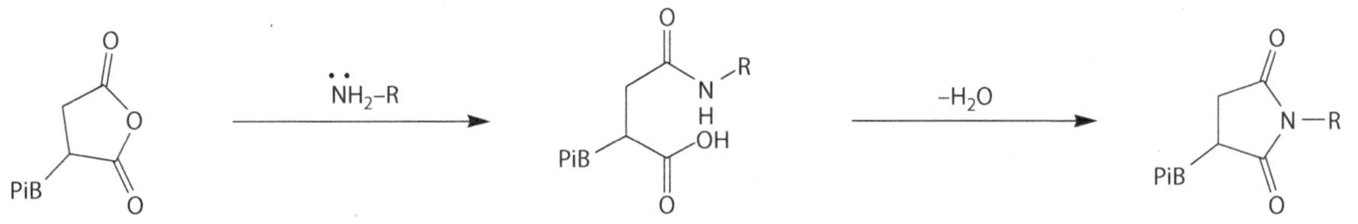

FIGURE 25.8 Imidation of a polyisobutenyl succinic anhydride.

By far the most common group of dispersants is polyisobutenyl succinimides. These are the preferred dispersants for tackling the black sludge problem that occurred in gasoline engines during the 1980s. During the 1990s and up to the present day, they have also been used in ever-increasing levels to reduce oil thickening, which is a result of the high levels of soot in diesel engine oils.

To prepare a polyisobutenyl succinimide dispersant, a hook is attached by reacting maleic anhydride (MA) with PIB to make polyisobutene succinic anhydride (PIBSA). PIBSA is then typically neutralized with a polyamine mixture to yield the succinimide dispersant (Figure 25.8).

The reaction between the PIB and the MA occurs via the unsaturated end group of the PIB. This can be achieved by directly reacting the PIB with MA at temperatures in excess of 200°C (direct alkylation or DA for short). Once the PIBSA is formed, it can react again to make the disuccinated product. The reaction of the PIB and MA depends on the unsaturated end groups of the PIB being reactive enough to add to the MA. As the reaction proceeds, the addition of MA to the PIB slows down as all the reactive end groups in the PIB are used up. With some types of PIBs, this will leave a relatively large amount of unreacted PIB. In recent years, this problem has been overcome by using PIBs with very high levels of terminal vinylidene groups, in some cases greater than 80%. These vinylidene groups are more reactive via the direct alkylation route, enabling DA PIBSAs to be prepared with relatively low levels of unreacted PIB (see Figure 25.9).

The key to the PIB–MA reaction is to ensure that the PIB is fully converted into PIBSA. Any unreacted PIB in the finished lubricant will result in a formulation with poor low-temperature viscometric properties. In addition, the dispersant will be less effective in preventing sludge formation and in dispersing soot since not all the oil-soluble PIB will contain active dispersant chemistry.

For these reasons, it is beneficial to maximize the conversion of the PIB to succinic anhydride. The addition of MA to the PIB is made easier by using chlorine, which catalyses the reaction via a PIB diene intermediate, which reacts more readily with the MA, yielding a highly converted PIB with minimal unreacted PIB. A schematic of the reaction is shown in Figure 25.10.

25.3.3 Polyisobutene Synthesis

PIB is produced by cationic polymerization of either pure isobutene or a C_4 stream from an oil refinery. The isobutene in the C_4 refinery stream reacts preferentially while other compounds such as n-butenes and butanes do not. Typical catalysts for these reactions include $AlCl_3$ and BF_3.

The molecular weight of the PIB is very important and can have a significant effect on the dispersant performance. Typically, the number average molecular weight (M_n) of PIB ranges from 500 to 3000 although there are instances where PIBs outside this range may be used. The higher–molecular weight PIB dispersants have viscosity modifier (VM) properties and aid the formulation of multigrade oils. In addition, they are much more effective in the handing of black sludge and soot. While lower–molecular weight dispersants have found use as sludge and soot dispersants, they are clearly less effective. Dispersants made using higher–molecular weight PIB (Mn > 2500) have an adverse effect on the low-temperature viscosity properties of the finished lubricant, particularly the cold crank viscosity, and for this reason they are not normally used. In addition, they are more difficult to react with MA due to their higher viscosity.

25.3.4 Dispersant Basicity

The amount of polyamines added to the PIBSA will determine the basicity of the dispersant. This is referred to as TBN and is measured by the ASTM D2896 method. The unit of measurement for TBN is mg KOH/g. The TBN of the dispersant will give a good indication of its structure. Higher levels of polyamines, that is, more base, will predominately yield a mono-succinimide structure, which is the most basic of the PIBSA dispersants (see Figure 25.11). When there is a molar excess of PIBSA, more of the bis- and tris- structures will predominate (see Figure 25.12).

25.3.5 Succinate Ester Dispersants

These dispersants are prepared from PIBSA and a polyol as shown in Figure 25.13. Ester-based dispersants are used to reduce sludge and piston deposits and function in a similar fashion to the succinimide-based dispersants.

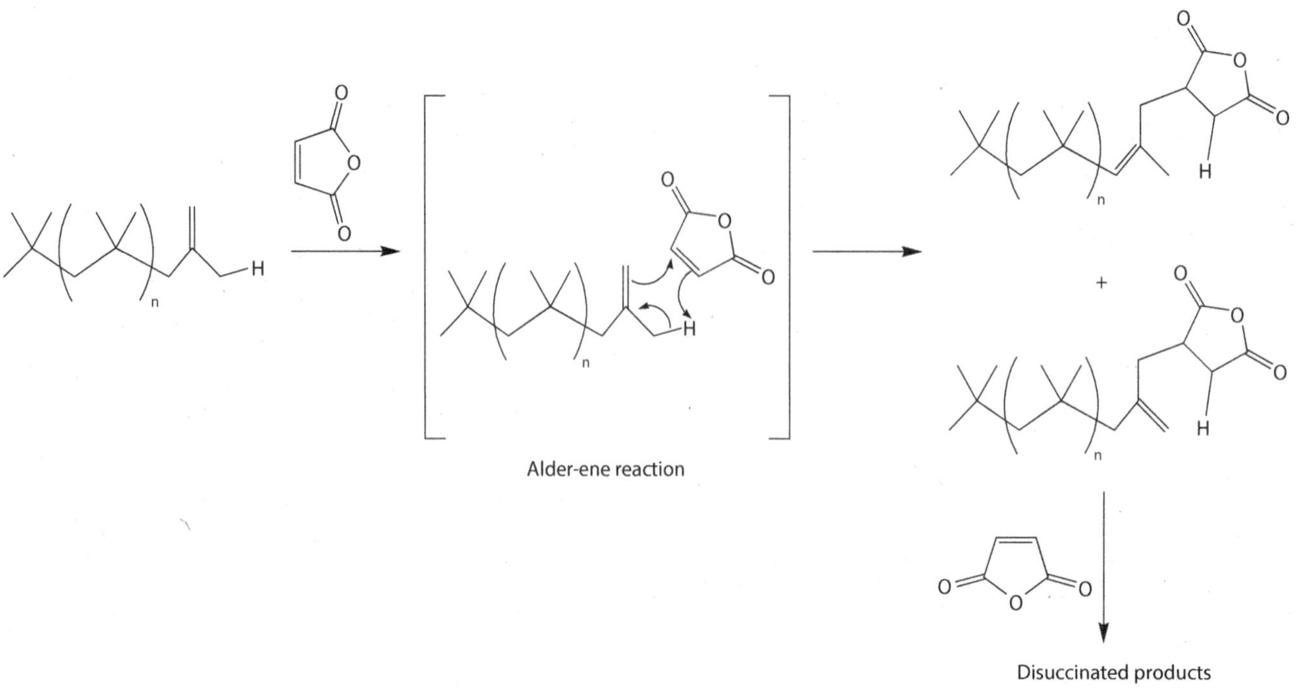

FIGURE 25.9   Direct alkylation reaction of polyisobutene with maleic anhydride.

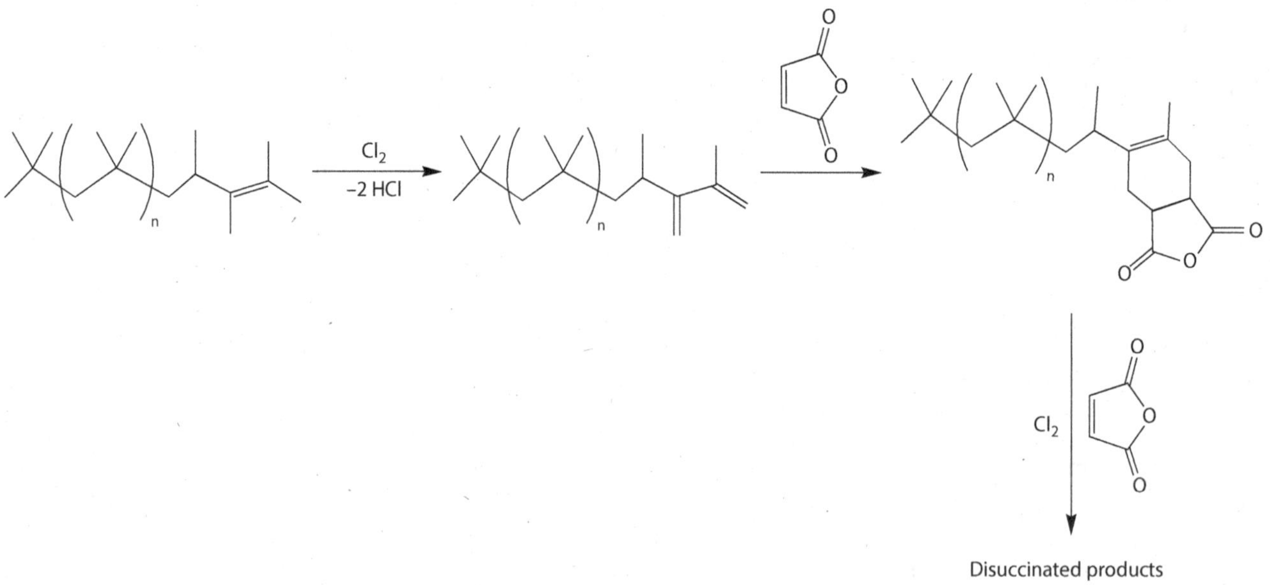

FIGURE 25.10   Chlorination/succination of PIB.

25.3.6   MANNICH DISPERSANTS

These types of dispersants are prepared by reacting a PIB phenol with a polyamine in the presence of formaldehyde. The resulting dispersant has some antioxidant properties (Figure 25.14). This family of dispersants is typically used in gasoline engine oils.

25.3.7   SOOT CONTAMINATION IN DIESEL ENGINE OILS

The soot contaminants in heavy-duty diesel engines usually appear when the engine is operating under very high loads or when the fuel injection is retarded (i.e., injected late in the cranking cycle). Under specific engine conditions, small soot particles, typically less than 200 nm, will agglomerate into a

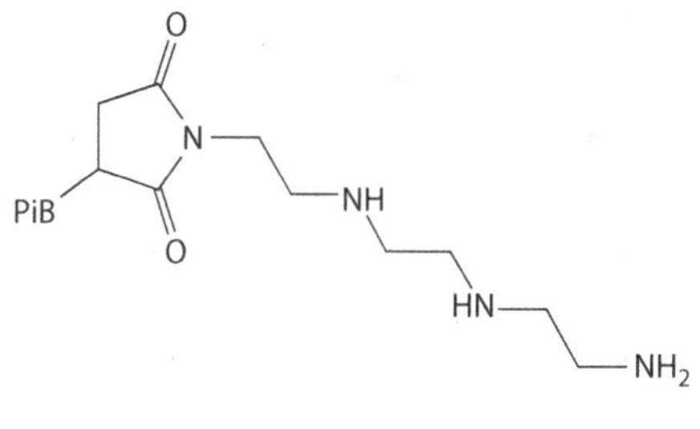

FIGURE 25.11 Example of mono-succinimide from TETA (triethylenetetramine).

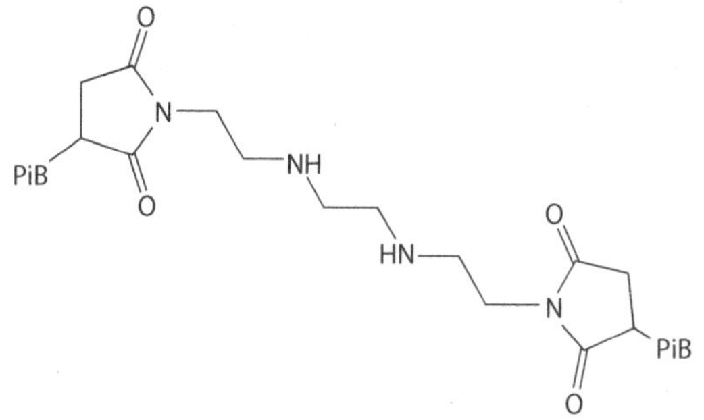

FIGURE 25.12 Example of bis-succinimide from TETA.

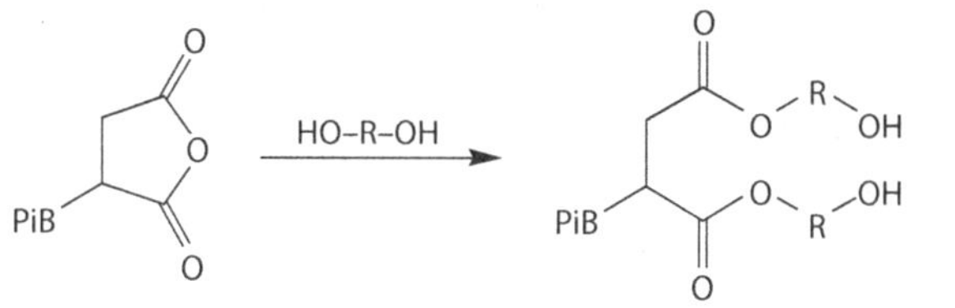

FIGURE 25.13 Example of a simple succinate ester dispersant.

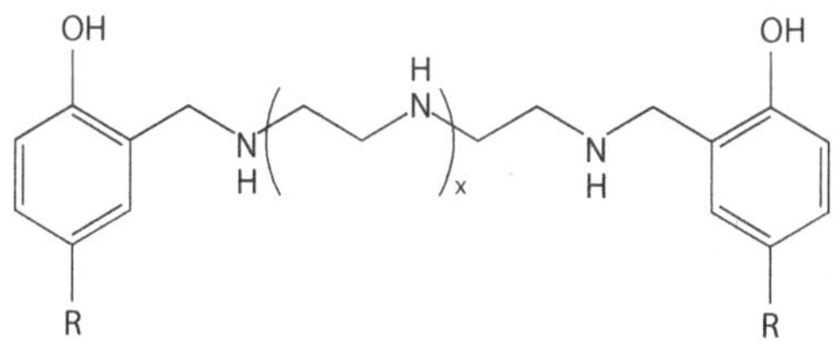

FIGURE 25.14 Typical Mannich dispersant.

larger macro structure, leading to a significant rise in oil viscosity. The structure of the dispersant is key to the lubricant's ability to minimize these structures from forming. Thus, with the correct choice of dispersant, it is possible to reduce the soot-related viscosity increase and also maintain cleanliness throughout the engine by preventing the large soot structures from settling out.

Over the past 10–20 years, the main reason for adding dispersants to diesel engine oils has been the increasing levels of soot, especially in modern low-emission engines. During the

1990s, the level of soot in diesel engine oil showed a higher trend as a result of new engine technologies designed to reduce NOx and particulate emissions. This trend accelerated in the United States as older engine designs were changed to meet the U.S. 1990 exhaust emission regulations and then later in Europe during 1992 for the Euro I regulations. During this period, retarded fuel injection was one of the main strategies employed by OEMs to reduce NOx emissions, which resulted in more unburned fuel entering the crankcase oil as soot particles. Combined with extended oil drain intervals, it is not unusual to see soot levels as high as 5% in certain types of duty cycles. This trend is set to continue as more legislation appears on the horizon aimed at reducing diesel exhaust emissions.

One of the most important factors in dispersing soot is the dispersant TBN. Typically as the base number of the succinimide dispersant tends to be higher, the dispersant molecular weight tends to lower values, since more of the mono-succinimide structure prevails with more primary nitrogen. This type of dispersant is the most efficient for use in diesel engines, especially in modern low-emission engines where there are high levels of oil soot. The effectiveness of the dispersant is determined by its ability to separate soot particles and thus prevent them from agglomerating. The mechanism whereby this occurs can be represented by the basic dispersant molecule attaching itself to an acidic site on the surface of the soot particle. Assuming that there are sufficient dispersant molecules attached to the soot particle, the PIB chains will prevent the soot particles from coming together. The stability of the soot dispersion is influenced by the molecular weight of the PIB and the level of reactive amines (primary and secondary) per molecule of PIB. A very simplistic representation of a soot–dispersant interaction is given in Figure 25.15. It is fair to say that a lot of energy has been spent finding the optimum dispersant for dispersing soot as the patent literature demonstrates. Each individual additive company will have their favored structures.

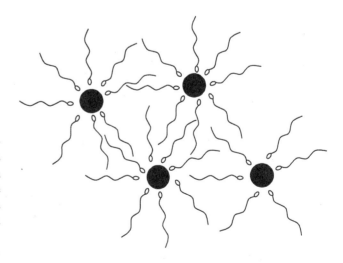

FIGURE 25.15 Schematic representation of dispersant–soot interaction.

25.3.8 SOOT THICKENING TESTS

As the soot levels of heavy-duty diesel engine oils increased, OEMs, particularly in North America, introduced engine tests to evaluate the oils' ability to disperse soot. One of the tests used to evaluate soot-mediated oil thickening for low-emission engines was Mack T-7. This is included in the Mack EO-K and API CF-4 specifications. As the engine technologies progressed to lower emission levels, more severe tests were introduced into API CJ-4 specifications. Figure 25.16 illustrates the progression of the engine and bench tests defined as the API C (heavy-duty diesel) lubricant categories.

For light-duty diesel applications, Peugeot introduced the XUD11 soot-thickening test for ACEA passenger car diesel oil approvals. This test, now in its second generation and called XUD11 BTE, is used to measure oil thickening at soot levels up to 4%. Because the level of soot is much higher compared to what is typically found in European passenger car diesels pre–Euro 2 (1996), the test makes a good measure of the oils' ability to minimize oxidative soot thickening for small European diesel engines. Criteria have also been applied to the Mercedes OM602A, OM364LA, and OM441LA diesel engine tests in respect of soot-related oil thickening and sludge control. Despite all these tests, the engines that have driven oil formulations to ever-higher levels of basic dispersant are still the Mack engines. A comparison of an API CG-4 formulation and an API CH-4 formulation in the Mack T-8 is shown in Figure 25.17. This demonstrates the progression of dispersant characteristics from API CG-4 to API CH-4.

Performance Parameter	CH-4	CI-4	CI-4 Plus	CJ-4	PC-11
Valve train wear	Cummins M11	Cummins M11 EGR	Cummins M11 EGR	Cummins ISB	Cummins ISB
Valve train wear, filter plugging, and sludge				Cummins ISM	Cummins ISM
Roller follower wear	RFWT	RFWT	RFWT	RFWT	RFWT
Oil oxidation	IIIE	IIIF	IIIF	IIIF or IIIG	Mack T13
Ring and liner wear	Mack T9	Mack T10	Mack T10	Mack T12	Mack T12
Soot dispersancy	Mack T8E (300 HS)	Mack T8E (300 HS)	Mack T11	Mack T11 (more severe limits)	Mack T11 (more severe limits)
Piston deposits and oil consumption	CAT 1P (Fe) CAT 1K	CAT 1R (Fe/all piston)	CAT 1R (Fe/all piston)	CAT C13 and CAT 1N Mack T12 (OC)	CAT C13 and CAT 1N Mack T12 (OC)
Corrosion	HTCBT	HTCBT	HTCBT	HTCBT	HTCBT
Used oil low-temperature pumpability		T-10 drain (75 h)	T-10 drain (75 h)	T-11 drain (180 h)	T-11 drain (180 h)
Elastomer capability		Yes	Yes	Yes	Yes
Volatility loss	20%	15%	15%	13%	13%
AT compatibility				Yes	Yes
Shear stability		Bosch, after 30 cycles	Bosch, after 90 cycles	Bosch, after 90 cycles	Bosch, after 90 cycles (more severe limit XW-40)
Adhesive wear					
Aeration				EOAT	CAT aeration

FIGURE 25.16   List of required tests for API C category: Evolution from API CH-4 to proposed PC 11.

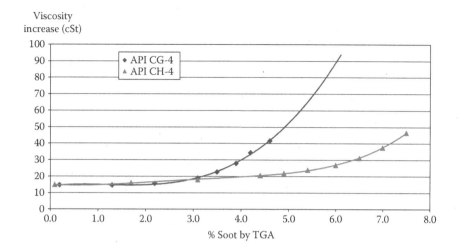

FIGURE 25.17 Comparison of an API CG-4 formulation versus an API CH-4 formulation performance in the Mack T-8 engine test.

25.3.9 SEAL TESTING

The balance between soot handling and seal compatibility has provided lubricant formulators with significant challenges over the past 10 years, especially as seal testing is a major part of the oil approval process in Europe.

The most difficult issue to contend with is that the highly basic dispersants, used in diesel oil formulations to disperse soot, are aggressive toward fluoroelastomer seals. There are many fluoroelastomer seal tests, one common example being the Volkswagen PV3344 test, which is a requirement for Volkswagen oil approvals. There are also seal test requirements for most of the European engine manufacturers, for example, Mercedes Benz, MAN, and MTU, and collectively through the Association of European Automotive Manufacturers (ACEA). A comparison of three formulations with different dispersants in a Mercedes Benz seal test is given in Table 25.1. This clearly shows that as the soot dispersancy is increased for API CH-4, the MB fluoroelastomer seal test worsens.

TABLE 25.1
Effect of Dispersant TBN on Mercedes Benz Fluoroelastomer Seals

Formulation	A	B	C
Max variation of AK6 viton			
Hardness (Shore A)	−3	0	1
Volume (%)	1	1	0.9
Tensile strength (%)	−8	−32	−46
Elongation rupture (%)	−15	−33	−38
	Pass	Pass	Borderline fail

Key to formulations:
A is the European passenger car gasoline/diesel formulation.
B is the North American heavy duty diesel formulation (API CG-4).
C is the North American heavy duty diesel formulation (API CH-4).

25.3.10 CORROSION

Another potential drawback with high-TBN dispersants is that they tend to be more aggressive toward Cu/Pb bearings. This has led some additive companies to treat their dispersants with boron compounds, for example, boric acid, to reduce Cu/Pb corrosion. In some cases, this can improve the antiwear properties but usually it reduces the effectiveness of the dispersant in dispersing soot. Clearly, a balance has to be found and formulators will use different dispersants to meet the various requirements.

25.3.11 SLUDGE

Under certain conditions, sludge will accumulate in an internal combustion engine. During the 1980s, this problem intensified throughout the world, particularly in gasoline engines in Germany, the United Kingdom, and the United States. The origins of the problem are most likely related to fuel quality, drive cycles, extended oil drain intervals, and the redirection of blow-by gas into the rocker cover. Sludge buildup, if left unchecked, can spread throughout the engine, leading to reduced oil flow through the filter and drain-back holes on the valve deck. In extreme situations, this will lead to engine seizure. The problems were not linked to poor-quality lubricants. However, as lubricant technology evolved it was found that newer lubricant formulations could help alleviate the problem.

Sludges can be split into low-temperature sludges and high-temperature sludges. Sludges formed at low temperatures will tend to be soft and easily removable from surfaces by wiping. As previously mentioned, the sludge-forming mechanism is thought to be accelerated by the transfer of blow-by gases into the oil. Blow-by gases contain water, acids, and partially burned hydrocarbons in the form of oxygenates and olefins. Olefins react further with nitrogen oxides to form oil-insoluble products. Fuels with high-end boiling points or with high aromatics also tend to contribute to sludge formation. Once formed, highway driving will exacerbate the problem as the engine heats up more, thereby causing the sludge to bake.

Sludge formation is reduced by the addition of basic succinimide dispersants, particularly the high–molecular weight types. The addition of boric acid to a succinimide dispersant to make a lower TBN dispersant will reduce the effectiveness of a dispersant in dispersing sludge [7]. While this indicates that basic dispersants are required to neutralize sludge precursors, it has been found that very high levels of basic nitrogen are not necessarily required for sludge dispersion in gasoline engines. For Sequence VE performance, high–molecular weight PIB bis-succinimide dispersants have demonstrated excellent performance. This would appear to indicate that the mechanism of sludge formation in a gasoline engine is different from the agglomeration of soot in a diesel engine, which does require high levels of basic nitrogen, especially in low-saturate mineral base stocks. While bis- and mono-succinimide dispersants are used in both gasoline and diesel engine oils, it would be fair to say that, more recently, diesel engine oils have tended to higher levels of basic dispersants, with correspondingly higher levels of PIB mono-succinimides compared to gasoline engine oils. One of the big challenges for formulators since API CG-4 has been to develop lubricants that meet the requirements of both gasoline and diesel engine oils, the so-called universal engine oil. This requires careful formulation with both bis- and mono-succinimide dispersants.

When formulating engine oil, it is also important to consider the interactions between the succinimide dispersant and other additives such as zinc dialkyldithiophosphates (ZDPs). A complex between ZDP and succinimide polyamine has been observed in laboratory tests, especially where there is a high basic nitrogen to ZDP ratio [8]. As the level of basic dispersant is increased, sludge and wear will improve. Above a critical concentration of dispersant, the antiwear properties of the ZDP will be dramatically reduced as the basic nitrogen forms a stable complex with the ZDP, effectively reducing its antiwear capability. It has been demonstrated that the ZDP–amine complex retards the rate of peroxide decomposition by ZDP [9], which in turn accelerates the formation of sludge. Sludge will usually increase in line with wear, so clearly a high level of ZDP with a sufficient dispersant level to disperse the sludge is required. High levels of ZDP permit high levels of dispersant to be used in the lubricant, thereby giving superior engine sludge and wear performance. However, phosphorous has been implicated in exhaust catalyst poisoning. Because of this, a maximum phosphorous limit of 0.1% w in the lubricant has been applied to recent API and ILSAC specifications. This means that where phosphorous limits are imposed for API SJ and ILSAC GF-3 lubricants, careful choice of dispersant and ZDP is required. The phosphorous limits are set to be reduced further for ILSAC GF-4, thereby presenting further challenges for the lubricant to meet the sludge and wear requirements in future engines.

25.3.12 Sludge Engine Tests

Although sludge tests have been around for many years, the most significant lubricant tests to address the sludge problems during the 1980s and 1990s were the Sequence VE and Mercedes M102E. More recently, the Sequence VG and the MB M111E have superceded these tests and have become the current benchmarks for measuring lubricant sludge–handling performance in Europe and North America. Table 25.2 lists the test conditions for the various gasoline sludge tests. The Sequence IIIE and IIIF are also included since high-temperature sludge is a rated parameter in these gasoline engine tests.

High-temperature sludges are also present in diesel engines and are a rated parameter for several engine tests, particularly the Cummins M11 for API CH-4. Dispersants based on high–molecular weight PIB are essential in minimizing sludge buildup in this test. Sludge is also a rated parameter for the Mercedes Benz OM364LA and OM441LA engine test although it is not as critical a parameter as the Cummins M11 engine test.

25.4 ANTIWEAR

25.4.1 Introduction

As the power of engines has risen, the need for additives to prevent wear has become more important. Initially, engines were lightly loaded and could withstand the load on the bearings and valve train. Corrosive protection of bearing metals was one of the early requirements for engine oils. Fortunately, the additives used to protect bearings usually had mild antiwear properties. These antiwear agents were compounds such as lead salts of long-chain carboxylic acids and were often used in combination with sulfur-containing materials. Oil-soluble sulfur–phosphorous and chlorinated compounds also worked well as antiwear agents. However, the most important advance in antiwear chemistry was made during the 1930s and 1940s with the discovery of ZDP [10–12]. These compounds were initially used to prevent bearing corrosion but were later found to have exceptional antioxidant and antiwear properties. The antioxidant mechanism of the ZDP was the key to its ability to reduce bearing corrosion. Since the ZDP suppresses the formation of peroxides, it prevents the corrosion of Cu/Pb bearings by organic acids.

25.4.2 Wear Mechanisms

Antiwear additives minimize wear in a mixed or partial lubricant film operating under boundary conditions (Figure 25.18). A classic example of boundary lubrication is in a nonconforming contact such as a cam on a follower. The partial lubricant film is most likely to occur when the oil is not viscous enough to separate the two surfaces completely.

The tendency to boundary lubrication increases as the temperature rises due to the viscosity–temperature dependence of the lubricant. Low-contact speeds, high-contact pressures, and rough surfaces will also contribute to more boundary lubrication. If these circumstances are taken to the extreme and minimal or no lubricant film exists, then maximum surface contact will exist (Figure 25.19). This is

TABLE 25.2
Passenger Car Lubricants: Sludge Engine Test Comparison

Test	Cylinder Config.	Disp. (cc)	Test Duration (h)	Test Operation	Duration (h)	Speed (rpm)	Power (kW)	Coolant Temp. (°C)	Oil Temp. (°C)	Fuel Type
Sequence IIIE (Buick)	V-6	3800	64	Steady speed	64	3000	50.6	115	149	Phillips GMR leaded gasoline
Sequence IIIF (Buick)	V-6	3800	80	Steady speed	80	3600	75.0	115	155	Howell EEE unleaded
Sequence VE (Ford)	I.L.-4 Slider follower	2290	288	Cyclic	2.00	2500	25.0	51.7	68.3	Phillips J unleaded gasoline
					1.25	2500	25.0	85.0	98.9	
					0.75	750	0.75	46.1	46.1	
Sequence VG (Ford)	V-8 Roller follower	4600	216	Cyclic	2	1200	69 kPa	57	68	Unleaded gasoline
					1.25	2900	66 kPa	85	100	
					0.75	700	Record	45	45	
Mercedes Benz M-111E Sludge	I.L.-4	1998	224	Cyclic	48 h cold 75 h WOT 100 h cyclic	Idle to 5500	Idle to max.	40–98	45–130	CEC RF-86-T-94 (ULG)
Nissan VG-20E		2000	200	Cyclic	200	800–3500	2–9 kgf m	38–100	50–117	Unleaded gasoline
Toyota 1G-FE		2000	48	Steady speed	48	4800	6 kgf m	120	149	Unleaded gasoline

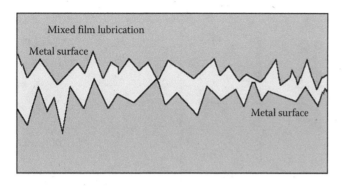

FIGURE 25.18 Schematic of mixed film lubrication regime.

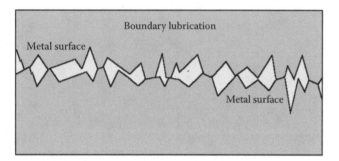

FIGURE 25.19 Schematic of boundary lubrication regime.

defined as an extreme pressure (EP) contact and is usually associated with very high temperatures and loads. Additives that prevent wear in an EP contact typically require higher activation temperatures and loads than an antiwear additive.

Antiwear and EP additives function by thermally decomposing to yield compounds that react with the metal surface. These surface-active compounds form a thin layer that preferentially shears under boundary lubrication conditions.

After ZDP was discovered, it rapidly became the most widespread antiwear additive used in lubricants. As a result, many interesting studies have been undertaken on ZDP with many mechanisms being proposed for the antiwear and antioxidant actions [13–17]. The performance of the ZDP is strongly influenced by the decomposition pathways. These pathways are thermolysis, oxidation, and hydrolysis, which in turn depend on the conditions under which the ZDP is working. In general, it can safely be assumed that the degradation products of the ZDP form a film on the metal surface, typically rich in phosphorus and oxygen and possibly polymeric. As mentioned earlier, this layer preferentially shears under boundary lubrication, thus reducing the wear on the metal surfaces. As this layer needs to be constantly replenished, the concentration of ZDP in the lubricant is critical. It is therefore not unusual to find up to 2% wt. of ZDP in a modern lubricant. This equates to a maximum level of about 0.15% phosphorous in the lubricant.

25.4.3 ZDP PREPARATION

ZDP is prepared by reacting a dialkyldithiophosphoric acid with zinc oxide. The first stage of the process involves the preparation of the acid. The acid is prepared from an alcohol and phosphorous pentasulfide:

$$4ROH + P_2S_5 \xrightarrow{H_2S} 2(RO)_2 P(S)SH$$
$$\text{Dialkyldithiophosphoric acid}$$

The acid is then neutralized with zinc oxide to yield ZDP at around 70°C–90°C. There are two types of ZDP structures, neutral and basic, both of which can be observed by ^{31}P NMR. The basic salt is observed at 104 ppm with the neutral salt at 101 ppm. The neutral salt exists as an equilibrium mixture of monomer, dimer, and oligomer. The basic salt consists of a central oxygen atom surrounded tetrahedrally by four zinc atoms and six dithiophosphate groups attached symmetrically to the six edges of the tetrahedron. In most industrial processes, the ZDP is left slightly basic for improved stability.

$$2(RO)_2P(S)SH + ZnO \xrightarrow{-H_2O} \left[\begin{matrix} RO & \diagdown \\ & P \diagup S \\ RO & \diagup \diagdown S \end{matrix}\right]_2 Zn$$

Neutral ZDP

$$3[(RO)_2P(S)S]_2Zn + ZnO \longrightarrow \left[\begin{matrix} RO & \diagdown \\ & P \diagup S \\ RO & \diagup \diagdown S \end{matrix}\right]_6 Zn_4O$$

Basic ZDP

25.4.4 ZDP DEGRADATION MECHANISMS

The type of alcohol used to prepare the ZDP will determine its thermal and oxidative stability. The most reactive ZDPs are derived from secondary alcohols, especially those that are lower in molecular weight. Solubility is a limiting factor at carbon numbers less than 5; therefore, most ZDPs will use alcohols with carbon numbers greater than 5. Alcohols with lower carbon number may be used if they are combined with a higher alcohol during the synthesis.

The type of alcohol used in the preparation will have a significant effect on the stability. In most cases, the thermal stability of the ZDP is as follows:

Aryl > Primary alkyl > Secondary alkyl

The least stable ZDPs tend to provide improved wear at lower engine oil temperatures. Therefore, the following applies for antiwear action:

Secondary alkyl > Primary alkyl > Aryl

Tables 25.3 through 25.5 indicate the performance of various ZDPs in a range of gasoline engine wear tests.

TABLE 25.3
Comparison of ZDP Antiwear Types, Performance in Sequence VE Wear Test

Alcohol Type	%Zn as ZDP	Sequence VE Wear	
		Average (µm)	Maximum (µm)
Mixed C$_3$ secondary C$_8$ primary	0.127	36	203
C$_8$ primary	0.124	121	495

TABLE 25.4
Comparison of ZDP Types, Performance in Sequence VD Wear Test

Alcohol Type	%Zn as ZDP	Sequence VD Wear
		Average (µm)
C$_6$ secondary	0.13	18
Mixed C$_6$ secondary C$_8$ primary	0.13	48

TABLE 25.5
Comparison of ZDP Types, Performance in Sequence IIID Wear Test

Alcohol Type	%Zn as ZDP	Sequence IIID Wear
		Average (µm)
Mixed C$_6$ secondary C$_8$ primary	0.13	25
C$_8$ primary	0.13	175

The mechanism by which secondary alcohol ZDPs thermally degrade is shown in Figure 25.20. The degradation mechanism proceeds rapidly as the temperature rises and is made easier as a hydrogen on the β position readily leaves to form the alkene. This mechanism may explain why the secondary ZDPs are much more active antiwear agents, particularly at lower temperatures.

In contrast, the primary alcohol ZDPs are more stable due to the absence of a tertiary hydrogen on the β carbon and therefore more useful for higher-temperature operation and wear such as those found in diesel engines. The mechanism of thermal degradation is via sequential alkyl transfers and relies on an intermolecular alkyl transfer (Figure 25.21).

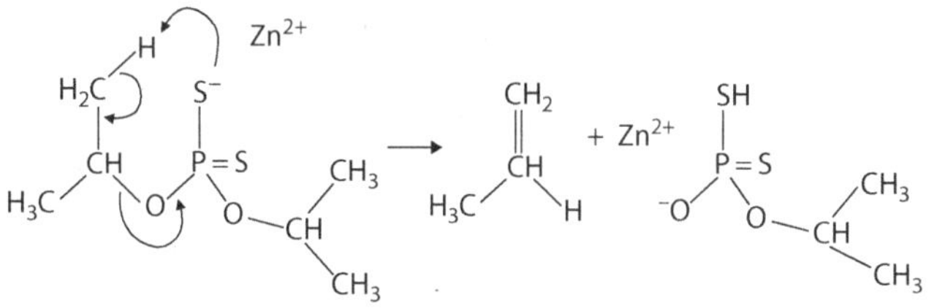

FIGURE 25.20 β-Elimination (secondary ZDP).

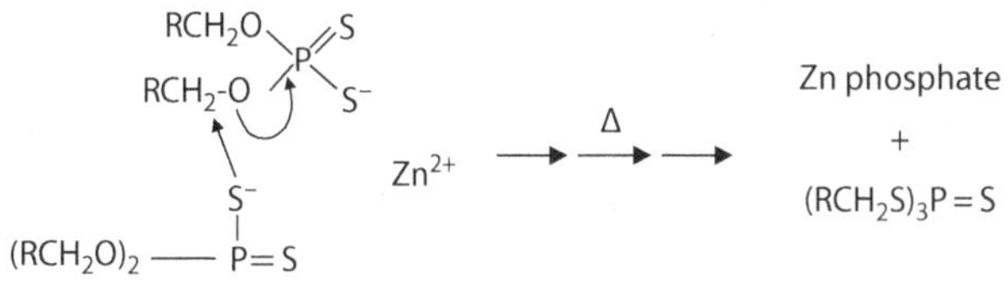

FIGURE 25.21 Sequential alkyl transfers (primary ZDP).

25.4.5 SEQUENTIAL ALKYL TRANSFERS (PRIMARY ZDP)

Other additives will affect the rate of thermal degradation of the ZDP. For instance, it is known that succinimide dispersants will complex with the ZDP, making it more resistant to thermal degradation. It is therefore important to recognize this when formulating an oil additive. Too much dispersant may tie up the ZDP, leaving it unable to form an effective antiwear film. A balance has to be found that will depend on the ZDP type and the dispersant structure.

As well as thermal degradation pathways for the ZDP, it also degrades at lower temperatures by oxidation (<100°C), yielding compounds that are beneficial antiwear agents (see Figure 25.22). This leads on to the mechanism of oxidative inhibition by ZDP that occurs via the thiophosphoryl disulfide intermediate (see Figure 25.23). A more detailed mechanism of the antioxidant function of the thiophosphoryl disulfide intermediate has been reported [18].

The effectiveness of ZDP in decomposing hydroperoxides has been linked to wear rates in a motored engine test [19]. Secondary ZDPs were better at decomposing peroxides than their primary counterparts. There is a plethora of papers in the literature with detailed analyses of the various degradation pathways for the ZDP molecule and its subsequent effect on the wear and oxidation properties of the lubricant [13,14,20–22]. While these give an excellent insight into the ZDP degradation mechanism, it must be remembered that the degradation pathways are strongly dependent on the test conditions, that is, temperature, amount and type of oxidants, and so on.

25.4.6 ANTIWEAR TESTS

The Sequence VE and Sequence IV-A gasoline engine tests (for API SJ and ILSAC GF-3, respectively) are designed to test the ability of a lubricant to prevent low-temperature sliding wear in the valve train of a gasoline engine. Secondary

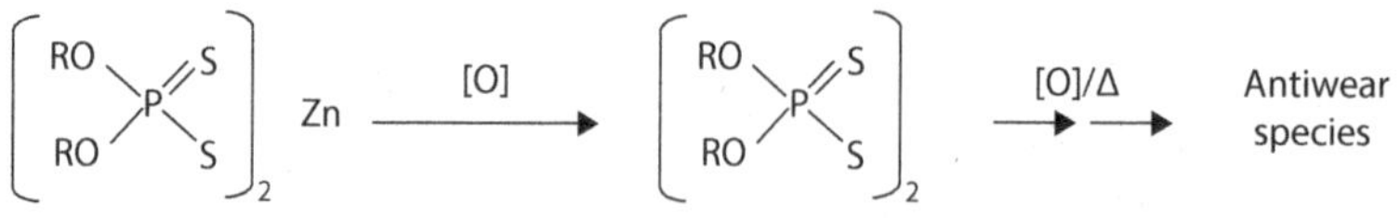

FIGURE 25.22 Thiophosphoryl disulfide formation. ([O] can be almost any oxidant such as O_2, ROOH, H_2O_2, Cu^{2+}, NOx, etc.).

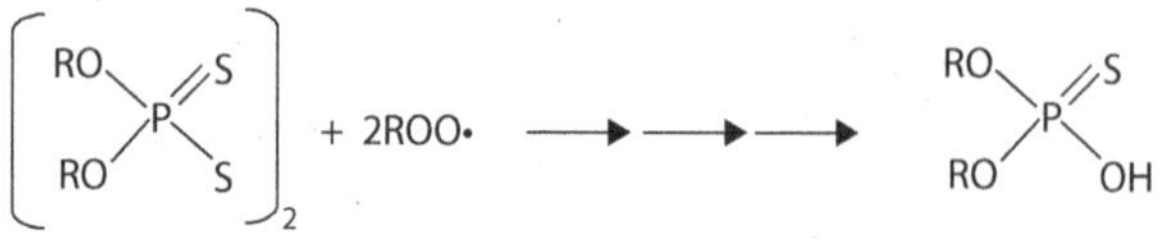

FIGURE 25.23 Thiophosphoryl disulfide decomposition of peroxy radical.

ZDPs will usually perform better than primary ZDPs in both tests. However, there are certain cases where it is possible to achieve a passing test result with primary ZDP, especially with formulations where there is a low level of detergent present. High-temperature gasoline wear is measured in the Sequence IIIE and Peugeot TU3 scuffing test and responds well to most types of ZDP.

Wear in a diesel engine is usually accelerated by the presence of soot. There are several tests that measure wear in diesel engines, and four of the most common ones are listed in Table 25.6.

In addition, field tests such as the Volvo VDS and Scania LDF specifications also specify limits for wear on critical engine components. Depending on the operating temperature of the oil, primary or secondary ZDPs will be used. For heavy-duty diesel engine oils, both primary and secondary ZDPs are widely used. In certain cases, an aryl ZDP may be used for maximum thermal stability.

25.4.7 OTHER ANTIWEAR AGENTS

Boron is widely used to provide antiwear properties in lubricants. A common method is to react boric acid with the amine moiety of a succinimide dispersant. Boron works by reacting at the metal surface to form boric acid, which has a layered structure with weak interlayer bonds [15]. On its own, it is unlikely to give the same level of antiwear performance as ZDP but when used in combination with other antiwear additives it will reduce wear.

TABLE 25.6
Diesel Engine Lubricants: Details of Wear Tests

Engine Test	Specification	Wear Measured
GM Roller Follower Wear Test	API CG-4, CH-4	Roller follower wear
Cummins M11	API CH-4	Valve bridge wear
Mercedes OM602A	MB Sheet 228.X, 229.X	Cam nose wear
	ACEA B and E sequences	Cylinder wear
Mitsubishi 4D34T	Global DHD-1, JASO DH-1	Cam nose wear

25.5 ANTIOXIDANTS

25.5.1 INTRODUCTION

Most lubricants, by virtue of being hydrocarbon-based, are susceptible to oxidation [23,24]. If oxidation is not controlled, lubricant decomposition will lead to oil thickening, sludge formation, and the formation of varnish, resin, and corrosive acids [25,26].

The oxidation reactions that occur in a lubricant at elevated temperatures in the presence of atmospheric oxygen may lead to declined lubricant performance, such as significant increase in kinematic viscosity observed in severe service conditions or during extended drain intervals (Figure 25.24).

In general, all types of base oils require the addition of antioxidants depending on the amount of unsaturation and "natural inhibition" present. The refined mineral base oils contain "natural inhibitors" in the form of sulfur and nitrogen compounds sufficient for many applications. The oxidative stability of such oils shows a distinct, relatively long induction period. On the other hand, hydrofined oils do not contain these natural inhibitors or contain them only in small quantities. Besides the sulfur and nitrogen compounds, other compounds such as aromatics or partially hydrogenated aromatics and phenolic oxidation products can be of importance; their inhibiting effect is lost during various refining processes. Traditional mineral oils, such as group I and II base stocks, show moderate oxidative resistance. Synthetic oils such as polyol esters and hydrogenated poly-α-olefins (PAOs) exhibit the highest oxidative stability due to their low unsaturated content. The most unstable oils include highly unsaturated vegetable triglycerides, such as corn, sunflower, canola, and peanut.

An overall scheme of lubricant oxidation starts with the conversion of hydrocarbon substrates to carbonyl compounds (Figure 25.25). Coupling of these polar compounds through Aldol and related reactions builds molecular weight. Eventually, at very high molecular weight (i.e., >1000), the oil is converted to insoluble sludge, which can precipitate on the engine hardware.

25.5.2 MECHANISM OF OXIDATION OF LUBRICATING OILS

The oxidation of petroleum hydrocarbons proceeds according to a radical chain mechanism via alkyl and peroxy radicals in three stages [27].

Initiation:

$$RH + O_2 \rightarrow R^\bullet + HOO^\bullet$$

FIGURE 25.24 Formulating with antioxidants can prevent premature oil thickening.

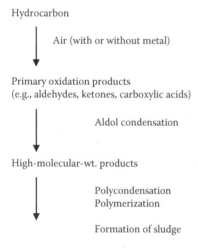

Hydrocarbon

↓ Air (with or without metal)

Primary oxidation products
(e.g., aldehydes, ketones, carboxylic acids)

↓ Aldol condensation

High-molecular-wt. products

↓ Polycondensation
Polymerization

Formation of sludge

FIGURE 25.25 Schematic steps in degradation of mineral oils.

The initiation step starts by the slow abstraction of a hydrocarbon proton by molecular oxygen to form alkyl and hydroperoxy free radicals. This process is also referred to as "auto-oxidation" and is favored by time, higher temperature, and transition metal (i.e., iron, nickel, copper, etc.) catalysis.

Propagation: Propagation starts by the rapid reaction of more oxygen with an alkyl free radical to form an alkyl peroxy radical, which is also capable of hydrocarbon abstraction to form a hydroperoxide and another alkyl radical. The alkyl radical may then react with more oxygen, starting the chain over [25].

$$R^\bullet + O_2 \rightarrow ROO^\bullet$$
$$ROO^\bullet + RH \rightarrow ROOH + R^\bullet$$

Peroxide decomposition: Alkyl hydroperoxides are quite reactive and may decompose, especially at higher temperature, to form additional radical species. These can undergo further abstraction and chain propagation reactions that increase the overall oxidation process. Alkyl peroxides and alkyl peroxy radicals also decompose to neutral oxidation

products such as alcohols, aldehydes, ketones, and carboxylic acids. Hydroperoxide decomposition to neutral oxidation products can be viewed as a chain termination step, since free additional radicals are not formed.

$$ROOH \rightarrow RO^\bullet + {}^\bullet OH$$
$$2ROOH \rightarrow RO^\bullet + ROO^\bullet + H_2O$$
$$RO^\bullet + ROOH \rightarrow \text{Various products}$$

Engine surface • Free radicals
or lubricant + ROOH → • Alcohols
• Inactive products

Termination (self and chain breaking): During termination stage, the radicals either self-terminate or terminate by reacting with oxidation inhibitors [23].

$$ROO^\bullet + ROO^\bullet \rightarrow \text{Inactive products}$$
$$ROO^\bullet + IH \rightarrow ROOH + I^\bullet$$

Alternative oxidation pathways are listed later.
Radical formation:

$$RH + O_2 \rightarrow ROOH \quad \text{Hydroperoxides}$$
$$ROOH \rightarrow RO^\bullet + {}^\bullet OH$$
$$ROH + NO_x \rightarrow R^\bullet + HNO_{x+1}{}^\bullet$$
$$2R^\bullet + O_2 + 2NOx \rightarrow RONO + RONO_2 \quad \text{Nitrite and nitrate esters}$$

Nitrogen oxides can react with alcohols (an oxidation product) to form alkyl free radicals, which react with NOx and oxygen to form nitrate and nitrate ester oxidation products.

Decomposition and rearrangement:

$$ROOH + SO_2 \rightarrow ROH + SO_3$$
$$ROOH + H_2SO_4 \rightarrow \text{Carbonyl compounds}$$
$$RO^\bullet \text{ and } ROO^\bullet \rightarrow \text{Oxidation products}$$

The action of sulfur dioxide and H_2SO_4 (from SO_3 + water) on alkyl hydroperoxides also leads to neutral oxidation products such as alcohols and carbonyl compounds. Aldehydes and ketones can react further and form polymers. Carboxylic acids can attack metallic hardware such as rings, valve train, and bearings, leading to extensive wear. Furthermore, they can form metal carboxylates, which further increase the oxidation rate. Wear metals can also enhance the rate of oxidation [26].

25.5.3 Oxidation Inhibitors

Oxidation inhibitors can be classified in the following manner:

1. Radical scavengers:
   a. *Nitrogen-containing inhibitors*: aryl amines
   b. *Oxygen-containing inhibitors*: phenols
   c. ZDPs
2. Hydroperoxide decomposers:
   a. *Sulfur-containing inhibitors*: sulfides, dithiocarbamates, sulfurized olefins
   b. *Phosphorous-containing inhibitors*: phosphites, ZDPs

Note that ZDPs function by both antioxidant mechanisms in addition to providing antiwear protection (covered in the earlier section). This dual (or tri-) functional ability of ZDPs to provide antiwear performance and antioxidancy by two different pathways explains why these additives are by far the most effective inhibitors, especially on a cost/performance basis.

Phenols or amines of specific structures function as radical acceptors by transfer of a hydrogen atom from the oxygen or nitrogen atom to the hydrocarbon or peroxy radical. The inhibitor radicals thus formed react through radical combination or electron transfer to give ionic compounds, or by additional reactions or formation of complexes that do not maintain the radical chain mechanism of the autoxidation reaction. In further reactions, very often ethers, betones, polyaromatic systems, and so on are formed.

25.5.4 Hindered Phenols and Arylamines

Hindered phenols and arylamines are two prominent examples of inhibitors that act as radical scavengers through hydrogen transfer. Figure 25.26 illustrates the mechanism of phenol performance.

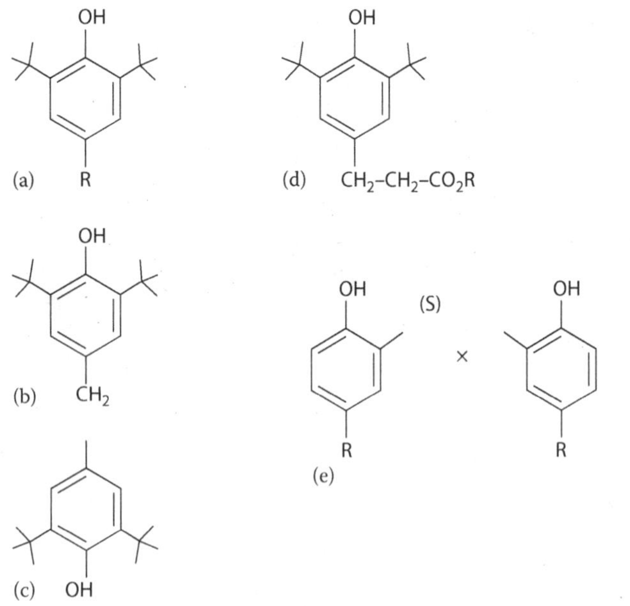

FIGURE 25.27 (a–d) Structures of sulfur free phenolic antioxidants and (e) structure of sulfur containing phenolic antioxidants. Major types of phenolic antioxidants.

Among various types of phenols, polyalkylphenols have significant performance advantage over nonalkylated compounds (Figure 25.27). The effect of substituents demonstrates the role of electron density and steric hindrance at the phenolic oxygen; an increase in the number of alkyl radicals and the introduction of electron donors in the o- and p-positions increases the efficiency, and the introduction of electron acceptors reduces it. Alfa-branching of alkyl groups in the o-position or increasing the alkyl group to C_4 in the p-position has a beneficial effect [27].

A new type of antioxidant involves the base-catalyzed addition of hindered phenol to Michael acceptors such as acrylate esters. Although more expensive than simple alkylphenols, the originators of this chemistry claim improved upper piston deposits' control in certain diesel engine tests and better seal compatibility than arylamines.

S-coupled (X = 1, 2) alkylphenols combine the antioxidant benefits of the phenol and sulfide groups.

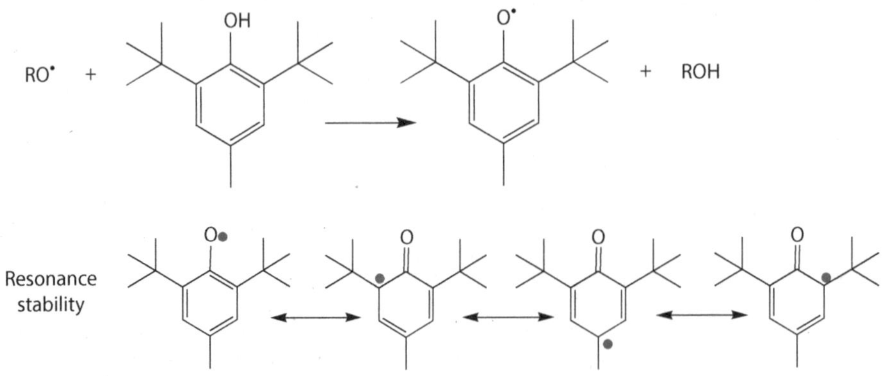

FIGURE 25.26 Mechanism of inhibition by hindered phenols.

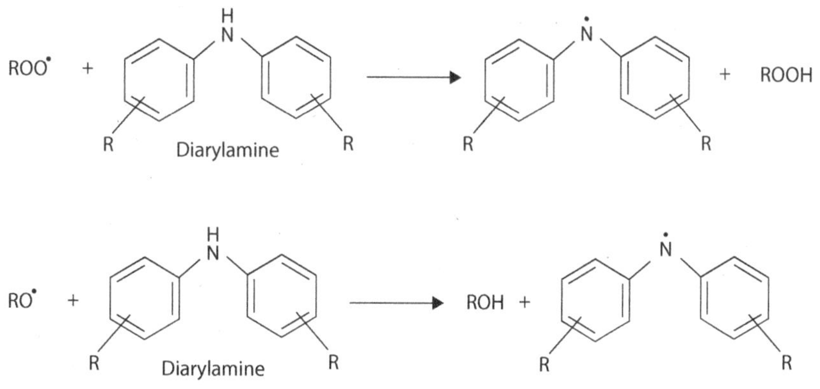

FIGURE 25.28 Mechanism of inhibition by arylamines.

The mechanism of oxidation inhibition (based on generation of nitrogen radicals) by arylamines is presented in Figure 25.28.

In general, alkylated arylamines are more effective antioxidants than alkylphenols because they are able to

- *Trap more equivalents of radicals*: four vs. two
- Better stabilize nitrogen or nitroxyl radicals (by two aryl rings instead one)
- Operate at both low- and high-temperature mechanisms (the latter of which regenerates the alkylated arylamine)

Oil-soluble amines such as diphenylamine, phenyl-alfanaphthylamine, and so on are the most common type of amine antioxidants used in lubricants. Diphenylamine can be alkylated with an aluminum chloride catalyst and a mixture of branched nonene olefins. A mixture of mono- and di-alkylate (Figure 25.29) is intentionally targeted by the reactant charge ratio to produce a liquid product. Diphenylamine is particularly suited to applications at elevated temperatures. Therefore, it is often used to lubricate supersonic aircraft engines and bearings. In these applications, arylamines prevent sludge formation in synthetic ester oils.

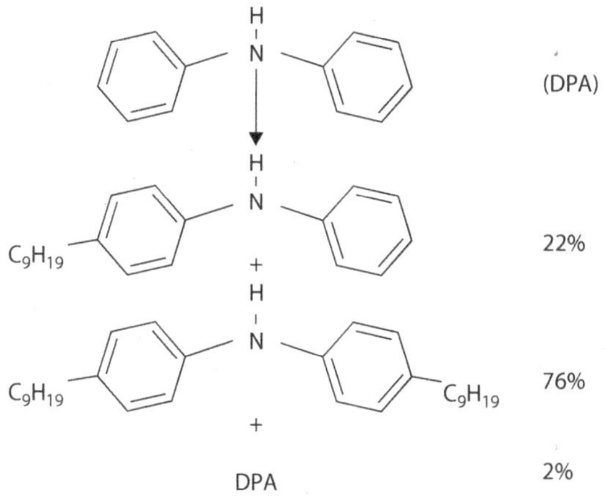

25.5.5 S- AND P-CONTAINING ANTIOXIDANTS

Sulfur, phosphorus, and compounds containing both elements decompose peroxides by reducing the hydroperoxide in the radical chain to alcohols; the sulfur or phosphorus atoms are correspondingly oxidized (Figure 25.30). Bivalent sulfur compounds (sulfides, etc.) yield sulfoxides and sulfones; trivalent phosphorus compounds (phosphates) are transformed into pentavalent ones (phosphates). Organic compounds of tetravalent sulfur act as peroxide decomposers, but those of the corresponding hexavalent, in line with theory, do not. Destruction of hydroperoxides is important so that these intermediates do not decompose into radicals, which can continue the chain oxidation process.

Phosphates are somewhat hydrolytically unstable, and due to limits on the phosphorous level in crankcase oils, these additives are generally limited to application in gear lubricants. ZDPs are more efficient phosphorous-containing antioxidants for crankcase oils, with the added benefit of antiwear performance.

ZDPs convert hydroperoxides to one equivalent of an alcohol and a carbonyl compound (Figure 25.31). The ZDP is regenerated intact and is available to convert another hydroperoxide molecule. Many cycles of hydroperoxide

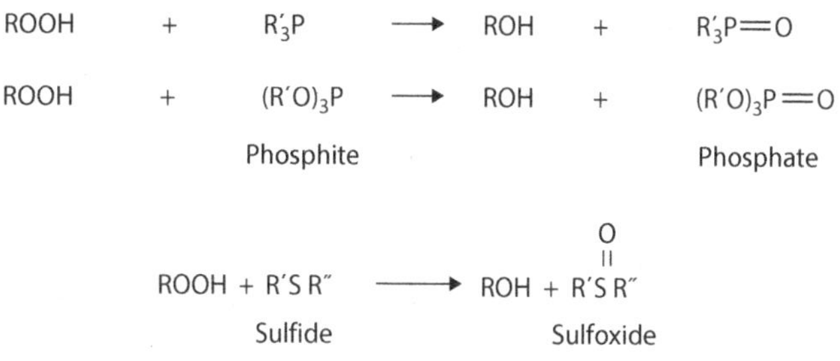

FIGURE 25.29 Alkyl aromatic amine as antioxidant.

FIGURE 25.30 Mechanism of S- and P-containing antioxidants.

FIGURE 25.31 ZDPs' hydroperoxide destruction.

decomposition are carried out by the ZDP, prior to its eventual breakdown.

25.5.6 Sulfur Compounds

Elemental sulfur is an efficient oxidation inhibitor; however, it shows a strong corrosion tendency.

From a practical approach, numerous dialkyl sulfides and polysulfides, diaryl sulfides, modified thiols, mercaptobenzimidazoles, thiophene derivatives, xanthogenates, zinc dialkyldithiocarbamates, thioglycols, thioaldehydes, and others have been examined as inhibitors. Dibenzyl disulfide must be mentioned among alkylaromatic S compounds.

Alkylphenol sulfides that are formed in the reaction of alkylphenols, such as butyl-, amyl- or octylphenol, with sulfur chloride are more active than compounds of the dibenzyl disulfide type, due to the position of the sulfur next to the OH group.

Modern compounds of this type contain mostly tert-butyl radicals besides methyl groups, such as 4,4′-thio-bis-(2-tert-butyl)-5-methylphenol. Sulfur–nitrogen compounds are also suited as oxidation inhibitors to lubricating oils (2-mercapto-benzimidazole, mercaptotriazines, reaction products of benzotriazole–alkylvinyl ethers or esters, and phenothiazine and its alkyl derivatives). Among sulfur-containing carboxylic acid esters, 3, 3′-thio-bis-(propionic-acid dodecyl ester) and bis-(3,5-di-tert-butyl-4-hydroxybenzyl)-malonicaid-bis-(3-thiapentadecyl) esters have been applied with success. These compounds have been replaced by the dialkyldithiophosphates due to their broad application spectrum. Sulfoxides, sometimes in combination with aromatic amines, have also been utilized.

25.5.7 Phosphorous Compounds

Red phosphorus possesses oxidation-inhibiting properties but cannot be used because of its corrosivity toward nonferrous metals and alloys. Triaryl and trialkyl phosphates have been proposed as thermally stable inhibitors; however, their applications are limited. Combined phosphoric acid–phenol derivatives such as 3,5-di-tert-butyl-4-hydroxybenzyl-phosphonic acid dialkyl esters or phosphonic acid piperazides have a better effect.

25.5.8 Sulfur–Phosphorus Compounds

Today, metal salts of thiophosphoric acids are used predominantly as oxidation inhibitors for crankcase oils. In principle, compounds that contain sulfur and phosphorus are significantly more efficient than inhibitors that contain only sulfur or phosphorus. Most widely used are ZDPs that are prepared by the reaction of P_2S_5 with the respective higher alcohols (e.g., hexyl, 2-ethylhexyl, octyl alcohols), followed by the reaction with zinc oxide. The temperature of the exothermic salt formation is kept at 20°C by successive addition of zinc oxide and cooling and is limited even at the end of the reaction to 80°C because of the thermal ability of the free dialkyldithiophosphoric acids. They react very corrosively with metals

and are toxic. Metal dialkyldithiophosphates are prepared in mineral oil solution. Their solubility in hydrocarbon oils increases with the increasing number of carbon atoms of the alkyl residues and is satisfactory with the diamyl compounds and higher. Longer-chain derivatives act as solubilizers for short-chain products. Metal dialkyl dithiophosphates act not only as antioxidants, but also as corrosion inhibitors and EP additives.

The efficiency of Zn dialkyldithiophosphates (with octyl or cetyl and, respectively, propyl, butyl, or octyl radicals in various combinations) decreases with increasing molecular mass of the alcohol substituents. The best results have been obtained with isopropyl and isoamyl radicals.

Reaction products of P_2S_5 with terpenes (dipentene, -pentene), polybutenes, olefins, and unsaturated esters belong to the same group; among these, the terpene and polybutene products have been proposed for crankcase oils. Metal dialkyldithiophosphates serve at the same time as detergents, EP additives, and anticorrosion agents, when the corrosion is caused by oxidation products; in this case, the metal is protected from attack by organic acids by the formation of sulfide or phosphate films. 2,5-Dimercaptothiadiazole derivatives have a similar effect.

25.5.9 Antioxidant Selection, Synergism, and Testing

Most of the crankcase engine antioxidant needs are met by the ZDP. Today's higher-temperature engines (Table 25.7) require supplemental ashless antioxidants to pass the Sequence IIIE engine. Therefore, modern engine oils use three or more different types of antioxidants, with the highest level (0.1%–0.15% wt.) being the ZDP.

The previous-generation engine to measure lubricant oxidation was the Sequence IIID, which required a maximum oil kinematic viscosity increase of 375 cSt at 80 h of operation. A fail result is shown for the base formulation at less than halfway through the test (30 h). Addition of 0.5% wt.

TABLE 25.7

Approximate Temperature of Internal Surfaces in a V-8 Engine

Area of Engine	Temperature Range (°C)
Exhaust valve head	650–730
Exhaust valve stem	635–675
Combustion chamber gases	2300–2500
Combustion chamber wall	204–260
Piston crown	204–426
Piston rings	149–315
Piston (wrist) pin	120–230
Piston skirt	93–204
Top cylinder wall	93–371
Bottom cylinder wall	Up to 149
Main bearings	Up to 177
Connecting rod bearings	93–204

arylamine antioxidant to base formulation easily meets the increased requirement in kinematic viscosity. This pass result could also be obtained by increasing the ZDP level, but the current limits on the phosphorous level prevent this option.

Another important antioxidant phenomenon is known as inhibitor synergism. Another way of presenting the definition shown here is by stating that "1/2 (or partial) levels of two compounds produces a greater benefit than a full level of either alone." Synergistic combinations of inhibitors can extend a lubricant's temperature and use ranges, thereby boosting its performance.

Either formulation containing 0.5% wt. of alkylphenol (rating = 63) or 0.5% wt. of arylamine as the inhibitor (rating = 60) alone does not provide a sufficient deposit control to pass this test. However, a mixture of two inhibitors at 0.25% wt. each offered an outstanding rating of 71, which is a pass result.

The efficiency of antioxidants in lubricating oils is tested on the laboratory scale through a battery of bench tests and engine tests under more severe conditions. This is particularly true for motor oils, where only the practical test in the engine can assess, for instance, the high-temperature efficiency of dithiophosphates and the frequently antagonistic effects of dispersants and oxidation inhibitors in a given additive combination.

As a rule, however, the final formulations of products are subjected to time-consuming field testing.

Here are some screen and engine oxidation performance tests.

> The ASTM D943 or turbine oil oxidation test is used for hydraulic oils. In this screen, molecular oxygen is blown through a quart of the oil at 95°C containing an iron and copper coil. The time needed to reach an oil TAN of 2.0 is recorded, with a pass result considered to be anywhere from 1000 to 3000 h, depending on the specific customer approval.
>
> ASTM D2272 (rotary bomb oxidation test, RBOT) and ASTM D4742 (thin-film oxygen uptake test, TFOUT) are two other industry-recognized bench tests. These screens use pressure to accelerate the oxidation process to simulate the lubricant's extended service or "real-world" conditions. The RBOT also finds utility in formulating industrial oils, while the TFOUT is used to measure the oxidation performance of passenger car and heavy-duty diesel engine oils.
>
> The Indiana stirred oxidation test has several versions, with temperature ranging from 150°C to 165°C. The beaker of oil is stirred vigorously for 24 h in the presence of iron and copper catalysts. Measurements include the oil's kinematic viscosity increase, TAN buildup, and amount of pentane-insoluble sludge generated.
>
> Pressure differential scanning of calorimetry is designed to predict the improvement the inhibitors add to an oil of poor oxidative stability.

TABLE 25.8
Historical Perspective: Changes in the V-8 Engine Over Decades

	1920	1960	1990
Engine capacity (L)	6	2	1.6
Brake horse power	50	70	130
Engine speed (rpm)	1200	5000	7000
Oil temperature (°C/°F)	60/140	90/194	130/266
Oil capacity (L)	14	4.5	3.5
Valve train	Side valves	Push rod, overhead valves	Twin overhead camshafts 4 valves
Fueling system	Single-choke carburetor	Multichoke carburetor	Fuel injected

Passenger car motor oil engines for oxidation performance consist of the following:

- Sequence IIIE and Sequence IIIF to measure the oil's viscosity buildup
- Sequence VE and Sequence VG to measure sludge control (Table 25.2)

The extreme temperature ranges reached in the Sequence VE engine are given in Table 25.7. The oil bulk temperature does not reach these high temperatures, but the fluid operates in close proximity to these parts. Since the oxidation process is accelerated by temperature, one can see why an oil's antioxidant system must be specially designed to be effective.

Typical passenger engine design has been undergoing dramatic changes since it has been introduced to the public in the early 1920s.

Table 25.8 summarizes several major design/fluid system changes observed [7]. The two critical parameters affecting lubricant stability and projected life are

1. The large increase in the engine oil temperature (2×)
2. The steady decrease in the amount of lubricant (~5×)

These two factors have put extra demands on the lubricant's antioxidant system, while recent limits on phosphorous levels have "capped" the use of ZDP.

25.6 VISCOSITY MODIFIERS

25.6.1 INTRODUCTION

Viscosity modifiers (VMs) are added to a lubricant formulation in order to reduce the viscosity–temperature dependence of base oils. This class of lubricant additives is the technology that enabled the development of multigrade lubricants in the 1960s. The primary feature of multigrades is that they allow the engine to start at low temperatures while providing sufficient viscosity at elevated temperatures to protect the engine against wear.

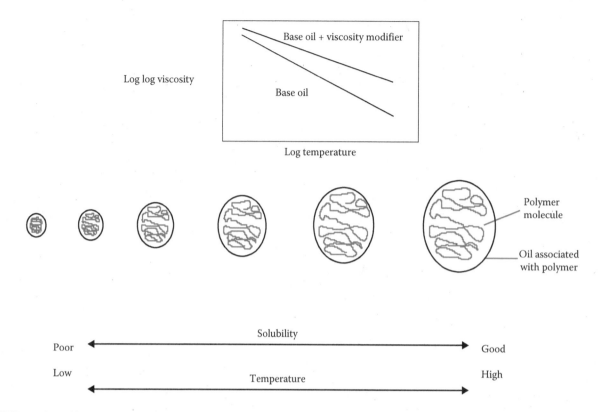

FIGURE 25.32 Effect of viscosity modifier on base oil viscosity–temperature dependence.

A lubricant oil formulation based on a group I mineral base stock without a VM additive will have a limited range of temperature operation since it has a relatively strong viscosity–temperature relationship. VMs are oil-soluble polymers that, when added to a base oil and additive mixture, will thicken the mixture at high temperatures while having a minimal thickening effect at lower temperatures (Figure 25.32). A very simplified mechanism to explain this phenomenon would be that the polymer–oil interaction at low temperature is minimal but increases as the temperature rises. This interaction of the polymer with the base oil at elevated temperatures increases the effective hydrodynamic volume of the polymer, thereby increasing the effective volume fraction of the VM. This, in turn, leads to an increase in lubricant viscosity. In order to make use of this phenomenon, a low-viscosity base oil mix is used to which is added the VM and additive mixture. Low-viscosity base oils permit the oil to flow freely at low temperatures while the VM will increase the viscosity at higher temperatures.

25.6.2 Viscosity Modifier Types

There are several types of VMs available to the oil formulator. Each has its strengths and weaknesses. Therefore, it is sensible to select the correct type for the intended application. Table 25.9 details some of the most common VMs in use today.

OCPs are prepared by either Ziegler–Natta or metallocene catalysis. Besides containing ethylene and propylene, diene monomers may be added to improve the handling characteristics of the solid form of the polymer. The relative amounts of each monomer will dictate the solubility of the copolymer. As the level of ethylene in the copolymer increases, polymer solubility decreases, especially at low temperatures. This reaction at low temperatures has the beneficial effect of lowering the polymer contribution to sub-ambient viscosity. But one must be careful not to go too far since at very high ethylene contents the OCP will drop out of solution or form gels at low temperature. This may only manifest after several months' storage at low temperatures. The monomer sequence distribution is also critical and can be carefully controlled to maximize the amount of ethylene in the copolymer and hence give minimal low-temperature thickening. OCPs are the most popular type of VM in use today due to their high thickening efficiency and relatively low cost.

Styrene–diene copolymers are prepared by an anionic polymerization of styrene with either butadiene or isoprene. The residual unsaturation in the backbone is removed by hydrogenation. Tapered block copolymers are prepared by charging both monomers together at the beginning of the reaction. To synthesize A–B block copolymers, the monomers are added sequentially. Anionic polymerization yields polymers with very narrow molecular weight distributions, especially when compared with those of OCPs and PMAs. A narrow molecular weight distribution yields the maximum shear-stability-to-thickening-efficiency ratio for a linear polymer. Because styrene–diene copolymers also possess excellent low-temperature properties and good thickening efficiency, they have found widespread use in engine oils. One of the few drawbacks is that, compared to OCPs, they are relatively

TABLE 25.9

Types of Viscosity Modifier Additives Utilized in Crankcase Oils

Viscosity Modifier Name	Abbreviation	Polymer Structure
Olefin copolymer	OCP	Linear copolymer, can contain long-chain branching
Poly(ethylene/propylene) (may contain diene termonomer)		
Polyalkylmethacrylate	PMA	Linear copolymer
Hydrogenated radial polyisoprene, can contain styrene as comonomer	HRI	Star polymer
Hydrogenated styrene–isoprene	HIS	Linear A-B block copolymer
Hydrogenated styrene–butadiene	HSB	Linear tapered block copolymer
Polyisobutylene	PIB	Linear homopolymer
Styrene–ester, alternating copolymer of styrene and alkylmaleate	SE	Linear copolymer

expensive to make due to the two-stage synthesis process and more expensive starting materials. Another issue is that A–B block copolymers undergo more temporary viscosity loss under high-temperature high–shear (HTHS) rate conditions (such as in lubricated journal bearings), especially when compared to OCPs.

Star polymers are prepared by an "arms-first" process, whereby isoprene and an optional comonomer are anionically polymerized to a predetermined molecular weight. The "arms" are then linked together by adding divinylbenzene, which forms an ill-defined gel core. The number of arms per star is defined by the amount of divinylbenzene relative to the polymer "arm" concentration. Finally, the star polymer is hydrogenated. These polymers have exceptional low-temperature properties but undergo more permanent viscosity loss under severe operating conditions when compared to OCPs at the same nominal shear stability index (SSI).

PMAs are prepared by free radical polymerization of alkylmethacrylate monomers. This produces a polymer with a relatively broad molecular weight distribution. To reduce molecular weight and hence increase shear stability of the polymer, the initiator concentration is increased. In addition, a chain transfer agent may be added during the polymerization process. This stops the polymer chains from growing and enables the production of a polymer with low molecular weight and exceptional shear stability. The composition of the monomers in the polymer backbone is chosen to optimize low-temperature properties, VI, and thickening efficiency. PMAs are frequently used because of their exceptional low-temperature viscometric properties. Because of this, they are used extensively as VMs in gear oils, automatic transmission fluids, and hydraulic oils, as well as pour point depressants. In recent years, OCPs have gained market share from PMAs in engine oils due to their relatively low cost and acceptable performance. PMAs can also be manufactured with dispersant functionality, which adds another performance attribute to an already highly versatile additive. The most direct method for preparing dispersant PMAs is to incorporate a nitrogen-containing monomer into the polymer during copolymerization. Alternately, various nitrogen-monomer grafting techniques can be used.

25.6.3 Dispersant Viscosity Modifiers

A variety of VMs can be prepared with the incorporation of dispersant properties. These dVMs have been used to provide improved sludge and soot handling performance to engine oil lubricants. They can either substitute for or add to the existing dispersant in the formulation. As a replacement for dispersant, they are required to provide equivalent performance. As a top treat to the existing additive, they will provide a boost in performance, especially in areas of sludge and soot handling. Examples include dispersant PMAs and dispersant OCPs.

25.6.4 Shear Stability of Engine Oils

Polymeric VMs are susceptible to mechanical and, in some cases, thermal shearing, leading to a loss in oil viscosity. The SSI of an engine oil VM is defined by the following equation. The SSI gives an indication of the mechanical stability of the VM:

$$SSI = \frac{m_i - m_f}{m_i - m_o} \times 100$$

where

m_i is the initial viscosity of lubricant with the viscosity modifier, cSt

m_f is the final viscosity of lubricant after shear*

m_o is the viscosity of lubricant without the viscosity modifier, cSt

Numerically, the SSI can be explained as follows: a VM with a low SSI number is more shear-stable while a VM with a high SSI number is less shear-stable. The SSI value of the VM will depend on its molecular weight. Higher–molecular weight VMs will have the greatest thickening efficiency for

---

* Note that the SSI value is dependent upon the particular shearing device used to measure shear stability. Common devices include the Kurt Orbahn fuel injector rig, CRC L-38 engine test, sonic shear device, and the KRL tapered bearing rig. The latter is primarily used to measure the SSI of driveline fluids such as highly shear-stable ATF and gear oils.

a given weight of polymer but will have the lowest shear stability (highest SSI). A value of 55 SSI represents the lowest shear stability acceptable for modern engine oils. While this is not unusual for North American passenger car motor oils, it is quite uncommon to find this shear stability being used in Europe, where minimum shear stability is specified in the ACEA and OEM specifications. A VM with 25 SSI or less is typical for European diesel and gasoline applications while a 35 SSI VM or less is preferred for North American diesel formulations.

25.6.5 VISCOSITY GRADE

A lubricant's viscosity grade, which defines the useful operating temperatures under which it may be used, is defined by a number of viscometric measurements that are given in the various SAE specifications. Polymeric VMs are crucial in allowing the manufacture of many viscosity grades. Most OEMs will use the viscosity grade to define the lubricant that works best in their equipment. For engine oil lubricants, the viscosity grade is defined by the SAE J300 specification.

In general, an OEM will aim to use the lowest-viscosity oil possible, in order to reduce energy losses due to friction. However, they will usually recommend a minimum oil viscosity with which their equipment will work comfortably. These limits may not always coincide with the SAE J300 specification, especially in the case of HTHS viscosity. European OEMs have historically been more conservative, and most have specified a minimum HTHS viscosity of 3.5cP for their engine oils irrespective of viscosity grade. This limit is still applied by European heavy-duty diesel OEMs since bearing wear is seen as a critical performance parameter. This would mean that an SAE 10W-30 oil would not only have to meet the SAE viscosity grade but would also have to be greater than 3.5cP in the HTHS viscosity test. North American diesel oils are similar to their European counterparts with 15W-40 making up the majority of the market. However, there are some 10W-30 oils in use for which there are no HTHS viscosity minimum limits other than that specified in SAE J300.

For some European passenger car OEMs, there has been some relaxation in the OEM limits applied to the HTHS viscosity as engine designs have changed to accommodate lower-viscosity oils. Lower-viscosity oils enable improvements in fuel economy. For example, Volkswagen has published oil specifications VW 503 and 506 for their gasoline and diesel engines. These specify an SAE 0W-30 with an HTHS of 2.9–3.4cP together with a fuel economy target. The reduction in HTHS viscosity for passenger car motor oils in Europe is moving in line with North America where most of the gasoline oils are 5W-30 with HTHS limits in line with the SAE J300 specification. In addition, some OEMs are now using 5W-20 oils for factory fill to ensure even better fuel economy. This is important since some VMs have poor temporary viscosity loss performance at high temperatures. If an HTHS of greater than 3.5cP is required, then it is easiest to formulate this oil with an OCP. If the HTHS target is less than 3.5cP, then most VMs will suffice.

25.6.6 VISCOSITY MODIFIER REQUIREMENTS

The necessary attributes for polymeric viscosity index improvers are as follows:

- Good oil thickening for cost-effectiveness
- Proper shear stability to stay in grade
- Temporary shear stability to meet HTHS requirements (especially where it is specified)
- Minimum low-temperature viscosity for cold cranking and pumping performance
- Minimal deposit formation at high temperature, for example, upper pistons and turbochargers

Correct selection of the VM will allow all the viscometric requirements to be achieved at minimum cost. This being done, it is then necessary to evaluate the formulation in the various engine and field tests. The performance of the VM in the engine and the service duty requirements will then define whether the VM is suitable for the intended application.

25.7 POUR POINT DEPRESSANTS

All mineral base oils contain some paraffinic components. These compounds have very good viscosity–temperature dependence but they are liable to form waxes at lower temperatures. While the effective volume of these waxes is low, they can still form a network of wax crystals that prevents the oil from flowing. One way to investigate this phenomenon is to measure an oil's ability to flow at low temperatures by way of its pour point. The pour point temperature of an oil is defined as the lowest temperature at which the oil is still capable of flowing and is measured according to ASTM D97. Additives used to reduce the pour point temperature may also benefit the low-temperature pumping viscosity.

Pour point depressants work at low temperatures, not by preventing wax crystals from being formed, but by minimizing the formation of wax networks and thereby reducing the amount of oil bound up in the network. Examples of pour point depressants include polyalkylmethacrylates, styrene ester polymers, alkylated naphthalenes, ethylene vinyl acetate copolymers, and polyfumarates. Treat rates are typically less than 0.5% (Figure 25.33).

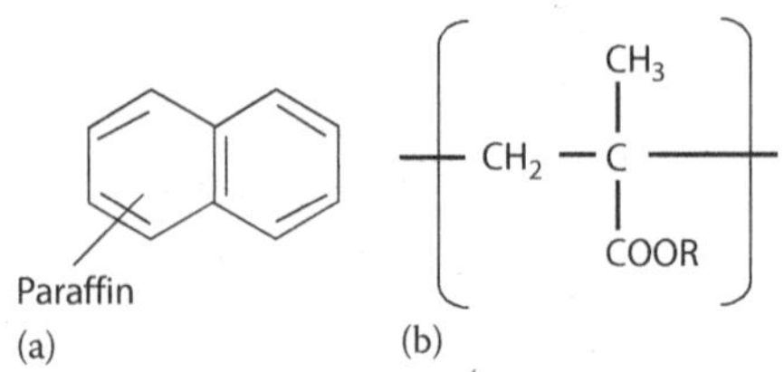

FIGURE 25.33 Examples of pour point depressants: (a) alkylated wax naphthalene and (b) polyalkylmethacrylate.

25.8 FOAM INHIBITORS/ANTIFOAMS

Entrained gas or foaming of a lubricant will reduce its effectiveness and needs to be minimized and at best stopped from building up during engine operation. In engine oils, the presence of foams can result in reduced oil pressure leading to engine damage, particularly for hydraulic lash adjusters or for hydraulically actuated unit injectors. Air entrainment is another problem. This may lead to cavitation of the oil film in bearings and possible failure. The performance of hydraulically actuated unit injectors is also sensitive to the amount of entrained air.

For simple oil formulations, a foam inhibitor may not be necessary. However, the stability of foams increases as more additives are added to the lubricant. This has necessitated the use of foam inhibitors such as those listed as follows:

* Dimethylsiloxane polymers
* Alkylmethacrylate copolymers
* Alkylacrylate copolymers

These compounds have borderline solubility in the lubricant and function by reducing the surface tension at the interface of the air bubble, thus allowing the bubble to burst more easily. To function effectively, they must be present as a fine dispersion.

Foam inhibitors are usually added to the lubricant at very low levels, typically less than 20 ppm. At higher levels, the solubility becomes an issue and is noticeable by an increase in the cloudiness of the lubricant and possible dropout of the foam inhibitor.

Foaming is measured according to ASTM D892, which uses air flowing through a porous ball to create foam in the test oil sample. The amount of foam and its stability is measured at 24°C and 94°C. ASTM D6082A is a variation of this test but measures foaming at 150°C. Additional engine tests such as the Navistar HEUI have also been introduced for diesel oils since API CG-4. This test measures a diesel oil's ability to minimize air entrainment, which could affect the performance of the unit injectors, which are hydraulically actuated. Gamma-ray detection techniques are also being used to monitor the entrained air in an engine by way of oil density. This method is used by BMW for their oil approval system.

REFERENCES

1. Klamann, D., *Lubricants and Related Products*, Verlag Chemie, Hamburg, Germany, 1984.
2. Mortier, R. M. and Orszulik, S. T., eds., *Chemistry and Technology of Lubricants*, Blackie Academic & Professional, London, U.K., 1997.
3. Smalheer, C. V. and Smith, R. K., *Lubricant Additives*, The Lezius-Hiles Co., Cleveland, OH, 1967.
4. Gergel, W. C., Detergents: What are they?, Presented at the *JSLE/ASLE Meeting*, Tokyo, Japan, June 10, 1975 (and references therein).
5. Kreuz, K. L., Gasoline engine chemistry as applied to lubricant problems, *Lubrication*, 55, 53–64, 1969.
6. Kreuz, K. L., Diesel engine chemistry as applied to lubricant problems, *Lubrication*, 56, 77–88, 1970.
7. Roby, S. et al., Deposit formation in gasoline engines–Dispersant effect on Seq.VE deposits. *Lubrication Engineering*, 50(12), 989–996, 1994.
8. Harrison, P. G., Brown, P., and McManus, J., 31P NMR study of the interaction of a commercial succinimide-type lubricating oil dispersant with zinc (II) bis (O, O′-di-iso-butyl-dithiophosphate). *Wear*, 156, 345–349, 1992.
9. Inoue, K. and Watanabe, H., Interactions of engine oil additives. *ASLE Transactions*, 26, 189–199, 1983.
10. Asseff, P. A., US Patent 2,261,047, Lubricant. Lubrizol, October 28, 1941.
11. Freuler, H. C., US Patent 2,364,283, Modified lubricating oil. Union Oil Co., December 5, 1944.
12. Schreiber, W., US Patent 2,370,080, Stabilized lubricant composition. Atlantic Refining Co., February 20, 1945.
13. Jones, R. and Coy, R. C., The thermal degradation and EP performance of zinc dialkyldithiophosphate additives in white oil. *ASLE Transactions*, 24(1), 77–90, 1981.
14. So, H. and Lin, Y. C., The theory of antiwear for ZDDP at elevated temperature in boundary lubrication condition. *Wear*, 177, 105–115, 1994.
15. Papey, A. G., Antiwear and extreme-pressure additives in lubricants. *Lubrication Science*, 10(3), 209–224, 1998.
16. Georges, J. M. et al., Presentation from *53rd Annual Meeting of STLE*, Detroit, MI, May 17–21, 1998.
17. Willermet, P. A. et al., Mechanism of formation of antiwear films from zinc dialkyldithiophosphates. *Tribology International*, 28, 163–175, 1995.
18. Eberan-Eberhost, C. et al., Ash-forming extreme pressure/anti-wear lubricant additives. Paper presented at the seminar *Additives for Lubricants*, TAE, Esslingen, Germany, December 11–13, 1991.
19. Habeeb, J. J. and Stover, W. H., The role of hydroperoxides in engine wear and the effect of zinc dialkyldithiophosphates. *ASLE Transactions*, 30(4), 419–426, 1986.
20. Korcek, S. et al., Mechanisms of antioxidant decay in gasoline engines: Investigations of zinc dialkyldithiophosphate additives. SAE Technical Paper 810014, 1981.
21. Jones, R. and Coy, R.C., The chemistry of the thermal degradation of zinc dialkyldithiophosphate additives. *ASLE Transactions*, 24(1), 91–97, 1981.
22. Kawanura, M. et al., Change of ZnDTP in engine oils during use. SAE Technical Paper 852076, 1985.
23. Ingold, K. U., Inhibition of autoxidation. *ACS Book - Oxidation of Organic Compounds*, Chapter 23, pp. 296–305, 1968.
24. Johnson, D., S. Korcek, and Zinbo, M., Inhibition of oxidation by ZDTP and ashless antioxidants in the presence of hydroperoxides at 160°C, *Fuels and Lubricants Meeting*, San Francisco, CA, October 31 to November 3, 1983, pp. 71–81.
25. Rasberger, M., Oxidative degradation and stabilisation of mineral oil based lubricants. *Chemistry and Technology of Lubricants*, Springer Science, Springer Netherlands, 98–143, 1997.
26. Abou El Naga, H. H. and Salem, A. E. M., Effect of worn metals on the oxidation of lubricating oils, *Wear* 96(3), 267–283, 1984.
27. Ingold, K. U., Inhibition of oil oxidation by 2,6-di-t-butyl-4-substituted phenols. *Journal of Physical Chemistry*, 64, 1636–1642, 1960.

26 Additives for Industrial Lubricant Applications

Leslie R. Rudnick

CONTENTS

26.1   INTRODUCTION

The objective of this chapter is to provide a review of additives and additive chemistry specifically for industrial lubricant applications. This chapter reviews the history of additives and some types of additives used to improve the performance of lubricants in industrial applications. This chapter also describes the chemistry and mechanisms of action of these additives and some of the important test methods used to evaluate formulated oils.

An additive is a chemical substance added to a lubricant to improve its properties and performance. Additives can function by physical and chemical interactions. Some additives remain essentially unchanged as a lubricant is used and others are depleted during use. Still others must chemically change to perform the desired function. Additives are used widely in many of the largest industries, including the food, soap and detergent, pharmaceutical, paint, plastics, fuel, and automotive industries. In fact, many of the antioxidants that are used in plastics are also used to stabilize automotive and industrial lubricants.

Today, most additive (and lubricant) companies have corporate websites that provide information. Some provide only cursory information and would do well to expend the effort to provide the scientist, formulator, and potential customer with more useful information. Other companies publish almost encyclopedic detail on all of their commercially available additives. Some include detailed chemistry, properties and performance characteristics, and examples of formulations. The Internet over the past several years has become a repository of information on additives and lubricants for the lubricant industry. A list of additive suppliers and the uniform resource locators for the corresponding website can be found in Chapter 35 and the additional material is available at www.crcpress.com/9781498731720. In addition to websites from the individual companies, industry organizations such as Noria and Society of Tribologists and Lubrication Engineers (STLE) provide information on lubrication, which includes discussion of additive interactions and depletion during use. Book publishers have also placed online the contents of entire books that can be purchased to download in their entirety or only the desired chapter(s). Several recent examples of books specifically related to additives and lubricant fluids have been published [1–3].

26.2 HISTORY OF ADDITIVES

Egyptians around 1500 BC may have used animal fats with or without water-based additives to move the huge stones and statues to the building sites of the early pyramids on lubricated wooden sledges [4]. Most certainly, there were earlier applications involving the lubrication of wheels and axles with mixtures of animal fats and natural oils. The earliest reported evidence of solid film lubrication was metal inserts in wooden implements found in the Middle Ages, around AD 500 [5].

The use of additives and additive industry has grown as a result of the larger use of lubricants and the more stringent requirements placed on lubricating fluids. The Industrial Revolution, 1700 to 1800, resulted in the greater need for both lubricants and lubricant additives. Addition of inorganic chemicals to water and the use of animal fats, vegetable oils, and fish oils were common.

The discovery of oil at Titusville, Pennsylvania, in 1859 resulted in not only the growth of a new petroleum-based industry but also a need for an additive industry to supply chemical components that would impart improved properties and

performance to the oils being used as lubricants at the time [6]. Early use of additives to improve lubricating oils was in metalworking fluids required by the rapid expansion of the railroad and automobile industries [7]. Animal fats, fish oils, vegetable oils, and blown rapeseed oils were added as friction modifiers. Rosin oil, mica, and wool yarn were among the early grease additives. Sulfur was added to metalworking oils as early as 1916, and shortly thereafter phosphorus additives came into use.

Petroleum refining continued to grow in scale and in process technology during the beginning of the twentieth century. At first, refining operations were aimed at improvements in the quality and quantity of fuel, heating oils, and asphaltic components for road paving. Between these ends of the spectrum, there were petroleum distillates that found their way into use as lubricants. These were high-boiling components and contained various amounts of sulfur and nitrogen that were later found to retard oxidation and wear when these fluids were used as lubricants. So one might consider the natural nonhydrocarbon components of petroleum as the first modern lubricant additives. Chemists at the major oil companies worked to isolate these materials and identify their structures so that they could be made as desired and added in optimum concentrations depending on the application requirements. There is an analogy here to the pharmaceutical industry where natural plant substances are extracted, isolated, and characterized as

---

TABLE 26.1
Chronological History of Additive Developments

Period	Application	Additive	Comments
BC	Wheels	Animal fats, oils	Chariots, carts
	Construction	Water	Pyramids, moving heavy stones
1500	Wire drawing	Wax	Au, Ag metals
1750	Industrial	Water	Industrial revolution, cooling
1800	Industrial, metalworking	Water plus inorganics	Rust prevention
1900	Industrial	Phosphorus, sulfur	Oxidation, wear protection grease
1920	Metalworking	Emulsifiers	Soluble oils
	Automotive, engine oils	Fatty acids	Friction reduction
1930	Engine oils	Isoparaffins, polymethacrylates	Pour point depressants
	Gears	Lead soaps	EP/wear protection
1940	Engine oils	Calcium carboxylates	Detergents
	Engine oils	Zinc dithiophosphates	Antiwear/antioxidant
	Engine oils	Sulfonates, phosphonates, phenates	Detergents, dispersants
1950	Metalworking	Soluble oils	Cooling capability
	Engine oils	Viscosity modifiers	Temperature–viscosity improvers
		Basic sulfonates	Detergents
		Overbasing, salicylates	Dispersants, acidity, oxidation control
1970	Industrial	Solid additives	EP/wear protection
		Phosphate glass	
		Boron nitride	
		Fluorocarbons	"Teflon-like" additives
		Low-toxicity	
		Biodegradable	"Environmentally friendly"
1990	Industrial	Low volatility	Reduced emissions
		Food-grade	Safety in food processing/handling
		Nano materials	Improved micro-lubrication

a prelude to direct chemical synthesis of these materials or improved analogs. Process engineers at the same time were developing processes to eliminate these materials so as to produce more stable, clearer lubricants. Synthetic chemists were also working to develop synthetic oils formed from basic chemicals derived from petroleum and other sources. All of these resulted in lubricant fluids that contained lower concentrations of naturally occurring antioxidants and antiwear (AW) and corrosion components. This made these oils more susceptible to degradation and less-effective lubricants, and therefore there arose a need to develop specific additives to protect these new fluids from all of the degradation processes that occur under the newer, more severe operating conditions. It should be noted that in general for petroleum-derived lubricants, the greater the extent of hydroprocessing, the lower the amount of heteroatom components, and therefore the more nonpolar and hydrocarbon-like the fluid.

Improvements in the performance of industrial lubricants since the late 1930s, for the most part, resulted from the development and use of synthetic additives. These developments occurred initially by researchers in major oil companies, but as the industry developed, many additive companies came into existence and flourished. A chronology of the development of some additives is given in Table 26.1.

Several reviews of lubricant additives and their applications have been published [3,8–12]. Companies interested in protecting their research investments and technology patented their new innovations, and by the 1950s, over 3000 additive patents had been issued. The rapid growth of the additive market in the 1900s has slowed in recent years because the world's finished lubricant market continues to remain flat. In 2006, it was predicted that global additive demand would grow at only 1%–1.2% per year until about 2015 [13].

26.3 FUNCTION OF ADDITIVES

As evidenced by the plethora of structures for additives described in the first 24 chapters of this book, the lubricant industry has had the advantage of the talents of many excellent organic chemists able to provide new organic and inorganic components as additives. Many of these structures, or slight modifications, are used to stabilize similar materials in other industries. Naturally occurring materials or synthetic versions are also used. For example, alpha-tocopherol, vitamin E, is used as an antioxidant both in food products and in lubricant products to maintain the stability of highly isoparaffinic base fluids that are used in food-grade and nonfood-grade industrial and automotive lubricants. Antioxidants, in particular, are employed to protect raw materials used in lubricant formulations as well as to protect the fully formulated oil during storage and use.

Although most industries have specific performance requirements related to their use of additives, there are some general requirements of additives that apply to all applications.

The overriding criterion for additives in lubricant formulations is that it be effective for the application. That said, there are certain criteria that make this possible.

These include the following:

1. *Solubility.* Additives must all be soluble in the formulated oil, and in some applications, this may require solubility in water. Additive solubility is critical in a lubricant throughout the temperature range that the lubricant will experience. This includes conditions of storage as well as the lowest and highest temperatures the oil will experience during use. Lubricants must be formulated such that additives remain in solution as the oil is used so that additives do not drop out and lower their effective concentration in the lubricant and so that they do not cause any deleterious effects by virtue of their insolubility.

2. *Stability.* Additives should have good stability to ensure acceptable performance over the useful life of the lubricant product. Lack of additive stability can affect the color and odor of the product. Breakdown of additive components caused by heat or light or moisture can result in reduced product stability and therefore performance. Many additives, especially arylamine antioxidants, are susceptible to darkening on exposure to light. This can affect product performance in terms of stability and also affect personal acceptance of the product in terms of quality.

3. *Volatility.* Low volatility is generally required for most industrial lubricant applications. This reduces the nonuniform loss of oil from the system and therefore thickening of the oil caused by the heavy components remaining in the system. Furthermore, this is often not considered when formulating; the additives that are chosen must be nonvolatile under the conditions of use to maintain the concentration of these additives in the lubricating fluid. This is especially true of lubricants that are to be used in high-temperature applications. In some particular applications, there is a need for a certain degree of volatility of the additive, for example, vapor-phase corrosion inhibitors must have a controlled rate of volatility.

4. *Compatibility.* Additives must be compatible with other components in the systems, including base oils and other additives. One of the most important considerations is compatibility with the base fluids being used in the formulation. This also involves compatibility with system components such as seals, gaskets, and hoses. Incompatibility is initially observed as a haze or cloudiness during formulation. This should be the first clue that something is insoluble or reacting. Bench tests that include the visual observation of the oil after testing can also provide information as to the compatibility of additives in the formulated oil.

5. *Odor.* Odor must be acceptable for the particular application. This is especially true in food-processing applications in which any odor from the base oil or additives can affect food product quality due to adsorption of odorous oil components, including additives. The same may be true for lubricants that

are used in the manufacture of textile or paper or other fibrous materials that can absorb odors.

6. *Activity.* Additives must have functional activity over the lifetime of the application. For example, a very reactive antioxidant that is depleted in a short time period will render a lubricant susceptible to oxidation sooner than a product formulated with long-term acting antioxidants as part of the additive chemistry in the product.

7. *Environmental compatibility.* Three issues to consider here are biodegradability of the additive, disposal constraints, and toxicity. These issues are important both from the standpoint of use and proper disposal and also in the event of a spill. Many industries are now using life cycle analysis to understand and control the use of all components that enter and leave their products and processes [14–16]. Green chemistry is also applied to the manufacture of additives to minimize the environmental impact of these processes and the disposal of waste products.

8. *Health and safety issues.* These issues are concerned with any of the aspects of human contact and transportation of the additives as raw materials or as components in finished products. Health and safety issues are particularly important for additives when used in food-grade lubricants for use in food-processing plants. Every additive that is allowed for use in food-grade H-1 (incidental contact) lubricants has an upper concentration limit. Lubricants formulated with quantities higher than these limits are not considered food grade because of the potential for unsafe levels of the additive to contact the food.

Before describing the chemistry of the various classes of additives used in the wide variety of industrial applications, it is appropriate to consider the additive types in terms of how they function as part of the formulated oil, that is, physical interaction or chemical interaction. Additives that function by physical interaction act through physical adsorption–desorption phenomena. These additives include pour point depressants, additives that impart lubricity, color stabilizers, additives that change in structural form with changes in temperature (viscosity index improver [VII]), additives that cause changes in surface or interfacial tension (antifoam and emulsifiers), those that cause the formation of structures that trap base fluid (tackifiers, thickeners, and fillers), antimisting additives, water repellants, or additives that affect the perceptive qualities of a lubricant (odorants, color stabilizers or dyes, and chemical marking agents) (Table 26.2).

Table 26.3 describes those additive types that act by chemically reacting with surfaces and are consumed or chemically changed during use. Chemical processes can also change the physical properties of the lubricant. For example, oxidation of the base fluid in a formulation results in degradation, producing

TABLE 26.2
Additives That Function by Physical Interaction

Additive	Function
Antifoam	Prevents the formation of stable foam
Antimisting	Reduces the tendency to mist or aerosol
Color stabilizer	Slows darkening of fluids
Demulsifier	Enhances the separation of water and oil by promoting drop-coalescence and gravity-induced phase separation
Dyes	Impart color, mask color, product identification
Emulsifier	Reduces interfacial tension and allows dispersion of water
Friction modifiers	Associate with the metal surface and improve sliding between surfaces
Odor control	Prevents or masks undesirable odors or maintains odor level
Pour point depressant	Lowers low-temperature fluidity by reducing the formation of wax crystals
Tackiness	Improves cohesion of fluid and nondrip quality
Thickener, solid filler	Converts oil into solid or semisolid lubricant
VII	Improves viscosity–temperature characteristics
Water repellant	Imparts water resistance to greases and other lubricants

TABLE 26.3
Additives That Function by Chemical Interaction

Additive	Function
Antibacteria (biocides)	Prevents or slows growth of bacteria in systems
Anticorrosion (corrosion inhibitors)	Protects surfaces against chemical attack
Antioxidant	Reduces the rate of lubricant oxidation and deterioration and increases oil and machine life
Antirust	Eliminates rusting due to water or moisture
AW	Reduces thin-film, boundary wear
Basicity control	Neutralizes acids from oxidation processes
Detergent	Maintains surface cleanliness
Dispersant	Suspends and disperses undesirable combustion, wear, and oxidation products
EP (temperature)	Prevents seizing and increases load-carrying ability
Friction modifiers	Reduce friction and increase lubricity
Metal deactivator	Counteracts catalytic effects of surfaces by passivating surfaces

smaller molecules that are more volatile than the original molecules. This has the effect of increasing the viscosity of the oil as the lower-molecular-weight components volatilize and can result in a decrease in oil volume if the process continues.

One should note that the chemical structure of the base fluid and that of the additives used in a lubricant formulation are all crucial to the properties and performance of the lubricant. The polarity of the base fluids and the additives is very important because each component of the mixture is competing for the metal surfaces. A polar additive that would normally adsorb and desorb reversibly from a metal surface in a nonpolar base fluid and has a reasonable steady-state concentration on the surface might have a much lower, even ineffective, concentration on the surface when in a formulation that contains a high concentration of polar base fluid such as ester or vegetable oil.

The AW performance of tricresyl phosphate (TCP) has been studied in 11 different base fluids. When molecular weight, dielectric constant, and solubility parameters were considered, correlations for the optimum concentration of the additive were obtained [17].

26.4 CLASSES OF ADDITIVES

26.4.1 ANTIOXIDANTS

Antioxidants chemically interact with free radicals and hydroperoxides to slow the chain reactions that result in oxidation of the base fluid. The two most common classes of compounds that are used as antioxidants are hindered phenols and aromatic nitrogen compounds. Some phenolic antioxidants commonly used in industrial applications include di-*tert*-butyl *p*-cresol (BHT) and 3,5-di-*tert*-butyl-4-hydroxy-hydrocinnamic acid, C7–C9 branched alkyl ester. Examples of arylamines include diphenylamine, dioctyl diphenylamine, styrenated diphenylamine, phenyl-alpha-naphthylamine (PANA), and polymerized versions of trimethylquinoline. These types of additives are described in detail in Chapter 1. In general, these types of antioxidants are used in gear oils, turbine oils, hydraulic oils, compressor oils, and various greases. Some additive molecules containing different chemical structures, such as phenothiazines, are sometimes used in ester formulations and can react many times with oxygen molecules before they are depleted. Other additives such as zinc dithiophosphates (ZDTPs) and ashless phosphorus-containing additives can also function as antioxidants in addition to other functions. These are described in Chapters 2 and 8, respectively.

26.4.2 ANTIWEAR AND EXTREME-PRESSURE (ANTISCUFFING) ADDITIVES

AW- and extreme-pressure (EP)-type (antiscuffing) additives react with metal surfaces to form a chemical film that is sheared more easily than the moving metal surfaces. Additives that perform these functions, ZDTPs, ashless phosphorus additives, ashless AW and EP additives, and sulfur carriers, are described in detail in Chapters 2, 7, 8, and 9, respectively. Two criteria differentiate between AW and EP additives. The differences depend on the load the additive film formed can carry and how it survives the high temperatures generated by the friction forces. Although bulk oil temperatures may be low, the temperatures at the metal-to-metal interfaces in the contact zone can exceed 300°C. At extreme loads, the frictional heat can exceed the melting temperature of the metal. In fact, during four-ball EP testing, the weld load is the load that caused the balls to actually weld to each other.

Some additives, for example, ZDTPs and alkylamines, perform multiple functions. These additives can function as antioxidants and AW additives or as AW and surface deactivators. ZDTPs are some of the most economical and effective additives. The largest application is in the formulation of automotive and diesel engine oils; however, they are used in some industrial hydraulic fluids. The temperature at which zinc dialkyldithiophosphates (ZDDPs) are most effective depends on the mix of hydrocarbon group, that is, primary alkyl, secondary alkyl, or aryl. Therefore, both the amount and structure of the ZDDP components affect the performance of the fully formulated oil.

ZDTP has come under scrutiny from an environmental viewpoint in automotive engine oils. Phosphorus and sulfur in the additive are suspected of being detrimental to the effectiveness of after-treatment catalysts in spark (gasoline) and compression ignition (diesel) engines. Research is ongoing to develop a cost-effective chemical alternative to ZDDPs. Alternative additives, evaluated to date, are not as effective and are significantly more expensive than ZDTPs. As combustion in the presence of lubricants and after-treatment catalysts are not factors in industrial applications, there is no problem with use of ZDDPs.

The adsorption and desorption of dibenzyl disulfide and dibenzyl sulfide on steel have been investigated. This involved the study of the rate of uptake of labeled additives onto stainless steel disks. The kinetics of both adsorption of these additives and the desorption from the metals was examined. Two processes, reversible physisorption and irreversible chemisorption, were identified, and the implications in AW and EP behavior were described [18].

26.4.3 FRICTION MODIFIERS

Friction modifiers or lubricious additives (see Chapter 5) absorb on the surface to form films that reduce the friction between the moving surfaces. These additives are generally polar molecules having a polar functional group (alcohol, aldehyde, ketone, ester, and carboxylic acid) and a nonpolar hydrocarbon *tail*. The polar portion of the molecule adsorbs onto the surface with the long hydrocarbon chains exposed to the moving surfaces, reducing the friction. They can also contain polar elements that adsorb and chemically react with the surface to form a protective film. Vegetable oils and animal-based fats have structures that make these materials excellent friction modifiers. In addition to liquid organic friction modifiers, there are various solid lubricants that reduce friction. These materials are reviewed in Chapter 6. A recent example from the literature reports on the influence of graphite, molybdenum disulfide, and perfluoropolyalkylethers on changes in the shear stress values in lubricating greases [19].

26.4.4 VISCOSITY MODIFIERS/VISCOSITY INDEX IMPROVERS

VIIs are very high-molecular-weight compounds (polymers) that function by physically changing form with changes in temperature. VIIs can be depleted chemically through thermal or oxidative breaking of bonds. Higher-molecular-weight versions can be mechanically sheared to form smaller, less-effective molecules. The stability of the VII depends on both the chemical structure and the size of the molecule. There are three main structural types of molecules used as VIIs—the nonpolar olefin copolymers (Chapter 11) and the polar poly-methacrylates and related compounds (Chapter 12) and hydrogenated styrene–diene viscosity modifiers (Chapter 13).

26.4.5 TACKINESS AND ANTIMISTING ADDITIVES

Tackiness and antimisting additives are used in several applications. In particular, tackiness is important in maintaining contact between the lubricant and a metal surface. Lubricants can have the tendency to migrate away from the surfaces that need lubrication depending on the lubricant viscosity and the geometry of the equipment. Antimisting is important in terms of reducing the tendency of lubricants to become airborne aerosols or fine mists. The materials that are used to reduce these phenomena are described in Chapter 15.

26.4.6 SEAL SWELL ADDITIVES

Seal swell additives are critical in maintaining contact between rubber or other types of seal materials and the metal surfaces in equipment. These seals keep a lubricant where it is needed and away from areas where it is not desired. Seal swell agents are adsorbed by the seal material and prevent the seals from cracking and deteriorating. These additives are described in Chapter 16.

26.5 DYES

Dyes are often used to identify fluids by application. Hydraulic fluids were often dyed red and coolants green during World War II to make it easier for technicians to select the correct lubricant and thus properly maintain equipment. MIL-L-Spec.5605 Red Oil was used as an aircraft hydraulic fluid in World War II in many aircraft. Dyes are also used for product branding and are used to identify that a particular supplier's oil has been used instead of a lower-cost-competitive oil.

26.6 BIOCIDES/ANTIMICROBIALS

Biocides are used to reduce the growth of bacteria in lubricants used in metalworking. These applications by nature use large volumes of lubricants that are exposed to air and the environment and consequently have the tendency to decompose by natural processes. To maintain lubricant quality and performance, biocides are added to these lubricants. These are described in Chapter 17.

26.7 SURFACTANTS

Surfactants are one of the most widely applied and diverse group of materials. They have been applied to lubrication for centuries and form the basis for a combination of performance features that include reducing friction and wear and cooling

during lubrication. As these materials can solubilize polar and nonpolar components, surfactants make possible many of the fluids used in metalworking and other areas of lubrication. Surfactants adsorb to the metal surface and provide a *cushion* or molecular boundary between moving surfaces to protect them from wear. The chemistry of surfactants is described in Chapter 18.

26.8 CORROSION INHIBITORS

Corrosion is potentially devastating to industrial equipment that comes in contact with water and other corrosive materials. Even water in the form of atmospheric humidity, which can enter equipment or react with exterior surfaces, can reduce the lifetime of expensive industrial machines. Corrosion inhibitors are summarized in Chapter 21.

26.9 ADDITIVES FOR FOOD-GRADE LUBRICANTS

Lubrication is obviously needed in the food production and preparation industries for the same reasons as are needed to lubricate machinery in other industries. An additional requirement, however, in the food industry is the need to be certain that the chemistries used are safe in the event that any of the lubricant makes contact with the food product. There are specific additives that are allowed and limiting concentrations of these additives is permitted in food-grade additives. Additives for food-grade lubricants are described in detail in Chapter 27.

26.10 MAGNETIC DISK DRIVE ADDITIVES

Magnetic disk drives are a critical component of our modern technology. These components are evolving in terms of the rate or rotation of the magnetic disk, and precision and smaller tolerances in the rotating spindle. Lubricants are designed to provide adequate lubrication, and additives must maintain performance at both the spindle motor and the head–disk interface. Additives protect against oxidation of the base fluid and control conductivity and other properties necessary for long-term performance. Additives for magnetic disk recording applications are described in Chapter 28.

26.11 ADDITIVES FOR GREASE

Additives for use in grease are in general similar to those used in other industrial lubricants, but their concentrations and combinations can be quite different. These additives are described in Chapter 29.

26.12 ADDITIVES USED IN BIODEGRADABLE LUBRICANTS

Bioderived and biodegradable lubricants are becoming more important due to increased interest in protecting the environment and due to regulations imposed by many governments around the world. For lubricants to meet the environmental criteria of biodegradability and toxicity, the additives must perform needed functions without being toxic themselves. These additives are sometimes the same as those currently in use in other applications, but more and more, they are specifically developed to meet the performance criteria and be nontoxic and biodegradable. The performance of some of these materials is described in Chapter 23. One example of newly synthesized additives directed at biodegradable esters is reported [20]. These additives were reported to form stable lubricating films on the surface of stainless steel balls used in four-ball wear testing.

Additives improve the properties and performance of lubricants by modifying chemical and physical characteristics of the oil. The mechanisms of action of additives depend on the chemical structure of the additive, the chemical structure of the base fluids, the temperature of use, and the mechanical pressures applied to the formulated oil. Many of these mechanisms are well understood and have been described in the literature. Some aspects of the additive chemistry of the more common additives are briefly reviewed in the following section.

26.13 ADDITIVE CHEMISTRY

An understanding of additive chemistry is necessary both for the synthesis of additives and to develop an understanding of how lubricants and additives react, perform, and degrade during use. Beginning in the 1940s, the severe operating conditions of military equipment during World War II led to the investigations of these areas in much greater detail than had previously been done. More severe operating conditions required greater thermal and oxidative stability, less friction, and better wear protection from the lubricants. As lubricant base fluids provide only part of the required properties and performance, the additives used in formulations needed to provide greater performance than ever before.

Research on oxidation mechanisms and oxidation inhibitor additives has been investigated by Ingold [21], Watson [22], Mahoney et al. [23], Chao et al. [24], Jensen et al. [25], Booser et al. [26–28], Dennison and Condit [29], and Zuidema [30], and others.

Antioxidants reduce oxidation by trapping free radicals or by decomposition of hydroperoxides generated during the oxidation process. As discussed earlier, substances containing amines or phenolic groups are effective antioxidants. Antioxidants that contain sulfur and combinations of phosphorus and sulfur are effective against hydroperoxides [31,32]. Some metal-containing compounds such as copper naphthenates can be either prooxidants or affect antioxidants depending on the formulation and the additive concentration [33,34].

A review of the effects of metals and the effect of hard-coated metals on the thermooxidative stability of branched perfluoroalkylether lubricants have also been reported [35,36].

Metal deactivators reduce the effect of metals and their salts on the rate of oxidation of the bulk fluid [37]. Some of these, such as ethylenediaminetetraacetic acid (EDTA), are complex with the metal particles suspended in the fluid rendering them inactive.

AW additives form protective films by a chemical or physical reaction with the surface [38–42]. They minimize the removal of metals through formation of lower-shear-strength films that reduce friction, decreasing the contact temperatures or by increasing the contact surface thereby reducing the effective load. The most effective AW additives contain sulfur or phosphorus or both. Studies of the performance and mechanisms of action of TCP and ZDDP have been well documented in the literature [43–48].

EP, or more correctly extreme-temperature additives, also contain phosphorus or sulfur compounds. Chlorinated compounds are also effective [49]. Use of phosphorus compounds started around the time of World War I, and addition of sulfur and chlorine to gear oils began in the 1930s [50–53]. EP gear oils in the 1960s and 1970s were formulated using combinations of sulfur-, phosphorus-, and chlorine-containing additives. The use of chlorine-containing AW and EP compounds is gradually declining due to environmental concerns with some suppliers developing new replacement products [54].

Rust and corrosion inhibitors adsorb on the surface forming a barrier hydrocarbon film against water and corrosion-causing materials. These are often basic amine-type compounds that have the ability to neutralize acids. They may also passivate the metal surface reducing the catalytic effect on oxidation of the oil [55,56]. When combined with antioxidants in an additive package, this combination is typically referred to as *rust and oxidation package* or *rust and oxidation oil (R&O)* and generally provides protection to industrial oils that are used under relatively mild operating conditions.

Detergents can neutralize acids produced by oxidation of the oil or, in the case of internal combustion engines, combustion. The combustion products get into the fluid as a result of blowby. Detergents can be phenates, sulfonates, and salicylates [38,57]. Typical dispersants are succinimides, succinates, and Mannich-type reaction products. Dispersants slow the formation of deposits and sludge by dispersing the precursors and insoluble particles in the fluid. These, like detergents, are large molecules with a polar end and a nonpolar portion. Dispersants and detergents are typically used in engine oils and not in industrial oils. These two categories of additives represent the largest category of additives sold. Normally, dispersants and detergents are not required in industrial gear oils, turbine oils, or industrial hydraulic fluids.

Foam inhibitors prevent formation of foam by changing surface tension that results in the collapse of gas bubbles as they form [58]. The inhibitors are high-molecular-weight polymers that are relatively insoluble in the oils. Antifoam additives are generally used at concentrations measured in parts per million and are usually blended into a more soluble solvent system to facilitate the accuracy of addition of the small quantities required. Some commercial antifoam additives supplied are already diluted in a suitable carrier fluid at concentrations that help to make addition of the material more accurate. Air entrainment is often mistaken for foaming in hydraulic systems. Additives can sometimes help with air entrainment, but usually, mechanical changes to the system to prevent air from entering the system or modifying the base fluid is a better approach. Polydimethylsiloxanes, polydiarylsiloxanes, mixed siloxanes, and polyglycol ethers are examples of these additives.

Friction modifiers are used to lower the energy requirements of a system by reducing the friction of the system. These additives are often long-chain hydrocarbon molecules with one end of the molecule containing a polar group that is adsorbed on the surface [59–61]. Chain length and branching of the carbon chain affect both the surface adsorption of these additives and the packing density on the metal surface. Oleic acid and other organic fatty acids, alcohols, and amides are typical structures of commercial friction modifiers.

In addition to liquids, there are also solids that act as friction modifiers. Graphite or molybdenum compounds are examples of solid friction modifiers. These materials provide reduction in friction and often help reduce wear. These materials are not soluble in hydrocarbon oils and are therefore suspended in the lubricating fluid.

Rheological properties of industrial oils are important performance criteria [62–65].

Viscosity at normal operating temperatures is an important lubricant specification. The lubricant film thickness under operating conditions needs to be considered to mitigate wear and damage to equipment. Viscosity is a function of temperature and pressure (and shear rate for non-Newtonian fluids). VIIs are used to modify the temperature–viscosity characteristics of an oil. VIIs are high-molecular-weight polymers that are compact molecules at low temperatures and expand in size as the temperature increases. Styrene-butadienes, acryloids, polymethacrylates, and various copolymers are the main VIIs used in industrial oils today. Furthermore, VIIs usually increase the viscosity of the lubricant and can be used in formulating a fluid to increase the viscosity of a base fluid.

Low-temperature viscosity is also an important specification for equipment that must start or operate at low ambient temperatures. Paraffinic petroleum–derived base oils become very viscous at low temperatures unless the high wax content is removed. Typically, isoparaffinic and naphthenic base oils are fluid down to temperatures below −40°C. Synthetic fluids generally have excellent low-temperature properties. Natural oils, such as vegetable oils, are solids or become very viscous by ~0°C.

Pour point depressant additives are often used to improve deficiencies in lubricants based on petroleum-derived paraffinic base fluids. These additives interfere with the wax crystalline network formed in these paraffinic oils at low temperatures. The effectiveness depends on both the type of fluid and the chemical structure of the additive used. Evaluation of concentration levels and fluid blending is required to achieve the maximum low-temperature properties for these additives. Several Group II and Group III base fluids contain high levels of isoparaffinic molecules and have acceptable pour points without the need of pour point depressants. Natural oils, on the contrary, may require

blending with synthetic fluids and pour point depressants to achieve the necessary viscosity levels at lower temperatures. Some commercially available additives are effective in mineral oils but have limited or no effect in natural oils. Acceptable pour point depends on the conditions of use. Obviously, fluids to be used in arctic conditions must have fluidity at the lowest temperature of use. Generally, this will be most important at start-up because some frictional heating will help bring a lubricant to a lower viscosity as the equipment continues to run. This is critical for equipment in remote locations such as space applications and remote measuring equipment. However, these fluids are generally designed with synthetic base fluids that have low pour points and low low-temperature viscosities.

System components such as seals and gaskets can shrink, swell, crack, or otherwise deteriorate if they are not compatible with the fluids used. Seal swell control is important to ensure fluid compatibility with the proper seal materials to prevent oil leaks. Organic phosphates have been used; however, the generally accepted approach, especially for synthetic fluids, is to blend the base fluids to achieve the correct degree of seal swell. For example, blending mixtures of synthetic hydrocarbons, such as poly-α-olefins (PAOs), and esters to achieve desired seal swell results is a common practice.

In the case of water base fluids such as hydraulic, metalworking, water glycols, antifreeze, and other similar fluids, the use of demulsifiers, emulsifiers, biocides, and corrosion prevention additives is often required.

Metalworking fluids can contain AW, EP, antimist, corrosion inhibitors, biocides, antifungicides, defoamers, couplers, and dyes.

26.14 INDUSTRIAL APPLICATIONS

Section 26.13 has described the form and function of additives and the chemistry involved in the use of these additives. The areas of industrial application are extremely diverse, both in terms of applications and in the severity of the conditions under which the additives must perform and maintain the integrity and quality of the lubricant.

Following is a list of some of the more important lubricants for industrial applications:

1. Air tool lubricants
2. Bearing lubricants
3. Cable oils
   a. Hollow cables
   b. Solid cables
4. Chain lubricants
5. Compressor lubricants
   a. Air compressors
   b. Reciprocating compressors
   c. Rotary screw compressors
      i. Oil-flooded compressors
   d. Rotary vane compressors
   e. Centrifugal compressors
   f. Axial flow compressors

g. Gas compressors
   i. Natural gas compressors
   ii. Ethylene compressors
   iii. Process gas compressors
h. Refrigeration compressors
   i. Reciprocating compressors
   ii. Rotary screw compressors
6. Condenser oils
7. Conveyor lubricants
8. Drilling fluids
9. Gear lubricants
   a. AW
   b. EP
   c. R&O
10. Grease
11. Heat transfer fluids
12. Hydraulic oils
   a. Gear pumps
   b. High-pressure hydraulics
   c. Piston pumps
   d. Vane pumps
13. Metalworking fluids
   a. Continuous casting
   b. Cold forging
   c. Cutting
   d. Coolants
   e. Drawing
   f. Extrusion
   g. Grinding
   h. Honing
   i. Lapping
   j. Punching
   k. Rolling
   l. Quenching
   m. Spark erosion oils
14. Mold release agents
15. Rubber release agents
16. Shock absorber oils
17. Switch gear oils
18. Textile lubricants
19. Turbines
   a. Gas turbines
      i. Aircraft turbines
   b. Steam turbines
20. Transformer fluids
21. Vacuum pump oils
22. Vibration damping oils
23. Wire rope protection fluids

Each of these applications has various subsets, and lubricants and the additives used will depend on the properties of the oil and the performance requirements for that oil. Some of these applications require oil-only formulations and, therefore, require oil-soluble additives. Other formulations, especially in some metalworking applications, use oil-in-water emulsions, and these formulations require different additive chemistries. Details on the types of additives that are used in each of these

areas can be found in various chapters of this book and in several excellent references [2,3,66].

Essentially all industrial lubricants include some additives, even if only a protective amount of antioxidant or a small amount of AW additive. Bench tests are typically used to screen the performance of additives. This is done because the cost of testing using bench tests is significantly less than the cost of field test or full-scale equipment tests. An additional reason is that tests can be performed using less lubricant, and test times permit more variations to be evaluated than can typically be done in full-scale equipment.

There are countless methods available for evaluating lubricant formulations. These range from measuring physical properties, for example, viscosity, pour point, flash point, fire point, specific gravity, and volatility, to structural analysis using infrared spectroscopy, elemental analysis, aniline point, and carbon residue. Further testing might include corrosion testing, hydrolytic stability, seal compatibility, and seal swell tests. There are also a wide variety of mechanical test methods such as the four-ball wear test and various Falex test methods. Most of the industry-accepted tests are those listed by the Institute of Petroleum, the American Standard Testing Materials (ASTM), the Deutsche Industrie Normung, the Japanese Industrial Standards, and the Coordinating European Council. A summary of many of these standard test methods can be found in Chapter 34.

In addition to the standard test methods, there are several laboratory test techniques that have been used to evaluate various aspects of lubricant fluids and formulations. Furthermore, there are many unpublished proprietary methods that have been designed and used by various companies in the lubricant industry. A few examples of analytical techniques that have been applied to lubricant evaluation will be described in the following few paragraphs.

Thermal methods, such as thermogravimetric analysis and pressure differential scanning calorimetry (PDSC), have been used to evaluate volatility, oxidation stability, and deposit-forming tendencies [67–74]. Noel and Cranton were among the first to realize the potential of using DSC for characterizing lubricant fluids [67–69].

A method has been developed using DSC to evaluate the remaining useful life of a lubricant. This work showed that one could use the data from DSC to essentially predict when to change the oil in turbine engines resulting in savings in the costs associated with equipment failure by not changing the oil soon enough and the costs associated with changing the oil too soon because of inaccurate data on the quality of the used oil [70].

Rhee [71] has reported on the development of a new oxidation stability test method for greases using PDSC. Advantages of using PDSC are that only small samples are required, the analyses are rapid and repeatable, and sensitivity is high.

The observation that lubricant base fluids exhibit differences in volatility depending on whether air is present or not has been demonstrated, and this difference has been described as the oxidation-mediated volatility [75]. Others have used DSC to evaluate the thermooxidative stability of formulated lubricants using thermal analytical methods [76].

Standard oxidation tests are routinely used to determine the quality of a lubricant formulation, and the results of these tests are frequently listed in the product specification or on the technical data sheets for the product. Two important industry tests are the rotating pressure vessel oxidation test, ASTM Method D2272, and the oxidation characteristics of inhibited mineral oils, ASTM Method D943. Both of these are examples of bulk oxidation tests [77]. More recently, developed oxidation tests are the thin-film oxygen uptake test and a thermooxidation engine oil simulation test. These, in contrast to the previously described tests, are thin-film oxidation tests [78,79]. All of these tests are used to evaluate the thermal and oxidative stability of base fluids and formulations containing additives expected to enhance performance.

Evaluation of friction and wear can be performed on various testers. ASLE (now STLE) in 1976 [80] compiled a list of 234 wear test methods. ASTM lists several friction and wear methods including four four-ball methods—ASTM Methods D2266, D2783, D4172, and D51283 [76]. Variations of the use of the four-ball test as a research tool are reported in the literature [81,82]. The Cameron-Plint and pin-on-disk tests are also widely used for the evaluation of materials, friction, and wear [83,84].

26.15 SUMMARY AND CONCLUSIONS

Additives perform a wide variety of functions and represent an important and necessary contribution to the overall properties and performance of industrial lubricants. Without additives, even the best base fluids are deficient in some features. Formulation of a product is always a compromise. There is always the issue of cost/performance. Many times, there is a great amount of research necessary to balance the various effects of the different additives used in combination with the base oils.

The arsenal of lubricant additives is large, and yet, new additives continue to be developed. This will help future lubricants meet the increasingly demanding conditions that equipment manufacturers require. Furthermore, environmental issues will require more benign and environment-friendly additive design in the future to meet the global regulations that are being proposed and implemented.

ACKNOWLEDGMENT

I acknowledge J. M. Perez for his original chapter on additives for industrial applications in *Lubricant Additives: Chemistry and Applications*, Marcel Dekker, New York, 1st Edition, 2003 from which this chapter has been expanded, updated, and revised.

REFERENCES

1. Booser, E. R. (Ed.), *Tribology Data Handbook*, CRC Press, Boca Raton, FL, 1997.
2. Rudnick, L. R. (Ed.), *Synthetics, Mineral Oils and Bio-Based Lubricants*, CRC Press, Boca Raton, FL, 2006.
3. Rudnick, L. R. (Ed.), *Lubricant Additives: Chemistry and Applications*, 1st edn., Marcel Dekker, New York, 2003; Rudnick, L. R. (Ed.), *Lubricant Additives: Chemistry and Applications*, 2nd edn., CRC Press, Boca Raton, FL, 2009.
4. Krim, J., Friction on the atomic scale, *Lubr. Eng.*, 8–13, 1997.
5. Dowson, D., *History of Tribology*, Longman Group Ltd., London, U.K., 1979.
6. Dickey, P. A., The first oil well, *J. Pet. Technol.*, 12–26, January 1959.
7. Larson, C. M. and Larson, R., Lubricant additives, in *Standard and Book of Lubrication Engineering*, Chapter 14, O'Conner, J. J. and Boyd, J. (Eds.), McGraw-Hill, New York, 1968.
8. Smalheer, C. V. and Mastin, A., Lubricant additives and their action, *J. Pet. Inst.*, 42(395), 337–346, 1956.
9. Zuidema, H. H., *The Performance of Lubricating Oils*, 2nd edn., Reinhold, New York, 1959.
10. Zisman, W. A. and Murphy, C. M., *Advances in Petroleum Chemistry and Refining*, Interscience, New York, 1960.
11. Stewart, W. T. and Stuart, F. A., in *Advances in Petroleum Chemistry and Refining*, Vol. 7, Chapter 1, Mckette, J. Jr. (Ed.), Wiley, New York, 1963.
12. Smalheer, C. V. and Kennedy, S., in *Lubricant Additives*, Bowen, S. N. and Harrison, R. G. (Eds.), The Lubrizol Corp., The Lezius-Hiles Co., Cleveland, OH, 1967 (Library of Congress Cat. Card No. 67-19868).
13. DeMarco, N., Additive supply: Tight and tighter, Lube Report, February 21, 2006.
14. de Wild-Scholten, M. J. and Alsema, E. A., Environmental life cycle inventory of crystalline silicon photovoltaic module production, *Materials Research Society Symposium Proceedings*, Vol. 895, 2006.
15. Brinkman, N., Wang, M., Weber, T., and Darlington, T., Well-to-wheels analysis of advanced fuel/vehicle systems—A North American Study of energy use, greenhouse gas emissions, and criteria pollutants, May 2005.
16. Papasavva, S., Hill, W. R., and Brown, R. O., GREEN-MAC-LCCP®: A Tool for assessing life cycle greenhouse emissions of alternative refrigerants, SAE Technical Paper 2008-01-0829, SAE, Warrendale, PA. April 2008.
17. Han, D. H. and Masuko, M., Comparison of antiwear additive response among several base oils of different polarities, *Preprint of the Society of Tribologists and Lubrication Engineers*, Las Vegas, NV, May 23–27, 1999, pp. 1–6.
18. Dacre, B. and Bovington, C. H., The adsorption and desorption of dibenzyl disulfide and dibenzyl sulfide on steel, *ASLE Trans.*, 25(2), 272–278.
19. Czarny, R. and Paszkowski, M., The influence of graphite, molybdenum disulfide and PTFE on changes in shear stress values in lubricating greases, *J. Synth. Lubr.*, 24(1), 19–30.
20. Wu, H. and Ren, T. H., Tribological performance of sulfonamide derivatives in lubricating oil additives in the diester, *J. Synth. Lubr.*, 23(4), 211–222.
21. Ingold, K. U., *J. Phys. Chem.*, 64, 1636, 1960.
22. Watson, J. I. and Smith, W. M., *Ind. Eng. Chem.*, 45, 197, 1953.
23. Mahoney, L. R., Korcek, S., Norbeck, J. M., and Jesen, R. K., Effects of structure on the thermoxidative stability of synthetic ester lubricants: Theory and predictive method development, *ACS Prepr. Div. Petrol. Chem.*, 27(2), 350–361, 1992.
24. Chao, T. S., Hutchinson, D. A., and Kjonaas, M., Some synergistic antioxidants for synthetic lubricants, *Ind. Eng. Chem. Prod. Res. Dev.*, 23(1), 21–27, 1984.
25. Jensen, R. K., Korcek, S., and Jimbo, M., Oxidation and inhibition of pentaerythritol esters, *J. Synth. Lubr.*, 1(2), 91–105, 1984.

26. Booser, E. R., PhD thesis, The Pennsylvania State University, State College, PA, 1948.
27. Booser, E. R. and Hammond, G. S., *J. Am. Chem. Soc.*, 76, 3861, 1954.
28. Booser, E. R., Hammond, G. S., Hamilton, C. E., and Sen, J. N., *J. Am. Chem. Soc.*, 77, 3233, 3380, 1955.
29. Dennison, G. H. Jr. and Condit, P. C., *Ind. Eng. Chem.*, 37, 1102, 1945.
30. Zuidema, H. H., *Chem. Rev.*, 38, 197, 1946.
31. Johnson, M. D., Korcek, S., and Zimbo, M., Inhibition of oxidation by ZDTP and ashless antioxidants in the presence of hydroperoxides at 160°C, SAE SP-558, pp. 71–81, 1983.
32. Scott, G., Synergistic effects of peroxide decomposing and chain-breaking antioxidants, *Chem. Ind.*, 271, 1963.
33. Mortier, R. M. and Orszulik, S. T., *Chemistry and Technology of Lubricants*, Blackie Academic and Professional, Suffolk, Great Britain, 1997.
34. Brooke, J. H. T., Iron and copper as catalysts in the oxidation of hydrocarbon lubricating oils, *Discuss. Faraday Soc.*, 10, 298–307, 1951.
35. Hellman, P. T., Zabinski, J. S., Geshwender, L., and Snyder, C. E. Jr., The effect of hard-coated metals on the thermo-oxidative stability of branched perfluoroalkylether lubricants, *J. Synth. Lubr.*, 24(1), 1–18.
36. Hellman, P. T., Geshwender, L., and Snyder, C. E. Jr., A review of the effect of metals on the thermo-oxidative stability of perfluoroalkylether lubricants, *J. Synth. Lubr.*, 23(4), 197–210.
37. Klaus, E. E. et al., U.S. Patent 7,776,524, 1993.
38. Rizvi, S. Q. A., Additives—chemistry and testing, in *STLE Tribology Data Handbook*, Chapter 12, Booser, E. R. (Ed.), CRC Press, Boca Raton, FL, 1997.
39. Roby, S. H., Yamaguchi, E. S., and Ryason, P. R., Lubricant additives for mineral based hydraulic oils, in *Handbook of Hydraulic Fluid Technology*, Chapter 15, Totten, G. E. (Ed.), Marcel Dekker, New York, 2000.
40. Liston, T. V., Lubricating additives—What they are and how do they function, *Lubr. Eng.*, 48(5), 389–397, 1992.
41. Spikes, H. A., Additive–additive and additive–surface interactions in lubricants, *Lubr. Sci.*, 2(1), 4–23, 1988.
42. Barton, D. B., Klaus, E. E., Tewksbury, E. J., and Strang, J. R., Preferential adsorption in the lubrication process of zinc dialkyldithiophosphate, *ASLE Trans.*, 16(3), 161–167, 1972.
43. Rounds, F. G., Some factors affecting the decomposition of three commercial zinc organic phosphates, *ASLE Trans.*, 18, 78–89, 1976.
44. Jahanmir, S., Wear reduction and surface layer formation by a ZDDP additive, *J. Trib.*, 109, 577–586, 1987.
45. Spedding, H. and Watkins, R. C., The antiwear mechanism of ZDDPs, Part I, *Trib. Int.*, 15, 9–12, 1982.
46. Watkins, R. C., The antiwear mechanism of ZDDPs, Part II, *Trib. Int.*, 15, 13–15, 1982.
47. Coy, R. C. and Jones, R. B., The thermal degradation and E.P. performance of zinc dialkyldithiophosphate additives in white oil, *ASLE Trans.*, 24, 77–90, 1981.
48. Perez, J. M., Ku, C. S., Pei, P., Hegeemann, B. E., and Hsu, S. M., Characterization of tricresylphosphate lubricating films using micro-Fourier transform infrared spectroscopy, *Trib. Trans.*, 33(1), 131–139, 1990.
49. Taylor, L., Dratva, A., and Spikes, H. A., Friction and wear behaviour of zinc dialkyl-dithiophosphate additives, *Trib. Trans.*, 43(3), 469–479, 2000.
50. L. B. Turner, Lubricating oil and method of manufacturing the same. U.S. Patent 2,124,598, Standard Oil, 1938.
51. F. L. Miller and C. F. Smith, High pressure lubricant. U.S. Patent 2,156,265, Standard Oil, 1939.
52. C. F. Prutton, Lubricating composition. U.S. Patent 2,153,482, Lubrizol, 1939.
53. C. F. Prutton and A. K. Smith, Lubricating composition. U.S. Patent 2,178,513, Lubrizol, 1939.
54. Tocci, L., Under pressure, *Lubes-n-Greases*, 3(12), 20–23, 1997.
55. Ford, J. F., Lubricating oil additives—A chemist's eye view, *J. Inst. Petrol.*, 54, 198, 1968.
56. Hamblin, P. C., Kristen, U., and Chasen, D., A review: Ashless antioxidants, copper deactivators and corrosion inhibitors. Their use in lubricating oils, in *International Colloquium on Additives for Lubricants and Operational Fluids*, Bartz, W. J. (Ed.), Techn. Akad., Esslingen, Germany, 1986.
57. O'Brien, J. A., Lubricating oil additives, in *STLE Handbook of Lubrication*, Vol. II, Booser, E. R. (Ed.), CRC Press, Boca Raton, FL, pp. 301–315, 1984.
58. O'Brien, J. A., Lubricating oil additives, in *STLE Handbook of Lubrication*, Vol. II, Booser, E. R. (Ed.), CRC Press, Boca Raton, FL, p. 309.
59. Haviland, M. L. and Goodwin, M. C., Fuel economy improvements with friction modified engine oils in EPA and road tests, SAE Technical Paper No. 790945, 1979.
60. Dinsmore, E. W. and Smith, A. H., Multi-functional friction modifying agents, *Lubr. Eng.*, 20, 353, 1964.
61. Dancy, J. H., Marshall, H. T., and Oliver, C. R., Determining frictional characteristics of engine oils by engine friction and viscosity measurements, SAE Technical Paper No. 800364, SAE, Warrendale, PA, 1980.
62. Johnston, W. G., A method to calculate pressure–viscosity coefficient from bulk properties of lubricants, *ASLE Trans.*, 24, 232, 1981.
63. Klaus, E. E. and Tewksbury, E. J., Liquid lubricants, in *STLE Handbook of Lubrication*, Vol. II, Booser, E. R. (Ed.), CRC Press, Boca Raton, FL, pp. 229–253, 1984.
64. Peter Wu, C. S., Melodick, T., Lin, S. C., Duda, J. L., and Klaus, E. E., The viscous behavior of polymer modified lubricating oils over a broad range of temperature and shear rate, *J. Trib.*, 112(July), 417, 1990.
65. Manning, R. and Hoover, Flow properties and shear stability cold flow properties, in *ASTM Manual on Fuels, Lubricants and Standards: Application and Interpretation*, Trotten, G. (Ed.), ASTM, Basel, Switzerland, 2001.
66. Klamen, D., *Lubricants and Related Products*, Verlag Chemie, Basel, Switzerland, 1984.
67. Noel, F., *J. Inst. Pet.*, 57, 354, 1971.
68. Noel, F., *Thermochim. Acta*, 4, 377, 1972.
69. Noel, F. and Cranton, C. E., in *Analytical Calorimetry*, Vol. 3, Porter, R. S. and Johnson, J. F. (Eds.), Plenum Publishing Company, New York, p. 305, 1974.
70. Kauffman, R. E. and Rhine, W. E., Development of a remaining useful life of a lubricant evaluation technique. Part I: Differential scanning calorimetric techniques, *Lubr. Eng.*, 154–161, 1988.
71. Rhee, I.-S., The development of a new oxidation stability test method for greases using a pressure differential scanning calorimeter, *NLGI Spokesman*, 7–123, 1991.
72. Walker, J. T. and Tsang, W., Characterization of lubricating oils by differential scanning calorimetry, SAE Technical Paper No. 801383, 1980.
73. Hsu, S. M. and Cummins, A. L., Thermogravimetric analysis of lubricants, SAE Tech Publication 831682, San Francisco, CA, SAE Special Technical Publication No. 558, 1983.

74. Perez, J. M., Zhang, Y., Pei, P., and Hsu, S. M., Diesel deposit forming tendencies of lubricants—Microanalysis methods, SAE Technical Paper No. 910750, Detroit, MI, 1991.

75. Rudnick, L. R., Buchanan, R. P., and Medina, F., *J. Synth. Lubr.*, 23, 11–26, 2006.

76. Zeeman, DSC cell—A versatile tool to study thermo-oxidation of aviation lubricants, *J. Synth. Lubr.*, 5(2), 133–148, 1988.

77. *Annual Book of ASTM Standards, Petroleum Products, Lubricants and Fossil Fuels*, American Society of Testing Materials, West Conshohocken, PA.

78. Ku, C. S., Pei, P., and Hsu, S. M., A modified thin-film oxygen uptake test (TFOUT) for the Evaluation of Lubricant Stability in ASTM Sequence IIIE, SAE Technical Paper No. 902121, Tulsa, OK, 1990.

79. Florkowski, D. W. and Selby, T. W., The development of a thermo-oxidation engine oil simulation test (TEOST), SAE Technical Paper No. 932837, SAE, Warrendale, PA, 1993.

80. *Friction and Wear Devices*, 2nd edn., ASLE (STLE), Park Ridge, IL, 1976.

81. Perez, J. M., A review of four-ball methods for the evaluation of lubricants, *Proceedings of the ASTM Symposium on Tribology of Hydraulic Pump Stand Testing*, ASTM, Houston, TX, 1995.

82. Klaus, E. E., Duda, J. L., and Chao, K. K., A study of wear chemistry using a micro sample four-ball wear test, *Trib. Trans.*, 34(3), 426–432, 1991.

83. Operating Instructions, PLINT TE 77 High Frequency Machine, Serial No. TE77/84 08/96, Wokingham, England.

84. Sheasby, J. S., Coughlin, T. A., Blahey, A. G., and Katcicjm, D. F., A reciprocating wear test for evaluating boundary lubrication, *Trib. Int.*, 23(5), 301–307, 1990.

27 Lubricants and Fluids for the Food Industry

Saurabh Lawate

CONTENTS

27.1 INTRODUCTION

Consider the following facts related to the food industry in the United States:

- About 1.5 million workers are employed by the food industry (nonfarm payrolls). This represents about 14% of the total number of people employed in the U.S. manufacturing sector (Figure 27.1).*
- U.S. households spend 12.6% of their income on food, which comes third after housing and transportation, which represented 33.3% and 17% respectively of the household expenditure in 2014.†

- The agricultural sector contributed about 4.8% of the gross domestic product (GDP) in the United States in 2014.‡
- The food-processing sector in the United States had 21,000 companies that generated revenues of $750 billion in 2015 representing ~37.5% of the global processed food sales of $2 trillion.[1]

The food industry in the United States is thus fairly sizable; it may be extrapolated that the same is applicable to most other countries, albeit with varying numbers.[2] In this chapter, the term "food industry" or "food-processing industry" will be used generically and may encompass any or all of the following stages: manufacturing, processing (including heating and cooling, slicing, dicing, mixing, cooking, bottle filling, carbonation, and blending), and packaging.

---

* Food manufacturing is assigned code 311 and is a subsector of the manufacturing sector as per the North American Industry Classification System (NAICS), formerly SIC codes.

† http://www.ers.usda.gov/data-products/chart-gallery/detail.aspx?chartId =40045&ref=collection&embed=True, United States Department of Agriculture, Economic Research Service (USDA-ERS) using data from the U.S. Bureau of Labor Statistics, Consumer Expenditure Survey, 2014.

‡ http://www.ers.usda.gov/data-products/chart-gallery/detail.aspx?chartId=40 037&ref=collection&embed=True, United Stated Department of Agriculture, Economic Research Service (USDA-ERS) using data from the U.S. Department of Commerce, Bureau of Economic Analysis, Value Added by Industry, 2014.

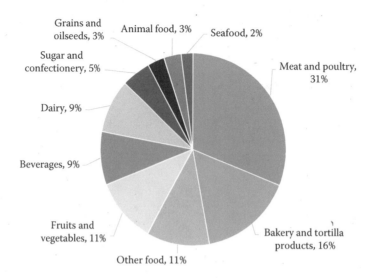

FIGURE 27.1 Employment breakdown in the food industry in the United States (2013). (From An overview of the state of the food processing industry, Pollack papers, http://www.pollock.com/an-overview-of-the-state-of-the-food-processing-industry, 2015.)

Specialized warehousing (refrigerated and atmosphere-controlled[3]) as well as refrigerated mobile transportation constitutes a major component in food manufacturing at all stages: pre-manufacturing (raw materials), during manufacture, and post manufacturing but pre-retail (finished goods). To put this in perspective, there are 4.17 billion cubic feet of refrigerated storage capacity in the United States, with five states contributing to about 37% of the total.[4] Despite its interesting nature, this aspect will not be discussed in this chapter.

The food industry uses raw materials, hard goods used for packaging, as well as utilities and services. Aside from this, a large variety of equipment is used in the food industry and it often relies on a similarly large variety of lubricants and fluids. The focus of this chapter will be on lubricants and fluids preceded by a general overview of the equipment used (Figure 27.2).

This chapter will be split into four sections:

1. Equipment used in the food industry
2. Lubricants and fluids used in the food industry*
3. Formulation considerations for lubricants and fluids
4. Trends and summary

27.2 EQUIPMENT USED IN THE FOOD INDUSTRY

The equipment used in the food industry ranges from air compressors, hydraulic pumps, hydraulic motor-driven systems, gear drives, and conveyors to cutters and slicers, mixers, homogenizers, filtration systems, and packaging lines. Additionally, heating and cooling systems are extensively used during the manufacture and processing of food (Table 27.1). Depending on the operation, a food-manufacturing plant may have all or some of the equipment mentioned earlier. A lot of the equipment is supplied by "original equipment makers" (OEMs). Like other industries, there are OEMs whose equipment is either directly used by the industry, for example, compressors, or the equipment is integrated into larger systems that are customized for the operation, for example, packaging lines which may involve hydraulic pumps and gear boxes. Some OEMs are exclusively dedicated to the food industry, for example, meat cutting and slicing equipment, while others service a variety of industries. (The reader may refer to Section 27.3.1 for details on OEM approvals, etc.)

It is beyond the scope of this book to go into further details of the equipment. For example, it is beyond the scope to describe that type of hydraulic systems (vane or piston),

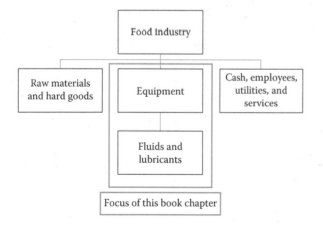

FIGURE 27.2 Food industry—essential components that keep it going.

---

* This chapter will exclude greases, pastes, and aerosols and will only include a discussion of products designed for incidental food contact (NSF HX-1, H1, and HT1)—see the relevant section for definitions of these categories.

TABLE 27.1

Types of Equipment and Its Application in the Food Industry[a]

Equipment Type	Application	Extent of Usage in Plants
Compressors	Used to compress air, CO_2, nitrogen, refrigeration gases (ammonia, HFC, etc.), and other specialized gases	100% of plants have air compressors; few exceptions exist
	Air compressors can be used for blow tanks, or moving solid renderings to various locations of the plant including dumpsters	A typical plant may have 3–15 rotary screw air compressors and anywhere from 5 to 40 rotary screw ammonia compressors
Hydraulic systems	Used for various types of work such as lifting or in equipment used for hide pulling	A typical plant has numerous hydraulic systems; some are small integrated into the machinery while some are large centralized systems which send oil out to multiple locations and it is returned to a centralized pump
Gear boxes	Used in conveyer systems and chain systems in slaughter houses. Other equipment is blood dryers	Wide usage especially where the product has to be continuously transported from one location to another (which is most plants)
Heat exchangers	Used in heating or cooling processes in plants; usually the fluid flowing through this is coming in hot and is cooled to the temperature needed or vice versa, for example, frozen food manufacturing	Most plants have one or multiple systems with heat exchangers

[a] Part of this information is based on private communication from Mr. Ben Duval, US Petrolon Industrial, Lincoln, NE, 2016.

or what type of gears (worm, planetary), or what type of air compressors (screw or reciprocating). Discussion of specific OEMs is also excluded since the "Food Industry Master" serves as an excellent source of this information.[5]

Most of the gear and hydraulic equipment used in the food industry is often lighter duty than what is found in heavy industries such as steel, mining, and so on. However, equipment reliability constitutes a major consideration in the food industry as downtime can be very costly. In many food-processing applications, "just in time" processing of food occurs. For example, chicken operations literally have trucks lined up and deliveries are timed. Any breakdown in operation creates major financial losses. Coupled to this, the food industry continues to focus on higher throughput, which means that machinery is moving at higher speeds and higher temperatures. This means that every

effort needs to be made to service and maintain equipment; ideally, longer service intervals automatically reduce the need for backup equipment or additional downtime. As a corollary, the lubricants and fluids used to maintain the equipment need to be of high quality and reliability as well.

27.2.1 ROTATING MACHINERY USED IN THE FOOD INDUSTRY

Rotating machinery includes pneumatic equipment such as air compressors or hydraulic systems and gear boxes. The rotating equipment generally converts electrical power to mechanical work and is almost ubiquitously used in the food industry. As a consequence, the food industry spends a sizable amount on power transmission equipment purchases (Figure 27.3).

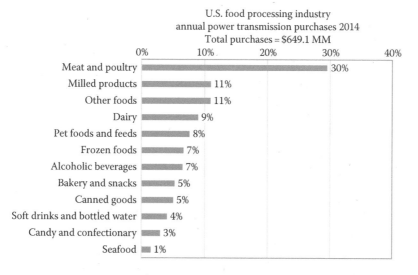

FIGURE 27.3 Annual power transmission purchases by the food-manufacturing industry in the United States. (From An overview of the state of the food processing industry, Pollack papers, http://www.pollock.com/an-overview-of-the-state-of-the-food-processing-industry, 2015.)

27.2.2 Other Equipment Used in the Food Industry

Numerous other pieces of equipment are used in the food industry such as cutters, slicers, mixers, sifters, homogenizers, can seamers, ovens, and associated systems, which include heating and cooling systems. Direct fired or natural gas–powered heaters and boilers are also common. Aside from this, there are several ancillary pieces of equipment such as filtration devices, waste handlers, and other specialized equipment. Last but not least, food-packaging lines constitute a big portion of equipment, which is often located at the point of exit in a plant. Food-packaging equipment can get extremely sophisticated with multiple filling lines often requiring the use of inter gas or carbonation for sodas and beverages, and sometimes for handling perishable items.

27.3 LUBRICANTS AND FLUIDS USED IN THE FOOD INDUSTRY

With the wide usage of equipment, it is not surprising that food plants use a fair amount of lubricants and fluids (Figure 27.4).

The total volume of fluids used in a plant is dependent on many factors starting with the size of the plant and the nature of the operation.

There are generally numerous lubricant fluids used. Typically hydraulic fluids, gear oils, and compressor oils constitute the bulk of the annual lubricating fluid volume. This is followed by specialized fluids such as (but not limited to) chain oils and can-seaming oils. Fill volumes and change intervals vary for fluids. Greases, pastes, and aerosols are also used but are excluded from the discussion.

Heating and cooling constitutes a key function in the industry. The process is facilitated by the use of heat transfer systems (utilizing heat transfer fluids) and refrigeration systems

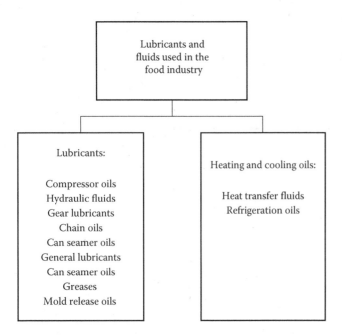

FIGURE 27.4 Types of lubricants and fluids used in the food industry.

TABLE 27.2
Type of Fluid and Typical Change Interval in a Food Plant

Type of Fluid	Typical Change Interval	Typical Volume in a Large Beef Processing Plant[a]
Air compressor oil	Depends on name plate oil life; typically 2,000, 4,000, 6,000, 8,000, 12,000 h	220–440 gal
Hydraulic fluid	At least annual or more frequent	10,000–15,000 gal
Industrial gear oil	Long-lasting	500–1,000 gal
Chain oil	Consumable	Varies
Vacuum pump oils	Depends on application	220–330 gal
Grease and miscellaneous lubricants, e.g., aerosols	Grease—often filled for life bearings	Varies
Heat transfer fluid	Long-lasting except for leaks; more frequent with commonly sold paraffinic oils	5000–30,000 gal
Ammonia (R-717) refrigeration oil	Unlikely to require change	440–660 gal

[a] Private communication from Mr. Ben Duval, President, US Petrolon Industrial, Lincoln, NE, April 2016.

(utilizing refrigeration compressor oils). These constitute a fairly significant category of non-lubricant fluids used in a plant. The highest volume of heat transfer fluids and refrigeration oil generally occurs at start-up while subsequent fills occur due to servicing or operational losses and/or upgrades. A list of types of fluids used in a food plant and typical change intervals is given in Table 27.2.

27.3.1 Special Considerations for Lubricants and Fluids Used in the Food Industry

27.3.1.1 OEM Approvals

One must be very mindful of OEM guidelines for the recommended fluids and lubricants used. Often, an OEM may offer its own branded fluid which is developed to best work in their equipment. Furthermore, the equipment warranty may be tied to the usage of the OEM-branded fluid. The choice of lubricants and fluids should be made according to the following order: OEM-branded fluids, OEM-approved or OEM-recommended fluids, a fluid known to be an equivalent.

27.3.1.2 Regulatory Compliance

While there is no mandated law that requires the use of specific fluids, there are several regulations and specific industry-accepted practices that apply to the food industry requiring the use of specific lubricants and fluids. There is also a growing trend overseas regarding applying these as well as local regulations.[6]

- Good manufacturing practices (GMP)
- Hazard analysis critical control point (HAACP)
- National Sanitation Foundation (NSF) registration for lubricants and fluids
- Compliance with other certifications/approvals (Kosher, Halal, others)
- ISO certifications (ISO 9001, ISO 21469)
- Responsible care certifications (RC140001)

If one decides to purchase fluids that meet all of these requirements, then the list of available suppliers and fluids becomes narrower. Due to this, the fluids and lubricants market for the food industry is often considered a niche market.

27.3.1.3 NSF Registration

Lubricants are one of the most widely used and serviced fluids in food manufacturing. While the term "food-grade lubricants" is widely used in the industry (for H1 lubricants), it is the author's opinion that this term should be avoided; it is misleading in that the end user may be led to believe that the lubricants meet the same safety standards as edible food and/or food ingredients, which is not the case. The term "incidental food contact lubricants" should be used instead and only for H1 lubricants. Furthermore, the United States Department of Agriculture (USDA) has strict guidelines on how much contamination of food by the lubricant or fluid is tolerated and the percentage of contamination is rather low.

From a historical perspective, the USDA first initiated guidelines for incidental food contact lubricants (this was mainly driven by meat- and poultry-processing plants) but abruptly discontinued the program in 1998, and it was then taken over by the National Sanitation Foundation (NSF), Ann Arbor, MI. An alternate attempt by the Underwriters Laboratory (UL), Chicago, IL, was initiated but for various reasons the NSF program is now the program of choice. While lubricants can be self-certified, most food manufactures now expect NSF registration from their suppliers. This stems from increased awareness of the program, mainly due to the diligent efforts of the NSF, which have been coupled with a highly efficient and reliable system of scrutiny and approvals. The online availability of the information at the NSF website (www.nsf.org) plus well-organized annual steering committee meetings has lent further value to the program.[7]

The industry is increasingly using lubricants and fluids registered with the NSF in Ann Arbor, MI, under the following categories:

NSF H1: Lubricants where indirect contact with food is possible

NSF H2: Lubricants where no contact with food is possible

NSF H3: Trolley and miscellaneous oils

NSF 3H: Release agents for direct food contact

NSF HT1: Heat transfer fluids where indirect contact with food is possible

NSF HX1: Category for components—base oils and additives

The distinction between the categories is shown in Table 27.3.

The NSF program for lubricants (H1–H3 categories) falls under its "Nonfood Compounds" program, which also includes "Water Treatment Compounds" (G1–G7), "Cleaners" (A1–A8), "Non-processing Area Products" (C1–C3), and some other registrations. Lubricants constituted 9889 registrations in 2015, which represented 54% of all the nonfood registrations. Since 2009, the lubricant category (H1 and HX-1) has seen a steady increase in registrations with approximately 6250 registrations in 2009 and 9889 registrations in 2015.[2]

27.3.1.3.1 NSF Registration Process

The NSF registration is initiated via a formal application process that requires full disclosure of components, the trade name of the product, and a representative label. The application is then reviewed at the NSF against a list of registered components and the limits of use set for these components. Should all criteria be met, an NSF registration number in the appropriate category (H1, H2, H3, HT1, HX-1) is granted. Should any component not be on the registered list, additional information is often required by the NSF and approval is not guaranteed. Rebrand registrations are also permitted.

27.3.1.3.1.1 Approval of Components and Base Oils Additive components and base oils used in the formulation of H1 lubricants must belong to a specific section in the

---

TABLE 27.3

Classification of Lubricants Used in the Food Industry

Lubricant/Fluid Classification	Application	Compliance
H1	Application where there is possibility of food contact	21 CFR Section 178.3570 or NSF review[a]
H2	Application where there is no possibility of food contact	Not defined but generally the lubricants must be free of any toxic heavy metals or ingredients; also subject to NSF review
H3	Trolley oils	None defined
HT1	Heat transfer fluids	21 CFR Section 178.3570 or NSF review
3H[a]	Release agents	21 CFR 172.878, GRAS substances and defoamers 21 CFR 173.340 (a)(1) and (a)(2)

[a] Greases are excluded but are classified as H1 or H2.

Code of Federal Regulations, Book 21. These sections include the following[8]:

- Incidental food contact lubricants (21 CFR Section 178.3570)
- Prior sanctions list (21 CFR Section 181)
- "Generally Recognized as Safe" (GRAS) (21 CFR Section 182)

As noted earlier, the NSF is the primary entity that registers H1 lubricants.[9] All ingredients and base oils that belong to the 21 CFR lists have now been classified as HX-1 or H1 respectively by the NSF.

Section 178.3570 specifically contains ingredients that can be used as lubricating additives or base oils for incidental food contact applications. To be listed in this section, the supplier has to petition the Food and Drug Administration (FDA). This review could result in an opinion letter from the FDA (which generally applies to chemical equivalents such as in cases where the sodium salt is listed but the potassium salt is not) or a full review for new ingredients. A full review can be very expensive and time-consuming. Over the years, a key global supplier headquartered in Germany has effectively developed, petitioned, and won approval for a large number of its additives. These are listed in Section 178.3570. On the base oil side, there are many suppliers who have successfully passed FDA scrutiny.

The prior sanctions list does not specifically include ingredients for use in incidental food contact lubricants. However, if ingredients on these lists will function as additives or base oils, they may be used in incidental food contact lubricants. The prior sanctions list does not change as its components have been grandfathered due to history of safe use.

GRAS is a well-established list that contains numerous ingredients which are mostly of value for food use but there are some ingredients on the GRAS list that have lubricating properties. For example, vegetable oils would fit into this category as they are listed on the GRAS list and can be used as lubricant base oils. Likewise, lecithin and certain fatty acids are GRAS-listed and may also be used in lubricants. In order to include ingredients on the GRAS list, the supplier can either self-affirm its GRAS status or petition the FDA.

27.3.1.3.1.2 Registration of the Fully Formulated Lubricants and Fluids Generally, lubricant suppliers to the food industry undertake the exercise of registering their products as H1 or H2 lubricants with the NSF. It must be noted that NSF registration is not mandatory and suppliers may self-certify their products.

27.3.1.4 Other Certifications/Approvals

The two key approvals are Kosher and Halal. Food-manufacturing facilities that make Kosher and Halal food usually expect that the ingredients as well as the fluids and lubricants they use would also have Kosher and Halal approvals.

For Kosher and Halal manufacturers of additives, base oils or finished lubes must generally meet two distinct levels of compliance:

- The plant and process where the product is made must generally be Kosher- or Halal-compliant.
- All the components used in the finished lubricant or fluids must themselves be Kosher- or Halal-certified.

Due to this, it is typical to seek out two letters from the Kosher- and Halal-certifying authorities: one for the facility and one for the products. Jewish dietary laws are termed "kosher."[10] Although there may be slight differences in interpretation by various kosher-certifying agencies, the ingredients and/or facility must meet the following criteria:

> The products cannot mix meat and milk products or contain pork and related products. Moreover, if there is an animal-derived product then the animal must be slaughtered according to kosher guidelines. From a procedural standpoint no product should be processed on equipment that is processing non-Kosher material. In such cases, special cleaning and 24-hour wait periods apply.

In the United States, there are about 300 agencies that offer kosher certification. The Orthodox Union in Manhattan, New York, is the largest and certifies about 4500 facilities in 68 countries. The process involves a fee and a mandatory visit to the facilities by a rabbi. In order to maintain certification, a company must be willing to pay the requisite fee and be amenable to unannounced inspections.

Muslim dietary laws require the imposition of "halal" (meaning lawful vs. "haram," which means unlawful), which means that the facility and/or ingredients must meet the following criteria[11]:

> Kosher approvals described earlier and additionally animal products must be made using Halal laws. Alcohol must be excluded from a product and so should pork-derived products. The Islamic Food and Nutrition Council of America (IFANCA), Chicago, Illinois issues Halal certificates, and the process involves a fee as well as annual inspections.

While neither of these is mandatory, the potential customer base that expects adherence to these certifications is significant. In 2010, there were about 13 million Jews and almost 1.6 billion Muslims worldwide, which translates to a sizable customer base.

27.3.1.5 Other Factors to Consider for Incidental Food Contact Lubricants

In addition to the considerations indicated earlier, the manufacturer may consider using "good manufacturing practices" (GMPs) or registration against specific "International Standards Organization" (ISO) standards such as 9001, 21469 or against the American Chemistry Council's Responsible Care RC14001 initiative.

GMP involves self-established protocols during manufacture that can ran range from cleaning protocols to analysis for

Ingredients	Finished lubricants and fluids[a]	Facility
Additives Other components Base oils	Lubricating oils Heat transfer fluids	Manufacturing and packaging sites

21 CFR sections	21 CFR Sections	Facility[a]

Incidental food contact
Lubricants
  178.3570
GRAS
  182
Prior sanctions
  181
Other
 	
NSF HX-1
Kosher
Halal
	Incidental food contact
	Lubricants
	  178.3570
	GRAS
	  182
	Prior sanctions
	  181
	Other
	
	NSF H-1
	Kosher
	Halal
	ISO 21469
	OEM approvals
		GMP
		Kosher
		Halal
		ISO 21469
		ISO 9001

FIGURE 27.5 Summary of criteria for incidental food contact lubricants. [a]*Note:* Greases, pastes, and aerosols are excluded from the discussion; as are H2 lubricants and 3H category fluids (release oils).

heavy metals and can be self-managed. The ISO standards, which are now common, involve annual audits and fees.

All the registrations involved add cost to the entire process. This is generally reflected in the price of these lubricants when compared to conventional lubricants. An excellent overview of challenges encountered in the development and manufacture of incidental food contact lubricants has been recently published.[12]

Figure 27.5 provides a summary of various criteria that must be met at the ingredient level, the finished fluid or grease level, or the plant level.

27.4 FORMULATION AND APPLICATION CONSIDERATIONS FOR LUBRICANTS AND FLUIDS

27.4.1 FORMULATION CONSIDERATIONS FOR LUBRICANTS

Lubricants are generally comprised of additives and base oils that are formulated together to meet specific performance requirements. These requirements may be dictated by the OEM (OEM-specific) or by industry standards (DIN, AGMA, NLGI). When such requirements do not exist, qualified lubricant or fluid suppliers, based on their knowledge, may offer products fit for the purpose.

27.4.1.1 Components Used in Incidental Food Contact Lubricants

Under the USDA program, the components that could be used in incidental food contact lubricants belonged either to the 21 CFR Section 178.3570, the GRAS list, or the prior sanctions list. Therefore, there was no specific category for

lubricant components. The finished lubricants were then registered by the USDA under the NSF H1 category. Since the NSF took over the program, a new category termed NSF HX-1 has been created for components, which include additives, thickeners, and base oils. Finished lubricants are still classified under the NSF H1 category.

27.4.1.1.1 Base Oils for Incidental Food Contact Lubricants

Whether the lubricant is an oil or grease, the base oil is often the majority ingredient in a formulation. Reference was made to the FDA review of components and in a review of mineral oils, the presence of aromatics in mineral base oils is not tolerated. This led to the approval and use of technical white mineral oil as a base oil. While there may be some speculation, it could be argued that the fact that poly-α-olefins (PAOs) and polybutenes/polyisobutylenes (PIBs) are free of aromatics may have formed one of the bases on which approval was granted. In addition to hydrocarbon base oils, polyalkylene glycols (PAGs) are also NSF-registered. More recently, the variety of esters and oil-soluble polyalkylene glycols (OSPs)[13] have been NSF-registered as NSF HX-1. The benefit of esters is the ability to solubilize additives and varnish while OSPs are not only compatible with both PAGs and hydrocarbon base oils but also result in cleaner lubricants.[14] Additionally, technology is available for vegetable oil–based incidental food contact lubricants and is covered in a U.S. patent.[15] Vegetable oils can be used by virtue of their GRAS status. This means that a wider range of base oils is now available to a formulator (Table 27.4) than it was at the time of writing of the first edition of this book.

The approval of new base oils such has esters has helped the development of improved lubricants. For example, a key

TABLE 27.4

Typical Properties of Base Oils

Base Oil	ISO Viscosity Range	Pour Point (°C)	Antioxidant Response[a]
Technical white mineral oil	32/46	−9	Excellent
Group II oils	32/46	−9	Excellent
PAO	32/46	<−30	Excellent
PIB	220 to >1000		Good
PAG	32/46	−12	Good
Vegetable oils	32 and 220	−15	Moderate to poor
Esters	46–1000	−45	Good
Silicones	46	−50	[b]
Perfluoro oils	32	~−12	[b]

[a] Subjective comparison.
[b] Most additives are not soluble in these base oils.

supplier of synthetic air compressor oils has recently developed a 12,000 h air compressor oil that is NSF H1-registered. The data on this product are shown in Table 27.5.[16]

27.4.1.1.2 Additives

27.4.1.1.2.1 Thickeners Typically, varying viscosity grades of lubricants are required. Common viscosity grades for various lubricants are listed in Table 27.6.

Polyisobutylenes (PIBs; Figure 27.6) are effective thickeners for mineral oil lubricants.

The thickening efficiency of a commercially available PIB is shown in Figure 27.7.[17] PIBs are available in various viscosity grades. Special handling of PIBs is required as the viscosity of the PIB itself increases. PIBs are soluble in hydrocarbon oils but may be insoluble in water-insoluble PAGs and certain esters. Vegetable oils have a limited viscosity range between ISO 32–46, which limits their usage in higher-viscosity applications, and PIB is generally incompatible.

27.4.1.1.2.2 Tackifiers Tackifiers are used to impart stringiness to a lubricant, which in turn improves its adherence to surfaces. An excellent general review on this topic is available.[18] Tackifiers find use in chain oils used in the food industries. Traditionally, tack has been qualitatively measured and is often tested by the "sticky finger test." Here the number of strings between two fingers is counted and a large number of strings indicates better tack. A recent publication mentions that substantial progress has been made toward quantifying tack using an open-siphon method that ultimately measures "free jet" length (Figure 27.8), which is correlated with the tackiness.[19] Tack was found to be a result of exudation of solvent from a swollen gel in dilute polymer solutions of the tackifier in oil. Tack depends on the MW and concentration of the polymer and its viscoelastic properties. The only NSF HX-1-registered tackifiers are high-molecular-weight polybutene tackifiers. These polymers have the ability

to "extend" and are also oxidatively stable due to their low level of unsaturation.

27.4.1.1.2.3 Pour Point Depressants While low-temperature properties are important in industrial and off-highway applications, most food manufacturing occurs in controlled conditions of temperature and low-temperature properties are not as critical. However, from the time of writing the book chapter for the first edition, there are now NSF HX-1-registered pour point depressants. These are typically maleic anhydride–styrene copolymers and, as seen in Figure 27.9, they are highly effective in lowering the pour point of technical white mineral oils.

27.4.1.1.2.4 Antioxidants Incidental food contact lubricants are used in applications, including hydraulic fluids, gear oils, compressor oils, chain oils, and other lubricants and fluids. These lubricants are subject to oxidative stresses much like conventional lubricants but present additional challenges. First, the approved hydrocarbon base oils have very low polarity and solubilizing power due to the removal of the aromatic content in the oil. As a result, by-products of oxidation such as sludge are not easily solubilized and this can affect equipment performance. Second, these lubricants can be subject to temperatures as high as 350°F in oven chain oils and bearing grease. Finally, only a limited choice of approved antioxidants is available. Fortunately, Group II base oils and technical white mineral oils and PAOs, which are commonly used base oils in these applications, respond well to antioxidants due to the fact that they have very low unsaturation. Many antioxidants are available for incidental food contact lubricants. These include hindered phenols to substituted amine antioxidants. Some natural antioxidants such as tocopherols may also be used. The structures of some of these antioxidants are shown in Figure 27.10.

Substituted amine antioxidants are very effective in mineral oils although some darkening may be expected. Hindered phenols such as butylated hydroxyl toluene (BHT) are also very effective and may not cause as much discoloration. However, BHT may be difficult to solubilize in a blend and therefore there is a clear need for more soluble antioxidants.

The introduction of an antioxidant package, at the time of printing of the first edition, offers a product that is in liquid form and is highly effective in a variety of oils. The results of this package in technical white mineral oil are shown in Figure 27.11.[17]

27.4.1.1.2.5 Antiwear and Extreme-Pressure Agents Antiwear agents are a crucial component of a lubricant. One of the most commonly used antiwear agents in incidental food contact lubricants is an amine phosphate salt shown in Figure 27.12. The amine phosphate salt also provides a degree of rust protection.

TABLE 27.5

Air Compressor Oil with >8000 h Life That Uses PAO and an NSF HX-1 Registered Ester

CPI®-4265-68-F

EXP-5158

S010-3864-12-736

OS343168A

Physical and Analytical Test Results

Properties	Test Method	LZ Test Code	ISO VG 46	ISO VG 68
Viscosity	ASTM D445			
40°C		D445_40	46.88	68.18
100°C		D445_100	7.71	10.56
Viscosity index	ASTM D2270	D2270	132	143
Density (g/mL)	ASTM D4052			
15.6°C		D4052_15.6	0.8551	0.8579
20.0°C		D4052_XX	0.9517	0.8552
Total acid number (mgKOH/g)	ASTM D974	D974	0.77	0.18
Neutralization number (mgKOH/g)	ASTM D664	D664		
TAN—inflection			0.7	0.1
TAN—buffer point (aqueous)			0.7	0.1
Flash point (°C)	ASTM D92	D92	262	288
Fire point (°C)	ASTM D92	D92	298	302
Pour point (°C)	ASTM D5950	D5950	−51	−42
ICP trace levels	ICP 7	1272		
Sulfur				
Color	ASTM D1500	D1500	L0.5	L0.5
Foaming tendency (mL)	ASTM D892	D892		
Sequence I				340/0
Sequence II				20/0
Sequence III				360/0
Demulsibility	ASTM D1401	D1402_54	42-38-0 (15)	40-38-2 (15)
Copper strip corrosion	ASTM D130	D130	1B	1B
Rust test	ASTM D665 A and B			
Part A		D665_A	Pass	Pass
Part B		D665_B	Pass	Pass
Air separation (min)	ASTM D3427			6.6
RPVOT (min)	ASTM D2272	D2272	—	2668
PDSC (min)	PDSC_FINOIL	PDSC_FINOIL	—	151.6

Performance Test Results

Properties	Test Method	LZ Test Code	Results
Foaming tendency (mL)	ASTM D892	D892	
Sequence I			340/0
Sequence II			20/0
Sequence III			360/0
Demulsibility	ASTM D1401	D1402_54	40-38-2 (15)
Copper strip corrosion	ASTM D130	D130	1B
Rust test	ASTM D665 A & B		
Part A		D665_A	Pass
Part B		D665_B	Pass
Air separation (min)	ASTM D3427		6.6
RPVOT (min)	ASTM D2272	D2272	2668
PDSC (min)	PDSC_FINOIL	PDSC_FINOIL	151.6
NOACK volatility	ASTM D5800	D5800_200	
Carbon residue	ASTM D189	D189	
Seal compatibility	ASTM D5662		
Fluoroelastomer			
Volume change (%)			1.00%
Durometer change (points)			−2
Tensile change (%)			3%
Elongation change (%)			−3%

(Continued)

TABLE 27.5 (*Continued*)
Air Compressor Oil with >8000 h Life That Uses PAO and an NSF HX-1 Registered Ester

Performance Test Results

Properties	Test Method	LZ Test Code	Results
Polyacrylate			
Volume change (%)			0.00%
Durometer change (points)			5
Tensile change (%)			25.60%
Elongation change (%)			−34.90%
Nitrile			
Volume change (%)			1.00%
Durometer change (points)			−1
Tensile change (%)			−7.00%
Elongation change (%)			−25.70%

Thermodynamic Test Results

Properties	Test Method	LZ Test Code	Results
Dielectric constant	ISO 17025	DIELEC_CONST	
20.3°C			2.48
96°C			2.33
149.3°C			2.24
Electrical conductivity (ps/cm)	ASTM D2624	COND-1	
21°C			0.08
Thermal conductivity (W/mK)	ASTM E1530	TC-OIL	
44.64°C			0.198
74.27°C			0.183
104.05°C			0.178
123.81°C			0.173
Heat capacity	ASTM E1269	CP-DSC	
40°C			2.220
50°C			2.259
60°C			2.297
70°C			2.336
80°C			2.375
90°C			2.412
100°C			2.450
110°C			2.490
120°C			2.532
130°C			2.568
140°C			2.607
Latent heat (kcal/mol)		LATENT_HEAT	10.6
Vapor pressure(torr)	ASTM D2897		
100°F			0.010
150°F			0.043
200°F			0.140
250°F			0.40
300°F			1.00
350°F			2.3

Storage Stability

Week Number	65°C @ Room Temp	Room Temp. @ Room Temp.	0°C @ 0°C	−18°C @ −18°C	0°C @ Room Temp.	−18°C @ Room Temp.
0		C				
1	C	C	C	C	C	C
2	C	C	C	C	C	C
3	C	C	C	C	C	C
4	C	C	C	C	C	C
8	C	C				

TABLE 27.6

Typical Viscosity Ranges for Incidental Food Contact Lubricants

Fluid Type	ISO Viscosity Range (Typical)
Hydraulic fluids	32–68 (46–68)
Gear oils	100–460 (220)
Air compressor oils	32–100 (46)

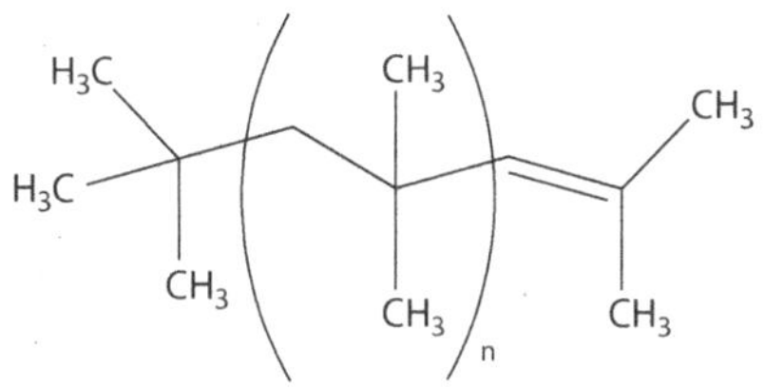

FIGURE 27.8 Free jet length resulting from tack. (From Breitner, A., 2015 nonfood compounds registration & ISO 21469 certification program update, NSF Steering Committee Meeting, 2015.)

FIGURE 27.6 Chemical structure of PIB. Increasing "n" increases the molecular weight of the PIB and its thickening efficiency.

Other antiwear agents include glyceryl esters as well as certain fatty acids.

A limited number of extreme-pressure (EP) agents is available. One of the most commonly used EP components is triphenylphosphorothiorane (TPPT), which is a white crystalline solid (Figure 27.13). This does not contain "active sulfur" like conventional EP agents that are not HX-1-registered. While the approval of ingredients with active sulfur is desirable to obtain the EP performance of conventional EP agents (typical Timken OK loads in the range of 60+ lb), it appears that the industry has quite successfully managed with the use of EP agents shown in Figure 27.13. Very often, Timken OK load values obtained for mineral oil and PAO-based incidental food contact lubricants treated with TPPT are in the 15–25 lb range. Soluble versions of

this EP agent containing alkyl substitutions on the aromatic rings are also available.

27.4.1.1.2.6 Rust and Corrosion Inhibitors and Metal Passivators Incidental food contact lubricants can be exposed to high-moisture conditions, which means that it is important that the lubricants have adequate rust and corrosion protection. Oleyl sarcosine is commonly used and is highly effective in mineral oils, PAO, and vegetable oils (Figure 27.14). A consistent pass test result can be obtained with this additive in both parts of the ASTM D665 rust test. The maximum treat rate permitted for this additive is 0.5% by weight. Substituted imidazolines are metal passivators (deactivator) (Figure 27.14).

27.4.1.1.3 Defoamers

Defoamers are used to prevent excessive foaming. Silicon defoamers are approved for use in incidental food contact

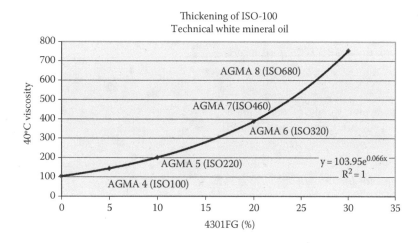

Thickening of ISO-100
Technical white mineral oil

$y = 103.95e^{0.066x}$
$R^2 = 1$

FIGURE 27.7 Increasing the ISO grade of incidental food contact lubricants.

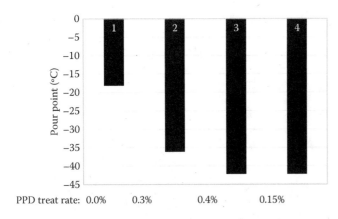

FIGURE 27.9 Commercial HX-1 registered pour point depressant in H1 registered white mineral oils. *Note:* Base oil kinematic viscosity at 40°C = 20 cSt.

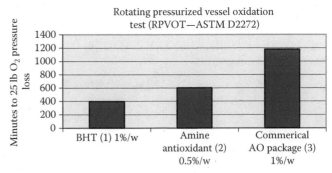

FIGURE 27.11 Performance of a commercial antioxidant package. (1) Shown at treat level comparable to the commercial antioxidant package, (2) shown at maximum permitted level, (3) shown at half the maximum permitted treat level. (From Raghaven, G.S.V. et al., Controlled atmosphere storage, in D. Heldman (ed.), *Encyclopedia of Agriculture, Food, and Biological Engineering*, CRC Press, 2003, pp. 148–150. United Nations Industrial Development Organization (UNIDO), Controlled atmosphere storage, paper 20, Retrieved from http://www.unido.org/fileadmin/32113_20ControlledAtmosphere Storage.2.pdf.)

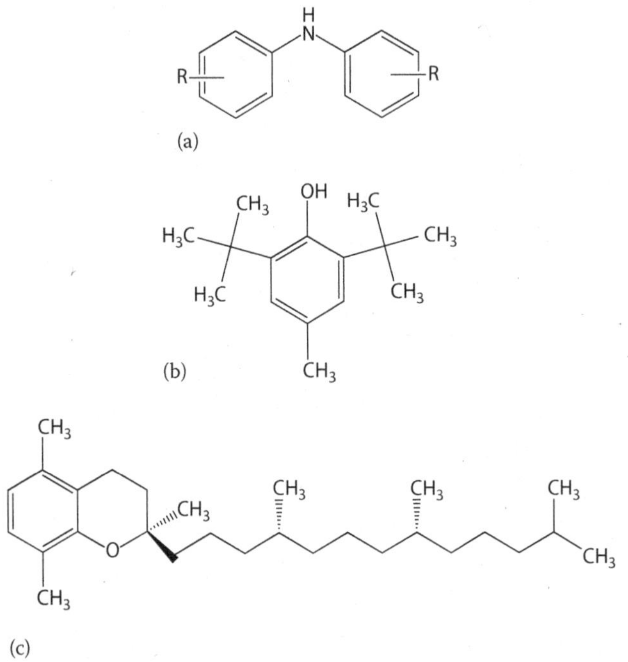

FIGURE 27.10 Chemical structures of antioxidants: (a) substituted diphenyl amine, (b) butylated hydroxyl toluene (BHT), (c) tocopherol E.

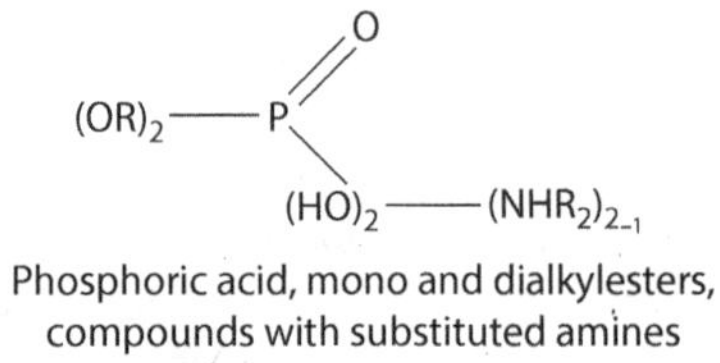

Phosphoric acid, mono and dialkylesters, compounds with substituted amines

FIGURE 27.12 Chemical structure of a commonly used antiwear agent.

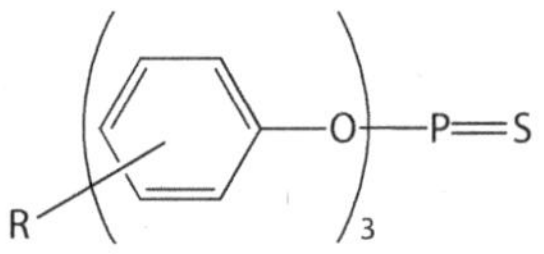

FIGURE 27.13 Chemical structures of commonly used EP agents—structure of triphenylphosphorothiorane (TPPT). R = H (solid), R = t-butyl; or nonyl (liquid).

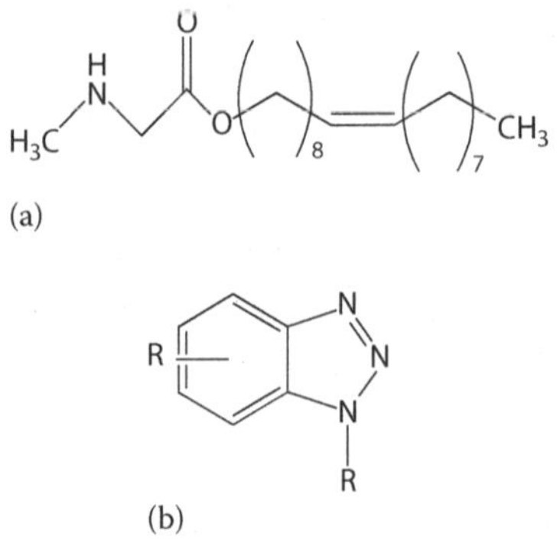

lubricants as long as the molecular weight of the defoamers is >2000 (Figure 27.15). They are highly effective in all approved base oils for this application.

27.4.1.1.4 Additive Packages for Incidental Food Contact Lubricants

As in conventional lubricants, additive packages simplify operation and can cut development time. One of the challenges in offering additive packages is to ensure that the limits specified for components must be adhered to. Hydraulic fluid additive packages are now offered by two leading additive suppliers.

FIGURE 27.14 Chemical structures of rust inhibitors and metal deactivators. (a) N-methyl-N-(1-oxo-9-octadecynyl) glycine—oleyl sarcosine, (b) substituted imidazoline.

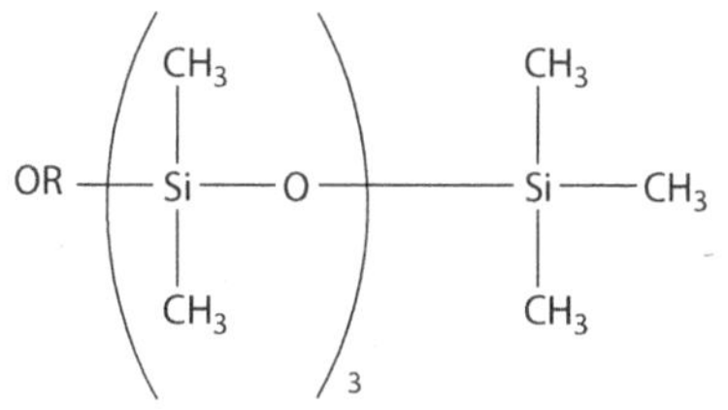

FIGURE 27.15 Chemical structure of a dimethylpolysiloxane defoamer.

One of the suppliers offers a product that may be used in hydraulic oils as well as gear oils, thereby providing additional logistics simplicity. Key performance results for this additive package are shown in Table 27.7 and pump performance tests are shown in Table 27.8.[17]

27.4.1.2 Formulation Challenges for Incidental Food Contact Lubricants

The formulation of incidental food contact lubricants is challenging because the formulator has to abide by many restrictions and yet is often expected to deliver the same performance that is expected from conventional lubricants.[20] These include the following:

- Limited list of approved ingredients compared to conventional lubricants—this can also create product differentiation challenges since all suppliers have the same list of ingredients to draw from
- Limitations on the amount of additive that can be used
- Limited solubility of additives in technical white mineral oil and PAO

TABLE 27.7

Performance of a Commercial Hydraulic and Gear Oil Additive Package

Test Method	Test Method ASTM or Other	ISO46 Hydraulic Fluid Typical Result	ISO 220 Gear Oil Typical Result
Performance tests			
Air release @ 50°C	D3427		
Air bubble separation in minutes		2.5	
Oxidation and seal performance			
RBOT (minutes to 25 lb oxygen pressure loss)	D2272	571	623
Seals 168 h/100°C	Hydraulic SRE-NBR28		
%Volume/hardness change			4.8/−5
%Tensile strength/elongation decrease			1.3/−4
Corrosion and wear performance			
Copper strip	D130		
3 h at 100°C		1B	1A
3 h at 121°C			1B
Rust test—part A/part B	D665	Pass/pass	Pass/pass
4-Ball wear test (167°F, 1200 rpm, 20 kg)			
Scar diameter (mm)		0.30	
Average friction coefficient		0.108	
4-Ball wear test (167°F, 1200 rpm, 40 kg)	D4172		
Scar diameter (mm)		0.37	0.37
Average friction coefficient		0.113	0.117

TABLE 27.8

Pump Test Performance of a Commercial Hydraulic and Gear Oil Package

Test Method	Initial/Final Flow (gpm)	Total Ring and Vane Weight Loss (mg)[a]	Ring Weight Loss (mg)[a]	Vane Weight Loss (mg)[a]
Eaton-Vickers V104-C 100 h at 150°F Conestoga	5.3/5.3	42	39	3
Eaton-Vickers 35VQ (50 h)		52	45	7

[a] The values indicated in the table are typical results and are not intended to be specifications.

An article summarizing the challenges that was published almost 8 years ago still continues to be valid.[12]

27.4.2 FORMULATION CONSIDERATIONS FOR OTHER FLUIDS USED IN THE FOOD INDUSTRY

27.4.2.1 Heat Transfer Fluids

Heat transfer fluids are often the highest-volume fluid used in a food plant, and their usage can account for tens of thousands of gallons for a single plant since these are circulated to various locations in a plant. For a high-quality fluid, when temperature range limits are monitored and followed, and oil condition is monitored, the initial fill can last for a decade. Replenishment is not common unless leaks or other reasons require a changeover. Typical off-cycle fill of heat transfer fluids often stems from system leaks and fires. Some years ago, frying processes utilized direct fired frying kettles. However, with certain unfortunate accidents and the industry's attention to safety, most of these systems now utilize high-temperature heat transfer fluids. High-temperature heat transfer fluids are mostly used for frying chicken and potato chips. Instead of direct fire, specialized heaters (supplied by specific OEMs), often fired by natural gas, are used to heat the heat transfer fluid, which is then circulated to its point of use in the plant.

Heat transfer fluids are one of the most commonly used fluids in this category aside from ammonia refrigeration oils. Heat transfer fluids may be classified by the chemistry or temperature-operating range (Figure 27.16).

The high temperature range within which a heat transfer fluid operates is often dictated by the chemistry—conventional paraffinic oils have the lowest high-temperature limit while silicones have the highest.

For high-temperature heat transfer fluids used at >450°F operating temperatures, NSF HT-1-registered paraffinic oils are widely promoted. However, despite claims, these often rapidly degrade as temperatures exceed 500°F. At temperatures exceeding 500°F, highly refined and custom-produced NSF HT-1-registered cycloparaffinic oils are a prudent choice that offers a very high cost/benefit ratio with outstanding thermal stability to 625°F. A comparison of key operational benefits of cycloparaffinic base oils versus paraffinic base oils is shown in Figure 27.17.[21]

For low-temperature applications, propylene glycol (PG)–based heat transfer fluids are the staple of most food plants. They are low cost and NSF HT1-registered. The low-temperature properties are simply modified by using deionized water—each dilution leads to a progressively higher operating temperature for the fluid (higher pour points or worse low-temperature behavior). Typically, the fluid contains an inhibitor package to prevent corrosion. Formate-based brines are a very good alternative offering much better operational viscosity at lower temperatures than glycols.[22] Care must be taken for glycols and the formate brines (especially the brines) that deionized water is used and *all* materials of construction are compatible to avoid corrosion. Silicones are the most expensive option with a wide temperature range.

27.4.2.2 Refrigeration Oils

Refrigeration is a very common operation in the food industry. It can be used to cool secondary fluids or a space or a process. Ammonia-based refrigeration systems are widely used; CO_2 systems used independently or more often as cascade systems with ammonia are common.[23] Both these systems are common for plant-wide operations, and you can have several miles of ammonia piped in a large food plant. Finally, hydrofluorocarbon (HFC)-based systems are also used but generally in more localized cooling applications and rarely piped through an entire operation. Each gas poses its own benefits and challenges. With certain regulations on the volume of ammonia in a plant, it is common to use a secondary loop heat transfer fluid that is cooled by ammonia.

All refrigeration systems use compressors and these require lubrication. Screw compressors are very common. Ammonia systems can use mineral oil or synthetic PAO but never polyolester (POE) refrigeration oils. Ammonia–CO_2 cascade systems[24] additionally use POE-based lubricants, but strictly on the CO_2 side of the cascade. HFC refrigeration gases utilize POE-based lubricants.

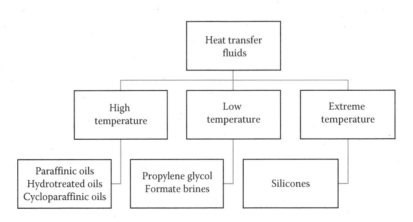

FIGURE 27.16 Heat transfer fluids used in the food industry.

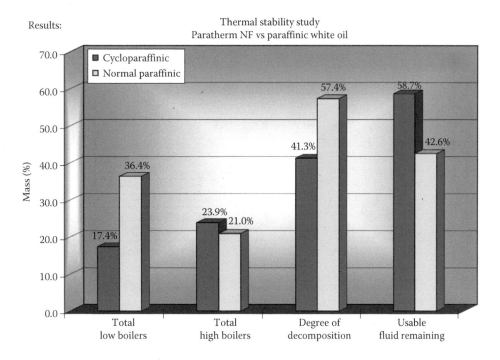

FIGURE 27.17 Comparison of thermal stability of heat transfer fluids. *Notes:* Value for *Total Low Boilers* includes gaseous decomposition products. Value for *Total High Boilers* includes residual materials not vaporized.

As stated earlier, ammonia gas (R-717) refrigeration systems are very common in food-manufacturing plants as well as in cold storage applications. This is because despite its toxicity, ammonia is a natural refrigerant, which is cheap, highly efficient, and has zero "global warming potential" (GWP). Ammonia systems can save as much as 10%–20% on electricity costs versus comparable HFC systems as per a reputed refrigeration system supplier.[25] While large distributed systems are common, low-charge ammonia systems in which a secondary cooling fluid is circulated are growing in use as they lower the total ammonia charge at a facility.[25,26] The design of closed-circuit ammonia refrigeration systems is covered by the new American National Standard, ANSI/IIAR 2-2014.[27]

For lubrication of ammonia refrigeration systems, the choice of base oils is critical for longevity and performance. Almost all of the systems utilize screw compressors and have an oil return system that is set up for immiscible oils. The ideal viscosity in most ammonia applications is ISO VG 68. Hydrotreated mineral oils (HTMOs) work best from a cost performance standpoint and can last up to 7000 h compared to naphthenic oils that last only a few thousand hours. Alkylbenzenes (ABs) are also used. PAO lubricants offer premium performance (7000+ h) coupled with excellent low-temperature properties and (like the other oils mentioned) are immiscible with ammonia (see PAO data in Figure 27.18). NSF H1-registered products are available only in the PAO category.

PAGs are the fluids of choice for DX-style evaporators where a miscible oil is desirable (Figure 27.19).[28]

POE refrigeration oils, which are commonly used for HFC refrigeration, must *never* be used with ammonia oils as they result in the formation of solid amides (Figure 27.20).[28] This can be a very costly cleanup procedure in addition to leading to significant downtime in critical operations.

27.4.3 Application Considerations for Users in the Food Industry

There are many suppliers of lubricants and fluids for the food industry. However, one must be careful in choosing a supplier and the fluid. Here are some guidelines that ideal suppliers should possess:

- Supplier reputation and longevity in the lubricant business as well as their ability to develop new product versus copied-me-too products
- Facility certification such as ISO 9001 and ISO 21469 and RC 14001 and Kosher and Halal
- NSF registration of products in the appropriate NSF H categories
- Ability to invest in new technology and manufacturing capabilities
- Ability to provide full data sets with performance tests included
- Ability to offer oil condition monitoring services
- OEM approvals if available

From a product standpoint, it is extremely important to realize that none of the above criteria (except the last bullet) are associated with performance of the lubricant in an application. The suitability of using a product in an application must still be independently verified by the user. OEM-approved products, products recommended by OEMs, and products supplied by independent parties with full performance data sets will be most reliable. Coupling this with oil condition monitoring

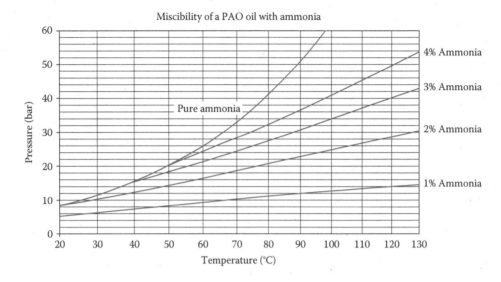

FIGURE 27.18   PAO-based ammonia oils make sense where immiscible oils are required.

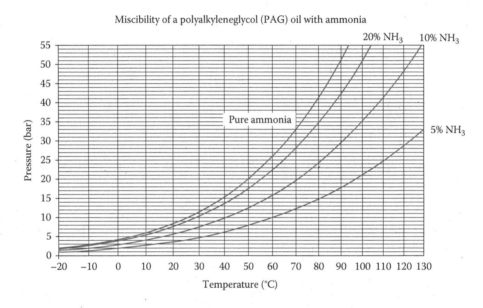

FIGURE 27.19   Why PAG-based ammonia oils make sense when miscibility is desired. *Note:* At typical operating temperatures of 60°C and pressures of 20–25 bar, PAG-based products will see a 20% dilution with ammonia resulting in a greatly reduced working viscosity making them unacceptable for use with regular ammonia systems. (Contrast that with a PAO lubricant shown in Figure 27.18.) PAG Oils are however used when a miscible oil is required as in DX style evaporators. At typical operating temperatures of 60°C and pressures of 20–25 bar, a PAO-based product will see a 2%–3% dilution with ammonia that provides a highly workable lubricant for ammonia systems. (Contrast that with a PAG lubricant shown in Figure 27.19.)

improves overall operational efficiency. Synthetic lubricants utilizing PAO, while seemingly expensive up front, generally provide the longest life, reduced downtime, and ultimately worry-free performance that translates into cost savings.

Finally, a very important point to note is that the presence of various additives and ingredients requires that global harmonized system (GHS) reporting guidelines are followed for the safety data sheets (SDS) for these products. Despite their incidental food contact status, the SDSs for these products can contain hazard labels, pictograms, signal words and risk phrases, as well as transportation, personal protection, and exposure guidelines.

27.5   OUTLOOK AND TRENDS

The growth of the food and beverage industry in the United States tracked at about 4.4% in 2014 and 4.1% in 2015.[29] Based on this, an approximately 4% growth may be anticipated in 2016. Developing countries constitute another avenue for growth and one can see sizable growth for manufactured and packaged food in many regions.[30] The industry continues to be under marginal pressures and therefore improved productivity is emphasized, and hence there is greater focus on compliance with regulations. This means

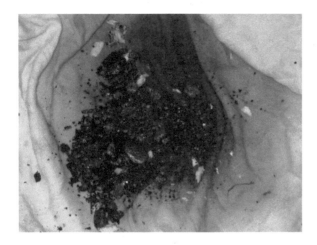

FIGURE 27.20 Formation of solid amides via the interaction of POE refrigeration oils with ammonia (R-717) refrigerant gas.

that higher-performing lubricants and fluids that are also NSF-compliant are more often in demand. This is evidenced in the growth of NSF-registered products[2] as well as growth in ISO 21469-certified facilities; as per the NSF, 15 companies now have 21 ISO 21469-registered facilities (10 in Europe and 11 in North America) covering 684 products.[2]

Product performance continues to improve as previously unregistered components such as esters and pour point depressants as well as oil-soluble PAGs are registered under the HX1 category. This is giving formulators extra latitude to push the performance of their products higher.

As more H1 lubricants and HT1 heat transfer fluids continue to be registered, the H2 category is beginning to come under scrutiny. Most plants are moving to all H1/HT1-registered products, and the H2 category is thus becoming a source of confusion. Furthermore, the performance difference between H1 and H2 lubricants continues to diminish and many H1 lubricants do not perform as well as their H2 counterparts. The use of these lubricants in a food plant and the need for this category are thus becoming questionable.[2] This may lead to the discontinuance of this category in the future.

The growth in sales of fluids and lubricants that are NSF-registered is expected to track between 2% and 4%. Finally, mergers and acquisitions activity continues in the food- and beverage-manufacturing sector,[29] as well as in the lubricants and fluids sector.[31,32] This can lead to more consolidated buying of fluids and lubricants, requiring suppliers to meet the prevailing standards and offer improved products.

ACKNOWLEDGMENTS

The author would like to acknowledge Amy Shifflett, OEM Account Manager, CPI Fluid Engineering, Midland, MI, for reviewing the chapter in great detail; David DeVore, President of Functional Products, Macedonia, OH, for a discussion on tackifiers; Ed Delate, General Manager, Paratherm Heat Transfer Fluids Division of the Lubrizol Corporation, King of Prussia, PA, for a general discussion related to the food industry; Yulia Reinhardt, Global Key Account Manager OEM, Mannheim, Germany, for a general discussion on the topic; and Ben Duval, President of U.S. Petrolon Industrial, Lincoln, NE, for a discussion related to the use of fluids in food plants.

REFERENCES

1. Pollock Group, An overview of the state of the food processing industry, Pollock papers, http://www.pollock.com/an-overview-of-the-state-of-the-food-processing-industry, Pollock Group, Grand Prairie, TX, 2015.
2. A. Breitner, 2015 nonfood compounds registration & ISO 21469 certification program update, *NSF Steering Committee Meeting*, Ann Arbor, MI, 2015.
3. G.S.V. Raghaven, Y. Gariepy, and C. Vigneault, Controlled atmosphere storage, In D. Heldman (ed.), *Encyclopedia of Agriculture, Food, and Biological Engineering*, pp. 148–150. CRC Press, Boca Raton, FL, 2003. United Nations Industrial Development Organization (UNIDO), Controlled atmosphere storage, paper 20, Retrieved from http://www.unido.org/fileadmin/import/32113_20ControlledAtmosphereStorage.2.pdf.
4. USDA-NASS, Capacity of refrigerated warehouses 2015 Summary, Report # ISSN: 1949-1638, United States Department of Agriculture-National Agricultural Statistics Service, Harrisburg, PA, January 2016.
5. P. Young, Food master food engineering section, A BNP media publication, Deerfield, IL, 2015.
6. A. Breitner, International regulations for food grade lubricants, *Lubes N'Greases Europe-Middle East-Africa*, R. Beercheck (ed.), 36–43, June 2014.
7. L. Rudnick (ed.), *Lubricant Additives—Chemistry and Application*, Additives Handbook, Marcel Dekker Publishing, Silver Springs, MD, 2003.
8. United States Food and Drug Administration (USFDA) Code of Federal Regulations, Book 21 Section 178.3570, United States.
9. NSF, http://www.nsf.org/services/by-industry/food-safety-quality/nonfood-compounds, Ann Arbor, Michigan.
10. Public Broadcasting Service, Belief and Practice, Kosher Certification, Episode No. 847, http://www.pbs.org/wnet/religionandethics/week847/belief.html., July 22, 2005.
11. http://www.ifanca.org/halal/.
12. S. Lawate and R. Profilet, Five hurdles to making food-grade lubes, *Compounding*, 57(10), 27–28, October 2007.
13. L. Tocci, A full plate, *Lubes and Greases*, 20(7), July 2014.
14. L. Inoue and M. Greaves, Building a better food grade lubricant, *Tribology and Lubrication Technology*, 50–52, November 2014.
15. S.S. Lawate, V.A. Carrick, and P.C. Naegely, U.S. Patent No. 5538654, Assignee Lubrizol Corporation, Wickliffe, OH, 1996.
16. Private communication from CPI Fluid Engineering Services, Midland, MI, 2016.
17. S. Lawate, Products for incidental food contact lubricants—Lubrizol 4300FG, Presented at the commercial forum, Society of Tribology and Lubrication Engineers (STLE) Annual Meeting, Philadelphia, PA, May 2017.
18. F. Litt, *Lubricant Additives—Chemistry and Application*, L. Rudnick (ed.), Additives Handbook, Marcel Dekker Publishing, New York, NY, pp. 355–361, 2003.
19. V.J. Levin, R.J. Stepan, and A.I. Leonov, Evaluating the tackiness of polymer containing lubricants by open-siphon methods: Experimental theory and observation, *TAE Conference*, Stuttgart, Germany, July 11, 2007.

20. S. Lawate, Environmentally friendly hydraulic fluids, In L. Rudnick (ed.), *Synthetic Base Oils*, Marcel Decker Publishing, New York, NY, 1999.

21. Private communication with Ryan Ritz of the Paratherm Heat Transfer Fluids, a Division of the Lubrizol Corporation, a Berkshire Hathaway company.

22. S. Lawate and D. Hamil, *IGHSPA Technical Conference and Expo*, Stillwater, OK, April 2004.

23. A. Pearson and A. Campbell, Using CO_2 in supermarkets, *ASHRAE Journal*, 24–28, February 2010.

24. A. Pearson, *The Use of Carbon Dioxide/Ammonia Cascade Systems for Low Temperature Food Refrigeration*, International Institute of Ammonia Refrigeration (IIAR), Nashville, TN, March 2000.

25. J. Spinner, Low charge ammonia systems to hit US market, *Food Production Daily*, May 1, 2014.

26. T.L. Chapp, Systems for low charge ammonia refrigeration, *Process Cooling*, 20–29, May 2016.

27. ANSI/IIAR Standard, Published, September 2015.

28. A. Shifflett, Ammonia lubricants: Ensuring the proper usage of ammonia rated products in your everyday maintenance, *RETA Conference*, Milwaukee, WI, September 2015.

29. N.P. Jachim, G.G. Ricco, T.F. Cummins, and M.A. Ewald, Food and beverage industry snapshot, *Stout, Risius and Ross (SRR) Newsletter*, Atlanta, GA, May 2015, http://www.srr.com/assets/pdf/food-beverage-industry-snapshot-spring-2015.pdf.

30. *International Year Book of Industrial Statistics*, United Nations Industrial Development Organization, Vienna, Austria and Edward Elgar Publishing Ltd. Cheltenham Glos, U.K., 2017.

31. Fuchs Petrolub AG acquires worldwide food grade lubricants from shell, *Food and Beverage Industry News*, www.foodmag.com.au/category/news; reported by Rita Mu, September 2010.

32. F+L Daily, FUCHS acquires Chevron's white oil and food machinery lubricants business, https://fuelsandlubes.com/fuchs-acquires-chevrons-white-oils-and-food-machinery-lubricants-business, 2016.

28 Lubricants for the Disk Drive Industry

Tom E. Karis

CONTENTS

28.1 INTRODUCTION

When thinking of a disk drive, one picture that comes to mind is that of digital data bits stored on a spinning disk housed inside a device such as a computer, a digital video recorder, or a music player. The outstanding precision and reliability of these high-speed rotating electromechanical devices is, perhaps, the leading example of microelectromechanical systems and nanotechnology at work today. For example, the magnetic recording read/write head floats on an air-lubricated bearing just about 1 nm away from the disk surface with a relative velocity that is typically 30 m/s. The shear rate across the gap is then about 30,000,000,000 1/s. As the data track width in the near future is decreased below 200 nm, the spindle motor on which the disks are mounted must have increasing stiffness with diminishing vibration amplitudes that are well below the track width to minimize servo seek time and track following. Ball bearing spindle motors used in the past have reached their limit, and current products incorporate fluid dynamic bearing spindle motors. In addition, when there is a high relative velocity between metallic and insulating components, electrostatic charge generation and dissipation must be controlled. Lubricants play a vital role in fulfilling these requirements for the disk drive industry. Fundamental understanding of the lubrication mechanism and lubricant chemistry is essential to the continued advancement of the technology.

This chapter focuses on lubricants for the magnetic recording disk and the spindle bearing motor. Emphasis is placed on the analytical tools that are common to lubricants in general. Similar techniques are applied to characterize the physical properties of lubricants that influence their performance. Rheological measurements are employed not only to characterize the viscosity but also to estimate the short-time dynamic response of disk lubricants through time–temperature superposition. Shear rheometry is exploited to characterize the yield stress of grease and the effect of blending on fluid dynamic bearing motor oils. Dielectric spectroscopy is utilized to probe the dipole relaxation of disk lubricant end groups. Dielectric permittivity and conductivity measurement are employed to develop the conductivity additives for ferrofluid used in motor seals and to investigate the effects of contamination on ball bearing grease electrochemistry. Another powerful technique that is highlighted in this chapter is Fourier transform infrared (FTIR) spectroscopy. This method can be used to study thin films in reflection or bulk samples in transmission. Examples are shown in which infrared spectroscopy is also applied to identify reaction products formed during electrochemical oxidation of ball bearing grease. Thermal analysis is employed to measure the vapor pressure of disk lubricants. A model is described that simulates evaporation of polydisperse lubricants from molecular weight distributions measured with gel permeation chromatography (GPC). Surface energy from contact angle measurements is combined with the chemical kinetic model for viscous flow and evaporation to predict the viscosity of molecularly thin films and to understand the factors that limit lubricant spin-off from rotating disks. The chemical kinetic model is also employed to combine vapor pressure and viscosity data in the quest for the molecular structure of a fluid bearing motor oil that has both low viscosity and vapor pressure. Not only are the techniques illustrated here with examples from the disk drive industry applicable to the lubrication industry in general, but they also will be particularly useful in adapting these methodologies to the tribology of micro- and nanoelectromechanical systems.

Nuclear magnetic resonance (NMR) spectra of several fluid dynamic bearing oils are included to highlight the chemical structural differences. Since accelerated motor life tests, and even tests on stabilized oils at elevated temperatures, take a very long time, a kinetic model that includes the synergistic effects of primary antioxidant (PriAOX) and secondary antioxidant (SecAOX) and metal catalyst on oil oxidation lifetime is presented to help estimate optimum additive concentrations. Examples of accelerated oil life testing are shown with an aromatic amine PriAOX and an alkyl dithiocarbamate SecAOX as well as the effect of a dissolved metal catalyst in isothermal accelerated oil oxidation life tests.

A case study illustrates the application of the oil oxidation and stabilization chemical reaction mechanisms and modeling for the development of an oil formulation with improved bearing lifetime for magnetic recording disk drive fluid dynamic bearing spindle motors. The guidelines provided by understanding the chemical mechanism enabled rapid convergence on an optimized formulation. Knowledge of the chemistry provides a significant benefit because it reduces the development cycle time and lowers the total cost to achieve improved motor lifetime.

In this edition, a method for measuring the amount of lubricant transferred from the disk to the slider by dip coating with a specially formulated oil is discussed in Section 28.2.2. Of particular interest is the ability to adjust the surface tension of oil with surfactants to optimize the detection sensitivity. Novel compounds for charge control in fluid dynamic bearings based on electron donor acceptor charge transport (electronic hopping) are discussed in Section 28.3.4. These provide an alternative to the usual ionic charge transport additives for charge control.

28.2 RECORDING DISK LUBRICANTS

The soft magnetic layers on the magnetic recording disk substrate are typically overcoated with a 1–2 nm thick carbonaceous film. Since the bare carbon overcoat has a relatively high surface energy, a low-surface-energy perfluoropolyether (PFPE) lubricant is applied on top of the overcoat. The most widely used PFPEs in the past were those having the Z-type backbone chain. These are random copolymers with the linear backbone chain structure

$$X - \left[\left(OCF_2\right)_m - \left(OCF_2CF_2\right)_n - \left(OCF_2CF_2CF_2\right)_p - \left(OCF_2CF_2CF_2CF_2\right)_q \right]_{x_o} - OX,$$

where

  X is the end group
  m, n, p, q, and x_0 are defined in Table 28.4.

A wide range of end groups are available to tailor the lubricant for optimum lubrication properties in magnetic recording disk drive systems [1]. The end groups for some of the commercially available lubricants are shown in Table 28.1. The adsorption energy of end groups (other than $-CF_3$) on the carbon overcoat surface is higher than that of the backbone

TABLE 28.1
Molecular Structure for Some of PFPE End Groups on the Z-Type PFPE Chain

Name	Structure
Z	$-CF_3$
Zdol	$-CF_2CH_2OH$
Ztetraol	$-CF_2CH_2OCH_2CHCH_2OH$ (with OH on central CH)
Zdiac	$-CF_2COOH$
Zdeal	$-CF_2COOCH_3$
Zdol TX	$-CF_2CH_2(OCH_2CH_2)_{1.5}OH$
AM-3001	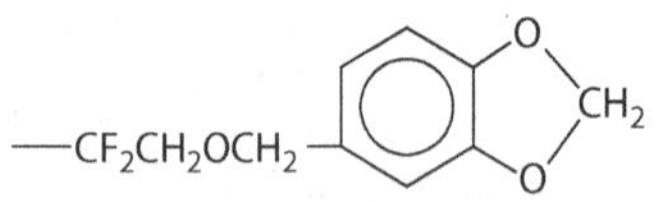 $-CF_2CH_2OCH_2-$ (benzodioxole ring structure)
A20H	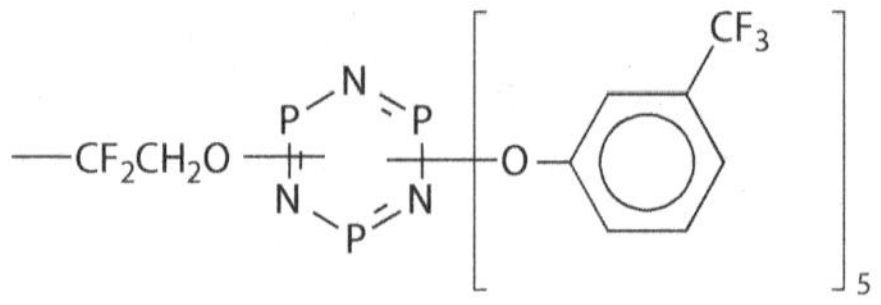 $-CF_2CH_2O-$ (cyclophosphazene ring with O-phenyl-CF$_3$, subscript 5)
ZDPA	$-CF_2CH_2N$ with $CH_2CH_2CH_3$ groups

Source: Adapted from Karis, T.E., Lubricants for the disk drive industry, in: Rudnick, L.R., ed., *Synthetic, Mineral Oil, and Bio-Based Lubricants Chemistry and Technology*, CRC Press, Taylor & Francis, Boca Raton, FL, 2006, pp. 623–654.
A20H has one Zdol end group.

chain [2–4]. The X1P-type end group on A20H [5,6] is sterically large in comparison to the chain monomers [7], and the X1P end group molecular weight of about 1000 Da is a significant contribution to the molecular weight of commonly used backbone chains of 2000–4000 Da [8]. Lower-molecular-weight end groups, also intended to passivate Lewis acid sites, are derived from Zdol with dipropylamine [9] and referred to as ZDPA (Table 28.1).

The molecular structures of the D and K series of PFPEs, also considered for magnetic recording disk lubricants at one time, are shown in Table 28.2. The repeat unit of the D chain is perfluoro n-propylene oxide. The D series includes Demnum with nonpolar end groups, Demnum SA with a hydroxyl end group, and Demnum SH with a carboxylic acid end group. Demnum tetraol, with the same end groups as Ztetraol in Table 28.1, is now the most widely used magnetic recording disk lubricant. The repeat unit of the K chain is perfluoro iso-propylene oxide. The K series includes Krytox with nonpolar end groups and Krytox COOH with a carboxylic acid end group.

Recently, PFPE magnetic recording disk lubricants have been developed with a multiplicity of hydroxyl end groups to increase their chemisorption to the overcoat (Table 28.3).

TABLE 28.2
Molecular Structure for D- and K-Type PFPEs

Name	Structure
Demnum S100	$CF_3CF_2CF_2O\left(CF_2CF_2CF_2O\right)_{x_0}CF_2CF_3$
Demnum SA	$CF_3CF_2CF_2O\left(CF_2CF_2CF_2O\right)_{x_0}CF_2CF_2CH_2-OH$
Demnum DPA	$CF_3CF_2CF_2O\left(CF_2CF_2CF_2O\right)_{x_0}CF_2CF_2CH_2N\begin{smallmatrix}CH_2CH_2CH_3\\CH_2CH_2CH_3\end{smallmatrix}$
Demnum SH	$CF_3CF_2CF_2O\left(CF_2CF_2CF_2O\right)_{x_0}CF_2CF_2\overset{\overset{O}{\|}}{C}-OH$
Krytox 143 AD	$CF_3CF_2CF_2O\left(\underset{}{\overset{CF_3}{\underset{\|}{CF}}}-CF_2O\right)_{x_0}CF_2CF_3$
Krytox COOH	$CF_3CF_2CF_2O\left(\overset{CF_3}{\underset{\|}{CF}}-CF_2O\right)_{x_0}\overset{CF_3}{\underset{\|}{CF}}-\overset{\overset{O}{\|}}{C}-OH$

Source: Adapted from Karis, T.E., Lubricants for the disk drive industry, in: Rudnick, L.R., ed., *Synthetic, Mineral Oil, and Bio-Based Lubricants Chemistry and Technology*, CRC Press, Taylor & Francis, Boca Raton, FL, 2006, pp. 623–654.

TABLE 28.3
Multidentate Disk Lubricant Structures Derived from Z-Type PFPEs

Name	Structure
ZDMD	$HOCH_2CF_2-Z-OCF_2CH_2OCH_2\underset{\underset{OH}{\|}}{CH}CH_2OCH_2CF_2-Z-OCF_2CH_2OH$
ZTMD	$HOCH_2\underset{\underset{OH}{\|}}{CH}CH_2OCH_2CF_2-Z-OCF_2CH_2OCH_2\underset{\underset{OH}{\|}}{CH}CH_2OCH_2\underset{\underset{OH}{\|}}{CH}CHCF_2CF_2CF_2\underset{\underset{OH}{\|}}{CH}CH_2OCH_2\underset{\underset{OH}{\|}}{CH}CH_2OCH_2CF_2-Z-OCF_2CH_2OCH_2\underset{\underset{OH}{\|}}{CH}CH_2OH$
LTA-30	$HOCH_2CF_2\left[OCF_2CF_2\right]_x OCF \begin{smallmatrix} CF_2O\left[CF_2CF_2O\right]_x CF_2CH_2OH \\ \\ CF_2O\left[CF_2CF_2O\right]_x CF_2CH_2OH \end{smallmatrix}$

Source: Adapted from Karis, T.E., Lubricants for the disk drive industry, in: Rudnick, L.R., ed., *Synthetics, Mineral Oil, and Bio-Based Lubricants Chemistry and Technology*, 2nd edn., CRC Press, Taylor & Francis, Boca Raton, FL, 2013, pp. 657–698.
The ZDMD is derived from Zdol 1000 or Zdol 2000. The ZTMD is derived from Ztetraol 1000. The LTA-30 has a molecular weight of 3000 Da.

Two Zdol chains are linked with epichlorohydrin to form a Zdol multidentate (ZDMD) with one hydroxyl group at each end and one in the middle [10]. Two Ztetraol chains are linked with a fluorinated di-epoxide to form a Ztetraol multidentate (ZTMD) with two hydroxyl groups at each end and four near the middle [11]. A novel three-arm star multidentate PFPE comprising polyperfluoroethylene oxide was derived by direct fluorination of the hydrocarbon ester with one hydroxyl group at the end of each arm (LTA-30) [12,13] (Table 28.3).

28.2.1 PROPERTIES

PFPEs are attractive as magnetic recording disk lubricants because of their low surface energy, low vapor pressure, wide liquid range, transparency, and lack of odor. PFPEs are related to polytetrafluoroethylene, but they have lower glass transition temperatures [14–16]. The first commercially available PFPEs had perfluoromethyl end groups and are referred to as nonpolar PFPEs. Subsequently, polar PFPEs with hydroxyl, carboxylic acid, and other polar end groups came into widespread use. The polar end group provides an additional means to adjust the fluid properties and the interaction with surfaces. PFPEs with polar end groups are predominantly used to lubricate present-day rigid magnetic recording media.

Their versatility has motivated considerable and detailed study of PFPEs. The bulk viscosity and glass transition temperature of nonpolar PFPEs have been extensively characterized by Sianesi et al. [14], Ouano and Appelt [17], Cantow et al. [18], Marchionni et al. [19–22], Cotts [23], and Ajroldi et al. [24]. Later investigations report the properties of PFPEs with polar end groups, for example, Danusso et al. [25], Tieghi et al. [26], Ajroldi et al. [27], Kono et al. [28], and Smith et al. [29].

The composition and molecular weight of several PFPE lubricants, measured by NMR spectroscopy [30], are given in Table 28.4.

28.2.1.1 Viscoelastic (Rheological)

Oscillatory shear and creep measurements were done with a Carri-Med CSL 500 (now TA Instruments) Stress Rheometer with the Extended Temperature Module and a 40 mm diameter parallel plate fixture. The dynamic strain amplitude was 5%, and this was within the range of linear viscoelasticity for these materials. The storage G' and loss modulus G'' were measured between 1 and 100 rad/s at each temperature. Typically, measurements were done each 20°C from −20°C to −100°C. Low-temperature measurements were performed to provide the high-frequency properties needed for calculations at the short timescales encountered in asperity contacts. The data measured at low temperature are transformed to high frequency through time–temperature superposition with Williams–Landel–Ferry (WLF) coefficients [31] that are derived from

TABLE 28.4
The Composition of Several PFPEs

Lubricant	m	n	p	q	m/n	O/C	x	M_n(Da)
Z03	0.530	0.405	0.057	0.008	1.31	0.754	73.4	6810
Zdiac	0.508	0.435	0.048	0.008	1.17	0.744	24.4	2310
Zdeal	0.567	0.426	0.003	0.004	1.33	0.782	22.8	2070
Ztetraol 2000	0.485	0.515	0	0	0.94	0.743	23.2	2300
Ztetraol 1000	0.523	0.477	0	0	1.10	0.762	14.2	1270
Ztx	0.475	0.517	0.007	0.001	0.92	0.736	22.7	2230
Zdol4KL819	0.612	0.383	0.003	0.0025	1.60	0.720	46.5	4000
Zdol4KL492	0.568	0.425	0.005	0.002	1.34	0.693	39.1	3600
Zdol4KL990	0.515	0.475	0.005	0.005	1.08	0.666	39.2	3600
Zdol4KBL598	0.492	0.508	0	0	0.97	0.658	47.2	4300
Zdol4KL905	0.469	0.526	0.0025	0.0025	0.89	0.650	41.5	3900
Zdol 2500	0.456	0.544	0	0	0.84	0.728	26.1	2420
Demnum S100	—	—	—	—	—	0.333	31.7	5230
Demnum SA2000	—	—	—	—	—	0.333	12.6	2080
Demnum SA2	—	—	—	—	—	0.333	18.6	3080
Demnum DPA	—	—	—	—	—	0.333	48.4	8100
Demnum SH	—	—	—	—	—	0.333	18.3	3040
Krytox 143 AD	—	—	—	—	—	0.333	39.8	6580
Krytox COOH	—	—	—	—	—	0.333	32.3	5370

Source: Adapted from Karis, T.E., Lubricants for the disk drive industry, in: Rudnick, L.R., ed., *Synthetic, Mineral Oil, and Bio-Based Lubricants Chemistry and Technology*, CRC Press, Taylor & Francis, Boca Raton, FL, 2006, pp. 623–654.

Note: O/C is the oxygen to carbon ration in the backbone chain. The degree of polymerization $x = x_o + 2$. M_n is the number average molecular weight. The Zdol4K series are different batches of Zdol 4000 from the manufacturer.

the rheological measurement data. The PFPEs were linearly viscoelastic in these test conditions. The dynamic properties were independent of strain amplitude, and no harmonic distortion of the sinusoidal angular displacement waveform was observed even at the lowest measurement temperatures. Time–temperature superposition was employed to obtain the master curves [32]. Viscosities for the lubricants at each temperature were calculated from the steady-state creep compliance. The glass transition temperatures, T_g, were measured using a modulated differential scanning calorimeter manufactured by TA Instruments model number 2920 MDSC V2.5F. The samples were cooled to $-150°C$ and heated to $20°C$ at $4°C/min$ with a $1.5°$ modulation over a period of 80 s. The differential heat flow and temperature phase shift were measured to determine the reversible and nonreversible components of the heat flow. The glass transition temperatures of several PFPE lubricants are listed in Table 28.5.

The temperature dependence of viscosity is shown in Figures 28.1 through 28.3 as the ratio of viscosity to the molecular weight η/M_n plotted as a function of distance from the glass transition temperature $T–T_g$. The ratio η/M_n is proportional to the segmental friction coefficient [32], and shifting the temperature by T_g takes into account the effect of T_g on the relaxation times. The smooth curves are from the regression fit of the shift factors in the WLF equation (28.1).

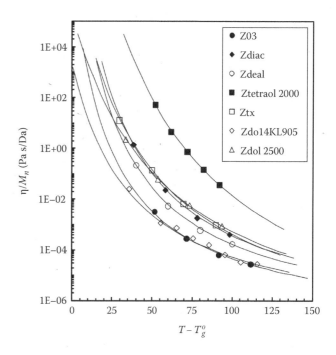

FIGURE 28.1 The ratio of viscosity to molecular weight as a function of distance from the glass transition temperature for the PFPE Z series. (Adapted from Karis, T.E., Lubricants for the disk drive industry, in: Rudnick, L.R., ed., *Synthetic, Mineral Oil, and Bio-Based Lubricants Chemistry and Technology*, CRC Press, Taylor & Francis, Boca Raton, FL, 2006, pp. 623–654.)

TABLE 28.5
The Glass Transition Temperature and the WLF Coefficients of Several PFPEs

Lubricant	T_g (°C)	C_1	C_2
Z03	−131.8	14.13	24.51
Zdiac	−118.4	18.14	25.90
Zdeal	−120.2	17.25	23.64
Ztetraol 2000	−112.2	23.22	45.81
Ztx	−109.9	15.67	42.75
Zdol4KL819	−126.7	11.73	38.46
Zdol4KL492	−123.3	16.27	49.82
Zdol4KL990	−119.7	15.98	52.22
Zdol4KBL598	−117.2	16.66	37.14
Zdol4KL905	−115.6	10.54	38.05
Zdol 2500	−113.6	13.62	59.72
Demnum S100	−111.2	13.06	62.76
Demnum SA2000	−114.1	13.75	43.89
Demnum SA2	−110.2	13.77	62.11
Demnum DPA	−110.7	12.13	78.52
Demnum SH	−110.1	13.27	63.56
Krytox 143 AD	−66.1	12.22	31.65
Krytox COOH	−61.4	11.97	40.79

Source: Adapted from Karis, T.E., Lubricants for the disk drive industry, in: Rudnick, L.R., ed., *Synthetic, Mineral Oil, and Bio-Based Lubricants Chemistry and Technology*, CRC Press, Taylor & Francis, Boca Raton, FL, 2006, pp. 623–654.

The reference temperature for C_1 and C_2 is T_g.

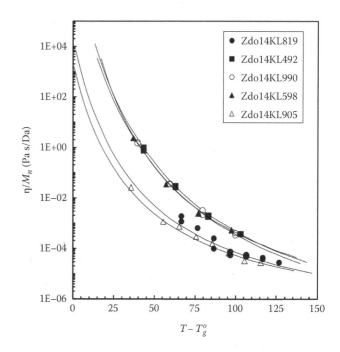

FIGURE 28.2 The ratio of viscosity to molecular weight as a function of distance from the glass transition temperature for the PFPE Zdol4K series, showing the effect of *O/C* ratio. The smooth curves are from the WLF equation. (Adapted from Karis, T.E., Lubricants for the disk drive industry, in: Rudnick, L.R., ed., *Synthetic, Mineral Oil, and Bio-Based Lubricants Chemistry and Technology*, CRC Press, Taylor & Francis, Boca Raton, FL, 2006, pp. 623–654.)

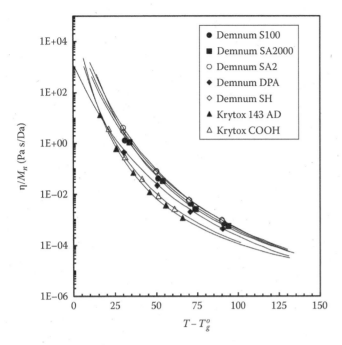

FIGURE 28.3 The ratio of viscosity to molecular weight as a function of distance from the glass transition temperature for the PFPE Demnum and Krytox series. The smooth curves are from the WLF equation. (Adapted from Karis, T.E., Lubricants for the disk drive industry, in: Rudnick, L.R., ed., *Synthetic, Mineral Oil, and Bio-Based Lubricants Chemistry and Technology*, CRC Press, Taylor & Francis, Boca Raton, FL, 2006, pp. 623–654.)

A subset of the Z series showing the effects of different end groups are shown in Figure 28.1. Most of the PFPEs shown in Figure 28.1 had an oxygen to carbon (*O/C*) ratio of about 0.65, except for the Zdeal, which had an *O/C* ratio of 0.694. The segmental friction coefficient was the lowest for nonpolar Z03 and the Zdol4KL905 (and Zdol4KL819 shown in Figure 28.2), and highest for the Ztetraol, with two hydroxyls on each end group. The segmental friction coefficients for Z chains with other types of end groups were in between the Z03 and Ztetraol. The friction coefficient for the Zdeal was slightly lower than that for the Zdiac, because the methyl ester probably blocks some of the hydrogen bonding. The Ztx, Zdiac, and Zdol 2500 had nearly the same segmental friction coefficient as one another.

The effect of the *O/C* ratio on the segmental friction coefficient for the Zdol4K series is shown in Figure 28.2. The lots with intermediate *O/C* ratio, Zdol4K L492, 990, and 598, were above Zdol4K L819 with (high) *O/C* = 0.72 and Zdol4KL905 with (low) *O/C* = 0.65, which were about the same as one another, even though their T_g are 11° apart. This surprising relationship may arise from a dependence of the segmental friction coefficient on the chain flexibility and the cohesive energy density that is different from the dependence of T_g on these properties.

The segmental friction coefficient for the D and K series, shown in Figure 28.3, was within the range of that observed for the Zdol4K series in Figure 28.2. The nonpolar Krytox and

the Krytox COOH were nearly the same as one another and were below most of the Demnum series. All of the Demnum series were nearly the same as one another. The addition of one polar end group had little effect on the segmental friction coefficient of the D and K series.

The storage and shear moduli, *G'* and *G''*, were measured and shifted along the temperature axis to obtain the master curves. The WLF coefficients [31] were calculated from the shift factors $a_{T_o}(T)$ by nonlinear regression analysis using the functional form

$$\log\left(a_{T_o}\right) = \frac{-C_1\left(T - T_o\right)}{C_2 + \left(T - T_o\right)} \qquad (28.1)$$

where the reference temperature $T_o = T_g$ and C_1 and C_2 are the WLF coefficients with respect to T_g. The WLF coefficients are listed in Table 28.5. Our range of coefficients $10.5 < C_1 < 23.5$ and $23.5 < C_2 < 79$ is consistent with that for nonpolar PFPEs Y and Z reported by Marchionni et al. [19].

Up to three Maxwell elements were derived from the master curves by nonlinear regression analysis from the linearly viscoelastic shear storage modulus, *G'*, and loss modulus, *G''*:

$$G' = \sum_i \frac{G_i\left(\omega a_{T_o}\tau_i\right)^2}{1 + \left(\omega a_{T_o}\tau_i\right)^2} \qquad (28.2)$$

and

$$G'' = \sum_i \frac{G_i\omega a_{T_o}\tau_i}{1 + \left(\omega a_{T_o}\tau_i\right)^2} \qquad (28.3)$$

where ω is the shear strain sinusoidal oscillation frequency. The shear rigidities G_i and corresponding relaxation times τ_i are listed in Table 28.6. The WLF coefficients, the shear rigidities, and the relaxation times provide the solid curves in Figures 28.4 through 28.6. The dynamic response for the Z series with different end groups is shown in Figure 28.4. The polar end group increases the relaxation times. Two relaxation times are observed in the Zdiac, Zdeal, and Zdol4KL905. Three relaxation times are observed in the Ztetraol and Zdol4KL1819. At ambient temperature, the Z03 has nearly the shortest characteristic time, τ_1, of all the PFPEs even though it has the highest M_n. Ztetraol has the longest τ_1 within the Z series. The response of the Zdol with a range of *O/C* ratio is shown in Figure 28.5. The *O/C* ratio has a significant effect on the dynamic response of the Zdol4K series. The dynamic response of the Demnum and Krytox is shown in Figure 28.6. The τ_1 for the Krytox is much longer than that for the Demnum.

The linear viscoelastic properties—the zero shear viscosity $\eta = G_1\tau_1$ and the equilibrium recoverable compliance $J_e^0 = \tau_1/\eta$—may be calculated from the dynamic properties listed in Table 28.6. The viscosity or relaxation time can be

TABLE 28.6

The Coefficients of the Maxwell Elements from the Master Curves at Reference Temperature T_g

Lubricant	G_1 (kPa)	τ_1 (s)	G_2 (kPa)	τ_2 (s)	G_3 (kPa)	τ_3 (s)	$\eta(-20\,^\circ\text{C})$ (Pa s) From Creep	From Dynamic
Z03	49.3	1.11E+07	—	—	—	—	0.2	0.2
Zdiac	28.4	5.17E+10	5.6	3.09E+09	—	—	1.0	1.0
Zdeal	31.1	6.31E+09	4.0	1.44E+08	—	—	0.4	0.4
Ztetraol 2000	36.6	4.02E+13	8.9	3.86E+12	5.5	3.16E+11	83	70
Ztx	55.6	7.42E+06	—	—	—	—	2.0	1.6
Zdol4KL819	4.0	3.56E+05	5.2	7.91E+04	14.3	8.13E+3	0.2	0.5
Zdol4KL492	43.4	1.51E+07	—	—	—	—	1.3	1.1
Zdol4KL990	49.7	5.77E+06	—	—	—	—	1.3	1.5
Zdol4KBL598	48.4	2.10E+08	—	—	—	—	2.3	1.4
Zdol4KL905	19.3	2.40E+03	19.7	2.21E+02	—	—	0.3	0.2
Zdol 2500	51.9	5.03E+04	—	—	—	—	2.2	2.0
Demnum S100	11.8	4.52E+04	38.0	9.91E+03	3.4	2.84E+2	3.7	1.5
Demnum SA2000	54.0	4.09E+05	—	—	—	—	1.1	1.5
Demnum SA2	35.1	5.49E+04	10.4	6.43E+03	—	—	2.8	2.1
Demnum DPA	42.0	1.87E+08	3.38	1.22E+07	—	—	3.6	2.5
Demnum SH	47.5	2.28E+04	6.8	1.60E+03	—	—	2.9	2.8
Krytox 143 AD	55.3	1.35E+05	3.1	4.30E+03	1.0	1.16E+2	81	69
Krytox COOH	44.9	3.08E+04	4.7	1.21E+03	2.3	7.17E+1	220	200

Source: Adapted from Karis, T.E., Lubricants for the disk drive industry, in: Rudnick, L.R., ed., *Synthetic, Mineral Oil, and Bio-Based Lubricants Chemistry and Technology*, CRC Press, Taylor & Francis, Boca Raton, FL, 2006, pp. 623–654.

The steady shear viscosity measured in creep, and the zero shear viscosity was calculated from the dynamic data at −20°C.

calculated at an arbitrary temperature T with the ratio of the shift factors from the WLF equation, for example, $\tau_1(T)$ or $\eta(T) = \eta(T_g)a_{T_g}g(T)$. The relaxation times for the Z series of lubricants calculated at 50°C are shown in Figure 28.7.

28.2.1.2 Dielectric

The lubricant dielectric properties provide complementary information to the rheological data. The concept is similar in that both energy storage and dissipation are characterized in response to a sinusoidal application of an electric field.

The permittivity and loss factor of different lubricant samples were measured using a TA Instruments Dielectric Analyzer (DEA) model 2970 with a single-surface ceramic sensor. Measurements were taken at an applied voltage of 1 V. The frequency sweep ranged from 0.1 to 10,000 Hz. Measurements were done at temperatures ranging from −100°C to 100°C. The data at various temperatures were shifted relative to reference temperature $T_o = 50\,^\circ\text{C}$ to provide the dielectric master curves for several magnetic recording disk lubricants, shown in Figure 28.8.

The dielectric properties are derived from the master curves with a discrete relaxation time (Debye) model [33] for the dielectric loss factor, ε'', and the dielectric permittivity, ε':

$$\varepsilon'' = \frac{\sigma}{\varepsilon_0 \omega a_{T_o}} + \sum_i \frac{\left(\varepsilon_{s,i} - \varepsilon_\infty\right)\omega a_{T_o}\tau_i}{1 + \left(\omega a_{T_o}\tau_i\right)^2} \quad (28.4)$$

and

$$\varepsilon' = \varepsilon_\infty + \sum_i \frac{\varepsilon_{s,i} - \varepsilon_\infty}{1 + \left(\omega a_{T_o}\tau_i\right)^2} \quad (28.5)$$

where
   ω is the sinusoidal oscillation frequency of the applied voltage
   τ_i are the dielectric relaxation times
   ε_0 is the absolute permittivity of free space $(8.85 \times 10^{-12}$ F/m)

The parameters in the discrete relaxation time series are determined by a regression fit to the dielectric master curves. There are multiple dielectric relaxation times for the Zdol and Ztetraol. Four relaxation times were employed to fit the data in Figure 28.8. These provide estimates for conductivity, σ, the dc relative permittivity, $\varepsilon'(0) = \sum_i \varepsilon_{s,i}$, and the limiting high-frequency permittivity, ε_∞. Note that the capacitive energy storage is proportional to the dc relative permittivity, and the refractive index $\mathbf{n}$ is related to the high-frequency relative permittivity by the Maxwell relation $\mathbf{n} \approx \sqrt{\varepsilon_\infty}$. For PFPEs, $\mathbf{n} \approx 1.3$ [30], which gives $\varepsilon_\infty \approx 1.7$. The dielectric properties and the four relaxation times and static relaxation amplitudes are listed in Table 28.7.

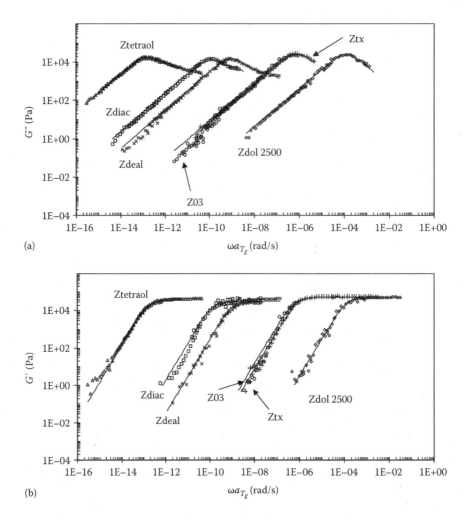

FIGURE 28.4 Shear loss (a) and storage (b) modulus master curves for the Z series. The smooth curves are from the discrete relaxation time series fit to the frequency–temperature shifted data. Reference temperature T_g. (Adapted from Karis, T.E., Lubricants for the disk drive industry, in: Rudnick, L.R., ed., *Synthetic, Mineral Oil, and Bio-Based Lubricants Chemistry and Technology*, CRC Press, Taylor & Francis, Boca Raton, FL, 2006, pp. 623–654.)

28.2.1.3 Thin Film Viscosity

The results mentioned earlier show that the viscosity increases exponentially as the measurement temperature approaches the glass transition temperature with bulk PFPE disk lubricants. This is because chain motions are progressively "frozen out" as the thermal energy becomes less than their activation energy. The lubricant viscosity also increases as the lubricant film thickness decreases, which helps to prevent the lubricant from flowing off the magnetic recording disks in the air shear [34]. Viscosity enhancement of thin films arises from a different mechanism than that found with decreasing temperature.

Dispersive interaction has a dramatic effect on the viscosity of the molecular layers closest to the surface, which can be explained in terms of the rate theory for viscous flow. Within the rate theory, a flow event comprises the transition of a flow unit from its normal or quiescent state, through a flow-activated state, to a region of lower free energy in an external stress field. For small molecules, the flow unit is the whole molecule, while for longer chains, the flow unit is a segment of the whole molecule. By analogy with chemical reaction rate theory, there is a flow-activation enthalpy, ΔH_{vis}, and entropy, ΔS_{vis}, for transition into the flow-activated state.

A flow unit is approximated by a particle in a box, with the energy being partitioned among rotational and translational degrees of freedom, which govern the transition probability. On this basis, the viscosity $\eta = (Nh_p/V_l)\exp(\Delta G_{vis}/RT)$, where N is Avogadro's number, h_p is the Planck constant, V_l is the molar volume, R is the universal gas constant, T is the temperature, and $\Delta G_{vis} = \Delta H_{vis} - T\Delta S_{vis}$ is the flow-activation Gibbs free energy. The flow-activation enthalpy $\Delta H_{vis} = \Delta E_{vis} + \Delta(pV)_{vis}$, where ΔE_{vis} is the flow-activation energy and $\Delta(pV)_{vis}$ is the pressure–volume work. At constant pressure, $\Delta(pV) = p\Delta V_{vis}$. For PFPE Z, the flow-activation volume $\Delta V_{vis} \approx 0.1$ nm^3 [18], which is equivalent to a spherical region ≈ 0.6 nm in diameter. At ambient pressure (100 kPa), $\Delta(pV)_{vis} \approx 6.2$ J/mol, so that near ambient conditions, $\Delta H_{vis} \approx \Delta E_{vis}$. Therefore, the viscosity is given by

$$\eta = \left(\frac{Nh_p}{V_l}\right)\exp\left[\frac{(\Delta E_{vis} - T\Delta S_{vis})}{RT}\right]. \qquad (28.6)$$

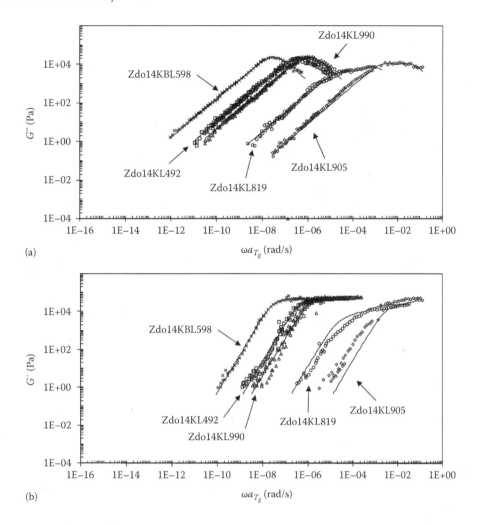

FIGURE 28.5 Shear loss (a) and storage (b) modulus master curves for the Zdol4K series. The smooth curves are from the discrete relaxation time series fit to the frequency–temperature shifted data. Reference temperature T_g. (Adapted from Karis, T.E., Lubricants for the disk drive industry, in: Rudnick, L.R., ed., *Synthetic, Mineral Oil, and Bio-Based Lubricants Chemistry and Technology*, CRC Press, Taylor & Francis, Boca Raton, FL, 2006, pp. 623–654.)

A regression fit to the bulk viscosity as a function of temperature [34], provided ΔE_{vis} = 34.7 kJ/mol and ΔS_{vis} = 9.87 J/mol-K. The flow-activation energy is close to that reported for bulk Zdol with a molecular weight of 3100 Da in References 35 and 36. A positive value for the flow-activation entropy of bulk Zdol means that the entropy of the flow unit increases on going into the flow-activated state.

Changes in the lubricant flow-activation energy and entropy near the solid surface cause changes in the viscosity with decreasing film thickness. The flow-activation energy near the solid surface is estimated from the thin film vaporization energy as follows. In an ideal gas, the chemical potential μ (or partial molar Gibbs free energy) is given by

$$d\mu = RTd \ln P, \tag{28.7}$$

where P is the partial pressure of the lubricant in the vapor phase. The chemical potential energy per unit volume in the lubricant film $\mu/V_l = \Pi$. The ratio of the film surface

vapor pressure to the vapor pressure of the bulk lubricant, $P^o (h)/P^o (\infty)$, is derived by integrating Equation 28.7:

$$\mu(h) - \mu(\infty) = RT \ln\left[\frac{P^o(h)}{P^o(\infty)}\right]. \tag{28.8}$$

The reference state is taken to be the chemical potential and vapor pressure of the bulk lubricant, $u(\infty) = 0$, and $P^o (\infty)$ is the vapor pressure of the bulk liquid.

In general, since surface energy is defined as the free energy per unit area, the total disjoining pressure (Π) for these fluids can be derived from the experimental surface energy (contact angle) data by

$$\Pi = -\frac{\partial}{\partial h}\left(\gamma^d + \gamma^p\right) \tag{28.9}$$

where

γ^d and γ^p are the dispersive and polar components of the surface energy, respectively

h is the film thickness

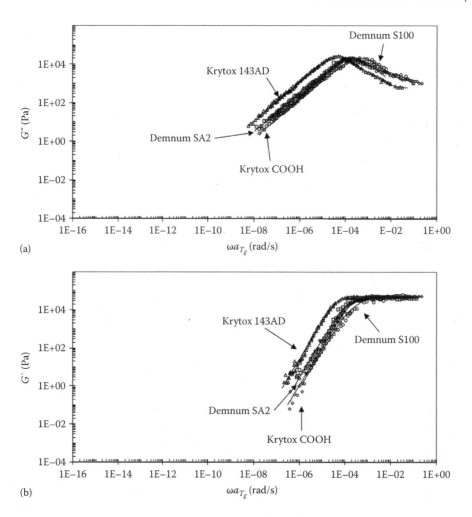

(a)

(b)

FIGURE 28.6 Shear loss (a) and storage (b) modulus master curves for the Demnum and Krytox series. The smooth curves are from the discrete relaxation time series fit to the frequency–temperature shifted data. Reference temperature T_g. (Adapted from Karis, T.E., Lubricants for the disk drive industry, in: Rudnick, L.R., ed., *Synthetic, Mineral Oil, and Bio-Based Lubricants Chemistry and Technology*, CRC Press, Taylor & Francis, Boca Raton, FL, 2006, pp. 623–654.)

The regression fit to the surface energy data, shown as the smooth curves in Figure 28.9a and b, was numerically differentiated to obtain the disjoining pressure [37,38]. The total disjoining pressure, including the individual contributions from the dispersive and polar components, is shown in Figure 28.10a. Notice that γ^d decreases monotonically with h, which is consistent with Equation 28.10. Below the film thicknesses of approximately 0.5 nm, Π at each molecular weight is dominated by γ^d, which increases rapidly with decreasing film thickness and is largely independent of molecular weight. γ^p, however, oscillates with film thickness and becomes larger in magnitude than γ^d as h increases. Oscillations in γ^p provide an additional contribution to Π for PFPE Zdol that produces alternating regions of stable and unstable film thickness [39]. The sum of the two contributions gives rise to oscillations in the total disjoining pressure. It is remarkable that the disjoining pressure from the surface energies as a function of film thickness, and Equation 28.8 relating the disjoining pressure to the degree of saturation, predicts the adsorption isotherms for low-molecular-weight Zdols, according to $P/P^0 = \exp(-\Pi V_l/RT)$, which

are shown in Figure 28.10b. There are two thermodynamically stable regions of film thickness for degrees of saturation corresponding to regions where $\Pi > 0$ and $\partial \Pi/\partial h < 0$. For thicknesses in between these regions, condensing Zdol molecules will either reevaporate or form islands at the next higher stable film thickness.

For the purpose of explaining the viscosity increase of thin films, surface chemical potential is approximated by the unretarded atom-slab dispersive interaction energy:

$$\mu = -\frac{V_l A}{6\pi h^3}. \tag{28.10}$$

The dispersion interaction coefficient A is also referred to as the Hamaker constant, and $A = 10^{-19}$ J for Zdol. Further discussion and review of the relationship between the dispersion force and the dielectric properties of the lubricant and overcoat are given in Reference 40.

As mentioned, the vaporization energy is the energy required to remove a molecule from the liquid without leaving behind a hole, which is the energy needed to form a hole

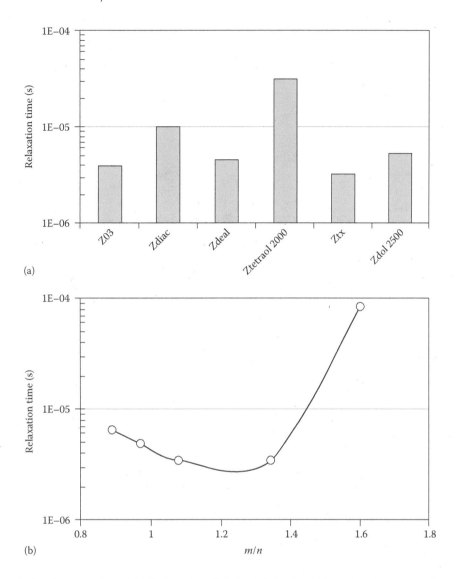

FIGURE 28.7 Longest relaxation time at 50°C for the Z series of lubricants (a) and Zdol4K series with a range of monomer chain composition (b) calculated from the dynamic rheological measurements with time–temperature superposition. (Adapted from Karis, T.E., Lubricants for the disk drive industry, in: Rudnick, L.R., ed., *Synthetic, Mineral Oil, and Bio-Based Lubricants Chemistry and Technology*, CRC Press, Taylor & Francis, Boca Raton, FL, 2006, pp. 623–654.)

the size of a molecule in the liquid. The free volume needed for a flow unit to transition into the flow-activated state is less than the size of the entire molecule. It is found that the ratio $n \equiv \Delta E_{vap,\infty}/\Delta E_{vis,\infty} > 3$, where $\Delta E_{vap,\infty}$ and $\Delta E_{vis,\infty}$ are the vaporization and flow-activation energy of the bulk liquid, respectively. Thus, the flow-activation energy near the surface is given approximately by

$$\Delta E_{vis} = \Delta E_{vis,\infty} - \frac{\mu}{n}. \qquad (28.11)$$

For linear chains longer than 5 or 10 carbon atoms, n increases due to the onset of segmental flow. In practice, n is experimentally determined from the measured values of the vaporization and flow-activation energy. For PFPE Zdol 4000, $\Delta E_{vap,\infty} = 166$ kJ/mol, giving $n \approx 4.8$. This is consistent with segmental flow.

In order to calculate the thin film viscosity with Equation 28.6, the flow-activation entropy near the surface is also needed. Experimental flow-activation entropy is calculated from the spin-off data [34] with Equations 28.6 and 28.11 as follows. The experimental η versus h is determined from the dh/dt during air shear–induced flow on a rotating disk. Equation 28.6 is then solved for ΔS_{vis} versus h using Equation 28.11 for ΔE_{vis}.

The flow-activation entropy and entropy are shown in Figure 28.11a. The flow-activation energy suddenly increases below about 0.8 nm due to the strong film thickness dependence of the dispersion force. The retarding effect of this increase on flow is compounded by the apparent effect of confinement on restricting the degrees of freedom in the flow transition state, as seen by the negative entropic contribution in Figure 28.11a. Below 2.3 nm, $T\Delta S_{vis} \approx -1.9$ kJ/mol, which corresponds to the critical configurational entropy change for

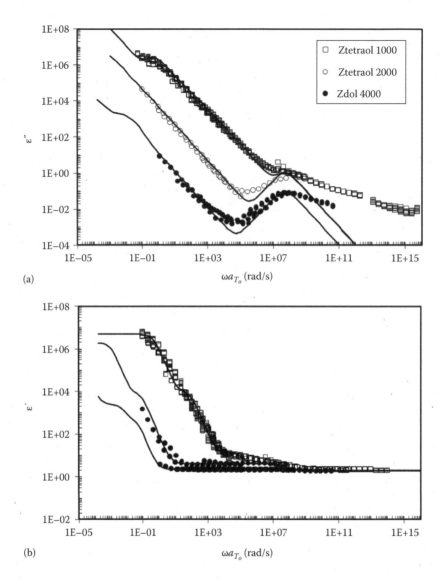

FIGURE 28.8 Dielectric loss factor (a) and relative permittivity (b) master curves for the Ztetraol 2000, Ztetraol 1000, and Zdol 4000. The smooth curves are from the discrete relaxation time series fit to the frequency–temperature shifted data. Reference temperature $T_o = 50\,^\circ$C. (Adapted from Karis, T.E., Lubricants for the disk drive industry, in: Rudnick, L.R., ed., *Synthetic, Mineral Oil, and Bio-Based Lubricants Chemistry and Technology*, CRC Press, Taylor & Francis, Boca Raton, FL, 2006, pp. 623–654.)

flow ($-R\ln 2 \approx -5.76$ J/mol K). The combined effects give rise to the observed increase in viscosity with film thickness shown in Figure 28.11b and enable the extrapolation of the viscosity to even thinner films where the spin-off is so slow that it takes years to measure.

The viscosity increases by a large amount with decreasing film thickness, which is much greater than the change in viscosity with temperature variations inside the disk drive. The bulk viscosity for several PFPE lubricants is shown in Figure 28.12. Since the increase in viscosity with decreasing thickness below about 0.8 nm is so much more than the increase with temperature between 0°C and 60°C, the operating temperature of disk drives should have no significant effect on lubricant spin-off from the disk by air shear, that is, excluding air shear force due to the head suspension assembly and the air bearing [41] and dispersion force between the lubricant and a low-flying slider [42].

28.2.1.4 Vapor Pressure

The vapor pressure of PFPE lubricants must be sufficiently low to prevent evaporation from the disk. The following method was developed to measure the vapor pressure. A model was derived to calculate the vapor pressure from the measured Zdol molecular weight distribution and evaporation rate. Molecular weight distributions were measured by GPC, as described in Karis et al. [30]. The vapor pressure of discrete molecular masses was calculated from the evaporation rate measured by isothermal thermogravimetric analysis (TGA) with a stagnant film diffusion model as in Karis and Nagaraj [43]. Polymers such as Zdol differ from the low-molecular-weight synthetic hydrocarbon oils in that polymers comprise a variety of different molecular weights. The molecular weight distribution must be taken into account in modeling the evaporation of polymers.

A numerical model was developed to simulate the evaporation of a polymer from an initial molecular weight distribution

TABLE 28.7

The Coefficients of the Debye Equation from the Dielectric Master Curves at Reference Temperature $T_o = 50\,°C$

Parameter	Lubricant		
	Zdol 4000	Ztetraol 2000	Ztetraol 1000
$\varepsilon'(0)$	3.3E3	1.1E4	1.1E7
σ (S/m)	1E−11	4E−8	6E−7
$\varepsilon_{s,1}$	3E3	1E4	1E7
τ_1 (s)	500	50	8
$\varepsilon_{s,2}$	3E2	1E3	1E6
τ_2 (s)	50	5	0.8
$\varepsilon_{s,3}$	30	100	100,000
τ_3 (s)	5	0.5	0.06
$\varepsilon_{s,4}$	2	4	5
τ_4 (s)	1E−8	1E−8	1E−8

Source: Adapted from Karis, T.E., Lubricants for the disk drive industry, in: Rudnick, L.R., ed., *Synthetic, Mineral Oil, and Bio-Based Lubricants Chemistry and Technology*, CRC Press, Taylor & Francis, Boca Raton, FL, 2006, pp. 623–654.

The high frequency $\varepsilon_\infty \approx 1.7$ from the index of refraction.

measured by GPC. The evaporation simulation is written in terms of mass flux and the discrete form of the molecular weight distribution $w_i(t)$ as

$$w_i\left(t^+\right) = w_i\left(t\right) - flux_i\left(t\right)\left\{\frac{A}{m_0}\right\}\Delta t, \qquad (28.12)$$

where

A is the surface area of the evaporating lubricant
m_0 is the initial mass of lubricant
Δt is the time step in the simulation

The mass flux of the ith molecular weight fraction M_i is given by stagnant film diffusion:

$$flux_i\left(t\right) = \frac{D_i}{\delta}\left(\frac{M_i}{RT}\right)P_i, \qquad (28.13)$$

where

D_i is the vapor-phase diffusion coefficient
δ is the diffusion length (calculated or measured with a liquid of known vapor pressure)

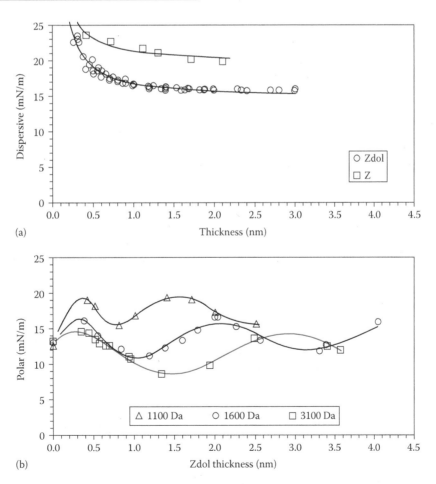

(a)

(b)

FIGURE 28.9 The components of the surface energy measured on CHx-overcoated thin film magnetic recording media with fractionated Zdol of narrow polydispersity index. (a) The dispersive component of the surface energy for PFPE Z and Zdol and (b) the polar component of the surface energy for PFPE Zdol. (Adapted from Karis, T.E., Lubricants for the disk drive industry, in: Rudnick, L.R., ed., *Synthetic, Mineral Oil, and Bio-Based Lubricants Chemistry and Technology*, CRC Press, Taylor & Francis, Boca Raton, FL, 2006, pp. 623–654.)

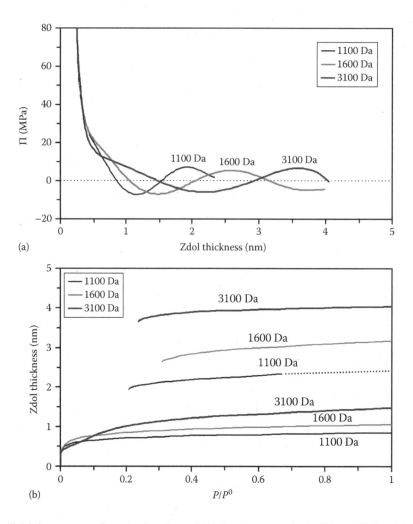

FIGURE 28.10 (a) The disjoining pressure from the fractionated Zdol surface energies in Figure 28.9 and (b) the corresponding Zdol adsorption isotherms at 60°C. (Adapted from Karis, T.E., Lubricants for the disk drive industry, in: Rudnick, L.R., ed., *Synthetic, Mineral Oil, and Bio-Based Lubricants Chemistry and Technology*, CRC Press, Taylor & Francis, Boca Raton, FL, 2006, pp. 623–654.)

The mass flux divided by the mass density yields the rate of film thickness change. The solution vapor pressure for the *i*th molecular fraction was approximated assuming an ideal solution according to Raoult's law $P_i = x_i P_i^0$, where x_i is the mole fraction of the *i*th molecular fraction.

The Hirschfelder approximation [44] is used for the vapor-phase diffusion coefficient:

$$D_i = 1.858 \times 10^{-4} \left(\frac{1}{M_i} + \frac{1}{M_{gas}} \right)^{1/2} \frac{T^{3/2}}{P \sigma_i^2 \Omega}, \quad (28.14)$$

where

M_{gas} is the molecular weight of the ambient atmosphere (air or nitrogen to suppress oxidation)

P is the ambient pressure

$\sigma_i = \left(\sigma_{lube}^i + \sigma_{gas}^i \right) / 2$ is the collision diameter

For nitrogen, $\sigma_{gas}^i = 0.315$ nm. The vapor-phase molecular diameter of the *i*th molecular fraction employed in estimating the binary mass diffusion coefficient is approximately given

by $\sigma_{lube}^i = 2 \times R_{g,i} \approx 0.05 \times \sqrt{M_i}$, where the molecular weight M_i is in daltons and the radius of gyration $R_{g,i}$ is in nanometers. This expression was derived from the radius of gyration for Zdol measured by light scattering [16,23,45]. The molecular diameters from this approximation for a range of ideal monodisperse Zdol molecular weights are listed in Table 28.8.

By analogy to hydrocarbon oils, the collision integral Ω for collision between molecules in the gas phase is a function of their binary Lennard–Jones interaction potential. The collision integral between Zdol molecules and nitrogen molecules was taken to be the same as that for collision between hydrocarbon molecules and nitrogen, $\Omega = 1.2$.

The Clapeyron equation is employed to calculate the pure component vapor pressure:

$$P_i^0 = P \exp \left\{ \frac{\Delta S_{vap}^i}{R} \right\} \exp \left\{ -1 \right\} \exp \left\{ \frac{-\Delta E_{vap}^i}{RT} \right\}, \quad (28.15)$$

where the vaporization entropy $\Delta S_{vap}^i = S_{vap}^i - S_{liq}$ is the difference between the entropy in the vapor state and that in the

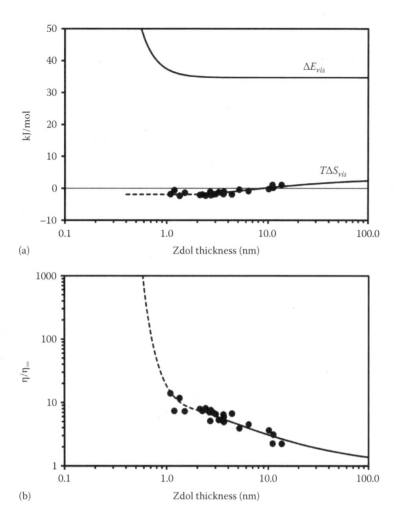

FIGURE 28.11 (a) Flow-activation energy and the entropic component of the flow-activation free energy and (b) dispersion-enhanced viscosity as a function of film thickness. The filled symbols are from the spin-off measurements, and the dashed region of the curve was calculated with constant flow-activation entropy below 2.3 nm. Fractionated Zdol molecular weight 4500 Da, temperature 50°C. (Adapted from Karis, T.E., Lubricants for the disk drive industry, in: Rudnick, L.R., ed., *Synthetic, Mineral Oil, and Bio-Based Lubricants Chemistry and Technology*, CRC Press, Taylor & Francis, Boca Raton, FL, 2006, pp. 623–654.)

liquid state. The Zdol liquid entropy is assumed to be independent of molecular weight. It was determined along with the activation energy by comparison of the simulated evaporation with isothermal TGA evaporation weight loss data. The liquid entropy for Zdol, $S_{liq} = 107$ J/mol K, is within the range obtained for the synthetic hydrocarbon oils. The vapor-phase translational entropy for oils is approximated with the Sackur–Tetrode equation:

$$S_{vap,trans}^i = R\left[\frac{5}{2} + \ln\left\{ \frac{(2\pi)^{3/2}(RT)^{5/2}}{h_p^3 \, N^4 P} \right\} + \frac{3}{2}\ln(M_i) \right]. \quad (28.16)$$

The vapor-phase rotational entropy is given by [46]

$$S_{vap,rot}^i = R\left[1 + \ln\left\{ \frac{1}{\pi q}\left(\frac{8\pi^3 IRT}{h_p^2} \right)^{a/2} \right\} \right], \quad (28.17)$$

where

a is the number of independent rotation axes
q is the degeneracy
I is the moment of inertia

An approximation for $S_{vap,rot}^i$ is given in Reference 43.

The vaporization entropy also includes the vibrational entropy. The available vibrational states are comparable between liquid and vapor for these high-molecular-weight hydrocarbons or polymeric oil molecules, whereas the translational and rotational states are much more restricted in the liquid phase.

Note that the ideal gas law is employed in deriving Equation 28.15 as follows: The vaporization enthalpy $\Delta H_{vap}^i = \Delta E_{vap}^i + \Delta(PV)_i$. The pressure–volume expansion work term has been replaced by $\Delta(PV)_i = RT$. The vaporization-activation energy, ΔE_{vap}^i, also depends on molecular weight because a longer molecule requires more energy to overcome the intermolecular interaction force between itself and its

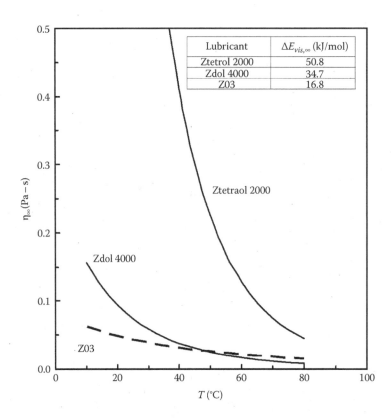

Lubricant	$\Delta E_{vis,\infty}$ (kJ/mol)
Ztetrol 2000	50.8
Zdol 4000	34.7
Z03	16.8

FIGURE 28.12 Bulk viscosity and flow-activation energy of several PFPE lubricants as a function of temperature. (Adapted from Karis, T.E., Lubricants for the disk drive industry, in: Rudnick, L.R., ed., *Synthetic, Mineral Oil, and Bio-Based Lubricants Chemistry and Technology*, CRC Press, Taylor & Francis, Boca Raton, FL, 2006, pp. 623–654.)

TABLE 28.8

The Gas-Phase Molecular Diameter for Ideal Monodisperse Zdol Fractions Calculated from the Radius of Gyration in Theta Solvent d (nm) $= 0.05\sqrt{M}$ (Da)

M (Da)	d (nm)	Degree of Polymerization	Contour Length (nm)	Equilibrium Thickness (nm)
500	1.12	3.54	2.71	0.66
750	1.37	6.29	4.06	0.74
1000	1.58	9.03	5.41	0.82
1350	1.84	12.88	7.29	0.93
1500	1.94	14.53	8.10	0.98
2000	2.24	20.02	10.79	1.14
3000	2.74	31.01	16.18	1.46
4000	3.16	42.00	21.56	1.79
4300	3.28	45.30	23.18	1.88
5400	3.67	57.38	29.10	2.24

Source: Adapted from Karis, T.E., Lubricants for the disk drive industry, in: Rudnick, L.R., ed., *Synthetic, Mineral Oil, and Bio-Based Lubricants Chemistry and Technology*, CRC Press, Taylor & Francis, Boca Raton, FL, 2006, pp. 623–654.

Also included are the degree of polymerization, the contour length measured along the chain from one end to the other, and equilibrium thickness. Equilibrium thickness is the maximum stable film thickness, or dewetting thickness, determined from the first zero crossing of the disjoining pressure with increasing film thickness.

neighbors in the surrounding liquid. Polar end groups contribute a fixed amount to the vaporization energy, which accounts for the non-zero intercept in the plot of vaporization energy as a function of molecular weight. As one might expect, there is a linear relation between the activation energy and the molecular weight for Zdol [47], $\Delta E_{vap}^i \approx \Delta E_{int} + \Delta E_{slope} \times M_i$. The slope, $\Delta E_{slope} = 0.029$ kJ/mol/Da, and intercept, $\Delta E_{int} = 50$ kJ/mol, of the vaporization–activation energy dependence on molecular weight for Zdol were determined by comparing the simulated evaporation data with that measured by isothermal TGA. The thermodynamic properties, vapor-phase diffusion coefficient, and vapor pressure for a range of ideal monodisperse Zdol molecular weights calculated as described earlier are listed in Table 28.9. The numerical values in Table 28.9 can be used in Equation 28.15 to calculate the vapor pressure of perfectly monodisperse molecular weight fractions.

The actual samples of commercial PFPE lubricant such as Zdol are polydisperse. Consequently, there is a wide range of partial pressures for a given sample, and the lowest-molecular-weight species in the distribution have the highest vapor pressure. In the case of Zdol 2000, since it is a copolymer of perfluoromethylene and perfluoroethylene oxide, the lowest-molecular-weight oligomers group together with similar molecular weights, hence similar vapor pressures. Figure 28.13a shows the molecular weight distribution of Zdol 2000 measured by GPC. The oscillations in the molecular weight distribution are visible up through 1000 Da. The mole fraction distribution is also shown, since it plays a key role in

TABLE 28.9

Vaporization Entropy ΔS_{vap}^i, Binary Diffusion Coefficient D_i, and Vapor Pressure p_{vap}^i for Perfectly Monodispersed Zdol Fractions Evaporating into Nitrogen at Ambient Pressure (10^5 Pa) at Three Different Temperatures, S_{liq} = 107 J/mol K,

$$\Delta E_{vap}^i \; kJ/mol = 50 \; kJ/mol + 0.029 \; kJ/mol \times M_i \,(Da)$$

M_i (Da)	ΔS_{vap}^i (J/mol K)	D (m²/s)	P_{vap}^i (Pa)
Temperature 35°C			
500	123	3.16E–06	1.13E+00
750	133	2.27E–06	2.26E–01
1000	140	1.78E–06	3.17E–02
1350	147	1.38E–06	1.48E–03
1500	150	1.26E–06	3.72E–04
2000	157	9.79E–07	3.07E–06
3000	167	6.82E–07	1.25E–10
4000	174	5.25E–07	3.61E–15
4300	176	4.91E–07	1.50E–16
5400	182	3.98E–07	1.16E–21
Temperature 45°C			
500	123	3.32E–06	2.76E+00
750	133	2.38E–06	6.01E–01
1000	141	1.87E–06	9.19E–02
1350	148	1.44E–06	4.87E–03
1500	151	1.32E–06	1.29E–03
2000	158	1.02E–06	1.27E–05
3000	168	7.15E–07	7.44E–10
4000	175	5.51E–07	3.05E–14
4300	177	5.15E–07	1.41E–15
5400	183	4.18E–07	1.61E–20
Temperature 60°C			
500	124	3.56E–06	9.51E+00
750	135	2.55E–06	2.34E+00
1000	142	2.00E–06	4.05E–01
1350	149	1.55E–06	2.55E–02
1500	152	1.41E–06	7.28E–03
2000	159	1.10E–06	9.19E–05
3000	169	7.66E–07	8.80E–09
4000	176	5.90E–07	5.91E–13
4300	178	5.52E–07	3.17E–14
5400	184	4.48E–07	6.26E–19

Source: Adapted from Karis, T.E., Lubricants for the disk drive industry, in: Rudnick, L.R., ed., *Synthetic, Mineral Oil, and Bio-Based Lubricants Chemistry and Technology*, CRC Press, Taylor & Francis, Boca Raton, FL, 2006, pp. 623–654.

Other parameters used are given in the text.

determining the actual vapor pressure. Qualitatively, the vapor pressure is increasing with decreasing molecular weight, but as the molecular weight becomes lower, there are fewer of these molecules in the solution, so Raoult's law acts to partly offset the increase in vapor pressure, causing the vapor pressure to decrease in the limit of low molecular weight. Hence, the shape of the partial pressure distribution superimposed on

the distribution in Figure 28.13a, calculated at 50°C. The partial pressure distribution for Zdol is shown with units on an expanded scale in Figure 28.13b. This shows the great detail provided by the GPC method, and also, the partial pressure peaks show the molecular weights that will evaporate with the highest rate, or distill out of the distribution. The total vapor pressure of polydisperse Zdol 2000 at 50°C is the sum of the partial pressures of each component, in this case, 0.2 Pa.

There are some other important properties of magnetic recording disk lubricants that will not be covered in this chapter, and several references on these are provided in the following. Lubricant spin-off and transfer to the slider are minimized by chemisorption to the overcoat [48]. Chemisorption [49], also referred to as bonding, is well described by Tyndall et al. [50]. Disk lubricants also serve to inhibit corrosion. The corrosion protection ability of Zol lubricants was related to surface energy by Tyndall et al. [50]. The most successful disk lubricant additive has been cyclic phosphazenes. However, cyclic phosphazene increases the lubricant mobility [51] and decreases the dewetting thickness [39]. More recently, an effort has been made to combine the desirable properties of both by incorporating a cyclic phosphazene end group onto Zdol. This lubricant is referred to as A20H, and it is described by Waltman et al. [6]. The A20H end group is shown in Table 28.1.

28.2.2 DISK LUBRICANT PICKUP ON MAGNETIC RECORDING SLIDERS

While the magnetic recording slider body flies over the magnetic recording disk at about 10 nm in the absence of read/write operations, and read/write operations are performed by thermally protruding the read/write head to within 1 nm of contact, it is inevitable that some of the liquid lubricant on the disk carbon overcoat transfers to the slider carbon overcoat [52,53]. The lubricant thickness on the disk is 0.8–1.2 nm thick, so that even a small portion of lubricant transfer from disk to slider can increase the slider flying height by several percent or more [54]. Thus, the magnetic signal becomes weaker with increased spacing gap [55], or more power is needed to protrude the slider to within the same distance from the surface during read/write operations (touchdown power). The lubricant typically accumulates on the trailing end of the slider and interferes with the aerodynamics of the air bearing [56]. There has been much focus on the mechanism of lubricant transfer from disk to slider throughout the past decade. The amount of the lubricant transferred to the slider in a given sweep test cycle increases as the lubricant film thickness on the disk approaches the equilibrium thickness (a.k.a. dewetting thickness) listed in Table 28.8. Lubricant transfer is thought to be through evaporation [57–59]. Alternatively, the lubricant may transfer at physical contacts due to dispersion energy gradient driving force [40,42].

Since the thickness of the lubricant on an approximately 0.8 × 1 mm slider is typically less than 1 nm, and the slider is only partially covered with the lubricant, there are limited options for quantitative measurement. One method that has the spatial resolution and chemical sensitivity to detect the molecularly thin

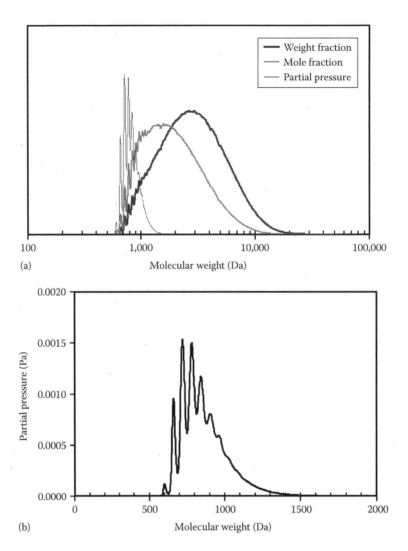

(a)

(b)

FIGURE 28.13 The molecular weight distribution, mole fraction distribution, and calculated partial pressure distribution (a) and the partial pressure distribution on an expanded scale with units (b) for Zdol 2000 at 50°C. (Adapted from Karis, T.E., Lubricants for the disk drive industry, in: Rudnick, L.R. ed., *Synthetic, Mineral Oil, and Bio-Based Lubricants Chemistry and Technology*, CRC Press, Taylor & Francis, Boca Raton, FL, 2006, pp. 623–654.)

films of the PFPE lubricant on the carbon-overcoated slider rails is time-of-flight secondary ion mass spectroscopy (TOF-SIMS) [60,61]. However, the TOF-SIMS measurement is not practical for large numbers of samples and usually has an unacceptably long turnaround time. This section describes a simple characterization method that can be performed using standard laboratory equipment and a microscope. It is referred to as lube pickup area measurement by oil decoration. The measurement is based on the contrast in the surface energy/wetting properties between the slider carbon overcoat and regions of the overcoat that have picked up the lubricant from the disk. The slider is lowered into the oil and then gradually withdrawn vertically out from the oil. Pure di(2-ethylhexyl)sebacate (DOS) oil remains covering the whole slider even when it is coated with the lubricant. It was found that an oil formulation in DOS base oil, referred to as blue oil, wets the carbon while it drains from lubricant-coated areas of the slider. For example, Figure 28.14a shows a slider that has no lubricant on the surface, which is fully covered with oil after dipping into the oil. A small amount of lubricant on

the trailing end surface is detected as the brighter region in the lower part of Figure 28.14b which is not wetted by the oil. The slider in Figure 28.14c is more than half covered with lubricant. The dark region is the residual oil and the lighter area is coated with lubricant and not wetted by the oil. The slider in Figure 28.14d is almost fully covered with lubricant. Only a few small residual oil droplet remained on the surface.

The presence of fluorinated disk lubricant on the nonwetted area of the slider was verified by TOF-SIMS. Figure 28.15a shows a slider after dipping into the blue oil has disk lubricant on the trailing end (right side of Figure 28.15a). A TOF-SIMS map of the CF3 ion on the same slider mirrors the lubricant-coated region as indicated by the oil (Figure 28.15b). The area of the slider coated by the lubricant can easily be determined by image analysis of the slider after dipping in the oil as lube pickup area in mm^2.

The conception of the previously mentioned blue oil formulation was intended to improve the contrast of the oil-wetted areas during the microscopic examination of the dipped sliders. Two blue colorants were added to DOS oil: (1) Silc-Pig® that

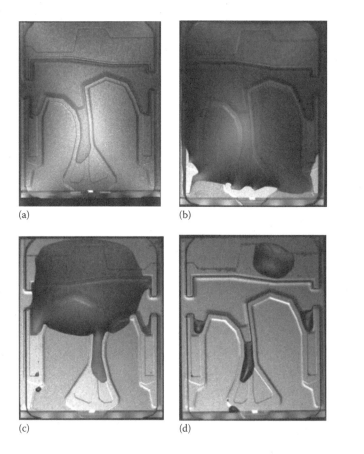

(a)

(b)

(c)

(d)

FIGURE 28.14 Magnetic recording slider after being dipped in blue oil formulation (a) unused slider is fully wetted by the oil, (b) used slider with some disk lubricant near the trailing end, (c) slider about half covered with disk lubricant, and (d) slider almost fully covered by disk lubricant. (Courtesy of Nguyen, T., Guo, X.-C., and Raman, V., Western Digital Corp., San Jose, CA.)

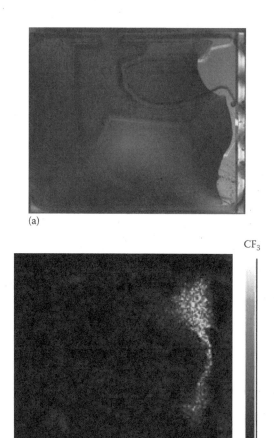

(a)

(b)

FIGURE 28.15 (a) Magnetic recording slider after being dipped in blue oil formulation. The lubricant-coated area is the bright area on the right, and (b) TOF-SIMS CF_3 ion image of the same slider showing lubricant in the bright area. (Courtesy of Nguyen, T., Chu, T., Spool, A., and Raman, V., Western Digital Corp., San Jose, CA.)

is used for coloring tin-cure silicone rubber compounds and (2) SO-Strong® that is used for coloring Smooth-On liquid urethane rubber, urethane plastic, or urethane foam (Smooth-On Corp., 5600 Lower Macungie Road, Macungie, PA 18062). Samples of the blue oil were prepared as follows: 17 g of DOS, 2 drops of SO-Strong blue tint (about 0.086 g), 0.17 g of blue Silc-Pig, mix with Vortex mixer and warming with heat gun alternately until no more change in appearance, and filter through 1 μm glass fiber filter into a clean vial. The result is approximately 1% Silc-Pig blue and 0.5% SO-Strong blue in DOS base oil.

28.2.2.1 Model Test Strips

In order to investigate how the draining of the blue oil provides contrast between the bare carbon and carbon overcoated with the PFPE disk lubricant, special test strips were prepared in the form of Wilhelmy plates. The Si plates were cut from 7 in. wafers, smooth on both sides, with dimensions $5 \times 20 \times 1.2$ mm. Both sides of the plates were coated with 4.67 nm of C_2H_4 diamond-like carbon (DLC). After the carbon overcoating, the lower 2–3 mm on some of the plates was dip coated with 1.28 nm of PFPE magnetic recording disk lubricant ZTMD. (ZTMD is defined in Table 28.3.) The location of the ZTMD on the lower end of the plate is shown in Figure 28.16a.

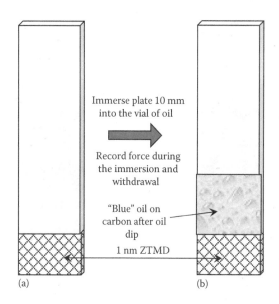

Immerse plate 10 mm into the vial of oil

Record force during the immersion and withdrawal

"Blue" oil on carbon after oil dip

1 nm ZTMD

(a)

(b)

FIGURE 28.16 Si strip coated with C_2H_4 DLC (diamond-like carbon overcoat, COC). Some of the plate ends were dip coated with ZTMD (cross-hatched region). (a) Before dipping into the oil and (b) after dipping into the oil during the Wilhelmy plate force microbalance dip coating test.

(The carbon and ZTMD thickness were measured with ESCA by X.-C. Guo, Western Digital Corp., Materials Analysis Lab, San Jose, CA.)

The carbon-overcoated Si plates were dipped in and out of the oil with a Cahn Radian 315 dynamic contact angle analyzer to an immersion depth of 10 mm at a traverse rate of 200 μm/s. A fresh plate was used for each test. The blue oil fully wetted and a thin film coated the carbon overcoat after withdrawal. When disk lubricant was present on the lower several mm of the plate, an oil film remained on the carbon above the lubricated region but drained from the lubricated region at the lower end of the plate (Figure 28.16b). A micrograph in Figure 28.17 shows the oil meniscus at the interface between the bare carbon and the ZTMD-coated portion at the lower end of the plate. The residual oil film thickness can be calculated from the interference fringes [62].

The focus of this investigation is the meniscus force on the plate as it is immersed and withdrawn from the oil. The force–distance curve for the bare carbon-overcoated plate immersion and withdrawal from DOS base oil is shown in Figure 28.18a. The meniscus pulls down on the plate as it enters the oil, and a higher force is measured as the meniscus is stretched and separates from the plate during withdrawal from the oil. The force–distance curve for the carbon-overcoated plate with ZTMD on the lower end is shown in Figure 28.18b. The meniscus force of the oil advancing onto the ZTMD-coated end is about 12 versus 35 mg for oil advancing onto the bare carbon overcoat in Figure 28.18a. Then the pull force during the oil advancing from the ZTMD to the bare carbon overcoat is about 22 mg (Figure 28.18b). The meniscus force at separation between the oil and the carbon overcoat is about 42 mg in Figure 28.18a, while the meniscus force at separation between the oil and the ZTMD-coated carbon is about 38 mg.

The meniscus force is related to surface tension by $\gamma \times PR \times \cos(\theta)$, where γ is the surface tension, PR is the

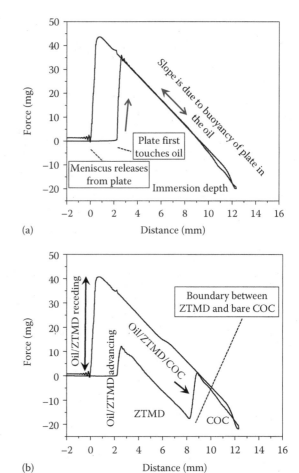

FIGURE 28.18 Force versus distance curves during dip coating in the microbalance. (a) Pure oil, no ZTMD on COC-coated plate. (b) Pure oil, ZTMD on end of COC-coated plate showing the definition of the advancing and receding meniscus force listed in Figure 28.20.

perimeter of the plates' lateral cross section, and θ is the contact angle. Assuming that the contact angle is close to zero at the moment of penetration and separation, this formula can be used to estimate the surface tension between the oil and the plate. The conversion factor to calculate the surface tension in mN/m from the pull force jump in mg is 0.8 mN/m/mg. Thus, the advancing surface tension between the oil and the carbon overcoat is 28 mN/m, and the receding surface tension between the oil and the carbon overcoat is 34 mN/m. The surface tension of the carbon overcoat is about 40 mN/m [49], while the surface tension of the DOS oil is 28.1 mN/m (Table 28.10). The surface tension change as the ZTMD enters the oil is 9.6 mN/m, and when the oil crosses from ZTMD to carbon, the additional jump in the pull force is 17.6 mN/m. The sum of the changes (9.6 + 17.6 = 27.2 mN/m) is close to the advancing pull force jump when the bare COC enters the oil. It is surprising that there is any force between the oil and the ZTMD because the surface energy of 1 nm thick ZTMD on carbon overcoat is 28.4 mN/m [63], which is nearly the same as the surface tension of the DOS oil. It is significant that the receding surface tension during separation between the oil and the carbon overcoat is 33.6 mN/m

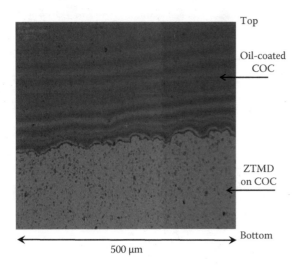

FIGURE 28.17 Interference fringes in the oil film that was prevented from draining over the ZTMD during withdrawal from the oil. Oil droplets are visible on the ZTMD.

TABLE 28.10

Parameters That Govern the Draining of the Pure and Blue Oils from Chemically Patterned Surfaces during Withdrawal

	Surface Tension (mN/m)	Capillary Number ($\times 10^{-4}$)	Bond Number ($\times 10^{-5}$)	Capillary Length (mm)	Maximum Thickness (μm)
Pure oil	28.1	1.44	0.70	1.78	4.71
0.01% FCA 1910	27.9	1.46	0.71	1.77	4.72
0.24% FCA 1910	23.9	1.70	0.87	1.64	4.84
0.3% FCA 1910[a]	22.5	1.80	0.95	1.59	4.89
0.005% Tegopren 6841[a]	22.5	1.80	0.95	1.59	4.89
0.01% Tegopren 6841	22.0	1.85	0.98	1.57	4.91
0.1% Tegopren 6841	20.8	1.95	1.05	1.53	4.95
0.1% Novec 4430	22.7	1.79	0.94	1.60	4.88
0.6% Novec 4430	21.9	1.85	0.98	1.57	4.91
Blue/32	21.2	1.92	1.03	1.54	4.94
Blue oil	21.0	1.93	1.04	1.53	4.95

The capillary number Ca is on the order of 10^{-4} (draining is controlled primarily by capillary force rather than viscous force) and the Bond number B_o is on the order of 10^{-5} (gravitational force is negligible relative to capillary force). Oil viscosity 20.3 mPa s, density 910 kg/m³, and pull rate 2×10^{-4} m/s.

[a] Surface tension was estimated from Figure 28.21.

while the receding surface tension between the oil and the ZTMD is only slightly less at 30.4 mN/m. This suggests that there is a molecular layer of oil that remains on the carbon and the ZTMD, so that the separation meniscus force is that between the oil and itself rather than that between the oil and the underlying material (carbon or ZTMD). This conjecture is supported by the observation of oil droplets that remained behind on the ZTMD after withdrawal. These dewetted droplets are visible in the lower portion of Figure 28.17.

The force–distance curve of the blue oil was qualitatively similar to that of the pure DOS oil. However, the advancing pull force onto the ZTMD and the receding meniscus force at separation from the ZTMD were less. A series of increasingly more dilute blue oils were prepared by successive dilution of the original blue oil in half with pure DOS oil down to 1/32 or about 3% of the original concentration. The UV/Vis spectrum measured for this set of oils is shown in Figure 28.19. The detector was partially saturated for the most concentrated blue oils. The minimum absorbance is in the blue region of wavelength from 435 to 500 nm, which accounts for the blue color of the oil. The force–distance curve was measured for each concentration using the test plates with ZTMD on the lower portion as shown in Figure 28.18a. There was little change in the meniscus force jumps over the whole range of concentration (Figure 28.20).

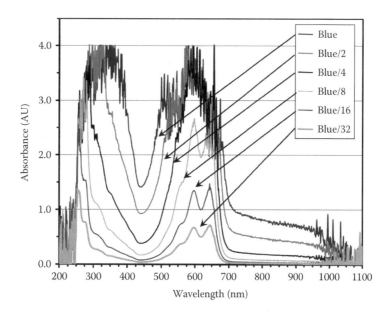

FIGURE 28.19 UV/Vis spectrum of blue oils successively diluted in half with pure DOS oil.

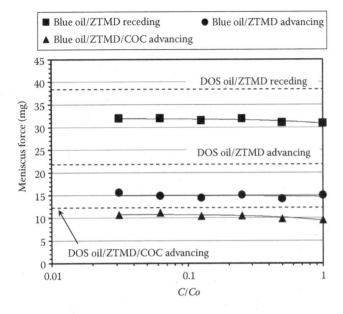

FIGURE 28.20 Meniscus force measured by the Wilhelmy plate microbalance test versus fraction of the initial blue oil concentration after dilution with pure oil. Dashed lines show values for pure DOS oil.

The dashed lines show the level of the corresponding pull force jumps for the pure DOS base oil. Advancing onto the ZTMD, the pull force jump was about 15 mg, or effective surface tension of 12 versus 17 mN/m for the pure oil. Advancing from the ZTMD to the carbon-overcoated region of the plate, the pull force jump was about 10 mg, or 8 versus 9.6 mN/m for the pure oil. Separating from the ZTMD on withdrawal, the pull force was about 31 mg with effective surface tension 25 versus 30 mN/m for the pure oil.

The surface tension of the blue oils measured separately using a bead-blasted stainless steel Wilhelmy plate [64] was about 21.8 mN/m (Table 28.10). Overall, the blue oil formulation decreased the surface tension between the ZTMD and the oil more than the surface tension between the carbon and the oil. The interaction surface energy between the ZTMD and the oil was 5 mN/m less than that of the pure oil, while the surface tension of the blue oil was 7 mN/m less than that of the pure DOS base oil. The difference between the surface tension reduction and the surface tension calculated from the ZTMD-coated plate is probably accounted for by the solid–liquid surface energy between the oil and the ZTMD of about 2 mN/m. Thus, it is reasonable that the blue oil wets the carbon-overcoated regions of the magnetic recording slider and drains from the ZTMD-coated regions rather than totally covering the slider like the pure DOS base oil because the SO-Strong and Silc-Pig colorants contain a surfactant that lowers the oil surface tension. Dilution of the original blue oil formulation up to 32× did not improve the slider decoration contrast.

28.2.2.2 Dip Coating Surface Chemistry

In principle, the surface tension of the decorating oil formulation can be adjusted to detect a certain lubricant film thickness on the slider carbon overcoat because the dispersive surface energy of the lubricated region increases toward that of the underlying carbon film with decreasing lubricant thickness [65]. Considering only dispersion force interaction, for a given oil formulation, lubricant type, and carbon overcoat, the equilibrium contact angle is determined by [66]

$$\gamma_s \left(1 - \cos\theta\right) \approx A^* / 12\pi \left(d_0 + h_s\right)^2 \qquad (28.18)$$

where

A^* is the effective Hamaker constant

d_0 is a constant that is often referred to as the "closest approach," which is on the order of a lubricant chain diameter [67]

h_s is the lubricant thickness on the slider

γ_s is the surface energy on the solid surface (slider)

In general, surface tension is an important property in determining how fluid drains from a surface. The slider decoration problem is analogous to the selective dip coating of chemically micropatterned surfaces [68]. The following nondimensional groups and length scales may provide some guidance for optimization of the oil formulation. The nondimensional capillary number compares the viscous force relative to the surface tension force

$$Ca = \frac{\eta U}{\gamma}, \qquad (28.19)$$

where

η is the oil viscosity

U is the withdrawal rate

γ is the surface tension of the oil

From Table 28.10, $Ca \ll 1$; hence, the surface tension force outweighs the viscous force. The nondimensional Bond number compares the gravitational body force relative to the surface tension force

$$B_o = \frac{\rho g h_\infty^2}{\gamma} \qquad (28.20)$$

where

ρ is the density

g is the gravitational acceleration

h_∞ is the maximum thickness of a film entrained on an infinite plate withdrawn vertically from a reservoir of a Newtonian liquid

From Table 28.10, $B_o \ll 1$; hence, the surface tension force outweighs the gravitational force. The characteristic length scale for an interface with surface tension and gravity determines the thickness of the entrained film

$$l_c = \sqrt{\frac{\gamma}{\rho g}}. \qquad (28.21)$$

The maximum film thickness on a plate withdrawn vertically is

$$h_\infty = 0.964 l_c Ca^{2/3}. \tag{28.22}$$

From Table 28.10, the maximum film thickness does not change much with over the maximum range of oil surface tension. As in [68], the geometry and angular orientation of the oleophobic and oleophilic surfaces strongly influence the residual oil film thickness.

28.2.2.3 Oil Surface Tension Modification

Previously, oil surface tension was modified to adjust the oil evaporation rate [64]. Two of the most promising surfactants were evaluated to enhance the resolution of the disk lubricant-coated areas of the magnetic recording slider, FCA 1910 and Novec 4430. The FCA 1910 is a foam control agent. Foam control agents, for example, alkyl polyacrylates and low-molecular-weight silicone oil, have been developed for use in lubricating oils [69]. Novec 4430 is a commercial nonionic polymeric fluorosurfactant that is based on perfluorobutane sulfonate and was provided by 3M Corp. Another surfactant, Tegopren 6841, a siloxane wax provided by Evonik Industries, was also evaluated.

The 0.1% and 0.6% Novec 4430 in DOS oil resulted in a droplet covering the center of the slider and an oil film showing interference fringes on the four corners of the slider. Even though the Novec 4430 solutions had intermediate surface tension values, they exhibited deviant draining behavior. This suggests that the molecular structure and composition of the surfactant play a role in determining the draining properties of the oil formulation in addition to the surface tension alone. For example, it is likely that the fluorinated portion of the Novec 4430 compound partially dissolved in the fluorinated portion of the PFPE of the disk lubricant on the slider with the polar and hydrocarbon portions of the Novec chain at the interface with the hydrocarbon ester oil.

The surface tension for several different concentrations of FCA 1910 and Tegopren 6841 in DOS oil is shown in Figure 28.21. For the same weight percent, Tegopren lowered the surface tension much more than the FCA. These formulations provided surface tension that was in between the pure DOS oil and the blue oil. Head sweep tests were done on carbon overcoat magnetic recording media lubricated with 1, 1.2, or 1.4 nm of the PFPE lubricant ZTMD. Increasingly, more ZTMD lubricant is transferred to the slider with increasing thickness because the ZTMD chains at the surface of the thicker layers are less tightly bound to the disk carbon overcoat. Figure 28.22 shows that the lube pickup area measured with the oil dip coating method increased with ZTMD thickness. This trend is expected as the thickness increases toward the dewetting thickness. Even though the surface tension of the blue oil is lower than the other two formulations, the lube pickup area was in between the other two. Moreover, the surface tension of the two surfactant formulations is nominally identical (Table 28.10), while the 0.3% FCA 1910 measured a significantly larger lube pickup area than the 0.005%

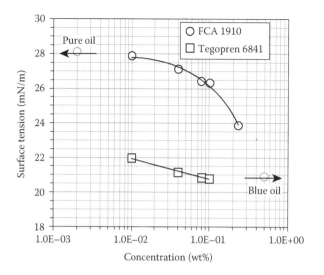

FIGURE 28.21 Surface tension for DOS oil containing various amounts of two different types of surfactants.

Tegopren 6841. Clearly, draining of the oils from the slider that is partially coated with a molecularly thin layer of PFPE cannot be accounted for by surface tension of the oil measured by the Wilhelmy plate method alone. A more complete model should take into account the interaction of surfactant with the carbon overcoat and lubricant. This is apparent from the microbalance curves shown in Figure 28.18 and the force values plotted in Figure 28.20. Specifically, the meniscus force when the ZTMD-coated end of the plate is withdrawn from the oil is much larger than the meniscus force when the ZTMD-coated end of the plate first touches the oil. Furthermore, there is a distinct increase in the meniscus force when the meniscus crosses from the ZTMD-coated end of the plate onto the carbon-overcoated region of the plate, but there is no meniscus force when the meniscus crosses from the carbon-overcoated region to the ZTMD-coated region during withdrawal.

28.3 SPINDLE MOTOR LUBRICANTS

Early magnetic recording disk drives used ball bearing and more recently fluid dynamic bearing spindle motors. The arrangement of the spindle motor and types of spindle motor bearings are shown in Figure 28.23.

28.3.1 BALL BEARING SPINDLE MOTOR BEARING GREASE

Ball bearing spindle motor bearings are typically lubricated with an NLGI grade 2 lithium grease. The grease composition, referred to as SRL, is a lithium grease comprising approximately 10% Li 12-hydroxy stearate, 17% di-2-ethyl-hexyl sebacate, and 70% pentaerythritol tetraesters, and the rest is a sulfonate rust inhibitor and an amine antioxidant. Lithium soap micelle fibers thicken the grease [70–72]. The grease base oil viscosity at 40°C is 22 mPa s, and the worked penetration is 245. A great variety of greases could potentially be used in these bearings, but in practice, the grease is limited

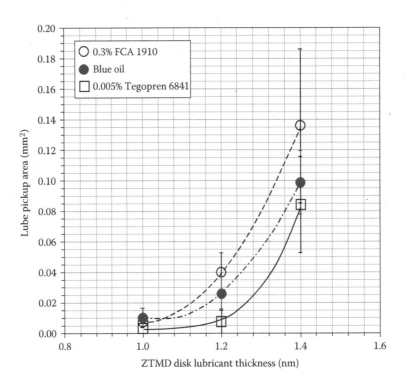

FIGURE 28.22 Lubricant pickup area versus ZTMD disk lubricant thickness after the head sweep test showing the results for DOS oil containing two different types of surfactants and the blue oil.

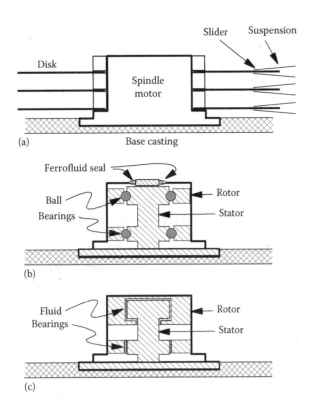

FIGURE 28.23 The arrangement of the magnetic recording disks and head suspension assembly on the spindle motor (a), schematic ball bearing spindle motor (b), and schematic fluid bearing spindle motor (c). (Adapted from Karis, T.E., Lubricants for the disk drive industry, in: Rudnick, L.R., ed., *Synthetic, Mineral Oil, and Bio-Based Lubricants Chemistry and Technology*, CRC Press, Taylor & Francis, Boca Raton, FL, 2006, pp. 623–654.)

by stringent requirements of low volatility, yield stress at temperature, low torque noise, and good thermal stability.

28.3.1.1　Yield Stress at Temperature

Typical magnetic recording disk drive ball bearing spindle motor grease rheological properties and yield stress are described, for example, by Karis et al. [73] and Cousseau et al. [74]. For practical purposes, the yield stress can be measured by gradually increasing the stress in a stress rheometer with a cone-plate fixture. The yield stress is detected when the cone begins to rotate. For example, the yield stress as a function of temperature for several grease candidates for use in ball bearing spindle motors is shown in Figure 28.24. There is a general trend of decrease with temperature, but all the greases maintain a measurable yield stress up through at least 80°C. The decrease of the grease yield stress with temperature is much less than that of the base oil viscosity.

Diluting the grease with additional base oil, or incorporation of contaminants in the grease, also affects the yield stress. Additional oil is often added to prelubricate the new bearing once it has been filled with grease. This is done to provide a lubrication film during initial start-up of the new bearing, before the base oil from the gel thickener of the grease has had time to diffuse throughout the surfaces of the balls and raceways. Prelube can either be the grease base oil itself or specially formulated prelube oil. The results with two types of prelube oils are also described here. Prelube oil A is diester oil with a sulfonate rust inhibitor and a hindered phenol antioxidant. Prelube oil B is mostly diester oil with several percent of a poly-α-olefin (PAO) oil, a sulfonate rust inhibitor, and a Zn dialkyldithiocarbamate (ZDTC) antiwear additive.

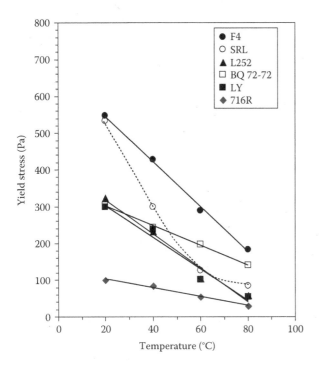

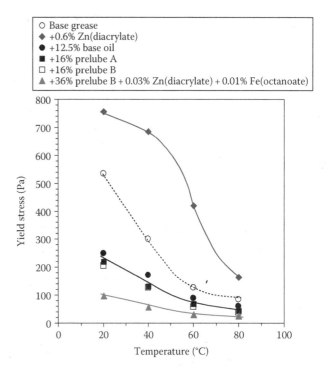

FIGURE 28.24 The yield stress of various candidate greases for ball bearing spindle motors as a function of temperature. (Adapted from Karis, T.E., Lubricants for the disk drive industry, in: Rudnick, L.R., ed., *Synthetic, Mineral Oil, and Bio-Based Lubricants Chemistry and Technology*, CRC Press, Taylor & Francis, Boca Raton, FL, 2006, pp. 623–654.)

FIGURE 28.25 The yield stress of base grease SRL showing the effect of additional base oil, prelube oils, and organometallic salt contamination on yield stress as a function of temperature. (Adapted from Karis, T.E., Lubricants for the disk drive industry, in: Rudnick, L.R., ed., *Synthetic, Mineral Oil, and Bio-Based Lubricants Chemistry and Technology*, CRC Press, Taylor & Francis, Boca Raton, FL, 2006, pp. 623–654.)

Grease may also be exposed to organometallic salts formed from various components within the bearing, bearing shields, or motor. Zn was incorporated as Zn(diacrylate), and Fe was incorporated as iron(III) 2-ethylhexanoate. The Zn(diacrylate) contaminant was intended to model products of bearing corrosion by the incomplete curing of a motor bearing adhesive [75].

Model grease containing prelube or contaminants was prepared in the laboratory by mixing in a custom-built lab-scale grease mill. The grease mill capacity was about 10 g of grease. The mill comprised two 32 mm diameter disks perforated with 35 circular holes, each 460 μm in diameter, inside a stainless steel tube. The perforated disks were separated by a 3.8 mm wide cavity. Grease was forced back and forth through the holes in the perforated disks by the reciprocating action of two opposing pneumatic cylinders driving Teflon pistons against the perforated plate within the steel tube. Air pressure was alternately applied to the cylinders using a cam and follower arrangement driven by a variable speed gear motor.

The yield stress of these model greases is shown in Figure 28.25. The yield stress was increased by Zn(diacrylate), while prelube oils decreased the yield stress.

For comparison with the yield stress versus temperature, the viscosity and density of the SRL grease base oil and two prelube oils are shown as a function of temperature in Figure 28.26. The viscosity and density of the base oil are somewhat higher than that of the prelube oils. Blends between the base oil and the prelube oil A or B will have intermediate viscosities.

The oil viscosity decreases much more than the yield stress with temperature. This implies that most of the yield stress change with temperature (Figure 28.25) is due to the gel network of the thickener. The reduction in yield stress on blending grease with prelube oil is probably due to dilution of a transient network in the gel thickener.

28.3.1.2 Hydrodynamic Film Thickness

The hydrodynamic film thickness of the oil provided by the grease must be sufficient to clear the asperities on the balls and race during operation at the specified load and velocity. The hydrodynamic film thickness is given by [76]

$$h = k\left(U\eta_0\right)^{0.67}\left(\alpha_p\right)^{0.53} \qquad (28.23)$$

where
h is the film thickness
k is a material and geometry parameter
U is the entrainment velocity
η_0 is the viscosity at atmospheric pressure
α_p is the pressure–viscosity coefficient

The film thickness between a steel ball and a plate was measured by Prof. H.A. Spikes, and his students, at Imperial College in London using ultrathin film interferometry [77]. The film thickness as a function of sliding speed for the grease

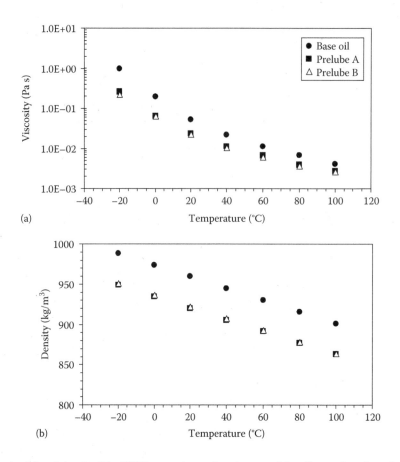

FIGURE 28.26 The viscosity (a) and density (b) of SRL grease base oil and two prelube oils as a function of temperature. (Adapted from Karis, T.E., Lubricants for the disk drive industry, in: Rudnick, L.R., ed., *Synthetic, Mineral Oil, and Bio-Based Lubricants Chemistry and Technology*, CRC Press, Taylor & Francis, Boca Raton, FL, 2006, pp. 623–654.)

base oil, and the prelube oils A and B, is shown in Figure 28.27. There is some variation in the power law slope between the oils, which slightly varies from the coefficients used in Equation 28.23. By comparison of a fluid with a known pressure–viscosity coefficient, they estimated the pressure–viscosity coefficients over a limited speed range between 0.1 and 1 m/s to be approximately 15 1/GPa for the base oil, 12 1/GPa for prelube oil A, and 10.5 1/GPa for prelube oil B. Additives and base oil type are known to affect the sliding/rolling traction [78]. The difference between prelube oils A and B is probably due to the minor fraction of PAO in prelube oil B.

28.3.1.3 Grease Electrochemistry

Some types of high-performance disk drive spindle motors incorporate ball bearings with silicon nitride ceramic balls for higher stiffness and lower vibration. It is critical that the bearings and grease provide smooth rotation so as not to excite resonances of the disk pack (Figure 28.23a). Electrostatic potential generated by bearings can induce a small current flow through the bearing, with a return path through the ferrofluid seal (Figure 28.23b). In order to investigate the effect of electrochemistry, grease containing various types of contaminants was sandwiched between two steel electrode plates. The plates were made of 25 mm diameter mirror-polished 304

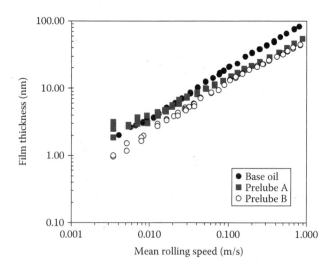

FIGURE 28.27 The film thickness as a function of rolling speed measured by ultrathin film interferometry. (Adapted from Karis, T.E., Lubricants for the disk drive industry, in: Rudnick, L.R., ed., *Synthetic, Mineral Oil, and Bio-Based Lubricants Chemistry and Technology*, CRC Press, Taylor & Francis, Boca Raton, FL, 2006, pp. 623–654.)

stainless steel disks with grease in between them on a160 μm thick filter paper. The plates were subjected to 25 V to simulate the passage of electrical current through the grease in the bearing. After several hundred hours, the plates were separated and examined for degraded grease as deposits on the plates. Film deposits were characterized by optical microscopy, FTIR spectroscopy in reflection, and x-ray photoelectron spectroscopy (XPS).

Figure 28.28 shows the current through the electrode plates plotted as a function of the voltage applied across the grease film with fresh grease between the plates. The conductance of the ferrofluid seal in the motor was about 77 nS (13 MΩ), so that the current through the bearing is typically 30–80 nA. In steady state, electrochemical cells were operated at 25 V, or between 100 and 3000 nA, depending on the type of grease contamination. Higher voltage was employed in the electrochemical cells to increase the rate of any electrochemical reactions that might take place. The initial conductance of the electrochemical cells, calculated from the linear region of the current–voltage plot, is listed in the second column of Table 28.11. The lowest conductance was obtained with the SRL grease alone. The highest conductances were found with the grease containing 16% prelube oil B and 300 ppm Zn and grease containing 16% prelube oil B. The grease conductance gradually varied with time during voltage application, as shown for several greases with and without contaminants in

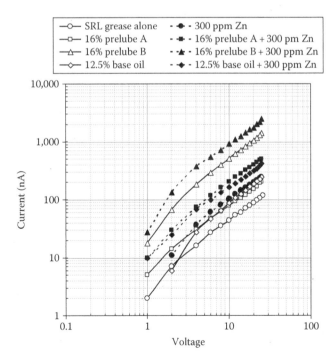

FIGURE 28.28 Initial current–voltage plot for SRL grease alone and SRL grease with the indicated additives and contaminants measured in the electrochemical cells. (Adapted from Karis, T.E., Lubricants for the disk drive industry, in: Rudnick, L.R. ed., *Synthetic, Mineral Oil, and Bio-Based Lubricants Chemistry and Technology*, CRC Press, Taylor & Francis, Boca Raton, FL, 2006, pp. 623–654.)

TABLE 28.11

Grease Electrochemical Cell Test Results for Grease on 160 μm Thick Filter Paper between 1 in. Diameter Electrode Plates

Sample	Initial Conductance (nS)	Time (h)	Film Deposit
SRL grease alone	4	960	Light
SRL grease +36% Prelube A +300 ppm Zn +100 ppm Fe	28	336	Medium
SRL grease +300 ppm Zn	7	336	Heavy
SRL grease +16% Prelube A	8	576	Heavy
SRL grease +16% Prelube A +300 ppm Zn	20	336	Light
SRL grease +16% Prelube B	52	576	Light
SRL grease +16% Prelube B +300 ppm Zn	93	336	Heavy
SRL grease +12.5% SRL base oil	9	336	Light
SRL grease +12.5% SRL base oil +300 ppm Zn	16	336	Heavy

Source: Adapted from Karis, T.E., Lubricants for the disk drive industry, in: Rudnick, L.R., ed., *Synthetic, Mineral Oil, and Bio-Based Lubricants Chemistry and Technology*, CRC Press, Taylor & Francis, Boca Raton, FL, 2006, pp. 623–654.

The initial conductance was calculated from the linear region of the current–voltage data measured between 1 and 25 V (Figure 28.28). Prelube is defined in the text. The right-hand column gives the appearance of the film deposit on the negative electrode plate after application of 25 V for the amount of time listed in the third column.

Figure 28.29. The pure grease, with no diluents or contaminants, maintained the lowest conductance.

After several hundred hours, the plates were separated and washed with chloroform by squirting from a pipette. When present, films were observed on the negative electrode plate. Although there was sometimes minor film formation or slight pitting on the positive plate, there was too little to quantify. Micrographs of the film deposits on several of the negative electrode plates are shown in Figure 28.30. These show the fibrous appearance. The film deposits were highly viscous. Film deposits were qualitatively ranked in terms of their severity, which is referred to as light, medium, and heavy, after the indicated electrolysis time, in Table 28.11. The lightest deposits were observed with the virgin grease, and the grease diluted with its own base oil. The heaviest deposit coincided with the highest conductance. However, even though they had nearly the lowest conductance, grease contaminated by 300 ppm of Zn as acrylate, or with 16% of the prelube oil A, also formed heavy deposits.

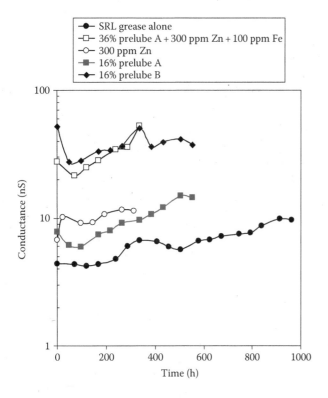

FIGURE 28.29 Conductance–time plot for SRL grease alone and SRL grease with additives and contaminants measured in the electrochemical cells. (Adapted from Karis, T.E., Lubricants for the disk drive industry, in: Rudnick, L.R., ed., *Synthetic, Mineral Oil, and Bio-Based Lubricants Chemistry and Technology*, CRC Press, Taylor & Francis, Boca Raton, FL, 2006, pp. 623–654.)

Reflection FTIR was performed on the residue on the plates after each test. Typical FTIR spectra are shown in Figure 28.31. The IR peaks were assigned to chemical groups according to the peak assignments in Table 28.12. Similar IR spectra and peak assignments are reported for mechanically stressed grease by Cann et al. [79]. The IR peak assignments, in conjunction with XPS measurements on the residue in Table 28.13, show electrochemical oxidation of the grease thickener. The ratio of carbonyl groups increased following the electrochemistry. For the pure thickener, the ratio of carbonyl to Li is 1.07, while aged grease and electrochemically oxidized grease have increased carbonyl due to oxidation. For black grease from a failed bearing, residue in a noisy bearing and a pin on disk wear test track also show increased carbonyl relative to the original thickener.

In summary, for the longest lifetime and best performance under all conditions, lithium grease should be kept free of metallic impurities and diluents. When electrochemical oxidation does occur, it forms a residue from the soap thickener on the raceway.

28.3.2 Ball Bearing Spindle Motor Ferrofluid Seal

As mentioned previously, the return path from the rotor to the stator for charge generated by the ball bearings is through the ferrofluid seal (Figure 28.23b). The ferrofluid is held in place by magnets in the seal housing, and the primary function of the ferrofluid seal is to prevent air flow through the motor into the disk drive enclosure. The typical ferrofluid is a suspension of 10–30 wt% subdomain magnetite particles 10 nm in diameter in a trimellitic/trimethylolpropane (TMP) ester oil with 10–20 wt% dispersing agent and up to 10 wt% antioxidant.

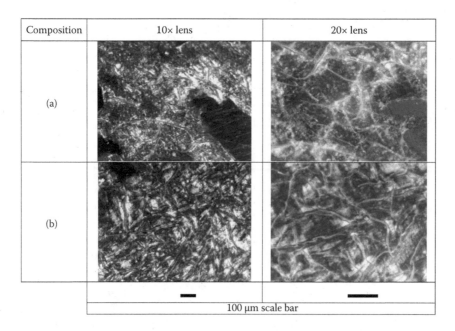

FIGURE 28.30 Optical micrographs showing the film deposits on the negative electrode plate following electrochemical oxidation of contaminated grease at two different levels of magnification. The residue was insoluble in chloroform and isopropanol, while the initial grease was easily removed from the plates by rinsing with chloroform. SRL grease + 300 ppm Zn for 336 h (a) and SRL grease + 16% Prelube A for 576 h (b). (Adapted from Karis, T.E., Lubricants for the disk drive industry, in: Rudnick, L.R., ed., *Synthetic, Mineral Oil, and Bio-Based Lubricants Chemistry and Technology*, CRC Press, Taylor & Francis, Boca Raton, FL, 2006, pp. 623–654.)

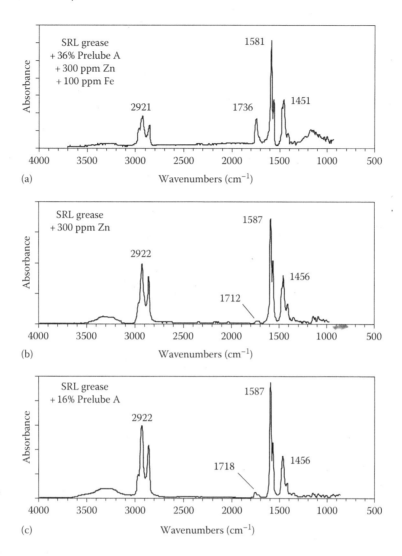

FIGURE 28.31 Reflection FTIR spectra of residue deposited on the negative electrode plates from grease containing various contaminants: (a) Prelube A, Zn, and Fe, (b) Zn, and (c) Prelube A. Oil was removed by washing with solvent before measurement. (Adapted from Karis, T.E., Lubricants for the disk drive industry, in: Rudnick, L.R., ed., *Synthetic, Mineral Oil, and Bio-Based Lubricants Chemistry and Technology*, CRC Press, Taylor & Francis, Boca Raton, FL, 2006, pp. 623–654.)

Ferrofluid is a mature technology, and these fluids are highly stable. The most recent effort to modify the properties of the ferrofluid was intended to increase the electrical conductivity so as to reduce the electrical potential between the rotor and the stator of the spindle motor. The development of conductivity additives for ferrofluids is described in the following.

Additives to increase the conductivity of the carrier oil were investigated. A number of conductivity-enhancing compounds were incorporated in a model carrier oil, trioctyltrimellitate (TOTM), and the conductivity was measured by DEA, as described in Section 28.2.1.2. The results of the initial screening are given in Table 28.14. Most of the additives reduced the conductivity. This probably indicates that the additives were associated with impurities, which were the primary charge carriers in the oil. The most promising initial results were obtained with a micellar solution of succinimide and dodecylbenzenesulfonic acid [80]. Variations of the organic acid and the succinimide/acid ratio were explored to optimize the conductivity of the TOTM carrier oil. The results are shown in Table 28.15. The mixture of succinimide and acid provided the highest conductivity to the oil. The most promising conductivity additives based on the tests in the model oil are shown in Figure 28.32. Even the best combination of conductivity additives in TOTM still had lower conductivity than any of the ferrofluids.

Dielectric spectroscopy was performed to determine the conductivity mechanism of the ferrofluid. The ferrofluid has three dielectric relaxation times, 260, 43, and 6.3 ms. These relaxation modes probably comprise the phoretic motion of the magnetite particles, phoretic motion of ions, and electronic charge hopping, respectively. The activation energy for conductivity is close to the viscous flow-activation energy, so the conductivity of the ferrofluid is mostly due to the phoretic motion of the magnetite particles. The relaxation times were unchanged by the conductivity additives.

TABLE 28.12
FTIR Peak Assignments

Absorbance	Wavenumber (cm⁻¹)
Broad dimer hydrogen-bonded carbonyl O–H stretch in 12-hydroxy stearic acid	3500–2500
Hydrogen-bonded O–H stretch in alcohol	3500–3200
Asymmetrical methylene C–H stretch	2928–2917
Aliphatic aldehyde or ester C=O stretch	1740
Aliphatic methyl ketone C=O stretch	1730
Aliphatic internal ketone C=O stretch	1725
Carboxylic acid dimer C=O stretch in 12-hydroxy stearic acid	1695
C–O–H in-plane bend in 2-hydroxy stearic acid	1470
Carboxylate anion, asymmetrical stretch	1589–1581
Carboxylate anion, symmetrical stretch	1456–1442
Ester C–C(=O)–O in base oil	1166

Source: Adapted from Karis, T.E., Lubricants for the disk drive industry, in: Rudnick, L.R., ed., *Synthetic, Mineral Oil, and Bio-Based Lubricants Chemistry and Technology*, CRC Press, Taylor & Francis, Boca Raton, FL, 2006, pp. 623–654.

The carboxylate anion is formed with Li or Zn and 12-hydroxy stearic acid. The ratio $(C=O)_{salt}/C(-H)_2$ was measured using the carboxylate anion, asymmetrical stretching, and asymmetrical methylene C–H stretching.

Since it became apparent that conventional additives used to enhance the conductivity of the carrier oil are of no benefit or decrease the conductivity of the ferrofluid, a different approach was needed. Ferrofluid is significantly more conductive than the carrier oil, due to the presence of the magnetite particles. The conductivity of a ferrofluid can only be enhanced by improving the efficiency of charge transfer between the suspended magnetite particles. This may be done by incorporating particles coated with conducting polymer, conducting polymer oligomers, or nanowires in the form of multiwall carbon nanotubes. Conducting polymer-coated carbon black particles (Eeonomer, Eeonyx Corp., 750 Belmont Way, Pinole, CA [81]) and multiwall nanotubes (BU200, Bucky USA, 9402 Alberene Dr., Houston, TX) were evaluated in the TOTM model oil. As shown in Table 28.16, the Eeonomers and nanotubes increased the oil conductivity by about five orders of magnitude. A stable mixed suspension of the nanotubes could not be adequately dispersed in the ferrofluid.

The conductivity of the ferrofluid increased with the Eeonomer concentration and was nearly independent of temperature between 5°C and 50°C, as shown in Figure 28.33. The conductivity as a function of temperature for Ferrofluid 1-containing Eeonomers is shown in Figure 28.34a. The

TABLE 28.13

The Ratio of Carboxylic Acid to Methylene from FTIR, and the Ratio of Total Carbonyl Carbon to Methylene Carbon and to Li, and Atomic Percent of Li and Zn from XPS, in Model Compounds, Electrochemically Deposited Films, Inner Race Deposits, and Black Grease from Failed Motor Bearings

	FTIR		XPS			
Film	Hydrogen-Bonded OH	$(C=O)_{salt}/C(-H)_2$	$(C=O)_{total}/C(-H)_2$	$(C=O)_{total}/Li$	Li (at%)	Zn (at%)
Li 12-hydroxy stearate	Yes	0.067 (exact)	0.067	1.07	4.3	0.09
Li 12-hydroxy stearate From grease (brown, stored 8 years)	Yes	0.035	0.079	1.5	3.6	—
Zn (12-hydroxy stearate)₂	—	—	0.063	—	<0.3	3.0
Zn (12-hydroxy stearate)₂ After 10 min at 130°C	Yes	0.049	—	—	—	—
eChem SRL grease +36% Prelube A +300 ppm Zn +100 ppm Fe	Yes	0.17, 0.05, 0.054	0.097	1.3	5.1	0.06
eChem SRL grease +300 ppm Zn	Yes	0.086, 0.082	0.11	1.4	4.7	0.12
eChem SRL grease +16% Prelube A	Yes	0.072, 0.077	0.11	1.4	5.3	—
Black grease	No	0.076	0.14	0.5	15.8	0.16
In-plane residue	No, Yes	0.10, 0.06	—	—	—	—
Ball track residue	No	0.12, 0.20, 0.31, 0.13	—	—	—	—

Source: Adapted from Karis, T.E., Lubricants for the disk drive industry, in: Rudnick, L.R., ed., *Synthetic, Mineral Oil, and Bio-Based Lubricants Chemistry and Technology*, CRC Press, Taylor & Francis, Boca Raton, FL, 2006, pp. 623–654.

The extinction coefficient ratio, $\varepsilon_{C(-H)_2}/\varepsilon_{C=O} = 0.0475$, was calculated from FTIR spectra of methylene $C(-H)_2$ asymmetric stretch and carboxylic acid (C=O–O) anion asymmetric stretch measured for Li 12-hydroxy stearate and Li stearate. Grease samples were rinsed with solvent to remove the oil. For Li 12-hydroxy stearate, the exact ratio of $C=O/C(-H)_2$ is 1/15 = 0.067, and the exact ratio of C(=O)/Li is 1; the exact atomic percent of Li = 4.8, and of Zn = 2.4. 300 ppm of Zn corresponds to 0.06 atomic percent of the lithium 12-hydroxy stearate. These results show evidence for electrochemical oxidation of the 12-hydroxy stearate thickener.

TABLE 28.14
Trial Matrix of Potential Conductivity Additives in Trioctyltrimellitate Model Carrier Oil

Additive	Concentration (wt%)	Comments	Conductivity (pS/m)
None	—	—	250–300
Disodium sebacate (Ciba DSSG)	1	Milky	5–7
Succinimide (2,5-pyrrolidinedione)	2	Mostly insoluble	80–100
Dodecylbenzenesulfonic acid	2	Clear solution	180–230
BHT (or Ionol)	2	Clear solution	50–70
Ciba Irgamet 30	2	Clear solution	170–180
Dicyclohexylammonium benzoate	2	Milky	200–260
PMC Cobratec 911S	2	Clear solution	130–150
Mellitic acid	2	Mostly insoluble	8–17
Mixtures			
Succinimide (2,5-pyrrolidinedione)	1.6	Hazy	860–990
Dodecylbenzenesulfonic acid	1		
Succinimide (2,5-pyrrolidinedione)	7.6	Hazy/gel	3000–5000
Dodecylbenzenesulfonic acid	4.8		

Source: Adapted from Karis, T.E., Lubricants for the disk drive industry, in: Rudnick, L.R., ed., *Synthetic, Mineral Oil, and Bio-Based Lubricants Chemistry and Technology*, CRC Press, Taylor & Francis, Boca Raton, FL, 2006, pp. 623–654.
Conductivity measured at 1 Hz and 50°C.
BHT, butylated hydroxytoluene.

TABLE 28.15
Optimization Matrix of Succinimide/Sulfonate Conductivity Additives in Trioctyltrimellitate Model Carrier Oil

Additive	Concentration (wt%)	Comments	Permittivity, 1 Hz		Conductivity (pS/m)	
			5°C	50°C	5°C	50°C
None	—	—	3.9	6.3	15	250–300
Succinimide	2	Mostly insoluble	—	4.0	—	80–100
DDBSA	2	Clear solution	—	4.0	—	180–230
Succinimide	1.6	Hazy	4.9	6.5	140	860–990
DDBSA	1					
Succinimide	7.6	Hazy/gel	—	26.0	—	3000–5000
DDBSA	4.8					
Succinimide	1.5	Brownish, hazy	3.0–3.2	4.8–5.6	160–220	4400
DNNSA	1.4					
Succinimide	1.5	Whitish, hazy,	—	2.9–3.2	—	5–8
DNNDSA	0.8	looks good				

Source: Adapted from Karis, T.E., Lubricants for the disk drive industry, in: Rudnick, L.R., ed., *Synthetic, Mineral Oil, and Bio-Based Lubricants Chemistry and Technology*, CRC Press, Taylor & Francis, Boca Raton, FL, 2006, pp. 623–654.
Succinimide is 2,5-pyrrolidinedione (99.1 g/mol). DDBSA is dodecylbenzenesulfonic acid (326.5 g/mol). DNNSA is dinonyl naphthalenesulfonic acid (460.71 g/mol). DNNDSA is dinonyl naphthalene disulfonic acid (540.79 g/mol). Amounts of DNNSA and DNNDSA adjusted for molecular weight and normality.

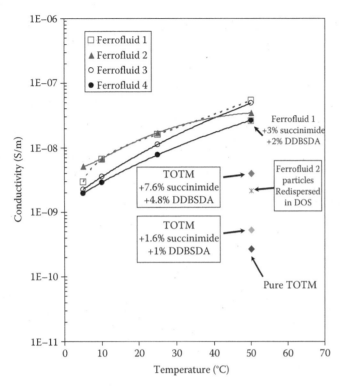

FIGURE 28.32 Conductivity of several types of ferrofluid, conductivity additives in model carrier oil, and Ferrofluid 1 combined with one of the most promising conductivity additives. (Adapted from Karis, T.E., Lubricants for the disk drive industry, in: Rudnick, L.R., ed., *Synthetic, Mineral Oil, and Bio-Based Lubricants Chemistry and Technology*, CRC Press, Taylor & Francis, Boca Raton, FL, 2006, pp. 623–654.)

TABLE 28.16

Conductivity Evaluation of Multiwall Nanotubes and Eeonomers in Trioctyltrimellitate Model Carrier Oil

Additive	Conductivity, σ (S/m)
None	2.5–3.0E–10
2% Eeonomer 200F	6.1E–4
	3.1E–4
2% Eeonomer 610F	2.0E–4
	2.3E–4
1% BS200 Multiwall Nanotubes	9.0E–4
	9.0E–4
10% BU200 Multiwall Nanotubes	8.8E–4
	8.6E–4

Source: Adapted from Karis, T.E., Lubricants for the disk drive industry, in: Rudnick, L.R., ed., *Synthetic, Mineral Oil, and Bio-Based Lubricants Chemistry and Technology*, CRC Press, Taylor & Francis, Boca Raton, FL, 2006, pp. 623–654.

The results of two separate measurements are shown for the mixtures.

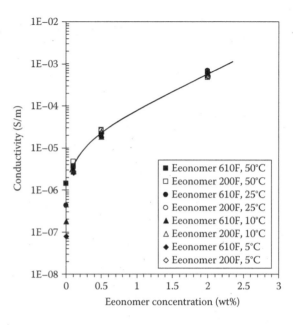

FIGURE 28.33 Conductivity of Eeonomers in Ferrofluid 1 as a function of Eeonomer concentration and temperature. (Adapted from Karis, T.E., Lubricants for the disk drive industry, in: Rudnick, L.R., ed., *Synthetic, Mineral Oil, and Bio-Based Lubricants Chemistry and Technology*, CRC Press, Taylor & Francis, Boca Raton, FL, 2006, pp. 623–654.)

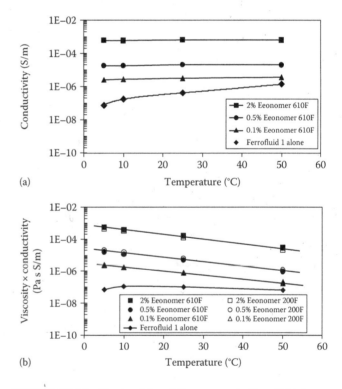

FIGURE 28.34 Conductivity of Eeonomers in Ferrofluid 1 as a function of Eeonomer temperature (a) and the product of conductivity × viscosity as a function of temperature (b) with additives in Ferrofluid 1. (Adapted from Karis, T.E., Lubricants for the disk drive industry, in: Rudnick, L.R., ed., *Synthetic, Mineral Oil, and Bio-Based Lubricants Chemistry and Technology*, CRC Press, Taylor & Francis, Boca Raton, FL, 2006, pp. 623–654.)

conduction mechanism seems to be limited by electronic transport through the Eeonomers, rather than by diffusive contacts between the particles, since the product of the conductivity and the viscosity of the ferrofluid-containing Eeonomers decreased with temperature (Figure 28.34b). The conductivity of the conducting polymer sheath was decreased by scattering due to lattice vibrations more than the transport between particles was increased by Brownian motion with increasing temperature.

While the Eeonomers appear to be the most effective means to increase the ferrofluid conductivity, the long-term stability may be degraded by reaction of the dodecylbenzene-sulfonic acid with the ester carrier oil. Furthermore, the conducting polymer sheath may also be crushed off the carbon black in milling during the initial dispersion phase of ferrofluid manufacturing.

28.3.3 FLUID BEARING MOTOR OIL

Disk drive motor bearing dynamics determine the precision of the spindle rotation. As the rotor spins relative to the stator, the spin axis traces out an orbit. The spin-axis motion has a component that is in phase and at the same frequency as the spindle rotation. This is repeatable runout. There is also a component of spin-axis motion that is random. This is nonrepeatable runout (NRRO). Increasing computer magnetic recording disk drive data storage density is potentially limited by NRRO. The spindle bearing is the primary contributor to NRRO. Improved NRRO improves the seek time and track following for a given track pitch and servomechanism. A disk drive with a low-NRRO spindle bearing accommodates higher track density, allowing higher areal data recording density.

Nearly all disk drives are now being built with fluid dynamic bearing motors (Figure 28.23c). These replace ball bearing spindle motors, because their lower NRRO allows the servomechanism to follow narrower data tracks. At the heart of a fluid dynamic bearing are a shaft in a sleeve for radial thrust and a thrust plate and a bushing for axial thrust. Embossed on one face of the radial and axial thrust members are grooves that force the oil inward, creating the internal pressure that provides the bearing stiffness [82]. The fluid dynamic bearing achieves lower levels of NRRO because a relatively thick film of lubricant separates the sleeve and the stator. The oil film thickness is typically 3–5 μm. The film provides viscous damping that attenuates NRRO below that achievable in ball bearings. The forces generated on the sleeve by the fluid dynamic bearing are of a lower frequency and amplitude than with ball bearings. The reduction in the excitation bandwidth and power allows the servo to seek and hold on a higher track density.

While the bearing must be designed to avoid cavitation [83] and air ingestion [84], two fundamental requirements of fluid dynamic bearing oil are low viscosity and low vapor pressure. The viscosity must be low enough to allow for spin-up in a reasonably short period of time at temperatures approaching 0°C. For example, the spin-up time for

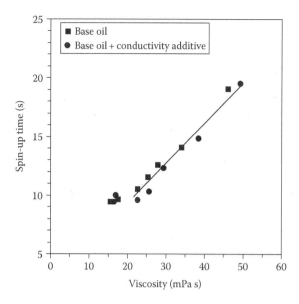

FIGURE 28.35 Spin-up time for prototype disk drives as a function of fluid bearing oil viscosity measured between 1°C and 25°C. The base oil was a diester and the conductivity additive was 2% Vanlube 9317 [64]. (Adapted from Karis, T.E., Lubricants for the disk drive industry, in: Rudnick, L.R., ed., *Synthetic, Mineral Oil, and Bio-Based Lubricants Chemistry and Technology*, CRC Press, Taylor & Francis, Boca Raton, FL, 2006, pp. 623–654.)

prototype disk drives with fluid bearing motors is shown in Figure 28.35. The viscosity dominates the spin-up time when it is higher than about 25 mPa s. For lower viscosities, the spin-up time is limited by inertia of the disk pack and peak motor power.

The oil vapor pressure must be low enough so that there is no appreciable evaporation over the 5- to 7-year lifetime of the motor at its maximum internal operating temperature of about 80°C. Well-formulated oil should contain antioxidants and other additives to inhibit oxidation and corrosion. A chemical kinetic model [85] is presented in Section 28.3.3.2, and the model is applied to development and testing of a new formulation [86] in Section 28.3.3.3. In addition, the fluid bearing oil must be sufficiently conductive to dissipate static charge accumulation due to air shear, electrical double-layer shear [87], and the work function difference between the slider and the disk [88]. The oil formulation must also have a low dc dielectric permittivity with minimal charge storage by ionic separation or dipole orientation. In this edition, novel charge control additives based on an electronic hopping transport mechanism are presented in Section 28.3.4.

Formulation starts with a base oil that has a sufficiently low vapor pressure and viscosity. A variety of base oils were investigated with the goal of providing guidelines for selection of the oil with the lowest possible viscosity and vapor pressure. The oils studied were diesters (Table 28.17), triesters (Table 28.18), and nonpolar hydrocarbons (Table 28.19). The first five model oils are diesters made from alkanedioic acids and alcohols with a range of diacid chain lengths and several alcohol molecular weights and

TABLE 28.17

Molecular Structures of the Diester Oils Investigated for Fluid Dynamic Bearing Motors

Acronym	Molecular Weight (g/mol)	Structure
DBS	314	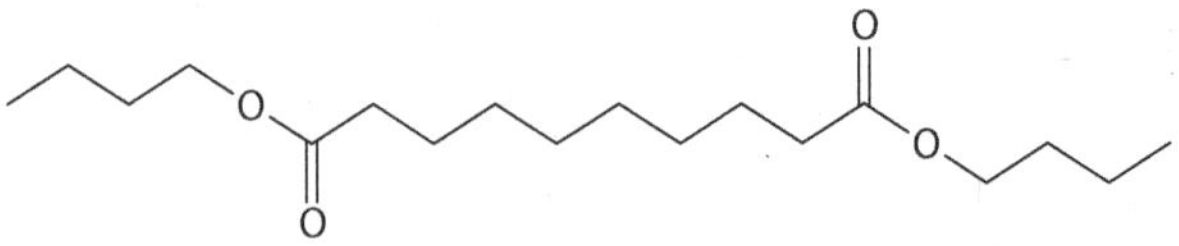
DOA	371	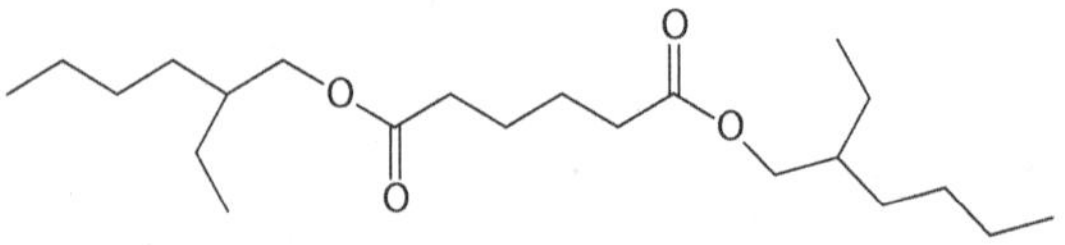
DOZ	413	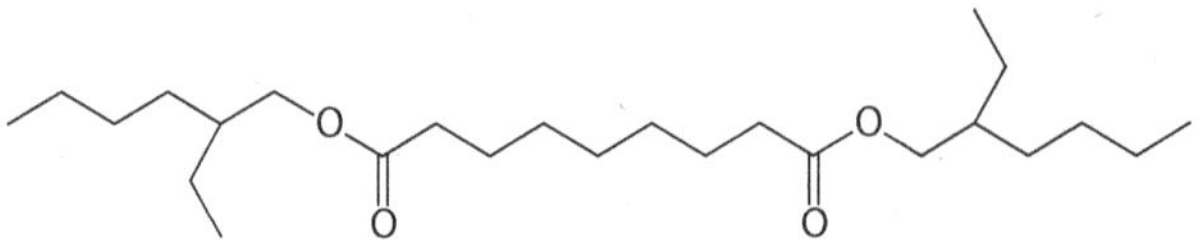
DOS	427	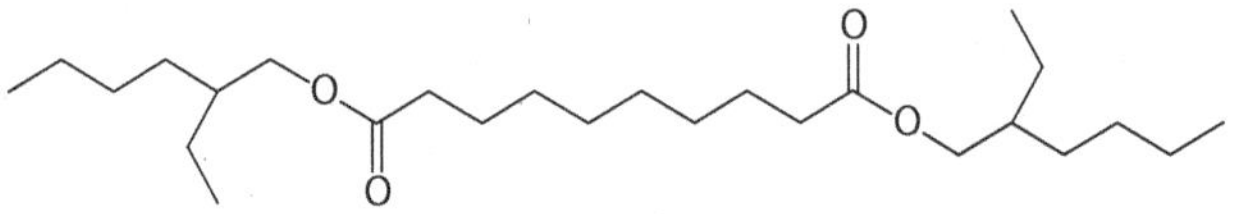
DIA	427	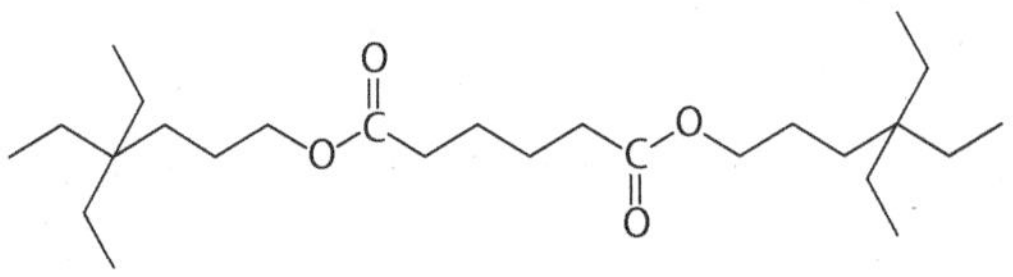

Source: Adapted from Karis, T.E., Lubricants for the disk drive industry, in: Rudnick, L.R., ed., *Synthetic, Mineral Oil, and Bio-Based Lubricants Chemistry and Technology*, CRC Press, Taylor & Francis, Boca Raton, FL, 2006, pp. 623–654.

branched isomers. Di(n-butyl)sebacate (DBS), di(2-ethylhexyl)adipate (DOA), and DOS were obtained from Aldrich Chemical Company (Saint Louis, MO). Di(2-ethylhexyl) azelate (DOZ) is Emery 2958 and di(4,4-diethylhexyl)adipate (DIA) is Emery 2970. The Emery oil samples were provided by Henkel Corporation (Emery Group, Cincinnati, OH). The next four model oils are triesters made from triols and alkanoic acids with a range of acid chain length. The triglycerides—glycerol tributanoate (TRIB), glycerol trihexanoate (TRIH), and glycerol trioctanoate (TRIO) (also referred to as tributyrin, tricaproin, and tricaprylin, respectively)—were obtained from Aldrich Chemical Company. About 8% of low-molecular-weight impurities were evaporated from Tricaproin by vacuum baking at 6.3 kPa for 96 h. The level of impurities in the other model ester oils was less than 0.2% during vacuum baking in the same conditions and did not show up in the NMR spectra. The trimethylolpropane C7 ester, trimethylolpropane TMP triheptanoate, is Emkarate 1510 from ICI Americas, Inc. (Wilmington, DE). The branched alkane, a low-molecular-weight PAO, was NYE 167A from Nye Lubricants. Some of the oils are isomers, having the same molecular weight and composition but differing only in molecular structure.

Representative proton NMR spectra for the diester oils are shown here for DBS and DOS in Figure 28.36. The diester base oils are most similar to the oils that are presently used in disk drive fluid dynamic bearing spindle motors. The NMR measurements were done on a Bruker Aspect 3000 Nuclear Magnetic Resonance Spectrometer attached to a 250 MHZ superconducting magnet with an automatic sample changer. The samples for proton NMR were 10% w/w oil in chloroform-d1. The solution spectra were measured in 5 mm NMR tubes from Kontes Scientific Glassware (Grade 6, 5 pack, #897235). The proton NMR measurements were done using a spectral width of 5 kHz, 45° pulse width of 3.0 μs, 1 s relaxation delay, and 512 scans.

The proton corresponding to each peak in the NMR spectra in Figure 28.36 is indicated by a letter, and an arrow shows the location of the corresponding proton on the molecular structure. Peak assignments corresponding to these structures are listed in Table 28.20. The peak area is proportional to the number of protons at each chemical shift. The peak areas were calculated from the NMR spectra with commercial software (NUTS NMR Data Processing Software, Version 4.27, Acorn NMR, Inc., 46560 Fremont Blvd., #418, Fremont, CA). The reference peak area (indicated by the * in the first columns of the peak assignment tables) was assigned to the expected area. The rest of the integrated areas are relative to the reference peak area. Both the integrated peak areas and the peak areas expected from

TABLE 28.18

Molecular Structures of the Triester Oils Investigated for Fluid Dynamic Bearing Motors

Acronym	Molecular Weight (g/mol)	Structure
TRIB	302	
TRIH	387	
TRIO	471	
TMP	471	

Source: Adapted from Karis, T.E., Lubricants for the disk drive industry, in: Rudnick, L.R., ed., *Synthetic, Mineral Oil, and Bio-Based Lubricants Chemistry and Technology*, CRC Press, Taylor & Francis, Boca Raton, FL, 2006, pp. 623–654.

the abundance of each type of proton in the model chemical structure agree within 10%. Small changes in the chemical shifts are due to differences in the chemical environment of the protons.

28.3.3.1 Viscosity and Vapor Pressure

The oil viscosity and vapor pressure are thermodynamically interrelated by the dispersion and dipolar forces between the molecules. The total intermolecular interaction energy must be overcome for a molecule to evaporate, while only part of the intermolecular interaction energy must be overcome during viscous flow. To first order, the dispersion force is proportional to the number and type of atoms and thus to molecular weight. For the relatively low-molecular-weight oils useful in fluid bearings, the flow is mostly by the whole molecule rather

than incrementally by segments of the molecule. Increasing the molecular weight, or polarity, to reduce the vapor pressure increases the viscosity. The relationship between oil vapor pressure and viscosity is thoroughly investigated with Eyring's chemical reaction rate theory of evaporation and flow in Karis and Nagaraj [43]. The essence of their findings is summarized here.

According to the Eyring equation (28.6), a plot of $\ln(\eta)$ versus $1/T$ is a straight line with slope $\Delta E_{vis}/R$ and intercept $\ln(Nh_p/V_l) - \Delta S_{vis}/R$. The viscosity of several low-molecular-weight ester and hydrocarbon oils is shown plotted as $\ln(\eta)$ versus $1/T$ in Figure 28.37. There is some systematic deviation from linearity, but overall the general trend is fit by the Eyring equation between −20°C and 100°C. The flow-activation entropy $\Delta S_{vis} = \Delta S_{vis}^{trans} + \Delta S_{vis}^{rot}$ is the sum of the translational

TABLE 28.19

Molecular Structures of the Nonpolar Oils Investigated for Fluid Dynamic Bearing Motors

Acronym	Molecular Weight (g/mol)	Structure
2,2-DPP	196	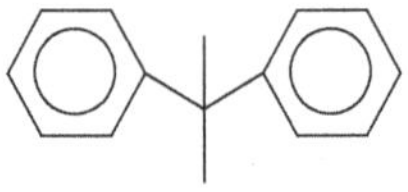
1,3-DPP	196	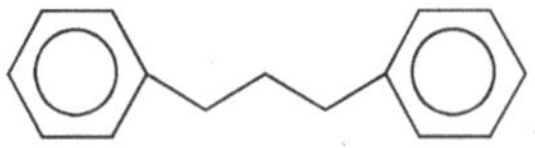
PAO	240	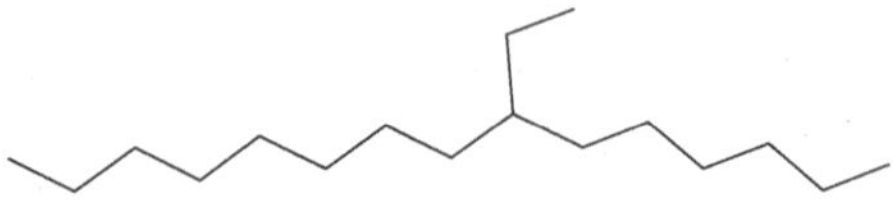
PRS	269	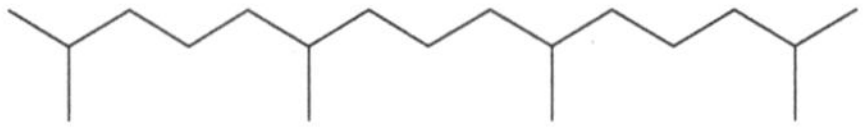
SQL	423	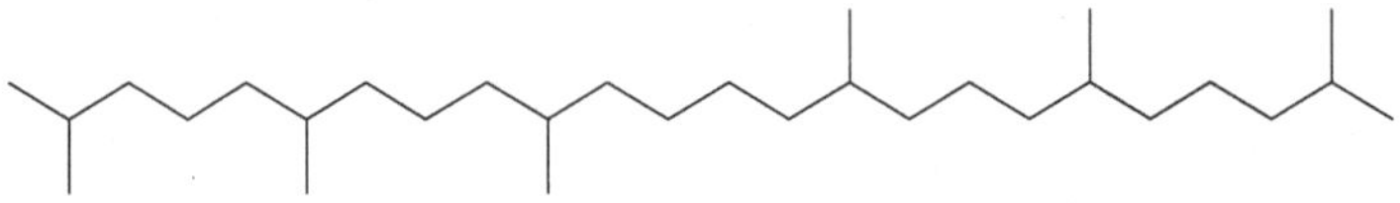

Source: Adapted from Karis, T.E., Lubricants for the disk drive industry, in: Rudnick, L.R., ed., *Synthetic, Mineral Oil, and Bio-Based Lubricants Chemistry and Technology*, CRC Press, Taylor & Francis, Boca Raton, FL, 2006, pp. 623–654.

Note: PRS is pristane, SQL is squalane, DPP is diphenyl propane, and PAO is poly-α-olefin.

and rotational contributions. The slope and intercept of the $\ln(\eta)$ versus $1/T$ plot are employed to calculate ΔS_{vis} and ΔE_{vis} for each of the oils. These thermodynamic properties for flow are listed in Table 28.21.

A similar analysis was performed for the vapor pressure of these oils. According to the Clapeyron equation (28.15), a plot of $\ln(P^0)$ versus $1/T$ is a straight line with slope $-\Delta E_{vap}/R$ and intercept $\ln(P) - 1 + \Delta S_{vap}/R$. The vapor pressure of several low-molecular-weight ester and hydrocarbon oils is shown plotted as $\ln(P^0)$ versus $1/T$ in Figure 28.38. The vaporization entropy $\Delta S_{vap} = \Delta S_{vap}^{trans} + \Delta S_{vap}^{rot}$ is approximately the sum of the translational and rotational contributions. The intercept of the $\ln(P^0)$ versus $1/T$ plot is then employed to calculate ΔS_{vap} and ΔE_{vap} for each of the oils. The translational component of the vaporization entropy is approximately equal to the vapor phase entropy given by Equation 28.16, and the rotational component is similarly calculated in the vapor phase with Equation 28.17 [43]. The vaporization entropy and the calculated vapor-phase translational and rotational entropy are employed to estimate the liquid-phase entropy as $S_{liq} \approx S_{vap} - \Delta S_{vap}$. These thermodynamic flow properties for vaporization are listed in Table 28.21.

Further insight into the relationship between the evaporation and flow properties of the fluid bearing motor base oil candidates is obtained by combining the results together within the framework of thermodynamics and the reaction rate model for evaporation and flow [89,90]. The ratio of the vaporization–activation energy to the flow-activation energy $n = \Delta E_{vap}/\Delta E_{vis}$ measures the partial decoupling of intermolecular forces between molecules during flow relative to the complete decoupling that takes place upon vaporization [91]. Another thermodynamic quantity that plays a key role in the flow process is the flow-activation rotational entropy. The flow-activation rotational entropy, ΔS_{vis}^{rot},

was calculated from the combined vapor pressure and viscosity versus temperature in Karis and Nagaraj [43]. If there is a way to lower the viscosity without increasing the vapor pressure, it seems that it can only be done by increasing the flow-activation entropy, ΔS_{vis}. For the oils shown here, ΔS_{vis} is proportional to ΔS_{vis}^{rot}, as shown in Figure 28.39a. However, ΔS_{vis}^{rot} decreases with n, as shown in Figure 28.39b. Thus, the flow-activation rotational entropy decreases with the amount of decoupling or asymmetry between vaporization and flow. For example, spherical molecules are more freely rotating in the flow-activated state, while for rodlike molecules rotation becomes less likely in the flow-activated state, as they surmount the energy barrier between adjacent molecules in the surrounding liquid.

Although the preceding discussion shows how the thermodynamic properties of vaporization and flow govern the oil vapor pressure and viscosity, little can be said about specifically what molecular structure and compositions can be synthesized to provide an improved fluid bearing motor. Recent developments on molecular dynamic simulation promise to bridge this gap between molecular structure and thermodynamics. For example, Bair et al. have demonstrated good agreement between high shear viscosity measurements and predictions from a molecular dynamic simulation [92–94]. Yet, the model requires input of a particular molecular structure, in their case, squalane, and numerically intensive computations are required to predict the viscosity. Even more complexity is involved if one tries to similarly predict the viscosity of more complex polar molecules such as diesters.

Oil blends, or viscosity-reducing additives, have also been considered. Blending oils changes the pressure–viscosity coefficient [76,95] and provides an intermediate viscosity [96–98]. The vaporization–activation energy and entropy and

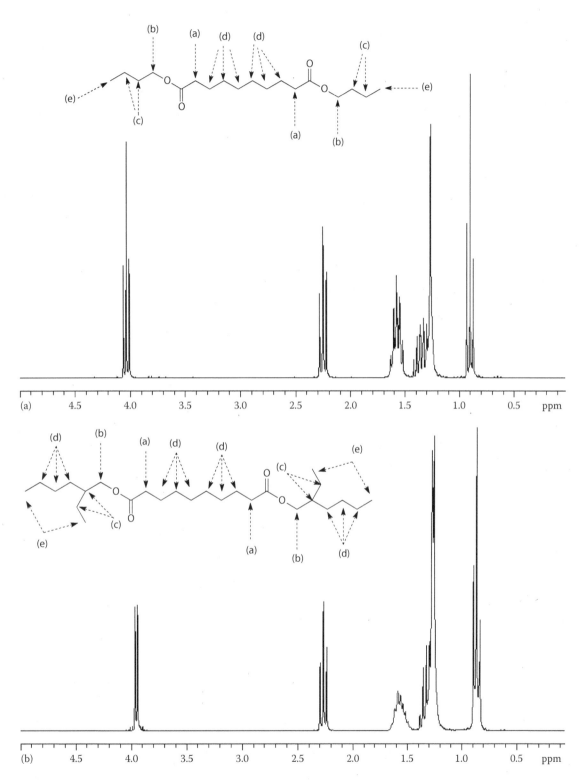

FIGURE 28.36 Proton NMR spectrum of diester fluid dynamic bearing oil: (a) DBS and (b) DOS. The peak assignments are indicated on the molecular structures. (Adapted from Karis, T.E., Lubricants for the disk drive industry, in: Rudnick, L., ed., *Lubricant Additives: Chemistry and Applications*, 2nd edn., CRC Press, Taylor & Francis, Boca Raton, FL, 2009, pp. 523–584.)

TABLE 28.20

Proton NMR Chemical Shifts and Peak Assignments for the Spectra in Figure 28.36

Proton	Chemical Shift (ppm)	Peak Area for DBS	
		Measured	Expected
(a)*	4.01	4	4
(b)	2.23	4.2	4
(c)	1.57	8.4	8
(d)	1.36	12.4	12
(e)	0.91	6.1	6

Proton	Chemical Shift (ppm)	Peak Area for DOS	
		Measured	Expected
(a)*	3.97	4	4
(b)	2.27	4.1	4
(c)	1.59	6.6	6
(d)	1.28	24.6	24
(e)	0.86	12.1	12

Source: Adapted from Karis, T.E., Lubricants for the disk drive industry, in: Rudnick, L., ed., *Synthetics, Mineral Oil, and Bio-Based Lubricants Chemistry and Technology*, 2nd edn., CRC Press, Taylor & Francis, Boca Raton, FL, 2013, pp. 657–698.

* indicates the reference peak used as the basis for the other peak areas.

flow-activation energy and entropy are shown for a blend of polar diester oil in nonpolar PAO oil in Figure 28.40. In this case, the vaporization–activation energy of the blend was less than that of the pure components. The vaporization–activation entropy of the blend was less than that of the pure components. The reduction in the vaporization–activation energy outweighed the effect of the increase in liquid entropy, and the vapor pressure of the blend was higher than that of the pure components. There is, however, an interesting dip in the properties near 95% DOS, which was reproducible. The activation energies of the blends are shown in Figure 28.40a and c, and the activation entropies of the blends are shown in Figure 28.40b and d. The flow-activation energy is ideally given by a linear combination of the weight fractions of blend components. The blend viscosity is actually less than the linear combination due to an increase in the flow-activation entropy near 50%. Taking the analysis one step further, the flow-activation rotational entropy has a sharp maximum near 95% DOS in Figure 28.41a. The ratio $n = \Delta E_{vap}/\Delta E_{vis}$ has a sharp minimum near 95% DOS, Figure 28.41b. This unusual behavior near 95% DOS suggests that synergistic effects are possible in oil blends. These probably arise from clustering of the nonpolar oil with the aliphatic chains of the diester. Perhaps this type of effect could be exploited to develop viscosity-reducing additives, which have a lower vapor pressure than the base oil.

However, for a blend to be useful, it would also need to be capable of forming an azeotrope and then it could only be used in motors at the azeotropic composition. Otherwise,

the evaporation of the oil component with the highest vapor pressure would gradually change the oil blend viscosity and vapor pressure with time.

28.3.3.2 Formulation Chemistry

Thermal stability is imparted to grease and oil by formulation with additives. Over two decades, in our lab we employed the principles of oil oxidation chemistry in developing formulations for ball bearing grease [99] and fluid dynamic bearing [100] spindle motors at the leading edge of the industry. The general outline of the chemical mechanisms for oxidation and stabilization, potential additives, and accelerated test results is published in Reference 86. This section presents the low-temperature elementary reactions for each type of oxidation pathway including inhibition by antioxidants. The rate equations are solved numerically in nondimensional form, and the model results are compared with thermal oxidation life test data to illustrate the synergistic effects of PriAOX and SecAOX and metal catalysis.

The following sets of elementary reactions employed to model the data are derived from schemes in References 86 and 101. Intrachain proton abstraction and self-termination are not included. Some of the non-rate-controlling reactants and products are not shown. The initiation reactions are shown in Figure 28.42a, and the propagation reactions are shown in Figure 28.42b. The dot denotes a carbon radical. Only interchain proton abstraction (no cleavage reaction) is considered because this is the dominant reaction pathway that we observed with ester oils. Proton abstraction from RH by thermal or mechanical excitation is the rate-limiting step in the initiation mechanism. Decomposition of the hydroperoxide ROOH is the rate-limiting step in the propagation mechanism.

PriAOX protonates radicals, as shown in Figure 28.43a. Examples of PriAOX are butylated hydroxytoluene, dioctyldiphenyl amine (DDA), and phenyl naphthylamine (PNA). SecAOX decomposes hydroperoxide, as shown in Figure 28.43b. Examples of SecAOX are zinc dialkyldithiophosphate and ZDTC. The products of hydroperoxide decomposition by these SecAOXs form an antiwear film on metal surfaces [102]. Dissolved catalytic metal ions M^n and M^{n+1} decompose hydroperoxide in a charge shuttle reaction [103] as shown in Figure 28.43c. The superscripts refer to the valence state of the metal ion. The rate constants in the elementary reactions were estimated by approximating isothermal oil oxidation life test results on a pentaerythritol tetraester oil [101] (Figure 28.44a). The values of the model rate constants are given in Table 28.22. Unless stated otherwise, model calculations and isothermal measurements were performed at 150°C. Since the activation energies for the individual elementary reactions are unknown, the activation energies for nonmetal-catalyzed reactions were approximated by the overall activation energy of 120 kJ/mol determined from the measured lifetimes of oil with and without antioxidants. The activation energy for metal-catalyzed elementary reactions was set to 10 kJ/mol in order to fit the overall activation energy of 70 kJ/mol determined from the

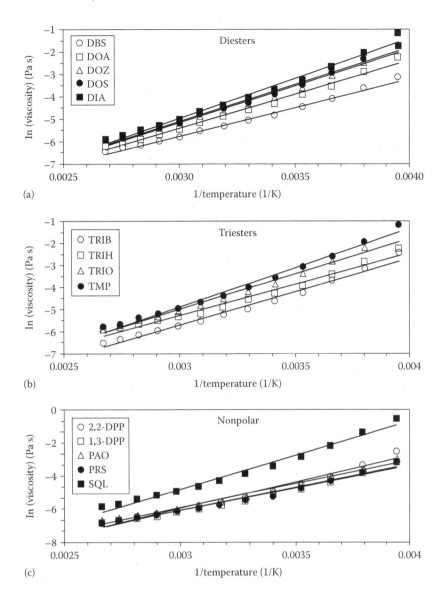

FIGURE 28.37 Natural logarithm of viscosity as a function of inverse temperature: (a) diesters, (b) triesters, and (c) hydrocarbon oils. The line is a fit to the Eyring equation. (Adapted from Karis, T.E., Lubricants for the disk drive industry, in: Rudnick, L.R., ed., *Synthetic, Mineral Oil, and Bio-Based Lubricants Chemistry and Technology*, CRC Press, Taylor & Francis, Boca Raton, FL, 2006, pp. 623–654.)

measured lifetimes of oil containing antioxidant and catalytic metal ions.

The model calculated oil lifetime is taken to be the peak in the ROOH concentration (Figure 28.45), which corresponds to the sudden consumption of oil molecules RH shown in Figure 28.45a. Note that the radical products actually continue on to further reactions, which are not included in the present model. Low-molecular-weight products and water evaporate at elevated temperature. Hydroxyl groups form hydrogen bonding that increases the oil viscosity.

The model calculated oil lifetime curves from the previously mentioned rate constants and activation energies are shown in Figure 28.44b. The slopes of the curves with pure oil and oil containing antioxidants are in reasonable agreement with the experimental data in Figure 28.44a. A more complicated temperature dependence is predicted for the oil

containing PriAOX and SecAOX and metal catalyst. The increase in the hydroperoxide decomposition rate by the metal catalyst [103] is apparently more significant at lower temperatures [104].

The model calculated depletion of PriAOX and SecAOX is shown in Figure 28.45b. The end of life for the oil corresponds to the simultaneous depletion of both the PriAOX and SecAOX. This suggests that oil life could be extended by replenishing the antioxidant prior to the end of the oil lifetime. The qualitative ability of the model to match the experimental effect of increasing metal ion concentration on the PriAOX depletion with time in oil containing PriAOX and SecAOX is shown in Figure 28.46. Dissolved catalytic metal ions decrease the oil life by increasing the rate of antioxidant consumption, hence decreasing the induction period of time until the onset of rapid oil oxidation.

TABLE 28.21

Evaporation and Flow Thermodynamic Properties Calculated from Vapor Pressure and Viscosity Data for Model Fluid Dynamic Bearing Oils

Oil	Evaporation			Flow			
	ΔE_{vap} (kJ/mol)	ΔS_{vap} (J/mol K)	S_{liq} (J/mol K)	ΔE_{vis} (kJ/mol)	ΔS_{vis} (J/mol K)	ΔS_{vis}^{rot} (J/mol K)	n
DBS	90.7	160	114	20.6	−4.2	−3.1	4.4
DOA	94.4	160	117	24.7	4.1	6.61	3.8
DOZ	84.5	125	154	26.7	6.8	19.8	3.2
DOS	108.6	175	105	26.7	5.8	2.6	4.1
DIA	94.1	143	137	28.9	11.2	18.1	3.3
TRIB	81.8	154	163	25.1	10.5	12.5	3.3
TRIH	93.3	156	167	23.2	−1.5	2.4	4.0
TRIO	116.3	185	143	26.2	3.5	−1.4	4.4
TMP	81.0	110	218	28.9	10.4	30.3	2.8
2,2-DPP	59.1	120	144	25.5	17.0	30.4	2.3
1,3-DPP	74.2	159	105	22.3	8.4	6.6	3.4
PAO	89.6	188	80	22.7	4.5	−4.2	3.9
PRS	78.5	166	104	22.8	4.6	2.6	3.5
SQL	82.8	124	156	32.9	21.2	36.6	2.5

Source: Adapted from Karis, T.E., Lubricants for the disk drive industry, in: Rudnick, L.R., ed., *Synthetic, Mineral Oil, and Bio-Based Lubricants Chemistry and Technology*, CRC Press, Taylor & Francis, Boca Raton, FL, 2006, pp. 623–654. Evaporation properties are referenced to 100°C and atmospheric pressure (100 kPa).

Since PriAOX and SecAOX interact with the complex oxidation mechanism in distinctly different ways, and depending on the values of the elementary reaction rate constants k_{pi} and k_s, there may be synergistic effects between the two. For example, the maximum inhibition of oxidation could be achieved when the peroxy radicals react with PriAOX to form hydroperoxide (Figure 28.43a) before they can abstract a proton (Figure 28.42a), and the hydroperoxide is decomposed to inert products by SecAOX (Figure 28.43b) before the hydroperoxide decomposes into alkoxy and hydroxy radicals (Figure 28.42b). The synergistic effect of PriAOX and SecAOX in oil containing ball bearings is shown in Figure 28.47a. The optimum mole ratio of primary/(primary + secondary) antioxidant for maximum oil oxidation lifetime is close to 0.2, or when 20 mol% of the antioxidant is primary and 80 mol% of the antioxidant is secondary. The model calculation for the effect of the ratio on oil life is shown in Figure 28.47b. There is a maximum (synergistic) oil life predicted both with and without catalytic metals in the oil. The optimum ratio is closer to 50 mol% of PriAOX and SecAOX. The presence of catalytic metal ions diminishes the benefit of the synergistic effect. Further refinement of the individual rate constants and possibly incorporation of more detailed elementary reactions could improve the model fit to the measured oil life.

The kinetic model also qualitatively predicts the oxygen consumption during oil oxidation, the effects of antioxidant concentration, oxygen pressure on induction time in isothermal pressure differential scanning calorimetry, exothermal peak temperature in nonisothermal pressure differential scanning calorimetry (NIPDSC), and the effect of heating rate on the exotherm temperature.

28.3.3.3 Formulation Example

This section illustrates the development of oil formulations for magnetic recording disk drive fluid dynamic bearing spindle motors. In this case, the sleeve is made of steel and the shaft is made from bronze. Carboxylic acid products of chain cleavage by intrachain proton abstraction dissolve catalytic copper ions into the oil [86]. Once catalytic metal ions are dissolved in the oil, there is rapid consumption of antioxidants due to hydroperoxide decomposition (Figure 28.43c), followed by oil loss through evaporation of low-molecular-weight cleavage products. Therefore, the fluid bearing motor oil was formulated with corrosion inhibitors.

Triazole corrosion inhibitors were selected for their compatibility with the diester type of oil used in these bearings [105,106] and with the magnetic recording head disk interface [107]. Two types of triazole are shown in Table 28.23. Tolyltriazoles (TTAs) were more easily dissolved into the oil and had a lower evaporation rate from the oil than benzotriazole (BTA).

An initial set of oil formulations was prepared and tested in continuously stirred oil at 150°C with the oil loss by evaporation prevented with a reflux condenser due to the long time duration for some of the tests. Some of the tests included steel in the form of ball bearings and/or leaded bronze powder to simulate the motor components. The compositions and test results are listed in Table 28.24. The antioxidant lifetime was measured by thin-layer chromatography, and the oil lifetime was the time at which significant peak broadening was observed in the proton NMR of the oil. In the absence of bronze powder, the oil lifetime was more than 2500 h. Addition of the bronze powder shortened the oil lifetime to 780 h. The TTA increased the antioxidant and oil

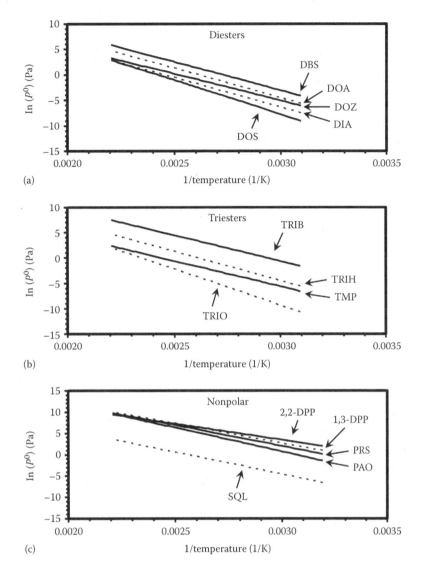

FIGURE 28.38 Natural logarithm of vapor pressure as a function of inverse temperature: (a) diesters, (b) triesters, and (c) hydrocarbon oils. The line is a fit to the Clapeyron equation. Ambient pressure $P = 0.1$ MPa. (Adapted from Karis, T.E., Lubricants for the disk drive industry, in: Rudnick, L.R., ed., *Synthetic, Mineral Oil, and Bio-Based Lubricants Chemistry and Technology*, CRC Press, Taylor & Francis, Boca Raton, FL, 2006, pp. 623–654.)

lifetime more and was more effective at low concentration, relative to the BTA corrosion inhibitor.

Another series of fluid dynamic bearing oil formulations was prepared in a lower-viscosity oil. The composition of these oils is given in Table 28.25. The key performance attribute for each of these formulations is listed in the comment column. The rotating bomb oxidation test (RBOT) induction time and NIPDSC exotherm temperature for these oil formulations are shown in Figure 28.48. Note that the RBOT contains Cu metal, while the NIPDSC test is performed in a (noncatalytic) Al pan. The induction time and exotherm temperature both decreased as the antioxidant concentration was decreased from 2% to 0.5%, F21 to F22 (but this improved the cold start stability of some motor types, depending on the number of poles in the stator). Addition of 0.05 wt% 5-methyl benzotriazole (5MeBTA) did not significantly increase the NIPDSC exotherm temperature (no Cu) (Figure 28.48b), but 0.05 wt%

5MeBTA did significantly increase the RBOT induction time (Figure 28.39a), F23 and F24. This result demonstrates that the 5MeBTA inhibited dissolution of catalytic Cu into the oil in the RBOT test, while the 5MeBTA did not participate in the oxidation/inhibition chemistry (without Cu present in the NIPDSC test). The ZDTC in F25 is a boundary lubrication additive [108–110], which also acts as a SecAOX. From the chemical reaction model in the previous section, a synergistic effect is expected with mixtures of PriAOX and SecAOX. The ZDTC included in F25 increased the exotherm temperature in Figure 28.48b; however, the ZDTC slightly decreased the RBOT induction time in Figure 28.48a. It was found that within 30 days at 100°C, the ZDTC formed a sediment with the epoxide and/or the 5MeBTA and these compounds were precipitated from solution. As long as the ZDTC remains in solution during the shorter time duration of the NIPDSC test, it is effective as a SecAOX.

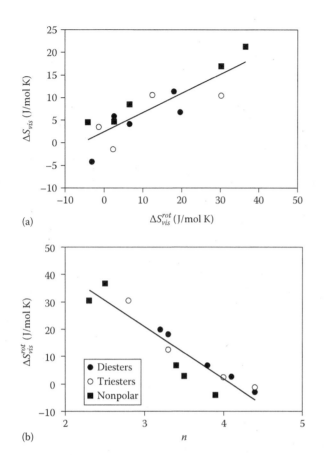

(a)

(b)

FIGURE 28.39 Flow-activation entropy versus flow-activation rotational entropy (a) and the flow-activation rotational entropy versus the ratio of the vaporization energy to the flow-activation energy (b) for the oils listed in Tables 28.16 through 28.18. (Adapted from Karis, T.E., Lubricants for the disk drive industry, in: Rudnick, L.R., ed., *Synthetic, Mineral Oil, and Bio-Based Lubricants Chemistry and Technology*, CRC Press, Taylor & Francis, Boca Raton, FL, 2006, pp. 623–654.)

Further analysis of the oil formulations was performed with the NIPDSC at heating rates of 1°C/min, 5°C/min, 10°C/min, and 20°C/min. According to the first-order model, the heat flow rate

$$\dot{Q} \propto -\frac{d[RH]}{dt} = Z[RH]\exp\left(-\frac{E_a}{RT}\right), \quad (28.24)$$

where

[RH] is the alkyl proton concentration
R is the gas constant
T is absolute temperature [111–113]

In the NIPDSC test with a linear temperature increase rate β,

$$\ln\left(\frac{\beta}{T_p^2}\right) = \ln\left(\frac{ZR}{E_a}\right) - \frac{E_a}{RT_p}. \quad (28.25)$$

A plot of the data reduced according to Equation 28.25 fits the first-order model, as shown in Figure 28.49, and provides the frequency factor Z and activation energy E_a for each oil sample. The lines in Figure 28.49 are from a linear regression fit to the data points. (*Note*: the lower right data point for F00 was excluded from the regression.) The activation energies are plotted in Figure 28.50a and frequency factors are shown in Figure 28.50b. In terms of the linear regression, the higher slope means a higher activation energy, which represents the effective activation energy for the system of oxidation reactions. The frequency factor shifts the curves parallel to one another at constant activation energy and is also referred to as the attempt frequency of the reaction. Overall, the frequency factor is correlated with the activation energy for this set of formulations (Figure 28.51). This is surprising, because they are generally considered to be independent quantities. The base oil formulation had the lowest Z and E_a. The conductivity additive in F20 had a similar effect on Z and E_a as the antioxidant in F21 and F22. Addition of the 5MeBTA in F23 and F24 slightly decreased Z and E_a. Incorporation of the SecAOX ZDTC in F25 greatly increased Z and E_a. For correlation between Z and E_a shown in Figure 28.51, the linear regression fit provides $\log_{10}(Z) = -0.4664 + 0.0966 \times E_a$ with $R^2 = 0.995$, which implies that the attempt frequency increases with the activation energy. This could be the case if there is an association between the additive molecules and the radical species in the oil. The large increase in both Z and E_a for the small amount of ZDTC in F25 further suggests that it may also associate within this complex, which inhibits the explosive chain propagation reaction from taking place until much higher temperatures because of the shorter distance over which diffusion is required for the PriAOX/SecAOX to quench a radical. We have seen qualitative evidence for such association among mixtures of PriAOX and SecAOX as shifts in the UV/Vis absorption frequency. In the severe case, the association leads to flocculation and sedimentation of these formulation components from solution.

In summary of the formulation test results on the fluid dynamic bearing oil formulations bench level testing, corrosion inhibitor improved the thermal stability of the oil in the presence of bronze powder (Table 28.24). Another set of formulations was prepared in a lower-viscosity base oil and tested with RBOT and NIPDSC. The most stable oil in the RBOT was F24, which contained the highest amount of antioxidant and corrosion inhibitor. The most stable oil in the NIPDSC test was F25, which was F24 with an added 0.5% of ZDTC. The original F25 suffered from sedimentation. A subset of F25 was optimized to eliminate sedimentation by including only the oligomeric antioxidant (2%) and ZDTC (0.5%). This formulation was compared with F24 on the high-frequency reciprocating rig (HFRR) for 1, 2, 4, and 6 h with two repetitions (PCS Instruments, London, UK). The test parameters were 10 Hz, 1 kg, 2 mm stroke, and 85°C with AISI 52100 steel, contact pressure 1.4 GPa, and Hertz contact diameter 116 μm. The HFRR test found that F24 had a

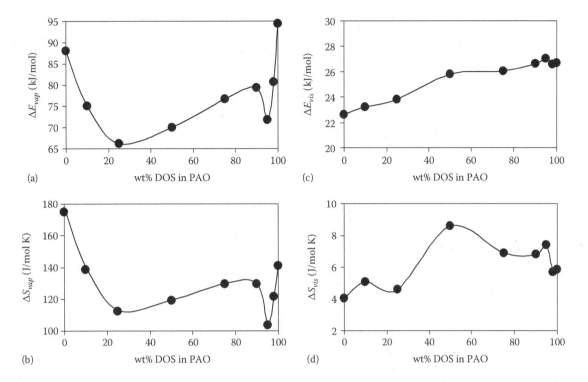

FIGURE 28.40 Vaporization–activation energy and entropy (a and b) and flow-activation energy and entropy (c and d) for blends of DOS in PAO. (Adapted from Karis, T.E., Lubricants for the disk drive industry, in: Rudnick, L.R., ed., *Synthetic, Mineral Oil, and Bio-Based Lubricants Chemistry and Technology*, CRC Press, Taylor & Francis, Boca Raton, FL, 2006, pp. 623–654.)

stable friction coefficient (0.14), developed 100% electrical contact resistance (ECR) within 1.5 h, and had a constant small wear scar cross section of about 800 μm² while the optimized F25 friction coefficient increased up to nearly 0.18, the ECR remained near 0, and the wear scar section increased up to 6000 μm². Apparently, the ZDTC chemistry aggravates the wear in this ester oil rather than forming an organo-zinc boundary lubrication film at the interface.

Several of the formulations were selected for accelerated component life testing in fluid dynamic bearing motors. The relative mean time before failure (MTBF) is shown in Figure 28.52. The MTBF was increased by 2× on going from F22 (0.5 wt% oligomeric antioxidant) to F21 (2 wt% oligomeric antioxidant). The MTBF was increased by more than 3× between F22 and F23 (F22 + 0.05% corrosion inhibitor). These results show that while the antioxidant prolongs the motor lifetime, the dominant contribution to improved motor life is through the corrosion inhibitor preventing dissolution of catalytic copper ions from the bronze shaft into the oil.

28.3.4 CHARGE CONTROL

In a rigid magnetic recording disk drive, there is an electrical potential difference between the recording head and the disk due in part to the work function difference between the slider body ceramic and the metallic magnetic recording layers or AlMg disk substrate. The work function is modified by the dielectric layers of carbon overcoating the slider and the disk and the dipolar layer of the disk lubricant that has polar end groups for chemisorption to the overcoat. Another contribution to the potential difference across the magnetic recording head disk interface is the fluid bearing motor. Charge separation occurs in the motor due to work function differences between the bearing metals and the oil coupled with streaming charge separation due to the relative velocity of the rotor and the stator. Fluid bearing motor oil conductivity additives are reviewed in Karis [64]. The sum of the component contributions to the interface potential difference typically adds up to approximately ±100 to ±500 mV and could be as high as ±1000 mV.

The presence of this electrical potential difference across the interface contributes to lubricant transfer from the disk to the slider [60,114], wear of the carbon overcoat on the disk and slider, and spacing gap reduction by the electrostatic force between the disk and the slider. Lubricant transfer and wear also cause the spacing gap to change, resulting in a change in the amplitude of the magnetic flux at the magnetic recording head. Over time, wear and accumulation can lead to performance degradation. The magnetic recording industry is implementing methods for measuring and controlling interface voltage by application of a slider bias to offset the inherent potential difference, for example, see Wang et al. (2016) [115].

Since charge separation in the fluid bearing motor oil contributes to the streaming potential generated across the bearing gap, any change in the ionic content or other charge transport

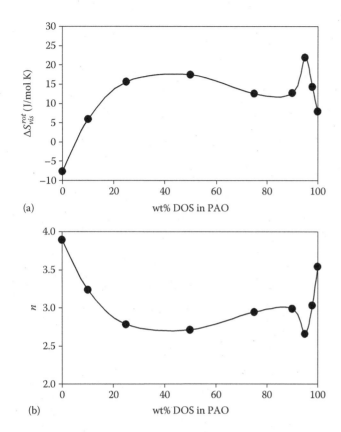

FIGURE 28.41 The rotational component of the flow-activation entropy (a) and the ratio of the vaporization–activation energy to the flow-activation energy, n, (b) for blends of DOS in PAO. (Adapted from Karis, T.E., Lubricants for the disk drive industry, in: Rudnick, L.R., ed., *Synthetic, Mineral Oil, and Bio-Based Lubricants Chemistry and Technology*, CRC Press, Taylor & Francis, Boca Raton, FL, 2006, pp. 623–654.)

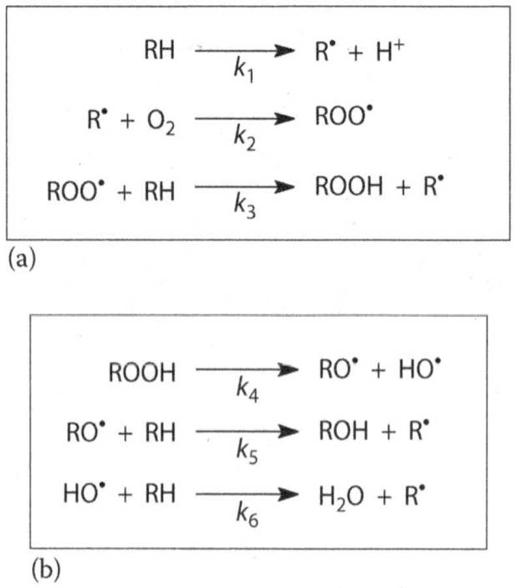

FIGURE 28.42 Oil oxidation initiation (a) and propagation (b) elementary reactions. RH stands for the hydrocarbon oil. H is a proton and R is the rest of the molecule. (Adapted from Karis, T.E., Lubricants for the disk drive industry, in: Rudnick, L.R., ed., *Synthetics, Mineral Oil, and Bio-Based Lubricants Chemistry and Technology*, 2nd edn., CRC Press, Taylor & Francis, Boca Raton, FL, 2013, pp. 657–698.)

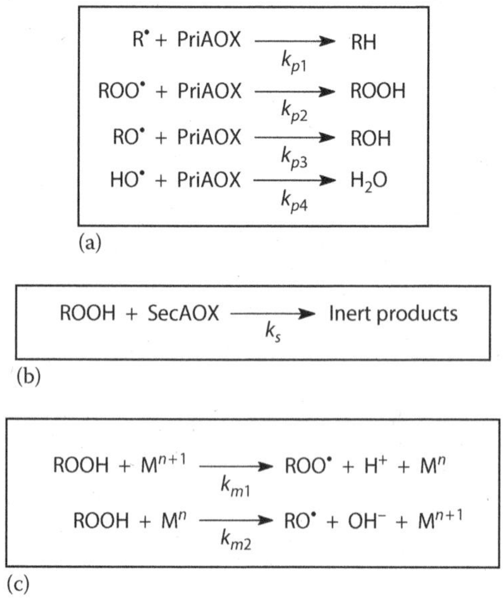

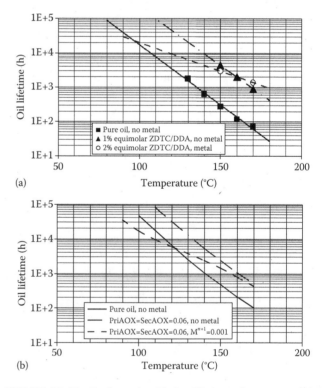

FIGURE 28.43 Oxidation inhibition by PriAOX (a), SecAOX (b), and metal M catalysis of hydroperoxide decomposition (c). (Adapted from Karis, T.E., Lubricants for the disk drive industry, in: Rudnick, L.R., ed., *Synthetics, Mineral Oil, and Bio-Based Lubricants Chemistry and Technology*, 2nd edn., CRC Press, Taylor & Francis, Boca Raton, FL, 2013, pp. 657–698.)

FIGURE 28.44 Isothermal oxidation lifetime of a pentaerythritol tetraester oil and oil containing equimolar amounts of PriAOX and SecAOX and metal catalyst (included as ball bearings) (a) and model calculated oil oxidation lifetime for equimolar PriAOX and SecAOX concentrations showing the qualitative effect of metal catalyst (b). (Adapted from Karis, T.E., Lubricants for the disk drive industry, in: Rudnick, L.R., ed., *Synthetics, Mineral Oil, and Bio-Based Lubricants Chemistry and Technology*, 2nd edn., CRC Press, Taylor & Francis, Boca Raton, FL, 2013, pp. 657–698.)

TABLE 28.22
Elementary Reaction Rate Constants for the Reactions in Figures 28.33 and 28.34

Rate Constant	Value
k_1	1E–10
k_2	5
k_3	1
k_4	0.05
k_5	1
k_6	5
kp_1	10
kp_2	1
kp_3	1
kp_4	1
ks	0.3
k_{m_1}	10
k_{m_2}	10

Source: Adapted from Karis, T.E., Lubricants for the disk drive industry, in: Rudnick, L., ed., *Synthetics, Mineral Oil, and Bio-Based Lubricants Chemistry and Technology*, 2nd edn., CRC Press, Taylor & Francis, Boca Raton, FL, 2013, pp. 657–698.

The units are nondimensional based on initial nondimensional oil proton concentration of RH = 1 and nondimensional dissolved oxygen concentration of 0.5. The rate constants were determined to provide qualitative agreement with oil oxidation life in hours at 150°C.

properties of the oil causes a change in the interface voltage over time. As antioxidant reacts to quench free radicals, it is decomposed into chemically altered fragments. For example, in one case the interface voltage changed from about −800 to −200 mV during 150 days on test. NMR analysis of the oil extracted from the tested motors found that the original 0.5% of a low-temperature hindered phenol antioxidant Octadecyl-(3,5-di-t-butyl-4-hydroxyphenyl)propionate was depleted from the oil. While the level of high-temperature antioxidant was unchanged after 150 days on test, it is expected that the high-temperature antioxidant comprising oligomers of N-(4-ocylphenyl)-1-naphthylamine and 4,4′-dioctyldiphenylamine should eventually deplete with motor run time. This antioxidant, also known as Vanlube 9317, provides an intermediate level of conductivity to the oil formulation. However, even oil containing Vanlube 9317 exhibits gradual changes over time due to the base oil oxidation and inhibition mechanism forming degradation products that remain dissolved in the oil. Figure 28.53 shows the relative permittivity and conductivity of Vanlube 9317 over a range of concentration in DOS oil versus aging time at 100°C in air. Both of these properties increase exponentially with time. The acceleration factor is approximately a factor of 2 for each 10°C, so that for a disk drive motor operating continuously at 50°C, 1500 h is 1.7 years.

The polyaromatic oligomers impart intermediate conductivity and charge storage to the oil, thus providing fluid bearing motors with a moderate voltage and a low peak discharge current [64]. Based on these results, it appeared that the polyaromatic oligomers provide charge control through a charge hopping mechanism, as opposed to bearing metal solubilization to transport ionic charge through the oil as with Stadis 450 [64]. The ionic mechanism can provide oil with higher conductivity, but the increased charge separation across the bearing gap leads to increased energy storage. The high conductivity and high charge separation combine to provide rapid charge dissipation referred to as peak discharge current when a contact forms across the rotor and the stator at the magnetic recording head disk interface. Thus, it is desirable to control the motor voltage with an engineered additive designed to be chemically stable with respect to oxidation while providing an electronic charge transport mechanism. Beyond the need for limiting flow electrification of ester oils in disk drive motors, there is also interest for improvement in the use of synthetic ester oils for electric power transformers [116,117].

28.3.4.1 Aryl Esters

One of the earliest observations of the electronic hopping (electron donor acceptor) conduction mechanism in lubricating oils is by Bronshtein et al. [80]. This was determined by comparing the conductivity versus temperature of sulfonate ionic additives with nonionic succinimide additives. Ionic mobility is limited by the viscous drag force on the ions, which decreases with decreasing viscosity, so that the product of the conductivity and the viscosity is relatively independent of temperature for the ionic conduction mechanism. On the other hand, electronic hopping conduction is independent of the base oil viscosity and increases more with temperature than the viscosity or ionic conductivity (higher activation energy). More recently, charge transport models from amorphous inorganic materials are being adapted to organic materials [118]. These models can be employed for engineering design of electron donor acceptor additives and to determine the required concentration to provide desired conductivity by treating the oil as a doped organic semiconductor.

One of the challenges to incorporating a polyaromatic charge control agent into an ester base oil is that aromatic compounds have limited solubility in the base oil. They also tend to crystallize and sediment due to their planer structure and strong intermolecular dispersion force. To mitigate this problem, polyaromatic compounds are solubilized by esterification with the organic diacid in the case of a diester such as DOS base oil, which uses sebacic acid. Experimental electronic hopping charge control agents for use in ester oils shown listed in Table 28.26 were synthesized in our lab for evaluation [119].

Aryl esters shown in Table 28.26 were synthesized with diacid core molecules sebacic acid and adipic acid. The symmetric aryl esters, with the same group attached to both ends of the diacid, dibenzyl sebacate, diphenethyl sebacate, and dinaphthyl sebacate, are solid at room temperature and are useful as soluble charge control additives. The asymmetric aryl esters (benzyl, 2-ethylhexyl) sebacate, (phenethyl, 2-ethylhexyl) sebacate, and (phenethyl, 2-ethylhexyl) adipate are liquids at room temperature. The asymmetric aryl esters

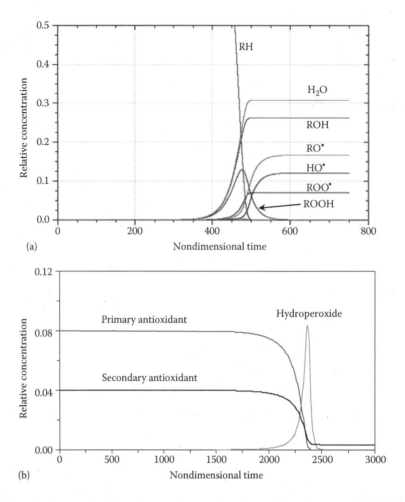

(a)

(b)

FIGURE 28.45 Oil oxidation products and reaction intermediates during the model isothermal oxidation of pure base oil (a) and the change of PriAOX and SecAOX and the hydroperoxide concentration peak with time for (PriAOX + SecAOX) = 0.12 and PriAOX/(PriAOX + SecAOX) = 0.67 (b). (Adapted from Karis, T.E., Lubricants for the disk drive industry, in: Rudnick, L.R., ed., *Synthetics, Mineral Oil, and Bio-Based Lubricants Chemistry and Technology*, 2nd edn., CRC Press, Taylor & Francis, Boca Raton, FL, 2013, pp. 657–698.)

may be useful as a conductive fluid lubricant without the necessity to dissolve them in a base oil. All of the aryl esters were completely soluble in the base oil DOS at least up to 5 wt%.

The electrical properties of the aryl esters and formulated oils were measured by saturating a 1 in. disk of filter paper (0.18 mm thick) with the liquid. A parallel plate test cell was formed by clamping the oil saturated filter paper between two mirror-polished SUS304 stainless steel disks (as in Section 28.3.1.3). The test cell was attached to an electrometer. The resistance of the cell was measured with a source voltage between 0.1 and 4 V. The charge flow into the cells after discharge by grounding the plates together was measured for 2 min after the application of a 1 V source. The conductivity was calculated from the resistance using the filter paper thickness and the nominal area corrected for porosity of the filter paper. The relative permittivity was calculated from a curve fit to derive the RC time constant. The effective area was determined by matching the relative permittivity of the base oil (DOS; Table 28.27, row 4, column 5).

Figure 28.54 shows the conductivity versus source voltage and charge versus time for several of the aryl-modified charge control oils compared with the pure base oil and two types of oils

that were tested in fluid bearing motors for disk drives, DOS oil in Table 28.17 and neopentyl glycol (NPG) oil, neopentyl glycol dicaprate (decanoic acid,1,1′-(2,2-dimethyl-1,3-propanediyl) ester, CAS 27841-06-1). The 0.5% Stadis 450 (Octel Starreon L.L.C., 8375 S. Willow St., Littleton, CO) in NPG base oil was highly conductive, but the accumulated charge was so high that the stored electrical energy is capable of damaging the magnetic recording interface. The conductivity of the NPG base oil is so low that a high voltage is built up across the bearing, which degrades the reliability of the head disk interface. The combination of 2% Vanlube 9317 (50 wt% reaction mixture of DDA and PNA in a tetraester oil carrier, R.T. Vanderbilt Corp.) [120] in NPG is known to provide both sufficient conductivity and acceptably low permeability for fluid bearing motors in magnetic recording disk drives. However, Vanlube 9317 is an antioxidant and is subject to chemical reaction with carbon radicals and oxidation products over time and thus is an unstable species to rely on for long-term stability of charge control and interface reliability.

The most promising chemically stable aryl charge control agent is dinaphthyl sebacate, which is shown at 1 and 5 wt% in DOS in Figure 28.54. The electrical properties of the 5 wt% formulation are sufficiently close to those of the 2% Vanlube

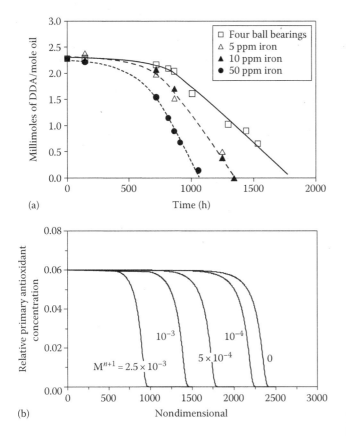

FIGURE 28.46 The effect of dissolved metal catalyst on PriAOX concentration during aging at 150°C in air measured by NMR spectroscopy with 1 wt% (ZDTC + DDA) in the ratio DDA/(ZDTC + DDA) = 0.57 in a pentaerythritol tetraester oil (a) and model calculated relative PriAOX concentration at various initial amounts of dissolved metal ion M^{n+1} where nondimensional PriAOX + SecAOX = 0.12 and the ratio PriAOX/(PriAOX + SecAOX) = 0.5. (Adapted from Karis, T.E., Lubricants for the disk drive industry, in: Rudnick, L.R., ed., *Synthetics, Mineral Oil, and Bio-Based Lubricants Chemistry and Technology*, 2nd edn., CRC Press, Taylor & Francis, Boca Raton, FL, 2013, pp. 657–698.)

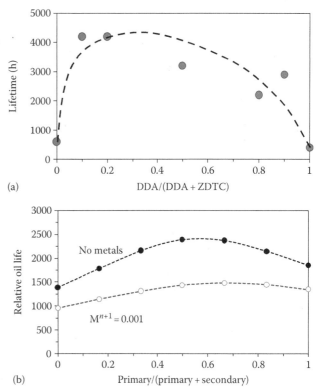

FIGURE 28.47 Oil oxidation lifetime as a function of the molar ratio of primary to primary + secondary antioxidant measured at 170°C in pentaerythritol tetraester oil containing ball bearings with total antioxidant amount 1 wt% (a) and calculated from the kinetic model with PriAOX + SecAOX = 0.12 with and without catalytic metals (b). (Adapted from Karis, T.E., Lubricants for the disk drive industry, in: Rudnick, L.R., ed., *Synthetics, Mineral Oil, and Bio-Based Lubricants Chemistry and Technology*, 2nd edn., CRC Press, Taylor & Francis, Boca Raton, FL, 2013, pp. 657–698.)

9317 in NPG. Therefore, the 5 wt% dinaphthyl sebacate can provide a more stable voltage and reliable magnetic recording interface. Also shown in Figure 28.54 is the pure (benzyl, 2-ethylhexyl) sebacate. This asymmetric aryl/alkyl ester may be useful alone as a formulated base oil where a highly conductive bearing oil or grease is desirable in disk drives or other electromechanical devices.

The electrical properties of conductivity and relative permittivity derived from the data such as that shown in Figure 28.54 are listed in Table 28.27. The first three rows in Table 28.27 show the values for the typical fluid bearing base oil NPG, an oil formulated with an oligomeric amine antioxidant Vanlube 9317 that provides initially acceptable motor voltage and charge characteristics in disk drives, and a highly conductive experimental oil formulated with Stadis 450. The motor voltage on the oil made with Vanlube 9317 drifts over long periods of time as the antioxidant reacts with oil oxidation products and carbon radicals. The oil made with Stadis 450 has

TABLE 28.23

Bronze Corrosion Inhibitors Tested in the Fluid Dynamic Bearing Oil Formulation

Benzotriazole (BTA)

TTA (mixture of 4- and 5MeBTA)

Source: Adapted from Karis, T.E., Lubricants for the disk drive industry, in: Rudnick, L., ed., *Synthetics, Mineral Oil, and Bio-Based Lubricants Chemistry and Technology*, 2nd edn., CRC Press, Taylor & Francis, Boca Raton, FL, 2013, pp. 657–698.

TABLE 28.24

Antioxidant Life and Oil Life in the Accelerated Oil Oxidation Test [121] Showing the Effect on Metals and Corrosion Inhibitors

Metals	Inhibitor		Lifetime (h)	
	Type	wt%	Antioxidant	Oil
None	None	—	—	>2500
Steel	None	—	—	>2500
Bronze	None	—	—	780
Bronze + steel	None	—	—	910
Bronze + steel	BTA	0.10	250	930
Bronze + steel	BTA	1.00	250	>1200
Bronze + steel	TTA	0.10	360	1120
Bronze + steel	TTA	0.25	360	970

Source: Adapted from Karis, T.E., Lubricants for the disk drive industry, in: Rudnick, L., ed., *Synthetics, Mineral Oil, and Bio-Based Lubricants Chemistry and Technology*, 2nd edn., CRC Press, Taylor & Francis, Boca Raton, FL, 2013, pp. 657–698.

The base formulation was 2% R.T. Vanderbilt Vanlube 9317 (oligomers of octylated diphenyl amine and octylated PNA) and 1% tricresyl phosphate in DOS oil. The metals were steel ball bearings and/or leaded bronze powder.

a high relative permittivity, which leads to high capacitance of the oil bearing. Since the conductivity is also high, this oil allows for rapid discharge of excessive electric energy into the head disk interface. The oil made with Stadis 450 is unacceptable for use in magnetic recording disk drive motor bearings.

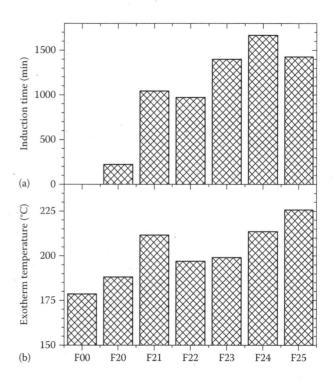

FIGURE 28.48 (a) RBOT induction time and (b) NIPDSC exotherm temperature for the oils listed in Table 28.25. RBOT test: 50 g oil, 5 mL water, polished Cu wire, 6.2 kg/cm^2 initial O$_2$ pressure, 100 rpm, at 150°C until O$_2$ pressure drops to 1.76 kg/cm^2 (JIS K 2514). NIPDSC test: 10–12 mg of oil in an open aluminum pan, O$_2$ pressure 3.4 MPa, heating rate 1°C/min. (Adapted from Karis, T.E., Lubricants for the disk drive industry, in: Rudnick, L.R., ed., *Synthetics, Mineral Oil, and Bio-Based Lubricants Chemistry and Technology*, 2nd edn., CRC Press, Taylor & Francis, Boca Raton, FL, 2013, pp. 657–698.)

TABLE 28.25

Composition of Oils Tested with the RBOT and the NIPDSC to Investigate the Effects of Corrosion Inhibitor and Secondary Antioxidant

Oil/Formulation	Conductivity Additive	Oligomers	5MeBTA	ZDTC	Comment
Base oil, F00	0	0	0	0	Base formulation
F20	1.2	0	0	0	High charging
F21	0	2	0	0	Oxidation stability, charge control
F22	0	0.5	0	0	Cold start, charge control
F23	0	0.5	0.05	0	Cold start, charge control, bronze
F24	0	2	0.05	0	Oxidation stability, charge control, bronze
F25	0	2	0.05	0.5	Oxidation stability, charge control, bronze, boundary lubrication

Source: Adapted from Karis, T.E., Lubricants for the disk drive industry, in: Rudnick, L.R., ed., *Synthetics, Mineral Oil, and Bio-Based Lubricants Chemistry and Technology*, 2nd edn., CRC Press, Taylor & Francis, Boca Raton, FL, 2013, pp. 657–698.

Concentrations are wt% of additive in the base formulation. The base oil formulation was NPG didecanoate with 0.5 wt% of a phenolic primary (low-temperature) antioxidant and 0.1 wt% of an epoxide. The oligomers were Vanlube 9317 (high-temperature PriAOX) and the ZDTC SecAOX was R.T. Vanderbilt Vanlube AZ [119]. One of the oils contained a mixed polymer conductivity additive DuPont Stadis 450. Values of the additives are given in wt%. See Table 28.23 for 5MeBTA.

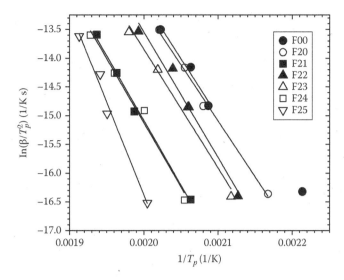

FIGURE 28.49 Reduced exothermic peak temperature plot according to Equation 28.25 from the NIPDSC test at various heating rates for the oils listed in Table 28.25. NIPDSC test: 10–12 mg of oil in an open aluminum pan, O_2 pressure 3.4 MPa, heating rates 1°C/min, 5°C/min, 10°C/min, or 20°C/min. (Adapted from Karis, T.E., Lubricants for the disk drive industry, in: Rudnick, L.R., ed., *Synthetics, Mineral Oil, and Bio-Based Lubricants Chemistry and Technology*, 2nd edn., CRC Press, Taylor & Francis, Boca Raton, FL, 2013, pp. 657–698.)

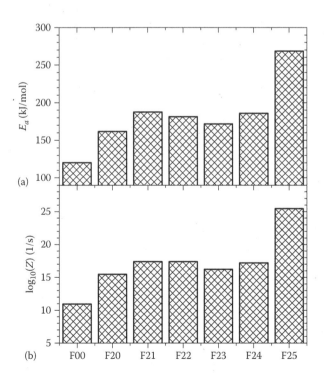

FIGURE 28.50 First-order model (a) activation energy and (b) frequency factor from the slope and intercept of the lines in Figure 28.49 for the oils listed in Table 28.25 with Equation 28.25. (Adapted from Karis, T.E., Lubricants for the disk drive industry, in: Rudnick, L.R., ed., *Synthetics, Mineral Oil, and Bio-Based Lubricants Chemistry and Technology*, 2nd edn., CRC Press, Taylor & Francis, Boca Raton, FL, 2013, pp. 657–698.)

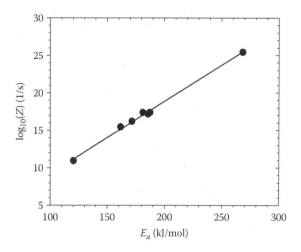

FIGURE 28.51 Correlation plot between the log of the frequency factor and the activation energy from the NIDPSC data measured for the oil formulations in Table 28.25. (Adapted from Karis, T.E., Lubricants for the disk drive industry, in: Rudnick, L.R., ed., *Synthetics, Mineral Oil, and Bio-Based Lubricants Chemistry and Technology*, 2nd edn., CRC Press, Taylor & Francis, Boca Raton, FL, 2013, pp. 657–698.)

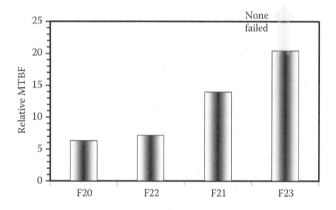

FIGURE 28.52 Relative MTBF from accelerated motor life test failure rates with several of the oil formulations listed in Table 28.25. (Adapted from Karis, T.E., Lubricants for the disk drive industry, in: Rudnick, L.R., ed., *Synthetics, Mineral Oil, and Bio-Based Lubricants Chemistry and Technology*, 2nd edn., CRC Press, Taylor & Francis, Boca Raton, FL, 2013, pp. 657–698.)

The next eight rows of Table 28.27 show the electrical properties of a DOS base oil with diaryl sebacate charge control additives. Diaryl sebacate charge control additives are solids at room temperature. They completely dissolve to form a clear solution in the base oil with mild heat and stirring in a Vortex mixer. The conductivity of dibenzyl sebacate solutions is not much improved over the DOS base oil. The conductivity of the 5 wt% diphenethyl sebacate is twice that of the DOS base oil. The conductivity of the 5% dinaphthyl sebacate is about eight times higher than that of the DOS base oil, while the permittivity is comparable to that of the 2 wt% Stadis 450 in NPG. This oil is an acceptable candidate for use in the magnetic recording disk drive motor bearing.

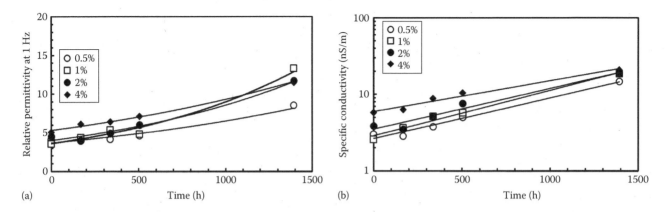

FIGURE 28.53 (a) Relative permittivity and (b) conductivity of Vanlube 9317 (wt%) in DOS oil as a function of aging time at 100°C in air. Shows the change in oil electrical properties due to oil oxidation product formation. Measured at ambient temperature 25°C.

The last three rows of Table 28.27 show the properties of the asymmetric aryl sebacates. These compounds are liquid at room temperature and appear to be slightly more viscous than the original DOS base oil. The conductivity of the (benzyl, 2-ethylhexyl) sebacate is 32 times higher than that of the DOS, while there was no measurable capacitance. The last two rows in Table 28.27 illustrate the effect of using a lower-molecular-weight diacid. Changing the core diacid from C8 to C6 increased the conductivity by 5× and the permittivity by only a factor of 2×. This is due to the increased volumetric density of the aryl group that acts as the charge transfer site. These asymmetric oils may be used in some types of bearings that permit a slightly higher viscosity and viscous power dissipation. A lower-viscosity asymmetric aryl ester that is liquid over the disk drive operating temperature may be obtained by optimization of the aryl group and acid molecular weight and structure.

TABLE 28.26
Molecular Structures of the Base Oil Di(2-ethylhexyl) Sebacate, Aryl Esters Synthesized on the Sebacate Core Molecule, and an Aryl Ester with a Smaller Adipate Core [119]

Dibenzyl sebacate

Diphenethyl sebacate

Dinaphthyl sebacate

(Benzyl, 2-ethylhexyl) sebacate

(Phenethyl, 2-ethylhexyl) sebacate

(Phenethyl, 2-ethylhexyl) adipate

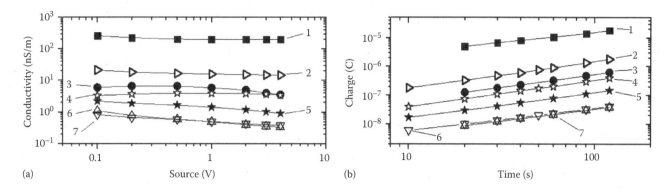

FIGURE 28.54 (a) Conductivity versus source voltage and (b) charge versus time with source at 1 V. 1, 0.5% Stadis 450 in NPG; 2, (benzyl, 2-ethylhexyl)sebacate; 3, 2% Vanlube 9317 in NPG; 4, 5% dinaphthyl sebacate in DOS; 5, 1% dinaphthyl sebacate in DOS; 6, NPG base oil; 7, DOS base oil [119]. Measured at ambient temperature 25°C.

TABLE 28.27

Conductivity and Relative Permittivity of Typical Fluid Bearing Base Oils, Formulated Oils, and Aryl Ester Charge Control Additives [119]

Additive Compound	wt%	Base Oil	Conductivity (nS/m)	Relative Permittivity
None	0	NPG	0.50	2.5
Vanlube 9317	2.0	NPG	5.9	73.0
Stadis 450	0.5	NPG	200	747.0
None	0	DOS	0.50	2.3
Dibenzyl sebacate	1.0	DOS	0.67	3.5
Dibenzyl sebacate	3.0	DOS	0.53	2.8
Dibenzyl sebacate	5.0	DOS	0.37	2.4
Diphenethyl sebacate	1.0	DOS	0.50	2.8
Diphenethyl sebacate	5.0	DOS	1.1	4.3
Dinaphthyl sebacate	1.0	DOS	1.4	15.6
Dinaphthyl sebacate	5.0	DOS	3.9	76.0
(Benzyl, 2-ethylhexyl) sebacate	100	None	16	—
(Phenethyl, 2-ethylhexyl) sebacate	100	None	14	349.0
(Phenethyl, 2-ethylhexyl) adipate	100	None	66	720.0

Base oil is either neopentyl glycol dicaprate (NPG) or di(2-ethylhexyl) sebacate (DOS). The conductivity in the table was measured at 1 V.

28.4 CONCLUSIONS AND FUTURE OUTLOOK

Magnetic recording disk drives employ the state-of-the-art lubrication science and technology to achieve unparalleled reliability. Continued growth in storage capacity relies more than ever before on fundamental understanding to provide key advancements in the critical areas of lubrication, nanolubrication of the head disk interface, and fluid bearing motor lubricants. The physical properties of the PFPE disk lubricants were described here. The bulk and thin film properties, in combination with polar end groups, provide many years of lubrication by virtue of viscosity enhancement and vapor pressure reduction of molecularly thin films. Advanced disk lubricants are being designed to fulfill increasingly stringent requirements of environmental variations [122–124] and low

friction at low-flying height [42]. Currently, advanced lubricants are being developed with a high level of chemisorptive attachment to the disk overcoat to prevent transfer of liquid lubricant to the slider and to enable reduced element spacing to the magnetic layer before the onset of adhesive interaction.

Measurement of lubricant transfer to the magnetic recording slider by dip coating with an oil formulation provides a rapid means for testing new disk lubricants, overcoats, and slider biasing technology for improved reliability of magnetic recording hard disk drives. While the oil coating method is practical, it is largely empirical and based on a poorly defined formulation. The physical chemistry that controls draining from micropatterned surfaces was reviewed. Surfactants were found to control the surface tension of the ester base oil, but more research is needed to determine the additional contribution of the surfactant molecular structure.

Grease-lubricated ball bearing spindle motors reached their pinnacle with 10,000 rpm drives. Over the course of developing sophisticated channeling greases for these high-performance, low-vibration, long-life bearings, much was learned about grease chemistry, physical properties, and the effects of contamination. Herein was described a comprehensive overview of grease yield stress and the effects of grease contamination on yield stress and electrochemical oxidation of an ester oil–based grease with a lithium soap thickener. The same principles apply to grease in general and therefore may potentially be useful in wide-ranging fields of application where smooth high-speed rotation is needed.

Spindle motors have evolved to fluid dynamic bearings, which provide much smoother rotation with comparable or better stiffness than ball bearings. Stiffness is provided by the relative motion between a shaft and a sleeve, one of which contains a pattern of grooves on its surface. The grooves pressurize a several-micron-thick film of oil. This chapter highlighted two key properties of the fluid bearing oil: viscosity and vapor pressure. Present fluid bearing oils have a low enough vapor pressure so that the evaporation loss of oil is small over the lifetime of the drive. However, there is a trade-off between the vapor pressure and the oil viscosity such that the lowest operating temperature of the drive is limited by the maximum power of the motor and the oil viscosity. Attempts to use lower-viscosity oils result in oils with a higher vapor pressure, and this appears to be a physical constraint of oils in general. Detailed thermodynamic analysis of the relationship between oil vapor pressure and viscosity was reviewed. The flow and vaporization–activation energies are linked to one another by interaction forces between the oil molecules. There is, however, in principle, some chance of reducing the viscosity without altering the activation energies if a molecular structure could be found, which has a large positive rotational component of the flow-activation entropy. Molecular structures examined have so far provided limited insight into how this may be accomplished. Another approach is to develop a blend of two different oils, combining mixtures of polar and nonpolar molecules. There is potential to develop an oil blend that forms an azeotrope and also has a lower viscosity and vapor pressure than pure component oils. Much fruitful work remains to be done here in the quest for a lower-viscosity and vapor pressure fluid bearing motor oil. The reward for success will be a reduction in spin-up time at cold temperature and a lower minimum operating temperature for drives with fluid bearing motors. Current work on advanced fluid bearing oil formulation is aimed at suppressing evaporation, aerosolization, and air ingestion at the air–oil interface meniscus, and cavitation in the oil film between the rotor and the stator.

Thermally stable oil and grease additive packages were developed with the assistance of a chemical kinetic model. The model provides interpretation of thermally accelerated oil oxidation life tests and predicts the interaction between PriAOX and SecAOX and dissolved catalytic metal ions. Oil oxidation tests at low temperature take a long time, while certain tests at higher temperatures such as isothermal and nonisothermal pressure differential scanning calorimetry can be conducted in a short period of time. The kinetic model links the short-time high-temperature test results with the long-time low-temperature tests and enables the test data to be used to predict ideal oil oxidation life at the used temperature.

The kinetic model provides guidance for the rapid development of new oil formulations. An example was given of the oil formulation for high-speed fluid dynamic motor bearing oils employed in magnetic recording disk drives. Motor lifetime was significantly improved with relatively few iterations of the formulation.

Novel charge control agents based on the electronic hopping transport mechanism have the potential to impart moderate levels of conductivity to ester oils. Polyaromatic compounds with delocalized electron clouds grafted onto the diacid to solubilize them in a diester oil provide nearly 8× higher conductivity than the pure base oil at a concentration of 5 wt%.

ACKNOWLEDGMENTS

The work presented in this chapter was carried out between 1982 and 2016 at the IBM Research Division Almaden Research Center and Hitachi Global Storage Technologies/HGST San Jose Research Center. Over these three decades of work on drive industry lubricants, a substantial team effort enabled the progress described in this chapter. Thanks are due to P. Kasai at the IBM Almaden Research Center for technical discussions on lubricant end groups and NMR peak assignments. Thermal analysis, dielectric spectroscopy, NIPDSC, and rheological and viscosity measurements were done by M. Carter, and NMR spectroscopy and interpretation were provided by J. Burns, in the Hitachi Global Storage Technologies San Jose Materials Analysis Laboratory. Also in the Hitachi Global Storage Technologies San Jose Materials Analysis Laboratory, the author thanks D. Pocker, for the XPS, and F. Eng for IR microscopy, on grease residues. Thanks are due to R. Siemens, who carried out the GPC measurements on PFPE lubricants. The author is grateful to T. Gregory for invaluable advice and technical discussions on PFPE vapor pressure and development of the polydisperse lubricant evaporation model. Many of the fractionated PFPE lubricants were graciously provided by R. Waltman in the San Jose Development Laboratory. Many thanks are due to the excellent summer students who contributed to this work: A. Brooks, J. Castro, A. Voss, and A. Greenfield. Much of the NMR characterization, viscosity, and density measurement and evaporation data analyses were performed by D. Hopper over about 5 years part-time and summers working with us. H.S. Nagaraj first introduced me to ball bearing grease and then fluid bearing oil for spindle motors. The grease mill was designed and built by J. Miller, who also prepared the modified grease samples and performed the electrochemical measurements on the greases. Invaluable insight into and technical discussions on surface energy were provided by G. Tyndall at the IBM Almaden Research Center. The author thanks K. Shida of Hitachi Global Storage Technologies and Nidec Corp. for assistance with some of the fluid dynamic bearing oil formulations and the RBOT measurements. Thanks to P. Brock at the IBM Almaden Research Center and N. Conley at the HGST San Jose Research Center for chemical synthesis and purification of the aryl esters. Finally, I thank my managers Q. Dai, C. Lee, B. Marchon, J. Lyerla, O. Melroy, and J. Kaufman for their encouragement and support throughout this work. Special thanks are also due to A. Hanlon and C. Hignite for their guidance and advice.

REFERENCES

1. B. Marchon, Thin-film media lubricants: Structure, characterization, and performance, in *Developments in Data Storage: Materials Perspective*, 1st edn., eds. S. N. Piramanayagam and T. C. Chong, John Wiley & Sons, Inc., Hoboken, NJ, Chapter 8, pp. 144–166 (2012).

2. T. E. Karis, Lubricants for the disk drive industry, in *Synthetic, Mineral Oil, and Bio-Based Lubricants Chemistry and Technology*, ed. L. R. Rudnick, CRC Press, Taylor & Francis, Boca Raton, FL, pp. 623–654 (2006).

3. J. Ruhe, G. Blackman, V. J. Novotny, T. Clarke, G. B. Street, and S. Kuan, Terminal attachment of perfluorinated polymers to solid surfaces, *J. Appl. Polym. Sci.*, 53, 825–836, (1994).

4. C. Kajdas and B. Bhushan, Mechanism of interaction and degradation of perfluoropolyethers with a DLC coating in thin-film magnetic rigid disks: A critical review, *J. Info. Stor. Proc. Syst.*, 1, 303–320 (1999).

5. H. Tani, H. Matsumoto, M. Shyoda, T. Kozaki, T. Nakakawaji, and Y. Ogawa, Magnetic recording medium, US Patent 2002/0006531A1, January 17, 2002.

6. R. J. Waltman, N. Kobayashi, K. Shirai, A. Khurshudov, and H. Deng, The tribological properties of a new cyclotriphosphazene-terminated perfluoropolyether lubricant, *Tribol. Lett.*, 16(1–2), 151–162 (2004).

7. P. H. Kasai, Degradation of perfluoropoly(ethers) and role of X-1P additives in disk files, *J. Info. Stor. Proc. Syst.*, 1, 23–31 (1999).

8. N. Tagawa, T. Tateyama, A. Mori, N. Kobayashi, Y. Fujii, and M. Ikegami, Spreading of novel cyclotriphosphazine-terminated PFPE films on carbon surfaces, in *Proceedings of the 2003 Magnetic Storage Symposium, Frontiers of Magnetic Hard Disk Drive Tribology and Technology*, eds. A. A. Polycarpou, M. Suk, and Y.-T. Hsia, ASME, New York, TRIB-Vol. 15, pp. 17–20 (2003).

9. P. H. Kasai and V. Raman, Perfluoropolyethers with dialkylamine end groups: Ultrastable lubricant for magnetic disks application, *Tribol. Lett.*, 12(2), 117–122 (2002).

10. H. Chiba, Y. Oshikubo, K. Watanabe, T. Tokairin, and E. Yamakawa, Tribological characteristics of newly synthesized multi-functional PFPE lubricants, *Proceedings of WTC2005 World Tribology Congress III* (September 12–16, 2005), Washington, DC.

11. B. Marchon, X.-C. Guo, T. Karis, H. Deng, Q. Dai, J. Burns, and R. Waltman, Fomblin multidentate lubricants for ultra-low magnetic spacing, *IEEE Trans. Magn.*, 42(10), 2504–2506 (2006).

12. K. Sonoda, D. Shirakawa, T. Yamamoto, and J. Itoh, The tribological properties of the new structure lubricant at the head-disk interface, *IEEE Trans. Magn.* 43(6), 2250–2252 (2007).

13. D. Shirakawa, K. Sonoda, and K. Ohnishi, A study on design and synthesis of new lubricant for near contact recording, *IEEE Trans. Magn.*, 43(6), 2253–2255 (2007).

14. D. Sianesi, V. Zamboni, R. Fontanelli, and M. Binaghi, Perfluoropolyethers: Their physical properties and behavior at high and low temperatures, *Wear*, 18(2), 85–100 (1971).

15. G. Marchionni, G. Ajroldi, M. C. Righetti, and G. Pezzin, Molecular interactions in perfluorinated and hydrogenated compounds: Linear paraffins and ethers, *Macromolecules*, 26(7), 1751–1757 (1993).

16. A. Sanguineti, P. A. Guarda, G. Marchionni, and G. Ajroldi, Solution properties of perfluoropolyether polymers, *Polymer*, 36(19), 3697–3703 (1995).

17. A. C. Ouano and B. Appelt, Poly(perfluoroethers): Viscosity, density and molecular weight relationships, *Org. Coat. Appl. Polym. Sci. Proc.*, 46, 230–236 (1981).

18. M. J. R. Cantow, E. M. Barrall, Jr., B. A. Wolf, and H. Geerissen, Temperature and pressure-dependence of the viscosities of perfluoropolyether fluids, *J. Polym. Sci. Polym. Phys.*, 25(3), 603–609 (1987).

19. G. Marchionni, G. Ajroldi, and G. Pezzin, Molecular weight dependence of some rheological and thermal properties of perfluoropolyethers, *Eur. Polym. J.*, 24(12), 1211–1216 (1988).

20. G. Marchionni, G. Ajroldi, P. Cinquina, E. Tampellini, and G. Pezzin, Physical properties of perfluoropolyethers: Dependence on composition and molecular weight, *Polym. Eng. Sci.*, 30(14), 829–834 (1990).

21. G. Marchionni, G. Ajroldi, M. C. Righetti, and G. Pezzin, Perfluoropolyethers: Critical molecular weight and molecular weight dependence of glass transition temperature, *Polym. Commun.*, 32(3), 71–73 (1991).

22. G. Marchionni, G. Ajroldi, and G. Pezzin, Viscosity of perfluoropolyether lubricants: Influence of structure, chain dimensions and molecular structure, Society of Automotive Engineers, SP936, pp. 87–96 (1992).

23. P. M. Cotts, Solution properties of a group of perfluoropolyethers: Comparison of unperturbed dimensions, *Macromolecules*, 27(22), 6487–6491 (1994).

24. G. Ajroldi, G. Marchionni, M. Fumagalli, and G. Pezzin, Mechanical relaxations in perfluoropolyethers, *Plast. Rubb. Composit. Process. Appl.*, 17(5) 307–315 (1992).

25. F. Danusso, M. Levi, G. Gianotti, and S. Turri, Some physical properties of two homologous series of perfluoro-polyoxyalkylene oligomers, *Eur. Polym. J.*, 30(5) 647–651 (1994).

26. G. Tieghi, M. Levi, and R. Imperial, Viscosity versus molecular weight and temperature of diolic perfluoropoly(oxyethylene-ran-oxymethylene) oligomers: Role of the end copolymer effect, *Polymer*, 39(5), 1015–1018 (1998).

27. G. Ajroldi, G. Marchionni, and G. Pezzin, The viscosity-molecular weight relationships for diolic perfluoropolyethers, *Polymer*, 40, 4163–4164 (1999).

28. R.-N. Kono, M. S. Jhon, H. J. Choi, and A. Kim, Effect of reactive end groups on the rheology of disk lubricant systems, *IEEE Trans. Magn.*, 35(5), 2388–2390 (1999).

29. R. Smith, P.-S. Chung, J. A. Steckel, M. S. Jhon, and L. T. Biegler, Force field parameter estimation of functional perfluoropolyether lubricants, *J. Appl. Phys.*, 109, 07B728 (2011).

30. T. E. Karis, B. Marchon, D. A. Hopper, and R. L. Siemens, Perfluoropolyether characterization by nuclear magnetic resonance spectroscopy and gel permeation chromatography, *J. Fluor. Chem.*, 118(1–2), 81–94 (2002).

31. M. L. Williams, R. F. Landel, and J. D. Ferry, The temperature dependence of relaxation mechanisms in amorphous polymers and other glass forming liquids, *J. Am. Chem. Soc.*, 77, 3701–3707 (1955).

32. J. D. Ferry, *Viscoelastic Properties of Polymers*, 3rd edn., John Wiley & Sons, Inc., New York (1980).

33. P. Debye, *Polar Molecules*, Lancaster Press, Inc., Lancaster, PA (1929).

34. T. E. Karis, B. Marchon, V. Flores, and M. Scarpulla, Lubricant spin-off from magnetic recording disks, *Tribol. Lett.*, 11(3–4), 151–159 (2001).

35. B. G. Min, J. W. Choi, H. R. Brown, D. Y. Yoon, T. M. O'Connor, and M. S. Jhon, Spreading characteristics of thin liquid films of perfluoropolyalkylethers on solid surfaces: Effects of chain-end functionality and humidity, *Tribol. Lett.*, 1, 225–232 (1995).

36. T. M. O'Connor, M. S. Jhon, C. L. Bauer, B. G. Min, D. Y. Yoon, and T. E. Karis, Surface diffusion and flow activation energies of perfluoropolyalkylethers, *Tribol. Lett.*, 1, 219–223 (1995).

37. G. W. Tyndall, T. E. Karis, and M. S. Jhon, Spreading profiles of molecularly thin perfluoropolyether films, *Tribol. Trans.*, 42(3), 463–470 (1999).

38. T. E. Karis and G. W. Tyndall, Calculation of spreading profiles for molecularly thin films from surface energy gradients, *J. Non Newt. Fluid Mech.*, 82, 287–302 (1999).

39. R. J. Waltman, A. Khurshudov, and G. W. Tyndall, Autophobic dewetting of perfluoropolyether films on amorphous nitrogenated carbon surfaces, *Tribol. Lett.*, 12(3), 163–169 (2002).

40. T. E. Karis, The role of surface science in magnetic recording tribology, in *Surfactants in Tribology*, Vol. 2, eds. G. Biresaw and K. L. Mittal, CRC Press, Taylor & Francis, Boca Raton, FL, pp. 457–505 (2011).

41. Q. Dai, F. Hendriks, and B. Marchon, Modeling the washboard effect at the head/disk interface, *J. Appl. Phys.*, 96(1), 696–703 (2004).

42. T. E. Karis, X.-C. Guo, and J.-Y. Juang, Dynamics in the bridged state of a magnetic recording slider, *Tribol. Lett.*, 30(2), 123–140 (2008).

43. T. E. Karis and H. S. Nagaraj, Evaporation and flow properties of several hydrocarbon oils, *Tribol. Trans.*, 43(4), 758–766 (2000).

44. J. O. Hirschfelder, R. B. Bird, and E. L. Spotz, The transport properties of gases and gaseous mixtures. II, *Chem. Rev.*, 44, 205–231 (1949).

45. M. Levi, S. Turri, and A. Sanguineti, On the unperturbed dimensions of perfluoropoly-(oxymethylene-co-oxyethylene)-acetals, *Polymer*, 40(15), 4273–4278 (1999).

46. A. W. Adamson, *A Textbook of Physical Chemistry*, Academic Press, New York, p. 228, (1973).

47. M. J. Stirniman, S. J. Falcone, and B. J. Marchon, Volatility of perfluoropolyether lubricants measured by thermogravimetric analysis, *Tribol. Lett.*, 6(3–4), 199–205 (1999).

48. R. J. Waltman, G. W. Tyndall, G. J. Wang, and H. Deng, The effect of solvents on the perfluoropolyether lubricants used on rigid magnetic recording media, *Tribol. Lett.*, 16 (3), 215–230 (2004).

49. T. E. Karis, Water adsorption on thin film magnetic recording media, *J. Colloid Interface Sci.*, 225(1), 196–203 (2000).

50. G. W. Tyndall, R. J. Waltman, and J. Pacansky, Effect of adsorbed water on perfluoropolyether-lubricated magnetic recording disks, *J. Appl. Phys.*, 90(12), 6287–6296 (2001).

51. H. Matsumoto, H. Tani, and T. Nakakawaji, Adsorption properties of lubricant and additive for high durability of magnetic disks, *IEEE Trans. Magn.*, 37(4), 3059–3061 (2001).

52. R. P. Ambekar, D. B. Bogy, Q. Dai, and B. Marchon, Critical clearance and lubricant instability at the head-disk interface of a disk drive, *Appl. Phys. Lett.*, 92, 033104-1–033104-3 (2008).

53. B. Marchon, T. Pitchford, Y.-T. Hsia, and S. Gangopadhyay, The head–disk interface roadmap to an area density of 4 Tbit/in², Article ID 521086, in *Advances in Tribology: Tribology of the Head Disk Interface*, eds. B. Marchon, N. Tagawa, B. Liu, T. Karis, and J.-Y. Juang, Hindawi Publishing Corp., London, UK (2013).

54. C. M. Mate, B. Marchon, A. N. Murthy, and S.-H. Kim, Lubricant-induced spacing increases at slider–disk interfaces, *Tribol. Lett.*, 37, 581–590 (2010).

55. X. Ma, J. Chen, H. J. Richter, H. Tang, and J. Gui, Contribution of lubricant thickness to head–media spacing, *IEEE Trans. Magn.*, 37(4), 1824–1826 (2001).

56. R. P. Ambekar and D. B. Bogy, Effect of slider lubricant pickup on stability at the head-disk interface, *IEEE Trans. Magn.*, 41(10), 3028–3030 (2005).

57. B. Marchon, T. Karis, Q. Dai, and R. Pit, A model for lubricant flow from disk to slider, *IEEE Trans. Magn.*, 39(5), 2447–2449 (2003).

58. Y. Ma and B. Liu, Lubricant transfer from disk to slider in hard disk drives, *Appl. Phys. Lett.*, 90, 143516-1–143516-3 (2007).

59. H. Kubotera and T. Imamura, Monte Carlo simulations of air shielding effect on lubricant transfer at the head disk interface, *Appl. Phys. Lett.*, 94, 243112-1–243112-3 (2009).

60. Y. Goto, N. Nakamura, A. Mizutani, H. Chiba, and K. Watanabe, Head disk interface technologies for high recording density and reliability, *Fujitsu Sci. Tech. J.*, 42(1), 113–121 (2006).

61. L. Zhu and F. Li, In situ measurement of the bonded film thickness of Z-Tetraol lubricant on magnetic recording media, *J. Appl. Phys.*, 108, 084907-1–084907-6 (2010).

62. A. L. Brown, Fluid film thickness measurement with Moire fringes, *Appl. Opt.*, 11(10), 2269–2277 (1972).

63. R. J. Waltman and X.-C. Guo, AFM force–distance curves for perfluoropolyether boundary lubricant films as a function of molecular polarity, *Tribol. Lett.*, 45, 275–289 (2012).

64. T. E. Karis, Surfactants for electric charge and evaporation control in fluid bearing motor oil, in *Surfactants in Tribology*, Vol. 3, eds. G. Biresaw and K. L. Mittal, CRC Press, Taylor & Francis, Boca Raton, FL, pp. 437–462 (2013).

65. R. J. Waltman and N. Kobayashi, Thin-film properties of some functionalized perfluoropolyether boundary lubricants based on the n-perfluoropropylene oxide monomer unit, *Tribol. Trans.*, 54, 168–177 (2011).

66. J. N. Israelachvili, van der Waals dispersion force contribution to works of adhesion and contact angles on the basis of macroscopic theory, *J. Chem. Soc. Faraday Trans.*, 69, 1729–1738 (1973).

67. B. Marchon and T. E. Karis, Poiseuille flow at a nanometer scale, *Europhys. Lett.*, 74(2), 294–298 (2006).

68. A. A. Darhuber, S. M. Troian, J. M. Davis, S. M. Miller, and S. Wagner, Selective dip-coating of chemically micropatterned surfaces, *J. Appl. Phys.*, 88(9), 5119–5120 (2000).

69. I. C. Callaghan, Antifoams for nonaqueous systems in the oil industry, in *Defoaming: Theory and Industrial Applications*, ed. P. R. Garrett, Surfactant Science Series, Vol. 45, Marcel Dekker, New York, pp. 119–150 (1993).

70. H. Kimura, Y. Imai, and Y. Yamamoto, Study on fiber length control for ester-based lithium soap grease, *Tribol. Trans.*, 44(3), 405–410 (2001).

71. M. A. Delgado, M. C. Sánchez, C. Valencia, J. M. Franco, and C. Gallegos, Relationship among microstructure, rheology and processing of a lithium lubricating grease, *Chem. Eng. Res. Design*, 83(9), 1085–1092 (2005).

72. L. Salomonsson, G. Stang, and B. Zhmud, Oil/thickener interactions and rheology of lubricating greases, *Tribol. Trans.*, 50(3), 302–309 (2007).

73. T. E. Karis, R.-N. Kono, and M. S. Jhon, Harmonic analysis in grease rheology, *J. Appl. Polym. Sci.*, 90, 334–343 (2003).

74. T. Cousseau, B. M. Graca, A. V. Campos, and J. H. O. Seabra, Influence of grease rheology on thrust ball bearings friction torque, *Tribol. Int.*, 46(1), 106–113 (2012).

75. D. J. MacLeod, A. Gredinberg, and G. P. Stevens, Method and apparatus for assembling disc drive motors utilizing multiposition preload/cure fixtures, US Patent 6,061,894 (2000).

76. A. R. LaFountain, G. J. Johnston, and H. A. Spikes, The elastohydrodynamic traction of synthetic base oil blends, *Tribol. Trans.*, 44(4), 648–656 (2001).

77. G. J. Johnston, R. Wayte, and H. A. Spikes, The measurement and study of very thin lubricant films in concentrated contacts, *Tribol. Trans.*, 34(2), 187–194 (1991).

78. M. T. Costello, Effects of basestock and additive chemistry on traction testing, *Tribol. Lett.*, 18(1), 91–97 (2005).

79. P. M. Cann, M. N. Webster, J. P. Doner, V. Wikstrom, and P. Lugt, Grease degradation in ROF bearing tests, *Tribol. Trans.*, 50(2), 187–197 (2007).

80. L. A. Bronshtein, Y. N. Shekhter, and V. M. Shkol'nikov, Mechanism of electrical conduction in lubricating oils (review), *Chem. Technol. Fuels Oils*, 15(5–6), 350–355 (1979).

81. J. K. Avlyanov, Stable polyaniline and polypyrrole nanolayers on carbon surface, *Synth. Met.*, 102(1–3), 1272–1273 (1999).

82. J. Zhu and K. Ono, A comparison study on the performance of four types of oil lubricated hydrodynamic thrust bearings for hard disk spindles, *J. Tribol. Trans. ASME*, 121(1), 114–120 (1999).

83. T. Hirayama, T. Sakurai, and H. Yabe, Analysis of dynamic characteristics of spiral-grooved journal bearing with considering cavitation occurrence, *JSME Int. J. Series C Mech. Syst. Mach. Elem. Manuf.*, 48(2), 261–268 (2005).

84. T. Asada, H. Saitou, Y. Asaida, and K. Itoh, Characteristic analysis of hydrodynamic bearings for HDDs, *IEEE Trans. Magn.*, 37(2), 810–814 (2001).

85. T. E. Karis, Lubricants for the disk drive industry, in *Lubricant Additives: Chemistry and Applications*, 2nd edn., ed. L. Rudnick, CRC Press, Taylor & Francis, Boca Raton, FL, pp. 523–584, (2009).

86. T. E. Karis, Lubricants for the disk drive industry, in *Synthetics, Mineral Oil, and Bio-Based Lubricants Chemistry and Technology*, 2nd edn., ed. L. Rudnick, CRC Press, Taylor & Francis, Boca Raton, FL, pp. 657–698 (2013).

87. J. Gavis and I. Koszman, Development of charge in low conductivity liquids flowing past surfaces: A theory of the phenomenon in tubes, *J. Colloid Interface Sci.*, 16, 375–391 (1961).

88. Z. Feng, C. Shih, V. Gubbi, and F. Poon, A study of tribo-charge/emission at the head-disk interface, *J. Appl. Phys.*, 85(8), 5615–5617 (1999).

89. H. Eyring, Viscosity, plasticity, and diffusion as examples of absolute reaction rates, *J. Chem. Phys.*, 4, 283–291 (1936).

90. S. Glasstone, K. L. Laidler, and H. Eyring, *The Theory of Rate Processes*, McGraw-Hill, Inc., New York, p. 477 (1941).

91. R. A. Ewell, The reaction rate theory of viscosity and some of its applications, *J. Appl. Phys.*, 9, 252–269 (1938).

92. S. Bair, C. McCabe, and P. T. Cummings, Comparison of non-equilibrium molecular dynamics with experimental measurements in the nonlinear shear-thinning regime, *Phys. Rev. Lett.*, 88(5), 58302-1–58302-4 (2002).

93. P. Kumar, M. M. Khonsari, and S. Bair, Full EHL simulations using the actual Ree–Eyring model for shear-thinning lubricants, *J. Tribol.*, 131(1), 011802 (2008).

94. C. Mary, D. Philippon, L. Lafarge, D. Laurent, F. Rondelez, S. Bair, and P. Vergne, New insight into the relationship between molecular effects and the rheological behavior of polymer-thickened lubricants under high pressure, *Tribol. Lett.*, 52(3), 357–369 (2013).

95. D. M. Heyes, E. R. Smith, D. Dini, H. A. Spikes, and T. A. Zaki, Pressure dependence of confined liquid behavior subjected to boundary-driven shear, *J. Chem. Phys.*, 136, 134705 (2012).

96. M. A. Eiteman and J. W. Goodrum, Rheology of the triglycerides tricaproin, tricaprylin, and tricaprin and of diesel fuel, *Trans. ASAE*, 36, 503–507 (1993).

97. D. Valeri and A. J. Meirelles, Viscosities of fatty acids, triglycerides, and their binary mixtures, *J. Am. Oil Chem. Soc.*, 74(10), 1221–1226 (1997).

98. G. B. Bantchev and G. Biresaw, Film-forming properties of castor oil–polyol ester blends in elastohydrodynamic conditions, *Lubr. Sci.*, 23(5), 203–219 (2011).

99. T. E. Karis and H. S. Nagaraj, Magnetic recording device, US Patent 6,194,360 (2001).

100. S. Wong, R. Kroeker, J. Burns, T. Karis, A. Hanlon, T.-C. Fu, and C. Hignite, Disk drive system with hydrodynamic bearing lubricant having charge-control additive comprising dioctyldiphenylamine and/or oligomer thereof, US Patent 7,212,376 (2007).

101. T. E. Karis, J. L. Miller, H. E. Hunziker, M. S. de Vries, D. A. Hopper, and H. S. Nagaraj, Oxidation chemistry of a pentaerythritol tetraester oil, *Tribol. Trans.*, 42(3), 431–442 (1999).

102. H. Spikes, The history and mechanisms of ZDDP, *Tribol. Lett.*, 17(3), 469–489 (2004).

103. N. K. Chakravarty, The influence of lead, iron, tin, and copper on the oxidation of lubricating oils. Part II. On the metal catalysed oxidation of oils, *J. Inst. Petrol.*, 49(479), 353–358 (1963).

104. A. Bondi, *Physical Chemistry of Lubricating Oils*, Reinhold Publishing Corp., New York, p. 289 (1951).

105. D. C. Byford, H. B. Silver, and H. Wood, Lubricants having improved anti-wear and anti-corrosion properties, US Patent 3,697,427 (1972).

106. J. M. Burns, M. D. Carter, T. E. Karis, K. Shida, and S. Y. Wong, Fluid dynamic bearing comprising a lubricating fluid having tolutriazole, US Patent 7,535,673 (2009).

107. R.-H. Wang, H. R. Wendt, C. A. Brown, S. Lum, S. McCoy, and T. Karis, Enhanced reliability of hard disk drive by vapor corrosion inhibitor, *IEEE Trans. Magn.*, 42(10), 2498–2500, (2006).

108. J. M. Palacios, Thickness and chemical composition of films formed by antimony dithiocarbamate and zinc dithiophosphate, *Tribol. Int.*, 19(1), 35–39, (1986).

109. J. M. Palacios, Films formed by antiwear additives and their incidence in wear and scuffing, *Wear*, 114(1), 41–49 (1987).

110. Y. A. Lozovoi, K. Meyer, G. N. Kuz'mina, and G. Bokhinek, Chemical modification of friction surfaces: Surface layers formed in boundary friction, *Sov. J. Fric. Wear*, 11(2), 270–277 (1990).

111. P. A. Redhead, Thermal desorption of gases, *Vacuum*, 12(4), 203–211 (1962).

112. A. Adhvaryu, S. Z. Erhan, Z. S. Liu, and J. M. Perez, Oxidation kinetic studies of oils derived from unmodified and genetically modified vegetables using pressurized differential scanning calorimetry and nuclear magnetic resonance spectroscopy, *Thermochim. Acta*, 364(1–2), 87–97 (2000).

113. R. Z. Lei, A. J. Gellman, and P. Jones, Thermal stability of fomblin Z and fomblin zdol thin films on amorphous hydrogenated carbon, *Tribol. Lett.*, 11(1), 1–15 (2001).

114. B.-K. Tan, B. Liu, Y. Ma, M. Zhang, and S.-F. Ling, Effect of electrostatic force on slider-lubricant interaction, *IEEE Trans. Magn.*, 43, 2241–2243 (2007).

115. Y. Wang, X. Wei, X. Liang, S. Yin, Y. Zi, and K-L. Tsui, An in situ measurement method for electric potential at head–disk interface using a thermal asperity sensor, *IEEE Trans. Magn.*, 52(1), Article#: 3300106 (2016).

116. Y. Zelu, T. Paillat, G. Morin, C. Perrier, and M. Saravolac, Study on flow electrification hazards with ester oils, *2011 IEEE International Conference on IEEE Trans. Magn., Dielectric Liquids (ICDL)*, Trondheim, Norway, pp. 1–4 (2011).

117. T. Toudja, F. Chetibi, A. Beldjilali, H. Moulai, and A. Beroual, Electrical and physicochemical properties of mineral and vegetable oils mixtures, *2014 IEEE 18th International Conference on Dielectric Liquids (ICDL)*, Bled, Slovenia pp. 1–4 (2014).

118. R. Schmechel, Gaussian disorder model for high carrier densities: Theoretical aspects and application to experiments, *Phys. Rev. B*, 66, 235206 (2002).

119. T.E. Karis, Charge control agent for fluid dynamic bearing motor lubricant, US Patent 9,368,150 (2016).

120. P. Bartle and C. Volkl, Thermo-oxidative stability of high-temperature stability polyol ester jet engine oils—A comparison of test methods, *J. Synth. Lubr.*, 17, 179–189 (2000).

121. T. E. Karis, and H.S. Nagaraj, Lubricant additives for magnetic recording disk drives, in *Lubricant Additives: Chemistry and Applications*, ed. L. Rudnick, Marcel Dekker, Inc., New York, NY, pp. 467–511 (2003).

122. T. E. Karis, B. Marchon, M. D. Carter, P. R. Fitzpatrick, and J. P. Oberhauser, Humidity effects in magnetic recording, *IEEE Trans. Magn.*, 41(2), 593–598 (2005).

123. T. E. Karis, W. T. Kim, and M. S. Jhon, Spreading and dewetting in nanoscale lubrication, *Tribol. Lett.*, 17(4), 1003–1017 (2004).

124. T. E. Karis and M. A. Tawakkul, Water adsorption and friction on thin film magnetic recording disks, *Tribol. Trans.*, 46(3), 469–478 (2003).

Section VII

Trends

29 Impacts of the Globally Harmonized System on Lubricant Manufacturers

Luc Séguin

CONTENTS

29.1 ORIGINS, OBJECTIVES, AND WORLD IMPLEMENTATION OF THE GLOBALLY HARMONIZED SYSTEM

The Globally Harmonized System (GHS) is a classification system that aims to classify chemicals according to their physical, health, and environmental hazards. It came about in the course of the 1992 United Nations Conference on Environment and Development (UNCED),* often called the Earth Summit. During this meeting, 172 governments—with 108 of them represented by heads of state or government—adopted three major agreements to guide future approaches to development. The aforementioned agreements are as follows: Agenda 21, a global plan of action to promote sustainable development; the resulting Rio Declaration on Environment and Development, a series of principles defining the rights and responsibilities of states; and the Statement of Forest Principles, which underpins the sustainable management of forests worldwide. In addition, two legally binding instruments were opened for signature at the Summit: the United Nations Framework Convention on Climate Change and the Convention on Biological Diversity. More specifically, it included the first official series of discussions on setting up a globally adopted classification system to identify the physical, health, and environmental hazards of chemicals, their proper labeling, and also the framework for the Safety Data Sheet (SDS), which must be provided with each shipment. This was a nonbinding agreement between all participant countries.

The general objective of the system is to ensure better protection of human health and of the environment during the handling of chemicals, including their transport and use. The Rio Summit Declaration stipulated that a worldwide classification system of all chemicals be released, if possible, by the year 2000. In fact, the very first consolidated release of the GHS (Revision 1.0), also known as *The Purple Book*,[†] was finally endorsed by the UN Economic and Social Council in July 2003. It has been continuously revised every 2 years, which ensures that it is kept current with the findings of modern science and with the legal progress made by each country adopting the system. We are now looking at the implementation of Revision 6.0[‡] that was released in November 2015.

Because the GHS is a nonbinding agreement, each country must willingly adopt the GHS within its own law and must do so based on a given enforceable revision. Moreover, the "building blocks" approach adopted for GHS implementation allows each country to determine what will be selected in terms of hazard categories within its territory, as well as the transition phases needed for full implementation. These factors are essential to ensure that the governmental authorities have full control over how the system is implemented in their territory. Moreover, it gives the authorities the ability to not adopt *all parts* of the UN GHS *Purple Book* and to even *add some new parts* to ensure a specific level of protection for their workers and their environment.

---

* http://www.unece.org/trans/danger/publi/ghs/implementation_e. html#transport.

† http://www.unece.org/trans/danger/publi/ghs/ghs_rev01/01files_e.html.
‡ http://www.unece.org/trans/danger/publi/ghs/ghs_rev06/06files_e. html#c38156.

We can therefore imagine the high potential for disharmony when Europe, North America, and Asia-Pacific countries, regardless of having a chemical classification system in place, look at required changes and the need for transition periods without any coordination from the United Nations.

We identified several reports and websites that may be used to follow the worldwide implementation of the GHS. The European Chemical Industry Council (CEFIC) published its Global GHS Implementation Task Force report on Overview of GHS implementations in May 2014.* This report lists 59 specific countries and Europe as a whole with their respective levels of implementation of the GHS. This includes, whenever applicable, their GHS legislation revision and/or phase-in transition period and approach. Looking at this document, it is important to realize that the European countries adopted the GHS under the REACH (Registration, Evaluation and Assessment of Chemicals)† regulation and, more specifically, under the Classification, Labelling and Packaging (CLP) regulation that governs the classification of chemicals in all member states, as well as the content of the GHS SDS and labels. REACH entered into force in June 2007 with a substance–mixture transition approach, wherein all pure substances were to be classified and labeled under the GHS by December 1, 2010, and shipped with an SDS strictly compliant with the 453/2010 format and content. Mixtures were to be properly classified, labeled, and sold with the same SDS format by June 1, 2015. The United States adopted the GHS under HAZCOM 2012 (Hazard Communication System)‡ in May of that year with a 3-year global transition approach. By June 1, 2015, all chemicals sold and found in workplaces must be classified, labeled, and sold with an SDS and a label based on the GHS, regardless of whether it is a pure substance or a mixture. Canada adopted the GHS via WHMIS 2015 and the Hazardous Product Regulation (HPR)§ in February 2015, with a transition period ending June 1, 2019. It will be phased in by initial suppliers and manufacturers, followed by blenders and distributors, and subsequently ending by ensuring that employers comply with the new system (classification of pure and mixtures, new SDS and label format/content).

In the course of these three major implementations, the label, SDS format and content, hazard categories, and ingredient thresholds for the classification of some health hazards have all been adopted differently by the respective authorities. The United States and Canada have moreover decided to make environmental hazards "optional." As such, they are among the few countries in the world adopting GHS with the option for their chemical manufacturers not to classify their products for the environment, Australia being the only other country also opting out on that specific hazard.

29.2 REGION-BY-REGION REVIEW OF THE MOST IMPORTANT GHS DISHARMONIZATIONS

It is not our intention to cover the hundreds of countries that adopted the GHS at the time this chapter was written but more to highlight the most important sources for disharmonizations, so that the reader can foresee where products can "easily or not easily" be shipped and sold. We will list the specifics that countries have decided to enforce and not to explain why they did so, leaving that for the reader's considerations. We will start this worldwide review with the Asia-Pacific countries, going to Africa and the Middle East, then Europe as a whole and with its own country specifics, and lastly with the Americas, where Canada was the last one to adopt the system among the G20 countries. We will make sure we insert at least one table per region, comparing several implementation criteria and our more specific comments on special ones needing a careful look.

29.2.1 GHS Adoption in the Asia-Pacific Countries

Looking at Table 29.1, the language in which the SDS and labels must be provided in this region of the world is obviously still one of the most important challenges today for North American and European exporters. The various enforced GHS revision for implementation and the widely spread implementation periods (going up to 2017 and 2019 in some cases) also creates a challenge in itself. The omission of some hazards must also be scrutinized against the exporters' product lines and families.

Finally, as we will also notice in Europe, the fact that some governmental authorities published mandatory (or optional) lists of GHS substance classifications all play against a harmonized mixture classification and therefore SDS and label content.

29.2.2 GHS Adoption in Africa and the Middle East

Obviously, this region of the world is far from being "easy to export" because of the vague GHS implementations or absence thereof. Africa, except for South Africa, is expected to adopt GHS by 2020 but no or minimal information circulates as to the details of its implementation. The author recommends to use the European implementation of GHS and to publish SDSs and labels in English, French, or Portuguese, as indicated in Table 29.2, until a specific country regulation is published.

One must also be aware that the Gulf Cooperation Council (GCC) was created back in 1981 and comprises Saudi Arabia, Bahrain, Kuwait, Qatar, Oman, and the United Arab Emirates. Since 2002, this organization has put forward the Common System for the Management of Hazardous Chemicals to coordinate GHS implementation among their member states and it is mainly based on the European Union's GHS implementation, documents must be first provided in the Arabic language, and the English language is an option.

* CEFIC-Global GHS Implementations—May 2014(1).pdf.
† http://ec.europa.eu/growth/sectors/chemicals/reach/index_en.htm.
‡ https://www.osha.gov/dsg/hazcom/ghs-final-rule.html.
§ WHMIS-2015-HPR-Gazette2-14903.pdf.

TABLE 29.1

Summary of GHS Implementation in Asia-Pacific Countries

Country	GHS Revision	Date/Implementation	Language	Specifics
Part I				
Australia	3	January 1, 2012, to December 31, 2016 (5 years)	English (UK)	Substances/mixtures GHS at same time Local emergency information required *Hazards not enforced*: Acute toxicity Cat 5 (oral-dermal-inhal.) Skin corrosion Cat 3 Serious eye damage Cat 2B Aspiration hazard Cat 2 Flammable gas Cat 2 Environmental hazards (acute/chronic) all categories
China	4	October 16, 2013, to November 1, 2014 (1 year)	Chinese (simplified)	A substance classification list has been published and must be used for all pure-mixture products used in the country. China new legislation requires the registration of substances imported into the territory (October 15, 2010).
India	X	Not yet implemented (2015)	Hindi English (UK)	India has not yet implemented GHS but it imports products being classified by the GHS under various revisions. We read that the government is being pressured to adopt a regulation within 1–2 years.
Indonesia	4	2009 to December 2017 (7 years)	Bahasa Indonesia	Substances to comply since 2010, mixtures to comply by December 31, 2016
Japan	4	2012 to December 2016 (5 years)	Japanese	Still allows the use of the old JIS 7250:2010 (MSDS)[a] and Z7251:2010 (Label) until the end of 2016 A list of 500–600 substances classified by the Japanese government must be used when substances are found as pure or in mixtures in the country.
Part II				
South Korea	3	2013	Korean	*Hazards not enforced*: Carcinogenic Cat 1B Hazard to the ozone layer The Korea Occupational Safety and Health Agency (KOSHA) published 6000 substance GHS classifications (nonmandatory, only for reference). The National Institute of Environmental Research (NIER) published 800 substance classifications that are mandatory. Some disharmonizations exist for some substance classifications between KOSHA and NIER.
Malaysia	3	April 2015	Malay and English	The bilingual SDSs and labels are the major specifics of the GHS implementation in this country. The Malaysia Environmentally Hazardous Substances Notification and Registration (EHSNR) Scheme requires the registration of substances imported in the territory (2011) and serves as the basis for substance-related legislation.
New Zealand	3	2006	English (UK)	This country adopted the GHS in its most disharmonized approach known by the author of this chapter. They mirror GHS hazard classification to transport classification (whenever applicable) and created an HSNO Hazard Classification being specific to the prescribed Hazardous Substances and New Organisms Act (HSNO Act). For example, explosives would have all HSNO codes starting with the reciprocal transport classes (1.1, 1.2, etc.), the flammable gases HSNO codes would all start with 2.1, and all health hazard HSNO codes start with 6.1. The very specific labeling system is clearly described in a document titled "Labelling of Hazardous Substances: Hazard and Precautionary Information" published in 2006 by the Environmental Risk Management Authority of New Zealand.

(Continued)

TABLE 29.1 (*Continued*)
Summary of GHS Implementation in Asia-Pacific Countries

Country	GHS Revision	Date/Implementation	Language	Specifics
Philippines	3	March 2015 to 2016 for substances and 2019 for mixtures	English	The GHS in this country also covers consumer products. A revision of all documents *must* be done every 5 years or earlier if a change in classification occurs. Official publications are available in English from the Department of Environment and National Resources (DENR) of the Republic of Philippines dated August–September 2015.
Part III				
Singapore	3	July 2015	English	*Hazards not enforced*: Acute toxicity category 5, Skin corrosion/irritation category 3, Aspiration hazard category 2, Acute hazard to aquatic life cat 2 and 3, and Chronic hazard to aquatic life cat 3 an 4.
Taiwan	4	January 2009/substances; January 2017/mixtures	Chinese (traditional)	This Chinese language differs drastically from the one spoken in mainland China (Chinese simplified). Taiwan shows apart in implementing Revision 4 (instead of 3 or 5) of the GHS.
Thailand	3	March 2013 substances; March 2017 for mixtures	Thai	Almost 3000 substances have been classified according to the GHS by the Thai authority. The database must be used and is web searchable at www.mot.go.th
Vietnam	3	March 2014 substances; March 2016 mixtures	Vietnamese	*Hazards not enforced*: Aspiration hazard STOT Short-term Category 3.

[a]　http://www.msdscompliance.com/data/ghs_implementation_status_35.html

TABLE 29.2
Summary of the GHS Implementation in Africa and the Middle East

Country	GHS Revision	Date/Implementation	Language	Specifics
Abu Dhabi	—	Not yet implemented	English	None known.
Angola	—	Expected by 2020	Portuguese	None known.
Botswana	—	Expected 2017	English	None known.
Chad	—	Expected by 2020	English	None known.
Congo	—	Expected by 2020	English	None known.
Egypt	—	—	Arabic	None known.
Israel	4	Mid-2015	Hebrew	Mainly following the European Union implementation scheme. Standard SI 2302 is published in Hebrew language only; several hundreds of pure substances are listed with their mandatory governmental GHS classifications.
Malawi	—	Expected by 2020	English	None known.
Mauritius	1	2004	English	None known.
Mozambique	—	Expected by 2020	French	None known.
Namibia	—	Expected by 2020	English	None known.
Saudi Arabia	—	Not yet implemented	Arabic/English	The Arabic language is mandatory and English is optional. The same rule applies for all of GCC members (see Section 29.2.2).
South Africa	4	Not confirmed	English	An annex of the South African Standard (SANS) 10234 a list of substance GHS classifications is to be used.
Tanzania	—	Expected by 2020	English	None known.
Zambia	—	Expected by 2020	English	None known.
Zimbabwe	—	Expected by 2020	English	None known.

29.2.3 GHS Adoption in European Countries

In Table 29.3 for European countries, a lot of details have been added for which the reader will have to access a specialized database to access the needed information and data. In fact, we refrained from adding "all possible additional information" required by these countries and covering aspects related to environmental requirements (waste codes for instance). In addition, we do not elaborate on the said "lists" and products' registration numbers in some of these countries; the reader must keep in mind this supplementary information *may* not apply to a specific product type or use and a local expert team will be needed to be able to fulfill needed requirements.

Let us keep in mind that Europe is *not* a country but really a union of still distinct authorities all classified under the "same GHS Revision" (four at the time this chapter is written) but with different strict country-specific requirements, such as local occupational exposure limits, and mandatory language requirements minimally for the labels but more and more often extended to all documentation, including Safety Data Sheets. Therefore, we do strongly recommend the readers to make sure they carefully implement their "European" SDSs in view of all applicable and strongly enforced requirements and not consider sending one "English" SDS and label per product, hoping it will be accepted at the border (outside the UK or other countries not speaking English as their main language). From experience, we strongly recommend the full enforcements of all requirements in France and Germany, these two countries being the ones mostly scrutinizing the SDSs and labels in circulation in their respective territory. The reader must also consider that the enforcement of the GHS in Europe has been done under the implementation of the REACH law. There is *no mention* of any REACH requirements in Table 29.3 as this law alone comprises several hundreds of pages. In very short terms, it makes it mandatory for importers of chemicals in Europe to register their products by substance present in several tons per year. The registration is associated with participation to the Substances Information Exchange Forum (SIEF) which is controlled by a Germany-based IT team, making sure that all importers of the same substance in Europe share and use the same data and classification. Non-negligible fees are linked to the mandatory participation to a SIEF and as one can imagine, exporting a mixture would result in multiple SIEF participations! Detailed references on REACH are numerous and freely available on the web. Another important requirement under REACH is that an "Only Representative" (OR) is also mandatory to participate in SIEF, and this OR can *only* be procured in one of the European countries; it can also be acquired from a local division of a non-European company. In summary, this huge market is highly regulated and the application of the regulation is fully enforced and more over since the obligation of SDSs and labels to be GHS and to be kept updated, via the possible issuance of substance new GHS mandatory classifications by the European Chemical Agency (ECHA) requires from all exporters in these countries, a *high level* of constant vigilance. The published ECHA substance classification is also another highly enforced obligation, whether or not it coincides with another country's

classification; the one being officially published must be used in all documents and for all classifications of the substance or of mixtures containing such substance.

29.2.4 GHS Adoption in the Americas

Looking at Table 29.4 for the Americas, one can "easily conclude" the existence of two blocks, the first block of fully enforced GHS countries (Argentina, Brazil, Canada, Mexico, and the United States) and the second block of countries, to enforce the regulation in the coming years. The countries adopting GHS in America are making huge efforts to align their implementations with one another, but there are still important disharmonizations that we discuss in the following sections.

29.2.4.1 America's GHS Hazards and Classifications

Among all countries in the world adopting GHS, the American countries, under the leadership of the United States' adoption, decided to lower the concentration cutoff to be considered in the calculation of some health hazards of mixtures from 1% to 0.1%. The classification of environmental hazards have been "left to a separate governmental agency" in Canada (Canadian Environmental Protection Agency [CEPA]) and the United States (Environmental Protection Agency [EPA]), which in both cases at the time this chapter was written have not yet started to review SDSs and labels for hazards, leaving it "optional to enforce if applicable" for each manufacturer-distributor in these two countries. Another "source of disharmonization" is the enforcement of the "Hazard Not Otherwise Classified—HNOC" under HAZCOM 2012 which are split in two (Heath and Physical) Hazard Not Otherwise Classified—HHNOC and PHNOC under WHMIS 2015 in Canada. On top of splitting them and changing their name and acronym, a more important difference comes with the way Canada enforces these hazards, only considering the ones causing "severe injuries and possibly death" of the users, requiring the full labeling of a category 1 hazardous product with the "Danger" signal word and applicable symbols, H and P phrases, while the definition is much smoother in the United States where we can indicate "any other known hazard(s)" by listing these specifically in the SDS but *not* creating a label for such products.

Other differences related to the hazards and classifications between Canada and the United States come from the revision under which the two countries adopted the GHS (Revision 3 in the United States and 5 in Canada). Some hazards (e.g., pyrophoric gases) must be covered under HNOC in the United States or are simply not considered (nonflammable aerosol) in the United States because these are not part of Revision 3, these being fully enforced in Canada being under Revision 5.

29.2.4.2 America's SDS and Label Content

Canada made it mandatory to show the "initial supplier" information in all SDSs and labels. The initial supplier would be an outside-of-Canada entity if this entity sells its product to the "end user," or would be a Canada-based entity if the products are sold in Canada via this organization (a local company entity

TABLE 29.3
Summary of the GHS Implementation in European Countries

Country	GHS Revision	Date/Implementation	Language	Specifics
Part I				
Austria	4	June 2015	German	Occupational exposure limits (OELs); Carcinogen, Mutagen, Reproductive Toxin (CMR) lists; Vbf Class; storage code; and limitations on the use of organic solvents
Belgium	4	June 2015	French (Brussels), Dutch (Outside Brussels)	OELs; CMR lists
Bulgaria	4	June 2015	Bulgarian	OELs; CMR lists
Croatia	4	June 2015	Croatian	OELs; CMR lists
Czech Republic	4	June 2015	Czech	OELs; CMR lists; Storage Code
Part II				
Denmark	4	June 2015	Danish	OELs; CMR lists; Danish Fire Class; Danish Cancer Risk Registry, the MAL Code; low boiling liquids; List of Undesirable Substances; others[a]
Estonia	4	June 2015	Estonian	OELs; CMR lists
Europe[b]	4	June 2015	See below	See below
Finland	4	June 2015	Finnish	OELs; CMR lists; Product Registration Code under NACE and UC62
France	4	June 2015	French	OELs; CMR lists, an occupational disease list; reinforced medical surveillance
Germany	4	June 2015	German	OELs, CMR lists, German Water Klasse (GWK); Storage Code; Hazardous Incidence Ordinance; AOX
Greece	4	June 2015	Greek	OELs; CMR lists
Hungarian	4	June 2015	Hungarian	OELs; CMR lists
Iceland	4	June 2015	Icelandic	OELs; CMR lists
Ireland	4	June 2015	English (UK)	OELs; CMR lists
Italy	4	June 2015	Italian	OELs; CMR lists; D.Lgs.152/06
Latvia	4	June 2015	Latvian	OELs; CMR lists
Lithuania	4	June 2015	Lithuanian	OELs; CMR lists
Netherlands	4	June 2015	Dutch	OELs; CMR lists
Norway	4	June 2015	Norwegian	OELs; CMR lists; product registration numbers
Poland	4	June 2015	Polish	OELs; CMR lists
Portugal	4	June 2015	Portuguese[c]	OELs; CMR lists
Part III				
Romania	4	June 2015	Romanian	OELs; CMR lists
Slovakia	4	June 2015	Slovak	OELs; CMR lists
Slovenia	4	June 2015	Slovenian	OELs; CMR lists
Spain	4	June 2015	Spanish[c]	OELs; CMR lists
Sweden	4	June 2015	Swedish	OELs; CMR lists; flammable liquid class
Switzerland	4	June 2015	German (Zurich) French (Geneva) Italian balance	OELs; CMR lists; Swiss LRV Klasse Ta-Luft; VOC content; product registration numbers
Turkey[d]	4	June 2015	Turkish	OELs; CMR lists
United Kingdom	4	June 2015	English (UK)	OELs; CMR lists

[a] This country adopted GHS with many other requirements then the one in the above table. The MAL law is the one requiring many of the extras.

[b] Europe is *not* a country but we can build a European SDS and label to be used in countries not having specific enforcement.

[c] The language indicated for these countries differs from the one spoken in American countries.

[d] Turkey enforced GHS with a specific requirement that the writers of SDSs and labels be certified in Turkey. This requirement alone sure has an impact on the commerce with other countries.

TABLE 29.4

Summary of the GHS Implementation in the Americas

Country	GHS Revision	Date/ Implementation	Language[a]	Specifics
Argentina	4	2013	Spanish	IRAM Standard No. 41400 (SDS) on a voluntary basis
Brazil	4	June 2015	Portuguese	ABNT (14725-3:2012 and 4:2012) are the regulations for the enforcement of GHS in the country at the time this chapter is written.
Canada	5	February 2015 to December 2018	English and French	Two languages as a mandatory requirement for all documents (labels and SDS) were enforced by the WHMIS 2015 law and the Hazardous Product Regulation (HPR). It is meant to be aligned with the US HAZCOM 2012 but in several ways; see Section 29.2.4.
Chile	—	Being worked out	Spanish	No confirmation on its GHS implementation plan
Colombia	—	Being worked out	Spanish	No confirmation on its GHS implementation plan
Costa Rica	—	Being worked out	Spanish	No confirmation on its GHS implementation plan
Ecuador	—	Being worked out	Spanish	No confirmation on its GHS implementation plan
Guatemala	—	Being worked out	Spanish	No confirmation on its GHS implementation plan
Honduras	—	Being worked out	Spanish	No confirmation on its GHS implementation plan
Mexico	5	October 2015 to December 2018	Spanish	A decree published in October 2015 changed the enforcement of GHS from a voluntary basis to a mandatory Revision 5 version with a 3 years transition period. See below.
Peru	—	Being worked out	Spanish	No confirmation on its GHS implementation plan.
Uruguay	—	Being worked out	Spanish	No confirmation on its GHS implementation plan.
United States	3	June–December 2015	English (US)	HAZCOM 2012 implemented the GHS with a 3-year transition period. See Section 29.2.4.

[a] Languages mentioned in the table differ from the same spoken in Europe.

or a local distributor). Some precautionary statements changed from Revision 3 to 5, so if applicable, they must match the correct revision in the country of destination of the products. Canada does not require chemicals listed as carcinogens by the OSHA, NTP, or IARC to be listed in Section 11 of the SDS, but the United States made it mandatory. Canada allows minimal information to be shown on small container labels; the United States requires the full label content regardless of the container size. Canada keeps it mandatory to label products containing between 0.1% and 1% of carcinogen ingredients, while the United States allows for nonlabeling of such products.

29.2.4.3 Overall Language Issue

The reader must consider the country specific mandatory publication of the SDSs and labels in four different languages (English, French, Spanish, and Portuguese) to be able to cover all of the countries on this continent from north to south. Portuguese being a requirement in Brazil only, it sure sets aside these documents from the others. English being a "big driver," it becomes the first language in which documents are normally written to be translated in French or Spanish afterward. The habit of translating an English document must be taken with extreme care because on top of the language difference, the countries all have their specific requirements, the major ones being reported in Table 29.5 for the ones that have fully enforced GHS by the time this chapter was written.

TABLE 29.5

Disharmonized Non-GHS Requirements in the Americas

Country	Specific Non-GHS Regulatory Requirements
Brazil	National OELs, when available
Canada	Provincial OELs, when available
	If the following sections of the SDS are filled:
	Transport of Dangerous Goods (TDG) to be shown in Section 14 classification
	Section 15
	DSL and NDSL (National Inventory status)
	National Pollutant Release Inventory (NPRI) (if applicable)
Mexico	National OELs, when available
	If the following sections of the SDS are filled:
	Mexican Transport NOMs to be shown in Section 14 classification
United States	ACGIH, OSHA, and NIOSH OELs, when available.
	If the following sections of the SDS are filled:
	Department of Transport (DOT) to be shown in Section 14 classification.
	Section 15
	Federal and State lists
	TSCA, CWA, CAA, SARA, and CERCLA
	States' Right-to-Know lists
	California Proposition 65 (if applicable)

Table 29.5 in two parts shows some other topics to be considered by the SDS and label writers looking to issue the same documents for more than one country. Although it is possible to add information that is country specific so a single document can be used, one must consider the need to translate, or not, some subsections for the benefit of some readers and this can lead to more complications than benefits. Canada clearly stipulates in the HPR that a single bilingual (English–French) SDS is totally compliant; nevertheless, the length of such document and the "relative complexity" to read the correct language portion needed in case of an emergency have lead the document to not be widely used, even 1 year after the enforcement of the WHMIS 2015. The unique language–country document is still widely used; we are also seeing a lot of US-Can-Mex SDSs being published but these in one language at a time. Labels published in two (English and French) or three (English, French, and Spanish) languages are often used.

29.3 THE LUBRICANT MANUFACTURING INDUSTRY AND ITS SPECIFIC GHS CHALLENGES

Of course, lubricant manufacturers deal with the same basic challenges as those faced by the rest of the chemical industry, relating to the correct classification of chemicals and the issuance of SDSs and labels that signify a country's compliance. We will now specifically address the ones related to the country-specific enforcement of trade secrets (Confidential Business Information [CBI]) in various lubricating additives when GHS health and environmental hazards are calculated in various countries.

Under the GHS, the common basic challenge faced by all chemical manufacturers is ensuring that the hazard classification of the chemicals (pure substances or mixtures) is made in accordance with the current GHS revision and the specific building blocks adopted in the country where it is being shipped.

Without addressing in full the complex GHS calculations involved in the determination of the correct hazard classes and categories of mixtures, there are several necessary considerations. First, the classification and class are based on the specific ingredient's concentration and the hazard's specific thresholds. These thresholds vary with some hazards from one country or continent to another. For example, carcinogenic ingredients must be considered starting at the 1% concentration level in Europe but down to 0.1% in the United States and Canada. Along the same lines, an eye irritation category 2 ingredient (or mixture) in Europe must be "converted" or reclassified according to category 2A or category 2B in North America. The product may also go from a nonenvironmentally hazardous chemical in the United States (and Canada) to a fully classified and category 1 "very toxic to aquatic life" when shipped anywhere outside these two countries.

These fundamental differences have a critical impact when looking at the correct classification, labeling, and SDS for a given product line that is to be sold in these different regions.

Considering a situation where a company wishes to sell the same product lines in all of the said countries and continents, there is a need for major strategic decisions to be made in regard to classification and the development of the mandatory documentation and its distribution worldwide.

More specifically, if a lubricating company wishes to have one classification for its entire line of products, it must include the applicable environmental hazards. Doing so would likely create huge competitive and/or financial disadvantages for North American branches and divisions. Doing otherwise would represent an important cost in the creation and management of labeling international versus domestic shipments, using different classifications, SDSs, and labels for each region.

In some cases regarding the same classification for a set of chemicals, for example, "skin and eye irritant, nonenvironmentally hazardous," the labeling of such a product would still fall under countries' specific requirements for shipments in North America, as opposed to those in Europe, for several reasons. The name of the hazardous ingredient(s) responsible for the hazard classification must appear on European labels, not in North America. In North America, the labels must be issued in English, French, and Spanish (Mexico, Central and South America—except Brazil) as opposed to a spoken language of the country of destination in Europe.

Furthermore, the Safety Data Sheets would be disharmonized between different countries for several reasons. First, the mandatory 16-section format adopted under HAZCOM 2012 in the United States and the WHMIS 2015 in Canada, with some slight adjustments from the HAZCOM 2012 format, versus the formats required under the European CLP-Reach; they differ in critical sections 1, 3, 8, 12, 13, 14, and 15. Section 1 must provide information for local contact(s) in an effort to group as many countries as possible within a single document. In Europe, all sections of the SDS must be subnumbered (1.1, 1.2, 1.3, 1.4, and 1.5), while this subnumbering is not required outside of Europe. Section 3 in Europe must show each disclosed ingredient together with its GHS hazard classification at 100% concentration level; this information is not mandatory or even suggested in the United States and Canada. In Europe, the disclosed ingredient(s) must be specifically listed because they have an Occupational Exposure Limit (OEL) in the country for which the SDS is written, and because it is hazardous, the disclosure of ingredient(s) with *only* an applicable OEL in the United States (and Canada) is not mandatory in that section (they must only be shown with these values in Section 8). Sections 12, 13, 14, and 15 are considered optional for the OSHA and Health Canada inspector compliance review. Sections 12 and 13 are deferred to the countries' specific Environmental Protection Agencies (CEPA in Canada and EPA in the United States), Section 14 covers the transportation of dangerous goods and should be revised by the United States and Canadian transportation authorities, and Section 15 should be revised by the member states (California's Proposition 65 and New Jersey, Pennsylvania, and other states' Right-to-Know lists) and against other federal agencies' lists (EPA, SARA, TSCA, CEPA, etc.). In Europe, these last two sections (14 and 15) would need to show the European country's specific transport classification and member states' lists (occupational diseases in France and German Water Klasse [GWK] in Germany) for instance. Finally, as previously indicated, all Safety Data Sheets must be published in a language spoken in the country of destination.

Another point of consideration is the additional complexity created by hazardous ingredient(s) contributing to the hazard classification of a given mixture being a "trade secret" or CBI. This situation varies for each manufacturer and each of their products in a given market and is governed under country-specific legislation and processes.

In the United States, a manufacturer claiming a trade secret ingredient has the right not to disclose its chemical identity (name and CAS number) and exact concentration (using unlimited ranges). In Europe, the use of a trade secret is strictly enforced by a request made in given member states and carried out in others; however, this is only possible for some low-hazard ingredients. In Canada, the authority (Health Canada) must receive a formal request and be paid a fee to issue a Trade Secret Registration Number to be shown in the documentation with the enforcement date. Other countries not specifically mentioned herein may forbid the use of trade secrets completely.

When using trade secret raw material in a finished product, the non-European country's SDS will likely not disclose enough information on the ingredient's specific hazard classification and its respective GHS hazard classification (at 100%) for the downstream user to perform the exact recalculation of its final mixture's hazards. If a European version of this product's SDS is available, supplemental information such as the specific ingredient disclosure and 100% concentration level (pure substance) GHS hazard classification are displayed, but the formulation of the product may not be the same for both continents unless specifically confirmed by the raw material manufacturer.

29.3.1 Strategies and Possible Solutions

In today's world, IT software systems and specialized service companies represent two distinct and possibly complementary options for these global manufacturing and distribution companies. They ensure the correct classification, documentation, and product distribution. Multiple cost analyses have been published to demonstrate when an investment is best recommended toward using software and an internal team of professionals as opposed to outsourcing a solution to ensure total and global compliance. Some large companies have even decided to go with a mixed approach, managing their domestic shipments using software while relying on outsourcing to manage their international requirements.

One can imagine that the challenge differs greatly if a company sells its products only in a single country or continent.

Unfortunately it is clear that the main objectives of the GHS—to have chemicals classified, labeled, and shipped under the same rules and with the same documentation—is definitely unachievable for multinational organizations. Therefore, each company must optimize its own approach by considering the markets for the distribution of its products, the volume of sales in each country, and the investment needed for an internal or an external solution.

Taking advantage of better harmonization between countries of the same continent (the Americas or Europe) is an important key to success. Assuming that the label content is the same, with language being the only real challenge, is more feasible than trying to fit all the countries of the world on a single pealing label.

Another important item to consider that is specific to the lubricant manufacturers is the additives that are normally mixtures of 2, 3, 5, or more ingredients that are often added in relatively small amounts (less than 10%, 5%, or even 1%) in a final lubricant product. One must imagine that before the GHS, the classification of the final lubricant was often not impacted by the additives because each and every (or most) of the hazardous ingredients would fall under the concentration threshold to be considered in the hazard assessment and classification. Now, with the summation rules for health and environmental hazards, along with the larger ranges of toxicity and ecotoxicity (Categories 1–5 depending on whether acute or chronic and on the country of destination), it is increasingly possible that the hazardous ingredients found in the additives would not only impact the classification of the lubricant, but may even make a nonhazardous pre-GHS lubricant have to be labeled as hazardous under the GHS.

Coupled with the said challenges is the fact that multiple additives sold in the United States are not displaying their hazardous ingredients because of trade secrets. In these cases, if a compliant SDS is provided, it would give the manufacturer/blender of the lubricant a GHS classification for the entire additive submixture without the specific hazard contribution of each of its ingredients. The hazard contribution would be kept as CBI. Using the classification of the additive as a whole would result in the over classification of the lubricant final product. The manufacturer may have no choice but to comply and be penalized for a more hazardous lubricant than a competitor using the same additive made by another provider showing all the GHS hazardous ingredients needing to be disclosed by law, with their respective name and CAS numbers.

This difficult situation could be handled better in respect to additive manufacturers and their desire not to disclose formulations, while allowing lubricant manufacturers to classify their final lubricant adequately. This could be done by adding, for each health- or environmentally hazardous ingredient with a generic name, a reasonably correct and narrow concentration range, along with the specific hazard classification of the ingredient at 100% level *(as is currently mandatory in Europe)*. Section 3 of the additive manufacturers' SDSs would protect them for not disclosing their formulations, while allowing the blenders of the lubricants to properly classify the resulting hazards of their finished products.

This simple addition in Section 3 of the HAZCOM 2012 (United States) and WHMIS 2015 (Canada) would enhance the harmonization of the format and content with Europe. It would also allow CBI or trade secrets to be used and allow for the correct calculation of all mixtures containing raw materials that fall under industry protection.

In conclusion, we strongly recommend that the additional information (hazards and concentration ranges) presented be added to Section 3 of GHS SDSs in circulation in all countries so we can better harmonize all documents while protecting the right for additive companies not to disclose the identity of critical ingredients in their formulations thus allowing the downstream users of these products to correctly classify their products under the GHS.

30 Long-Term Trends in Industrial Lubricant Additives

*Fay Linn Lee and John W. Harris**

CONTENTS

---

* In memory of John W. Harris, his Shell colleagues—Howard B. Mead (United-Kingdom), Cliff Henderson, and Fay Linn Lee (United States)—have contributed this chapter update on behalf of John. In the technical field of industrial lubricants and greases, John was recognized as an international expert by Shell Oil colleagues and by his peers in professional technical societies, including the American Society for Testing and Materials (ASTM), the National Lubricating Grease Institute (NLGI), the European Lubricating Grease Institute, and the Society of Tribologists and Lubrication Engineers. He wrote and presented numerous technical papers and patents and contributed to the first edition of this book. As a technical advisor in Shell Global Solutions, John was known as Dr. AeroShell Grease for his contributions in developing and patenting aircraft lubricants. Dr. John W. Harris passed away on November 21, 2005.

30.1 INTRODUCTION

The formulation of the various types of lubricating oils used in industry varies significantly depending on their application. The types of industrial oil lubricants considered in this chapter, in decreasing order of volumes consumed by industry, are as follows:

1. Circulating oils, including turbine oils
2. Hydraulic oils
3. Gear oils
4. Compressor oils

Trends in lubricating grease formulations are also considered.

Performance requirements relevant to each type of lubricant are listed in Table 30.1. As may be expected, the greater the number of performance properties required of a lubricant, the more complex the formulation.

30.2 ADDITIVES FOR CIRCULATING OILS AND TURBINE OILS

30.2.1 PERFORMANCE REQUIREMENTS FOR CIRCULATING OILS AND TURBINE OILS

The following performance requirements are typically required for circulating oils and, in particular, steam and gas turbine oils:

1. Provide bearing lubrication.
2. Remove heat through circulation.
3. Serve as hydraulic fluid for governor and other control equipment.
4. Lubricate reducing gears.
5. Protect against corrosion.
6. Allow rapid separation of water.
7. Resist foaming.
8. Resist oxidation and sludge formation.

As gas turbine oils typically operate at higher temperatures than steam turbine oils, requirement 8 is even more critical for gas turbine oils.

Various military and civilian agencies and turbine builders issue specifications for turbine oils [1,2]. Performance tests that usually appear in these specifications are listed in Table 30.2.

30.2.2 ADDITIVES

The types of additives commonly used in turbine oils are listed in Table 30.2 with the performance properties they impart and the related test methods.

30.2.2.1 Antioxidants

Antioxidants are used to inhibit the attack of turbine oil by oxygen and to reduce the formation of adverse oxidation products such as corrosive organic acids and insoluble sludge. In turbine oils, this is often accomplished with synergistic blends of hindered phenols and diaryl amines, which inhibit oxidation by donating a hydrogen atom to reactive peroxy radical intermediates formed during the oxidation process, thus terminating a chain reaction that would otherwise propagate the degradation of more oil molecules [3–7]. Antioxidants that deactivate peroxy radicals are termed *primary antioxidants*. Volatile antioxidants are generally avoided because they can be depleted by evaporation during use.

Another type of antioxidant inhibits oil oxidation by decomposing reactive hydroperoxide intermediates to less-reactive chemical species. Otherwise, these hydroperoxides would decompose to reactive radicals, resulting in the propagation of the oxidation process. Antioxidants that act as peroxide decomposers are termed *secondary antioxidants* [5]. Types of additives that fall into this class include thioethers, zinc dialkyldithiophosphates (ZDDPs), zinc dialkyldithiocarbamates, and organic phosphites.

Industrial oil formulators exploit the synergistic relationship between hindered phenols and diaryl amines and between primary antioxidants and secondary antioxidants [8]. Manufacturers of antioxidants have also developed additives that contain both the phenol moiety and the thioether moiety in the same molecule so that one additive can potentially act as both primary antioxidant and secondary antioxidant [5].

TABLE 30.1

Performance Requirements for Industrial Oils

Type of Oil	Performance Property				
	Rust and Oxidation Inhibition	Ability to Resist Foaming	Ability to Separate Water	Antiwear Properties	EP
Circulating and turbine oils	X	X	X	—	—
Hydraulic oils	X	X	X	X	—
Compressor oils	X	X	X	X	—
Gear oil	X	X	X	X	X
Lubricating grease	X	—	—	X	X

TABLE 30.2
Additives Used in Turbine Oils

Additive Function	Types of Additives	Related Performance Tests
Antioxidant	Diaryl amines	ASTM[a] D943, standard test method for oxidation characteristics of inhibited mineral oils
	Hindered phenols	ASTM D2272, standard test method for oxidation stability of steam turbine oils by rotating pressure vessel
	Organic sulfides	
		ASTM D4310, standard test method for the determination of the sludging and corrosion tendencies of inhibited mineral oils
		ASTM D6186, standard test method for oxidation induction time of lubricating oils by pressure differential scanning calorimetry (PDSC)
Rust inhibitor	Alkylsuccinic acid derivatives	ASTM D665, standard test method for rust-preventing characteristics of inhibited mineral oil in the presence of water
	Ethoxylated phenols	
	Imidazoline derivatives	
Foam inhibitor	Polydimethylsiloxanes/polyacrylates	ASTM D892, standard test method for foaming characteristics of lubricating oils
Metal deactivator	Triazoles	ASTM D130, standard test method for detection of copper corrosion from petroleum products by the copper strip tarnish test
	Benzotriazoles	
	2-Mercaptobenzothiazoles	
	Tolutriazole derivatives	
Mild antiwear/EP additive	Alkylphosphoric acid esters and salts	ASTM D4172, standard test method for wear preventive characteristics of lubricating fluid (four-ball method)
		ASTM D5182, standard test method for evaluating the scuffing load capacity of oils (FZG visual method)
Demulsifier	Polyalkoxylated phenols	ASTM D1401, standard test method for water separability of petroleum oils and synthetic fluids
	Polyalkoxylated polyols	
	Polyalkoxylated polyamines	

[a] American Society for Testing and Materials, *Annual Book of ASTM Standards: Petroleum Products, Lubricants and Fossil Fuels*, Vols. 5.01–5.04, issued annually.

30.2.2.2 Rust Inhibitors

Turbine oils are formulated to prevent rust in areas in contact with the oil. Typical turbine oil specifications require that the oil pass corrosion tests using both demineralized water and saltwater, ASTM D665, procedures A and B, respectively. Anticorrosion additives are generally polar in nature and preferentially form water-resistant films on iron or steel surfaces [5]. Such additives should not be readily leached from the turbine oil by water entrained in the circulating oil and should not adversely affect other desirable performance properties such as fast water separation and low-foaming tendencies. Types of rust inhibitors that may be used in turbine oils are listed in Table 30.2.

Rust inhibitor–related compatibility problems can occur if certain turbine oils are contaminated with industrial oils that contain organometallic performance additives. For example, antiwear hydraulic oils may contain zinc-based antiwear additives and metal sulfonate rust inhibitors, and interactions between acidic rust inhibitors in the turbine oil and metal-containing additives in the antiwear hydraulic oil can result in the formation of insoluble salts, which can plug filters. Similarly, if a new turbine component is installed in a unit and metal sulfonate rust preservatives on the component parts have not been completely washed off, a similar filter-plugging situation can occur.

30.2.2.3 Metal Deactivators

Metal deactivators prevent the corrosion of copper and copper alloys such as bronze and brass. They function by preferentially adsorbing onto the surface of copper alloys, forming a protective layer that inhibits attack by acids formed during the oxidation of the bulk lubricant [5,9]. Conversely, the metal deactivator can act as a copper passivator to protect the oil from the catalytic effects of the copper alloy, which could accelerate bulk oil oxidation. Table 30.2 lists the types of metal deactivators that may be used in turbine oils.

30.2.2.4 Foam Inhibitor

Antifoam additives of the type listed in Table 30.2 are added to turbine oils to control foaming tendencies. One method for measuring foaming tendencies is ASTM D892. Antifoam additives are not completely soluble in oil and form dispersions of tiny droplets of low surface tension, which accelerate the coalescence of air bubbles in oil. Typical dose rates for silicone antifoam additives are in the 10–20 ppm range, whereas typical dose rates for polyacrylates are in the 50–200 ppm range. Because antifoam additives are only partially soluble in oil, they can eventually separate from the oil, resulting in decreased foam control.

It is important that the antifoam additives do not adversely affect the air release properties of the turbine oil, or otherwise, air entrainment in the oil can result in the sluggish operation of hydraulic systems and other problems. Air release is measured by the method ASTM D3427. Typically, values of less than 5–10 min are desirable. Overdosing with antifoam additive can adversely affect air release properties.

30.2.2.5 Demulsifiers

Demulsifiers facilitate water separation from oil. They concentrate at the water–oil interface and promote the coalescence of water droplets, resulting in faster separation from water so that it can settle to the bottom of the oil reservoir and be drained off. Types of demulsifiers are listed in Table 30.2.

30.2.3 TURBINE OIL MONITORING

Despite the use of performance additives, turbine oil properties may deteriorate in use, and guidelines for condemnation values for various physical properties of turbine oils have been developed. Such guidelines are described in ASTM D4378. Turbine manufacturers may also have their own set of monitoring guidelines for turbine oils.

30.2.4 BASE OIL EFFECTS ON ADDITIVES

Modern engine oil requirements have driven changes in the mineral oil base stocks available for formulating lubricants, including industrial lubricants. Three of the types of base oils defined by the American Petroleum Institute (API) are described as follows [10]:

1. Group I base stocks contain less than 90% saturates and greater than 0.03% sulfur and have a viscosity index (VI) greater than or equal to 80 and less than 120.
2. Group II base stocks contain greater than or equal to 90% saturates and less than or equal to 0.03% sulfur and have a VI greater than or equal to 80 and less than 120.
3. Group III base stocks contain greater than or equal to 90% saturates and less than or equal to 0.03% sulfur and have a VI greater than or equal to 120.

A step-change improvement in the oxidation performance of mineral oil–based turbine oils has been achieved by using group II base oils in place of group I base oils. Group II base oils contain significantly lower levels of aromatic compounds, which are susceptible to oxidation, than group I base oils. However, group II base oils also contain lower levels of sulfur components, some of which can act as secondary oxidation inhibitors. It has been reported that the oxidation resistance of group II base oil blends can be further enhanced by adding sulfur-containing secondary oxidation inhibitors along with optimized combinations of hindered phenols and diaryl amine primary antioxidants [8]. Owing to the lower solvency of group II base oils, resulting from the lower levels of aromatic compounds, not all the additive formulations that performed satisfactorily in group I base stocks display adequate solubility in group II base stocks [11].

30.2.5 ENVIRONMENTAL EFFECTS OF TURBINE OILS

As turbine oil additives are metal-free and are used at low rates, typically less than 1%, the environmental impact of turbine oil additives is generally fairly minimal. In some cases, local wastewater treatment requirements may be such that turbine oil users prefer turbine oils free from detectable phenolic antioxidants.

30.2.6 TURBINE OIL TRENDS

As the oxidation life of turbine oils is extended, new test methods will be needed to evaluate field performance. New turbine oils formulated with group II base oils typically have ASTM D943 lives between 10,000 and 30,000 h, making them impractical to characterize by this method. The ASTM D943 method primarily monitors acid formation as the results of oxidation. As sludge formation can affect the operation of critical equipment, such as servo valves, more emphasis will be placed on measuring the tendency of turbine oils to form sludge and varnish and on choosing additives that minimize the formation of insoluble oxidation products. Wider availability of group III base oils in the future may lead to the development of turbine oil formulations with even longer performance lives.

30.3 ADDITIVES FOR HYDRAULIC OILS

30.3.1 PERFORMANCE REQUIREMENTS

There are three basic types of hydraulic pumps: gear pumps, vane pumps, and piston pumps. Each is designed to perform specific tasks. Generally, vane pumps operating under high pressure have the greatest requirements for wear protection by hydraulic fluids [12].

The main functions of hydraulic oils are as follows:

1. Transmit power.
2. Protect and lubricate system components.
3. Remove heat through circulation.
4. Protect the component parts from oxidation and corrosion.
5. Provide wear protection.
6. Provide mild extreme-pressure (EP) performance.

Hydraulic oils are expected to meet the following performance requirements:

1. Consistent performance and appropriate viscosity and compressibility
2. Corrosion protection
3. Wear protection and oxidation and thermal stability
4. Hydrolytic stability
5. Long life
6. Filterability
7. Compatibility with system components
8. Good demulsibility or emulsibility, depending on application

The ASTM has issued a standard specification for mineral oil–based hydraulic oil, ASTM D6158. Many pump manufacturers have issued hydraulic oil specifications, which include satisfactory pump performance under severe conditions. A number of these specifications have been summarized in the literature [13], as have military specifications for hydraulic oils [2]. Performance tests that often appear in hydraulic oil specifications are listed in Table 30.3.

30.3.2 Additives

The types of additives commonly used in hydraulic oils are listed in Table 30.3.

30.3.2.1 Antioxidants

In addition to the types of antioxidants described previously for turbine oils, hydraulic oils may use organometallic antioxidants such as ZDDP [15,16] additives. The ZDDP additives are multifunctional, acting as secondary antioxidants and antiwear additives. The types of antioxidants that can be used in hydraulic oils are listed in Table 30.3.

30.3.2.2 Rust Inhibitors

Rust inhibitors used in hydraulic oils may contain metals, such as metal sulfonates, and may be overbased to provide the capability to neutralize acidic oxidation products. Various rust inhibitors that may be used in hydraulic oils are listed in Table 30.3.

30.3.2.3 Antiwear Additives

Antiwear additives are needed to inhibit frictional wear of moving parts such as the sliding vanes of vane pumps. These additives react with metal surfaces, forming sacrificial films with lower shear strength than the metal itself. The sacrificial films help to reduce friction, and although a small amount of metal is removed with the sacrificial film as it is

TABLE 30.3
Additives Used in Industrial Hydraulic Oils

Additive Function	Types of Additives	Related Performance Tests
Antioxidant	Hindered phenols	ASTM D943, standard test method for oxidation characteristics of inhibited mineral oils
	Diaryl amines	
	Phenothiazine	
	Metal dialkyldithiocarbamates	ASTM D2272, standard test method for oxidation stability of steam turbine oils by rotating pressure vessel
	Ashless dialkyldithiophosphates	
	Metal dialkyldithiophosphates	ASTM D4310, standard test method for the determination of the sludging and corrosion tendencies of inhibited mineral oils
		ASTM D6186, standard test method for oxidation induction time of lubricating oils by pressure differential scanning calorimetry (PDSC)
Rust inhibitor	Alkylsuccinic acid derivatives	ASTM D665, standard test method for rust-preventing characteristics of inhibited mineral oil in presence of water
	Ethoxylated phenols	
	Fatty amines	
	Salts of fatty acids and amines	
	Salts of phosphate esters and amines	
	Metal sulfonates	
	Ammonium sulfonates	
	Imidazoline derivatives	
Metal deactivator	Benzotriazoles	ASTM D130, standard test method for detection of copper corrosion from petroleum products by the copper strip tarnish test
	2-Mercaptobenzothiazoles	
	Thiadiazoles	
	Tolutriazole derivatives	
Antiwear additive	Alkylphosphoric acid esters and salts	ASTM D4172, standard test method for wear preventive characteristics of lubricating fluid (four-ball method)
	Dialkyldithiophosphates	
	Metal dialkyldithiocarbamates [14]	ASTM D7043, standard test method for indicating wear characteristics of non-petroleum and petroleum hydraulic fluids in a constant volume vane pump various pump tests as specified by pump manufacturers
	Phosphate esters	
	Dithiophosphate esters	
	Derivatives of 2,5-dimercapto- 1,3,4-thiadiazoles	
	Molybdenum carboxylates	ASTM D5182, standard test method for evaluating the scuffing load capacity of oils (FZG visual method)
Demulsifiers	Polyalkoxylated phenols	ASTM D1401, standard test method for water separability of petroleum oils and synthetic fluids
	Polyalkoxylated polyols	
	Polyalkoxylated polyamines	ASTM D2711, standard test method for demulsibility characteristics of lubricating oils
PPDs	Poly(alkyl methacrylates)	ASTM D97, standard test method for pour point of petroleum product
		ASTM D5949, standard test method for pour point of petroleum products
Foam inhibitors	Polydimethylsiloxanes	ASMT D892, standard test method for foaming characteristics of lubricating oils
Shear stable VI improvers	Poly(alkyl methacrylates)	ASTM D5621, standard test method for sonic shear stability of hydraulic fluids
	Olefin copolymers	
	Styrene diene copolymers	

rubbed off by the moving parts, the rate of metal removal is much lower than if the sacrificial film was not present.

The types of antiwear additives used in hydraulic oils are listed in Table 30.3. Most hydraulic oil formulations contain ZDDP antiwear additives. The composition of the films that ZDDP additives deposit on metal surfaces has been studied and determined to be complex [17]. The popularity of these antiwear additives is due to their cost-effectiveness as they are multifunctional, acting both as antiwear additives and as antioxidants. Although ZDDP additives provide excellent antiwear properties, they can also cause corrosion and deposit formation under conditions that promote their hydrolysis or thermal decomposition. By judicious selection of the types of alcohols used to form the phosphate ester portion of the ZDDP molecule, additive developers can maximize hydrolytic and thermal stability while maintaining good antiwear properties. For example, ZDDP antiwear additives derived from primary alcohols have better thermal stability than the corresponding ZDDP antiwear additives derived from secondary alcohols.

Environmentally aware hydraulic oils are formulated with zinc-free, ashless antiwear additives [12,18]. These additives are generally organic compounds containing sulfur and phosphorus. As with the ZDDP antiwear additives, the ashless antiwear additives must be designed in such a way as to maximize hydrolytic and thermal stability as well as improve antiwear performance. It has been claimed that ashless antiwear oils can exhibit better filterability properties than antiwear oils containing ZDDP [12,19]. Because the possibility exists that ashless and zinc-containing antiwear oils may be inadvertently mixed in commercial applications, it is best to formulate ashless antiwear oils in such a way that they will be compatible with antiwear additives containing ZDDP so that no precipitates that could plug filters are formed [12].

Test methods have been developed to evaluate the thermal and hydrolytic stability of antiwear hydraulic oils. Test method ASTM D2070 evaluates the thermal stability of an oil by heating the test oil in the presence of copper and iron rods for 168 h at 135°C. At the end of the test, the metal rods are rated by visual examination and the sludge is isolated by filtration and weighed. Test method ASTM D2619 evaluates hydrolytic stability by heating and agitating the test oil with water and a copper strip in a sealed glass bottle for 48 h at 93°C. At the end of the test, the viscosity change of the oil and the acidity changes of the oil and the water phases are measured, the weight loss of the copper strip is determined, and any resulting insoluble material is isolated and weighed.

30.3.2.4 Metal Deactivators

Owing to the potential corrosiveness of the decomposition products of the antiwear additives, the presence of an effective metal deactivator is even more important in antiwear hydraulic oils than in turbine oils. Table 30.3 lists the types of metal deactivators that can be used in hydraulic oils.

30.3.2.5 Foam Inhibitor

Because air is compressible, foam or entrained air interferes with the primary function of a hydraulic oil, which is to transfer power. Therefore, the type and amount of silicone or polyacrylate antifoam additive must be carefully chosen to maximize foam inhibition while maintaining good air release properties. Some original equipment manufacturers are extremely concerned about the presence of silicone materials in their plants as it is believed that even a trace of silicone on a metal surface can affect paint coating properties. Therefore, these manufacturers require that only nonsilicone antifoam additives be used in lubricating oils employed in their plants.

The choice of antifoam additive may also be affected by the practice of hydraulic oil condition monitoring. Often, optical particle counters based on light scattering techniques are used to monitor oil cleanliness. As antifoam additives are finely dispersed, partially insoluble fluids in oil, they can falsely be registered as particles by some optical particle counters. In the future, one criterion for choosing antifoam additives may be their ability to disperse sufficiently so as not to interfere with the function of optical particle counters.

30.3.2.6 Viscosity Index Improver

VI improvers are used to widen the useful operating temperature range of hydraulic oils. They are long-chain polymers that function by uncoiling or dissociating at elevated temperatures, resulting in thickening of the oil. Owing to this property, oils containing VI improvers will display higher measured kinetic viscosities at elevated temperatures than comparable oils that have the same 40°C oil viscosity but do not contain VI improver. The types of polymers used as VI improvers are listed in Table 30.3.

An important property of VI improvers for hydraulic oils is shear stability. If a VI improver lacks shear stability, it will quickly be degraded by the hydraulic pump and will lose its ability to increase the viscosity of the base oil at elevated temperatures. This will result in less-than-optimum lubricant film thickness and accelerated pump wear. Generally, the shear stability of VI improvers increases as the molecular weight decreases. However, the thickening capability of the VI improver also decreases as the molecular weight decreases. A trend in the industry is to use VI improvers with better shear stability for better pump protection. Therefore, lower-molecular-weight VI improvers are used although they must be used at higher percentages by weight.

One method for determining the shear stability of hydraulic oil containing a VI improver, ASTM D5621, is based on sonic irradiation. In this method, a sample of test oil is irradiated in a sonic oscillator for a specified time interval and the change in viscosity after shear is determined. It is claimed that the amount of shear measured by this method is similar to the amount of shear imparted to a hydraulic oil in a hydraulic vane pump test [20].

30.3.2.7 Pour Point Depressants

Pour point depressant (PPD) additives are chemicals that help mineral oil–based hydraulic oils remain pumpable at low temperatures by lowering the pour point of the oil. The PPD additives are polymers that interfere with the crystallization of wax from mineral oil at low temperatures. By retarding the formation of a wax crystal network in oil at low temperatures, PPDs can reduce the pour points of some paraffinic oils by as much as 30°C–40°C. The effect is less marked in nonparaffinic oils, which have lower wax contents.

30.3.3 ENVIRONMENTAL CONCERNS

Hydraulic oils have received considerable attention regarding their impact on the environment. Hydraulic oil lines carry oil under moderate to high pressure; therefore, rupture of a line in a forest or near a waterway can result in the release of a significant amount of hydraulic fluid into an environmentally sensitive area. Many agencies have drafted methods for evaluating the environmental impact of hydraulic oils and for defining what can be classified as a rapidly biodegradable hydraulic oil. For example, the ASTM has issued the following standards and methods related to the evaluation of the impact of hydraulic oils on the environment:

1. D 5864, Standard Test Method for Determining the Aerobic Aquatic Biodegradation of Lubricants or Their Components
2. D 6006, Standard Guide for Assessing Biodegradability of Hydraulic Fluids
3. D 6046, Standard Classification of Hydraulic Fluids for Environmental Impact
4. D 6081, Standard Practice for Aquatic Toxicity Testing of Lubricants: Sample Preparation and Results Interpretation

Generally, biodegradable hydraulic oils are based on natural or synthetic esters. Additives that are effective with these types of oils [21] and that are environmentally aware must be chosen. Regulations governing environmentally aware hydraulic oils generally require that they be metal-free, readily biodegradable, display low ecotoxicity in water and soil, and exhibit a low tendency for bioaccumulation [22,23].

30.3.4 FUTURE TRENDS

An ongoing trend toward reducing the cost of hydraulic equipment by making it smaller obviates the need for oils that function at higher temperatures and with shorter oil sump residence times. Also, an environmentally driven trend to minimize waste oil disposal will require lubricating oils with longer functional lives. These trends will require oils with improved oxidation resistance, water separation, and air release properties. Some of these improvements will be accomplished by upgrading base oil quality, first from group I to group II base oils and then to group III base oils. The remainder of the improved performance will be accomplished with improved additives that are also environmentally friendly. As base oils are upgraded, changes in additives may also be required to ensure additive–base oil compatibility.

30.4 ADDITIVES FOR GEAR OILS

30.4.1 PERFORMANCE REQUIREMENTS FOR GEAR OILS

Most industrial operations use equipment driven by one type of gear or another, for example, rock drills, crushers, hoists, mobile equipment, grinders, mixers, and speed increases/reducers.

There are many combinations of gear types and materials; therefore, the requirements for gear oils are varied [24,25]. Some of the common types of gears are listed in Table 30.4 and illustrated in Figures 30.1 through 30.7. The relative amount of sliding and rolling motions between interacting gears can be quite different between gear types. For example, worm gears interact by a predominantly sliding motion,

TABLE 30.4
Common Gear Types

Gear Type	Configuration of Gear	Comments
Spur gear (Figure 30.1)	Straight teeth cut parallel to the axis of the gear	For moderate loads and moderate speeds
Helical gear (Figure 30.2)	Teeth cut at an angle to the axis of the gear	For higher speeds and higher loads than spur gears
		Produce axial thrust
Herringbone gear (Figure 30.3)	Teeth cut in a "V" shape	Similar to helical gear, but the "V" shape of the gears eliminates the axial load
Straight bevel gear (Figure 30.4)	Straight teeth cut on an angular surface of a truncated one	For transfer of motion of shafts where shaft centerlines intersect and shafts are at right angles to each other
		For moderate loads and moderate speeds
Spiral bevel gear (Figure 30.5)	Similar to straight bevel gear except the teeth are cut at an angle on a radial line	For higher speeds and higher loads than straight bevel gear
		Quieter operation than straight bevel gear
Hypoid gear (Figure 30.6)	Similar to spiral bevel gear, but teeth cut to accommodate shaft centerlines on different planes	For high loads and quiet running
		Interaction of meshing gear teeth is strictly sliding
		Tooth loads can be high, requiring gear oil with good EP performance
Worm gear (Figure 30.7)	Gear wheel turning at right angles to an offset worm (screw) gear	Interaction of meshing gear teeth is strictly sliding
		Tooth loads can be kept low with proper gear tooth design

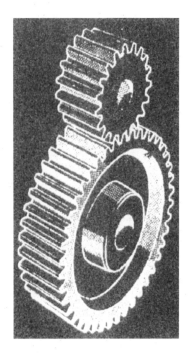

FIGURE 30.1 Spur gears. (Courtesy of Lubrizol Corp, Wickliffe, OH.)

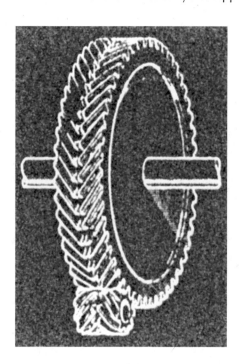

FIGURE 30.3 Herringbone gear. (Courtesy of Lubrizol Corp., Wickliffe, OH.)

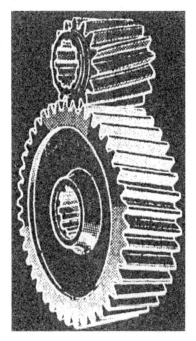

FIGURE 30.2 Helical gear. (Courtesy of Lubrizol Corp, Wickliffe, OH.)

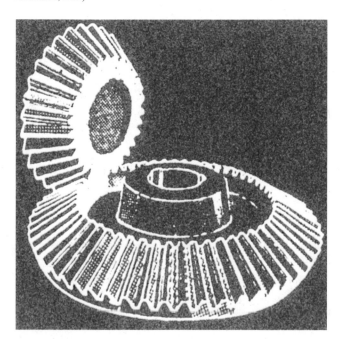

FIGURE 30.4 Straight bevel gear. (Courtesy of Lubrizol Corp., Wickliffe, OH.)

whereas spur gears operate by a combination of rolling and sliding interactions. For straight gears, such as spur gears, only one tooth from each gear supports load at one time. Angled and curved gears are designed to allow more than one gear tooth on each gear to simultaneously support load, resulting in a greater load-carrying capacity.

Because the amount of sliding and rolling action and the load-carrying capacity of the various gear types can vary significantly, their gear oil performance requirements can vary substantially. For example, lightly loaded spur gears require

an oil with only rust and oxidation inhibitors, whereas heavily loaded hypoid gears require oils with high levels of EP additives. In the case of worm gears, almost all their action is sliding, the teeth are generally not heavily loaded, and the smaller worm gears may be made of bronze for better sliding wear resistance. For these types of gears, an oil compounded with a friction modifier, such as acidless tallow, may be sufficient.

The American Gear Manufacturers Association (AGMA) provides a series of gear oil specifications for industrial enclosed gear drives and industrial open gearing. The AGMA

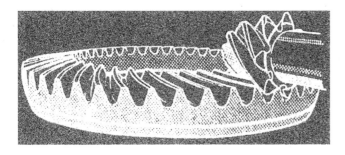

FIGURE 30.5 Spiral bevel gear. (Courtesy of Lubrizol Corp., Wickliffe, OH.)

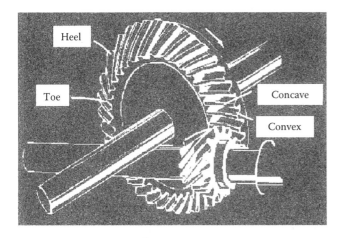

FIGURE 30.6 Hypoid gear. (Courtesy of Lubrizol Corp., Wickliffe, OH.)

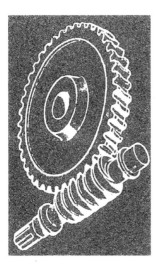

FIGURE 30.7 Worm gear. (Courtesy of Lubrizol Corp., Wickliffe, OH.)

classifications provide for three types of oil in various viscosity grades. Oils are classified as those containing rust and oxidation inhibitors only, those that also contain EP additives, and those containing fatty material along with rust and oxidation inhibitors. The fatty material, such as acidless tallow, provides mild EP and friction reduction for predominantly sliding applications such as worm gears. The types of performance tests that appear in gear oil specifications are listed in Table 30.5.

30.4.2 ADDITIVES FOR GEAR OILS

The types of additives used in industrial gear oils are listed in Table 30.5.

Many of the types of additives used in hydraulic oils are also used in gear oils. Additives that differ from those used in hydraulic oils are discussed next.

30.4.2.1 Extreme-Pressure Additives

Organic sulfur and phosphorus compounds and inorganic borates are primary components in gear oil EP additives [33]. The roles of these various elements in gear oil packages have been studied and reviewed [34]. The phosphorus in sulfur/phosphorus EP additives yields low wear rates under normal operating conditions, whereas the sulfur prevents seizure under high loads. Modern gear oils often contain borates as well. Finely divided inorganic borates dispersed in gear oil can further enhance load-carrying capacities at high and low speeds, possibly by forming a chemically bound borate film on the iron surfaces [34].

Combinations of organic phosphorus- and chlorine-containing compounds have also been used as EP additives for gear oils. However, it is believed that low melt temperatures of iron chlorides are a limiting factor as well as the potential for organic chlorides to decompose in the presence of water and yield hydrochloric acid [34].

30.4.2.2 Friction Modifiers

For applications involving primarily sliding contact, such as worm gears, friction modifiers such as acidless tallow may be used. In these applications, friction modifiers can be more effective than sulfur-containing EP additives, which might promote the corrosion of bronze worm gears.

30.4.2.3 Viscosity Index Improvers

Because gear oils can be subjected to even greater mechanical shear than hydraulic oils, the polymers used in VI improvers in gear oils must possess exceptional shear stability. It is preferable to use low-molecular-weight VI improvers in gear oils as low-molecular-weight polymers display greater shear stability than high-molecular-weight polymers. The use of lower-molecular-weight VI improvers increases formulation costs as they must be used at higher percentages than high-molecular-weight VI improvers to achieve the same VI increase. Despite the higher costs associated with the use of lower-molecular-weight VI improvers, this approach is necessary to maintain gear oil viscosity grade and lubricating film thickness during operation.

30.4.2.4 Antimisting Additives

Mist lubrication is an effective means of lubricating gears and other elements; however, care must be taken to minimize stray mist, which could cause health problems for those working in the vicinity. Polymeric additives, such as polyisobutenes, can be used to promote the coalescence of oil drops and reduce the formation of stray mist [30–32].

TABLE 30.5
Additives Used in Industrial Gear Oils

Additive Function	Types of Additives	Related Performance Tests
Antioxidant	Hindered phenols	ASTM D6186, standard test method for oxidation induction time of
	Diaryl amines	lubricating oils by pressure differential scanning calorimetry (PDSC)
	Phenothiazine	ASTM D2893, standard test method for oxidation characteristics of
	Metal dialkyldithiocarbamates	extreme pressure lubricating oils
	Ashless dialkyldithiocarbamates	ASTM D665, standard test method for rust-preventing characteristics of
	Metal dialkyldithiophosphates	inhibited mineral oil in presence of water
Rust inhibitor	Alkylsuccinic acid derivatives	
	Ethoxylated phenols	
	Fatty amines	
	Salts of fatty acids and amines	
	Salts of phosphate esters and amines	
	Metal sulfonates	
	Ammonium sulfonates	
	Substituted imidazolines	
Metal deactivator	Benzotriazoles	ASTM D130, standard test method for detection of copper corrosion from
	2-Mercaptobenzothiazoles	petroleum products by the copper strips tarnish test
	Thiadiazoles	
	Tolutriazole derivatives	
Antiwear additive	Alkylphosphoric acid esters and salts	ASTM D4172, standard test method for wear preventive characteristics of
	Dialkyldithiophosphates	lubricating fluid (four-ball method)
	Metal dialkyldithiocarbamates	
	Phosphate esters	
	Dithiophosphate esters	
	Derivatives of 2,5-dimercapto-1,3,4-thiadiazoles	
	Molybdenum carboxylates	
Soluble EP additive	Sulfurized esters [26]	ASTM D5182, standard test method for evaluating the scuffing load
	Sulfurized olefins [26]	capacity of oils (FZG visual method)
	Diaryl disulfides [26]	ASTM D2782, standard test method for measurement of extreme-pressure
	Dialkyldithiophosphate esters	properties of lubricating fluids (Timken method)
	Lead naphthenate	ASTM D2783, standard test method for measurement of extreme-pressure
	Bismuth naphthenate [27]	properties of lubricating fluids (four-ball method)
	Antimony dialkyldithiocarbamate	
	Antimony dialkyldithiophosphate	
	Ammonium salts of phosphate esters	
	Chlorinated waxes	
	High-molecular-weight complex esters [28]	
	Borate esters [29]	
Solid EP additive	Molybdenum disulfide	See the above test method for EP properties and wear preventive properties
	Graphite	
Tackifier	Polyisobutylenes	No ASTM standard test method
Demulsifiers	Polyalkoxylated phenols	ASTM D1401, standard test method for water separability of petroleum
	Polyalkoxylated polyols	oils and synthetic fluids
	Polyalkoxylated polyamines	ASTM D2711, standard test method for demulsibility characteristics of
		lubricating oils
Friction modifiers	Acidless tallow	ASTM D5183, standard test method for determination of coefficient of
		friction of lubricants using the four-ball wear test machine
PPDs	Poly(alkyl methacrylates)	ASTM D97, standard test method for pour point of petroleum products
		ASTM D5949, standard test method for pour point of petroleum products
Foam inhibitors	Polydimethylsiloxanes	ASTM D892, standard test method for foaming characteristics of
		lubricating oils
Shear stable VI improvers	Poly(alkyl methacrylates)	Severe shear tests are under development
	Olefin copolymers	
	Styrene diene copolymers	
Antimisting additives	Polyisobutylenes [30–32]	ASTM D3705, standard test method for misting properties of lubricating
		fluids

30.4.3 ADDITIVES FOR OPEN GEAR LUBRICANTS

For large, slow-moving gears that cannot practically be enclosed in oil-tight casings, an adhesive, high-viscosity material is required to stick to the gears. Such open gear lubricants are generally applied by spraying or brushing. In cold climates, a synthetic base oil may be required to provide a high-viscosity base fluid that can be sprayed at low temperatures [35,36]. Such lubricants generally contain the following core components:

1. High-viscosity base oil
2. Oil-soluble EP additives
3. Solid EP additives
4. Tackifier
5. Anticorrosion additive

30.4.4 ENVIRONMENTAL CONCERNS

In the past, gear oils have contained organic derivatives of lead for EP performance, and open gear lubricants have contained chlorinated solvents to improve pumpability. Over the last decade, environmental considerations have essentially resulted in the elimination of these components. Newer open gear lubricant formulations also tend to avoid the use of asphalt on the basis of environmental and performance considerations [35,36].

For applications in which industrial gear oils are used in environmentally sensitive areas such as forests and near waterways, biodegradable gear oils based on natural and synthetic esters have been developed [37].

30.4.5 FUTURE TRENDS

A failure phenomenon known as micropitting has surfaced as a problem in certain applications such as wind turbine gearboxes [38]. Although the failure mode is not completely understood, it is believed to be related to the surface finish of the gears. Test methods have been developed to measure the ability of gear oils to protect against micropitting [39]. There is more to learn in terms of understanding micropitting and minimizing its occurrence.

As in the case of hydraulic pumps, gearbox efficiencies are improved to provide lower operating costs for the end users. This will result in higher gear speeds and loads and higher oil operating temperatures. The challenge for gear oil formulators will be to use base oils and additives that provide

1. Improved thermal stability and cleanliness
2. Improved high-temperature EP performance
3. Improved oxidation resistance
4. Extended demulsibility life
5. Lower friction
6. Improved foam resistance
7. Improved surface fatigue protection

30.5 ADDITIVES FOR COMPRESSOR OILS

30.5.1 PERFORMANCE REQUIREMENTS FOR COMPRESSOR OILS

It is difficult to generalize the lubrication requirements of compressors as requirements depend on a number of parameters such as the type of compressor, the properties of the gas being compressed, and designed discharge temperatures and pressures [40,41].

Compressors can be classified into two types, positive displacement and dynamic, depending on the process by which they compress gas. Compressors based on positive displacement trap the gas, reduce its volume, and then discharge the compressed gas. Dynamic compressors accelerate the gas and convert velocity to increased pressure. Positive-displacement compressors are useful for high-pressure applications with one limiting factor being the discharge gas temperature. Dynamic compressors are useful for high-volume, low-pressure applications.

Positive-displacement compressors can be further classified into reciprocating and rotary compressors. Reciprocating compressors utilize single- or double-acting pistons. Rotary types include rotary vane and rotary screw compressors.

Principal sites for lubrication in reciprocating compressors include the crankshaft and associated bearings, connecting rods, wrist pins, pistons, cylinders, piston rings, and valves. Rotary screw compressor lubrication sites include the rotors, bearings, and shaft seals. Rotary vane compressors require the lubrication of bearings and shafts [41].

Dynamic centrifugal compressors require the lubrication of bearings, gear reducers, and seals [41].

Critical lubrication requirements of the various types of compressors include the following:

1. Oxidation resistance
2. Thermal stability
3. Low carbon deposits at high temperatures (high-pressure reciprocating compressor)
4. Compatibility with gas being compressed
5. Viscosity retention (resistance to dilution by gas being compressed)
6. Antiwear properties (rotary vane)
7. Gas seal provided
8. General lubrication
9. Heat removal
10. Corrosion prevention
11. Compatibility with compressor components
12. Good demulsibility
13. General cleanliness
14. Low-foaming tendencies
15. Low flammability

30.5.2 COMPRESSOR OIL FORMULATION

High-pressure reciprocating compressors require that lubricants should be stable at high temperatures because discharge temperatures can range from 350°F to 500°F. Lubricants for such reciprocating compressors must not generate heavy carbon deposits at hot valve areas as this increases the risk for fires and explosions. Consequently, lubricants for these applications are usually based on synthetic base oils with low carbon-forming tendencies such as diesters and polyol esters.

Although the requirements for compressor oils vary significantly depending on a number of parameters and a substantial portion of compressor oils are based on synthetic fluids, many formulations still rely on the types of additives listed in Tables 30.2 and 30.3 for turbine oils and hydraulic oils. Rotary vane compressors require lubricants that minimize vane wear and vane-sticking problems. In addition to the types of additives already mentioned for turbine and hydraulic oils, detergent additives may be used in some rotary vane compressor oils to help prevent deposit formation and vane sticking.

The formulations of compressor oils can vary significantly depending on the type of compressor and the type of gas being compressed. For example, mineral oils are generally not used for compressing natural gas because natural gas is soluble in mineral oil. The dilution of mineral oil–based compressor oil with natural gas would decrease the viscosity of the oil below an effective lubricating level. In this case, a polyglycol-based oil with limited solubility for natural gas is more suitable.

30.5.3 Compressor Oil Trends

Although mineral oil–based lubricants may be suitable for normal duty with nonreactive gases, a variety of synthetic-based lubricants are used for severe-duty, high-temperature, and high-pressure applications and for gases that are reactive with mineral oils. As a general trend, ester-based compressor lubricants are often used for reciprocating compressors in severe service duty due to their lower tendency to form carbon deposits that can lead to fires in reciprocating compressors. For rotary screw and rotary vane compressors, synthetic poly-α-olefins (PAOs) are growing in popularity due to their good thermal stability and lower cost relative to ester-based fluids. The formulation of compressors oils for severe service is so dependent on equipment design and the chemical nature of the gas being compressed that the compressor manufacturers generally designate the compressor oil formulations that are suitable for their equipment.

30.6 ADDITIVES FOR LUBRICATING GREASES

30.6.1 Performance Requirements for Lubricating Greases

A lubricating grease consists of a lubricating oil, performance additives, and a thickener, which is a fine dispersion of an insoluble solid that is capable of forming a matrix that retains the oil in a semisolid state. Owing to their semisolid consistency, lubricating greases hold some advantages over lubricating oils. A grease is more easily sealed into a bearing than an oil; a grease does not require a circulating system; and in wet or dusty environments, a grease can act as a seal against

contaminants. For these reasons, greases are often the preferred lubricant for ball and roller bearings in electric motors, household appliances, mobile equipment, wheel and chassis lubrication, and some industrial plain and roller bearings.

Owing to the wide variety of applications in which a lubricating grease may be used, there are many different specifications for lubricating greases. Many multipurpose greases meet the LB-GC requirements of ASTM D4950, Standard Classification and Specification for Automotive Service Greases, which is a classification system that was developed by a cooperative effort of the ASTM, the NLGI, and the Society of Automotive Engineers (SAE). The ASTM D4950 specification covers lubricating greases suitable for the periodic relubrication of chassis systems and wheel bearings of passenger cars, trucks, and other vehicles. In the ASTM D4950 classification system, the letters LB signify service typical of lubrication for chassis components and universal joints in passenger cars, trucks, and other vehicles under mild to severe duty. The letters GC signify service typical of lubrication for wheel bearings operating in passenger cars, trucks, and other vehicles under mild to severe duty. It is possible to formulate lubricating greases that meet both the LB and GC specifications, and such greases may display the NLGI GC-LB certification mark [42].

Many military grease specifications have been written [2], and greases meeting particular military specifications have been designated for critical applications such as aviation applications. Additionally, many original equipment manufacturers, such as automobile manufacturers, issue specifications for greases that are specific for the equipment that they build.

The main ingredient that differentiates a grease from an oil is the grease thickener. Most grease thickeners are soaps composed of the reaction product of a fatty acid derivative and a metallic hydroxide. Other salts may be combined with the soap thickeners to impart higher grease dropping points (temperature at which the grease liquefies). Such greases are called *complex greases*. A smaller proportion of lubricating greases are manufactured with nonsoap thickeners such as organoclays or polyurea compounds. Some of the more common thickeners are as follows:

1. Lithium salt of hydrogenated castor oil fatty acid
2. Calcium stearate (hydrous)
3. Calcium stearate (anhydrous)
4. Lithium complex
5. Aluminum complex
6. Calcium complex
7. Organoclay
8. Polyurea
9. Silica

30.6.2 Additive

Performance additives for greases are similar to those used in lubricating oils; however, it is important that the grease additives also be compatible with the grease thickener.

Lack of compatibility between an additive and a grease thickener could lead to the disruption of the thickener matrix that holds the oil in a semisolid state. This would result in an uncontrolled softening or hardening of the grease or a decrease in the grease dropping point.

Various grease thickeners behave differently with respect to their compatibility with performance additives. Generally, lithium soap thickener is compatible with a much wider variety of grease additives than organoclay thickener. Consequently, an anticorrosion additive such as calcium sulfonate may perform quite well in a lithium soap-thickened grease, whereas it could cause significant softening of an organoclay-thickened grease. Therefore, performance additives must be carefully matched with the type of grease thickener.

The levels at which additives are used in lubricating greases are generally higher than in lubricating oils. Several factors can contribute to the need for higher additive addition rates in lubricating greases as compared with lubricating oils:

1. The grease thickener may compete with the performance additive for adsorption sites on the metal surfaces.
2. The grease thickener may adsorb additive so that less is available for performing the desired additive function.
3. Compared with lubrication by circulating oil, grease lubrication does not provide a mechanism for heat removal, and the reserve of lubricant is much less.

Another difference between lubricating oils and greases, in terms of additive use, is that lubricating greases can more readily utilize solid additives. Although attempts have been made to suspend various dispersions of solid additives in lubricating oils, such lubricants generally suffer from a tendency to separate or plug filters. However, the semisolid consistency of greases is such that they can easily keep fine solid lubricants, such as graphite and molybdenum disulfide, suspended indefinitely.

The types of additives used in greases are listed in Table 30.6 along with corresponding test methods used to evaluate the performance properties these additives impart to the grease. Some of the additives mentioned in Table 30.6 are used much less frequently than in the past with the increasing HSE concerns over the use of heavy metals, chlorinated compounds, and components that may react to produce potential carcinogens. National Health Safety and Environment (HSE) regulations as well as equipment manufacturer requirements are making the selection and treat rate of additive components a challenging process. See Section 30.6.3 for more details.

30.6.2.1 Rust and Oxidation Inhibitors
Greases for lightly loaded and high-speed applications rely on elastohydrodynamic lubrication and generally require only rust and oxidation inhibitors and antiwear additives. The types of additives used to achieve these performance properties are similar to those used in lubricating oils (Table 30.6).

As mentioned earlier, the additives may be used at higher percentages in lubricating grease than are typically used in lubricating oils.

Unlike oil formulators, grease formulators are not restricted to performance additives that give clear and bright solutions in oil. Even insoluble additives, such as sodium nitrite and sodium sebacate, have been used as rust inhibitors in greases. A drawback of solid additives is that they can contribute to higher noise levels in grease-lubricated bearings [51].

30.6.2.2 Antiwear and Extreme-Pressure Additives
For mobile equipment and industrial bearings in which heavier loads are involved and the oil film may not be sufficient to prevent metal-to-metal contact, EP and antiwear additives, such as those listed in Table 30.6, are used. These additives include most of the types used in EP gear oil formulations.

As indicated by the EP test methods listed in Table 30.6, there are several methods for measuring the EP properties of greases. These test methods differ in the way that they evaluate antiwear and EP performance, and, therefore, it is desirable to develop grease formulations that perform well in all these test methods when designing multipurpose industrial greases. This is usually accomplished by using more than one additive and taking advantage of synergistic effects between the additives [52]. It has been reported that the performance of some EP additives may be affected by the temperature at which they are added to the grease. For example, the Timken OK load (ASTM D2509) of a lithium base grease appeared to increase from 35 to 55 lb when the temperature of addition of the sulfurized olefin EP additive was raised from 90°C to 120°C [53].

Applications in which shock loading may occur, such as the bearings of large shovels and certain conveyor bearings, the rapid application of load could cause the scoring of metal surfaces before soluble EP additives could react to form a sacrificial layer. Solid lubricants are used in such applications to provide a physical separation of metal surfaces during shock loading. Such lubricants are usually finely dispersed solids, which have the ability to form films on metal surfaces that decrease sliding friction. Various solid lubricants that have been used in lubricating greases are listed in Table 30.6. Perhaps, the two most commonly used solid lubricants in greases are graphite and molybdenum disulfide. Synergistic EP effects between these two solid lubricants have been exploited in greases and open gear lubricants for a number of years [54]. It has also been claimed that the EP/antiwear performance of graphite can be enhanced by the treatment of graphite with polar agents such as alkali molybdates and tungstates and alkali earth sulfates and phosphates. Treatment of graphite with these polar inorganic materials helps the graphite to adhere to metal surfaces more strongly, even in the absence of trace water, which is normally necessary for untreated graphite to adhere to metal, and act as a friction-reducing lubricant [55].

Specialized semifluid grease containing solid lubricants have been developed for the lubrication of open gears that cannot be lubricated by means of an oil reservoir [56].

TABLE 30.6
Additives Used in Lubricating Greases

Additive Function	Types of Additives	Related Performance Tests
Antioxidant	Hindered phenols	ASTM D5483, standard test method for oxidation induction time of lubricating
	Diaryl amines	greases by pressure differential scanning calorimetry
	Phenothiazine	ASTM D942, standard test method for oxidation stability of lubricating grease by
	Oligomers of trimethyldihydroquinoline	the oxygen pressure vessel method
	Metal dialkyldithiocarbamates	
	Ashless dialkyldithiocarbamates	
	Metal dialkyldithiophosphates	
Rust inhibitor	Alkylsuccinic acid derivatives	ASTM D1743, standard test method for corrosion preventive properties of
	Ethoxylated phenols	lubricating greases
	Nitrogen-containing heterocyclic compounds	ASTM D5969, standard test method for corrosion preventive properties of
	Fatty amines	lubricating greases in the presence of dilute synthetic sea water environments
	Salts of fatty acids and amines	ASTM D6138, standard test method for determination of corrosion preventive
	Salts of phosphate esters and amines	properties of lubricating grease under dynamic wet conditions (Emcor test)
	Metal sulfonates [43,44]	
	Ammonium sulfonates	
	Substituted imidazolines	
	Lead naphthenate	
	Bismuth naphthenate	
	Sodium nitrite	
	Sodium sebacate	
Metal deactivator	Disalicylidenepropanediamine	ASTM D4048, standard test method for detection of copper corrosion from
	Derivatives of 2,5-dimercapto-1,3,4-thiadizole	lubricating greases
Antiwear additive	Alkylphosphoric acid esters and salts	ASTM D5707, standard test method for measuring friction and wear properties
	Dialkyldithiophosphates	of lubricating grease using a high-frequency, linear-oscillating (SRV) test
	Metal dialkyldithiocarbamates	machine
	Phosphate esters	ASTM D2266, standard test method for wear preventive characteristics of
	Dithiophosphate esters	lubricating grease (four-ball method)
	Derivatives of 2,5-dimercapto-1,3,4-thiadiazoles	
	Molybdenum carboxylates	
Soluble EP additive	Sulfurized esters [26]	ASTM D5706, standard test method for determining extreme properties of
	Sulfurized olefins [26]	lubricating greases using a high-frequency, linear-oscillation (SRV) test machine
	Diaryl disulfides [26]	ASTM D2596, standard test method for measurement of extreme pressure
	Organic sulfur/phosphorus compounds	properties of lubricating grease (four-ball method)
	Lead naphthenate	ASTM D2509, standard test method for measurement of load-carrying capacity
	Bismuth naphthenate [27,45]	of lubricating grease (Timken method)
	Antimony dialkyldithiocarbamate	
	Antimony dialkyldithiophosphate	
	Ammonium salts of phosphate esters	
	Chlorinated waxes	
	High-molecular-weight complex esters	
	Chlorinated waxes	
	High-molecular-weight complex esters	
	Borate esters [46]	
Solid EP additive	Metal borates	See the above test methods for EP properties and friction and wear properties
	Molybdenum disulfide [47]	
	Perfluorinated polyolefins	
	Graphite [48]	
	Calcium carbonate	
	Calcium phosphate	
	Calcium acetate	
	Boron nitride	
	Boric acid	
	Metal powders	
	Phosphate glasses [49]	

(Continued)

TABLE 30.6 (*Continued*)
Additives Used in Lubricating Greases

Additive Function	Types of Additives	Related Performance Tests
Tackifier	Polyisobutylenes	No standard test method
Polymers for water resistance	Atactic polypropylene	ASTM D4049, standard test method for determining the resistance of lubricating grease to water spray
	Polyethylene	
	Functionalized polyolefins [50]	ASTM D1264, standard test method for determining the water washout characteristics of lubricating greases
	Alkylsuccinimides	
	Polyisobutenes	
	Styrene–butadiene block copolymers	

These greases are required to produce a sturdy protective film on heavily loaded gears that may also be subjected to shock loading. The grease is usually applied at intervals through a spray nozzle and therefore must be readily pumpable and sprayable. In the past, pumpability has been achieved by the addition of a solvent such as 1,1,1-trichloroethane or a hydrocarbon solvent. More recently, solvent-free open gear greases have been developed using synthetic base oils to achieve low-temperature performance and pumpability [56]. Open gear greases generally contain combinations of soluble EP additives and solid lubricants to achieve high EP performance. These lubricants may also contain high levels of polymers to promote adhesion and water resistance. Open gear greases are usually gray to black in color due to the presence of graphite and molybdenum disulfide; however, light-colored open gear greases have been developed through the use of colorless solid lubricants such as phosphate glass [49]. With the increasing cost of raw materials, alternatives to molybdenum disulfide are used where technically suitable.

Thread compounds are another example of specialty greases that contain high levels of solid lubricants. These compounds are generally composed of grease combined with a very high level of metal powder and other solid lubricants. For example, API Bulletin 5A2 [57] describes a thread compound composed of 36 wt% grease, 30.5 wt% lead powder, 12.2 wt% zinc powder, 3.3 wt% copper powder, and 18 wt% natural graphite. Thread compounds may also be formulated with anticorrosion additive to protect pipe threads during storage.

30.6.2.3 Polymers for Water Resistance

Mobile equipment is often used in applications in which grease-lubricated bearings may be subjected to water, as are bearings in certain industrial applications such as in steel and paper mills. It has been found that the addition of various polymers to lubricating greases improves their water resistance. Some of the types of polymers that have been used to impart water resistance to lubricating greases are listed in Table 30.6. Although most of these polymers improve water resistance by raising the grease base fluid viscosity, at least one functionalized polymer actually reacts with alkali during the formation of the soap thickener and becomes a part of the grease thickener matrix [58].

30.6.2.4 Grease Thickeners as Performance Additives

A recent development in the evolution of lubricating grease technology has been the use of solids that act both as grease thickeners and as performance additives. For example, a number of patents have been written concerning the preparation and use of calcium sulfonate complex greases in which calcium carbonate dispersed by calcium sulfonate acts as a thickener, an anticorrosion additive, and an EP additive [59,60]. Similar multifunctional claims have been made for greases thickened with specially designed titanium complex thickeners [61,62].

30.6.3 Environmental Concerns

Environmental concerns and regulations have influenced modern grease formulations, and this trend will likely accelerate in the future. The disposal of lubricants that contain restricted substances is onerous and expensive. Therefore, there has been a shift away from additives that appear on restricted substance lists or that may appear on future restricted substance lists. Examples of formulation changes instigated primarily by environmental concerns include the following:

1. Substitution of alternative EP and anticorrosion additives to replace lead naphthenates and lead dialkyldithiocarbamates
2. Removal of heavy-metal powders from thread compounds [63–65]
3. Removal of chlorinated solvents from open gear lubricants and greases [56]
4. Substitution of alternative EP and antiwear additives to replace chlorinated paraffins [28,66]
5. Substitution of alternative anticorrosion additives to replace sodium nitrite

Concern was raised in California over the possible toxicity of the solid lubricant molybdenum disulfide. This could have been a problem for the disposal of used grease containing molybdenum disulfide. However, after toxicological review, the California Department of Toxic Substances Control ruled that wastes containing molybdenum in the form of molybdenum disulfide are excluded from identification as hazardous waste [67].

With the increasingly global nature of the lubricant business, regulations such as those in the European Union covering Registration, Evaluation and Authorisation of Chemicals (REACh) are affecting the lubricants and additive industries worldwide. The European Union's new chemical legislation came into effect on June 1, 2007. Although the details of its implementation are still to be finalized, it is likely to have a significant impact on suppliers, producers, and consumers of more than 30,000 chemicals.

Various original equipment manufacturers, such as automobile manufacturers, provide long lists of restricted substances that are not allowed in materials, such as lubricants, which they purchase. There is not much doubt that the list of restricted substances will grow in the future. Concerns about environmental persistence will likely increase the demand for biodegradable greases. Although the market for this type of grease is currently quite modest, the U.S. Army has written a specification for biodegradable multipurpose grease that should help to advance the technology [68]. Such greases usually contain natural or synthetic ester base oils that are biodegradable and additives of low toxicity that, if not readily biodegradable themselves, at least do not impede the ability of microorganisms to ultimately biodegrade organic portions of the grease to carbon dioxide, water, and biomass. Studies have been performed to determine the effectiveness of various grease additives in these types of base greases [69,70].

30.6.4 LUBRICATING GREASE TRENDS

As in the case of other industrial lubricants, the trend for lubricating greases is to develop products with improved high-temperature performance leading to longer bearing lives. Grease formulators are also searching for anticorrosion additives that provide better protection against saltwater corrosion and that do not significantly affect the performance of the grease thickener [44].

Because the interaction of grease additives and grease thickener must be considered for the development of optimum grease formulations, it follows that the importance of various grease additives may change as the popularity of the various grease thickeners changes. Although the versatility of lithium soap and lithium complex greases causes them to be the most widely used grease types, polyurea greases are growing in popularity as they tend to give significantly longer bearing lives in certain sealed bearing applications. It is expected that the popularity of polyurea- and diurea-thickened greases will continue to grow, and, as they do, more emphasis will be placed on developing additives that perform optimally with these types of thickeners.

REFERENCES

1. I Macpherson, DD McCoy. Hydraulic and turbine oil specifications. In ER Booser, ed., *Tribology Data Handbook*. Boca Raton, FL: CRC Press, 1997, pp. 279–281.
2. BD McConnell. Military specifications. In ER Booser, ed., *Tribology Data Handbook*. Boca Raton, FL: CRC Press, 1997, pp. 182–241.
3. ET Denisov, IV Kyudyakor. Mechanisms of action and reactivities of the free radicals of inhibitors. *Chem Rev* 87: 1313–1357, 1987.
4. M Rasberger. Oxidative degradation and stabilization of mineral oil based lubricants. In RM Mortier, ST Orszulik, eds., *Chemistry and Technology of Lubricants*. New York: VCH Publishers Inc., 1992, pp. 108–109.
5. PC Hamblin, U Kristen, DC Ardsley. Ashless antioxidants, copper deactivators and corrosion inhibitors: Their use in lubricating oils. *J Lubr Sci* 2(4): 287–318, 1989.
6. JR Thomas, CA Tolman. Oxidation inhibition by diphenylamine. *J Am Chem Soc* 84: 2930–2935, 1962.
7. JR Thomas. The identification of radical products from the oxidation of diphenylamine. *Am Chem Soc* 82: 5955–5956, 1960.
8. VJ Gatto, MA Grina. Effects of base oil type, oxidation test conditions and phenolic antioxidant structure on the detection and magnitude of hindered phenol/diphenylamine synergism. *Lubr Eng* 55(1): 11–20, 1999.
9. Y Luo, B Zhong, J Zhang. The mechanism of copper-corrosion inhibition by thiadiazole derivatives. *Lubr Eng* 51(4): 293–296, 1995.
10. American Petroleum Institute. Appendix E—API base oil interchangeability guidelines for passenger car motor oils and diesel engine oils, API 1509, 14th ed., Addendum 1, 1998, p. E-1.
11. AS Galliano-Roth, NM Page. Effect of hydroprocessing on lubricant base stock composition and product performance. *Lubr Eng* 50(8): 659–664, 1994.
12. D Clark et al. New generation of ashless top tier hydraulic fluids. *Lubr Eng* 56(4): 22–31, 2000.
13. I Macpherson, DD McCoy. Hydraulic and turbine oil specifications. ER Booser (ed.), In *Tribology Data Handbook*. Boca Raton, FL: CRC Press, 1997, pp. 274–278.
14. DK Tuli et al. Synthetic metallic dialkyldithiocarbamates as antiwear and extreme-pressure additives for lubricating oils: Role of metal on their effectiveness. *Lubr Eng* 51(4): 298–302, 1995.
15. MD Sexton. Mechanisms of antioxidant action. Part 3. The decomposition of 1-methyl-1-phenylethyl hydroperoxide by OO′-dialkyl(aryl)phosphorodithioate complexes of cobalt, nickel, and copper. *J Chem Soc Perkin Trans* 2: 1771–1776, 1984.
16. AJ Bridgewater, JR Dever, MD Sexton. Mechanisms of antioxidant action. Part 2. Reactions of zinc bis-[OO-dialkyl(aryl) phosphorodithioates] and related compounds with hydroperoxides. *J Chem Soc Perkin Trans* 2: 1006–1086, 1980.
17. JM Georges et al. Mechanism of boundary lubrication with zinc dithiophosphate. *Wear* 53: 9–34, 1979.
18. WB Bowden, JY Chien, KA Colapret. Scrapping downtime in a scrapyard—A field test of ashless hydraulic oils in a high pressure application. *Lubr Eng* 54(11): 7–11, 1998.
19. LR Kray, JH Chan, KM Morimoto. New direction for paper machine lubricants, ashless vs. ash type additive formulations. *Lubr Eng* 51(10): 834–838, 1995.
20. RL Stambaugh, RJ Kopko, TF Roland. Hydraulic pump performance—A basis for fluid viscosity classification, SAE Paper no. 901633. Society of Automotive Engineers, Warrendale, PA, 1990.
21. S Asadauskas, JM Perez, JL Duda. Oxidative stability and antiwear properties of high oleic vegetable oils. *Lubr Eng* 52(12): 877–892, 1996.
22. RL Goyan, RE Melley, WC Ong. Biodegradable lubricants. *Lubr Eng* 54(7): 10–17, 1998.

23. DE Tharp, KD Erdman, GH Kling. Development of the BF-1 biodegradable hydraulic fluid requirements. *Lubr Eng* 54(7): 21–23, 1998.

24. VW Castleton. Development and use of multipurpose gear lubricants in industry. *Lubr Eng* 51(7): 593–598, 1995.

25. RW Cain, PS Greenfield, GF Hermann. Industrial gear oil: Past, present and future. *NLGI Spokesman J* 64(8): 22–29, 2000.

26. O Rohr. Sulfur: The ashless additive, sulfur as a key element in EP greases and in general lubricants. *NLGI Spokesman* 58(5): 20–27, 1994.

27. RW Hein. Evaluation of bismuth naphthenate as an EP additive. *Lubr Eng* 56(11): 45–51, 2000.

28. PR Miller, H Patel. Using complex polymeric esters as multifunctional replacements for chlorine and other additives. *Lubr Eng* 53(2): 31–33, 1997.

29. J Yao, J Dong. Improvement of hydrolytic stability of borate esters used as lubricant additives. *Lubr Eng* 51(6): 475–478, 1995.

30. JP Maxwell. Practical application of oil mist lubrication for gear driven machine tool heads. *Lubr Eng* 49(6): 435–437, 1993.

31. K Bajaj. Oil-mist lubrication of high-temperature paper machine bearings. *Lubr Eng* 50(7): 564–568, 1994.

32. RS Marano et al. Polymeric additives as mist suppressants in metal cutting fluids. *Lubr Eng* 53(10): 25–34, 1997.

33. M Masuko, T Hanada, H Okabe. Distinction in antiwear performance between organic sulfide and organic phosphate as EP additives for steel under rolling with sliding partial EHD conditions. *Lubr Eng* 50(12): 972–976, 1994.

34. SM Kim et al. Boundary lubrication of steel surfaces with borate, phosphorus, and sulfur containing lubricants at relatively low and elevated temperatures. *Trib Trans* 43(4): 569–578, 2000.

35. DB Barret, IR Bjel. High viscosity gels for heavy duty open gear drives a break from tradition. *Lubr Eng* 49(11): 820–827, 1993.

36. G Daniel et al. Lubricants for open gearing. *NLGI Spokesman* 63(6): 20–30, 1999.

37. BR Hohn, K Michaelis, R Dobereiner. Load carrying capacity properties of fast biodegradable gear lubricants. *Lubr Eng* 55(11): 15–38, 1999.

38. J Muller. The lubrication of wind turbine gearboxes. *Lubr Eng* 49(11): 839–843, 1993.

39. P Oster, H Winter. Influence of the lubricant on pitting and micro pitting resistance of case carburized gears—Test procedures. AGMA paper 87-FTM-9, Alexandria, VA, 1987.

40. W Scales. Air compressor lubrication. In ER Booser, ed., *Tribology Data Handbook*. Boca Raton, FL: CRC Press, 1997, pp. 242–247.

41. KC Lilje, RE Rajewski, EE Burton. Refrigeration and air conditioning lubricants. In ER Booser, ed., *Tribology Data Handbook*. Boca Raton, FL: CRC Press, 1997, pp. 342–354.

42. *Lubricating Grease Guide*, 4th edn. Kansas, MO: The National Lubricating Grease Institute, 1996, pp. 3.28–3.36.

43. R Baker. A comparison of rust inhibitors in ASTM D 1743 and IP 220 tests. *NLGI Spokesman* 57(9): 20–23, 1993.

44. ME Hunter, F Baker. Corrosion rust and beyond. *NLGI Spokesman* 63(1): 14–20, 1999.

45. O Rohr. Bismuth, the new and "green" extreme pressure-element and the best replacement for lead. *NLGI Spokesman* 61(6): 10–17, 1997.

46. JP Donor. U.S. Patent 5,242,610, Grease composition, 1993.

47. TJ Risdon. EP additive response in greases containing MoS_2. *NLGI Spokesman* 63(8): 10–19, 1999.

48. WM Kenan. Characteristics of graphite lubricants. *NLGI Spokesman* 56(11): 14–17, 1993.

49. L Tocci. Transparent grease bears a heavy load. *Lubes-n-Greases* 3(4): 52, 1997.

50. HS Pink, T Hutchings, JF Stadler. U.S. Patent 5,110,490, Water resistant grease composition, 1992.

51. CE Ward, CE Littlefield. Practical aspects of grease noise testing. *NLGI Spokesman* 58(5): 8–12, 1994.

52. S Beret. Assessment of grease EP performance and EP testing. *NLGI Spokesman* 58(8): 7–22, 1994.

53. PR Todd, MR Sivik, GW Wiggins. An investigation into the performance of grease additives as related to the temperature of addition: Part 1. *NLGI Spokesman* 64(3): 14–18, 2000.

54. FG Fischer, AD Cron, RG Huber. Graphite and molybdenum disulphide-synergisms. *NLGI Spokesman* 46(6): 190, 1982.

55. R Holinski, M Jungk. New solid lubricants as additives for greases—"Polarized graphite." *NLGI Spokesman* 64(6): 23–27, 2000.

56. N Samman, SN Lau. Grease-based open gear lubricants: Multi-service product development and evaluation. *Lubr Eng* 55(4): 40–48, 1999.

57. *API Bulletin 5A2 on thread compounds for casing, tubing, and line pipe*, 5th edn. Washington, DC: American Petroleum Institute, 1972, p. 9.

58. C Scharf, HF George. The enhancement of grease structure through the use of functionalized polymer systems. *NLGI Spokesman* 59(11): 8, 1996.

59. WD Olsen, RJ Muir, T Eliades, T Steib. U.S. Patent 5,308,514 sulfonate greases, 1994.

60. W Macwood, R Muir. Calcium sulfonate greases one decade later. *NLGI Spokesman* 63(5): 24–37, 1999.

61. A Mukar et al. A new generation high performance Ti-complex grease. *NLGI Spokesman* 58(1): 25–30, 1994.

62. A Kumar et al. Titanium complex greases for girth gear application. *NLGI Spokesman* 63(6): 15–19, 1999.

63. WD Stringfellow, NL Jacobs, RV Hendricks. Environmentally acceptable specialty lubricants, an interpretation of the environmental regulations of the North Sea. *NLGI Spokesman* 57(9): 14–19, 1993.

64. N Adamson, JM Pinoche, T Noel. A novel tubing/casing thread compound offering both corrosion protection and performance with environmental acceptability. *NLGI Spokesman* 62(7): 10–22, 1998.

65. NL Jacobs. U.S. Patent 5,180,509, Metal-free lubricant composition containing graphite for use in threaded connections, 1993.

66. RJ Fensterheim. Where in the world are chlorinated paraffins heading. *Lubes-n-Greases* 4(4): 14, 1998.

67. J Dumdum. Government regulations update. *NLGI Spokesman* 58(4): 36, 1994.

68. I Rhee. 21st century military biodegradable greases. *NLGI Spokesman* 64(1): 8–17, 2000.

69. A Fessenbecker, I Roehrs, R Pegnoglou. Additives for environmentally acceptable lubricant. *NLGI Spokesman* 60(6): 9–25, 1996.

70. N Kato et al. Lubrication life of biodegradable greases with rapeseed oil base. *Lubr Eng* 55(8): 19–25, 1999.

31 Long-Term Additive Trends in Aerospace Applications

Carl E. Snyder, Jr., Lois J. Gschwender, and Shashi Kant Sharma

CONTENTS

31.1 UNIQUE REQUIREMENTS FOR AEROSPACE APPLICATIONS

Aerospace, especially military aerospace systems, requires faster, highly maneuverable, and higher-temperature systems. The higher temperatures typically experienced in aerospace applications are common for all lubricants, but we use hydraulic fluids as an example. The high temperatures experienced in aerospace hydraulic systems are the result of three related factors. First, for military aircraft, the operational requirements for the aircraft (e.g., high speed and high degree of maneuverability) impose more severe demands on the hydraulic systems. Second, weight is always a critical issue for aircraft that drives hydraulic system components that are small and operate under more strenuous conditions (e.g., higher speed, higher loads) and that imposes more severe operating conditions on the hydraulic fluid (i.e., higher temperatures, higher shear rates). Third, the high emphasis on weight savings for aircraft applications results in smaller fluid volume systems. All these factors result in higher temperatures being imposed on the hydraulic fluids. While high maneuverability is usually not an issue for commercial aerospace applications, the other two factors are equally important for commercial and military applications. In addition to the higher temperatures imposed on aerospace hydraulic fluids, they must also be capable of operating at extremely low temperatures. Aircraft hydraulic and lubrication systems are required to operate at temperatures as low as −54°C. Typical operational temperature ranges for the hydraulic fluids used in aerospace are shown in Figure 31.1. Similar factors make unique demands on other aerospace lubricants, such as gas turbine engine oils and greases. Aerospace applications include both spacecraft and aircraft applications. The major emphasis of this chapter is aircraft applications. Hydraulic fluids and gas turbine engine oils are discussed in this chapter because they represent the larger volume usage, but similar factors and issues exist for the other aerospace lubricants and fluids such as greases and coolants. Liquid and grease lubricants for space applications are also discussed because the space environment creates some unique and challenging problems for lubricant additives. Often, customary additives are not adequate for aerospace applications, and more development is needed to make acceptable aerospace lubricants [1–8].

31.2 WHY ARE ADDITIVES NEEDED?

Additives are required for aerospace fluids and lubricants to enhance or impart a required operational capability to a specific fluid or lubricant not provided by the base fluid alone. When a lubricant (e.g., lubricating oil, hydraulic fluid, and grease) is stressed in its operating environment, it can degrade such that the performance of the system or components in which it is used is adversely affected. In order to enhance the performance of a lubricant, or to mitigate its degradation during use, additives are required. Some of the additive types used in aerospace applications are listed below. Depending on the application, a formulation may contain one or more of these and other additives:

1. Boundary lubrication additives (antiwear, lubricity, extreme pressure)
2. Antioxidants
3. Anticorrosion additives
4. Metal deactivators
5. Viscosity index (VI) improvers

A lubricant has to be effective in both the fluid-film lubrication and the boundary lubrication regimes. In fluid-film lubrication, a fluid film separates the interacting surfaces, while in boundary lubrication, more direct contact occurs between the two surfaces. Fluid-film lubrication can be further divided into two broad categories—hydrodynamic lubrication and elastohydrodynamic

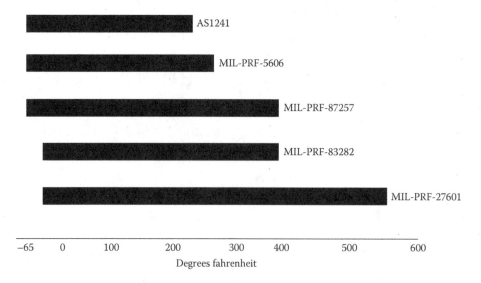

AS1241

MIL-PRF-5606

MIL-PRF-87257

MIL-PRF-83282

MIL-PRF-27601

−65   0   100   200   300   400   500   600
Degrees fahrenheit

FIGURE 31.1   Temperature ranges for aerospace hydraulic fluids.

lubrication. In fluid-film lubrication, the physical properties of the lubricant such as viscosity, pressure viscosity, and traction determine the performance of the lubricated contact. While fluid-film lubrication is the desired mode of operation, the boundary lubrication regime cannot be avoided. Even in fluid-film lubrication, boundary lubrication occurs during the start-up and shutdown, and during occasional asperity interaction during operation. Therefore, the material (both the surface and the lubricant) properties important for the boundary lubrication regime are equally important for the fluid-film lubrication. Boundary lubrication additives are generally needed for reducing friction and wear of the components (Table 31.1) [9].

Antioxidants are required to prolong the useful life of fluids and lubricants. Most fluids and lubricants are hydrocarbon based and are susceptible to oxidative degradation via a free-radical mechanism. Antioxidants react with the initially produced free radicals to prevent the oxidation process from proceeding from the slow initiation phase of the process to the rapid autocatalytic phase.

Corrosion inhibitors are usually synonymous with rust inhibitors for most lubricants. However, as we go to more exotic metals and with the emergence of halogenated lubricants, corrosion can occur with nonferrous metals and by attack of very aggressive fluid degradation products. For example, degradation products from perfluoropolyalkylethers react with a variety of metals to cause what one could call corrosion [10,11]. A better definition of corrosion inhibitors would be additives that inhibit the reaction of moisture or fluid degradation products with metals. They are widely used for that function in many classes of fluids and lubricants.

Metal deactivators are additives that passivate the surface of metals. The passivation can result in either reducing the metal's catalytic effect on lubricant degradation or reducing the effect of the lubricant on the surface of the metal. They are widely used in hydraulic fluids and lubricants.

VI improving additives are discussed in Section 31.5.

TABLE 31.1

Typical Additives Used in Aerospace Fluids and Lubricants

| | | Fluid/Lubricant | | | |
| | | Hydraulic Fluid | | Space Lubricant | |
Additive	Gas Turbine Engine Oil	Hydrocarbon	Phosphate Ester	Hydrocarbon	PFPAE
Antioxidant	Aromatic amines, for example, PANA	Hindered phenol	[a]	Hindered phenol	[a]
Lubricity/antiwear	Aryl phosphate	Aryl phosphate	[a]	Aryl phosphate	[a]
VI improver	N/A	Polymethyl methacrylate	[a]	N/A	N/A
Rust inhibitor	Metal sulfonate or organic acid	Metal sulfonate	[a]	N/A	NaNO$_2$[b]
Antifoam	Silicone	Silicone	[a]	N/A	N/A
Metal deactivator	Benzotriazole	Benzotriazole	[a]	N/A	N/A

N/A, not typically used for this application.

[a]   Proprietary.

[b]   Greases only.

31.3 HOW ADDITIVES WORK

In general, certain performance-improving additives are optimized for specific base oils, but in some instances, the same additive can be effective in a variety of classes of base fluids. However, in some cases, additives that have been effective at improving a specific property or performance parameter in one or several classes of base fluids are not effective in others. For example, substituted aryl phosphates have been very effective antiwear additives for steel-on-steel rubbing contacts in a large number of chemical base stocks, for example, mineral oils, synthetic hydrocarbons, esters, and even more exotic base fluids such as the polyphenyl ethers and the sila-hydrocarbons. They are not effective additives, however, for steel-on-steel contacts in polydimethyl siloxanes or chlorotri-fluoroethylene lubricants. The reason for specificity of some additives is beyond the scope of this chapter, but it is important to recognize this phenomenon as we consider future trends for additive applications. As we attempt to develop advanced lubricants using new and novel base oils, we must recognize that new additive technology may be required. How different additives enhance the performance of a lubricant is discussed ahead. Additives may be less effective as metals are improved such as high chromium and high carbide steels that have lower levels of iron on the metal surface. This is further discussed in Section 31.7.

In boundary lubrication, the interacting surfaces react with the lubricant components to form protective physisorbed, chemisorbed, or reaction films [12]. When metal-to-metal contact occurs between two surfaces in a lubricated contact, the asperities on the surfaces shear, thereby exposing a fresh metal surface. The appropriate lubricant component reacts with the exposed fresh metal to form protective surface films. The surface films thus formed are generally low friction and wear resistant and protect the surfaces from early failure/wear. During operation, these surface films can wear off and regenerate. Antiwear or lubricity additives are generally added to the base oil to enhance the formation of the protective surface films during boundary lubrication. Thus, in boundary lubrication, the chemistry of the lubricant along with the material properties of the interacting surfaces determines the performance of the lubricated contact.

Many researchers have expended significant effort at determining the detailed mechanism by which specific additives work. In some cases, for example, the free-radical interceptor mechanism of most effective antioxidants, the mechanisms, have been intensively studied, worked out, and generally well understood [13]. In other cases, however, although extensive research has been conducted, for example, the mechanism of action of tricresylphosphate, there are a number of theories, many of which are supported by chemical data, but no one theory is universally accepted.

When one reviews the published data and, in the case of many researchers, their unpublished data, a few common facts can be found for many types of performance-improving additives. First, most additives appear to be effective through a "competitive reaction" mechanism. In other words, the additive is reacting with some undesirable species that occurs in the lubrication system that, if unchecked, will lead to premature failure of the lubricant or component. In the case of an antioxidant in a hydrocarbon lubricant, the additive is reacting with the initial free radicals formed early in the oxidation process to retard or prevent their attack on the lubricant molecules in an autocatalytic process that results in rapid deterioration of the lubricant. In the case of the antiwear additives in hydrocarbon lubricants, the additive reacts with the newly formed high-energy, fresh metal surfaces formed by metal-to-metal contact, thereby passivating these surfaces by forming, in most cases, lubricious reaction products that prevent the catastrophic wear of the component. The chemistry of that reaction product will vary depending on the chemistry of the additive and the surface. The mechanism of wear protection, however, may vary with lubricants and surfaces of different chemistries, as we discuss later in this chapter.

Another common factor with many lubricant/hardware systems is that the additives are generally unreactive and stable during the "normal" functioning of the lubricant and only become effective and reactive on demand. For example, antioxidants in hydrocarbons become effective and reactive only when free radicals are formed in the lubricant due to oxidation of the lubricant molecules. Similarly, effective antiwear additives in hydrocarbon lubricants generally do not react until the fresh metal surfaces are formed. Whether this reaction is induced by the reactivity of the fresh metal surface or the momentary high temperatures that occur during the metal-to-metal contact is an item still under debate, but it is generally accepted that the additive, until this reactivity is triggered by a specific event, is inactive and stable. Something occurs to lower the activation energy and allows the additive to be effective.

Other additives are less affected by "events" that trigger their action. For example, VI improvers [14] and rust inhibitors are generally effective during the entire life of the lubricants and do not require some specific (unusual) event to trigger their actions. But as we consider future additive requirements for future classes of base fluids and surfaces requiring lubrication, new additive technology must be developed.

An additive may work by more than one mechanism. For example, many metal-alkylated sulfonate rust inhibitors form a physical barrier on metal surfaces. The polar portions of the molecules are adsorbed on the metal, effectively blocking water from the corrosion-prone surfaces [8]. The molecules also form micelles or groups of molecules loosely in rings with their polar portions toward the middle. These act as little water traps to surround and tie up water molecules and, again, to prevent them from getting to the metal surfaces.

31.4 ADDITIVES FOR SPACE APPLICATIONS

Lubricants for outer-space applications are particularly challenging. First of all, the volume of space lubricants used is so small that major oil companies have little interest in investing in research, development, and commercialization, even if the lubricants command a high price. Fortunately, several companies deal with specialty lubricants and are willing to supply to this niche market.

The most important additive for a space lubricant is an antiwear additive [7]. Spacecraft mechanisms generally operate on a very small quantity of lubricant. With lubricant depletion due primarily to evaporation or lubricant creep, the lubricated components experience a higher degree of boundary lubrication. Effective antiwear additives are required for prolonged life of these components. As a general rule, perfluoropolyalkylether (PFPAE) lubricants are not used in applications where the bearing contact stress is over 100,000 psi because they may undergo tribocorrosion-induced failure. An antioxidant is often used to protect a spacecraft's hydrocarbon oil and its antiwear additive for the time the formulation is on the earth, waiting for launch, but is not essential in space, where little oxygen is available.

The major technical challenges for spacecraft lubricant additives are low volatility and good solubility. Spacecraft lubricants were originally mineral oils/greases, but these are gradually being replaced with synthetics including hydrocarbon poly-α-olefins (PAOs) and multialkylated cyclopentane (Pennzane®), and PFPAEs. Even lower volatility and improved low-temperature viscosity silahydrocarbon fluids were developed but not yet implemented [15]. Soluble and effective additives are readily available for terrestrial hydrocarbon oils and, with some looking, commercially available low-volatility additives may also be found. In greases, solubility is less important, but it is generally believed that at least partially soluble additives are advantageous in grease performance over nonsoluble additives.

For many years there were no soluble additives for PFPAE lubricants. Conventional additives are not soluble in PFPAE fluids. Solubilizing substituents, similar to the backbone structure of PFPAE fluids, are required in the additive structure to make it soluble in PFPAE fluids. However, PFPAE manufacturers do offer some soluble additives for their fluids, for use in the manufacturers' own product lines. These additives are not typically available by themselves. The U.S. Air Force has conducted research and development of soluble additives for PFPAE fluids as part of a High-Performance Turbine Engine Technology program [16]. Additives are needed for PFPAE fluids because of their unique behavior especially in boundary lubrication in low-humidity environments [17]. A phenomenon called *tribocorrosion* can occur, which is a combination of fluid decomposition and wear metal corrosion. This happens at much lower temperatures than are predicted by static, nonwear stability tests. The additives act to interrupt the cycle of fluid decomposition and metal interaction, thereby minimizing the tribocorrosion and improving wear performance [18].

Besides antiwear additives, antirust additives are needed to prevent steel components from rusting when used with PFPAE fluids or greases. Conventional hydrocarbon lubricants tend to leave a protective film on the surface of steels, while PFPAE fluids do not, thus leaving low-chrome steel components prone to rusting after use with a PFPAE fluid.

This is due to the high solubility of oxygen into the PFPAE fluids, which allows oxygen to get to the iron surfaces, leading to rust. PFPAE fluids and greases containing an antirust additive can provide less rust protection for steels than hydrocarbon products without antirust additives. However, PFPAE fluids and greases formulated with these special antirust additives do demonstrate improved protection against rusting than unformulated fluids and greases.

Several important new classes of PFPAE additives were developed in the Air Force program, [16] most notably a perfluoroalkyldiphenylether, a perfluoroalkyltriphenylether, and a tri(perfluoropolyalkyletherphenyl)phosphine (PH3) [19]. These provide improved antiwear and high-temperature stability to the PFPAE fluids by interacting with and neutralizing fluid decomposition products that, without the additives, attack metal surfaces and cause an autocatalytic decomposition of the fluid.

As the molecular weight of any lubricant increases, the solubility for polar additives decreases. This is especially true for the very nonpolar PAOs. Long-chain alkyl groups on the additives improve additive solubility but often reduce effectiveness. Additives, in general, are polar compounds, which is what makes them effective. All of this makes for a very short list of effective, soluble, and low-volatility additives for spacecraft.

The spacecraft lubricant additives do not need to be especially thermally stable since the spacecraft mechanisms do not operate at high temperatures. They may, however, operate at extreme cold temperatures, and the additives must stay in solution in oils.

Test methods for spacecraft lubricants are not well standardized and often are peculiar to a particular satellite company. Because there are few second chances in spacecraft lubricants once in operation, usually a "life test" is conducted for years on earth under high vacuum to simulate actual operation.

31.5 ADDITIVES FOR HYDRAULIC FLUIDS

Hydraulic fluids are operated in closed systems, although the fluid reservoir may be open to air. Hydraulic fluid additives may include antioxidant, antiwear, rubber swell, VI improver, and minor amounts of metal deactivator and dye. (This is a simpler list than that of engine oil, where thermal-oxidative stability is critical.) The application temperature often dictates the additives used. Hydraulic fluids for military applications are defined by the military specifications including MIL-PRF-5606H [20] (mineral oil based fluid) and MIL-PRF-83282D [21] and MIL-PRF-87257 [22] (both synthetic hydrogenated PAO-based fluids). Hydraulic fluids for commercial aircraft that are phosphate ester based are defined by SAE AS 1241 [23].

Antioxidants are usually ashless (containing no metals) and are believed to serve primarily to protect the fluid in storage rather than in an application. A test such as ASTM D4636, Corrosiveness and Oxidation Stability of Hydraulic Oils, Aircraft Turbine Engine Lubricants, and Other Highly Refined Oils is often used to evaluate the antioxidant effectiveness.

Aryl phosphate antiwear additives in hydraulic fluids are widely used and are very effective for steel-on-steel rubbing surfaces. The ASTM D4172, Wear Preventive Characteristics of Lubricating Fluid (Four-Ball Wear Method), is widely used to evaluate antiwear additives, but other methods are also used.

Rubber swell additives are used with synthetics such as PAOs that, by themselves, provide little swell or even shrinkage to nitrile seals, the most widely used seal material in military aircraft. Nitrile seals were originally developed to swell an appropriate volume with naphthenic mineral oil. When synthetic PAOs were introduced, they had to be adjusted to be a drain-and-fill replacement for mineral oils, so rubber swell additives were added to the synthetic formulations. Rubber swell additives are usually polar esters actually adsorbed into the elastomeric compounds, creating a positive seal to the hydraulic system. Seal tests may include swell per FED-STD-791 method 3603 and other physical property determinations.

VI improvers are used mainly with naphthenic mineral and phosphate ester oils to increase the VI (lower the amount of change in viscosity with temperature). They are long-chain polymers such as polymethylmethacrylates, but other chemical classes are also used. In theory, they are coiled at low temperature and strung out at higher temperature and therefore impede flow less at lower temperatures but more at higher temperatures. The VI improvers do impede flow in bulk, such as past seals, and provide additional fluid-film thickness in low-shear environments but do little to improve the fluid-film thickness in the elastohydrodynamic regime in bearings [14]. VI improvers are broken into smaller pieces in high-shear and high-stress environments, present in both elastohydrodynamic and boundary lubrication regimes, and therefore become less effective. A shear stability test, for example, as in the MIL-PRF-5606 hydraulic fluid specification [20], is used to determine a formulation's resistance to shear under a specific set of conditions, but this test does not duplicate the severity of real lubricated contacts.

A minor amount (0.05%) of metal deactivator is needed with PAO-based hydraulic fluids for them to pass all the metal weight changes in ASTM D4636. A small amount of dye is used in hydraulic fluids for leak detection. Hydrocarbon hydraulic fluids are usually dyed red, while phosphate ester hydraulic fluids are dyed purple.

31.6 ADDITIVES FOR GAS TURBINE ENGINE OILS

Aircraft engine oils must have a critical balance of performance-improving additives, especially if used at high temperatures. Engine oils are often sprayed onto hot bearings with considerable airflow present to help cool the bearing. This gives rise to hot spots and thin oil films with lots of contact with oxygen. Coking of the fluid and additives, when they oxidize and turn to carbonaceous deposits, is detrimental to the cooling efficiency of the engine and can be abrasive. Also, coke can clog the breathing and oil delivery tubes in an engine.

Engine oils are very oxidatively stable polyol ester based with one or more antioxidant and antiwear, antifoam, and metal deactivator additives. All the additives must have a low propensity to form coke. Often, the optimized oil represents a compromise between low coking and optimum performance as provided by specific additives. For example, different aryl phosphate antiwear additives exhibit various degrees of effectiveness, and it may not be possible to select the best antiwear additive because its presence in the oil causes too much coke formation. As a rule, the coking tendencies of formulations are greater with the more effective aryl phosphates and greater with higher percentages of additives.

An antifoam additive, a specific molecular-weight-range dimethyl silicone, may be used.

The following tests are generally used to evaluate the effectiveness of the additives:

> *Oxidative stability*: ASTM D4636
> *Antiwear performance*: ASTM D1947, Ryder Gear Test and Modified ASTM D5182, FZG Test
> *Anticoking performance*: Fed-Std 791, Method 3450, Bearing Deposition
> *Foaming*: Fed-Std 791, Method 3213, Foaming Characteristics of Aircraft Turbine Engine Oils (Static Foam Test)

In addition to these tests, each new engine oil qualified to a military specification must pass an actual engine test as described in the specifications.

31.7 FUTURE TRENDS

Three major factors will drive the need for new additive research and development for aerospace applications. The first one, which has been discussed in detail in the Introduction, is the need for additives effective over a wide temperature range, going from as low as $-54°C$ to over $300°C$. The second has already been discussed to some extent, and that is the new classes of the base oils developed to meet new, more demanding requirements. The class discussed earlier, the PFPAE, sometimes called perfluoroethers or perfluoropolyethers, is only one example. Another class that demonstrated unique requirements is the chlorofluorocarbon oligomers that served as the base fluid for a $-54°C$ to $175°C$ nonflammable aerospace hydraulic fluid. Although most additives typically used in hydrocarbon-based lubricants are quite soluble in the chlorofluorocarbon base stock, most of them have no beneficial effect. The successful development of an effective formulation using that unique base fluid required years of research. The final formulation that met the military specification MIL-H-53119 [24] utilized a carefully balanced additive package composed of a lubricity additive and a rust-inhibiting additive. While these are two new and unique base fluids with which the authors have firsthand experience, other new base stocks could be developed for future applications that also require new additive research and development.

The third factor that will prompt the development of new performance-improving additives for future lubricating systems is the change in the chemistry of the surfaces with which the lubricants will come in contact and be expected to lubricate and provide long life. The main type of additive influenced by this factor is the lubricity additive. Two examples are new metal alloys and surface modifications such as hard coatings. In both cases, the iron, which in the past has been available to interact with the additive, is not available or is not readily available, and the chemical reactions that activate the additive are halted. On the other hand, the surface modifications may provide a barrier for any undesirable interaction between the lubricant and the substrate material. New bearing materials such as Cronidur-30®, Rex20® (CRU20®), and CCS42L® [25] have been developed that are more corrosion resistant and/or harder than the traditional tool-steel bearing materials such as M50 and 52100. Higher hardness of these new bearing materials gives them higher load-carrying capability. For high-speed precision bearings, hybrid bearings, utilizing the lower-density silicon nitride rolling elements, are gaining popularity.

The lack of reactivity of these materials with the environment (including the lubricant/additive system) that makes them effective at corrosion resistance also inhibits the formation of the beneficial surface films for enhanced boundary lubrication performance. The state-of-the-art boundary additives, such as tricresylphosphate, are not optimized for the newer chemical composition bearing steels. Additive development will be required for optimized performance of bearings utilizing hard coatings on the balls and/or races. These hard, wear-resistant coatings, such as titanium carbonitride (TiCN)– [26] and titanium carbide (TiC)–coated balls, have been shown to extend the lifetime of bearings in critical applications, but in most cases, they have been used either with no liquid-grease lubricant or with lubricants optimized for steel-to-steel contacts. Again, the chemistry of the lubricity additives was optimized to react with fresh steel (iron) asperities caused by contact in boundary lubrications. With the changed chemistry of the surface of the bearing material from iron-based alloys to a hard coating, the typical lubricity additives may not effectively provide the significant increase in lifetime as is typically experienced with the steel bearings. In many cases, improved lifetimes are experienced with hybrid bearings that incorporate a coated or ceramic ball and a standard steel race. However, with a fully ceramic bearing or one that uses coated surfaces on both the ball and the race, the maximum enhancement in performance cannot be experienced unless new additives are developed. These new additives will beneficially interact with these new bearing surfaces to produce the lubricious coatings that are believed to be the main mechanism resulting in the currently experienced wear improvement in steel bearings with the state-of-the-art lubricity additives.

REFERENCES

1. CE Snyder Jr, LJ Gschwender. Development of a nonflammable hydraulic fluid for aerospace applications over a −54°C to 135°C temperature range. *Lubr Eng* 36:458–465, 1980.
2. CE Snyder Jr et al. Development of high temperature (−40° to 288°C) hydraulic fluids for advanced aerospace applications. *Lubr Eng* 38:173–178, 1982.
3. GA Beane IV et al. Advanced lubricants for aircraft turbine engines, SAE Technical Paper Series Paper 851834, 1985.
4. CE Snyder Jr et al. Development of a shear stable viscosity index improver for use in hydrogenated polyalphaolefin-based fluids. *Lubr Eng* 42:547–557, 1986.
5. LJ Gschwender, CE Snyder Jr, GA Beane IV. Military aircraft 4 cSt gas turbine engine oil development. *Lubr Eng* 42:654–659, 1987.
6. LJ Gschwender et al. Advanced high-temperature air force turbine engine oil program. In WR Herguth, TM Warne, eds., *Turbine Lubrication in the 21st Century*. ASTM STP 1407. West Conshohocken, PA: American Society for Testing and Materials, 2001.
7. LJ Gschwender et al. Improved additives for multiply alkylated cyclopentane-based lubricants. *J Synth Lubr* 25:1, 31–41, 2008.
8. LJ Gschwender et al. Computational chemistry of soluble additives for perfluoropolyalkylether liquid lubricants. *Trib Trans* 39:368–373, 1996.
9. CE Snyder Jr et al. Liquid lubricants and lubrication. In B Bhushan, ed., *CRC Handbook of Modern Tribology*, Vol. I. Boca Raton, FL: CRC Press, 2000 pp. 361–382.
10. PT Hellman, L Gschwender, CE Snyder Jr. A review of the effect of metals on the thermo-oxidative stability of perfluoropolyalkylether lubricants. *J Synth Lubr* 23:197–210, October–December 2006.
11. PT Hellman, JS Zabinski, L Gschwender, CE Snyder Jr., AL Korenyi-Both. The effect of hard coated metals on the thermo-oxidative stability of a branched perfluoropolyalkylether lubricant. *J Synth Lubr* 24:1–18, January–March 2007.
12. J Liang, B Cavdar, SK Sharma. A study of boundary lubrication thin films produced from a perfluoroplkylether fluid on M-50 surfaces. I. Film species characterization and mapping studies. *Trib Lett* 3:107–112, 1997.
13. RK Jensen et al. Liquid-phase autoxidation of organic compounds at elevated temperatures. Absolute rate constant for intermolecular hydrogen abstraction in hexadecane autoxidation at 120–190°C. *Int J Chem Kinet* 26:673–680, June 1994.
14. SK Sharma, N Forster, LJ Gschwender. Effect of viscosity index improvers on the elastohydrodynamic lubrication characteristics of a chlorotrifluoroethylene and a polyalphaolefin fluid. *Trib Trans* 36:555–564, 1993.
15. WR Jones et al. The tribological properties of several silahydrocarbons for use in space mechanisms. National Aeronautics and Space Administration, Washington, DC, NASA/TM–2001–211196, November 2001.
16. LJ Gschwender, CE Snyder Jr, GW Fultz. Soluble additives for perfluoropolyalkylethers liquid lubricants. *Lubr Eng* 49:702–708, 1993.
17. LS Helmick et al. The effect of humidity on the wear behavior of bearing steels with $R_fO(n-C_3F_6O)_xR_f$ perfluoropolyalkylether fluids and formulations. *Trib Trans* 40:393–402, 1997.
18. SK Sharma, LJ Gschwender, CE Snyder Jr. Development of a soluble lubricity additive for perfluoropolyalkylether fluids. *J. Synth Lubr* 7:15–23, 1990.
19. C Tamborski, CE Snyder Jr. Perfluoroalkylether substituted aryl phosphines and their synthesis. U.S. Patent 4,011,267. US Patent Office, Washington, DC. March 1977.
20. MIL-PRF-5606H: Military specification, hydraulic fluid, petroleum base; aircraft, missile and ordinance. September 12, 2011. http://quicksearch.dla.mil.
21. MIL-PRF-83282D: Performance specification, hydraulic fluid, fire resistant, synthetic hydrocarbon base, metric. NATO Code Number 537, September 30, 1997. http://quicksearch.dla.mil

22. MIL-PRF-87257: Military specification, hydraulic fluid, fire resistant; low temperature synthetic hydrocarbon base, aircraft and missile. April 22, 2004. http://quicksearch.dla.mil.

23. AS1241: Fire resistant phosphate ester hydraulic fluid for aircraft. SAE, Society of Automotive Engineers, Warrendale, PA, 1992.

24. MIL-H-53119: Military specification, hydraulic fluid, non-flammable, chlorotrifluoroethylene base. March 1, 1991. http://quicksearch.dla.mil.

25. JJC Hoo, WB Green, eds., *Bearing Steels: Into the 21st Century.* ASTM STP 1327. American Society for Testing and Materials, ASTM International, West Conshohocken, PA, 1998.

26. AK Rai et al. Performance evaluation of some pennzane-based greases for space applications. Published in the *Proceedings of the 33rd Aerospace Mechanisms Symposium*, Jet Propulsion Laboratory, Pasadena, CA, NAS/CP-1999-209259, 1999, pp. 213–220.

32 Eco Requirements for Lubricant Additives and Base Stocks

Thomas Rühle and Matthias Fies

CONTENTS

32.1 BIOLUBRICANTS

The public awareness of environmental pollution has grown globally for decades influenced by environmental disasters like, for example, the nuclear disaster of Fukushima, the Deepwater Horizon explosion, and increasing air pollution in Asian mega cities.

The growing global industrialization caused the global climate change by global greenhouse gas (GHG) emissions.

The major scientific agencies of the United States, Europe, and many independent scientific organizations agree that climate change is occurring and that humans are contributing to it [1]. At the 2015 United Nations Climate Change Conference in Paris, 195 countries agreed by consensus to the final global pact, the Paris Agreement, to reduce emissions as part of the method for reducing greenhouse gas [2].

Another trend is the growing social recognition of environmental protection. Many globally active companies added sustainability claims to their company's strategy [3] and accept the need to balance economic, environmental, and social requirements to ensure a more sustainable development.

The development of energy-efficient engines will have a direct impact on greenhouse gas emissions, and the development of a suitable lubricant that offers reduced friction and decreases wear is part of this development.

Beside the efforts to reduce energy consumption, there is a growing requirement to protect nature from unintended lubricant discharge into the environment. According to literature, 30%–50% of all lubricants sold worldwide end up released into the environment [4], either intentionally through total loss applications like chain saw lubricants or unintentionally by spills or inadequate disposal [5]. Estimates for the loss of hydraulic fluids are as high as 70%–80% [6,7]. These lubricants can contaminate the soil, the surface- and groundwater, and the air.

32.2   DEFINITION OF BIOLUBRICANTS

Until 2015, a general, noncontentious, and well-accepted description and definition of biodegradability, renewability, and toxicity valid for all kinds of lubricants was missing.

Despite great interest in environmentally compatible lubricants, the lack of standards and technical language describing these fluids and greases has impeded the growth of the market for these types of lubricants. Only for some groups of lubricants are standards and definitions available, for example, ISO 15380 [8] and some ecolabels for hydraulic fluids.

Biolubricants' base oils can be made from both biomass and fossil resources. Within the different standards, there was a common agreement for the conditions of biodegradability and toxicity for biolubricants. However, the content of renewable raw material for the formulation of biolubricants was not defined and varies between 0% and 70%, depending on the application. With regard to greenhouse gas emissions, renewability of a lubricant is an important factor. Manufacturing a lubricant from biologically produced carbon-containing materials rather than using mineral oils enables the reduction of the amount of new carbon released into the atmosphere. For this reason, the use of bio-based lubricants has a positive impact on the environment. Våg et al. [9] undertook a comparative life cycle assessment study into base oils used in the manufacture of hydraulic fluids and calculated that the energy consumption per functional unit during the production of a mineral oil–based fluid (45,000 MJ) is greater than that for a synthetic ester (22,000 MJ).

The use of bio-based material is also recommended to conserve the limited resources of mineral oil. To quantify the term bio-based, methods such as ASTM D6866-4 [10] using radiocarbon and isotope ratio mass spectrometry analysis to determine the bio-based content are used.

Both relevant aspects of biolubricants, being biodegradable and bio-based, have been summarized by the European Committee for Standardization (CEN) in the standard "Liquid petroleum products—Bio-lubricants—Criteria and requirements of bio-lubricants and bio-based lubricants" (Technical Report 16227, prEN 16807). It has to be emphasized that both terms, biolubricants and bio-based lubricants, are equivalently treated in the same standard. This standard defines minimum requirements for renewability, biodegradability, toxicity, and technical performance.

For the United States, the U.S. Environmental Protection Agency (EPA) defines the condition for Environmentally Acceptable Lubricants (EAL) with limits for biodegradability, minimal toxicity, and a minimal tendency for bioaccumulation [11]. Environmentally acceptable lubricants are described as those that are rapidly biodegradable, have a low ecotoxicity, and do not harm water. Of course, these fluids could also be bio-based, but this is not mandatory—in principle; some products derived from petrochemicals could achieve the aforementioned environmental criteria.

The situation gets even more complex by the fact that terms, such as "biodegradability" are not described by one single definition. Biodegradation can be measured in a number of ways and is usually done indirectly by the measurement of the amount of oxygen consumed or the amount of carbon dioxide produced by microorganisms. Typically, tests standardized by the Organisation for Economic Co-operation and Development (OECD) are used, such as the following:

301: Ready Biodegradability
301 A: Dissolved Organic Carbon (DOC) Die-Away Test B: CO_2 Evolution Test
301 C: Modified Ministry of International Trade and Industry (MITI), Japan Test (I)
301 D: Closed Bottle Test
301 E: Modified OECD Screening Test F: Manometric Respirometry Test
302 A: Inherent Biodegradability: Modified Semi-continuous Activated Sludge (SCAS) Test
302 B: Inherent Biodegradability: Zahn-Wellens/Eidgenössische Materialprüfungs-und Versuchsanstalt für Industrie, Bauwesen und Gewerbe (EMPA) Test
302 C: Inherent Biodegradability: Modified MITI Test (II)
303: Simulation Test—Aerobic Sewage Treatment A: Activated, B: Biofilms
304 A: Inherent Biodegradability in Soil
305: Bioconcentration: Flow-through Fish Test
306: Biodegradability in Seawater
307: Aerobic and Anaerobic Transformation in Soil
308: Aerobic and Anaerobic Transformation in Aquatic Sediment Systems
309: Aerobic Mineralization in Surface Water—Simulation Biodegradation Test
310: Ready Biodegradability—CO_2 in Sealed Vessels (Headspace Test)
311: Anaerobic Biodegradation of Organic Compounds in Digested Sludge—Method by Measurement of Gas Production
312: Leaching in Soil Columns [12]

The determination of the ecotoxicity is normally based on the extrapolation of results obtained for aquatic organisms (algae, daphnia, and fish) using defined assessment factors. Such an evaluation results in a concentration limit. A tested substance is then believed to have no negative impact on an ecosystem, as long as the concentration of the substance is below the set limit.

Additionally, many countries have unique regulations that are not harmonized, for example, the German "water hazard classes" [13,14].

With different definitions applicable for a biolubricant, it is important to clarify the exact requirements and specifications to be met before starting to formulate such a fluid. A good biolubricant must not only comply with the environmental requirements but also has to meet necessary technical requirements for the intended application. Ideally, a biolubricant should have at least the same technical performance compared with a mineral oil–based fluid and should meet all relevant original equipment manufacturer (OEM) specifications and national and international norms.

32.3 MARKET OF BIOLUBRICANTS

According to different market reports, the global synthetic lubricants demand was 37 million tons in 2015 and is expected to reach 42 million tons in 2018, growing at a combined annual growth rate (CAGR) of 2.5%. Bio-based lubricant demand was 600 ktons in 2015 and is expected to reach 900 ktons in 2020, growing at a CAGR of 6.6% [15]. Biolubricants have a market share of app. 1.5% of the global lubricant demand. North America and Europe are the most important markets for biolubricants accounting for 85%–90% of the global market [16].

The increased supply of high-performing, cost-competitive biodegradable base oils in the context of government regulations is driving biolubricants market growth globally. Moreover, this market, supported by industrial interest in developing innovative green formulations for various end users, is forecast to outpace the growth of finished lubricants. However, growth rates of biolubricants vary by region, and despite relatively strong progress, penetration of biolubricants is limited. The reasons for this relatively limited uptake of biolubricants include high prices as well as different performances that need technical support for the usage of biolubricants. However, in Europe the market share of biolubricants for special applications is high, reaching almost 100% for chainsaw and demolding oils and lubricating grease and lower levels for two-stroke, hydraulic, and compressor lubricants used in excavators and forklifts operating in environmentally sensitive areas.

32.4 DRIVERS FOR THE USE OF BIOLUBRICANTS

Increasing environmental concerns and rising consumer awareness, especially in the U.S. and Western European markets has been the primary driving force behind the demand for bio-based lubricants. As cost and performance are the major market drivers for lubricants, the success of biolubricants in the market was supported by regulatory agencies promoting these products with tax benefits and preferred purchasing. More and more governmental regulations mandate the use of biolubricants for selected applications.

According to literature, between 30% and 50% of all lubricants sold worldwide end up released into the environment either intentionally through total loss applications like, for example, chain saw lubricants or unintentionally by spills or inadequate disposal. Estimates for the loss of hydraulic fluids are as high as 70%–80%. These lubricants can contaminate the soil, the surface- and groundwater, and the air. The U.S. Environmental Protection Agency reports operational discharges in marine applications, for example, stern tube or hydraulic leakages on marine vessels, of 36.9–61 million liters of lubricating oil into marine port waters annually [17]. To protect the environment against pollution by lubricants, the best approach is to prevent the loss of the lubricant and appropriate recycling systems. If this is not possible, environmentally acceptable fluids have to be used. Beside the ecological benefits of biolubricants, health and safety considerations make the adoption of biolubricants favorable; typically, higher flashpoints at similar viscosities lead to a lower flammability risk, and with lower hazard ratings, the potential for harm to a person from inhalation or skin contact is reduced.

32.5 LEGISLATION

With legislation, the government regulates the use of certain ingredients positively or negatively—either there is an obligation or mandate to use special components or a ban on some ingredients or chemical classes. Biolubricant products are available for many applications but it is up to the consumer to make the purchasing choice. Biolubricants are more expensive and, without regulations, the choice is usually for the cheaper mineral oil–based product. To enforce the use of biolubricants by law, however, is still somewhat of an exception, more common being the obligation to use. The first obligation to use biodegradable lubricants was mandated in Portugal in 1991 [18]. This obligation required for outboard two-stroke engine oils a degree of biodegradability of 66% (according to CEC-L-33-T-82) [19], but stopped short of further specifying the base fluid to be used. Austria made another approach in 1992, when it banned mineral oil–based chainsaw oils and mandated to end users the use of oils that are at least 90% biodegradable [20]. In Sweden, the Swedish Standard is mandatory, for example, for hydraulic equipment in the forestry application [21].

32.5.1 REACH

In 2006 the European Parliament agreed on a regulation of the European Union for the Registration, Evaluation, Authorization and Restriction of Chemicals (REACH) [22]. REACH will improve the protection of human health and the environment from the risks that can be posed by chemicals, while enhancing the competitiveness of the EU chemicals industry. It also promotes alternative methods for the hazard assessment of substances in order to reduce the number of animal tests. REACH requires from companies a registration, evaluation, and authorization of chemical substances that are manufactured or imported in Europe in a quantity of more than 1 metric ton per year and will be fully in place in 2018. The law will affect the petroleum additives, lubricant, and grease industries in Europe. It will also affect companies that import equipment or machinery that contains chemicals in Europe. Biolubricants, as more environmentally friendly lubricants, could benefit from REACH legislation.

32.5.2 CLP/GHS

In 2008, the different European laws concerning chemical safety and systems of classification and labelling have been aligned to the Classification, Labelling and Packaging (CLP) Regulation [23]. The CLP ensures that hazards presented by chemicals are clearly communicated to workers and consumers in the European Union through the classification and labelling of chemicals. Before placing chemicals on the market, the industry must establish the potential risks to human health and the environment of such substances and mixtures, classifying them in line with the identified hazards. The hazardous chemicals also have to be labelled according to a standardized system so that workers and consumers know about their effects before they handle them. Thanks to this process, the hazards of chemicals are communicated through standard statements and pictograms on labels and safety data sheets. For example, when a supplier identifies a substance as "acute toxicity category 1 (oral)," the labelling will include the hazard statement "fatal if swallowed," the word "Danger," and a pictogram with a skull and crossbones. The method of classifying and labelling chemicals is based on the United Nations Globally Harmonized System of Classification and Labelling of Chemicals (GHS) and was fully implemented in many countries in 2015. The GHS was designed as one universal standard for all countries to follow and to replace all the diverse classification systems. Before it was created and implemented, there were many different regulations on hazard classification in use in different countries. While those systems may have been similar in content and approach, they resulted in multiple standards, classifications, and labels for the same hazard and made a worldwide approach necessary. The adoption of the GHS is facilitating international trade by increasing consistency between the laws in different countries that currently have different hazard communication requirements.

32.5.3 OSPAR

Since 1972, the Convention for the Protection of the Marine Environment of the North-East Atlantic (OSPAR Convention) has worked to identify threats to the marine environment and has organized, across its maritime area, programs and measures to ensure effective national action to combat them [24]. The OSPAR Convention is the current legislative instrument regulating international cooperation on environmental protection in the North-East Atlantic. It combines and updates the 1972 Oslo Convention on dumping waste at sea and the 1974 Paris Convention on land-based sources of marine pollution. Work carried out under the convention is managed by the OSPAR Commission, which is made up of representatives of the governments of the 15 signatory nations, and representatives of the European Commission. The OSPAR convention now regulates European standards on marine environmental conditions and creates a list of chemical substances that are banned for formulations. The OSPAR List of Substances of Possible Concern is a dynamic working list and will be regularly revised, as new information on persistence, toxicity, and liability to bioaccumulate becomes available.

CLP/GHS and OSPAR regulations are used in the European standards for biolubricants.

32.6 NATIONAL AND INTERNATIONAL LABELLING SCHEMES

Companies can apply at different, mostly nongovernmental, organizations to have their products and services certified with a specially created label for environmental friendly lubricants (European Ecolabel, German Blue Angel, Nordic Swan, Swedish Standard, etc.). The labelling schemes are not intended to grant subsidies, but to provide a commonly recognized promotional tool for purchasing decisions and to increase public awareness and create sensitivity for environmentally friendly products and services. Various ecolabels have different requirements and result in different degrees of eco-friendliness. Basic requirements on biodegradability and nontoxicity are common to all standards; however, a minimum percentage of renewable carbon is not required with every standard. Currently, several labels exist worldwide.

32.6.1 GERMAN BLUE ANGEL

This ecolabel is issued by the Federal Environmental Agency (UBA) and the German Institute for Quality Assurance and Labeling (RAL). The general Blue Angel label was founded in 1978 [25], and the first voluntary environmental standard specifically for lubricants  (RAL-UZ 48 [26]) came into effect in 1988. The German ecolabel Blue Angel now contains three standards for the use of lubricants:

- RAL-UZ 48 Readily Biodegradable Chain Lubricants for Power Saws
- RAL-UZ 64 Readily Biodegradable Lubricants and Forming Oils [27]
- RAL-UZ 79 Readily Biodegradable Hydraulic Fluids [28]

The version RAL-UZ 178 [29] dates from August 2014. According to VDMA 24568 [30], not only technical performance requirements, but also the environmental effect of the used lubricants play an important role. As can be deduced from the naming of the standards or norms, biodegradability plays an important role for this label.

For example, hydraulic fluids must have a biodegradability of 70% according to OECD 301—B, C, D, or F [28]. In addition, hydraulic fluids must not contain, among others, substances with restrictions due to their intrinsic properties according to European chemical law (REACH, CLP). Those products are carcinogenic, reprotoxic, and mutagenic substances and any substances that are classified in the hazard class "Acute Toxicity" Category 1, 2, 3, or 4 in accordance with the guidelines in Regulation (EC) No 1272/2008.

Substances classified in the hazard class "Skin Irritation, Eye Irritation or Sensitisation" can only be added to the final product up to a maximum total concentration that is smaller than the concentration that would lead to classification in the hazard class "Skin Irritation 2" or "Eye Irritation 2."

Furthermore, substances that are classified in the hazard class "Hazardous to the Aquatic Environment - Acute Hazard" category cannot be used in the formulation. Beside the ecotoxicity requirements, the Blue Angel label explicitly includes technical performance requirements. To be awarded, hydraulic fluids need to fulfill at least the minimum technical requirement of ISO 15380 [31].

Broadly, this German ecolabel focuses on aquatic toxicity, biodegradation, bioaccumulation, the content of dangerous components, and the technical performance for the different categories of lubricants. Other characteristics such as renewability, CO_2 balance, or life cycle analysis are not considered in detail.

32.6.2 Swedish Standard

The Nordic region is the second largest biolubricant consumer in Europe, with Sweden being the largest consumer among Nordic countries. Sweden's biolubricants demand is driven by a high percentage of forestry acreage (70%) and enforcement by forestry associations. Sweden and Finland remain mostly focused on the forestry sector for biolubricants, while Denmark is focused on the off-shore and marine segments and Norway on transformer oil applications. Biolubricants in the Nordic region are forecast to post moderate growth for the foreseeable future, driven mainly by anticipated new public procurement directives for the public works and construction sector, the gradual conversion to green transformer oils in other Nordic countries based on best practices from Norway, and the marine and off-shore lubricant demand in Denmark [16]. The dominant standard in the region is the Swedish Standard, which applies to hydraulic fluids and greases and covers biodegradability, toxicity, and renewable content. The environmental requirements are focused on individual chemical substances regarding sensitizing properties, acute toxicity, and biodegradability [32]. The Swedish standard SS 155434 is a legal requirement for hydraulic equipment in forestry application. Since 2015 the Standard "Hydraulic Fluids – Requirements and Test Methods – SS15 54 34" has been available in edition No. 5, where the demands on biodegradability have been sharpened [33].

Depending on the applicable method, the biodegradability of the base fluid must be at least 60% (ISO 9439 [34], ISO

9408 [35], ISO 10707 [36], ISO 10708 [37]) or more than 70% (ISO 7827 [38]). The biodegradability of the oil and the toxicity of the oils and the additives are considered. To be awarded with this label, technical application requirements offering nearly the same performance as mineral oil–based lubricants must be fulfilled.

32.6.3 European Ecolabel

The European Ecolabel (EEL) is the primary pan-European label used to define biolubricants. In 1992, the European Union introduced a generally accepted label to help the consumer identify environmentally friendly products and services [39]. It represents a voluntary scheme with a clear objective to market more environmentally friendly products and services and has been developed in cooperation with industry and nongovernmental organizations (NGOs). In 2011 the regulation was adapted to the CLP standard. The EEL has stringent renewable content and biodegradation criteria. About 45%–70% renewable content is required based on the product application. Also, each component of the lubricant that exceeds 0.1% has to be evaluated for biodegradation. Products covered under the EEL include hydraulic fluids, tractor fluids, grease, chainsaw oils, wire rope oils, two-stroke engine oils, and gear oils. The EEL helps to identify products and services that have a reduced environmental impact throughout their life cycle from the extraction of raw material through to production, use, and disposal. Recognized throughout Europe, the EEL is a voluntary label promoting environmental excellence which helps you distinguish environmentally friendly products of high quality.

Components must be evaluated under ecotoxicological perspectives for acceptability with respect to the following criteria:

- Exclusion or limit of substances and mixtures that are toxic or substances rated as carcinogen, mutagen, or reprotoxic (CMR)
- Exclusion of specific substances listed on OSPAR [24], organic halogen and nitrite compounds and selected metals or metallic compounds.
- Additional aquatic toxicity requirements
- Biodegradability and bioaccumulative potential
- Renewable raw materials, depending on application between 45% up to 70%.
- Minimum technical performance
- Information appearing on the EEL

The EEL lubricant cannot bear any R-phrase or H-statement related to health and environment. Evaluation of aquatic toxicity, biodegradation, bioaccumulation, and so on must be undertaken for all ingredients separately, or for the finished formulation. This implies that for all additives, a fully documented toxicological dossier has to be submitted to meet the requirements stated by the EEL. There is the possibility for lubricant raw material suppliers to add products that fulfill the EEL criteria on a Lubricant Substance Classification list (LuSC-list) [40]. Formulators can use these substances or brands without requesting the underlying documents for EEL classification. The technical performance requirements depend on the intended application; hydraulic fluids, for example, should fulfill ISO 15380. In practice, this means that tests such as the Baader test (DIN 51554-3), the steel corrosion test (ASTM D665 A), the copper corrosion test (ASTM D130), and the Forschungsstelle für Zahnräder und Getriebebau (Technical University Munich) (FZG) test (DIN 51354-2) must be passed. Some criteria only have to be reported or can be agreed with the end user—for example, the dry TOST test (ASTM D943 modified). As ISO 15380 differentiates technical performance according to different base oil types, lubricants with differing qualities may be developed.

32.6.4 CEN

Europe has many different and regional standards, and some like, for example, the EEL refers to specific biolubricant families. The CEN defines a European standard for biolubricants by harmonizing the existing regional standards [41]. The CEN is an association that brings together the national standardization bodies of 33 European countries. CEN is one of three European standardization organizations (together with CENELEC and ETSI) that have been officially recognized by the EU and by the European Free Trade Association (EFTA) as being responsible for developing and defining voluntary standards at the European level.

This European Standard CEN–PREN 16807 (Liquid petroleum products–bio-products–bio-lubricants–criteria and requirements of bio-lubricants and bio-based–lubricants) defines the term biolubricant and minimum requirements for all kinds of biolubricants [42]. This European Standard also briefly describes relevant test method needs with respect to the characterization of biolubricants. It presents recommendation for related standards in the field of biodegradability, product functionality, and the amount of different renewable raw materials and/or different bio-based contents used during the manufacturing of such biolubricants forming one product group. The minimum requirements for biolubricants are

* Biodegradability—≥60% according to OECD 301 for oils and ≥50% for lubricating greases
* Toxicity—not labelled as "Dangerous to the Environment" according CLP directive
* Renewability—≥25% bio-based carbon according to ASTM D6866 (^{14}C method)
* Technical performance—fit for purpose

Table 32.1 compares the basic requirements for the important European labels.

TABLE 32.1

Comparison of European Directives: Product Groups and Minimum Renewable Content

Directive	Product Group	Minimum Renewable Content (%)
German Blue Angel	Chainsaw lubricants, lubricants and forming oils, and hydraulic and gear oils	Not specified
Swedish Standard	Hydraulic fluids and greases	Not specified
European Ecolabel	Hydraulic fluids, greases, chainsaw oils, two-stroke oils, and concrete release agents	45–70
CEN PR 16807	All types of lubricants, including engine oils and metalworking fluids	25

32.6.5 CANADIAN ECOLABEL: EcoLogo

Manufacturers, suppliers, importers, or dealers of environmental preferable products participating the Environmental Choice Program (ECP) are awarded with the Canadian ecolabel, EcoLogo [43]. Guidelines for different product categories shall be fulfilled. These categories include,

among others, automotive engine oil (recycled), bicycle chain oil (biodegradable), chainsaw lubricants (biodegradable and nontoxic), marine engine oil (inboard and outboard), and industrial lubricants (re-refined oil, synthetic oil, and vegetable oil–based). For each category, a different set of criteria must be fulfilled to qualify for the Canadian EcoLogo. For example, for synthetic industrial lubricants, biodegradability according to CEC-L33-T82 is required. The lubricant should not contain more than 0.1% petroleum oil (or additives containing petroleum oil) and any of the following metals: lead, zinc, chromium, magnesium, and vanadium [44]. The lubricant should not be labelled a Class D, poisonous and infectious material [45]. Technical performance is tested with various ASTM test methods—for example, ASTM D97 for pour point, ASTM D665 for rust, ASTM D525 for oxidation stability, and ASTM D892 for foam. In general, the Canadian EcoLogo concentrates on aquatic toxicity and biodegradability and technical performance, while the CO_2 balance and renewability not in the focus. The Canadian EcoLogo is one of the few labels that include a life cycle analysis in its considerations for awarding products.

32.6.6 UNITED STATES: BIOPREFERRED AND VGP

The BioPreferred program from the U.S. Department of Agriculture is designed to promote the use of biodegradable lubricants [46]. The program's goal is to encourage the use of bio-based products, thereby lowering the nation's reliance on petroleum, increasing the use of renewable agricultural resources, and reducing adverse environmental and health effects. The program includes mandatory purchasing requirements for federal agencies and their contractors, a voluntary labelling initiative, and an online catalog of certified bio-based products. The USDA Certified Biobased Product label provides information about the product's bio-based content and assures consumers that an independent third party has confirmed the manufacturer's claims with regard to renewable content. The BioPreferred Program added bio-based motor oil in 2013.

The U.S. Environment Protection Agency's (EPA) updated Vessel General Permit (VGP) program took effect in 2013 [47]. The VGP requires the use of EALs for all oil-to-sea interfaces for vessels unless technically infeasible. The goal is to reduce the environmental impact of lubricant discharges on the aquatic ecosystem by increasing the use of environmentally acceptable lubricants for vessels operating in U.S. waters. This regulation should promote the use of bio-based or biodegradable synthetic lubricants.

The minimum requirement for environmental acceptable lubricants are

- Biodegradability—$\geq 60\%$ according to test methods OECD 301 A-F, 306, and 310.
- Minimal Toxicity—A substance must pass either OECD 201, 202, and 203 for acute toxicity testing or OECD 210 and 211 for chronic toxicity testing.
- Not bioaccumulative—The partition coefficient in the marine environment is log K_{OW} <3 or >7 using test methods OECD 117 and 107.

Products meeting the permit's definitions of an EAL include those labelled by the following voluntary labelling programs: the Blue Angel, European Ecolabel, Nordic Swan, Swedish Standards SS 155434 and 155470, Convention for the Protection of the Marine Environment of the North-East Atlantic (OSPAR) requirements, and the EPA's Design for the Environment (DfE). For all applications where lubricants are likely to enter the sea, environmentally acceptable lubricant formulations including vegetable oils, biodegradable synthetic esters, or biodegradable polyalkylene glycols as oil bases instead of mineral oils must be used.

32.6.7 Japanese Eco Mark

The Eco Mark label is issued by the Japan Environment Association (JEA) [48]. It is awarded to products that minimize environmental load and aid environmental protection across their entire life cycle (from production to disposal). Since 1989, the JEA has administered the Eco Mark Program with the goal of disseminating environmental information on products and encouraging consumers to choose environmentally sound products. The symbol itself represents the desire to protect the earth with our own hands, using the phrase "Friendly to the Earth" at the top of the symbol. This goal will be accomplished by authorizing the Eco

Mark to de displayed on products that reduce the environmental load caused by everyday activities, thereby contributing to the preservation of the environment. In principle, products must meet the following criteria: creates less environmental load in other ways, thus contributing significantly to environmental conservation. With the Eco Mark labelling, consumers who wish to be in environment-friendly living condition can make the choice of products more easily.

32.6.8 Other Countries

Many other countries developed ecolabelling programs [49] with the aim of providing accurate environmental information to consumers and to induce firms to develop and produce environment-friendly products in line with consumers' purchasing preferences by affixing an ecolabel on products causing relatively less pollution or using less resources in the production and consumption processes among products for the same purpose. Most of these programs do not consider lubricants, yet. Examples include the following:

China Environmental Labeling plan (China): This government-run ecolabel (the Ten Circle Mark) is issued by the China Environmental United Certification Center under the jurisdiction of the State Environmental Protection Administration.

Environmental Choice (New Zealand): The Environmental Choice ecolabel was introduced by the national government of New Zealand and is issued by New Zealand Ecolabelling Trust.

Green Mark (Taiwan): The Green Mark was introduced in Taiwan by the Environmental Protection Administration and is issued by the Environment and Development Foundation.

Korea Ecolabel (South Korea): The Korea ecolabel is issued by the Korea Environment Industry & Technology Institute that was established in accordance with the Development of and Support for Environmental Technology Act.

Thai Green Label: The Green Label is an environmental certification awarded to specific products that are shown to have minimum detrimental impact on the environment in comparison with other products serving the same function.

32.7 BASE FLUIDS FOR BIOLUBRICANTS

32.7.1 Triglycerides

Triglycerides are the main components of fats and oils found in plants and animals. In technical terms triglycerides are esters composed of glycerin and three (typically natural) fatty acids (see Figure 32.1). The most important fatty acids contained in such oils are the saturated acids—palmitic acid (C16:0) and stearic acid (C18:0)—and the unsaturated acids—oleic acid (C18:1), linoleic acid (C18:2), and linolenic acid (C18:3). The exact fatty acid composition of natural oil depends very

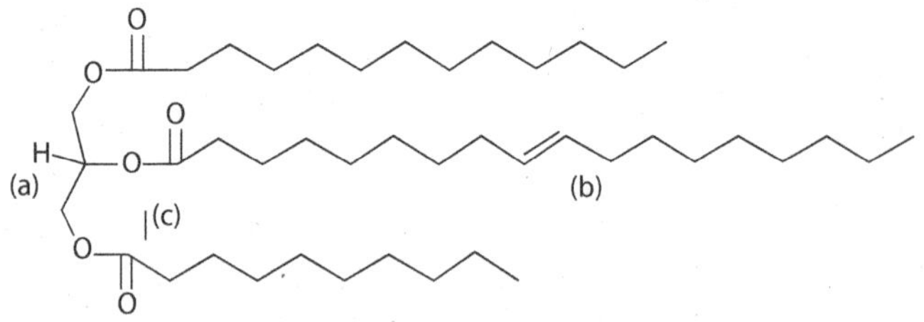

FIGURE 32.1 Example of a triglyceride. (a) β Hydrogen, (b) unsaturation, and (c) ester group.

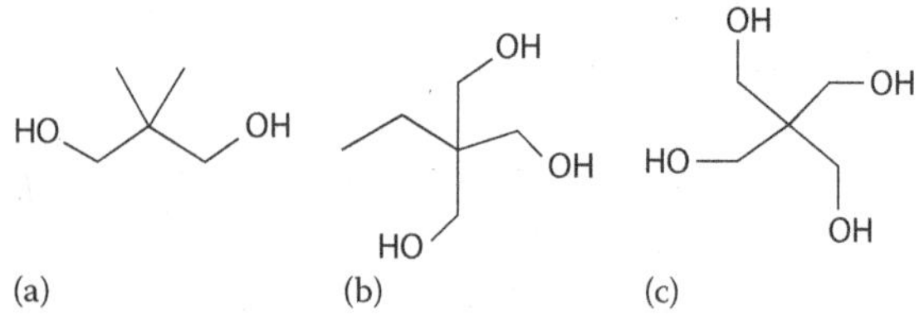

FIGURE 32.2 Neo-alcohols. (a) NPG, (b) TMP, and (c) PE.

much on the source from whence it is derived. The main fatty acid in a rapeseed oil, for example, is oleic acid (C18:1), and in a soybean oil, linoleic acid (C18:2) is dominant. Modern agricultural technology such as breeding new plants or the use of genetically modified organisms allows further possibilities to tailor the fatty acid composition. An example is the development of high oleic sunflower oils composed of almost 90% oleic acid (C18:1). Triglycerides have three drawbacks in regard to the technical performance as a lubricant.

First, due to the hydrogen atom in a β position (a), triglycerides can decompose relatively easily through a low-energy, six-membered ring intermediate. The thermal stability of natural esters is hence very low. Second, the presence of unsaturated double bonds (b) in the fatty acid moiety negatively affects the oxidation stability of these compounds. Typically, the bonding energy of a hydrogen atom in an allylic position is much lower ($\Delta H(C–H) = 345 \pm 5$ kJ/mol), compared with a hydrogen atom in a fully saturated hydrocarbon ($\Delta H(C–H) = 411 \pm 2$ kJ/mol) [50].

A hydrogen atom in an allylic position can therefore be abstracted easily in a radical reaction initiating the oxidation process. Even worse is the oxidative stability of polyunsaturated fatty acids. According to Falk and Meyer-Pittroff [51], the difference in the reactivity of oleic acid (C18:1), linoleic acid (C18:2), and linolenic acid (C18:3) is approximately 1:10:100.

The third drawback of a triglyceride is the low sterical demand of the ester groups (c) when glycerin is used as an alcohol. This allows water to attack the carbonyl group and hydrolyze the ester, releasing free fatty acids.

Despite the fact that unsaturated fatty acids are prone to oxidation, the unsaturation also provides a benefit to the natural oils with respect to their low-temperature flow properties. As a rule of thumb, it can be stated that a natural triglyceride needs at least one cis-configured double bond to show a suitable low-temperature performance. A fully saturated triglyceride would be a solid fat at room temperature and not be usable for a lubricant formulation.

32.7.2 SYNTHETIC ESTERS

To overcome the deficits of natural esters, the glycerol moiety can be replaced by sterically demanding alcohols not bearing a hydrogen atom in β position (see Figure 32.2). Those synthetic esters are either manufactured by a transesterification process of fats and oils or esterification of acids with alcohols.

Usually alcohols such as trimethylolpropane (TMP), neopentyl glycol (NPG), or pentaerythritol (PE) are used [52–54].

The corresponding acids for the esters can be either bio-based fatty acids or petrochemical acids. The resulting polyol esters show improved hydrolytic and thermal stability; the oxidation stability depends on the degree of saturation of the acid. From the saponification of fats and oils, different bio-based acids are available with carbon numbers from C-8 to C-22, fully saturated and highly unsaturated, respectively.

Alternatively, synthetic esters can be manufactured using a polycarboxylic acid in combination with alcohols. Commonly used polycarboxylic acids are, for example, the bio-based azelaic, sebacic, and dimeric fatty acid. Petrochemical-based adipic acid is also used for this application. The corresponding alcohols may be bio-based or petrochemical-based as well. However, for performance reasons branched petrochemical alcohols are mainly used to achieve a low pour point. Such esters can be classified as semisynthetic and most of them may still have a high degree of bio-based material. Most of these esters have good biodegradability and a good ecotoxicological profile. Esters stemming from petrochemicals do not meet bio-based requirements anymore. However, modern concepts like mass balance offer an opportunity to compensate the petrochemical carbon portion by bio-based [55]. There is a broad range of acids and alcohols commercially available. Synthetic esters can hence be tailored to the expected or required performance. The use of branched alcohols or fatty acids allows, for example, for the production of esters with a good oxidative stability without compromising on low-temperature performance. A high degree of branching also has a positive effect on the seal compatibility of such products.

32.7.3 POLYALKYLENE GLYCOLS

Polyalkylene glycols are manufactured from petrochemical epoxides, such as ethylene oxide and propylene oxide yielding a polymeric structure.

Depending on the starting material used, polyalkylene glycols can be either water soluble (due to an excess of ethylene glycol groups) or water insoluble (due to an excess of propylene glycol groups) (see Figure 32.3). However, polyalkylene glycols are not miscible with mineral oil–based lubricants.

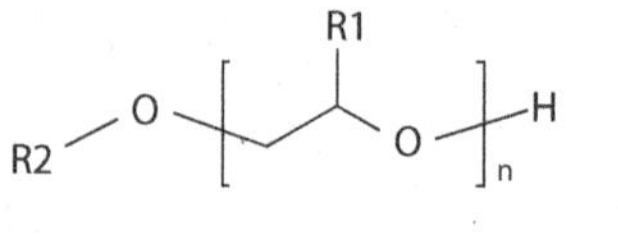

FIGURE 32.3 R1 is typically hydrogen or a methyl group. R2 can be a n-butyl group.

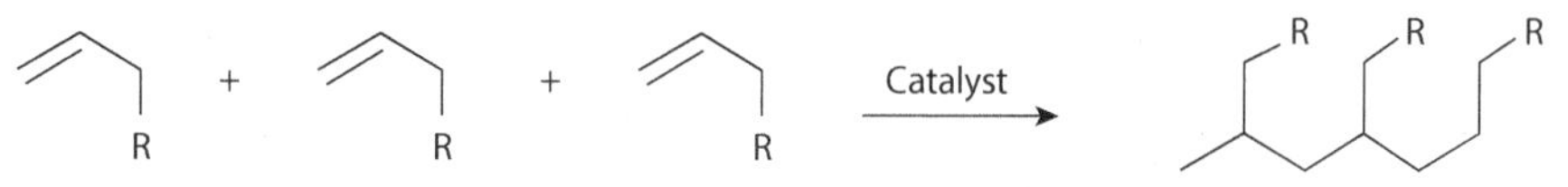

FIGURE 32.4 R1 and R2 are typically linear C-7 or C-9 groups.

Cross-contamination needs to be avoided and changing from mineral oil–based lubricants to PAG-based lubricants requires careful flushing of the equipment. The property of the products can be designed by the copolymerization of varying rations of ethylene oxide and propylene oxide. The properties also depend on the starting alcohol of the polymerization process. Very commonly, n-butanol is used as the starting material. However, some n-butanol-started polyalkylene glycols may have respiratory toxicity [56]. Some of the low-molecular-weight polyalkylene gylcols fulfill the requirements of high biodegradability; the higher molecular weight products are not readily biodegradable. Currently, polyalkylene glycols are produced from petrochemicals and do not meet the bio-based requirement. However, there are different ways to produce polyalkylene glycols from natural resources, such as lactic acid or sugar, but these are not common industrial processes [57]. The mass balance concept is an alternative option for bio-based polyalkyene glycols.

32.7.4 POLY-α-OLEFINS

Poly-α-olefins (PAOs) are produced from petrochemical resources and are hence not bio-based. PAOs are polymerized from olefins, in which the double bounds are in α position (see Figure 32.4). For lubricants, typical starting materials used are 1-decen and 1-dodecen and, despite the name, only oligomers such as trimers and tetramers are used.

The technical performance of PAOs is good, especially in regard to the oxidative stability. As PAOs are very nonpolar, the additive solubility can be a concern, and therefore, PAOs are often used in mixtures with polar esters to overcome this deficiency. Such mixtures can have very good technical performance, but the variabilities and permutations are so diverse

that an in-depth description would be out of the scope of this chapter. PAOs do not meet the eco standard biodegradability criterion. The low-viscosity types of PAOs, which are claimed to be readily biodegradable [58], are not generally used for the formulation of lubricants. PAOs are therefore not embraced by the eco standard definition of biolubricants and will consequently not be dealt with in this chapter.

32.7.5 CONCLUSION

There are different types of base fluids in current commercial use that are marketed with the claim that they are biolubricants. However, only vegetable oils and a high number of synthetic esters easily meet the eco standard biodegradability criteria. This is also true for some low-molecular-weight PAGs. The PAOs and mineral oil–based oil do not satisfy the eco standard. Synthetic esters may have different amounts of bio-based carbon atoms, which could be critical for some eco standards. To formulate a biolubricant, the choice of the correct base fluid is critical, and depending on the specific requirements, each of the described fluids could be optimally selected. The advantages and disadvantages should be balanced and matched with the lubrication needs of the application. A comparison of selected parameters for lubricant base stocks is listed in Table 32.2.

32.8 ADDITIVES FOR BIOLUBRICANTS

32.8.1 INTRODUCTION

The selection of the right additives for bio lubricants is a challenging task, since the additives have to fulfill certain toxicological and ecological requirements. These aspects

TABLE 32.2
Comparison of Selected Parameters for Lubricant Base Stocks

Parameter	Trigyleride	Synthetic Ester	Mineral Oil	PAG	PAO
Renewability content	High	Variable	Not	Not	Not
Readily biodegradable	Good	Good	Not	Variable	Not
Water solubility	Low	Low	Low	Variable	Low
Low-temperature performance	Weak	Variable	Weak	Good	Good
Oxidation resistance	Weak	Good	Variable	Good	Good
Hydrolytic stability	Low	Medium	Good	Good	Good
Seal material compatibility	Good	Limited	Good	Limited	Good
Paint and varnish compatibility	Good	Limited	Good	Limited	Good
Additive solubility	Good	Good	Variable	Good	Limited
Lubricity of base oil	Good	Good	Limited	Good	Limited
Corrosion resistance	Poor	Limited	Good	Limited	Good

have been discussed [59], addressing the toxicological and environmental impacts of important lubricant additive classes. Most additives have a log K_{ow}* between 3 and 8; in other words, the material is presumed to have ability to bioconcentrate in aquatic organisms to potentially harmful levels. Due to their limited water solubility, their biodegradation rate is low [59].

Considering the toxicology for different additive classes, the concerns can be summarized as follows [59,60]:

- Calcium sulfonates cause skin irritation in guinea pigs and also few humans show a weak sensation.
- Calcium phenates can affect male reproductive organs in rabbits.
- Zinc dialkyl dithiophosphates have a human eye irritation potential and are mutagenic due to the presence of zinc.
- Oxidation inhibitors based on phenols can be eye and skin irritants.
- A number of boron compounds, principally boric acid, boric oxide, sodium borate, and sodium perborate referred to as "borates" have recently been proposed for classification as Reprotoxic Category 2 [61].
- Some additives used in the past, for example, lead naphthenates, chlorinated naphthenates, and tri(o-cresyl) phosphates have been discontinued.

From a technical perspective, nearly all conventional lubricant additives can be used. But considering human health and impact on the environment, there are some restrictions. Environmentally acceptable components should be heavy metal–free, since heavy metals negatively affect the ecotoxicology and biodegradability. A limited number of additives is available for which the full ecotoxicological and degradability data have been assessed. The EEL is a widely accepted label and a number of additives are approved according to its requirements. For each additive, a maximum treat rate is allowed for different application categories. Table 32.3 shows all additive types approved according to the EEL along with the five different application categories they can be used for. Including additive packages, basically six different additive classes, each consisting of different chemistries, are suitable for the formulation of EALs.

---

* *Octanol-Water Partition Coefficient* (K_{OW})—a "coefficient representing the ratio of the solubility of a compound in octanol (a non-polar solvent) to its solubility in water (a polar solvent). The higher the K_{OW}, the more nonpolar the compound. Log K_{OW} is generally used as a relative indicator of the tendency of an organic compound to adsorb to soil. Log K_{OW} values are generally inversely related to aqueous solubility and directly proportional to molecular weight."—U.S. Environmental Protection Agency, 2009.

Due to EAL's requirement of biodegradability and low toxicity, ester base stocks are often used, since many of them fulfill these requirements. The additive formulation strategy used for mineral oils (Grp. I–III) or PAOs cannot always be applied to these base stocks, since esters show some special features like high polarity or a lack of hydrolytic stability, which makes it sometimes necessary to adapt the additive formulation to the specific ester base stock. Many esters have quite a high hydrolytic stability like, for example, polyol esters, but some esters have a rather low hydrolytic stability which might result in a higher acid number, which itself propagates the aging of the base fluid and the steel corrosion in the machine.

Plenty of esters have a considerable degree of unsaturation, which affects their oxidative stability dramatically. Normally, a higher degree of unsaturation results in a lower oxidative stability [63].

32.8.2 Antioxidants

In practice, the oxidative ageing of the lubricant is the dominating process, which influences the lifetime of a lubricant significantly [64]. The oxidative degradation can cause the formation of acids, varnish, deposits, and sludge and an increase in the viscosity. It is generally accepted that the oxidative degradation follows a cyclic free radical mechanism through alkyl and peroxide radicals as shown in Figure 32.5.

To prevent oxidation, antioxidants are added to the formulation. Radical scavengers, which inhibit the propagation of the radical chain mechanism, are described as primary antioxidants. Sterically hindered phenols and secondary aromatic amines are typical examples of primary antioxidants.

Aminic and phenolic antioxidants stabilize a lubricant through similar reaction mechanisms. It has been found that both antioxidant classes act synergistically [65]. Hence often, combinations of aminics and phenolics are applied. Unlike phenolic antioxidants, aminic antioxidants are active at higher temperatures (>120°C) [66]. Therefore, it is essential to know the operation temperature of the intended application. A chemically unmodified vegetable oil, which is used typically only in low-temperature applications, preferably would be stabilized with a higher amount of phenolic antioxidants (>1%), whereas for a high-temperature application, for example, a PAO-based base fluid, would be optimized with the addition of a synergistic mixture of aminic and phenolic antioxidants. As primary antioxidants can be regenerated, they are catalytically active in the stabilization of a lubricant.

Secondary antioxidants react stoichiometrically with peroxides and deactivate these aggressive chemical intermediates which are formed during the chemical decomposition. Examples of secondary antioxidants are phosphorus- and sulfur-containing compounds, in which the heteroatom has a lower valence state, thus it still can be readily oxidized. As secondary antioxidants deactivate peroxides, they prevent the chain propagation reaction initiated when a peroxide is divided into two radicals. Standard secondary antioxidants

TABLE 32.3

Environmental and Toxicological Assessment of Important Lubricant Additive Classes for the European Ecolabel

	EEL Categories				
Additives	I: Hydraulic Fluids and Tractor Transmission Oils	II: Greases and Stern Tube Greases	III: Stern Tube Oils Chainsaw Oils, Concrete Release Agents, Wire Rope Lubricants and Other Total Loss Lubricants	IV: Two-Stroke Oils	V: Industrial and Marine Gear Oils
Antioxidants					
Phenolic antioxidants	√	√	√	√	√
Aminic antioxidants	√	√	√	√	√
Antioxidant blends	√	√	√	√	√
Secondary antioxidants	√	√	√	√	√
Antiwear additives					
Alkylphosphites	√	√	√	√	√
Thiophosphonates	√	√	√	√	√
Dithiophosphates	√	√	√	√	√
Amine dithiophosphates	√	√	√	√	√
Alkylphosphates	√	√	√	√	√
EP additives					
Sulfurized esters	√	√	√	√	√
Sulfurized triglycerides	√	√	√	√	√
Sulfurized olefins	√	√	√	√	√
Corrosion inhibitors					
Succinic acid esters	√	√	√	√	√
Sarcosine	√	√	√	√	√
Amine neutralized phosphoric acid ester	√	√	√	√	√
Metal Deactivators					
Benzotriazoles	√	√	√	√	√
Tolytriazoles	√	√	√	√	√
Thiadiazoles	√	√	√	√	√
Industrial Additive Packages					
Circulating oil package	√	√	√	√	√
Turbine oil packages	√	√	√	√	√

Source: IVAM-research and consultancy on sustainability, Amsterdam (Netherlands), Lubricant Substance Classification list (LuSClist), Version date: 09 July 2012, http://www.ivam.uva.nl/wpcontent/uploads/2014/11/LuSC-list-30062015-no-track.pdf (download December 12, 2015).

are zinc dialkyldithiophosphates (ZDTPs). However, as zinc compounds are classified to be ecotoxic and labelled "N" based on toxicity to algae [67], they cannot be used to formulate environmentally acceptable fluids. For such fluids, ashless phosphites and thioesters are much better suited. As a consequence of the different additive mode of action, a combination of primary and secondary antioxidants is very favorable and shows synergistic effects.

As an alternative to the aforementioned antioxidants, the use of metal dithiocarbamates is described in the literature [68]. These additives show very good performance in different tests, by far exceeding the more commonly used ZDTPs [69]. Unfortunately, the metal-free dithiocarbamates, using an ammonium salt or using a methylene bridge, do not show this performance-boosting effect. The use of metal-containing dithiocarbamates—such as bismuth and zinc dithiocarbamates—cannot be utilized to formulate an environmentally

acceptable fluid. Metal-free dithiocarbamate borate esters show excellent load-carrying and antiwear properties in rapeseed oil, but at the same time, they increase friction significantly, especially at higher loads [70]. By using thiazol-based dithiocarbamate derivatives, other authors find a significant reduction of the friction coefficient and wear in rapeseed oil [71].

For the oxidative stabilization of vegetable oils, which mostly have a significant degree of unsaturation, a number of antioxidants can be used [72]: butylated hydroxyanisole (BHA), Butylated mono-tert-butyl hydroxyl quinone (TBHQ), butylated hodroxytoluol (BHT), propyl gallate (PG), tocopherol, mixtures of ZDTP, and dithiocarbamates (DTC) with primary and secondary antioxidants (AO) performance.

Unsaturated oils are more susceptible to oxidation than saturated oils because a radical is easily formed by the removal of a hydrogen of the methylene group next to a double bond. The free radical readily forms a peroxiradical with oxygen,

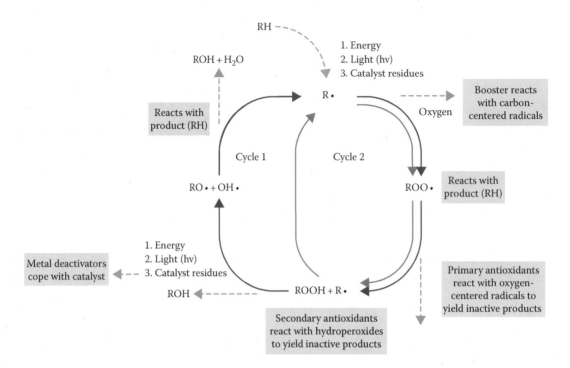

FIGURE 32.5 Schematic mechanism of oxidative degradation and the influence of antioxidants.

which removes another hydrogen atom and thereby propagates the oxidation process [72].

Organic sulfides in unsaturated esters can increase the wear rate, but antioxidants prevent the magnified wear by these sulfides. Peroxides generated during the rubbing process seem to play a significant role in the magnified wear by organic sulfides in vegetable oils and are quenched by antioxidants [73]. Similar effects can be found in vegetable oil formulations containing ZDTP. In a rapeseed oil, an impact of the peroxy value (POV) on the wear scar is observed. Peroxides generated during the oxidation of the base stock react with zinc dithiophosphates (ZDTP) and form disulfides which show a poor antiwear performance. The peroxides themselves cause oxidative wear. The use of antioxidants reduces the POV and therefore the wear [74].

Polyesters basically contain two different kinds of methylene groups: alpha-methylene groups from the polyol fragment and methylene groups neighbored to the carboxyl group. The first readily react with oxygen to form peroxidic radicals, whereas the latter are less reactive and quite stable, but their ether analogues are very reactive and form unstable peroxides even at room temperature [75].

In sebacate esters, a polymer phenolic AO, Poly-(p-methoxyphenole) (PMOP) (see Figure 32.6), was found to have a better AO performance compared to classical AOs like BHT or BHA and also are better than its monomer p-methoxyphenole (MOP) (see Figure 32.7) [76]. The better performance can be explained by a higher AO efficiency of the remaining -OH groups and the higher thermal stability due to the higher molecular weight of the polymer.

Besides the improved oxidative stability, the additive also provides a lower friction coefficient and wear scar compared to the pure base stock. Furthermore, the copper corrosion was

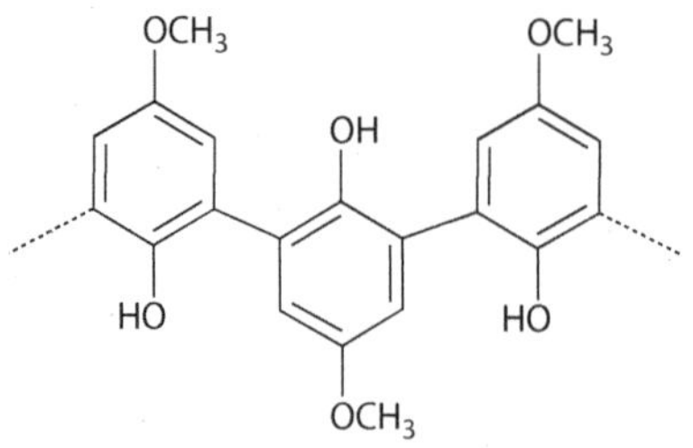

FIGURE 32.6 Structure of poly(p-methoxyphenol)—PMOP. (Own drawing according to Miao, C. et al., *Tribol. Int.*, 88, 95, 2015.)

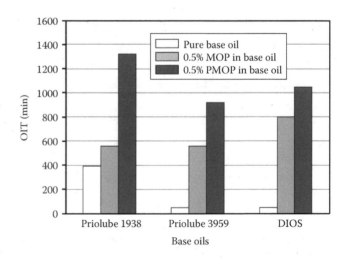

FIGURE 32.7 Oxidation induction time (OIT) of PMOP in different base oils compared to its monomer MOP (DIOS-di-iso-octyl-sebacate, Priolube 3959—diester, Priolube 1938—trimellithate ester). (Taken from Miao, C. et al., *Tribol. Int.*, 88, 95, 2015.)

improved from a 2b in the pure base stock to a 1b for the PMOP containing base oil [76].

Another approach for unsaturated ester base stocks like TMP-oleate (trimethylol propane oleate) is the use of a classical AO like BHT in combination with Tolyltriazole (TTZ) as a metal deactivator and a compound which regenerates the depleted antioxidant. A formulation containing Na-ascorbate was superior compared to the formulation without Na-ascorbate in the RPVOT, what is explained by the effect of regeneration of the BHT by Na-ascorbate [77].

Polymerized soybean oil has a lower oxidative stability compared to its monomer due to the tertiary carbon atom which is generated during the polymerization [78]. By using thermogravimetry (TG) and high pressure differential scanning calorimetry (HPDSC), it was shown that the use of BHT or zincdithiocarbamte—ZnDTC—can improve the oxidation temperature again by 70°C–80°C.

Another strategy to stabilize base stocks is to build in a phenolic block into the ester chain. Antioxidant-modified esters of dipentaerythritol with a phenolic block in the ester chain where investigated by thermogravimetry and Fourier transform infrared spectroscopy (FT-IR). A significant AO effect was found for these base stocks. Unfortunately, the authors didn't benchmark the fluid against a physical mixture of an ester and a phenolic AO in order to identify the intrinsic benefit of this solution [79].

For polyolesters, it was found that phenyl-a-naphthylamine (PANA) seems to be a very efficient Antioxidant [80], which was confirmed by a study of the authors by using rotating pressure vessel oxidation test (RPVOT) (Figure 32.8a) [81]. We also found out that besides PANA, DPA is quite an efficient AO solution for other ester types (see Figure 32.8b and c). Combinations of aminic and phenolic antioxidants didn't show any significant synergistic effects.

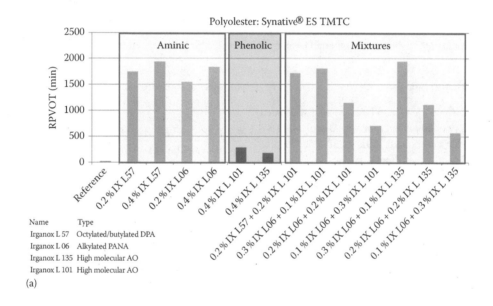

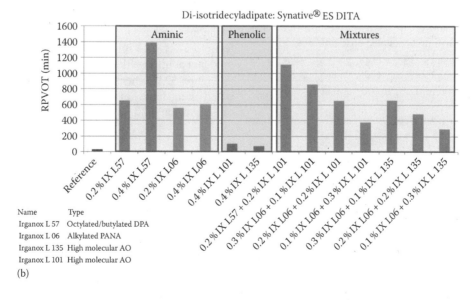

FIGURE 32.8 (a) Oxidation behavior of different AOs in (a) a polyol ester base stock, (b) an adipate ester base stock. *(Continued)*

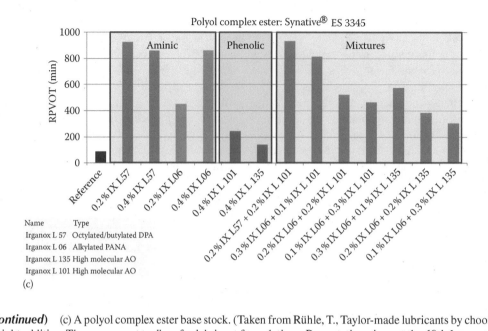

Name	Type
Irganox L 57	Octylated/butylated DPA
Irganox L 06	Alkylated PANA
Irganox L 135	High molecular AO
Irganox L 101	High molecular AO

(c)

FIGURE 32.8 (*Continued*) (c) A polyol complex ester base stock. (Taken from Rühle, T., Taylor-made lubricants by choosing the right base stocks AND the right additive: The component toolbox for lubricant formulations, Presentation given at the *19th International Colloquium Tribology*, Technische Akademie Esslingen, Ostfildern, Germany, January 21–23, 2014.)

32.8.3 METAL DEACTIVATORS

In various applications, the lubricant can be in contact with parts made of yellow metals, typically copper, or with dissolved metal contaminants. As copper compounds catalytically decompose peroxides generating radicals, these metal species are very efficient pro-oxidatives and need to be deactivated. Dissolved copper ions are deactivated with chelating agents, whereas the metal parts are protected and deactivated with film-forming metal-passivating agents. The advantage of film- forming agents is the fact that they prevent the progressive oxidation of the metal surface, which would release copper ions to the fluid. Complexing agents, on the contrary, could even promote the transfer of copper ions from the metal surface into the fluid. A typical example of a chelating agent would be disalicyliden-propylene-diamine, in which film-forming agents are very often derivatives of benzotriazol [82]. Both groups of metal deactivators are used only in very low treat rates, below 0.1%, and are active in all suitable base fluids used for the formulation of biolubricants. A critical property of some metal deactivators is the low solubility; especially when PAO is used as a base fluid, this property must be carefully checked.

32.8.4 CORROSION INHIBITORS

Another important class of additives used to protect metal surfaces is corrosion inhibitors; in this chapter, the term *corrosion inhibitor* is used synonymously with *antirust additives* and describes additives that are used to protect ferrous metals against the attack of moisture, oxygen, and other aggressive lubricant ingredients. Corrosion inhibitors possess as a general molecule design a long alkyl chain to improve the solubility in lubricants and a polar group to provide the surface activity. Corrosion inhibitors are absorbed physically or chemically on the metal surface and form a dense protective layer. Examples of corrosion inhibitors include metal sulfates,

amines, carboxylic acid derivatives, and amine-neutralized alkyl phosphoric acid partial esters. Because of their surface activity, corrosion inhibitors tend to compete with polar base fluids for the metal surface; therefore, it can be quite complex to improve the corrosion stability of an ester or a PAG.

A potential problem associated with corrosion inhibitors is the fact that acids, bases, and metals can catalyze the hydrolysis of esters. To prevent the hydrolysis of ester-based fluids, the treat rate of the corrosion inhibitors should be as low as possible, and in addition, preferably neutral additives, such as the neutralized acids, should be applied. It has been described that the response of corrosion inhibitors in different esters can show significant variation [83]. For example, to stabilize a rapeseed oil, twice the amount of a corrosion inhibitor was necessary than was required for a TMP-oleate. In addition, it has been found that at higher treat rates, the use of corrosion inhibitors can have a detrimental effect on the corrosion protection. They also might compete with other surface-active lubricant additives like, for example, antiwear additives.

Figure 32.9 schematically shows the competitive effect based on the model of equilibrium adsorption for the example of a corrosion inhibitor in an ester base stock.

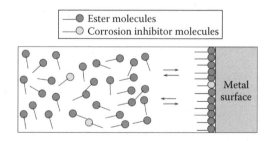

FIGURE 32.9 Competitive adsorption of corrosion inhibitor and ester base stock molecules.

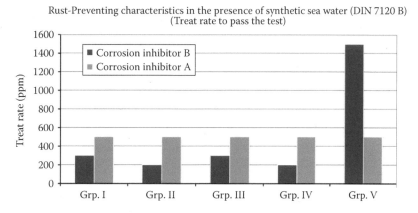

FIGURE 32.10 Salt water corrosion test of two different corrosion inhibitors in Grp. I–V base stocks. (According to Rühle, T. et al., A new corrosion inhibitor with unique performance advantages for industrial lubricant applications, *Proceedings of the OilDoc Conference & Exhibition*, January 22–24, 2013, Rosenheim, Germany, 2013.)

Data from the authors' lab confirmed this kind of interaction (see Figure 32.10) [84] for two different corrosion inhibitors in an adipate ester and in Grp. I–IV base stocks, whereas Corrosion Inhibitor B based on a carboxylic acid shows a considerably low treat rate to pass the salt water corrosion test according to DIN 7120 B in Grp. I–IV base stocks, the treat rate in the adipate ester is quite high. Corrosion Inhibitor A, based on a dicarboxylic acid half ester, shows the opposite behavior: the required treat rate is rather low in the ester but higher in Grp. I–IV base stocks compared to Corrosion Inhibitor B.

This effect is explained in Figure 32.11 by using the model of the simple Langmuir adsorption isotherm, considering also the additive solubility in an ester oil (according to [85]). This figure represents only the corrosion data in the Grp. V base stock data from Figure 32.10.

C_q is defined as the optimum surface coverage to provide a sufficient CI performance. Corrosion Inhibitor A shows a rather poor solubility in an ester base stock (and good solubility in Grp. I–IV base stocks) and thereby reaches the required surface coverage C_q at a lower treat rate TR_A. Curve B represents the curve for the Corrosion Inhibitor B that shows a good solvency in the ester base stock. As a consequence, the concentration C_q is reached at a higher treat level TR_B (see also [86]). This is exactly the result shown in Figure 32.10, where the required treat rate TR_B to pass the corrosion test in the ester base stock is higher than TR_A. This model can also be used in order to explain the behavior of both corrosion inhibitors in less polar base stocks like Grp. I–IV. Due to the rather poor solubility of Corrosion Inhibitor B in these base stocks, the situation is just the opposite as shown in Figure 32.12. Here, the required treat rate TR_B to pass the corrosion test is lower than TR_A, since the solubility of Corrosion Inhibitor B in these base stocks is not as good as that of Corrosion Inhibitor A. That means that the treat rate in order to reach the optimum surface coverage C_q is

Langmuir isotherm: $q = \dfrac{K_L q_{max} C_{eq}}{1 + K_L C_{eq}}$

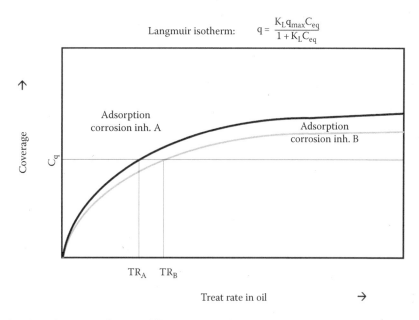

FIGURE 32.11 Langmuir adsorption model for two different corrosion inhibitors in an adipate ester base stock. (Corrosion inhibitor A: *high* adsorption affinity, corrosion inhibitor B: *low* adsorption affinity.) (According to Rühle, T., Hydraulic oils—New requirements and the answers of the lubricant industry, UNITI Mineralöltechnik No. 1, 2014, pp. 1–48.)

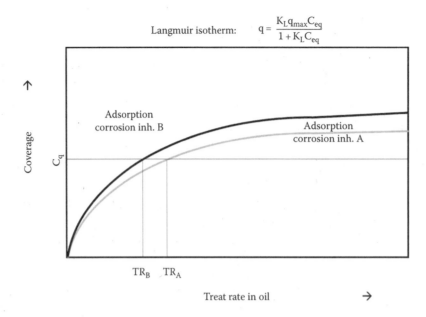

Langmuir isotherm: $q = \dfrac{K_L q_{max} C_{eq}}{1 + K_L C_{eq}}$

FIGURE 32.12 Langmuir adsorption model for two different corrosion inhibitors in mineral oils. (Corrosion Inhibitor A: *low* adsorption affinity, Corrosion Inhibitor B: *high* adsorption affinity.) (According to Rühle, T., Hydraulic oils—New requirements and the answers of the lubricant industry, UNITI Mineralöltechnik No. 1, 2014, pp. 1–48.)

higher for Corrosion Inhibitor A than for Corrosion Inhibitor B in Grp. I–IV base stocks [85,86].

32.8.5   EXTREME-PRESSURE AND ANTIWEAR ADDITIVES

Extreme-pressure and antiwear (EP/AW) additives are needed in lubricants to protect metal surfaces against mechanical damage in case the lubricant film is ruptured under critical conditions. These additives not only absorb onto the metal surface, but also react further in the case of a thermal fragmentation or tribofragmentation reaction, building a chemical reaction layer on the metal. The formed layers make the tribological contact softer, preventing a direct metal-to-metal contact of moving parts. AW additives are used for applications with medium loads. One model to explain the mode of action of these additives is to assume that polymer condensation and polymerization reactions are initiated, which provide a protecting layer on the metal. This layer is sheared off and renewed constantly during the operation [87]. Under very high loads, the performance of AW additives becomes insufficient and designated EP additives are needed. These additives undergo real reactions with the metal surface to form tribolayers consisting of iron phosphite or iron sulfides or iron chlorides. Here, the metal is a part of the protective tribolayer.

As EP/AW additives take part in real chemical reactions, they require certain activation energy to display the intended performance. The activation energy is provided partly due to the operation temperature, but more largely due to the local temperature spikes and mechanical forces in the friction zone. It is hence important to know these conditions for the selection of the correct EP/AW additives. For low-temperature applications, more active additives might be necessary, whereas for higher temperatures, it has to be ensured that the additives at least partially stay in grade

in the bulk fluid. ZDTPs are the most common AW additives, but they are not suitable for environmentally friendly fluids due to the content of a heavy metal, zinc. They can't be used in biolubricants, since they bioaccumulate in the food chain [88] and show aquatic toxicity as well as soil toxicity [89,90].

For biolubricants, ashless additives such as triarylphosphates or trialkylphosphates, neutralized alkyl phosphoric acids, and ashless dialkyl-dithiophosphates are used. All these additives are very reactive in nonpolar PAOs as they can easily be absorbed on the metal surface. In polar esters and PAGs, however, the situation is more complex, as these base fluids interact strongly with metal surfaces and consequently provide an already good protection against friction and wear, at least under mild conditions. For more demanding applications, in which the protective reaction layer provided from additives is required, the high surface activity of the base fluids is detrimental, as the additives are hindered to approach the metal surface in the first stage. Hence, higher treat rates of additives need to be applied, and preferentially more polar additives have to be used. A further problem is the fact that even ashless sulfur- and phosphorus-containing compounds do not necessarily have excellent ecotoxicological profiles. Of course, this statement cannot be generalized; however, so far, only a very limited number of additives are available that would be officially approved for the use in stringent environmentally friendly lubricants such as EEL fluids (see Table 32.3).

If the application is very demanding and high loads need to be managed, the sometimes still-used chlorinated products should definitely be avoided for toxicological reasons. Sulfur carriers, on the contrary, could be used for biolubricants. For example, sulfurized esters, which can have a good toxicological profile as well as a high degree of biodegradability, have been tested and found to show good performance [25] (see also Table 32.3).

TABLE 32.4

Chemical Structure and Dipole Moments of Octadecanoic Acid (OA) and 2-Carboxyoctadecanoic (2-COA), and Their Wear and Friction Behavior in a Polyether

Chemical Name	Stearic Acid	2-Carboxy Stearic Acid
Chemical structure		
Dipole moment (D)	2.2	2.5
Wear index		
2 mM conc.	−0.1	−1.0
5 mM conc.		−1.0
20 mM conc.	−0.19	
Friction index		
2 mM conc.	~−0.35	~−0.62
5 mM conc.		~−0.63
20 mM conc.	~−0.55	

Source: Table compiled according to Minami, I. and Mori, S., *Lubric. Sci.*, 19, 127, 2007.

Note: Friction index = $\dfrac{(\mu \text{ of additive solution}) - (\mu \text{ of additive free oil})}{(\mu \text{ of additive free oil})}$.

The effectiveness of antiwear (AW) additives is strongly dependent on the chemical structure of the ester, such as length and branching of the carbon chain [91]. Polar groups affect AW but not friction, whereas chain length is the main factor for friction. It was also found that some additives work better in unsaturated, some better in saturated esters. Many sulfur-containing additives function better in unsaturated esters. The epoxidation of unsaturated esters leads to improved oxidation stability, but the AW performance might suffer [91].

An important factor is the polarity of the ester base stock. The efficiency of AW additives can be increased by decreasing the polarity of the base stock, since the solubility of the additive in the base stock and therefore the affinity of this additive towards the surface is lower; in principle, this is the same effect which was already discussed in the section about the corrosion inhibitor and can be applied to any kind of surface-active lubricant additive [92].

One option to increase the affinity of molecules toward the surface is to increase the polarity of the molecule, for example, by just adding a second, vicinal carboxylic group to a carboxylic acid like, for example, octadecanoic acid (OA) in order to obtain 2-carboxyoctadecanoic acid (2-COA). The second one has a dipole moment of 2.5 D versus a dipole moment of 2.2 D for the single acid (see Table 32.4). In a polar base stock like a polyether, this small change has a significant positive impact on the wear as well as on the friction coefficient (Table 32.4) [86].

There are other ways to tune the polarity of an ester on a molecular level. One possibility for fatty acid–based products is to decrease oleophilic moieties, which basically has the consequences of lower molecular weight and lower viscosities. Another possibility is the introduction of polar groups. C=C double bonds also show a higher contribution to dipole moment

than C–C single bonds [92]. Hydroxylphosphonates, for example, reduce wear significantly in polar synthetic esters, where conventional AW additives can't be applied. The substituents and thereby the dipole moment affect the AW properties [92]. Whereas saturated alkyl and allyl derivatives show excellent AW properties, allyl derivatives show poor results.

Since antiwear additives and corrosion inhibitors are both surface active and show a similar physisorption behavior in esters, we use the model already discussed in the chapter about the corrosion inhibitors in order to explain this behavior (see Section 32.8.4). Investigating the wear behavior of a dibutyl phosphate with a fixed treat rate in esters with different polarities measured by the so-called polarity index,* quite a strong positive linear correlation between the polarity index and wear index† can be found [86]. This is because AW additives show higher solubility in esters with a higher polarity index and thereby less adsorption on the surface resulting in a higher wear rate indicated by the wear index.

On the other hand, it was shown, for P-containing AW additives in polar esters, that too much adsorption capacity of the AW additive might cause corrosive wear. Using hydrolytic stable AW additives reduces corrosive wear to a certain extent [86].

Traditional AW additives in some synthetic esters lubricants show a rather low performance. Waare et al. investigated different additive packages consisting of EP (dithiocarbamate–DTC and triphenyl phosphorothionate—TPPT), yellow metal deactivator –YMD (thiadoazole and triazole), and AW (amine phosphate) in

---

* Polarity index = $\dfrac{(\text{Number of ester groups}) \times 10{,}000}{(\text{Total carbon number}) \times (\text{Formula weight})}$.

† Wear index = $\dfrac{(\text{WSD of additive solution}) - (\text{WSD of additive free oil})}{(\text{WSD of additive free oil})}$.

mineral oils and synthetic TMP-polyolesters. They tested four different additive combinations: (1) amine phosphate (AW) + thiadiazole (YMD) + triphenylphosphothionate (EP), (2) amine phosphate (AW) + ashless DTC (EP) + tolyltriazole (YMD), (3) a combination like in (1) but with a lower treat rate TPPT and no YMD, and (4) a combination like in (2) but with a lower treat rate of DTC and no YMD. The synthetic ester–based lubricants, with all combinations, resulted in much higher surface wear than for the mineral oil based formulations. The EP additive was much more effective than the AW additives in both base fluids [93].

Organic sulfides under boundary conditions in TMP-esters and rapeseed oil magnify wear in unsaturated esters, whereas a large wear was observed during short rubbing times and low concentrations of the sulfide. Surface analytical tests revealed some organic oxides and unreacted additive in the worn surface most probably generated from organic peroxides by the oxidation of the unsaturated esters. Phenolic antioxidants may prevent wear to some extent [73].

Sulfur-containing triazine derivatives show a high thermal stability and are able to reduce the friction coefficient and the wear scar and also provide corrosion protection and extreme-pressure performance in rapeseed oils under a wide range of test conditions at treat levels around 1.5%. They are supposed to adsorb on the surface and generate surface-protecting layers [94].

Also some rather "exotic" AW additives like nanoparticles or ionic liquids have been tested. Two different ionic liquids—phosphonium phosphates—have been benchmarked against amine phosphate (see Figure 32.13) [95]. The base stocks used were safflower oil, TMP, and polyolester. A combination of phenolic and aminic AOs has been used. The antioxidant properties are similar for all three formulations, but the wear scar diameter (WSD) and the coefficient of friction (COF) measured by a pin-on-disc apparatus was lower for the lubricants with

the ionic liquid additives in all three base stocks even at lower treat rates compared to the aminophosphates. These results can be explained by film-forming properties and in the boundary regime, despite the fact that the viscosity of the ionic liquids is much lower than for the amine phosphate.

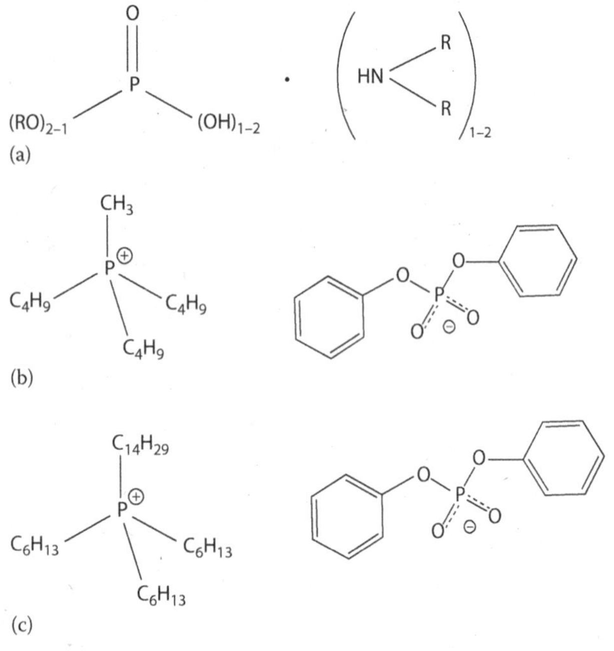

Borane and fluorine-containing imidazolium ionic liquids have been tested in oleate esters for the aluminum–steel contacts [96]. Whereas all compounds reduced wear at high temperatures, only the highly polar compound reduced wear at room temperature.

Nanoparticles have been discussed for the use in lubricants already since a long time. *Reeves et al.* used boron nitride (BN) particles as AW additives for biolubes [97]. The base stock used was canola oil. BN particles have a layered structure and cause a reduction of the COF in many lubricants. Applying 2.5% and 5% BN in canola oil and testing these formulations on a pin-on-disk, it was found that the wear scar and COF can be reduced significantly compared to the pure base oil, whereas nanoparticles of 70 nm size show the best effect and microparticles of 5 μm size show an increase in friction wear. This can be explained by a model including the surface roughness of the metal. Particles larger than the surface asperities carry a portion of the load between the asperities, but at the same time increasing friction and being abrasive causing higher wear. Nanoparticles are filled into the asperities creating a protective transfer film and thereby reducing friction and wear. Since the nanoparticles are significantly smaller than the hills and valleys created by the surface roughness, they can therefore be adsorbed as a monolayer on the surface, protecting the surface. When large particles with a similar or larger size than the "hills" and "valleys" are used, they can not fill up the valley anymore, becoming rather abrasive instead of protecting the surface. This principle is shown in Figure 32.14.

32.8.6 POUR POINT DEPRESSANTS AND VISCOSITY INDEX IMPROVERS

Pour point depressants (PPDs) and viscosity index improvers (VIIs) are added to improve the rheology of a lubricant. At reduced temperatures, paraffinic compounds in lubricants can form wax crystals that agglomerate and crystallize, solidifying the lubricant. To prevent this crystallization, a PPD is added. PPDs are specially designed polymers, which have a bifunctional molecular structure. Some parts of the molecules are similar to paraffin wax crystals, enabling cocrystallization with the waxes, whereas other parts of the polymer have a structure very dissimilar to the wax crystals, preventing any further agglomeration and growth of the wax matrices. The effect is a better flow behavior and a depression of the pour point. As PAGs, PAO, and most synthetic esters already have good low-temperature behavior, the addition of a PPD is not very common. Natural esters, however, can crystallize relatively easily—even in some cases at ambient temperatures—and therefore the demand for a PPD can be high. The amount of PPD used varies according to the ester type, concentration of the polymer in the fluid, type of polymer, and degree of pour point correction desired.

VIIs are added to minimize the extreme viscosity variation of a given fluid with change in temperature, especially important for applications working in different temperature regimes.

FIGURE 32.13 Amine phosphate structure (a) and the structure of two different ionic liquids (b and c used as antiwear additives). (Own drawings according to Khemchandani, B. et al., *Tribol. Int.*, 77, 171, 2014.)

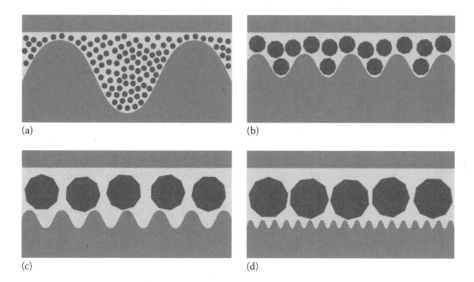

FIGURE 32.14 Solid antiwear particles of different size used on a metal surface with constant surface roughness (particle sizes: (a)—70 nm, (b)—500 nm, (c)—1500 nm, and (d)—5000 nm). (Own drawing according to Reeves, C.J. et al., *Tribol. Lett.*, 51, 437, 2013.)

VIIs have a chemical structure similar to PPD, but the polymers used have a much higher molecular weight. As base fluids for biolubricants have in general already quite a high viscosity index, a further improvement of this property is not very common and only necessary for special applications.

32.8.7 ANTIHYDROLYSIS AGENTS

Antihydrolysis agent is a special class of additives, which is used only for ester-based fluids. As described earlier, the ester bond can be easily cleaved reacting with water, forming acids and alcohol. As this reaction is catalyzed by acids, the reaction is to be seen as autocatalytic, because each molecule of

acid formed during the hydrolysis can itself catalyze a new hydrolysis reaction. In the absence of catalyst, this process is kinetically very slow. Additionally, the formed acid might attack and dissolve bearing metals like lead, zinc, and tin.

In order to stabilize an ester, special carbodiimides can be added, which work as acid scavengers and hence are able to remove the catalyst and therefore stop the degradation process to a certain extent [98]. The disadvantage of such acid scavengers is that they also might react with acids intentionally added as a corrosion inhibitor or an EP/AW additive; hence, the potential for an antagonistic effect is high.

Figure 32.15 shows the effect of a carbodiimide on the hydrolytic stability of a TMP-oleate ester by recording the

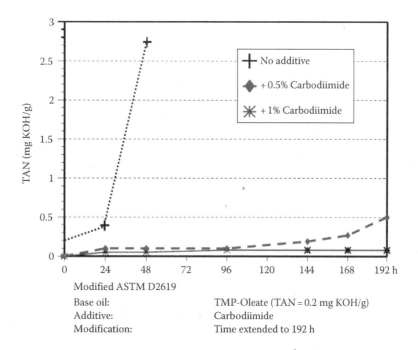

FIGURE 32.15 Effect of carbodiimides on the hydrolytic stability of a TMP-oleate ester. (Taken from Pazdzior, D. and Fessenbecker, A., Additive solutions for industrial lubricants, Presentation given at the *Lubricants Russia Conference*, Moscow, Russia, 2005.)

total acid number as a function of time. The additization of 1% carbodiimide into the TMP-oleate leads to a significant improvement of the hydrolytic stability. It also can be seen that in the non-aged formulation (0 hours), the acid number of the TMP-oleate with carbodiimide is already significantly lower, which basically results in a smaller autocatalytic effect and therefore a better ageing behavior.

32.8.8 ANTIFOAM AGENTS, DEMULSIFIERS, AND EMULSIFIERS

To improve the surface properties of biolubricants, antifoam agents, demulsifiers, and emulsifiers can be added. In typical industrial applications, the amount of additive is extremely low (10–100 ppm) and hence will not be discussed in detail here.

32.9 SUMMARY

Currently, biolubricants are still regarded as niche products. However, in certain applications such as hydraulic fluids, the use of biolubricants is steadily growing. The VGP legislation which was introduced by the U.S. government after the disaster in the Gulf of Mexico also put some pressure on the market players to come up with solutions. Government incentives and labelling schemes are aiming to create market drivers, public promotion, and compensation for potential economical disadvantages. Absolutely mandatory is, of course, the suitable technical performance of the biolubricants, as well as the lowest possible harm to the environment. To achieve this, various base oils can be used. To ensure performance requirements are met, additives are needed, which are tailored for the base fluid used and well suited for the intended application. These additives need to be carefully formulated, making use of synergistic effects and avoiding any antagonism. Especially for ester base stock, not all traditional lubricants show a sufficient performance. The choice of suitable surface-active additives like corrosion inhibitors or AW additives is strongly dependent on the properties of the ester like dipole moment, polarity, degree of unsaturation, et cetera. Since some esters have a lack of hydrolytic stability, the use of antihydrolysis agents might be required, additives which usually are not necessary in mineral oils.

So far, it is not yet globally accepted, but with a growing regional importance, biolubricants are required to be ecologically friendly, with criteria such as renewability of feedstocks, low aquatic toxicity, and a high rate of biodegradability becoming demanded more often. These environmental factors are not limited to the base oil, but are also valid in a certain degree for the additives; hence, in future, the demand for environmentally acceptable, high-performance additives will be growing steadily.

REFERENCES

1. United States Environmental Protection Agency (EPA), Climate change in the United States: Benefits of global actions, 2015, Available at: www.epa.gov/cira. Last accessed on May 2, 2017.
2. European Commission, Climate strategies and targets, 2017. Available at: www.ec.europa.eu/clima/policies/international/negotiations/paris/index_en.htm. Last accessed on May 2, 2017.
3. BASF Publication, UN sustainable development goals and our contribution, 2016. Available at www.basf.com/en/company/sustainability/employees-and-society/goals.html. Last accessed on May 2, 2017.
4. D. Horner, Recent trends in environmentally friendly lubricants. *Journal of Synthetic Lubrication*, 18, 327–348, 2002.
5. C.J. Reeves, P.L. Menezes, T.-C. Jen, M.R. Lovell, Evaluating the tribological performance of green liquid lubricants and powder additive based green liquid lubricants. *STLE Annual Meeting and Exhibition*, STLE, St. Louis, MO, 2012.
6. K. Carnes, Offroad hydraulic fluids beyond biodegradability. *Tribology & Lubrication Technology*, 60, 32–40, 2004.
7. S. Miller, C. Scharf, M. Miller, Utilizing new crops to grow the biobased market, in trends, in *New Crops and New Uses*, J. Janick, A. Whipkey (eds.), ASHS Press, Alexandria, VA, pp. 26–28, 2000.
8. DIN ISO 15380:2011, Lubricants, industrial oils and related products (class L)—Family H (Hydraulic systems)—Specifications for categories HETG, HEPG, HEES and HEPR. ASTM International, West Conshohocken, PA, 2011.
9. C. Vag, A. Marby, M. Kopp, L. Furberg, T. Norrby, A comparative life-cycle assessment of the manufacture of base fluids for lubricants. *Journal of Synthetic Lubrication*, 19(1), 39–57, 2002.
10. ASTM D 6866-04, Standard test methods for determining the biobased content of natural range materials using radiocarbon and isotope ration mass spectrometry analysis. ASTM International, West Conshohocken, PA, 2004.
11. United States Environmental Protection Agency, Environmental Acceptable Lubricants; EPA 800-R-11-002; Office of Wastewater Management; Washington, DC, November 2011.
12. Organisation for Economic Co-operation and Development (OECD), OECD guidelines for the testing of chemicals. Available at: www.oecd.org/chemicalsafety/testing/oecd-guidelinesforthetestingofchemicals.html. Last accessed on May 2, 2017.
13. Umweltbundesamt, LTWS-Schriftenreihe Nr. 12, Katalog wassergefahrdender Stoffe, Umweltbundesamt, Berlin, Germany, 1996.
14. Bundesministerium für Jusitz, Bundesrepublik Deutschland, Allgemeine Verwaltungsvorschrift zum Wasserhaushaltsgesetz uber die Einstufung wassergefahrdender Stoffe in Wassergefahrdungsklassen (Verwaltungsvorschrift wassergefahrdende Stoffe—VwVwS). Published at BAnz. Nr. 98a; 29, May 17, 1999.
15. A. Gaikward, Synthetic & bio-based lubricants market—Global industry analysis, market size, share, trends, analysis, growth and forecast, 2012–2018. Transparency Market Research Pvt. Ltd., Published January 23, 2013.
16. N. Aslanian, *Lubes'n'Greases Europe Middle East-Africa*, (68), 6–8, February 2015.
17. United States Environmental Protection Agency (EPA), Economic and benefits analysis of the final Vessel General Permit (VGP), Office of Wastewater Management, Washington, DC, 2013. Available at: www3.epa.gov/npdes/pubs/vgp_ea2013.pdf. Last accessed on May 2, 2017.
18. M. Schneider, P. Smith, Government-industry forum on non-food uses of crops (GIFNFC 7/7) case study: Plant oil based lubricants in total loss & potential loss applications, Final Report, p. 20, May 16, 2002.

19. A. Scherlofsky, EU Eco-label marketing for products. Project 2006. Work on the implementation of the EU Eco-Label Scheme in the areas of marketing, product group development and stakeholder representation. LOT 7. Marketing of products. Wien: Report on behalf of DG Environment. http://ec.europa.eu/environment/ecolabel/pdf/marketing/lot_7/final_report.zip. Accessed April 28, 2008.

20. Bundesgesetzblatt für die Republik Österreich, Verordnung über das Verbot bestimmter Schmiermittelzusatze und Verwendung von Kettensägeölen, BGBl. Nr. 647, 1990. Available at: www.ris.bka.gv.at/Dokumente/BgblPdf/1990_647_0/1990_647_0.pdf. Last accessed on May 2, 2017.

21. Bundesministerium fur Verbaucherschutz, Ernahrung und Landwirtschaft (Hrsg.), Bericht uber schnell abbaubare Schmierstoffe und Hydraulikflussigkeiten, June 2002.

22. The European Union, Regulation 1907/2006/EG; (REACH), 2007.

23. European Chemicals Agency, Regulation (EC) No 1272/2008 on the classification, labelling and packaging of substances and mixtures (CLP Regulation). Available at: www.echa.europa.eu/regulations/clp/legislation. Last accessed on May 2, 2017.

24. OSPAR, Available at: www.ospar.org/. Last accessed on May 2, 2017.

25. Ral German Institute for Quality Assurance and Certification. Available at: http://www.blauer-engel.de/deutsch/navigation/body_blauer_engel.htm.

26. RAL German Institute for Quality Assurance and Certification, Basic criteria for award of the environmental label: Readily biodegradable chain lubricants for power saw, RAL-UZ 48, edn., April 2007.

27. RAL German Institute for Quality Assurance and Certification, Basic criteria for award of the environmental label: Readily biodegradable lubricants and forming oils, RAL-UZ 64, edn., April 2007.

28. RAL German Institute for Quality Assurance and Certification, Basic criteria for award of the environmental label: Readily biodegradable hydraulic fluids. RAL-UZ 79, edn., April 2007.

29. Umweltbundesamt der Bundesrepublik Deutschland, Biodegradable lubricants and hydraulic fluids. RAL-UZ 178, edn. 2013. Available at: www.blauer-engel.de/en/our-label-environment. Last accessed on May 2, 2017.

30. Verein deutscher Maschinen- und Anlagenbau (VDMA), Biologisch schnell abbaubare Druckflussigkeiten, Technische Mindestanforderungen. VDMA 24568; ICS 75.120; 1994, March 2014.

31. DIN ISO 15380, Mineralole und Brennstoffe Bd. 1/14. Erg. Lieferung, Oktober 2004.

32. H. Gustafsson, Environmental requirements for lubricants in Swedish standards, Abstract to European Lubricating Grease Institute, 15th Annual Meeting, Vienna, Austria, April 2003.

33. RISE, Research Institutes of Sweden, Hydraulic fluids which meet environmental requirements according to Swedish Standard SS 15 54 34, 2015. Available at: www.sp.se/en/index/services/Hydraulic%20fluids/Sidor/default.aspx. Last accessed on May 2, 2017.

34. DIN EN ISO 9439 Norm, 2000-10, Wasserbeschaffenheit—Bestimmung der vollstandigen aeroben biologischen Abbaubarkeit organischer Stoffe im wa.rigen Medium—Verfahren mit Kohlenstoffdioxid-Messung (ISO 9439:1999); Deutsche Fassung EN ISO 9439:2000.

35. DIN EN ISO 9408 Norm, 1999-12, Wasserbeschaffenheit—Bestimmung der vollstandigen aeroben biologischen Abbaubarkeit organischer Stoffe im wa.rigen Medium uber die Bestimmung des Sauerstoffbedarfs in einem geschlossenen Respirometer (ISO 9408:1999), Deutsche Fassung EN ISO 9408:1999.

36. DIN EN ISO 10707 Norm, 1998-03, Wasserbeschaffenheit—Bestimmung der vollstandigen aeroben biologischen Abbaubarkeit organischer Stoffe in einem wa.rigen Medium—Verfahren mittels Bestimmung des biochemischen Sauerstoffbedarfs (geschlossener Flaschentest) (ISO 10707:1994), Deutsche Fassung EN ISO 10707:1997.

37. ISO 10708 Norm, 1997-02, Wasserbeschaffenheit—Bestimmung der vollstandigen aeroben biologischen Abbaubarkeit organischer Verbindungen in einem wa.rigen. Medium—Bestimmung des biochemischen Sauerstoffbedarfs mit dem geschlossenen Flaschentest in zwei Phasen.

38. DIN EN ISO 7827 Norm, 1996-04, Wasserbeschaffenheit—Bestimmung der vollstandigen aeroben biologischen Abbaubarkeit organischer Stoffe in einem wa.rigen. Medium—Verfahren mittels Analyse des gelosten organischen Kohlenstoffs (DOC) (ISO 7827:1994), Deutsche Fassung EN ISO 7827:1995.

39. European Commission, Regulation (EU) 880/92, 1992. Available at: www.ec.europa.eu/environment/ecolabel/eu-ecolabel-for-consumers.html. Last accessed on May 2, 2017.

40. European Union Ecolabel application pack for lubricants, 2014. Available at www.ec.europa.eu/environment/ecolabel/documents/lusclist.pdf. Last accessed on May 2, 2017.

41. European Committee for Standardization (CEN), Available at: www.cen.eu. Last accessed on May 2, 2017.

42. European Committee for Standardization (CEN), Liquid petroleum products—Bio-lubricants—Criteria and requirements of bio-lubricants and bio-based lubricants, 2016; EN 16807:2016.

43. The EcoLogoM Program, TerraChoice Environmental Marketing. Available at www.ecologo.org/. Last accessed on May 2, 2017.

44. Environment Choice Program, Certification Criteria Document (CCD-069), Synthetic Industrial Lubricants, November 10, 1996, edits June 2007.

45. Government of Canada; Department of Justice, Hazardous products act, SOR/2015-17,s.21. Available at: http://laws-lois.justice.gc.ca/eng/acts/H-3/. Last accessed on May 2, 2017.

46. United States Department of Agriculture, 2002. Available at www.biopreferred.gov/BioPreferred/faces/Welcome.xhtml. Last accessed on May 2, 2017.

47. United States Environmental Protection Agency (EPA), Vessel general permit for discharges incidental to the normal operation of vessels (VGP), Office of Wastewater Management; Washington, DC, 2013. Available at www3.epa.gov/npdes/pubs/vgp_permit2013.pdf. Last accessed on May 2, 2017.

48. Japan Environmental Association (JEA), Eco Mark Office, Available at www.ecomark.jp/english/. Last accessed on May 2, 2017.

49. Thailand Environment Institute (TEI), 1994. Available at: http://www.tei.or.th/greenlabel/index.html. Last accessed on May 2, 2017.

50. D.R. Lide, Handbook of Chemistry and Physics, Vol. 86, CRC Press, Boca Raton, FL, pp. 9.64–9.72, 2005.

51. O. Falk, R. Meyer-Pittroff, The effect of fatty acid composition on biodiesel oxidative stability. Journal of Lipid Science and Technology, 106(12), 837–843, 2004.

52. E. Kosukainen, Y.Y. Linko, M. Lamasa, T. Tevvakangas, P. Linko, Plant-oil-based lubricants and hydraulic fluids. Journal of the American Oil Chemists' Society, 75, 1557–1563, 1998.

53. D. Kodali, Bio-based lubricants—Chemical modification of vegetable oils. Inform, 14, 121–123, 2003.

54. H. Wagner, R. Luther, T. Mang, Lubricant base fluids based on renewable raw materials—Their catalytic manufacture and modification. *Applied Catalysis A: General*, 221, 429–442, 2001.

55. C. Kormann, A. Kicherer, A mass balance approach to link sustainable renewable resources in chemical synthesis with market demand, in *Sustainability Assessment of Renewables—Based Products: Methods and Case Studies*, J. Dewulf, R. Alvarenga, and S. De Meester (eds.), John Wiley & Sons Ltd., 2015.

56. ECETOC: TR 055, Pulmonary toxicity of polyalkylene glycols, December 1997.

57. P. Harmsen, M. Hackmann, *Green Building Blocks for Biobased Plastics*, 2013.

58. F. Joel, Carpenter. *Journal of Synthetic Lubrication*, 12(1), 13–20, April 1995.

59. C.M. Cisson, G.A. Rausina, P.M. Stonebreaker, Human health and environmental hazard characterization of lubricating oil additives. *Lubrication Science*, 8-2, 144–177 (January 1996).

60. J.C.J. Bart, E. Gucciardi, S. Cavallaro, *Biolubricants—Science and Technology*, Woodhead Publishing Oxford, U.K., 2013, pp. 351–395.

61. European Commission, Assessment of the risk to consumers from borates and the impact of potential restrictions on their marketing and use, Final report, Directorate-General Enterprise and Industry by Risk & Policy Analysts Limited, London, U.K., November 2008.

62. IVAM-research and consultancy on sustainability, Amsterdam (Netherlands), Lubricant Substance Classification list (LuSC-list), Version date: 09 July 2012, http://www.ivam.uva.nl/wp-content/uploads/2014/11/LuSC-list-30062015-no-track.pdf (download December 12, 2015).

63. M. Miller, Additives for bioderived and biodegredable lubricants, L. Rudnick (ed.) in *Lubricant Additives—Chemistry and Applications*, CRC Press, Boca Raton, FL, 2009, p. 445.

64. T. Mang, W. Dresel, *Lubricants and Lubrication*, Wiley VCH, Weinheim, Germany, 2001.

65. R.K. Jensen, S. Korcek, M. Zinbo, J.L. Gerlock, Regeneration of amines in catalytic inhibition and oxidation. *The Journal of Organic Chemistry*, 17, 5396–5400, 1995.

66. P. Hamblin, D. Chasen, U. Kristen, Paper at the *Fifth International Colloquium, Additives for Lubricants Operational Fluids*, Technische Akademie Esslingen (TAE), Germany, Jan 14th–16th 1986.

67. Amendment to the Technical Progress 2004/73EG of the Material Directive, European Union, 2004.

68. R. Becker, A. Knorr, Antioxidantien für pflanzliche ö. *Tribologie Schmierungstechnik*, 42(5), 272–276, 1995.

69. R. Becker, A. Knorr, An evaluation of antioxidants for vegetable oils at elevated temperatures. *Lubrication Science*, 8(2), 95–116, 1996.

70. Y. Sun, L. Hu, Q. Xue, Tribological properties and action mechanism of N,N, dialkyl dithiocarbamte-derived S-hydroxylethyl borate esters as additives in rapeseed oil. *Wear*, 266, 917–924, 2009.

71. W. Huang, B. Hou, P. Zhang, J. Dong, Trobological performance and action mechanism of S-[2-(acetamido)thiazol-1-yl] dialkyl dithiocarbamate as additive in rapeseed oil. *Wear*, 256, 1106–1113, 2004.

72. N.J. Fox, G.W. Stachowiak, Vegetable oil-based lubricants—A review of oxidation. *Tribology International*, 40, 1035–1046, 2007.

73. I. Minami, T. Kubo, D. Shimamoto, D. Takahashi, H. Nanao, S. Mori, Investigation of wear mechanism by organic sulphides in vegetable oils. *Lubrication Science*, 19, 113–126, 2007.

74. I. Minami, K. Mimura, Synergystic effect of antiwear additives and antioxidants in vegetable oils. *Journal of Synthetic Lubrication*, 21-3, 193–205, 2004.

75. V.N. Bakunin, O.P. Parenago, A mechanism of thermooxidative degradation of polyol ester lubricants. *Journal of Synthetic Lubrication*, 9, 127–143, 1992.

76. C. Miao, L. Zhang, K. Zheng, Y. Cui, S. Zhang, L. Yu, P. Zhang, Synthesis of poly(p-methoxyphenol) and evaluation of its antioxidation behavior as an antioxidant in several ester oils. *Tribology International*, 88, 95–99, 2015.

77. H. Murrenhoff, X. Zhang, C. Göhler, Ein Antioxidantien-System für Esteröle. O+P 11-12/2004.

78. Z. Liu, B.K. Sharma, S.Z. Erhan, A. Biswas, R. Wang, T.P. Schuman, Oxidation and low temperature stability of polymerized syobean oil-based lubricants. *Thermochimica Acta*, 601, 9–16, 2015.

79. L. Zhang, G. Cai, Y. Wang, W. Eli, Synthesis and characterization of antioxidant-modified esters of dipentaerythitol as lubricating base oil. *Lubrication Science*, 25, 329–337, 2013.

80. P. Mousavi, D. Wang, C.S. Grant, W. Oxenham, P.J. Hauser, Effects of antioxidants on the thermal degradation of a polyol ester lubricant using GPC. *Industrial & Engineering Chemistry Research*, 45, 15–22, 2006.

81. T. Rühle, Taylor-made lubricants by choosing the right base stocks AND the right additive: The component toolbox for lubricant formulations, Presentation given at the *19th International Colloquium Tribology*, Technische Akademie Esslingen, Ostfildern, Germany, January 21–23, 2014.

82. J. Waynick, The development and use of metal deactivators in the petroleum industry: A review. *Energy Fuels*, 15(6), 1325–1331, 2001.

83. A. Fessenbecker, J. Korff, Additive für ökologisch unbedenklichere Schmierstoffe. *Tribologie Schmierungstechnik*, 42(1), 26–30, 1995.

84. T. Rühle, M. Hof, S. Seibel, A new corrosion inhibitor with unique performance advantages for industrial lubricant applications. *Proceedings of the OilDoc Conference & Exhibition*, January 22–24, 2013, Rosenheim, Germany, 2013.

85. T. Rühle, Hydraulic oils—New requirements and the answers of the Lubricant industry, UNITI Mineralöltechnik No. 1, pp. 1–48 , 2014.

86. I. Minami, S. Mori, Concept of molecular design towards additive technology for advanced lubricants. *Lubrication Science* 19, 127–149, 2007.

87. K. Meyer, Schichtbildungsprozesse und Wirkungsmechanismus schichtbildender Additive für Schmierstoffe. *Zeitschrift für Chemie*, 24(12), 425–435, 1984.

88. A.G. Heath, *Water Pollution and Fish Physiology*, CRC Press, Boca Raton, FL, 1995, p. 81.

89. A.G. Heath, *Water Pollution and Fish Physiology*, CRC Press, Boca Raton, FL, 1995, pp. 150–151.

90. T. Kunito, K. Saeki, S. Goto, H. Hayashi, H. Oyaizu, S. Matsumoto, Copper and zinc fractions affecting microorganisms in long-term sludge amended soils. *Bioresource Technology*, 79, 135–146, 2001.

91. W. Castro, D.E. Weller, K. Cheenkachorn, J.M. Perez, The effect of chemical structure of basefluids on antiwear effectiveness of additives. *Tribology International*, 38, 321–326, 2005.

92. I. Minami, K. Hirao, M. Memita, S. Mori, Investigation of anti-wear additives for low viscous esters: Hydroxyl phosphonates. *Tribology International*, 40, 626–631, 2007.

93. P. Waara, J. Hannu, T. Norrby, A. Byheden, Additive influence on wear and friction performance of environmentally adapted lubricants. *Tribology International*, 34, 547–556, 2001.

94. X. Zeng, H. Wu, H. Yi, T. Ren, Tribological behavior of three novel triazine derivatives as additives in rapeseed oil. *Wear*, 262, 718–726, 2007.

95. B. Khemchandani, A. Somers, P. Howlett, A.K. Jaiswal, E. Syanna, M. Forsyth, A biocompatible ionic liquid as an antiwear additive for biodegradable lubricants. *Tribology International*, 77, 171–177, 2014.

96. A.-E. Jimenéz, M.-D. Bermúdez, Imidazolium ionic liquids as additives of the synthetic ester propylene glycol dioleate in aluminium-steel lubrication. *Wear*, 256, 787–798, 2008.

97. C.J. Reeves, P.L. Menezes, M.R. Lovell, T.-C. Jen, The size effect off boron nitride particles on the trobological performance of biolubricants for energy conservation and sustainability. *Tribology Letters*, 51, 437–452, 2013.

98. D. Pazdzior, A. Fessenbecker, Additive solutions for industrial lubricants. Presentation given at the *Lubricants Russia Conference*, Moscow, Russia, 2005.

Section VIII

Methods and Resources

33 Lubricant Industry–Related Terms and Acronyms

Leslie R. Rudnick

CONTENTS

The plethora of acronyms related to the field of lubrication continues to grow. These acronyms and abbreviations come from a variety of diverse industries and disciplines, including original equipment manufacturers, component suppliers, lubricant additive and fluid suppliers and producers, and professional societies directly and peripherally involved in the lubricant industry. Each class of lubricants, synthetic and conventional, has its set of abbreviations reserved to describe differences in structure or performance characteristics. Terms and acronyms for lubricant additives are numerous and generally reflect the chemical structure or the type of additives. In some cases, the acronym reflects the function of the additive. Acronyms created at different times by different industries have resulted in identical abbreviations that refer to different things.

This chapter lists many of the important terms generally used in the lubricant industry. A complete list would require far more space than can be devoted in this book.

33.1 TERMS AND ACRONYMS

21 CFR 178.3570	The section of the Code of Federal Regulations that deals with lubricants with incidental food contact.
2T/2-cycle	A term applied to lubricants for two-cycle engines (i.e., motorcycles, outboard marine motors, and weed whackers).
3P2E	Three-ring polyphenyl ether.
4P3E	Four-ring polyphenyl ether.
4T	A term applied to lubricants for four-cycle engines.
5P4E	Five-ring polyphenyl ether.
6P5E	Six-ring polyphenyl ether.
AAM	Alliance of Automobile Manufacturers.
AAMA	American Automobile Manufacturers Association.
AAR	Association of American Railroads.
AB	Alkylbenzene.
ABIL	Agriculture-based industrial lubricants.
ABMA	American Bearing Manufacturers Association—a nonprofit association of American manufacturers of antifriction bearings, spherical plain bearings, or major components thereof. The purpose of ABMA is to define national and international standards for bearing products and maintain bearing industry statistics.

ABOT	Aluminum Beaker Oxidation Test for Ford MERCON ATF approval.
ABSA	Alkylbenzenesulfonic acid; precursor to overbased calcium sulfonate.
ACC	American Chemistry Council.
ACEA	Association des Constructeurs Européens d'Automobiles (European Automobile Manufacturers' Association).
ACERT	Advanced Combustion Emissions Reduction Technology (Caterpillar).
ACIL	American Council of Independent Laboratories—ACIL is the national trade association representing independent, commercial engineering and scientific laboratory, testing, consulting, and R&D firms.
ACS	American Chemical Society.
Additive	Chemical compound or formulation of several chemical compounds added to a base oil to alter its physical, chemical, and performance properties.
AEL	Allowable exposure limit.
AEOT	Engine oil aeration Test.
AES	Average engine sludge.
AEV	Average engine varnish.
A/F	Air–fuel ratio.
AFNOR	Association Française de Normalisation.
AFOA	American Fats and Oils Association.
AFR	Air/fuel ratio.
AFV	Alternative fuel vehicle.
AGELFI	Cooperative Research Organization of AGIP, ELF, and FINA oil companies.
AGMA	American Gear Manufacturers Association—an organization for the establishment and promotion of industrial gear lubricant standards.
AGO	Automotive gas oil.
AHEM	Association of Hydraulic Equipment Manufacturers.
AIAM	Association of International Automobile Manufacturers.
AIChE	American Institute of Chemical Engineers.
AIT	Autoignition temperature (ASTM D2155)—the lowest temperature at which a gas or vaporized liquid will ignite in the absence of an ignition source.
AL	Atmospheric lifetime.
ALTNER	Alternative Energy Programs of the European Commission.
AMA	Automobile Manufacturers Association.

(Continued)

ANFAVEA	Brazil Automobile Manufacturers Association.
ANIQ	Mexican equivalent to the CMA.
ANSI	American National Standards Institute.
Antioxidant	A chemical component added to lubricants to reduce the tendency for oxidation-related degradation of the oil.
Antiwear additive	Additives that can deposit multilayer films thick enough to supplement marginal hydrodynamic films and prevent asperity contact, or preferentially wear rather than allow contact between asperities that result in wear.
AO	Antioxidant.
AOCA	American Oil Change Association—provides a link between the motoring public and auto maintenance specialists.
AOCS	American Oil Chemists' Society—a global forum for the science and technology of fats, oils, surfactants, and related materials.
APE	Association of Petroleum Engineers (United States).
API	American Petroleum Institute—society organized to further the interest of the petroleum industry.
API GL-4	Designates the type of service characteristic of gears, particularly hypoid, in passenger cars and other automotive-type equipment operated under high-speed, low-torque and low-speed, and high-torque conditions; largely replaced by performance standard API GL-5.
API GL-5	Designates the type of service characteristic of gears, particularly hypoid, in passenger cars and other automotive-type equipment operated under high-speed, shock load; high-speed, low-torque; and low-speed, high-torque conditions.
AQIRP	Auto/Oil Air Quality Improvement Research Program.
ARB	Air Resources Board (California).
ASA	American Soybean Association.
ASEAN	Association of Southeast Asian Nations.
Ashless	Additive containing no metallic elements.
ASLE	American Society of Lubrication Engineers (now STLE).
ASM	American Society for Materials.
ASME	American Society of Mechanical Engineers.
ASTM	American Society for Testing and Materials.
A/T	Conventional shifting automatic transmission.
ATA	American Trucking Associations.
ATC	Additive Technical Committee (European Petroleum Additive Industry Association, European CMA).
ATD	Allison Transmission Division; division of General Motors.
ATF	Automatic transmission fluid
ATIEL	Association Technique de l'Industrie Européenne des Lubrifiants (European Oil Marketers Association).
AT-PZEV	Advanced technology partial zero-emission vehicle.
Auto-Oil Program	A joint activity between the European Union (EU), the European Oil Industry (EUROPIA), and the European motor industry (ACEA).
AW	Antiwear agent—minimizes wear by reacting with a metal surface to provide a protective layer.
b-CVT	Belt CVT.
BIA	Boating Industry Association—industry body organized to specify lubricants for marine application (now NMMA).
Biodegradability	The ability of a chemical compound to be broken down by living organisms.

BFPA	British Fluid Power Association.
bhp-hr	Brake horsepower-hour.
BHRA	British Hydromechanics Research Association.
BLF	British Lubricants Federation.
Block grease	A very firm grease produced as a block that is applied to large open plain bearings, which operate at low speed and high temperatures.
BNA	Bureau des Normes de l'Automobile (France).
BNP	Bureau de Normalisation du Pétrole.
BOCLE	Ball-on-cylinder lubricity evaluator.
BOFT	Bearing oil film thickness.
BOI	Base oil interchange.
BOIG	Base Oil Interchange Guidelines.
BOTD	Ball on three disks.
Boundary	A regime of lubrication where there is partial contact between the metal components and partial separation of the surfaces by the lubricant fluid film.
BPD	Biocidal Products Directive.
BPT	Borderline pumping temperature (as defined in ASTM D3829).
Br	Bromine number (ASTM D1158).
Brookfield viscosity	Viscosity in centipoise, as determined on the Brookfield viscometer, is the torque resistance on a spindle rotating in the fluid being tested. Although Brookfield viscosities are most frequently associated with low-temperature properties of gear oils and transmission fluids, they are in fact determined for many other types of lubricants.
BRT	Ball rust test—the new bench test to replace Sequence IID engine test to measure rust and corrosion at low temperatures.
BSFC	Brake-specific fuel consumption.
BSI	British Standards Institution.
BTC	British Technical Council of the Motor and Petroleum Industries (CEC).
BTU	British thermal unit.
C-3	Specification by Allison Division of General Motors covering transmission application.
CA (API)	Service typical of diesel engines operated in mild to moderate duty with high-quality fuels and occasionally has included gasoline engines in mild service. Oils designed for this service provide protection from bearing corrosion and from ring-belt deposits in some naturally aspirated diesel engines when using fuels of such quality that they impose no unusual requirements for wear and deposit protection. They were widely used in the late 1940s and 1950s but should not be used in any engine unless specifically recommended by the equipment manufacturer.
CAA	Clean Air Act.
CAAA	Clean Air Act Amendment.
CAFÉ	Corporate Average Fuel Economy.
CARB	California Air Resources Board.
Carbon residue	Percentage of coked material remaining after a sample of lubricating oil has been exposed to high temperatures under ASTM Method D189 (Conradson) or D524 (Ramsbottom).
Caterpillar IP	A single-cylinder engine test designed to measure piston deposit control of an engine oil.

(Continued)

CB (API)	Service typical of diesel engines operated in mild to moderate duty, but with lower-quality fuels that necessitate more protection for wear and deposits. Occasionally has included gasoline engines in mild service. Oils designed for this service provide necessary protection from bearing corrosion and from ring belt deposits in some naturally aspirated diesel engines with higher-sulfur fuels. Oils designed for this service were introduced in 1949.
CBO	Conventional base oil.
CBOT	Chicago Board of Trade.
CC (API)	Service typical of certain naturally aspirated, turbocharged, or supercharged diesel engines used when highly effective control of wear and deposits is vital or using fuels of a wide quality range including high-sulfur fuels. Oils designed for this service were introduced in 1955 and provide protection from bearing corrosion and from high-temperature deposits in these diesel engines.
	Oil meeting the performance requirements measured in the following diesel and gasoline engine tests: The 1-G2 diesel engine test has been correlated with indirect injection engines used in heavy-duty operation, particularly with regard to piston and ring groove deposits. The L-38 gasoline engine test requirement provides a measurement of copper–lead bearing weight loss and piston varnish under high-temperature operating conditions.
CCD	Combustion chamber deposits.
CCR	Conradson carbon residue,
CCS	Cold-cranking simulator.
CD-II (API)	Service typical of two-stroke cycle engines requiring highly efficient control over wear and deposits. Oils designed for this service also meet all performance requirements of API service category CD.
	Oil meeting the performance requirements measured in the following diesel and gasoline engine tests: The I-G2 diesel engine test has been correlated with indirect injection engines used in heavy-duty operation, particularly with regard to piston and ring groove deposits. The 6V-53T diesel engine test has been correlated with vehicles equipped with two-stroke cycle diesel engines in high-speed operation before 1985, particularly with regard to ring and liner distress. The L-38 gasoline engine test requirement provides a measurement of copper–lead bearing weight loss and piston varnish under high-temperature operating conditions.
CDP	Cresyl diphenyl phosphate.
CE (API)	Service typical of many turbocharged or supercharged high-performance diesel engines, operated under both low-speed/high-load and high-speed/high-load conditions. Oils designed for this service have been available since 1984 and provide improved control of oil consumption, oil thickening, and piston assembly deposits and wear relative to the performance potential offered by oils designed for category CD service.
	Oil meeting the performance requirements of the following diesel and gasoline engine tests: The 1-G2 diesel engine test has been correlated with indirect injection engines used in heavy-duty service, particularly with regard to piston and ring groove deposits. The T-6, T-7, and NTC-400 are direct injection diesel engine tests. The T-6 has been correlated with vehicles equipped with engines used in high-speed operation before 1980, particularly with regard to deposits, oil consumption, and wear. The T-7 test has been correlated with vehicles equipped with engines used in lugging operation before 1984, particularly with regard to oil thickening. The NTC-400 diesel engine test has been correlated with vehicles equipped with engines in highway operation before 1983, particularly with regard to oil consumption, deposits, and wear. The L-38 gasoline engine test requirement provides a measurement of copper–lead bearing weight loss under high-temperature operating conditions.
CEC	California Energy Commission.
	Conseil Européen de Coordination pour les Développement des Essais de Performance des Lubrifiants et des Combustibles pour Moteurs (Coordinating European Council of Motor and Petroleum Industries: test standardization like ASTM).
CEFIC	Conseil Européen des Federations de l'Industrie Chimique (European Chemical Industry Council).
CEN	European Standardization Council.
CEPA	Canadian Environmental Protection Act.
CERCLA	Comprehensive Environmental Response, Compensation, and Liability Act.
CF (API)	API service category CF denotes service typical of indirect injected diesel engines and other diesel engines that use a broad range of fuel types including those using fuel with higher sulfur content (e.g., 0.5 wt%).
	Effective control of piston deposits, wear, and copper-containing bearing corrosion is essential for these engines, which may be naturally aspirated, turbocharged, or supercharged. Oils designated for this service have been in existence since 1994. Oils designated for this service may also be used when API service category CD is recommended.
CF-2 (API)	API service category CF-2 denotes service typical of two-stroke cycle engines requiring highly effective control over cylinder and ring-face scuffing and deposits. Oils designated for this service have been in existence since 1994 and may also be used when API service category CD-II is recommended. These oils do not necessarily meet the requirements of CF or CF-4 unless passing test requirements for these categories.
CF-4 (API)	This category was adopted in 1990 and describes oils for use in high-speed, four-stroke diesel engines. API CF-4 oils exceed the requirements of the CE category, providing improved control of oil consumption and piston deposits.

(Continued)

	Oil meeting the performance requirements in the following diesel and gasoline engine tests: the 1K diesel engine test, which has been correlated with direct injection engines used in heavy-duty service before 1990, particularly with regard to piston and ring groove deposits in the T-6, T-7, NTC-400, and L-38 engines; see "CE (API)" for explanation.
CFC	Chlorofluorocarbon.
CFPP	Cold filter plugging point.
CFR	Coordinating Fuel and Equipment Research Committee.
CFV	Clean fuel vehicle.
CG	Conventional gasoline.
CG-4 (API)	API service category CD-4 describes oils for use in high-speed four-stroke cycle diesel engines used in both heavy-duty on-highway (<0.05 wt% sulfur fuel) and off-highway (<0.5 wt% sulfur fuel) applications. CG-4 oils provide effective control over high-temperature piston deposits, wear corrosion, foaming, oxidation stability, and soot accumulation. These oils are especially effective in engines designed to meet 1994 exhaust emission standards and may also be used in engines requiring API service categories CD, CE, and CF-4. Oils designated for this service have been in existence since 1994.
CGSB	Canadian General Standards Board—a consensus organization of producers, users, and general interest groups, which develops standards for test methods and products for Canada.
CH-4 (PC-7)	New (proposed) classification for the generation of heavy-duty engine oils.
CI	Cetane index.
CIA	Chemical Industries Association (part of the CEFIC).
CIDI	Compression ignition direct injection (diesel).
CIMAC	International Council on Combustion Engines.
CLCA	Comité de Liaison de la Construction de l'Automobile.
CLEPA	Comité de Liaison de la Construction d'Equipements et de Pièces d'Automobiles.
CLR	Cooperative Lubricants Research.
CMA	Chemical Manufacturers Association—a standardizing body composed of additive manufacturers (United States).
CMAQ	Congestion Mitigation and Air Quality Improvement Program.
CMMO	Chemically modified mineral oil.
CMVO	Chemically modified vegetable oil.
CN	Cetane number.
CNG	Compressed natural gas.
CNHTC	China National Heavy Duty Truck Group Corporation.
CNPC	China National Petroleum Corporation.
CO	Carbon monoxide.
CO_2	Carbon dioxide.
CONCAWE	Conservation of Clean Air and Water in Europe.
Corrosion inhibitor	A lubricant additive used to protect surfaces against chemical attack from contaminants in the lubricating fluid or grease. These additives generally operate by reacting chemically and forming a film on the metal surfaces.
cP	Centipoise = mPa · s (SI unit).
CPC	Chinese Petroleum Corporation.

CPPI	Canadian Petroleum Products Institute.
CRC	Coordinating Research Council—an American standardizing body for performance testing.
CSA	Canadian Standards Association.
cSt	Centistokes.
CSTCC	Continuously slipping torque converter clutch.
Cummins M11	A heavy-duty engine test to measure crosshead wear.
CUNA	Commissione Tecnica di Unificazione Nell'Autoveicolo (CEC).
CVMA	Canadian Vehicle Manufacturers Association.
CVS	Constant volume sampling.
CVT	Continuously variable transmission.
DAP	Detroit Advisory Panel (API).
DASMIN	Deutsche Akkreditierungastelle Mineralol (German).
DBC	Dibutyl carbonate.
DBPP	Dibutyl phenyl phosphate.
DCT	Dual-clutch transmission.
DDC	Detroit Diesel Corporation.
DEC	Diethyl carbonate.
DEER	Diesel engine emissions reduction.
Demulsibility	A measure of the ability of an oil to separate from water to that measured by the test time required for a specified oil–water emulsion to break using ASTM D1401.
DEO	Diesel engine oil.
DEOAP	Diesel Engine Oil Advisory Panel (API/EMA).
DETA	Diethylene triamine.
Detergent	Oil additive that prevents deposits from forming on engine surfaces and may remove previously formed deposits.
DEXRON®-II	General Motors trademark specification for ATF.
DEXRON®-III	General Motors trademark specification for ATF, issued in 1993.
DEXRON®-IIIG	General Motors trademark specification for ATF, issued in 1998.
DFA	Diesel fuel additive.
DGMK	Deutsche Gesellschaft fur Mineralölwissenschaft und Kohlechemie.
DH-I	AJASO diesel engine oil category—a category mainly for Japanese-made heavy-duty diesel engine providing wear and soot-handling properties, and thermal-oxidative stability.
DHYCA	Direction des Hydrocarbarres et Carburants (French Ministry of Industry).
DI	Direct injection (normally diesel).
DI	Driveability index.
DII	Diesel ignition improver.
DIN	Deutsches Institut für Normung (German Standards Institute).
Dispersant	Oil additive that keeps engines clean by holding in suspension the insoluble products for oil oxidation and fuel combustion formed during engine operation.
DIOC	Diisooctyl carbonate.
DiPE	Dipentaerythritol.
DKA	Deutscher Koordinierungsausschuss—the German National Body in the Coordinating European Council.
DMC	Dimethyl carbonate.
DME	Dimethyl ether.

(Continued)

DNA	Deutsche Normenausschuss.
DOA	Dioctyl adipate.
DOC	Diesel oxidation catalyst.
DOCP	Dispersant olefin copolymer viscosity modifier or viscosity index improver.
DOD	Department of Defense.
DOE	Department of Energy.
DOHC	Dual overhead cam.
DOP	Dioctyl phthalate.
DOS	Di-2-ethylhexylsebacate.
DOT	Department of Transportation (United States).
DPF	Diesel particulate filter.
DPMA	Dispersant polymethacrylate viscosity modifier.
Dropping point	The temperature at which a grease changes from a semisolid to a fluid under the test conditions. This temperature can be considered a measure of the upper use limit for the grease.
DSC	Differential scanning calorimetry—used to measure onset oxidation temperatures of oils.
DTBP	Di-*tert*-butyl phenol.
DVM	Dispersant viscosity modifier.
EC	European Community.
	European Commission.
	Environmental Council (Japan).
	Energy conserving.
	Environment Canada.
ECCC	Electronically controlled computer clutch.
	EC-II (API).
	Energy Conserving-II—designation that an engine oil provides 2.7% fuel economy improvement versus a reference oil in the Sequence VI test.
ECE	Economic Commission for Europe.
ECHA	European Chemicals Agency—control point in the REACh system established in Helsinki.
ECTC	Engine Coolants Technical Committee (CEC).
EDC	Electronic diesel control.
EEB	European Environmental Bureau.
EEC	European Economic Community.
	Electronic Emission Controls.
EELQMS	European Engine Lubricants Quality Management System.
EFI	Electronic fuel injection.
EFTC	Engine Fuels Technical Committee (CEC).
EGR	Exhaust gas recirculation.
EHD	Elastohydrodynamic (lubrication).
EHDPP	2-Ethylhexyl diphenyl phosphate.
EHEDG	European Hygienic Engineering Design Group.
EIA	Energy Information Administration (U.S. DOE).
EINECS	European Inventory of Existing Commercial Chemical Substances.
ELGI	European Lubricating Grease Institute.
ELINCS	European List of Notified Chemical Substances.
ELTC	Engine Lubricants Technical Committee (CEC).
ELV	End-of-life vehicle.
EMA	Engine Manufacturers Association (United States); heavy-duty diesel.
EMPA	Swiss Federal Laboratories for Materials Science and Technology.

Emulsion	A mechanical mixture of two mutually insoluble fluids. Some metalworking fluids are designed to remain as a stable emulsion by incorporation of an emulsifier.
ENGVA	European Natural Gas Vehicle Association.
EOFT	Engine Oil Filterability Test (GM).
EO-J/EO-K/ EO-K/2	Mack Truck Company heavy-duty diesel specifications.
EO-L	Mack Truck Company heavy-duty diesel specification, issued in 1993.
EOLCS	Engine Oil Licensing and Certification System (API-1520).
EO-M	Mack Truck Company heavy-duty diesel specifications, issued in 1998.
EP additive	Extreme-pressure additive—an additive designed to prevent metal–metal adhesion or welding when the degree of surface control is sufficiently high that the normal protective (oxide) films are removed and other surface-active species in the oil are not reactive enough to deposit a protective film. EP additives function by reacting with the metal surface to form a metal compound, for example, iron sulfide.
EPA	Environmental Protection Agency (United States).
EPACT	The Energy Policy Act of 1992 (EPAct).
EPDM	Ethylenepropylene diene–based elastomeric seal material.
EPEFE	European Program on Emission, Fuels, and Engine Technologies (an advisory group consisting of 14 motor companies [ACEA] and 32 oil companies [EUROPIA]).
EPM	Ethylene–propylene-based elastomeric seal material.
ERC	European Registration Centre.
ESCS	Engine Service Classification System.
ESI	Extended service interval.
ESIS	European Chemical Substances Information System.
ETC	European Transient Cycle.
ETLP	Engine Tests of Lubricants Panel (IP).
EU	European Union.
EUC	Elementary Urban Cycle.
EUDC	Extra-Urban Driving Cycle.
EULIM	European Union of Independent Lubricant Manufacturers.
EUROPIA	European Petroleum Industry Association.
EV	Electric vehicle.
EVA	Ethylene vinyl alcohol.
FATG	Fuel Additives Task Group (CMA).
FBP	Final boiling point.
FCAAA	Federal Clean Air Act Amendments.
FCC	Fluid catalytic cracking.
FCEV	Fuel cell electric vehicle.
FDA	Food and Drug Administration.
FE	Fuel economy.
FEI	Fuel economy improvement (fuel efficiency increase).
FERC	Federal Energy Regulatory Commission (United States).
FF	Factory fill.
FFV	Flexible-fuel vehicle.
FIE	Fuel injection equipment.
FIMS	Field ionization mass spectrometry.

(Continued)

Fire point	ASTM D92—a laboratory test to measure the lowest temperature at which a sample will sustain burning for 5 s.
FISITA	Federation Internationale des Sociétes d'Ingénieurs des Techniques de l'Automobile.
Flash point	ASTM D92—a laboratory test to measure the tendency of a sample to form a flammable mixture with air, the lowest temperature at which a test flame causes the vapor of a fluid to ignite.
Four-ball test	Two test procedures based on the same principle: 1. Four-Ball EP Test (ASTM D2596) 2. Four-Ball Wear Test (ASTM D2266) The three lower balls are clamped together to form a cradle upon which the fourth ball rotates in a vertical axis. The balls are immersed in the lubricant under investigation. The test is used to determine the relative wear-preventing properties of lubricants operating under boundary lubrication conditions. The test is carried out at a specified speed, temperature, and load. At the end of the specified period, the average diameter of wear scar on the three balls is reported. The Four-Ball EP Test is designed to evaluate performance under much higher loads. In this test, the top ball is rotated at a specified speed (1700 ± 60 rpm), but temperature is not controlled. The loading is increased at specified intervals until the rotating ball seizes and welds to the other balls. At the end of each interval, the average scar diameter is recorded. Two values are generally reported—load wear index and weld point.
FSIS	Food Safety Inspection Service.
FT	Fischer–Tropsch.
FTC	Federal Trade Commission (United States).
FTIR	Fourier transform infrared spectroscopy.
FTM	Federal Test Method (United States).
FTP	Federal Test Procedure (EPA).
FZG	Forschungsstelle für Zahnrader und Getriebebau (Research Institute for Gears and Gearboxes).
GAO	Government Accountability Office (United States).
GATC	Gross additive treat cost.
GC	Gas chromatography. In the ASTM D4950 Standard Classification and Specification for Automotive Service Greases, the letters GC designate service typical of the lubrication of wheel bearings operating in passenger cars, trucks, and other vehicles under mild to severe duty.
GDI	Gasoline direct injection.
GDTC	Gross delivered treating cost.
GEO	Gas engine oil—lubricant used for natural gas engines.
GEPE	Group des Experts pour la Pollution et l'Energie (Group of Experts for Pollution and Energy).
GF-2	ILSAC PCMO oil classification standard, effective from August 1997.
GF-3	ILSAC PCMO oil classification after GF-2, proposed in 2000.
GFC	Groupement Français de Coordination (CEC).
GHS	Globally Harmonized System.
GI	Gelation index (as defined in ASTM D5133).
GL-4/5	Gear service characteristics (API).
GM	General Motors.
GMO	Glycerol monooleate.
GO-H	Gear lubricant specified by Mack Truck Company.
GRAS	Generally recognized as safe.
GRPE	Groupe des Rapporteurs pour la Pollution et l'Énergie.
GSA	General Services Administration (United States).
GTL	Gas-to-liquid.
GWP	Global warming potential.
H-1	The USDA classification that applies to lubricants with incidental food contact and to ingredients used to make these lubricants. Under NSF, H-1 applies to lubricants, and HX-1 applies to ingredients for H-1 lubricants.
H-2	The USDA classification that applies to lubricants in food-processing plants with no food contact and to ingredients used to make these lubricants. Under NSF, H-2 applies to lubricants, and HX-2 applies to ingredients for H-2 lubricants.
HACCP	Hazard Analysis and Critical Control Point implement procedures for USDA regulatory requirements.
HAP	Hazardous air pollutant.
HC	Hydrocarbon.
HCB	Hydrocracker bottoms.
HCCI	Homogeneous charge combustion ignition.
HCFC	Hydrochlorofluorocarbon.
HD	Heavy duty.
HDD	Heavy-duty diesel.
HDDEO	Heavy-duty diesel engine oil.
HDDO	Heavy-duty diesel oil.
HDEO	Heavy-duty engine oil.
HDEOCP	Heavy-Duty Engine Oil Classification Panel.
HDMO	Heavy-duty motor oil.
HEFCAD	High-energy, friction characteristics and durability; part of DEXRON-II qualification program.
HEUI	Hydraulically operated electronically controlled unit injectors.
HFC	Hydrofluorocarbon.
HFE	Hydrofluoroether.
HF-O	Denison specification for heavy-duty hydraulic antiwear fluids.
HFRR	High-frequency reciprocating rig.
HOOT	Hot oil oxidation test.
HOPOE	Highly optimized polyol ester.
HPV	High production volume.
HRMS	High-resolution mass spectrometry.
HSPOE	High-stability polyol ester.
HTHS	High-temperature, high-shear viscosity measured at 150°C and 10^6 s^{-1} (ASTM D4683, CEC L-36-A-90), (ASTM D4741), or (ASTM D5481).
HTHSRV	High-temperature, high-shear rate viscosity.
HVI	High-viscosity index.
Hydrodynamic lubrication	A lubrication regime characterized by a full-fluid film between two moving surfaces. As oil is moved between the moving parts, the action causes a high pressure in the lubricant fluid, and this separates the moving parts.
HX-1	NSF classification that applies to ingredients for H-1 lubricants.

(Continued)

HX-2	NSF classification that applies to ingredients for H-2 lubricants.
IBP	Initial boiling point.
IC	Internal combustion.
IchemE	Institution of Chemical Engineers (United Kingdom).
ICOA	International Castor Oil Association.
ICOMIA	International Council of Marine Industry Associations.
IDDPP	Isodecyl diphenyl phosphate.
IDI	Indirect injection (diesel).
IEA	International Energy Agency.
IENICA	Interactive European Network for Industrial Crops and their Applications.
IFP	Institut Français du Petrole.
IGL	Investigation Group Lubricants (CEC).
ILMA	Independent Lubricant Manufacturers Association—an association of oil companies, also called as compound blenders.
ILSAC	International Lubricants Standardization and Approval Committee.
I/M	Vehicle Inspection and Maintenance Program.
IMECHE	Institution of Mechanical Engineers.
IOP	Institute of Physics—Tribology Group.
IP	Institute of Petroleum (United Kingdom).
IPPP	Isopropylphenyl phenyl phosphate.
IR	Infrared (spectroscopy).
ISO	International Standards Organization.
ISOT	Indiana Stirred Oxidation Test (adopted as JIS K 2514).
ISTEA	Intermodel Surface Transportation Efficiency Act.
IVD	Intake valve deposit.
IVT	Infinitely variable transmission.
JALOS	Japan Lubricating Oil Society.
JAMA	Japan Automobile Manufacturers Association, Inc.
JARI	Japan Automobile Research Institute.
JASIC	Japan Automobile Standards Internationalization Center.
JASO	Japan Automotive Standards Organization.
JAST	Japanese Society of Tribologists.
JATA	Japan Automobile Transport Technology Association.
JCAP	Japan Clean Air Program.
JD	John Deere—a farm implement manufacturer.
JIC	J. I. Case—a farm implement manufacturer.
JIS	Japanese Industrial Standards.
JISC	Japanese Industrial Standards Committee.
JPI	Japan Petroleum Institute.
JSAE	Japan Society of Automotive Engineers.
KTH	Royal Institute of Technology, Sweden.
KV	Kinematic viscosity.
LB	In the ASTM D4950 Standard Classification and Specification for Automotive Service Greases, the letters LB designate service typical of lubrication of chassis components and universal joints in passenger cars, trucks, and other vehicles under mild to severe duty.
LCO	Light cycle oil.
LCST	Lower critical solution temperature.
LDV	Light-duty vehicle.
LeRC	Lewis Research Center, NASA.
LEV	Low-emission vehicle.

LMOA	Locomotive Maintenance Officers Association.
LNG	Liquefied natural gas.
LOFI	Lubricant oil flow improver.
LPG	Liquefied petroleum gas.
LPV	Low production volume.
LRI	Lubricant Review Institute—a body associated with SAE that qualifies heavy-duty engine oil and gear lubricants for the industry and the U.S. military.
LSC	Lubricant Standards Committee.
LSD	Low-sulfur diesel.
M11	Cummins heavy-duty engine test (CH-4) to measure crosshead wear.
M111	A Mercedes-Benz gasoline engine oil fuel economy test (CEC-L-54-T-96).
Mack T-8	Engine test used by Mack Truck Company in specifying heavy-duty diesel lubricants, extended to 300 h.
Mack T-9	Engine test used by Mack Truck Company in specifying heavy-duty diesel lubricants; more severe (higher-soot loading) test required for CH-4 specifications.
Mannich	A name reaction that produces unique fuel and lubricant detergent/dispersants.
MB	Mercedes-Benz.
MB OM364A	A Mercedes-Benz heavy-duty diesel engine oil test to measure bore polish, piston cleanliness, and turbo deposits (CEC-L-52-T-97).
MB OM602A	A Mercedes-Benz diesel engine oil test to measure cam wear (CEC-L-51-T-95).
MERCON® V	Ford Motor Company trademark ATF specification.
MIL	Military standards specification.
MIL-L-210F	Latest in a series of heavy-duty diesel lubricants designated by the U.S. military for over-the-road and off-highway applications.
MIL-L-2105C	U.S. military specifications for mobile equipment gear lubricant, now superseded.
MIL-L-2105D	U.S. military specification for mobile equipment gear lubricants; active through 1995.
MIL-PRF-2105E	U.S. military gear specification, issued on August 22, 1995.
MIRA	Motor Industry Research Association (United Kingdom).
MITI	Ministry of International Trade and Industry (Japan).
MOD	Ministry of Defence.
MOE	Ministry of Energy (United Kingdom).
MOFT	Minimum oil film thickness.
MOL	Ministry of Health, Labour, and Welfare (Japan).
MON	Motor octane number.
MOT	Ministry of Transport (United Kingdom).
MOU	Memorandum of understanding.
MRV	Minirotary viscometer—a measure of oil pumpability at various temperatures (ASTM D3829 and ASTM D4683).
MSDS	Material Safety Data Sheets.
MT-1	Manual transmissions specifications, issued in 1995 (API).
MTAC	Multiple test acceptance criteria.
MTBE	Methyl *tert*-butyl ether.
MTF	Manual transmission fluid.
MVEG	Motor Vehicle Emissions Group (Europe).

(Continued)

MVMA	Motor Vehicle Manufacturers Association (U.S. passenger cars).
MWF	Metalworking fluid(s).
NA	Normally (or naturally) aspirated (diesel engine).
NAAQS	National Ambient Air Quality Standards (United States).
NACE	National Association of Corrosion Engineers (United States).
NAEGA	North American Export Grain Association.
NAFTA	North American Free Trade Agreement.
NATC	Net additive treat cost.
NCM	National Comite Motorproeven (Netherlands) (CEC).
NCPA	National Cottonseed Products Association.
NCWM	National Conference on Weights and Measures.
NDOCP	Nondispersant olefin copolymer.
NDTC	Net delivered treat cost.
NDVM	Nondispersant viscosity modifier.
NEB	National Energy Board (NA).
	National Environment Board (Thailand).
NEDO	New Energy and Industrial Technology Development Organization (Japan).
NEFI	New England Fuel Institute.
NEL	National Engineering Laboratory (United Kingdom).
NESCAUM	Northeast States for Coordinated Air Use Management.
NESHAP	National Emission Standards for Hazardous Air Pollutants.
NFPA	National Fluid Power Association.
NGEO	Natural gas engine oil.
NGFA	National Grain and Feed Association.
NI	Nonpolarity index.
NIOP	National Institute of Oilseed Products.
NIST	National Institute of Standards and Technology (United States).
NLEV	National low-emission vehicle.
NLGI	National Lubricating Grease Institute (United States).
NLP	No-longer polymers.
NMHC	Nonmethane hydrocarbon.
NMMA	National Marine Manufacturers Association (United States) (formerly Boating Industry Association [BIA]).
NMOG	Nonmethane organic gases (includes alcohols).
NMR	Nuclear magnetic resonance.
NOACK	NOACK volatility; DIN 51851 (ASTM D5800).
NOPA	National Oilseed Processors Association.
NORA	National Oil Recyclers Association (United States).
NOX	Nitrogen oxide.
NPA	National Petroleum Authority (United States).
NPG	Neopentylglycol.
NPI	Nonpolarity index.
NPRA	National Petrochemical and Refiners Association (United States).
NRC	Natural Resources Canada.
NRCC	National Research Council of Canada.
NREL	National Renewable Energy Laboratory.
NRL	Naval Research Laboratory.
NSF	National Sanitation Foundation—a nongovernmental, nonprofit corporation assumed the role formerly held by the USDA. NSF creates registry numbers for approved lubricant products.
NTC-400	Cummins diesel engine test.
NUTEK	Swedish National Board for Industrial and Technical Development.
OCP	Olefin copolymer viscosity modifier or viscosity index improver.
ODI	Oil drain interval.
ODP	Ozone depleting potential.
ODS	Ozone-depleting substance.
OEM	Original equipment manufacturer (e.g., GM and Ford).
OICA	Organization Internationale des Constructeurs d'Automobiles (was BPICA).
OMB	Office of Management and Budget (United States).
OMS	Office of Mobile Sources (U.S. EPA).
ON	Octane number.
OPEC	Organization of the Petroleum Exporting Countries.
OPEST	Oil Protection and Emission System Test.
ORD	Octane requirement decrease.
	Office of Research and Development (U.S. EPA).
ORI	Octane requirement increase.
ORNL	Oak Ridge National Laboratory.
OSHA	Occupational Safety and Health Administration (United States).
OTA	Office of Technology Assessment (United States).
OTAG	Ozone Transport Assessment Group.
OTC	Ozone Transport Commission.
Oxidation	One of several modes of oil degradation. The process generally involves the addition of oxygen to the lubricant structure, followed by cleavage or polymerization, resulting in unfavorable oil properties and performance.
Package	Formulation of various chemical compounds.
PADD	Petroleum Administration for Defense District.
PAG	Polyalkene glycols, polyalkylene glycols.
PAH	Polycyclic aromatic hydrocarbons.
PAHO	Pan American Health Organization.
PAJ	Petroleum Association of Japan.
PAO	Poly-α-olefin (PAO); base stocks of various viscosity classifications.
PAPTG	Product Approval Protocol Task Group (CMA0).
PBT	Persistent, bioaccumulative, and toxic.
PC	Proposed classification.
PCD	Passenger car diesel.
PCEO	Passenger car engine oil.
PCMO	Passenger car motor oil.
PCTFE	Polychlorotrifluoroethylene.
PDSC	Pressure differential scanning calorimetry.
PDVSA	Petroleos de Venezuela.
PE	Pentaerythritol.
PEA	Polyether amine.
PEC	Petroleum Energy Center.
PFPAE	Perfluoropolyalkylether.
PFPE	Perfluoropolyether.
Phenolic	Antioxidant based on 2,6-di-*tert*-butyl phenol chemistry or other alkylated phenolics.
PIB	Polyisobutene.
PIBSA	Polyisobutenyl succinic anhydride; precursor for ashless dispersants.
PIO	Polyinternalolefins.
PM	Particulate matter.
PM$_{2.5}$	Particulate matter <2.5 µm diameter.

(Continued)

PM_{10}	Particulate matter <10 μm diameter.
PMA	Polymethacrylate.
PMAA	Petroleum Marketers Association of America.
PMC	Pensky-Martens Closed Cup—flash point test.
PNA	Polynuclear aromatic.
PNGV	Partnership for a New Generation of Vehicles (United States).
POFA	Polymerized fatty acids.
Poise	The CGS unit of absolute viscosity (dyne s/cm^2) as measured by the shear stress required to move one layer of fluid along another over a total thickness of 1 cm at a shear rate of 1 cm/s. Absolute viscosity values are independent of density and are directly related to the resistance to flow.
Pour point	This is a widely used low-temperature flow indicator and is 3°C above the temperature at which normally a liquid petroleum product maintains fluidity. It is a significant factor in cold-weather start-up, but must be considered along with pumpability, the ease with which an oil forms a honeycomb or crystals at low temperatures near the pour point. A conventional measure of the lower-temperature limit for low-temperature flow of a lubricating fluid.
PPD	Pour point depressant—an additive used to lower the pour point of an oil by modifying the structure of wax crystals.
PPE	Polyphenyl ether.
ppm	Parts per million.
PSA TU3M	A Peugeot gasoline engine test to measure cam wear and scuffing (CEC-L-38-A-94). A Peugeot gasoline engine test to measure high-temperature ring sticking and piston varnish (CEC-L-55-T-95).
PSA XUD 11	A Peugeot diesel engine test to measure medium-temperature dispersivity (soot-induced oil thickening) (CEC L-56-T-95).
PTFE	Polytetrafluoroethylene.
PTIT	Petroleum Institute of Thailand.
PTT	Petroleum Authority of Thailand.
PVA	Polymethacrylate viscosity index improver or viscosity modifier.
PVC	Pressure viscosity coefficient.
PVE	Polyvinyl ether.
QPL	Qualified Products List (U.S. military).
RBOT	Rotating bomb oxidation test.
RCRA	Resource Conservation and Recovery Act.
RFG	Reformulated gasoline.
RI	Radial isoprene (star polymer) viscosity modifier or viscosity index improver.
RME	Rapeseed methyl ester.
R&O	Rust and oxidation—inhibited lubricant for use in circulating systems, compressors, hydraulic systems, and gear cases.
ROCOT	Rotating compressor oxidation test.
RON	Research octane number.
ROSE	Rose Foundation—Recycling Oil Saves the Environment.
RSI	Registration Systems, Inc. (CMA monitoring agency).
Rust inhibitor	A lubricant additive for protecting ferrous (iron and steel) components from rusting caused by water contamination or other harmful materials formed by oil degradation. Some rust inhibitors operate similarly to corrosion inhibitors by reacting chemically to form an inert film on metal surfaces. Other rust inhibitors absorb water by incorporating it into a water-in-oil emulsion so that only the oil touches the metal surfaces.
RVP	Reid vapor pressure.
SA (API)	Service typical of older engines operated under such mild conditions that the protection afforded by compounded oils is not required. This category should not be used in any engine unless specifically recommended by the equipment manufacturer.
SAE	Society of Automotive Engineers.
SAIC	Shanghai Automotive Industry Corporation (Group).
SAIT	South African Institute of Tribology.
SARA	Superfund Amendments and Reauthorization Act.
SB	Styrene–butadiene viscosity modifier or viscosity index improver.
SB (API)	Service typical of older gasoline engines operated under such mild conditions that only minimum protection afforded by compounding is desired. Oils designed for this service have been used since 1930s and provide only antiscuff capability and resistance to oil oxidation and bearing corrosion. They should not be used in any engine unless specifically recommended by the equipment manufacturer.
SC (A[O])	Service typical of gasoline engines in 1964–1967 models of passenger cars and some trucks operating under engine manufacturers' warranties in effect during those model years. Oils designed for this service provide control of high- and low-temperature deposits, wear, rust, and corrosion in gasoline engines.
SCAQM	South Coast (California) Air Quality Management.
Scuffing	Wear caused by the localized welding and fracture of rubbing surfaces.
SD (API)	Service typical of gasoline engines in 1968–1970 models of passenger cars and some trucks operating under engine manufacturers' warranties in effect during those model years. Also may apply to certain 1971 and later models as specified (or recommended) in the owners' manuals. Oils designed for this service provide more protection against high- and low-temperature engine deposits, wear, rust, and corrosion in gasoline engine than oils that are satisfactory for API engine service category SC and may be used when API engine service category SC is recommended.
SE (API)	Service typical of gasoline engines in passenger cars and some trucks beginning with 1972 and certain 1971 models operating under engine manufacturers' warranties. Oils designed for this service provide more protection against oil oxidation, high-temperature engine deposits, rust, and corrosion in gasoline engines than oils that are satisfactory for API Engine service categories SD or SC and may be used when either of these classifications is recommended.

(Continued)

Sequence IIIE, F	Engine test to simulate expressway driving to measure oxidation and wear characteristics of formulated oils at high temperature.	SIA	Société des Ignenieurs de l'Automobile.
		SIAM	Society of Indian Automobile Manufacturers.
Sequence tests	A series of (ASTM) industry-standardized tests used to determine the quality of crankcase lubricants (i.e., Sequence IID, IIIE, VE, and VIA).	SIB	Sulfurized isobutylene—a basic EP additive for gear lubricant packages.
		SIEF	Substance Information Exchange Forum.
		SIGMA	Society of Independent Gasoline Marketers of America.
Sequence UL-38	Engine test similar to L-38 except fuel used is unleaded; test measures bearing wear and deposits.	SIP	Styrene–isoprene copolymer.
		SIP(s)	State Implementation Plan(s).
Sequence VE, F	Engine test to simulate stop-and-go driving and to measure sludge, varnish, cam wear, and oil screen plugging.	SIS	Sveriges Standardiseringskommission.
		SJ (API)	Service requirement adopted for gasoline engines in 1996 for use in current and earlier passenger car, sport utility vehicle, van, and light truck under vehicle manufacturers' recommended maintenance procedures.
Sequence VIA, B	Engine test designed to measure fuel efficiency properties of an engine oil (VIA for ILSAC GF-2 and VIB for ILSAC GF-3).		
SF (API)	Service typical of gasoline engines in passenger cars and some trucks beginning with the 1980 model year operating under manufacturers' recommended maintenance procedures. Oils developed for this service provide increased oxidation stability and improved antiwear performance relative to oils that meet the minimum requirements for API service category SE. The oils also provide protection against engine deposits, rust, and corrosion. Oils meeting API service classification SF may be used where API service categories SE, SD, or SC are recommended.		Oils developed for the category may be used where API SH and earlier categories are recommended.
			Engine oils that meet the category may use the API Base Oil Interchangeability Guidelines and the API Guidelines for SAE Viscosity-Grade Engine Testing.
		SMDS	Shell Middle Distillate Synthesis.
		SME	Society of Manufacturing Engineers.
		SMM&T	Society of Motor Manufacturers and Traders Ltd. (United Kingdom).
		SMR	Svenska Mekanisters Riksforenig (National Organization in Sweden [CEC]).
	Oils meeting the performance requirements measured in the following gasoline engine tests: the IID gasoline engine test has been correlated with vehicles used in short-trip service before 1978, particularly with regard to rusting. The IIIE gasoline engine test has been correlated with vehicles used in high-temperature service before 1988, particularly with regard to oil thickening and valve train wear. The VE gasoline engine test has been correlated with vehicles used in stop-and-go service before 1988, particularly with regard to sludge and valve train wear. The L-38 gasoline engine test requirement provides a measurement of copper–lead bearing weight loss and piston varnish under high-temperature operating conditions. The 1-H2 diesel engine test requirement provides a measurement of high-temperature deposits.	SMRP	Society for Maintenance and Reliability Professionals.
		SNAP	Significant New Alternatives Policy.
		SNCF	Societe Nationale des Chemins de fer Francais.
		SNPRM	Supplemental notice of proposed rule making.
		SNV	Schweizerische Normenvereinigung (National Organization, Switzerland) (CEC).
		SOF	Soluble oil fraction.
		SOT	Spin orbit tribometer.
		SSI	Shear stability index.
		SSU or SUS	Saybolt Seconds Universal or Saybolt Universal Second—a measure of viscosity, or a fluid's resistance to flow.
		STLE	Society of Tribologists and Lubrication Engineers.
		STOU	Super tractor oil universal—a term commonly applied to a lubricant that can be used for all applications of a tractor, including hydraulic, engine, wet brakes, and gear.
SH (API)	Service requirement for gasoline engines and dated in 1993 for use in current and earlier passenger car, van, and light-truck operation under vehicle manufacturers' recommended maintenance procedures.		
		SULEV	Super ultra-low-emission vehicle.
		SUV	Sport utility vehicle.
		SwRI	Southwest Research Institute.
	Oils developed for this category may be used where SG oils are recommended. SH oil provides improved control of engine deposits, oil oxidation, wear, corrosion, and rust relative to oils developed for previous categories.	T-8	Engine test used by Mack Truck Company in specifying heavy-duty diesel lubricants, extended to 300 h.
		T-9	Engine test used by Mack Truck Company in specifying heavy-duty diesel lubricants, more severe (higher-soot loading) test required for CH-4 specifications.
	Engine oils that meet SH designation may use the API Base Oil Interchangeability Guidelines and API Guidelines for SAE Viscosity-Grade Engine Testing.	TAD	Technische Vereinigung fur Mineralöladditive in Deutschland EV (subgroup of the ATC).
SHC	Synthetic hydrocarbon.	TAN	Total acid number—a measure of a lubricant's acidity.
SHPD	Super high-performance diesel.	TBEP	Tributoxyethyl phosphate.
SHPDO	Super high-performance diesel oil.	TBN	Total base number—a measure of the acid-neutralizing property of a lubricating oil.
SI	Styrene–isoprene viscosity modifier or viscosity index improver spark ignition.		
		TBP	Tributyl phosphate.
		TBPP	*tert*-Butylphenyl phenyl phosphate.
	Système International—units of measurement (m, kg, s, K).	TBS	Tapered bearing simulator (ASTM D4683).

(Continued)

TCP	Tricresyl phosphate.
t-CVT	Toroidal CVT.
TCW-3	Water-cooled two-cycle engine oil specification.
TEOST	Thermal engine oil stability test.
TFB	Swedish Transport Research Board.
TFMO	Thin-film micro-oxidation test.
TFOUT	Thin-film oxygen uptake test.
TGA	Thermogravimetric analysis.
THCT	Turbo hydramatic cycling test; part of DEXRON® qualification program.
ThOD	Theoretical oxygen demand.
THOT	Turbo hydramatic oxidation test; part of DEXRON qualification program.
TiBP	Triisobutyl phosphate.
TISI	Thailand Industrial Standards Institute.
TLEV	Transitional low-emission vehicle.
TLTC	Transmission Lubricants Technical Committee (of CEC).
Timken OK load	This is a measure of the extreme-pressure properties of a lubricant. Lubricated by the product under investigation, a standard steel roller rotates against a block. Timken OK load is the heaviest load that can be carried without scoring.
TMC	ASTM Test Monitoring Center.
TMP	Trimethylolpropane.
TO-2	Caterpillar specification for transmission oil, including durability and frictional property testing.
TO-3	Caterpillar specification, including TO-2 plus fluoroelastomer seal compatibility.
TO-4	Caterpillar specification, including TO-2 plus fluoroelastomer seal compatibility, nearest version of Caterpillar transmission and drive train fluid requirements.
TOCP	Tri-orthocresyl phosphate.
TOST	Turbine oil stability test.
TOTM	Trioctyl trimellitate.
TOU	Tractor oil universal—an oil that goes in all parts of the tractor, except for the engine.
TPP	Triphenyl phosphate.
Tribology	The science and technology of interacting surfaces in relative motion and the practice related thereto.
TSCA	Toxic Substances Control Act.
TÜV	Technischer Uberwachungs Verein.
TVMD	See "TAD."
TXP	Trixylenyl phosphate.
UCBO	Unconventional base oil.
UCST	Upper critical solution temperature.
UDC	Urban Driving Cycle (Europe).
UEIL	Union Européen des Independants en Lubrifiants (European Union of Independent Lubricant Manufacturers).
UFIP	Union Francaise des Industries Petrolieres.
UIC	Union des Industries Chimiques.
UKPIA	United Kingdom Petroleum Industry Association.
ULEV	Ultralow-emission vehicle.
ULSD	Ultralow-sulfur diesel.
USB	United Soybean Board.
USCAR	United States Council for Automotive Research.
USDA	United States Department of Agriculture.

USX	Formerly U.S. Steel; dominant maker of steel and specifier of industrial lubricants for the steel and related industries.
UTAC	L'Union Technique de l'Automobile Oil—used in all lubricating places, except the crankcase on a farm tractor.
UTTO	Universal Tractor Transmission Oil—used in all lubricating places, except the crankcase on a farm tractor.
VCI	Verband der Chemischen Industrie e. V.
VDA	Verband der Automobilindustrie.
VDI	Verein Deutscher Ingenieure.
VDS, VDS2	Volvo Long Drain Lubricant Specification.
VGO	Vacuum gas oil.
VGRA	Viscosity-grade read across.
VHVI	Very high viscosity index.
VI	Viscosity index—a measure of the rate of change in viscosity with temperature (higher VI means less change) (ASTM D2270).
VII	Viscosity index improver—additive that increases the viscosity index beyond that which can be obtained using ordinary refining methods.
Viscosity	Measure of resistance of flow of a liquid; dynamic viscosity is the ratio between the applied shear and the rate of shear, expressed in poise or centipoise; kinematic viscosity is the ratio of the viscosity (dynamic viscosity) to the density of the liquid, expressed in stokes or centistokes.
Viton®	A fluoroelastomeric compound from DuPont, presently used by Caterpillar and Mercedes-Benz for use in transmission application.
VM	Viscosity modifier.
VMI	Viscosity modifier interchange.
VOC	Volatile organic compound.
VOF	Volatile organic fraction.
Volatility	A measure of the amount of material evaporated from a sample under a particular set of conditions, usually expressed as a percentage of the original sample.
VTC	Viscosity-temperature coefficient.
VVT	Variable valve timing.
VW 1431 TCIC	A Volkswagen indirect injection diesel engine oil test to measure ring sticking and piston cleanliness (CEC-L-46-T-93).
VW DI	A Volkswagen direct injection diesel engine oil test to measure ring sticking and piston cleanliness (CEC-L-78-T-97).
WAFI	Wax antisettling flow improver.
WASA	Wax antisettling additive.
WCM	Wax crystal modifier.
WSPA	Western States Petroleum Association.
XDP	Xylenyl diphenyl phosphate.
ZDDP	Zinc dialkyldithiophosphate—a widely used antiwear and antioxidant agent for motor oils and industrial fluids; also referred to as ZDTP, ZDP, and "zinc."
ZDP/ZDTP	Zinc dithiophosphate.
ZEV	Zero-emission vehicle.
ZF	German transmission manufacturer, Zahnradfabrik Friedrichshafen A.G.

33.2   SOME FEDERAL SUPPLY CHAIN ABBREVIATIONS

Air Force/11	Air Force Aeronautical Systems Center (ASC), Wright-Patterson AFB, OH.
Air Force/68	Air Force San Antonio Air Logistics Center (SAALC), Kelly AFB, TX.
AOAP	Auto-Oil Advisory Panel.
Army/AR	Army Tank-Automotive and Armaments Command, Armament Research Development and Engineering Center (ARDEC), Picatinny, NJ.
Army/AT	Army Tank-Automotive and Armaments Command, Tank-Automotive Research Development and Engineering Center (TARDEC), Warren, MI.
Dexos 1	General Motor's trademarked global gasoline engine oil specification.
Dexos 2	General Motor's trademarked global diesel engine oil specification.
DfE	Design for the Environment.
DSCR/GS	Defense Logistics Agency's Defense Supply Center Richmond (DSCR), Richmond, VA.
FAT	No QPL exists, but a first article test (FAT) is required, or may be optional.
FAU	"First allowable use" of a designated category within the API donut logo.
GF-6	A new gasoline engine oil category that addresses fuel economy.
LSPI	Low-speed preignition.
HETG	Environmentally acceptable hydraulic fluid (triglycerides, e.g., vegetable oils).
HEPG	Environmentally acceptable hydraulic fluid (polyglycol).
HEES	Environmentally acceptable hydraulic fluid (synthetic esters).
HEPR	Environmentally acceptable hydraulic fluid (PAOs and related hydrocarbons).

Navy/AS	Naval Air Systems Command (NAVAIR), Patuxent River, MD.
Navy/AS^2	Naval Air Systems Command (NAVAIR), Lakehurst, NJ.
Navy/SH	Naval Sea Systems Command (NAVSEA), Arlington, VA.
Navy/YD	Naval Facilities Engineering Command (NAVFAC), Alexandria, VA.
OSPAR	Oslo and Paris Commissions for the protection of the marine environment of the Northeast Atlantic. OSPAR is the mechanism by which 15 governments and the EU cooperate to protect the Northeast Atlantic marine environment. The 15 governments include Belgium, Denmark, Finland, France, Germany, Iceland, Ireland, Luxembourg, the Netherlands, Norway, Portugal, Spain, Sweden, Switzerland, and the United Kingdom.
OECD	Organisation for Economic Co-operation and Development.
PLONOR list	OSPAR list of substances/preparations used and discharged offshore Which are considered to pose little or no risk to the environment.
QPL	Qualified Products List.

Global Registrations

Australia	AICS
Canada	DSL
China	IECSC
Europe	REACh
Japan	MITI, ENCS
Korea	KECS
New Zealand	New Zealand
Philippines	PICCS
Taiwan	ECN
USA	TSCA

34 Testing Methods for Additive/Lubricant Performance

Leslie R. Rudnick

CONTENTS

This chapter contains a selection of many of the most commonly used test methods and specifications selected from the United States, Europe, and Japanese lubricant testing methods. These include methods, standards, and specifications from the American Society for Testing and Materials (ASTM), Federal Test Method (FTM), Military (Spec) (MIL), Conseil Européen de Coordination pour les Développement des Essais de Performance des Lubrifiants et des Combustibles pour Moteurs (CEC), Deutsches Institut für Normung (DIN), Japan Petroleum Institute (JPI), and Federal Supply Class 9150 Product Commodities. Some cross-references are also given.

ASTM publishes an *Annual Book of ASTM Standards. Petroleum Products, Lubricants, and Fossil Fuels*, Volumes 5.01–5.04, in which a complete list of ASTM methods pertaining to lubricants may be found.

34.1 SUMMARY OF STANDARD TEST METHODS AND SPECIFICATIONS

34.1.1 AMERICAN SOCIETY FOR TESTING AND MATERIALS

D56	Standard test method for flash point by tag closed cup tester
D86	Standard test method for distillation of petroleum products
D88	Standard test method for viscosity in Saybolt universal seconds
D91	Standard test method for precipitation number of lubricating oils
D92	Standard test method for flash and fire points by Cleveland open cup tester
D93	Standard test method for flash point by Pensky–Martens closed cup tester
D94	Standard test method for saponification number of petroleum products
D95	Standard test method for water in petroleum products and bituminous materials by distillation
D97	Standard test method for pour point of petroleum products
D130	Standard test method for the detection of copper corrosion from petroleum products by the copper strip tarnish test
D150-98 (2004)	Standard test methods for AC loss characteristics and permittivity (dielectric constant) of solid electrical insulation
D156	Standard test method for Saybolt color of petroleum products (Saybolt chromometer method)
D189	Standard test method for Conradson carbon residue of petroleum products
D217	Standard test method for cone penetration of lubricating grease
D257-99	Standard test methods for DC resistance or conductance of insulating materials
D287	Standard test method for API gravity of crude petroleum and petroleum products (hydrometer method)
D323	Standard test method for Vapor Pressure of Petroleum Products (Reid Method)
D445	Standard test method for kinematic viscosity of transparent and opaque liquids (and the calculation of dynamic viscosity)
D482	Standard test method for ash from petroleum products
D524	Standard test method for Ramsbottom carbon residue of petroleum products
D525-05	Standard test method for oxidation stability of gasoline (induction period method)
D566	Standard test method for dropping point of lubricating grease
D567	Standard test method for calculating viscosity index

(Continued)

D611	Standard test method for (1993)el aniline point and mixed aniline point of petroleum products and hydrocarbon solvents
D664	Standard test method for acid number of petroleum products by potentiometric titration
D665	Standard test method for rust-preventing characteristics of inhibited mineral oil in the presence of water
D873-12	Standard test method for Oxidation Stability of Aviation Fuels (Potential Residue Method)
D874	Standard test method for sulfated ash from lubricating oils and additives
D877-02e1	Standard test method for dielectric breakdown voltage of insulating liquids using disk electrodes
D892	Standard test method for foaming characteristics of lubricating oils
D893	Standard test method for insolubles in used lubricating oils
D942	Standard test method for oxidation stability of lubricating grease by the oxygen bomb method
D943	Standard test method for oxidation characteristics of inhibited mineral oils
D972	Standard test method for evaporation loss of lubricating greases and oils
D974	Standard test method for acid and base number by color-indicator titration
D1091	Standard test methods for phosphorus in lubricating oils and additives
D1092	Standard test method for measuring apparent viscosity of lubricating greases
D1093	Standard test method for acidity of hydrocarbon liquids and their distillation residues
D1159	Standard test method for bromine numbers of petroleum distillates and commercial aliphatic olefins by electrometric titration
D1160	Standard test method for distillation of petroleum products at reduced pressure
D1177	Standard test method for Freezing Point of Aqueous Engine Coolants
D1209	Standard test method for color of clear liquids (platinum–cobalt scale) (APHA color)
D1217	Standard test method for density and relative density (specific gravity) of liquids by Bingham pycnometer
D1238	Standard test method for flow rates of thermoplastics by extrusion plastimeter (melt index) (or ISO 1133–1991)
D1264	Standard test method for determining the water washout characteristics of lubricating greases
D1268	Standard test method for leakage tendencies of automotive wheel bearing greases
D1296	Standard test method for odor of volatile solvents and diluents
D1298	Standard test method for (1990)el density, relative density (specific gravity), or API gravity of crude petroleum and liquid petroleum products by hydrometer method
D1319	Standard test method for Hydrocarbon Types in Liquid Petroleum Products by Fluorescent Indicator Adsorption
D1331	Standard test method for surface and interfacial tension of solutions of surface-active agents

D1358	Standard test method for (1995)el spectrophotometric diene value of dehydrated castor oil and its derivatives
D1401	Standard test method for water separability of petroleum oils and synthetic fluids
D1403	Standard test methods for cone penetration of lubricating grease using one-quarter and one-half scale cone equipment
D1478	Standard test methods for low-temperature torque of ball bearing grease
D1480	Standard test method for density and relative density (specific gravity) of viscous material by Bingham pycnometer
D1481	Standard test method for density and relative density (specific gravity) of viscous material by Lipkin bicapillary pycnometer
D1500	Standard test method for ASTM color of petroleum products (ASTM color scale)
D1552-15	Standard test method for Sulfur in Petroleum Products by High Temperature Combustion and IR Detection
D1646	Standard test methods for rubber viscosity, stress relaxation, and prevulcanization characteristics (Mooney viscometer)
D1662	Standard test method for active sulfur in cutting oils
D1742	Standard test method for oil separation from lubricating grease during storage
D1743	Standard test method for determining corrosion preventive properties of lubricating greases
D1744	Standard test method for determination of water in liquid petroleum products by Karl Fischer reagent
D1747	Standard test method for refractive index of viscous materials
D1748	Standard test method for rust protection by metal preservatives in the humidity cabinet
D1796	Standard test method for Water and Sediment in Fuel Oils by the Centrifuge Method (Laboratory Procedure)
D1831	Standard test method for roll stability of lubricating grease
D1947	Standard test method for load-carrying capacity of petroleum oil and synthetic fluid gear lubricants
D2007	Standard test method for characteristic groups in rubber extender and processing oils and other petroleum-derived oils by the clay–gel absorption chromatographic method
D2070	Standard test method for thermal stability of hydraulic oils
D2155	Discontinued from 1981, replaced by E 659
D2161	Standard practice for conversion of kinematic viscosity to Saybolt universal viscosity or to Saybolt Furol viscosity
D2265	Standard test method for dropping point of lubricating grease over wide temperature range
D2266	Standard test method for wear preventive characteristics of lubricating grease (four-ball method)
D2270	Standard test method for standard practice for calculating viscosity index from kinematic viscosity at 40 and 100°C
D2272	Standard test method for oxidation stability of steam turbine oils by rotating bomb
D2273	Standard test method for trace sediment in lubricating oils
D2274	Standard test method for Oxidation Stability of Distillate Fuel Oil (Accelerated Method)

(Continued)

D2440	Standard test method for Oxidation Stability of Mineral Insulating Oil
D2500	Standard test method for cloud point of petroleum oils
D2502	Standard test method for estimation of molecular weight (relative molecular mass) of petroleum oils from viscosity calculation
D2509	Standard test method for measurement of load-carrying capacity of lubricating grease (Timken method)
D2512-95 (2002)	Standard test method for compatibility of materials with liquid oxygen (impact sensitivity threshold and pass–fail techniques)
D2549-02	Standard test method for separation of representative aromatics and nonaromatics fractions of high-boiling oils by elution chromatography
D2595	Standard test method for evaporation loss of lubricating greases over wide temperature range
D2596	Standard test method for measurement of extreme-pressure properties of lubricating grease (four-ball method)
D2602	Replaced by D5293, standard test method for apparent viscosity of engine oils at low temperature using the cold-cranking simulator
D2603-01	Test method for sonic shear stability of polymer-containing oils
D2619-95 (2002)e1	Standard test method for hydrolytic stability of hydraulic fluids (beverage bottle method)
D2620	Discontinued from 1993, replaced by D5293
D2622	Standard test method for sulfur in petroleum products by x-ray spectrometry
D2625	Standard test method for determining endurance life and load-carrying capacity of dry solid film lubricants (Falex method)
D2670	Standard test method for measuring wear properties of fluid lubricants (Falex pin and Vee block method)
D2710	Standard test method for bromine index of petroleum hydrocarbons by electrometric titration
D2711	Standard test method for demulsibility characteristics of lubricating oils
D2714	Standard test method for calibration and operation of the Falex block-on-ring friction and wear testing machine
D2766	Standard test method for specific heat of liquids and solids
D2782	Standard test method for measurement of extreme-pressure properties of lubricating fluids (Timken method)
D2783	Standard test method for measurement of extreme-pressure properties of lubricating fluids (four-ball method) (load wear index)
D2786-91 (2001)e1	Standard test method for hydrocarbon types analysis of gas–oil saturates fractions by high ionizing voltage mass spectrometry
D2878	Standard test method for estimating apparent vapor pressures and molecular weights of lubricating oils
D2879	Standard test method for vapor pressure–temperature relationship and initial decomposition temperature of liquids by isoteniscope
D2887	Standard test method for boiling range distribution of petroleum fractions by gas chromatography
D2893	Standard test method for oxidation characteristics of extreme-pressure lubricating oils
D2896	Standard test method for base number of petroleum products by potentiometric perchloric acid titration
D2982	Standard test method for detecting glycol-base antifreeze in used lubricating oils
D2983	Standard test method for low-temperature viscosity of automotive fluid lubricants measured by Brookfield viscometer
D3120	Standard test method for trace quantities of sulfur in light liquid petroleum hydrocarbons by oxidative microcoulometry
D3228	Standard test method for total nitrogen in lubricating oils and fuel oils by modified Kjeldahl method
D3232	Standard test method for measurement of consistency of lubricating greases at high temperatures
D3233	Standard test method for measurement of extreme pressure properties of fluid lubricants (Falex Pin and Vee block methods)
D3238	Standard test method for calculation of carbon distribution and structural group analysis of petroleum oils by the ndM method
D3244	Standard test method for standard practice for utilization of test data to determine conformance with specifications
D3336	Standard test method for life of lubricating greases in ball bearings at elevated temperatures
D3427-03	Standard test method for air release properties of petroleum oils
D3525	Standard test method for gasoline diluent in used gasoline engine oils by gas chromatography
D3527	Standard test method for life performance of automotive wheel bearing grease
D3603	Standard test method for Rust-Preventing Characteristics of Steam Turbine Oils in the Presence of Water (Horizontal Disk Method)
D3704	Standard test method for wear preventive properties of lubricating greases using the (Falex) block on ring test machine in oscillating motion
D3705	Standard test method for misting properties of lubricating fluids
D3711	Standard test method for deposition tendencies of liquids in thin films
D3829	Standard test method for predicting the borderline pumping temperature of engine oil
D3850-94 (2000)	Standard test method for rapid thermal degradation of solid electrical insulating materials by thermogravimetric method (TGA)
D3945	Standard test method for shear stability of polymer-containing fluids using diesel injector nozzle (deactivated 1998, replaced by D6278)
D4047	Standard test method for phosphorus in lubricating oils and additives by quinoline phosphomolybdate method
D4048	Standard test method for detection of copper corrosion from lubricating greases
D4049	Standard test method for determining the resistance of lubricating grease to water spray
D4057	Standard test method for standard practice for manual sampling of petroleum and petroleum products

(Continued)

D4172	Standard test method for wear preventive characteristics of lubricating fluid (four-ball method)
D4294	Standard test method for sulfur in petroleum products by energy-dispersive x-ray fluorescence spectroscopy
D4310	Standard test method for determination of the sludging and corrosion tendencies of inhibited mineral oils
D4485	Standard test method for standard specification performance of automotive engine oils
D4624	Standard test method for measuring apparent viscosity by capillary viscometer at high temperature and high shear rates
D4627	Standard test method for iron chip corrosion for water-dilutable metalworking fluids
D4628	Standard test method for analysis of barium, calcium, magnesium, and zinc in unused lubricating oils by atomic absorption spectrometry
D4629	Standard test method for trace nitrogen in liquid petroleum hydrocarbons by syringe/inlet oxidative combustion and chemiluminescence detection
D4636	Standard test method for corrosiveness and oxidation stability of hydraulic oils, aircraft turbine engine lubricants, and other highly refined oils
D4683	Standard test method for measuring viscosity at high shear rate and high temperature by tapered bearing simulator
D4684	Standard test method for determination of yield stress and apparent viscosity of engine oils at low temperature (MRV TP-1 cycle)
D4693	Standard test method for low-temperature torque of grease-lubricated wheel bearings
D4739	Standard test method for base number determination by potentiometric titration
D4741	Standard test method for measuring viscosity at high temperature and high shear rate by tapered-plug viscometer
D4742	Standard test method for oxidation stability of gasoline automotive engine oils by thin-film oxygen uptake (TFOUT)
D4781-03	Standard test method for mechanically tapped packing density of fine catalyst particles and catalyst carrier particles
D4857	Standard test method for determination of coefficient of friction of lubricants using the four-ball wear test machine
D4871-88	Standard guide for universal oxidation/thermal stability test apparatus
D4898	Standard test method for insoluble contamination of hydraulic fluids by gravimetric analysis
D4927	Standard test method for elemental analysis of lubricant and additive components, barium, calcium, phosphorus, sulfur, and zinc, by wavelength-dispersive x-ray fluorescence spectroscopy
D4950	Standard classification and specification for automotive service greases
D4951	Standard test method for determination of additive elements in lubricating oils by inductively coupled plasma atomic emission spectrometry
D5119	Standard test method for evaluation of automotive engine oils in CRC L-38 spark ignition engine
D5133	Standard test method for low temperature, low shear rate, viscosity/temperature dependence of lubricating oils using a temperature-scanning technique (scanning Brookfield test with gelation index calculation)
D5182	Standard test method for evaluating the scuffing load capacity of oils (FZG visual method)
D5183	Standard test method for evaluating coefficient of friction of lubricants using the four-ball wear test machine
D5185	Standard test method for determination of additive elements, wear metals, and contaminants in used lubricating oils and determination of selected elements in base oils by inductively coupled plasma atomic emission spectrometry (ICP-AES)
D5293	Standard test method for apparent viscosity of engine oils between −5 and −30°C using the cold-cranking simulator
D5302	Standard test method for evaluation of automotive engine oils in the Sequence VE spark ignition engine
D5306-92(2002)e1	Standard test method for linear flame propagation rate of lubricating oils and hydraulic fluids
D5480	Standard test method for engine oil volatility by gas chromatography
D5483	Standard test method for oxidation induction time of lubricating greases by pressure differential scanning calorimetry
D5533	Standard test method for evaluation of automotive engine oils in the Sequence IIIE spark ignition engine
D5570	Standard test method for evaluating the thermal stability of manual transmission lubricants in a cycle durability test
D5620	Standard test method for evaluating thin-film lubricants in a drain and dry mode using Falex pin and Vee block test machine
D5621	Standard test method for sonic shear stability of hydraulic fluids
D5704	Standard test method for evaluation of thermal and oxidative stability of lubricating oils used for manual transmissions and final drive axles
D5706	Standard test method for determining extreme-pressure properties of lubricating greases using a high-frequency, linear-oscillation (SRV) test machine
D5707	Standard test method for measuring friction and wear properties of lubricating grease using a high-frequency, linear-oscillation (SRV) test machine
D5800	Standard test method for evaporation loss of lubricating oils by the NOACK method
D5862	Standard test method for evaluation of engine oils in two-stroke cycle turbo-supercharged 6V92TA diesel engine
D5864	Standard test method for determining the aerobic aquatic biodegradation of lubricants or their components
D5949	Standard test method for pour point of petroleum products
D5968-04	Standard test method for evaluation of corrosiveness of diesel engine oil at 121°C
D5969	Standard test method for corrosion preventive properties of lubricating greases in the presence of dilute synthetic seawater environments

(Continued)

D6006	Standard guide for assessing biodegradability of hydraulic fluids
D6022	Standard test method for calculation of permanent shear stability index
D6046	Standard classification of hydraulic fluids for environmental impact
D6079 04e1	Standard test method for evaluating lubricity of diesel fuels by the high-frequency reciprocating rig (HFRR)
D6080 97(2002)	Standard practice for defining the viscosity characteristics of hydraulic fluids
D6081	Standard practice for aquatic toxicity testing of lubricants: sample preparation and results interpretation
D6082	Standard test method for high temperature foaming characteristics of lubricating oils
D6121	Standard test method for evaluation of load-carrying capacity of lubricants under conditions of low speed and high torque used for final hypoid drive axles
D6138	Standard test method for determination of corrosion preventive properties of lubricating greases under dynamic wet conditions (Emcor test)
D6158	Standard specification for mineral oil hydraulic oils
D6186	Standard test method for oxidation induction time of lubricating oils by pressure differential scanning calorimetry (PDSC)
D6203	Standard test method for thermal stability of way lubricants
D6278	Standard test method for shear stability of polymer-containing fluids using a European diesel injector apparatus (see also D3945)
D6417	Standard test method for estimation of engine oil volatility by capillary gas chromatography
D6425	Standard test method for measuring friction and wear properties of extreme-pressure (EP) lubricating oils using SRV test machine
D6546-00	Standard Test Methods for and Suggested Limits for Determining Compatibility of Elastomer Seals for Industrial Hydraulic Fluid Applications
D6546-15	Standard Test Methods for and Suggested Limits for Determining Compatibility of Elastomer Seals for Industrial Hydraulic Fluid Applications
D6557	Standard test method for evaluation of rust preventive characteristics of automotive engine oils
D6594-04a	Standard test method for evaluation of corrosiveness of diesel engine oil at 135°C
D6595	Standard test method for determination of wear metals and contaminants in used lubricating oils or used hydraulic fluids by rotating disk electrode atomic emission spectrometry
D7097	Standard test method for Determination of Moderately High Temperature Piston Deposits by Thermo-Oxidation Engine Oil Simulation Test—TEOST MHT
E 537-02	Standard test method for the thermal stability of chemicals by differential scanning calorimetry
E 659	Standard test method for (1994)el autoignition temperature of liquid chemicals
E 1064	Standard test method for water in organic liquids by coulometric Karl Fischer titration

G72-82(1996) e1	Standard test method for autogenous ignition temperature of liquids and solids in a high-pressure oxygen-enriched environment
G133/95	Standard test method for linearly reciprocating ball-on-flat sliding wear
STP 315H	Multicylinder test sequence for evaluating automotive engine oils
STP 509A	Single-cylinder engine test for evaluating the performance of crankcase lubricants

34.1.2 CEC Test Methods

L-01-A-79	Test for diesel engine crankcase oils using the Petter AVI single-cylinder laboratory diesel engine
L-02-A-78	Oil oxidation and bearing corrosion test using the Petter W1 single-cylinder gasoline engine
L-07-A-85	Load-carrying capacity test for transmission lubricants using the FZG test rig
L-11-T-72	The coefficient of friction of automatic transmission fluids using the DKA friction machine
L-12-A-76	Evaluation of piston cleanliness in the MWM KD 12 E test engine (method B more severe)
L-14-A-93	Evaluation of the shear stability of lubricating oils containing polymers using the Bosch diesel fuel injector pump rig
L-18-A-80	Procedure for measurement of low-temperature apparent viscosity by means of the Brookfield viscometer (liquid bath method)
L-19-T-77	Evaluation of the lubricity of two-stroke engine oils (using the Motobecane engine AV7L 50 cm³)
L-20-A-79	Evaluation of two-stroke engine lubricants with respect to engine deposit formation oils (using the Motobecane engine AV7L 50 cm³)
L-21-T-77	The evaluation of two-stroke engine lubricants: Sequence I—Piston antiseizure Sequence II—General performance Sequence III—Preignition (using a Piaggio Vespa 180 SS engine)
L-24-A-78	Engine cleanliness under severe conditions using the Petter AVB supercharged diesel engine
L-25-A-78	Engine oil viscosity stability test (using a Peugeot 204 engine)
L-28-T-79	The evaluation of outboard engine lubricant performance (using Johnson and Evinrude marine outboard engines)
L-29-T-81	Ford Kent test procedure for evaluating the influence of the lubricating oil on piston ring sticking and deposit formation (using a Ford Kent engine)
L-30-T-81	Cam and tappet pitting test procedure (using MIRA cam and tappet test machine)
L-31-T-81	Predicting the borderline pumping temperature of engine oils using the Brookfield viscometer
L-33-A-93	Biodegradability of two-stroke cycle outboard engine oils in water

(Continued)

L-33-A-94	Biodegradability of two-stroke cycle outboard engine oils in water
L-34-T-82	Preignition tendencies of engine lubricants (using a Fiat 132C engine)
L-35-T-84	Motor oil evaluation in a turbocharged passenger car diesel engine (using a VW ATL 1.6 L)
L-36-96	The evaluation of oil-elastomer compatibility (laboratory test)
L-36-A-90	The measurement of lubricant dynamic viscosity under conditions of high shear (using a Ravenfield viscometer)
L-36-A-97	High temperature high shear (HTHS)
L-37-T-85	Shear stability test for polymer-containing oils (using the FZG test rig)
L-38-A-94	Valve train scuffing (using a PSA TU3 engine) wear test
L-39-T-87	Oil/elastomer compatibility test
L-40-A-93	Lubricating oil evaporative losses (using NOACK evaporative tester)
L-41-T-88	Evaluation of sludge-inhibition qualities of motor oils in a gasoline engine (using a Mercedes-Benz M102E engine)
L-42-A-92	Evaluation of bore polish, piston cleanliness, liner wear, and sludge in a DI turbo-charged diesel engine (using Mercedes-Benz OM364A engine)
L-46-T-93	VW intercooled turbo-diesel ring stitching and piston cleanliness test
L-51-T-95	OM 602A neon test
L-51-T-98	The evaluation of engine crankcase lubricants with respect to low-temperature lubricant thickening and wear under severe operating conditions (MB-OM602A engine) "A" status granted basis cam-wear only
L-53-T-95	M111 black sludge test
L-54-T-96	Fuel economy effects of engine lubricants (MB M111 E20)
L-55-T-95	TU3 MH high-temperature deposits, ring sticking, and oil thickening test
L-56-T-95	XUD11 ATE medium-temperature dispensarity test
L-56-T-98	Oil dispersion test at medium temperature for automobile diesel engines (XUD11BTE engine)

34.1.3 General Motors

| 9099P | Engine oil filterability test (EOFT) (to be modified for GF-3) |

34.1.4 Society of Automotive Engineers

J183	Engine oil performance and engine service classification (other than *energy conserving*)
J300	Engine oil viscosity classification standard
J357	Physical and chemical properties of engine oils
J1423	Classification of energy-conserving engine oil for passenger cars, vans, and light-duty trucks

AS5780	Core Requirement Specification for Aircraft Gas Turbine Engine Lubricants; September 2000
AS5780B	Specification for Aero and Aero-Derived Gas Turbine Engine Lubricants; February 2013
AMS3384	Rubber, Fluorocarbon Elastomer (FKM), 70–80 Hardness, Low Temperature Sealing T_g −22°F (−30°C), for Elastomeric Shapes or Parts in Gas Turbine Engine Oil, Fuel, and Hydraulic Systems; August 2013
AMS7287	Fluorocarbon Elastomer (FKM) High Temperature/ HTS Oil Resistant/Fuel-Resistant Low Compression Set/70–80 Hardness, Low Temperature T_g −22°F (−30°C) for Seals in Oil/Fuel/Specific Hydraulic Systems; August 2012
AMS7379	*Rubber*: Fluorocarbon Elastomer (FKM) 70–80 Hardness, Low-Temperature Sealing T_g −40°F (−40°C) for Elastomeric Seals in Aircraft Engine Oil, Fuel, and Hydraulics Systems; July 2008
AMS-P-83461	Packing, Preformed, Petroleum Hydraulic Fluid Resistant, Improved Performance at 275°F (135°C); April 1998
AMS-R-83485	Rev. A Rubber, Fluorocarbon Elastomer, Improved Performance at Low Temperatures; May 1998

34.1.5 Miscellaneous Test Methods

CEM	Electric motor test (grease)
DIN 51350	Part 2 weld load
DIN 51350	Part 3 wear scar
DIN 51352-1	Testing of lubricants; determination of aging characteristics of lubricating oils; increase in Conradson carbon residue after aging by passing air through the lubricating oil
DIN 51381	Air release properties per temperature
DIN 51554-1	Testing of mineral oils; test of susceptibility to aging according to Baader; purpose, sampling, aging
DIN 51587	Testing of lubricants; determination of the aging behavior of steam turbine oils and hydraulic oils containing additives
DIN 51802 (IP-220)	Emcor rust test
DIN 51817	Testing of Lubricants—Determination of oil separation from greases under static conditions
DIN 51851	ASTM D5100 NOACK volatility
DIN 53169	pH at 20°C
Emcor rust test (grease)	
FE-8	Test (grease)
FTM 321.2	Oil Separation from lubricating grease (static technique)
FTM 350	Evaporation loss
FTM 352	Wick ignition
FTM 791B	Cone bleed
FTM 791C (Method 3470.1)	Homogeneity and miscibility

(Continued)

FTM 3009	Contamination, particulate (oils)
FTM 3010	Gravimetric contamination and ash residue by filtration
FTM 3011	Particulate contamination by HIAC counter
FTM 3012	Particulate contamination of oils by filtration
FTM 3403	Compatibility of turbine lubricating oils
FTM 3411	Thermal stability and corrosivity
FTM 3432	Compatibility of FKM elastomer
FTM 3433	Navy S silicone rubber, swelling and tensile strength
FTM 3456	Channel point of lubricating oils
FTM 3480	Volatility
FTM 3603	Swelling of rubber NBR-L
FTM 3604	Swelling of rubber NBR-H
FTM 4001.2	Salt spray corrosion (see ASTM B 117)
FTM 5306	Corrosiveness of cutting fluid
FTM 5307	Oxidation and corrosion stability
FTM 5308	Oxidation and corrosion stability
FTM 5309	Corrosion, copper, 24 h
FTM 5322	Corrosiveness (bimetallic couple)
GE electric motor test (grease)	
JPI-55-55-99	Hot tube test
MIL-G-22050	Gasket and packing material, rubber, for use with polar fluids, steam, and air at moderately high temperatures
MIL-G-81322	Grease, aircraft wide temperature range
MIL-H-22072C(AS)	Hydraulic fluid catapult
MIL-H-27601B	Hydraulic fluid, petroleum base, high temperature, flight vehicle MIL-H-46170, water sensitivity
MIL-H-46170B	Hydraulic fluid, rust inhibited, fire resistant, synthetic hydrocarbon base
MIL-H-53119	Corrosion rate evaluation procedure (CREP) for CTFE hydraulic fluids MIL-H-83282 high-temperature stability (sealed ampule)
MIL-H-5606A	Hydraulic Fluid, Petroleum Base, Aircraft and Ordnance, 21 February 1957
MIL-H-5606G	Hydraulic Fluid, Petroleum Base; Aircraft; Missile and Ordnance—Notice 1; 29 March 1996
MIL-H-83282	Linear flame propagation rate
MIL-H-83282C	Hydraulic fluid, fire resistant, synthetic hydrocarbon base, aircraft MIL-H-83306 hydraulic fluid, fire resistant, phosphate ester base, aircraft
MIL-H-87257	High-temperature stability (purged with nitrogen)
MIL-H-87257	Hydraulic Fluid, Fire Resistant; Low Temperature, Synthetic Hydrocarbon Base, Aircraft and Missile; 2 March 1992
MIL-L-23699E	Lubricating Oil, Aircraft Turbine Engine, Synthetic Base, NATO Code Number O-156; 25 August 1994

MIL-O-5606	Oil; Hydraulic, Aircraft, Petroleum Base; 31 January 1950
MIL-P-25732	Cold-resistant acrylonitrile-butadiene rubber
MIL-P-25732C	Packing, Preformed, Petroleum Hydraulic Fluid Resistant, Limited Service at 275°F (135°C); 25 February 1980
MIL-P-25732C	Packing, Preformed, Petroleum Hydraulic Fluid Resistant, Limited Service at 275°F (135°C)—Notice 1; 15 November 1989
MIL-P-83461B	Packing, Preformed, Petroleum Hydraulic Fluid Resistant, Improved Performance at 275°F (135°C); 25 February 1980
MIL-PRF-2104	Lubricating oil, internal combustion engine, combat/tactical service MIL-PRF-2105 lubricating oil, gear multipurpose
MIL-PRF-5606H	Hydraulic Fluid, Petroleum Base; Aircraft Missile, and Ordnance; 7 June 2002
MIL-PRF-7808L	Lubricating Oil, Aircraft Turbine Engine, Synthetic Base; 2 May 1997
MIL-PRF-10924	Grease, automotive, and artillery
MIL-PRF-23699F	Lubricating Oil, Aircraft Turbine Engine, Synthetic Base, NATO Code Number O-156; 21 May 1997
MIL-PRF-23699G	Lubricating Oil, Aircraft Turbine Engine, Synthetic Base, NATO Code Numbers: O-152, O-154, O-156, and O-167; 13 March 2014
MIL-PRF-46170	Hydraulic fluid, rust inhibited, fire resistant, synthetic hydrocarbon base
MIL-PRF-63460	Lubricant, cleaner and preservative for weapons and weapons systems (metric)
MIL-PRF-81322	Grease, aircraft, general purpose, wide temperature range
MIL-PRF-83282B	Hydraulic fluid, fire resistant, synthetic hydrocarbon base, metric, NATO code number H-537, 10 February 1982
MIL-PRF-83282D	Hydraulic fluid, fire resistant, synthetic hydrocarbon base, metric, NATO code number H-537, 30 September 1997
MIL-PRF-87252	Coolant fluid, hydrolytically stable, dielectric
MIL-PRF-87257 B	Hydraulic Fluid, Fire Resistant; Low Temperature, Synthetic Hydrocarbon Base, Aircraft and Missile; 22 April 2004
MIL-R-83248	Rubber, fluorocarbon elastomer, high-performance fluid, and compression set resistant
MIL-R-83485	(USAF) Rubber, Fluorocarbon Elastomer, Improved Performance At Low Temperatures; 8 September 1976
MIL-STD-1246	Cleanliness levels SKF R2F test (simulates paper mill applications)
USS	Low-temperature mobility test (grease)

34.1.6 FEDERAL SUPPLY CLASS 9150 PRODUCT COMMODITIES

Document Summarized Title and Description		QPL	Custodian	NATO Code
MIL-PRF-23699F	Synthetic aircraft turbine engine oil	Yes	Navy/AS	O-156/O-154
MIL-PRF-23827C	Aircraft and instrument grease	Yes		G-354
MIL-PRF-81322F	Aircraft wide temperature range grease	Yes		G-395
MIL-PRF-81329D	Solid film lubricant	No (FAT)		S-1737
MIL-PRF-83282D	Synthetic fire-resistant hydraulic fluid	Yes		H-537
MIL-PRF-85336B	All-weather lubricant for weapons	Yes		
MIL-L-19701B	Semifluid lubricant for weapons	Yes		
MIL-G-21164D	Molybdenum disulfide grease	Yes		G-353
MIL-L-23398D	Solid film lubricant, air cured	Yes		S-748
MIL-G-23549C	General-purpose grease	Yes		
MIL-G-25013E	Aircraft bearing grease	Yes		G-372
MIL-G-25537C	Aircraft helicopter bearing grease	Yes		G-366
MIL-H-81019D	Hydraulic fluid for ultralow temperatures	Yes		
MIL-S-81087C[a]	Silicone fluid antiwear grease	Yes		H-536
MIL-G-81827A	Aircraft high loading and antiwear grease	Yes		
MIL-L-81846	Instrument ball bearing lubricating oil	No		
MIL-G-81937A	Ultraclean instrument grease	Yes		
DOD-L-85645A[a]	Dry thin-film lubricant	No		
DOD-G-85733	High-temperature catapult grease	Yes		
DOD-L-85734	Synthetic helicopter transmission lubricant	Yes		
VV-D-1078B	Silicone fluid damping fluid	No		S-1714–1732
SAE J1899	Aircraft piston engine oil, ashless dispersant	Yes		O-123/O-128
SAE J1966	Aircraft piston engine oil, nondispersant	Yes		O-113/O-117
SAE AMS-G-4343	Pneumatic systems grease	No		G-392
SAE AMS-G-6032	Plug valve grease	Yes		G-363
MIL-H-22072C	Hydraulic fluid for catapults	Yes	Navy/AS[b]	H-579
A-A-59290	Arresting gear hydraulic fluid	No	Navy/AS[b]	
MIL-PRF-9000H	Diesel engine oil	Yes	Navy/SH	O-278
MIL-PRF-17331H	Steam turbine lubricating oil	Yes		O-250
MIL-PRF-17672D	Hydraulic fluid	Yes		H-573
MIL-PRF-24139A	Multipurpose grease	Yes		
DOD-PRF-24574	Lubricating fluid for oxidizing mixtures	Yes		
MIL-L-15719A	High-temperature electrical bearing grease	Yes		
MIL-T-17128C	Transducer fluid	No		
MIL-G-18458B	Exposed gear and rope grease	Yes		
MIL-H-19457D	Fire-resistant hydraulic fluid (FAT)	No		H-580
MIL-L-24131B	Graphite and alcohol lubricant	Yes		
MIL-L-24478C	Molybdenum disulfide and alcohol lubricant	No		
DOD-G-24508A	Multipurpose grease	Yes		
DOD-G-24650	Food processing equipment grease	No		
DOD-G-24651	Food processing equipment lubricating oil	No		
VV-L-825C	Lubricating oil for refrigerant compressors	No		O-282/O-290
A-A-50433	Seawater-resistant grease	No		
A-A-50634	Lubricating oil for compressors using HFC-134A	No		
A-A-59004A	Antigalling compound	No		
MIL-PRF-6081D	Jet engine lubricating oil	Yes	Air Force/11	O-132/O-133
MIL-PRF-6085D	Aircraft instrument lubricating oil	Yes		O-147
MIL-PRF-6086E	Aircraft gear petroleum lubricating oil	Yes		O-153/O-155
MIL-PRF-7808L	Aircraft turbine synthetic engine oil	Yes		O-148/O-163
MIL-PRF-7870C	Low-temperature lubricating oil	Yes		O-142
MIL-PRF-8188D	Corrosion preventive engine oil (FSC 6850)	Yes		C-638
MIL-PRF-27601C	Hydraulic fluid	Yes		

(Continued)

Document Summarized Title and Description		QPL	Custodian	NATO Code
MIL-PRF-27617F	Aircraft and instrument grease	Yes		G-397–399/ G-1350
MIL-PRF-32014	Aircraft and missile high-speed grease	No		
MIL-PRF-83261B	Aircraft extreme-pressure grease	No		
MIL-PRF-83363C	Helicopter transmission grease	No		G-396
MIL-PRF-87100A	Aircraft turbine synthetic engine oil	Yes		
MIL-PRF-87252C	Dielectric coolant fluid (FSC 9160)	Yes		S-1748
MIL-PRF-87257A	Synthetic fire-resistant hydraulic fluid	Yes		H-538
MIL-H-5606G[a]	Petroleum hydraulic fluid for aircraft/ordnance	Yes		H-515
DOD-L-25681D	Silicone fluid with molybdenum	No (disulfide)	Air Force/68	S-1735
MIL-L-87177A	Synthetic corrosion preventive lubricant	No (FAT)	Air Force/70	
MIL-PRF-2104G	Combat/tactical diesel	Yes (engine oil)	Army/AT	O-236/O-237/O-1236
MIL-PRF-2105E	Multipurpose gear oil	Yes		O-186/O-226/O-228
MIL-PRF-3150D	Preservative oil	Yes		O-192
MIL-PRF-6083F	Operational and preservative hydraulic fluid	Yes		C-635
MIL-PRF-10924G	Automotive/artillery grease	Yes		G-403
MIL-PRF-12070E	Fog oil	No		F-62
MIL-PRF-21260E	Preservative and break-in engine oil	Yes		C-640/C-642
MIL-PRF-32033	Preservative and water-displacing oil	Yes		0-190
MIL-PRF-46002C	Vapor corrosion inhibitor (VCI)	No (FAT) preservative oil		
MIL-PRF-46010F	Solid-film lubricant	Yes		S-1738
MIL-PRF-46147C	Solid-film lubricant	Yes		
MIL-PRF-46167C	Arctic engine oil	Yes		O-183
MIL-PRF-46170C	Synthetic fire-resistant hydraulic fluid	Yes		H-544
MIL-PRF-46176B	Silicon brake fluid	Yes		H-547
MIL-PRF-53074A	Steam cylinder lubricating oil	No		O-258
MIL-PRF-53131A	Precision bearing synthetic lubricating oil	Yes		
VV-G-632B	General-purpose industrial grease	No		
VV-G-671F	Graphite grease	No		G-412
A-A-52039B	Automotive engine oil API service SH	No		
A-A-52036A	Commercial heavy-duty diesel engine oil	No		
A-A-59354	Hydraulic fluid for machines	No		
SAE J1703	Conventional brake fluid	No		H-542
MIL-PRF-63460D	Cleaner-lubricant-preservative for weapons	Yes	Army/AR	S-758
MIL-L-11734C	Synthetic lubricant for mechanical fuse	No		
Systems MIL-L-14107C	Low-temperature weapons lubricant	Yes		O-157
MIL-L-45983	Heat-cured solid film lubricant	No		
MIL-L-46000C	Semifluid weapons lubricant	Yes		O-158
MIL-G-46003A	Rifle grease	Yes		
MIL-L-46150	Semifluid high-loading weapons lubricant	Yes		
MIL-PRF-3572B	Colloidal graphite in oil	No		DSCR/GS
MIL-DTL-17111C	Power transmission fluid	No		(FAT) H-575
MIL-PRF-26087C	Reciprocating compressor lubricating oil	No		
MIL-L-3918A[a]	Instrument lubricating oil for jewel bearings	No		
MIL-L-46014[a]	Spindle lubricating oil	No		
MIL-L-83767B[a]	Vacuum pump lubricating oil	No		
VV-C-846B	Emulsifiable oil-type cutting fluids	No		
A-A-50493A	Penetrating oil	No		
A-A-59113	Machine tools/slideways lubricating oil	No		
A-A-59137	Breech block lubricating oil (naval ordnance)	No		
A-A-59173	Silicone grease	No		
A-A-59197	Fatty oil for metalworking lubricants	No		
SAE AS1241C	Fire-resistant phosphate ester hydraulic fluid	No		

[a] These specifications had been designated as "Inactive for New Design" and are no longer used except for replacement purposes. Their QPLs will be maintained until the products are no longer required.

[b] See Navy/AS2 under Abbreviations section.

34.1.7 Specifications Having Cross-Reference among JIS, ASTM, and Others

ASTM or Others	JIS	Title and Contents
F 312	B 9930	Hydraulic fluid—determination of particulate contamination by the particle count method
F 313	B 9931	Fluid contamination—determination of contaminants by the gravimetric method
D117	C 2101	Testing method of electrical insulating oils
D923		Sampling
D4559		Evaporation
D1218/21807		Refractive index and specific dispersion
D974		Total acid number
D1275		Corrosive sulfur
D1533		Water content
D2112/2440		Oxidation stability
D877/1816		Dielectric strength
D924		Dielectric loss tangent and relative dielectric constant
D1169		Volume resistivity
	K 2249	Crude petroleum and petroleum products—determination of the density and petroleum measurement tables based on a reference temperature (15°C)
D1298/E100		I-shaped float method
D4052/5002		Oscillating method
ISO 3833		Warden pycnometer method
D941		I-shaped pycnometer method
D70		Hubbard pycnometer method
D1250		Density, mass, and volume conversion table
D140/4057/4177	K 2251	Crude petroleum and petroleum products—sampling
D1093	K 2252	Testing method for reaction of petroleum products
	K 2254	Petroleum products—determination of distillation characteristics
D86, E133		Test method for distillation of petroleum products at atmospheric pressure
D1160		Test method for distillation of petroleum products at reduced pressure
D2287		Test method for boiling range distillation of petroleum products by gas chromatography
	K 2255	Petroleum products—gasoline—determination of lead content
D3341		Iodine monochloride method
D3237		Atomic absorption spectroscopy method
D661	K 2256	Testing methods for aniline point and mixed aniline point of petroleum products
D323	K 2258	Testing method for vapor pressure of crude oil and petroleum products (Reid method)
D381	K 2261	Petroleum products—motor gasoline and aviation fuels—determination of existent gum—jet evaporation method
	K 2265	Crude oil and petroleum products—determination of flash point

ASTM or Others	JIS	Title and Contents
D56		Tag closed test
D3828/3278		Small-scale closed test
D93		Pensky–Martens closed cup test
D92		Cleveland open cup test
	K 2269	Testing methods for pour point and cloud point of crude oil and petroleum products
D97		Pour point
D2500		Cloud point
	K 2270	Crude petroleum and petroleum products—determination of carbon residue
D189		Conradson method
D4530		Micro method
	K 2272	Testing methods for ash and sulfated ash of crude oil and petroleum products
D482		Ash
D874		Sulfated ash
	K 2275	Crude oil and petroleum products—determination of water content
D95/4006		Distillation method
D4377/1744		Karl Fischer volumetric method
		Karl Fischer coulometric method
DIN 9114		Hydride reaction method
	K 2276	Petroleum products—testing methods for aviation fuels
D873		Oxidation stability (potential residue)
D2386		Freezing point
D1094		Water tolerance
D235/4952		Doctor test
D3227		Determination of mercaptan sulfur (potentiometric method)
D1740		Luminometer number test
D1840		Determination of naphthalene (ultraviolet spectroscopy)
FS 1151.2		Explosive vapor test
D3242		Total acid number
D3948		Water separation index (micro separometer)
D2550		Water separation index (water separometer)
D3241		Thermal stability (JFTOT)
D2276/5452		Particulate contaminant
IP 227		Copper corrosion
D2624		Electric conductivity
D3343		Hydrogen content
	K 2279	Crude petroleum and petroleum products—determination and estimation of heat of combustion
D4529/4868		Net heat of combustion
D4868		Gross heat of combustion
	K 2280	Petroleum products—fuels—determination of octane number, cetane number, and calculation of cetane index
D2699		Research octane number
D2700		Motor octane number
D909		Supercharge octane number

(Continued)

ASTM or Others	JIS	Title and Contents	ASTM or Others	JIS	Title and Contents
D613		Cetane number		K 2514	Lubricating oils—determination of oxidation stability
D4737		Calculation method for cetane index using four variable equation	ISOT		Oxidation stability of lubricants for internal combustion engine
D1368		Small amount of lead in n-heptane and isooctane (dithizone method)	D943		Turbine oil oxidation stability test (TOST)
D2268		Purity of n-heptane and isooctane (capillary gas chromatography)	D2272		Rotating pressure vessel oxidation test (RBOT)
	K 2283	Crude petroleum and petroleum products—determination of kinematic viscosity and calculation of viscosity index from kinematic viscosity	D3397		Total acid number (semimicro method)
			IP 280		Turbine oil oxidation stability (oil-soluble catalyst method)
D445/446		Kinematic viscosity	D892	K 2518	Petroleum products—lubricating oils—determination of foaming characteristics
D2270		Viscosity index		K 2619	Lubricating oils—testing methods for load-carrying capacity
D341		Estimated relation between kinematic viscosity and temperature	D2619		Soda four-ball test (four-ball test modified by Dr. Soda)
D525	K 2287	Testing methods for oxidation stability of gasoline (induction period method)	D2782		Timken
IP 309	K 2288	Gas oil—determination of cold filter plugging point		K 2520	Petroleum products—lubricating oils—determination of demulsibility characteristics
	K 2301	Fuel gas and natural gas—methods for chemical analysis and testing	D1401		Demulsibility test
D1145		Sampling of gas sample	IP 19		Steam emulsion number
D1945/1946		Chemical analysis (gas chromatography)		K 2536	Liquid petroleum products—testing method of components
ISO 6326-1		Analysis of total sulfur	D1319/2001/2427		Fluorescent indicator adsorption analysis (FIA)
ISO 6326-1		Analysis of hydrogen sulfide	D2267/4420/5580		Determination of aromatics by gas chromatography
ISO 6327		Analysis of water (dew point method)			
D900/1826		Heat of combustion (Junkers gas calorimeter)	D1322	K 2537	Petroleum product—aviation turbine fuels and kerosene—determination of smoke point
D3588		Heat of combustion (calculation method)		K 2540	Testing method for thermal stability of lubricating oils
D1070		Specific gravity (pycnometer method)		K 2541	Crude oil and petroleum products—determination of sulfur content
D3588		Specific gravity (calculation method)			
D4057	K 2420	Method of sampling for aromatic hydrocarbon and tar products	D2785/ISO 4260		Oxyhydrogen combustion method
	K 2501	Petroleum products and lubricants—determination of neutralization number	D3120		Microcoulometric titration
D974		Color indicator titration (TAN, strong acid number, strong base number)	D1551		Quartz tube test
			D4294/ISO 8754		Energy-dispersive x-ray fluorescence spectroscopy
D664		Potentiometric titration (TAN, strong acid number)	D129		General bomb method
D4739		Potentiometric titration (TBN, strong base number)	D1266		Lamp method
D2896		Potentiometric titration (TBN, perchloric acid method)	D2622		Wavelength-dispersive x-ray fluorescence spectroscopy
	K 2503	Testing method of lubricating oil for aircraft		K 2580	Petroleum products—determination of color
D91/2273		Precipitation number	D156		Saybolt
D94		Saponification number	D1500		ASTM
FS 3006.3		Contamination		K 2601	Testing methods for crude petroleum
FS 204.1		Diluted pour point	D3828		Flash point
ISO 6617		Oxidation stability	D96/4007/1796		Water and sediment
FS 5308.7		Corrosiveness and oxidation stability	IP 77		Salt content (titration)
D665	K 2510	Testing method for rust-preventing characteristics of lubricating oil	D3230		Salt content (coulometric)
D130	K 2513	Petroleum products—corrosiveness to copper—copper strip test	D2892		Distillation at atmospheric pressure

(Continued)

ASTM or Others	JIS	Title and Contents
D1159/2710	K 2605	Petroleum distillates and commercial aliphatic olefins—determination of bromine number—electrometric method
	K 2609	Crude petroleum and petroleum products—determination of nitrogen content
D3228		Macro-Kjedahl method
D3431		Microcoulometric titration
D4629/5762		Chemiluminescence method

Note: ISOT = Indiana Stirring Oxidation Test.

Air Force/11	Air Force Aeronautical Systems Center (ASC), Wright-Patterson AFB, OH
Air Force/68	Air Force San Antonio Air Logistics Center (SAALC), Kelly AFB, TX
Air Force/70	Hill Air Force Base Logistics Center, UT
Army/AT	Army Tank-Automotive and Armaments Command, Tank-Automotive Research Development and Engineering Center (TARDEC), Warren, MI
Army/AR	Army Tank-Automotive and Armaments Command, Armament Research, Development and Engineering Center (ARDEC), Picatinny, NJ
DSCR/GS	Defense Logistics Agency's Defense Supply Center Richmond (DSCR), Richmond, VA

34.2 ABBREVIATIONS USED

QPL	Qualified products listing
FAT	No QPL exists, but a First Article Test (FAT) is required or may be optional
Navy/AS	Naval Air Systems Command (NAVAIR), Patuxent River, MD
Navy/AS2	Naval Air Systems Command (NAVAIR), Lakehurst, NJ
Navy/SH	Naval Sea Systems Command (NAVSEA), Arlington, VA
Navy/YD	Naval Facilities Engineering Command (NAVFAC), Alexandria, VA

ACKNOWLEDGMENTS

This chapter was compiled over the years with the generous help of several colleagues in the lubricant community: Ed Zaweski and Hiroshi Yamaochi (Amoco Chemicals—both retired), Alan Plomer (BP Belgium), and Darryl Spivey (BP, Naperville, IL).

I also thank Piet Purmer (Shell Chemical Company), Dick Kuhlman (Afton), Ed Snyder (AFRL/MLBT), Don Campbell, and Bob Rhodes (Shell—retired).

35 Internet Resources for the Additive/Lubricant Industry

Leslie R. Rudnick

CONTENTS

35.1 ALPHABETICAL LISTING

2V Industries Inc., www.2vindustries.com

49 North, www.49northlubricants.com

76 Lubricants Company, www.tosco.com

A.W. Chesterton Company, www.chesterton.com

A/R Packaging Corporation, www.arpackaging.com

Acculube, www.acculube.com

Accumetric LLC, www.accumetric.com

Accurate Lubricants & Metalworking Fluids Inc. (dba Acculube), www.acculube.com

Acheson Colloids Company, www.achesonindustries.com

Acme Refining, Division of Mar-Mor Inc., www.acmerefining.com

Acme-Hardesty Company, www.acme-hardesty.com

Adco Petrol Katkilari San Ve. Tic. AS, www.adco.com.tr

Advanced Ceramics Corporation, www.advceramics.com

Advanced Lubrication Technology Inc. (ALT), www.altboron.com

Aerospace Lubricants Inc., www.aerospacelubricants.com

AFD Technologies, www.afdt.com

AG Fluoropolymers USA Inc., www.fluoropolymers.com

Airflow Systems Inc., www.airflowsystems.com

Airosol Company Inc., www.airosol.com

Akzo Nobel, www.akzonobel.com

Alco-Metalube Company, www.alco-metalube.com

Alfa Laval Separation, www.alfalaval.com

Alfa Romeo, www.alfaromeo.com

Alithicon Lubricants, Division of Southeast Oil & Grease Company Inc., www.alithicon.com

Allegheny Petroleum Products Company, www.oils.com

Allen Filters Inc., www.allenfilters.com

Allen Oil Company, www.allenoil.com

Allied Oil & Supply Inc., www.allied-oil.com

Allied Washoe, www.alliedwashoe.com

Amalie Oil Company, www.amalie.com

Amber Division of Nidera Inc., www.nidera-us.com

Amcar Inc., www.amcarinc.com

Amerada Hess Corporation, www.hess.com

American Agip Company Inc., www.americanagip.com

American Bearing Manufacturers Association, www.abma-dc.org

American Board of Industrial Hygiene, www.abih.org

American Carbon Society, www.americancarbonsociety.org

American Chemical Society (ACS), www.acs.org

American Council of Independent Laboratories (ACIL), www.acil.org

American Gear Manufacturers Association (AGMA), www.agma.org

American International Chemical Inc., www.aicma.com

American Lubricants Inc., www.americanlubricantsb-flo.com

American Lubricating Company, www.americanlubricating.com

American Machinist, www.americanmachinist.com

American National Standards Institute (ANSI), www.ansi.org

American Oil & Supply Company, www.aosco.com

American Oil Chemists Society (AOCS), www.aocs.org

American Petroleum Institute (API), www.api.org

American Refining Group Inc., www.amref.com

American Society for Horticultural Science (ASHS), www.ashs.org

American Society for Testing and Materials (ASTM), www.astm.org

American Society of Agricultural Engineering (ASAE), www.asae.org

American Society of Agronomy (ASA), www.agronomy.org

American Society of Mechanical Engineers International (ASME), www.asme.org

American Synthol Inc., www.americansynthol.com

Amoco, www.amoco.com

Amptron Corporation, www.superslipperystuff.com/organisation.htm

Amrep Inc., www.amrep.com

AMSOIL Inc., www.amsoil.com

Ana Laboratories Inc., www.analaboratories.com

Analysts Inc., www.analystinc.com

Anderol Specialty Lubricants, www.anderol.com

Andpak Inc., www.andpak.com

ANGUS Chemical Company, www.dowchemical.com

Anti Wear 1, www.dynamicdevelopment.com

API Links, www.api.org/links

Apollo America Corporation, www.apolloamerica.com

Aral International, www.aral.com

Arch Chemicals Inc., www.archbiocides.com

ARCO, www.arco.com

Argonne National Laboratory, www.et.anl.gov

Arizona Chemical, www.arizonachemical.com

Asbury Carbons Inc.—Dixon Lubricants, www.asbury.com

Asbury Graphite Mills Inc., www.asbury.com

Asheville Oil Company Inc., www.ashevilleoil.com

Ashia Denka, www.adk.co.jp/en/chemical/index.html

Ashland Chemical, www.ashchem.com

Ashland Distribution Company, www.ashland.com

Asian Oil Company, www.asianoilcompany.com

Associated Petroleum Products, www.associatedpetroleum.com

Associates of Cape Cod Inc., www.acciusa.com

Atlantis International Inc., www.atlantis-usa.com

Atlas Oil Company, www.atlasoil.com

Audi, www.audi.com

Ausimont, www.ausiusa.com

Automotive Aftermarket Industry Association (AAIA), www.aftermarket.org

Automotive News, www.autonews.com

Automotive Oil Change Association (AOCA), www.aoca.org

Automotive Parts and Accessories Association (APAA), www.apaa.org

AutoWeb, www.autoweb.com

AutoWeek Online, www.autoweek.com

Avatar Corporation, www.avatarcorp.com

Badger Lubrication Technologies Inc., www.badgerlubrication.com

Baker Petrolite, www.bakerhughes.com/bakerpetrolite/

BALLISTOL USA, www.ballistol.com

Bardahl Manufacturing Corporation, www.bardahl.com

Baron USA Inc., www.baronusa.com

BASF Corp., www.basf.com

Battenfeld Grease and Oil Corporation of New York, www.battenfeld-grease.com

Bayer Corp., www.bayer.com

Behnke Lubricants Inc./JAX, www.jax.com

Behnke Lubricants/JAX, www.jaxusa.com

Bell Additives Inc., www.belladditives.com

Bel-Ray Company Inc., www.belray.com

Benz Oil Inc., www.benz.com

Berenfield Containers, www.berenfield.com

Bericap NA, www.bericap.com

Bestolife Corporation, www.bestolife.com

BF Goodrich, www.bfgoodrich.com

BG Products Inc., www.bgprod.com

Bharat Petroleum, www.bharatpetroleum.com

Big East Lubricants Inc., www.bigeastlubricants.com

Bijur Lubricating Corporation, www.bijur.com

Bio-Rad Laboratories, www.bio-rad.com

BioTech International Inc., www.info@biotechintl.com

Bismuth Institute, www.bismuth.be

Blackstone Laboratories, www.blackstone-labs.com

Blaser Swisslube, www.blaser.com

BMW (International), www.bmw.com/bmwe

BMW (United States), www.bmwusa.com

BMW Motorcycles, www.bmw-motorrad.com

Boehme Filatex Inc., www.boehmefilatex.com

BP Amoco Chemicals, www.bpamocochemicals.com

BP Lubricants, www.bplubricants.com

BP, www.bppetrochemicals.com

Brenntag Northeast Inc., www.brenntag.com

Brenntag, www.brenntag.com

Briner Oil Company, www.brineroil.com

British Lubricants Federation Ltd., www.blf.org.uk

British Petroleum (BP), www.bp.com

Britsch Inc., www.britschoil.com

Brno University of Technology, Faculty of Mechanical Engineering, Elastohydrodynamic Lubrication Research Group, www.fme.vutbr.cz/en

Brugarolas SA, www.brugarolas.com/html/eng/actividades_industriales.php

Buckley Oil Company, www.buckleyoil.com

Buckman Laboratories Inc., www.buckman.com

Buick (GM), www.buick.com

Burlington Chemical, www.burco.com

BVA Oils, www.bvaoils.com

Cabot Corporation (fumed metal oxides), www.cabot-corp.com/cabosil

Cadillac (GM), www.cadillac.com

California Air Resources Board, www.arb.ca.gov

Callahan Chemical Company, www.calchem.com

Caltex Petroleum Corporation, www.caltex.com

Calumet Lubricants Company, www.calumetlub.com

Calvary Industries Inc., www.calvaryindustries.com

CAM2 Oil Products Company, www.cam2.com

Cambridge, chemfinder.cambridgesoft.com

Cambridge University, Department of Materials Science and Metallurgy, Tribology, www.msm.cam.ac.uk/tribo/tribol.htm

Cambridge University, Department of Engineering, Tribology, www.mech.eng.cam.ac.uk/Tribology/

Canner Associates Inc., www.canner.com

Cannon Instrument Company, www.cannon-ins.com

Capital Enterprises (Power-Up Lubricants), www.nnl690.com

Car and Driver Magazine Online, www.caranddriver.com

Cargill—Industrial Oil & Lubricants, www.techoils.cargill.com

Car-Stuff, www.car-stuff.com

Cary Company, www.thecarycompany.com

Castle Products Inc., www.castle-comply.com

Castrol Heavy Duty Lubricants Inc., www.castrolhdl.com

Castrol Industrial North America Inc., www.castrolindustrialna.com

Castrol International, www.castrol.com

Castrol North America, www.castrolusa.com

CAT Products Inc., www.run-rite.com

Caterpillar, www.cat.com

Caterpillar, www.caterpillar.com

Center for Innovation Inc., www.centerforinnovation.com

Center for Tribology Inc. (CETR), www.cetr.com

Centurion Lubricants, www.centurionlubes.com

CEPSA (Spain), www.cepsa.es

Champion Brands LLC, www.championbrands.com

Chart Automotive Group Inc., www.chartauto.com

Chattem Chemicals Inc., www.chattemchemicals.com

Chem Connect, www.chemconnect.com

Chem-EcoI Ltd., www.chem-ecol.com

Chemetall Foote Corporation, www.chemetall.com

Chemical Abstracts Service, www.cas.org

Chemical Resources, www.chemcenter.org

Chemical Week Magazine, www.chemweek.com

Chemicolloid Laboratories Inc., www.colloidmill.com

Chemlube International Inc., www.chemlube.com

Chemsearch Lubricants, www.chemsearch.com

Chemtool Inc./Metalcote, www.chemtool.com

Chevrolet (GM), www.chevrolet.com

Chevron Chemical Company, www.chevron.com

Chevron Oronite, www.chevron.com

Chevron Phillips Chemical Company LP, www.cpchem.com

Chevron Phillips Chemical Company, www.chevron.com

Chevron Products Company Lubricants & Specialties Products, www.chevron.com/lubricants

Chevron Products Company, www.chevron.com

Chevron Texaco, www.chevrontexaco.com

Chevron, www.chevron.com

Christenson Oil, www.christensonoil.com

Chrysler (Mercedes Benz), www.chrysler.com

Ciba Specialty Chemicals Corporation, www.cibasc.com

CITGO Petroleum Corporation, www.citgo.com

Citroen (France), www.citroen.com

Citroen (United Kingdom), www.citroen.co.uk/fleet

Clariant Corporation, www.clariant.com

Clark Refining and Marketing, www.clarkusa.com

CLC Lubricants Company, www.clclubricants.com

Climax Molybdenum Company, www.climaxmolybdenum.com

Coastal Hydraulic Engineering Ltd, www.coastalhydraulics.com/OilsLubricants.htm

Coastal Unilube Inc., www.coastalunilube.com

Cognis, www.cognis-us.com

Cognis, www.cognis.com

Cognis, www.na.cognis.com

College of Petroleum and Energy Studies CPS Home Page, www.colpet.ac.uk/index.html

College of Petroleum and Energy Studies, www.colpet.ac.uk

Colorado Petroleum Products Company, www.colopetro.com

Colorado School of Mines Advanced Coating and Surface Engineering Laboratory (ACSEL), www.mines.edu/research/acsel/acsel.html

Commercial Lubricants Inc., www.comlube.com

Commercial Ullman Lubricants Company, www.culc.com

Commonwealth Oil Corporation, www.commonwealthoil.com

Como Industrial Equipment Inc., www.comoindustrial.com

Como Lube & Supplies Inc., www.comolube.com

Computational Systems Inc., www.compsys.com/index.html

Concord Consulting Group Inc., www.concordcg.com

Condat Corporation, www.condatcorp.com

Conklin Company Inc., www.conklin.com

Conoco, www.conoco.com

Containment Solutions Inc., www.containmentsolutions.com

Coolants Plus Inc., www.coolantsplus.com

Co-ordinating European Council (CEC), www.cectests.org

Coordinating Research Council (CRC), www.crcao.com

Cortec Corporation, www.cortecvci.com

Cosby Oil Company, www.cosbyoil.com

Cosmo Oil, www.cosmo-oil.co.jp

Country Energy, www.countryenergy.com

CPI Engineering Services, www.cpieng.com

CRC Industries Inc., www.crcindustries.com

Creanova Inc., www.creanovainc.com

Crescent Manufacturing, www.crescentmfg.net

Croda Inc., www.croda.com

Crompton Corporation, www.cromptoncorp.com

Crop Science Society of America (CSSA), www.crops.org

Cross Oil Refining and Marketing Inc., www.crossoil.com

Crowley Chemical Company Inc., www.crowleychemical.com

Crystal Inc-PMC, www.pmc-group.com

CSI, www.compsys.com

Cummins Engine Company, www.cummins.com

Custom Metalcraft Inc., www.custom-metalcraft.com

Cyclo Industries LLC, www.cyclo.com

D&D Oil Company Inc., www.amref.com

D.A. Stuart Company, www.d-a-stuart.com

D.B. Becker Company Inc., www.dbbecker.com

D.W. Davies & Company Inc., www.dwdavies.com

D-A Lubricant Company, www.dalube.com

Daimler Chrysler, www.daimlerchrysler.com

Danish Technological Institute (DTI) Tribology Centre, www.dti.dk

Darmex Corporation, www.darmex.com

Darsey Oil Company Inc., www.darseyoil.com

David Weber Oil Company, www.weberoil.com

Davison Oil Company Inc., www.davisonoil.com

Dayco Inc., www.dayco.com

DeForest Enterprises Inc., www.deforest.net

Delkol, www.delkol.co.il

Delphi Automotive Systems, www.delphiauto.com

Dennis Petroleum Company Inc., www.dennispetroleum.com

Department of Defense (DOD), www.defenselink.mil

Des-Case Corporation, www.des-case.com

Detroit Diesel, www.detroitdiesel.com

Deutsches Institute Fur Normung e. V. (DIN), www.din.de

Dexsil Corporation, www.dexsil.com

Dialog, www.dialog.com

Diamond Head petroleum Inc., www.diamondheadpetroleum.com

Diesel Progress, www.dieselpub.com

Digilube Systems Inc., www.digilube.com

Dingo Maintenance Systems, www.dingos.com

Dion & Sons Inc., www.dionandsons.com

Diversified Petrochemical Services, www.chemhelp.com

Division of Machine Elements Home Page Niigata University, Japan, tmtribol.eng.niigata-u.ac.jp/index_e.html

Dixon Lubricants & Special Products Group, Division of Asbury Carbons, www.dixonlube.com

Dodge, www.dodge.com

Don Weese Inc., www.schaefferoil.com

Dover Chemical, www.doverchem.com

Dow Chemical Company, www.dow.com

Dow Corning Corporation, www.dowcorning.com

Dumas Oil Company, www.esn.net/dumasoil

DuPont-Dow Elastomers, www.dupont-dow.com

DuPont Krytox Lubricants, www.lubricants.dupont.com

DuPont, www.dupont.com/intermediates

Duro Manufacturing Inc., www.duromanufacturing.com

Dutton-Lainson Company, www.dutton-lainson.com

Dylon Industries Inc., www.dylon.com

E. I. DuPont de Nemours and Company, www.dupont.com/intermediates

Eagle, www.eaglecars.com

Eastech Chemical Inc., www.eastechchemical.com

Eastern Oil Company, www.easternoil.com

Easy Vac Inc., www.easyvac.com

Ecotech Div., Blaster Chemical Companies, www.pbblaster.com

Edjean Technical Services Inc., www.edjetech.com

EidgenSssische Technische Hochschule (ETH), Zurich Laboratory for Surface Science and Technology (LSST), www.surface.mat.ethz.ch

EKO, www.eko.gr

Elco Corporation, The, www.elcocorp.com

El Paso Corporation, www.elpaso.com

Elementis Specialties-Rheox, www.elemetis-specialties.com

Elf Lubricants North America Inc., www.keystonelubricants.com

Eljay Oil Company Inc., www.eljayoil.com

EMERA Fuels Company Inc., www.emerafuels.com

Emerson Oil Company Inc., www.emersonoil.com

Energy Connection, The, www.energyconnect.com

Engel Metallurgical Ltd., www.engelmet.com

Engen Petroleum Ltd., www.engen.co.za

Engineered Composites Inc., www.engineeredcomposites.net

ENI, www.eni.it

Environmental and Power Technologies Ltd., www.cleanoil.com

Environmental Lubricants Manufacturing Inc. (ELM), www.elmusa.com

Environmental Protection Agency (EPA), www.epa.gov

Equilon Enterprises LLC-Lubricants, www.equilonmotivaequiva.com

Equilon Enterprises LLC-Lubricants, www.shellus.com

Equilon Enterprises LLC-Lubricants, www.texaco.com

Ergon Inc., www.ergon.com

Esco Products Inc., www.escopro.com

Esslingen, Technische Akademie, www.tae.de

Ethyl Corporation, www.ethyl.com

Ethyl Petroleum Additives, www.ethyl.com

ETNA Products Inc., www.etna.com

Etna-Bechem Lubricants Ltd., www.etna.com

European Automobile Manufacturers Association (ACEA), www.acea.be

European Oil Companies Organization of E. H. and S. (CONCAWE), www.concawe.be

European Patent Office, www.epo.co.at/epo/

EV1, www.gmev.com

Evergreen Oil, www.evergreenoil.com

Exxon, www.exxon.com

ExxonMobil Chemical Company, www.exxonmobil chemical.com

ExxonMobil Corp., www.exxonmobil.com

ExxonMobil Industrial Lubricants, www.mobilindustrial.com

ExxonMobil Lubricants & Petroleum Specialties Company, www.exxonmobil.com

F. Bacon Industriel Inc., www.f-bacon.com

F.L.A.G. (Fuel, Lubricant, Additives, Grease) Recruiting, www.flagsearch.com

Fachhochschule Hamburg, Germany, www.haw-hamburg.de/fh/forum/f12/indexf.html/tribologie/etribology.html

Falex Corporation, www.falex.com

Falex Tribology NV, www.falexint.com

Fanning Corporation, The, www.fanncorp.com

Far West Oil Company Inc., www.farwestoil.com

Farmland Industries Inc., www.farmland.com

Federal World, www.fedworld.gov

Ferrari, www.ferrari.com

Ferro/Keil Chemical, www.ferro.com

FEV Engine Technology Inc., www.fev.com

Fiat, www.fiat.com

Fina Oil and Chemical Company, www.totalpetrochemicalsusa.com

Findett Corporation, www.findett.com

Finish Line Technologies Inc., www.finishlineusa.com

FINKE Mineralolwerk, www.finke-mineraloel.de

Finnish Oil and Gas Federation, www.oil.fi

Flamingo Oil Company, www.pinkbird.com

Flo Components Ltd., www.flocomponents.com

Flowtronex International, www.flowtronex.com

Fluid Life Corporation, www.fluidlife.com

Fluid Systems Partners US Inc., www.fsp-us.com

Fluid Technologies Inc., www.fluidtechnologies.com

Fluidtec International, www.fluidtec.com

Fluitec International, www.fluitec.com

FMC Blending & Transfer, www.fmcblending-transfer.com

FMC Lithium, www.fmclithium.com

FMC, www.fmc.com

Ford Motor Company, www.ford.com

Fortum (Finland), www.fortum.com

Forward Corporation, www.forwardcorp.com

Freightliner, www.freightliner.com

Frontier Performance Lubricants Inc., www.frontierlubricants.com

Fuchs Lubricants Company, www.fuchs.com

Fuchs, www.fuchs-oil.de

Fuel and Marine Marketing (FAMM), www.fammllc.com

Fuels and Lubes Asia Publications Inc., www.flasia.inf

Fuki America Corporation, www.fukiamerica.com

Functional Products, www.functionalproducts.com

G-C Lubricants Company, www.gclube.com

G & G Oil Company of Indiana Inc., www.ggoil.com

G. R. O'Shea Company, www.groshea.com

G. T. Autochemilube Ltd., www.gta-oil.co.uk

Galactic, www.galactic.com

Gamse Lithographing Company, www.gamse.com

Gard Corporation, www.gardcorp.com

Gas Tops Ltd., www.gastops.com

Gasco Energy, www.gascoenergy.com

Gateway Additives, www.lubrizol.com

Gear Technology Magazine, www.geartechnology.com

General Motors (GM), www.gm.com

Generation Systems Inc., www.generationsystems.com

Geo. Pfau's Sons Company Inc., www.pfauoil.com

Georgia Tech Tribology, www.me.gatech.edu

Georgia-Pacific Resins Inc.—Actrachem Division, www.gapac.com

Georgia-Pacific Resins Inc.—Actrachem Division, www.gp.com

Global Electric Motor Cars, LLC, www.gemcar.com

Globetech Services Inc., www.globetech-services.com

Glover Oil Company, www.gloversales.com

GMC, www.gmc.com

Gold Eagle Company, www.goldeagle.com

Golden Bear Oil Specialties, www.goldenbearoil.com

Golden Gate Petroleum, www.ggpetrol.com

GoldenWest Lubricants, www.goldenwestlubricants.com

Goldschmidt Chemical Corporation, www.goldschmidt.com

Gordon Technical Service Company, www.gtscofpa.com

Goulston Technologies Inc., www.goulston.com

Graco Inc., www.graco.com

Granitize Products Inc., www.granitize.com

Greenland Corporation, www.greenpluslubes.com

Grignard Company LLC, www.purelube.com

Gulf Oil, www.gulfoil.com

H & W Petroleum Company Inc., www.hwpetro.com

H. L. Blachford Ltd., www.blachford.ca

H.N. Funkhouser & Company, www.hnfunkhouser.com

Haas Corporation, www.haascorp.com

Hall Technologies Inc., www.halltechinc.com

Halocarbon Products Corporation, www.halocarbon.com

Halron Oil Company Inc., www.halron.com

Hammonds Fuel Additives Inc., www.hammondscos.com

Hampel Oil Distributors, www.hampeloil.com

Hangsterfer's Laboratories Inc., www.hangsterfers.com

Harry Miller Corp., www.harrymillercorp.com

Hasco Oil Company Inc., www.hascooil.com

Hatco Corporation, www.hatcocorporation.com

Haynes Manufacturing Company, www.haynesmfg.com

HCI/Worth Chemical Corporation, www.hollandchemical.com

Hedwin Corporation, www.hedwin.com

HEF, France, www.hef.fr

Henkel Surface Technologies, www.henkel.com

Hercules Inc., Aqualon Division, www.herc.com

Herguth Laboratories Inc., www.herguth.com

Heveatex, www.heveatex.com

Hexol Lubricants, www.hexol.com

High Performance Lubricants, www.hplubricants.com

Hindustan Petroleum Corporation Ltd., www.hind-petro.com

Hino Motor Ltd., www.hino.co.jp

Hi-Port Inc., www.hiport.com

Hi-Tech Industries Inc., www.hi-techind.com

Holland Applied Technologies, www.hollandapt.com

Honda (Japan), www.worldhonda.com

Honda (United States), www.honda.com

Hoosier Penn Oil Company, www.hpoil.com

Hoover Materials Handling Group Inc., www.hooveribcs.com

Houghton International Inc., www.houghtonintl.com

How Stuff Works, www.howstuffworks.com/engine.htm

Howes Lubricator, www.howeslube.com

Huntsman Corporation, www.huntsman.com

Huskey Specialty Lubricants, www.huskey.com

Hydraulic Repair & Design Inc., www.h-r-d.com

Hydrocarbon Asia, www.hcasia.safan.com

Hydrocarbon Processing Magazine, www.hydrocarbonprocessing.com

Hydrosol Inc., www.hydrosol.com

Hydrotex Inc., www.hydrotexlube.com

Hy-Per Lube Corporation, www.hyperlube.com

Hysitron Incorporated: Nanomechanics, www.hysitron.com

Hyundai, www.hyundai-motor.com

I.S.E.L. Inc., www.iselinc.com

ICIS-LOR Base Oils Pricing Information, www.icislor.com

Idemitsu, www.idemitsu.co.jp

ILC/Spectro Oils of America, www.spectro-oils.com

Illinois Oil Products Inc., www.illinoisoilproducts.com

Imperial College, London ME Tribology Section, www.me.ic.ac.uk/tribology/

Imperial Oil Company Inc., www.imperialoil.com

Imperial Oil Ltd., www.imperialoil.ca

Imperial Oil Products and Chemicals Division, www.imperialoil.ca

Independent Lubricant Manufacturers Association (ILMA), www.ilma.org

Indian Institute of Science, Bangalore, India, Department of Mechanical Engineering, www.mecheng.iisc.ernet.in

Indian Oil Corporation, www.iocl.com

Indiana Bottle Company, www.indianabottle.com

Industrial Packing Inc., www.industrialpacking.com

Infineum USA LP, www.infineum.com

Infiniti, www.infiniti.com

Ingenieria Sales SA de CV, www.isalub.com

Inolex Chemical Company, www.inolex.com

Innovene, www.innovene.com

Insight Services, www.testoil.com

Institut National des Sciences Appliquees de Lyon, France, Laboratoire de Mechanique des Contacts, www.insa-lyon.fr/Laboratoires/LMC/index.html

Institute of Materials Inc. (IOM), www.savantgroup.com

Institute of Mechanical Engineers (ImechE), www.imeche.org.uk

Institute of Petroleum (IP), www.energyinst.org.uk

Institute of Physics (IOP), Tribology Group, www.iop.org

Instruments for Surface Science, www.omicroninstruments.com/index.html

Interline Resources Corporation, www.interlineresources.com

Internal Energy Agency (IEA), www.iea.org

International Group Inc., The (IGI), www.igiwax.com

International Lubricants Inc., www.lubegard.com

International Organization for Standardization (ISO), www.iso.ch

International Products Corp., www.ipcol.com

Intertek Testing Services-Caleb Brett, www.itscb.com

Invicta a.s., www.testoil.com

Irving Oil Corporation, www.irvingoil.com

ISO Translated into Plain English, connect.ab.ca/praxiom

Israel Institute of Technology (Technion), meeng.technion.ac.il/Labs/energy.htm#tribology

Isuzu, www.isuzu.com

ITW Fluid Products Group, www.itwfpg.com

J & H Oil Company, www.jhoil.com

J & S Chemical Corporation, www.jschemical.com

J.H. Calo Company Inc., www.jhcalo.com

J.R. Schneider Company Inc., www.jrschneider.com

J.A.M. Distributing Company, www.jamdistributing.com

J.B. Chemical Company Inc., www.jbchemical.com

J.B. Dewar Inc., www.jbdewar.com

J.D. Streett & Company Inc., www.jdstreett.com

J.N. Abbott Distributor Inc., www.jnabbottdist.com

Jack Rich Inc., www.jackrich.com

Jaguar, www.jaguarcars.com

Japan Association of Petroleum Technology (JAPT), www.japt.org

Japan Automobile Manufacturers Association (JAMA), www.japanauto.com

Japan Energy Corporation, www.j-energy.co.jp/english

Japan Energy, www.j-energy.co.jp

Japanese Society of Tribologists (JAST) (in Japanese), www.jast.or.jp

Jarchem Industries Inc., www.jarchem.com

Jasper Engineering & Equipment, www.jaspereng.com

Jeep, www.jeep.com

Jenkin-Guerin Inc., www.jenkin-guerin.com

Jet-Lube (United Kingdom) Ltd., www.jetlube.com

John Deere, www.deere.com

Johnson Packings & Industrial Products Inc., www.johnsonpackings.com

Journal of Tribology, engineering.dartmouth.edu/thayer/research/index.html

K.C. Engineering Ltd., www.kceng.com

K.l.S.S. Packaging Systems, www.kisspkg.com

Kafko International Ltd., www.kafkointl.com

Kanazawa University, Japan, Tribology Laboratory, web.kanazawa-u.ac.jp/~tribo/labo5e.html

Kath Fuel Oil Service, www.kathfuel.com

Kawasaki, www.kawasaki.com

Kawasaki, www.khi.co.jp

Keck Oil Company, www.keckoil.com

Kelsan Lubricants USA LLC, www.kelsan.com

Kem-A-Trix Specialty Lubricants & Compounds, www.kematrix.com

Kendall Motor Oils, www.kendallmotoroil.com

Kennedy Group, The, www.kennedygrp.com

King Industries Specialty Chemicals, www.kingindustries.com

Kittiwake Developments Limited, www.kittiwake.com

Kleenoil Filtration Inc., www.kleenoilfiltrationusa.com/index.htm

Kleentek-United Air Specialists Inc., www.uasinc.com

Kline & Company Inc., www.klinegroup.com

Klüber Lubrication North America LP, www.kluber.com

Koehler Instrument Company, www.koehlerinstrument.com

KOST Group Inc., www.kostusa.com

Kruss USA, www.krussusa.com

Kuwait National Petroleum Company, www.knpc.com.kw

Kyodo Yushi USA Inc., www.kyodoyushi.co.jp

Kyushu University, Japan, Lubrication Engineering Home Page, www.mech.kyushu-u.ac.jp/index.html

Lambent Technologies, www.petroferm.com

Lamborghini, www.lamborghini.com

Laub/Hunt Packaging Systems, www.laubhunt.com

Lawler Manufacturing Corporation, www.lawler-mfg.com

Leander Lubricants, www.leanderlube.com

Leding Lubricants Inc., www.automatic-lubrication.com

Lee Helms Inc., www.leehelms.com

Leffert Oil Company, www.leffertoil.com

Leffler Energy Company, www.leffler.com

Legacy Manufacturing, www.legacymfg.com

Les Industries Sinto Racing Inc., www.sintoracing.com

Lexus, www.lexususa.com

Liftomtic Inc., www.liftomatic.com

Lilyblad Petroleum Inc., www.lilyblad.com

Lincoln-Mercury, www.lincolnmercury.com

Linpac Materials Handling, www.linpacmh.com

Liqua-Tek Inc., www.hdpluslubricants.com

Liquid Controls Inc., A Unit of IDEX Corporation, www.lcmeter.com

Liquid Horsepower, www.holeshot.com/chemicals/additives.html

LithChem International, www.lithchem.com

Loos & Dilworth Inc.—Automotive Division, www.loosanddilworth.com

Loos & Dilworth Inc.—Chemical Division, www.loosanddilworth.com

Lormar Reclamation Service, www.lormar.com

Los Alomos National Laboratory, www.lanl.gov/worldview/

Lowe Oil Company/Champion Brands LLC, www.championbrands.com

LPS Laboratories, www.lpslabs.com

LSST Tribology and Surface Forces, www.surface.mat.ethz.ch

LSST Tribology Letters, www.surface.mat.ethz.ch/people/senior-scientists/zstefan/ sitepublications

LubeCon Systems Inc., www.lubecon.com

Lubelink, www.lubelink.com

Lubemaster Corporation, www.lubemaster.com

LubeNet, www.lubenet.com

LubeRos—A Division of Burlington Chemical Company Inc., www.luberos.com

Lubes and Greases, www.lngpublishing.com

LuBest, Division of Momar Inc., www.momar.com

Lubricant Additives Research, www.silverseries.com

Lubricant Technologies, www.lubricanttechnologies.com

Lubricants Network Inc., www.lubricantsnetwork.com

Lubricants USA, www.finalube.com

Lubricants World, www.lubricantsworld.com

Lubrication Engineers Inc., www.le-inc.com

Lubrication Engineers of Canada Ltd., www.lubeng.com

Lubrication Systems, www.lsc.com

Lubrication Technologies Inc., www.lube-tech.com

Lubrication Technology Inc., www.lubricationtechnology.com

Lubrichem International Corporation, www.lubrichem.net

Lubrifiants Distac Inc., www.inspection.gc.ca/english/ppc/reference/n2e.shtml

Lubri-Lab Inc., www.lubrilab.com

LUBRIPLATE Div., Fiske Bros. Refining Company, www.lubriplate.com

Lubriport Labs, www.ultralabs.com/lubriport

Lubriquip Inc., www.lubriquip.com

Lubritec, www.lubritec.com

Lubrizol Corporation, The, www.lubrizol.com

Lubromation Inc., www.lubromation.com

Lub-Tek Petroleum Products Corporation, www.lubtek.com

Lucas Oil Products Inc., www.lucasoil.com

LukOil (Russian Oil Company), www.lukoil.com

Lulea University of Technology, Department of Mechanical Engineering, www.luth.se/depts/mt/me/

Lyondell Lubricants, www.lyondelllubricants.com

Machines Production Web Site, www.machpro.fr

Mack Trucks, www.macktrucks.com

MagChem Inc., www.magchem.com

Magnalube, www.magnalube.com

Maine Lubrication Service Inc., www.mainelube.com

Manor Technology, www.manortec.co.uk

Manor Trade Development Corporation, www.amref.com

Mantek Lubricants, www.mantek.com

Marathon Ashland Petroleum LLC, www.mapllc.com

Marathon Oil Company, www.marathon.com

MARC-IV, www.marciv.com

Marcus Oil & Chemical, www.marcusoil.com

Markee International Corporation, www.markee.com

Marly, www.marly.com

Maryn International Ltd., www.maryngroup.com

Maryn International, www.poweruplubricants.com

Master Chemical Corporation, www.masterchemical.com

Master Lubricants Company, www.lubriko.com

Maxim Industrial Metalworking Lubricants, www.maximoil.com

Maxima Racing Lubricants, www.maximausa.com

Mays Chemical Company, www.mayschem.com

Mazda, www.mazda.com

McCollister & Company, www.mccollister.com

McGean-Rohco Inc., www.mcgean-rohco.com

McGee Industries Inc., www.888teammclube.com

McIntyre Group Ltd., www.mcintyregroup.com

McLube Division/McGee Industries Inc., www.888teammclube.com

Mechanical Engineering Magazine, www.memagazine.org/index.html

Mechanical Engineering Tribology Website, widget.ecn.purdue.edu/<metrib/

Mega Power Inc., www.megapowerinc.com

Mercedes-Benz (Germany), www.mercedes-benz.de

Metal Forming Lubricants Inc., www.mflube.com

Metal Mates Inc., www.metalmates.net

Metalcote/Chemtool Inc., www.metalcote.com

Metalworking Lubricants Company, www.metalworkinglubricants.com

Metalworking Lubricants, www.maximoil.com

Metorex Inc., www.metorex.fi

Mettler Toledo, www.mt.com

MFA Oil Company, www.mfaoil.com

Michel Murphy Enterprises Inc., www.michelmurphy.com

Micro Photonics Inc., www.microphotonics.com

Mid-Michigan Testing Inc., www.tribologytesting.com

Mid-South Sales Inc., www.mid-southsales.com

Mid-Town Petroleum Inc., www.midtownoil.com

Migdal's Lubricant Web Page, members.aol.com/sirmigs/lub.htm

Milacron Consumable Products Division, www.milacron.com

Milatec Corporation, www.militec.com

Mitsubishi Motors, www.mitsubishi-motors.co.jp

Mobil, www.mobil.com

MOL Hungarian Oil & Gas, www.mol.hu

Molyduval, www.molyduval.com

Molyslip Atlantic Ltd., www.molyslip.co.uk

Monlan Group, www.monlangroup.com

Monroe Fluid Technology Inc., www.monroefluid.com

Moore Oil Company, www.mooreoil.com

Moraine Packaging Inc., www.hdpluslubricants.com

Moroil Technologies, www.moroil.com

Motiva Enterprises LLC, www.motivaenterprises.com

Motorol Lubricants, www.motorolgroup.com

Motul USA Inc., www.motul.com

Mozel Inc., www.mozel.com

Mr. Good Chem Inc., www.mrgoodchem.com

Murphy Oil Corporation, www.murphyoilcorp.com

Muse, Stancil & Company, www.musestancil.com

Nalco Chemical Company, www.nalco.com

NanoTribometer System, www.ume.maine.edu/LASST

Naptech Corporation, www.satec.com

NASA Lewis Research Center (LeRC) Tribology & Surface Science Branch, www.lerc.nasa.gov/Other_Groups/SurfSci

National Centre of Tribology, United Kingdom, www.acat/net

National Fluid Power Association (NFPA), www.nfpa.com

National Institute for Occupational Safety and Health, www.cdc.gov/homepage.html

National Institute of Standards and Technology, webbook.nist.gov/chemistry

National Lubricating Grease Institute (NLGI), www.nlgi.org

National Metal Finishing Resource Center, www.nmfrc.org

National Oil Recyclers Association (NORA), www.recycle.net/Associations/rs000141.html

National Petrochemical & Refiners Association (NPRA), www.npradc.org

National Petroleum News, www.npnweb.com

National Petroleum Refiners Association (NPRA), www.npra.org

National Resource for Global Standards, www.nssn.org

National Tribology Services, www.natrib.com

Naval Research Lab Tribology Section—NRL Code 6176, stm2.nrl.navy.mil/~wahl/6176.htm

NCH, www.nch.com

Neale Consulting Engineers Limited, www.tribology.co.uk

Neo Synthetic Oil Company Inc., www.neosyntheticoil.com

Newcomb Oil Company, www.newcomboil.com

Niagara Lubricant Company Inc., www.niagaralubricant.com

Nissan (Japan), www.nissan.co.jp

Nissan (United States), www.nissandriven.com

Nissan (United States), www.nissanmotors.com

NOCO Energy Corporation, www.noco.com

Noco Lubricants, www.noco.com

Nordstrom Valves Inc., www.nordstromaudco.com

Noria —OilAnalysis.Com, www.oilanalysis.com

Northern Technologies International Corporation, www.ntic.com

Northwestern University, Tribology Lab, cset.mech. northwestern.edu/member.htm

Nyco SA, www.nyco.fr

Nye Lubricants, www.nyelubricants.com

Nynas Naphthenics, www.nynas.com

O'Rourke Petroleum, www.orpp.com

Oak Ridge National Laboratory (ORNL) Tribology Test Systems, www.html.ornl.gov/mituc/tribol.htm

Oakite Products Inc., www.oakite.com

OATS (Oil Advisory Technical Services), www.oats. co.uk

Occidental Chemical Corporation, www.oxychem.com

Occupational Safety and Health Administration (OSHA), www.osha.gov

Ocean State Oil Inc., www.oceanstateoil.com

Oden Corporation, www.odencorp.com

Oil Analysis (Noria), www.oilanalysis.com

Oil Center Research Inc., www.oilcenter.com

Oil Depot, www.oildepot.com

Oil Directory.com, www.oildirectory.com

Oil Distributing Company, www.oildistributing.com

Oil Online, www.oilonline.com

Oil-Chem Research Corporation, www.avblend.com

Oilkey Corporation, www.oilkey.com

Oil-Link Oil & Gas Online, www.oilandgasonline.com

Oilpure Technologies Inc., www.oilpure.com

Oilspot.com, www.oilspot.com

OKS Speciality Lubricants, www.oks-india.com

OMGI, www.omgi.com

OMICRON Vakuumphysik GmbH, www.omicronin-struments.com/index.html

Omni Specialty Packaging, www.nuvo.cc

OMS Laboratories Inc., members.aol.com/labOMS/ index.html

Opel, www.opel.com

Orelube Corporation, www.orelube.com

Oronite, www.oronite.com

O'Rourke Petroleum Products, www.orpp.com

Ottsen Oil Company Inc., www.ottsen.com

Owens-Illinois Inc., www.o-i.com

Oxford Instruments Inc., www.oxinst.com

Paper Systems Inc., www.paper-systems.com

Paramount Products, www.paramountproducts.com

PARC Technical Services Inc., www.parctech.com

Parent Petroleum, www.parentpetroleum.com

PATCO Additives Division—American Ingredients Company, www.patco-additives.com

Pathfinder Lubricants, www.pathfinderlubricants.ca

Patterson Industries Ltd. (Canada), www.patterson-industries.com

PBM Services Company, www.pbmsc.com

PdMA Corporation, www.pdma.com

PDVSA (Venezuela), www.pdvsa.com

PED Inc., www.ped.vianet.ca

PEMEX (Mexico), www.pemex.com

Pennine Lubricants, www.penninelubricants.co.uk

Pennwell Publications, www.pennwell.com

Pennzoil, www.pennzoil.com

Pennzoil-Quaker State Company, www.pennzoilquaker-state.com

PENRECO, www.penreco.com

Penta Manufacturing Company/Division of Penta International Corporation, www.pentamfg.com

Performance Lubricants & Race Fuels Inc., www. performanceracefuels.com

Perkin Elmer Automotive Research, www.perkinelmer. com/ar

Perkins Products Inc., www.perkinsproducts.com

Pertamina (Indonesia), www.pertamina.com

Petro Star Lubricants, www.petrostar.com

PetroBlend Corporation, www.petroblend.com

Petrobras (Brazil), www.petrobras.com.br

Petro-Canada Lubricants, www.htlubricants.com

Petrofind.com, www.petrofind.com

Petrogal (Portugal), www.petrogal.pt

Petrolab Corporation, www.petrolab.com

Petrolabs Inc., pages.prodigy.net/petrolabsinc

Petroleum Analyzer Company LP (PAC), www.petro-leum-analyzer.com

Petroleum Authority, Thailand, www.nectec.or.th

Petroleum Authority of Thailand, www.nectec.or.th/ users/htk/SciAm/12PTT.html

Petroleum Marketers Association of America (PMAA), www.pmaa.org

Petroleum Packers Inc., www.pepac.com

Petroleum Products Research, www.swri.org/4org/d08/ petprod/

Petroleumworld.com, www.petroleumworld.com

Petro-Lubricants Testing Laboratories Inc., www.pltlab. com

PetroMin Magazine, www.petromin.safan.com

PetroMoly Inc., www.petromoly.com

Petron Corporation, www.petroncorp.com

Petroperu (Peru), www.petroperu.com

Petrotest, www.petrotest.net

Peugeot, www.peugeot.com

Pfaus Sons Company Inc., www.pfauoil.com

Pflaumer Brothers Inc., www.pflaumer.com

Philips Industrial Electronics Deutschland, www. philips-tkb.com

Phillips Petroleum Company/Phillips 66, www. phillips66.com/phillips66.asp

Phoenix Petroleum Company, www.phoenixpetroleum. com

Pico Chemical Corporation, www.picochemical.com

Pilot Chemical Company, www.pilotchemical.com

Pinnacle Oil Inc., www.pinnoil.com

Pipeguard of Texas, www.pipeguard-texas.com

Pitt Penn Oil Company, www.pittpenn.com

Plastic Bottle Corporation, www.plasticbottle.com

Plastican Inc., www.plastican.com

Plews/Edelmann Division, Stant Corporation, www.stant.com

PLI LLC, www.memolub.com

Plint and Partners: Tribology Division, www.plint.co.uk/trib.htm

Plymouth, www.plymouthcars.com

PMC Specialties Inc., www.pmcsg.com

PolimeriEuropa, www.polimerieuropa.com

PoIySi Technologies Inc., www.polysi.com

Polaris Laboratories, LLC, www.polarislabs.com

Polar Company, www.polarcompanies.com

Polartech Ltd., www.polartech.co.uk

Pontiac (GM), www.pontiac.com

Power Chemical, www.warcopro.com

Practicing Oil Analysis Magazine, www.practicingoilanalysis.com

Precision Fluids Inc., www.precisionfluids.com

Precision Industries, www.precisionind.com

Precision Lubricants, www.precisionlubricants.com

PREDICT/DLI—Innovative Predictive Maintenance, www.predict-dli.com

Predictive Maintenance Corporation, www.pmaint.com

Predictive Maintenance Services, www.theoillab.com

Premo Lubricant Technologies, www.premolube.com

Prime Materials, www.primematerials.com

Primrose Oil Company Inc., www.primrose.com

Probex Corporation, www.probex.com

Products Development Manufacturing Company, www.veloil.com

ProLab TechnoLub Inc., www.prolab-technologies.com

ProLab-Bio Inc., www.prolab-lub.com

Prolong Super Lubricants, www.prolong.com

ProTec International Inc., www.proteclubricants.com

Pulsair Systems Inc., www.pulsair.com

Purac America Inc., www.purac.com

Purdue University Materials Processing and Tribology Research Group, www.ecn.purdue.edu/<farrist/lab.html

Pure Power Lubricants, www.gopurepower.com

QMI, www.qminet.com

Quaker Chemical Corporation, www.quakerchem.com

Quaker State, www.qlube.com

R.A. Miller & Company Inc., www.ramiller.on.ca

R.T. Vanderbilt Company Inc., www.rtvanderbilt.com

R.E. Carroll Inc., www.recarroll.com

R.E.A.L. Services, www.realservices.com

R.H. Foster Energy LLC, www.rhfoster.com

Radian Inc., www.radianinc.com

Radio Oil Company Inc., www.radiooil.com

Ramos Oil Company Inc., www.ramosoil.com

Rams-Head Company, www.doall.com

Ransome CAT, www.ransome.com

Ravenfield Designs Ltd., www.ravenfield.com

Reade Advanced Materials, www.reade.com

Red Giant Oil Company, www.redgiantoil.com

Red Line Oil, www.redlineoil.com

Reed Oil Company, www.reedoil.com

Reelcraft Industries Inc., www.realcraft.com

Reit Lubricants Company, www.reitlube.com

Reitway Enterprises Inc., www.reitway.com

Reliability Magazine, www.pmaint.com/tribo/docs/oil_anal/tribo_www.html

Renewable Lubricants Inc., www.renewablelube.com

Renite Company, www.renite.com

Renite Company—Lubrication Engineers, www.renite.com

Renkert Oil, www.renkertoil.com

Rensberger Oil Company Inc., www.rensbergeroil.com

Rexam Closures, www.closures.com

Rhein Chemie Corporation, www.bayer.com

Rhein Chemie Rheinau GmbH, www.rheinchemie.com

Rheotek (PSL SeaMark), www.rheotek.com

Rheox Inc., www.rheox.com

Rhodia, www.rhodia.com

Rhone-Poulenc Surfactants & Specialties, www.rpsurfactants.com

Ribelin, www.ribelin.com

RiceChem, A Division of Stilling Enterprises Inc., www.ricechem.com

RichardsApex Inc., www.richardsapex.com

Riley Oil Company, www.rileyoil.com

RO-59 Inc., members.aol.com/ro59inc

Rock Valley Oil & Chemical Company, www.rockvalleyoil.com

Rocol Ltd., www.rocol.com

Rohm & Haas Company, www.rohmhaas.com

RohMax Additives GmbH, www.rohmax.com

Ross Chem Inc., www.rosschem.com

Rowleys Wholesale, www.rowleys.com

Royal Institute of Technology (KTH), Sweden Machine Elements Home Page, www.damek.kth.se/mme

Royal Lubricants Inc., www.royallube.com

Royal Manufacturing Company Inc., www.royalube.com

Royal Purple Inc., www.royalpurple.com

Russell-Stanley Corporation, www.russell-stanley.com

RWE-DEA (Germany), www.rwe-dea.de

RyDol Products, www.rydol.com

Saab, www.saab.com

Saab Cars USA, www.saabusa.com

Safety Information Resources on the Internet, www.siri.org/links1.html

Safety-Kleen Corporation, www.safety-kleen.com

Safety-Kleen Oil Recovery, www.ac-rerefined.com

Saitama University, Japan Home Page of Machine Element Laboratory, www.mech.saitama-u.ac.jp/youso/home.html

San Joaquin Refining Company, www.sjr.com

Sandia National Laboratories Tribology, www.sandia.gov/materials/sciences/

Sandstrom Products Company, www.sandstromproducts.com

Sandy Brae Laboratories Inc., www.sandy/brae.com

Santie Oil Company, www.santiemidwest.com

Santotrac Traction Lubricants, www.santotrac.com

Santovac Fluids Inc., www.santovac.com

Sasol (South Africa), www.sasol.com

SATEC Inc., www.satec.com

Saturn (GM), www.saturncars.com

Savant Group of Companies, www.savantgroup.com

Savant Inc., www.savantgroup.com

Saxton Industrial Inc., www.schaefferoil.com

Scania, www.scania.se

Schaeffer Manufacturing, www.schaefferoil.com

Schaeffer Oil and Grease, www.schaefferoil.com

Schaeffer Specialized Lubricants, www.schaefferoil. com

Scully Signal Company, www.scully.com

Sea-Land Chemical Company, www.sealandchem.com

Selco Synthetic Lubricants, www.synthetic-lubes.com

Senior Flexonics, www.flexonics-hose.com

Sentry Solutions Ltd., www.sentrysolutions.com

Sexton & Peake Inc., www.sexton.qpg.com

SFR Corporation, www.sfrcorp.com

SGS Control Services Inc., www.sgsgroup.com

Shamrock Technologies Inc., www.shamrocktechnologies.com

Share Corp., www.sharecorp.com

Shell, www.shellus.com

Shell (United States), www.shellus.com

Shell Chemicals, www.shellchemical.com

Shell Global Solutions, www.shellglobalsolutions.com

Shell International, www.shell.com/royal-en

Shell Lubricants (United States), www.shell-lubricants. com

Shell Oil Products US, www.shelloilproductsus.com

Shepherd Chemical Company, www.shepchem.com

Shrieve Chemical Company, www.shrieve.com

Silvas Oil Company Inc., www.silvasoil.com

Silverson Machines Inc., www.silverson.com

Simons Petroleum Inc., www.simonspetroleum.com

Sinclair Oil Corporation, www.sinclairoil.com

Sinopec (China Petrochemical Corporation), www. sinopec.com.cn

SK Corporation (Houston Office), www.skcorp.com

SKF Quality Technology Centre, www.qtc.skf.com

Sleeveco Inc., www.sleeveco.com

Slick 50 Corporation, www.slick50.com

Snyder Industries, www.snydernet.com

Sobit International Inc., www.sobitinc.com

Society of Automotive Engineers (SAE), www.sae.org

Society of Environmental Toxicology and Chemistry (SETAC), www.setac.org

Society of Manufacturing Engineers (SME), www.sme. org

Society of Tribologists and Lubrication Engineers (STLE), www.stle.org

Soltex, www.soltexinc.com

Sourdough Fuel, www.petrostar.com

Southern Illinois University, Carbondale Center for Advanced Friction Studies, www.frictioncenter.com

Southwest Grease Products, www.stant.com/brochure. cfm?brochure=155&location_id=119

Southwest Research Institute (SwRI) Engine Technology Section, www.swri.org/default.htm

Southwest Research Institute, www.swri.org

Southwest Spectro-Chem Labs, www.swsclabs.com

Southwestern Graphite, www.asbury.com

Southwestern Petroleum Corporation (SWEPCO), www.swepco.com

Southwestern Petroleum Corporation, www.swepcousa. com

SP Morell & Company, www.spmorell.com

Spacekraft Packaging, www.spacekraft.com

Spartan Chemical Company Inc. Industrial Products Group Division, www.spartanchemical.com

Spartan Oil Company, www.spartanonline.com

Specialty Silicone Products Inc., www.sspinc.com

Spectro Oils of America, www.goldenspectro.com

Spectro Oils of America, www.spectro-oils.com

SpectroInc. Industrial Tribology Systems, www. spectroinc.com

Spectronics Corporation, www.spectroline.com

Spectrum Corp., www.spectrumcorporation.com

Spencer Oil Company, www.spenceroil.com

Spex CertiPrep Inc., www.spexcsp.com

SQM North America Corporation, www.sqmna.com

St. Lawrence Chemical, www.stlawrencechem.com

Star Brite, www.starbrite.com

State University of New York, Binghamton Mechanical Engineering Laboratory, www.me.binghamton.edu

Statoil (Norway), www.statoil.com

Steel Shipping Containers Institute, www.steelcontainers.com

Steelco Industrial Lubricants Inc., www.steelcolubricants. com

Steelco Northwest Distributors, www.steelcolubricants. com

STP Products Inc., www.stp.com

Stratco Inc., www.stratco.com

Suburban Oil Company Inc., www.suburbanoil.com

Summit Industrial Products Inc., www.klsummit.com

Summit Technical Solutions, www.lubemanagement. com

Sunnyside Corporation, www.sunnysidecorp.com

Sunoco Inc., www.sunocoinc.com

Sunohio, Division of ENSR, www.sunohio.com

Superior Graphite Company, www.superiorgraphite. com

Superior Lubricants Company Inc., www.superior lubricants.com

Superior Lubrication Products, www.s-l-p.com

Surtec International Inc., www.surtecinternational.com

Swiss Federal Laboratories for Materials Testing and Research (EMPA) Centre for Surface Technology and Tribology, www.empa.ch

Synco Chemical Corporation, www.super-lube.com

SynLube Inc., www.synlube.com

Synthetic Lubricants Inc., www.synlube-mi.com

Syntroleum Corporation, www.syntroleum.com

T.S. MoIy-Lubricants Inc., www.tsmoly.com

T W Brown Oil Company Inc., www.brownoil.com/soypower.html

Taber Industries, www.taberindustries.com

TAI Lubricants, www.lubekits.com

Tannas Company, www.savantgroup.com

Tannis Company, www.savantgroup.com/tannas.sht

Technical Chemical Company (TCC), www.technical-chemical.com

Technical University of Delft, Netherlands Laboratory for Tribology, www.tudelft.nl

Technical University, Munich, Germany, www.fzg.mw.tu-muenchen.de

Tek-5 Inc., www.tek-5.com

Terresolve Technologies, www.terresolve.com

Texaco Inc., www.texaco.com

Texas Refinery Corporation, www.texasrefinery.com

Texas Tech University, Tribology, www.ttu.edu

Textile Chemical Company Inc., www.textilechem.com

The Maintenance Council, www.trucking.org

Thermal-Lube Inc., www.thermal-lube.com

Thermo Elemental, www.thermoelemental.com

Thomas Petroleum, www.thomaspetro.com

Thornley Company, www.thornleycompany.com

Thoughtventions Unlimited Home Page, www.tvu.com/%7Ethought/

Tiodize Company Inc., www.tiodize.com

Titan Laboratories, www.titanlab.com

TMC, www.truckline.com

Tokyo Institute of Technology, Japan Nakahara Lab. Home Page, www.mep.titech.ac.jp/data/labs_staffE.html

Tomah Products Inc., www.tomah3.com

Tom-Pac Inc., www.tom-pac.com

Top Oil Products CompanyLtd.., www.topoil.com

Torco International Corporation, www.torcoracingoils.com

Tosco, www.tosco.com

Total, www.total.com

Total, www.totalfinaelf.com/ho/fr/index.htm

Totalfina Oleo Chemicals, www.totalfina.com

Tower Oil & Technology Company, www.toweroil.com

Toyo Grease Manufacturing (M) SND BHD, www.toyointernational.com

Toyota (Japan), www.toyota.co.jp

Toyota (United States), www.toyota.com

TransMontaigne, www.transmontaigne.com

Trans Mountain Oil Company, www.transmountainoil.com

Tribologist.com, www.wearcheck.com/sites.html

Tribology Group, www.msm.cam.ac.uk/tribo/tribol.htm

Tribology International, www.elsevier.nl/inca/publications/store/3/0/4/7/4/

Tribology Letters, www.kluweronline.com/issn/1023-8883

Tribology Research Review 1992–1994, www.me.ic.ac.uk/department/review94/trib/tribreview.html

Tribology Research Review 1995–1997, www.me.ic.ac.uk/department/review97/trib/tribreview.html

Tribology/Tech-Lube, www.tribology.com

Tribos Technologies, www.tribostech.com

Trico Manufacturing Corporation, www.tricomfg.com

Trilla Steel Drum Corporation, www.trilla.com

Troy Corporation, www.troycorp.com

Tsinghua University, China, State Key Laboratory of Tribology, www.pim.tsinghua.edu.cn

TTi's Home Page, www.tti-us.com

Turmo Lubrication Inc., www.lubecon.com

TXS Lubricants Inc., www.txsinc.com

U.S. Data Exchange, www.usde.com

U.S. Department of Energy (DOE), www.energy.gov

U.S. Department of Transportation (DOT), www.dot.gov

U.S. Energy Information Administration, www.eia.doe.gov

U.S. Oil Company Inc., www.usoil.com

U.S. Patent Office and Trademark Office, www.uspto.gov

UEC Fuels and Lubrication Laboratories, www.uec-usx.com

Ultimate Lubes, www.ultimatelubes.com

Ultra Additives Inc., www.ultraadditives.com

Ultrachem Inc., www.ultracheminc.com

Unilube Systems Ltd., www.unilube.com

Unimark Oil Company, www.gardcorp.com

Union Carbide Corporation, www.unioncarbide.com

Uniqema, www.uniqema.com

Uniroyal Chemical Company Inc., www.uniroyalchemical.com

UniSource Energy Inc., www.unisource-energy.com

Unist Inc., www.unist.com

Unit Pack Company Inc., www.unitpack.com

United Color Manufacturing Inc., www.unitedcolor.com

United Lubricants, www.unitedlubricants.com

United Oil Company Inc., www.duralene.com

United Soybean Board, www.unitedsoybean.org

Universal Lubricants Inc., www.universallubes.com

University of Applied Sciences, Hamburg, Germany, www.haw-hamburg.de

University of California, Berkeley Bogey's Tribology Group, cml.berkeley.edu/tribo.html

University of California, San Diego Center for Magnetic Recording Research, orpheus.ucsd.edu/cmrr/

University of Illinois, Urbana-Champaign Tribology Laboratory, www.mie.uiuc.edu

University of Kaiserslautern, Germany Sektion Tribologie, www.uni-kl.de/en/

University of Leeds, M.Sc. (Eng.) Course in Surface Engineering and Tribology, leva.leeds.ac.uk/tribology/msc/tribmsc.html

University of Leeds, United Kingdom, Research in Tribology, leva.leeds.ac.uk/tribology/research.html

University of Ljubljana, Faculty of Mechanical Engineering, Center for Tribology and Technical Diagnostics, www.ctd.uni-lj.si/eng/ctdeng.htm

University of Maine Laboratory for Surface Science and Technology (LASST), www.ume.maine.edu/LASST/

University of Newcastle upon Tyne, United Kingdom, Ceramics Tribology Research Group, www.ncl.ac.uk/materials/materials/resgrps/certrib.html

University of Northern Iowa, www.uni.edu/abil

University of Notre Dame Tribology/Manufacturing Laboratory, www.nd.edu/<ame

University of Pittsburg, School of Engineering, Mechanical Engineering Department, www.engrng.pitt.edu

University of Sheffield, United Kingdom, Tribology Research Group, www.shef.ac.uk/mecheng/tribology/

University of Southern Florida, Tribology, www.eng.usf.edu

University of Texas at Austin, Petroleum & Geosystems Engineering, Reading Room, www.pe.utexas.edu/Dept/Reading/petroleum.html

University of Tokyo, Japan, Mechanical Engineering Department, www.mech.t.u-tokyo.ac.jp/english/index.html

University of Twente, Netherlands Tribology Group, www.wb.utwente.nl/en/index.html

University of Western Australia Department of Mechanical and Material Engineering, www.mech.uwa.edu.au/tribology/

University of Western Ontario, Canada Tribology Research Centre, www.engga.uwo.ca/research/tribology/Default.htm

University of Windsor, Canada, Tribology Research Group, venus.uwindsor.ca/research/wtrg/ index.html

Unocal Corporation, www.unocal.com

Uppsala University, Sweden Tribology Group, www.angstrom.uu.se/materials/index.htm

U.S. Department of Agriculture (USDA), www.usda.gov

U.S. Department of Energy (DOE), www.energy.gov

U.S. Department of Defense (DOD), www.dod.gov

USX Engineers & Consultants, www.uec.com/labs/ctns

USX Engineers and Consultants: Laboratory Services, www.uec.com/labs/

Vacudyne Inc., www.vacudyne.com

Valero Marketing & Supply, www.valero.com

Valvoline Canada, www.valvoline.com

Valvoline, www.valvoline.com

Van Horn, Metz & Company Inc., www.vanhornmetz.com

Vauxhall, www.vauxhall.co.uk

Vesco Oil Corporation, www.vesco-oil.com

Victoria Group Inc., The, www.victoriagroup.com

Viking Pump Inc., A Unit of IDEX Corporation, www.vikingpump.com

Vikjay Industries Inc., www.vikjay.com

Virtual Oil Inc., www.virtualoilinc.com

Viswa Lab Corporation, www.viswalab.com

Vogel Lubrication System of America, www.vogel-lube.com

Volkswagen (Germany), www.vw-online.de

Volkswagen (United States), www.vw.com

Volvo (Sweden), www.volvo.se

Volvo Cars of North America, www.volvocars.com

Volvo Group, www.volvo.com

Vortex International LLC, www.vortexfilter.com

VP Racing Fuels Inc., www.vpracingfuels.com

Vulcan Oil & Chemical Products Inc., www.vulcanoil.com

Vulsay Industries Ltd., www.vulsay.com

Wallace, www.wallace.com

Wallover Oil Company, www.walloveroil.com

Walthall Oil Company, www.walthall-oil.com

Warren Distribution, www.wd-wpp.com

Waugh Controls Corporation, www.waughcontrols.com

WD-40 Company, www.wd40.com

Web-Valu Intl., www.webvalu.com

Wedeven Associates Inc., members.aol.com/wedeven/

West Central Soy, www.soypower.net

West Penn Oil Company Inc., www.westpenn.com

Western Michigan University Tribology Laboratory, www.mae.wmich.edu/labs/Tribology/Tribology.html

Western Michigan University, Department of Mechanical and Aeronautical Engineering, www.mae.wmich.edu

Western States Oil, www.lubeoil.com

Western States Petroleum Association, www.wspa.org

Whitaker Oil Company Inc., www.whitakeroil.com

Whitmore Manufacturing Company, www.whitmores.com

Wilks Enterprise Inc., www.wilksir.com

Winfield Brooks Company Inc., www.tapfree.com

Winzer Corp., www.winzerusa.com

Witco (Crompton Corporation), www.witco.com

Wolf Lake Terminals Inc., www.wolflakeinc.com

Worcester Polytechnic Institute, Department of Mechanical Engineering, www.me.wpi.edu/Research/labs.html

Worldwide PetroMoly Inc., www.petromoly.com

WSI Chemical Inc., www.wsi-chem-sys.com

WWW Tribology Information Service, www.shef.ac.uk/<mpe/tribology/

WWW Virtual Library: Mechanical Engineering, www.vlme.com/

Wynn Oil Company, www.wynnsusa.com

X-1R Corporation, The, www.x1r.com

Yahoo Lubricants, dir.yahoo.com/business_and_economy/shopping_and_services/automotive/supplies/lubricants/

Yahoo Tribology, ca.yahoo.com/Science/Engineering/
    Mechanical_Engineering/Tribology/
Yocum Oil Company Inc., www.yocumoil.com
YPF (Argentina), www.ypf.com.ar
Yuma Industries Inc., www.yumaind.com
Zimmark Inc., www.zimmark.com
Zinc Corporation of America, www.zinccorp.com

35.2 INTERNET LISTINGS BY CATEGORY

35.2.1 LUBRICANT FLUIDS

2V Industries Inc., www.2vindustries.com
49 North, www.49northlubricants.com
76 Lubricants Company, www.tosco.com
A/R Packaging Corporation, www.arpackaging.com
Acculube, www.acculube.com
Accurate Lubricants & Metalworking Fluids Inc. (dba
    Acculube), www.acculube.com
Acheson Colloids Company, www.achesonindustries.
    com
Acme Refining, Division of Mar-Mor Inc., www.
    acmerefining.com
Acme-Hardesty Company, www.acme-hardesty.com
Advanced Ceramics Corporation, www.advceramics.
    com
Advanced Lubrication Specialties Inc., www.advanced-
    lubes.com
Aerospace Lubricants Inc., www.aerospacelubricants.
    com
African Lubricants Industry, www.mbendi.co.za/aflu.
    htm
AG Fluiropolymers USA Inc., www.fluoropolymers.
    com
Airosol Company Inc., www.airosol.com
Akzo Nobel, www.akzonobel.com
Alco-Metalube Company, www.alco-metalube.com
Alithicon Lubricants, Division of Southeast Oil &
    Grease Company Inc., www.alithicon.com
Allegheny Petroleum Products Company, www.oils.
    com
Allen Oil Company, www.allenoil.com
Allied Oil & Supply Inc., www.allied-oil.com
Allied Washoe, www.alliedwashoe.com
Alpha Grease & Oil Inc., www.alphagrease.thomas
    register.com/olc/alphagrease/
ALT Inc., www.altboron.com
Amalie Oil Company, www.amalie.com
Amber Division of Nidera Inc., www.nidera-us.com
Amcar Inc., www.amcarinc.com
Amerada Hess Corporation, www.hess.com
American Agip Company Inc., www.americanagip.com
American Eagle Technologies Inc., www.frictionrelief.
    com
American Lubricants Inc., www.americanlubricants
    bflo.com
American Lubricating Company, www.alcooil.com

American Oil & Supply Company, www.aosco.com
American Petroleum Products, www.americanpetro-
    leum.com
American Refining Group Inc., www.amref.com
American Synthol Inc., www.americansynthol.com
Amptron Corporation, www.supersllpperystuff.com/
    organisation.htm
Amrep Inc., www.amrep.com
AMSOIL Inc., www.amsoil.com
Anderol Specialty Lubricants, www.anderol.com
Anti Wear 1, www.dynamicdevelopment.com
Apollo America Corporation, www.apolloamerica.com
Aral International, www.aral.com
Arch Chemicals Inc., www.archbiocides.com
ARCO, www.arco.com
Arizona Chemical, www.arizonachemical.com
Asbury Carbons Inc.—Dixon Lubricants, www.asbury.
    com
Asbury Graphite Mills Inc., www.asbury.com
Asheville Oil Company Inc., www.ashevilleoil.com
Ashland Chemical, www.ashchem.com
Ashland Distribution Company, www.ashland.com
Aspen Chemical Company, www.aspenchemical.com
Associated Petroleum products, www.associated-
    petroleum.com
Atlantis International Inc., www.atlantis-usa.com
Atlas Oil Company, www.atlasoil.com
ATOFINA Canada Inc., www.atofinacanada.com
Ausimont, www.ausiusa.com
Avatar Corporation, www.avatarcorp.com
Badger Lubrication Technologies Inc., www.badger
    lubrication.com
BALLISTOL USA, www.ballistol.com
Battenfeld Grease and Oil Corporation of New York,
    www.battenfeld-grease.com
Behnke Lubricants/JAX, www.jaxusa.com
Behnke Lubricants Inc./JAX, www.jax.com
Bell Additives Inc., www.belladditives.com
Bel-Ray Company Inc., www.belray.com
Benz Oil Inc., www.benz.com
Berry Hinckley Industries, www.berry-hinckley.com
Bestolife Corporation, www.bestolife.com
BG Products Inc., www.bgprod.com
Big East Lubricants Inc., www.bigeastlubricants.com
Blaser Swisslube, www.blaser.com
Bodie-Hoover Petroleum Corporation, www.bodie-
    hoover.com
Boehme Filatex Inc., www.boehmefilatex.com
BoMac Lubricant Technologies Inc., www.boma-
    clubetech.com
Boncosky Oil Company, www.boncosky.com
Boswell Oil Company, www.boswelloil.com
BP Amoco Chemicals, www.bpamocochemicals.com
BP Lubricants, www.bplubricants.com
BP, www.bptechchoice.com
BP, www.bppetrochemicals.com
Brascorp North America Ltd., www.brascorp.on.ca

Brenntag Northeast Inc., www.brenntag.com

Brenntag, www.brenntag.com

Briner Oil Company, www.brineroil.com

British Petroleum (BP), www.bp.com

Britsch Inc., www.britschoil.com

Brugarolas SA, www.brugarolas.com/english.htm

Buckley Oil Company, www.buckleyoil.com

BVA Oils, www.bvaoils.com

Callahan Chemical Company, www.calchem.com

Caltex Petroleum Corporation, www.caltex.com

Calumet Lubricants Company, www.calumetlub.com

Calvary Industries Inc., www.calvaryindustries.com

CAM2 Oil Products Company, www.cam2.com

Canner Associates Inc., www.canner.com

Capital Enterprises (Power-Up Lubricants), www.nnl690.com

Cargill-Industrial Oil & Lubricants, www.techoils.cargill.com

Cary Company, www.thecarycompany.com

CasChem Inc., www.cambrex.com

Castle Products Inc., www.castle-comply.com

Castrol Heavy Duty Lubricants Inc., www.castrolhdl.com

Castrol Industrial North America Inc., www.castrolindustrialna.com

Castrol International, www.castrol.com

Castrol North America, www.castrolusa.com

CAT Products Inc., www.run-rite.com

Centurion Lubricants, www.centurionlubes.com

Champion Brands LLC, www.championbrands.com

Charles Manufacturing Company, www.tsmoly.com

Chart Automotive Group Inc., www.chartauto.com

Chem-EcoI Ltd., www.chem-ecol.com

Chemlube International Inc., www.chemlube.com

Chempet Corporation, www.rockvalleyoil.com/chempet.htm

Chemsearch Lubricants, www.chemsearch.com

Chemtool Inc./Metalcote, www.chemtool.com

Chevron Chemical Company, www.chevron.com

Chevron Oronite, www.chevron.com

Chevron Phillips Chemical Company LP, www.cpchem.com

Chevron Phillips Chemical Company, www.chevron.com

Chevron Products Company, Lubricants & Specialties Products, www.chevron.com/lubricants

Chevron Products Company, www.chevron.com

Christenson Oil, www.christensonoil.com

Ciba Specialty Chemicals Corporation, www.cibasc.com

Clariant Corporation, www.clariant.com

Clark Refining and Marketing, www.clarkusa.com

Clarkson & Ford Company, www.clarkson-ford.com

CLC Lubricants Company, www.clclubricants.com

Climax Molybdenum Company, www.climaxmolybdenum.com

Coastal Unilube Inc., www.coastalunilube.com

Cognis, www.cognislubechem.com

Cognis, www.cognis-us.com

Cognis, www.cognis.com

Cognis, www.na.cognis.com

Colorado Petroleum Products Company, www.colopetro.com

Commercial Lubricants Inc., www.comlube.com

Commercial Oil Company Inc., www.commercialoilcompany.com

Commercial Ullman Lubricants Company, www.culc.com

Commonwealth Oil Corporation, www.commonwealthoil.com

Como Lube & Supplies Inc., www.comolube.com

Condat Corporation, www.condatcorp.com

Conklin Company Inc., www.conklin.com

Coolants Plus Inc., www.coolantsplus.com

Cortec Corporation, www.cortecvci.com

Cosby Oil Company, www.cosbyoil.com

Country Energy, www.countryenergy.com

CPI Engineering Services, www.cpieng.com

CRC Industries Inc., www.crcindustries.com

Crescent Manufacturing, www.crescentmfg.net

Crompton Corporation, www.cromptoncorp.com

Crown Chemical Corporation, www.brenntag.com

Cyclo Industries LLC, www.cyclo.com

D & D Oil Company Inc., www.amref.com

D. A. Stuart Company, www.d-a-stuart.com

D. W. Davies & Company Inc., www.dwdavies.com

D-A Lubricant Company, www.dalube.com

Darmex Corporation, www.darmex.com

Darsey Oil Company Inc., www.darseyoil.com

David Weber Oil Company, www.weberoil.com

Davison Oil Company Inc., www.davisonoil.com

Dayco Inc., www.dayco.com

D.B. Becker Co. Inc., www.dbbecker.com

Degen Oil and Chemical Company, www.eclipse.net/<degen

Delkol, www.delkol.co.il

Dennis Petroleum Company Inc., www.dennispetroleum.com

Diamond Head Petroleum Inc., www.diamondheadpetroleum.com

Diamond Shamrock Refining Company LP, www.udscorp.com

Digilube Systems Inc., www.digilube.com

Dion & Sons Inc., www.dionandsons.com

Dixon Lubricants & Special Products Group, Division of Asbury Carbons, www.dixonlube.com

Don Weese Inc., www.schaefferoil.com

Dow Chemical Company, www.dow.com

Dow Corning Corp., www.dowcorning.com

Dryden Oil Company Inc., www.castrol.com

Dryson Oil Company, www.synergynracing.com

DSI Fluids, www.dsifluids.com

Dumas Oil Company, www.esn.net/dumasoil

DuPont Krytox Lubricants, www.lubricants.dupont.com

DuPont, www.dupont.com/intermediates

E.I. DuPont de Nemours and Company, www.dupont.com/intermediates

Eastech Chemical Inc., www.eastechchemical.com

Eastern Oil Company, www.easternoil.com

Ecotech Div., Blaster Chemical Companies, www.pbblaster.com

EKO, www.eko.gr

El Paso Corporation, www.elpaso.com

Elf Lubricants North America Inc., www.keystonelubricants.com

Eljay Oil Company Inc., www.eljayoil.com

EMERA Fuels Company Inc., www.emerafuels.com

Emerson Oil Company Inc., www.emersonoil.com

Engen Petroleum Ltd., www.engen.co.za

Enichem Americas Inc., www.eni.it/english/mondo/americhe/usa.html

Environmental Lubricants Manufacturing Inc. (ELM), www.elmusa.com

Equilon Enterprises LLC, www.equilon.com

Equilon Enterprises LLC-Lubricants, www.equilonmotivaequiva.com

Equilon Enterprises LLC-Lubricants, www.shellus.com

Equilon Enterprises LLC-Lubricants, www.texaco.com

Esco Products Inc., www.escopro.com

ETNA Products Inc., www.etna.com

Etna-Bechem Lubricants Ltd., www.etna.com

Evergreen Oil, www.evergreenoil.com

Exxon, www.exxon.com

ExxonMobil Chemical Company, www.exxonmobilchemical.com

ExxonMobil Industrial Lubricants, www.mobilindustrial.com

ExxonMobil Lubricants & Petroleum Specialties Company, www.exxonmobil.com

F&R Oil Company Inc., www.froil.com

F. Bacon Industriel Inc., www.f-bacon.com

Far West Oil Company Inc., www.farwestoil.com

Fina Oil and Chemical Company, www.fina.com

Findett Corp., www.findett.com

Finish Line Technologies Inc., www.finishlineusa.com

FINKE Mineralolwerk, www.finke-mineraloel.de

Finnish Oil and Gas Federation, www.oil.fi

Flamingo Oil Company, www.pinkbird.com

Forward Corporation, www.forwardcorp.com

Frontier Performance Lubricants Inc., www.frontierlubricants.com

Fuchs Lubricants Company, www.fuchs.com

Fuchs, www.fuchs-oil.de

Fuel and Marine Marketing (FAMM), www.fammllc.com

Fuki America Corporation, www.fukiamerica.com

G-C Lubricants Company, www.gclube.com

G & G Oil Company of Indiana Inc., www.ggoil.com

G.T. Autochemilube Ltd., www.gta-oil.co.uk

Gard Corp., www.gardcorp.com

Geo. Pfau's Sons Company Inc., www.pfauoil.com

Georgia-Pacific Pine Chemicals, www.gapac.com

Glover Oil Company, www.gloversales.com

GOA Company., www.goanorthcoastoil.com

Gold Eagle Company, www.goldeagle.com

Golden Bear Oil Specialties, www.goldenbearoil.com

Golden Gate Petroleum, www.ggpetrol.com

GoldenWest Lubricants, www.goldenwestlubricants.com

Goldschmidt Chemical Corporation, www.goldschmidt.com

Goulston Technologies Inc., www.goulston.com

Great Lakes Chemical Corporation, www.glcc.com

Granitize Products Inc., www.granitize.com

Greenland Corporation, www.greenpluslubes.com

Grignard Company LLC, www.purelube.com

Groeneveld Pacific West, www.groeneveldpacificwest.com

Gulf Oil, www.gulfoil.com

H & W Petroleum Company Inc., www.hwpetro.com

H.L. Blachford Ltd., www.blachford.ca

H.N. Funkhouser & Company, www.hnfunkhouser.com

Halocarbon Products Corporation, www.halocarbon.com

Halron Oil Company Inc., www.halron.com

Hampel Oil Distributors, www.hampeloil.com

Hangsterfer's Laboratories Inc., www.hangsterfers.com

Harry Miller Corp., www.harrymillercorp.com

Hasco Oil Company Inc., www.hascooil.com

Hatco Corporation, www.hatcocorporation.com

Haynes Manufacturing Company, www.haynesmfg.com

HCI/Worth Chemical Corp., www.hollandchemical.com

Henkel Surface Technologies, www.henkel.com

Henkel Surface Technologies, www.thomasregister.com/henkelsurftech

Hexol Canada Ltd., www.hexol.com

Hexol Lubricants, www.hexol.com

High Performance Lubricants, www.hplubricants.com

Holland Applied Technologies, www.hollandapt.com

Hoosier Penn Oil Company, www.hpoil.com

Houghton International Inc., www.houghtonintl.com

Howes Lubricator, www.howeslube.thomasregister.com

Huls America, www.CreanovaInc.com

Huls America, www.huls.com

Huskey Specialty Lubricants, www.huskey.com

Hydrosol Inc., www.hydrosol.com

Hydrotex Inc., www.hydrotexlube.com

Hy-Per Lube Corporation, www.hyperlube.com

I.S.E.L. Inc., www.americansynthol.com

ILC/Spectro Oils of America, www.spectro-oils.com

Illinois Oil Products Inc., www.illinoisoilproducts.com

Imperial Oil Company Inc., www.imperialoil.com

Imperial Oil Ltd., www.imperialoil.ca

Imperial Oil Products and Chemicals Division, www.imperialoil.ca

Ingenieria Sales SA de CV, www.isalub.com

Innovene, www.innovene.com

Inolex Chemical Company, www.inolex.com

International Lubricants Inc., www.lubegard.com

International Products Corporation, www.ipcol.com

IQA Lube Corporation, www.iqalube.com

Irving Oil Corp, www.irvingoil.com

ITW Fluid Products Group, www.itwfpg.com

J & H Oil Company, www.jhoil.com

J & S Chemical Corporation, www.jschemical.com

J.A.M. Distributing Company, www.jamdistributing.com

J.B. Chemical Company Inc., www.jbchemical.com

J.B. Dewar Inc., www.jbdewar.com

J.D. Streett & Company Inc., www.jdstreett.com

J.N. Abbott Distributor Inc., www.jnabbottdist.com

Jack Rich Inc., www.jackrich.com

Jarchem Industries Inc., www.jarchem.com

Jasper Engineering & Equipment, www.jaspereng.com

Jenkin-Guerin Inc., www.jenkin-guerin.com

Jet-Lube (United Kingdom) Ltd., www.jetlube.com

Johnson Packings & Industrial Products Inc., www.johnsonpackings.com

Kath Fuel Oil Service, www.kathfuel.com

Keck Oil Company, www.keckoil.com

Kelsan Lubricants USA LLC, www.kelsan.com

Kem-A-Trix Specialty Lubricants & Compounds, www.kematrix.com

Kendall Motor Oil, www.kendallmotoroil.com

Kluber Lubrication North America LP, www.kluber.com

KOST Group Inc., www.kostusa.com

Kyodo Yushi USA Inc., www.kyodoyushi.co.jp

Lambent Technologies, www.petroferm.com

LaPorte, www.laporteplc.com

Leander Lubricants, www.leanderlube.com

Lee Helms Inc., www.leehelms.com

Leffert Oil Company, www.leffertoil.com

Les Industries Sinto Racing Inc., www.sintoracing.com

Lilyblad Petroleum Inc., www.lilyblad.com

Liqua-Tek Inc., www.hdpluslubricants.com

Liquid Horsepower, www.holeshot.com/chemicals/additives.html

LithChem International, www.lithchem.com

Loos & Dilworth Inc.—Automotive Division, www.loosanddilworth.com

Loos & Dilworth Inc.—Chemical Division, www.loosanddilworth.com

Lowe Oil Company/Champion Brands LLC, www.championbrands.com

LPS Laboratories, www.lpslabs.com

LubeCon Systems Inc., www.lubecon.com

Lubemaster Corporation, www.lubemaster.com

LubeRos—A Division of Burlington Chemical Company Inc., www.luberos.com

LuBest, Division of Momar Inc., www.momar.com

Lubricant Technologies, www.lubricanttechnologies.com

Lubricants USA, www.finalube.com

Lubrication Engineers Inc., www.le-inc.com

Lubrication Engineers of Canada, www.lubeng.com

Lubrication Technologies Inc., www.lube-tech.com

Lubrication Technology Inc., www.lubricationtechnology.com

Lubrichem International Corporation, www.lubrichem.net

Lubrifiants Distac Inc., www.inspection.gc.ca/english/ppc/reference/n2e.shtml

Lubri-Lab Inc., www.lubrilab.com

LUBRIPLATE Div., Fiske Bros. Refining Company, www.lubriplate.com

Lubritec, www.ensenada.net/lubritec/

Lucas Oil Products Inc., www.lucasoil.com

Lyondell Lubricants, www.lyondelllubricants.com

MagChem Inc., www.magchem.com

Magnalube, www.magnalube.com

Maine Lubrication Service Inc., www.mainelube.com

Manor Trade Development Corporation, www.amref.com

Mantek Lubricants, www.mantek.com

Markee International Corporation, www.markee.com

Marly, www.marly.com

Maryn International Ltd., www.maryngroup.com

Maryn International, www.poweruplubricants.com

Master Chemical Corporation, www.masterchemical.com

Master Lubricants Company, www.lubriko.com

Maxco Lubricants Company, www.maxcolubricants.com

Maxim Industrial Metalworking Lubricants, www.maximoil.com

Maxima Racing Lubricants, www.maximausa.com

McCollister & Company, www.mccollister.com

McGean-Rohco Inc., www.mcgean-rohco.com

McGee Industries Inc., www.888teammclube.com

McLube Division l/McGee Industries Inc., www.888teammclube.com

Mega Power Inc., www.megapowerinc.com

Metal Forming Lubricants Inc., www.mflube.com

Metal Mates Inc., www.metalmates.net

Metalcote/Chemtool Inc., www.metalcote.com

Metalworking Lubricants Company, www.metalworkinglubricants.com

Metalworking Lubricants, www.maximoil.com

MFA Oil Company, www.mfaoil.com

Mid-South Sales Inc., www.mid-southsales.com

Mid-Town Petroleum Inc., www.midtownoil.com

Milacron Consumable Products Division, www.milacron.com

Millennium Lubricants, www.millenniumlubricants.com

Miller Oil of Indiana Inc., www.milleroilinc.com

Mohawk Lubricants Ltd., www.mohawklubes.com

Molyduval, www.molyduval.com

Molyslip Atlantic Ltd., www.molyslip.co.uk

Monroe Fluid Technology Inc., www.monroefluid.com

Moore Oil Company, www.mooreoil.com

Moraine Packaging Inc., www.hdpluslubricants.com

Morey's Oil Products Company, www.moreysonline.com

Moroil Technologies, www.moroil.com

Motiva Enterprises LLC, www.motivaenterprises.com

Motorol Lubricants, www.motorolgroup.com

Motul USA Inc., www.motul.com

Mr. Good Chem Inc., www.mrgoodchem.com

NCH, www.nc.com

Neo Synthetic Oil Company Inc., www.neosyntheticoil.com

Niagara Lubricant Company Inc., www.niagaralubricant.com

NOCO Energy Corporation, www.noco.com

Noco Lubricants, www.noco.com

Nyco SA, www.nyco.fr

Nye Lubricants, www.nyelubricants.com

Nynas Naphthenics, www.nynas.com

O'Rourke Petroleum, www.orpp.com

Oakite Products Inc., www.oakite.com

OATS (Oil Advisory Technical Services), www.oats.co.uk

Occidental Chemical Corporation, www.oxychem.com

Ocean State Oil Inc., www.oceanstateoil.com

Oil Center Research Inc., www.oilcenter.com

Oil Center Research International LLC, www.oilcenter.com

Oil Depot, www.oildepot.com

Oil Distributing Company, www.oildistributing.com

Oil-Chem Research Corporation, www.avblend.com

Oilkey Corporation, www.oilkey.com

Oilpure Technologies Inc., www.oilpure.com

OKS Speciality Lubricants, www.oks-india.com

Omega Specialties, www.omegachemicalsinc.com

Omni Specialty Packaging, www.nuvo.cc

OMO Petroleum Company Inc., www.omoenergy.com

Orelube Corp., www.orelube.com

Oronite, www.oronite.com

O'Rourke Petroleum Products, www.orpp.com

Ottsen Oil Company Inc., www.ottsen.com

Paramount Products, www.paramountproducts.com

Parent Petroleum, www.parentpetroleum.com

PATCO Additives Division-American Ingredients Company, www.patco-additives.com

Pathfinder Lubricants, www.pathfinderlubricants.ca

PBM Services Company, www.pbmsc.com

Pedroni Fuel Company, www.pedronifuel.com

Pennine Lubricants, www.penninelubricants.co.uk

Pennzoil, www.pennzoil.com

Pennzoil-Quaker State Company, www.pennzoilquakerstate.com

PENRECO, www.penreco.com

Penta Manufacturing Company/Division of Penta International Corporation, www.pentamfg.com

Perkins Products Inc., www.perkinsproducts.com

Petro Star Lubricants, www.petrostar.com

PetroBlend Corporation, www.petroblend.com

Petro-Canada Lubricants, www.htlubricants.com

Petroleum Packers Inc., www.pepac.com

PetroMoly Inc., www.petromoly.com

Petron Corp., www.petroncorp.com

Pfaus Sons Company Inc., www.pfauoil.com

Pflaumer Brothers Inc., www.pflaumer.com

Phoenix Petroleum Company, www.phoenixpetroleum.com

Pico Chemical Corporation, www.picochemical.com

Pinnacle Oil Inc., www.pinnoil.com

Pitt Penn Oil Company, www.pittpenn.com

Plews/Edelmann Div., Stant Corp., www.stant.com

PolySi Technologies Inc., www.polysi.com

Polar Company, www.polarcompanies.com

PolimeriEuropa, www.polimerieuropa.com

Power Chemical, www.warcopro.com

Power-Up Lubricants, www.mayngroup.com

Precision Fluids Inc., www.precisionfluids.com

Precision Industries, www.precisionind.com

Precision Lubricants Inc., www.precisionlubricants.com

Prime Materials, www.primematerials.com

Primrose Oil Company Inc., www.primrose.com

Probex Corporation, www.probex.com

Products Development Manufacturing Company, www.veloil.com

ProLab TechnoLub Inc., www.prolab-technologies.com

ProLab-Bio Inc., www.prolab-lub.com

Prolong Super Lubricants, www.prolong.com

ProTec International Inc., www.proteclubricants.com

Pure Power Lubricants, www.gopurepower.com

QMI, www.qminet.com

Quaker Chemical Corporation, www.quakerchem.com

Quaker State, www.qlube.com

R.E. Carroll Inc., www.recarroll.com

Radio Oil Company Inc., www.radiooil.com

Ramos Oil Company Inc., www.ramosoil.com

Rams-Head Company, www.doall.com

Ransome CAT, www.ransome.com

Red Giant Oil Company, www.redgiantoil.com

Red Line Oil, www.redlineoil.com

Reed Oil Company, www.reedoil.com

Reit Lubricants Company, www.reitlube.com

Reitway Enterprises Inc., www.reitway.com

Renewable Lubricants Inc., www.renewablelube.com

Renite Company, www.renite.com

Renite Company—Lubrication Engineers, www.renite.com

Renkert Oil, www.renkertoil.com

Rensberger Oil Company Inc., www.rensbergeroil.com

RichardsApex Inc., www.richardsapex.com

Riley Oil Company, www.rileyoil.com

RO-59 Inc., http://members.aol.com/ro59inc

Rock Valley Oil & Chemical Company, www.rockvalleyoil.com

Rocol Ltd., www.rocol.com

Rowleys Wholesale, www.rowleys.com

Royal Lubricants Inc., www.royallube.com

Royal Manufacturing Company Inc., www.royalube.com

Royal Purple Inc., www.royalpurple.com

RyDol Products, www.rydol.com

Safety-Kleen Oil Recovery, www.ac-rerefined.com

Sandstrom Products Company, www.sandstromproducts.com

Santie Oil Company, www.santiemidwest.com

Santotrac Traction Lubricants, www.santotrac.com

Santovac Fluids Inc., www.santovac.com

Saxton Industries Inc., www.saxton.thomasregister.com

Saxton Industries Inc., www.schaefferoil.com

Schaeffer Manufacturing, www.schaefferoil.com

Schaeffer Oil and Grease, www.schaefferoil.com

Schaeffer Specialized Lubricants, www.schaefferoil.com

Selco Synthetic Lubricants, www.synthetic-lubes.com

Sentry Solutions Ltd., www.sentrysolutions.com

Service Supply Lubricants LLC, www.servicelubricants.com

SFR Corporation, www.sfrcorp.com

Share Corporation, www.sharecorp.com

Shell Global Solutions, www.shellglobalsolutions.com

Shell Lubricants (United States), www.shell-lubricants.com

Shrieve Chemical Company, www.shrieve.com

Simons Petroleum Inc., www.simonspetroleum.com

SK Corporation (Houston Office), www.skcorp.com

Slick 50 Corporation, www.slick50.com

Smooth Move Company, www.theprojectsthatsave.com

Sobit International Inc., www.sobitinc.com

Soltex, www.soltexinc.com

Sourdough Fuel, www.petrostar.com

Southwest Grease Products, www.stant.com/brochure.cfm?brochure=155&location_id=119

Southwestern Graphite, www.asbury.com

Southwestern Petroleum Corporation, www.swepcousa.com

Spartan Chemical Company Inc. Industrial Products Group Division, www.spartanchemical.com

Spartan Oil Company, www.spartanonline.com

Specialty Silicone Products Inc., www.sspinc.com

Spectro Oils of America, www.goldenspectro.com

Spectro Oils of America, www.spectro-oils.com

Spectrum Corporation, www.spectrumcorporation.com

Spencer Oil Company, www.spenceroil.com

St. Lawrence Chemicals, www.stlawrencechem.com

Steelco Industrial Lubricants Inc., www.steelcolubricants.com

Steelco Northwest Distributors, www.steelcolubricants.com

STP Products Inc., www.stp.com

Suburban Oil Company Inc., www.suburbanoil.com

Summit Industrial Products Inc., www.klsummit.com

Sunnyside Corporation, www.sunnysidecorp.com

Superior Graphite Company, www.superiorgraphite.com/sgc.nsf

Superior Lubricants Company Inc., www.superiorlubricants.com

Superior Lubrication Products, www.s-l-p.com

Surtec International Inc., www.surtecinternational.com

Synco Chemical Corporation, www.super-lube.com

SynLube Inc., www.synlube.com

Synthetic Lubricants Inc., www.synlube-mi.com

Syntroleum Corporation, www.syntroleum.com

T.S. MoIy-Lubricants Inc., www.tsmoly.com

T.W. Brown Oil CompanyInc., www.brownoil.com/soypower.html

TAI Lubricants, www.lubekits.com

Technical Chemical Company (TCC), www.technical-chemical.com

Tek-5 Inc., www.tek-5.com

Terrresolve Technologies, www.terresolve.com

Texas Refinery Corporation, www.texasrefinery.com

Textile Chemical Company Inc., www.textilechem.com

Thermal-Lube Inc., www.thermal-lube.com

Thornley Company, www.thornleycompany.com

Tiodize Co. Inc., www.tiodize.com

Tom-Pac Inc., www.tom-pac.com

Top Oil Products Company, Ltd., www.topoil.com

Torco International Corporation, www.torcoracingoils.com

Totalfina Oleo Chemicals, www.totalfina.com

Tower Oil & Technology Company, www.toweroil.com

Toyo Grease Manufacturing (M) SND BHD, www.toyogrease.com

TransMontaigne, www.transmontaigne.com

Transmountain Oil Company, www.transmountainoil.com

TriboLogic Lubricants Inc., www.tribologic.com

Tribos Technologies, www.tribostech.com

Trico Manufacturing Corporation, www.tricomfg.com

Tricon Specialty Lubricánts, www.tristrat.com

Turmo Lubrication Inc., www.lubecon.com

TXS Lubricants Inc., www.txsinc.com

U.S. Industrial Lubricants Inc., www.usil.cc

U.S. Oil Company Inc., www.usoil.com

Ultrachem Inc., www.ultracheminc.com

Unimark Oil Company, www.gardcorp.com

Union Carbide Corporation, www.unioncarbide.com

Uniqema, www.uniqema.com

Uniroyal Chemical Company Inc., www.uniroyalchemical.com

UniSource Energy Inc., www.unisource-energy.com

Unist Inc., www.unist.com

United Lubricants, www.unitedlubricants.com

United Oil Company Inc., www.duralene.com

United Oil Products Ltd., http://ourworld.compuserve.com/homepages/Ferndale_UK

United Soybean Board, www.unitedsoybean.org

Universal Lubricants Inc., www.universallubes.com

Unocal Corporation, www.unocal.com

Valero Marketing & Supply, www.valero.com
Valvoline Canada, www.valvoline.com
Valvoline, www.valvoline.com
Vesco Oil Corporation, www.vesco-oil.com
Vikjay Industries Inc., www.vikjay.com
Virtual Oil Inc., www.virtualoilinc.com
Vogel Lubrication System of America, www.vogel-lube.com
VP Racing Fuels Inc., www.vpracingfuels.com
Vulcan Oil & Chemical Products Inc., www.vulcanoil.com
Wallover Oil Company, www.walloveroil.com
Walthall Oil Company, www.walthall-oil.com
Warren Distribution, www.wd-wpp.com
WD-40 Company, www.wd40.com
West Central Soy, www.soypower.net
Western States Oil, www.lubeoil.com
Whitaker Oil Company Inc., www.whitakeroil.com
Whitmore Manufacturing Company, www.whitmores.com
Wilcox and Flegel Oil Company, www.wilcoxandflegel.com
Winfield Brooks Company Inc., www.tapfree.com
Winzer Corporation, www.winzerusa.com
Witco (Crompton Corporation), www.witco.com
Wolf Lake Terminals Inc., www.wolflakeinc.com
Worldwide PetroMoly Inc., www.petromoly.com
Wynn Oil Company, www.wynnsusa.com
X-1R Corp., The, www.x1r.com
Yocum Oil Company Inc., www.yocumoil.com
Yuma Industries Inc., www.yumaind.com

35.2.2 ADDITIVES

Acheson Colloids Company, www.achesonindustries.com
Acme-Hardesty Company, www.acme-hardesty.com
ALT, www.altboron.com
AFD Technologies, www.afdt.com
AG Fluoropolymers USA Inc., www.fluoropolymers.com
Akzo Nobel, www.akzonobel.com
Amalie Oil Company, www.amalie.com
Amber Division of Nidera Inc., www.nidera-us.com
American International Chemical Inc., www.aicma.com
Amitech, www.amitech-usa.com
ANGUS Chemical Company, www.dowchemical.com
Anti Wear 1, www.dynamicdevelopment.com
Arch Chemicals Inc., www.archbiocides.com
Arizona Chemical, www.arizonachemical.com
Asbury Carbons Inc.—Dixon Lubricants, www.asbury.com
Ashland Distribution Company, www.ashland.com
Aspen Chemical Company, www.aspenchemical.com
ATOFINA Chemicals Inc., www.atofina.com
ATOFINA Canada Inc., www.atofinacanada.com

Baker Petrolite, www.bakerhughes.com/bakerpetrolite/
Bardahl Manufacturing Corporation, www.bardahl.com
BASF Corporation, www.basf.com
Bayer Corporation, www.bayer.com
Bismuth Institute, www.bismuth.be
BoMac Lubricant Technologies Inc., www.bomaclubetech.com
BP Amoco Chemicals, www.bpamocochemicals.com
Brascorp North America Ltd., www.brascorp.on.ca
British Petroleum (BP), www.bp.com
Buckman Laboratories Inc., www.buckman.com
Burlington Chemical, www.burco.com
Cabot Corporation (fumed metal oxides), www.cabot-corp.com/cabosil
Callahan Chemical Company, www.calchem.com
Calumet Lubricants Company, www.calumetlub.com
Cargill-Industrial Oil & Lubricants, www.techoils.cargill.com
Cary Company, www.thecarycompany.com
CasChem Inc., www.cambrex.com
Center for Innovation Inc., www.centerforinnovation.com
Certified Laboratories, www.certifiedlaboratories.com
Chattem Chemicals Inc., www.chattemchemicals.com
Chemetall Foote Corporation, www.chemetall.com
Chemsearch Lubricants, www.chemsearch.com
Chevron Oronite, www.chevron.com
Ciba Specialty Chemicals Corporation, www.cibasc.com
Clariant Corp., www.clariant.com
Climax Molybdenum Company, www.climaxmolybdenum.com
Cognis, www.cognislubechem.com
Cognis, www.cognis-us.com
Cognis, www.cognis.com
Cognis, www.na.cognis.com
Commonwealth Oil Corporation, www.commonwealthoil.com
Cortec Corporation, www.cortecvci.com
Creanova Inc., www.creanovainc.com
Croda Inc., www.croda.com
Crompton Corporation, www.cromptoncorp.com
Crowley Chemical Company Inc., www.crowleychemical.com
Crown Chemical Corporation, www.brenntag.com
Crystal Inc.-PMC, www.pmc-group.com
Cummings-Moore Graphite Company, www.cumograph.com
D.A. Stuart Company, www.d-a-stuart.com
D.B. Becker Company Inc., www.dbbecker.com
DeForest Enterprises Inc., www.deforest.net
Degen Oil and Chemical Company, www.eclipse.net/<degen
Dover Chemical, www.doverchem.com
Dow Chemical Company, www.dow.com
Dow Corning Corporation, www.dowcorning.com

DuPont-Dow Elastomers, www.dupont-dow.com

Dylon Industries Inc., www.dylon.com

E.I. DuPont de Nemours and Company, www.dupont.com/intermediates

E.W. Kaufmann Company, www.ewkaufmann.com

Elco Corporation, The, www.elcocorp.com

Elementis Specialties, www.elementis-na.com

Elementis Specialties-Rheox, www.rheox.com

Elf Atochem Canada, www.atofinachemicals.com

Environmental Lubricants Manufacturing Inc. (ELM), www.elmusa.com

Ethyl Corporation, www.ethyl.com

Ethyl Petroleum Additives, www.ethyl.com

Fanning Corporation, The, www.fanncorp.com

Ferro/Keil Chemical, www.ferro.com

FMC Lithium, www.fmclithium.com

FMC, www.fmc.com

Functional Products, www.functionalproducts.com

G.R. O'Shea Company, www.groshea.com

Gateway Additives, www.lubrizol.com

Geo. Pfau's Sons Company. Inc., www.pfauoil.com

Georgia-Pacific Pine Chemicals, www.gapac.com

Georgia-Pacific Resins Inc.—Actrachem Division, www.gapac.com

Georgia-Pacific Resins Inc.—Actrachem Division, www.gp.com

Goldschmidt Chemical Corporation, www.goldschmidt.com

Great Lakes Chemical Corporation, www.glcc.com

Grignard Company LLC, www.purelube.com

Hall Technologies Inc., www.halltechinc.com

Hammonds Fuel Additives Inc., www.hammondscos.com

Heveatex, www.heveatex.com

Holland Applied Technologies, www.hollandapt.com

Huntsman Corporation, www.huntsman.com

Infineum USA LP, www.infineum.com

International Lubricants Inc., www.lubegard.com

J.H. Calo Company Inc., www.jhcalo.com

J.B. Chemical Company Inc., www.jbchemical.com

Jarchem Industries Inc., www.jarchem.com

Keil Chemical Division; Ferro Corporation, www.ferro.com

King Industries Specialty Chemicals, www.kingindustries.com

Lambent Technologies, www.petroferm.com

LaPorte, www.laporteplc.com

Lockhart Chemical Company, www.lockhartchem.com

Loos & Dilworth Inc.—Chemical Division, www.loosanddilworth.com

LubeRos—A Division of Burlington Chemical Company Inc., www.luberos.com

Lubricant Additives Research, www.silverseries.com

Lubricants Network Inc., www.lubricantsnetwork.com

Lubri-Lab Inc., www.lubrilab.com

Lubrizol Corporation, The, www.lubrizol.com

Lubrizol Metalworking Additive Company, www.lubrizol.com

Mantek Lubricants, www.mantek.com

Marcus Oil and Chemical, www.marcusoil.com

Master Chemical Corporation, www.masterchemical.com

Mays Chemical Company, www.mayschem.com

McIntyre Group Ltd., www.mcintyregroup.com

Mega Power Inc., www.megapowerinc.com

Metal Mates Inc., www.metalmates.net

Metalworking Lubricants Company, www.metalworkinglubricants.com

Militec Corporation, www.militec.com

Nagase America Corporation, www.nagase.com

Naptech Corporation, www.satec.com

Northern Technologies International Corporation, www.ntic.com

Oil Center Research Inc., www.oilcenter.com

OKS Speciality Lubricants, www.oks-india.com

Omega Specialties, www.omegachemicalsinc.com

OMG Americas Inc., www.omgi.com

OMGI, www.omgi.com

Oronite, www.oronite.com

PATCO Additives Division-American Ingredients Company, www.patco-additives.com

Pflaumer Brothers Inc., www.pflaumer.com

Pilot Chemical Company, www.pilotchemical.com

PMC Specialties Inc., www.pmcsg.com

Polartech Ltd., www.polartech.co.uk

Precision Fluids Inc., www.precisionfluids.com

Purac America Inc., www.purac.com

R.T. Vanderbilt Company Inc., www.rtvanderbilt.com

R.H. Foster Energy LLC, www.rhfoster.com

Reade Advanced Materials, www.reade.com

Rhein Chemie Corporation, www.bayer.com

Rhein Chemie Rheinau GmbH., www.rheinchemie.com

Rheox Inc., www.rheox.com

Rhodia, www.rhodia.com

Rhone-Poulenc Surfactants & Specialties, www.rpsurfactants.com

RiceChem, A Division of Stilling Enterprises Inc., www.ricechem.com

Rohm & Haas Company, www.rohmhaas.com

RohMax Additives GmbH, www.rohmax.com

Ross Chem Inc., www.rosschem.com

Santotrac Traction Lubricants, www.santotrac.com

Santovac Fluids Inc., www.santovac.com

Sea-Land Chemical Company, www.sealandchem.com

Shamrock Technologies Inc., www.shamrocktechnologies.com

Shell Chemicals, www.shellchemical.com

Shepherd Chemical Company, www.shepchem.com

Soltex, www.soltexinc.com

SP Morell & Company, www.spmorell.com

Spartan Chemical Company Inc. Industrial Products Group Division, www.spartanchemical.com

SQM North America Corporation, www.sqmna.com

St. Lawrence Chemicals, www.stlawrencechem.com

Stochem Inc., www.stochem.com

Thornley Company, www.thornleycompany.com
Tiodize Company Inc., www.tiodize.com
Tomah Products Inc., www.tomah3.com
Troy Corporation, www.troycorp.com
Ultra Additives Inc., www.ultraadditives.com
Uniqema, www.uniqema.com
Uniroyal Chemical Company Inc., www.uniroyalchemical.com
United Color Manufacturing Inc., www.unitedcolor.com
United Lubricants, www.unitedlubricants.com
Valhalla Chemical, www.valhallachem.com
Van Horn, Metz & Company Inc., www.vanhornmetz.com
Virtual Oil Inc., www.virtualoilinc.com
Wynn Oil Company, www.wynnsusa.com
Zinc Corporation of America, www.zinccorp.com

35.2.3 OIL COMPANIES

Adco Petrol Katkilari San Ve. Tic. AS, www.adco.com.tr
Amoco, www.amoco.com
Aral International, www.aral.com
Asian Oil Company, www.nilagems.com/asianoil/
Bharat Petroleum, www.bharatpetroleum.com
BP, www.bp.com
CEPSA (Spain), www.cepsa.es
Chevron Texaco, www.chevrontexaco.com
Chevron, www.chevron.com
CITGO Petroleum Corporation, www.citgo.com
Coastal Corporation, www.elpaso.com
Conoco, www.conoco.com
Cosmo Oil, www.cosmo-oil.co.jp
Cross Oil Refining and Marketing Inc., www.crossoil.com
Ecopetrol (Columbian Petroleum Company), www.ecopetrol.com.co
ENI, www.eni.it
Ergon Inc., www.ergon.com
ExxonMobil Corp., www.exxonmobil.com
Fortum (Finland), www.fortum.com
Gasco Energy, www.gascoenergy.com
Hindustan Petroleum Corporation Ltd., www.hindpetro.com
Idemitsu, www.idemitsu.co.jp
Indian Oil Corporation, www.indianoilcorp.com
Interline Resources Corporation, www.interlineresources.com
Japan Energy Corporation, www.j-energy.co.jp/eng/index.html
Japan Energy, www.j-energy.co.jp
Kuwait National Petroleum Company, www.knpc.com.kw
Lukoil (Russian Oil Company), www.lukoil.com
Marathon Ashland Petroleum LLC, www.mapllc.com
Marathon Oil Company, www.marathon.com
Mobil, www.mobil.com
MOL Hungarian Oil & Gas, www.mol.hu

Murphy Oil Corporation, www.murphyoilcorp.com
PDVSA (Venezuela), www.pdvsa.com
PEMEX (Mexico), www.pemex.com
Pertamina (Indonesia), www.pertamina.com
Petrobras (Brazil), www.petrobras.com.br
Petrogal (Portugal), www.petrogal.pt
Petroleum Authority of Thailand, www.nectec.or.th/users/htk/SciAm/12PTT.html
Petroperu (Peru), www.petroperu.com
Phillips Petroleum Company/Phillips 66, www.phillips66.com/phillips66.asp
RWE-DEA (Germany), www.rwe-dea.de
San Joaquin Refining Company, www.sjr.com
Sasol (South Africa), www.sasol.com
Shell (United States), www.shellus.com
Shell International, www.shell.com/royal-en
Shell Oil Products US, www.shelloilproductsus.com
Sinclair Oil Corp., www.sinclairoil.com
Sinopec (China Petrochemical Corporation), www.sinopec.com.cn
Statoil (Norway), www.statoil.com
Sunoco Inc., www.sunocoinc.com
Texaco Inc., www.texaco.com
Tosco, www.tosco.com
Total, www.total.com
Total, www.totalfinaelf.com/ho/fr/index.htm
YPF (Argentina), www.ypf.com.ar

35.2.4 UNIVERSITY SITES

Brno University of Technology, Faculty of Mechanical Engineering, Elastohydrodynamic Lubrication Research Group, fyzika.fme.vutbr.cz/ehd/
Cambridge University, Department of Materials Science and Metallurgy, Tribology, www.msm.cam.ac.uk/tribo/tribol.htm
College of Petroleum and Energy Studies CPS Home Page, www.colpet.ac.uk/index.html
College of Petroleum and Energy Studies, www.colpet.ac.uk
Colorado School of Mines Advanced Coating and Surface Engineering Laboratory (ACSEL), www.mines.edu/research/acsel/acsel.html
Danish Technological Institute (DTI) Tribology Centre, www.tribology.dti.dk
Departments of Mechanical Engineering Luleå Technical University, Sweden, www.luth.se/depts/mt/me/
Division of Machine Elements Home Page Niigata University, Japan, tmtribol.eng.niigata-u.ac.jp/index_e.html
Ecole Polytechnique Federale de Lausanne, Switzerland, igahpse.epfl.ch
EidgenSssische Technische Hochschule (ETH), Zurich Laboratory for Surface Science and Technology (LSST), www.surface.mat.ethz.ch
Esslingen, Technische Akademie, www.tae.de

Fachhochschule Hamburg, Germany, www.haw-hamburg.de/fh/forum/f12/indexf.html/tribologie/etribology.html

Georgia Tech Tribology, www.me.gatech.edu/research/tribology.html

Imperial College, London ME Tribology Section, www.me.ic.ac.uk/tribology/

Indian Institute of Science, Bangalore, India, Department of Mechanical Engineering, www.mecheng.iisc.ernet.in

Institut National des Sciences Appliquées de Lyon, France, Laboratoire de Mécanique des Contacts, www.insa-lyon.fr/Laboratoires/LMC/index.html

Iowa State University, Tribology Laboratory, www.eng.iastate.edu/coe/me/research/labs/tribology_lab.html

Israel Institute of Technology (Technion), meeng.technion.ac.il/Labs/energy.htm#tribology

Kanazawa University, Japan, Tribology Laboratory, web.kanazawa-u.ac.jp/<tribo/ labo5e.html

Kyushu University, Japan, Lubrication Engineering Home Page, www.mech.kyushu-u.ac.jp/index.html

Lulea University of Technology, Department of Mechanical Engineering, www.luth.se/depts/mt/me/

Northwestern University, Tribology Lab, cset.mech.northwestern.edu/member.htm

Ohio State University, Center for Surface Engineering and Tribology, Gear Dynamics and Gear Noise Research Laboratory, gearlab.eng.ohio-state.edu

Pennsylvania State University, The, www.me.psu.edu/research/tribology.html

Purdue University, Materials Processing and Tribology Research Group, www.ecn.purdue.edu/<farrist/lab.html

Purdue University, Mechanical Engineering Tribology Web Site, widget.ecn.purdue.edu/<metrib/

Royal Institute of Technology (KTH), Sweden Machine Elements Home Page, www.damek.kth.se/mme

Saitama University, Japan Home Page of Machine Element Laboratory, www.mech.saitama-u.ac.jp/youso/home.html

Sandia National Laboratories Tribology, www.sandia.gov/materials/sciences/

Shamban Tribology Laboratory Kanazawa University, Japan, web.kanazawa-u.ac.jp/<tribo/labo5e.html

Southern Illinois University, Carbondale Center for Advanced Friction Studies, www.frictioncenter.com

State University of New York, Binghamton Mechanical Engineering Laboratory, www.me.binghamton.edu/me_labs.html

Swiss Federal Laboratories for Materials Testing and Research (EMPA) Centre for Surface Technology and Tribology, www.empa.ch

Swiss Tribology Online, Nanomechanics and Tribology, dmxwww.epfl.ch/WWWTRIBO/home.html

Technical University of Delft, Netherlands Laboratory for Tribology, www.ocp.tudelft.nl/tribo/

Technical University of Munich, Germany, www.fzg.mw.tu-muenchen.de

Technische Universitat Ilmenau, Faculty of Mathematics and Natural Sciences, www.physik.tu-ilmenau.de/index_e.html

Texas Tech University, Tribology, www.osci.ttu.edu/ME_Dept/Research/tribology.htmld/

The Pennsylvania State University, www.me.psu.edu/research/tribology.html

Tokyo Institute of Technology, Japan Nakahara Lab. Home Page, www.mep.titech.ac.jp/Nakahara/home.html

Trinity College, Dublin Tribology and Surface Engineering, www.mme.tcd.ie/Groups/ Tribology/

Tsinghua University, China, State Key Laboratory of Tribology, www.pim.tsinghua.edu.cn/index_cn.html

University of Akron Tribology Laboratory, www.ecgf.uakron.edu/<mech

University of Applied Sciences, Hamburg, Germany Department of Mechanical Engineering Tribology, www.fh-hamburg.de/fh/fb/m/tribologie/e_index.html

University of Applied Sciences, Hamburg, Germany, www.haw-hamburg.de/fh/fb/m/tribologie/e_index.html

University of California, Berkeley Bogey's Tribology Group, cml.berkeley.edu/tribo.html

University of California, San Diego, Center for Magnetic Recording Research, orpheus.ucsd.edu/cmrr/

University of Florida, Mechanical Engineering Department, Tribology Laboratory, grove.ufl.edu/<wgsawyer/

University of Illinois, Urbana-Champaign Tribology Laboratory, www.mie.uiuc.edu

University of Kaiserslautern, Germany Sektion Tribologie, www.uni-kl.de/en/

University of Leeds, M.Sc. (Eng.) Course in Surface Engineering and Tribology, leva.leeds.ac.uk/tribology/msc/tribmsc.html

University of Leeds, United Kingdom, Research in Tribology, leva.leeds.ac.uk/tribology/research.html

University of Ljubljana, Faculty of Mechanical Engineering, Center for Tribology and Technical Diagnostics, www.ctd.uni-lj.si/eng/ctdeng.htm

University of Maine Laboratory for Surface Science and Technology (LASST), www.ume.maine.edu/LASST/

University of Maine, NanoTribometer System, www.ume.maine.edu/LASST

University of Newcastle upon Tyne, United Kingdom, Ceramics Tribology Research Group, www.ncl.ac.uk/materials/materials/resgrps/certrib.html

University of Northern Iowa, www.uni.edu/abil

University of Notre Dame Tribology/Manufacturing Laboratory, www.nd.edu/<ame

University of Pittsburg, School of Engineering, Mechanical Engineering Department, www.engrng.pitt.edu/<mewww

University of Sheffield, United Kingdom, Tribology Research Group, www.shef.ac.uk/mecheng/tribology/

University of Southern Florida, Tribology, www.eng.usf.edu/<hess/

University of Texas at Austin, Petroleum & Geosystems Engineering, Reading Room, www.pe.utexas.edu/Dept/Reading/petroleum.html

University of Tokyo, Japan, Mechanical Engineering Department, www.mech.t.u-tokyo.ac.jp/english/index.html

University of Twente, Netherlands Tribology Group, www.wb.utwente.nl/vakgroep/tr/ tribeng.htm

University of Western Australia Department of Mechanical and Material Engineering, www.mech.uwa.edu.au/tribology/

University of Western Ontario, Canada Tribology Research Centre, www.engga.uwo.ca/research/tribology/Default.htm

University of Windsor, Canada Tribology and Wear Research Group, zeus.uwindsor.ca/research/wtrg/index.html

University of Windsor, Canada, Tribology Research Group, venus.uwindsor.ca/research/wtrg/index.html

Uppsala University, Sweden Tribology Group, www.angstrom.uu.se/materials/index.htm

Western Michigan University Tribology Laboratory, www.mae.wmich.edu/labs/Tribology/Tribology.html

Western Michigan University, Department of Mechanical and Aeronautical Engineering, www.mae.wmich.edu

Worcester Polytechnic Institute, Department of Mechanical Engineering, www.me.wpi.edu/Research/labs.html

35.2.5 GOVERNMENT SITES/INDUSTRY SITES

American Bearing Manufacturers Association, www.abma-dc.org

American Board of Industrial Hygiene, www.abih.org

American Carbon Society, www.americancarbonsociety.org

American Chemical Society (ACS), www.acs.org

American Council of Independent Laboratories (ACIL), www.acil.org

American Gear Manufacturers Association (AGMA), www.agma.org

American National Standards Institute (ANSI), www.ansi.org

American Oil Chemists Society (AOCS), www.aocs.org

American Petroleum Institute (API), www.api.org

American Society of Agricultural Engineering (ASAE), www.asae.org

American Society of Agronomy (ASA), www.agronomy.org

American Society for Horticultural Science (ASHS), www.ashs.org

American Society for Testing and Materials (ASTM), www.astm.org

American Society of Mechanical Engineers International (ASME), www.asme.org

Argonne National Laboratory, www.et.anl.gov

Automotive Aftermarket Industry Association (AAIA), www.aftermarket.org

Automotive Oil Change Association (AOCA), www.aoca.org

Automotive Parts and Accessories Association (APAA), www.apaa.org

Automotive Service Industry Association (ASIA), www.aftmkt.com

British Lubricants Federation Ltd., www.blf.org.uk

California Air Resources Board, www.arb.ca.gov

Center for Tribology Inc. (CETR), www.cetr.com

Co-ordinating European Council (CEC), www.cectests.org

Coordinating Research Council (CRC), www.crcao.com

Crop Science Society of America (CSSA), www.crops.org

Department of Defense (DOD), www.dodssp.daps.mil/dodssp.htm

Deutsches Institute Fur Normung e. V. (DIN), www.din.de

Environmental Protection Agency (EPA), www.fedworld.gov

European Automobile Manufacturers Association (ACEA), www.acea.be

European Oil Companies Organization of E. H. and S. (CONCAWE), www.concawe.be

Federal World, www.fedworld.gov

Independent Lubricant Manufacturers Association (ILMA), www.ilma.org

Industrial Maintenance & Plant Operation (IMPO), www.mcb.co.uk/cgi-bin/mcb_serve/table1.txt&ilt&stanleaf.htm

Institute of Materials Inc. (IOM), www.savantgroup.com

Institute of Mechanical Engineers (ImechE), www.imeche.org.uk

Institute of Petroleum (IP), http://212.78.70.142

Institute of Physics (IOP), Tribology Group, www.iop.org

Internal Energy Agency (IEA), www.iea.org

International Organization for Standardization (ISO), www.iso.ch

Japan Association of Petroleum Technology (JAPT), www.japt.org

Japan Automobile Manufacturers Association (JAMA), www.japanauto.com

Japanese Society of Tribologists (JAST) (in Japanese), www.jast.or.jp

Los Alomos National Laboratory, www.lanl.gov/worldview/

NASA Lewis Research Center (LeRC) Tribology & Surface Science Branch, www.lerc.nasa.gov/Other_Groups/SurfSci

National Centre of Tribology, United Kingdom, www.aeat.com/nct/

National Fluid Power Association (NFPA), www.nfpa.com

National Institute for Occupational Safety and Health, www.cdc.gov/homepage.html

National Institute of Standards and Technology, webbook.nist.gov/chemistry

National Lubricating Grease Institute (NLGI), www.nlgi.org

National Metal Finishing Resource Center, www.nmfrc.org

National Oil Recyclers Association (NORA), www.recycle.net/Associations/rs000141.html

National Petrochemical & Refiners Association (NPRA), www.npradc.org

National Petroleum Refiners Association (NPRA), www.npra.org

National Research Council of Canada Lubrication Tribology Services, 132.246.196.24/en/fsp/service/lubrication_trib.htm

Naval Research Lab Tribology Section—NRL Code 6176, stm2.nrl.navy.mil/<wahl/6176.htm

Oak Ridge National Laboratory (ORNL) Tribology Test Systems, www.ms.ornl.gov/htmlhome

Occupational Safety and Health Administration (OSHA), www.osha.gov

Petroleum Authority of Thailand, www.nectec.or.th/users/htk/SciAm/12PTT.html

Petroleum Marketers Association of America (PMAA), www.pmaa.org

Society of Automotive Engineers (SAE), www.sae.org

Society of Environmental Toxicology and Chemistry (SETAC), www.setac.org

Society of Manufacturing Engineers (SME), www.sme.org

Society of Tribologists and Lubrication Engineers (STLE), www.stle.org

Southwest Research Institute (SwRI) Engine Technology Section, www.swri.org/4org/d03/engres/engtech/

Southwestern Petroleum Corporation (SWEPCO), www.swepco.com

U.S. Department of Agriculture (USDA), www.usda.gov

U.S. Department of Defense (DOD), www.dod.gov

U.S. Department of Energy (DOE), www.energy.gov

U.S. Department of Transportation (DOT), www.dot.gov

U.S. Energy Information Administration, www.eia.doe.gov

U.S. Patent Office, www.uspto.gov

U.S. Data Exchange, www.usde.com

Western States Petroleum Association, www.wspa.org

35.2.6 Testing Labs/Equipment/Packaging

A.W. Chesterton Company, www.chesterton.com

A/R Packaging Corporation, www.arpackaging.com

Accumetric LLC, www.accumetric.com

Airflow Systems Inc., www.airflowsystems.com

Alfa Laval Separation, www.alfalaval.com

Allen Filters Inc., www.allenfilters.com

Ana Laboratories Inc., www.analaboratories.com

Analysts Inc., www.analystinc.com

Anatech Ltd., www.anatechltd.com

Andpak Inc., www.andpak.com

Applied Energy Company, www.appliedenergyco.com

Aspen Technology, www.aspentech.com

Associates of Cape Cod Inc., www.acciusa.com

Atico-Internormen-Filter, www.atico-internormen.com

Baron USA Inc., www.baronusa.com

Berenfield Containers, www.berenfield.com

Bericap NA, www.bericap.com

BF Goodrich, www.bfgoodrich.com

Bianco Enterprises Inc., www.bianco.net

Bijur Lubricating Corporation, www.bijur.com

Biosan Laboratories Inc., www.biosan.com

Bio-Rad Laboratories, www.bio-rad.com

BioTech International Inc., www.info@biotechintl.com

Blackstone Laboratories, www.blackstone-labs.com

Cannon Instrument Company, www.cannon-ins.com

Certified Laboratories, www.certifiedlaboratories.com

Chemicolloid Laboratories Inc., www.colloidmill.com

Como Industrial Equipment Inc., www.comoindustrial.com

Computational Systems Inc., www.compsys.com/index.html

Containment Solutions Inc., www.containmentsolutions.com

CSI, www.compsys.com

Custom Metalcraft Inc., www.custom-metalcraft.com

Delphi Automotive Systems, www.delphiauto.com

Des-Case Corporation, www.des-case.com

Dexsil Corporation, www.dexsil.com

Diagnetics, www.entek.com

Digilube Systems Inc., www.digilube.com

Dingo Maintenance Systems, www.dingos.com

DSP Technology Inc., www.dspt.com

Duro Manufacturing Inc., www.duromanufacturing.com

Dutton-Lainson Company, www.dutton-lainson.com

Dylon Industries Inc., www.dylon.com

Easy Vac Inc., www.easyvac.com

Edjean Technical Services Inc., www.edjetech.com

Engel Metallurgical Ltd., www.engelmet.com

Engineered Composites Inc., www.engineeredcomposites.net

Environmental and Power Technologies Ltd., www.cleanoil.com

Evans Industries Inc., www.evansind.com

Falex Corporation, www.falex.com

Falex Tribology NV, www.falexint.com

FEV Engine Technology Inc., www.fev-et.com

Flo Components Ltd., www.flocomponents.com

Flowtronex International, www.flowtronex.com

Fluid Life Corporation, www.fluidlife.com

Fluid Systems Partners US Inc., www.fsp-us.com

Fluid Technologies Inc., www.fluidtechnologies.com

Fluids Analysis Lab, www.butler-machinery.com/oil. html

Fluidtec International, www.fluidtec.com

Fluitec International, www.fluitec.com/

FMC Blending & Transfer, www.fmcblending-transfer. com

Framatome ANP, www.framatech.com

Fuel Quality Services Inc., www.fqsgroup.com

G.R. O'Shea Company, www.groshea.com

G.T. Autochemilube Ltd., www.gta-oil.co.uk

Galactic, www.galactic.com

Gamse Lithographing Company, www.gamse.com

Gas Tops Ltd., www.gastops.com

Generation Systems Inc., www.generationsystems.com

Georgia-Pacific Resins Inc.—Actrachem Division, www.gapac.com

Gerhardt Inc., www.gerhardths.com

Globetech Services Inc., www.globetech-services.com

Graco Inc., www.graco.com

Gulfgate Equipment Inc., www.gulfgateequipment.com

Hedwin Corporation, www.hedwin.com

Hercules Inc., Aqualon Division, www.herc.com

Herguth Laboratories Inc., www.herguth.com

Hi-Port Inc., www.hiport.com

Hi-Tech Industries Inc., www.hi-techind.com

Hoover Materials Handling Group Inc., www.hooveribcs. com

Horix Manufacturing Company, www.sgi.net/horix

Hydraulic Repair & Design Inc., www.h-r-d.com

Hysitron Incorporated: Nanomechanics, www.hysitron. com

Indiana Bottle Company, www.indianabottle.com

Industrial Packing Inc., www.industrialpacking.com

Insight Services, www.testoil.com

Instruments for Surface Science, www.omicron-instruments.com/index.html

Interline Resources Corporation, www.interlineresources. com

International Group Inc., The (IGI), www.igiwax.com

Intertek Testing Services-Caleb Brett, www.itscb.com

Invicta AS, www.testoil.com

J & S Chemical Corporation, www.jschemical.com

J.R. Schneider Company Inc., www.jrschneider.com

JAX-Behnke Lubricants Inc., www.jax.com

Johnson Packings & Industrial Products Inc., www. johnsonpackings.com

K.C. EngineeringLtd.., www.kceng.com

K.l.S.S. Packaging Systems, www.kisspkg.com

Kafko International Ltd., www.kafkointl.com

Kennedy Group, The, www.kennedygrp.com

Kittiwake Developments Limited, www.kittiwake.com

Kleenoil Filtration Inc., www.kleenoilfiltrationinc.com

Kleentek-United Air Specialists Inc., www.uasinc.com

Koehler Instrument Company, www.koehlerinstrument.com

Kruss USA, www.krussusa.com

Laub/Hunt Packaging Systems, www.laubhunt.com

Lawler Manufacturing Corporation, www.lawler-mfg. com

Leding Lubricants Inc., www.automatic-lubrication.com

Legacy Manufacturing, www.legacymfg.com

Liftomtic Inc., www.liftomatic.com

Lilyblad Petroleum Inc., www.lilyblad.com

Linpac Materials Handling, www.linpacmh.com

Liqua-Tek/Moraine Packaging, www.globaldialog. com/<mpi

Liquid Controls Inc., A Unit of IDEX Corporation, www.lcmeter.com

Lormar Reclamation Service, www.lormar.com

LubeCon Systems Inc., www.lubecon.com

Lubricant Technologies, www.lubricanttechnologies. com

Lubrication Engineers of Canada, www.lubeng.com

Lubrication Systems, www.lsc.com

Lubrication Technologies Inc., www.lube-tech.com

Lubriport Labs, www.ultralabs.com/lubriport

Lubriquip Inc., www.lubriquip.com

Lubrizol Corporation, The, www.lubrizol.com

Lubromation Inc., www.lubromation.com

Lub-Tek Petroleum Products Corporation, www.lubtek. com

Machines Production Web Site, www.machpro.fr

Manor Technology, www.manortec.co.uk

Metalcote/Chemtool Inc., www.metalcote.com

Metorex Inc., www.metorex.fi

Mettler Toledo, www.mt.com

Michel Murphy Enterprises Inc., www.michelmurphy. com

Micro Photonics Inc., www.microphotonics.com

Mid-Michigan Testing Inc., www.tribologytesting.com

Monlan Group, www.monlangroup.com

Motor Fuels/Combustibles Testing, www.empa.ch/ englisch/fachber/abt133/index.htm

Mozel Inc., www.mozel.com

Nalco Chemical Company, www.nalco.com

Naptec Corporation, www.satec.com

National Tribology Services, www.natrib.com

NCH, www.nc.com

Newcomb Oil Company, www.newcomboil.com

Nordstrom Valves Inc., www.nordstromaudco.com

Oden Corporation, www.oden.thomasregister.com

Oden Corporation, www.odencorp.com

Oil Analysis (Noria), www.oilanalysis.com

OMICRON Vakuumphysik GmbH, www.omicron-instruments.com/index.html

OMS Laboratories Inc., members.aol.com/labOMS/ index.html

Owens-Illinois Inc., www.o-i.com

Oxford Instruments Inc., www.oxinst.com

Paper Systems Inc., www.paper-systems.com

PARC Technical Services Inc., www.parctech.com

Patterson Industries Ltd. (Canada), www.patterson industries.com

PCS Instruments, www.pcs-instruments.com

PdMA Corporation, www.pdma.com

PED Inc., www.ped.vianet.ca

Perkin Elmer Automotive Research, www.perkinelmer. com/ar

Perkins Products Inc., www.perkinsproducts.com

Perma USA, www.permausa.com

Petrolab Corporation, www.petrolab.com

Petrolabs Inc., http://pages.prodigy.net/petrolabsinc

Petroleum Analyzer Company LP (PAC), www. petroleum-analyzer.com

Petroleum Products Research, www.swri.org/4org/d08/ petprod/

Petro-Lubricants Testing Laboratories Inc., www.pltlab. com

Petrotest, www.petrotest.net

Pflaumer Brothers Inc., www.pflaumer.com

Philips Industrial Electronics Deutschland, www. philips-tkb.com

Pipeguard of Texas, www.pipeguard-texas.com

Plastic Bottle Corporation, www.plasticbottle.com

Plastican Inc., www.plastican.com

Plews/Edelmann Division, Stant Corporation, www. stant.com

PLI LLC, www.memolub.com

Plint and Partners: Tribology Division,www.plint.co.uk/ trib.htm

Polaris Laboratories, LLC, www.polarislabs.com

PREDICT/DLI—Innovative Predictive Maintenance, www.predict-dli.com

Predictive Maintenance Corporation, www.pmaint.com

Predictive Maintenance Services, www.theoillab.com

Premo Lubricant Technologies, www.premolube.com

Pulsair Systems Inc., www.pulsair.com

Qorpak, www.qorpak.com

R & D/Fountain Industries, www.fountainindustries.com

R.A. Miller & Company Inc., www.ramiller.on.ca

R.E.A.L. Services, www.realservices.com

Radian Inc., www.radianinc.com

Ramos Oil Company Inc., www.ramosoil.com

Ravenfield Designs Ltd., www.ravenfield.com

Reelcraft Industries Inc., www.realcraft.com

Rexam Closures, www.closures.com

Rheotek (PSL SeaMark), www.rheotek.com

Ribelin, www.ribelin.com

Russell-Stanley Corporation, www.russell-stanley.com

Safety-Kleen Corporation, www.safety-kleen.com

Saftek: Machinery Maintenance Index, www.saftek. com/boiler/machine/mmain.htm

Sandy Brae Laboratories Inc., www.sandy/brae.com

SATEC Inc., www.satec.com

Savant Group of Companies, www.savantgroup.com

Savant Inc., www.savantgroup.com

Saxton Industries, www.saxton.thomasregister.com

Scully Signal Company, www.scully.com

Senior Flexonics, www.flexonics-hose.com

Service Supply Lubricants LLC, www.service-lubricants.com

Sexton & Peake Inc., www.sexton.qpg.com

Silvas Oil Company Inc., www.silvasoil.com

Silverson Machines Inc., www.silverson.com

Sinclair Oil Corporation, www.sinclairoil.com

SKF Quality Technology Centre, www.qtc.skf.com

Sleeveco Inc., www.sleeveco.com

Snyder Industries, www.snydernet.com

Southwest Research Institute, www.swri.org

Southwest Spectro-Chem Labs, www.swsclabs.com

Spacekraft Packaging, www.spacekraft.com

Specialty Silicone Products Inc., www.sspinc.com

Spectro Inc. Industrial Tribology Systems, www. spectroinc.com

Spectronics Corporation, www.spectroline.com

Spex CertiPrep Inc., www.spexcsp.com

Star Brite, www.starbrite.com

Steel Shipping Containers Institute, www.steelcontain-ers.com

Stratco Inc., www.stratco.com

Sunohio, Division of ENSR, www.sunohio.com

Superior Lubricants Company Inc., www.superior lubricants.com

Taber Industries, www.taberindustries.com

Tannas Company, www.savantgroup.com

Tannis Company, www.savantgroup.com/tannas.sht

Thermo Elemental, www.thermoelemental.com

Thomas Petroleum, www.thomaspetro.com

Thoughtventions Unlimited Home Page, www.tvu. com/%7Ethought/

Titan Laboratories, www.titanlab.com

TriboLogic Lubricants Inc., www.tribologic.com

Trico Manufacturing Corporation, www.tricomfg.com

Trilla Steel Drum Corporation, www.trilla.com

TTi's Home Page, www.tti-us.com

UEC Fuels and Lubrication Laboratories, www.uec-usx. com

Ultimate Lubes, www.ultimatelubes.com

Unilube Systems Ltd., www.unilube.com

Unit Pack Company Inc., www.unitpack.com

USX Engineers & Consultants, www.uec.com/labs/ctns

USX Engineers and Consultants: Laboratory Services, www.uec.com/labs/

Vacudyne Inc., www.vacudyne.com

Van Horn, Metz & Company Inc., www.vanhornmetz. com

Viking Pump Inc., A Unit of IDEX Corporation, www. vikingpump.com

Viswa Lab Corporation, www.viswalab.com

Vortex International LLC, www.vortexfilter.com

Vulsay Industries Ltd., www.vulsay.com

Wallace, www.wallace.com
Waugh Controls Corporation, www.waughcontrols.com
Wearcheck International, www.wearcheck.com
Wedeven Associates Inc., members.aol.com/wedeven/
West Penn Oil Company Inc., www.westpenn.com
Western States Oil, www.lubeoil.com
Wilks Enterprise Inc., www.wilksir.com
WSI Chemical Inc., www.wsi-chem-sys.com
Zimmark Inc., www.zimmark.com

35.2.7   CAR/TRUCK MFG

Alfa Romeo, www.alfaromeo.com
Audi, www.audi.com
BMW (International), www.bmw.com/bmwe
BMW (United States), www.bmwusa.com
BMW Motorcycles, www.bmw-motorrad.com
Buick (GM), www.buick.com
Cadillac (GM), www.cadillac.com
Caterpillar, www.cat.com
Caterpillar, www.caterpillar.com
Chevrolet (GM), www.chevrolet.com
Chrysler (Mercedes Benz), www.chrysler.com
Citroen (France), www.citroen.com
Citroen (United Kingdom), www.citroen.co.uk/fleet
Cummins Engine Company, www.cummins.com
Daimler Chrysler, www.daimlerchrysler.com
Detroit Diesel, www.detroitdiesel.com
Dodge, www.dodge.com
Eagle, www.eaglecars.com
EV1, www.gmev.com
Ferrari, www.ferrari.com
Fiat, www.fiat.com
Ford Motor Company, www.ford.com
General Motors (GM), www.gm.com
Global Electric Motor Cars, LLC, www.gemcar.com
Hyundai, www.hyundai-motor.com
Infiniti, www.infiniti.com
Isuzu, www.isuzu.com
Jaguar, www.jaguarcars.com
Jeep, www.jeep.com
John Deere, www.deere.com
Kawasaki, www.kawasaki.com
Kawasaki, www.khi.co.jp
Lamborghini, www.lamborghini.com
Lexus, www.lexususa.com
Lincoln-Mercury, www.lincolnmercury.com
Mack Trucks, www.macktrucks.com
Mazda, www.mazda.com
Mercedes-Benz (Germany), www.mercedes-benz.de
Mitsubishi Motors, www.mitsubishi-motors.co.jp
Nissan (Japan), www.nissan.co.jp
Nissan (United States), www.nissandriven.com
Nissan (United States), www.nissanmotors.com
Opel, www.opel.com
Peugeot, www.peugeot.com
Plymouth, www.plymouthcars.com

Pontiac (GM), www.pontiac.com
Saab Cars USA, www.saabusa.com
Saab, www.saab.com
Saturn (GM), www.saturncars.com
Scania, www.scania.se
Toyota (Japan), www.toyota.co.jp
Toyota (United States), www.toyota.com
Vauxhall, www.vauxhall.co.uk
Volkswagen (Germany), www.vw-online.de
Volkswagen (United States), www.vw.com
Volvo (Sweden), www.volvo.se
Volvo Cars of North America, www.volvocars.com
Volvo Group, www.volvo.com

35.2.8   PUBLICATIONS/REFERENCES/ RECRUITING/SEARCH TOOLS, ETC.

American Machinist, www.penton.com/cgi-bin/super-directory/details.pl?id=317
API Links, www.api.org/links
Automotive & Industrial Lubricants Guide, www.wearcheck.com
Automotive and Industrial Lubricants Guide by David Bradbury, www.escape.ca/<dbrad/index.htm
Automotive and Industrial Lubricants Tutorial, www.escape.ca/<dbrad/index.htm
Automotive News, www.autonews.com
Automotive Service Industry Association (ASIA), www.aftmkt.com/asia
Automotive Services Retailer, www.gcipub.com
AutoWeb, www.autoweb.com
AutoWeek Online, www.autoweek.com
Bearing.Net, www.wearcheck.com
Cambridge, chemfinder.cambridgesoft.com
Car and Driver Magazine Online, www.caranddriver.com
Car-Stuff, www.car-stuff.com
Center for Innovation Inc., www.centerforinnovation.com
Chem Connect, www.chemconnect.com
Chemical Abstracts Service, www.cas.org
Chemical Resources, www.chemcenter.org
Chemical Week Magazine, www.chemweek.com
Concord Consulting Group Inc., www.concordcg.com
Dialog, www.dialog.com
Diesel Progress, www.dieselpub.com
Diversified Petrochemical Services, www.chemhelp.com
European Patent Office, www.epo.co.at/epo/
F.L.A.G. (Fuel, Lubricant, Additives, Grease) Recruiting, www.flagsearch.com
Farmland Industries Inc., www.farmland.com
Fuel Quest, www.fuelquest.com/cgi-bin/fuelqst/corporate/fq_index.jsp
Fuels and Lubes Asia Publications Inc., www.flasia.com.ph
Gear Technology Magazine, www.geartechnology.com/mag/gt-index.html

Haas Corporation, www.haascorp.com

HEF, France, www.hef.fr

How Stuff Works, www.howstuffworks.com/engine.htm

Hydrocarbon Asia, www.hcasia.safan.com

Hydrocarbon Online, www.wearcheck.com

Hydrocarbon Processing Magazine, www.hydrocarbon processing.com

ICIS-LOR Base Oils Pricing Information, www.icislor.com/

Industrial Lubrication and Tribology Journal, www.mcb.co.uk/ilt.htm

Industrial Maintenance and Engineering Links (PLI, LLC), www.memolub.com/link.htm

International Tribology Conference, Yokohama 1995, www.mep.titech.ac.jp/Nakahara/jast/itc/itc-home.htm

ISO Translated into Plain English, connect.ab.ca/praxiom

Journal of Fluids Engineering, borg.lib.vt.edu/ejournals/JFE/jfe.html

Journal of Tribology, engineering.dartmouth.edu/thayer/research/index.html

Kline & Company Inc., www.klinegroup.com

LSST Tribology and Surface Forces, bittburg. ethz.ch/LSST/Tribology/default.html

LSST Tribology Letters, http://bittburg.ethz.ch/LSST/Tribology/letters.html

Lubelink, www.lubelink.com

LubeNet, www.lubenet.com

Lubes and Greases, www.lngpublishing.com

Lubricants Network Inc., www.lubricantsnetwork.com

Lubricants World, www.lubricantsworld.com

Lubrication Engineering Magazine, www.stle.org/le_magazine/le_index.htm

MaintenanceWorld, www.wearcheck.com

MARC-IV, www.marciv.com

Mechanical Engineering Magazine, www.memagazine.org/index.html

Migdal's Lubricant Web Page, http://members.aol.com/sirmigs/lub.htm

Muse, Stancil & Company, www.musestancil.com

National Petroleum News, www.petroretail.net/npn

National Resource for Global Standards, www.nssn.org

Neale Consulting Engineers Limited, www.tribology.co.uk

Noria—OilAnalysis.Com, www.oilanalysis.com

Oil Directory.com, www.oildirectory.com

Oil Online, www.oilonline.com

Oil-Link Oil & Gas Online, www.oilandgasonline.com

Oilspot.com, www.oilspot.com

Pennwell Publications, www.pennwell.com

Petrofind.com, www.petrofind.com

PetroleumWorld.com, www.petroleumworld.com

PetroMin Magazine, www.petromin.safan.com

Practicing Oil Analysis Magazine, www.practicin-goilanalysis.com

Predictive Maintenance Corporation: Tribology and the Information Highway, www.pmaint.com/tribo/docs/oil_anal/tribo_www.html.ref

Reliability Magazine, www.pmaint.com/tribo/docs/oil_anal/tribo_www.html

Safety Information Resources on the Internet, www.siri.org/links1.html

Savant Group of Companies, www.savantgroup.com

SGS Control Services Inc., www.sgsgroup.com

Shell Global Solutions, www.shellglobalsolutions.com

SubTech (Petroleum Service & Supply Information), www.subtech.no/petrlink.htm

Summit Technical Solutions, www.lubemanagement.com

Sunohio, Division of ENSR, www.sunohio.com

Tannas Company, www.savantgroup.com

Test Engineering Inc., www.testeng.com

The Energy Connection, www.energyconnect.com

The Maintenance Council, www.trucking.org

TMC, www.truckline.com

Tribologist.com, www.wearcheck.com

Tribology Consultant, hometown.aol.com/wearconsul/wear/wear.htm

Tribology International, www.elsevier.nl/inca/publications/store/3/0/4/7/4/

Tribology Letters, www.kluweronline.com/issn/1023-8883

Tribology Research Review 1992–1994, www.me.ic.ac.uk/department/review94/trib/tribreview.html

Tribology Research Review 1995–1997, www.me.ic.ac.uk/department/review97/trib/tribreview.html

Tribology/Tech-Lube, www.tribology.com

Truklink (truck fleet information), www.truklink.com

United Soybean Board, www.unitedsoybean.org

Victoria Group Inc., The, www.victoriagroup.com

Wear Chat: WearCheck Newsletter, www.wearcheck.com

Wear, www.elsevier.nl/inca/publications/store/5/0/4/1/0/7/

World Tribologists Database, greenfield.fortunecity.com/fish/182/tribologists.htm

WWW Tribology Information Service, www.shef.ac.uk/<mpe/tribology/

WWW Virtual Library: Mechanical Engineering, www.vlme.com

Yahoo Lubricants, dir.yahoo.com/business_and_economy/shopping_and_services/automotive/supplies/lubricants/

Yahoo Tribology, ca.yahoo.com/Science/Engineering/Mechanical_Engineering/Tribology/

Appendix A: Grease Compatibility Chart

	Aluminum Complex	Barium	Bentonite Clay	Calcium	Calcium 12-Hydroxy	Calcium Complex	Calcium Sulfonate	Lithium	Lithium 12-Hydroxy	Lithium Complex	Polyurea	Sodium
Aluminum complex	C	I	I	I	C	I	B	I	I	C	I	I
Barium	I	C	I	I	C	I	B	I	I	I	I	I
Bentonite clay	I	I	C	C	C	I	B	I	I	I	I	I
Calcium	I	I	C	C	C	B	I	C	B	B	I	I
Calcium 12-hydroxy	C	C	C	C	C	B	NA	C	C	C	B	I
Calcium complex	I	I	I	B	B	C	C	I	I	C	B	I
Calcium sulfonate	B	B	B	I	NA	C	C	C	C	C	B	I
Lithium	I	I	I	C	C	I	C	C	C	C	I	B
Lithium 12-hydroxy	I	I	I	B	C	I	C	C	C	C	I	I
Lithium complex	C	I	I	B	C	C	C	C	C	C	I	B
Polyurea	I	I	I	I	B	B	I	I	I	I	C	I
Sodium	I	I	I	I	I	I	I	B	I	B	I	C

C, usually compatible; B, borderline compatibility (probably incompatible); I, incompatible; NA, no information on compatibility.

Appendix B: Color Scale Comparison Chart

Comparison of Color Test Values Used for Lubricant Evaluation			
Hazen 1 Units D1209	Saybolt 2 Color D156	ASTM 3 Color D1500	Gardner 4 Color D1544
5	30		
10	27		
15	25		
20			
20–30	23		
30			
30–40	21		
40			
50			
60			
70	16		
100	14		
150	1		
200	–6		1
250			2
			3
>250		0.5	4
			5
			6
		1	7
			8
			9
		1.5	10
		2	11
		2.5	12
		3	13
		3.5	14
		4	15
		4.5	16

ASTM D1209 Platinum–cobalt color test method.
ASTM D156 Saybolt colorimetric test method.
ASTM D1500 ASTM visual color scale method.
ASTM D1544 Gardner color scale method.

Appendix C: ISO Viscosity Ranges

	ASTM D2422 ISO Viscosity Grades for Industrial Lubricants			
Viscosity Grade ISO VG	Midpoint Viscosity at 40°C, cSt	Midpoint Viscosity at 40°C, SUS Approximate Equivalents	Viscosity Limits at 40°C, cSt Minimum	Viscosity Limits at 40°C, cSt Maximum
2	2	32	1.98	2.42
3	3	36	2.88	3.52
5	5	40	4.14	5.06
7	7	50	6.12	7.48
10	10	60	9	11
15	15	75	13.5	16.5
22	22	105	19.8	24.2
32	32	150	28.8	35.2
46	46	215	41.4	50.6
68	68	315	61.2	74.8
100	100	465	90	110
150	150	700	135	165
220	220	1000	198	242
320	320	1500	288	352
460	460	2150	414	506
680	680	3150	612	748
1000	1000	4650	900	1100
1500	1500	7000	1350	1650

Index